Springer Monographs in Mathematics

This series publishes advanced monographs giving well-written presentations of the "state-of-the-art" in fields of mathematical research that have acquired the maturity needed for such a treatment. They are sufficiently self-contained to be accessible to more than just the intimate specialists of the subject, and sufficiently comprehensive to remain valuable references for many years. Besides the current state of knowledge in its field, an SMM volume should ideally describe its relevance to and interaction with neighbouring fields of mathematics, and give pointers to future directions of research.

Šárka Nečasová • Stanislav Kračmar •
Jiří Neustupa • Patrick Penel

Mathematical Theory of a Fluid Flow Around a Rotating and Translating Body

Springer

Šárka Nečasová
Institute of Mathematics
Academy of Sciences
Praha 1, Czech Republic

Jiří Neustupa
Institute of Mathematics
Academy of Sciences
Praha 1, Czech Republic

Stanislav Kračmar
Department of Mathematics
Czech Technical University
Praha 2, Czech Republic

Patrick Penel
University of Toulon
Le Beausset, France

ISSN 1439-7382 ISSN 2196-9922 (electronic)
Springer Monographs in Mathematics
ISBN 978-3-031-84122-4 ISBN 978-3-031-84120-0 (eBook)
https://doi.org/10.1007/978-3-031-84120-0

Mathematics Subject Classification: 76D05, 76D07, 35Q35

This Springer imprint is published by the registered company Springer Nature Switzerland AG
The registered company address is: Gewerbestrasse 11, 6330 Cham, Switzerland

The book is dedicated to four extraordinary and excellent personalities, who have left us, but who remain with us in our hearts: Jindřich Nečas, Mariarosaria Padula, Miroslav Krbec, and Antonín Novotný.

Š. N. would like to express very deep thanks Š. N. would like to thank her friend , mentor, and collaborator Professor Giovanni Paolo Galdi for introducing her to the wonderful field of mathematical modeling of flows around rotating and translating bodies, which is the subject of the book. Moreover, she would like to thank her friend and mentor Professor Ronald B. Guenther. Furthermore, she would like to thank her family: her mother Zdeňka, sister Jindra and children Martin, Jan, and Lucie for their great support.

S. K. would like to thank his wife Dagmar for her support and patience.

P. P. is very much thankful to Czech colleagues and friends for 26 years of cooperation in excellent working conditions between Toulon and Prague. He would also like to dedicate the book to Jessie, the youngest of his grandchildren.

Š. N. and S. K. would like to express their thanks to Paul Deuring for his support to this project and wonderful collaboration. Moreover, Š. N. would like to express her gratitude to Reinhard Farwig for introducing her to the wonderful world of weighted spaces and applications in harmonic analysis.

Prague and Toulon, November 2024

Stanislav Kračmar,
Šárka Nečasová,
Jiří Neustupa,
Patrick Penel

Acknowledgements

Š. Nečasová, S. Kračmar, and J. Neustupa have been supported by the Czech Science Foundation (GAČR), project No. 19-04243S. Š. Nečasová and J. Neustupa have also been supported by the Institute of Mathematics of the Czech Academy of Sciences, RVO 67985840, and S. Kračmar has been supported by the Czech Technical University, RVO 12000. Moreover, the last version of the book was supported by the Czech Science Foundation (GAČR), project No. 22-01591S. Finally, Š.N. has also been supported by the Praemium Academica of Š. Nečasová.

Contents

Part I Introduction and Preliminaries

1 Introduction 3

2 Preliminaries 17
2.1 Basic Notation and Weighted Function Spaces 17
2.2 The Bogovskii Operator 22
2.3 Some Auxiliary Results 23
2.4 The Littlewood–Paley Decomposition 27
2.5 Fundamental Solution of the Nonsteady "Rotating" Oseen Problem 28
2.5.1 Basic Properties of the Fundamental Solution $\Gamma = \Gamma_{jk}$ 35
2.5.2 Further Properties of Γ_{jk}: Pointwise Estimates 38
2.6 T-Time-Periodic Functions 47

Part II Linear Theory

3 The Steady Stokes Type Problem 53
3.1 A Strong Solution in L^q 53
3.1.1 Introduction 53
3.1.2 Two Differential Forms of the Problem 55
3.1.3 The Main Results 59
3.1.4 Various Remarks 61
3.1.5 Proofs of Theorems 3.1 and 3.3 Using the Associated Ordinary Angular Differential Equation 64
3.1.6 The Associated Time-Periodic Stokes Problem and the Proof of Theorem 3.1 73
3.1.7 The Nonhomogeneous Problem and Proof of Theorem 3.2 77
3.2 A Weak Solution in the Whole Space $\mathbb{R}^3$ and in an Exterior Domain 77

3.2.1 The Whole Space Problem ... 79
3.2.2 The Problem in an Exterior Domain ... 82
3.3 The Steady Stokes Type Problem in Weighted Spaces ... 88
3.3.1 A Strong Solution in the Whole Space ... 88
3.3.2 Proof of Theorem 3.8 ... 91
3.3.3 A Strong Solution in an Exterior Domain ... 104
3.4 Very Weak Solutions to the Stokes Type Problem ... 115
3.4.1 The Homogeneous Case: Very Weak Solutions ... 115
3.4.2 The Case of Nonhomogeneous Dirichlet Conditions: Generalized Very Weak Solution ... 119
3.4.3 Proof of Theorem 3.13 ... 123
3.5 Notes to the Stokes Type Problem ... 127
3.5.1 Concluding Remarks to Previous Sections ... 127
3.5.2 Asymptotic Behavior ... 127
3.5.3 2D Case ... 130

4 The Steady Oseen Type Problem ... 133
4.1 A Strong Solution in L^q ... 133
4.1.1 The Main Result and Corollaries ... 134
4.1.2 Proof of Theorem 4.1 Using the Associated Ordinary Angular Differential Equation ... 136
4.1.3 Proofs of Lemmas 4.2 and 4.3 ... 145
4.1.4 Proof of Corollary 4.1 ... 148
4.1.5 The Associated Time-Periodic Oseen Problem and Proof of Theorem 4.1 ... 152
4.2 L^q-Theory of Weak Solutions of the Oseen Type Problem in the Whole Space and in an Exterior Domain ... 159
4.2.1 Introduction ... 159
4.2.2 The Main Result ... 161
4.2.3 Details of Proof—Boundedness of $\partial_3 U_0$ ... 183
4.3 Further Properties of the Oseen Type Problem ... 185
4.3.1 Further Properties of the Rotating Oseen in the Whole Space $\mathbb{R}^3$ ... 187
4.3.2 Further Properties of the Rotating Oseen in the Exterior Domain ... 191
4.3.3 The Regularity of Very Weak Solutions ... 194
4.3.4 Uniqueness of the Very Weak Solution ... 200
4.4 L^q-Weighted Theory of the Strong Solution to Oseen Type Problem ... 208
4.4.1 Preliminaries ... 211
4.4.2 Proofs ... 221
4.5 L^q-Weighted Theory of the Weak Solution to the Oseen Type Problem ... 232
4.5.1 Proofs ... 238

4.6 L^q-Weighted Theory of the Very Weak Solutions to the Oseen Type Problem ... 243
4.7 Notes to the Oseen Type Problem ... 246

5 Representation Formula and the Asymptotic Behavior of the Steady Oseen Type Problem ... 247
5.1 Introduction ... 247
5.1.1 Fundamental Solution of the Stationary Problem ... 249
5.1.2 Mathematical Preliminaries to the Representation Formula ... 264
5.2 Derivation of the Representation Formula ... 267
5.3 Asymptotic Behavior ... 271
5.3.1 Some Volume Potentials ... 271
5.3.2 Asymptotic Profile ... 275
5.3.3 Asymptotic Profile of the Gradient of the Velocity Field ... 293
5.3.4 Decay Estimates of the Second Derivatives of the Velocity ... 295
5.4 Leray Solution ... 296
5.4.1 Introduction ... 296
5.4.2 Auxiliary Results ... 297
5.4.3 Representation Formula for the Leray Solution ... 298
5.4.4 Asymptotic Profile ... 302
5.5 Notes on Chap. 5 ... 304

6 Artificial Boundary Conditions for the Oseen Type Problem ... 307
6.1 Introduction ... 307
6.2 Preliminaries ... 308
6.3 Auxiliary Results to Asymptotic Behavior of the Pressure ... 309
6.4 Decay Estimate of the Pressure ... 310
6.5 Formulation of the Problem with Artificial Boundary Conditions ... 320

7 The Oseen-Type Problem: A Weak Solution in Anisotropic L^2-Spaces ... 327
7.1 Introduction ... 327
7.2 Friedrichs–Poincaré Inequality ... 330
7.3 Pressure in $\mathbb{R}^3$ and Uniqueness Theorems ... 332
7.3.1 The Pressure in $\mathbb{R}^3$ ... 333
7.3.2 Uniqueness in $\mathbb{R}^3$ ... 333
7.3.3 Uniqueness in an Exterior Domain $\mathcal{D} \subset \mathbb{R}^3$ ... 335
7.4 Existence of a Solution in the Whole Space ... 336
7.4.1 The Problem in B_R ... 336
7.5 Existence of Solution in Exterior Domains ... 342
7.5.1 Homogenous Dirichlet Boundary Conditions ... 342
7.5.2 Appendix A ... 349
7.5.3 Appendix B ... 349
7.6 Notes on Chap. 7 ... 350

8 Stokes and Oseen Type Operators: Spectral Theory and Generated Semigroups 351
8.1 Associated Stokes and Oseen Type Operators and Their Basic Properties 351
8.2 A Brief Survey of Basic Notions from Spectral Theory of Operators in a Banach Space 364
8.3 Spectral Theory of the Stokes Type Operator 368
8.3.1 L^2-Theory 368
8.3.2 L^q-Theory 377
8.4 Spectral Theory of the Oseen Type Operator 389
8.4.1 L^2-Theory 390
8.4.2 L^q-Theory 401
8.5 Semigroups Generated by the Stokes and Oseen Type Operators 416
8.5.1 Basic Information on the Semigroups That Involve the Rotational Effect 416
8.5.2 On the Uniform Growth Bound of the Semigroups 418

Part III Navier–Stokes Type Equations

9 A Brief Introduction to the Existence Theory for the Stationary Navier–Stokes Type Problem Around Translating and Rotating Body 427
9.1 L^2 Theory 427
9.1.1 Navier–Stokes Type Problem Without the Oseen Term 427
9.1.2 The Navier–Stokes Type Problem with the Oseen Term 432
9.2 L^p-Theory 438

10 Pointwise Decay for the Stationary Navier–Stokes Type Problem 441
10.1 Introduction 441
10.2 Auxiliary Results 446
10.3 L^p-Estimates of the Volume Potential $\mathfrak{R}(f)$ 457
10.4 L^p-Estimates of the Velocity and Its Gradient 463
10.5 Pointwise Estimates of the Velocity 469
10.6 Pointwise Estimates of the Gradient of the Velocity 475

11 Asymptotic Behavior of the Solutions of the Navier–Stokes Type Problem 479
11.1 Introduction 479
11.2 Discussion and Main Results 480
11.3 Historical Remarks 487
11.4 Notation and Auxiliary Results 488
11.5 Representation Formula for the Navier–Stokes System 494
11.6 Proof of Theorem 11.2 498

11.7 Preliminaries to Fourier Transforms ... 507
11.8 Proof of Theorem 11.3 and Corollary 11.1 ... 510
11.9 Proof of Theorem 11.4 ... 513
11.10 Notes on Chap. 11 ... 514

12 A Brief Introduction to the Existence Theory for the Nonstationary Navier–Stokes Type Problem Around Translating and Rotating Body ... 517
12.1 Local in Time Existence of a Strong Solution ... 518
12.1.1 Introduction ... 518
12.1.2 Preliminaries ... 520
12.1.3 The Local Transformation ... 521
12.1.4 The System in the Cylindrical Domain ... 523
12.1.5 Proof of Theorem 12.1 ... 525
12.1.6 Proof of the Local Existence and Uniqueness Result ... 529
12.1.7 Proof the Global Existence and Uniqueness in 2D ... 532
12.2 Global in Time Existence of a Strong Solution ... 538
12.2.1 Global Existence and Convergence to Steady State ... 539
12.3 Notes on Chap. 12 ... 549

Part IV Appendix

13 Appendix ... 553
13.1 The Case When $\mathbf{u}_\infty$ Is Not Vanished or Not Parallel to the Angular Velocity $\boldsymbol{\omega}$ ... 553
13.2 Estimates of the Oseen Kernels ... 559
13.2.1 Introduction and Basic Notation ... 559
13.2.2 Oseen Fundamental Solution ... 560
13.2.3 L^∞-Estimates of a Convolution in $\mathbb{R}^N$ ... 569
13.2.4 L^∞-Estimates for Weakly Singular Oseen Kernels ... 582
13.2.5 L^p-Estimates for Weakly Singular Oseen Kernels ... 595
13.3 Singular Integrals ... 602

References ... 613

Index ... 623

List of Symbols

$\mathbb{N}$	The set of all natural numbers
$\mathbb{N}_0$	$\mathbb{N}_0 = \mathbb{N} \cup \{0\}$
$\mathbb{Z}$	The set of integer numbers
$\mathbb{Z}_+$	The set of positive integer numbers
$\mathbb{R}$	The set of all real numbers
x	A scalar
A^c	The complement $\mathbb{R}^3 \backslash A$ of set A
c, c_i, C, C_j	Generic constants
$\mathbb{R}^n$	The n-dimensional Euclidian space
$\{\mathbf{e}_1, \ldots, \mathbf{e}_n\}$	The canonical basis
$\mathbf{x} \equiv (x_i)_{1 \leq i \leq n}$	A vector
$\cdot$	The inner product
$\wedge$	The vector product
$B_R(\mathbf{x}) =$	$\{\mathbf{y} \in \mathbb{R}^n;\ \lvert\mathbf{x} - \mathbf{y}\rvert < R\}$ for some $R > 0$
$B^R =$	$\mathbb{R}^n \setminus \overline{B_R}$
$\mathcal{D}$	An exterior domain in $\mathbb{R}^n$
$\mathcal{D}^c =$	$\mathbb{R}^n \setminus \mathcal{D}$ interpreted as a body $\mathcal{B}$ or an obstacle
$\partial\mathcal{D}$	The boundary of the body $\mathcal{B}$
$\mathbf{n}$	The outer unit normal vector field on $\partial\mathcal{D}$
$\mathbf{u}_\infty := u_\infty \mathbf{e}_3$	The time-independent translational velocity of body $\mathcal{B}$, assumed to be parallel to $\mathbf{e}_3$
$\boldsymbol{\omega} := \omega \mathbf{e}_3$	The time–independent angular velocity of body $\mathcal{B}$, assumed to be parallel to $\mathbf{e}_3$
$\mathbf{u}, \mathbf{v}$	Various vectorial functions (denoted by boldface letters)
p	The scalar function (or distribution), expressing the pressure
$\nu > 0$	The kinematic coefficient of viscosity
$Q_\omega(t)$	The matrix of rotation in terms of ω, $Q_\omega(t)^T = Q_\omega(t)^{-1}$

$\partial_i f$	$\frac{\partial f}{\partial x_i}$
$\partial_t \varphi,\ \dot{\varphi}$	$\frac{\partial \varphi}{\partial t}$
$\partial^\alpha,\ D^\alpha$	The differential operator $\frac{\partial^{\lvert\alpha\rvert}}{\partial_{x_1}^{\alpha_1}\ldots\partial_{x_n}^{\alpha_n}}$ with the multi-index $\alpha = (\alpha_1, \ldots, \alpha_n)$
∇	The nabla operator, denoting the gradient
div, Div	The divergence operators
div $\mathbf{v}$	$= \partial_i v_i$ with the implicit Einstein summation convention
Δ	The Laplace operator, $\Delta \mathbf{v} = \partial_{ii} \mathbf{v}$
$\mathcal{R}$	The "rotational" operator, $\mathcal{R}\mathbf{v} := \mathbf{e}_3 \wedge \mathbf{x} \cdot \nabla \mathbf{v} - \mathbf{e}_3 \wedge \mathbf{v}$
$S_{\omega\nu}$	The "rotational" Stokes operator, $S_{\omega\nu}\mathbf{v} := -\nu\Delta\mathbf{v} - \omega\mathcal{R}\mathbf{v}$
L	A simplified notation for $S_{\omega\nu}$ or $S_{1,1}$
$\mathfrak{H}$	a normed space with the norm denoted by $\|.\|_{\mathfrak{H}}$; the norm of $v \equiv (v_1, \ldots, v_n) \in \mathfrak{H}^n$ is defined by the formula $$\|v\|_{\mathfrak{H}}^{(n)} := \left(\sum_{j=1}^{n} \|v_j\|_{\mathfrak{H}}^2\right)^{1/2},$$ however, we write only $\|.\|_{\mathfrak{H}}$ instead of $\|.\|_{\mathfrak{H}}^{(n)}$ for simplicity
$\|\cdot\|_q$	The norm in the Lebesque spaces $L^q(\cdot)$ and $\mathbf{L}(\mathcal{D})^q$ (over an appropriate domain), $1 \le q < \infty$, where $\mathbf{L}^q = (L^q)^n$ for $n \in \mathbb{N}$
$\|\cdot\|_{m,q}$	The norm in the Sobolev spaces $H^{m,q}(\mathcal{D})$ and $\mathbf{H}^{m,q}(\mathcal{D})$ (over an appropriate domain) for $m \in \mathbb{N}$, $1 \le q < \infty$; where $\mathbf{H}^{m,q} = (H^{m,q})^n$ for $n \in \mathbb{N}$
$C_0^\infty(\mathcal{D})$	The linear space of infinitely differentiable functions with compact support in $\mathcal{D}$
$\mathbf{C}_0^\infty(\mathcal{D}) =$	$(C_0^\infty(\mathcal{D}))^n$, $n \ge 1$
$H_0^{m,q}(\mathcal{D})$	$:= \overline{C_0^\infty(\mathcal{D})}^{\|\cdot\|_{m,q}}$ the completion in the indicated topology
$L^1_{loc}(\mathbb{R}^n)\vert_{\pi_{k-1}}$	the set of classes of functions belonging to $L^1(\mathbb{R}^n)$ modulo π_{k-1}, which is a polynomial of degree less than or equal to $k-1$, $k \ge 1$
$\widehat{\mathbf{H}}^{k,q}(\mathbb{R}^n)$	The linear space of $\mathbf{v} \in L^1_{\mathrm{loc}}(\mathbb{R}^n)\vert_{\pi_{k-1}}$, satisfying $\partial^\alpha \mathbf{v} \in L^q(\mathbb{R}^n)$, where $\lvert\alpha\rvert = k$, $k \ge 1$ the homogeneous Sobolev spaces
$\lvert\cdot\rvert_{k,q}$	The norm in $\widehat{\mathbf{H}}^{k,q}(\mathbb{R}^n)$ or in $\widehat{\mathbf{H}}_0^{k,q}(\mathcal{D}) = \overline{C_0^\infty(\mathcal{D})}$
X_σ	A space X with divergence free functions
X'	The dual space to X

$L^{q'} = (L^q)'$	$\frac{1}{q} + \frac{1}{q'} = 1$, The dual space to L^q
$H^{-1,q'} = (H_0^{1,q})'$	The dual space to $H_0^{1,q}$
$\mathcal{H}_0^{-1,q} = (\widehat{H}^{1,q'})'$	The dual space to $\widehat{H}^{1,q'}$
$\mathbf{T}_w^{k,q}(\partial D) := (\mathbf{H}_w^{k,q}(D))\|_{\partial D}$	Vector space of boundary trace equipped with the norm of the quotient space $$\|\mathbf{h}\|_{\mathbf{T}_w^{k,q}(\partial D)} := \inf\{\mathbf{u} \in \mathbf{H}_w^{k,q}(D)\,,\ \mathbf{u}\|_{\partial D} = \mathbf{h}\}.$$
$\mathbf{T}_w^{0,q}(\partial D) := (\mathbf{H}_w^{0,q}(D))\|_{\partial D}$	Vector space of boundary trace equipped with the norm of the quotient space $$\|\mathbf{h}\|_{\mathbf{T}_w^{0,q}(\partial D)} := \inf\{\mathbf{u} \in \mathbf{H}_w^{0,q}(D)\,,\ \mathbf{u}\|_{\partial D} = \mathbf{h}\}.$$
$\mathbf{T}_w^{-1,q'}(\partial D) := (\mathbf{T}_{w'}^{1,q}(\partial D))'$	It is a dual space to $(\mathbf{T}_{w'}^{1,q}(\partial D))$
$\mathcal{S}(\mathbb{R}^n)$	Schwartz' space of rapidly decreasing functions
$\mathcal{S}'(\mathbb{R}^n)$	The space of tempered distributions
$\mathcal{F}$	The Fourier transform on $\mathcal{S}(\mathbb{R}^n)$
$\mathcal{F}\mathbf{v}$ or $\hat{\mathbf{v}}$	Notations of Fourier transformation of $\mathbf{v}$
A_q	The Muckenhoupt class of weight functions in $L^1_{\text{loc}}(\mathbb{R}^n)$ $1 \le q < \infty$
w	Systematically a weight function
$\|\cdot\|_{q,w}$, $\|\cdot\|_{k,q,w}$	The norms in $\mathbf{L}^q_w(\mathcal{D})$ resp. $\mathbf{H}^{k,q}_w(\mathcal{D})$ the weighted Lebesque or Sobolev spaces
$\widehat{H}^{k,q}_{w,0}(\mathcal{D}) :=$	$\overline{C_0^\infty(\mathcal{D})}^{\|\nabla^k\cdot\|_{q,w}}$
$H^{k,q}_{w,0}(\mathcal{D}) :=$	$\overline{C_0^\infty(\mathcal{D})}^{\|\cdot\|_{k,q,w}}$
$\mathbf{L}^2_{\alpha,\beta}$, $V_{\alpha,\beta}$	Special notation when $w = \eta^\alpha_\beta$ (see Chap. 4)
$K(z,t)$	Fundamental solution of the heat equation
$\Gamma(y,z,t)$	Fundamental solution of nonsteady Oseen problem with rotating terms
Γ_a	Gamma function
${}_1F_1(a,c,\lambda)$	Kummer function
Σ_s	Fundamental solution of the steady Stokes type of problem
Σ_{os}	Fundamental solution of the steady Oseen type of problem
(E,Q)	Fundamental solution of the unsteady Oseen system
$N(\mathscr{T})$	Null space of operator $\mathscr{T}$
$R(\mathscr{T})$	Range of operator $\mathscr{T}$
$d(\mathcal{B})$	The diameter of body $\mathcal{B}$
ρ	$\rho := \|\boldsymbol{\omega}\|^2\, d(\mathcal{B})^4/\nu^2$
τ	$\tau := k\,\|\boldsymbol{\omega}\|\, d(\mathcal{B})/\nu$
Re	$\text{Re} = \|\boldsymbol{\omega}\|\, d(\mathcal{B})/\nu$

$\boldsymbol{L}^q_\sigma(\mathcal{D})$	A subspace of $\boldsymbol{L}^q(\mathcal{D})$, that are divergence–free in the sense of distributions and their normal component on $\partial\mathcal{D}$ is equal to zero in the sense of traces
$\boldsymbol{G}^q(\mathcal{D})$	A subspace of $\boldsymbol{L}^q(\mathcal{D})$, $\boldsymbol{G}^q(\mathcal{D}) := \{\nabla p \in \boldsymbol{L}^q(\mathcal{D});\ p \in W^{1,q}_{loc}(\mathcal{D})\}$
Π^q_σ	Helmholtz projection in $\boldsymbol{L}^q(\mathcal{D})$
$\boldsymbol{W}^{1,q}_{0,\sigma}(\mathcal{D}_R)$	A subspace of $\boldsymbol{W}^{1,2}(\mathcal{D}_R)$, $\boldsymbol{W}^{1,q}_{0,\sigma}(\mathcal{D}) := \boldsymbol{W}^{1,2}_0(\mathcal{D}) \cap \boldsymbol{L}^q_\sigma(\mathcal{D})$ $\mathscr{A}^\omega_k$ $\mathscr{A}^\omega_k\mathbf{u}\& := -\Pi_\sigma{}^q S_{\nu,\omega}\mathbf{u} + k\Pi_\sigma{}^q\partial_3\mathbf{u}$
$\mathscr{A}^\omega_k$	Operator $D(\mathscr{A}^\omega_k) := \{\mathbf{u} \in \boldsymbol{W}^{2,q}(\mathcal{D}) \cap \boldsymbol{W}^{1,q}_{0,\sigma}(\mathcal{D});\ (\boldsymbol{\omega}\wedge \mathbf{x})\cdot\nabla\mathbf{u} \in \boldsymbol{L}^q(\mathcal{D})\}$
$\mathscr{A}^0_0$	The Stokes operator
$\mathscr{A}^0_k$	The Oseen operator
nul	The nullity of an operator
def	The deficiency of an operator
ind	The index of an operator
nul′	An approximate nullity of an operator
Sp	The spectrum of an operator
Sp_p	The point spectrum of an operator
Sp_r	The residual spectrum of an operator
Sp_c	The continuous spectrum of an operator
Sp_ess	The essential spectrum of an operator

Part I
Introduction and Preliminaries

Chapter 1
Introduction

Dynamics of fluids, i.e., liquids and gases, is an important part of continuum mechanics. Fundamental classical concepts and results in fluid dynamics are connected with the names of I. Newton, L. Euler, A. L. Cauchy, J. L. Lagrange, D. Bernoulli, C. Huygens, etc. Their works was later extended by G. Kirchhoff, H. Helmholtz, G. G. Stokes, C. L. Navier, and many others. Great progress in fluid dynamics was achieved using differential and integral calculus, differential equations, and the theory of functions of complex variable. It was later followed by the development of numerical computational methods. A deeper analysis of corresponding mathematical models has been enabled by advances in theory of ordinary and partial differential equations, which are mainly based on methods and tools of functional analysis. Most of the mathematical models of fluid flows consist of a system of partial differential equations and appropriate boundary and/or initial conditions. Theoretical investigation especially focuses on studies of existence of solutions (which are usually considered to be elements of appropriate function spaces), uniqueness, regularity, stability, the way the solution depends on various parameters, branching and bifurcations of solutions, etc. Results, achieved in this field, significantly contribute to deeper understanding of the models, the role of various individual terms in the models, verification of the models, and their adequacy and correctness. Moreover, many numerical methods whose aim is to find concrete solutions are directly based on, or are influenced by, theoretical results.

While the stress tensor typically depends on the deformation (or, more precisely, on the tensor of deformation) in solid bodies, one can characterize fluids as continuous media, where the stress tensor $\mathbb{T}$ depends on the velocity of deformation, i.e., the rate of deformation. More precisely, $\mathbb{T}$ depends on the rate of deformation tensor $\mathbb{D}$, which coincides with the symmetrized gradient of velocity. Relations between the tensors $\mathbb{T}$ and $\mathbb{D}$ are given by the so-called constitutive equations. The constitutive equations, roughly speaking, express the response of the fluid motion to acting stresses or vice versa. The large diversity of possible fluids, appearing in nature or being produced by modern technologies, leads to a large variety of

Š. Nečasová et al., *Mathematical Theory of a Fluid Flow Around a Rotating and Translating Body*, Springer Monographs in Mathematics,
https://doi.org/10.1007/978-3-031-84120-0_1

the constitutive equations. In this text, we restrict ourselves to the Newtonian viscous incompressible fluid (we briefly say "viscous incompressible fluid"), where $\mathbb{T} = 2\mu\mathbb{D}$. (Here, μ is the so-called dynamic coefficient of viscosity.) We also assume that the fluid is homogeneous in the sense that its density ρ is a positive constant. Moreover, we also assume that the system of physical units is chosen so $\rho = 1$. Then we can also write $\mathbb{T} = 2\nu\mathbb{D}$, where $\nu := \mu/\rho$ is the so-called kinematic coefficient of viscosity.

The used partial differential equations mostly reflect the general principles of mechanics, applied to the moving fluid: conservation of mass, momentum, energy, and the second law of thermodynamics. In our case, the law of conservation of mass is expressed by the equation of continuity, and the law of conservation of momentum is expressed by the vectorial Navier–Stokes equation. (The plural form "Navier–Stokes equations" usually involves both the equations of conservation of momentum and mass.) The conservation of energy and the second law of thermodynamics play important roles, especially in the theory of compressible fluids. The modern theory of mathematical models in the fluid dynamics preferably studies the equations in the so-called variational forms rather than in their "classical forms." (This is closely connected with the notion of a "weak solution," in contrast to the "classical solution.") It is for mathematical convenience: the variational forms, usually based on certain integral equations, which represent the most natural framework for mathematical formulation of the main conservation laws. This corresponds to the well-known fact that the conservation laws, applied to a sample of fluid, taking at some time an arbitrarily chosen volume, have been originally mostly formulated by means of certain integral identities and not directly by differential equations.

Of enormously many particular problems of fluid dynamics, investigation of flows of a viscous incompressible fluid around rigid bodies belongs to widely studied fundamental questions of theoretical and experimental fluid dynamics. Naturally, a body moving in a fluid influences the fluid flow and vice versa. This is a special case of the so-called fluid-structure interaction. The fluid–structure interaction generally denotes any interaction between the moving fluid and material boundary of the flow field. Results, obtained in this discipline, play a crucial role in mechanical and civil engineering (e.g., in the design of aircrafts, turbines, bridges, and buildings), and they also remarkably contribute to understanding of processes and effects in other fields, like, e.g., the blood flow (aneurysms in large arteries, artificial heart valves, etc.). Mathematical theory of the fluid–structure interaction has been rapidly developed, especially in recent years. To simplify the models, many authors neglect the influence of the fluid on the boundary and assume that the motion of the boundary is a priori known. An important special case is a model of a fluid flow around a compact rigid body, which moves in the fluid with constant, known translational and rotational velocities.

Systematic and rigorous mathematical studies of flows around translating bodies were initiated by pioneering works of C. W. Oseen and J. Leray in the 20s and 30s of the twentieth century, and they were later followed by a large number of other mathematicians and physicists. A lot of effort has been devoted to qualitative analysis of corresponding mathematical models, studying problems in the whole space or in an exterior domain, seeking solutions in various function spaces (with or without weights), etc. Important results in this field are connected with the names of J. Leray, J. Serrin, R. Finn, J. Heywood, O. A. Ladyzhenskaya, V. A. Solonnikov, G. P. Galdi, C. Simader, H. Sohr, H. Kozono, Y. Shibata, V. Šverák, R. Farwig, and many others. Rotation of the body, its expression in mathematical models, and effect on the qualitative theory have been intensively studied since the last decades of the twentieth century.[1]

This book is devoted to qualitative analysis of mathematical models of motion of a viscous Newtonian incompressible fluid around a compact rigid body $\mathcal{B}$, translating and rotating in the fluid with the given time-independent translational and angular velocities $k\mathbf{e}_3$ and $\boldsymbol{\omega} \equiv \omega\mathbf{e}_3$, respectively, where $\mathbf{e}_3$ is the unit vector in the direction of the x_3-axis and $k, \omega \in \mathbb{R}$. Thus, the body translates along the x_3-axis and rotates about the same axis. Throughout the book, we assume that $\omega \neq 0$; otherwise the motion of body $\mathcal{B}$ would be purely translational.

Before we proceed with the derivation of the corresponding mathematical model, let us note that the study of motion of a body in a fluid, or vice versa, flow of a fluid around a fixed or moving body, is generally much wider discipline than just the study of a flow past a rotating body. The motion is significantly influenced by the type of the fluid. (For example, the body can execute, under the same acting force, a completely different motion in a Newtonian fluid and in a viscoelastic fluid.) Furthermore, the motion of a body in a fluid depends on the way the body is propelled. (The body can, e.g., move due to the self-propulsion so that it generates a certain velocity profile on its surface, or due to an acting external force, or it moves because it appropriately changes its shape.) All these cases are described in greater detail, together with fundamentals of the corresponding theory, in the work [96] by G. P. Galdi (a chapter in a book). Here, the author discusses several interesting examples, like, e.g., the motion of a mechanical fish or a fall of one or more particles in a Newtonian or a more complicated viscous fluid. The mechanical fish is just a cylinder with a tail which flaps to and fro. It moves forward in water due to the motion of the tail. However, repeating the same motion, the fish does not move forward in a very viscous liquid, like, e.g., the corn syrup. As to a particle or a system of particles, falling in a fluid due to the gravity, it is shown in [96] that the final position, the particle (or the system of particles) attains in the fluid relative to the direction of the gravity force, substantially depends not only on the geometry of the particles but also on the type of the fluid. Another field, rapidly developing in the last several decades, is the mathematical studies of mutual interactions of bodies, flowing in the fluid, their interactions with boundaries of the flow field, types of strokes, motion of the bodies and the fluid before or after the strokes, etc. Summarizing these notes, we want to tell the readers that the mathematical theory of movement of bodies in a fluid is a really extensive area in which many problems

[1] Of many works on this theme, we refer, e.g., to the papers [57, 60, 66, 70, 72, 96, 97, 109, 125–127, 147–150, 175–177].

have already been solved, but apparently many more are still open. You will see that the topic of this book, which is the flow of an incompressible Newtonian fluid around a rotating body, enables one to create a relatively consistent mathematical theory, which, despite the number of known results, generates a number of open questions.

Derivation of the Mathematical Model It is well known that the motion of a compact rigid body $\mathcal{B}$ can be decomposed to the rotation and translation. Thus, there exist a unitary 3×3 matrix $Q(t)$ and a vector $\mathbf{a}(t)$ in $\mathbb{R}^3$ so that if $\mathbf{x}$ is a position (at time $t \geq 0$) of a point of body $\mathcal{B}$, whose position at time $t = 0$ was $\mathbf{x}'$, then

$$\mathbf{x} = Q(t)\mathbf{x}' + \mathbf{a}(t). \tag{1.1}$$

Denote by $\mathcal{D}(t)$ the exterior of body $\mathcal{B}$ at time t, and assume that $\mathcal{D}(t)$ is filled with a viscous incompressible fluid.

Note that $Q(t)$ and $\mathbf{x}'$ can be formally identified with a second order and a first order tensor, respectively. Then $Q(t)\mathbf{x}'$ denotes the inner tensorial product of $Q(t)$ and $\mathbf{x}'$. Similarly, for example, $\dot{Q}(t)\, Q(t)^{-1}$ is the inner product of two second order tensors, etc.

Motion of the fluid in the space-time region $\{(\mathbf{x}, t) \in \mathbb{R}^3 \times I;\ \mathbf{x} \in \mathcal{D}(t);\ t \in I\}$, where I is an interval on the time-axis and $\mathbf{x} \equiv (x_1, x_2, x_3)$, is described by the system of the Navier–Stokes equations

$$\partial_t \mathbf{v} + \mathbf{v} \cdot \nabla \mathbf{v} = -\nabla p + \nu \Delta \mathbf{v} + \mathbf{f}, \tag{1.2}$$

$$\operatorname{div} \mathbf{v} = 0, \tag{1.3}$$

where $\mathbf{v}$ is the velocity field, p is the pressure, and $\mathbf{f}$ represents an external specific body force, acting on the fluid. Equation (1.2) expresses the law of conservation of momentum, and Eq. (1.3) (the so-called equation of continuity) expresses the conservation of mass. Parameter ν is the kinematic coefficient of viscosity. We suppose that ν is a positive constant.

We assume that the first reference frame for the description of motion of body $\mathcal{B}$, and the fluid is chosen so that the fluid is at rest in infinity from the point of view of the observer. This leads to the condition

$$\mathbf{v}(\mathbf{x}, t) \to \mathbf{0} \quad \text{for } |\mathbf{x}| \to \infty. \tag{1.4}$$

The assumption on adherence of the fluid to body $\mathcal{B}$ on the surface of $\mathcal{B}$ and formula (1.1) provide the boundary condition

$$\mathbf{v}(\mathbf{x}, t) = \dot{Q}(t)\, Q(t)^{-1}[\mathbf{x} - \mathbf{a}(t)] + \dot{\mathbf{a}}(t) \quad \text{for } \mathbf{x} \in \partial\mathcal{D}(t),\ t \in I. \tag{1.5}$$

The disadvantage of this description is the variability of the spatial domain $\mathcal{D}(t)$, filled by the moving fluid, in dependence on time. This is why we transform the considered problem to the fixed domain $\mathcal{D}(0)$, i.e., to the exterior of body $\mathcal{B}$ at time $t = 0$. Inverting formula (1.1), we obtain

$$\mathbf{x}' = \eta_t(\mathbf{x}) := Q(t)^{-1}[\mathbf{x} - \mathbf{a}(t)]. \tag{1.6}$$

η_t is a volume-preserving diffeomorphism of $\mathbb{R}^3$ onto itself. Particularly, η_t is a volume-preserving diffeomorphism of $\mathcal{D}(t)$ onto $\mathcal{D}(0)$, and η_t^{-1} is a volume-preserving diffeomorphism of $\mathcal{D}(0)$ onto $\mathcal{D}(t)$.

The velocity field $\mathbf{v}$ in domain $\mathcal{D}(t)$ transforms to the velocity $\mathbf{v}'$ in domain $\mathcal{D}(0)$ in accordance with the general rule on transformation of a time-dependent divergence-free vector field under the action of a time-dependent volume-preserving diffeomorphism η_t:

$$\mathbf{v}(\mathbf{x}, t) \mapsto \mathbf{v}'(\mathbf{x}', t) = \frac{\mathrm{d}}{\mathrm{d}\vartheta} \eta_{t+\vartheta}(\mathbf{X}_\vartheta(\mathbf{x}))\big|_{\vartheta=0}, \tag{1.7}$$

where $\mathbf{x}$ and $\mathbf{x}'$ are coupled through (1.6) and $\mathbf{X}_\vartheta$ is a solution of the initial-value problem

$$\frac{\mathrm{d}}{\mathrm{d}\vartheta}\mathbf{X}_\vartheta(\mathbf{x}) = \mathbf{v}(\mathbf{X}_\vartheta(\mathbf{x}),\ t+\vartheta), \quad \mathbf{X}_\vartheta(\mathbf{x})\big|_{\vartheta=0} = \mathbf{x}. \tag{1.8}$$

Thus, denoting by dot the derivative with respect to t, we calculate that

$$\mathbf{v}'(\mathbf{x}', t) = \dot{\eta}_t(\mathbf{x}) + \nabla\eta_t(\mathbf{x})\mathbf{v}(\mathbf{x}, t). \tag{1.9}$$

Expressing η_t and η_t^{-1} from (1.6), we obtain

$$\begin{aligned} \dot{\eta}_t(\mathbf{x}) &= \dot{\eta}_t\left(\eta_t^{-1}(\mathbf{x}')\right) = \dot{Q}(t)^{-1} Q(t)\mathbf{x}' - Q(t)^{-1}\dot{\mathbf{a}}(t), \\ \nabla\eta_t(\mathbf{x})\mathbf{v}(\mathbf{x}, t) &= \nabla\eta_t(\eta_t^{-1}(\mathbf{x}'))\mathbf{v}(\mathbf{x}, t) = Q(t)^{-1}\mathbf{v}(\mathbf{x}, t). \end{aligned}$$

Substituting this to (1.9), we get

$$\mathbf{v}'(\mathbf{x}', t) = \dot{Q}(t)^{-1}Q(t)\mathbf{x}' - Q(t)^{-1}\dot{\mathbf{a}}(t) + Q(t)^{-1}\mathbf{v}(\mathbf{x}, t). \tag{1.10}$$

The terms $\dot{Q}(t)^{-1}Q(t)\cdot\mathbf{x}'$ and $Q(t)^{-1}\dot{\mathbf{a}}(t)$ represent the contributions coming purely from the rotation and translation, respectively. The disadvantage of transformation (1.10) is that although $\mathbf{v}$ tends to zero for $|\mathbf{x}| \to \infty$ (see condition (1.4)), $\mathbf{v}'$ does not behave in the same way for $|\mathbf{x}'| \to \infty$ due to the presence of the terms $\dot{Q}(t)^{-1}Q(t)\mathbf{x}'$ and $-Q(t)^{-1}\dot{\mathbf{a}}(t)$ on the right-hand side of (1.10). It is therefore useful to subtract both the terms $\dot{Q}(t)^{-1}Q(t)\mathbf{x}'$ and $-Q(t)^{-1}\dot{\mathbf{a}}(t)$ (which are obviously divergence-free) from $\mathbf{v}'$ and to consider the transformed velocity in the form

$$\begin{aligned}\widetilde{\mathbf{v}}'(\mathbf{x}',t) &:= \mathbf{v}'(\mathbf{x}',t) - \dot{Q}(t)^{-1}Q(t)\mathbf{x}' + Q(t)^{-1}\dot{\mathbf{a}}(t)\\ &= Q(t)^{-1}\mathbf{v}(\mathbf{x},t) \;=\; Q(t)^{-1}\mathbf{v}\big(Q(t)\mathbf{x}' + \mathbf{a}(t),t\big).\end{aligned} \tag{1.11}$$

Inverting this formula, we get

$$\mathbf{v}(\mathbf{x},t) = Q(t)\widetilde{\mathbf{v}}'(\mathbf{x}',t). \tag{1.12}$$

By analogy, we define

$$\begin{aligned}p'(\mathbf{x}',t) &:= p\big(Q(t)\mathbf{x}' + \mathbf{a}(t),\, t\big) = p(\mathbf{x},t),\\ \mathbf{f}'(\mathbf{x}',t) &:= Q(t)^{-1}\,\mathbf{f}\big(Q(t)\mathbf{x}' + \mathbf{a}(t),\, t\big) = Q(t)^{-1}\,\mathbf{f}(\mathbf{x},t)\end{aligned}$$

for $\mathbf{x} \in \mathcal{D}(t)$ and $\mathbf{x}' = \eta_t(\mathbf{x}) \in \mathcal{D}(0)$. Reversing these formulas, substituting to the Navier–Stokes system (1.2), (1.3), and multiplying the first equation from the left by the matrix $Q(t)^{-1}$, we obtain

$$\begin{aligned}\partial_t\widetilde{\mathbf{v}}' + Q(t)^{-1}\,\dot{Q}(t)\widetilde{\mathbf{v}}' + \dot{Q}(t)^{-1}\,Q(t)\mathbf{x}'\cdot\nabla'\widetilde{\mathbf{v}}' - Q(t)^{-1}\dot{\mathbf{a}}(t)\cdot\nabla'\widetilde{\mathbf{v}}' + \widetilde{\mathbf{v}}'\cdot\nabla\widetilde{\mathbf{v}}'&\\ = -\nabla' p' + \nu\Delta'\widetilde{\mathbf{v}}',& \qquad (1.13)\\ \operatorname{div}'\widetilde{\mathbf{v}}' = 0.& \qquad (1.14)\end{aligned}$$

Since $Q(t)^{-1}\,Q(t)$ is the identity matrix and its derivative with respect to t therefore equals the zero matrix, we deduce that $\dot{Q}(t)^{-1}\,Q(t) = -Q(t)^{-1}\,\dot{Q}(t)$. Using this formula and the fact that $Q(t)$ is a unitary matrix, one can verify that the matrix $Q(t)^{-1}\,\dot{Q}(t)$ is skew-symmetric. Hence there exist numbers $\omega_1(t)$, $\omega_2(t)$, $\omega_3(t)$, such that

$$Q(t)^{-1}\,\dot{Q}(t) = \begin{pmatrix} 0, & -\omega_3(t), & \omega_2(t)\\ \omega_3(t), & 0, & -\omega_1(t)\\ -\omega_2(t), & \omega_1(t), & 0\end{pmatrix}. \tag{1.15}$$

Consequently, denoting $\boldsymbol{\omega}(t) := \big(\omega_1(t),\, \omega_2(t),\, \omega_3(t)\big)$, we obtain the formulas $Q(t)^{-1}\,\dot{Q}(t)\mathbf{v}' = \boldsymbol{\omega}(t)\wedge\mathbf{v}'$ and $\dot{Q}(t)^{-1}\,Q(t)\mathbf{x}'\cdot\nabla'\mathbf{v}' = -\big(\boldsymbol{\omega}(t)\wedge\mathbf{x}'\big)\cdot\nabla'\mathbf{v}'$. Thus, Eq. (1.13) takes the form

$$\begin{aligned}\partial_t\widetilde{\mathbf{v}}' + \boldsymbol{\omega}(t)\wedge\widetilde{\mathbf{v}}' - \big(\boldsymbol{\omega}(t)\wedge\mathbf{x}'\big)\cdot\nabla'\widetilde{\mathbf{v}}' - Q(t)^{-1}\dot{\mathbf{a}}(t)\cdot\nabla'\widetilde{\mathbf{v}}' + \widetilde{\mathbf{v}}'\cdot\nabla\mathbf{v}'&\\ = -\nabla' p' + \nu\Delta'\widetilde{\mathbf{v}}' + \mathbf{f}'.&\end{aligned} \tag{1.16}$$

Assuming further that $\boldsymbol{\omega}(t)$ and $Q(t)^{-1}\dot{\mathbf{a}}(t)$ are time-independent vectors, Eq. (1.16) becomes autonomous. One can choose the system of coordinates so that $\boldsymbol{\omega}(t) \equiv \boldsymbol{\omega} = (0, 0, \omega)^T = \omega\mathbf{e}_3$. Using (1.15) and the condition that $Q(t)$ equals the identity matrix for $t = 0$, we calculate that

$$Q(t) = \begin{pmatrix} \cos\omega t, & \sin\omega t, & 0 \\ -\sin\omega t, & \cos\omega t, & 0 \\ 0, & 0, & 1 \end{pmatrix}. \tag{1.17}$$

(In order to stress the dependence on ω, we also often use the notation $Q_\omega(t)$ instead of $Q(t)$ throughout the text.)

Denoting the constant vector $Q(t)^{-1}\dot{\mathbf{a}}(t)$ by $(k_1, k_2, k_3)^T$ and using the form (1.17) of $Q(t)$, we get

$$\dot{\mathbf{a}}(t) = \begin{pmatrix} k_1 \cos\omega t + k_2 \sin\omega t \\ -k_1 \sin\omega t + k_2 \cos\omega t \\ k_3 \end{pmatrix}^T.$$

Assuming at first that $\omega \neq 0$ and using the initial condition $\mathbf{a}(0) = \mathbf{0}$, we obtain

$$\begin{aligned} \mathbf{a}(t) &= \begin{pmatrix} (k_1/\omega) \sin\omega t - (k_2/\omega) \cos\omega t + k_2/\omega \\ (k_1/\omega) \cos\omega t + (k_2/\omega) \sin\omega t - k_1/\omega \\ k_3 t \end{pmatrix}^T \\ &= \big(k_2/\omega, -k_1/\omega, k_3 t\big)^T + Q(t) \cdot \big(-k_2/\omega, k_1/\omega, 0\big). \end{aligned}$$

Substituting this into (1.1), we obtain

$$\mathbf{x} = Q(t) \cdot \big[\mathbf{x}' - \big(k_2/\omega, -k_1/\omega, 0\big)\big] + \big(k_2/\omega, -k_1/\omega, k_3 t\big). \tag{1.18}$$

Thus, the body rotates about the line $x_1' = k_2/\omega,\ x_2' = -k_1/\omega$ and moves along this line with the translational velocity $(0, 0, k_3)$. We may assume without loss of generality that $k_1 = k_2 = 0$; otherwise we shift the system of coordinates. We may also write, for simplicity, just k instead of k_3. Thus, $\mathbf{a}(t) = (0, 0, kt)$, $\dot{\mathbf{a}}(t) = (0, 0, k)$, and (1.18) yields

$$\mathbf{x} = Q(t)\mathbf{x}' + kt\,\mathbf{e}_3. \tag{1.19}$$

If $\omega = 0$, then $Q(t)$ reduces to the identity matrix, which means that body $\mathcal{B}$ moves just with the translational velocity $\dot{\mathbf{a}}(t) = (0, 0, k)$.

Substituting now for $\boldsymbol{\omega}(t)$, $\mathbb{Q}(t)$, and $\mathbf{a}(t)$ in Eq. (1.16) and writing just $\boldsymbol{\omega}$ instead of $\boldsymbol{\omega}(t)$, we obtain

$$\begin{aligned}\partial_t\widetilde{\mathbf{v}}' + \boldsymbol{\omega}\wedge\widetilde{\mathbf{v}}' - (\boldsymbol{\omega}\wedge\mathbf{x}')\cdot\nabla'\widetilde{\mathbf{v}}' - k\,\partial_3'\widetilde{\mathbf{v}}' + \widetilde{\mathbf{v}}'\cdot\nabla\mathbf{v}' \\ = -\nabla' p' + \nu\Delta'\widetilde{\mathbf{v}}' + \mathbf{f}'.\end{aligned} \tag{1.20}$$

The boundary condition (1.5) transforms to

$$\widetilde{\mathbf{v}}'(\mathbf{x},t) = \boldsymbol{\omega}\wedge\mathbf{x}' + k\,\mathbf{e}_3 \qquad \text{for } \mathbf{x}\in\partial\mathcal{D}(t),\ \ t\in I, \tag{1.21}$$

and condition (1.4) leads to

$$\widetilde{\mathbf{v}}'(\mathbf{x}',t) \to \mathbf{0} \quad \text{for } |\mathbf{x}'|\to\infty. \tag{1.22}$$

In order to preserve a simple notation, we further omit the primes and tildes everywhere in the equations and conditions (1.14), (1.20)–(1.22), and we write only $\mathcal{D}$ instead of $\mathcal{D}(0)$. Thus, the spatial variable and the transformed velocity, pressure, and specific body force are further denoted only by $\mathbf{x}$, $\mathbf{v}$, p, and $\mathbf{f}$ instead of $\mathbf{x}'$, $\widetilde{\mathbf{v}}'$, p', and $\mathbf{f}'$. The system (1.20), (1.14) thus takes the simpler form

$$\partial_t\mathbf{v} + \boldsymbol{\omega}\wedge\mathbf{v} - (\boldsymbol{\omega}\wedge\mathbf{x} + k\mathbf{e}_3)\cdot\nabla\mathbf{v} + \mathbf{v}\cdot\nabla\mathbf{v} = -\nabla p + \nu\Delta\mathbf{v} + \mathbf{f}, \tag{1.23}$$

$$\operatorname{div}\mathbf{v} = 0 \tag{1.24}$$

for $(\mathbf{x},t)\in\mathcal{D}\times I$.

Note that our assumptions do not exclude the special case when $\mathcal{B} = \emptyset$, and consequently $\mathcal{D}(t) = \mathcal{D}(0) = \mathcal{D} = \mathbb{R}^3$ for all $t \geq 0$.

Unlike the momentum equation (1.2), Eq. (1.23) contains new terms: $\boldsymbol{\omega}\wedge\mathbf{v}$ (the Coriolis force) and $(\boldsymbol{\omega}\wedge\mathbf{x} + k\mathbf{e}_3)\cdot\nabla\mathbf{v}$. Especially the term $(\boldsymbol{\omega}\wedge\mathbf{x})\cdot\nabla\mathbf{v}$ plays an important role in the theory of Eqs. (1.23) and (1.24): It is not subordinate to $\nu\Delta\mathbf{v}$ due to the presence of $\mathbf{x}$, which is not bounded in $\mathcal{D}$, and it therefore substantially changes the character and properties of Eqs. (1.23) and (1.24) in comparison with (1.3) and (1.2).

By analogy with the theory of the Navier–Stokes system (1.2), (1.3), the important part of the mathematical theory of the system (1.23), (1.24) consists of studies of linearized models. We successively consider four possible linear models in the next chapters. In particular, we have in mind these models (= problems):

- **The Stokes type problem in the whole $\mathbb{R}^3$:**

$$\partial_t\mathbf{v} + \boldsymbol{\omega}\wedge\mathbf{v} - (\boldsymbol{\omega}\wedge\mathbf{x})\cdot\nabla\mathbf{v} = -\nabla p + \nu\Delta\mathbf{v} + \mathbf{f} \quad \text{in } \mathbb{R}^3\times I, \tag{1.25}$$

$$\operatorname{div}\mathbf{v} = 0 \quad \text{in } \mathbb{R}^3\times I, \tag{1.26}$$

$$\mathbf{v}\to\mathbf{0} \quad \text{for } |\mathbf{x}|\to\infty \tag{1.27}$$

- **The Stokes type problem in the exterior domain $\mathcal{D}$:**

$$\partial_t \mathbf{v} + \boldsymbol{\omega} \wedge \mathbf{v} - (\boldsymbol{\omega} \wedge \mathbf{x}) \cdot \nabla \mathbf{v} = -\nabla p + \nu \Delta \mathbf{v} + \mathbf{f} \quad \text{in } \mathcal{D} \times I, \tag{1.28}$$

$$\operatorname{div} \mathbf{v} = 0 \quad \text{in } \mathcal{D} \times I, \tag{1.29}$$

$$\mathbf{v} = \boldsymbol{\omega} \wedge \mathbf{x} + k\mathbf{e}_3 \quad \text{for } \mathbf{x} \in \partial\mathcal{D}, \tag{1.30}$$

$$\mathbf{v} \to \mathbf{0} \quad \text{for } |\mathbf{x}| \to \infty \tag{1.31}$$

- **The Oseen type problem in the whole $\mathbb{R}^3$:**

$$\partial_t \mathbf{v} + \boldsymbol{\omega} \wedge \mathbf{v} - (\boldsymbol{\omega} \wedge \mathbf{x} + k\mathbf{e}_3) \cdot \nabla \mathbf{v} = -\nabla p + \nu \Delta \mathbf{v} + \mathbf{f} \quad \text{in } \mathbb{R}^3 \times I, \tag{1.32}$$

$$\operatorname{div} \mathbf{v} = 0 \quad \text{in } \mathbb{R}^3 \times I, \tag{1.33}$$

$$\mathbf{v} \to \mathbf{0} \quad \text{for } |\mathbf{x}| \to \infty \tag{1.34}$$

- **The Oseen type problem in the exterior domain $\mathcal{D}$:**

$$\partial_t \mathbf{v} + \boldsymbol{\omega} \wedge \mathbf{v} - (\boldsymbol{\omega} \wedge \mathbf{x} + k\mathbf{e}_3) \cdot \nabla \mathbf{v} = -\nabla p + \nu \Delta \mathbf{v} + \mathbf{f} \quad \text{in } \mathcal{D} \times I, \tag{1.35}$$

$$\operatorname{div} \mathbf{v} = 0 \quad \text{in } \mathcal{D} \times I, \tag{1.36}$$

$$\mathbf{v} = \boldsymbol{\omega} \wedge \mathbf{x} + k\mathbf{e}_3 \quad \text{for } \mathbf{x} \in \partial\mathcal{D}, \tag{1.37}$$

$$\mathbf{v} \to \mathbf{0} \quad \text{for } |\mathbf{x}| \to \infty \tag{1.38}$$

- **The Navier–Stokes type problem in the whole $\mathbb{R}^3$:**

$$\partial_t \mathbf{v} + \boldsymbol{\omega} \wedge \mathbf{v} - (\boldsymbol{\omega} \wedge \mathbf{x} + k\mathbf{e}_3) \cdot \nabla \mathbf{v} + \mathbf{v} \cdot \nabla \mathbf{v} = -\nabla p + \nu \Delta \mathbf{v} + \mathbf{f} \quad \text{in } \mathbb{R}^3 \times I, \tag{1.39}$$

$$\operatorname{div} \mathbf{v} = 0 \quad \text{in } \mathbb{R}^3 \times I, \tag{1.40}$$

$$\mathbf{v} \to \mathbf{0} \quad \text{for } |\mathbf{x}| \to \infty \tag{1.41}$$

- **The Navier–Stokes type problem in the exterior domain $\mathcal{D}$:**

$$\partial_t \mathbf{v} + \boldsymbol{\omega} \wedge \mathbf{v} - (\boldsymbol{\omega} \wedge \mathbf{x} + k\mathbf{e}_3) \cdot \nabla \mathbf{v} + \mathbf{v} \cdot \nabla \mathbf{v} = -\nabla p + \nu \Delta \mathbf{v} + \mathbf{f} \quad \text{in } \mathcal{D} \times I, \tag{1.42}$$

$$\operatorname{div} \mathbf{v} = 0 \quad \text{in } \mathcal{D} \times I, \tag{1.43}$$

$$\mathbf{v} = \boldsymbol{\omega} \wedge \mathbf{x} + k\mathbf{e}_3 \qquad \text{for } \mathbf{x} \in \partial\mathcal{D}, \tag{1.44}$$

$$\mathbf{v} \to \mathbf{0} \qquad \text{for } |\mathbf{x}| \to \infty \tag{1.45}$$

All these problems can also be considered with the more general equation

$$\operatorname{div} \mathbf{v} = g \quad \text{in } \mathcal{D} \times I \tag{1.46}$$

instead of the simple condition $\operatorname{div} \mathbf{v} = 0$. Studies of all these problems are also interesting and useful in the two-dimensional case, i.e., either in $\mathbb{R}^2$ or in an exterior domain $\mathcal{D}$ in $\mathbb{R}^2$. (Here, however, the terms $\boldsymbol{\omega} \wedge \mathbf{x}$ and $\boldsymbol{\omega} \wedge \mathbf{v}$ must be replaced by $\omega\mathbf{x}^{\perp} \equiv \omega(-x_2, x_1)$ and $\omega\mathbf{v}^{\perp} \equiv \omega(-v_2, v_1)$.) Steady-state variants of all the problems are also of special use and interest. Note that the steady-state solution of any of the aforementioned problems in fact corresponds to a time-periodic solution of a certain corresponding problem in the original Cartesian coordinates, considered at the beginning before the transformations.

It should also be noted that:

- In addition to the transformation (1.6), (1.10), the flow around a rotating body has been studied by means of a certain local transformation, introduced by Inoue and Wakimoto in [135] (see Chap. 12 or [26, 47, 207, 208] for further details), and in a more general setting in time-dependent domains in [166, 180, 181],
- The fact that the axis of rotation is parallel to the direction of the translational motion of body $\mathcal{B}$ naturally follows from the requirement that the transformed momentum equation (1.16) be autonomous. This means, on the other hand, that if the axis of rotation is not parallel to the translational motion of $\mathcal{B}$, then Eq. (1.16) is not autonomous. This case is discussed in the Appendix. More details on this theme can also be found in [65].
- It is sometimes advantageous to study the equations of motion in the so-called dimensionless form.

  For example, the dimensionless form of the Navier–Stokes type problem in the case of a pure rotation of the body (i.e., the case when $k = 0$) is

$$\left.\begin{aligned} \mathrm{Re}(\mathbf{v} \cdot \nabla\mathbf{v} - \boldsymbol{\mu} \wedge \mathbf{x} \cdot \nabla\mathbf{v} + \boldsymbol{\mu} \wedge \mathbf{v}) &= \Delta\mathbf{v} - \nabla p \\ \operatorname{div} \mathbf{v} &= 0 \\ \lim_{|\mathbf{x}| \to \infty} \mathbf{v}(\mathbf{x}) &= 0 \end{aligned}\right\} \text{ in } \mathcal{D} \tag{1.47}$$

$$\mathbf{v}(\mathbf{x}) = \boldsymbol{\mu} \wedge \mathbf{x}, \ \ \mathbf{x} \in \partial\mathcal{D}, \tag{1.48}$$

  where $\mathrm{Re} = |\boldsymbol{\omega}|\, d(\mathcal{B})/\nu$ is the appropriate Reynolds number, $d(\mathcal{B})$ is the diameter of body $\mathcal{B}$, $\boldsymbol{\mu} = \boldsymbol{\omega}/|\boldsymbol{\omega}|$. (See [96].) Note that the dimensionless form is also often defined in another way. This concerns, e.g., the case with a nonzero translation (i.e., $k \neq 0$), where an appropriate change of coordinates leads to this dimensionless form of the steady Oseen type problem:

$$\begin{aligned} \Delta \mathbf{v} + \tau \frac{\partial \mathbf{v}}{\partial x_3} + \sqrt{\rho}\,(\boldsymbol{e}_3 \wedge \mathbf{x} \cdot \mathbf{v} - \boldsymbol{e}_3 \wedge \mathbf{v}) &= -\nabla p + \tau \mathbf{f} && \text{in } \mathcal{D}, \\ \operatorname{div} \mathbf{v} &= 0 && \text{in } \mathcal{D}, \\ \mathbf{v} &= \rho \boldsymbol{e}_3 \wedge \mathbf{x} + \tau \mathbf{e}_3 && \text{for } \mathbf{x} \in \partial\mathcal{D}, \\ \mathbf{v} &\to \mathbf{0} && \text{for } |\mathbf{x}| \to \infty, \end{aligned} \tag{1.49}$$

where the dimensionless parameters ρ and τ are defined as follows: $\rho := |\boldsymbol{\omega}|^2 \, d(\mathcal{B})^4/\nu^2$, $\tau := k \, |\boldsymbol{\omega}| \, d(\mathcal{B})/\nu$, see [98]. (Note that ρ and τ are called the Taylor number and the Reynolds number, respectively. However, this Reynolds number differs from number Re, used in (1.47).) By analogy, the dimensionless form of the full Navier–Stokes type problem (with nonzero rotation and translation) is

$$\begin{aligned} \Delta \mathbf{v} + \tau \frac{\partial \mathbf{v}}{\partial x_3} + \tau \cdot (\mathbf{v} \cdot \nabla)\mathbf{v} + \sqrt{\rho}\,(\boldsymbol{e}_3 \wedge \mathbf{x} \cdot \mathbf{v} - \boldsymbol{e}_3 \wedge \mathbf{v}) &= -\,\nabla p + \tau \mathbf{f} \\ & \qquad \text{in } \mathcal{D} \\ \operatorname{div} \mathbf{v} &= 0 \quad \text{in } \mathcal{D}, \\ \mathbf{v} &\to \mathbf{0} \quad \text{for } |\mathbf{x}| \to \infty \end{aligned} \tag{1.50}$$

$$\mathbf{v} = \rho \boldsymbol{e}_3 \wedge \mathbf{x} + \tau \mathbf{e}_3. \tag{1.51}$$

By analogy with theory of the Navier–Stokes and related equations, qualitative results on the linearized equations (i.e., the Oseen type or Stokes type equations) play a fundamental role in the theory of the full nonlinear system of equations, describing the fluid flow around a rotating body. This is why we pay a lot of attention to the linearized problems in this book. We present and explain four different approaches (methods), which can all be used to obtain results on existence, uniqueness, asymptotic behavior, and possibly also other properties of solutions of the linearized problems.

- The first method, introduced by Hishida, Farwig, and Müller, is based on the *Paley–Littlewood decomposition*, respectively, the theory of the Paley–Littlewood decomposition with Muckenhoupt weights, which is developed in the book.
- The second approach, which was developed by Galdi and Kyed, considers steady equations in the rotating frame, uses the transformation back to the external observer's frame, and afterward applies methods and tools from the theory of time-periodic solutions.

- The third method, based on the *variational approach,* originally introduced for the Oseen problem by Kračmar and Penel, is applied to the Oseen type problem and uses the L^2 Muckenhoupt-weighted spaces.
- The last approach, introduced by Deuring, Kračmar, and Nečasová, uses the fundamental solution, constructed by Thomann and Guenther for the unsteady Oseen problem. Applying this approach, we obtain the fundamental solution for the Oseen type problem (i.e., for the Oseen problem with terms, caused by the rotation), and from this we further derive the fundamental solution for the nonlinear Navier–Stokes problem with "rotating" terms. This fundamental solution is used in Chaps. 9 and 10 in order to obtain new information on the asymptotic behavior of a solution, which is one of the main new results, presented in the book.

The book is divided into three parts:

Part I has two chapters. The basic notation and mathematical description of studied problems is introduced in Chap. 1. A series of basic lemmas are recalled in Chap. 2.

Part II is devoted to the linear Stokes type and Oseen type of problems.

Chapter 3 deals with the theory of the steady Stokes type problem. We present the L^q-theory for general $q \in (1, \infty)$. The explanation begins with the problem in the whole $\mathbb{R}^3$. Both strong and weak solutions are considered. In order to get appropriate a priori L^q-estimates, we apply the Paley–Littlewood theory and also give an alternative proof based on the known form of the fundamental solution of the corresponding unsteady Stokes type problem. Then we show the existence of weak and strong solutions in the exterior domain, applying the localization technique. We also investigate the existence of very weak solutions. A weighted space approach (with Muckenhoupt's weights) is also involved.[2]

In Chap. 4 we investigate the theory of the Oseen type problem. We begin with the study of the existence of a strong solution in the whole space $\mathbb{R}^3$, and we give existence in L^q setting. We again apply the Paley–Littlewood theory in order to obtain a priori estimates in the L^q-spaces. By analogy with Chap. 3, we also present an alternative proof based on the fundamental solution of the unsteady Oseen type problem. Then we study the weak solution of the problem in the whole $\mathbb{R}^3$ and in the exterior domain as well. The solution in the weighted L^2—and L^q—spaces (with the one-sided Muckenhoupt weights) is also considered for the strong and weak sense. Finally the very weak solution is considered. Moreover, the very weak solution in approach of Kim is also studied.[3]

Chapter 5 is devoted to derivation of the representation formula of the steady Oseen type problem. As a first step, a truncated exterior domain $B_R \backslash \mathcal{D}^c$ is considered. Then, considering the limit transition $R \to \infty$, we derive the representation formula for the case of an exterior domain. Volume potentials, which arise in the

[2] The related literature: [62, 67, 68, 68, 70, 74, 75, 102, 126, 127, 152, 177], etc.

[3] See [57, 62, 72, 76, 102, 147, 149, 153, 175, 176, 194, 195], etc.

problem, and the asymptotic velocity profile are studied in next sections. Using the representation formula for the Oseen type problem, an analogous formula is also derived for the Navier–Stokes type problem, see Chap. 11. Here, the asymptotic profile of the gradient of the velocity field and the second derivatives of the velocity are also investigated. Finally, we study the so-called Leray solution (representation formula and asymptotic profile).[4]

Chapter 6 deals with the Oseen type problem in the exterior domain $\mathcal{D}$. One of the goals is to obtain a pointwise estimate of the pressure. Furthermore, we study the Oseen type problem in a truncated domain $\mathcal{D}_R := B_R \cap \overline{\mathcal{D}}$ under certain artificial boundary conditions on the truncated boundary ∂B_R. In order to obtain a truncation error estimate, we compare this solution with the solution in the whole exterior domain $\overline{\mathcal{D}}$.

The variational approach to the Oseen type problem in the L^2-space with a Muckenhoupt weight is a subject of Chap. 7. We focus on the problem in an exterior domain $\mathcal{D}$ in $\mathbb{R}^3$ and on the whole space case $\mathcal{D} = \mathbb{R}^3$.

Chapter 8 deals with studies of basic functional analytic properties of a linear operator, associated with the Stokes type and Oseen type problems (closedness, generation of a semigroup, spectrum, resolvent estimates, etc.) in the whole space and in an exterior domain.

Part III is devoted to the nonlinear Navier–Stokes type problem, describing a stationary viscous incompressible flow around a translating and rotating body.

Chapter 9 gives a brief introduction to the existence theory of Navier–Stokes type problem in the case or pure rotation and in the case when rotation and translation is considered.

The goal of Chap. 10 is to derive optimal rates of decay (for $|\mathbf{x}| \to \infty$) for the velocity and its gradient on the basis of a representation formula, involving a fundamental solution constructed by R. B. Guenther, E. A. Thomann, JMFM 8 (2006), 77–98, for the corresponding linearized system and further developed in the work by Deuring et al. [35–38, 40], see also Chap. 5.

Asymptotic properties of solutions in infinity are studies in Chap. 11. Before going into detail, it is given a brief historical introduction to this problem to point out the fundamental contributions of *R. Finn, K. I. Babenko, and G. P. Galdi* in this theory. Finally, the contribution by Deuring, Kračmar, and Nečasová is explained. It is shown that the velocity can be split into a constant vector, multiplied by the first column of the fundamental solution of the Oseen system, plus a remainder, decaying pointwise near infinity at a rate which is higher than the decay rate of the Oseen tensor. Moreover, we derive optimal rates of decay for velocity, gradient of velocity, and pressure.[5]

---

[4] Results from this chapter can also be found in Lecture Notes, [145]. See also results [38–43, 67, 68, 100, 160, 162], etc.

[5] Here, we also discuss differences between our approach and that of Galdi and Kyed. More details on these themes can be found in [39–41, 63, 66, 67, 69, 97, 100, 102, 108–110, 125, 163, 193].

Chapter 12 gives a brief introduction to the local and global existence theory of nonsteady Navier–Stokes type problem. Moreover, the convergence to the steady case is shown.

Finally, in the Appendix we describe the case when rotation is not parallel to the velocity at infinity. Since the approach by Deuring, Kračmar, and Nečasová used estimates of the Oseen kernels and singular integrals from the works of Kračmar, Novotný, and Pokorný [119, 155], we also included to the Appendix.

Chapter 2
Preliminaries

2.1 Basic Notation and Weighted Function Spaces

First of all we recall the notation of some function spaces, together with some properties that will be needed in the next chapters. The vector functions and the spaces of vector functions are denoted by boldface letters.

Let $\mathcal{D}$ be a domain in $\mathbb{R}^n$.

- The linear space $C_0^\infty(\mathcal{D})$ consists of all infinitely differentiable functions that have compact supports in $\mathcal{D}$. For $q \in [1, \infty)$, the *Lebesgue space* $L^q(\mathcal{D})$ is the normed linear space of all (equivalence classes of) real Lebesgue-measurable functions u defined on $\mathcal{D}$ such that

$$\|u\|_q := \left(\int_{\mathcal{D}} |u|^q \right)^{1/q} < \infty. \tag{2.1}$$

- For $q \in [1, \infty)$ and $m \in \mathbb{N}$, the *Sobolev space* $H^{m,q}(\mathcal{D})$ is the space of all $u \in L^q(\mathcal{D})$ such that $D^\alpha u \in L^q(\mathcal{D})$ for all $0 \leq |\alpha| \leq m$. It is a normed linear space with the norm

$$\|u\|_{m,q} := \left(\sum_{0 \leq |\alpha| \leq m} \|D^\alpha u\|_q^q \right)^{1/q}, \tag{2.2}$$

where $\alpha = (\alpha_1, \ldots, \alpha_n) \in \mathbb{N}_0^n$ is the *multi-index* and

$$D^\alpha u := \frac{\partial^{|\alpha|} u}{\partial x_1^{\alpha_1} \ldots \partial x_n^{\alpha_n}}, \quad |\alpha| := \sum_{i=1}^{n} \alpha_i.$$

Š. Nečasová et al., *Mathematical Theory of a Fluid Flow Around a Rotating and Translating Body*, Springer Monographs in Mathematics,
https://doi.org/10.1007/978-3-031-84120-0_2

- $H_0^{m,q}(\mathcal{D})$ is a subspace of $H^{m,q}(\mathcal{D})$, defined as the completion of $C_0^\infty(\mathcal{D})$ in the norm (2.2).
- We define the dual spaces

$$H^{-m,q'}(\mathbb{R}^3) = (H^{m,q}(\mathbb{R}^3))',$$
$$H^{-m,q'}(\mathcal{D}) = (H_0^{m,q}(\mathcal{D}))'$$

with the norms denoted by

$$\|\cdot\|_{-1,q,\mathbb{R}^3} \quad \text{and} \quad \|\cdot\|_{-1,q,\mathcal{D}}, \text{ respectively,}$$

where $1/q + 1/q' = 1$.
- In the case when $\mathcal{D} = \mathbb{R}^n$ the *homogeneous Sobolev space* $\widehat{H}^{k,q}(\mathbb{R}^n)$ is defined as the normed linear space of all $u \in L^1_{\mathrm{loc}}(\mathbb{R}^n)|_{\pi_{k-1}}$ $\partial^\alpha u \in L^q(\mathbb{R}^n)$ for all $\alpha \in \mathbb{N}_0^n$, such that $|\alpha| = k$. Here, $L^1_{\mathrm{loc}}(\mathbb{R}^n)|_{\pi_{k-1}}$ denotes the space of equivalent classes of functions that differ at most by a polynomial of degree less than or equal to $k-1$. (The class of such polynomials is denoted by π_{k-1}. See [93] for more details.) The space $\widehat{H}^{k,q}(\mathbb{R}^n)$ is equipped with the norm

$$|u|_{k,q} := \left(\sum_{|\alpha|=k} \|D^\alpha u\|_q^q \right)^{1/q}.$$

Note that in the case $k = 1$, we have $\widehat{H}^{1,q}(\mathbb{R}^n) = \overline{C_0^\infty(\mathbb{R}^n)}^{\|\nabla\cdot\|_q}$ (the completion of $C_0^\infty(\mathbb{R}^n)$ in the norm $\|\nabla \cdot\|_q$). Thus space can be characterized as

$$\{v \in L^q_{\mathrm{loc}}(\mathbb{R}^3);\ \nabla v \in L^q(\mathbb{R}^3)\}|_{\pi_0}.$$

- In the case $\mathcal{D} \neq \mathbb{R}^n$, we can define the space $\widehat{H}_0^{1,q}(\mathcal{D}) := \overline{C_0^\infty(\mathcal{D})}^{\|\nabla\cdot\|_q}$ with the same norm $|u|_{k,q}$. This space can be identified with $\{v \in L^q_{\mathrm{loc}}(\overline{\mathcal{D}});\ \nabla v \in L^q(\mathcal{D}),\ v|_{\partial\mathcal{D}} = 0\}$.
- The dual spaces

$$\widehat{H}^{-1,q}(\mathbb{R}^3) = (\widehat{H}^{1,q'}(\mathbb{R}^3))',$$
$$\widehat{H}^{-1,q}(\mathcal{D}) = (\widehat{H}_0^{1,q'}(\mathcal{D}))'$$

with norms $|\cdot|_{-1,q,\mathbb{R}^3}$ and $|\cdot|_{-1,q,\mathcal{D}}$, respectively.

Remark 2.1

- Let us mention that in the case of a bounded domain the Sobolev spaces $H^{k,q}(\mathcal{D})$ and the homogeneous Sobolev spaces $\widehat{H}^{k,q}(\mathcal{D})$ coincide.

- Applying Sobolev's imbedding the homogeneous Sobolev spaces $\widehat{H}_0^{1,q}(\mathcal{D})$ can be characterized in the following way:

$$\widehat{H}_0^{1,q}(\mathcal{D}) = \begin{cases} \{v \in L^{3q/(3-q)}(\mathcal{D});\, \nabla v \in L^q(\mathcal{D}),\, v|_{\partial\mathcal{D}} = 0\} & \text{for } 1 < q < 3\,(= n), \\ \{v \in L^q_{\mathrm{loc}}(\overline{\mathcal{D}});\, \nabla v \in L^q(\mathcal{D}),\, v|_{\partial\mathcal{D}} = 0\} & \text{for } 3 \le q < \infty. \end{cases}$$

The spaces $\widehat{H}^{1,q}(\mathbb{R}^3)$ can be characterized by means of the next lemma (see [144, Proposition 2.4]).

Lemma 2.1

- *For* $1 < r < n$, *we have* $\widehat{H}^{1,r}(\mathbb{R}^n) = \{u \in L^s(\mathbb{R}^n)\colon \nabla u \in L^r(\mathbb{R}^n)\}$, *where* $s = 3r/(3-r)$.
- *Let* $r \ge n$. *Suppose* $\{u_k\} \subset C_0^\infty(\mathbb{R}^n)$ *is a Cauchy sequence in the norm of* $\widehat{H}^{1,r}(\mathbb{R}^n)$. *Then there is a sequence* $\{w_k\} \subset C_0^\infty(\mathbb{R}^n)$, *which is a Cauchy sequence in the norm of* $\widehat{H}_0^{1,r}(\mathbb{R}^n)$ *and a function* $u \in L^r_{\mathrm{loc}}(\mathbb{R}^n)$ *with* $\nabla u \in L^r(\mathbb{R}^n)$ *satisfying*

$$\begin{aligned} \|\nabla u_k - \nabla w_k\|_{L^r(\mathbb{R}^n)} \to 0, \\ w_k \to u \quad \text{in } L^r_{\mathrm{loc}}(\mathbb{R}^n), \\ \nabla w_k \to \nabla u \quad \text{in } L^r(\mathbb{R}^n) \quad \text{as } k \to \infty. \end{aligned} \tag{2.3}$$

Such u *is unique up to an additive constant. In this case, we have the inclusion* $\widehat{H}_0^{1,r}(\mathbb{R}^n) \subset \{[u] \in L^r_{\mathrm{loc}}(\mathbb{R}^n)|\pi_0\colon \nabla u \in L^r(\mathbb{R}^n)\}$, *where* $[u] = \{w \in L^r_{\mathrm{loc}}(\mathbb{R}^n)\colon w - u \in \mathbb{R}\}$.

Let $B_r(\mathbf{x})$ be the open ball centered point at $\mathbf{x}$ with radius $r > 0$. For sufficiently large $r > 0$, we set $\mathcal{D}_r = \mathcal{D} \cap B_r$.

- A weight function $w \in L^1_{\mathrm{loc}}(\mathbb{R}^n)$ belongs to the *Muckenhoupt class* A_q, $1 \le q < \infty$, if there exists a constant C such that

$$A_q(w) := \sup_{\mathcal{Q}}(|\mathcal{Q}|^{-1} w(\mathcal{Q}))(|\mathcal{Q}|^{-1} w'(\mathcal{Q}))^{q-1} \tag{2.4}$$

is finite if $1 < q < \infty$, and

$$A_1(w) := \sup_{\mathcal{Q} \ni \mathbf{x}_0} (|\mathcal{Q}|^{-1} w(\mathcal{Q})) \le C w(\boldsymbol{x}_0) \quad \text{for a.a. } \mathbf{x}_0 \in \mathbb{R}^n, \tag{2.5}$$

where $w' := w^{-1/(q-1)}$ and $w(\mathcal{Q}) = \int_{\mathcal{Q}} w(\mathbf{x})\mathrm{d}\mathbf{x}$, the supremum is taken over all cubes $\mathcal{Q}$ in $\mathbb{R}^n$. In the first case, the supremum is taken over all cubes $\mathcal{Q}$ in $\mathbb{R}^n$ with edges parallel to the coordinate axes, in the second case over those of such cubes which contain $\mathbf{x}_0$; here $|\mathcal{Q}|$ denotes the n-dimensional Lebesgue measure of $\mathcal{Q}$.

- The infimum of the constants C satisfying (2.4) in case $1 < q < \infty$ or (2.5) in case $q = 1$ is called the A_q-*constant* of the weight.

Remark 2.2 It is easily seen from (2.4) that $|\mathcal{Q}|^{-1}(w(\mathcal{Q}))^{1/q}(w'(\mathcal{Q}))^{1/q'} \le C$, and as a consequence we get a dual weight satisfying $w' \in A_{q'}$, $1/q+1/q' = 1$. By Stein [205], for $1 < q < \infty$ and $w \in A_q$, there exists s such that $1 \le s < q$ and $w \in A_s$.

Remark 2.3 We note that in the first case we may restrict ourselves to cubes $\mathcal{Q}$ centered at $\mathbf{x}$. Since the A_q weights satisfy the *doubling property*, that is, $w(2\mathcal{Q}) \le cw(\mathcal{Q})$, where $w(\mathcal{Q}) = \int_{\mathcal{Q}} w(x)\mathrm{d}x$ and $2\mathcal{Q}$ is the cube with the same center as $\mathcal{Q}$ but with double side length, one can consider balls instead of cubes; the observation on centers of cubes applies to balls as well (see, e.g., [115]).

- For $q \in (1,\infty)$, $w \in A_q$, we define the weighted Lebesgue spaces $L^q_w(\mathcal{D}) := \{f \in L^1_{\mathrm{loc}}(\overline{\mathcal{D}}), \int_{\mathcal{D}} |f|^q w \mathrm{d}x < \infty\}$ with the norm $\|f\|_{q,w} := \left(\int_{\mathcal{D}} |f|^q w \mathrm{d}x\right)^{1/q}$.
- For $q \in (1,\infty)$, $w \in A_q$, we define the *weighted Sobolev space* $H^{k,q}_w(\mathcal{D}) := \{f \in L^1_{\mathrm{loc}}(\overline{\mathcal{D}}), \nabla^j f \in L^q_w(\mathcal{D}), j \le k\}$ equipped with the norm $\|u\|_{k,q,w} := \sum_{j=0}^k \|\nabla^j u\|_{q,w}$.
- We define the *homogeneous weighted Sobolev spaces* $\widehat{H}^{k,q}_w(\mathcal{D}) := \{f \in L^1_{\mathrm{loc}}(\overline{\mathcal{D}}), \nabla^k f \in L^q_w(\mathcal{D})\}$, with the norm $\|u\|_{\widehat{H}^{k,q}_w} := \|\nabla^k u\|_{q,w}$.
- Finally, to take the advantage of the density properties, the spaces $\widehat{H}^{k,q}_{w,0}(\mathcal{D}) := \overline{C^\infty_0(\mathcal{D})}^{\|\cdot\|_{\widehat{H}^{k,q}_w}}$, $H^{k,q}_{w,0}(\mathcal{D}) := \overline{C^\infty_0(\mathcal{D})}^{\|\cdot\|_{H^{k,q}_w}}$ with norms $\|\nabla^k u\|_{q,0}$, $\|u\|_{k,q,0}$. In all cases, we will add the subscript σ to denote the divergence-free counterpart. We can also consider the respective dual spaces.

Remark 2.4 If $\mathcal{D}$ is a bounded domain, let $1 < s \le q < \infty$ and $w \in A_s$. It follows from Hölder's inequality that the weighted Lebesgue spaces are embedded into unweighted ones as follows:

$$L^q_w(\mathcal{D}) \hookrightarrow L^r(\mathcal{D}) \quad \text{for every } r < \frac{q}{s},\ r \le q. \tag{2.6}$$

On the other hand, the Sobolev embedding

$$H^{k,r}_w(\mathcal{D}) \hookrightarrow L^q_w(\mathcal{D}) \tag{2.7}$$

holds under Lipschitz regularity for $\mathcal{D}$ and under the condition $1/r \le 1/q + k/3s$ with $k < 3$. That property is shown by Fröhlich in [88].

Remark 2.5 It has been shown by Fröhlich in [89] that as in the unweighted case the embeddings

$$\mathbf{H}_w^{k+1,q}(\mathcal{D}) \hookrightarrow \mathbf{H}_w^{k,q}(\mathcal{D}) \tag{2.8}$$

are compact in a bounded Lipschitz domain $\mathcal{D}$.

The proof of (2.8) is based on the application of the mollifiers instead of the translation. It is a consequence of the property of the translation operators $\tau_{\mathbf{h}}: \mathbf{u} \mapsto \mathbf{u}(\cdot + \mathbf{h})$ for $\mathbf{h} \in \mathbb{R}^n$. These operators do not map $\mathbf{L}_w^q$ to $\mathbf{L}_w^q$ in general, and the classical criterion of Kolmogorov characterizing precompact subsets of $\mathbf{L}^q(\mathcal{D})$ cannot be valid for the weighted spaces $\mathbf{L}_w^q(\mathcal{D})$ for arbitrary $w \in A_q$. Thus, the usual proof of the compact embedding $H^{1,q}(\mathcal{D}) \hookrightarrow\hookrightarrow L^q(\mathcal{D})$ does not transfer to the weighted case, and it is necessary to apply an idea of Alt [3] using the mollifiers instead of translations.

Remark 2.6 It is also very important to choose constants $C = C(w)$ in estimates depending on the weight function $w \in A_q$ uniformly whenever the A_q-constant $A_q(w)$ is bounded from above, i.e., $A_q(w) \leq c \leq \infty$. We call this property of the *constant A_q-consistency* . More precisely:

- A mapping $C: A_q \mapsto \mathbb{R}_+$ is called *A_q-consistently increasing*, iff

  $$\sup\{C(w) : w \in A_q, A_q(w) \leq c\} < \infty \quad \text{for all } c \in \mathbb{R}_+.$$

  A mapping $C: A_q \mapsto \mathbb{R}_+$ is called *A_q-consistently decreasing*, iff $1/C$ is A_q-consistently increasing.
- We define *the Schwartz space $\mathcal{S}(\mathbb{R}^n)$ of "rapidly decreasing functions"* as the set of $C^\infty(\mathbb{R}^n)$ functions $\mathbf{u}$ on $\mathbb{R}^n$ such that for any $k \in \mathbb{N}_0$ we have

  $$\||\mathbf{u}|\|_{k,S} \equiv \sup_{x \in \mathbb{R}^n} (1+|\mathbf{x}|)^k |\partial^\alpha \mathbf{u}(\mathbf{x})| \leq \infty \text{ for all } |\alpha| \leq k, \text{ for all } \mathbf{x} \in \mathbb{R}^n. \tag{2.9}$$

- If $\mathbf{u} \in \mathcal{S}(\mathbb{R}^n)$, let

  $$\mathcal{F}\mathbf{u}(\boldsymbol{\xi}) = \widehat{\mathbf{u}}(\boldsymbol{\xi}) = \frac{1}{(2\pi)^{n/2}} \int_{\mathbb{R}^n} \mathrm{e}^{-ix\cdot\boldsymbol{\xi}} \mathbf{u}(\mathbf{x}) \mathrm{d}\mathbf{x}, \quad \boldsymbol{\xi} \in \mathbb{R}^n,$$

  be *the Fourier transform of* $\mathbf{u}$. Its inverse will be denoted by $\mathcal{F}^{-1}$.
- We define *the centered Hardy–Littlewood maximal operator*

  $$\mathcal{M}\mathbf{u}(\mathbf{x}) = \sup_{\mathcal{Q} \ni x} \frac{1}{|\mathcal{Q}|} \int_{\mathcal{Q}} |\mathbf{u}(\mathbf{y})| \mathrm{d}\mathbf{y}, \quad \mathbf{x} \in \mathbb{R}^n,$$

  for $\mathbf{u} \in \mathbf{L}^1_{\mathrm{loc}}(\mathbb{R}^n)$, where again $\mathcal{Q}$ runs over the set of all cubes centered at $\mathbf{x}$.

2.2 The Bogovskii Operator

We shall discuss the solvability of the equation $\operatorname{div} u = g$ in $\mathcal{D}$, $u = 0$ on $\partial\mathcal{D}$ in various bounded and unbounded domains. For more details, see, e.g., [15, 16, 93, 144].

Assumptions on domain $\mathcal{D}$

Let $1 < q < +\infty$. Let $\mathcal{D} \subset \mathbb{R}^n$, $n \geq 2$, be a domain with boundary $\partial\mathcal{D} \in C^{1,1}$, and suppose one of the following cases:

(i) $\mathcal{D}$ is bounded.
(ii) $\mathcal{D}$ is an exterior domain, i.e., a domain having a compact nonempty complement.

In the bounded situation, Bogovskii [15, 16] has constructed a bounded linear operator $\mathfrak{B}\colon L_0^q(\mathcal{D}) \to H_0^{1,q}(\mathcal{D})^n$ the so called *Bogovskii operator* such that $u = \mathfrak{B}g$ is a solution to

$$\begin{aligned} \operatorname{div} u &= g \quad \text{in } \mathcal{D}, \\ u &= 0 \quad \text{on } \partial\mathcal{D} \end{aligned} \tag{2.10}$$

satisfying

$$\|\mathfrak{B}g\|_{1,q} \leq c\|g\|_q,$$

where $L_0^q(\mathcal{D})$ denotes a class of functions $g(x)$ from $L^q(\mathcal{D})$, such that $\int_{\mathcal{D}} g(x)\,\mathrm{d}x = 0$. Moreover, $\mathfrak{B}$ maps $W_0^{1,q}(\mathcal{D}) \cap L_0^q(\mathcal{D})$ into $H_0^{2,q}(\mathcal{D})^n$, see [15]. As to domains with less smooth boundaries, let us note that the problem of solvability of the equation $\operatorname{div} u = g$ has been solved in a *star-shaped domain*, see Galdi [93].

It should also be mentioned that in many situations, the Bogovskii operator is applied in Sobolev spaces of negative order. For $s < 0$ we define

$$(H_0^{-s,p'}(\mathcal{D}))' := H^{s,p}(\mathcal{D}), \quad (H^{-s,p'}(\mathcal{D}))' := H_0^{s,p}(\mathcal{D}),$$

where $1/p + 1/p' = 1$. See [117] for more details.

In connection with the Bogovskii operator, we also mention comments on Sobolev spaces of negative order in works of Galdi [93] and Farwig and Sohr [78]. Farwig and Sohr [79] observed that if $\mathcal{D}$ is the domain satisfying one of the assumptions (i) and (ii), then $\mathfrak{B}$ is a bounded operator:

- From $\widehat{H}^{1,q}(\mathcal{D}) \cap L_0^q(\mathcal{D})$ in $H_0^{1,q}(\mathcal{D})^n \cap H^{2,q}(\mathcal{D})^n$, if $\mathcal{D}$ is bounded
- From $\widehat{H}^{-1,q}(\mathcal{D})$ in $L^q(\mathcal{D})^n$, also if $\mathcal{D}$ is bounded
- From $H^{1,q}(\mathcal{D}) \cap \widehat{H}^{-1,q}(\mathcal{D})$ in $H_0^{1,q}(\mathcal{D})^n \cap H^{2,q}(\mathcal{D})^n$, if $\mathcal{D}$ is unbounded

and $u = \mathfrak{B}g$ solves always $\operatorname{div}\mathfrak{B}g = g$ with $\mathfrak{B}g|_{\partial\mathcal{D}} = 0$ under the condition $\int_{\mathcal{D}} g(x)\mathrm{d}x = 0$ and satisfies the estimates

$$\|u\|_q \leq c\|g\|_{-1,q},$$
$$\|u\|_{2,q} \leq c(\|\nabla g\|_q + \|g\|_{-1,q}),$$

where $c = c(\mathcal{D}, q) > 0$ is a constant.

2.3 Some Auxiliary Results

Here, we recall several important lemmas.

Lemma 2.2

(1) Let μ be a nonnegative regular Borel measure such that the centered Hardy-Littlewood maximal operator

$$\mathcal{M}\mu(\mathbf{x}) = \sup_{\mathcal{Q}\ni\mathbf{x}} \frac{1}{|\mathcal{Q}|}\int_{\mathcal{Q}} \mathrm{d}\mu$$

is finite for almost all $\mathbf{x} \in \mathbb{R}^n$; here $\mathcal{Q}$ runs over the set of all cubes in $\mathbb{R}^n$ centered at $\mathbf{x}$ (with edges parallel to the coordinate axes) and $|\mathcal{Q}|$ denotes the Lebesgue measure of $\mathcal{Q}$. Then $(\mathcal{M}\mu)^\gamma \in A_1$ for all $\gamma \in [0, 1)$.[1]

(2) Let $w_1, w_2 \in A_1$ and $0 < \theta < 1$. Then $w_1^{1-\theta} w_2^{\theta} \in A_1$.

(3) For all $1 < q < \tau$ one has $A_1 \subset A_q \subset A_\tau$. [2]

(4) Let $1 < q < \infty$ and $w \in A_q$. Then there are $w_1, w_2 \in A_1$ such that

$$w = \frac{w_1}{w_2^{q-1}}.$$

Conversely, a weight $w = w_1 w_2^{1-q}$ belongs to A_q if $w_1, w_2 \in A_1$.

Proof For part (1) see [115, Theorem II 3.4].

The claims (2) and (3) and the second part of (4) are simple consequences of Hölder's inequality. The first part of (4), the factorization of A_q-weights, can be found, e.g., in [205], [V 5.3, Proposition 9]. □

[1] Let us note that definition of A_1 can be found before, see 2.5.

[2] Let us note that definition of A_q can be found before, see 2.4.

Theorem 2.1 *Let $1 < q < \infty$ and $w \in A_q$. Then the following statements hold true:*

(i) *Operator $\mathcal{M}$, considered on $\mathcal{S}(\mathbb{R}^n)$, can be extended to a bounded operator from $L^q_w(\mathbb{R}^n)$ to $L^q_w(\mathbb{R}^n)$.*
(ii) *Let $m \in C^n(\mathbb{R}^n \setminus \{0\})$ satisfy the pointwise* Hörmander–Mikhlin multiplier *condition*

$$|\boldsymbol{\xi}|^{|\alpha|}|D^\alpha m(\boldsymbol{\xi})| \le c_\alpha \quad \text{for all} \quad \boldsymbol{\xi} \in \mathbb{R}^n \setminus \{\mathbf{0}\}$$

and all multi-indices $\alpha \in \mathbb{N}_0^n$ with $|\boldsymbol{\alpha}| \le n$. Then the multiplier operator $\mathbf{u} \mapsto \mathcal{F}^{-1}(m\widehat{\mathbf{u}})$, $\mathbf{u} \in \mathcal{S}(\mathbb{R}^n)$, can be extended to a bounded linear operator from L^q_w to L^q_w.

Proof

(i) See [115, Theorem IV 2.8].
(ii) See [115, Theorem IV 3.9] or [158, Theorem 4]. Note that the pointwise condition on m implies the integral condition in [115, 158].

□

Further, we recall some basic lemmas.

Lemma 2.3 (Kozono and Sohr [144]) *Let $\mathcal{D} \subset \mathbb{R}^n$ $(n \ge 2)$ be any domain, and let $1 < q < \infty$. For all $\mathbf{f} \in \widehat{\mathbf{H}}^{-1,q}(\mathcal{D})$, there is $F \in L^q(\mathcal{D})$ such that*

$$\nabla \cdot \mathbf{F} = \mathbf{f}, \quad \|\mathbf{F}\|_{q,\mathcal{D}} \le C\|\mathbf{f}\|_{-1,q,\mathcal{D}} \tag{2.11}$$

with some $C > 0$. As a result, the space $\{\nabla\mathbf{F}, \mathbf{F} \in \mathbf{C}_0^\infty(\mathcal{D})^n\}$ is dense in $\widehat{\mathbf{H}}^{-1,q}(\mathcal{D})$.

The next lemma concerns the Stokes problem with the inhomogeneous divergence condition

$$-\Delta\mathbf{u} + \nabla p = \mathbf{f}, \quad \operatorname{div}\mathbf{u} = g \quad \text{in } \mathcal{D}; \ \mathbf{u}|_{\partial\mathcal{D}} = 0. \tag{2.12}$$

Lemma 2.4 (Cattabriga [22]) *Let $\mathcal{D}$ be a bounded domain with smooth boundary $\partial\mathcal{D}$ and let $1 < q < \infty$. Suppose that*

$$\mathbf{f} \in \mathbf{W}^{-1,q}(\mathcal{D})^3, \quad g \in L^q(\mathcal{D}), \quad \int_{\mathcal{D}} g(\mathbf{x})\mathrm{d}\mathbf{x} = 0.$$

Then the problem (2.12) *possesses a unique (up to an additive constant for p) weak solution $\{\mathbf{u}, p\} \in \mathbf{W}_0^{1,q}(\mathcal{D})^3 \times L^q(\mathcal{D})$ subject to the estimate*

$$\|\nabla\mathbf{u}\|_{q,\mathcal{D}} + \|p - \bar{p}\|_{q,\mathcal{D}} \le C(\|\mathbf{f}\|_{-1,q,\mathcal{D}} + \|g\|_{q,\mathcal{D}}), \tag{2.13}$$

where $\bar{p} = 1/|\mathcal{D}| \int_{\mathcal{D}} p(\mathbf{x})\mathrm{d}\mathbf{x}$.

Lemma 2.5 (Weighted Poincaré Inequality [89]) *Let $\mathcal{D} \subset \mathbb{R}^n$ be a bounded Lipschitz domain, $q \in (1,\infty)$, and $w \in A_q$. Let h be a continuous seminorm on $\mathbf{W}^{1,1}(\mathcal{D})$ such that $h(c) = 0$ implies $c = 0$ for constant functions c.*

1. *There exists an A_q-consistently increasing constant $c \in \mathbb{R}$ such that*

$$\|\mathbf{u}\|_{q,w} \leq c\|\nabla\mathbf{u}\|_{q,w} \quad \text{for every } \mathbf{u} \in \widehat{\mathbf{H}}^{1,q}_w(\mathcal{D}) \text{ with } h(\mathbf{u}) = 0 \tag{2.14}$$
$$\text{and for every } \mathbf{u} \in \widehat{\mathbf{H}}^{1,q}_w(\mathcal{D}) \text{ with } \mathbf{u}|_{\Gamma_{co}} = 0,$$

where Γ_{co} is a connected component of $\partial\mathcal{D}$.

2. *There exists an A_q-consistently constant c such that*

$$\|\mathbf{u}\|_{2,q,w} \leq c\|\nabla^2\mathbf{u}\|_{q,w}$$

for every $\mathbf{u} \in \widehat{\mathbf{H}}^{2,q}_w(\mathcal{D})$ with $\mathbf{u}|_{\Gamma_{co}} = 0$.

Proof Both Assertions 1 and 2 are proved in [89] for $\mathbf{u}|_{\partial\mathcal{D}} = 0$. However, the same approach can also be used to show a slightly more general assertion. □

Due to the presence of "the rotating term" $-(\boldsymbol{\omega}\wedge\mathbf{x})\nabla\mathbf{v}$ and $\boldsymbol{\omega}\times\mathbf{v}$, we must restrict the class of weight functions. Namely, we consider (in the Stokes case) weights w on $\mathbb{R}^n$, $n = 2$, $n = 3$, that are independent of the angular variable θ in cylindrical coordinate system attached to the axis of revolution $(0, 0, 1)^T$ and assume that $\mathbf{w}$ satisfies the following condition depending on $q \in (1,\infty)$:

$$\begin{aligned} 2 < q < \infty\colon\ & w^r \in A_{rq/2} && \text{for some } r \in [1,\infty), \\ q = 2\colon\ & w^r \in A_r, \text{ or } w^{-r} \in A_r && \text{for some } r \in [1,\infty), \\ 1 < q < 2\colon\ & w^r \in A_{rq/2} && \text{for some } r \in (2/q, 2/(2-q)]. \end{aligned} \tag{2.15}$$

Lemma 2.6 *Let $(T_\alpha)_{\alpha\in\mathbb{N}}$ be a family of operators such that $T_\alpha\colon \mathbf{L}^q_w(\mathbb{R}^n) \to \mathbf{L}^q_w(\mathbb{R}^n)$ is continuous for every $q \geq 2$ and every $w \in A_{q/2}$ and such that $\|T_\alpha\|_{\mathcal{L}(\mathbf{L}^q_w)}$ is uniformly bounded in α. Moreover, we assume for the adjoint operators T^*_α that*

$$T^*_\alpha\mathbf{v} = T_\alpha\mathbf{v} \quad \text{for every } \mathbf{v} \in \mathbf{L}^q_w(\mathbb{R}^n) \cap \mathbf{L}^{q'}_{w'}(\mathbb{R}^n) \tag{2.16}$$

for every $\alpha \in \mathbb{N}$.

Then $T_\alpha\colon \mathbf{L}^p_v(\mathbb{R}^n) \to \mathbf{L}^p_v(\mathbb{R}^n)$ is continuous for every $\mathbf{v}$ fulfilling (2.15), and $\|T_\alpha\|_{\mathcal{L}(\mathbf{L}^p_v)}$ is uniformly bounded in α.

Proof As in the proof of [72, Proposition 3.1], one employs the duality argument and (2.16) to show that $T_\alpha\colon \mathbf{L}^p_v(\mathbb{R}^n) \to \mathbf{L}^p_v(\mathbb{R}^n)$ is continuous for $p \in (1,2)$, if the weight function $\mathbf{v}$ fulfills $\mathbf{v}^{2/(2-q)} \in A_{q/(2-q)}$. The operator norm of T_α, i.e., $\|T_\alpha\|_{\mathcal{L}(\mathbf{L}^{q'}_{v'})}$, is uniformly bounded in α.

Now one uses the same interpolation procedure as in the proof in [72, Theorem 1.2. (i)], to conclude the continuity of T_α for an arbitrary weight function w fulfilling (2.15). Since the complex interpolation does not increase the continuity constants, the proof of the lemma is finished. □

Corollary 2.1 *Let the weight function* $\mathbf{w}$ *be chosen according to* (2.15)*. Then* $\mathbf{w} \in A_q$.

Proof Let φ be nonnegative, radial, and radially decreasing with $\int \varphi(\mathbf{x})d\mathbf{x} = 1$. Moreover, we set $\varphi_\varepsilon(\mathbf{x}) := \varphi(\mathbf{x}/\varepsilon)/\varepsilon^n$ and define $T_\varepsilon f := f * \varphi_\varepsilon$.

By Stein [205, Proposition V.2.1], the operator $T_\varepsilon\colon \mathbf{L}^r_v(\mathbb{R}^n) \to \mathbf{L}^r_v(\mathbb{R}^n)$ is continuous and $(T_\varepsilon)_{\varepsilon>0}$ is uniformly bounded in $\mathcal{L}(\mathbf{L}^r_v(\mathbb{R}^n))$ if and only if $\mathbf{v} \in A_r$; in particular, this is true for $r \geq 2$ and $\mathbf{v} \in A_{r/2} \subset A_r$.

By Lemma 2.6 we obtain that $T_\varepsilon : \mathbf{L}^q_w(\mathbb{R}^n) \to \mathbf{L}^q_w(\mathbb{R}^n)$ is continuous for q and w according to (2.15). Thus again by [205, Proposition V.2.1] we obtain that $\mathbf{w} \in A_q$. □

The following lemma is a weighted analog to well-known Bogowski's theorem on the solution of the equation $\operatorname{div}\mathbf{u} = \mathbf{f}$. It is proved in [189].

Lemma 2.7

(i) Let $\mathcal{D} \subseteq \mathbb{R}^n$ $(n \geq 2)$ be a bounded domain with Lipschitz boundary $\partial\mathcal{D}$, and let $k \in \mathbb{N}_0$, $1 < r < \infty$, and $w \in A_r$. Then for every $\mathbf{f} \in \widehat{\mathbf{H}}^{k,q}_{w,0}(\mathcal{D})$ with $\int_\mathcal{D} \mathbf{f}(\mathbf{x})\,\mathrm{d}\mathbf{x} = 0$, there exists $\mathbf{u} \in \widehat{\mathbf{H}}^{k+1,q}_{w,0}(\mathcal{D})$ such that

$$\operatorname{div}\mathbf{u} = \mathbf{f} \quad \textit{and} \quad \|\mathbf{u}\|_{\widehat{\mathbf{H}}^{k+1,q}_w(\mathcal{D})} \leq c\|\mathbf{f}\|_{\widehat{\mathbf{H}}^{k,q}_w(\mathcal{D})}, \tag{2.17}$$

where $c = c(\mathcal{D}, k, r, w) > 0$.

(ii) Let $\mathcal{D} = \mathbb{R}^3$ or $\mathcal{D}$ be an exterior domain $C^{1,1}$ in $\mathbb{R}^3$. For every $g \in \widehat{H}^{-1,q}_w(\mathcal{D})$ there exists a function $\mathbf{u} \in L^q_w(\mathcal{D})$ that solves

$$\operatorname{div}\mathbf{u} = g \textit{ and } \|\mathbf{u}\|_{q,w} \leq c\|g\|_{\widehat{H}^{-1,q}_w} \tag{2.18}$$

with $c = c(q, w)$, independent of g.

Observation 2.1

In Lemma 2.7 (ii) we consider $\mathbf{u} = 0$ on $\partial\mathcal{D}$. In the case $\mathbf{u} = \boldsymbol{h}$ on $\partial\mathcal{D}$, the following estimate is satisfied:

$$\|\mathbf{u}\|_{q,w} \leq c\left(\|g\|_{\widehat{H}^{-1,q}_w} + \|h\|\|_{T^{0,q}_w(\partial\mathcal{D})}\right),$$

where $c = c(\mathcal{D}, q, w)$.

For the definition of $T^{0,q}_w(\partial\mathcal{D})$ see Sect. 3.4.

In the case of weighted Sobolev spaces the solvability of the Stokes equations in bounded domains has been shown by Fröhlich in [89]. More precisely one has the following theorem.

Theorem 2.2 *Let $\mathcal{D} \subset \mathbb{R}^n$ $(n \geq 2)$ be a bounded domain with boundary of the class $C^{1,1}$, let $1 < r < \infty$, and $\mathbf{w} \in A_r$. Then for every $\mathbf{f} \in \mathbf{L}^r_w(\mathcal{D})$ and $g \in \widehat{H}^{1,r}_w(\mathcal{D})$ with $\int_{\mathcal{D}} g = 0$, there is a unique $\{\mathbf{u}, p\} \in \widehat{\mathbf{H}}^{2,r}_w(\mathcal{D}) \times \widehat{H}^{1,r}_w(\mathcal{D})$ with $\int_{\mathcal{D}} p \, \mathrm{d}x = 0$ such that*

$$-\Delta \mathbf{u} + \nabla p = \mathbf{f}, \quad \operatorname{div} \mathbf{u} = g \quad \text{in } \mathcal{D} \text{ and } \mathbf{u}|_{\partial\mathcal{D}} = 0 \tag{2.19}$$

in the sense of distributions in $\mathcal{D}$. This solution $\{\mathbf{u}, p\}$ satisfies the inequality

$$\|\nabla^2 \mathbf{u}\|_{\mathbf{L}^r_w(\mathcal{D})} + \|\nabla p\|_{L^r_w(\mathcal{D})} \leq C\left(\|\mathbf{f}\|_{\mathbf{L}^r_w(\mathcal{D})} + \|g\|_{\widehat{H}^{1,r}_w(\mathcal{D})}\right) \tag{2.20}$$

with a constant $C = C(\mathcal{D}, r, w)$.

2.4 The Littlewood–Paley Decomposition

First of all we introduce the dyadic blocks of frequencies:

Definition 2.1 Let $\varphi \in C_0^\infty(\mathbb{R}^n)$ be a nonnegative function so that $|\xi| \leq \frac{1}{2} \Rightarrow \varphi(\xi) = 1$ and $|\xi| \geq 1 \Rightarrow \varphi(\xi) = 0$. Let ψ be defined as $\psi(\xi) = \varphi(\xi/2) - \varphi(\xi)$. Let S_j and Δ_j be defined as the Fourier multipliers $\mathcal{F}(S_j\mathbf{f}) = \varphi(\xi/2^j)\mathcal{F}\mathbf{f}$ and $\mathcal{F}(\Delta_j\mathbf{f}) = \psi(\xi/2^j)\mathcal{F}\mathbf{f}$. The distribution $\Delta_j\mathbf{f}$ is called the jth dyadic block of *the Littlewood–Paley decomposition* of $\mathbf{f}$. Then for all $N \in \mathbb{Z}$ and all $f \in \mathcal{S}'(\mathbb{R}^n)$, we have $\mathbf{f} = S_N\mathbf{f} + \sum_{j \geq N} \Delta_j\mathbf{f}$ in $\mathcal{S}'(\mathbb{R}^n)$, this equality is called the Littlewood–Paley decomposition of the distribution $\mathbf{f}$.

Theorem 2.3 (Littlewood–Paley Decomposition of $\mathbf{L}^p(\mathbb{R}^n)$) *Let $\mathbf{f} \in \mathcal{S}'(\mathbb{R}^n)$ and $1 < p < \infty$. Then the following assertions are equivalent:*

(i) $\mathbf{f} \in \mathbf{L}^p(\mathbb{R}^n)$.

(ii) $\mathbf{S}_0 \in \mathbf{L}^p(\mathbb{R}^n)$ *and* $\left(\sum_{j \in \mathbb{N}} |\Delta_j\mathbf{f}|^2\right)^{1/2} \in \mathbf{L}^p(\mathbb{R}^n)$.

(iii) $\mathbf{f} = \sum_{j \in \mathbb{Z}} \Delta_j\mathbf{f}$ *and* $\left(\sum_{j \in \mathbb{Z}} |\Delta_j\mathbf{f}|^2\right)^{1/2} \in \mathbf{L}^p(\mathbb{R}^n)$.

Moreover, the following norms are equivalent on L^p:

$$\|\mathbf{f}\|_p, \quad \|S_0\mathbf{f}\|_p + \left\|\left(\sum_{j \in \mathbb{N}} |\Delta_j\mathbf{f}|^2\right)^{1/2}\right\|_p \quad \text{and} \quad \left\|\left(\sum_{j \in \mathbb{Z}} |\Delta_j\mathbf{f}|^2\right)^{1/2}\right\|_p.$$

Proof See [168]. □

Now, we recall a well-known theorem on complex interpolation of L^q_w-spaces, see [14]. Note that we use a different definition of the weighted space $\mathbf{L}^q_w$ than in [14], but the validity of lemma is unchanged.

Lemma 2.8 *Let* $1 \le p_1, p_2 < \infty$, *let* $0 < w_1, w_2$ *be two weight functions,* $\delta \in (0, 1)$, *and*

$$\frac{1}{p} = \frac{1-\delta}{p_1} + \frac{\delta}{p_2}, \qquad w^{1/p} = w_1^{(1-\delta)/p_1} w_2^{\delta/p_2}.$$

Then

$$[L^{p_1}_{w_1}, L^{p_2}_{w_2}]_\delta = L^p_w$$

in the sense of the complex interpolation.

2.5 Fundamental Solution of the Nonsteady "Rotating" Oseen Problem

In this subsection we recall the main results proved by Guenther and Thomann, see [124], on the problem of the nonsteady "rotating" Oseen problem.

We consider the following system of equations:

$$\begin{aligned} \partial_t \boldsymbol{u} - \Delta \boldsymbol{u} - (\boldsymbol{U} + \boldsymbol{\omega} \wedge \boldsymbol{x}) \cdot \nabla \boldsymbol{u} + \boldsymbol{\omega} \wedge \boldsymbol{u} + \nabla p = \boldsymbol{f} \quad & \text{in } \mathcal{D} \times (0, \infty), \\ \operatorname{div} \boldsymbol{u} = 0 \quad & \text{in } \mathcal{D} \times (0, \infty), \\ \boldsymbol{u} = \boldsymbol{u}_{\partial \mathcal{D}} \quad & \text{on } \partial \mathcal{D} \times (0, \infty), \\ \boldsymbol{u}(\boldsymbol{x}, t) \to 0 \quad & \text{as } |\boldsymbol{x}| \to \infty, \end{aligned} \tag{2.21}$$

in a time-independent exterior domain $\mathcal{D}$.

We define the Oseen type operator

$$\mathcal{L} \boldsymbol{v} = \mathcal{L}_{y,t} \boldsymbol{v} = -\Delta \boldsymbol{v} - (\boldsymbol{U} + \boldsymbol{\omega} \wedge \boldsymbol{y}) \cdot \nabla \boldsymbol{v} + \boldsymbol{\omega} \wedge \boldsymbol{v}. \tag{2.22}$$

Then the fundamental tensor of (2.21) comprises a (3×3)-matrix of distributions $\boldsymbol{\Gamma}(\boldsymbol{y}, \boldsymbol{z}, t)$ and a three-dimensional vector of distributions $\boldsymbol{Q}(y, z, t)$ such that for any vector $a \in \mathbb{R}^3$ the distributions

$$\begin{aligned} \boldsymbol{v}_z(\boldsymbol{y}, t) &= \boldsymbol{\Gamma}(\boldsymbol{y}, \boldsymbol{z}, t - s)\boldsymbol{a}, \ t \ge s, \quad \boldsymbol{v}_z(\boldsymbol{y}, t) = 0, \ t < s, \\ \pi_z(y, t) &= \boldsymbol{Q}(\boldsymbol{y}, \boldsymbol{z}, t - s)\boldsymbol{a}, \ t \ge s, \quad \pi_z(\boldsymbol{y}, t) = 0, \ t < s, \end{aligned}$$

solve the system

$$\begin{aligned} \frac{\partial \boldsymbol{v}_{z,s}}{\partial t} + \mathcal{L} \boldsymbol{v}_{z,s} + \nabla \pi_{z,s} &= \delta_s(t) \delta_z(\boldsymbol{y}) \boldsymbol{a}, \\ \operatorname{div} \boldsymbol{v}_{z,s} &= 0, \end{aligned} \tag{2.23}$$

in the sense of distributions. Therefore for all test functions $\boldsymbol{\varphi} \in C_0^\infty(\mathbb{R}^3 \times \mathbb{R})^3$

$$\langle \partial_t \boldsymbol{v}_{z,s} + \mathcal{L}\boldsymbol{v}_{z,s} + \nabla \pi_{z,s}, \boldsymbol{\varphi} \rangle = \boldsymbol{\varphi}(z, s)\boldsymbol{a} = \langle \delta_z \otimes \delta_s, \boldsymbol{\varphi}\boldsymbol{a} \rangle$$

and $\operatorname{div} \boldsymbol{v}_{z,s} = 0$ for all $t > s$. Here $\delta_s(t)$, $\delta_z(\boldsymbol{y})$ denote the point masses concentrated at $t = s$, $\boldsymbol{y} = z$.

We will recall the main theorems from the work of Guenther and Thomann (see [124]) and sketch the main ideas of the proof.

Let K denote the usual *fundamental solution to the heat equation*, that is,

$$K(z, t) := (4\pi t)^{-3/2} \mathrm{e}^{-|z|^2/(4t)} \quad \text{for } z \in \mathbb{R}^3, \ t \in (0, \infty).$$

The estimate of the fundamental solution of heat equation K was proved in [202].

Lemma 2.9 *For $\boldsymbol{\alpha} \in \mathbb{N}_0^3$, $l \in \mathbb{N}_0$, there is $C(l, \boldsymbol{\alpha}) > 0$ such that*

$$|\partial_{\boldsymbol{x}}^{\boldsymbol{\alpha}} \partial_t^l K(\boldsymbol{x}, t)| \le C(l, \alpha)(|\boldsymbol{x}|^2 + t)^{-3/2-|\alpha|/2-l} \quad \textit{for } \boldsymbol{x} \in \mathbb{R}^3, \ t \in (0, \infty).$$

An easy computation involving the relation $\Gamma_a(1/2) = \sqrt{\pi}$ (where Γ_a denotes here the *Gamma function*) yields:

Lemma 2.10

$$\int_0^\infty K(\boldsymbol{x}, t)\, \mathrm{d}t = (4\pi |\boldsymbol{x}|)^{-1} \quad \text{for } x \in \mathbb{R}^3 \backslash \{0\}.$$

Theorem 2.4 *The fundamental tensor $\boldsymbol{\Gamma}(y, z, t)$, $\boldsymbol{Q}(y, z, t)$ of the problem* (2.21) *in the whole space can be written in the form (where $z(t) = \mathrm{e}^{-t\boldsymbol{\Omega}} z - t\boldsymbol{U}$ such that $\boldsymbol{\Omega u} = \boldsymbol{\omega} \wedge \boldsymbol{u}$)*

$$\begin{aligned}\boldsymbol{\Gamma}(y, z, t) =& K(y - z(t), \tau) \left\{ \left[\boldsymbol{I} - \frac{(y - z(t)) \otimes (y - z(t))}{|y - z(t)|^2} \right] \right. \\ & \left. - {}_1F_1\left(1, \frac{5}{2}, \frac{|y - z(t)|^2}{4\tau}\right) \left[\frac{1}{3}\boldsymbol{I} - \frac{(y - z(t)) \otimes (y - z(t))}{|y - z(t)|^2} \right] \right\} \mathrm{e}^{-t\boldsymbol{\Omega}}, \\ \boldsymbol{Q}(y, z, t) =& -\frac{1}{4\pi} \nabla_y \frac{1}{|y - z(t)|} \delta_0 \equiv \boldsymbol{Q}^*(y)\delta_0,\end{aligned}$$

where ${}_1F_1(1, c, u)$ denotes the so-called Kummer function. *In particular, for every initial value $\boldsymbol{u}_0 \in \mathcal{S}(\mathbb{R}^3)^3$,*

$$\lim_{(y,t)\to(y_0,0^+)} \int_{\mathbb{R}^3} \boldsymbol{\Gamma}(y, z, t)\boldsymbol{u}_0(z)\, \mathrm{d}z = \Pi_\sigma \boldsymbol{u}_0(y), \quad y \in \mathbb{R}^3,$$

where Π_σ denotes the Helmholtz projection *in $L^2(\mathbb{R}^3)$.*

Here and in the rest of this text, we use the abbreviation

$$z(t) := \mathrm{e}^{-t\boldsymbol{\Omega}} z - t\boldsymbol{U} = \mathrm{e}^{-t\boldsymbol{\Omega}} z + \tau t \boldsymbol{e}_1 \quad \text{for } z \in \mathbb{R}^3,\ t \in [0, \infty), \tag{2.24}$$

with the matrix Ω defined as

$$\boldsymbol{\Omega} = \varrho \begin{pmatrix} 0 & 0 & 0 \\ 0 & 0 & -1 \\ 0 & 1 & 0 \end{pmatrix}, \tag{2.25}$$

and $\rho > 0$. We recall the definition of the Kummer function later, see (2.31).

In the following we will also use the cylindrical coordinates $r, \theta, x_3 \in [0, \infty) \times [0, 2\pi) \times \mathbb{R}$ for x such that $(\boldsymbol{\omega} \wedge \boldsymbol{x}) \cdot \nabla \boldsymbol{u} = \partial_\theta \boldsymbol{u}$, where ∂_θ denotes the angular derivative with respect to θ. Obviously $-\Delta$ commutes with ∂_θ. Let $\nabla' = (\partial_1, \partial_2)$.

Let us define the function space $\mathcal{J}_T^{q,s}$, $1 < q, s < \infty$, of initial values with the norm

$$\|\boldsymbol{u}_0\|_{\mathcal{J}_T^{q,s}} = \left(\int_0^T (\|\mathrm{e}^{-t\mathscr{A}_0^0} \Pi_\sigma^q \boldsymbol{u}_0\|_q^s + \|\mathscr{A}_0^0 \mathrm{e}^{-t\mathscr{A}_0^0} \Pi_\sigma^q \boldsymbol{u}_0\|_q^s)\,\mathrm{d}t \right)^{1/s},$$

where Π_σ^q is the Helmholtz projection on $L^q(\mathbb{R}^3)^3$ and $\mathscr{A}_0^0 = -\Pi_\sigma^q \Delta$ is the Stokes operator (for more details on the Stokes operator, see Sect. 8.1). The following theorem states that the equation under consideration is well posed in this space.

Theorem 2.5 *Let $0 < T < \infty$, and assume that for some $1 < q, s < \infty$ the data $\boldsymbol{u}_0 \in L_\sigma^q(\mathbb{R}^3)$ and $\boldsymbol{f} \in L^s(0, T; L^q(\mathbb{R}^3))$ satisfy*

$$\boldsymbol{f}, \partial_\theta \boldsymbol{f}, \in L^s(0, T; L^q(\mathbb{R}^3)),$$

and $\boldsymbol{u}_0, \partial_\theta \boldsymbol{u}_0 \in \mathcal{J}_T^{q,s}$. Then the unique solution $(\boldsymbol{v}, \nabla p) \in L^s(0, T; (L^q(\mathbb{R}^3))$ of

$$\frac{\partial \boldsymbol{v}}{\partial t} + \mathcal{L}\boldsymbol{v} + \nabla p = \boldsymbol{f} \quad \text{in } \mathbb{R}^3 \times (0, \infty), \tag{2.26}$$
$$\nabla \cdot v = 0 \quad \text{in } \mathbb{R}^3 \times (0, \infty)$$

with the initial data $\boldsymbol{v}(0, \boldsymbol{y}) = \boldsymbol{u}_0(\boldsymbol{y})$ is given by

$$\boldsymbol{v}(\boldsymbol{y}, t) = \int_0^t \int_{\mathbb{R}^3} \boldsymbol{\Gamma}(\boldsymbol{y}, z, t - s) \boldsymbol{f}(z, s)\,\mathrm{d}z\mathrm{d}s + \int_{\mathbb{R}^3} \boldsymbol{\Gamma}(\boldsymbol{y}, z, t) \boldsymbol{u}_0(z)\mathrm{d}z, \tag{2.27}$$

$$p(\boldsymbol{y}, t) = \int_0^t \int_{\mathbb{R}^3} \boldsymbol{Q}(\boldsymbol{y}, z, t) \boldsymbol{f}(z, s)\mathrm{d}z\mathrm{d}s = \int_{\mathbb{R}^3} \boldsymbol{Q}^*(\boldsymbol{y} - z) \boldsymbol{f}(z, t)\mathrm{d}z. \tag{2.28}$$

Moreover, $\boldsymbol{v}$, p *satisfy the a priori estimate*

$$\begin{aligned}\|\boldsymbol{v};\ \nabla\boldsymbol{v};\ \nabla^2\boldsymbol{v};\boldsymbol{v}_t;\ \partial_\theta\boldsymbol{v};\ \nabla p\|_{L^s(0,T;L^q)} \qquad (2.29)\\ \le C(1+T)[\|\boldsymbol{u}_0;\ \partial_\theta\boldsymbol{u}_0\|_{\mathcal{J}_T^{q,s}}+\|\boldsymbol{f};\ \partial_\theta\boldsymbol{f}\|_{L^s(0,T;L^q)}],\end{aligned}$$

where the constant C *depends on* q, s, *and* $\boldsymbol{\Omega}$, *but not on* T.

Remark 2.7 The terms, already appearing in the estimate of $\boldsymbol{v}_t$ are present due to the fact that the operator $\mathcal{L}$ does not generate an analytic semigroup and will not satisfy the standard maximal regularity estimate, see [62, 74, 116, 126].

Corollary 2.2

(i) *The fundamental solution* $\boldsymbol{\Gamma}$ *from Theorem 2.4 is unique.*
(ii) *For any* $y, z \in \mathbb{R}^3$ *and* $s < \tau < t$, *one has the semigroup property*

$$\int_{\mathbb{R}^3}\boldsymbol{\Gamma}(y,z',t-\tau)\boldsymbol{\Gamma}(z',z,\tau-s)\mathrm{d}z' = \boldsymbol{\Gamma}(y,z,t-s). \qquad (2.30)$$

(iii) *For* $\boldsymbol{u} \in \mathcal{S}(\mathbb{R}^3)$,

$$\lim_{(y,t)\to(y^0,0^+)}\int_{\mathbb{R}^3}\boldsymbol{\Gamma}(y,z,t)\boldsymbol{u}(z)\mathrm{d}z = P\boldsymbol{u}(y^0).$$

(iv) *The (backward in time) adjoint problem*

$$(-\partial_s+\mathcal{L}^*)\boldsymbol{w}+\nabla\pi = \boldsymbol{g},\ \nabla\cdot\boldsymbol{w}=0 \quad on\ \mathbb{R}^3\times(0,T),\ \boldsymbol{w}(T)=0,$$

with the operator $\mathcal{L}^*\boldsymbol{w} = -\Delta\boldsymbol{w}+(\boldsymbol{U}+\boldsymbol{\omega}\wedge\boldsymbol{y})\cdot\nabla\boldsymbol{w}-\boldsymbol{\omega}\wedge\boldsymbol{w}$ *has the fundamental solution*

$$\boldsymbol{\Gamma}'(z,y,s) = \boldsymbol{\Gamma}(y,z,t).$$

Recall that Π_σ denotes the Helmholtz projection in $L^2(\mathcal{D})$. Considering it in $\mathbb{R}^3$, we can write

$$\Pi_\sigma = I+\mathcal{R} = I+\nabla\,\mathrm{div}(-\Delta)^{-1},$$

where $\mathcal{R}$ is the (3×3)-matrix operator with entries $(\mathcal{R}_i\mathcal{R}_j)_{i,j}$.
The Kummer function ${}_1F_1(1,c,u)$ is defined by

$${}_1F_1(1,c,u) := \sum_{n=0}^{\infty}\frac{\Gamma_a(c)}{\Gamma_a(n+c)}u^n \quad \text{for } u\in\mathbb{R},\ c\in(0,\infty). \qquad (2.31)$$

As basic results for Kummer functions, we mention the following facts:

Lemma 2.11 *For $a, c > 0$ the following results hold:*

(L1)

$$ {}_1F_1(1, c, \lambda) = \sum_{n=0}^{\infty} \frac{1}{(c)_n} \lambda^n, $$

where $(c)n = \Gamma_a(c+n)/\Gamma_a(c)$ is the Pochhammer symbol.

(L2)

$$ \frac{\mathrm{d}}{\mathrm{d}\lambda} {}_1F_1(a, c, \lambda) = \frac{a}{c} {}_1F_1(a+1, c+1, \lambda). $$

(L3) *There exists a constant $C > 0$ such that for all $\lambda > 0$*

$$ |\mathrm{e}^{-\lambda}({}_1F_1(1, c, \lambda) - 1)| \leq C \frac{\lambda}{(1+\lambda)^c}. $$

(L4)

$$ \begin{aligned} \frac{\mathrm{d}}{\mathrm{d}\lambda}(\mathrm{e}^{-\lambda}({}_1F_1(1, c, \lambda) - 1)) =& \frac{1}{c}\mathrm{e}^{-\lambda}\, {}_1F_1(1, c+1, \lambda) \\ & - \frac{\lambda}{c+1}\mathrm{e}^{-\lambda}\, {}_1F_1(1, c+2, \lambda), \\ \frac{\mathrm{d}^2}{\mathrm{d}\lambda^2}(\mathrm{e}^{-\lambda}({}_1F_1(1, c, \lambda) - 1)) =& \frac{-2}{c+1}\mathrm{e}^{-\lambda}\, {}_1F_1(1, c+2, \lambda) \\ & + \frac{\lambda}{c+2}\mathrm{e}^{-\lambda}\, {}_1F_1(1, c+3, \lambda). \end{aligned} $$

In particular, there exists a constant $C > 0$ such that for all $\lambda > 0$ and for $j = 1, 2$

$$ \left| \frac{\mathrm{d}^j}{\mathrm{d}\lambda^j}(\mathrm{e}^{-\lambda}({}_1F_1(1, c, \lambda) - 1)) \right| \leq C \frac{1}{(1+\lambda)^{c+j-1}}. $$

Proof **(L1)** and **(L2)** can be found in [124, page 82]. For the proof of **(L3)** we use the Gamma function Γ_a, the asymptotic result

$$ \mathrm{e}^{-\lambda}\, {}_1F_1(1, c, \lambda) \sim \Gamma_a(c) \frac{1}{\lambda^{c-1}} \quad \text{as} \quad \lambda \to \infty, \tag{2.32} $$

see [124, page 82], and that ${}_1F_1(1, c, 0) = 1$. **(L4)** $_1$ follows from the formula

$$\frac{\mathrm{d}}{\mathrm{d}\lambda}(\mathrm{e}^{-\lambda}\,{}_1F_1(1, c, \lambda)) = \frac{1-c}{c}\mathrm{e}^{-\lambda}\,{}_1F_1(1, c+1, \lambda),$$

see [124, Lemma 2.1], and the identity

$${}_1F_1(1, c, \lambda) - 1 = \frac{1}{c}\lambda\,{}_1F_1(1, c+1, \lambda),$$

see [124, equation (4.9)]. The second equation in (4) is proved analogously. The estimates follows from (2.32). □

First, ignoring the pressure term and the solenoidality condition in (2.21), we consider the linear operator

$$\widetilde{\mathcal{L}}\boldsymbol{w} = \widetilde{\mathcal{L}}_{y,t}\boldsymbol{w} = -\Delta\boldsymbol{w} - (\boldsymbol{U} + \boldsymbol{\omega} \wedge \boldsymbol{y}) \cdot \nabla\boldsymbol{w} + \boldsymbol{\omega} \wedge \boldsymbol{w}. \tag{2.33}$$

Proposition 2.1 *Assume $\boldsymbol{w}_0 \in \mathcal{S}(\mathbb{R}^3)$. Then the solution of the initial value problem*

$$\frac{\partial \boldsymbol{w}}{\partial t} + \widetilde{\mathcal{L}}\boldsymbol{w} = 0 \quad \text{in } (0, \infty), \quad \boldsymbol{w}(\cdot, s) = \boldsymbol{w}_0 \tag{2.34}$$

is given by

$$\boldsymbol{w}(y, t) = \int_{\mathbb{R}^3} \widetilde{\boldsymbol{\Gamma}}(y, z, t)\boldsymbol{w}_0(z)\,\mathrm{d}z, \tag{2.35}$$

where

$$\widetilde{\boldsymbol{\Gamma}}(y, z, t) = K(y - z(t), t)\mathrm{e}^{-t\boldsymbol{\Omega}}. \tag{2.36}$$

Proof By two elementary transformations we will reduce the problem (2.34) to the simpler problem

$$\frac{\partial \boldsymbol{v}}{\partial t} - (\boldsymbol{\omega} \wedge \boldsymbol{y}) \cdot \nabla\boldsymbol{v} - \Delta\boldsymbol{v} = 0 \quad \text{in } (0, \infty),\ \boldsymbol{v}(0) = \boldsymbol{w}_0. \tag{2.37}$$

First let $\boldsymbol{w}^*(t) = \exp(t\boldsymbol{\Omega})\boldsymbol{w}$. Then

$$\frac{\partial \boldsymbol{w}^*}{\partial t} - (\boldsymbol{U} + \boldsymbol{\omega} \wedge \boldsymbol{y}) \cdot \nabla\boldsymbol{w}^* - \Delta\boldsymbol{w}^* = 0 \quad \text{in } (0, \infty),\ \boldsymbol{w}^*(0) = \boldsymbol{w}_0. \tag{2.38}$$

Next, let

$$\boldsymbol{v}(y, t) = \boldsymbol{w}^*(y - \boldsymbol{U}t, t). \tag{2.39}$$

Then we get the problem

$$\frac{\partial \boldsymbol{v}}{\partial t} = (\boldsymbol{\omega} \wedge \boldsymbol{y}) \cdot \nabla \boldsymbol{v} + \Delta \boldsymbol{v}. \tag{2.40}$$

Now the solution v to (2.40) can be written in the form

$$\begin{aligned}\boldsymbol{v}(\boldsymbol{y}, t) &= \int_{\mathbb{R}^3} K(z, t)\boldsymbol{w}_0(\mathrm{e}^{t\boldsymbol{\Omega}}\boldsymbol{y} - z)\,\mathrm{d}z \\ &= \frac{1}{(4\pi t)^{3/2}} \int_{\mathbb{R}^3} \exp\left(-\frac{|\mathrm{e}^{t\boldsymbol{\Omega}}\boldsymbol{y} - z|^2}{4t}\right) \boldsymbol{w}_0(z)\,\mathrm{d}z,\end{aligned}$$

see, e.g., DaPrato and Lunardi in [27]. Hence we get

$$\begin{aligned}\boldsymbol{w}(\boldsymbol{y}, t) &= \mathrm{e}^{-t\boldsymbol{\Omega}}\boldsymbol{w}^*(\boldsymbol{y}, t) = \mathrm{e}^{-t\boldsymbol{\Omega}}\boldsymbol{v}(\boldsymbol{y} + t\boldsymbol{U}, t) \\ &= \frac{1}{(4\pi t)^{3/2}}\mathrm{e}^{-t\boldsymbol{\Omega}} \int_{\mathbb{R}^3} \exp\left(-\frac{|\mathrm{e}^{t\,\boldsymbol{\Omega}}(\boldsymbol{y} + t\boldsymbol{U}) - z|^2}{4t}\right) w_0(z)\,\mathrm{d}z.\end{aligned}$$

□

To obtain the fundamental solution of the linearized problem (2.21) taking into account the incompressibility condition, we have to adapt Proposition 2.1, cf. [124]. Using the Helmholtz projection Π_σ, it is easy to see that for every fixed $\boldsymbol{a} \in \mathbb{R}^3$

$$\boldsymbol{\Gamma}(\boldsymbol{y}, z, t)\boldsymbol{a} = \Pi_\sigma(\widetilde{\boldsymbol{\Gamma}}(\boldsymbol{y}, z, t)\boldsymbol{a}), \quad \boldsymbol{Q}(\boldsymbol{y}, z, t)\boldsymbol{a} = -\frac{1}{4\pi}\boldsymbol{a}\nabla\frac{1}{|\boldsymbol{y} - z|}\delta_0(t)$$

is the fundamental tensor for the linear equation (2.34); here, Π_σ acts on the variable $\boldsymbol{y}$. In particular, for $t > 0$

$$\left(\frac{\partial}{\partial t} + \mathcal{L}\right)(\boldsymbol{\Gamma}\boldsymbol{a}) + \nabla \boldsymbol{Q}\boldsymbol{a} = 0, \quad \nabla \cdot (\boldsymbol{\Gamma}\boldsymbol{a}) = 0. \tag{2.41}$$

Since

$$\boldsymbol{\Gamma}(\boldsymbol{y}, z, t)\boldsymbol{a} = \Pi_\sigma\widetilde{\boldsymbol{\Gamma}}(\boldsymbol{y}, z, t)\boldsymbol{a} = (I+\mathcal{R})\widetilde{\boldsymbol{\Gamma}}(\boldsymbol{y}, z, t)\boldsymbol{a} = [(I+\mathcal{R})K(\boldsymbol{y}-z(t), t]\mathrm{e}^{-t\boldsymbol{\Omega}}\boldsymbol{a}$$

and $\mathcal{R}_i\mathcal{R}_j\boldsymbol{f} = \partial/\partial y_i\partial/\partial y_j(-\Delta)^{-1}\boldsymbol{f}$, we get

$$\boldsymbol{\Gamma}(\boldsymbol{y}, z, t)\boldsymbol{a} = [K(\boldsymbol{y} - z(t), t)I + \mathrm{Hess}\,\psi(\boldsymbol{y}, z, \tau)]\mathrm{e}^{-t\boldsymbol{\Omega}}\boldsymbol{a}; \tag{2.42}$$

here $\psi(y, z, t)$ is the solution of the equation $-\Delta_y \psi(\boldsymbol{y}, t) = K(\boldsymbol{y} - \boldsymbol{z}(t), t)$, i.e.,

$$\psi(\boldsymbol{y}, \boldsymbol{z}, t) = \frac{1}{4\pi} \frac{1}{(4\pi t)^{3/2}} \int_{R^3} \frac{1}{|\boldsymbol{y} - \boldsymbol{x}|} \exp\left(-\frac{|\boldsymbol{x} - \boldsymbol{z}(t)|^2}{4t}\right) \mathrm{d}\boldsymbol{x}, \tag{2.43}$$

and Hess $\psi(\boldsymbol{y}, \tau) = ((\partial/\partial y_i)(\partial/\partial y_j))\psi(\boldsymbol{y}, \tau)$ denotes the Hessian of ψ.

To compute ψ and its Hessian, we follow [124] and introduce the

$$\operatorname{Erf}(s) = \frac{2}{\sqrt{\pi}} \int_0^s \mathrm{e}^{-u^2}\, \mathrm{d}u = \frac{2s}{\sqrt{\pi}} \mathrm{e}^{-s^2}\, {}_1F_1\left(1, \frac{3}{2}, s^2\right).$$

Lemma 2.12 *For all $t > 0$*

$$\psi(\boldsymbol{y}, \boldsymbol{z}, t) = \frac{1}{4\pi|\boldsymbol{y} - \boldsymbol{z}(t)|} \operatorname{Erf}\left(\frac{|\boldsymbol{y} - \boldsymbol{z}(t)|}{\sqrt{4t}}\right) \tag{2.44}$$

and

$$\begin{aligned}\frac{\partial^2}{\partial y_i \partial y_j} \psi(\boldsymbol{y}, \boldsymbol{z}, t) =& K(\boldsymbol{y} - \boldsymbol{z}(t), t) \left(-\frac{1}{3}\, {}_1F_1\left(1, \frac{5}{2}, \frac{|\boldsymbol{y} - \boldsymbol{z}(t)|^2}{4t}\right) \delta_{ij}\right. \\ &\left. + \frac{(y_i - z_i(t))(y_j - z_j(t))}{|\boldsymbol{y} - \boldsymbol{z}(t)|^2} \left[{}_1F_1\left(1, \frac{5}{2}, \frac{|\boldsymbol{y} - \boldsymbol{z}(t)|^2}{4t}\right) - 1\right]\right).\end{aligned}$$

Proof See [124, Lemma 3.1, Proposition 3.2]. □

From Proposition 2.1 and Lemma 2.12 it follows for all $\boldsymbol{a} \in \mathbb{R}^3$ that $(\partial_t + \mathcal{L})(\boldsymbol{\Gamma a}) + \nabla(\boldsymbol{Q a}) = 0$ for $t > 0$ and $\operatorname{div}(\boldsymbol{\Gamma a}) = 0$. It remains to show, for every initial value $\boldsymbol{u}_0 \in \mathcal{S}(\mathbb{R}^3)^3$ with the Helmholtz decomposition $\boldsymbol{u}_0 = \boldsymbol{h} + \nabla q$, that

$$\lim_{t \to 0+} \int_{\mathbb{R}^3} \boldsymbol{\Gamma}(y, z, t)\boldsymbol{u}_0(z)\mathrm{d}z + \nabla_y \int_{\mathbb{R}^3} \boldsymbol{Q}^*(y, z)\boldsymbol{u}_0(z)\mathrm{d}z = \boldsymbol{u}_0(y). \tag{2.45}$$

It can be found in detail, see [124].

2.5.1 *Basic Properties of the Fundamental Solution $\boldsymbol{\Gamma} = \boldsymbol{\Gamma}_{jk}$*

We will use the following notation in this paragraph:

$$\boldsymbol{w} = \boldsymbol{y} - \boldsymbol{z}(t), \quad \hat{\boldsymbol{w}} = \frac{\boldsymbol{w}}{|\boldsymbol{w}|},$$

$$l(\hat{\boldsymbol{w}}) = \hat{\boldsymbol{w}} \otimes \hat{\boldsymbol{w}},$$

$$\lambda = \frac{|\boldsymbol{y} - \boldsymbol{z}(t)|^2}{4t},$$

$$\mathcal{G}(\lambda) = \mathrm{e}^{-\lambda}\ {}_1F_1\left(1, \frac{5}{2}, \lambda\right),$$

$$\mathcal{M}(\boldsymbol{y}, \boldsymbol{z}, t) = \frac{1}{3}\frac{1}{(4\pi t)^{3/2}}\mathcal{G}(\lambda)[I - 3 \cdot l(\hat{\boldsymbol{w}})],$$

so that

$$\boldsymbol{\Gamma}(\boldsymbol{y}, \boldsymbol{z}, t) = \left[K(\boldsymbol{y} - \boldsymbol{z}(t), t)\left(I - l(\hat{\boldsymbol{w}})\right) - \mathcal{M}(\boldsymbol{y}, \boldsymbol{z}, t)\right]\mathrm{e}^{-t\boldsymbol{\Omega}}.$$

Proposition 2.2 *The fundamental solution* $\boldsymbol{\Gamma}$ *has (in each component of the* (3×3)*-matrix) the following asymptotic properties:*

(i) *For any vectors* $\boldsymbol{y}, \boldsymbol{z} \in \mathbb{R}^3$, $\boldsymbol{y} \neq \boldsymbol{z}$,

$$\boldsymbol{\Gamma}(\boldsymbol{y}, \boldsymbol{z}, t) \sim -\frac{1}{4\pi}\frac{1}{|\boldsymbol{y} - \boldsymbol{z}|^3}\left[I - 3\frac{(\boldsymbol{y} - \boldsymbol{z}) \otimes (\boldsymbol{y} - \boldsymbol{z})}{|\boldsymbol{y} - \boldsymbol{z}|^2}\right] \quad \text{as } t \to 0.$$

(ii) *For any vectors* $\boldsymbol{y}, \boldsymbol{z} \in \mathbb{R}^3$ *and for large* t*, which means* $t \to \infty$*, we consider two cases:*

- *When* $\boldsymbol{U} = 0$*, then*

$$\mathrm{e}^{t\boldsymbol{\Omega}}\boldsymbol{\Gamma}(\boldsymbol{y}, \boldsymbol{z}, t) \sim \frac{2}{3}\frac{1}{(4\pi t)^{3/2}}I, \ t \to \infty$$

- *When* $\boldsymbol{U} \neq 0$,

$$\boldsymbol{\Gamma}(\boldsymbol{y}, \boldsymbol{z}, t) \sim -\frac{1}{4\pi}\frac{1}{|t\boldsymbol{U}|^3}\left[I - 3\frac{\boldsymbol{U} \otimes \boldsymbol{U}}{|\boldsymbol{U}|^2}\right], \ t \to \infty.$$

(iii) *Let* $\boldsymbol{y}^0, \boldsymbol{z}, \boldsymbol{s} \in \mathbb{R}^3$, $|\boldsymbol{s}| = 1$*, be fixed, and let* $\boldsymbol{y} = \boldsymbol{y}^0 + \rho\boldsymbol{\zeta}$, $\rho > 0$*. Then for* $t > 0$

$$\boldsymbol{\Gamma}(\boldsymbol{y}, \boldsymbol{z}, t) \sim -\frac{1}{4\pi}\frac{1}{|\boldsymbol{y} - \boldsymbol{z}(t)|^3}[I - 3\boldsymbol{\zeta} \otimes \boldsymbol{\zeta}] \quad \text{as } \rho \to \infty.$$

Proof

(i) Since $\boldsymbol{y} \neq \boldsymbol{z}$, the term $\lambda \to \infty$ as $t \to 0$. Hence the leading term in $\boldsymbol{\Gamma}$ is determined by $\mathcal{M}$, where by Lemma 2.11 (3), $\mathcal{G}(\lambda) \sim \Gamma_a(5/2)\lambda^{-3/2} = \frac{3}{4}\sqrt{\pi}\lambda^{-3/2}$. This proves (i).

(iii) In this case $\lambda \to \infty$ and the leading term in $\boldsymbol{\Gamma}$ is determined by $\mathcal{M}$, cf. (i). Since $l(\hat{\boldsymbol{w}}) \sim \boldsymbol{\zeta} \otimes \boldsymbol{\zeta}$ as $\rho \to \infty$ for $t > s$ fixed, we get (iii).

(ii) In the case $\boldsymbol{U} = 0$, then $u = |\boldsymbol{y} - \mathrm{e}^{-t\boldsymbol{\Omega}}\boldsymbol{z}|^2/4t \to 0$ and

$$\mathrm{e}^{-u}[1 - {}_1F_1(1, 5/2, u)] \to 0;$$

therefore the first part of (ii) is proven.

In the second case $\boldsymbol{z}(t) = \mathrm{e}^{-t\boldsymbol{\Omega}}\boldsymbol{z} - t\boldsymbol{U}$, and we get for large t that

$$\lambda = \frac{|\boldsymbol{y} - \boldsymbol{z}(t)|^2}{4t} \sim \frac{t|\boldsymbol{U}|^2}{4},$$
$$\widehat{\boldsymbol{w}} = \frac{\boldsymbol{y} - \boldsymbol{z}(t)}{|\boldsymbol{y} - \boldsymbol{z}(t)|} \sim \frac{\boldsymbol{U}}{|\boldsymbol{U}|}.$$

Since by Lemma 2.11 (3) the leading term in Γ is determined by M, we have

$$\begin{aligned}\mathrm{e}^{t\boldsymbol{\Omega}}\Gamma(\boldsymbol{y}, \boldsymbol{z}, t) &\sim -\frac{1}{3}\frac{\Gamma(5/2)}{(4\pi t)^{3/2}}\left(\frac{4}{|t\boldsymbol{U}|^2}\right)^{3/2}\left[I - 3\frac{\boldsymbol{U}\otimes\boldsymbol{U}}{|\boldsymbol{U}|^2}\right] \\ &= -\frac{1}{4\pi}\frac{1}{|t\boldsymbol{U}|^3}\left[I - 3\frac{\boldsymbol{U}\otimes\boldsymbol{U}}{|\boldsymbol{U}|^2}\right]\end{aligned}$$

as $t \to \infty$.

□

We will give a summary of previous results, Proposition 2.2 and Theorems 2.4 and 2.5 with results from [124]:

Theorem 2.6 ([124, Proposition 4.1], Proposition 2.2 (i)) *Let $j, k \in \{1, 2, 3\}$, $\boldsymbol{y}, \boldsymbol{z} \in \mathbb{R}^3$ with $\boldsymbol{y} \neq \boldsymbol{z}$. Then*

$$\Gamma_{jk}(\boldsymbol{y}, \boldsymbol{z}, t) \to -(4\pi)^{-1}(\delta_{jk}|\boldsymbol{y} - \boldsymbol{z}|^{-3} - 3(\boldsymbol{y} - \boldsymbol{z})_j(\boldsymbol{y} - \boldsymbol{z})_k|\boldsymbol{y} - \boldsymbol{z}|^{-5})$$

for $t \downarrow 0$.

Theorem 2.7 ([124, Theorem 1.3], Theorems 2.4 and 2.5) *Let $\boldsymbol{y}, \boldsymbol{z} \in \mathbb{R}^3$ with $\boldsymbol{y} \neq \boldsymbol{z}$, $t \in (0, \infty)$, $j, k \in \{1, 2, 3\}$. Then*

$$\begin{aligned}\partial_t\Gamma_{jk}(\boldsymbol{y}, \boldsymbol{z}, t) - \Delta_z\Gamma_{jk}(\boldsymbol{y}, \boldsymbol{z}, t) + (\boldsymbol{U} + \boldsymbol{\omega} \wedge \boldsymbol{z})\nabla_z\Gamma_{jk}(\boldsymbol{y}, \boldsymbol{z}, t) \\ - [\boldsymbol{\omega} \wedge (\Gamma_{js}(\boldsymbol{y}, \boldsymbol{z}, t))_{1\le s\le 3}]_k = 0. \qquad (2.46)\end{aligned}$$

We further set

$$E_{4j}(\boldsymbol{x}) := (4\pi)^{-1}x_j|\boldsymbol{x}|^{-3} \quad (1 \le j \le 3,\ \boldsymbol{x} \in \mathbb{R}^3\backslash\{0\}).$$

E_{4j} is the pressure part of the fundamental tensor of the stationary Oseen problem. It follows

$$Q_j(\boldsymbol{y}, \boldsymbol{z}, t) = E_{4j}(\boldsymbol{y} - \boldsymbol{z}(t))\delta_0(t).$$

Theorem 2.8 ([124, Proposition 4.1], Theorems 2.4 and 2.5) *Let $j, k \in \{1, 2, 3\}$, $\boldsymbol{y}, \boldsymbol{z} \in \mathbb{R}^3$. Then*

$$\Gamma_{jk}(\boldsymbol{y}, \boldsymbol{z}, t) \to -\partial_k E_{4j}(\boldsymbol{y} - \boldsymbol{z}) \quad \textit{for } t \downarrow 0, \quad \textit{if } \boldsymbol{y} \neq \boldsymbol{z}. \tag{2.47}$$

2.5.2 Further Properties of Γ_{jk}: Pointwise Estimates

We derive some additional properties of the fundamental solution of the evolution problem in order to derive the fundamental solution of the stationary problem. We will need the following technical lemmas. Let us introduce the term $s_\tau(\boldsymbol{x})$ as

$$s_\tau(\boldsymbol{x}) := 1 + \tau\ (|\boldsymbol{x}| - x_1) \quad \text{for } \boldsymbol{x} \in \mathbb{R}^3.$$

Lemma 2.13 ([58, Lemma 4.3]) *Let $\beta \in (1, \infty)$. Then*

$$\int_{\partial B_r} s_\tau(\boldsymbol{x})^{-\beta} \mathrm{d}o_x \leq C(\beta, \tau)\, r \quad \textit{for } r \in (0, \infty).$$

Proof $s_\tau(\boldsymbol{x}) \geq \min(1, \tau)\, s_1(\boldsymbol{x})$, $s_\tau(\boldsymbol{x})^{-\beta} \leq \min(1, \tau)^{-\beta} \int_{\partial B_r} s_1(\boldsymbol{x})^{-\beta} \mathrm{d}o_x$. Introducing the spherical coordinates $r > 0$, $\varphi \in [0, 2\pi)$, $\theta \in [-\pi/2, \pi/2]$, with $x_1 = r\sin\theta$, $x_2 = r\cos\varphi\cos\theta$, $x_3 = r\sin\varphi\cos\theta$, and $v = r(1 - \sin\theta)$, we get

$$\begin{aligned}
&\min(1, \tau)^{-\beta} \int_{\partial B_r} s_1(\boldsymbol{x})^{-\beta} \mathrm{d}o_x \\
&= \min(1, \tau)^{-\beta} \int_{-\frac{\pi}{2}}^{\frac{\pi}{2}} \int_0^{2\pi} (1 + r(1 - \sin\theta))^{-\beta} r^2 \cos\theta \,\mathrm{d}\varphi \mathrm{d}\theta \\
&= 2\pi r \min(1, \tau)^{-\beta} \int_0^{2r} (1 + v)^{-\beta} \mathrm{d}v \leq 2\pi r \min(1, \tau)^{-\beta} \int_0^{\infty} (1 + v)^{-\beta} \mathrm{d}v \\
&= 2\pi(\beta - 1)^{-1} \min(1, \tau)^{-\beta} r.
\end{aligned}$$

□

Lemma 2.14 ([34, Lemma 4.8]) *For* $\boldsymbol{x}, \boldsymbol{y} \in \mathbb{R}^3$, *we have*

$$s_\tau(\boldsymbol{x}-\boldsymbol{y})^{-1} \le C(\tau)(1+|\boldsymbol{y}|)s_\tau(\boldsymbol{x})^{-1}.$$

Proof First, let us note $s_\tau(\boldsymbol{x}) \ge 1$ for all $\boldsymbol{x} \in \mathbb{R}^3$.

1. Assuming $|\boldsymbol{x}| - x_1 \le 4\,|\boldsymbol{y}|$, we get
   $s_\tau(\boldsymbol{x}) = 1+\tau(|\boldsymbol{x}|-x_1) \le 1+4\tau|\boldsymbol{y}| \le 4\max(1,\tau)(1+|\boldsymbol{y}|)$
   $\le 4\max(1,\tau)(1+|\boldsymbol{y}|)\,s_\tau(\boldsymbol{x}-\boldsymbol{y})$. So, the inequality is valid with the constant $C(\tau) = 4\max(1,\tau)$.
2. Now, assuming $|\boldsymbol{x}| - x_1 \ge 4\,|\boldsymbol{y}|$, we have
   $2\,s_\tau(\boldsymbol{x}) \le 2\,s_\tau(\boldsymbol{x}) + 2\tau(|\boldsymbol{y}|+y_1) = 2(1+\tau(|\boldsymbol{x}|-x_1)) - 2\tau(|\boldsymbol{y}|-y_1) + 4\,\tau|\boldsymbol{y}|$
   $\le 2(1+\tau(|\boldsymbol{x}-\boldsymbol{y}|-(x_1-y_1))+(1+\tau(\boldsymbol{x}-x_1)) = 2s_\tau(\boldsymbol{x}-\boldsymbol{y})+s_\tau(\boldsymbol{x})$. So, we get $s_\tau(\boldsymbol{x}) \le 2s_\tau(\boldsymbol{x}-\boldsymbol{y})$, and the inequality is valid with the constant $C(\tau)=2$.

Summarizing, we proved the inequality of the lemma with $C(\tau) = 4\max(1,\tau)$. □

Lemma 2.15 ([32, Lemma 2]) *Let* $S \in (0,\infty)$, $\boldsymbol{x} \in B_S$, $t \in (0,\infty)$. *Then*

$$|\boldsymbol{x} - \tau t\mathbf{e}_1|^2 + t \ge C(S,\tau)(|\boldsymbol{x}|^2+t).$$

Proof Assuming $c \in (0,1)$, $\boldsymbol{x} = [x_1, x_2 x_3]$, we define quadratic polynomial $P_2(t) = P_2(t;\boldsymbol{x},\tau,c) = |\boldsymbol{x}-\tau t\mathbf{e}_1|^2 + t - c^2(|\boldsymbol{x}|^2+t)$.

We will prove $P_2(t) \ge 0,\ t \in (0,\infty)$.

$P_2(t) = (1-c^2)(|\boldsymbol{x}|^2 - (\boldsymbol{x}\cdot\boldsymbol{e}_1)^2) + (\tau t - (\boldsymbol{x}\cdot\boldsymbol{e}_1))^2 - c^2(\boldsymbol{x}\cdot\boldsymbol{e}_1)^2 + (1-c^2)t$
$= (1-c^2)(x_2^2+x_3^2) + (\tau t - x_1)^2 - c^2x_1^2 + (1-c^2)t$. The minimum $P_2(t)$ on interval $(-\infty,+\infty)$ occurs at point $t_{\min} = x_1\tau^{-1} - (1-c^2)2^{-1}\tau^{-2}$ and $P_2(t_{\min}) = (1-c^2)(x_2^2+x_3^2) + (1-c^2)^2 4^{-2}\tau^{-2} - c^2x_1^2 + (1-c^2)t_{\min}$.

1. If $t_{\min} \le 0$, then $P_2(t)$ is increasing on $(0,+\infty)$, and $P_2(0) > 0$. Hence, $P_2(t) > 0$ for $t \in (0,+\infty)$, so the inequality of the lemma is valid for arbitrary constant $C = c^2 \in (0,1)$.
2. If $t_{\min} > 0$, then $P_2(t) > (1-c^2)^2 4^{-2}\tau^{-2} - c^2x_1^2 \ge 0$, where the last inequality is satisfied for $x_1 \le |\boldsymbol{x}| \le 2^{-2}(1-c^2)c^{-1}\tau^{-1} = S$. From the last formula we get the explicit expression for $c = c(S,\tau) = (\sqrt{4\tau^2S^2+1}+2\tau S)^{-1}$.

Summarizing, we get that the inequality of the lemma is valid for $C(S,\tau) = c(S,\tau)^2 = (\sqrt{4\tau^2S^2+1}+2\tau S)^{-2}$. □

Lemma 2.16 *Let* $S \in (0,\infty)$, $\boldsymbol{x} \in B_S^c$. *Then* $|\boldsymbol{x}| \ge C(S)s_\tau(\boldsymbol{x})$.

Proof $|\boldsymbol{x}| \ge S/2 + |\boldsymbol{x}|/2 \ge S/2 + (|\boldsymbol{x}|-x_1)/4 \ge \min\{S/2,\ 1/(4\tau)\}s_\tau(x)$. □

We define some further functions which will be useful in the following:

$$\mathcal{S}(u) := \sum_{n=1}^{\infty} \Gamma_a\left(\frac{5}{2}\right) \Gamma_a\left(\frac{5}{2}+n\right)^{-1} u^{n-1} \quad \text{for } u \in \mathbb{R},$$

$$\Lambda_{jk}(\boldsymbol{x},t) := K(\boldsymbol{x},t)\Big(\delta_{jk} - x_j x_k |\boldsymbol{x}|^{-2} - {}_1F_1\left(1, \frac{5}{2}, \frac{|\boldsymbol{x}|^2}{4t}\right)\left(\frac{\delta_{jk}}{3} - x_j x_k |\boldsymbol{x}|^{-2}\right)\Big)$$

for $\boldsymbol{x} \in \mathbb{R}^3 \backslash \{0\}$, $t \in (0,\infty)$, $j,k \in \{1,2,3\}$. Note that

$$(\Gamma_{jk}(\boldsymbol{y},\boldsymbol{z},t))_{1 \le j,k \le 3} = (\Lambda_{rs}(\boldsymbol{y}-\boldsymbol{z}(t),t))_{1 \le r,s \le 3}\, \mathrm{e}^{-t\boldsymbol{\Omega}} \tag{2.48}$$

for $\boldsymbol{y}, \boldsymbol{z} \in \mathbb{R}^3$, $t \in (0,\infty)$ with $\boldsymbol{y} \ne \boldsymbol{z}(t)$.[3] The ensuing estimates of Kummer functions are standard (see [171] for (2.49)).

Theorem 2.9 *Let $s \in (0,\infty)$. Then there is $C(s) > 0$ such that*

$$\left|\frac{\mathrm{d}^k}{\mathrm{d}u^k}\left(\mathrm{e}^{-u}\, {}_1F_1\left(1, \frac{5}{2}, u\right)\right)\right| \le C(s) u^{-3/2-k} \quad \textit{for } u \in [s,\infty),\ k \in \{0,1,2\}, \tag{2.49}$$

$$\left|\frac{\mathrm{d}^k}{\mathrm{d}u^k}\, {}_1F_1\left(1, \frac{5}{2}, u\right)\right| + |\mathcal{S}^{(k)}(u)| \le C(s), \quad \left|1 - {}_1F_1\left(1, \frac{5}{2}, u\right)\right| \le C(s)|u| \quad \textit{for } u \in [-s,s],\ k \in \{0,1,2\}. \tag{2.50}$$

Of course, in the preceding theorem one might admit any $k \in \mathbb{N}$, but we will need only values k from $\{0,1,2\}$.

The ensuing lemma is easy to verify.

Lemma 2.17 *For $\boldsymbol{y}, \boldsymbol{z} \in \mathbb{R}^3$, $t \in (0,\infty)$, put*

$$\widetilde{\boldsymbol{\Lambda}}(\mathbf{x},t) := K(\mathbf{x},t)\left[\left(1 - {}_1F_1\left(1, \frac{5}{2}, \frac{|\boldsymbol{x}|^2}{4t}\right)\frac{1}{3}\right)\delta_{rs}\ \mathcal{S}\Big(\frac{|\boldsymbol{x}|^2}{4t}\Big)\frac{(\boldsymbol{x})_r(\boldsymbol{x})_s}{4t}\right]_{1\le r,s\le 3},$$

$$\widetilde{\boldsymbol{\Gamma}}(\boldsymbol{y},\boldsymbol{z},t) := K(\boldsymbol{y}-\boldsymbol{z}(t),t)\Big[\left(1 - {}_1F_1\left(1, \frac{5}{2}, \frac{|\boldsymbol{y}-\boldsymbol{z}(t)|^2}{4t}\right)\frac{1}{3}\right)\delta_{rs}\ \mathcal{S}\Big(\frac{|\boldsymbol{y}-\boldsymbol{z}(t)|^2}{4t}\Big)\frac{(\boldsymbol{y}-\boldsymbol{z}(t))_r(\boldsymbol{y}-\boldsymbol{z}(t))_s}{4t}\Big]_{1\le r,s\le 3}\mathrm{e}^{-t\boldsymbol{\Omega}},$$

---

[3] By $\mathcal{S}$ we mean just in this subsection $\mathcal{S} := \sum_{n=1}^{\infty} \Gamma_a\left(\frac{5}{2}\right)\Gamma_a\left(\frac{5}{2}+n\right)^{-1} u^{n-1}$ and not the space of the Schwartz space as in the rest of the book.

where $z(t)$ was introduced in (2.24). Then

$$\Lambda_{jk}(\boldsymbol{x},t)=\widetilde{\Lambda}_{jk}(\boldsymbol{x},t) \quad \text{for } \boldsymbol{x}\in\mathbb{R}^3,\ t\in(0,\infty) \text{ with } \boldsymbol{x}\neq\boldsymbol{0}, \tag{2.51}$$

$$\Gamma_{jk}(\boldsymbol{y},\boldsymbol{z},t)=\widetilde{\Gamma}_{jk}(\boldsymbol{y},\boldsymbol{z},t) \quad \text{for } \boldsymbol{y},\boldsymbol{z}\in\mathbb{R}^3,\ t\in(0,\infty) \text{ with } \boldsymbol{y}\neq\boldsymbol{z}(t). \tag{2.52}$$

Since $\widetilde{\Lambda}_{jk}\in C^\infty(\mathbb{R}^3\times(0,\infty))$, $\widetilde{\Gamma}_{jk}\in C^\infty(\mathbb{R}^3\times\mathbb{R}^3\times(0,\infty))$, we may conclude.

Corollary 2.3 *The functions Λ_{jk} and Γ_{jk} can be extended continuously to a function from $C^\infty(\mathbb{R}^3\times(0,\infty))$ and $C^\infty(\mathbb{R}^3\times\mathbb{R}^3\times(0,\infty))$, respectively.*

Next we state that the function $\Gamma_{jk}(\boldsymbol{y},\boldsymbol{z},t)$ and its first order derivatives with respect to $\boldsymbol{z}$ may be estimated in a similar way as $K(\boldsymbol{y}-\boldsymbol{z}(t),t)$ and $\partial y_l K(\boldsymbol{y}-\boldsymbol{z}(t),t)$, respectively ($1\le l\le 3$).

Lemma 2.18 *For $x\in\mathbb{R}^3$, $t\in(0,\infty)$, $\boldsymbol{\beta}\in\mathbb{N}_0^3$ with $|\boldsymbol{\beta}|\le 1$, the inequality*

$$|\partial_x^{\boldsymbol{\beta}}\Lambda_{jk}(\boldsymbol{x},t)|\le C(|\boldsymbol{x}|^2+t)^{-3/2-|\beta|/2}$$

holds, where C is a positive constant.

By (2.48), *this means that*

$$|\partial_y^{\beta}\Gamma_{jk}(\boldsymbol{y},\boldsymbol{z},t)|+|\partial_z^{\beta}\Gamma_{jk}(\boldsymbol{y},\boldsymbol{z},t)|\le C(|\boldsymbol{y}-\boldsymbol{z}(t)|^2+t)^{-3/2-|\beta|/2}$$

for $\boldsymbol{y},\boldsymbol{z}\in\mathbb{R}^3$ and for t, $\boldsymbol{\beta}$ as above.

Proof Take $\boldsymbol{x}$, t as in the lemma, and suppose in addition that $\boldsymbol{x}\neq 0$. Let $n\in\{1,2,3\}$. We will consider two cases. Firstly, we suppose that $|\boldsymbol{x}|^2\le t$. It follows with (2.52), Lemma 2.9, and Theorem 2.9 with $s=1/4$ that

$$\begin{aligned}|\widetilde{\Lambda}_{jk}(\boldsymbol{x},t)|&\le CK(\boldsymbol{x},t)\left(1+\left|{}_1F_1\left(1,\frac{5}{2},\frac{|\boldsymbol{x}|^2}{4t}\right)\right|+\left|\mathcal{S}\left(\frac{|\boldsymbol{x}|^2}{4t}\right)\right|\frac{|\boldsymbol{x}|^2}{t}\right)\\&\le CK(\boldsymbol{x},t)\le C(|\boldsymbol{x}|^2+t)^{-3/2},\end{aligned} \tag{2.53}$$

$$\begin{aligned}|\partial_{x_n}\widetilde{\Lambda}_{jk}(\boldsymbol{x},t)|\le C\sum_{l=1}^{3}\Bigg[&|\partial_{x_l}K(\boldsymbol{x},t)|\left(1+\left|{}_1F_1\left(1,\frac{5}{2},\frac{|\boldsymbol{x}|^2}{4t}\right)\right|\right)\\&+K(\boldsymbol{x},t)\left|\frac{\mathrm{d}}{\mathrm{d}u}\,{}_1F_1\left(1,\frac{5}{2},u\right)\right|_{u=|x|^2/(4t)}\frac{|\boldsymbol{x}|}{t}\\&+|\partial_{x_l}K(\boldsymbol{x},t)|\left|\mathcal{S}\left(\frac{|\boldsymbol{x}|^2}{4t}\right)\right|\frac{|\boldsymbol{x}|^2}{t}\\&+K(\boldsymbol{x},t)\left|\mathcal{S}'\left(\frac{|\boldsymbol{x}|^2}{4t}\right)\right|\frac{|\boldsymbol{x}|^3}{t^2}+K(\boldsymbol{x},t)\left|\mathcal{S}\left(\frac{|\boldsymbol{x}|^2}{4t}\right)\right|\frac{|\boldsymbol{x}|}{t}\Bigg]\end{aligned} \tag{2.54}$$

$$
\begin{aligned}
&\le C\Big[(|\boldsymbol{x}|^2+t)^{-2}+(|\boldsymbol{x}|^2+t)^{-3/2}\frac{|\boldsymbol{x}|}{t}+(|\boldsymbol{x}|^2+t)^{-2}\frac{|\boldsymbol{x}|^2}{t}\\
&\quad+(|\boldsymbol{x}|^2+t)^{-3/2}\frac{|\boldsymbol{x}|^3}{t^2}+(|\boldsymbol{x}|^2+t)^{-3/2}\frac{|\boldsymbol{x}|}{t}\Big]\\
&\le C((|\boldsymbol{x}|^2+t)^{-2}+(|\boldsymbol{x}|^2+t)^{-3/2}t^{-1/2})\le C(|\boldsymbol{x}|^2+t)^{-2}.
\end{aligned}
$$

Secondly, we consider the case $|\boldsymbol{x}|^2 \ge t$. By Lemma 2.9 and (2.49) with $s = 1/4$,

$$
\begin{aligned}
|\Lambda_{jk}(\boldsymbol{x},t)| &\le C(K(\boldsymbol{x},t)+t^{-3/2}|\mathrm{e}^{-u}\,{}_1F_1(1,5/2,u)|_{u=|\boldsymbol{x}|^2/(4t)}) \qquad (2.55)\\
&\le C\left((|\boldsymbol{x}|^2+t)^{-3/2}+t^{-3/2}\left(\frac{t}{|\boldsymbol{x}|^2}\right)^{3/2}\right)\\
&\le C((|\boldsymbol{x}|^2+t)^{-3/2}+|\boldsymbol{x}|^{-3})\le C(|\boldsymbol{x}|^2+t)^{-3/2},
\end{aligned}
$$

$$
\begin{aligned}
|\partial_{x_n}\Lambda_{jk}(\boldsymbol{x},t)| \le& C\sum_{l=1}^{3}\Big[|\partial_{x_l}K(\boldsymbol{x},t)|+K(\boldsymbol{x},\,t)|\boldsymbol{x}|^{-1} \qquad (2.56)\\
&+t^{-3/2}\left|\frac{\mathrm{d}}{\mathrm{d}u}\left(\mathrm{e}^{-u}\,{}_1F_1\left(1,\frac{5}{2},u\right),\right)\right|_{u=|\boldsymbol{x}|^2/4t}\frac{|\boldsymbol{x}|}{t}\\
&+t^{-3/2}|\mathrm{e}^{-u}\,{}_1F_1\left(1,\frac{5}{2},u\right)|_{u=|\boldsymbol{x}|^2/4t}|\boldsymbol{x}|^{-1}\Big]\\
\le& C\Big[(|\boldsymbol{x}|^2+t)^{-2}+(|\boldsymbol{x}|^2+t)^{-3/2}|\boldsymbol{x}|^{-1}\\
&+t^{-5/2}\left(\frac{t}{|\boldsymbol{x}|^2}\right)^{5/2}|\boldsymbol{x}|+t^{-3/2}\left(\frac{t}{|\boldsymbol{x}|^2}\right)^{3/2}|\boldsymbol{x}|^{-1}\Big]\\
\le& C((|\boldsymbol{x}|^2+t)^{-2}+(|\boldsymbol{x}|^2+t)^{-3/2}|\boldsymbol{x}|^{-1}+|\boldsymbol{x}|^{-4})\le C(|\boldsymbol{x}|^2+t)^{-2}.
\end{aligned}
$$

From (2.53)–(2.56) the assertion of lemma holds. □

Concerning the matrix $\boldsymbol{\Omega}$, we observe:

Lemma 2.19 *Let* $\boldsymbol{x} \in \mathbb{R}^3$, $t \in \mathbb{R}$. *Then*

$$
(\mathrm{e}^{t\boldsymbol{\Omega}}\boldsymbol{x})_1 = x_1, \quad \mathrm{e}^{t\boldsymbol{\Omega}}\mathbf{e}_1 = \mathbf{e}_1.
$$

Proof The assertion immediately follows from the relation

$$\boldsymbol{\Omega} = \varrho \begin{pmatrix} 0 & 0 & 0 \\ 0 & 0 & -1 \\ 0 & 1 & 0 \end{pmatrix}.$$

□

Due to Lemma 2.19, we get the following lemma:

Lemma 2.20

$$(\Gamma_{jk}(\boldsymbol{y}, \boldsymbol{z}, t))_{1 \le j,k \le 3} = \mathrm{e}^{-t\boldsymbol{\Omega}} (\Lambda_{rs}(\mathrm{e}^{t\boldsymbol{\Omega}} \boldsymbol{y} - \tau t \mathbf{e}_1 - \boldsymbol{z}, t))_{1 \le r,s \le 3} \tag{2.57}$$

for $\boldsymbol{y}, \boldsymbol{z} \in \mathbb{R}^3$, $t \in (0, \infty)$.

Finally we prove an analog of Lemma 2.18 for the second derivatives of Λ_{jk}. We put $\eta(\boldsymbol{x}) := \boldsymbol{x} \otimes \boldsymbol{x} |\boldsymbol{x}|^{-2}$ for $\boldsymbol{x} \in \mathbb{R}^3 \backslash \{0\}$.

Lemma 2.21 *Let* $j, k, l, m \in \{1, 2, 3\}$, $\boldsymbol{x} \in \mathbb{R}^3$, $t \in (0, \infty)$. *Then*

$$|\partial_{x_l} \partial_{x_m} \Lambda_{jk}(\boldsymbol{x}, t)| \le C(|\boldsymbol{x}|^2 + t)^{-5/2}.$$

Proof In view of Lemma 2.16, we may suppose $\boldsymbol{x} \neq 0$. We have

$$|\partial^{\boldsymbol{\beta}} \eta_{jk}(\boldsymbol{x})| \le C |\boldsymbol{x}|^{-|\boldsymbol{\beta}|} \quad (\boldsymbol{\beta} \in \mathbb{N}_0^3 \text{ with } |\boldsymbol{\beta}| \le 2). \tag{2.58}$$

Moreover, for an arbitrary function $G \in C^2(\mathbb{R})$, the relations

$$\partial_{x_r} \left(G \left(\frac{|\boldsymbol{x}|^2}{4t} \right) \right) = G' \left(\frac{|\boldsymbol{x}|^2}{4t} \right) \frac{x_r}{2t}$$

$$\partial x_l \partial x_m \left(G \frac{|\boldsymbol{x}|^2}{4t} \right) = G' \frac{|\boldsymbol{x}|^2}{4t} \frac{\delta_{lm}}{2t} + G'' \left(\frac{|\boldsymbol{x}|^2}{4t} \right) \frac{x_l x_m}{4t^2}$$

hold ($l, m, r \in \{1, 2, 3\}$), so that

$$\left| \partial_{x_r} \left(G \left(\frac{|\boldsymbol{x}|^2}{4t} \right) \right) \right| \le C \left| G' \left(\frac{|\boldsymbol{x}|^2}{4t} \right) \right| \left(\frac{|\boldsymbol{x}|^2}{t} \right)^{1/2} t^{-1/2} \tag{2.59}$$

$$\left| \partial x_l \partial x_m \left(G \left(\frac{|\boldsymbol{x}|^2}{4t} \right) \right) \right| \le C \left(\left| G' \left(\frac{|\boldsymbol{x}|^2}{4t} \right) \right| + \left| G'' \left(\frac{|\boldsymbol{x}|^2}{4t} \right) \right| \frac{|\boldsymbol{x}|^2}{t} \right) t^{-1}.$$

Further, we distinguish as before two cases: the cases $|\boldsymbol{x}|^2 \leq t$ and $|\boldsymbol{x}|^2 > t$. First, suppose that $|\boldsymbol{x}|^2 \leq t$. Then $|\boldsymbol{x}|^2/4t \leq 1/4$, $t = t/2 + t/2 \geq (|\boldsymbol{x}|^2 + t)/2$, so that for $G \in C^2(\mathbb{R})$, with Lemma 2.9 and (2.59)

$$\begin{aligned}
&\left|\partial_{x_r}\left(K(\boldsymbol{x},t)G\left(\frac{|\boldsymbol{x}|^2}{4t}\right)\right)\right| \\
&\leq \left|\partial_{x_r}K(\boldsymbol{x},t)\,G\left(\frac{|\boldsymbol{x}|^2}{4t}\right)\right| + K(\boldsymbol{x},t)\left|\partial_{x_r}\left(G\left(\frac{|\boldsymbol{x}|^2}{4t}\right)\right)\right| \\
&\leq C\Big((|\boldsymbol{x}|^2+t)^{-2}\left\|G|\left[0,\frac{1}{4}\right]\right\|_\infty + (|\boldsymbol{x}|^2+t)^{-3/2}\left\|G'|\left[0,\frac{1}{4}\right]\right\|_\infty \frac{|\boldsymbol{x}|^2}{t}t^{-1/2}\Big) \\
&\leq C(|\boldsymbol{x}|^2+t)^{-2}\sum_{\nu=0}^{1}\left\|G^{(\nu)}|\left[0,\frac{1}{4}\right]\right\|_\infty,
\end{aligned} \tag{2.60}$$

$$\begin{aligned}
&\left|\partial_{x_l}\partial_{x_m}\left(K(\boldsymbol{x},t)G\left(\frac{|\boldsymbol{x}|^2}{4t}\right)\right)\right| \leq \left|\partial_{x_l}\partial_{x_m}(K(\boldsymbol{x},t))\,G\left(\frac{|\boldsymbol{x}|^2}{4t}\right)\right| \\
&\quad + \sum_{(r,s)\in\{(l,m),(m,l)\}}\left|\partial_{x_r}K(\boldsymbol{x},t)\partial_{x_s}\left(G\left(\frac{|\boldsymbol{x}|^2}{4t}\right)\right)\right| \\
&\quad + K(\boldsymbol{x},t)\left|\partial_{x_l}\partial_{x_m}\left(G\left(\frac{|\boldsymbol{x}|^2}{4t}\right)\right)\right| \\
&\leq C\Big((|\boldsymbol{x}|^2+t)^{-5/2}\left\|G|\left[0,\frac{1}{4}\right]\right\|_\infty + (|\boldsymbol{x}|^2+t)^{-2}\left\|G'|\left[0,\frac{1}{4}\right]\right\|_\infty (|\boldsymbol{x}|^2/t)^{1/2}t^{-1/2} \\
&\quad + (|\boldsymbol{x}|^2+t)^{-3/2}t^{-1}\Big(\left\|G'|\left[0,\frac{1}{4}\right]\right\|_\infty + \left\|G''|\left[0,\frac{1}{4}\right]\right\|_\infty \frac{|\boldsymbol{x}|^2}{t}\Big)\Big) \\
&\leq C(|\boldsymbol{x}|^2+t)^{-5/2}\sum_{\nu=0}^{2}\left\|G^{(\nu)}|\left[0,\frac{1}{4}\right]\right\|_\infty.
\end{aligned} \tag{2.61}$$

Let us denote in this paragraph $\mathcal{R}(u) := 1 - \; {}_1F_1(1, 5/2, u)/3$ for $u \in \mathbb{R}$, and we have

$$\boldsymbol{\Lambda}(\boldsymbol{x},t) = \widetilde{\boldsymbol{\Lambda}}(\boldsymbol{x},t) = K(\boldsymbol{x},t)\left[\mathcal{R}\Big(\frac{|\boldsymbol{x}|^2}{4t}\Big)\delta_{rs}\,\mathcal{S}\Big(\frac{|\boldsymbol{x}|^2}{4t}\Big)\frac{(\boldsymbol{x})_r(\boldsymbol{x})_s}{4t}\right]_{1\leq r,s\leq 3}.$$

Then it follows with Theorem 2.9, Corollary 2.3, Lemma 2.9, (2.60), and (2.61) that

$$
\begin{aligned}
&|\partial_{x_l}\partial_{x_m}\Lambda_{jk}(\boldsymbol{x},t)| \\
&\le \left|\partial_{x_l}\partial_{x_m}\left(K(\boldsymbol{x},t)\mathcal{R}\left(\frac{|\boldsymbol{x}|^2}{4t}\right)\right)\right| + \left|\partial_{x_l}\partial_{x_m}\left(K(\boldsymbol{x},t)\mathcal{S}\left(\frac{|\boldsymbol{x}|^2}{4t}\right)\right)(4t)^{-1}x_jx_k\right| \\
&\quad + \sum_{(r,s)\in\{(l,m),\,(m,l)\}}\left|\partial_{x_r}\left(K(\boldsymbol{x},t)\mathcal{S}\left(\frac{|\boldsymbol{x}|^2}{4t}\right)\right)(4t)^{-1}\partial_{x_s}(x_jx_k)\right| \\
&\quad + \left|K(\boldsymbol{x},t)\mathcal{S}\left(\frac{|\boldsymbol{x}|^2}{4t}\right)(4t)^{-1}\partial_{x_l}\partial_{x_m}(x_jx_k)\right| \\
&\le C\Bigg(\sum_{\nu=0}^{2}\left\|\mathcal{R}^{(\nu)}|\left[0,\frac{1}{4}\right]\right\|_\infty(|\boldsymbol{x}|^2+t)^{-5/2} \\
&\quad + \sum_{\nu=0}^{2}\left\|\mathcal{S}^{(\nu)}|\left[0,\frac{1}{4}\right]\right\|_\infty(|\boldsymbol{x}|^2+t)^{-5/2}\frac{|\boldsymbol{x}|^2}{t} \\
&\quad + \sum_{\nu=0}^{1}\left\|\mathcal{S}^{(\nu)}|\left[0,\frac{1}{4}\right]\right\|_\infty(|\boldsymbol{x}|^2+t)^{-2}|\boldsymbol{x}|/t + \left\|\mathcal{S}|\left[0,\frac{1}{4}\right]\right\|_\infty(|\boldsymbol{x}|^2+t)^{-3/2}t^{-1}\Bigg) \\
&\le C\Bigg((|\boldsymbol{x}|^2+t)^{-5/2}\left(1+\frac{|\boldsymbol{x}|^2}{t}\right) \\
&\quad + (|\boldsymbol{x}|^2+t)^{-2}\left(\frac{|\boldsymbol{x}|^2}{t}\right)^{1/2}t^{-1/2} + (|\boldsymbol{x}|^2+t)^{-3/2}t^{-1}\Bigg).
\end{aligned}
$$

Applying that $t \ge (|\boldsymbol{x}|^2+t)/2$ and $|\boldsymbol{x}|^2/t \le 1/4$, we arrive at the inequality

$$
|\partial_{x_l}\partial_{x_m}\Lambda_{jk}(\boldsymbol{x},t)| \le \mathfrak{C}(|\boldsymbol{x}|^2+t)^{-5/2} \quad \text{in the case } |\boldsymbol{x}|^2 \le t. \tag{2.62}
$$

Let $|\boldsymbol{x}|^2 > t$, so that $|\boldsymbol{x}|^2/(4t) \ge 1/4$, $(|\boldsymbol{x}|^2/t)^{-1} \le 1$, and $|\boldsymbol{x}|^2 = (|\boldsymbol{x}|^2+|\boldsymbol{x}|^2)/2 \ge (|\boldsymbol{x}|^2+t)/2$. Using notation $\mathcal{G}(u) := e^{-u}\,{}_1F_1(1,5/2,u)$ for $u \in \mathbb{R}$. This means that

$$
K(\boldsymbol{x},t)\,{}_1F_1\left(1,\frac{5}{2},\frac{|\boldsymbol{x}|^2}{4t}\right) = (4\pi t)^{-3/2}\mathcal{G}\left(\frac{|\boldsymbol{x}|^2}{4t}\right).
$$

Hence

$$
\Lambda_{jk}(\boldsymbol{x},t) = K(\boldsymbol{x},t)(\delta_{jk}-\eta_{jk}(\boldsymbol{x})) - (4\pi t)^{-3/2}\mathcal{G}\left(\frac{|\boldsymbol{x}|^2}{4t}\right)\left(\frac{\delta_{jk}}{3}-\eta_{jk}(\boldsymbol{x})\right).
$$

Thus, referring to (2.58)–(2.59), Theorem 2.9, and Lemma 2.9, we find

$$
\begin{aligned}
&|\partial_{x_l}\partial_{x_m}\Lambda_{jk}(\boldsymbol{x},t)| \le |\partial_{x_l}\partial_{x_m}K(\boldsymbol{x},t)(\delta_{jk}-\eta_{jk}(\boldsymbol{x}))| \\
&+ \sum_{(r,s)\in\{(l,m),\,(m,l)\}} |\partial_{x_r}K(\boldsymbol{x},t)\partial_{x_s}\eta_{jk}(\boldsymbol{x})| + |K(\boldsymbol{x},t)\partial_{x_l}\partial_{x_m}\eta_{jk}(\boldsymbol{x})| \\
&+ (4\pi t)^{-3/2}\left|\partial_{x_l}\partial_{x_m}\left(\mathcal{G}\left(\frac{|\boldsymbol{x}|^2}{4t}\right)\right)\left(\frac{\delta_{jk}}{3}-\eta_{jk}(\boldsymbol{x})\right)\right| \\
&+ (4\pi t)^{-3/2}\sum_{(r,s)\in\{(l,m),\,(m,l)\}}\left|\partial_{x_r}\left(\mathcal{G}\left(\frac{|\boldsymbol{x}|^2}{4t}\right)\right)\partial_{x_s}\eta_{jk}(\boldsymbol{x})\right| \\
&+ (4\pi t)^{-3/2}\left|\mathcal{G}\left(\frac{|\boldsymbol{x}|^2}{4t}\right)\partial_{x_l}\partial_{x_m}\eta_{jk}(\boldsymbol{x})\right| \\
\le& C\Bigg((|\boldsymbol{x}|^2+t)^{-5/2} + (|\boldsymbol{x}|^2+t)^{-2}|\boldsymbol{x}|^{-1} + (|\boldsymbol{x}|^2+t)^{-3/2}|\boldsymbol{x}|^{-2} \\
&+ t^{-5/2}\left[\left|\mathcal{G}'\left(\frac{|\boldsymbol{x}|^2}{4t}\right)\right| + \left|\mathcal{G}''\left(\frac{|\boldsymbol{x}|^2}{4t}\right)\right|\frac{|\boldsymbol{x}|^2}{t}\right] \\
&+ t^{-2}\left|\mathcal{G}'\left(\frac{|\boldsymbol{x}|^2}{4t}\right)\right|\left(\frac{|\boldsymbol{x}|^2}{t}\right)^{1/2}|\boldsymbol{x}|^{-1} + t^{-3/2}\left|\mathcal{G}\left(\frac{|\boldsymbol{x}|^2}{4t}\right)\right||\boldsymbol{x}|^{-2}\Bigg) \\
\le& C\Bigg((|\boldsymbol{x}|^2+t)^{-5/2} + \sum_{\nu=0}^{2}\sup\left\{\mathcal{G}^{(\nu)}(u)u^{3/2+\nu} : u\in\left[\frac{1}{4},\infty\right)\right\} \\
&\times\left[t^{-5/2}\left(\frac{|\boldsymbol{x}|^2}{t}\right)^{-5/2} + t^{-2}\left(\frac{|\boldsymbol{x}|^2}{t}\right)^{-2}|\boldsymbol{x}|^{-1} + t^{-3/2}\left(\frac{|\boldsymbol{x}|^2}{t}\right)^{-3/2}|\boldsymbol{x}|^{-2}\right]\Bigg) \\
\le& C((|\boldsymbol{x}|^2+t)^{-5/2} + |\boldsymbol{x}|^{-5}) \le C(|\boldsymbol{x}|^2+t)^{-5/2}. \qquad (2.63)
\end{aligned}
$$

The lemma follows from (2.62) and (2.63) □

The preceding lemma and Lemma 2.18 give us a preliminary estimate of Γ_{jk}:

Corollary 2.4 *Let* $j,k\in\{1,2,3\}$, $\boldsymbol{\alpha},\boldsymbol{\beta}\in\mathbb{N}_0^3$ *with* $|\boldsymbol{\alpha}+\boldsymbol{\beta}|\le 2$, $\boldsymbol{y},\boldsymbol{z}\in\mathbb{R}^3$, $t\in(0,\infty)$. *Then*

$$|\partial_{\boldsymbol{y}}^{\boldsymbol{\alpha}}\partial_{\boldsymbol{z}}^{\boldsymbol{\beta}}\Gamma_{jk}(\boldsymbol{y},\boldsymbol{z},t)| \le C(|\boldsymbol{y}-\boldsymbol{z}(t)|^2+t)^{-3/2-|\boldsymbol{\alpha}+\boldsymbol{\beta}|/2}.$$

Proof By the definition of Γ_{jk}, we have

$$\Gamma_{jk}(\boldsymbol{y},\boldsymbol{z},t) = \sum_{s=1}^{3}\Lambda_{js}(\boldsymbol{y}-\boldsymbol{z}(t),t)(\mathrm{e}^{-t\boldsymbol{\Omega}})_{sk}.$$

Differentiating the preceding equation and applying Lemmas 2.18 and 2.21 yield Corollary 2.4. □

2.6 T-Time-Periodic Functions

In this part, we introduce some spaces of T*-time-periodic functions*. For more details see [21, 104, 164]. The space of infinitely differentiable T-time periodic functions with compact support in the spatial variable is defined as follows:

$$C^{\infty}_{0,\mathrm{per}}(\mathcal{D}\times\mathbb{R}) := \{f \in C^{\infty}(\mathcal{D}\times\mathbb{R}) | f(\mathbf{x}, t+T) = f(\mathbf{x}, t), \quad f \in C^{\infty}_{0}(\mathcal{D}\times[0, T]\}.$$

Using the norm $\|\cdot\|_{k,j,q}$ on $C_{0,\mathrm{per}}(\mathcal{D}\times\mathbb{R})$, we define the Sobolev and Lebesgue spaces of time-periodic functions:

$$\begin{aligned}
W^{k,j,q}_{0,\mathrm{per}}(\mathcal{D}\times\mathbb{R}) &= \overline{C^{\infty}_{0,\mathrm{per}}(\mathcal{D}\times\mathbb{R})}^{\|\cdot\|_{k,j,q}}\\
W^{k,j,q}_{\mathrm{per}}(\mathcal{D}\times\mathbb{R}) &= \overline{C^{\infty}_{0,\mathrm{per}}(\mathbb{R}\times\overline{\mathcal{D}})}^{\|\cdot\|_{k,j,q}}\\
L^{q}_{\mathrm{per}}(\mathcal{D}\times\mathbb{R}) &:= \overline{C^{\infty}_{0,\mathrm{per}}(\mathcal{D}\times\mathbb{R})}^{\|\cdot\|_{q}}\\
L^{q}_{\mathrm{per}}(x;\mathbb{R}) &:= \{f:\mathbb{R}\to x \text{ measurable } (f(\cdot) = f(\cdot+T) \text{ a.e.}\\
&\qquad \|f\|_{L^{q}(0,T);x} < \infty\}\\
C^{\infty}_{0,\sigma,\mathrm{per}}(\mathcal{D}\times\mathbb{R}) &:= \{u \in C^{\infty}_{0,\mathrm{per}}(\mathcal{D}\times\mathbb{R}) | \operatorname{div} u = 0\}\\
D^{1,q}_{\mathrm{per}}(\mathcal{D}\times\mathbb{R}) &:= \overline{C^{\infty}_{0,\mathrm{per}}(\overline{\mathcal{D}})\times\mathbb{R}}^{\langle\cdot\rangle_{1,0,q}}.
\end{aligned}$$

[4] We define two simple complementary projections $\mathcal{P}$ and $\mathcal{P}_{\perp}$ defined for sufficiently regular functions by

$$\begin{aligned}
\mathcal{P}u(t,\mathbf{x}) &:= \frac{1}{T}\int_{0}^{T} u(s,\mathbf{x})\mathrm{d}s\\
\mathcal{P}_{\perp}u(t,\mathbf{x}) &:= u(t,\mathbf{x}) - \mathcal{P}(t,\mathbf{x}).
\end{aligned}$$

$\mathcal{P}$ and $\mathcal{P}_{\perp}$ decompose a time-periodic vector field u into a time-independent part Pu and time-periodic part $P_{\perp}$ with vanishing time-average over the period. Vector fields of the former type are called as steady states and vector fields of the latter type as oscillatory. The projections are used to decompose the Lebesgue and Sobolev spaces

[4] For definition of $\langle\cdot\rangle_{1,0,q}$, see the end of this subsection.

of T-time-periodic functions. The following property is satisfied

$$\mathcal{P}L^q_{\text{per}}(\mathcal{D} \times \mathbb{R}) = L^q(\mathcal{D}).$$

$\mathcal{P}_\perp$ induces the subspaces of oscillatory functions:

$$L^q_{\text{per},\perp}(\mathcal{D} \times \mathbb{R}) := \mathcal{P}_\perp L^q_{\text{per}}(\mathcal{D} \times \mathbb{R}) = \{f \in L^q_{\text{per}}(\mathcal{D} \times \mathbb{R}) | \mathcal{P}f = 0\}$$
$$W^{2,1,q}_{\text{per},\perp}(\mathcal{D} \times \mathbb{R}) := P_\perp W^{2,1,q}_{\text{per}}(\mathcal{D} \times \mathbb{R}) = \{u \in W^{2,1,q}_{\text{per}}(\mathcal{D} \times \mathbb{R}) | \mathcal{P}u = 0\}.$$

$\mathcal{P}$ and $\mathcal{P}_\perp$ are continuous projections on $L^q_{\text{per}}(\mathcal{D}\times\mathbb{R})$ and $W^{1,2,q}_{\text{per}}(\mathcal{D}\times\mathbb{R})$. We define

$$D^{1,q}_{\text{per},\perp}(\mathcal{D} \times \mathbb{R}) := \mathcal{P}_\perp \mathcal{D}^{1,q}_{\text{per}}(\mathcal{D} \times \mathbb{R}).$$

We define the *locally compact Abelian group* $G := \mathbb{R}^n \times \mathbb{R}/T\mathbb{Z}$, equipped with the quotient topology induced by the canonical quotient mapping

$$q \colon \mathbb{R}^n \times \mathbb{R} \to \mathbb{R}^n \times \frac{\mathbb{R}}{T}\mathbb{Z}, \quad q(\mathbf{x}, t) := (\mathbf{x}, [t]).$$

The restriction $\Pi := q|_{(0,T)\times\mathbb{R}^n}$ is used to identity G with the domain $\mathbb{R}^n \times [0, T]$; π is a continuous bijection. Using π, we identity the *Haar measure* dg on G as the product of the Lebesgue measure on $[0, T)$ and the Lebesgue measure on $\mathbb{R}^n$. The Haar measure is unique up to a normalization factor,

$$\int_G u(g)\,\mathrm{d}\mathbf{y} = \frac{1}{T}\int_0^+ \int_{\mathbb{R}^n} u \circ \Pi(\mathbf{x}, t)\,\mathrm{d}\mathbf{x}\mathrm{d}t.$$

The symbol Π in integrals of G-defined functions with respect to $\mathrm{d}x\mathrm{d}t$ shall be omitted. We denote

$$C^\infty(G) = \{u \colon G \mapsto \mathbb{R} | u \circ q \in C^\infty(\mathbb{R}^n \times \mathbb{R})\},$$

the space of smooth functions on G. For $u \in C^\infty(G)$ derivatives are defined by

$$\forall\,(\alpha, \beta) \in \mathbb{N}^n_0 \times \mathbb{N}_0 \colon \partial_{\mathbf{x}} u^\alpha \partial_t^\beta := [\partial_x^\alpha \partial_t^\beta (u \circ q)] \circ \pi^{-1}.$$

The subspace $C_0^\infty(G) := \{u \in C^\infty(G) |\operatorname{supp} u$ is compact) denotes the compactly supported smooth functions. We introduce the space of tempered distributions on *G-Schwartz–Bruhat space* of generalized Schwartz function, which are given by

$$\mathcal{S}(G) := \{u \in C^\infty(G) | \forall\,(\alpha, \beta, \gamma) \in \mathbb{N}^n_0 \times \mathbb{N}^n_0 \times \mathbb{N}^n_0\,;\, \rho_{\alpha,\beta,\gamma}(u) < \infty\},$$
$$\rho_{\alpha,\beta,\gamma}(u) := \sup_{(t,\mathbf{x})\in G} |x^\alpha \partial_{\mathbf{x}}^\beta \partial_t^\gamma u(t, \mathbf{x})|.$$

Together with the seminorm topology of the family $\{\rho_{\alpha,\beta,\gamma} \mid (\alpha,\beta,\gamma) \in \mathbb{N}_0^n \times \mathbb{N}_0^n \times \mathbb{N}_0\}$, $\mathcal{S}(G)$ defines a topological vector space. The dual space $\varphi'(G)$ equipped with the weak* topology is referred to as the space of tempered distribution on G. For a tempered distribution $u \in \varphi'(G)$, distributional derivatives $\partial^\beta \partial_{\mathbf{x}}^\alpha u \in \varphi'(G)$ are defined as in classical definition by duality. For each $(k, \boldsymbol{\xi}) \in \mathbb{Z} \times \mathbb{R}^n$ with $X\colon G \to K$, $\chi(t, \mathbf{x}) := \mathrm{e}^{i\mathbf{x}\boldsymbol{\xi}+i(2\pi/T)kt}$ on G, we can characterize the dual group $\widehat{G} = \mathbb{Z} \times \mathbb{R}^n$; the Haar measure on $\widehat{G}$ is simply the product of the counting measure on $\mathbb{Z}$ and the Lebesgue measure on $\mathbb{R}^n$.

$$C^\infty(\widehat{G}) := \{w \in C(\widehat{G})\colon \forall k \in \mathbb{Z} : w(k, .) \in C^\infty(\mathbb{R}^n)\}$$

the space of smooth functions on $\widehat{G}$.

The Schwartz–Bruhat space on the dual group $\widehat{G}$ is given by

$$\mathcal{S}(\widehat{G}) := \{w \in C^\infty(\widehat{G})\colon \forall(\alpha,\beta,\gamma) \in \mathbb{N}_0^n \times \mathbb{N}_0^n \times \mathbb{N}_0\colon \widehat{\rho}_{\alpha,\beta,\gamma}(w) < \infty\},$$

$$\widehat{\rho}_{\boldsymbol{\alpha},\boldsymbol{\beta},\boldsymbol{\gamma}}(w) := \sup_{(k,\boldsymbol{\xi})\in\widehat{G}} |\boldsymbol{\xi}^\alpha \partial_{\boldsymbol{\xi}}^\beta k^\gamma w(k, \boldsymbol{\xi})|,$$

and it is equipped with the canonical seminorm topology.

The Fourier transform associated with the locally compact Abelian group G is denoted by $\mathcal{F}_G$.

$$\mathcal{F}_G\colon L^1(G) \to C(\widehat{G})$$

$$\mathcal{F}_G(u)(k, \boldsymbol{\xi}) := \frac{1}{T}\int_0^T \int_{\mathbb{R}^n} u(t, \mathbf{x})\mathrm{e}^{-i\mathbf{x}\boldsymbol{\xi}-i2\pi/Tkt}\mathrm{d}\mathbf{x}\mathrm{d}t,$$

and the inverse Fourier transform is defined

$$\mathcal{F}_G^{-1} L^1(\widehat{G}) \to C(G),$$

$$\mathcal{F}_G^{-1}(w)(t, \mathbf{x}) := \sum_{k\in\mathbb{Z}} \int_{\mathbb{R}^n} w(k, \boldsymbol{\xi})\mathrm{e}^{i\mathbf{x}\boldsymbol{\xi}+i(2\pi/T)kt}\mathrm{d}\boldsymbol{\xi}.$$

It is easy to see that $\mathcal{F}_G : \mathcal{S}(G) \to \mathcal{S}(\widehat{G})$ is a homeomorphism with $\mathcal{F}_G^{-1}$, provided that the Lebesgue measure $\mathrm{d}\xi$ is appropriately normalized. By duality $\mathcal{F}_G$ extends to a homeomorphism $\varphi'(G) \to \varphi'(\widehat{G})$. We can define also Lebesgue and Sobolev spaces with respect to the domain

$$L^q(G) := \overline{C_0^\infty(G)}^{\|\cdot\|_q}, \quad W^{1,2,q}(G) := \overline{C_0^\infty(G)}^{\|\cdot\|_{1,2,q}}, \quad D^{1,2}(G) := \overline{C_0^\infty(G)}^{\langle\cdot\rangle_{0,1,2}},$$

where $\|\cdot\|_q$ denotes the L^q norm with respect to the Haar measure $\mathrm{d}g$, the norm $\|\cdot\|_{2,1,q}$

$$\|u\|_{2,1,q} = \left(\sum_{|\alpha|\leq 2, |\beta|\leq 1} \|\partial_{\mathbf{x}}^{\alpha} u\|^q_{L^q(\mathcal{D}\times(0,T))} + \|\partial_t^{\beta} u\|^q_{L^q((0,T)\times x)} \right)^{1/q},$$

and the norm

$$\langle p\rangle_{0,1,q} := \|p\|_q + \int_0^T \left| \int_{\mathcal{D}_{R_0}} p(t,\mathbf{x})\,\mathrm{d}\mathbf{x} \right| \mathrm{d}t$$

defines a norm on $C^{\infty}_{0,\mathrm{per}}(\overline{\mathcal{D}} \times \mathbb{R})$. We define

$$\begin{aligned} W^{1,2,q}_{\perp}(G) &:= \mathcal{P}_{\perp} W^{1,2,q}(G) \\ D^{1,q}_{\perp}(G) &:= \mathcal{P}_{\perp} D^{1,q}(G) \\ L^q_{\perp}(G) &:= \mathcal{P}_{\perp} L^q(G). \end{aligned}$$

Part II
Linear Theory

Chapter 3
The Steady Stokes Type Problem

3.1 A Strong Solution in L^q

3.1.1 Introduction

The Stokes type equations are the linearized form of the stationary case of flow (describing by the Navier–Stokes equations) around a rotating body (see Equations (1.25) and (1.26) from the Introduction section, with $\mathbf{u}_\infty = \mathbf{0}$ and $\operatorname{div} \mathbf{u} = 0$). In this section we consider the problem in the whole space $\mathbb{R}^3$ in the nonhomogeneous case. Let $\mathbf{f}$ be a given vector function in $\mathbb{R}^3$ and g be a given scalar function in $\mathbb{R}^3$. We seek a velocity field $\mathbf{u} = (u_1, u_2, u_3)$ and a scalar function p representing the associated pressure which satisfy the following equations:

$$
\begin{aligned}
S_{\omega,\nu}\mathbf{u} + \nabla p = \mathbf{f} \quad & \text{in } \mathbb{R}^3, \\
\operatorname{div} \mathbf{u} = g \quad & \text{in } \mathbb{R}^3, \\
\lim_{|\mathbf{x}|\to\infty} \mathbf{u}(\mathbf{x}) = \mathbf{0}, &
\end{aligned}
\tag{3.1}
$$

where

$$
\begin{aligned}
\omega &= |\boldsymbol{\omega}|, \\
S_{\omega,\nu}\mathbf{u} &= -\nu\Delta\mathbf{u} - \omega\mathcal{R}\mathbf{u}, \\
\mathcal{R}\mathbf{u} &= \mathbf{e}_3 \wedge \mathbf{x} \cdot \nabla\mathbf{u} - \mathbf{e}_3 \wedge \mathbf{u}.
\end{aligned}
\tag{3.2}
$$

The constant $\nu > 0$ is the coefficient of kinematic viscosity. The vector $\boldsymbol{\omega}$ is the angular velocity of the body which is rotating about the x_3-axis, so $\boldsymbol{\omega} = \omega\mathbf{e}_3$. The most general exterior force $\mathbf{f}$ will be in the form $\mathbf{f} = \mathbf{f}_0 + \operatorname{div} F$ with F is a matrix

Š. Nečasová et al., *Mathematical Theory of a Fluid Flow Around a Rotating and Translating Body*, Springer Monographs in Mathematics,
https://doi.org/10.1007/978-3-031-84120-0_3

function in $\mathbb{R}^3$. In the case $g = 0$, the fluid is homogeneous, introducing the so-called incompressibility constraint. Let us consider this case. Concerning the non-homogeneous case, more details can be found see Section 3.3.4. In the case $\omega = 0$, we will denote by $(\mathbf{u}_0, p_0)$ a solution to problem (3.1) with operator $S_{0,\nu}$. The steady Stokes type problem reduces to "classical" steady Stokes problem which always arises in theoretical studies of fluid flows described by the Navier–Stokes equations.

Applying $\text{div}(\cdot)$ to $(3.1)_1$, we get $\Delta p_0 = \text{div}\,\mathbf{f}$. There is a standard argument which authorizes the elimination of the pressure in Eq. $(3.1)_1$: $\mathbf{F}_0 := \mathbf{f} - \nabla p_0$ becoming the new exterior force in the second member. Obviously p_0 is defined up to an additive constant.

A sufficiently rapid decay of the velocity at infinity is required for integrability properties. We have the energy equality

$$\int_{\mathbb{R}^3} (S_{0,\nu}\mathbf{u}_0 + \nabla p_0)(\mathbf{x}) \cdot \mathbf{u}_0(\mathbf{x})d\mathbf{x} = \nu \int_{\mathbb{R}^3} |\nabla \mathbf{u}_0(\mathbf{x})|^2 d\mathbf{x} = \int_{\mathbb{R}^3} \mathbf{f}(\mathbf{x}) \cdot \mathbf{u}_0(\mathbf{x})d\mathbf{x}$$

and the following integral (or variational) identity

$$\nu(\nabla \mathbf{u}_0, \nabla \boldsymbol{\varphi}) - (p_0, \nabla \cdot \boldsymbol{\varphi}) = (\mathbf{f}, \boldsymbol{\varphi}).$$

To obtain it we integrate by parts the product of equation $S_{0,\nu}\mathbf{u}_0 + \nabla p_0 = \mathbf{f}$ by any smooth function with a compact support, $\boldsymbol{\varphi} \in \mathbf{C}_0^\infty(\mathbb{R}^3)$. We use the notation $(\mathbf{f}, \boldsymbol{\varphi}) := \int_{\mathbb{R}^3} \mathbf{f}(\mathbf{x}) \cdot \boldsymbol{\varphi}(\mathbf{x})\, d\mathbf{x}$.

The classical meaning imposes $\mathbf{u}_0 \in \mathbf{C}_0^2$, $p_0 \in C_0^1$, and $\mathbf{f} \in \mathbf{C}_0^0$. A weak representation of the previous integral identity with $\langle \mathbf{f}, \boldsymbol{\varphi} \rangle$ instead of $(\mathbf{f}, \boldsymbol{\varphi})$ is to be understood in the duality between $\widehat{\mathbf{H}}^{-1,2}(\mathbb{R}^3)$ and $\widehat{\mathbf{H}}_0^{1,2}(\mathbb{R}^3)$.

If Π_σ denotes the Helmholtz projection from $\mathbf{L}^2$ onto $\mathbf{L}^2_\sigma$, then $\mathscr{A}_0^0 = -\nu \Pi_\sigma \Delta$ is the Stokes operator with the domain $D(\Pi_\sigma \Delta) = \widehat{\mathbf{H}}^{2,2}(\mathbb{R}^3) \cap \widehat{\mathbf{H}}_0^{1,2}(\mathbb{R}^3)$. Therefore the integral identity holds in the extended sense for $(\mathbf{u}_0, p_0)$ with $\mathbf{u}_0 \in D(\Pi_\sigma \Delta)$, $p_0 \in L^2_{\text{loc}}(\mathbb{R}^3)$, $\nabla p_0 \in L^2(\mathbb{R}^3)$.

The general framework consists in subspaces of $\mathbf{L}^q(\mathbb{R}^3)$, looking for locally integrable $\mathbf{u}_0$ in $\mathbf{H}^{1,q}(\mathbb{R}^3)$ or $\mathbf{H}^{2,q}(\mathbb{R}^3)$. We are interesting to study a regular solution $(\mathbf{u}_0, p_0)$, so-called strong solution.

In the case $\omega \neq 0$, the rotation operator $\mathcal{R}$ contains two terms, a term of order one with respect to differentiability and a term of zeroth order. The latter represents the Coriolis force $\omega \mathbf{e}_3 \wedge \mathbf{u}$. The first one is a so-called drift term $\omega \mathbf{e}_3 \wedge \mathbf{x} \cdot \nabla$ compromising the behavior at large distances of $\mathbf{u}$: It cannot be subordinate to the Laplace operator (even if ω would be small). So it represents the main difficulties of the present study. Nevertheless we have the property that

$$\text{div}(\mathcal{R}\mathbf{u}) = \mathbf{e}_3 \wedge \mathbf{x} \cdot \nabla\, \text{div}\,\mathbf{u} = 0;$$

then the equation for the pressure p is always $\Delta p = \text{div}\,\mathbf{f}$.

Our attention will be focused on exact formulae to solve explicitly this Stokes type problem. Before turning to this construction and related $\mathbf{L}^q$-estimates of the strong solution $(\mathbf{u}, p)$ in terms of $\mathbf{f}$, we are interested in changes of variables giving new representations of the rotation operator. These changes of variables are dictated by the anisotropy of differential operator, the axis of rotation being $\mathbb{R}\mathbf{e}_3$.

3.1.2 Two Differential Forms of the Problem

Let us derive two different differential formulations of the rotation operator $\mathcal{R}$ in two different systems of coordinates parameterized either with an artificial time or with an angle. So it will be possible to overcome the cited difficulties and to rewrite the steady Stokes type problem in two different ways. Hereafter we will observe the following identities:

$$\mathcal{R}Q_1(t)\mathbf{u}(Q_1(t)^{\mathrm{T}}\cdot) = \partial_t[Q_1(t)\mathbf{u}(Q_1(t)^{\mathrm{T}}\cdot)] \tag{3.3}$$

for any vector function $\mathbf{u}(\cdot)$ and for any $t \in \mathbb{R}$, having the meaning of a time

$$\mathcal{R}\mathbf{w}(\cdot) = Q_1(t)\partial_t[Q_1(t)^{\mathrm{T}}\mathbf{w}(Q(t)\cdot)] = [\partial_t\mathbf{w} - \mathbf{e}_3 \wedge \mathbf{w}](\cdot) \tag{3.4}$$

or t having now the meaning of an angle, and the second identity holds for any vector function $\mathbf{w}(\cdot)$ described in the cylindrical coordinates $(|\mathbf{x}'|\cos t, |\mathbf{x}'|\sin t, x_3)$ for $t = \arctan(x_2/x_1)$.

The first viewpoint is to go back to the chosen rotating frame and to identify in $(3.1)_1$ $\mathcal{R}$ with a differentiation with respect to this artificial time t.

Indeed using the matrix of rotation (see (1.17) from the Introduction section)

$$Q(t) = Q_\omega(t) := \begin{pmatrix} \cos\omega t & \sin\omega t & 0 \\ -\sin\omega t & \cos\omega t & 0 \\ 0 & 0 & 1 \end{pmatrix}, \tag{3.5}$$

as the reader can immediately verify, we get

$$[\partial_t Q(t)]\mathbf{v} = \omega\mathbf{e}_3 \wedge Q(t)\mathbf{v}, \tag{3.6}$$

and then, introducing the new variable

$$\mathbf{y} = Q(t)^{\mathrm{T}}\mathbf{x}, \tag{3.7}$$

we can calculate that

$$\partial_t[Q(t)\mathbf{u}(Q(t)^{\mathrm{T}}\mathbf{x})]$$
$$= -\omega\Big(\mathbf{e}_3 \wedge Q(t)^{\mathrm{T}}\mathbf{x}\cdot\nabla[Q(t)\mathbf{u}(Q(t)^{\mathrm{T}}\mathbf{x})] + \mathbf{e}_3 \wedge [Q(t)\mathbf{u}(Q(t)^{\mathrm{T}}\mathbf{x})]\Big).$$

Therefore, we obtain

$$\partial_t[Q(t)\mathbf{u}(Q(t)^{\mathrm{T}}\mathbf{x})] = -\omega\mathcal{R}[Q(t)\mathbf{u}(Q(t)^{\mathrm{T}}\mathbf{x})]. \tag{3.8}$$

As a consequence the new velocity field and the new pressure

$$\widetilde{\mathbf{u}} = \widetilde{\mathbf{u}}(\mathbf{y},t) := Q(t)\mathbf{u}(Q(t)^{\mathrm{T}}\mathbf{x}) \tag{3.9}$$
$$\widetilde{p} = \widetilde{p}(\mathbf{y},t) := p(Q(t)^{\mathrm{T}}\mathbf{x}),$$

which are $(2\pi/\omega)$-periodic with respect to t, satisfy a t-time-periodic Stokes problem

$$\begin{aligned} \partial_t\widetilde{\mathbf{u}} - \nu\Delta\widetilde{\mathbf{u}} + \nabla\widetilde{p} &= \widetilde{\mathbf{f}} && \text{in } \mathbb{R}^3\times\mathbb{R} \\ \operatorname{div}\widetilde{\mathbf{u}} &= 0 && \text{in } \mathbb{R}^3\times\mathbb{R} \\ \widetilde{\mathbf{u}}\left(\mathbf{y}, t+\frac{2\pi}{\omega}\right) &= \widetilde{\mathbf{u}}(\mathbf{y},t) && \text{in } \mathbb{R}^3\times\mathbb{R} \\ \lim_{|\mathbf{y}|\to\infty}\widetilde{\mathbf{u}}(\mathbf{y},\cdot) &= \mathbf{0} && \text{in } \mathbb{R}, \end{aligned} \tag{3.10}$$

with an appropriate volume force $\widetilde{\mathbf{f}}$, also $(2\pi/\omega)$-periodic with respect to t, in the form $\widetilde{\mathbf{f}}(\mathbf{y},t) := Q(t)\mathbf{f}(Q(t)^{\mathrm{T}}\mathbf{y})$.

A good condition is directly associated with this system

$$\lim_{t\to 0+}\widetilde{\mathbf{u}}(\cdot,t) = \mathbf{u}(\cdot), \tag{3.11}$$

in a sense to be clarified, say $\lim_{t\to 0+}\|\widetilde{\mathbf{u}}(\cdot,t)-\mathbf{u}\|_6 = 0$.

Remark 3.1 The pertinent advantage of the previous system lies in the absence of the growing term as $|\mathbf{y}|\to\infty$.

Remark 3.2 Having the uniqueness of $\widetilde{\mathbf{u}}(\cdot,\cdot)$, periodic solution, we can conversely apply the inverse transformation $\mathbf{v}(\cdot,t) = Q(t)^{\mathrm{T}}\widetilde{\mathbf{u}}(Q(t)\cdot,t)$, $p(.) = \widetilde{p}(Q(t)\cdot,t)$ to get formally t-independent solution, of our original steady Stokes type problem.

Remark 3.3 We can immediately observe that $\int_0^{2\pi/\omega}\partial_t\widetilde{\mathbf{u}}(\cdot,t)\mathrm{d}t = 0$. We define the projection $\mathcal{P}$ by the formula $\mathcal{P}\mathbf{u}(\mathbf{y}) := (\omega/2\pi)\int_0^{2\pi/\omega}\widetilde{\mathbf{u}}(\mathbf{y},t)\mathrm{d}t$, which is the time-average over the period of $\widetilde{\mathbf{u}}$. We also can denote it by $\mathcal{P}\widetilde{\mathbf{u}}$. The respective time-mean values of $\widetilde{p}$ and $\widetilde{\mathbf{f}}$ will be $\mathcal{P}\widetilde{p}$ and $\mathcal{P}\widetilde{\mathbf{f}}$. Then we get

$$
\begin{aligned}
-\nu\Delta\mathcal{P}\widetilde{\mathbf{u}} + \nabla\pi &= \mathcal{P}\widetilde{\mathbf{f}} \\
\operatorname{div}(\mathcal{P}\widetilde{\mathbf{u}}) &= 0 \\
\lim_{|\mathbf{y}|\to\infty} \mathcal{P}\widetilde{\mathbf{u}}(\mathbf{y}) &= \mathbf{0},
\end{aligned}
\tag{3.12}
$$

where $\pi = \pi(\mathbf{y}) := (\omega/2\pi)\int_0^{2\pi/\omega} \widetilde{p}(\mathbf{y}, t)\mathrm{d}t$. From the identity (3.3), we also have $\mathcal{RP}\widetilde{\mathbf{u}} = \mathbf{0}$.

The functions $\widetilde{\mathbf{u}}$, $\widetilde{p}$, $\widetilde{\mathbf{f}}$ can be obviously expanded in Fourier series. Denoting all their coefficients by $\mathbf{u}_k$, p_k, $\mathbf{f}_k$, we observe that $\mathcal{P}\widetilde{\mathbf{u}} = \mathbf{u}_0$ and $\pi = p_0$ (up to an additive constant) solve the obtained Stokes equations. Further, we have the following property: $\mathcal{P}\mathbf{u}_0 = \mathcal{P}^2\widetilde{\mathbf{u}} = \mathcal{P}\widetilde{\mathbf{u}} = \mathbf{u}_0$.

From both previous remarks, we at once deduce the velocity field of interest, $\widetilde{\mathbf{u}} - \mathbf{u}_0$, which is $2\pi/\omega$-time-periodic invariant by the transformation we have denoted by $\widetilde{\cdot}$ and directly relied to $\mathbf{u}$

$$
\begin{aligned}
Q(t)^{\mathrm{T}}[\widetilde{\mathbf{u}}(\cdot, t) - \mathbf{u}_0](Q(t)\mathbf{x}) &:= Q(t)^{\mathrm{T}}[Q(t)\widetilde{\mathbf{u}}(Q(t)^{\mathrm{T}}Q(t)\mathbf{x}, t) - \mathbf{u}_0(Q(t)\mathbf{x})] \\
&= \widetilde{\mathbf{u}}(\mathbf{y}, t) - \mathbf{u}_0(\mathbf{y}).
\end{aligned}
$$

This quantity $\widetilde{\mathbf{u}} - \mathbf{u}_0$ the *effective oscillatory velocity part* is directly relied to $\mathbf{u}$ in the limit $t \to 0+$ by the property (3.11), and it satisfies the system of equations

$$
\left.
\begin{aligned}
\partial_t(\widetilde{\mathbf{u}} - \mathbf{u}_0) - \nu\Delta(\widetilde{\mathbf{u}} - \mathbf{u}_0) + \nabla(\widetilde{p} - p_0) &= \widetilde{\mathbf{f}} - \mathcal{P}\widetilde{\mathbf{f}} && \text{in } \mathbb{R}^3 \times \mathbb{R}, \\
\operatorname{div}(\widetilde{\mathbf{u}} - \mathbf{u}_0) &= 0 && \text{in } \mathbb{R}^3 \times \mathbb{R}, \\
\widetilde{\mathbf{u}}\left(\mathbf{y}, t + \tfrac{2\pi}{\omega}\right) - \mathbf{u}_0(\mathbf{y}) &= \widetilde{\mathbf{u}}(\mathbf{y}, t) - \mathbf{u}_0(\mathbf{y}) && \text{in } \mathbb{R}^3 \times \mathbb{R}, \\
\lim\nolimits_{|\mathbf{x}|\to\infty} \widetilde{\mathbf{u}}(\mathbf{y}, \cdot) - \mathbf{u}_0(\mathbf{y}\,) = \mathbf{0} & && \text{in } \mathbb{R},
\end{aligned}
\right\}
\tag{3.13}
$$

where the time-mean value of $\widetilde{\mathbf{f}} - \mathcal{P}\widetilde{\mathbf{f}}$ is obviously zero. This approach was introduced by G. P. Galdi and M. Kyed [102] and Galdi [98].

The obtained system is one of the possible formulations of our problem from which we shall investigate the solution in $\boldsymbol{L}^q$-setting. We call such an approach as the *associated time-periodic Stokes problem*. We will establish the required L^q-estimates for $\mathbf{u}$ in the limit as $t \to 0+$. On the other hand, from Remark 3.1, we also need to verify that $\widetilde{\mathbf{u}}$, $\mathbf{u}_0$, and $\mathbf{u}$ have the appropriate asymptotic properties at infinity in space.

In order to explain how the second identity (3.4) is derived, an alternative viewpoint lies in another choice of coordinates, the choice of cylindrical coordinates (r, t, x_3), $t \in (0, 2\pi)$ having now the meaning of an angle, and where $r = |\mathbf{x}'| = \sqrt{x_1^2 + x_2^2}$, $x_3 \in \mathbb{R}$. As the reader can immediately verify, we have

$$\mathbf{e}_3 \wedge \mathbf{x} \cdot \nabla = (-x_2, x_1, 0)^{\mathrm{T}} \cdot \nabla = -x_2\partial_1 + x_1\partial_2 = \partial_t. \tag{3.14}$$

Then for $\mathbf{u} = \mathbf{u}(|\mathbf{x}'|, t, x_3)$ we get the new identity

$$\mathcal{R}\mathbf{u} = \partial_t \mathbf{u} - \mathbf{e}_3 \wedge \mathbf{u} \tag{3.15}$$

and the partial differential equation

$$-\omega\partial_t \mathbf{u} - \nu\Delta\mathbf{u} + \omega\mathbf{e}_3 \wedge \mathbf{u} = \mathbf{F} \quad \text{in } \mathbb{R}^+ \times (0, 2\pi) \times \mathbb{R}, \tag{3.16}$$

the complements being $\operatorname{div}\mathbf{u} = 0$, $\mathbf{u}(\cdot, 0, \cdot) = \mathbf{u}(\cdot, 2\pi, \cdot)$ and $\lim_{|\mathbf{x}| \to \infty} \mathbf{u} = \mathbf{0}$. Note that div and Δ now denote the corresponding operators in cylindrical coordinates.

Using the Fourier transform from space variable $\mathbf{x} = (x_1, x_2, x_3)$ to $\boldsymbol{\xi} = (\xi_1, \xi_2, \xi_3)$ and denoting by ∂_θ the angular derivative to be applied to $\widehat{\mathbf{u}}(\boldsymbol{\xi}) = \widehat{\mathbf{u}}((|\boldsymbol{\xi}'|, \theta, \xi_3)$, we get

$$\partial_\theta = \mathbf{e}_3 \wedge \boldsymbol{\xi} \cdot \nabla_\xi = -\xi_2\partial_{\xi_1} + \xi_1\partial_{\xi_2}.$$

Then we obtain in $\mathbb{R}^3_\xi$ or in $\mathbb{R}^+_{|\xi'|} \times (0, 2\pi) \times \mathbb{R}_{\xi_3}$ the following ordinary differential equation:

$$-\omega\partial_\theta\widehat{\mathbf{u}} + \nu|\boldsymbol{\xi}|^2\widehat{\mathbf{u}} + \omega\mathbf{e}_3 \wedge \widehat{\mathbf{u}} = \widehat{\mathbf{F}}$$

of course with $\boldsymbol{\xi} \cdot \widehat{\mathbf{u}}(\boldsymbol{\xi}) = 0$. But a rotation around the ξ_3-axis is required to consider the term $\mathbf{e}_3 \wedge \widehat{\mathbf{u}}$: Clearly the variable $Q_1(\theta)\boldsymbol{\xi}$ corresponds to the variable $Q_1(t)\mathbf{x}$. We take temporarily $\omega = 1$. As a consequence the velocity field $\widehat{\mathbf{v}}(\boldsymbol{\xi}) = \widehat{\mathbf{v}}(|\boldsymbol{\xi}'|, \theta, \xi_3) := Q_1(\theta)^{\mathrm{T}}\mathbf{u}(\widehat{Q_1(t)}\mathbf{x})(\boldsymbol{\xi})$ satisfies the equation

$$-\partial_\theta\widehat{\mathbf{v}} + \nu|\boldsymbol{\xi}|^2\widehat{\mathbf{v}} = Q_1(\theta)^{\mathrm{T}}\widehat{\mathbf{F}}, \quad \text{in the variable} \quad Q_1(\theta)\boldsymbol{\xi}, \tag{3.17}$$

all functions being 2π-periodic with respect to θ; moreover $\boldsymbol{\xi} \cdot \widehat{\mathbf{v}}(\boldsymbol{\xi}) = 0$. The obtained system constitutes another formulation of our problem.

Naturally $\widehat{\mathbf{v}}$ can be expressed as

$$\widehat{\mathbf{v}}(\boldsymbol{\xi}) = \int_0^{2\pi} \mathrm{e}^{-\nu|\boldsymbol{\xi}|^2\theta} Q_1(\theta)^{\mathrm{T}}\widehat{\mathbf{F}}(Q_1(\theta)\boldsymbol{\xi})\mathrm{d}\theta. \tag{3.18}$$

This approach was introduced by Farwig, Hishida, and Müller, see [70]. By the inverse Fourier transform, we will receive an explicit integral representation of the solution of our initial Stokes type problem. We refer it as the *associated ordinary angular differential equation* $\mathbf{u}$, which we shall analyze in $\mathbf{L}^q$ framework. The main tools will be the application of the Littlewood–Paley theory and an appropriate decomposition of the integral operator. It will be explained in Sect. 3.1.5, and for more details, see [70].

Remark 3.4 The obtained *associated ordinary angular differential equation* is very interesting because it leads to an exact formula of the solution to our steady Stokes type model, it is clear in the case of the whole space, and it is a perfect analogy with the classical steady Stokes model ($\omega = 0$). (See Galdi or Sohr.) We will see that the obtained representation is in fact more complex.

Remark 3.5 In the first approach with the associated time-periodic Stokes problem, we finally are not far from a similar situation: Via an inverse appropriate space-time Fourier transform, we will also get an explicit integral representation of $\mathbf{w} = \widetilde{\mathbf{u}} - \mathcal{P}\widetilde{\mathbf{u}}$ in terms of $\mathbf{F} := \widetilde{\mathbf{f}} - \mathcal{P}\widetilde{\mathbf{f}} - \nabla(\widetilde{p} - p_0)$.

3.1.3 The Main Results

Let us recall the problem we deal with

$$S_{\omega,\nu}\mathbf{u} + \nabla p = \mathbf{f} \quad \text{in } \mathbb{R}^3 \qquad (3.19)$$
$$\operatorname{div} \mathbf{u} = g \quad \text{in } \mathbb{R}^3,$$

rewritten in the form

$$S_{\omega,\nu}\mathbf{u} = \mathbf{F} = f - \nabla p \quad \text{in } \mathbb{R}^3, \qquad (3.20)$$
$$\operatorname{div} \mathbf{u} = g \quad \text{in } \mathbb{R}^3.$$

Obtained via the reformulated problems, either the *associated time-periodic Stokes problem* or the *associated ordinary angular differential equation*, we have the following results:

Theorem 3.1 (Homogeneous Case)
(A)

1. *Let* $\mathbf{F} \in L^q(\mathbb{R}^3)$, $1 < q < \infty$, *and* $g = 0$. *Then (3.20) has a solution* $\mathbf{u} \in \mathbf{L}^1_{\text{loc}}(\mathbb{R}^3)$ *satisfying the estimate*

$$\|\nu\nabla^2\mathbf{u}\|_q + \|\omega\mathcal{R}\mathbf{u}\|_q \le c\|\mathbf{F}\|_q. \qquad (3.21)$$

2. *Let* $\mathbf{F} \in \mathbf{L}^{q_1}(\mathbb{R}^3) \cap \mathbf{L}^{q_2}(\mathbb{R}^3)$, $1 < q_1, q_2 < \infty$, *and let* $\mathbf{u}_1$ *and* $\mathbf{u}_2$ *be the solutions as given by* (**A**) *corresponding to* $q = q_1$ *and* $q = q_2$, *respectively. Then there*

are $\alpha, \beta, \gamma, \delta \in \mathbb{R}$ such that $\mathbf{u}_1$ coincides with $\mathbf{u}_2$ up to an affine linear vector field $\alpha\omega\mathbf{e}_3 + \beta\omega\mathbf{e}_3 \wedge \mathbf{x} + (\gamma x_1, \gamma x_2, \delta x_3)^{\mathrm{T}}$ and any solution remains a solution if one adds such a term.

(B)

1. *Let $\mathbf{f} \in \boldsymbol{L}^q(\mathbb{R}^3)$, $1 < q < \infty$; then there exists a solution $(\mathbf{u}, p) \in \widehat{\boldsymbol{H}}^{2,q}(\mathbb{R}^3) \times \widehat{H}^{1,q}(\mathbb{R}^3)$ of (3.19) satisfying the estimate*

$$\|\nu\nabla^2\mathbf{u}\|_q + \|\omega\mathcal{R}\mathbf{u}\|_q + \|\nabla p\|_q \le c_1\|\mathbf{f}\|_q. \tag{3.22}$$

Moreover, if $1 < q < 3$, then

$$\|\nabla\mathbf{u}\|_{3q/(3-q)} \le c_2\|\mathbf{f}\|_q, \tag{3.23}$$

and if $1 < q < 3/2$, then

$$\|\mathbf{u}\|_{3q/(3-2q)} \le c_3\|\mathbf{f}\|_q. \tag{3.24}$$

The constants c_1, c_2, c_3 are independent of $|\omega|$.

2. *Let $\mathbf{f} \in \boldsymbol{L}^{q_1}(\mathbb{R}^3) \cap \boldsymbol{L}^{q_2}(\mathbb{R}^3)$, $1 < q_1 < q_2 < \infty$, and let $(\mathbf{u}_1, p_1)$ and $(\mathbf{u}_2, p_2)$ be two such solutions; then p_1 equals p_2 up to a constant and $\mathbf{u}_1$ equals $\mathbf{u}_2$ up to an affine linear vector filed of the form $\alpha\omega\mathbf{e}_3 + \beta\omega\mathbf{e}_3 \wedge \mathbf{x} + (\gamma x_1, \gamma x_2, -2\gamma x_3)^{\mathrm{T}}$, $\alpha, \beta, \gamma \in \mathbb{R}$, and any solution remains the solution if one adds such terms.*

Theorem 3.2 (Nonhomogeneous Case) *Let $\mathbf{f} \in \boldsymbol{L}^q(\mathbb{R}^3)$, and let $g \in \widehat{H}^{1,q}_{\mathrm{loc}}$ such that $(\omega\mathbf{e}_3 \wedge \mathbf{x})g$, $\nabla g \in \boldsymbol{L}^q(\mathbb{R}^3)$, $1 < q < \infty$. Then there exists a solution $(\mathbf{u}, p) \in \widehat{\boldsymbol{H}}^{2,q}(\mathbb{R}^3) \times \widehat{H}^{1,q}(\mathbb{R}^3)$ of (3.19) satisfying the estimate*

$$\|\nu\nabla^2\mathbf{u}\|_q + \|\omega\mathcal{R}\mathbf{u}\|_q + \|\nabla p\|_q \le c(\|\mathbf{f}\|_q + \|\nu\nabla g + (\omega\mathbf{e}_3 \wedge \mathbf{x})g\|_q), \tag{3.25}$$

where $(3.19)_2$ has to be understood in the sense $\nabla \operatorname{div}\mathbf{u} = \nabla g$.

Theorem 3.3 (2D Case)
(A1)

1. *Let $\mathbf{F} \in \boldsymbol{L}^q(\mathbb{R}^2)$, $1 < q < \infty$. Then (3.20) has a solution $\mathbf{u} \in \boldsymbol{L}^1_{\mathrm{loc}}(\mathbb{R}^2)$ satisfying the estimate*

$$\|\nu\nabla^2\mathbf{u}\|_q + \|\omega((\mathbf{e}_3 \wedge \mathbf{x}) \cdot \nabla\mathbf{u} - \mathbf{e}_3 \wedge \mathbf{u})\|_q \le c\|\mathbf{F}_1\|_q. \tag{3.26}$$

2. *Let $\mathbf{F} \in \mathbf{L}^{q_1}(\mathbb{R}^3) \cap \mathbf{L}^{q_2}(\mathbb{R}^3)$, $1 < q_1, q_2 < \infty$, and let $\mathbf{u}_1$ and $\mathbf{u}_2$ be the solutions as given by* **(A1)** *corresponding to $q = q_1$ and $q = q_2$, respectively. Then there are $\beta, \gamma \in \mathbb{R}$ such that $\mathbf{u}_1$ coincides with $\mathbf{u}_2$ up to an affine linear vector field $\beta\omega\mathbf{e}_3 \wedge \mathbf{x} + (\gamma x_1, \gamma x_2)^{\mathrm{T}}$ and any solution remains a solution if one adds such a term.*

(B1)

1. *Let* $\mathbf{f} \in \boldsymbol{L}^q(\mathbb{R}^2)$, *and let* $g \in \widehat{H}^{1,q}_{\mathrm{loc}}(\mathbb{R}^2)$ *such that* $(\omega \mathbf{e}_3 \wedge \mathbf{x})g,\ \nabla g \in \boldsymbol{L}^q(\mathbb{R}^2)$, $1 < q < \infty$. *Then there exists a solution* $(\mathbf{u}, p) \in \widehat{\boldsymbol{H}}^{2,q}(\mathbb{R}^2) \times \widehat{H}^{1,q}(\mathbb{R}^2)$ *of* (3.19) *satisfying the estimate*

$$\|\nu\nabla^2\mathbf{u}\|_q+\|\omega((\mathbf{e}_3\wedge\mathbf{x})\cdot\nabla\mathbf{u}-\mathbf{e}_3\wedge\mathbf{u})\|_q+\|\nabla p\|_q \leq c(\|\mathbf{f}\|_q+\|\nu\nabla g+(\omega\mathbf{e}_3\wedge\mathbf{x})g\|_q),$$

where $(3.19)_2$ *has to be understood in the sense* $\nabla \operatorname{div} \mathbf{u} = \nabla g$. *Its equivalence class in* $\widehat{\boldsymbol{H}}^{2,q}(\mathbb{R}^2) \times \widehat{H}^{1,q}(\mathbb{R}^2)$ *is unique.*

2. *Moreover, if* $(\mathbf{u}_1, p_1)$ *and* $(\mathbf{u}_2, p_2)$ *are two such solutions, then* p_1 *equals* p_2 *up to a constant and* $\mathbf{u}_1$ *equals* $\mathbf{u}_2$ *up to the linear term* $\beta(-x_2, x_1)^{\mathrm{T}}$, $\beta \in \mathbb{R}$.

3.1.4 Various Remarks

Before dealing with the proof of Theorem 3.1 and deriving the $\mathbf{L}^q$ estimates, we present several remarks, where we comment some properties of the solution.

Remark 3.6 We consider a priori estimate (3.21) in $\mathbf{L}^2$-setting

$$\|\nu\nabla^2\mathbf{u}\|_2 + \|\omega((\mathbf{e_3} \wedge \mathbf{x}) \cdot \nabla\mathbf{u} - \mathbf{e_3} \wedge \mathbf{u})\|_2 \leq c\|\mathbf{f}\|_2.$$

Making use of the Plancherel identity, the Fubini theorem, and the Cauchy-Schwartz inequality, we obtain

$$\|\partial_j\partial_k\mathbf{u}\|_2 \leq \|\xi_j\xi_k\widehat{\mathbf{u}}\|_2 \leq \||\xi|^2\widehat{\mathbf{u}}\|_2 = \|\Delta\mathbf{u}\|_2,$$

where

$$\begin{aligned}
\|\Delta\mathbf{u}\|_2^2 &= \int_{\mathbb{R}^3} |\xi|^4 \left| \int_0^\infty \mathrm{e}^{-\nu|\xi|^2 t} Q_1(t)^{\mathrm{T}}\widehat{\mathbf{F}}(Q_1(t)\xi)\mathrm{d}t \right|^2 \mathrm{d}\xi \\
&\leq \int_{\mathbb{R}^3} \left(\int_0^\infty |\xi|^2\mathrm{e}^{-\nu|\xi|^2 t}\mathrm{d}t\right)\left(\int_0^\infty |\xi|^2 \cdot \mathrm{e}^{-\nu|\xi|^2 t}|\widehat{\mathbf{F}}(Q_1(t)\xi)|^2\mathrm{d}t\right)\mathrm{d}\xi \\
&= \frac{1}{\nu}\int_0^\infty \int_{\mathbb{R}^3} |\xi|^2\mathrm{e}^{-\nu|\xi|^2 t}|\widehat{\mathbf{F}}(\xi)|^2\mathrm{d}\xi\mathrm{d}t = \frac{1}{\nu^2}\|\mathbf{F}\|_2^2.
\end{aligned} \tag{3.27}$$

Remark 3.7 (**L^q-Integrability of u**) The results indicate in Theorem 3.1 that there is no reason for the $\mathbf{L}^2$- or $\mathbf{L}^q$-integrability of $\mathbf{u}$, the same comment for $\mathbf{e}_3 \wedge \mathbf{u}$. But, having a solution $(\mathbf{u}, p)$ given by the theorems, the $\mathbf{L}^q$-integrability of $\mathbf{u}$ is valid in the particular case canceling the mean value of the integral of $\mathbf{f}$, i.e., its time-average over the period or its angle-average over $[0, 2\pi)$. Let us ignore temporarily the pressure and assume

$$\mathbf{f}_m(|\mathbf{x}'|, x_3) = \frac{1}{2\pi}\int_0^{2\pi} Q_1(t)^T \mathbf{f}(|\mathbf{x}'|, t, x_3)dt = 0,$$

and denote

$$\mathbf{u}_m(|\mathbf{x}'|, x_3) = \frac{1}{2\pi}\int_0^{2\pi} Q_1(t)^T \mathbf{u}(|\mathbf{x}'|, t, x_3)dt,$$

which now solves $S_{1,\nu}\mathbf{u}_m = (S_{1,\nu}\mathbf{u})_M = \mathbf{f}_m$, which implies that $\mathbf{u}_m = 0$. Let us prove it for the *associated ordinary angular differential equation*. We recall the identity

$$\mathcal{R}\mathbf{u} = Q_1(t)\partial_t(Q_1(t)^{\mathrm{T}}\mathbf{u}) = \partial_t\mathbf{u} - \mathbf{e}_3 \wedge \mathbf{u}.$$

Applying the Wirtinger inequality, we get

$$\|\mathbf{u}\|_q^q = \int_0^\infty r\int_0^{2\pi} |Q_1(t)^{\mathrm{T}}\mathbf{u}(r,t)|^q \mathrm{d}t\mathrm{d}r \leq C\|\partial_t(Q_1(t)^{\mathrm{T}}\mathbf{u}\|_q^q) \leq C\|(\partial_t - \mathbf{e}_3 \wedge \mathbf{u})\|_q^q.$$

On the other hand, we observe the differential equation (3.16) becoming

$$S_{\omega,\nu}\mathbf{u}_m(|\mathbf{x}'|, x_3) = (S_{\omega,\nu}\mathbf{u})_m = \mathbf{f}_m(|\mathbf{x}'|, x_3),$$

$Q_1(t)^{\mathrm{T}}$ in front of $\mathbf{u}$ or $\mathbf{f}$ plays no role, so $\mathbf{u}_m(|\mathbf{x}'|, x_3)$ would be $\mathbf{0}$ in case of $\mathbf{f}_m = \mathbf{0}$.

Remark 3.8 (**L^q-Integrability of $\widetilde{\mathbf{u}} - \mathbf{u}_0$**) For the *associated time-periodic Stokes problem*, the same property naturally holds. In the meaningful model describing the effective oscillatory velocity field $\widetilde{\mathbf{u}} - \mathbf{u}_0$, we already observed that the time-mean value of the second member $\widetilde{\mathbf{f}} - \mathcal{P}\widetilde{\mathbf{f}}$ was obviously zero: Therefore we always have the control of the $\mathbf{L}^q$-integrability of the *effective oscillatory velocity part*, a second remarkable property.

Remark 3.9 The question of uniqueness of solution from Theorem 3.1 is connected to that estimate of $\|\nabla^2\mathbf{u}\|_q$ which is indifferent to a polynomial of degree 1 added to $\mathbf{u}$. Therefore in the homogeneous Sobolev space $\widehat{\boldsymbol{H}}^{2,q}(\mathbb{R}^3)$, the equivalence class of $\mathbf{u}$ solution is unique. Indeed, having a solution $(\mathbf{u}, p)$ given by Theorem 3.1, if $(\mathbf{u}_1, p_1)$ from the space $\widehat{\boldsymbol{H}}^{2,r}(\mathbb{R}^3) \times \widehat{H}^{1,r}(\mathbb{R}^3)$, for some r, $1 < r < \infty$, is another solution to (3.1), then $(\mathbf{u}_1, p_1)$ can be written in the form

$$\mathbf{u}_1 = \mathbf{u} + \alpha\omega\mathbf{e}_3 + \beta\omega\mathbf{e}_3 \wedge \mathbf{x} + \gamma(x_1, x_2, -2x_3)^{\mathrm{T}},$$
$$p_1 = p + \eta$$

for some $\alpha, \beta, \gamma, \eta \in \mathbb{R}$.

Proof Let us prove part B (2) of Theorem 3.1. Assume the existence of the solution $(\mathbf{u}, p) \in \widehat{\boldsymbol{H}}^{2,q}(\mathbb{R}^3) \times \widehat{H}^{1,q}(\mathbb{R}^3)$ to (3.19), and let $(\mathbf{u}_1, p_1) \in \widehat{\boldsymbol{H}}^{2,r}(\mathbb{R}^3) \times \widehat{H}^{1,r}(\mathbb{R}^3)$ for some r be another solution. Then $(\overline{\mathbf{w}}, \overline{\eta}) := (\mathbf{u} - \overline{\mathbf{u}}, p - \overline{p})$ satisfies (3.22) with a homogeneous right-hand side. Due to the divergence-free properties, we get $\Delta\overline{\eta} = 0$. Since $\nabla\overline{\eta} \in L^q(\mathbb{R}^3) + L^r(\mathbb{R}^3)$, it follows that $\overline{\eta}$ is a real number.

Because Δ and $\mathcal{R}$ commute, we get that $(\Delta\overline{\mathbf{w}}, \Delta\overline{\eta})$ also satisfy (3.19) with a homogeneous right-hand side. Since $\Delta\overline{\mathbf{w}} = \boldsymbol{L}^q(\mathbb{R}^3) + \boldsymbol{L}^r(\mathbb{R}^3)$, from the estimate (3.22), we get $\Delta\overline{\mathbf{w}} = 0$. Then $\overline{\mathbf{w}} = A\mathbf{x} + \boldsymbol{b}$ for some $\boldsymbol{b} \in \mathbb{R}^3$ and some $A \in \mathbb{R}^{3\times 3}$ with $(A\boldsymbol{x}) = TrA = 0$. Denoting $\widetilde{\mathbf{w}} = Q(t)\overline{\mathbf{w}}(Q(t)^{\mathrm{T}}\mathbf{x})$, we get

$$\partial_t\widetilde{\mathbf{w}} = \nu\Delta\widetilde{\mathbf{w}} - \nabla\widetilde{\eta} = \mathbf{0} \ = \omega\mathcal{R}Q(t)\widetilde{\mathbf{w}},$$

whence $Q(t)(AQ(t)^{\mathrm{T}}\mathbf{x}+\boldsymbol{b})$ is t-independent. Also $\mathcal{R}\widetilde{\mathbf{w}} = \mathbf{0}$, i.e., $\widetilde{\mathbf{w}}$ lies in the kernel of operator $\mathcal{R}$, then using (3.8) with $AQ(t)^T\mathbf{x} + \boldsymbol{b}$ instead of $\mathbf{u}(Q(t)^T\mathbf{x})$ we receive

$$\mathbf{0} = (\mathbf{e}_3\wedge AQ(t)^T\mathbf{x})\cdot\nabla(Q(t)AQ(t)^{\mathrm{T}}\mathbf{x}+Q(t)\boldsymbol{b}) - \mathbf{e}_3\wedge(Q(t)AQ^{\mathrm{T}}(t)\mathbf{x}) - \mathbf{e}_3\wedge Q(t)\boldsymbol{b},$$

so

$$\mathbf{0} = (\mathbf{e}_3 \wedge AQ^T(t)^T\mathbf{x}) \cdot \nabla(AQ(t)^{\mathrm{T}}\mathbf{x}) - \mathbf{e}_3 \wedge AQ^{\mathrm{T}}(t)\mathbf{x} - \mathbf{e}_3 \wedge \boldsymbol{b}$$

for all $(\mathbf{x}, t) \in \mathbb{R}^3 \times \mathbb{R}$.

It implies that $\boldsymbol{b} = \alpha\boldsymbol{e}_3$ for some $\alpha \in \mathbb{R}$ and secondly that $AQ(t)^{\mathrm{T}}\mathbf{x}$ is t-independent. Taking successively $t = 0$, $\mathbf{x} = \mathbf{e}_1$, $\mathbf{x} = \mathbf{e}_2$, $\mathbf{x} = \mathbf{e}_3$, we obtain $A\mathbf{x} = \beta\boldsymbol{e}_3 \wedge \mathbf{x} + \gamma(x_1, x_2, -2x_3)^{\mathrm{T}}$ for some $\beta, \gamma \in \mathbb{R}$. □

Remark 3.10 To extend Remark 3.7, another natural question asks for the control of a solution $(\mathbf{u}, p)$ given by the theorems, in the completed norm

$$\|\nabla^2\mathbf{u}\|_q + \|\nabla p\|_q + \|\mathbf{u}\|_{L^q(B_m)} + \|p\|_{L^q(B_m)},$$

where

$$B_m = B_m(\mathbf{0}) = \{\mathbf{x} \in \mathbb{R}^3 \text{ s.t. } |\mathbf{x}| \le m\}, \quad m \in \mathbb{N}.$$

For that, simply choose vectors $\mathbf{a}, \mathbf{b}, \mathbf{c}, \mathbf{d} \in \mathbb{R}^3$ such that $\|\mathbf{u} + x_1\mathbf{a} + x_2\mathbf{b} + x_3\mathbf{c} + \mathbf{d}\|_{L^q(B_m)} = 0$ and change $\mathbf{u}$ by $\mathbf{u} + \mathbf{w}$, where $\mathbf{w} := x_1\mathbf{a} + x_2\mathbf{b} + x_3\mathbf{c} + \mathbf{d}$ and eventually p by $p + \eta$, $\eta \in \mathbb{R}$.

Remark 3.11 (**C_0^∞ Right-Hand Side**) Let $(\mathbf{u}_n, p_n) \in \widehat{\boldsymbol{H}}^{2,q}(\mathbb{R}^3) \times \widehat{H}^{1,q}(\mathbb{R}^3)$ be the solution to (3.19) with $\mathbf{f}_n \in C_0^\infty(\mathbb{R}^3)$, $1 \le n \le \infty$, such that $\lim_{n\to\infty}\mathbf{f}_n = \mathbf{f}$ in $\boldsymbol{L}^q(\mathbb{R}^3)$. We know that $(\mathbf{u}_n, p_n)$ satisfies the estimate (3.22) and that $\lim_{n\to\infty} S_{\omega,\nu}\boldsymbol{u}_n + \nabla p_n = \lim_{n\to\infty}\boldsymbol{f}_n = \boldsymbol{f}$ in $\boldsymbol{L}^q(\mathbb{R}^3)$.

Let us denote by X_m the following Banach spaces ($m \in \mathbb{N}$)

$$X_m := \{(\boldsymbol{u}, p) \in \boldsymbol{L}^1_{\text{loc}}(\mathbb{R}^3) \times L^1_{\text{loc}}(\mathbb{R}^3);\ \|(\boldsymbol{u}, p)\|_{X_m} < \infty\},$$

where $\|(\boldsymbol{u}, p)\|_{X_m} := \|\nabla^2 \boldsymbol{u}\|_q + \|\nabla p\|_q + \|\boldsymbol{u}\|_{L^q(B_m)} + \|p\|_{L^q(B_m)}$.

From (3.22) and Remark 3.9, for appropriate $\{\mathbf{w}_n\}_{n=1}^{\infty}$ and $\{\eta_n\}_{n=1}^{\infty}$, we know that $\{(\boldsymbol{u}_n - \mathbf{w}_n,\ p_n - \eta_n)\}_{n=1}^{\infty}$ is a Cauchy sequence in X_m; then, passing to the limit, there exist $(\boldsymbol{u}, p) \in \bigcap_{m \in \mathbb{N}} X_m$ satisfying the estimate (3.22) and

$$\lim_{n\to\infty} (\boldsymbol{u}_n - \mathbf{w}_n, p_n - \eta_n) = (\boldsymbol{u}, p) \quad \text{in } X_m,\ \forall m \in \mathbb{N}.$$

It implies that

$$\lim_{n\to\infty} S_{\omega,\nu}(\boldsymbol{u}_n - \mathbf{w}_n) + \nabla(p_n - \eta_n) = S_{\omega,\nu}\boldsymbol{u} + \nabla p \quad \text{in } \mathcal{D}'(\mathbb{R}^3),$$

and

$$\lim_{n\to\infty} S_{\omega,\nu}\mathbf{w}_n = \mathbf{f} - S_{\omega,\nu}\boldsymbol{u} - \nabla p.$$

Consequently, $\boldsymbol{f} - (S_{\omega,\nu}\boldsymbol{u} + \nabla p)$ is equal to $S_{\omega,\nu}\boldsymbol{w}$ for some first order polynomial $\mathbf{w}$; in other words $S_{\omega,\nu}(\mathbf{u} + \mathbf{w}) + \nabla p = \mathbf{f}$ and $(\mathbf{u} + \mathbf{w}, p) \in \widehat{\boldsymbol{H}}^{2,q}(\mathbb{R}^3) \times \widehat{H}^{1,q}(\mathbb{R}^3)$.

Note that, adding $\|\nabla \mathbf{u}\|_{3q/(3-q)}$ or $\|\mathbf{u}\|_{3q/(3-2q)}$ to the X_m-norm, respectively, when $1 < q < 3$ or $1 < q < 3/2$, we can easily get the estimates (3.23) and (3.24).

Henceforth the proof of Theorems 3.1 can be established in the case $\mathbf{f} \in C_0^{\infty}(\mathbb{R}^3)$ or $\mathbf{f} \in \mathcal{S}(\mathbb{R}^3)$. Corresponding sequences $\{\mathbf{f}_n\}_{n\in\mathbb{N}}$ will converge to $\mathbf{f} \in \mathbf{L}^q(\mathbb{R}^3)$, and the inequalities (3.22–3.24) will be preserved.

3.1.5 Proofs of Theorems 3.1 and 3.3 Using the Associated Ordinary Angular Differential Equation

The aim of this subsection is to derive L^q-estimates (3.21–3.24), and also to observe deeply the integral representation formula of $\mathbf{u}$ which follows from (3.18). The method was introduced by Farwig, Hishida, and Müller [70], and we will follow their approach.

From (3.18) we get the integral representation formula

$$\widehat{\mathbf{u}}(\xi) = \int_0^{\infty} \mathrm{e}^{-\nu|\xi|^2 t} Q_1(t)^{\mathrm{T}} \widehat{\mathbf{F}}(Q(t)\boldsymbol{\xi}) \mathrm{d}t, \tag{3.28}$$

for $\mathbf{u}$ solution to our problem $S_{1,\nu}\mathbf{u} = \mathbf{F}$ with $\operatorname{div}\mathbf{u} = 0$, and $\omega = 1$.

Then using the heat kernel K it is clear that

$$\mathbf{u}(\mathbf{x}) = \int_0^\infty Q_1(t)^{\mathrm{T}} K * \mathbf{F}(Q_1(t)\cdot)(\mathbf{x})dt. \tag{3.29}$$

Now, denoting the fundamental solution of the stationary Stokes type problem by

$$\boldsymbol{\Sigma}_{st}(\mathbf{x}, \mathbf{y}) = \int_0^\infty Q_1(t)^{\mathrm{T}} K(Q_1(t)\mathbf{x} - \mathbf{y})dt,$$

it is clear that (3.29) can be rewritten

$$\mathbf{u}(\mathbf{x}) = \int_{\mathbb{R}^3} \boldsymbol{\Sigma}_{st}(\mathbf{x}, \mathbf{y})\mathbf{F}(\mathbf{y})d\mathbf{y},$$

and conforming the same integral form for $\Delta\mathbf{u}(\mathbf{x})$, we have

$$\Delta\mathbf{u}(\mathbf{x}) = \int_0^\infty \Delta_x \boldsymbol{\Sigma}_{st}(\mathbf{x}, \mathbf{y})\mathbf{F}(\mathbf{y})d\mathbf{y}.$$

Our reader can verify that

$$\Delta_x \boldsymbol{\Sigma}(\mathbf{x}, \mathbf{y}) = \int_0^\infty Q_1(t)^{\mathrm{T}} \frac{\mathrm{e}^{-|Q_1(t)\mathbf{x}-\mathbf{y}|^2/(4\nu t)}}{(4\pi\nu t)^{3/2}} \cdot \left(-\frac{3}{2\nu t} + \frac{|Q_1(t)\mathbf{x} - \mathbf{y}|^2}{(2\nu t)^2}\right) \mathrm{d}t.$$

Remark 3.12 We shall continue the study of $\Delta_x \boldsymbol{\Sigma}_{st}(\mathbf{x}, \mathbf{y})$ but apply the Littlewood–Paley theory. The kernel of $\Delta_x \boldsymbol{\Sigma}_{st}((\mathbf{x}, \mathbf{y})$ does not define a classical Calderon–Zygmund integral operator. This is a direct consequence of Proposition 2.1 from [70], pages 302–302, where it is established the estimate

$$\Delta_x \boldsymbol{\Sigma}_{st}(\mathbf{x}, \mathbf{y}) \geq \frac{\beta}{|\mathbf{x} - \mathbf{y}|},$$

when $|\mathbf{x}|, |\mathbf{y}| \to \infty$ for some real number $\beta > 0$.

Remark 3.13 Let us mention that the integrals in (3.28) are not converge in $\mathcal{S}'$ in the case of $\mathbb{R}^2$. In this case we have to modify the representation of the solution $\boldsymbol{u}$. We define

$$\begin{aligned}\langle \boldsymbol{u}, \boldsymbol{\varphi}\rangle = \langle \widehat{\boldsymbol{u}}, \check{\boldsymbol{\varphi}}\rangle &= \int_{|\boldsymbol{\xi}|\geq 1} \int_0^\infty \mathrm{e}^{-\nu s^2 t} Q_1(t)^{\mathrm{T}} \widehat{\boldsymbol{F}}(Q_1(t)\boldsymbol{\xi}) \cdot \check{\boldsymbol{\varphi}}(\boldsymbol{\xi})\, \mathrm{d}t\, \mathrm{d}\boldsymbol{\xi} \\ &+ \int_{|\boldsymbol{\xi}|<1} \int_0^\infty \mathrm{e}^{-\nu s^2 t} Q_1(t)^{\mathrm{T}} \widehat{\boldsymbol{F}}(Q_1(t)\boldsymbol{\xi}) \cdot (\check{\boldsymbol{\varphi}}(\boldsymbol{\xi}) - \check{\boldsymbol{\varphi}}(\mathbf{0}))\, \mathrm{d}t\, \mathrm{d}\boldsymbol{\xi}\end{aligned}$$

for all $\boldsymbol{\varphi} \in \mathcal{S}(\mathbb{R}^2)$. Here $\check{}$ denotes the inverse Fourier transform.

Forgetting all constants and assuming $\nu = 1$, $\omega = 1$, we see that the kernel model for $\Delta_x \boldsymbol{\Sigma}_{st}(\mathbf{x}, \mathbf{y})$ is

$$\mathcal{K}(\mathbf{x}, \mathbf{y}) = \int_0^\infty Q_1(t)^{\mathrm{T}} t^{-3/2} \psi\left(\frac{Q_1(t)\mathbf{x} - \mathbf{y}}{\sqrt{t}}\right) \frac{\mathrm{d}t}{t}, \tag{3.30}$$

where $\psi \in \mathcal{S}(\mathbb{R}^3)$,

$$\widehat{\psi}(\boldsymbol{\xi}) = (2\pi)^{-3/2} |\boldsymbol{\xi}|^2 \mathrm{e}^{-\nu|\boldsymbol{\xi}|^2} \tag{3.31}$$

$$\widehat{\psi}_t = \widehat{\psi}(\sqrt{t}\boldsymbol{\xi}) = (2\pi)^{-3/2} t |\boldsymbol{\xi}|^2 \mathrm{e}^{-\nu t|\boldsymbol{\xi}|^2} \tag{3.32}$$

therefore we can introduce $\psi_t(.)$ such that $\psi_t(\mathbf{x}) := t^{\ 3/2} \psi(\frac{\mathbf{x}}{\sqrt{t}})$.

We have in view to use the double inequality which is the basic and crucial result of the Littlewood–Paley theory, i.e.,

$$c_1 \|\mathbf{F}\|_q \leq \|\mathfrak{S}\mathbf{F}(\cdot)\|_q \leq c_2 \|\mathbf{F}\|_q, \tag{3.33}$$

where the square operator $\mathfrak{S}$ is given by

$$\mathfrak{S}\mathbf{F}(\mathbf{x}) := \left(\int_0^\infty |\varphi_s * \mathbf{F}(\mathbf{x})|^2 \frac{\mathrm{d}s}{s}\right)^{1/2}.$$

The family $\{\varphi_s\}_{s>0}$ lies in $\mathcal{S}(\mathbb{R}^3)$ and has special features

(P1) $\varphi_s(\mathbf{x}) = s^{-3/2} \varphi(\mathbf{x}/\sqrt{s})$, with $\varphi \in C_0^\infty(\frac{1}{2}, 2)$ and $\widehat{\varphi}(\boldsymbol{\xi}) = (2\pi)^{-3/2} \int_{\mathbb{R}^3} \mathrm{e}^{-i\mathbf{x}\cdot\boldsymbol{\xi}} \varphi(\mathbf{x})\mathrm{d}x$.

(P2) $\operatorname{supp} \widehat{\varphi_s} \subset \mathcal{A}(1/2\sqrt{s}, 2/\sqrt{s})$ and $\int_{\mathbb{R}^3} \varphi_s(\mathbf{x})\mathrm{d}\mathbf{x} = 0$.

(P3) Radial symmetry $\widehat{\varphi}(\boldsymbol{\xi}) = \widehat{\varphi}(|\boldsymbol{\xi}|)$, and $\int_0^\infty \widehat{\varphi_s}(\boldsymbol{\xi})^2 \mathrm{d}s/s = 1$.

(P4) $\varphi_s * \psi_j = \varphi_s * \chi_j * \psi \equiv 0$ unless $t \in \mathcal{A}(s, j) := [2^{2j-4}s, 2^{2j+4}s]$.

The double inequality (3.33) holds with constants $c_1, c_2 > 0$, depending on q, but independent of $\mathbf{F}$.

To continue, let us provide a specific partition of unity $\sum_{j\in\mathbb{Z}} \chi(2^{-j}r) = 1$ with $\chi \in C_0^\infty(\frac{1}{2}, 2), 0 \leq \chi(\cdot) \leq 1$. We shall decompose $\mathcal{K}$ and the integral operator $\mathcal{T}\colon \mathbf{F} \to \mathcal{T}\mathbf{F}$ defined by

$$\mathcal{T}\mathbf{F}(\mathbf{x}) := \int_{\mathbb{R}^3} \mathcal{K}(\mathbf{x}, \mathbf{y})\mathbf{F}(\mathbf{y})\mathrm{d}\mathbf{y}.$$

So formally $\mathcal{K} = \sum_{j\in\mathbb{Z}} \mathcal{K}_j$ and $\mathcal{T} = \sum_{j\in\mathbb{Z}} \mathcal{T}_j$ where the family $\{\widehat{\chi_j}\}_{j\in\mathbb{Z}}$

$$\psi^j = \chi_j * \psi_j\,, \widehat{\chi_j}(\boldsymbol{\xi}) := \widehat{\chi}(2^{-j}|\boldsymbol{\xi}|) \tag{3.34}$$

depends on frequencies of varying annuli

$$\operatorname{supp} \widehat{\chi_j} \subset \mathcal{A}(2^{j-1}, 2^{j+1}) = \{\boldsymbol{\xi} \in \mathbb{R}^3 : 2^{j-1} < |\boldsymbol{\xi}| < 2^{j+1}\}. \tag{3.35}$$

We also have $\sum_{j\in\mathbb{Z}} \widehat{\chi_j}(\boldsymbol{\xi}) = 1$. Let us define $\psi_j \in \mathcal{S}(\mathbb{R}^3))$ by

$$\widehat{\psi_j} = \widehat{\chi_j} \cdot \widehat{\psi}, \quad \text{or equivalently} \quad \psi_j = (2\pi)^{-3/2} \chi_j * \psi. \tag{3.36}$$

So, $\psi = \sum_{j\in\mathbb{Z}} \psi_j$ and

$$\psi_t^j(.) = t^{-3/2} \chi_j * \psi\left(\frac{\cdot}{\sqrt{t}}\right), \tag{3.37}$$

and

$$\widehat{\psi_t^j}(\boldsymbol{\xi}) = \widehat{\chi_j}(\boldsymbol{\xi}) \widehat{\psi}(\sqrt{t}\boldsymbol{\xi}), \tag{3.38}$$

$$\mathcal{K}_j(\mathbf{x}, \mathbf{y}) = \int_0^\infty Q_1(t)^{\mathrm{T}} \psi_j(Q_1(t)\mathbf{x} - \mathbf{y}) \frac{\mathrm{d}t}{t}, \tag{3.39}$$

$$\mathcal{T}_j \mathbf{F}(\mathbf{x}) = \int_{\mathbb{R}^3} \mathcal{K}_j(\mathbf{x}, \mathbf{y}) \mathbf{F}(\mathbf{y}) \mathrm{d}\mathbf{y} = \int_0^\infty Q_1(t)^{\mathrm{T}} \psi_j * \mathbf{F}(Q_1(t)\mathbf{x}) \frac{\mathrm{d}t}{t} \tag{3.40}$$

such that w.r.t. the operator norm topology on L^q, we can write $\mathcal{T} = \sum_{j=-\infty}^{\infty} \mathcal{T}_j$.

To prove Theorem 3.1 we ask for the estimate $\|\mathcal{T}_j \mathbf{F}\|_q$ inside the double inequality (3.33), and we shall need the next three lemmas which we will prove in the next subsection:

- Concerning behavior of ψ_t^j, see Lemma 3.1.
- An estimate of

$$\mathcal{M}^j \mathbf{g}(\mathbf{x}) = \sup_{r>0} \int_{A_r} (|\psi_t^j| * |\mathbf{g}|)(Q(t)^{\mathrm{T}} \mathbf{x}) \frac{\mathrm{d}t}{t},$$

  see Lemma 3.2.
- An estimate of the maximal operator

$$\widetilde{\mathcal{M}}_\theta \mathbf{g}(\mathbf{x}) = \sup_{r>0} \int_{A_r} |\mathbf{g}(Q(t)^{\mathrm{T}} \mathbf{x})| \frac{\mathrm{d}t}{t},$$

  see Lemma 3.3.

Lemma 3.1 *The functions* ψ_t^j *have the following properties:*

(1) *For* $j \in \mathbb{Z}$ *and* $t > 0$,

$$\operatorname{supp} \widehat{\psi}_t^j \subset A\left(\frac{2^{j-1}}{\sqrt{t}}, \frac{2^{j+1}}{\sqrt{t}}\right).$$

(2) *For* $m > \frac{1}{2}n$, *let* $h(x) = (1 + |\mathbf{x}|^2)^{-m}$ *and* $h_t(\mathbf{x}) = t^{-n/2}h(\mathbf{x}/\sqrt{t})$.

Then there exists a constant $C > 0$ *independent of* $j \in \mathcal{Z}$ *such that*

$$|\psi^j(\mathbf{x})| \le C2^{-2|j|}h_{2^{-2j}}(\mathbf{x}) \quad \textit{for all} \quad \mathbf{x} \in \mathbb{R}^n. \tag{3.41}$$

Moreover

$$\|\psi^j\|_1 \le C2^{-2|j|}. \tag{3.42}$$

Lemma 3.2 *For* $j \in \mathbb{Z}$ *we denote by* $\mathcal{M}^j$ *the maximal operator*

$$\mathcal{M}^j g(\mathbf{x}) = \sup_{r>0} \int_{A_r} (|\psi_t^j| * |g|)(Q(t)^{\mathrm{T}}\mathbf{x})\frac{\mathrm{d}t}{t},$$

where $A_r = [r/16, 16r]$. *Then for* $q \in (2, \infty)$ *the operator* $\mathcal{T}_j$ *satisfies the estimate*

$$\|\mathcal{T}_j\mathbf{f}\|_q \le C\|\psi^j\|_1^{1/2}\|\|\mathcal{M}^j\|\|_{(q/2)'}^{1/2}\|\mathbf{f}\|_q$$

with a constant $C > 0$ *independent of* $j \in \mathbb{Z}$. *The term* $\|\mathcal{M}^j\|_{(q/2)'}$ *denotes the operator norm of the sublinear operator* $\mathcal{M}^j$ *on* $L^{(q/2)'}(\mathbb{R}^n)$, *where* $1/(q/2)' + 1/(q/2) = 1$.

Lemma 3.3 *By* $\mathcal{M}$ *we denote the classical Hardy–Littlewood maximal operator on* $\mathbb{R}^n$, *i.e.,*

$$\mathcal{M}g(x) = \sup_{r>0} \frac{1}{|B_r(\mathbf{x})|} \int_{B_r(\mathbf{x})} |g(\mathbf{y})|\mathrm{d}\mathbf{y},$$

and let $\widetilde{\mathcal{M}}_\theta g$ *denote the "angular" maximal operator*

$$\widetilde{\mathcal{M}}_\theta g(\mathbf{x}) = \sup_{r>0} \int_{A_r} |g(Q(t)^{\mathrm{T}}\mathbf{x})|\frac{\mathrm{d}t}{t},$$

where $A_r = [r/16, 16r]$. *Then* $\mathcal{M}^j$ *in Lemma* 3.2 *satisfies the estimates*

$$\mathcal{M}^j g(\mathbf{x}) \le C2^{-2|j|}\mathcal{M}(\widetilde{\mathcal{M}}_\theta g)(\mathbf{x}) \quad \textit{for a.a.} \quad \mathbf{x} \in \mathbb{R}^n,$$

$$\|\mathcal{M}^j g\|_q \le C2^{-2|j|}\|g\|_q, \quad 1 < q < \infty.$$

Proof We will show the first part of Theorem 3.1. Applying the estimate $\|\mathcal{T}_j\mathbf{F}\|_q$ inside the double inequality (3.33) first for $\|S\mathcal{T}_j\mathbf{F}(\cdot)\|_q$ and secondly by $\|S\mathbf{F}(\cdot)\|_{q/2}$ itself, we have

$$\varphi_s * \mathcal{T}_j\mathbf{F}(\mathbf{x}) = \int_0^\infty Q_1(t)^{\mathrm{T}}(\varphi_s * (\psi_j * \mathbf{F}))(Q_1(t)\mathbf{x})\frac{\mathrm{d}t}{t},$$

and from the associativity of convolutions and using the Cauchy–Schwartz inequality, we get

$$|\varphi_s * \mathcal{T}_j\mathbf{F}(\mathbf{x})|^2 \le C\|\psi_j\|_1 \int_{\mathcal{A}(s,j)} (|\psi_j| * |\varphi_s * \mathbf{F}|^2)(Q_1(t)\mathbf{x})\frac{\mathrm{d}t}{t}.$$

Thus, we find a nonnegative function $g \in L^{(q/2)'}(\mathbb{R}^3)$ with the unit norm such that for

$$\mathbf{I} := \left\| \int_0^\infty |\varphi_s * \mathcal{T}_j\mathbf{F}(\mathbf{x})|^2 \frac{\mathrm{d}s}{s} \right\|_{q/2} = \int_0^\infty \int_{\mathbb{R}^3} |\varphi_s * \mathcal{T}_j\mathbf{F}(\mathbf{x})|^2 g(\mathbf{x})\mathrm{d}\mathbf{x}\frac{\mathrm{d}s}{s},$$

we obtain

$$\mathbf{I} \le C\|\psi_j\|_1 \int_0^\infty \frac{\mathrm{d}s}{s} \int_{\mathbb{R}^3} \mathrm{d}\mathbf{x} \left\{ |\varphi_s * \mathbf{F}|^2)(\mathbf{x}) \left[\int_{\mathcal{A}(s,j)} (|\psi_j| * g(Q_1(-t)\mathbf{x})\frac{\mathrm{d}t}{t} \right] \right\},$$

where the inner integral on $\mathcal{A}(s, j)$ is bounded uniformly for $s > 0$.

Finally $\|\mathcal{T}_j\mathbf{F}\|_q \le \mathbf{I} \le Cc_2\|\mathbf{F}\|_q$.

Let $q \in (2, \infty)$. Then from Lemmas 3.1–3.3 we get

$$\|\mathcal{T}_j\mathbf{f}\|_q \le C2^{-|j|} \cdot 2^{-|j|} \cdot \|\mathbf{f}\|_q.$$

It follows that $\sum_{j\in\mathbb{Z}} \mathcal{T}_j$ converges in the L^q-operator norm and $\mathcal{T} = \sum_{j\in\mathbb{Z}} \mathcal{T}_j$ is bounded on $\boldsymbol{L}^q(\mathbb{R}^n)$ for $q > 2$.

Similarly, we get the convergence also for the dual operator $\mathcal{T}^*\mathbf{f}$. Then the operator $\mathcal{T} = \Sigma\mathcal{T}_j$ is bounded for $q < 2$, $q = 2$ which has been already proven. Equations (3.23) and (3.24) follow from Sobolev imbedding. The last part will be proven in the next section.

Let us describe how to solve the pressure. As we explain before applying the div to the equation the problem is reduced to solving the equation

$$\Delta p = \operatorname{div}\mathbf{f} + \nu\Delta g + \partial_\theta g = \operatorname{div} F,$$

where $\mathbf{F} = \mathbf{f} + \nu\nabla g + (\omega \wedge \mathbf{x})g$ satisfying the estimate

$$\|F\|_q \le c(\|\mathbf{f}\|_q + \|\nu\nabla g + (\omega \wedge \mathbf{x})g).$$

Using the fact that Δ is an isomorphism from $\widehat{\boldsymbol{H}}^{1,q'}$ to $(\widehat{\boldsymbol{H}}^{1,q'})^*$, it implies the existence of a unique $p \in \widehat{H}^{1,q}(R^n)$ solving $\Delta p = \operatorname{div} F$, which satisfies

$$\|\nabla p\|_q \leq \|F\|_q.$$

□

Let us formulate Remark 3.6 as a proposition:

Proposition 3.1 *For* $\mathbf{f} \in L^2(\mathbb{R}^n)$, $n = 2$, *or* $n = 3$, *the solution* $\mathbf{u}$ *of* (3.1) *given by* (3.29) *satisfies the estimate*

$$\nu\|\nabla^2\mathbf{u}\|_2 + \|\omega((\mathbf{e}_3 \wedge \mathbf{x}) \cdot \nabla\mathbf{u} - \mathbf{e}_3 \wedge \mathbf{u})\|_2 \leq C\|\mathbf{f}\|_2. \tag{3.43}$$

3.1.5.1 Proofs of Lemmas 3.1, 3.2, and 3.3

First we shall prove Lemma 3.1.

Proof of Lemma 3.1 (1) follows from (3.32), (3.34), (3.36). Concerning (2) we will first prove

$$|2^{j|\alpha|}\partial^\alpha\widehat{\psi}^j(\boldsymbol{\xi})| \leq C_\alpha 2^{-2|j|}\eta(2^{-j}|\boldsymbol{\xi}|) \tag{3.44}$$

for all $\boldsymbol{\xi} \in \mathbb{R}^n$, $j \in \mathbb{Z}$, all multi-indices $\alpha \in \mathbb{N}_0^n$ and with a function $\eta \in C_0^\infty(\frac{1}{4}, 4)$, $0 \leq \eta \leq 1$. From the definition of $\widehat{\chi}_j$, (3.34)–(3.35), and the pointwise estimates

$$|\partial^\beta\widehat{\psi}(\boldsymbol{\xi})| \leq C_{\beta,N}\begin{cases} |\boldsymbol{\xi}|^{\max(0,2-|\beta|)}, & |\boldsymbol{\xi}| < 1, \\ |\boldsymbol{\xi}|^{-N}, & |\boldsymbol{\xi}| \geq 1, \end{cases} \tag{3.45}$$

for every $N \in \mathbb{N}$, $\beta \in \mathbb{N}_0^n$, applying Leibniz's formula we obtain the estimate

$$\begin{aligned} |2^{j|\alpha|}\partial^\alpha\widehat{\psi}^j(\boldsymbol{\xi})| &\leq C\sum_{0\leq\beta\leq\alpha} 2^{j|\alpha|}|\partial^{\alpha-\beta}\widetilde{\chi}(2^{-j}|\boldsymbol{\xi}|)|\partial^\beta\widehat{\psi}(\boldsymbol{\xi})| \\ &\leq C\sum_{0\leq\beta\leq\alpha} 2^{j|\beta|}\eta(2^{-j}|\boldsymbol{\xi}|)|\partial^\beta\widehat{\psi}(\boldsymbol{\xi})|. \end{aligned} \tag{3.46}$$

For $j \geq 0$ only $|\boldsymbol{\xi}| \sim 2^j$ has to be considered, and we get (3.44). For $j < 0$ and $|\boldsymbol{\xi}| \sim 2^j < 1$, the right-hand side of the last inequality is bounded by

$$C\sum_{0\leq\beta\leq\alpha}\eta(2^{-j}|\boldsymbol{\xi}|)2^{j\max(|\beta|,2)} \leq C2^{-2|j|}\eta(2^{-j|\boldsymbol{\xi}|}). \tag{3.47}$$

It implies (3.44).

Finally, we would like to estimate $\psi^j(\boldsymbol{\xi})$. We will use the following identity:

$$(1+|2^j\boldsymbol{x}|^2)^m\psi^j(\mathbf{x}) = \frac{1}{(2\pi)^{n/2}}\int_{\mathbb{R}^n}(1-2^{2j}\Delta)^m\widehat{\psi}_j(\boldsymbol{\xi})\mathrm{e}^{-ix\xi}\mathrm{d}s.$$

From (3.44)

$$|(1-2^{2j}\Delta)^m\widehat{\psi}^j(\boldsymbol{\xi})| \le C_{m,N}2^{-2|j|}\eta(2^{-j}|\boldsymbol{\xi}|)$$

for all $j \in \mathbb{Z}$ and $\boldsymbol{\xi} \in \mathbb{R}^n$. Therefore

$$\|(1-2^{2j}\Delta)^m\widehat{\psi}^j\|_1 \le C_m 2^{nj-2|j|},$$

and it follows that $|(1+|2^jx|^2)^m\psi^j(\mathbf{x})| \le C2^{nj-2|j|}$ which it proves (2) of Lemma 3.1. □

Now, we give a proof of Lemma 3.2.

Proof We will apply the Littlewood–Paley decomposition (3.33) for estimation of $\|\mathcal{T}_j\mathbf{f}\|_q$. Moreover, we find a function $0 \le g \in L^{(q/2)'}(\mathbb{R}^n)$ with $\|g\|_{(q/2)'} = 1$ such that

$$\|\mathcal{T}_j\mathbf{F}\|_q^2 \le \frac{1}{C^2}\left\|\int_0^\infty|\varphi_s * \mathcal{T}_j\mathbf{F}(\cdot)|^2\frac{\mathrm{d}s}{s}\right\|_{q/2} = \frac{1}{C^2}\int_0^\infty\int_{\mathbb{R}^n}|\varphi_s * \mathcal{T}_j\mathbf{F}|^2 g\,\mathrm{d}\mathbf{x}\frac{\mathrm{d}s}{s}.$$

From (3.36) and (3.39)

$$\varphi_s * \mathcal{T}_j\mathbf{F}(\mathbf{x}) = \int_0^\infty Q(t)^{\mathrm{T}}(\varphi_s * \psi_t^j * \mathbf{f})(Q(t)\mathbf{x})\frac{\mathrm{d}t}{t}.$$

Moreover, we know that $\varphi_s * \psi_t^j = 0$ unless $t \in A(s,j) := [2^{2j-4}s, 2^{2j+4}s]$. Using $\int_{t\in A(s,j)}\mathrm{d}t/t = \log 2^8$ for every $j \in \mathbb{Z}$, $s > 0$, the inequality of Cauchy–Schwartz and the associativity of convolutions yield

$$\begin{aligned}|\varphi_s * \mathcal{T}_j\mathbf{f}(x)|^2 &\le C\int_{A(s,j)}\left|(\psi_t^j * (\varphi_s * \mathbf{f}))(Q(t)\mathbf{x})\right|^2\frac{\mathrm{d}t}{t}\\ &\le C\|\psi^j\|_1\int_{A(s,j)}(|\psi_t^j| * |\varphi_s * \mathbf{f}|^2)(Q(t)\mathbf{x})\frac{\mathrm{d}t}{t}.\end{aligned}$$

Thus

$$\|\mathcal{T}_j\mathbf{F}\|_q^2 \le C\|\psi^j\|_1\int_{\mathbb{R}^n}\int_0^\infty|\varphi_s * \mathbf{F}|^2(\mathbf{x})\int_{A(s,j)}(|\psi_t^j| * g)(Q(-t)\mathbf{x})\frac{\mathrm{d}t}{t}\frac{\mathrm{d}s}{s}\mathrm{d}\boldsymbol{x}.$$

The inner integral on $A(s, j)$ is bounded by $\mathcal{M}^j g(\boldsymbol{x})$ uniformly for $s > 0$. Using the Hölder inequality and (3.33), we show that

$$\|\mathcal{T}_j \mathbf{F}\|_q^2 \leq C\|\psi^j\|_1 \left(\int_{\mathbb{R}^n} \left(\int_0 |\varphi_s * \mathbf{F}|^2 \frac{\mathrm{d}s}{s}\right)^{q/2} \mathrm{d}\mathbf{x}\right)^{2/q} \|\mathcal{M}^j g\|_{(q/2)'}$$
$$\leq C\widetilde{C}\|\psi^j\|_1 \|\mathbf{f}\|_q^2 \||\mathcal{M}^j\||_{(q/2)'} \|g\|_{(q/2)'}.$$

Since $\|g\|_{(q/2)'} = 1$, the proof is finished. □

We will give a proof of Lemma 3.3.

Proof From Lemma 3.1 (2) we have

$$|\psi_t^j(\mathbf{x})| \leq C2^{-2|j|} h_{t2^{-2j}}(\mathbf{x}),$$

and it implies

$$\mathcal{M}^j g(\mathbf{x}) \leq C2^{-2|j|} \sup_{r>0} \int_{A_r} (h_{t2^{-2j}} * |g|)(Q(t)^{\mathrm{T}}\mathbf{x}) \frac{\mathrm{d}t}{t}.$$

Then there exists a constant $C > 0$ independent of r, j such that $h_{t2^{-2j}} \leq ch_{r2^{-2j}}$ for all $t \in A_r$. It implies

$$\mathcal{M}^j g(\mathbf{x}) \leq C2^{-2|j|} \sup_{r>0} h_{r2^{-2j}} * \int_{A_r} |g|(Q(t)^{\mathrm{T}}\mathbf{x}) \frac{\mathrm{d}t}{t} \leq C2^{-2|j|} \sup_{t>0} h_t * \widetilde{\mathcal{M}}_\theta g(\mathbf{x}).$$

We have to use that h is a nonnegative, radially decreasing function and that $\int h_t \mathrm{d}x \equiv C_0 > 0$ for all $t > 0$. From Stein [205] it follows

$$\sup_{t>0} h_t * \widetilde{\mathcal{M}}_\theta g(\boldsymbol{x}) \leq C_0 \mathcal{M}(\widetilde{M}_\theta g)(\mathbf{x}).$$

For $q \in (1, \infty)$ the maximal operator $\mathcal{M}$ is bounded on $L^q(\mathbb{R}^n)$. Concerning $\widetilde{\mathcal{M}}_\theta$ we consider for given $g \in L^q(\mathbb{R}^n)$ the restriction

$$g_r(\theta) = g(r, \theta), \quad g_{r,x_3}(\theta) = g(r, \theta, x_3), \tag{3.48}$$

where we use polar or cylindrical coordinates. For $n = 2$, $g_r(\theta) \in L^q(0, 2\pi)$ for a.a. $r > 0$ by Fubini's theorem and with the classical one-dimensional Hardy–Littlewood maximal operator $\mathcal{M}_1$ on $L^q(0, 2\pi)$

$$|\widetilde{\mathcal{M}}_\theta g(r, \theta)| \leq C(\mathcal{M}_1 g_r)|t| \quad \text{for a.a.} \quad r > 0. \tag{3.49}$$

We obtain

$$\|\widetilde{\mathcal{M}}_\theta g\|_q^q \le C\int_0^\infty r\|\mathcal{M}_1 g_r\|_{L^q(0,2\pi)}^q \mathrm{d}r \le C\int_0^\infty r\|g_r\|_{L^q(0,2\pi)}^q \mathrm{d}r = c\|g\|_q^q$$

due to the L^q-boundedness of $\mathcal{M}_1$. For $n = 3$ the proof is similar. □

3.1.6 The Associated Time-Periodic Stokes Problem and the Proof of Theorem 3.1

The aim of this subsection is to present an alternative approach for deriving a priori L^q-estimates to the problem (3.1) by method introduced by Galdi and Kyed [102, 104].

For the completeness we also state the estimates in L^2 framework.

Lemma 3.4 ([93], Theorem VIII.5.1) *Let* $\mathbf{F}$ *be a second order tensor field such that* $div\,\mathbf{F} \in L^2(\mathbb{R}^3)$ *with* $\|\mathbf{F}\|_{2,R} < \infty$, *and let* $\mathbf{g} \in L^q(\mathbb{R}^3)$ $q \in (3,\infty)$ *with support contained in* $\mathcal{D}_\rho$ *for some* $\rho > 0$. *(By* $\mathcal{D}_\rho$ *we denote* $\mathcal{D} \cap B_\rho$.*) Then the problem* (3.1) *with* $\boldsymbol{f} \equiv div\,\mathbf{F} + \mathbf{g}$ *has at least one solution such that*

$$\left.\begin{aligned} &\mathbf{u} \in \boldsymbol{H}^{2,2}_{\mathrm{loc}}(\mathbb{R}^3) \cap \widehat{\boldsymbol{H}}^{2,2}(\mathbb{R}^3) \cap \widehat{\boldsymbol{H}}^{1,2}(\mathbb{R}^3),\\ &\|\mathbf{u}\|_{1,R} < \infty\\ &p \in H^{1,2}_{\mathrm{loc}}(\mathbb{R}^3) \cap \widehat{H}^{1,2}(\mathbb{R}^3) \cap L^r(\mathbb{R}^3), \quad r > 3/2. \end{aligned}\right\} \tag{3.50}$$

This solution satisfies the following estimates:

$$\left.\begin{aligned} &|\mathbf{u}|_{2,2} + |\mathbf{u}|_{1,2} + |p|_{1,2} \le c_1(\|div\,\mathbf{F}\|_2 + [[\mathbf{F}]]_2 + \|\mathbf{g}\|_q)\\ &\|\mathbf{u}\|_{1,R} + \|p\|_r \le C_2([[\mathbf{F}]]_2 + \|\mathbf{g}\|_q), \end{aligned}\right\} \tag{3.51}$$

where $c_1 = c_1(p, q, B)$, $C_2 = C_2(\rho, r, q, B)$ *whenever* $R \in [0, B]$, *for some* $B > 0$, *where by* $[[\boldsymbol{U}]]_2$ *we set* $[[\boldsymbol{U}]]_2 = \sup_{\boldsymbol{x}\in\mathbb{R}^3}[(1+|\boldsymbol{x}|)^2|\boldsymbol{U}(x)|]$. *Moreover, assume in particular that* $\mathbf{F} = 0$ *and* $\mathbf{g}$ *satisfies the further assumption* $\mathbf{g} \in \boldsymbol{H}^{1,\infty}(\Omega_\rho)$. *Then* $\mathbf{u}$ *satisfies also*

$$\|\nabla p\|_2 + \|\nabla^2\mathbf{u}\|_2 < \infty, \tag{3.52}$$

and there is constant $C_3 = C_3(\rho)$ *such that*

$$\|\nabla p\|_2 + \|\nabla^2\mathbf{u}\|_2 \le C_3\|\mathbf{g}\|_{1,\infty}. \tag{3.53}$$

Proof The existence and uniqueness in L^2 framework can be found in the book by Galdi [98, Theorem VIII.2.1, VIII.1.1, VIII.1.2 and Lemma VIII.3.2]. The asymptotic properties of $\mathbf{u}$ were derived in [93, Theorem VIII.5.1]. □

We shall study the L^q-estimates, presenting the announced alternative proof of Theorem 3.1.

Let us recall the equations (see (3.13)), $\widetilde{\mathbf{u}} - \mathbf{u}_0$ being the effective oscillatory velocity part

$$\begin{aligned} \partial_t(\widetilde{\mathbf{u}} - \mathbf{u}_0) - \nu\Delta(\widetilde{\mathbf{u}} - \mathbf{u}_0) + \nabla(\widetilde{p} - p_0) &= \widetilde{\mathbf{f}} - \mathcal{P}\mathbf{f} \quad \text{in } \mathbb{R}^3 \times \mathbb{R}, \\ \operatorname{div}(\widetilde{\mathbf{u}} - \mathbf{u}_0) &= 0 \quad \text{in } \mathbb{R}^3 \times \mathbb{R}, \\ \widetilde{\mathbf{u}}(\mathbf{y}, t + 2\pi/\omega) - \mathbf{u}_0(\mathbf{y}) &= \widetilde{\mathbf{u}}(\mathbf{y}, t) - \mathbf{u}_0(\mathbf{y}) \quad \text{in } \mathbb{R}^3 \times \mathbb{R}, \\ \lim_{|\mathbf{y}|\to\infty} \widetilde{\mathbf{u}}(\mathbf{y}, \cdot) - \mathbf{u}_0(\mathbf{y}) &= \mathbf{0} \quad \text{in } \mathbb{R}. \end{aligned} \tag{3.54}$$

We follow Galdi and Kyed [104] and introduce the group $\mathbf{G} := \mathbb{R}^3 \times \mathbb{R}/((2\pi/\omega)\mathbb{Z})$ where the $2\pi/\omega$-time periodicity is always insured, intrinsic in $\mathbf{G}$. The transfer of topological and differentiable structures must be obtained in all spaces of functions or distributions defined in $\mathbf{G}$. So, $\mathbf{C}_0^\infty(\mathbf{G})$, $\mathbf{L}^q(\mathbf{G})$, $\mathbf{H}_0^{m,q}(\mathbf{G})$, $\widehat{\boldsymbol{H}}^{2,q}(\mathbf{G})$, and $\mathcal{S}(\mathbf{G})$ are well defined. Finally we can use the Fourier transform

$$\mathcal{F}_{\mathbf{G}}[\mathbf{w}](\boldsymbol{\xi}, k) := \frac{\omega}{2\pi}\int_0^{2\pi}\int_{\mathbb{R}^3} \mathbf{w}(\mathbf{y}, t)\mathrm{e}^{-\mathrm{i}\mathbf{y}\cdot\boldsymbol{\xi}}\mathrm{e}^{-\mathrm{i}(2\pi/\omega)kt}\mathrm{d}\mathbf{y}\mathrm{d}t$$

and the corresponding inverse transform

$$\mathcal{F}_{\mathbf{G}}^{-1}[\varphi](\mathbf{y}, t) := \sum_{k\in\mathbb{Z}}\int_{\mathbb{R}^3} \varphi(\boldsymbol{\xi}, k)\mathrm{e}^{\mathrm{i}\mathbf{y}\cdot\boldsymbol{\xi}}\mathrm{e}^{\mathrm{i}(2\pi/\omega)kt}\mathrm{d}\boldsymbol{\xi}.$$

In fact, $\mathbf{G}$ is identified with $\mathbb{R}^3 \times [0, (2\pi/\omega))$ restricting the second component to the interval of the period. Then we have the property

$$\int_{\mathbf{G}} \mathbf{w}(g)dg := \frac{\omega}{2\pi}\int_0^{2\pi}\int_{\mathbb{R}^3} \mathbf{w}(\mathbf{y}, t)\mathrm{d}\mathbf{y}\mathrm{d}t.$$

Now, let $\widetilde{\mathbf{w}}(\mathbf{y}, t) = \widetilde{\mathbf{u}}(\mathbf{y}, t) - \mathbf{u}_0(\mathbf{x})$, $\widetilde{\pi}(\mathbf{y}, t) = \widetilde{p}(\mathbf{y}, t) - p_0(\mathbf{y}, t)$, and $\widetilde{\mathbf{f}}_0 := (I - \mathcal{P})\widetilde{\mathbf{f}}$, and our system reads

$$\begin{aligned} \partial_t\widetilde{\mathbf{w}} - \nu\Delta\widetilde{\mathbf{w}} + \nabla\widetilde{\pi} &= \widetilde{\mathbf{f}}_0 \quad \text{in } \mathbf{G}, \\ \operatorname{div}\widetilde{\mathbf{w}} &= 0 \quad \text{in } \mathbf{G}. \end{aligned} \tag{3.55}$$

Using $\mathcal{F}_{\mathbf{G}}$, we get a solution $(\widetilde{\mathbf{w}}, \widetilde{\pi})$ in the sense of $\mathcal{S}'(\mathbf{G})$

$$\mathcal{F}_{\mathbf{G}}[\widetilde{\mathbf{f}}_0](\xi, k) = (1 - \delta_{\mathbb{Z}}(k))\mathcal{F}_{\mathbf{G}}[\widetilde{\mathbf{f}}] \tag{3.56}$$

$$\left(\nu|\boldsymbol{\xi}|^2 + i\frac{2\pi}{\omega}k\right)\mathcal{F}_{\mathbf{G}}[\widetilde{\mathbf{w}}] = (I - \boldsymbol{\xi}\otimes\boldsymbol{\xi}/|\boldsymbol{\xi}|^2)\mathcal{F}_{\mathbf{G}}[\widetilde{\mathbf{f}}_0]$$
$$|\boldsymbol{\xi}|^2\mathcal{F}_{\mathbf{G}}[\widetilde{\pi}] = i\boldsymbol{\xi}\cdot\mathcal{F}_{\mathbf{G}}[\widetilde{\mathbf{f}}_0],$$

where $1-\delta_{\mathbb{Z}}(k)$ is the symbol of $I-\mathcal{P}$, and $I-\boldsymbol{\xi}\otimes\boldsymbol{\xi}/|\boldsymbol{\xi}|^2$ is the usual symbol of the projection Π^q_σ from $\mathbf{L}^q(\mathbb{R}^3)$ onto $\mathbf{L}^q_\sigma(\mathbb{R}^3)$ continuous on $\mathbf{L}^q(\mathbb{R}^3)$ for any q and continuous on $\mathbf{L}^q(G)$ in this context. For the pressure, it is standard $\widetilde{\pi} = \mathcal{F}^{-1}_{\mathbb{R}^3}[i\boldsymbol{\xi}/|\boldsymbol{\xi}|^2\cdot\mathcal{F}_{\mathbb{R}^3}[\widetilde{\mathbf{f}}]]$, so we have the estimate $\|\nabla\widetilde{\pi}\|_q \le C\|\widetilde{\mathbf{f}}_0\|_q$.

Note that the quantity $\left(\nu|\boldsymbol{\xi}|^2 + i\frac{2\pi}{\omega}k\right)$ vanishes only at $(0,\mathbf{0})$, and then the Fourier multipliers $M, kM, |\boldsymbol{\xi}|M, |\boldsymbol{\xi}|^2M$ such that

$$M(k,\boldsymbol{\xi}) := \frac{1-\delta_{\mathbb{Z}}(k)}{\nu|\boldsymbol{\xi}|^2 + i\frac{2\pi}{\omega}k}$$

belong to $L^\infty(\mathbb{Z}\times\mathbb{R}^3)\cap C^\infty(\mathbb{Z}\times\mathbb{R}^3)$ and can be viewed as $L^q(\mathbb{R}^4)$-multipliers. Precisely, multiplier theory is available in a Euclidean setting and not in the general setting of group multipliers, and therefore it is necessary to use the transference principle. Using a transference principle for Fourier multipliers on local compact Abelian groups gives possibility to study the properties of M via corresponding multipliers defined on $\mathbb{R}^4$. For more details see [28, 49]. From the transference principle follows that M is an $L^q(G)$ multiplier if there is a continuous homeomorphism $\varphi\colon \hat{G}\to\mathbb{R}^4$ such that $M = m\circ\varphi$ for some $L^q(\mathbb{R}^4)$ multiplier m. Moreover, the norm of $L^q(G)$ operator corresponding to M with the norm of the $L^q(\mathbb{R}^4)$ corresponding to m. We introduce $\chi\in C_0^\infty(\mathbb{R};\mathbb{R})$, $\chi(\eta)=1$ for $|\eta|\le 1/2$, and $\chi(\eta)=0$ for $|\eta|\ge 1$. Define $m\colon \mathbb{R}\times\mathbb{R}^3\to\mathbb{C}$, $m(\eta,\boldsymbol{\xi})\colon (1-\chi((T/2\pi)\eta)/(|\nu|\boldsymbol{\xi}|^2+i\eta))$. Further let $\varphi\colon\widehat{G}\to\mathbb{R}^{n+1}$, $\varphi(k,\boldsymbol{\xi}) = ((2\pi/T)k,\pi)$. φ is a continuous homeomorphism and $M = m\circ\varphi$. Using a standard multiplier theorem, we can get that $m\in L^q(\mathbb{R}^4)$. Since m is with nonvanishing denominator away from $(0,0)$, then

$$(\eta,\boldsymbol{\xi}) \to \xi_1^{\epsilon_1}\cdots\xi_n^{\epsilon_n}\eta^{\epsilon_{n+1}}\partial_1^{\epsilon_1}\cdots\partial_n^{\epsilon_n}\partial_\eta^{\epsilon_{n+1}}m(\eta,\xi) \tag{3.57}$$

stay bounded as $|(\eta,\boldsymbol{\xi})|\to\infty$. From Marcinkiewic's multiplier theorem follows that m is an $L^q(\mathbb{R}^{n+1})$ multiplier.

It follows for all $\widetilde{\mathbf{f}}\in\mathbf{L}^q(\mathbf{G})$,

$$\|\mathcal{F}^{-1}_{\mathbf{G}}[M(k,\boldsymbol{\xi})\mathcal{F}_{\mathbf{G}}[\widetilde{\mathbf{f}}](k,\boldsymbol{\xi})]\|_q \le C_\omega\|\widetilde{\mathbf{f}}\|_q.$$

Since $I-\boldsymbol{\xi}\otimes\boldsymbol{\xi}/|\boldsymbol{\xi}|^2$ is a symbol of the Helmholtz–Weyl projection, it means $\Pi^q_\sigma\colon L^q(\mathbb{R}^n)\to L^q_\sigma(\mathbb{R}^n)$, where Π^q_σ is a continuous projection on $L^q(G)$.

Now defining $(\widetilde{\mathbf{w}}, \widetilde{\pi})$ by (3.56), we can see that $(\widetilde{\mathbf{w}}, \widetilde{\pi})$ is a solution in the sense of distribution $\mathcal{S}'(G)$, and moreover

$$\|\widetilde{\mathbf{w}}\|_q = \|\mathcal{F}_{\mathbf{G}}^{-1}[M(k, \boldsymbol{\xi})\mathcal{F}_{\mathbf{G}}[\Pi_\sigma^q \widetilde{\mathbf{f}}](k, \boldsymbol{\xi})]\|_q \le C_\omega \|\Pi_\sigma^q \widetilde{\mathbf{f}}_0\|_q \le C_\omega \|\widetilde{\mathbf{f}}\|_q$$

and analogously estimates for $\partial_t \widetilde{\mathbf{w}}$, $\nabla \widetilde{\mathbf{w}}$, $\nabla^2 \widetilde{\mathbf{w}}$. Therefore we summarize the results.

Theorem 3.4 *Let* $\mathbf{f} \in L^q(\mathbb{R}^3)$. *Then there exists a unique effective oscillatory velocity* $\widetilde{\mathbf{w}}$ *and its associated pressure* $\widetilde{\pi}$, *which is a solution to problem* (3.55) *that satisfies the estimate*

$$\|\nu \nabla^2 \widetilde{\mathbf{w}}\|_q + \|\widetilde{\mathbf{w}}\|_q + \|\partial_t \widetilde{\mathbf{w}}\|_q + \|\nabla \widetilde{\pi}\|_q \le c C_\omega \|\widetilde{\mathbf{f}}\|_q,$$

$$(\widetilde{\mathbf{w}}, \widetilde{\pi}) \in W^{1,2,q}_{\mathrm{per},\perp}(\mathbb{R} \times \mathbb{R}^3) \times D^{1,q}_{\mathrm{per},\perp}(\mathbb{R} \times \mathbb{R}^3).$$

As a consequence, there exists a locally integrable solution $(\mathbf{u}, p)$, $\mathbf{u} = \mathbf{u}_0 + \widetilde{\mathbf{w}}$, $p = p_0 + \widetilde{\pi}$, *to* (3.1) *that satisfies the inequality*

$$\|\nu \nabla^2 \mathbf{u}\|_q + \|\nabla p\|_q + \|\omega \mathcal{R} \mathbf{u}\|_q \le c C_\omega \|\mathbf{f}\|_q. \tag{3.58}$$

If for some $r \in (1, \infty)$ $(\widetilde{\mathbf{w}}_1, \widetilde{\pi}_1) \in W^{1,2,4}_{\mathrm{per},\perp}(\mathbb{R} \times \mathbb{R}^3) \times D^{1,4}_{\mathrm{per},\perp}(\mathbb{R} \times \mathbb{R}^3)$ *is another solution, then* $\widetilde{\mathbf{w}}_1 = \widetilde{\mathbf{w}}$ *and* $\widetilde{\pi} = \widetilde{\pi}_1 + \mathrm{d}(t)$ *for some* T*-periodic function* $\mathrm{d}\colon \mathbb{R} \to \mathbb{R}$.

Proof It remains to prove the uniqueness property. Let us assume the existence of another solution $(\widetilde{\mathbf{w}}_1, \widetilde{\pi}_1)$. Applying the Helmholtz projection and the Fourier transformation, we get

$$\left(i\frac{2\pi}{T}k + \nu|\boldsymbol{\xi}|^2 \right) \mathcal{F}_G(\widetilde{\mathbf{w}} - \widetilde{\mathbf{w}}_1) = 0.$$

The polynomial $i(2\pi/T)k + \nu|\boldsymbol{\xi}|^2$ vanishes only in $(k, \boldsymbol{\xi}) = (0, 0)$, which implies that $\operatorname{supp} \mathcal{F}_G(\widetilde{\mathbf{w}} - \widetilde{\mathbf{w}}_1) \subset \{(0, 0)\}$. Also $\Pi_\sigma^q(\widetilde{\mathbf{w}} - \widetilde{\mathbf{w}}_1) = 0$ and also $\delta_Z(k)\mathcal{F}_G(k, \boldsymbol{\xi}) = 0$. Then $(0, 0) \notin \operatorname{supp} \mathcal{F}_G(\widetilde{\mathbf{w}} - \widetilde{\mathbf{w}}_1)$. It implies $\operatorname{supp} \mathcal{F}_G(\widetilde{\mathbf{w}} - \widetilde{\mathbf{w}}_1) = \emptyset$. So $\operatorname{supp} \mathcal{F}_G(\widetilde{\mathbf{w}} - \widetilde{\mathbf{w}}_1) = 0$, and thus $\widetilde{\mathbf{w}} = \widetilde{\mathbf{w}}_1$. Moreover, $p(t, x) = \pi(t, \mathbf{x}) + d(t)$, where $d(t)$ is some T-periodic function $d : R \to R$. □

Remark 3.14 Let us remark the important property of the fundamental solution in the time-periodic system. From previous observation it follows that the projection $\mathcal{P}$ can be written as $\mathcal{P}\mathbf{f} := \mathcal{F}_G^{-1}[\delta_Z \mathcal{F}_G[\mathbf{f}]]$. $\mathcal{P}$ can be extended to a projection in the sense of distribution $\mathcal{P} : \mathcal{S}'(G) \mapsto \mathcal{S}'(G)$ and similar to $\mathcal{P}_\perp$. The fundamental solution of time-periodic case can be decomposed to the well-known fundamental solution of the steady case and an oscillatory part. For more details, see [104].

3.1.7 *The Nonhomogeneous Problem and Proof of Theorem 3.2*

From the equations

$$\begin{aligned} S_{\omega,\nu}\mathbf{u} + \nabla p &= \mathbf{f} \quad \text{in } \mathbb{R}^3, \\ \operatorname{div}\mathbf{u} &= g \quad \text{in } \mathbb{R}^3, \\ \lim_{|\mathbf{x}|\to\infty} \mathbf{u}(\mathbf{x}) &= \mathbf{0}, \end{aligned} \tag{3.59}$$

we infer that the pressure p is given by the solution of

$$\begin{aligned} \Delta p &= \operatorname{div}\mathbf{f} + \nu\Delta g + \omega \operatorname{div}(\mathbf{e}_3 \wedge x \cdot g) \\ &:= \operatorname{div}\mathbf{G}, \end{aligned} \tag{3.60}$$

with

$$\mathbf{G} = \mathbf{f} + \nu\nabla g + \omega\mathbf{e}_3 \wedge x \cdot g.$$

Then we get the estimate

$$\|\nabla p\|_q \le c\|\mathbf{G}\|_q \le c\|\mathbf{f}\|_q + \|\nu\nabla g + \omega\mathbf{e}_3 \wedge x \cdot g\|_q, \tag{3.61}$$

where $c = c(q)$.

We can always solve the nonhomogeneous continuity equation $\operatorname{div}\mathbf{G}_0 = g$ and apply the results to our system (3.59) with new unknown $\mathbf{u} - \mathbf{G}_0$ but now with a homogeneous continuity equation and with an external force expressed by $\mathbf{F}_0 = \mathbf{f} + \nu\Delta\mathbf{G}_0 + \omega\mathcal{R}\mathbf{G}_0$. Summarizing,

$$\begin{aligned} S_{\nu,\omega}(\mathbf{u} - \mathbf{G}_0) + \nabla p &= \mathbf{F}_0 \quad \text{in } \mathbb{R}^3 \\ \operatorname{div}(\mathbf{u} - \mathbf{G}_0) &= 0 \quad \text{in } \mathbb{R}^3. \end{aligned} \tag{3.62}$$

3.2 A Weak Solution in the Whole Space $\mathbb{R}^3$ and in an Exterior Domain

The aim of the next subsection is to study the weak solutions of the Stokes type problem in the whole space $\mathbb{R}^3$ and in an exterior domain $\mathcal{D} \subset \mathbb{R}^3$ of the class $C^{1,1}$. The main tool in this part will be the proof of the existence of the weak solution applying the harmonic analysis, as in Sect. 3.1.5. The method of localization is applied in the case of existence of solution in the exterior domain. Here, we follow the work of Hishida, see [127]. The a priori estimates and the existence can be derived on the approach by Galdi and Kyed, see [102].

We investigate the following boundary value problem:

$$\left.\begin{aligned} S_{\omega,\nu}\mathbf{u} + \nabla p &= \mathbf{f} \text{ in } \mathcal{D} \\ \operatorname{div}\mathbf{u} &= 0 \text{ in } \mathcal{D} \\ \boldsymbol{u} &= 0 \text{ on } \partial\mathcal{D} \end{aligned}\right\} \tag{3.63}$$

or $\lim_{|\mathbf{x}|\to\infty}\mathbf{u}(\mathbf{x}) = 0$ when $\mathcal{D} = \mathbb{R}^3$. We recall the notation $S_{\omega,\nu} = -\nu\Delta - \omega\mathcal{R}$, where $\mathcal{R} = \mathbf{e}_3 \wedge \mathbf{x}\cdot\nabla - \mathbf{e}_3 \wedge \cdot$.

Definition of the Weak Solution
Let $\mathcal{D} \subseteq \mathbb{R}^3$ and $1 < q < \infty$. Given $\mathbf{f} \in \widehat{\boldsymbol{H}}^{-1,q}(\mathcal{D})$, we call $\{\mathbf{u}, p\} \in \widehat{\boldsymbol{H}}_0^{1,q}(\mathcal{D}) \times L^q(\mathcal{D})$ a weak solution to (3.63) if

1. $\operatorname{div}\mathbf{u} = 0$ *in* $L^q(\mathcal{D})$.
2. $\mathcal{R}\mathbf{u} \in \widehat{\boldsymbol{H}}^{-1,q}(\mathcal{D})$.
3. $\{\mathbf{u}, p\}$ *satisfies* (3.63) *in the sense of distributions, that is,*

$$\nu(\nabla\mathbf{u}, \nabla\boldsymbol{\varphi}) - \omega\langle\mathcal{R}\mathbf{u}, \boldsymbol{\varphi}\rangle - \langle p, \operatorname{div}\boldsymbol{\varphi}\rangle = \langle\mathbf{f}, \boldsymbol{\varphi}\rangle, \quad \forall\boldsymbol{\varphi} \in \boldsymbol{C}_0^\infty(\mathcal{D}), \tag{3.64}$$

where $\langle\cdot,\cdot\rangle$ stands for various duality pairings.

□

The next two theorems represent the main results. For simplicity we will consider the case $\nu = 1, \omega = 1$.

Theorem 3.5 *Let $1 < q < \infty$, and suppose that*

$$\mathbf{f} \in \widehat{\boldsymbol{H}}^{-1,q}(\mathbb{R}^3),\ g \in L^q(\mathbb{R}^3),\ (\omega\mathbf{e}_3 \wedge \mathbf{x})\, g \in \widehat{\boldsymbol{H}}^{-1,q}(\mathbb{R}^3).$$

Then the problem

$$\left.\begin{aligned} S_{1,1}\mathbf{u} + \nabla p &= \mathbf{f}, \\ div\, \mathbf{u} &= g \end{aligned}\right\} \text{ in } \mathbb{R}^3 \tag{3.65}$$

possesses a weak solution $\{\mathbf{u}, p\} \in \widehat{\boldsymbol{H}}^{1,q}(\mathbb{R}^3) \times L^q(\mathbb{R}^3)$ satisfying the estimate

$$\begin{aligned} &\|\nabla\mathbf{u}\|_{q,\mathbb{R}^3} + \|p\|_{q,\mathbb{R}^3} + \|\mathcal{R}\mathbf{u}\|_{-1,q,\mathbb{R}^3} \\ &\qquad \le C(\|\mathbf{f}\|_{-1,q,\mathbb{R}^3} + \|g\|_{q,\mathbb{R}^3} + \|(\mathbf{e}_3 \wedge \mathbf{x})g\|_{-1,q,\mathbb{R}^3}) \end{aligned} \tag{3.66}$$

with a constant $C > 0$. Solution $\mathbf{u}$ is unique in the class $\widehat{\boldsymbol{H}}^{1,q}(\mathbb{R}^3)$ up to an additive constant vector.

Theorem 3.6 *Let* $3/2 < q < 3$. *For every* $\mathbf{f} \in \widehat{\boldsymbol{H}}^{-1,q}(\mathcal{D})$, *there exists a unique weak solution* $\{\mathbf{u}, p\} \in \widehat{\boldsymbol{H}}_0^{1,q}(\mathcal{D}) \times L^q(\mathcal{D})$ *of the problem* (3.63) *satisfying the estimate*

$$\|\nabla \mathbf{u}\|_q + \|p\|_q + \|\mathcal{R}\mathbf{u}\|_{-1,q} \le C\|\mathbf{f}\|_{-1,q} \tag{3.67}$$

with a constant $C > 0$. □

3.2.1 The Whole Space Problem

The aim of this part is to give the proof of Theorem 3.5. We begin with the elliptic problem without constraint div $\mathbf{u} = 0$.

Theorem 3.7 *Let* $1 < q < \infty$, $\mathbf{f} \in \widehat{\boldsymbol{H}}^{-1,q}(\mathbb{R}^3)$. *Then the equation*

$$S_{1,1}\mathbf{u} \equiv -\Delta\mathbf{u} - \mathcal{R}\mathbf{u} = \mathbf{f} \quad \textit{in } \mathbb{R}^3 \tag{3.68}$$

has a weak solution $\mathbf{u} \in \widehat{\boldsymbol{H}}^{1,q}(\mathbb{R}^3)$ *satisfying the estimate*

$$\|\nabla \mathbf{u}\|_{q,\mathbb{R}^3} + \|\mathcal{R}\mathbf{u}\|_{-1,q,\mathbb{R}^3} \le C\|\mathbf{f}\|_{-1,q,\mathbb{R}^3} \tag{3.69}$$

with some $C > 0$. *A solution* $\mathbf{u}$ *is unique in* $\widehat{\boldsymbol{H}}^{1,q}(\mathbb{R}^3)$ *up to an additive constant vector.* □

Proof of Theorem 3.7 The essential step in the proof is to show that

$$\|\nabla \mathbf{u}\|_{q,\mathbb{R}^3} \le C\|\boldsymbol{F}\|_{q,\mathbb{R}^3} \tag{3.70}$$

with $\mathbf{f} = \operatorname{div} \boldsymbol{F}$ and $\boldsymbol{F} = (F)_{\mu\nu} \in \boldsymbol{C}_0^\infty(\mathbb{R}^3)$. As in the previous part we can formally write the solution to Eq. (3.68) in the form

$$T\boldsymbol{F}(\boldsymbol{x}) = \nabla\mathbf{u}(\boldsymbol{x}) = -\int_{\mathbb{R}^3} \nabla_x \nabla_y \Sigma_{st}(\mathbf{x}, \mathbf{y}) : F(\mathbf{y})\mathrm{d}\mathbf{y},$$

$$(\nabla_y \Sigma_{st}(\mathbf{x}, \mathbf{y}) : F(\mathbf{y}))_l = \sum_{1\le\mu,\nu\le 3} \partial y_\nu \, (\Sigma_{st})_{l\,\mu}(\mathbf{x}, \mathbf{y}) F_{\mu\,\nu}(\mathbf{y}) \quad (1 \le \ell \le 3).$$

Applying the Palley–Littlewood decomposition as in the previous subsection and using Lemmas 3.1–3.3, we obtain

$$\begin{aligned}\|T_{lm}^{(j)}\boldsymbol{F}\|_{q,\mathbb{R}^3} \le& C2^{-2|j|}\|\boldsymbol{F}\|_{q,\mathbb{R}^3},\\ T_{lm}\boldsymbol{F}(\mathbf{x}) =& \partial_{x_m}\mathbf{u}_l(\mathbf{x})\\ =& \sum_{\mu,\nu,k}\int_0^\infty Q_1(t)_{lm}^{\mathrm{T}}Q_1(t)_{kn}(H_{k\nu,t}*F_{\mu\nu})(Q_1(t)\mathbf{x})\frac{\mathrm{d}t}{t},\end{aligned}$$

where $\boldsymbol{H} = (H_{k\nu})_{1\le k,\nu\le 3}$ is the Hessian matrix of K (heat kernel), that is,

$$\begin{aligned}H_{k\nu}(\mathbf{x}) &= \frac{\partial^2 K_t(\mathbf{x})}{\partial x_\nu \partial x_k},\\ H_{k\nu,t}(\mathbf{x}) &= t^{-3/2}H_{k\nu}(\mathbf{x}/\sqrt{t}).\end{aligned}$$

Let $\mathbf{f} \in \widehat{\boldsymbol{H}}^{-1,q}(\mathbb{R}^3)$. By Lemma 3.5 there is $\boldsymbol{F} \in L^q(\mathbb{R}^3)$ such that

$$\operatorname{div}\boldsymbol{F} = \mathbf{f},\ \ \|\boldsymbol{F}\|_{q,\mathbb{R}^3} \le C\|\mathbf{f}\|_{-1,q,\mathbb{R}^3}.$$

This gives us the estimate

$$\|\nabla\mathbf{u}\|_{q,\mathbb{R}^3} \le C\|\mathbf{f}\|_{-1,q,\mathbb{R}^3}$$

together with

$$\|\mathcal{R}\mathbf{u}\|_{-1,q,\mathbb{R}^3} = \|\mathbf{f} + \Delta\mathbf{u}\|_{-1,q,\mathbb{R}^3} \le \|\mathbf{f}\|_{-1,q,\mathbb{R}^3} + \|\nabla\mathbf{u}\|_{q,\mathbb{R}^3}.$$

It implies the estimate (3.69). □

It remains to prove the uniqueness. Using the duality argument, we consider the adjoint equation

$$S_{1,1}^*\mathbf{v} \equiv -\Delta\mathbf{v} + \mathcal{R}\mathbf{u} = \operatorname{div}\boldsymbol{F} \tag{3.71}$$

with $\boldsymbol{F} \in \boldsymbol{C}_0^\infty(\mathbb{R}^3)$.

The solution $\mathbf{v}(\mathbf{x})$ satisfies

$$\mathbf{v}(\mathbf{x}) = \int_0^\infty Q_1(t)(\mathrm{e}^{t\Delta}\operatorname{div}\boldsymbol{F})(Q_1^{\mathrm{T}}(t)\mathbf{x})\mathrm{d}t. \tag{3.72}$$

As in the case of (3.68), we have $\mathbf{v} \in \widehat{\boldsymbol{H}}^{1,r}(\mathbb{R}^3)\ \forall r \in (1, \infty)$ with $\|\nabla \mathbf{v}\|_{r,\mathbb{R}^3} \leq C\|\boldsymbol{F}\|_{r,\mathbb{R}^3}$. Let $\mathbf{u} \in \widehat{\boldsymbol{H}}^{1,q}(\mathbb{R}^3)$ be a weak solution of $S_{1,1}\mathbf{u} = 0$ in $\widehat{\boldsymbol{H}}^{-1,q}(\mathbb{R}^3)$. Taking $\mathbf{v}$ as a test function, we get

$$\langle S_{1,1}\mathbf{u}, \mathbf{v}\rangle = 0. \tag{3.73}$$

By analogy to (3.70), we take $\mathbf{u} \in \widehat{\boldsymbol{H}}^{-1,q/q-1}(\mathbb{R}^3)$ as a test function:

$$\langle \mathbf{u}, S^*_{1,1}\mathbf{v}\rangle = \langle \mathbf{u},\ \mathrm{div}\,\mathbf{F}\rangle. \tag{3.74}$$

Then

$$\langle \mathbf{u},\ \mathrm{div}\,\boldsymbol{F}\rangle = 0. \tag{3.75}$$

Because $\boldsymbol{F} \in \boldsymbol{C}_0^\infty(\mathbb{R}^3)$ is arbitrary, we obtain $\mathbf{u} = 0$ in $\widehat{\boldsymbol{H}}^{1,q}(\mathbb{R}^3)$. Then $\mathbf{u}$ is a constant vector, and it should be a constant vector. To complete the proof of Theorem 3.5, we need the following lemma.

Lemma 3.5 *Let* $\mathbf{v} \in \mathcal{S}'(\mathbb{R}^3)$ *be the solution of*

$$-\Delta\mathbf{v} - (\mathbf{e}_3 \wedge \mathbf{x}) \cdot \nabla\mathbf{v} = 0 \quad \text{in } \mathbb{R}^3. \tag{3.76}$$

Then $\mathrm{supp}\,\widehat{\mathbf{v}} \subset \{\mathbf{0}\}$. *Especially, if* $\mathbf{v} \in \boldsymbol{L}^q(\mathbb{R}^3)$ *for* $1 \leq q < \infty$, *then* $\mathbf{v} = \mathbf{0}$. □

Proof Applying the Fourier transform we get

$$|\boldsymbol{\xi}|^2\widehat{\mathbf{v}} - (\mathbf{e}_3 \wedge \boldsymbol{\xi}) \cdot \nabla\widehat{\mathbf{v}} = 0. \tag{3.77}$$

For any $\boldsymbol{\varphi} \in C_0^\infty(\mathbb{R}^3 \setminus \{0\})$, the adjoint equation

$$|\boldsymbol{\xi}|^2\chi + (\mathbf{e}_3 \wedge \boldsymbol{\xi}) \cdot \nabla\chi = \varphi \text{ in } \mathbb{R}^3, \tag{3.78}$$

$$\chi(\boldsymbol{\xi}) = \int_0^\infty \mathrm{e}^{-|\boldsymbol{\xi}|^2 t}\varphi(Q_1(t)^{\mathrm{T}}\boldsymbol{\xi})\mathrm{d}t \tag{3.79}$$

is a solution $\boldsymbol{\chi} \in C_0^\infty(\mathbb{R}^3 \setminus \{0\})$. Hence, we have

$$\langle\widehat{\mathbf{v}}, \boldsymbol{\varphi}\rangle = \langle\widehat{\mathbf{v}}, |\boldsymbol{\xi}|^2\chi + (\mathbf{e}_3 \wedge \boldsymbol{\xi}) \cdot \nabla\boldsymbol{\chi}\rangle = \langle|\boldsymbol{\xi}|^2\widehat{\mathbf{v}} - (\mathbf{e}_3 \wedge \boldsymbol{\xi}) \cdot \nabla\widehat{\mathbf{v}}, \boldsymbol{\chi}\rangle = 0, \tag{3.80}$$

which completes the proof. □

Proof of Theorem 3.5 We have to get the estimate involving the pressure term p. The relation

$$\operatorname{div} \mathcal{R}\mathbf{u} = \mathbf{e}_3 \wedge \mathbf{x} \cdot \nabla \operatorname{div} \mathbf{u} = \operatorname{div} [(\mathbf{e}_3 \wedge \mathbf{x}) \operatorname{div} \mathbf{u}] \tag{3.81}$$

gives the pressure term by $\mathbf{f}$ and g as

$$p = -\operatorname{div} (-\Delta)^{-1}[\mathbf{f} + \nabla g + (\mathbf{e}_3 \wedge \mathbf{x})g].$$

It implies

$$\|p\|_{q,\mathbb{R}^3} \le C\|\mathbf{f} + \nabla g + (\mathbf{e}_3 \wedge \mathbf{x})g\|_{-1,q,\mathbb{R}^3}, \tag{3.82}$$

from which

$$\|\mathbf{f} - \nabla p\|_{-1,q,\mathbb{R}^3} \le C(\|\mathbf{f}\|_{-1,q,\mathbb{R}^3} + \|\nabla g + (\mathbf{e}_3 \wedge \mathbf{x})g\|_{-1,q,\mathbb{R}^3}). \tag{3.83}$$

Thus Theorem 3.7 provides a solution $\mathbf{u} \in \widehat{H}^{1,q}(\mathbb{R}^3)$ of

$$-\Delta\mathbf{u} - \mathcal{R}\mathbf{u} = \mathbf{f} - \nabla p. \tag{3.84}$$

Since

$$-\Delta(\operatorname{div} \mathbf{u}) - (\omega\mathbf{e}_3 \wedge \mathbf{x}) \cdot \nabla(\operatorname{div} \mathbf{u}) = \nabla \cdot \mathbf{f} - \Delta p = -\Delta g - (\mathbf{e}_3 \wedge \mathbf{x})\nabla g,$$

Lemma 2.3 gives $\operatorname{div} \mathbf{u} = g$ in $L^q(\mathbb{R}^3)$. The estimate (3.69) together with (3.82), (3.83) implies (3.66). □

Remark 3.15 As in the case of strong solution we can also derive a priori estimate (3.66) from the time-periodic Stokes problem (3.13); for more details see [102, 104]. □

3.2.2 The Problem in an Exterior Domain

First we assume the existence and uniqueness of a weak solution to the boundary value problem (3.63) (we always consider $\nu = 1, \omega = 1$), and we would like to obtain L^q-estimates. Later the existence will be justified in two steps, the intermediate one with $\mathbf{f} = \operatorname{div} \mathbf{F}$, where $\mathbf{F} \in C_0^\infty(\mathcal{D})$ follows from Remark 3.11. Applying the localization technique as in [144], we immediately observe that (3.63) combines both systems in the whole space and in a bounded domain.

The main lemma says:

Lemma 3.6 *Let* $3/2 < q < \infty$. *Given* $\mathbf{f} \in \widehat{\boldsymbol{H}}^{-1,q}(\mathcal{D})$, *let* $\{\mathbf{u}, p\} \in \widehat{\boldsymbol{H}}_0^{1,q}(\mathcal{D}) \times L^q(\mathcal{D})$ *be a weak solution to the problem* (3.63). *Choose* $\rho > \rho_0 > 0$ *so large that* $\mathbb{R}^3 \setminus \mathcal{D} \subset B_{\rho_0}$, *and take* $\psi \in C_0^\infty(B_\rho; [0,1])$ *such that* $\psi = 1$ *on* B_{ρ_0}. *Then*

$$\begin{aligned} &\|\nabla\mathbf{u}\|_q + \|p\|_q + \|\mathcal{R}\mathbf{u}\|_{-1,q} \\ &\le C\left(\|\mathbf{f}\|_{-1,q} + \|\mathbf{u}\|_{q,\mathcal{D}_\rho} + \|p\|_{-1,q,\mathcal{D}_\rho} + \left|\int_{\mathcal{D}_\rho}\psi(\boldsymbol{x})p(\boldsymbol{x})\mathrm{d}\boldsymbol{x}\right|\right) \end{aligned} \tag{3.85}$$

with some $C > 0$, *where* $\mathcal{D}_\rho = \mathcal{D} \cap B_\rho$. □

Proof Choose $\rho > \rho_0 > 0$ so large that $\Omega \subset B_{\rho_0} = \{\boldsymbol{x} \in \mathbb{R}^3 : |\boldsymbol{x}| < \rho_0\}$, and take a cutoff function $\psi \in C_0^\infty(B_\rho; [0,1])$ such that $\psi = 1$ on B_{ρ_0} and $\operatorname{supp}(\nabla\psi) \subset \{x : \rho_0 < |\boldsymbol{x}| < \rho\}$. Introducing now $\boldsymbol{U} = (1-\psi)\mathbf{u}$, $\boldsymbol{V} = \psi\mathbf{u}$, $\sigma = (1-\psi)p$ and $\tau = \psi p$, we get

$$u = \boldsymbol{U} + \boldsymbol{V}, \tag{3.86}$$

$$p = \sigma + \tau. \tag{3.87}$$

Then $\{\boldsymbol{U}, \sigma\}$ is a weak solution of

$$\left.\begin{aligned} -\Delta\boldsymbol{U} - \mathcal{R}\mathbf{u} + \nabla\sigma &= \boldsymbol{Z}_1, \\ \operatorname{div}\boldsymbol{U} &= -\boldsymbol{u}\cdot\nabla\psi \end{aligned}\right\} \quad \text{in } \mathbb{R}^3, \tag{3.88}$$

where

$$\boldsymbol{Z}_1 = (1-\psi)\mathbf{f} + 2\nabla\psi\cdot\nabla\mathbf{u} + [\Delta\psi + (\mathbf{e}_3 \wedge \mathbf{x})\cdot\nabla\psi]\mathbf{u} - (\nabla\psi)p. \tag{3.89}$$

The weak formulation of (3.88) can be written as

$$\langle\nabla\boldsymbol{U}, \nabla\boldsymbol{\varphi}\rangle - \langle(\mathbf{e}_3 \wedge \mathbf{x})\cdot\nabla\boldsymbol{U} - \mathbf{e}_3 \wedge \boldsymbol{U}, \boldsymbol{\varphi}\rangle - \langle\sigma, \nabla\cdot\boldsymbol{\varphi}\rangle = \langle\boldsymbol{Z}_1, \boldsymbol{\varphi}\rangle, \tag{3.90}$$

for all $\boldsymbol{\varphi} \in \boldsymbol{C}_0^\infty(\mathbb{R}^3)$, since $\{\mathbf{u}, p\}$ satisfies (3.64) with $\boldsymbol{\varphi} = (1-\psi)\boldsymbol{\varphi}$.

Further, $(\boldsymbol{V}, \tau)$ is a weak solution of the problem

$$-\Delta\boldsymbol{V} + \nabla\tau = \boldsymbol{Z}_2,\ \operatorname{div}\boldsymbol{V} = \mathbf{u}\cdot\nabla\psi \text{ in } \mathcal{D}_\rho;\ \ \boldsymbol{V}|_{\partial\mathcal{D}_\rho} = 0, \tag{3.91}$$

with

$$\boldsymbol{Z}_2 = \psi[\mathbf{f} + (\mathbf{e}_3 \wedge \mathbf{x})\cdot\nabla\mathbf{u} - \mathbf{e}_3 \wedge \mathbf{u}] - 2\nabla\psi\cdot\nabla\mathbf{u} - (\Delta\psi)\mathbf{u} + (\nabla\psi)p. \tag{3.92}$$

From Theorem 3.5 and Lemma 2.4, respectively, it follows that

$$\begin{aligned}\|\nabla \boldsymbol{U}\|_{q,\mathbb{R}^3} + \|\sigma\|_{q,\mathbb{R}^3} \leq & C(\|\boldsymbol{Z}_1\|_{-1,q,\mathbb{R}^3} + \|\mathbf{u}\cdot\nabla\psi\|_{q,\mathbb{R}^3} \\ & + \|(\mathbf{e}_3\wedge\mathbf{x})(\mathbf{u}\cdot\nabla\psi)\|_{-1,q,\mathbb{R}^3})\end{aligned} \tag{3.93}$$

and

$$\begin{aligned}&\|\nabla \boldsymbol{V}\|_{q,\mathcal{D}_\rho} + \|\tau\|_{q,\mathcal{D}_\rho} \\ &\leq C\left(\|\boldsymbol{Z}_2\|_{-1,q,\mathcal{D}_\rho} + c\|\mathbf{u}\cdot\nabla\psi\|_{q,\mathcal{D}_\rho} + \frac{1}{|\mathcal{D}_\rho|^{1-1/q}}\left|\int_{\mathcal{D}_\rho}\tau(\mathbf{x})\mathrm{d}x\right|\right).\end{aligned} \tag{3.94}$$

Let $\varphi \in C_0^\infty(\mathbb{R}^3)$. Then we obtain

$$\begin{aligned}|\langle(1-\psi)\mathbf{f},\boldsymbol{\varphi}\rangle| &\leq \|\mathbf{f}\|_{-1,q}\|\nabla\boldsymbol{\varphi}[(1-\psi)\boldsymbol{\varphi}]\|_{q/(q-1)} \\ &\leq \|\mathbf{f}\|_{-1,q}\left(\|\nabla\boldsymbol{\varphi}\|_{q/(q-1)} + c\|\nabla\boldsymbol{\varphi}\|_{q/(q-1),\mathcal{D}_\rho}\right) \\ &\leq C\|\mathbf{f}\|_{-1,q}\|\nabla\boldsymbol{\varphi}\|_{q/(q-1),\mathbb{R}^3}.\end{aligned} \tag{3.95}$$

We have used the condition $q > 3/2$, so that $q/(q-1) < 3$. Applying the Sobolev inequality, we further get

$$\|\boldsymbol{\varphi}\|_{q/(q-1),\mathcal{D}_\rho} \leq |D_\rho|^{1/3}\|\boldsymbol{\varphi}\|_{r,\mathcal{D}_\rho} \leq c\|\boldsymbol{\varphi}\|_{r,\mathbb{R}^3} \leq c\|\nabla\boldsymbol{\varphi}\|_{q/(q-1),\mathbb{R}^3}, \tag{3.96}$$

where $1/r = (q-1)/q - 1/3$. In the same way we obtain

$$\left|\langle 2\nabla\psi\cdot\nabla\mathbf{u} + [\Delta\psi + (\mathbf{e}_3\wedge\mathbf{x})\cdot\nabla\psi)]\mathbf{u},\boldsymbol{\varphi}\rangle\right| \leq c\|\mathbf{u}\|_{q,\mathcal{D}_\rho}\|\nabla\boldsymbol{\varphi}\|_{q/(q-1),\mathbb{R}^3}, \tag{3.97}$$

and

$$|\langle(\nabla\psi)p,\boldsymbol{\varphi}\rangle| \leq c\|p\|_{-1,q,\mathcal{D}_\rho}\|\nabla\boldsymbol{\varphi}\|_{q/(q-1),\mathbb{R}^3}, \tag{3.98}$$

$$|\langle(\mathbf{e}_3\wedge\mathbf{x})\cdot\mathbf{u}\cdot\nabla\psi),\boldsymbol{\varphi}\rangle| \leq c\|\mathbf{u}\|_{q,\mathcal{D}_\rho}\|\nabla\boldsymbol{\varphi}\|_{q/(q-1),\mathbb{R}^3}. \tag{3.99}$$

From (3.95)–(3.99) and (3.93) we get

$$\|\nabla\boldsymbol{U}\|_{q,\mathbb{R}^3} + \|\sigma\|_{q,\mathbb{R}^3} \leq C\left(\|\boldsymbol{f}\|_{-1,q} + \|\mathbf{u}\|_{q,\mathcal{D}_\rho} + \|p\|_{-1,q,\mathcal{D}_\rho}\right). \tag{3.100}$$

Similarly, we see that

$$|\langle\boldsymbol{Z}_2,\varphi\rangle| \leq C\left(\|\boldsymbol{f}\|_{-1,q} + \|\mathbf{u}\|_{q,\mathcal{D}_\rho} + \|p\|_{-1,q,D_\rho}\right)\|\nabla\boldsymbol{\varphi}\|_{q/(q-1),\mathcal{D}_\rho} \tag{3.101}$$

for all $\boldsymbol{\varphi} \in \boldsymbol{C}_0^\infty(\mathcal{D}_\rho)$.

The estimate (3.101) together with (3.95) implies that

$$\begin{aligned}&\|\nabla V\|_{q,\mathcal{D}_\rho} + \|\tau\|_{q,\mathcal{D}_\rho} \\ &\qquad \le C\left(\|\mathbf{f}\|_{-1,q} + \|\mathbf{u}\|_{q,\mathcal{D}_\rho} + \|p\|_{-1,q,\mathcal{D}_\rho} + \left|\int_{\mathcal{D}_\rho} \psi(x)p(\mathbf{x})\mathrm{d}x\right|\right).\end{aligned} \tag{3.102}$$

From (3.100) and (3.102) we obtain

$$\|\nabla\mathbf{u}\|_q + \|p\|_q \le C\left(\|\mathbf{f}\|_{-1,q} + \|\mathbf{u}\|_{q,\mathcal{D}_\rho} + \|p\|_{-1,q,\mathcal{D}_\rho} + \left|\int_{\mathcal{D}_\rho} \psi(\mathbf{x})p(\mathbf{x})\mathrm{d}\mathbf{x}\right|\right). \tag{3.103}$$

This inequality, together with

$$\|\mathcal{R}\mathbf{u}\|_{-1,q} = \|\mathbf{f} + \Delta\mathbf{u} - \nabla p\|_{-1,q} \le \|\mathbf{f}\|_{-1,q} + \|\nabla\mathbf{u}\|_q + \|p\|_q, \tag{3.104}$$

yields (3.85). □

Now, we are interested in the adjoint problem

$$\left.\begin{aligned} -\Delta\mathbf{v} + \mathcal{R}\mathbf{v} - \nabla\pi &= \mathbf{f}, \\ -\operatorname{div}\mathbf{v} &= 0 \end{aligned}\right\} \quad \text{in } \mathcal{D}, \tag{3.105}$$
$$\mathbf{v} = 0 \quad \text{on } \partial\mathcal{D}.$$

Lemma 3.7 *Let* $\mathbf{F} \in C_0^\infty(\mathcal{D})$. *Then the problem* (3.63) *with* $f = \operatorname{div}\mathbf{F}$ *has a weak solution* $\{\mathbf{u}, p\}$ *in the class*

$$\mathbf{u} \in \widehat{\boldsymbol{H}}_0^{1,r}(\mathcal{D}), \quad p \in L^r(\mathcal{D}) \quad \text{for all } r,\ \frac{3}{2} < r < \infty. \tag{3.106}$$

The same assertion holds for the adjoint problem (3.105) *as well.* □

Proof Applying the standard L^2-techniques, we show the existence of a distributional solution

$$\mathbf{u} \in \widehat{\boldsymbol{H}}_0^{1,2}(\mathcal{D}) \hookrightarrow \boldsymbol{L}^6(\mathcal{D}), \quad p \in L^2_{\mathrm{loc}}(\overline{\mathcal{D}}).$$

Indeed, when $q = 2$, we take $\boldsymbol{\varphi} = \mathbf{u}$ in (3.64) to get

$$\|\nabla\mathbf{u}\|_2^2 = \langle \operatorname{div}\mathbf{F}, \mathbf{u}\rangle,$$

since $\int_{\mathcal{D}_R}[(\mathbf{e}_3\wedge\mathbf{x})\cdot\nabla\mathbf{u}]\cdot\mathbf{u}\mathrm{d}x = \frac{1}{2}\int_{\partial\mathcal{D}_R}\mathbf{u}\mathrm{d}\cdot(\mathbf{e}_3\wedge\mathbf{x})(\mathbf{u})^2\mathrm{d}\sigma = 0$ for every sufficiently large $R > 0$, where $\mathcal{D}_R = \mathcal{D}\cap B_R$ and $\mathbf{n}$ is the unit exterior normal vector to the boundary $\partial\mathcal{D}_R$; here note that $\mathbf{x}\cdot(\mathbf{e}_3\wedge\mathbf{x}) = 0$ on ∂B_R. Thus, we have the a priori estimate

$$\|\nabla\mathbf{u}\|_2 \le \|\mathbf{F}\|_2.$$

Applying the Lax–Millgram theorem together with de Rham lemma (see [93]), we get a distributional solution $\{\mathbf{u}_R, p_R\} \in \boldsymbol{W}^{1,2}_{0,\sigma}(\mathcal{D}_R)\times L^2(\mathcal{D}_R)$ to the same equations in each bounded domain $\mathcal{D}_R$ where $\boldsymbol{H}^{1,2}_{0,\sigma}(\mathcal{D}_R) = \overline{\boldsymbol{C}^\infty_0(\mathcal{D}_R)}^{\|\nabla\cdot\|_{2,\mathcal{D}_R}}$. The bilinear form

$$(\mathbf{u},\boldsymbol{\varphi}) \to \langle\nabla\mathbf{u},\nabla\boldsymbol{\varphi}\rangle - \langle(\mathcal{R}\mathbf{u},\boldsymbol{\varphi}\rangle$$

on $\boldsymbol{H}^{1,2}_{0,\sigma}(\mathcal{D}_R)\times H^{1,2}_{0,\sigma}(\mathcal{D}_R)$ is continuous and coercive. Moreover $\|\nabla\mathbf{u}_R\|_2 \le \|\mathbf{F}\|_2$ is valid on $\mathcal{D}$ in a way that $\mathbf{u}_R$ is extend by 0 in $\mathcal{D}\setminus\mathcal{D}_R$. There exist $\mathbf{u}\in\widehat{\boldsymbol{H}}^{1,2}_0(\mathcal{D})$ and a sequence $\{R_n\}\uparrow^\infty$ such that $\mathbf{u}_{R_n} \rightharpoonup \mathbf{u}$ weakly in $\widehat{\boldsymbol{H}}^{1,2}_0(\mathcal{D})$ as $n\to\infty$. We find $\langle S_{1,1}\mathbf{u} - \operatorname{div}\mathbf{F},\boldsymbol{\varphi}\rangle = 0$ for all $\boldsymbol{\varphi}\in\boldsymbol{C}^\infty_0(\mathcal{D})$ with $\nabla\cdot\boldsymbol{\varphi} = 0$. By de Rham's theorem, there is a distribution p such that $-\nabla p = S_{1,1}\mathbf{u} -$ div $\mathbf{F}$. We know that the right-hand side belongs to $\boldsymbol{H}^{-1,2}(\mathcal{D}_R)$ for every large $R>0$, we get $p\in L^2(\mathcal{D}_R)$, and thus $p\in L^2_{\mathrm{loc}}(\overline{\mathcal{D}})$. Applying the cutting-off technique we get in the same way as in the proof of Lemma 3.6

$$\mathbf{u}\in\widehat{\boldsymbol{H}}^{1,r}_0(\mathcal{D}),\quad p\in L^r(\mathcal{D})\quad \text{for } 3/2 < r\le 6.$$

By localization we obtain (3.106) for the problem (3.64). Similarly for problem (3.105) the regularity result (3.106) is valid. □

Proposition 3.2 (Uniqueness) *Let* $1<q<3$. *Suppose that* $\{\mathbf{u}, p\}\in\widehat{\boldsymbol{H}}^{1,q}_0(\mathcal{D})\times L^q(\mathcal{D})$ *is a weak solution to the problem* (3.64) *with* $\mathbf{f}=0$. *Then* $\{\mathbf{u},p\} = \{\mathbf{0},0\}$. □

Proof Consider the adjoint problem (3.105) with $\mathbf{f} = \nabla\cdot\mathbf{F}$, where $\mathbf{F}\in C^\infty_0(\mathcal{D})$. By Lemma 3.7 there is a weak solution $\{\mathbf{v},\pi\}$ of the class (3.106) to the adjoint problem (3.105). Since $q/(q-1) > 3/2$, we can set $\boldsymbol{\varphi} = \mathbf{v}$ in (3.64) with $\nu=\omega=1$ and $\mathbf{f}=0$:

$$(\nabla\mathbf{u},\nabla\mathbf{v}) - \langle\mathcal{R}\mathbf{u},\mathbf{v}\rangle = 0. \tag{3.107}$$

Similarly we test (3.105) by $\mathbf{u}$ to get

$$\langle\nabla\mathbf{v},\nabla\mathbf{u}\rangle + \langle\mathcal{R}\mathbf{v},\mathbf{u}\rangle = \langle\operatorname{div}\mathbf{F},\mathbf{u}\rangle = \langle\mathbf{f},\mathbf{u}\rangle = \langle\operatorname{div}\mathbf{F},\mathbf{u}\rangle. \tag{3.108}$$

From (3.107) and (3.108) we obtain $\langle\operatorname{div}\mathbf{F},\mathbf{u}\rangle = 0$ for all $\mathbf{F}\in C^\infty_0(\mathcal{D})$. From Lemma 2.3 we get $\langle\mathbf{f},u\rangle = 0$ for all $\mathbf{f}\in\widehat{\mathbf{H}}^{-1,q/(q-1)}(\mathcal{D})$, which yields $\mathbf{u}=\mathbf{0}$ in $\widehat{\boldsymbol{H}}^{1,q}_0(\mathcal{D})$ and also $p=0$ in $L^q(\mathcal{D})$. □

From Lemma 3.6 with Proposition 3.2 we get the following a priori estimate.

Proposition 3.3 (A Priori Estimate) *Let* $3/2 < q < 3$. *Suppose that* $\{\mathbf{u}, p\} \in \widehat{\boldsymbol{H}}_0^{1,q}(\mathcal{D}) \times L^q(\mathcal{D})$ *is a weak solution to the problem* (3.64) *with* $\mathbf{f} \in \widehat{\boldsymbol{H}}^{-1,q}(\mathcal{D})$. *Then the estimate*

$$\|\nabla \mathbf{u}\|_q + \|p\|_q + \|\mathcal{R}\mathbf{u}\|_{-1,q} \leq C\|\mathbf{f}\|_{-1,q} \tag{3.109}$$

holds. □

Proof We prove the proposition by contradiction. Let us consider $\mathbf{f}_k \in \widehat{\boldsymbol{H}}^{-1,q}(\mathcal{D})$. Let $\{\mathbf{u}_k, p_k\} \in \widehat{\boldsymbol{H}}_0^{1,q}(\mathcal{D}) \times L^q(\mathcal{D})$ be a corresponding weak solution such that

$$\|\nabla \boldsymbol{u}_k\|_q + \|p_k\|_q + \|\mathcal{R}\mathbf{u}_k\|_{-1,q} = 1 \tag{3.110}$$

with

$$\|\mathbf{f}_k\|_{-1,q} \to 0$$

for $k \to \infty$.

Then

$$\|\mathbf{u}_k\|_{1,q,\mathcal{D}_\rho} \leq \|\nabla \mathbf{u}_k\|_{q,\mathcal{D}_\rho} + C\|\mathbf{u}_k\|_{q*,\mathcal{D}_\rho} \leq c\|\nabla \mathbf{u}_k\|_q \leq c, \tag{3.111}$$

$$\|p_k\|_{q,\mathcal{D}_\rho} \leq \|p_k\|_q \leq C \quad \text{with} \quad \frac{1}{q^*} = \frac{1}{q} - \frac{1}{3}, \tag{3.112}$$

$\mathcal{D}_\rho = \mathcal{D} \cap B_\rho$. There are subsequences, which we denote $\mathbf{u}_k$, p_k again, and they are weakly converging in $\boldsymbol{H}^{1,q}(\mathcal{D}_\rho)$ and $L^q(\mathcal{D}_\rho)$. From the Rellich compactness theorem, they converge strongly in $\boldsymbol{L}^q(\mathcal{D}_\rho)$ and $H^{-1,q}(\mathcal{D}_\rho)$. From (3.85) it follows that $\{\boldsymbol{u}_k, p_k\}$ and $\{\mathcal{R}\mathbf{u}\}$ are Cauchy sequences in $\widehat{\boldsymbol{H}}_0^{1,q}(\mathcal{D}) \times L^q(\mathcal{D})$ and in $\widehat{\boldsymbol{H}}^{-1,q}(\mathcal{D})$; hence there exists $\{\mathbf{u}, p\} \in \widehat{\boldsymbol{H}}_0^{1,q}(\mathcal{D}) \times L^q(\mathcal{D})$ so that

$$\|\nabla \mathbf{u}_k - \nabla \mathbf{u}\|_q + \|p_k - p\|_q \to 0, \tag{3.113}$$

$$\|\mathcal{R}\mathbf{u}_k - \mathcal{R}\mathbf{u}\|_{-1,q} \to 0$$

for $k \to \infty$.

Then it follows that $\{\mathbf{u}, p\}$ is a weak solution of (3.64) with $\mathbf{f} = 0$. Since $q < 3$, Proposition 3.2 implies that $\{\mathbf{u}, p\} = \{0, 0\}$ which is in contradiction with

$$\|\nabla \mathbf{u}\|_q + \|p\|_q + \|\mathcal{R}\mathbf{u}\|_{-1,q} = 1.$$

This completes the proof. □

Proof of Theorem 3.4 The uniqueness follows from Proposition 3.2. Using Lemma 2.3 with $\mathbf{f} \in \widehat{\boldsymbol{H}}^{-1,q}(\mathcal{D})$, we take $\mathbf{F}_k \in \boldsymbol{C}_0^\infty(\mathcal{D})$ so that

$$\| \operatorname{div} \mathbf{F}_k - \mathbf{f}\|_{-1,q} \to 0$$

as $k \to \infty$. By Lemma 3.6 we have a solution $\{\mathbf{u}_k, p_k\}$ of the class (3.106) to the problem (3.64) with the force $\operatorname{div} \mathbf{F}_k$. We take $r = q$ in (3.106) (we know that $q > 3/2$). From Proposition 3.3, using (3.106), we can show the existence of $\{\mathbf{u}, p\} \in \widehat{\boldsymbol{H}}_0^{1,q}(\mathcal{D}) \times L^q(\mathcal{D})$ so that the same convergence properties as in (3.113) hold. The pair $\{\mathbf{u}, p\}$ is a weak solution to (3.64) with the estimate (3.67). Thus, it implies the assertion of the theorem. □

3.3 The Steady Stokes Type Problem in Weighted Spaces

The goal of this section is to find appropriate estimates in weighted spaces, which give more precise information on the behavior at large distance. The weight functions are taken from the class of Muckenhoupt weights, $w \in A_q$. We work in the weighted Lebesgue and Sobolev spaces, i.e., we integrate with respect to the measure $w(\mathbf{x})\,d\mathbf{x}$. We refer to Chap. 2, Sect. 2.1 for duality pairs $L^q_w - L^{q'}_{w'}$, $1 \le q \le \infty$. Another approach would be to prove behavior at large distance of our solutions but in unweighted spaces and to show that it corresponds to some weights.

In the following two subsections, we consider the steady Stokes problem in the whole space, secondly in an exterior domain. We follow work by Farwig, Krbec, and Nečasová [73] and by Nečasová and Schumacher [177].

3.3.1 A Strong Solution in the Whole Space

The content of this subsection is mainly based on the paper by Farwig, Krbec, Nečasová [73]. We investigate the homogeneous Stokes system including, as in the previous parts, the two additional terms that follow from the transformation to the coordinate system attached to the rotating body. The problem considered is

$$\begin{aligned} S_{\omega,\nu}\mathbf{u} + \nabla p &= \mathbf{f} \quad \text{in } \mathbb{R}^n, \\ \operatorname{div} \mathbf{u} &= 0 \quad \text{in } \mathbb{R}^n, \\ \mathbf{u} &\to 0 \quad \text{as } |x| \to \infty, \end{aligned} \tag{3.114}$$

where $n = 2$ or $n = 3$; in the two-dimensional case $\omega\mathbf{e}_3 \wedge \mathbf{x}=(-x_2, x_1)$ for $\mathbf{x} = (x_1, x_2)$ and $\omega\mathbf{e}_3 \wedge \mathbf{u} =(-u_2, u_1)$ for $\mathbf{u} = (u_1, u_2)$.

The aim of this subsection is to find the *a priori estimate* to the problem (3.114) in the weighted L^q-spaces with the appropriate weights.

Theorem 3.8

(i) *Let* $\mathbf{w}$ *be a weight of Muckenhoupt type in* $\mathbb{R}^n$, $n = 2$ *or* $n = 3$, *and assume that* w *is independent of the angular variable* θ *in a cylindrical coordinate system attached to the* $\mathbf{e}_3$*-axis of revolution* $(0, 0, 1)^T$. *Moreover, let* w *satisfy the condition* (2.15).

Given $\mathbf{f} \in \boldsymbol{L}^q_w(\mathbb{R}^n)$, *there exists a solution* $(\mathbf{u}, p) \in \boldsymbol{L}^1_{\text{loc}}(\mathbb{R}^n) \times L^1_{\text{loc}}(\mathbb{R}^n)$ *of* (3.114) *satisfying the estimate*

$$\|\nu\nabla^2\mathbf{u}\|_{q,w} + \|\omega\mathcal{R}\mathbf{u}\|_{q,w} + \|\nabla p\|_{q,w} \leq c\|\mathbf{f}\|_{q,w} \tag{3.115}$$

with a constant $c = c(q, w) > 0$.

(ii) *Let* $\mathbf{f} \in \boldsymbol{L}^{q_1}_{w_1}(\mathbb{R}^n) \cap \boldsymbol{L}^{q_2}_{w_2}(\mathbb{R}^n)$, $n = 2$ *or* $n = 3$, *be such that both* (q_1, w_1) *and* (q_2, w_2) *satisfy the condition* (2.15), *and let* $\mathbf{u}_1, \mathbf{u}_2 \in \boldsymbol{L}^1_{\text{loc}}(\mathbb{R}^n)$ *together with the corresponding pressure functions* $p_1, p_2 \in \boldsymbol{L}^1_{\text{loc}}(\mathbb{R}^n)$ *be solutions of* (3.114) *satisfying* (3.115) *for* (q_1, w_1) *and* (q_2, w_2), *respectively. Then there are* $\alpha, \beta, \gamma \in \mathbb{R}$ *such that* $\mathbf{u}_1$ *coincides with* $\mathbf{u}_2$ *up to an affine linear field* $\alpha\omega\mathbf{e}_3 + \beta\omega\mathbf{e}_3 \wedge \mathbf{x} + \gamma(x_1, x_2, -2x_3)^T$. *For* $n = 2$, *obviously* $\mathbf{u}_1 = \mathbf{u}_2$ *up to the field* $\beta(-x_2, x_1)^T$. □

3.3.1.1 Preliminaries to Theorem 3.8

We basically consider the problem (3.114) as the associated differential equation

$$-\nu\Delta\mathbf{u} - \partial_\theta\mathbf{u} + \omega\mathbf{e}_3 \wedge \mathbf{u} = \mathbf{F} := \mathbf{f} - \nabla p \quad \text{in} \quad \mathbb{R}^n, \tag{3.116}$$

$\mathbf{u}$ being 2π-periodic with respect to θ, its solenoidality being assumed by the solenoidality of $\mathbf{F}$. The pressure p satisfied the equation $\Delta p = \operatorname{div}\mathbf{f}$. Applying the Hörmander–Mikhlin multiplier theorem in weighted L^q-spaces (Theorem 2.1 (ii)) yields for every weight $\mathbf{w} \in A_q(\mathbb{R}^n)$ the estimate

$$\|\nabla p\|_{q,w} \leq c\|\mathbf{f}\|_{q,w}\,, \tag{3.117}$$

where $c = c(q, w) > 0$; in particular $\nabla p \in L^q_w$.

As before in the proof of Theorem 3.1, we have an integral representation of $\mathbf{u}$ solution, and it suffices to derive an L^q_w-estimate for $\Delta\mathbf{u}$. The main tools are a weighted version of the Littlewood–Paley theory and related decompositions of the integral operators which we denote by $\mathcal{T}$ and $\Delta_x \Sigma_{st}(\mathbf{x}, .)$, where Σ_{st} is the fundamental solution. Precisely,

$$\Sigma_{st}(\mathbf{x}, \mathbf{y}) = \int_0^\infty Q(t)^{\mathrm{T}} K(Q(t)\mathbf{x} - \mathbf{y}) \mathrm{d}t, \tag{3.118}$$

$$\mathbf{u}(\mathbf{x}) = \int_{\mathbb{R}^3} \Sigma_{st}(\mathbf{x}, \mathbf{y}) F(\mathbf{y}) \mathrm{d}\mathbf{y}, \tag{3.119}$$

$$-\Delta \mathbf{u}(x) = \int_{\mathbb{R}^3} -\Delta_x \Sigma_{st}(\mathbf{x}, \mathbf{y}) \mathbf{F}(\mathbf{y}) \mathrm{d}\mathbf{y}. \tag{3.120}$$

Assuming for the simplicity $\nu = 1, \omega = 1$, we recall the kernel model for $\Delta_x \Sigma_{st}(\mathbf{x}, \mathbf{y})$

$$\mathcal{K}(\mathbf{x}, \mathbf{y}) = \int_0^\infty Q_1(t)^T t^{-3/2} \psi\Big(\frac{Q_1(t)\mathbf{x} - \mathbf{y}}{\sqrt{t}}\Big) \frac{\mathrm{d}t}{t}$$

and finally the integral operator $\mathcal{T}$ is defined by

$$\mathcal{T}\mathbf{F}(\mathbf{x}) = \int_{\mathbb{R}^n} \mathcal{K}(\mathbf{x}, \mathbf{y}) \mathbf{F}(\mathbf{y}) \mathrm{d}\mathbf{y}. \tag{3.121}$$

As before, we need an analogous double inequality (3.33) but in the weighted case for all $w \in A_q(\mathbb{R}^n)$, $1 < q < \infty$. There is a classical result, see [186], Proposition 1.9, Theorem 1.10, and also [158, 206], and we have

$$c_1 \|\mathbf{F}\|_{q,w} \leq \|\mathfrak{S}\mathbf{F}\|_{q,w} \leq c_2 \|\mathbf{F}\|_{q,w}, \tag{3.122}$$

with constants $c_1, c_2 > 0$, depending on q, w but independent of $\mathbf{F}$ where the square operator $\mathfrak{S}$ is defined by

$$\mathfrak{S}\mathbf{F}(\mathbf{x}) := \left(\int_0^\infty |\varphi_s * \mathbf{F}(\cdot)|^2 \frac{\mathrm{d}s}{s} \right)^{1/2}, \tag{3.123}$$

with $\{\varphi_s\}_{s>0}$ a family of functions in $S(\mathbb{R}^n)$ satisfying properties (P1),(P2), (P3), and (P4). We recall these properties:

(P1) $\varphi_s(\mathbf{x}) = s^{-3/2} \varphi(\mathbf{x}/\sqrt{s})$, with $\varphi \in C_0^\infty(\frac{1}{2}, 2)$,
and $\widehat{\varphi}(\boldsymbol{\xi}) = (2\pi)^{-3/2} \int_{\mathbb{R}^3} \mathrm{e}^{-i\mathbf{x}\cdot\boldsymbol{\xi}} \varphi(\mathbf{x}) \mathrm{d}x$.
(P2) $\operatorname{supp} \widehat{\varphi_s} \subset \mathcal{A}(1/2\sqrt{s}, 2/\sqrt{s})$ and $\int_{\mathbb{R}^3} \varphi_s(\mathbf{x}) \mathrm{d}\mathbf{x} = 0$.
(P3) Radial symmetry $\widehat{\varphi}(\boldsymbol{\xi}) = \widehat{\varphi}(|\boldsymbol{\xi}|)$, and $\int_0^\infty \widehat{\varphi_s}(\boldsymbol{\xi})^2 \mathrm{d}s/s = 1$.
(P4) $\varphi_s * \psi_j = \varphi_s * \chi_j * \psi \equiv 0$ unless $t \in \mathcal{A}(s, j) := [2^{2j-4}s, 2^{2j+4}s]$.

As in the previous part, we note that

$$\|\mathfrak{S}\mathbf{F}(\mathbf{x})\|_{q,w} = \left\| \int_0^\infty |\varphi_s * \mathbf{F}(\cdot)|^2 \frac{\mathrm{d}s}{s} \right\|_{q/2,w} = \int_0^\infty \int_{\mathbb{R}^n} |\varphi_s * \mathcal{T}_j \mathbf{F}(x)|^2 g(x) \mathrm{d}x \, \frac{\mathrm{d}s}{s}; \tag{3.124}$$

using a duality argument (also for weights) there exists a function $0 \le \mathbf{g} \in L_v^{(q/2)'} = \left(L_w^{(q/2)}\right)^*$ with $\|\mathbf{g}\|_{(q/2)',v} = 1$. Remember that w' is the weight $w^{-\frac{(g/2)'}{q/2}}$ which is $w^{-2/(q-2)}$ and belongs to $A_{(q/2)'} = A_{\frac{q}{q-2}}$, and $q > 2$. In case $q < 2$ we apply the adjoint operator.

Now, providing the same specific partition of the unity as in (3.34)–(3.36), we get

$$\mathcal{K}_j(\mathbf{x}, \mathbf{y}) = \int_0^\infty Q_1(t)^{\mathrm{T}} t^{-3/2} \psi^j \Big(\frac{Q_1(t)\mathbf{x} - \mathbf{y}}{\sqrt{t}} \Big) = \int_0^\infty Q_1(t)^{\mathrm{T}} \psi_t^j (Q_1(t)\mathbf{x} - \mathbf{y}) \frac{\mathrm{d}t}{t},$$

$$\mathcal{T}_j \mathbf{F}(\mathbf{x}) = \int_{\mathbb{R}^n} \mathcal{K}_j(\mathbf{x}, \mathbf{y}) \mathbf{F}(\mathbf{y}) \mathrm{d}\mathbf{y} = \int_0^\infty Q_1(t)^{\mathrm{T}} (\psi_t^j * \mathbf{F})(Q_1(t)\mathbf{x}) \frac{\mathrm{d}t}{t},$$

$j \in \mathbb{Z}$. We have in view to apply the square operator $\mathfrak{S}$ to $\mathcal{T}_j \mathbf{F}$ and to prove Theorem 3.8.

3.3.2 *Proof of Theorem 3.8*

To prove Theorem 3.8 we present the next proposition. We will start with the weight $w \in A_{q/2}, q > 2$, $w \in A_1$, or $w^{-1} \in A_1$ in the case $q = 2$, $w^{2/q-2} \in A_{q/(q-2)}, 1 < q < 2$. The case of more general weights based on complex interpolation of L^q_w-spaces is postponed to the end of this section.

Step 1 As in Lemmas 3.2 and 3.3, we introduce two maximal operators

$$\mathcal{M}^j g(\mathbf{x}) = \sup_{s>0} \int_{\mathcal{A}_s} \left(|\psi_t^j| * |g| \right) (Q_1(t)^{\mathrm{T}} \mathbf{x}) \frac{\mathrm{d}t}{t}, \tag{3.125}$$

and

$$\mathcal{M}_\theta g(x) = \sup_{s>0} \int_{\mathcal{A}_s} |g(Q_1(t)^{\mathrm{T}} \mathbf{x})| \, \frac{\mathrm{d}t}{t}, \tag{3.126}$$

where $\mathcal{A}_s = [1/16s, 16s]$.

Remark 3.16 Let us explain the reason of introducing the "angular" maximal operator. In the proof we need to estimate $\|\mathcal{M}^j g\|_{(q/2)',v}$. Since $\frac{1}{2}q \in (1,\infty)$ is arbitrary, we have to consider $\|\mathcal{M}^j\|_{L^p_\rho}$ for arbitrary $p \in (1,\infty)$ and θ-independent weights $\rho \in A_p$, and because of this reason we introduce the "angular" maximal operator $\mathcal{M}_\theta g(x)$. □

Then we will prove the auxiliary estimate

$$\|\mathcal{T}_j \mathbf{f}\|_{q,w} \leq c\|\psi^j\|_1^{1/2}\|\mathcal{M}^j\|^{1/2}_{L_v^{(q/2)'}}\|\mathbf{f}\|_{q,w}, \quad j \in \mathbb{Z}, \tag{3.127}$$

where v denotes the weight

$$v = w^{-(q/2)'/(q/2)} = w^{-\frac{2}{q-2}} \in A_{(q/2)'} = A_{q/(q-2)}. \tag{3.128}$$

Moreover we get

$$\mathcal{M}^j g(x) \leq c2^{-2|j|}\mathcal{M}(\mathcal{M}_\theta g)(\mathbf{x}) \quad \text{for a.a. } \mathbf{x} \in \mathbb{R}^n, \tag{3.129}$$

$$\|\mathcal{M}^j g\|_{(p),\rho} \leq c2^{-2|j|}\|g\|_{p,\rho} \qquad \text{for } 1 < p < \infty,\ \rho \in A_p. \tag{3.130}$$

Precisely, to prove (3.129) we apply the pointwise estimate $|\psi_t^j(\mathbf{x})| \leq c2^{-2|j|}h_{t2^{-2j}}(\mathbf{x})$, see Lemma 3.1, (3.41), and (3.42). We get

$$\mathcal{M}^j g(\mathbf{x}) \leq c2^{-2|j|}\sup_{s>0}\int_{A_s}(h_{t2^{-2j}} * |g|)(Q_1(t)^{\mathrm{T}}\mathbf{x})\,\frac{\mathrm{d}t}{t}.$$

Moreover, there exists a constant $c > 0$ independent of $s > 0$, $j \in \mathbb{Z}$, such that $h_{t2^{-2j}} \leq c h_{s2^{-2j}}$ for all $t \in A_s$. Consequently

$$\mathcal{M}^j g(\mathbf{x}) \leq c2^{-2|j|}\ \sup_{s>0} h_{s2^{-2j}} * \int_{A_s}|g|(Q_1(t)^{\mathrm{T}}\mathbf{x})\,\frac{\mathrm{d}t}{t} \leq c2^{-2|j|}\sup_{t>0} h_t * \mathcal{M}_\theta\, g(\mathbf{x}).$$

Since h is nonnegative, radially decreasing, and $\|h_t\|_1 = \|h\|_1 = c_0 > 0$ for all $t > 0$, a well-known convolution estimate, see [205, II §2.1], yields the pointwise estimate (3.129). Note that up to now we have not used any special properties of the weight $\rho \in A_p$.

Concerning (3.130), we will show for general L^p_ρ. Note that for arbitrary but fixed radial distance $r = (x_1^2 + x_2^2)^{1/2}$ and $x_3 \in \mathbb{R}$ for $n = 3$,

$$\mathcal{M}_\theta\, g(r, x_3, \theta) \leq \sup_{s>0}\frac{16}{s}\int_{-16s}^{16s}|g|(r, x_3, \theta)|\mathrm{d}t \leq c\mathcal{M}_1 g(r, x_3, \theta),$$

where $\mathcal{M}_1$ is the classical maximal operator on $L^1_{\mathrm{loc}}(\mathbb{R})$. Since $\boldsymbol{\rho}(\mathbf{x}) = \boldsymbol{\rho}(r, x_3, \theta)$ is independent of θ, Fubini's theorem and the boundedness of $\mathcal{M}_1$ on L^p-spaces of

2π-periodic functions imply that

$$\|\mathcal{M}_\theta\, g\|_{p,\rho}^p \le c \int_{\mathbb{R}} \int_0^\infty \|\mathcal{M}_1 g(r, x_3, \theta)\|_{L_{\rho(r,x_3,\cdot)}(0,2\pi)}^p \, r^2 \mathrm{d}r \mathrm{d}x_3$$
$$\le c \int_{\mathbb{R}} \int_0^\infty \|g(r, x_3, \theta)\|_{L_{\rho(r,x_3,\cdot)}(0,2\pi)}^p r^2 \mathrm{d}r \mathrm{d}x_3 = c\|g\|_{p,\rho}^p.$$

Finally, since $\mathcal{M}$ is bounded on L^p_ρ, see Theorem 2.1 (i), (3.129) yields (3.130)

Step 2 Let $q > 2$ and $w \in A_{q/2}$. We estimate $\|S\mathcal{T}_j \mathbf{F}\|_{q,w}$ systematically by the similar way as in Sect. 3.3.1.1, and we get

$$\begin{aligned}
\left\|\mathfrak{S}\mathcal{T}_j\mathbf{F}\right\|_{q/2,w} &= \int_{\mathbb{R}^n} \mathrm{d}x \int_0^\infty |\varphi_s * \mathcal{T}_j\mathbf{F}(x)|^2 g(x)\frac{\mathrm{d}s}{s} \\
&= \int_{\mathbb{R}^n} \mathrm{d}x \int_0^\infty \frac{\mathrm{d}s}{s} Q_1(t)^{\mathrm{T}}(\varphi_s * \psi_t^j * \mathbf{F})(Q_1(t)\mathbf{x}) g(Q_1(t)\mathbf{x})\frac{\mathrm{d}t}{t} \\
&= \int_{\mathbb{R}^n} \mathrm{d}x \int_0^\infty \frac{\mathrm{d}s}{s} \int_{\mathcal{A}(s,j)} Q_1(t)^{\mathrm{T}} |(\psi_t^j * (\varphi_s * \mathbf{F}))(Q_1(t)\mathbf{x})|^2 \\
&\quad \times g(Q_1(t)\mathbf{x})\frac{\mathrm{d}t}{t}
\end{aligned} \tag{3.131}$$

because $\varphi_s * \psi_t^j = \psi_t^j * \varphi_s = 0$ unless $t \in \mathcal{A}(s, j) := [2^{2j-4}s, 2^{2j+4}s]$.
Thus,

$$\begin{aligned}
&\|\mathfrak{S}\mathcal{T}_j\mathbf{F}\|_{q,w} \\
&\le c\|\psi^j\|_1 \left(\int_{\mathbb{R}^n} \left(\int_0^\infty |\varphi_s * \mathbf{F}(x)|^2 \frac{\mathrm{d}s}{s} \right)^{q/2} w(\mathbf{x})\mathrm{d}x \right)^{2/q} \|\mathcal{M}^j g(x)\|_{(q/2)',w'};
\end{aligned} \tag{3.132}$$

here we used successively Cauchy–Schwarz's inequality and Hölder inequality. Finally we obtain

$$\|\mathcal{T}_j\mathbf{F}\|_{q,w} \le \|\mathfrak{S}\mathcal{T}_j\mathbf{F}\|_{q,w} \le c2^{-2|j|}\|\mathbf{F}\|_{q,w}, \tag{3.133}$$

for all $j \in \mathbb{Z}$, for $q > 2$ and $w \in A_{q/2}$ and as a consequence the convergence of series $\sum_{j\in\mathbb{Z}} \mathcal{T}_j = \mathcal{T}$ in L^q_w.

Step 3 Now let $q = 2$ and $\mathbf{w} \in A_1$. In this case the Littlewood–Paley decomposition (3.122) in L^2_w implies that

$$\|\mathcal{T}_j\mathbf{F}\|_{2,w}^2 \le c \int_0^\infty \int_{\mathbb{R}^n} |\varphi_s * \mathcal{T}_j\mathbf{F}|^2(\mathbf{x})\, g(\mathbf{x})\mathrm{d}\mathbf{x}\, \frac{\mathrm{d}s}{s},$$

where

$$g \in L^\infty_v, \quad v = \frac{1}{w} \quad \text{and} \quad \|g\|_{\infty,v} = \operatorname{ess\,sup}_{\mathbb{R}^n} |g\, v| = 1,$$

cf. (3.128)–(3.131). By the same reasoning as before, we arrive at (3.127). That is,

$$\|\mathcal{T}_j \mathbf{F}\|_{2,w} \le c 2^{-|j|} \|\mathcal{M}^j g\|^{1/2}_{\infty,v} \|\mathbf{F}\|_{2,w}. \tag{3.134}$$

Since $|g(x)| \le \frac{1}{v(\mathbf{x})} = w(\mathbf{x})$ for a.a. $\mathbf{x} \in \mathbb{R}^n$ and ψ_t^j is radially symmetric, the operator $\mathcal{M}^j$ satisfies the pointwise estimate

$$\begin{aligned}\mathcal{M}^j g(\mathbf{x}) \le \mathcal{M}^j w(\mathbf{x}) &= \sup_{s>0} \int_{A_s} \left(|\psi_t^j| * w(Q_1(t)^{\mathrm{T}} \cdot) \right)(\mathbf{x}) \frac{\mathrm{d}t}{t} \\ &\le c 2^{-2|j|} \sup_{t>0} h_t * \mathcal{M}_\theta\, w(\mathbf{x}) \le c 2^{-2|j|} \mathcal{M}\left(\mathcal{M}_1\, w\right)(\mathbf{x}).\end{aligned}$$

Exploiting as before that w is θ-independent, we know that $\mathcal{M}_1 w(r, x_3) = w(r, x_3)$. Moreover, the assumption $w \in A_1$ implies that $\mathcal{M} w \le c w$. Hence, $\mathcal{M}^j g(\mathbf{x}) \le c 2^{-2|j|} w(\mathbf{x})$ and consequently

$$\|\mathcal{M}^j g(\mathbf{x})\|_{\infty,v} \le c 2^{-2|j|}. \tag{3.135}$$

Now (3.134) and (3.135) lead to the operator bound $\|\mathcal{T}_j\|_{L^2_w} \le c 2^{-2|j|}$ and the boundedness of $\mathcal{T}$ on L^2_w.

Step 4 The remaining estimates $1 < q < 2$ are proved by duality arguments. Obviously the adjoint operator to $\mathcal{T}$ is defined by

$$\mathcal{T}^* \mathbf{F}(x) = \int_0^\infty (-\Delta) Q_1(t) K * \mathbf{F}(Q_1(t)^{\mathcal{T}} \mathbf{x}) \mathrm{d}t = \int_{\mathbb{R}^n} \mathcal{K}^*(\mathbf{x}, \mathbf{y}) \mathbf{F}(\mathbf{y}) \mathrm{d}\mathbf{y},$$

where the kernel $\mathcal{K}^*$ has the same structure as $\mathcal{K}$. Hence $\mathcal{T}^*$ is bounded on L^q_w for $q \ge 2$ and $w \in A_{q/2}$. Now let $1 < q < 2$ and $w^{2/(2-q)} \in A_{q/(2-q)} = A_{(q'/2)'}$, or equivalently $w' = w^{-q'/q} \in A_{(q'/2)}$. Then

$$|\langle \mathcal{T}\mathbf{F}, \mathbf{G} \rangle| = |\langle \mathbf{F}, \mathcal{T}^* \mathbf{G} \rangle| \le \|\mathbf{F}\|_{q,w} \|\mathcal{T}^* \mathbf{G}\|_{q',w'} \le c \|\mathbf{F}\|_{q,w} \|\mathcal{T}\mathbf{G}\|_{q',w'}$$

since $q' > 2$ and $w' = w^{-q'/q} \in A_{(q'/2)}$.

Finally let $q = 2$ and $1/\mathbf{w} \in A_1$. As before, since $1/\mathbf{w} \in A_1$,

$$|\langle \mathcal{T}\mathbf{F}, \mathbf{G} \rangle| \le \|\mathbf{F}\|_{2,w} \|\mathcal{T}^* \mathbf{G}\|_{2,1/w} \le c \|\mathbf{F}\|_{2,w} \|G\|_{2,1/w}.$$

We formulate the results obtained in Steps 1–4 as a proposition.

Proposition 3.4 *Assume that the weight* **w** *satisfies*

$$\begin{array}{ll} w \in A_{q/2} & \text{if } q > 2, \\ w \in A_1 \text{ or } w^{-1} \in A_1 & \text{if } q = 2, \\ w^{2/(2-q)} \in A_{q/(2-q)} & \text{if } 1 < q < 2. \end{array} \tag{3.136}$$

Then the linear operator $\mathcal{T}$ defined by (3.121) *satisfies the estimate*

$$\|\mathcal{T}\mathbf{F}\|_{q,w} \le c\|\mathbf{F}\|_{q,w}, \quad \mathbf{F} \in \boldsymbol{L}^q_w, \tag{3.137}$$

with a constant $c = c(q, w) > 0$ *independent of* **F**. □

Step 5 To prove Theorem 3.8 we need to extend Proposition 3.4 applying the complex interpolation of L^q_w-spaces, see [14], Lemma 2.8.

Let us start with Part (i) of Theorem 3.8:

Let $q \in (1, \infty)$ and $w \in A_q$ such that the $\boldsymbol{L}^q_w$—estimate of ∇p holds, see (3.117). Hence it suffices to consider **u** defined by (3.116)–(3.119). Choose arbitrary q_1, q_2 with

$$1 < q_1 < q < q_2 < \infty \quad \text{and} \quad q_1 \le 2 \le q_2 \tag{3.138}$$

and $\delta \in (0, 1)$ satisfying

$$\frac{1}{q} = \frac{1-\delta}{q_1} + \frac{\delta}{q_2}, \tag{3.139}$$

as well as weights w_1, w_2 such that

$$w_1^{2/(2-q_1)} \in A_{q_1/(2-q_1)} \quad \text{and} \quad w_2 \in A_{q_2/2},$$

cf. (3.136) in Proposition 3.4. By Lemma 2.2 there exist $u_1, v_1, u_2, v_2 \in A_1$ with

$$w_1^{2/(2-q_1)} = \frac{u_1}{v_1^{2(q_1-1)/(2-q_1)}} \quad \text{and} \quad w_2 = \frac{u_2}{v_2^{(q_2-2)/2}}.$$

Since the linear operator $\mathcal{T}$ is bounded on both $L^{q_1}_{w_1}$ and $L^{q_2}_{w_2}$, Lemma 2.8 shows that it is bounded on $L^q_{\widetilde{w}}$ as well, where

$$\widetilde{w} = w_1^{q(1-\delta)/q_1} \cdot w_2^{q\delta/q_2} = \frac{u_1^{q(1-\delta)(2-q_1)/2q_1}}{v_1^{q(1-\delta)(q_1-1)/q_1}} \cdot \frac{u_2^{q\delta/q_2}}{v_2^{q\delta(q_2-2)/2q_2}}.$$

Note that the sum of the exponents of the numerators equals $1 - \frac{1}{2}q(1-\delta) = \frac{1}{2}(2-q(1-\delta))$. Therefore, taking the $\frac{1}{2}(2-q(1-\delta))$th root of the previous identity and choosing $u_1 = u_2$ and $v_1 = v_2$, we arrive with an elementary calculation at

$$\widetilde{w}^{2/(2-q(1-\delta))} = \frac{u_1}{v_1^{q/(2-q(1-\delta))-1}},$$

which by Lemma 2.2 is a weight in $A_{q/(2-q(1-\delta))}$. Since $u_1, v_1 \in A_1$ are arbitrary, we proved the boundedness of $\mathcal{T}$ on L^q_w for arbitrary $\mathbf{w} = \widetilde{w}$ if

$$w^\tau \in A_{\tau q/2}, \quad \tau = \frac{2}{2-q(1-\delta)}.$$

Further we have to find all admissible τ subject to the restrictions given by (3.138) and (3.139). For this reason consider the easier term

$$\gamma = 2\left(1 - \frac{1}{\tau}\right) = q(1-\delta) = q\frac{1/q - 1/q_2}{1/q_1 - 1/q_2}.$$

Case 1 $1 < q < 2$, in which $1 < q_1 < q$ and $q_2 \geq 2$. Due to monotonicity properties of γ as a function of $1/q_1$ and of $1/q_2$, it suffices to check γ at the corners of the rectangle $(1/q, 1) \times (0, \frac{1}{2}]$. The corresponding function values are q, 1 and $2-q$. Hence the range of γ equals the interval $(2-q, q)$ yielding for $\tau = 2/(2-\gamma)$ the condition

$$\frac{2}{q} < \tau < \frac{2}{2-q}.$$

Note that the limiting value $\tau = 2/(2-q)$ is allowed due to Proposition 3.4. Finally the condition $w^\tau \in A_{\tau q/2}$, $2/q < \tau \leq 2/(2-q)$ easily implies that $w \in A_q$: By Lemma 2.2 there exist $v_1, v_2 \in A_1$ such that

$$w = v_1^{1/\tau} \cdot v_2^{1/\tau - q/2}, \tag{3.140}$$

where $v_1^{1/\tau} \in A_1$ and $q/2 - 1/\tau \leq q - 1$ yielding $v_2^{(\frac{q}{2}-\frac{1}{\tau})/(q-1)} \in A_1$.

Case 2 $q > 2$, in which $1 < q_1 \leq 2$ and $q_2 > q$. In this case the values of γ at the corners of the rectangle $[\frac{1}{2}, 1) \times (0, 1/q)$ in the $(1/q_1, 1/q_2)$-plane are 0, 1, and 2. Hence, $1 < \tau < \infty$ and we observe that $\tau = 1$ is admissible due to Proposition 3.4. Finally note that the condition $w^\tau \in A_{\tau q/2}$ also implies $w \in A_q$: There exist $v_1, v_2 \in A_1$ such that w satisfies (3.140), where again $q/2 - 1/\tau + 1 \leq q$ for all $\tau \in (1, \infty)$.

Case 3 $q = 2$. In this case it suffices to interpolate between $L^2_{w_1}$ and $L^2_{w_2}$, where $w_1 \in A_1$ and $1/w_2 \in A_1$, see Proposition 3.4. Then $\mathcal{T}$ is bounded on L^2_w for all

$$w = \frac{w_1^{1-\delta}}{w_2^{\delta}}, \quad 0 < \delta < 1.$$

Then $w^{1/(1-\delta)} = w_1/w_2^{\delta/(1-\delta)}$, or with $\tau = 1/(1-\delta) \in (1,\infty)$,

$$w^{\tau} = \frac{w_1}{w_2^{\tau-1}} \in A_{\tau} = A_{\tau q/2}.$$

Part (ii) of Theorem 3.8 (Uniqueness):

(ii) Note that $L^{q_i}_{w_i}(\mathbb{R}^n) \subset S'(\mathbb{R}^n)$, $i = 1,2$; indeed, $w_i \in L^1_{\mathrm{loc}}(\mathbb{R}^n)$ and $\int_{|\mathbf{x}|\geq 1} w_i(\mathbf{x})|\mathbf{x}|^{-nq_i}\,d\mathbf{x} < \infty$, see [115, IV.3 (3.2)]. Since Eq. (3.114) is linear, it suffices to consider $\mathbf{f} = \mathbf{0}$ and a solution $\mathbf{u} \in S'(\mathbb{R}^n)^n$ of (3.114). In the proof of [70], Theorem 3.1 (2), (3), it was shown that under these assumptions $\mathbf{u}$ is a polynomial and that $\mathbf{u}(x) = \alpha\omega\mathbf{e}_3 + \beta\omega\mathbf{e}_3 \wedge \mathbf{x} + \gamma(x_1, x_2, -2x_3)^{\mathrm{T}}$, $\alpha, \beta, \gamma \in \mathbb{R}$ ($\mathbf{u}(\mathbf{x}) = \beta(-x_2, x_1)$ if $n = 2$).

3.3.2.1 Examples of Weights and Corollaries

As an example we consider the weights of the type (radial type) $|\mathbf{x}|^{\alpha}$ and $(1+|\mathbf{x}|)^{\alpha}$, $\alpha \in \mathbb{R}$, as well as weights of the type $(1+|\mathbf{x}|)^{\alpha}(1+r)^{\beta}$, $\alpha, \beta \in \mathbb{R}$, where $r = \sqrt{x_1^2 + x_2^2}$ is the radial distance of $\mathbf{x} = (x_1, x_2, x_3)$ from the axis of revolution.

Corollary 3.1 (i) *The a priori estimate* (3.115) *holds for weights*

$$w(\mathbf{x}) = |\mathbf{x}|^{\alpha} \quad \textit{and} \quad w(\mathbf{x}) = (1+|\mathbf{x}|)^{\alpha},$$

where

$$\begin{aligned} 2 \leq q < \infty: &\quad -n < \alpha < \frac{nq}{2}, \\ 1 < q < 2: &\quad -\frac{nq}{2} < \alpha < n(q-1). \end{aligned}$$

(ii) *The a priori estimate* (3.115) *holds for the anisotropic, axially symmetric weight* $w(\mathbf{x}) = (1+|\mathbf{x}|)^{\alpha}(1+r)^{\beta}$ *on* $\mathbb{R}^3$ *provided that*

$$\begin{aligned} 2 \leq q < \infty: &\quad -2 < \beta < q \quad &\textit{and} \quad -3 < \alpha+\beta < \frac{3q}{2}, \\ 1 < q < 2: &\quad -q < \beta < 2(q-1) \quad &\textit{and} \quad -\frac{3q}{2} < \alpha+\beta < 3(q-1). \end{aligned}$$

Remark 3.17 For $q = 2$ the conditions on α, β in Corollary 3.1 are optimal, see Lemma 3.8 (ii). For α close to n and $\beta = 0$ the estimate of $\nabla^2\mathbf{u}$ in $\boldsymbol{L}^2_w(\mathbb{R}^n)$ indicates that $\nabla^2\mathbf{u}(\mathbf{x})$ could decay as $|\mathbf{x}|^{-n}$ when $|\mathbf{x}| \to \infty$. In Corollary 3.1 (ii) the estimate $\int |\nabla\mathbf{u}|^2(1+|\mathbf{x}|)^\alpha(1+r)^\beta d\mathbf{x} < \infty$ for β close to 2 and α close to 1 indicates that for a suitable $\mathbf{f}$ such that $\int |\mathbf{f}|^2(1+|\mathbf{x}|)^\alpha(1+r)^\beta d\mathbf{x} < \infty$, but $\int |\mathbf{f}|^2(1+|\mathbf{x}|)^{\alpha+\beta} d\mathbf{x} = \infty$, the term $\nabla^2 u(\mathbf{x})$ may anisotropically decay as $|\mathbf{x}|^{-2}(1+r)^{-1}$. □

Actually Theorem 3.8 holds also for θ-dependent weights w provided that $w(\mathbf{x}) = w(r, x_3, \theta)$ satisfies an additional *one-dimensional Muckenhoupt condition* with respect to the angular variable $\theta \in [0, 2\pi)$ and with an $A_q(\theta)$-constant independent of (r, x_3), see Corollary 3.2 and its proof in Sect. 3.3.1.1 below. As an example we consider the anisotropic weight functions

$$w(\mathbf{x}) = \eta^\alpha_\beta(\mathbf{x}) = (1 + |\mathbf{x}|)^\alpha(1 + s(\mathbf{x}))^\beta, \quad s(\mathbf{x}) = r - x_1,$$

introduced in [58] to analyze the Oseen equations.

Corollary 3.2 *The a priori estimate* (3.115) *holds for the anisotropic, θ-dependent weights*

$$w(\mathbf{x}) = \sigma^\alpha_\beta(\mathbf{x}) = |\mathbf{x}|^\alpha s(\mathbf{x})^\beta \quad \textit{and} \quad w(\mathbf{x}) = \eta^\alpha_\beta(\mathbf{x})$$

provided that

$$\begin{aligned} &2 \le q < \infty: \quad -\frac{1}{2} < \beta < \frac{q}{4} \quad &&\textit{and } -n < \alpha + \beta < \frac{nq}{2},\\ &1 < q < 2: \quad -\frac{q}{4} < \beta < \frac{q-1}{2} \quad &&\textit{and } -\frac{nq}{2} < \alpha + \beta < n(q-1). \end{aligned}$$

Proof of Corollary 3.1 (i) Let $w(\mathbf{x}) = |\mathbf{x}|^\alpha$ or $(1 + |\mathbf{x}|)^\alpha$. As is well known, see, e.g., [205, page 218], $w \in A_q$ iff $-n < \alpha < n(q-1)$, $1 < q < \infty$, and $\mathbf{w} \in A_1$ iff $-n < \alpha \le 0$. If $q > 2$, then the condition $w^\tau \in A_{\tau q/2}$ in (2.15) implies that $-n < \alpha\tau < n(\tau q/2 - 1)$, $\tau \in [1, \infty)$, or equivalently that $-n < \alpha < nq/2$ by choosing $\tau = 1$ or τ as large as possible. If $1 < q < 2$, then $w^\tau \in A_{\tau q/2}$ in (2.15) is equivalent to $-n < \alpha\tau < n(\tau q/2 - 1)$ for $\tau \in (2/q, 2/(2-q)]$, yielding $-nq/2 < \alpha < n(q-1)$ by choosing $\tau = 2/(2-q)$ or $\tau \to 2/q$. Finally, for $q = 2$, (2.15) immediately admits all weights w satisfying $-n < \alpha < n$.

(ii) By Lemma 3.8 below $w(\mathbf{x}) = (1+|\mathbf{x}|)^\alpha(1+|(x_1, x_2)|)^\beta \in A_q$, $1 < q < \infty$, if $-2 < \beta < 2(q-1)$ and $-3 < \alpha+\beta < 3(q-1)$. If $q > 2$, the condition $w^\tau \in A_{\tau q/2}$, $\tau \in [1, \infty)$, implies that $-2 < \beta < q$ and $-3 < \alpha+\beta < 3q/2$; for the proof choose $\tau = 1$ or $\tau \to \infty$. For $q < 2$ the condition $\mathbf{w}^\tau \in A_{\tau q/2}$, $\tau \in (2/q, 2/(2-q)]$, yields the restrictions $-q < \beta < 2(q-1)$ and $-3q/2 < \alpha + \beta < 3(q-1)$ for α, β by choosing $\tau = 2/(2-q)$ or $\tau \to 2/q$. For $q = 2$ we obviously get the bounds $-2 < \beta < 2$ and $-3 < \alpha + \beta < 3$. □

Lemma 3.8 *Let w be the weight $w(\mathbf{x}) = (1+|\mathbf{x}|)^{\alpha}(1+r)^{\beta}$, where $r = |x'| = \sqrt{x_1^2+x_2^2}$ denotes the distance of $\mathbf{x} = (x_1, x_2, x_3)$ to the axis of revolution. Then the following statements hold true:*

(i) *For all $\beta \in (-2, 0]$ and $-3 < \alpha + \beta \le 0$, we have $w \in A_1$.*
(ii) *Given $1 < q < \infty$, the weight w lies in A_q if $-2 < \beta < 2(q-1)$ and $-3 < \alpha+\beta < 3(q-1)$.*

□

Proof (i) For $b \in (-1, 2]$ define the regular Borel measure

$$\mu(A) = \int_{A\cap\mathbb{R}} |y_3|^b \mathrm{d}\mathbf{y}$$

on $\mathbb{R}^3$, where $A \cap \mathbb{R}$ stands for $A \cap \{(0,0,y_3) \in \mathbb{R}^3 : y_3 \in \mathbb{R}\}$; for a similar ansatz, see [58]. We claim that for the maximal operator $\mathcal{M}$, see Lemma 2.2,

$$\mathcal{M}\mu(\mathbf{x}) = \sup_{\mathcal{Q}\ni\mathbf{x}} \frac{\mu(\mathcal{Q})}{|\mathcal{Q}|} \sim \frac{|\mathbf{x}|^b}{r^2}. \tag{3.141}$$

To prove the equivalence in (3.141) we consider an arbitrary $\mathbf{x} \in \mathbb{R}^n$ with $x_3 > 0$; let $R > 0$ denote the half of the side length of the cube $\mathcal{Q}$ centered at $\mathbf{x}$ in the definition of $\mathcal{M}\mu(\mathbf{x})$. For simplicity assume that $\mathcal{Q}$ is closed. Then we consider three cases:

Case 1 $0 < R < x_3/2$. Then $y_3 \sim x_3$ for all $(0,0,y_3) \in Q \cap \mathbb{R}$ and

$$\frac{\mu(\mathcal{Q})}{|\mathcal{Q}|} = \frac{1}{8R^3}\int_{\mathcal{Q}\cap\mathbb{R}} y_3^b \mathrm{d}y_3 \sim \frac{x_3^b}{R^2} \le \frac{x_3^b}{r^2}$$

since $r = |x'| \le R$ such that $\mathcal{Q} \cap \mathbb{R} \ne \emptyset$ is possible.

Case 2 $\frac{1}{2}x_3 \le R \le x_3$. Then

$$\frac{\mu(\mathcal{Q})}{|\mathcal{Q}|} \sim \frac{1}{R^3}\int_{x_3-R}^{x_3+R} |y_3|^b \mathrm{d}y_3 \sim R^{b-2} \sim x_3^{b-2}.$$

Case 3 $R > x_3$. Now

$$\frac{\mu(\mathcal{Q})}{|\mathcal{Q}|} \sim \frac{1}{R^3}\int_{-R}^{R} |y_3|^b \mathrm{d}y_3 \sim R^{b-2} \le x_3^{b-2}$$

since $b > -1$. This case also shows that $b \le 2$ is needed to get $\mathcal{M}\mu(\mathbf{x})$ finite for a.a. $\mathbf{x} \in \mathbb{R}^3$.

Summarizing the previous three cases we now consider $\mathbf{x} \in \mathbb{R}^3$ with either $r < x_3$ or $r \geq x_3$. If $r < x_3$, then $x_3 \sim |\mathbf{x}|$, and due to the first case

$$\mathcal{M}\mu(\mathbf{x}) \sim \frac{x_3^b}{r^2} \sim \frac{|\mathbf{x}|^b}{r^2}.$$

Finally, if $r \geq x_3$, then the third case applies with $R = r \sim |\mathbf{x}|$. Hence,

$$\mathcal{M}\mu(\mathbf{x}) \sim r^{b-2} \sim \frac{|\mathbf{x}|^b}{r^2}.$$

Now (3.141) is proved.

By Lemma 2.2, $w(\mathbf{x}) = (|\mathbf{x}|^b/r^2)^\gamma$, $0 \leq \gamma < 1$, $-1 < b \leq 2$, is an A_1-weight. In other words, $w(\mathbf{x}) = |\mathbf{x}|^\alpha r^\beta \in A_1$ for all $\beta \in (-2, 0]$ and $\alpha \in (\beta/2, -\beta]$. This set of admissible (α, β) defines a half open triangle in the (α, β)-plane with vertices $(0, 0)$, $(-1, -2)$, and $(2, -2)$. Eventually, since also $|\mathbf{x}|^\gamma \in A_1$ for $-3 < \gamma \leq 0$, Lemma 2.2 implies that $\mathbf{w} \in A_1$ for all (α, β) in the open parallelogram

$$P\colon \; -2 < \beta < 0, \quad -3 < \alpha + \beta < 0$$

plus the line segments $-2 < \beta \leq 0$, $\alpha = -\beta$, and $\beta = 0$, $-3 < \alpha + \beta \leq 0$.

To prove the same result for nondegenerate weights, note that $\alpha = 0$ is allowed, i.e., $r^\beta \in A_1$ for all $\beta \in (-2, 0]$. Moreover, since the sum of two A_1-weights and also the minimum of two A_1-weights is an A_1-weight as well, we conclude that $w(x) = (1+|\mathbf{x}|)^\alpha r^\beta \in A_1$ for the same α, β as before. Note that the same result will hold when the axis of revolution is parallel to the third unit vector but passes through $(x_0', 0)$, $|x_0'| = r \leq 1$. Obviously the corresponding A_1-constant is independent of x_0'. Hence for all cubes Q centered at $\mathbf{x} \in \mathbb{R}^3$,

$$\begin{aligned}
&\frac{1}{|Q|}\int_Q \left(\int_{|x_0'|\leq 1} (1 + |\mathbf{y}|)^\alpha |y' - x_0'|^\beta \mathrm{d}x_0' \right) \mathrm{d}\mathbf{y} \\
&= \int_{|x_0'|\leq 1} \left(\frac{1}{|Q|} \int_Q (1 + |\mathbf{y}|)^\alpha |y' - x_0'|^\beta \mathrm{d}\mathbf{y} \right) \mathrm{d}x_0' \\
&\leq \int_{|x_0'|\leq 1} (1 + |\mathbf{x}|)^\alpha |x' - x_0'|^\beta \mathrm{d}x_0'.
\end{aligned}$$

Since $\beta > -2$, we conclude that

$$\int_{|x_0'|\leq 1} (1 + |\mathbf{x}|)^\alpha \, |x' - x_0'|^\beta \mathrm{d}x_0' \sim (1 + |\mathbf{x}|)^\alpha (1 + |x'|)^\beta$$

is an A_1-weight for all $\beta \in (-2, 0]$, $-3 < \alpha + \beta \leq 0$.

(ii) Consider $w_j(\mathbf{x}) = (1+|\mathbf{x}|)^{\alpha_j}(1+r)^{\beta_j} \in A_1$, $j = 1, 2$, where α_j, β_j run through all of the parallelogram P. By Lemma 2.2

$$\frac{w_1(x)}{w_2(\mathbf{x})^{q-1}} = (1+|\mathbf{x}|)^{\alpha_1-(q-1)\alpha_2}(1+r)^{\beta_1-(q-1)\beta_2} \in A_q.$$

Now it can easily be seen that $w(\mathbf{x}) = (1+|\mathbf{x}|)^{\alpha}(1+r)^{\beta} \in A_q$ for all α, β satisfying $-2 < \beta < 2(q-1)$, $-3 < \alpha + \beta < 3(q-1)$. □

Proof of Corollary 3.2 As to the weight $w = \eta^{\alpha}_{\beta}$, we may proceed as in the proof of Theorem 3.8 based on Proposition 3.4. Hence the operators T_j satisfy the estimate (3.134), and the maximal operators $\mathcal{M}^j$ satisfy the pointwise estimate (3.129) just as before. However, we have to modify the proof of (3.130) by analyzing $\mathcal{M}_\theta$ more carefully.

For the moment let $\rho(\mathbf{x}) := \rho_1(r)\rho_2(r,\theta)$ be an A_p-weight on $\mathbb{R}^2$, $1 < p < \infty$, such that $\rho_2(r,\theta)$ is 2π-periodic with respect to θ and satisfies a modified Muckenhoupt condition for 2π-periodic weights, i.e.,

$$\sup_{r>0}\ \sup_{0<|b-a|\le 2\pi} I_{a,b}(\rho_2) < \infty, \tag{3.142}$$

where

$$I_{a,b}(\rho_2) = \left(\frac{1}{b-a}\int_a^b \rho_2(r,\theta)\mathrm{d}\theta\right)\cdot\left(\frac{1}{b-a}\int_a^b \rho_2(r,\theta)^{-1/(p-1)}\mathrm{d}\theta\right)^{p-1},$$

see [173, Corollary 4], without the additional parameter r. Then consider for given $g \in L^1_{\mathrm{loc}}(\mathbb{R}^n)$ the restriction $g_r(\theta) = g(r,\theta)$. Obviously, $\mathcal{M}_\theta$ is estimated by the one-dimensional Hardy–Littlewood maximal operator $\mathcal{M}_1^{\mathrm{per}}$ for 2π-periodic functions, see [173], i.e., for all $\theta \in (0, 2\pi)$

$$\mathcal{M}_\theta g(r,\theta) \le c(\mathcal{M}_1^{\mathrm{per}} g_r)(\theta), \quad r > 0.$$

Then by Muckenhoupt [173, Corollary 4], using polar coordinates and Fubini's theorem,

$$\begin{aligned}\|\mathcal{M}_\theta g\|^p_{p,\rho} &\le c\int_0^\infty r\rho_1(r)\|\mathcal{M}_1^{\mathrm{per}} g_r\|^p_{L^p(0,2\pi,\rho_2(r,\cdot))}\mathrm{d}r\\ &\le cB_p\int_0^\infty r\rho_1(r)\|g_r\|^p_{L^p(0,2\pi,\rho_2(r,\cdot))}\mathrm{d}r = c\|g\|^p_{p,\rho}.\end{aligned}$$

In the case $n = 3$ we proceed analogously for a weight $\rho(\mathbf{x}) := \rho_1(r, x_3)\times \rho_2(r, x_3, \theta)$, inserting the additional variable x_3 in condition (3.142) and performing an additional integration with respect to $x_3 \in \mathbb{R}$ in the last estimate.

Applying this general procedure we have to check for which $\alpha, \beta \in \mathbb{R}$ the weights $\rho = \sigma_\beta^\alpha$ and $\rho = \eta_\beta^\alpha$ satisfy the condition (3.142). First let $\rho(x) = \sigma_\beta^\alpha(\mathbf{x}) = r^{\alpha+\beta}(1-\cos\theta)^\beta$ yielding $\rho_2(r,\theta) = \rho_2(\theta) = (1-\cos\theta)^\beta$. Since $\rho_2(\theta) \sim \theta^{2\beta}$ near the origin and $\rho_2(\theta) \sim (2k\pi - \theta)^{2\beta}$ near $2k\pi$, $k \in \mathbb{Z}$, we may restrict ourselves to small $|a|$, $|b|$, $b > 0$, in (3.142). If $0 = a < b$, then

$$I_{0,b}(\rho_2) \sim \frac{1}{b}\int_0^b \theta^{2\beta}\mathrm{d}\theta \left(\frac{1}{b}\int_0^b \theta^{-2\beta/(p-1)}\mathrm{d}\theta\right)^{p-1} = c_\beta$$

provided $-\frac{1}{2} < \beta < \frac{1}{2}(p-1)$. For $0 < a < \frac{1}{2}b$, the term $I_{a,b}(\rho_2)$ may be compared with $I_{0,b}(\rho_2)$, and for $0 < \frac{1}{2}b \le a < b$ both integrands in $I_{a,b}(\rho_2)$ may be compared with the constants $b^{2\beta}$ and $b^{-2\beta/(p-1)}$, respectively. Finally, if $a < 0 < b$, $|a| \le b$, then $I_{a,b}(\rho_2) \sim I_{0,b}(\rho_2)$. Hence we have proved

$$\rho = \sigma_\beta^\alpha : \quad \sup_{\substack{0<|b-a|\le 2\pi \\ r>0}} I_{a,b}(\rho_2) < \infty \quad \text{for } -\frac{1}{2} < \beta < \frac{1}{2}(p-1).$$

Next we consider the weight $\boldsymbol{\rho} = \boldsymbol{\eta}_\beta^\alpha$ on $\mathbb{R}^2$. In this case

$$\rho_2(r,\theta) = (1 + r(1-\cos\theta))^\beta \sim \begin{cases} (1+r)^\beta, & \theta \sim (2k+1)\pi, \\ (1 + r(2k\pi - \theta)^2)^\beta, & \theta \sim 2k\pi, \end{cases}$$

is r-dependent. Then, if $0 = a < b < \pi$, using the change of variables $\vartheta = \sqrt{r}\theta$ and the notation $B = b\sqrt{r}$,

$$\begin{aligned} I_{0,b}(\rho_2) &\sim \frac{1}{b}\int_0^b (1+r\theta^2)^\beta \mathrm{d}\theta \left(\frac{1}{b}\int_0^b (1+r\theta^2)^{-\beta/(p-1)}\mathrm{d}\theta\right)^{p-1} \\ &\sim \frac{1}{B}\int_0^B (1+\vartheta)^{2\beta}\mathrm{d}\vartheta \left(\frac{1}{B}\int_0^B (1+\vartheta)^{-2\beta/(p-1)}\mathrm{d}\vartheta\right)^{p-1} \\ &\sim \frac{1}{B^p}((1+B)^{2\beta+1} - 1)\left(1 - (1+B)^{1-2\beta/(p-1)}\right)^{p-1}. \end{aligned} \tag{3.143}$$

Now it is easy to see that the last term is uniformly bounded in $B \in (0,\infty)$ provided that $-\frac{1}{2} < \beta < \frac{1}{2}(p-1)$. In this way, omitting further cases, we proved

$$\rho = \eta_\beta^\alpha, n = 2: \quad \sup_{\substack{0<|b-a|\le 2\pi \\ r>0}} I_{a,b}(\rho_2(r,\cdot)) < \infty \quad \text{for } -\frac{1}{2} < \beta < \frac{1}{2}(p-1).$$

Finally we investigate the weight $\rho = \eta_\beta^\alpha$ on $\mathbb{R}^3$. Since $|\mathbf{x}| = \sqrt{r^2 + x_3^2}$, we get $\rho(\mathbf{x}) = \rho_1(r)\rho_2(r, x_3, \theta)$, where

$$\begin{aligned}\rho_2(r, x_3, \theta) &= 1 + \sqrt{r^2 + x_3^2} - r + r(1 - \cos\theta) \\ &\sim 1 + \sqrt{r^2 + x_3^2} - r + r\begin{cases}1, & \theta \sim (2k+1)\pi, \\ (2k\pi - \theta)^2, & \theta \sim 2k\pi.\end{cases}\end{aligned}$$

If $0 = a < b < \pi$, we proceed as in the previous case and get with $R = 1 + \sqrt{r^2 + x_3^2} - r \geq 1$ and $B = b\sqrt{r/R}$ that

$$\begin{aligned}I_{0,b}(\rho_2) &\sim \frac{1}{b\sqrt{r}} \int_0^{b\sqrt{r}} (\sqrt{R} + \theta)^{2\beta} \mathrm{d}\theta \cdot \left(\frac{1}{b\sqrt{r}} \int_0^{b\sqrt{r}} (\sqrt{R} + \theta)^{-2\beta/(p-1)} \mathrm{d}\theta \right)^{p-1} \\ &= \frac{1}{B} \int_0^B (1 + \theta)^{2\beta} \mathrm{d}\theta \left(\frac{1}{B} \int_0^B (1 + \theta)^{-2\beta/(p-1)} \mathrm{d}\theta \right)^{p-1}.\end{aligned}$$

Comparing this with (3.143), we see that we have proved

$$\rho = \eta_\beta^\alpha, n = 3: \quad \sup_{\substack{0<|b-a|\leq 2\pi \\ r>0,\, x_3\in\mathbb{R}}} I_{a,b}(\rho_2(r, x_3, \cdot)) < \infty \quad \text{for } -\frac{1}{2} < \beta < \frac{1}{2}(p-1).$$

Even in this final case, (3.142) is satisfied (with an obvious modification for $x_3 \in \mathbb{R}$).

To complete the proof, we use Proposition 3.4 and the way it is proven. To far we have shown that the linear operator T, see (3.121), is bounded if (3.136) is satisfied—with an obvious modification of the class A_p. To be more precise, A_p is replaced by the class of weights

$$\widetilde{A_p} = \left\{ w = \sigma_\beta^\alpha : \ -n < \alpha + \beta < n(p-1), -\frac{1}{2} < \beta < \frac{1}{2}(p-1) \right\}$$

or

$$\widetilde{A_p} = \left\{ w = \eta_\beta^\alpha : \ -n < \alpha + \beta < n(p-1), -\frac{1}{2} < \beta < \frac{1}{2}(p-1) \right\};$$

for the bound on $\alpha + \beta$ we used the well-known result

$$\sigma_\beta^\alpha, \eta_\beta^\alpha \in A_p \quad \text{iff} -\frac{n-1}{2} < \beta < \frac{n-1}{2}(p-1), \quad -n < \alpha + \beta < n(p-1),$$

see [58] for the case $n = 3$, $p = 2$ and [119] for the general case. If $p = 1$, then $\alpha + \beta = 0$ and $\beta = 0$ are allowed as well in the previous characterizations. Now we use Lemma 2.8 on interpolation within each class $\widetilde{A}_p$. Looking at the proof of Theorem 3.8 we get a result analogous to (2.15) with A_p replaced by $\widetilde{A}_p$. Finally we repeat the proof of Corollary 3.1 (ii). The only differences are due to the new upper and lower bounds on β and on $\alpha + \beta$. □

3.3.3 A Strong Solution in an Exterior Domain

The contents of this subsection is mainly based on work by Nečasová and Schumacher [177]. In this subsection, we study the modified Stokes system in an exterior domain $\mathcal{D} \subset \mathbb{R}^3$ of class $C^{1,1}$

$$\begin{aligned} S_{\omega,\nu}\mathbf{u} + \nabla p &= f \quad \text{in } \mathcal{D}, \\ \operatorname{div} \boldsymbol{u} &= 0 \quad \text{in } \mathcal{D}, \\ \boldsymbol{u}|_{\partial\mathcal{D}} &= 0. \end{aligned} \tag{3.144}$$

In particular, we are interested in solutions $\mathbf{u}$ which decay at infinity, i.e., they fulfill

$$\boldsymbol{u}(\boldsymbol{x}) \to \mathbf{0} \quad \text{for } |\boldsymbol{x}| \to \infty. \tag{3.145}$$

The aim of this part is to study strong solutions to (3.144)–(3.145) in weighted function spaces.

As a preparation, we start to deal with the whole-space problem (3.146), where $n = 2$ or $n = 3$ with inhomogeneous data. However, in an exterior domain $\mathcal{D}$ we consider only the case of three space dimensions.

The goal of this subsection is to find appropriate estimates in weighted spaces in an exterior domain. For later use it is important that we can deal with inhomogeneous divergences g in the whole space.

3.3.3.1 The Inhomogeneous Case in $\mathbb{R}^n$

First, we are interested in the whole-space problem

$$\left.\begin{aligned} -\nu\Delta\boldsymbol{u} - \omega\mathcal{R}\boldsymbol{u} + \nabla p &= f \\ \operatorname{div}\boldsymbol{u} \qquad\qquad &= g \end{aligned}\right\} \text{ in } \mathbb{R}^n. \tag{3.146}$$

Theorem 3.9 *Let $n = 2$ or $n = 3$.*

(i) *Given $\boldsymbol{f} \in \boldsymbol{L}^q_w(\mathbb{R}^n)$ and $g \in W^{1,1}_{\text{loc}}(\mathbb{R}^n)$ with $\nu\nabla g + (\boldsymbol{\omega} \wedge \boldsymbol{x})g \in \boldsymbol{L}^q_w(\mathbb{R}^n)$, there exists a solution $(\boldsymbol{u}, p) \in \boldsymbol{L}^1_{\text{loc}}(\mathbb{R}^n) \times L^1_{\text{loc}}(\mathbb{R}^n)$ of (3.146) that satisfies the estimate*

$$\|\nu\nabla^2\boldsymbol{u}\|_{q,w} + \|\boldsymbol{\omega}\mathcal{R}\boldsymbol{u}\|_{q,w} + \|\nabla p\|_{q,w} \le C(\|\boldsymbol{f}\|_{q,w} + \|\nu\nabla g + (\boldsymbol{\omega} \wedge \boldsymbol{x})g\|_{q,w}), \tag{3.147}$$

where $C = C(q, w) > 0$.

(ii) *Let $\boldsymbol{f} \in \boldsymbol{L}^{q_1}_{w_1}(\mathbb{R}^n) \cap \boldsymbol{L}^{q_2}_{w_2}(\mathbb{R}^n)$ such that (q_1, w_1) and (q_2, w_2) satisfy the condition* (2.15)*, and let $\boldsymbol{u}_1, \boldsymbol{u}_2 \in \boldsymbol{L}^1_{\text{loc}}(\mathbb{R}^n)$ together with corresponding pressure functions $p_1, p_2 \in \boldsymbol{L}^1_{\text{loc}}(\mathbb{R}^n)$ be solutions of* (3.146) *satisfying* (3.147) *for (q_1, w_1), (q_2, w_2), respectively. Then there are $\alpha, \beta, \gamma \in \mathbb{R}$ such that u_1 coincides with u_2 up to an affine linear field $\alpha\boldsymbol{\omega} + \beta\boldsymbol{\omega} \wedge \boldsymbol{x} + \gamma(x_1, x_2, -2x_3)^{\mathrm{T}}$ if $n = 3$ and $\boldsymbol{u}_1 = \boldsymbol{u}_2$ up to the field $\beta(-x_2, x_1)^{\mathrm{T}}$ if $n = 2$.*

Moreover, if $\nabla^2\boldsymbol{u} \in \boldsymbol{L}^q_w(\mathbb{R}^n)$ satisfies (3.146) *with $\boldsymbol{f} = \boldsymbol{0}$, $g = 0$, then $\boldsymbol{u}$ is a polynomial of the above form. In particular, this implies that every solution $\boldsymbol{u} \in \boldsymbol{L}^1_{\text{loc}}(\mathbb{R}^n)$, $\nabla^2\boldsymbol{u} \in \boldsymbol{L}^q_w(\mathbb{R}^n)$ to* (3.146) *satisfies the estimate* (3.147).

□

Proof Let us just briefly mention the difference from the homogeneous case is the estimate of pressure.

A straightforward calculation shows that the unknown pressure p is given by the solution of

$$\Delta p = \operatorname{div} \boldsymbol{f} + \nu\Delta g + \partial_\theta g = \operatorname{div} \boldsymbol{G} \quad \text{in } \mathbb{R}^n$$

with $\boldsymbol{G} = \boldsymbol{f} + \nu\nabla g + (\omega\mathbf{e}_3 \wedge \boldsymbol{x})g$. Then the Hörmander–Mikhlin multiplier theorem in weighted L^q-spaces (Theorem 2.1 (ii)) yields for every weight $w \in A_q(\mathbb{R}^n)$ the estimate

$$\|\nabla p\|_{q,w} \le c\|\boldsymbol{G}\|_{q,w} \le c(\|\boldsymbol{f}\|_{q,w} + \|\nu\nabla g + (\boldsymbol{\omega} \wedge \boldsymbol{x})g\|_{q,w}), \tag{3.148}$$

where $c = c(q, w) > 0$; in particular $\nabla p \in \boldsymbol{L}^q_w(\mathbb{R}^n)$.

Then again as in the previous cases due to the geometry of the problem, it is reasonable to introduce (polar or) cylindrical coordinates $(r, x_3, \theta) \in (0, \infty) \times \mathbb{R} \times [0, 2\pi)$, which finally give us the formal solution

$$\boldsymbol{u}(\boldsymbol{x}) = \int_0^\infty Q_\omega(t)^T K * \boldsymbol{F}(Q_\omega(t)\cdot)(\boldsymbol{x})\mathrm{d}t = \int_0^\infty Q_\omega(t)^T (K * \boldsymbol{F})(Q_\omega(t)\boldsymbol{x})\mathrm{d}t. \tag{3.149}$$

To prove Theorem 3.9 note that due to the formula

$$\widehat{\partial_j \partial_k \boldsymbol{u}}(\boldsymbol{\xi}) = -\xi_j \xi_k \widehat{\boldsymbol{u}} = \frac{\xi_j \xi_k}{|\boldsymbol{\xi}|^2} \left(\widehat{-\Delta \boldsymbol{u}}(\boldsymbol{\xi}) \right), \quad 1 \le j,\ k \le n,$$

and Theorem 2.1 (ii), it suffices to find an estimate of $\|\Delta \boldsymbol{u}\|_{q,w}$, where

$$-\Delta \boldsymbol{u}(\boldsymbol{x}) = \int_{\mathbb{R}^n} \mathcal{K}(\boldsymbol{x},\boldsymbol{y}) \boldsymbol{F}(\boldsymbol{y}) \mathrm{d}\boldsymbol{y}, \quad \mathcal{K}(\boldsymbol{x},\boldsymbol{y}) = -\Delta_x \Sigma_{st}(\boldsymbol{x},\boldsymbol{y}).$$

To obtain the required estimates, we note that by [72, Proposition 3.1] and Corollary 2.1 the operator

$$\mathcal{T}\colon \boldsymbol{L}^q_w(\mathbb{R}^n) \to \boldsymbol{L}^q_w(\mathbb{R}^n), \quad \boldsymbol{F} \mapsto \int_{\mathbb{R}^n} \mathcal{K}(\boldsymbol{x},\boldsymbol{y}) \boldsymbol{F}(\boldsymbol{y}) \mathrm{d}\boldsymbol{y}$$

is continuous for weight functions that fulfill (2.15). Combining this with (3.148) and (3.116), it implies

$$\|\nabla^2 \boldsymbol{u}\|_{q,w} \le c \|\Delta \boldsymbol{u}\|_{q,w} \le c \|\boldsymbol{F}\|_{q,w} \le c \left(\|\boldsymbol{f}\|_{q,w} + \|\nu \nabla g + (\boldsymbol{\omega} \wedge \boldsymbol{x}) g\|_{q,w} \right),$$

and the proof of (i) is finished.

For the proof of (ii), we consider $\boldsymbol{u} - \boldsymbol{u}_k$, $k = 1, 2$, where $\mathbf{u}$ is the solution constructed in (i). Then $\boldsymbol{u}$ fulfills the estimate (3.147) for both (q_1, w_1) and (q_2, w_2). Moreover, $\boldsymbol{u} - \boldsymbol{u}_k$ is a solution with respect to homogeneous data. Thus the uniqueness result follows from the corresponding fact in the case of a homogeneous divergence shown in [72, Theorem 1.2 (ii)], in the proof of which it is shown that any strong solution to (3.146) with respect to zero data is equal to a polynomial of the asserted form. □

3.3.3.2 Exterior Domain in $\mathbb{R}^3$

Let us now focus on the problem (3.144) in the exterior domain $\mathcal{D}$. For simplicity, we consider the viscosity coefficient ν equal to 1. Moreover, throughout this section we assume that $1 < q < \infty$ and that w is a weight function that fulfills (2.15). Note that by Corollary 3.1 every such weight function is contained in the Muckenhoupt class A_q.

Moreover we consider the null space of the problem which is given by

$$\begin{aligned} \mathbf{K}_{\text{null}} := \Big\{ &\boldsymbol{u} \in \widehat{\boldsymbol{H}}^{2,q}_w(\mathcal{D})\colon \operatorname{div} \boldsymbol{u} = 0, \boldsymbol{u}|_{\partial\mathcal{D}} = 0,\ S_{\omega,\nu} \boldsymbol{u} + \nabla p = 0, \\ &\text{for some } p \in \widehat{H}^{1,q}_w(\mathcal{D}) \Big\}. \end{aligned} \tag{3.150}$$

Remark 3.18 The null space $\mathbf{K}_{\text{null}}$ is independent of q and $\mathbf{w}$. We will show it in the end of this subsection. □

The following theorem summarizes our results in the exterior domain.

Theorem 3.10 *For every* $\mathbf{f} \in \boldsymbol{L}^q_w(\mathcal{D})$*, there exists a strong solution* $\{\mathbf{u}, p\} \in \boldsymbol{L}^1_{\text{loc}}(\mathcal{D}) \times L^1_{\text{loc}}(\mathcal{D})$ *to (3.144) satisfying the estimate*

$$\|\nabla^2 \boldsymbol{u}\|_{q,w} + \|\omega \mathcal{R} \boldsymbol{u}\|_{q,w} + \|\nabla p\|_{q,w} \leq c\|\boldsymbol{f}\|_{q,w}, \tag{3.151}$$

with some $c > 0$*. The solution* $\boldsymbol{u}$ *is unique in* $\widehat{\boldsymbol{H}}^{2,q}_w(\mathcal{D})/\mathbf{K}_{\text{null}}$, $\dim \mathbf{K}_{\text{null}} = 3$. □

Corollary 3.3 *Assume that there exists* $r > 1$ *such that*

$$\|\varphi\|_{r,\mathcal{D}} \leq \|\nabla^2 \varphi\|_{q,w,\mathcal{D}}$$

for every $\varphi \in C_0^\infty(\mathcal{D})$*. This holds in particular in the case* $w = 1$ *and* $q \in (1, \frac{3}{2})$*. Let* $\boldsymbol{f} \in \boldsymbol{L}^q_w$*; then there exists a strong solution* $\boldsymbol{u} \in \widehat{\boldsymbol{H}}^{2,q}_w(\mathcal{D})$ *of* (3.144) *having the property* $\lim_{|x| \to \infty} \boldsymbol{u}(x) = 0$. □

Proof of Theorem 3.10 The idea of the proof is to split this problem to an unbounded part that is treated by the use of the results from the whole space problem in Theorem 3.9 and to a part on a bounded domain that can be estimated using the estimates of solutions to the Stokes equations in bounded domains in Theorem 2.4. The assertion about the dimension of $\mathbf{K}_{\text{null}}$ is proved in Proposition 3.5 and Proposition 3.6, and the proof of the existence is given in Lemma 3.10. See Proposition 3.7 for proof of the a priori estimate. □

Lemma 3.9 *Given* $f \in \boldsymbol{L}^q_w(D)$*, let* $\{\boldsymbol{u}, p\} \in \widehat{\boldsymbol{H}}^{2,q}_w(\mathcal{D}) \times \widehat{H}^{1,q}_w(\mathcal{D})$ *be a strong solution to the problem* (3.144)*. Fix* $\rho > \rho_0 > 0$ *so large that* $\mathbb{R}^3 \setminus \mathcal{D} \subset B_{\rho_0}$*, and take* $\psi \in C_0^\infty(B_\rho; [0, 1])$ *such that* $\psi = 1$ *on* B_{ρ_0}*. Then*

$$\begin{aligned}\|\nabla^2 \boldsymbol{u}\|_{q,w,\mathcal{D}} &+ \|\nabla p\|_{q,w,\mathcal{D}} + \|\omega \mathcal{R}\mathbf{u}\|_{q,w,\mathcal{D}} \\ &\leq C \left(\|\boldsymbol{f}\|_{q,w,\mathcal{D}} + \|\boldsymbol{u}\|_{1,q,w,\mathcal{D}_\rho} + \|p\|_{q,w,\mathcal{D}_\rho} \right),\end{aligned} \tag{3.152}$$

where by B_ρ *we denote the ball centered at* 0 *with radius* ρ *and* $\mathcal{D}_\rho = B_\rho \cap \mathcal{D}$. □

Proof By the use of the cutoff function ψ, we decompose the solution $\{\mathbf{u}, p\}$ as follows:

$$\begin{aligned}&\boldsymbol{u} = \boldsymbol{U} + \boldsymbol{V},\ \boldsymbol{U} = (1 - \psi)\boldsymbol{u},\ \ \boldsymbol{V} = \psi \boldsymbol{u}, \\ &p = \sigma + \tau,\ \ \sigma = (1 - \psi)p,\ \ \tau = \psi p.\end{aligned}$$

Then $\{\boldsymbol{U}, \sigma\}$ is a strong solution of

$$\left.\begin{aligned} S_{\omega,1}\boldsymbol{U} + \nabla\sigma &= \boldsymbol{Z}_1, \\ \nabla\cdot\boldsymbol{U} = -\boldsymbol{u}\cdot\nabla\psi &=: g_1 \end{aligned}\right\} \quad \text{in } \mathbb{R}^3, \tag{3.153}$$

where $\operatorname{supp} g_1 \subset \mathcal{D}_\rho$ and

$$\boldsymbol{Z}_1 = (1-\psi)\boldsymbol{f} + 2\nabla\psi\cdot\nabla\boldsymbol{u} + [\Delta\psi + (\omega\mathbf{e}_3\wedge\boldsymbol{x})\cdot\nabla\psi]\boldsymbol{u} - (\nabla\psi)p. \tag{3.154}$$

Similarly, $\{\boldsymbol{V}, \tau\}$ is a strong solution to the Stokes problem in $\mathcal{D}_\rho$

$$\begin{aligned} -\Delta\boldsymbol{V} + \nabla\tau &= \boldsymbol{Z}_2 \quad \text{in } \mathcal{D}_\rho, \\ \nabla\cdot\boldsymbol{V} = \boldsymbol{u}\cdot\nabla\psi &=: g_2 \quad \text{in } \mathcal{D}_\rho, \\ \boldsymbol{V}|_{\partial D_\rho} &= \boldsymbol{0}, \end{aligned} \tag{3.155}$$

where

$$\boldsymbol{Z}_2 = \psi[\boldsymbol{f} + \omega\mathcal{R}\boldsymbol{u}] - 2\nabla\psi\cdot\nabla\boldsymbol{u} - (\Delta\psi)\boldsymbol{u} + (\nabla\psi)p. \tag{3.156}$$

From Theorem 3.9 it follows that

$$\begin{aligned} &\|\nabla^2\boldsymbol{U}\|_{q,w,\mathbb{R}^3} + \|\omega\mathcal{R}\boldsymbol{U}\|_{q,w,\mathbb{R}^3} + \|\nabla\sigma\|_{q,w,\mathbb{R}^3} \\ &\qquad \le C(\|\boldsymbol{Z}_1\|_{q,w,\mathbb{R}^3} + \|\nu\nabla\boldsymbol{g}_1 + (\omega\mathbf{e}_3\wedge\boldsymbol{x})\, g_1\|_{q,w,\mathbb{R}^3}), \end{aligned} \tag{3.157}$$

and from Theorem 2.4 one obtains

$$\|\nabla^2\boldsymbol{V}\|_{q,w,\mathcal{D}_\rho} + \|\nabla\tau\|_{q,w,\mathcal{D}_\rho} \le C\left(\|\boldsymbol{Z}_2\|_{q,w,\mathcal{D}_\rho} + \|g_2\|_{1,q,w,\mathcal{D}_\rho}\right).$$

Now it remains to estimate $\boldsymbol{Z}_1$ and $\boldsymbol{Z}_2$.

$$\begin{aligned} \|\boldsymbol{Z}_1\|_{q,w,\mathbb{R}^3} \le& \|(1-\psi)\boldsymbol{f}\|_{q,w,\mathbb{R}^3} + 2\|\nabla\psi\cdot\nabla\boldsymbol{u}\|_{q,w,\mathbb{R}^3} \\ &+ \|[\Delta\psi + (\omega\mathbf{e}_3\wedge\boldsymbol{x})\cdot\nabla\psi]\boldsymbol{u}\|_{q,w,\mathbb{R}^3} + \|(\nabla\psi)p\|_{q,w,\mathbb{R}^3} \\ \le& c(\|\boldsymbol{f}\|_{q,w,\mathcal{D}} + \|\nabla\boldsymbol{u}\|_{q,w,\mathcal{D}_\rho} + \|\boldsymbol{u}\|_{q,w,\mathcal{D}_\rho} + \|p\|_{q,w,\mathcal{D}_\rho}) \\ \le& c(\|\boldsymbol{f}\|_{q,w,\mathcal{D}} + \|\nabla\boldsymbol{u}\|_{q,w,\mathcal{D}_\rho} + \|p\|_{q,w,\mathcal{D}_\rho}), \end{aligned} \tag{3.158}$$

using the weighted Poincaré inequality (2.14) together with $\boldsymbol{u}|_{\partial D} = 0$. Moreover,

$$\begin{aligned} \|\boldsymbol{Z}_2\|_{q,w,\mathcal{D}_\rho} \le& c(\|\boldsymbol{f}\|_{q,w,\mathcal{D}_\rho} + \|\omega\mathcal{R}\boldsymbol{u}\|_{q,w,\mathcal{D}_\rho} \\ &+ \|\nabla\boldsymbol{u}\|_{q,w,\mathcal{D}_\rho} + \|\boldsymbol{u}\|_{q,w,\mathcal{D}_\rho} + \|p\|_{q,w,\mathcal{D}_\rho}) \end{aligned}$$

$$\leq c(\|f\|_{q,w,\mathcal{D}_\rho} + \|\omega \mathcal{R}u\|_{q,w,\mathcal{D}_\rho} + \|\nabla u\|_{q,w,\mathcal{D}_\rho} + \|p\|_{q,w,\mathcal{D}_\rho}) \tag{3.159}$$

and the following inequalities

$$\begin{aligned} \|g_2\|_{q,w,\mathcal{D}_\rho} + \|\nabla g_2\|_{q,w,\mathcal{D}_\rho} &\leq c\|\nabla g_2\|_{q,w,\mathcal{D}_\rho} \leq c\|\nabla u\|_{q,w,\mathcal{D}_\rho}, \\ \|g_1\|_{q,w,\mathcal{D}_\rho} + \|\nabla g_1\|_{q,w,\mathcal{D}_\rho} &\leq c\|\nabla g_1\|_{q,w,\mathcal{D}_\rho} \leq c\|\nabla u\|_{q,w,\mathcal{D}_\rho}, \\ \|(\omega \mathbf{e}_3 \wedge x) \cdot \nabla u - \omega \mathbf{e}_3 \wedge u\|_{q,w} &\leq \|f + \Delta u - \nabla p\|_{q,w,\mathcal{D}} \\ &\leq \|f\|_{q,w,\mathcal{D}} + \|\Delta u\|_{q,w} + \|\nabla p\|_{q,w,\mathcal{D}}. \end{aligned} \tag{3.160}$$

Then from (3.153)–(3.160) it follows

$$\|\nabla^2 u\|_{q,w} + \|\nabla p\|_{q,w} + \|\omega \mathcal{R}u\|_{q,w} \leq C(\|f\|_{q,w} + \|\nabla u\|_{q,w,\mathcal{D}_\rho} + \|p\|_{q,w,\mathcal{D}_\rho}).$$

□

Proposition 3.5 (Uniqueness) *Suppose that*

$$\{u, p\} \in \widehat{H}_w^{2,q}(\mathcal{D}) \times \widehat{H}_w^{1,q}(\mathcal{D})$$

is a strong solution to the problem (3.144) *with* $f = \mathbf{0}$. *Then there exists a polynomial of the form*

$$\pi(\mathbf{x}) = \alpha\omega \mathbf{e}_3 + \beta\omega \mathbf{e}_3 \wedge x + \gamma(x_1, x_2, -2x_3)^{\mathrm{T}}$$

such that $u - \pi \mapsto 0$ *in the sense of pointwise convergence when* $|x| \to \infty$ *and* u *is uniquely determined by this polynomial. In particular,* $u = \mathbf{0}$ *if* $\pi = \mathbf{0}$.

Accordingly, the gradient of the pressure function ∇p *is uniquely determined by* π. □

Proof **Step 1.** As in the proof of Lemma 3.9, one obtains that $\{U, \sigma\}$ with $U := (1-\psi)u$ and $\sigma = (1-\psi)p$ is a strong solution to the whole-space problem (3.146) with f replaced by

$$Z_1 := 2\nabla\psi \cdot \nabla u + [\Delta\psi + (\omega \mathbf{e}_3 \wedge x) \cdot \nabla\psi]u - (\nabla\psi)p.$$

Applying Sobolev imbedding (2.7), we get $Z_1 \in H_w^{1,q}(\mathcal{D}_\rho) \hookrightarrow L_w^r(\mathcal{D}_\rho)$, where $1/r \geq 1/q - 1/3q$. Then by Theorem 3.9 (ii) we obtain $\nabla^2 U \in L_w^r(\mathcal{D})$. Note that the solutions in the whole space case are unique only up to a polynomial; however, if this polynomial is multiplied by a cutoff function, one still obtains a smooth function with compact support which is in $\boldsymbol{L}^r_w(\mathcal{D})$.

Now we apply the same procedure to the part with the bounded support $\boldsymbol{V} := \psi \boldsymbol{u}$ and $\tau = \psi p$. From the uniqueness of solutions to the Stokes equations, we thus obtain $\nabla^2 \boldsymbol{V} \in \boldsymbol{L}^r_w(\mathcal{D})$, and combined with the above we obtain for the solution u that $\nabla^2 \boldsymbol{u} \in \boldsymbol{L}^r_w(\mathcal{D})$.

A finite number of repetitions of this procedure gives us $\boldsymbol{Z}_1, \boldsymbol{Z}_2 \in \boldsymbol{L}^r_w(\mathcal{D}_\rho)$ for every $r \geq q$. In particular by Remark 2.4, (2.6) we may choose r so large that

$$L^r_w(\mathcal{D}_\rho) \hookrightarrow L^{6/5}(\mathcal{D}_\rho).$$

Then again the above procedure yields $\nabla^2 \boldsymbol{u} \in \boldsymbol{L}^{6/5}(\mathcal{D})$. Analogously one obtains $\nabla p \in L^{6/5}(\mathcal{D})$.

Step 2. Now the weighted version of Bogowski's Theorem 2.7 in $\mathcal{D}_\rho$ yields that there exists $\boldsymbol{v} \in H_0^{2,6/5}(\mathcal{D}_\rho)$ with $\operatorname{div} \boldsymbol{v} = \operatorname{div} \psi \boldsymbol{u}$. We set $\boldsymbol{U}_1 := (1 - \psi)\boldsymbol{u} + \boldsymbol{v}$.

Then $\operatorname{div} \boldsymbol{U}_1 = 0$; the Sobolev embedding theorem yields

$$S_{\omega,1} \boldsymbol{U}_1 + \nabla((1 - \psi)p) \in \boldsymbol{L}^{6/5}(\mathbb{R}^3) \hookrightarrow \widehat{\boldsymbol{H}}^{-1,2}(\mathbb{R}^3).$$

Step 3. By Theorem 3.6 there exists a weak solution $\{\widetilde{\boldsymbol{U}}_1, \tilde{p}\} \in \widehat{\boldsymbol{H}}^{1,2}(\mathbb{R}^3) \times L^2(\mathbb{R}^3)$ to the problem

$$S_{\omega,1} \widetilde{\boldsymbol{U}}_1 + \nabla \tilde{p} = S_{\omega,1} \boldsymbol{U}_1 + \nabla(1 - \psi)p, \quad \operatorname{div} \widetilde{\boldsymbol{U}}_1 = 0 \quad \text{in } \mathbb{R}^n, \quad \lim_{|x| \to \infty} \widetilde{\boldsymbol{U}}_1(x) = \boldsymbol{0} \tag{3.161}$$

with an appropriate pressure function $\tilde{p}$. This solution $\widetilde{\boldsymbol{U}}_1$ is contained in $\widehat{\boldsymbol{H}}_0^{1,2}(\mathbb{R}^3) \hookrightarrow \boldsymbol{L}^6(\mathbb{R}^3)$ and is given by the formula (3.149). Since the same formula yields a strong solution to the problem (3.146), we obtain $\nabla^2 \widetilde{\boldsymbol{U}}_1 \in \boldsymbol{L}^{6/5}(\mathbb{R}^3)$.

By the uniqueness of solutions to the whole space problem (see Theorem 3.9), one obtains

$$(\boldsymbol{U}_1 - \widetilde{\boldsymbol{U}}_1)(\mathbf{x}) = \boldsymbol{\pi}(\mathbf{x}) = \alpha\omega \mathbf{e}_3 + \beta\omega \mathbf{e}_3 \wedge \boldsymbol{x} + \gamma(x_1, x_2, -2x_3)^{\mathrm{T}}$$

with appropriate coefficients α, β, γ.

Step 4. Now let $g \in \boldsymbol{H}^{1,2}(\mathcal{D})$ be chosen such that

$$\boldsymbol{g}|_{\partial\mathcal{D}} = \boldsymbol{\pi}|_{\partial\mathcal{D}}, \quad \operatorname{div} \boldsymbol{g} = 0 \quad \text{and} \operatorname{supp} \boldsymbol{g} \subset \overline{\mathcal{D}_\rho}. \tag{3.162}$$

We show that such a function g exists. Note that

$$\int_{\mathcal{D}} \operatorname{div}(\psi \boldsymbol{\pi}) \cdot 1 \mathrm{d}\boldsymbol{x} = \int_{\partial \mathcal{D}_\rho} \psi \boldsymbol{\pi} \cdot \boldsymbol{N} \mathrm{d}S = -\int_{\partial(\mathbb{R}^n \setminus \mathcal{D})} \boldsymbol{\pi}(-\boldsymbol{N}) \mathrm{d}S =$$

$$-\int_{\mathbb{R}^n \setminus \mathcal{D}} \operatorname{div} \boldsymbol{\pi} \, \mathrm{d}\boldsymbol{x} = 0.$$

Now Bogowski's Theorem 2.7 gives us a function $\boldsymbol{\zeta} \in \boldsymbol{W}_0^{1,2}(\mathcal{D}_\rho)$ with $\operatorname{div}\boldsymbol{\zeta} = \operatorname{div}(\psi\boldsymbol{\pi})$. Then $\boldsymbol{g} := \psi\boldsymbol{\pi} - \boldsymbol{\zeta}$ fulfills (3.162).

Then one has for every $\boldsymbol{\varphi} \in \widehat{\boldsymbol{H}}_0^{1,2}(\mathcal{D})$

$$\begin{aligned} |\langle S_{\omega,1}\boldsymbol{g}, \boldsymbol{\varphi}\rangle_{\mathcal{D}}| &\leq |\langle \nabla \boldsymbol{g}, \nabla\boldsymbol{\varphi}\rangle| + |\langle \omega \mathbf{e}_3 \wedge \boldsymbol{g} - (\omega\mathbf{e}_3 \wedge \boldsymbol{x}) \cdot \nabla \boldsymbol{g}, \boldsymbol{\varphi}\rangle| \\ &\leq \|\nabla \boldsymbol{g}\|_{2,\mathcal{D}} \|\nabla\boldsymbol{\varphi}\|_{2,\mathcal{D}} + \|\boldsymbol{\omega} \wedge \boldsymbol{g} - (\boldsymbol{\omega} \wedge \boldsymbol{x}) \cdot \nabla \boldsymbol{g}\|_{2,\mathcal{D}_\rho} \|\boldsymbol{\varphi}\|_{2,\mathcal{D}_\rho} \\ &\leq c\|\boldsymbol{g}\|_{H^{1,2}(\mathcal{D})} \|\nabla\boldsymbol{\varphi}\|_{2,\mathcal{D}}. \end{aligned}$$

using the Poincaré inequality in $\mathcal{D}_\rho$ and the fact that $\boldsymbol{\varphi}|_{\partial\mathcal{D}} = 0$. We have shown that $S_{\omega,1}\boldsymbol{g} \in \widehat{\boldsymbol{H}}^{-1,2}(\mathcal{D})$. Thus by Theorem 3.6 there exists $\boldsymbol{G} \in \widehat{\boldsymbol{H}}_0^{1,2}(\mathcal{D})$ solving weakly the problem

$$S_{\omega,1}\boldsymbol{G} + \nabla p^* = S_{\omega,1}\boldsymbol{g} \quad \text{in } \mathcal{D}, \quad \operatorname{div}\boldsymbol{G} = 0 \tag{3.163}$$

for an appropriate pressure function p^*.

Step 5. Now we set

$$\widetilde{\boldsymbol{V}} := \widetilde{\boldsymbol{U}_1} - \boldsymbol{v} + \boldsymbol{g} - \boldsymbol{G} + \psi\boldsymbol{u} = \boldsymbol{u} + \boldsymbol{g} - \boldsymbol{G} - \boldsymbol{\pi}.$$

Then one has $\widetilde{\boldsymbol{V}} \in \widehat{\boldsymbol{H}}_0^{1,2}(\mathcal{D})$, and it solves the problem

$$S_{\omega,1}\widetilde{\boldsymbol{V}} + \nabla p = S_{\omega,1}\boldsymbol{u} + \nabla p = \boldsymbol{0}, \quad \operatorname{div}\widetilde{\boldsymbol{V}} = 0,$$

where the pressure function p consists of the pressure terms in (3.161) and (3.163), so $p = \tilde{p} - p^*$. Thus, by the uniqueness of weak solutions in [127], we obtain that $\boldsymbol{V} = \boldsymbol{0}$, which in turn yields

$$\boldsymbol{u} = \boldsymbol{\pi} - \boldsymbol{g} + \boldsymbol{G}.$$

Since the functions $\boldsymbol{g}$ and $\boldsymbol{G}$ depend linearly on $\boldsymbol{\pi}$, the function $\boldsymbol{u}$ is taken from some space with dimension 3. The behavior of $\boldsymbol{u}$, i.e., the growth of $\boldsymbol{\pi}$ at infinity determines the particular element.

For the pressure p corresponding to $\boldsymbol{u}$ one has $\nabla p = -S_{\omega,1}\boldsymbol{u} = -S_{\omega,1}\boldsymbol{g} - -S_{\omega,1}\boldsymbol{G} = \nabla p^*$ with ∇p^* defined as in (3.163) depending linearly on $\boldsymbol{\pi}$. □

Lemma 3.10 *Let $\boldsymbol{f} \in \boldsymbol{L}^q_w(\mathcal{D})$; then* (3.144) *has a strong solution in the class $\boldsymbol{u} \in \widehat{\boldsymbol{H}}_w^{2,q}(\mathcal{D})$. The corresponding pressure $p \in \widehat{H}_w^{1,q}(\mathcal{D})$.* □

Proof Let $\boldsymbol{f}_0$ be the extension of $\boldsymbol{f}$ by $\boldsymbol{0}$ on $\mathbb{R}^3 \setminus \mathcal{D}$, and let $(\boldsymbol{u}_1, p_1) \in \widehat{\boldsymbol{H}}_w^{2,q}(\mathbb{R}^3) \times \widehat{H}_w^{1,q}(\mathbb{R}^3)$ be a solution of

$$S_{\omega,1}\boldsymbol{u}_1 + \nabla p_1 = \boldsymbol{f}_0 \quad \text{and} \quad \operatorname{div}\boldsymbol{u}_1 = 0 \quad \text{in } \mathbb{R}^3,$$

which exists by Theorem 3.9 and having the appropriate behavior at infinity. Moreover, as in the proof of Proposition 3.5 one shows that there exists a function $\boldsymbol{g} \in \boldsymbol{H}^{2,q}_w(\mathcal{D}_\rho)$ with compact support in $\mathcal{D}_\rho \cup \partial\mathcal{D}$ and $\boldsymbol{g}|_{\partial\mathcal{D}} = \boldsymbol{u}_1|_{\partial\mathcal{D}}$ and $\operatorname{div} \boldsymbol{g} = 0$.

Then by Remark 2.4 there exists $s > 1$ such that $\boldsymbol{g}$ belongs to the unweighted space $\boldsymbol{H}^{2,s}(\mathcal{D})$. We show that $S_{\omega,1}\boldsymbol{g} \in \widehat{\boldsymbol{H}}^{-1,r}(\mathcal{D})$ for some $r \in (3/2, 3)$,

$$|\langle S_{\omega,1}\boldsymbol{g}, \boldsymbol{\psi}\rangle| \leq \|S_{\omega,1}\boldsymbol{g}\|_{L^s(\mathcal{D}_\rho)}\|\boldsymbol{\psi}\|_{L^{s'}(\mathcal{D}_\rho)} \leq c\|S_{\omega,1}\boldsymbol{g}\|_{L^s(\mathcal{D}_\rho)}\|\nabla\boldsymbol{\psi}\|_{L^{r'}(\mathcal{D}_\rho)}$$

for every $\boldsymbol{\psi} \in \widehat{\boldsymbol{H}}^{1,r'}_0(\mathcal{D})$, where r is chosen such that $W^{1,r'}(\mathcal{D}_\rho) \hookrightarrow L^{s'}(\mathcal{D}_\rho)$. This requires $1/r' \leq 1/s' + 1/3$, and thus $r \in (3/2, 3)$ is possible.

By Theorem 3.6 there exists a weak solution $\boldsymbol{G} \in \widehat{\boldsymbol{H}}^{1,r}_0(\mathcal{D})$ to the exterior problem

$$S_{\omega,1}\boldsymbol{G} + \nabla p_2 = S_{\omega,1}\boldsymbol{g}, \quad \boldsymbol{G}|_{\partial\mathcal{D}} = 0, \quad \operatorname{div} \boldsymbol{G} = 0.$$

Then $\boldsymbol{u} := \boldsymbol{u}_1 - \boldsymbol{g} + \boldsymbol{G}$, $p = p_1 + p_2$ solves the problem

$$S_{\omega,1}\boldsymbol{u} + \nabla p = \boldsymbol{f}, \quad \boldsymbol{u}|_{\partial\mathcal{D}} = 0, \quad \operatorname{div} \boldsymbol{u} = 0$$

in the sense of distributions in $\mathcal{D}$. It remains to show that $\nabla^2\boldsymbol{u} \in \boldsymbol{L}^q_w(\mathcal{D})$ and $(\omega\mathbf{e}_3 \wedge \boldsymbol{x}) \cdot \nabla\boldsymbol{u} - \omega\mathbf{e}_3 \wedge \boldsymbol{u} \in \boldsymbol{L}^q_w(\mathcal{D})$. This follows from a similar cutoff procedure as in the proof of Lemma 3.9 as follows.

Set $r_1 := r$. As a consequence of its construction, function $\boldsymbol{u}$ is in $\widehat{\boldsymbol{H}}^{2,q}_w(\mathcal{D}) + \widehat{\boldsymbol{H}}^{2,r_1}(\mathcal{D})$. Using the same notation as in the proof of Lemma 3.9, we obtain

$$\boldsymbol{Z}_1 \in \boldsymbol{L}^q_w(\mathbb{R}^n) + \boldsymbol{L}^{r_1}(\mathcal{D}_\rho) \quad \text{and} \quad \boldsymbol{Z}_2 \in \boldsymbol{L}^q_w(\mathcal{D}_\rho) + \boldsymbol{L}^{r_1}(\mathcal{D}_\rho) \quad \text{for } i = 1, 2,$$

i.e., we may write $\boldsymbol{Z}_1 = \boldsymbol{Z}^o_1 + \boldsymbol{Z}^i_1$, $\boldsymbol{Z}_2 = \boldsymbol{Z}^o_2 + \boldsymbol{Z}^i_2$ with $\boldsymbol{Z}^o_1 \in L^q_w(\mathbb{R}^n)$, $\boldsymbol{Z}^i_1 \in L^{r_1}(\mathbb{R}^n)$ with $\operatorname{supp} \boldsymbol{Z}^i_1 \subset \mathcal{D}_\rho$, $\boldsymbol{Z}^o_2 \in L^q_w(\mathcal{D}_\rho)$, and $\boldsymbol{Z}^i_2 \in L^{r_1}(\mathcal{D}_\rho)$. Let $\boldsymbol{u}^o_1 \in H^{2,q}_w(\mathbb{R}^n)$ and $\boldsymbol{u}^i_1 \in \boldsymbol{H}^{2,r_1}(\mathbb{R}^n)$ be the solutions to

$$S_{\omega,1}\boldsymbol{u}^o_1 + \nabla p^o_1 = \boldsymbol{Z}^o_1 \quad \text{and} \quad S_{\omega,1}\boldsymbol{u}^i_1 + \nabla p^i_1 = \boldsymbol{Z}^i_1$$

with vanishing divergence. Then $(1-\psi)\boldsymbol{u} = \boldsymbol{u}^o_1 + \boldsymbol{u}^i_1$ up to a polynomial of degree ≤ 1. Analogously let $\boldsymbol{v}^o_1 \in \boldsymbol{H}^{2,q}_w(\mathcal{D}_\rho)$ and $\boldsymbol{v}^i_1 \in \boldsymbol{H}^{2,r_1}(\mathcal{D}_\rho)$ be the solutions to the Stokes equations

$$-\Delta\boldsymbol{v}^o_1 + \nabla p^o_1 = \boldsymbol{Z}^o_2 \quad \text{and} \quad -\Delta\boldsymbol{v}^i_1 + \nabla p^i_1 = \boldsymbol{Z}^i_2$$

with vanishing divergence and boundary values. Choose a cutoff function $\widetilde{\psi} \in C^\infty_0(B_\rho)$ with $\widetilde{\psi} \equiv 1$ on $\operatorname{supp}\psi$. Then one has by the uniqueness of solutions to the

Stokes equation $\psi \boldsymbol{u} = \widetilde{\psi}\psi \boldsymbol{u} = \widetilde{\psi}\boldsymbol{v}_1^o + \widetilde{\psi}\boldsymbol{v}_1^i \in \widehat{\boldsymbol{H}}_w^{2,q}(\mathcal{D}) + \widehat{\boldsymbol{H}}^{2,r_1}(\mathcal{D})$. Combining the above yields $\boldsymbol{u} \in \widehat{\boldsymbol{H}}_w^{2,q}(\mathcal{D}) + \widehat{\boldsymbol{H}}^{2,r_1}(\mathcal{D})$.

This in turn shows that $\boldsymbol{Z}_i \in \boldsymbol{L}_w^q(\mathcal{D}) + \boldsymbol{H}^{1,r_1}(\mathcal{D}_\rho) \hookrightarrow \boldsymbol{L}_w^q(\mathcal{D}) + \boldsymbol{L}^{r_2}(\mathcal{D}_\rho)$ with $1/r_2 \leq 1/r_1 + 1/n$ and again $\boldsymbol{u} \in \widehat{\boldsymbol{H}}_w^{2,q}(\mathcal{D}) + \widehat{\boldsymbol{H}}^{2,r_2}(\mathcal{D})$. If one repeats this procedure finitely many times, one obtains $\boldsymbol{u} \in \widehat{\boldsymbol{H}}_w^{2,q}(\mathcal{D}) + \widehat{\boldsymbol{H}}^{2,r_k}(\mathcal{D})$ with r_k arbitrarily large. In particular, it can be chosen so large that $L^{r_k}(\mathcal{D}_\rho) \hookrightarrow L_w^q(\mathcal{D})$. Then one has $\boldsymbol{u} \in \widehat{\boldsymbol{H}}_w^{2,q}(\mathcal{D})$ as asserted. □

Proposition 3.6 *We consider the null space* $\mathbf{K}_{null}$ *defined in* (3.150). *Then* $\dim \mathbf{K}_{\text{null}} = 3$. □

Proof It has been already shown that $\dim \mathbf{K}_{\text{null}} \leq 3$, see Lemma 3.5. It remains to show that $\dim K \geq 3$. Let $\boldsymbol{\pi} = \alpha\boldsymbol{\omega} + \beta\boldsymbol{\omega} \wedge \boldsymbol{x} + \gamma(x_1, x_2, -2x_3)^{\mathrm{T}}$. Then as in the proof of Lemma 3.10 we find that there exists $\boldsymbol{g} \in \widehat{\boldsymbol{H}}_w^{2,q}(\mathcal{D})$ such that

$$\boldsymbol{g}|_{\partial D} = \boldsymbol{\pi}|_{\partial D}, \quad \operatorname{div} \boldsymbol{g} = 0, \quad \operatorname{supp} \boldsymbol{g} \text{ is compact in } \overline{\mathcal{D}} \text{ and } L\boldsymbol{g} \in \boldsymbol{H}^{-1,2}(\mathcal{D}).$$

Then, by Theorem 3.6, there exists $\boldsymbol{v} \in \boldsymbol{H}_0^{1,2}(\mathcal{D})$ solving $S_{\omega,1}\boldsymbol{v} + \nabla p = S_{\omega,1}\boldsymbol{g}$ for some p, $\operatorname{div} \boldsymbol{u} = 0$. Thus if we set $\boldsymbol{u} := \boldsymbol{v} - \boldsymbol{g} + \boldsymbol{\pi}$, we find as in the proof of Lemma 3.10 that actually $\nabla^2 \boldsymbol{u} \in \boldsymbol{L}_w^q(\mathcal{D})$, and thus

$$\boldsymbol{u} \in \mathbf{K}_{\text{null}} \quad \text{with} \quad \boldsymbol{u}(\boldsymbol{x}) - \boldsymbol{\pi}(\boldsymbol{x}) \xrightarrow{|\boldsymbol{x}| \to \infty} 0.$$

Thus for every α, β, γ there exists $\boldsymbol{u} \in \mathbf{K}_{\text{null}}$ with $\boldsymbol{u}(\boldsymbol{x}) - \boldsymbol{\pi}(\boldsymbol{x}) \xrightarrow{|x| \to \infty} 0$, which means $\dim \mathbf{K}_{\text{null}} \geq 3$. □

Similarly to the results in case of the classical Stokes equations in [93, V.4.], the solutions to (3.144) can be estimated in the factor space modulo the null space $\mathbf{K}_{\text{null}}$.

Proposition 3.7 *Suppose* $\{\boldsymbol{u}, p\} \in \widehat{\boldsymbol{H}}_w^{2,q}(\mathcal{D}) \times \widehat{H}_w^{1,q}(\mathcal{D})$ *is a strong solution to* (3.144) *with* $\boldsymbol{f} \in L_w^q(D)$. *Then the estimate*

$$\inf_{S_{\omega,1}\boldsymbol{v} + \nabla p^* = 0} (\|\boldsymbol{u} + \boldsymbol{v}\|_{\widehat{H}_w^{2,q}} + \|\omega \mathcal{R}(\boldsymbol{u} + \boldsymbol{v})\|_{q,w} + \|\nabla p + \nabla p^*\|_{q,w}) \leq c\|\boldsymbol{f}\|_{q,w}$$

holds, where the infimum is taken over all $\boldsymbol{v} \in \mathbf{K}_{\text{null}}$. *In particular, since the null space* $\mathbf{K}_{\text{null}}$ *has dimension* 3, *at least one representative of* $\boldsymbol{u} + \mathbf{K}_{\text{null}}$ *fulfills the estimate* (3.151). □

Proof We assume the contrary. Then there exist sequences $(\boldsymbol{f}_k) \subset \boldsymbol{L}_w^q(\mathcal{D})$ and corresponding strong solutions $(\boldsymbol{u}_k, p_k)_k \subset \widehat{\boldsymbol{H}}_w^{2,q}(\mathcal{D}) \times \widehat{H}_w^{1,q}(\mathcal{D})$ such that

$$\inf_{S_{\omega,\nu}\boldsymbol{v} + \nabla p^* = 0} (\|\nabla^2(\boldsymbol{u}_k + \boldsymbol{v})\|_{q,w,\mathcal{D}} + \|\omega \mathcal{R}(\boldsymbol{u}_k + \boldsymbol{v})\|_{q,w,\mathcal{D}} + \|\nabla p_k + \nabla p^*\|_{q,w,\mathcal{D}}) = 1 \tag{3.164}$$

and $\|\boldsymbol{f}_k\|_{q,w,D} \to 0$ as $k \to \infty$. Since $\mathbf{K}_{\text{null}}$ has a finite dimension, we may assume that

$$\begin{aligned}&\|\nabla^2 \boldsymbol{u}_k\|_{q,w,\mathcal{D}} + \|\omega \mathcal{R} \boldsymbol{u}_k\|_{q,w,\mathcal{D}} + \|\nabla p_k\|_{q,w,\mathcal{D}} \\ &\quad = \inf_{S_{\omega,1}\boldsymbol{v}+\nabla p^*=0} (\|\nabla^2(\boldsymbol{u}_k + \boldsymbol{v})\|_{q,w} \\ &\qquad + \|\omega\mathcal{R}(\boldsymbol{u}_k+\boldsymbol{v})\|_{q,w,\mathcal{D}} + \|\nabla p_k + \nabla p^*\|_{q,w,\mathcal{D}}).\end{aligned}$$

As in Proposition 3.9 we put $\mathcal{D}_\rho = \mathcal{D} \cap B_\rho$ and obtain

$$\|\nabla \boldsymbol{u}_k\|_{1,q,\mathcal{D}_\rho,w} \le \|\boldsymbol{u}_k\|_{2,q,w,\mathcal{D}_\rho} \le c\|\nabla^2 \boldsymbol{u}_k\|_{q,w,\mathcal{D}_\rho} \le c\|\nabla^2 \boldsymbol{u}_k\|_{q,w,\mathcal{D}} \le 1,$$

where we have used $\boldsymbol{u}_k|_{\partial D} = 0$ and the weighted version of Poincaré's inequality for second derivatives in Lemma 2.5. Similarly, if we subtract an appropriate constant from each p_k, we may assume that

$$\|p_k\|_{1,q,w,\mathcal{D}_\rho} \le \|\nabla p_k\|_{q,w,\mathcal{D}} \le 1.$$

We choose subsequences, which we again denote by $(\boldsymbol{u}_k)$, (p_k), such that

$$\begin{aligned} \boldsymbol{u}_k &\rightharpoonup \boldsymbol{u} \quad \text{in } \boldsymbol{H}_w^{2,q}(\mathcal{D}_\rho), \\ p_k &\rightharpoonup p \quad \text{in } H_w^{1,q}(\mathcal{D}_\rho). \end{aligned}$$

Now we apply the fact that by (2.8) the embeddings $\boldsymbol{H}_w^{2,q}(\mathcal{D}_\rho) \hookrightarrow \boldsymbol{H}_w^{1,q}(\mathcal{D}_\rho)$ and $\boldsymbol{H}_w^{1,q}(\mathcal{D}_\rho) \hookrightarrow \boldsymbol{L}_w^q(\mathcal{D}_\rho)$ are compact and obtain the strong convergence of $(\boldsymbol{u}_k)$ in $\boldsymbol{H}_w^{1,q}(\mathcal{D}_\rho)$ and p_k in $\boldsymbol{L}_w^q(\mathcal{D}_\rho)$. From (3.152) it follows that $(\boldsymbol{u}_k, p_k)_k$ and $((\boldsymbol{\omega} \wedge \boldsymbol{x}) \cdot \nabla \boldsymbol{u}_k - \boldsymbol{\omega} \wedge \boldsymbol{u}_k)_k$ are Cauchy sequences in $\widehat{\boldsymbol{H}}_w^{2,q}(\mathcal{D}) \times \widehat{H}_w^{1,q}(\mathcal{D})$ and $\boldsymbol{L}_w^q(\mathcal{D})$. This implies that there exists $\{\boldsymbol{u}, p\} \in \widehat{\boldsymbol{H}}_w^{2,q}(\mathcal{D}) \times \widehat{H}_w^{1,q}(\mathcal{D})$ such that

$$\|\nabla^2 \boldsymbol{u}_k - \nabla^2 \boldsymbol{u}\|_{q,w} + \|\nabla p_k - \nabla p\|_{q,w} \xrightarrow{k\to\infty} 0, \tag{3.165}$$

$$\|[\omega\mathcal{R}\boldsymbol{u}_k - \omega\mathcal{R}\boldsymbol{u}]\|_{q,w} \xrightarrow{k\to\infty} 0. \tag{3.166}$$

It follows from (3.165) and (3.166) that the pair $\{\boldsymbol{u}, p\}$ is a strong solution to (3.144) with $\boldsymbol{f} = 0$. This contradicts our assumption that

$$\inf_{S_{\omega,\nu}\boldsymbol{v}+\nabla p^*=0} (\|\nabla^2(\boldsymbol{u}+\boldsymbol{v})\|_{q,w} + \|\omega\mathcal{R}(\boldsymbol{u}+\boldsymbol{v})\|_{q,w} + \|\nabla p + \nabla p^*\|_{q,w}) = 1.$$

Thus, (3.151) is valid. □

Proof of Corollary 3.3 To see this approximate $\boldsymbol{f}$ by a sequence $(\boldsymbol{f}_n) \subset \boldsymbol{C}_0^\infty(\mathcal{D})$. Then as before one uses the weighted Poincaré inequality in $B_n \cap \mathcal{D}$, supp $\boldsymbol{f}_n \subset B_n$, to show that $\boldsymbol{f}_n \in \boldsymbol{H}^{-1,2}(\mathcal{D})$. Thus, by Theorem 3.6 there exist corresponding weak solutions $(\boldsymbol{u}_n)$ to (3.144). In particular $\boldsymbol{u}_n \in \boldsymbol{L}^r(\mathcal{D})$. As in the proof of Lemma 3.10 one shows that these solutions also satisfy $\nabla^2 \boldsymbol{u}_n \in \boldsymbol{L}^q_w(\mathcal{D})$. Thus, by the a priori estimate that we have shown in Proposition 3.7, we find that there exists some $\boldsymbol{u} \in \widehat{\boldsymbol{H}}^{2,q}_w(\mathcal{D})$ such that $\boldsymbol{u}_n \xrightarrow{n\to\infty} \boldsymbol{u}$ in $\widehat{\boldsymbol{H}}^{2,q}_w(\mathcal{D})/\mathbf{K}_{\text{null}}$. Since $\mathbf{K}_{\text{null}}$ is finite dimensional, there exists $\boldsymbol{v}_n$ with $\|\nabla^2 \boldsymbol{v}_n\|_{q,w} = \|\boldsymbol{u}_n\|_{\widehat{H}^{2,q}_w/\mathbf{K}_{\text{null}}}$. Thus the sequence of solutions $\boldsymbol{v}_n$ converges in the Banach space

$$(\widehat{\boldsymbol{H}}^{2,q}_w(\mathcal{D}) \cap \boldsymbol{L}^r(\mathcal{D})) + \mathbf{K}_{\text{null}},$$

and we obtain that there exists $\boldsymbol{\eta} \in \mathbf{K}_{\text{null}}$ such that $\boldsymbol{u} - \boldsymbol{\eta} \in \boldsymbol{L}^r(\mathcal{D})$. Then $\boldsymbol{u} - \boldsymbol{\eta}$ is the solution which decays at infinity. □

Proof of Remark 3.18 We would like to show that K does not depend on q and w. Let us consider $\boldsymbol{u} \in \mathbf{K}_{\text{null}}$ with corresponding q and w. We use the cutoff function ψ such that it is smooth, with compact support and equal to zero outside of an exterior domain $\mathcal{D}$. We denote $U = \psi \boldsymbol{u}$. Applying the operator $S_{\omega,1}$ to $\boldsymbol{U}$, we get $S_{\omega,1}\boldsymbol{U} = S_{\omega,1}(\psi \boldsymbol{u}) = \psi S_{\omega,1}\boldsymbol{u} + 2\nabla \boldsymbol{u} \nabla \psi + \boldsymbol{u} \Delta \psi - (\boldsymbol{\omega} \wedge \boldsymbol{x}) \cdot \nabla \psi)\boldsymbol{u}$. Again as in the proof of Proposition 3.5 we apply the Sobolev embedding, see Remark 2.4 that $\boldsymbol{Z}_1 = 2\nabla \boldsymbol{u} \nabla \psi + \boldsymbol{u} \Delta \psi - ((\boldsymbol{\omega} \wedge \boldsymbol{x}) \cdot \nabla \psi)\boldsymbol{u} \in \boldsymbol{H}^{1,q}_w(\operatorname{supp} \psi) \hookrightarrow \boldsymbol{L}^r_w(\operatorname{supp} \psi)$, where $1/r \geq 1/q - 1/3q$. Then from Theorem 2.2 we get that $\nabla^2 \boldsymbol{U} \in \boldsymbol{L}^r_w(\mathbb{R}^3)$, which implies that $\nabla^2 \boldsymbol{u} \in \boldsymbol{L}^r_w(\mathcal{D})$. Repeating this procedure, we may choose r so large that for every $r \geq q$, $\boldsymbol{L}^r_w(\operatorname{supp} \psi) \hookrightarrow \boldsymbol{L}^2(\operatorname{supp} \psi)$. It implies (again applying Theorem 2.2) that $\nabla^2 \boldsymbol{u} \in \boldsymbol{L}^2(\mathcal{D})$ with the same null space $\mathbf{K}_{\text{null}}$ but with corresponding $q = 2$, $w = 1$. It follows that the null space $\mathbf{K}_{\text{null}}$ does not depend on w and q. □

3.4 Very Weak Solutions to the Stokes Type Problem

The content of this section is mainly based on the work by Kračmar et al. [152].

3.4.1 The Homogeneous Case: Very Weak Solutions

In this subsection we investigate the Stokes problem in the whole space or in the exterior domain in a very weak sense. We consider

$$\begin{aligned} S_{\omega,\nu}\mathbf{u} + \nabla p &= \mathbf{f} \quad \text{in } \mathcal{D} \subseteq \mathbb{R}^3, \\ \operatorname{div} \mathbf{u} &= g \quad \text{in } \mathcal{D}, \\ \mathbf{u} &= \mathbf{0} \quad \text{on } \partial\mathcal{D}, \text{ if } \mathcal{D} \neq \mathbb{R}^3. \end{aligned} \tag{3.167}$$

Before starting our analysis let us give some introduction to the problem of a very weak solution. A very weak solution is based on the duality method. It was introduced for elliptic equations as early as 1966–1968 by Lions–Magenes, see [169], and it can also be found in the work by Nečas, see [174].

The concept of very weak solutions for Stokes or Navier–Stokes equations was introduced by Giga in 1981, see [120], and by Amrouche and Girault in 1994 in a domain class $C^{1,1}$, see [7]. More recently this concept was extended by Amrouche and Rodriguez–Bellido, see [9], and Galdi, Simader, Farwig, Kozono, and Sohr, see, e.g., [64, 112], to a setting in classical L^q-spaces. The only existence of inhomogeneous divergence of the velocity in some L^r-spaces is required with a convenient choice of exponents r and q. For the nonsteady case we can refer to the work of Amann in 2000–2002 in the setting of Besov spaces, see [5, 6].

The notion of a very weak solution has been generalized by Schumacher, see [190, 191], to a setting in weighted Lebesque and Bessel potential spaces using arbitrary Muckenhoupt weights. Further, a very weak solution may be generalized so that traces on the boundary can be defined, see [190].

Let us introduce the following spaces of functions and functionals

$$\begin{aligned}
\mathbf{Y}^{2,q'}_{w'}(\mathcal{D}) &:= \{\mathbf{u} \in \widehat{\mathbf{H}}^{2,q'}_{w'}(\mathcal{D}),\ \mathbf{u}|_{\partial\mathcal{D}} = \mathbf{0}\},\\
\widehat{\mathbf{H}}^{2,q'}_{w',\mathbf{K}_{\rm null}}(\mathcal{D}) &:= \widehat{\mathbf{H}}^{2,q'}_{w'}(\mathcal{D})/\mathbf{K}'_{\rm null},\\
\mathcal{H}^{-1,q}_{w,0}(\mathcal{D}) &:= \left(\widehat{H}^{1,q'}_{w'}(\mathcal{D})\right)',
\end{aligned}$$

and we will add if necessary the subscript σ in the case of divergence-free vector fields:

$$\begin{aligned}
\mathbf{Y}^{2,q'}_{w',\sigma}(\mathcal{D}) &:= \{\mathbf{u} \in \mathbf{Y}^{2,q'}_{w'}(\mathcal{D})|\ \operatorname{div}\mathbf{u} = 0\},\\
\mathbf{Y}^{-2,q}_{w,\sigma}(\mathcal{D}) &:= (\mathbf{Y}^{2,q'}_{w',\sigma}(\mathcal{D}))'.
\end{aligned}$$

In the lowest regularity context the data are given by functionals which now are in general no distributions. The corresponding functional spaces must be precisely defined.

Let $\mathcal{D} \subseteq \mathbb{R}^3$ be always an exterior domain of type $C^{1,1}$ or the whole space and w always a q-weight function according to (2.15).

Let us recall the null space $\mathbf{K}_{\rm null} = \mathbf{K}_{null,q,w}$ of the same system, which is given by

$$\mathbf{K}_{\rm null} = \left\{\mathbf{v} \in \widehat{\mathbf{H}}^{2,q}_{w,\sigma}(\mathcal{D})\colon S_{\omega,\nu}\mathbf{v} + \nabla\pi = \mathbf{0}, \mathbf{v}|_{\partial\mathcal{D}} = \mathbf{0},\ \text{for some } \pi \in \widehat{H}^{1,q}_{w}(\mathcal{D})\right\}. \tag{3.168}$$

Similarly we denote by $\mathbf{K}'_{\rm null} = \mathbf{K}_{{\rm null},q',w'}$ the null space associated with the dual system, $S'_{\omega,\nu}\mathbf{v} - \nabla\pi = \mathbf{0}$, $\operatorname{div}\mathbf{v} = 0$, $\mathbf{v}|_{\partial\mathcal{D}} = \mathbf{0}$. Both null spaces $\mathbf{K}_{\rm null}$ and $\mathbf{K}'_{\rm null}$

have with respect to the velocity fields the finite dimension 3, independently of q and w. Let us also define

$$\mathcal{K}_{\text{null}} = \mathcal{K}_{\text{null},q,w} = \left\{[\mathbf{v}, \pi]\colon \mathbf{v} \in \mathbf{K}_{q,w} \text{ and } \pi \text{ is the pressure associate with } \mathbf{v}\right\}, \tag{3.169}$$

i.e.,

$$\mathcal{K}_{\text{null}} = \left\{[\mathbf{v}, \pi] \in \widehat{\mathbf{H}}^{2,q}_{w,\sigma}(\mathcal{D}) \times \widehat{H}^{1,q}_{w}(\mathcal{D})\colon S_{\omega,\nu}\mathbf{v} + \nabla\pi = \mathbf{0}, \mathbf{v}|_{\partial\mathcal{D}} = \mathbf{0}\right\}.$$

Definition 3.1 Let $1 < q < \infty$ and w be a radially symmetric q-weight function according to (2.15). Let $\mathbf{F} \in \widehat{\mathbf{H}}^{-2,q}_{w}(\mathcal{D})$ and $g \in \mathcal{H}^{-1,q}_{w,0}(\mathcal{D})$ be given. Then a function $\{\mathbf{u}, p\} \in \mathbf{L}^{q}_{w}(\mathcal{D}) \times \mathcal{H}^{-1,q}_{w,0}(\mathcal{D})$ is called the very weak solution to the system (3.167), if the equation

$$\langle \mathbf{u}, S'_{\omega,\nu}\boldsymbol{\varphi} - \nabla\psi\rangle_w = \langle \mathbf{F}, \boldsymbol{\varphi}\rangle_w + \langle G, \psi\rangle_w \tag{3.170}$$

holds for all test functions $\boldsymbol{\varphi} \in \mathbf{Y}^{2,q'}_{w',\sigma}(\mathcal{D})$ with $\mathcal{R}\boldsymbol{\varphi} \in \mathbf{L}^{q'}_{w'}(\mathcal{D})$ and $\psi \in \widehat{H}^{1,q'}_{w'}(\mathcal{D})$. Operator $\mathcal{R}$ always denotes $-(\mathbf{e}_3 \wedge \mathbf{x}) \cdot \nabla + \mathbf{e}_3 \wedge \cdot$. The associated pressure p is obtained by

$$\langle p, \operatorname{div}\boldsymbol{\varphi}\rangle_W = \langle u, S'_{\omega,\nu}\boldsymbol{\varphi}\rangle - \langle \mathbf{F}, \boldsymbol{\varphi}\rangle_w.$$

The pairing $\langle\cdot,\cdot\rangle_w$ denotes convenient dualities between $\mathbf{L}^{q}_{w}$ and $\mathbf{L}^{q'}_{w'}$, between $\mathbf{Y}^{-2,q}_{w,\sigma}$ and $\mathbf{Y}^{2,q'}_{w',\sigma}$, and between $\mathcal{H}^{-1,q}_{w,0}$ and $\widehat{H}^{1,q'}_{w'}$. □

Adding $S'_{\omega,\nu}\boldsymbol{\Phi} - \nabla\Psi = \mathbf{0}$ with $\{\boldsymbol{\Phi}, \Psi\} \in \mathcal{K}'_{\text{null}} \in$ (3.170), it follows that the compatibility condition for the existence of a very weak solution is $\langle \mathbf{F}, \boldsymbol{\Phi}\rangle_w + \langle G, \Psi\rangle_w = 0$ for all $[\boldsymbol{\Phi}, \psi] \in \mathcal{K}'_{\text{null}}$.

Remark 3.19 In the case of the homogenous Dirichlet boundary condition, functionals $\mathbf{F}$, G mean the functionals corresponding to the given functions $\mathbf{f}$, g. In this part we follow the way introduced by Schumacher [190, 191], which means that first we work in a formal way and in next part we introduce more regularity to get the sense of trace. It also gives the possibility to explain the dependence of $\mathbf{f}$, g (3.167) and $\mathbf{F}$, G (3.170) in the case of the nonhomogeneous Dirichlet boundary conditions. It is a little bit different approach than the works of Amann [5, 6] or Farwig et al. [64, 71, 79]. From [190, Remark 5.1] it follows that in the unweighted case the corresponding norm introduced through the Stokes operator [112] is equivalent to the norm introduced in [190] in the case of a bounded domain, and the results coincide. □

Now we are in a position to state the existence and uniqueness theorem for very weak solutions.

Theorem 3.11 *Let* $1 < q < \infty$ *and* w *be a radially symmetric* q*-weight function according to* (2.15). *Then, for every* $[\mathbf{F}, G] \in \Big(\widehat{\mathbf{H}}_w^{-2,q}(\mathcal{D}) \times \mathcal{H}_{w,0}^{-1,q}(\mathcal{D})\Big) \cap \mathcal{K}_{\text{null}}'^{\perp}$, *there exists a unique very weak solution* $\mathbf{u} \in \mathbf{L}_w^q(\mathcal{D})$ *in the sense of Definition* 3.1 *satisfying the estimate*

$$\|\mathbf{u}\|_{q,w} \le c\left(\|\mathbf{F}\|_{\mathbf{Y}_w^{-2,q}} + \|G\|_{\mathcal{H}_{w,0}^{-1,q}}\right), \tag{3.171}$$

where $c = c(\nu, \omega, \mathcal{D}, w, q)$. *Moreover, there exists a pressure functional* $p \in \mathcal{H}_{w,0}^{-1,q}(\mathcal{D})$ *such that* $(\mathbf{u}, p)$ *solves*

$$\langle \mathbf{u}, S'_{\omega,\nu}\boldsymbol{\varphi}\rangle_w - \langle p, \operatorname{div}\boldsymbol{\varphi}\rangle_w = \langle \mathbf{F}, \boldsymbol{\varphi}\rangle_w \textit{ for all } \boldsymbol{\varphi} \in \mathbf{Y}_{w'}^{2,q'}(\mathcal{D}),\ \mathcal{R}\boldsymbol{\varphi} \in \mathbf{L}_{w'}^{q'}(\mathcal{D}).$$

$\mathbf{u}$ *and* p *fulfill the inequality*

$$\|\mathbf{u}\|_{q,w} + \|p\|_{\mathcal{H}_{w,0}^{-1,q}} \le c\left(\|\mathbf{F}\|_{\mathbf{Y}_w^{-2,q}} + \|G\|_{\mathcal{H}_{w,0}^{-1,q}}\right), \tag{3.172}$$

$c = c(\nu, \omega, \mathcal{D}, w, q)$. □

Proof We know that w' is a radially symmetric q'-weight function. For every $\mathbf{v} \in \mathbf{L}_{w'}^{q'}(\mathcal{D})$, in accordance with Theorem 3.10, there exists a couple in $\widehat{\mathbf{H}}_{w',\sigma}^{2,q'}(\mathcal{D}) \times \widehat{H}_{w'}^{1,q'}(\mathcal{D})$, a strong solution to

$$S'_{\omega,\nu}\boldsymbol{\varphi} - \nabla\psi = \mathbf{v}, \quad \operatorname{div}\boldsymbol{\varphi} = 0, \quad \boldsymbol{\varphi}|_{\partial\mathcal{D}} = \mathbf{0}, \tag{3.173}$$

i.e., $[\boldsymbol{\varphi}, \psi] \in \mathbf{Y}_{w',\sigma}^{2,q'}(\mathcal{D}) \times \widehat{H}_{w'}^{1,q'}(\mathcal{D})$. Due to Proposition 3.7, let $[\boldsymbol{\varphi}_v, \psi_v]$ be a representative of $[\boldsymbol{\varphi}, \psi] + \mathcal{K}'_{\text{null}}$ which satisfies the estimate

$$\|\nu\nabla^2\boldsymbol{\varphi}_v\|_{q',w'} + \|\omega\mathcal{R}\boldsymbol{\varphi}_v\|_{q',w'} + \|\nabla\psi_v\|_{q',w'} \le c\|\mathbf{v}\|_{q',w'}.$$

We now define $\mathbf{u} := \mathbf{u}(\mathbf{F}, G)$ setting

$$\langle \mathbf{u}, \mathbf{v}\rangle_w := \langle \mathbf{F}, \boldsymbol{\varphi}_v\rangle_w + \langle G, \psi_v\rangle_w,$$

where $[\mathbf{F}, G] \in \Big(\widehat{\mathbf{H}}_w^{-2,q}(\mathcal{D}) \times \mathcal{H}_{w,0}^{-1,q}(\mathcal{D})\Big) \cap \mathcal{K}_{\text{null}}'^{\perp} \subset \mathbf{Y}_w^{-2,q}(D) \times \mathcal{H}_{w,0}^{-1,q}(\mathcal{D})$ are given. Then one has

$$\begin{aligned}
|\langle \mathbf{u}, \mathbf{v}\rangle_w| \le& |\langle \mathbf{F}, \boldsymbol{\varphi}_v\rangle_w + \langle G, \psi_v\rangle_w| \le \|\mathbf{F}\|_{\mathbf{Y}_w^{-2,q}}\|\nabla^2\boldsymbol{\varphi}_v\|_{q',w'} \\
&+ \|G\|_{\mathcal{H}_{w,0}^{-1,q}}\|\nabla\psi_v\|_{q',w'} \\
\le& c\left(\|\mathbf{F}\|_{\mathbf{Y}_w^{-2,q}} + \|G\|_{\mathcal{H}_{w,0}^{-1,q}}\right)\|\mathbf{v}\|_{q',w'}.
\end{aligned}$$

Thus $\mathbf{u} = \mathbf{u}(\mathbf{F}, G)$ belongs to $(\mathbf{L}_{w'}^{q'}(\mathcal{D}))' = \mathbf{L}_w^q(\mathcal{D})$ and fulfills the estimate stated.

It remains to show that this function $\mathbf{u}(\mathbf{F}, G)$ is a very weak solution of our problem. To see this, let $\boldsymbol{\varphi}$, ψ be the test functions with $\boldsymbol{\varphi} \in \mathbf{Y}^{2,q'}_{w',\sigma}(\mathcal{D})$, $\mathcal{R}\boldsymbol{\varphi} \in \mathbf{L}^{q'}_{w'}(\mathcal{D})$, and $\psi \in \widehat{H}^{1,q'}_{w'}(\mathcal{D})$ independently. We can choose $\mathbf{v} := S'_{\omega,\nu}\boldsymbol{\varphi} - \nabla\psi \in \mathbf{L}^{q'}_{w'}(\mathcal{D})$; then

$$\langle \mathbf{u}(\mathbf{F}, G), S'_{\omega,\nu}\boldsymbol{\varphi} - \nabla\psi\rangle_w = \langle \mathbf{u}(\mathbf{F}, G), \mathbf{v}\rangle_w = \langle \mathbf{F}, \boldsymbol{\varphi}\rangle_w + \langle G, \psi\rangle_w.$$

Therefore $\mathbf{u}(\mathbf{F}, G)$ is a very weak solution to our problem.

Note that the relation $\langle \mathbf{u}(\mathbf{F}, G), S'_{\omega,\nu}(\boldsymbol{\varphi}+\boldsymbol{\Phi})-\nabla(\psi+\Psi)\rangle_w = \langle \mathbf{F}, \boldsymbol{\varphi}\rangle_w+\langle G, \psi\rangle_w$ holds for all $\{\boldsymbol{\Phi}, \Psi\} \in \mathcal{K}'_{\text{null}}$. The compatibility condition being satisfied by $\mathbf{F}$ and G, we get $\langle \mathbf{F}, \boldsymbol{\varphi}\rangle_w + \langle G, \psi\rangle_w = \langle \mathbf{F}, \boldsymbol{\varphi} + \boldsymbol{\Phi}\rangle_w + \langle G, \psi + \Psi\rangle_w$; it means that the previous relation is in accord with the given definition of the very weak solution.

Applying the standard de Rham theorem or Bogovski lemma, see Lemma 2.7, we get the existence of pressure which satisfies the estimate (3.172).

To prove the uniqueness of $\mathbf{u} = \mathbf{u}(\mathbf{F}, G)$, let $\mathbf{U} = \mathbf{U}(\mathbf{F}, G) \in \mathbf{L}^q_w(\mathcal{D})$ be another very weak solution to the system (3.167). We have

$$\langle \mathbf{U}, S'_{\omega,\nu}\boldsymbol{\varphi} - \nabla\psi\rangle_w = \langle \mathbf{F}, \boldsymbol{\varphi}\rangle_w + \langle G, \psi\rangle_w \tag{3.174}$$

for all test functions $\boldsymbol{\varphi} \in \mathbf{Y}^{2,q'}_{w',\sigma}(\mathcal{D})$ with $\mathcal{R}\boldsymbol{\varphi} \in \mathbf{L}^{q'}_{w'}(\mathcal{D})$ and $\psi \in \widehat{H}^{1,q'}_{w'}(\mathcal{D})$. Thus

$$\langle \mathbf{U} - \mathbf{u}, S'_{\omega,\nu}\boldsymbol{\varphi} - \nabla\psi\rangle_w = 0.$$

Here the quantity $S'_{\omega,\nu}\boldsymbol{\varphi} - \nabla\psi$ is an arbitrary element from $\mathbf{L}^{q'}_{w'}(\mathcal{D})$, and it follows that $\mathbf{U} - \mathbf{u} = \mathbf{0}$ in $\mathbf{L}^q_w(\mathcal{D})$. □

3.4.2 *The Case of Nonhomogeneous Dirichlet Conditions: Generalized Very Weak Solution*

3.4.2.1 Introduction

We are interested in the case of nonhomogeneous Dirichlet boundary conditions $\mathbf{u}|_{\partial\mathcal{D}} = \boldsymbol{\omega} \wedge \mathbf{x} := \mathbf{h}$, where $\mathcal{D} \subset \mathbb{R}^3$ is always an exterior domain of type $C^{1,1}$. Let w be always a q-weight function according to (2.15). Now

$$\begin{aligned} S_{\omega,\nu}\mathbf{u} + \nabla p &= \mathbf{f} && \text{in } \mathcal{D}, \\ \operatorname{div}\mathbf{u} &= g && \text{in } \mathcal{D}, \\ \mathbf{u}|_{\partial\mathcal{D}} &= \boldsymbol{\omega} \wedge \mathbf{x} := \mathbf{h}. \end{aligned} \tag{3.175}$$

Remark 3.20 Let us mention that $\omega \mathbf{e}_3 \wedge \mathbf{x}$ is solenoidal. The boundary condition follows from the fact that we consider a purely rotating body. □

To define the very weak solution in the case of nonhomogeneous boundary conditions—the so-called **generalized very weak solution,** a special class of the functionals $\{\mathbf{F}, G\} = \{\mathbf{F}(\mathbf{f}, \mathbf{h}), G(g, \mathbf{h})\}$ is needed. This will be described later.

Definition 3.2 Let $1 < q < \infty$ and w be a radially symmetric q-weight function according to (2.15). Let $\{\mathbf{F}, G\} = \{\mathbf{F}(\mathbf{f}, \mathbf{h}), G(g, \mathbf{h})\} \in \widehat{\mathbf{H}}_w^{-2,q}(\mathcal{D}) \times \mathcal{H}_{w,0}^{-1,q}(\mathcal{D})$ be given. We call $\{\mathbf{u}, p\} \in \mathbf{L}_w^q(\mathcal{D}) \times \mathcal{H}_{w,0}^{-1,q}(\mathcal{D})$ a generalized very weak solution to the system (3.175), if the equation

$$\langle \mathbf{u}, S'_{\omega,\nu}\boldsymbol{\varphi} - \nabla\psi\rangle_w = \langle \mathbf{F}(\mathbf{f}, h), \boldsymbol{\varphi}\rangle_w + \langle G(g, h), \psi\rangle_w, \tag{3.176}$$

where

$$\langle \mathbf{F}, \boldsymbol{\varphi}\rangle_w = \langle \mathbf{f} - \widetilde{\mathbf{F}}_h, \boldsymbol{\varphi}\rangle_w$$
$$\langle G, \psi\rangle_w = \langle g - \widetilde{G}_h, \psi\rangle_w,$$

holds for all test functions $\boldsymbol{\varphi} \in \mathbf{Y}_{w',\sigma}^{2,q'}(\mathcal{D})$ with $\mathcal{R}\boldsymbol{\varphi} \in \mathbf{L}_{w'}^{q'}(\mathcal{D})$ and $\psi \in \widehat{H}_{w'}^{1,q'}(\mathcal{D})$. The associated pressure p is obtained by

$$\langle p, \operatorname{div}\boldsymbol{\varphi}\rangle_W = \langle u, S'_{\omega,\nu}\boldsymbol{\varphi}\rangle - \langle \mathbf{F}, \boldsymbol{\varphi}\rangle_w.$$

The pairing $\langle\cdot,\cdot\rangle_w$ denotes convenient dualities between $\mathbf{L}_w^q$ and $\mathbf{L}_{w'}^{q'}$, between $\mathbf{Y}_{w,\sigma}^{-2,q}$ and $\mathbf{Y}_{w',\sigma}^{2,q'}$, and between $\mathcal{H}_{w,0}^{-1,q}$ and $\widehat{H}_{w'}^{1,q'}$. □

For more details on the definition of traces $\mathbf{T}_w^{k,q}(\partial D)$, see [152].

The next two theorems represent the main results of this part.

Theorem 3.12 *Let the data* $\mathbf{F}$, G *be regular enough to justify that, in the decompositions given by Proposition* 3.8, *smooth functions* $\mathbf{f} \in \mathbf{L}_w^q(\mathcal{D})$ *and* $g \in \widehat{H}_w^{1,q}(\mathcal{D})$ *exist. Let* $\mathbf{u}$ *be given as a very weak solution to problem* (3.175) *with* $\mathbf{h} \in \mathbf{T}_w^{2,q}(\partial\mathcal{D})$, *where* $\mathbf{F}$ *and* G *satisfy the compatibility conditions. Then, this very weak solution* $\mathbf{u}$ *to the nonhomogeneous modified Stokes system, whose data are* $\mathbf{F}$ *and* G, *coincides with the strong solution of the nonhomogeneous modified Stokes system data* $\mathbf{f}$, g, *and* $\mathbf{h}$. *In particular,* $\mathbf{u} \in \widehat{\mathbf{H}}_w^{2,q}(\mathcal{D})$ *and*

$$\|\nu\nabla^2\mathbf{u}\|_{q,w} + \|\boldsymbol{\omega}\mathcal{R}\mathbf{u}\|_{q,w} \le c(\|\mathbf{f}\|_{q,w} + \|g\|_{\widehat{H}_w^{1,q}} + \|\mathbf{h}\|_{T_w^{2,q}}), \tag{3.177}$$

where $c = c(D, q, w)$. □

Theorem 3.13 *Let* $1 < q < \infty$ *and* w *be a* q*-weight function that fulfills* (2.15). *Assume that*

$$[\mathbf{F}, G] \in \left(\widehat{\mathbf{H}}_w^{-2,q}(\mathcal{D}) \times \mathcal{H}_{w,0}^{-1,q}(\mathcal{D})\right) \cap \mathcal{K}_{\text{null}}'^{\perp}, \quad \textit{and} \quad \mathbf{h} \in \mathbf{T}_w^{0,q}(\partial\mathcal{D}).$$

Then there exists a unique very weak solution $\mathbf{u} \in \mathbf{L}^q_w(\mathcal{D})$ *to the problem* (3.175). *The associated pressure is obtained in* $\mathcal{H}^{-1,q}_{w,0}(\mathcal{D})$ *by*

$$\langle p, \operatorname{div} \boldsymbol{\varphi} \rangle_w = \langle \mathbf{u}, S'_{\omega,\nu} \boldsymbol{\varphi} \rangle_w - \langle \mathbf{F}, \boldsymbol{\varphi} \rangle_w, \quad \forall\, \boldsymbol{\varphi} \in \mathbf{Y}^{2,q'}_{w'}(\mathcal{D}).$$

These functions $\mathbf{u}$ *and* p *satisfy the estimate*

$$\|\mathbf{u}\|_{q,w} + \|p\|_{\mathcal{H}^{-1,q}_{w,0}} \leq c \left(\|\mathbf{F}\|_{\widehat{H}^{-2,q}_w} + \|G\|_{\mathcal{H}^{-1,q}_{w,0}} + \|\mathbf{h}\|_{T^{0,q}_w} \right),$$

where $c = c(\nu, \omega, \mathcal{D}, w, q)$. □

Properties of F, G

It is easily seen that if $\mathbf{h} \in \mathbf{T}^{0,q}_w(\partial\mathcal{D})$, then the functionals "supported by the boundary"

$$\widetilde{\mathbf{F}}_h : \boldsymbol{\varphi} \to \nu \langle \mathbf{h}, (\mathbf{N} \cdot \nabla) \boldsymbol{\varphi}|_{\partial\mathcal{D}} \rangle_{w,\partial\mathcal{D}}$$

and

$$\widetilde{G}_h : \psi \to \langle \mathbf{h}, \psi|_{\partial\mathcal{D}}\, \mathbf{N} \rangle_{w,\partial\mathcal{D}}$$

belong to $\left(\widehat{\mathbf{H}}^{2,q'}_{w'}(\mathcal{D})\right)'$ and $\left(\widehat{H}^{1,q'}_{w'}(\mathcal{D})\right)'$, respectively. In order to solve the system (3.175) "very weakly," we must introduce appropriate functionals $\mathbf{F} = \mathbf{F}(\mathbf{f}, \mathbf{h})$ and $G = G(g, \mathbf{h})$ with firstly the properties $\mathbf{F}(\mathbf{f}, \mathbf{0}) = \mathbf{f}$ and $G(g, \mathbf{0}) = g$ and secondly such that, for each $\boldsymbol{\varphi} \in \mathbf{Y}^{2,q'}_{w',\sigma}(\mathcal{D})$, $\mathcal{R}\boldsymbol{\varphi} \in \mathbf{L}^{q'}_{w'}(\mathcal{D})$ and $\psi \in \widehat{H}^{1,q'}_{w'}(\mathcal{D})$, the relations

$$\langle \mathbf{u}, S'_{\omega,\nu} \boldsymbol{\varphi} \rangle_w = \langle \mathbf{F}, \boldsymbol{\varphi} \rangle_w \quad \text{and} \quad - \langle \mathbf{u}, \nabla \psi \rangle_w = \langle G, \psi \rangle_w \tag{3.178}$$

give us a very weak solution.

Applying formally Green's formula, we have

$$\langle \mathbf{u}, S'_{\omega,\nu} \boldsymbol{\varphi} \rangle_w = \langle S_{\omega,\nu} \mathbf{u}, \boldsymbol{\varphi} \rangle_w - \nu \langle \mathbf{u}|_{\partial\mathcal{D}}, (\mathbf{N} \cdot \nabla) \boldsymbol{\varphi} \rangle_{w,\partial\mathcal{D}}. \tag{3.179}$$

Using the boundary condition "$\mathbf{u}|_{\partial\mathcal{D}} = \mathbf{h}$," we derive the definition of $\mathbf{F}$

$$\langle \mathbf{u}, S'_{\omega,\nu} \boldsymbol{\varphi} \rangle_w = \langle \mathbf{F}, \boldsymbol{\varphi} \rangle_w := \langle \mathbf{f} - \widetilde{\mathbf{F}}_h, \boldsymbol{\varphi} \rangle_w. \tag{3.180}$$

Applying now formally the Stokes–Ostrogradski formula, we similarly derive the definition of G

$$- \langle \mathbf{u}, \nabla \psi \rangle_w = \langle \operatorname{div} \mathbf{u}, \psi \rangle_w - \langle \mathbf{u}|_{\partial\mathcal{D}}, \psi|_{\partial\mathcal{D}}\, \mathbf{N} \rangle_{w,\partial\mathcal{D}} = \langle G, \psi \rangle_w := \langle g - \widetilde{G}_h, \psi \rangle_w. \tag{3.181}$$

Remark 3.21 The functionals $\mathbf{F}$ and G are a priori so irregular, and it is difficult to distinguish between boundary values and forces. □

The following assertion is important in this respect.

Proposition 3.8 *The decompositions* $\mathbf{F} = \mathbf{f} - \widetilde{\mathbf{F}}_h$ *and* $G = g - \widetilde{G}_h$ *are uniquely defined by the data* $\mathbf{F} \in \widehat{\mathbf{H}}_w^{-2,q}(\mathcal{D})$, $G \in \mathcal{H}_{w,0}^{-1,q}(\mathcal{D})$, *and* $\mathbf{h} \in \mathbf{T}_w^{0,q}(\partial\mathcal{D})$. □

Proof Let us assume that there exist two decompositions

$$\langle \mathbf{F}, \boldsymbol{\varphi}\rangle_w = \langle \mathbf{f}_i, \boldsymbol{\varphi}\rangle_w - \nu\langle \mathbf{h}_i, (\mathbf{N}\cdot\nabla)\boldsymbol{\varphi}\rangle_{w,\partial\mathcal{D}} \quad \text{for } i = 1, 2.$$

So we have

$$\langle \mathbf{f}_1 - \mathbf{f}_2, \boldsymbol{\varphi}\rangle_w - \nu\langle \mathbf{h}_1 - \mathbf{h}_2, (\mathbf{N}\cdot\nabla)\boldsymbol{\varphi}|_{\partial\mathcal{D}}\rangle_{w,\partial\mathcal{D}} = 0 \quad \text{for } \boldsymbol{\varphi} \in \mathbf{Y}_{w'}^{2,q'}(\mathcal{D}).$$

The previous functional $\langle \mathbf{f}_1 - \mathbf{f}_2, \cdot\rangle_w$ vanishes on $\mathbf{C}_0^\infty(\mathcal{D})$: It follows that $\mathbf{f}_1 - \mathbf{f}_2 = \mathbf{0}$ as a distribution on the domain D. Consequently, for every $\boldsymbol{\varphi}$, $\langle \mathbf{h}_1 - \mathbf{h}_2, (\mathbf{N}\cdot\nabla)\boldsymbol{\varphi}|_{\partial D}\rangle_{w,\partial\mathcal{D}} = 0$. The mapping defining the normal trace of $\nabla\boldsymbol{\varphi}$ on $\partial\mathcal{D}$ from $\widehat{\mathbf{H}}_{w'}^{2,q'}(\mathcal{D})$ to $\mathbf{T}_{w'}^{1,q'}(\partial\mathcal{D})$ is surjective; hence $\mathbf{h}_1 - \mathbf{h}_2 = \mathbf{0}$.

Similarly, one can verify the uniquely defined decomposition for G. The functionals $\mathbf{F}$ and G can be in smaller spaces, like $\widehat{\mathbf{H}}_w^{-1,q}(\mathcal{D})$ or maybe $\mathbf{L}_w^q(\mathcal{D})$, and $L_w^q(\mathcal{D})$ or maybe $H_w^{1,q}(\mathcal{D})$, respectively. □

Proof of Theorem 3.12 There exists an extension of $\mathbf{h} \in \mathbf{T}_w^{2,q}(\partial\mathcal{D})$, say $\mathbf{v}_{1h}$, which satisfies $\mathbf{v}_{1h} \in \widehat{\mathbf{H}}_w^{2,q}(\mathcal{D})$, $\mathbf{v}_{1h}|_{\partial\mathcal{D}} = \mathbf{h}$, and $\|\mathbf{v}_{1h}\|_{\widehat{H}_w^{2,q}} \le c\|\mathbf{h}\|_{T_w^{2,q}}$. Then we can solve $\operatorname{div}\mathbf{b}_1 = g - \operatorname{div}\mathbf{v}_{1h}$, and by Lemma 2.7 (ii) and Observation 2.1 there exists $\mathbf{b}_1 \in \widehat{\mathbf{H}}_{w,0}^{2,q}(\mathcal{D})$ such that $\|\mathbf{b}_1\|_{\widehat{H}_w^{2,q}} \le c(\|g\|_{\widehat{H}_w^{1,q}} + \|\mathbf{h}\|_{T_w^{2,q}})$, $c = c(\mathcal{D}, q, w)$.

By Theorem 3.10, there exists a strong solution $[\mathbf{u}_1, p_1] \in \widehat{\mathbf{H}}_w^{2,q}(\mathcal{D}) \times \widehat{H}_w^{1,q}(\mathcal{D})$ to the following homogeneous modified Stokes system:

$$\begin{aligned} S_{\omega,\nu}\mathbf{u}_1 + \nabla p_1 &= -S_{\omega,\nu}(\mathbf{b}_1 + \mathbf{v}_{1h}) \quad \text{in } \mathcal{D},\\ \operatorname{div}\mathbf{u}_1 &= 0 \qquad \text{in } \mathcal{D},\\ \mathbf{u}_1|_{\partial D} &= \mathbf{0}, \end{aligned} \tag{3.182}$$

which satisfies the estimate (3.151).

It is now clear that

$$\begin{aligned} S_{\omega,\nu}(\mathbf{u}_1 + \mathbf{b}_1 + \mathbf{v}_{1h}) + \nabla p_1 &= \mathbf{0} \quad \text{in } \mathcal{D},\\ \operatorname{div}(\mathbf{u}_1 + \mathbf{b}_1 + \mathbf{v}_{1h}) &= g \quad \text{in } \mathcal{D},\\ (\mathbf{u}_1 + \mathbf{b}_1 + \mathbf{v}_{1h})|_{\partial\mathcal{D}} &= \mathbf{h}; \end{aligned} \tag{3.183}$$

therefore $\mathbf{v}_1 := \mathbf{u}_1 + \mathbf{b}_1 + \mathbf{v}_{1h} \in \widehat{\mathbf{H}}_w^{2,q}(\mathcal{D})$ solves strongly the previous modified Stokes system (3.183) and satisfies the inequality

$$\|\nu\nabla^2\mathbf{v}_1\|_{q,w} + \|\boldsymbol{\omega}\mathcal{R}\mathbf{v}_1\|_{q,w} \le c\left(\|g\|_{\widehat{H}_w^{1,q}} + \|\mathbf{h}\|_{T_w^{2,q}}\right).$$

Finally, we have another strong solution $(\mathbf{v}_2, p_2)$ of the modified Stokes system

$$\begin{aligned} S_{\omega,\nu}\mathbf{v}_2 + \nabla p_2 &= \mathbf{f} \quad \text{in } \mathcal{D},\\ \operatorname{div}\mathbf{v}_2 &= 0 \quad \text{in } \mathcal{D},\\ \mathbf{v}_2|_{\partial D} &= \mathbf{0} \end{aligned} \tag{3.184}$$

with $\|\nu\nabla^2\mathbf{v}_2\|_{q,w} + \|\boldsymbol{\omega}\mathcal{R}\mathbf{v}_2\|_{q,w} \le c\|\mathbf{f}\|_{q,w}$.

Thus $\mathbf{u} := \mathbf{v}_1 + \mathbf{v}_2$ and $p := p_1 + p_2$ solve strongly the system

$$\begin{aligned} S_{\omega,\nu}\mathbf{u} + \nabla p &= \mathbf{f} \quad \text{in } \mathcal{D},\\ \operatorname{div}\mathbf{u} &= g \quad \text{in } \mathcal{D},\\ \mathbf{u}|_{\partial\mathcal{D}} &= \mathbf{h} \end{aligned} \tag{3.185}$$

with the estimate

$$\|\nu\mathbf{u}\|_{\widehat{H}_w^{2,q}} + \|\boldsymbol{\omega}\mathcal{R}\mathbf{u}\|_{\widehat{H}_w^{1,q}} \le c\left(\|\mathbf{f}\|_{q,w} + \|g\|_{\widehat{H}_w^{1,q}} + \|\mathbf{h}\|_{T_w^{2,q}}\right), \quad c = c(\mathcal{D}, w, q). \tag{3.186}$$

By means of the Stokes–Ostrogradski and Green formulas, we can easily verify that the strong solutions $\mathbf{v}_1$ and $\mathbf{v}_2$ are also very weak solutions of related modified Stokes models: Indeed, the relations $-\langle\mathbf{v}_1, \nabla\psi\rangle_w = \langle g - \widetilde{G}_h, \psi\rangle_w = \langle G, \psi\rangle_w$, $0 = \langle\mathbf{v}_1, S'_{\omega,\nu}\boldsymbol{\varphi}\rangle_w + \langle\widetilde{\mathbf{F}}_h, \boldsymbol{\varphi}\rangle_w$, $-\langle\mathbf{v}_2, \nabla\psi\rangle_w = 0$, and $\langle\mathbf{f}, \boldsymbol{\varphi}\rangle_w = \langle\mathbf{v}_2, S'_{\omega,\nu}\boldsymbol{\varphi}\rangle_w$ hold. Adding the respective formulas, we conclude that

$$\langle\mathbf{u}, S'_{\omega,\nu}\boldsymbol{\varphi}\rangle_w = \langle\mathbf{F}, \boldsymbol{\varphi}\rangle_w \quad \text{and} \quad -\langle\mathbf{u}, \nabla\psi\rangle_w = \langle G, \psi\rangle_w.$$

These equations hold for $\boldsymbol{\varphi} \in \mathbf{Y}^{2,q'}_{w',\sigma}$ with $\mathcal{R}\boldsymbol{\varphi} \in \mathbf{L}^{q'}_{w'}$ and $\psi \in \widehat{H}^{1,q'}_{w'}$. Then $\mathbf{u}$ is also a very weak solution to the initial system (3.175). □

3.4.3 Proof of Theorem 3.13

Remark 3.22 Under the assumptions, $\mathbf{f} \in \mathbf{L}_w^q(\mathcal{D})$ and $g \in \widehat{H}_w^{1,q}(\mathcal{D})$, we have proved a result on the regularity of the obtained very weak solution (which is unique as soon as $[\mathbf{F}, G] \in \left(\widehat{\mathbf{H}}_w^{-2,q}(\mathcal{D}) \times \mathcal{H}_{w,0}^{-1,q}(\mathcal{D})\right) \cap \mathcal{K}'^{\perp}_{\text{null}}$, see Theorem 3.11). □

In agreement with Proposition 3.8, assuming

$$[\mathbf{F}, G] \in \left(\widehat{\mathbf{H}}_w^{-2,q}(\mathcal{D}) \times \mathcal{H}_{w,0}^{-1,q}(\mathcal{D})\right) \cap \mathcal{K}_{\text{null}}'^{\perp}$$

and $\mathbf{h} \in \mathbf{T}_w^{0,q}(\partial D)$, it is clear that $\mathbf{v}_1$ and $\mathbf{v}_2$ as very weak solutions of the related modified Stokes systems (3.183) and (3.184) from the proof of Theorem 3.12 satisfy the following estimates:

$$\|\mathbf{v}_1\|_{q,w} + \|p_1\|_{\mathcal{H}_{w,0}^{-1,q}} \leq c(\|g\|_{\mathcal{H}_{w,0}^{-1,q}} + \|\mathbf{h}\|_{T_w^{0,q}}) \leq c(\|G\|_{\mathcal{H}_{w,0}^{-1,q}} + \|\mathbf{h}\|_{T_w^{0,q}})$$

$$\|\mathbf{v}_2\|_{q,w} + \|p_2\|_{\mathcal{H}_{w,0}^{-1,q}} \leq c\|\mathbf{f}\|_{\widehat{H}_w^{-2,q}} \leq c\|\mathbf{F}\|_{\widehat{H}_w^{-2,q}},$$

where $c = c(\nu, \omega, \mathcal{D}, w, q)$.

So we have proved Theorem 3.13.

Remark 3.23 From Theorem 3.12 it follows that in the case that data $\mathbf{F}$, G are regular enough then our very weak solution coincides with the strong solution, and this gives us right to define the notion of the trace on ∂D.

Another possibility can be done following the ideas of Schumacher [190]:

We can define the neighborhood of the boundary in the following way: Let $R > 0$ be such that $\mathcal{D}^c \subset B_R$. Let $U = \mathcal{D} \cap B_R$. So, we have $\partial U = \partial B_R \cup \partial D$. We will turn our attention to the restriction to a fixed Lipschitz domain U.

We choose $1 < r < \infty$ and a weight function $\tilde{w} \in A_r$ satisfying condition (2.4) such that

$$H_{w'}^{1,q'}(U) \hookrightarrow L_{\tilde{w}'}^{r'}(U) \hookrightarrow L_{w'}^{g'}(U). \tag{3.187}$$

Then

$$\mathbf{Y}_{w'}^{2,q'}(U) \subset \widehat{\mathbf{H}}_{\tilde{w}}^{1,r}(U) \quad \text{and} \quad L_{\tilde{w}}^{r'}(U) \subset \mathcal{H}_{w,0}^{-1,q}(U).$$

Using these embeddings it can be shown the next theorem.

Theorem 3.14 *Let* $\mathbf{F} \in (\mathbf{Y}_{w'}^{2,q'}(U))'$ *and* $G \in \mathcal{H}_{w,0}^{-1,q}(U)$ *allow a decomposition into*

$$\langle \mathbf{F}, \boldsymbol{\varphi}\rangle_w = \langle \mathbf{f}, \boldsymbol{\varphi}\rangle_w - \nu\langle \mathbf{h}, \mathbf{N}\cdot\nabla\boldsymbol{\varphi}\rangle_{w,\partial\mathcal{D}} \tag{3.188}$$

$$\textit{for all } \boldsymbol{\varphi} \in \mathbf{Y}_{w'}^{2,q'}(U),\ \boldsymbol{\varphi} = \mathbf{0} \textit{ on } \partial\mathcal{D} \cup \partial \mathrm{B}_R,$$

$$\langle G, \psi\rangle_w = \langle g, \psi\rangle_w - \langle \mathbf{h}, \mathbf{N}\cdot\psi\rangle_{w,\partial\mathcal{D}} \quad \textit{for all } \psi \in \widehat{H}_{w'}^{1,q'}(U)$$

with $\mathbf{h} \in \mathbf{T}_w^{0,q}(\partial U)$, $\mathbf{f} \in \mathbf{H}_{\tilde{w}}^{-1,r}(U)$, $g \in L_{\tilde{w}}^r(U)$, *where* $1 < r < \infty$ *and* $\tilde{w} \in A_r$ *are chosen according to* (3.187)$_1$.

Then:

- *The decomposition is uniquely determined by* $\mathbf{F}$ *and* G.
- *Every strong solution to the "rotating" Stokes problem corresponding to the data* $\mathbf{h} \in \mathbf{T}^{2,q}_w(\partial U)$, $\mathbf{f} \in \mathbf{L}^q_w(U)$, *and* $g \in \widehat{H}^{1,q}_w(U)$ *is a very weak solution corresponding to the data* $\mathbf{F}$ *and* G *and satisfying* (3.188).
- *Let* $\mathbf{h} \in \mathbf{T}^{2,q}_w(\partial U)$, $\mathbf{f} \in \mathbf{L}^q_w(U)$, *and* $g \in \widehat{H}^{1,q}_w(U)$ *with* $\int_{\mathcal{D}} g d\mathbf{x} = \int_{\partial\mathcal{D}} \mathbf{N} \cdot G dS$; *then the very weak solution* $\mathbf{u}$ *with respect to* $\mathbf{F}$ *and* G *is a strong solution with respect to* $\mathbf{f}$, g, *and* $\mathbf{h}$.

□

Note to the Problem of Introducing the Trace Operator

We will give more details on the approach introduced in the paper of Schumacher [190] and mentioned in Remark 3.23. Denoting by

$$\widetilde{\Delta \mathbf{u}} \text{ the extension of } \Delta \mathbf{u}|_{C_0^\infty(U)} \text{ to the dual space of } \widehat{\mathbf{H}}^{1,r'}_{\tilde{w}}(U), \tag{3.189}$$

i.e., $\mathcal{H}^{-1,r}_{\tilde{w},0}(U)$, we define the space

$$\widetilde{W}^{q,r}_{w,\tilde{w}}(U) := \{\mathbf{u} \colon \mathbf{u} \in \mathbf{L}^q_w(U),\ \widetilde{\Delta \mathbf{u}} \in \mathcal{H}^{-1,r}_{\tilde{w},0}(U)\},$$

which is a Banach space (see [190, Lemma 5.2]) equipped with the norm

$$\|\mathbf{u}\|_{\widetilde{W}^{q,r}_{w,\tilde{w}}} = \|\mathbf{u}\|_{q,r} + \|\widetilde{\Delta \mathbf{u}}\|_{\mathcal{H}^{-1,r}_{\tilde{w},0}(U)}.$$

Lemma 3.11 *Let* U *be a* $C^{1,1}$*-domain,* $1 < q < \infty$, *and* $w \in A_q$. *For every* $\mathbf{l} \in \mathbf{T}^{1,q}_w(\partial U)$ *with* $\mathbf{N} \cdot \mathbf{l} = 0$, *there exist functions* $\boldsymbol{\varphi}_l \in \widehat{\boldsymbol{H}}^{2,q}_w(U)$, *such that*

$$\operatorname{div} \boldsymbol{\varphi}_l = 0, \quad \boldsymbol{\varphi}_l|_{\partial U} = 0, \quad \text{and} \quad \mathbf{N} \cdot \nabla \boldsymbol{\varphi}_l = \mathbf{l}. \tag{3.190}$$

Moreover, $\boldsymbol{\varphi}_l$ *depends linearly on* $\mathbf{l}$ *and can be chosen to fulfill the estimate*

$$\|\nabla^2 \boldsymbol{\varphi}_l\|_{q,w} \le c \|\mathbf{l}\|_{\mathbf{T}^{1,q}_w(\partial U)} \tag{3.191}$$

with a constant $c = c(U, q, w)$. □

Proof See [190, Lemma 5.5]. □

Using this lemma, we define the tangential component of $\mathbf{u} \in \widetilde{W}^{q,r}_{w,\tilde{w}}$ as follows. If $\mathbf{u} \in \widetilde{W}^{q,r}_{w,\tilde{w}}$ and $\boldsymbol{\varphi} \in \mathbf{Y}^{2,q'}_{w',\sigma}$, we can write

$$\langle \Delta_\sigma \mathbf{u}, \boldsymbol{\varphi} \rangle := \lim_{i \to \infty} \langle \mathbf{u}, \Delta \boldsymbol{\varphi}_i \rangle, \tag{3.192}$$

where $(\boldsymbol{\varphi}_i)_i \in \mathbf{C}^\infty_{0,\sigma}$ converges to $\boldsymbol{\varphi}$ in $\mathbf{H}^{1,r'}_{\tilde{w}',0,\sigma}$, see [190, Lemma 5.1].

Lemma 3.12 *Let* $\mathbf{l} \in \mathbf{T}^{1,q'}_{w'}(\partial U)$ *and* $\mathbf{u} \in \widetilde{W}^{q,r}_{w,\tilde{w}}(U)$. *There exists a continuous linear operator*

$$\left.\begin{array}{l}\gamma\colon \widetilde{W}^{q,r}_{w,\tilde{w}}(U) \mapsto \mathbf{T}^{0,q}_{w}(\partial U) \text{ such that}\\ \langle \gamma(\mathbf{u}), \mathbf{l}\rangle_{w,\partial U} = \langle \mathbf{u}, \Delta\boldsymbol{\varphi}_l\rangle_w - \langle \Delta_\sigma \mathbf{u}, \boldsymbol{\varphi}_l\rangle_w \quad \text{if } \mathbf{N}\cdot\mathbf{l} = 0,\\ \langle \gamma(\mathbf{u}), \mathbf{l}\rangle_{w,\partial U} = 0 \quad \text{if } \mathbf{l} = \tilde{h}\mathbf{N},\end{array}\right\} \tag{3.193}$$

where $\boldsymbol{\varphi}_l$ *is given by the previous lemma, for some scalar-valued* $\tilde{h} \in T^{1,q'}_{w'}$. *Moreover, this tangential trace is independent of the choice of the extension* $\boldsymbol{\varphi}_l$ *and coincides with the tangential component of the usual restriction* $\mathbf{u} \in \mathbf{C}^\infty(\overline{U})$ *on* ∂U.

Proof See [190, Theorem 5.2]. □

For $\mathbf{u} \in \widetilde{W}^{q,r}_{w,\tilde{w}}(U)$ the tangential trace $\gamma(\mathbf{u})_\tau$ is given for all $\boldsymbol{\varphi} \in \mathbf{Y}^{2,q'}_{w',\sigma}(U)$ by

$$\langle \gamma(\mathbf{u})_\tau, \mathbf{N}\cdot\nabla\boldsymbol{\varphi}\rangle_{w,\partial U} = \langle \mathbf{u}, \Delta\boldsymbol{\varphi}\rangle_w - \langle \Delta\mathbf{u}, \boldsymbol{\varphi}\rangle_w.$$

Let $\mathbf{u} \in E^{q,r}_{w,\tilde{w}}(U) := \{\mathbf{v}\colon \mathbf{v} \in \mathbf{L}^q_w(U), \operatorname{div}\mathbf{v} \in \mathbf{L}^r_{\tilde{w}}(U)\}$. We define the normal component of the trace $\gamma(\mathbf{u})_\mathbf{N}$ as in the classical situation using Green's formula

$$\langle \gamma(\mathbf{u})_\mathbf{N}, \mathbf{N}\cdot\mathbf{v}\rangle_{w,\partial U} := \langle \operatorname{div}\mathbf{u}, \mathbf{v}\rangle_w + \langle \mathbf{u}, \nabla\mathbf{v}\rangle_w. \tag{3.194}$$

This gives us a functional in $\mathbf{T}^{0,q}_w$ because for every $\boldsymbol{\xi} \in \mathbf{T}^{1,q'}_{w'}(\partial U)$ there exists $\mathbf{v} \in \mathbf{H}^{1,q}_w(U)$ such that

$$\mathbf{v}|_{\partial U} = \boldsymbol{\xi} \quad \text{and} \quad \|\mathbf{v}\|_{1,q',w'} \le c\|\boldsymbol{\xi}\|_{\mathbf{T}^{1,q'}_{w'}}.$$

Since there exists $r \in (1,\infty)$ such that $\mathbf{H}^{1,q}_w(U) \hookrightarrow \mathbf{H}^{1,r}(U)$, then that mapping $\mathbf{u} \to \mathbf{N}\cdot\mathbf{u}|_{\partial U}\colon E^{q,r}_{w,\tilde{w}}(U) \to \mathbf{T}^{0,q}_w(\partial U)$ is continuous. So, for $\mathbf{u} \in \widetilde{W}^{q,r}_{w,\tilde{w}}(U) \cap E^{q,r}_{w,\tilde{w}}(U)$ we can write $\mathbf{u}|_{\partial U} = \mathbf{h}$ if

$$\langle \gamma(\mathbf{u}), \mathbf{l}\rangle = \langle \mathbf{h}, \mathbf{l}\rangle \quad \text{for all } \mathbf{l} \in \mathbf{T}^{1,q'}(\partial U) \text{ with } \mathbf{l}\cdot\mathbf{N} = 0 \text{ and } \mathbf{u}\cdot\mathbf{N}|_{\partial U} = \mathbf{h}\cdot\mathbf{N}.$$

For $\mathbf{u} \in \widetilde{W}^{q,r}_{w,\tilde{w}}(U) \cap E^{q,r}_{w,\tilde{w}}(U)$, we finally define the operator $\mathbf{u} \to \mathbf{h}$.

3.5 Notes to the Stokes Type Problem

3.5.1 Concluding Remarks to Previous Sections

- As we have already mentioned in the Introduction of Chap. 3, Sect. 3.1, we explain two different approaches in L^q setting for the strong solution:
  - An approach introduced by Farwig et al. [70]
  - An approach introduced by Galdi and Kyed [100] for the strong solution in L^q setting □
- After that we have investigated in Sect. 3.2 the weak solution in L^q setting using approach by Hishida [127] in the whole space, and by localization techniques we have extended into the exterior.
- Section 3.3 is devoted to the theory of the strong solution in L^q setting with Muckenhoupt weight w. Such a theory was proven by Farwig, Krbec, and Nečasová in the whole space [73] and by Nečasová and Schumacher [177] also in the exterior domain.
- The very weak solution is studied in Sect. 3.4 as a dual to the strong solution is investigated in the whole space and also in the exterior domain [152].

There are also results by Abada et al. on the case of the whole space, see [1]. Moreover, concerning time-periodic Stokes type case see the work of Eiter [54].

3.5.2 Asymptotic Behavior

Let us briefly mention results achieved by Farwig and Hishida [68] concerning the asymptotic behavior of the Stokes problem. A very important part is an explicit representation of the fundamental solution, which is related to the profile of flow, and it follows immediately from the theory introduced by Farwig et al. [70]. We will give a summary from the previous sections.

We introduce the fundamental solution of the usual Stokes problem

$$
\begin{aligned}
E_{st}(\mathbf{x}) &= \frac{1}{8\pi}\left(\frac{1}{|\mathbf{x}|}\mathcal{I} + \frac{\mathbf{x}\otimes\mathbf{x}}{|\mathbf{x}|^3}\right),\\
P_{st}(\mathbf{x}) &= \nabla\left(\frac{-1}{4\pi|\mathbf{x}|}\right).
\end{aligned}
\tag{3.195}
$$

We call the pair of a 3×3 matrix Σ_{st} and a column vector Q_{st} as a fundamental solution of the system

$$
S_{\omega,1}\mathbf{u} + \nabla p = \mathbf{f}, \quad \operatorname{div}\mathbf{u} = 0 \text{ in } \mathbb{R}^3 \tag{3.196}
$$

if the volume potentials

$$\mathbf{u}(\mathbf{x}) = \int_{\mathbb{R}^3} \Sigma_{st}(\mathbf{x}, \mathbf{y})\mathbf{f}(\mathbf{y})d\mathbf{y}, \; p(\mathbf{x}) = \int_{\mathbb{R}^3} Q_{st}(\mathbf{x}, \mathbf{y})\mathbf{f}(\mathbf{y})d\mathbf{y} \tag{3.197}$$

solve (3.196) for all $\mathbf{f} \in C_0^\infty(\mathbb{R}^3)$.

As we have already explained in the beginning of Chap. 3 that applying the div on $S_{\omega,1}\mathbf{u} + \nabla p = \mathbf{f}$, we get $\Delta p =$ div $\mathbf{f}$. It implies that pressure can be solved as

$$Q_{st} = \nabla_y \frac{1}{4\pi |\mathbf{x} - \mathbf{y}|} = P_{st}(\mathbf{x} - \mathbf{y}). \tag{3.198}$$

This gives us right to look for the fundamental solution Σ_{st} in the form $\Sigma_{st}^0 + \Sigma_{st}^1$, where Σ_{st}^0 is the fundamental solution of the operator $S_{\omega,1}$ and

$$\Sigma_{st}^1 = \int_{\mathbb{R}^3} \operatorname{div}_{\mathbf{z}} \Sigma_{st}^0(\mathbf{x}, \mathbf{z}) \otimes \nabla_{\mathbf{y}} \frac{1}{4\pi |\mathbf{z} - \mathbf{y}|} d\mathbf{z}.$$

Proposition 3.9 *Let* $\omega = \omega e_3$. *The pair* $\Sigma_{st}(\mathbf{x}, \mathbf{y})$, $Q_{st}(\mathbf{x} - \mathbf{y})$ *with* $\Sigma_{st}(\mathbf{x}, \mathbf{y}) = \Sigma_{st}^0(\mathbf{x}, \mathbf{y}) + \Sigma_{st}^1(\mathbf{x}, \mathbf{y})$ *is a fundamental solution of* (3.196), *where*

$$\Sigma_{st}^0(\mathbf{x}, \mathbf{y}) = \int_0^\infty Q_\omega^T(t) K(Q_\omega(t)\mathbf{x} - \mathbf{y}, t) dt,$$

$$\Sigma_{st}^1(\mathbf{x}, \mathbf{y}) = - \int_0^\infty \int_0^s \nabla_{\mathbf{x}} \nabla_{\mathbf{y}} K(Q_\omega(t)\mathbf{x} - \mathbf{y}, s) \, dt ds,$$

$$Q_{st} = \nabla_y \frac{1}{4\pi |\mathbf{x} - \mathbf{y}|} = P_{st}(\mathbf{x} - \mathbf{y}).$$

As a second step for the exterior domain problem:

$$S_{\omega,1}\mathbf{u} + \nabla p = \mathbf{f}, \quad \operatorname{div} \mathbf{u} = 0 \quad \text{in } \mathcal{D}, \tag{3.199}$$

using a potential theory the following potential representation is shown.

Proposition 3.10 *Let* $\mathbf{f} \in C_0^\infty(\mathcal{D})$. *Then the solution* $(\mathbf{u}, p)$ *of the class* $\mathbf{u} \in L^{3,\infty}(\mathcal{D})$, $\nabla \mathbf{u}$, $p \in L^{3/2,\infty}(\mathcal{D})$[1] *to* (3.199) *with* $\mathbf{u}|_{\partial\mathcal{D}} = 0$ *can be represented as*

$$\mathbf{u}(\mathbf{x}) = \int_{\mathcal{D}} \Sigma_{st}(\mathbf{x}, \mathbf{y})\mathbf{f}(\mathbf{y})d\mathbf{y} + \int_{\partial\mathcal{D}} \Sigma_{st}(\mathbf{x}, \mathbf{y})(\boldsymbol{n} \cdot T(\mathbf{u}, p))(\mathbf{y})d\sigma_{\mathbf{y}}, \tag{3.200}$$

[1] $L^{3,\infty}(\Omega)$ are the Lorentz spaces, which can be defined as a real interpolation $L_{q,r}(\Omega) = (L^1(\Omega), L^\infty(\Omega))_{1-1/q,r}$; for more details see Bergh and Loöfström [14].

$$
\begin{aligned}
p(\mathbf{x}) = & \int_{\mathcal{D}} Q_{st}(\mathbf{x}-\mathbf{y})\mathbf{f}(\mathbf{y})\mathrm{d}\mathbf{y} + \int_{\partial\mathcal{D}} \frac{\boldsymbol{n}(\nabla p - \mathbf{f})(\mathbf{y})}{4\pi|\mathbf{x}-\mathbf{y}|}\mathrm{d}\sigma_{\mathbf{y}} \\
& - \int_{\partial\mathcal{D}} \boldsymbol{n} Q_{st}(\mathbf{x}-\mathbf{y})p(\mathbf{y})\mathrm{d}\sigma_{\mathbf{y}},
\end{aligned}
\tag{3.201}
$$

with $T = \nabla\mathbf{u} + (\nabla\mathbf{u})^T - p\mathcal{I}$ *and* $\boldsymbol{n}$ *the exterior unit normal to the boundary* $\partial\mathcal{D}$. *Moreover, the formula* (3.201) *is valid also in the case* $\mathbf{u}|_{\partial\mathcal{D}} = \omega \wedge \mathbf{x}$. □

Theorem 3.15 *Let* $\boldsymbol{\omega} = \omega e_3$. *Given* $\mathbf{f} = \operatorname{div}\mathbf{F}$ *with* $\mathbf{F} \in C_0^\infty$, *let* $(\mathbf{u}, p)$ *be the solution of class* $\mathbf{u} \in L^{3,\infty}(\mathcal{D})$, $\nabla\mathbf{u},\ p \in L^{3/2,\infty}(\mathcal{D})$ *to* (3.199) *with* $\mathbf{u}|_{\partial\mathcal{D}} = \omega \wedge \mathbf{x}$, $\mathbf{u} \to 0$ *as* $|\mathbf{x}| \to \infty$. *Then the solution has the representation formulas*

$$
\mathbf{u} = \mathbf{U}_{1st}(\mathbf{x}) + \mathbf{U}_{2nd}(\mathbf{x}) + \left(1 + \frac{1}{\omega}\right) O\left(\frac{1}{|\mathbf{x}|^3}\right),
\tag{3.202}
$$

$$
p(\mathbf{x}) = P_{1st}(\mathbf{x}) + O\left(\frac{1}{|\mathbf{x}|^3}\right)
\tag{3.203}
$$

for $|\mathbf{x}| \to \infty$ *where*

$$
U_{1st}(\mathbf{x}) = E_{st} \begin{pmatrix} 0 \\ 0 \\ \int_{\partial\mathcal{D}} (\boldsymbol{n} \cdot (T + \mathbf{F}))_3 \,\mathrm{d}\sigma_{\mathbf{y}} \end{pmatrix},
\tag{3.204}
$$

$$
\begin{aligned}
U_{2nd}(\mathbf{x}) = & \frac{1}{8\pi|\mathbf{x}|^3} \begin{pmatrix} \alpha & -\beta & 0 \\ \beta & \alpha & 0 \\ 0 & 0 & \alpha \end{pmatrix} \begin{pmatrix} x_1 \\ x_2 \\ x_3 \end{pmatrix} - \frac{3(\mathbf{x} \otimes \mathbf{x})}{8\pi|\mathbf{x}|^5} \begin{pmatrix} \frac{\alpha'}{2} x_1 \\ \frac{\alpha'}{2} x_2 \\ \alpha_3 x_3 \end{pmatrix} \\
= & \frac{\beta(\mathbf{e}_3 \wedge \mathbf{x})}{8\pi|\mathbf{x}|^3} + \left(\alpha_3 - \frac{\alpha'}{2}\right) \frac{|\mathbf{x}'|^2 - 2x_3^2}{|\mathbf{x}|^2} \frac{\mathbf{x}}{8\pi|\mathbf{x}|^3},
\end{aligned}
\tag{3.205}
$$

and

$$
P_{1st}(\mathbf{x}) = \int_{\partial\mathcal{D}} \{(\boldsymbol{n} \cdot (\Delta\mathbf{u}))\mathbf{y} - p\boldsymbol{n} + \boldsymbol{n} \cdot \mathbf{F}\}\mathrm{d}\sigma_{\mathbf{y}} \cdot P_{st}(\mathbf{x}),
\tag{3.206}
$$

where

$$
\begin{aligned}
\alpha = & -\int_{\partial\mathcal{D}} \mathbf{y} \cdot (\boldsymbol{n} \cdot (T + \mathbf{F}))\,\mathrm{d}\sigma_{\mathbf{y}} + \int_{\mathcal{D}} tr\,\mathbf{F}\,\mathrm{d}\mathbf{y} = \alpha' + \alpha_3 \\
\alpha' = & -\int_{\partial\mathcal{D}} \mathbf{y}' \cdot (\boldsymbol{n} \cdot (T + \mathbf{F}))'\,\mathrm{d}\sigma_{\mathbf{y}} + \int_{\mathcal{D}} (\mathbf{F}_{11} + \mathbf{F}_{22})\,\mathrm{d}\mathbf{y},
\end{aligned}
$$

$$\alpha_3 = -\int_{\partial\mathcal{D}} y_3 \cdot (\boldsymbol{n} \cdot (T + \mathbf{F}))_3 \, \mathrm{d}\sigma_{\mathbf{y}} + \int_{\mathcal{D}} \mathbf{F}_{33} \, \mathrm{d}\mathbf{y},$$

$$\beta = \mathbf{e}_3 \cdot \int_{\partial\mathcal{D}} \mathbf{y} \wedge (\boldsymbol{n} \cdot (T + \mathbf{F})) \, \mathrm{d}\sigma_{\mathbf{y}} + \int_{\mathcal{D}} (\mathbf{F}_{12} - \mathbf{F}_{21}) \, \mathrm{d}\mathbf{y}.$$

Remark 3.24

- From Theorem 3.15 follows that the leading term is $E_{st}(\mathbf{x})\mathbf{e}_3$ which means the third column vector of the classical Stokes fundamental solution.
- When $\omega = 0$, the usual Stokes asymptotic representation is satisfied

$$\mathbf{u}(\mathbf{x}) = \int_{\partial\mathcal{D}} \boldsymbol{n}(T + \mathbf{F}) \, \mathrm{d}\sigma_y E_{st}(\mathbf{x}) + O\left(\frac{1}{|\mathbf{x}|^2}\right).$$

- The remaining term of the velocity profile decays like $\frac{1}{|\mathbf{x}|^3}$.
- The first part of the coefficient β is the third component of $\int_{\partial\mathcal{D}} \mathbf{y} \times (\boldsymbol{n} \cdot (T + \mathbf{F})) \, \mathrm{d}\sigma_{\mathbf{y}}$, which stands for the total torque exerted on the boundary $\partial\mathcal{D}$.

□

For more details, see [68].

Let us also mention results by Hishida and Shibata [129, 130] concerning L_p–L_q theory and decay estimates of the Stokes type operator.

3.5.3 2D Case

Let us consider the exterior problem of the Stokes type problem in a two-dimensional case. From Sects. 3.1 and 3.3 we got L^q-estimates with or without weights in two- or three-dimensional cases, and for three-dimensional case, see Sect. 3.5.2, we know the asymptotic behavior. The two-dimensional case has completely different behavior. It is known that for "classical" Stokes case the problem is related to the Stokes paradox which says that the problem

$$\begin{aligned} -\Delta\mathbf{u} + \nabla p &= \mathbf{0}, \\ \operatorname{div}\mathbf{u} &= 0, \\ \mathbf{u}|_{\partial\mathcal{D}} &= \mathbf{0}, \\ \mathbf{u} &\to \mathbf{u}_\infty \quad \text{as} \quad |\mathbf{x}| \to \infty \end{aligned} \tag{3.207}$$

has no solution unless $\mathbf{u}_\infty = \mathbf{0}$. Chang and Finn in [23] proved that the Stokes paradox can be explained in terms of the total net force exerted by the fluid to the obstacle

$$M = \int_{\partial\Omega} T(\mathbf{u}, p)\boldsymbol{\nu}\, \mathrm{d}\sigma.$$

Precisely, they showed that the flow satisfying (3.207) can be bounded at infinity only if the net force M vanishes. Other forms of the Stokes paradox are described in works of Galdi [98], and in Chap. 5. Hishida shows that there is no longer Stokes' paradox due to the rotating effect.

Chapter 4
The Steady Oseen Type Problem

The aim of the chapter is to give the comprehensive overview of strong, weak, and very weak solutions for the Oseen type problem. The existence and uniqueness of the strong solution in L^2 framework can be found in the work of Galdi, see [96] and Galdi, Silvestre, see [109]. As in the Stokes type of problem Chap. 3 there are two possible differential formulations of the rotating operator $\mathcal{R}$ in two different systems of coordinates parametrized either with an artificial time or with the angle. Using the parametrization by angle, the existence of the strong solution in L^q-spaces was shown by Farwig in [60], [61]. Later the results have been extended to the weighted L^q-spaces by Farwig, Krbec, Nečasová, see [72]. Parametrization by time and the proof of the strong solution have been shown by Kyed and Galdi, see [102]. Weak solutions in L^q framework have been shown by Kračmar, Nečasová, Penel in the whole space, see [149], and also in the exterior domain, see [151], using the parametrization by angle. The case of weak solution using the parametrization by time has been shown by Kyed and Galdi, see [102]. Further, the weighted approach in L^q-spaces of the weak and very weak solution was shown by Kračmar, Krbec, Nečasová, Penel and Schumacher, see [152], [153].

4.1 A Strong Solution in L^q

In this section we present analogously as in Chap. 3 two different methods of proving the existence and a priori estimates of a strong solution in L^q.

We will explain the method introduced by Farwig et al., see [70] to the Stokes type problem, and extended by Farwig, see [61] to the Oseen type problem.

Š. Nečasová et al., *Mathematical Theory of a Fluid Flow Around a Rotating and Translating Body*, Springer Monographs in Mathematics,
https://doi.org/10.1007/978-3-031-84120-0_4

4.1.1 The Main Result and Corollaries

Let us recall the problem in $\mathbb{R}^3$

$$\begin{aligned} -\nu\Delta\mathbf{u}+k\partial_3\mathbf{u}-\omega(\mathbf{e}_3\wedge\mathbf{x})\cdot\nabla\mathbf{u}+\omega\mathbf{e}_3\wedge\mathbf{u}+\nabla p &=\mathbf{f}, \text{ in } \mathbb{R}^3 \\ \operatorname{div}\mathbf{u} &=g, \text{ in } \mathbb{R}^3 \end{aligned} \tag{4.1}$$

which can be written

$$\begin{aligned} S_{\nu,\omega}\boldsymbol{u}+\nabla p+k\partial_3\boldsymbol{u} &=\boldsymbol{f}, \\ \operatorname{div}\boldsymbol{u} &=g, \\ S_{\nu,\omega}\boldsymbol{u} &=\nu\Delta\boldsymbol{u}-\omega\mathcal{R}\boldsymbol{u}, \\ \mathcal{R}\boldsymbol{u} &=(\boldsymbol{e}_3\wedge\boldsymbol{x}\cdot\nabla\boldsymbol{u}-\boldsymbol{e}_3\wedge\boldsymbol{u}). \end{aligned} \tag{4.2}$$

Let us denote the operator

$$\mathcal{L}_{\nu,\omega,k} := S_{\nu,\omega}\boldsymbol{u}-k\partial_3\boldsymbol{u}.$$

We state the main result of Subsection 4.1 (see [61]).

Theorem 4.1

(A1) *Let* $1<q<\infty$, $\mathbf{f}\in \boldsymbol{L}^q(\mathbb{R}^3)$, *and* $g\in \widehat{H}^{1,q}(\mathbb{R}^3)$ *such that* $|(x_1,x_2)|g\in L^q(\mathbb{R}^3)$. *Furthermore, let* $\nu>0$, $k\in\mathbb{R}$.
Then the problem (4.1) has a solution $(\mathbf{u},p)\in \widehat{\boldsymbol{H}}^{2,q}(\mathbb{R}^3)\times\widehat{H}^{1,q}(\mathbb{R}^3)$ *satisfying the a priori estimates*

$$\|\nu\nabla^2\mathbf{u}\|_q+\|\nabla p\|_q\le C(\|\mathbf{f}\|_q+\|\nu\nabla g+\omega(\mathbf{e}_3\wedge\mathbf{x})g-kg\boldsymbol{e}_3\|_q), \tag{4.3}$$

$$\begin{aligned} &\|k\partial_3\mathbf{u}\|_q+\|\omega(\mathbf{e}_3\wedge\mathbf{x})\cdot\nabla\mathbf{u}-\omega\mathbf{e}_3\wedge\mathbf{u}\|_q \\ &\qquad\le C\left(1+\frac{k^4}{\nu^2|\omega|2}\right)(\|\mathbf{f}\|_q+\|\nu\nabla g+\omega(\mathbf{e}_3\wedge\mathbf{x})g-kg\boldsymbol{e}_3\|_q) \end{aligned} \tag{4.4}$$

with a constant $C>0$ *independent of* ν, k *and* ω.

(A2) *Let* $(\mathbf{u},p)\in\widehat{\boldsymbol{H}}^{2,q}(\mathbb{R}^3)\times\widehat{H}^{1,q}(\mathbb{R}^3)$ *be solution of (4.1) with* (A1) *assumptions, suppose that* $\mathbf{f}\in\boldsymbol{L}^r(\mathbb{R}^3)$, $g\in W^{1,r}(\mathbb{R}^3)$, $|(x_1,x_2)|g\in L^r(\mathbb{R}^3)$ *and let* $(\mathbf{u}_1,p_1)\in\widehat{\boldsymbol{H}}^{2,r}(\mathbb{R}^3)\times\widehat{H}^{1,r}(\mathbb{R}^3)$ *be another solution of (4.1). Then* $p-p_1$ *is constant, and* $\mathbf{u}-\mathbf{u}_1$ *equals* $\alpha\mathbf{e}_3+\beta\omega\boldsymbol{e}_3\wedge\mathbf{x}$, $\alpha,\beta\in\mathbb{R}$.

□

Corollary 4.1

(B1) *Let* $1 < q < 4$, $\mathbf{f} \in \boldsymbol{L}^q(\mathbb{R}^3)$, $g \in \widehat{H}^{1,q}(\mathbb{R}^3)$ *such that* $|(x_1, x_2)|g \in L^q(\mathbb{R}^3)$ *and let* $(\mathbf{u}, p) \in \widehat{\boldsymbol{H}}^{2,q}(\mathbb{R}^3) \times \widehat{H}^{1,q}(\mathbb{R}^3)$ *be the solution of (4.1). Then there exists* $\beta \in \mathbb{R}$ *such that*

$$\nabla'(\mathbf{u} - \beta\omega\mathbf{e}_3 \wedge \mathbf{x}) \in \boldsymbol{L}^r(\mathbb{R}^3) \quad \text{for all } r > 1, \tag{4.5}$$

$$\frac{1}{r} \in \frac{1}{q} - \left[\frac{1}{4}, \frac{1}{3}\right],$$

$(\nabla' = (\partial_1, \partial_2))$. *Furthermore,*

$$\|\nabla'(\mathbf{u} - \beta\omega\mathbf{e}_3 \wedge \mathbf{x})\|_r \le C(\|\mathbf{f}\|_q + \|\nu\nabla g + (\omega\mathbf{e}_3 \wedge \mathbf{x})g - kg\boldsymbol{e}_3\|_q \tag{4.6}$$

with a constant $c = c(\nu, k, \omega, r) > 0$.

Moreover to assumption (B1) *we assume that* $1 < q < 2$. *Then there exist* $\alpha, \beta \in \mathbb{R}$ *such that*

$$\mathbf{u} - \beta\omega\mathbf{e}_3 \wedge \mathbf{x} - \alpha\boldsymbol{e}_3 \in \boldsymbol{L}^s(\mathbb{R}^3) \quad \text{for all } s > 1, \tag{4.7}$$

$$\frac{1}{s} \in \frac{1}{q} - \left[\frac{1}{2}, \frac{2}{3}\right].$$

Further,

$$\|\mathbf{u} - \beta\omega\mathbf{e}_3 \wedge \mathbf{x} - \alpha\boldsymbol{e}_3\|_s \le C(\|\mathbf{f}\|_q + \|\nu g + (\omega\mathbf{e}_3 \wedge \mathbf{x})g - kg\boldsymbol{e}_3\|_q) \tag{4.8}$$

with a constant $C = C(\nu, k, \omega; s) > 0$.

□

Theorem 4.2

(C1) *Let* $1 < q < 2$, $\nu > 0$, $k > 0$, $\mathbf{f} \in L^q(\mathbb{R}^3)$. *Then (4.1) has a solution* $\mathbf{u} \in \widehat{H}^{2,q}(\mathbb{R}^3)$ *satisfying the a priori estimates*

$$\left\| \frac{\nabla\mathbf{u}}{|\mathbf{x}'|} \right\|_q \le \frac{C}{\nu}\|\mathbf{f}\|_q \tag{4.9}$$

and

$$\left\| \frac{\omega((\mathbf{e}_3 \wedge \mathbf{x}) \cdot \nabla\mathbf{u} - \mathbf{e}_3 \wedge \mathbf{u})}{1 + |\mathbf{x}'|} \right\|_q \le c\left(1 + \frac{k}{\nu}\right)\|\mathbf{f}\|_q \tag{4.10}$$

with a constant $c > 0$ *independent of* k, ν, ω. *By* $|\mathbf{x}'| = \sqrt{\mathbf{x}_1^2 + \mathbf{x}_2^2}$, $\mathbf{x}' = (\mathbf{x}_1, \mathbf{x}_2)$.

(C2) *If* $1 < q < 3$, *then*

$$\left\| \frac{\nabla \mathbf{u}}{|\mathbf{x}|} \right\|_q \leq \frac{1}{\nu} \|\mathbf{f}\|_q \tag{4.11}$$

and

$$\left\| \frac{\omega((\mathbf{e}_3 \wedge \mathbf{x}) \cdot \nabla \mathbf{u} - \mathbf{e}_3 \wedge \mathbf{u})}{1 + |\mathbf{x}|} \right\|_q \leq C \left(1 + \frac{k}{\nu}\right) \|\mathbf{f}\|_q \tag{4.12}$$

with a constant $c > 0$ *independent of* k, ν, ω.

(C3) *For all* $1 < q < \infty$ *the third component* $\mathbf{u}_3$ *of the solution* $\mathbf{u}$ *satisfies the a priori estimate*

$$\left\| \frac{k \partial_3 \mathbf{u}_3}{1 + |\mathbf{x}'|} \right\|_q + \left\| \frac{\omega(\mathbf{e}_3 \wedge \mathbf{x}) \cdot \nabla \mathbf{u}_3}{1 + |\mathbf{x}'|} \right\|_q \leq c \left(1 + \frac{k}{\nu}\right) \|\mathbf{f}\|_q. \tag{4.13}$$

□

4.1.2 Proof of Theorem 4.1 Using the Associated Ordinary Angular Differential Equation

We will describe the method used by Farwig [61], applied to the Stokes type problem and extended to the Oseen type problem.

As in the Stokes type case, see Chap. 3, we have to eliminate the pressure. Applying div to (4.1) we get

$$\Delta p = \operatorname{div} \mathbf{f} + \nu \Delta g + (\omega \mathbf{e}_3 \wedge \mathbf{x}) \cdot \nabla g - k\, \partial_3 g, \tag{4.14}$$

i.e.,

$$(\nabla p, \nabla \boldsymbol{\varphi}) = (\mathbf{f}, \nabla \boldsymbol{\varphi}) + (\nu \nabla g + (\omega \mathbf{e}_3 \wedge \mathbf{x}) g, \nabla \varphi) - (k\, g, \partial_3 \boldsymbol{\varphi}) \tag{4.15}$$

for all test functions $\boldsymbol{\varphi} \in C_0^\infty(\mathbb{R}^3)$.

From the assumptions of Theorem 4.1 we get that the right-hand side of (4.15) defines a functional F on $\widehat{H}^{1,q'}(\mathbb{R}^3)$ satisfying the estimate

$$\|F\|_{\widehat{H}^{-1,q}(\mathbb{R}^3)} \leq \|\mathbf{f}\|_q + \|\nu \nabla g + (\omega \mathbf{e}_3 \wedge \mathbf{x}) g - k\, g e_3\|_q. \tag{4.16}$$

Moreover $-\Delta$ defines an isomorphism from $\widehat{H}^{1,q}(\mathbb{R}^3)$ onto $\widehat{H}^{-1,q}(\mathbb{R}^3)$. It implies an existence of a unique $p \in \widehat{H}^{1,q}(\mathbb{R}^3)$ solving (4.14) and satisfying the a priori estimate

$$\|\nabla p\|_q \le \|\mathbf{f}\|_q + \|\nu \nabla g + (\omega \mathbf{e}_3 \wedge \mathbf{x}) g - k g \boldsymbol{e}_3\|_q. \tag{4.17}$$

This gives us a right to look for $\mathbf{u}$ in (4.1) as a solution of the problem

$$-\nu \Delta \mathbf{u} + k\partial_3 \mathbf{u} - \omega(\mathbf{e}_3 \wedge \mathbf{x}) \cdot \nabla \mathbf{u} + \omega \mathbf{e}_3 \wedge \mathbf{u} = \mathbf{f} - \nabla p. \tag{4.18}$$

From unicity we get that $\operatorname{div} \mathbf{u} = g$.

Now, we are looking for an explicit solution of (4.1), and we write $\mathbf{f}$ instead of $\mathbf{f} - \nabla p$.

Dividing (4.1) by ω, we get

$$-\frac{\nu}{\omega}\Delta \mathbf{u} + \frac{k}{\omega}\partial_3 \mathbf{u} - (\boldsymbol{e}_3 \wedge \mathbf{x}) \cdot \nabla \mathbf{u} + \boldsymbol{e}_3 \wedge \mathbf{u} = \frac{1}{\omega}\mathbf{f}. \tag{4.19}$$

Introducing cylindrical coordinates $(r, \theta, x_3) \in \mathbb{R}_+^- \times [0, 2\pi) \times \mathbb{R}$, $r = \sqrt{x_1^2 + x_2^2}$ for $r = (x_1, x_2, x_3)^{\mathrm{T}}$ observe that

$$\partial_\theta \mathbf{u} = (\boldsymbol{e}_3 \wedge \mathbf{x}) \cdot \nabla \mathbf{u}.$$

We apply the Fourier transform $\mathcal{F}$ which we also denote as $= \hat{}$ to (4.19) using the cylindrical coordinates $(s, \varphi, \xi_3) \in \mathbb{R}_+ \times [0, 2\pi) \times \mathbb{R}$, $s = \sqrt{\xi_1^2 + \xi_2^2}$ for $\boldsymbol{\xi} = (\xi_1, \xi_2, \xi_3)^{\mathrm{T}}$ as well and using $\widehat{\partial_\theta \mathbf{u}} = \partial_\varphi \widehat{\mathbf{u}}$.

It implies that $\widehat{\mathbf{u}}$ satisfies the equation

$$\frac{1}{\omega}(\nu|\boldsymbol{\xi}|^2 + \mathrm{i}k\xi_3)\widehat{\mathbf{u}} - \partial_\varphi \widehat{\mathbf{u}} + \boldsymbol{e}_3 \wedge \widehat{\mathbf{u}} = \frac{1}{\omega}\widehat{\mathbf{f}}. \tag{4.20}$$

Denoting $\widehat{\mathbf{v}}(\varphi) = Q_{e_3}^{\mathrm{T}}(\varphi)\widehat{\mathbf{u}}(t, \varphi, \boldsymbol{\xi}_3)$ $\widehat{\mathbf{v}}$ solves the equation

$$\frac{1}{\omega}(\nu|\boldsymbol{\xi}|^2 + \mathrm{i}k\xi_3)\widehat{\mathbf{v}} - \partial_\varphi \widehat{\mathbf{v}} = \frac{1}{\omega} Q_{e_3}^{\mathrm{T}}(\varphi)\widehat{\mathbf{f}}, \tag{4.21}$$

which is the inhomogeneous linear ordinary differential equation of the first order with respect to φ. Such equation has a unique 2π-periodic solution

$$\widehat{\mathbf{v}}(\varphi) = \frac{1/\widetilde{\omega}}{1 - \mathrm{e}^{-2\pi(\nu|\boldsymbol{\xi}|^2 + \mathrm{i}k\xi_3)/\omega}} \int_0^{2\pi} \mathrm{e}^{-(\nu|\boldsymbol{\xi}|^2 + \mathrm{i}k\xi_3)t/\omega} Q_{\boldsymbol{e}_3}^{\mathrm{T}}(\varphi + t)\widehat{f}(Q_{\boldsymbol{e}_3}(t)\boldsymbol{\xi})\mathrm{d}t, \tag{4.22}$$

where $\boldsymbol{\xi} = (s, \varphi, \xi_3)$.

Then

$$\widehat{\mathbf{u}}(\xi) = \frac{1}{1-\mathrm{e}^{-2\pi(\nu|\boldsymbol{\xi}|^2+\mathrm{i}k\xi_3)/\omega}} \int_0^{2\pi/\omega} \mathrm{e}^{-(\nu|\boldsymbol{\xi}|^2+\mathrm{i}k\xi_3)t} Q_\omega^{\mathrm{T}}(t)\widehat{\mathbf{f}}(Q_\omega(t)\boldsymbol{\xi})\mathrm{d}t. \tag{4.23}$$

Using the geometric series and applying $\frac{2\pi}{\omega}$ periodicity of the map

$$t \mapsto Q_\omega^{\mathrm{T}}(t)\widehat{\mathbf{f}}(Q_\omega(t)\boldsymbol{\xi}),$$

$$\widehat{\mathbf{u}}(\boldsymbol{\xi}) = \int_0^\infty \bar{\mathrm{e}}^{\nu|\boldsymbol{\xi}|^2+\mathrm{i}k\xi_3)t} Q_\omega^{\mathrm{T}}(t)\widehat{\mathbf{f}}(Q_\omega(t)\boldsymbol{\xi})\mathrm{d}t.\widehat{\mathbf{f}}(Q_\omega(t)\boldsymbol{\xi})\mathrm{d}t. \tag{4.24}$$

The term $\mathrm{e}^{itk\xi_3} \in S(\mathbb{R}^3)$ is the Fourier transformation of the shift operator

$$\mathbf{f} \mapsto \mathbf{f}(\cdot - kt\boldsymbol{e}_3) \quad \text{in } S'(\mathbb{R}^3).$$

We can write $\widehat{u}(\boldsymbol{\xi})$ as

$$\widehat{\mathbf{u}}(\boldsymbol{\xi}) = \int_0^\infty \bar{\mathrm{e}}^{-\nu|\boldsymbol{\xi}|^2 t} Q_\omega^{\mathrm{T}}(t)(\hat{\mathbf{f}}(Q_\omega(t)\cdot -kt\boldsymbol{e}_3))(\boldsymbol{\xi})\mathrm{d}t. \tag{4.25}$$

Since $\mathrm{e}^{-\nu|\boldsymbol{\xi}|^2 t}$ is the Fourier transform of the heat kernel

$$K(\mathbf{x}) = \frac{1}{(4\pi\nu t)^{3/2}}\mathrm{e}^{-|\mathbf{x}|^2/4\nu t},$$

this gives us

$$\mathbf{u}(\mathbf{x}) = \int_0^\infty K * Q_\omega^{\mathrm{T}}(t)\mathbf{f}(Q_\omega(t)\cdot -kt\boldsymbol{e}_3)(\mathbf{x})\mathrm{d}t. \tag{4.26}$$

The fundamental solution which we derived (4.26) coincides up to the Helmholtz correction with the steady version of the fundamental solution derived by Deuring et al., see [35, 36] from unsteady variant derived by Guenther, Thomann, see [124].

To prove an a priori estimate of (4.26), it is sufficient to estimate $\Delta\mathbf{u}$ on $\mathbf{f}$. Given $\mathbf{f} \in \boldsymbol{S}(\mathbb{R}^3)$ from (4.25), we have

$$\begin{aligned} -\widehat{\nu\Delta\mathbf{u}}(\boldsymbol{\xi}) =& \nu|\boldsymbol{\xi}|^2\widehat{\mathbf{u}}(\boldsymbol{\xi}) = \int_0^\infty \widehat{\boldsymbol{\psi}}_{\nu t}(\boldsymbol{\xi}) Q_\omega^{\mathrm{T}}(t)\widehat{\mathbf{f}}(Q_\omega(t)\cdot -kt\boldsymbol{e}_3)(\boldsymbol{\xi})\frac{\mathrm{d}t}{t} \\ &= \int_0^\infty \widehat{\boldsymbol{\psi}}_t(\boldsymbol{\xi}) Q_{\omega/\nu}^{\mathrm{T}}(t)\widehat{\mathbf{f}}\left(Q_{\omega/\nu}(t)\cdot -\frac{k}{\nu}t\boldsymbol{e}_3\right)(\boldsymbol{\xi})\frac{\mathrm{d}t}{t}, \end{aligned}$$

where

$$\widehat{\boldsymbol{\psi}}(\boldsymbol{\xi}) = \frac{1}{(2\pi)^{3/2}}|\boldsymbol{\xi}|^2 e^{-|\boldsymbol{\xi}|^2} \quad \text{and} \quad \widehat{\boldsymbol{\psi}}_t(\boldsymbol{\xi}) = \widehat{\boldsymbol{\psi}}(\sqrt{t}\boldsymbol{\xi})$$

for $t > 0$ are the Fourier transforms of $\boldsymbol{\psi} \in \mathcal{S}(\mathbb{R}^3)$ and $\boldsymbol{\psi}_t(\mathbf{x}) = t^{-3/2}\boldsymbol{\psi}(\mathbf{x}/\sqrt{t})$, $t > 0$, respectively. Decompose $\widehat{\boldsymbol{\psi}}_t$ so that $\widetilde{\chi} \in C_0^\infty(\frac{1}{2}, 2)$, $0 \le \widetilde{\chi} \le 1$, and $\sum_{j=-\infty}^{\infty} \widetilde{\chi}(2^{-j}r) = 1$ for all $r > 0$. Moreover

$$\widehat{\chi}_j(\xi) = \widetilde{\chi}(2^{-j}|\boldsymbol{\xi}|), \quad \chi_j \in S(\mathbb{R}^3)$$

with

$$\operatorname{supp} \widehat{\chi}_j \subset A(2^{j-1}, 2^{j+1}) := \{\boldsymbol{\xi} \in \mathbb{R}^3 \colon 2^{j-1} \le |\boldsymbol{\xi}| \le 2^{j+1}\}.$$

Finally, define $\psi^j \in S(\mathbb{R}^3)$ by

$$\psi^j = (2\pi)^{-3/2}\chi_j * \psi$$

or equivalently

$$\widehat{\psi}^j = \widehat{\chi}_j \cdot \widehat{\psi}, \quad j \in Z, \quad \psi = \sum_{j=-\infty}^{+\infty} \psi^j.$$

Using ψ^j, we introduce the operator

$$T_j\mathbf{f}(\mathbf{x}) = \int_0^\infty \psi_t^j * Q_{\omega/\nu}^{\mathrm{T}}(t)\mathbf{f}\left(Q_{\omega/\nu}(t) \cdot -\frac{k}{\nu}t\boldsymbol{e}_3\right)(\mathbf{x})\frac{\mathrm{d}t}{t}. \tag{4.27}$$

We have to prove that the series $\sum_{j=-\infty}^{\infty} T_j$ converges in the operator norm topology on $L^q(\mathbb{R}^3)^3$.

Before proving a priori estimates, we introduce the following lemmas. For completeness, we also recall Lemma 3.1.

Lemma 4.1 (Lemma 3.1)

The function ψ_t^j, $j \in \mathbb{Z}$, $t > 0$, has the following properties:

(P1) $\operatorname{supp} \psi_t^j \subset A(2^{j-1}/\sqrt{t}, 2^{j+1}/\sqrt{t})$.

(P2) $\|\psi^j\|_1 \le C2^{-2|j|}$ *and* $|\psi_j(\mathbf{x})| \le c2^{-2|j|}h_{2^{-2|j|}}(\mathbf{x})$ *for all* $\mathbf{x} \in \mathbb{R}^3$, *where* $h(\mathbf{x}) = (1 + |\mathbf{x}|^2)^{-2}$, $h_t(\mathbf{x}) = t^{-3/2}h(\mathbf{x}/\sqrt{t})$, *and* $c > 0$ *is a constant independent of* $j \in \mathbb{Z}$ *and of* $\mathbf{x} \in \mathbb{R}^3$.

□

Lemma 4.2 *For $j \in \mathbb{Z}$, let $\mathcal{M}_j$ denote the maximal operator*

$$\mathcal{M}_j g(\mathbf{x}) = \sup_{r>0} \int_{A_r} (|\psi_t^j * |g|) \left(Q^t_{\omega/\nu}(t)\mathbf{x} + \frac{k}{\nu} t\boldsymbol{e}_3 \right) \frac{dt}{t}, \tag{4.28}$$

where $A_r = [r/16, 16r]$. Then for $q \in (2, \infty)$ the operator T_j satisfies the estimate

$$\|T_j \mathbf{f}\|_q \leq c 2^{-|j|} \|\mathcal{M}_j\|^{1/2}_{(q/2)'} \|\mathbf{f}\|_q \tag{4.29}$$

with a constant $c > 0$ independent of $j \in \mathbb{Z}$. The term $\|\mathcal{M}_j\|_{(q/2)'}$ denotes the operator norm of the sublinear operator $\mathcal{M}_j$ on $L^{(q/2)'}(\mathbb{R}^3)$, the dual of $\boldsymbol{L}^{q/2}(\mathbb{R}^3)^3$.

□

Lemma 4.3 *The maximal operator $\mathcal{M}_j$ satisfies on $L^p(\mathbb{R}^3)^3$, $1 < p < \infty$, the operator norm estimate*

$$\|\mathcal{M}_j\|_p \leq c 2^{-2|j|}, \quad j \in \mathbb{Z},$$

where $c = c(p)$ is independent of $j \in \mathbb{Z}$. □

We continue in the proof of Theorem 4.1(A1).

Proof From Lemmas 4.2, 4.3, operator T_j satisfies

$$\|T_j \mathbf{f}\|_q \leq c 2^{-2|j|} \|\mathbf{f}\|_q, \quad j \in \mathbb{Z}, \quad \mathbf{f} \in \mathcal{S}(\mathbb{R}^3), \quad q \in (2, \infty), \quad c(q) = c > 0.$$

$T = \sum_{j=-\infty}^{\infty} T_j$ converges in the operator norm on $\boldsymbol{L}^q(\mathbb{R}^3)$ and $\|T\mathbf{f}\|_q \leq c\|\mathbf{f}\|_q$, i.e., for every $\mathbf{f} \in \mathcal{S}(\mathbb{R}^3)$ Eq. (4.18) with $p = 0$ has a solution $\mathbf{u} \in \mathcal{S}'(\mathbb{R}^3) \cap \widehat{\boldsymbol{H}}^{2,q}(\mathbb{R}^3)$ satisfying the inequality

$$\|\nu \nabla^2 \mathbf{u}\|_q \leq c\|\nu \Delta \mathbf{u}\|_q \leq C\|\mathbf{f}\|_q, \quad \mathbf{f} \in \mathcal{S}(\mathbb{R}^3). \tag{4.30}$$

Now, we would like to prove the inequality for $\partial_3 \mathbf{u}$. Using a representation of $k\partial_3 \mathbf{u}$ from (4.23), denoting

$$k' = \frac{k}{\omega}, \quad \nu' = \frac{\nu}{\omega}, \quad D(\boldsymbol{\xi}) = 1 - \mathrm{e}^{-2\pi(\nu'|\boldsymbol{\xi}|^2 + \mathrm{i}k'\xi_3)},$$

we have

$$\widehat{k\partial_3 \mathbf{u}}(\boldsymbol{\xi}) = \frac{\mathrm{i}k'\xi_3}{D(\boldsymbol{\xi})} \int_0^{2\pi} \mathrm{e}^{-(\nu'|\boldsymbol{\xi}|^2 + \mathrm{i}k'\xi_3)t} Q^{\mathrm{T}}_{\boldsymbol{e}_3}(t) \widehat{\mathbf{f}} Q_{\boldsymbol{e}_3}(t)\boldsymbol{\xi}) \mathrm{d}t \quad \text{for } \mathbf{f} \in \mathcal{S}(\mathbb{R}^3). \tag{4.31}$$

We define a cutoff function $\eta \in C_0^\infty(B_1(\mathbf{0}))$ with $\eta(\boldsymbol{\xi}) = 1$ for $\boldsymbol{\xi} \in B_{1/2}(\mathbf{0})$ and recall the effect of the multiplicative term $e^{-ik't\xi_3}$ in (4.23)–(4.25). We introduce $m_0(\boldsymbol{\xi}), m_1(\boldsymbol{\xi}), I_0(\boldsymbol{\xi}), I_1(\boldsymbol{\xi})$ such that

$$\widehat{k\partial_3\mathbf{u}(\boldsymbol{\xi})} = m_0(\boldsymbol{\xi})\widehat{I_0}(\boldsymbol{\xi}) + m_1(\boldsymbol{\xi})\widehat{I_1}(\boldsymbol{\xi}), \tag{4.32}$$

$$m_0(\boldsymbol{\xi}) = \frac{ik'\xi_3\eta_{\nu'}(\boldsymbol{\xi})}{D(\boldsymbol{\xi})}, \quad m_1(\boldsymbol{\xi}) = \frac{k'}{\sqrt{\nu'}}\frac{1-\eta_{\nu'}(\boldsymbol{\xi})}{D(\boldsymbol{\xi})}, \tag{4.33}$$

where $\eta_{\nu'}(\boldsymbol{\xi}) = \eta(\sqrt{\nu'}\boldsymbol{\xi})$, with

$$I_0(\mathbf{x}) = \int_0^{2\pi} K_t' * Q_{e_3}^{\mathrm{T}}(t)\mathbf{f}(Q_{e_3}(t)\cdot -k't\boldsymbol{e}_3)(\mathbf{x})\mathrm{d}t, \tag{4.34}$$

$$I_1(\mathbf{x}) = \sqrt{\nu'}\int_0^{2\pi} \partial_3 K_t' * Q_{e_3}^{\mathrm{T}}(t)\mathbf{f}(Q_{e_3}(t)\cdot -k't\boldsymbol{e}_3)(\mathbf{x})\mathrm{d}t. \tag{4.35}$$

$K_t'(\cdot)$ denotes the heat kernel, where ν' is replaced by ν. Since $\|K_t'\|_1 = 1$, then $|\partial_3 K_t'(\mathbf{x})| \le c/\sqrt{\nu' t}K_t'(\frac{1}{2}\mathbf{x})$ and $\|\mathbf{f}(Q_{e_3}(t)\cdot -k't\boldsymbol{e}_3)\|_p = \|\mathbf{f}\|_p$. Applying Young's inequality yields

$$\|I_0\|_q \le 2\pi\|\mathbf{f}\|_q, \quad \|I_1\|_q \le c\|\mathbf{f}\|_q, \tag{4.36}$$

where $c > 0$ is independent of k, ω, and ν. Moreover, m_0 and m_1 satisfy the pointwise estimates

$$\max_{j=0,1}\max_{\alpha}\sup_{\boldsymbol{\xi}\neq 0} |\boldsymbol{\xi}^\alpha D_{\boldsymbol{\xi}}^\alpha m_j(\boldsymbol{\xi})| \le c\left(1 + \left(\frac{k'}{\sqrt{\nu'}}\right)^4\right) = c\left(1 + \frac{k^4}{\nu^2|\boldsymbol{\omega}|^2}\right) \tag{4.37}$$

with a constant $C > 0$ independent of ν, ω and k; $\alpha \in \mathbb{N}^3$ is the set of all multi-indices $\alpha \in \{0, 1\}^3$.

Proof of (4.37).

(i) For m_1, can be seen easily that $m_1(\boldsymbol{\xi}) = 0$ unless $\nu'|\boldsymbol{\xi}|^2 \ge \frac{1}{2}$, which gives us a uniform pointwise lower bound of denominator $D(\boldsymbol{\xi})$, it means

$$\begin{aligned}\xi_3\partial_3 m_1(\boldsymbol{\xi}) = \frac{k'\xi_3}{\sqrt{\nu'}}\Bigg(&-\frac{\sqrt{\nu'}}{D(\boldsymbol{\xi})}(\partial_3\eta)(\sqrt{\nu'}\boldsymbol{\xi}) - (1-\eta_{\nu'}(\boldsymbol{\xi})) \\ &\times\frac{2\pi}{D(\boldsymbol{\xi})^2}(2\nu'\xi_3 + ik')e^{-2\pi(\nu'|\boldsymbol{\xi}|^2 + ik'\xi_3)}\Bigg)\end{aligned} \tag{4.38}$$

and then

$$|\xi_3 \partial_3 m_1(\boldsymbol{\xi})| \leq c \left(\frac{k'}{\sqrt{\nu'}} + \left(\frac{k'}{\sqrt{\nu'}} \right)^2 \right) \leq c \left(1 + \left(\frac{k'}{\sqrt{\nu'}} \right)^2 \right).$$

(ii) Now for m_0, observing that $m_0(\boldsymbol{\xi}) = 0$ unless. $\nu'(\boldsymbol{\xi})^2 \leq 1$..,we note that since $k'\xi_3$ may take arbitrary values, we have to be more precise:

For each $\boldsymbol{\xi}$....and using Taylor's expansion to estimate $D(\boldsymbol{\xi})$ in the denominator, with $|k'\xi_3 - n| \leq 1/2$ we get

Firstly, let us mention that $m_0(\boldsymbol{\xi}) = 0$ unless $\nu'(\boldsymbol{\xi})^2 \leq 1$.

Since $k'\xi_3$ may take arbitrary values, we have to be more precise. For each $\xi \in \mathbb{R}^3$ there exists an $n \in \mathbb{Z}$ such that $|k'\xi_3 - n| \leq \frac{1}{2}$ and using Taylor's expansion $1 - e^{-z}$,

$$D(\boldsymbol{\xi}) \sim 2\pi(\nu'|\boldsymbol{\xi}|^2 + i(k'\xi_3 - n)).$$

If $n = 0$, i.e., $|k'\xi_3| \leq \frac{1}{2}$, we get

$$|m_0(\boldsymbol{\xi})| \leq c \frac{|ik'\xi_3|}{|\nu'|\boldsymbol{\xi}|^2 + ik'\xi_3|} \leq c.$$

If $n \neq 0$ and $|k'\xi_3 - n| \leq \frac{1}{4}$, we obtain

$$|m_0(\boldsymbol{\xi})| \leq c \frac{|ik'\xi_3|}{|\nu'|\boldsymbol{\xi}|^2 + i(k'\xi_3 - n)|} \leq c \frac{|k'\xi_3|}{\nu'|\boldsymbol{\xi}|^2} \leq c \left| \frac{k'}{\nu'\xi_3} \right| \leq c \left(\frac{k'}{\sqrt{\nu'}} \right)^2$$

using $|k'\xi_3| \geq \frac{3}{4}$.

Finally, if $\frac{1}{4} < |k'\xi_3 - n| \leq \frac{1}{2}$ and $n \neq 0$, then $|D|$ has a uniform positive lower bound and the following estimate holds

$$|m_0(\boldsymbol{\xi})| \leq c|k'\xi_3| \leq c \frac{|k'|}{\sqrt{\nu}}$$

since $\nu'\xi_3^2 \leq 1$.

It implies that we have the bounds for m_0 such that

$$\|m_0\|_\infty \leq c \left(1 + \left(\frac{k'}{\sqrt{\nu'}} \right)^2 \right). \tag{4.39}$$

(iii) The differentiation of $\xi_3 \partial_3 m_0(\xi)$ gives the term

$$2\pi \frac{ik'\xi_3}{D(\boldsymbol{\xi})^2} (2\nu'\xi_3^2 + ik'\xi_3) e^{-(2\pi(\nu'|\boldsymbol{\xi}|^2 + ik'\xi_3))},$$

which is estimated by

$$c\left(1+\left(\frac{k'}{\sqrt{\nu'}}\right)^4\right).$$

In $\xi_1\partial_1$ and $\xi_2\partial_2$ we are not getting further powers of $k'/\sqrt{\nu'}$, and it follows that (4.37) holds.

Applying Marcinkiewicz's multiplier theorem and (4.33)–(4.36), for every $q \in (1, \infty)$ the a priori estimate

$$\|k\partial_3\mathbf{u}\|_q \le c\left(1+\frac{k^4}{\nu^2|\omega|^2}\right)\|\mathbf{f}\|_q, \quad \mathbf{f} \in \boldsymbol{\mathcal{S}}(\mathbb{R}^3), \tag{4.40}$$

holds with a constant $c > 0$ independent of $\mathbf{f}$, k, ν, and ω.

Now, our goal is to extend (4.30) and (4.37) for arbitrary $\mathbf{f} \in \boldsymbol{L}^q(\mathbb{R}^3)$, $q > 2$ and to get a vector field $\mathbf{u} \in \boldsymbol{L}^1_{\text{loc}}(\mathbb{R}^3)$ with $\nabla^2\mathbf{u}$, $\partial_3\mathbf{u} \in \boldsymbol{L}^q(\mathbb{R}^3)$ solving

$$\mathcal{L}_{\nu,\omega,k}\mathbf{u} := \mathbf{f}.$$

We will choose a sequence $(\boldsymbol{f}_j) \subset \boldsymbol{\mathcal{S}}(\mathbb{R}^3)$ such that $\mathbf{f}_j \to \mathbf{f}$ in $\boldsymbol{L}^q(\mathbb{R}^3)$ as $j \to \infty$. Let $(\mathbf{u}_j) \subset \boldsymbol{L}^1_{\text{loc}}(\mathbb{R}^3)$ denote the corresponding solutions of $\mathcal{L}_{\nu,\omega,k}\mathbf{u}_j = \mathbf{f}_j$ satisfying

$$\sup_j(\|\nu\nabla^2\mathbf{u}_j\|_q + \|k\,\partial_3\mathbf{u}_j\|_q) < \infty. \tag{4.41}$$

It implies the existence of constant vectors $c_j, d_{1j}, d_{2j} \in \mathbb{R}^3$ such that

$$\int_{B_1}(\partial_1\mathbf{u}_j - d_{1j})\mathrm{d}\mathbf{x} = \int_{B_1}(\partial_2\mathbf{u}_j - d_{2j})\mathrm{d}\mathbf{x} = \int_{B_1}(\mathbf{u}_j - (c_j + d_{1j}x_1 + d_{2j}x_2))\mathrm{d}\mathbf{x} = 0.$$

From Poincare's inequality and the a priori estimate, (4.41) implies that for all $m \in \mathbb{N}$

$$\sup_j(\|\nabla^2(\mathbf{u}_j - \boldsymbol{r}_j)\|_q + \|\partial_3(\mathbf{u}_j - \boldsymbol{r}_j)\|_q + \|\mathbf{u}_j - \boldsymbol{r}_j\|_{L^q(B_m)}) \le c_M$$

for some constant $c_M > 0$, where $\boldsymbol{r}_j(\mathbf{x}) = c_j + d_{1j}x_1 + d_{2j}x_2$.

Using the concepts of weak convergence and compact imbedding, we get a subsequence denoted again $(\mathbf{u}_j - \boldsymbol{r}_j)$ and $\widetilde{\mathbf{u}} \in \boldsymbol{L}^1_{\text{loc}}(\mathbb{R}^3)$ such that

$$\begin{aligned}&\|\nabla^2(\mathbf{u}_j - \boldsymbol{r}_j) - \nabla^2\widetilde{\mathbf{u}}\|_q + \|\partial_3(\mathbf{u}_j - \boldsymbol{r}_j) - \partial_3\widetilde{\mathbf{u}}\|_q \to 0,\\ &\mathbf{u}_j - \boldsymbol{r}_j \to \widetilde{\mathbf{u}} \quad \text{in } L^q(B_m) \text{ for } m \in \mathbb{N}\end{aligned} \tag{4.42}$$

as $j \to \infty$.

In particular, $\mathcal{L}_{\nu,\omega,k}(\mathbf{u}_j - \boldsymbol{r}_j) \to \mathcal{L}_{\nu,\omega,k}\widetilde{u}$ in the sense of distributions and since $\mathcal{L}_{\nu,\omega,k}\mathbf{u}_j = \mathbf{f}_j$ also $\mathcal{L}_{\nu,\omega,k}\boldsymbol{r}_j \to \mathcal{L}\widetilde{\mathbf{u}} - \mathbf{f}$ in $\mathcal{D}'(\mathbb{R}^3)$. The space of linear polynomials π_1 and $L(\pi_1^3) \subset \pi_1^3$ is finite dimensional, $\mathcal{L}_{\nu,\omega,k}\boldsymbol{r}_j \to \mathcal{L}\boldsymbol{r}$ for some $\boldsymbol{r} \in \pi_1^3$ as $j \to \infty$. Hence $\mathbf{u} = \widetilde{\mathbf{u}} + \boldsymbol{r}$, we get $\mathcal{L}_{\nu,\omega,k}\mathbf{u} = \mathbf{f}$, and by (4.42)

$$\|\nu\nabla^2 u\|_q \le c\|\mathbf{f}\|_q, \quad \|k\partial_3\mathbf{u}\|_q \le c\left(1 + \frac{k^4}{\nu^2|\omega|^2}\right)\|f\|_q, \tag{4.43}$$

for $q > 2$.

It remains two cases $q = 2$ and $q \in (1,2)$. Using a duality argument, we prove (4.43) for $q \in (1,2)$.

We define the adjoint T^* by

$$T^*\mathbf{g}(\mathbf{x}) = \int_0^\infty (\boldsymbol{\psi}_t * Q_{\omega/\nu}(t)\mathbf{g})\left(Q^{\mathrm{T}}_{\omega/\nu}(t)\mathbf{x} + \frac{k}{\nu}te_3\right)\frac{\mathrm{d}t}{t}, \quad \mathbf{g} \in \boldsymbol{S}(\mathbb{R}^3).$$

Using Lemmas 4.2 and 4.3 we get the estimates in $L^{q'}$ for all $j \in Z$ and $q' > 2$, and via the convergence of the series we obtain the estimate for T^*

$$\|T^*\mathbf{g}\|_{q'} \le c\|\mathbf{g}\|_{q'}.$$

It implies $\|T\mathbf{f}\|_q \le c\|\mathbf{f}\|_q \quad \forall q \in (1,2),\ \mathbf{f} \in \mathcal{S}(\mathbb{R}^3)$. Thus (4.37) holds for all $q \in (1,\infty)$ and (4.43) holds for $q \in (1,2)$. Case $q = 2$ can be proved by Plancherel's Theorem. □

Proof of Theorem 4.1 (A2): We consider a solution $(\mathbf{u}, p) \in S'(\mathbb{R}^3)$ of (4.1) when $\mathbf{f} = 0$ and $g = 0$. The goal is to show that $\mathbf{u} = \alpha e_3 + \beta\omega e_3 \wedge \mathbf{x}$ for suitable $\alpha, \beta \in \mathbb{R}$ and that p is a constant. In Fourier spaces (4.1) yields the equations

$$\begin{aligned} \nu|\boldsymbol{\xi}|^2\widehat{\mathbf{u}} + \mathrm{i}k\xi_3\widehat{\mathbf{u}} - \omega(\mathbf{e}_3 \wedge \boldsymbol{\xi}) \cdot \nabla_\xi\widehat{\mathbf{u}} + \omega(\mathbf{e}_3 \wedge \widehat{\mathbf{u}}) + \mathrm{i}\boldsymbol{\xi}\widehat{p} &= 0 \\ \mathrm{i}\xi \cdot \widehat{\mathbf{u}} &= 0, \end{aligned} \tag{4.44}$$

hence $|\xi|^2\widehat{p} = 0$. Thus $\operatorname{supp}\widehat{p} \subset \{0\}$ and p is a polynomial. Since ∇p, we assume to belong in $L^q(\mathbb{R}^3) + L^r(\mathbb{R}^3)$ for $1 < q, r < \infty$, p must be a constant. To analyze $\widehat{\mathbf{u}}$ we apply the cylindrical coordinates (s, φ, ξ_3) for ξ, and we denote $\widehat{v}(\varphi) := Q^{\mathrm{T}}_{\mathbf{e}_3}(\varphi)\widehat{\mathbf{u}}(s, \varphi, \xi_3)$. Since $\widehat{p} = \widehat{p}_0\delta_0$, where $\widehat{p}_0 \in \mathbb{C}$, δ_0 denote Dirac's δ distribution,

$$\frac{1}{\omega}(\nu|\boldsymbol{\xi}|^2 + \mathrm{i}k\xi_3)\widehat{\mathbf{v}} - \partial\varphi\widehat{\mathbf{v}} = -\frac{\mathrm{i}\widehat{p}_0}{\omega}\delta_0 \quad \text{in } S(\mathbb{R}^3). \tag{4.45}$$

Next we consider an arbitrary test function $\boldsymbol{\psi} \in C_0^\infty(\mathbb{R}^3 \setminus \{0\}\}^3$, and let

$$\boldsymbol{\psi}_0(s, \varphi, \xi_3) = \mathrm{e}^{-(\nu'|\boldsymbol{\xi}|^2 + \mathrm{i}k'\xi_3)\varphi}\int_{-\infty}^{\varphi} \mathrm{e}^{(-\nu'|\boldsymbol{\xi}|^2|\mathrm{i}k'\xi_3)\varphi'}\boldsymbol{\psi}(s, \varphi, \xi_3),$$

where $\nu' = \nu/\omega$, $k' = k/\omega$.

It implies

$$(\nu'|\boldsymbol{\xi}|^2 + \mathrm{i}k'\xi_3)\boldsymbol{\psi}_0 + \partial_\varphi \boldsymbol{\psi}_0 = \boldsymbol{\psi}.$$

Consequently,

$$\langle \widehat{\mathbf{v}}, \boldsymbol{\psi}\rangle = \langle \widehat{\mathbf{v}}, (\nu'|\boldsymbol{\xi}|^2 + \mathrm{i}k'\xi_3 + \partial_\varphi)\boldsymbol{\psi}_0\rangle = \langle (\nu'|\boldsymbol{\xi}|^2 + \mathrm{i}k\,\xi_3 - \partial_\varphi)\widehat{\mathbf{v}}, \boldsymbol{\psi}_0\rangle = 0$$

due to (4.45) and since $\operatorname{supp}\psi_0 \subset \mathbb{R}^3 \setminus \{0\}$. It implies $\widehat{\mathbf{v}} \subset \{0\}$ and also $\operatorname{supp}\widehat{\mathbf{u}} \subset \{0\}$. This gives us that $\mathbf{u}$ is polynomial. From assumption $\nabla^2\mathbf{u} \in L^q(\mathbb{R}^3) + L^r(\mathbb{R}^3)$, $\nabla^2\mathbf{u} = 0$ and $\mathbf{u}$ is linear polynomial $\mathbf{u}(\mathbf{x}) = \boldsymbol{a} + B\mathbf{x}$, where $\boldsymbol{a} \in \mathbb{R}^3$, $B \in \mathbb{R}^{3,3}$. Then an elementary calculation shows that $\boldsymbol{a} = \alpha\mathbf{e}_3$, $\alpha \in \mathbb{R}$, $B_{ij} = 0$ except $B_{21} = -B_{12} = \beta \in \mathbb{R}$. Hence, $\mathbf{u}(x) = \alpha\mathbf{e}_3 + \beta\omega\boldsymbol{e}_3 \wedge \mathbf{x}$. □ □

4.1.3 Proofs of Lemmas 4.2 and 4.3

Let us begin with proof of Lemma 4.2.

Proof of Lemma 4.2 To estimate $\|T_j\mathbf{f}\|_q$, we will apply as in Chap. 3 the Littlewood–Paley decomposition of $T_j\mathbf{f}$, and precisely we replace it by the equivalent L^q-norm

$$\left\| \left(\int_0^\infty |\varphi_s * T_j\mathbf{f}(\cdot)|^2 \frac{\mathrm{d}s}{s} \right)^{1/2} \right\|_q, \tag{4.46}$$

where $\varphi \in S(\mathbb{R}^3)$, $\varphi_s(\mathbf{x}) = s^{-3/2}\varphi(\mathbf{x}/\sqrt{s})$ for $s > 0$, $\widehat{\varphi_s}(\boldsymbol{\xi}) = \widetilde{\varphi}(\sqrt{s}|\boldsymbol{\xi}|)$, and where $\widetilde{\varphi} \in C_0^\infty(\frac{1}{2}, 2)$ satisfies $0 \le \widetilde{\varphi} \le 1$, and $\int_0^\infty \widetilde{\varphi}(s)^2 \mathrm{d}s/s = \frac{1}{2}$. Thus, we can find $0 \le g \in L^{(q/2)'}(\mathbb{R}^3)$ with $\|g\|_{(q/2)'} = 1$ such that

$$\begin{aligned}
\|T_j\mathbf{f}\|_q^2 \le c \left\| \int_0^\infty |\varphi_s * T_j\mathbf{f}(\cdot)|^2 \frac{\mathrm{d}s}{s} \right\|_{q/2} &= \int_{\mathbb{R}^3} g(\mathbf{x}) \int_0^\infty |\varphi_s * T_j\mathbf{f}(\mathbf{x})|^2 \frac{\mathrm{d}s}{s}\mathrm{d}\mathbf{x} \\
&= \int_0^\infty \left(\int_{\mathbb{R}^3} |\varphi_s * T_j\mathbf{f}(\mathbf{x})|^2 g(\mathbf{x})\mathrm{d}x \right) \frac{\mathrm{d}s}{s}.
\end{aligned}$$

Using (4.27) and the radial symmetry of φ_s and ψ_t^j,

$$\begin{aligned}
\varphi_s * T_j f(\mathbf{x}) &= \int_0^\infty Q^{\mathrm{T}}_{\omega/\nu}(t)\boldsymbol{\varphi}_s * \psi_t^j * \left[\mathbf{f}\left(Q_{\omega/\nu}(t) \cdot -\frac{k}{\nu}t\boldsymbol{e}_3 \right) \right](\mathbf{x}) \frac{\mathrm{d}t}{t} \qquad (4.47) \\
&= \int_{A(s,j)} Q^{\mathrm{T}}_{\omega/\nu}(t)(\varphi_s * \psi_t^j * \mathbf{f}) \left(Q_{\omega/\nu}(t)\mathbf{x} - \frac{k}{\nu}t\boldsymbol{e}_3 \right) \frac{\mathrm{d}t}{t}
\end{aligned}$$

since $\varphi_s * \psi_t^j = 0$ unless $t \in A(s,j) := [2^{2j-4}s, 2^{2j+4}s]$. Let us remark that $\int_{A(s,j)} \mathrm{d}t/t = \log 2^8$ for every $j \in U$ and $s > 0$. Applying the Cauchy–Schwartz, the associativity of convolutions, the inequality

$$|\psi_t^j * (\varphi_s * \mathbf{f})(\mathbf{y})|^2 \leq \|\psi_t^j\|_1 (\psi_t^j * |\varphi_s * \mathbf{f}|^2)(\mathbf{y}),$$

and Lemma 4.1, it implies that

$$\begin{aligned} |\varphi_s * T_j \mathbf{f}(\mathbf{x})|^2 \leq & c \int_{A(s,j)} \left[(\psi_t^j * (\varphi_s * \mathbf{f})) \left(Q_{\omega/\nu}(t)\mathbf{x} - \frac{k}{\nu} t e_3 \right) \right]^2 \frac{\mathrm{d}t}{t} \\ \leq & c_j \int_{A(s,j)} (|\psi_t^j| * |\varphi_s * \mathbf{f}|^2) \left(Q_{\omega/\nu}(t)\mathbf{x} - \frac{k}{\nu} t e_3 \right) \frac{\mathrm{d}t}{t}, \end{aligned} \tag{4.48}$$

where $c_j = c2^{-2|j|}$. Then

$$\begin{aligned} \|T_j \mathbf{f}\|_q^2 \leq & c_j \int_0^\infty \int_{A(s,j)} \int_{\mathbb{R}^3} |\psi_t^j| * |\varphi_s * \mathbf{f}|^2 \left(Q_{\omega/\nu}(t)\mathbf{x} - \frac{k}{\nu} t e_3 \right) g(\mathbf{x}) \mathrm{d}\mathbf{x} \frac{\mathrm{d}t}{t} \frac{\mathrm{d}s}{s} \\ = & c_j \int_j^\infty \int_{A(s,j)} \int_{\mathbb{R}^3} |\psi_t^j| * |\varphi_s * \mathbf{f}|^2(\mathbf{x}) \mathbf{g} \left(Q^{\mathrm{T}}_{\omega/\nu}(t)\mathbf{x} + \frac{k}{\nu} t e_3 \right) \mathrm{d}\mathbf{x} \frac{\mathrm{d}t}{t} \frac{\mathrm{d}s}{s} \\ = & c_j \int_{\mathbb{R}^3} \int_0^\infty |\varphi_s * \mathbf{f}|^2(\mathbf{y}) \int_{A(s,j)} |\psi_t^j| * \mathbf{g} \left(Q^{\mathrm{T}}_{\omega/\nu}(t)\mathbf{y} + \frac{k}{\nu} t e_3 \right) \frac{\mathrm{d}t}{t} \frac{\mathrm{d}s}{s} \mathrm{d}y \\ \leq & c_j \int_{\mathbb{R}^3} \left(\int_0^\infty |\varphi_s * \mathbf{f}|^2(\mathbf{y}) \frac{\mathrm{d}s}{s} \right) \mathcal{M}_j \mathbf{g}(\mathbf{y}) \mathrm{d}\mathbf{y}. \end{aligned}$$

□

Remark 4.1 Let us mention that the inequality

$$\int_{\mathbb{R}^3} |\psi^t| * v(\mathbf{x}) h(\mathbf{x}) \mathrm{d}\mathbf{x} = \int_{\mathbb{R}^3} v(\mathbf{y}) |\psi_t^j| * h(\mathbf{y}) \mathrm{d}\mathbf{y}$$

has been used. Finally, using the Hölder inequality

$$\|T_j \mathbf{f}\|_q^2 \leq c 2^{2-|j|} \left\| \int_0^\infty |\varphi_s * \mathbf{f}|^2(\cdot) \frac{\mathrm{d}s}{s} \right\|_{q/2} \|\mathcal{M}_j \mathbf{g}\|_{(q/2)}$$

and the Paley–Littlewood decomposition of $\mathbf{f}$, and (4.46) for $T_j\mathbf{f}$, it gives us claim of Lemma 4.2. □

Now we prove Lemma 4.3.

Proof of Lemma 4.3 Using $h_{t2^{-2|j|}}(\mathbf{x}) \leq c h_{s2^{-2|j|}}(\mathbf{x})$ for all $j \in \mathbb{Z}$, $t \in A_s \in [s/16, 16s]$, $s > 0$, $\mathbf{x} \in \mathbb{R}^3$, and c being independent of j, t, s, and $\mathbf{x}$ and by

Lemma 4.1 we get

$$\begin{aligned}\mathcal{M}_j\mathbf{g}(\mathbf{x}) \leq & c2^{-2|j|}\sup_{s>0} h_{s2^{-2|j|}} * \int_{A_s} |\mathbf{g}| \left(Q^{\mathrm{T}}_{\omega/\nu}(t) + \frac{k}{\nu}t\boldsymbol{e}_3\right)\frac{\mathrm{d}t}{t} \\ \leq & c2^{-2|j|}\sup_{r>0} h_r * \sup_{s>0}\frac{1}{s}\int_{A_s} |g|| \left(Q^{\mathrm{T}}_{\omega/\nu}(t)\mathbf{x} + \frac{k}{\nu}t\boldsymbol{e}_3\right)\mathrm{d}t.\end{aligned}$$

We apply the Hardy–Littlewood maximal operator $\mathcal{M}$ on $L^p(\mathbb{R}^3)$ defined as

$$\mathcal{M}g(\mathbf{x}) = \sup_{s>0}\frac{1}{|B_s(\mathbf{x})|}\int_{B_s(\mathbf{x})} |g(\mathbf{y})|\mathrm{d}\mathbf{y}$$

and the "helical" maximal operator

$$\mathcal{M}_{\mathrm{hel}}\mathbf{g}(\theta, x_3) := \sup_{s>0}\frac{1}{s}\int_{A_S} |\mathbf{g}| \left(\theta - \frac{\omega}{\nu}t, x_3 + \frac{k}{\nu}t\right)\mathrm{d}t$$

for functions $\mathbf{g}$ depending on (θ, x_3) and 2π-periodic in θ. $0 \leq h \in L^1(\mathbb{R}^3)$ is radially symmetric and strictly decreasing,

$$\sup_{r>0} h_r * \mathbf{u}(\mathbf{x}) \leq c\mathcal{M}\mathbf{u}(\mathbf{x}).$$

Then

$$\mathcal{M}_j\mathbf{g}(\mathbf{x}) \leq c2^{-2|j|}\mathcal{M}(\mathcal{M}_{\mathrm{hel}}g_r(\cdot,\cdot))(\mathbf{x}),$$

where $g_r(\theta, x_3) = g(r, \theta, x_3) = g(\mathbf{x})$ is considered as function of θ, x_3 only and

$$\|\mathcal{M}_j g\|_p \leq c2^{-2|j|}\|\mathcal{M}_{\mathrm{hel}}g_r(\cdot,\cdot)\|_{L^p(\mathbb{R}^3)} \tag{4.49}$$

because of the L^p-continuity of $\mathcal{M}$. We would like to estimate $\mathcal{M}_{\mathrm{hel}}g_r(\cdot,\cdot)$ in $L^p(\mathbb{R}^3)$, fix $r > 0$, and use the 2π-periodicity of g_r with respect to θ, and then we get

$$\begin{aligned}&\int_{\mathbb{R}}\int_0^{2\pi} |\mathcal{M}_{\mathrm{hel}}g_r(\theta, x_3)|^p\mathrm{d}\theta\mathrm{d}x_3 \\ &\leq \int_{\mathbb{R}}\int_0^{2\pi}\left|\sup_{s>0}\frac{1}{s}\int_{-16s}^{16s} |g_r| \left(\theta - \frac{\omega}{\nu}\left(x_3 + \frac{k}{\nu}t\right), x_3 + \frac{k}{\nu}t\right)\mathrm{d}t\right|^p \mathrm{d}t\mathrm{d}x_3 \\ &= \int_{\mathbb{R}}\int_0^{2\pi}\left|\sup_{s>0}\frac{1}{s}\int_{-16s}^{16s} \gamma_{r,\theta}\left(x_3 + \frac{k}{\nu}t\right)\mathrm{d}t\right|^p \mathrm{d}\theta\mathrm{d}x_3,\end{aligned}$$

where $\gamma_{r,\theta}(y_3) = |g_r|(\theta - \frac{\omega}{k}y_3, y_3)$. Thus the Hardy–Littlewood maximal operator on $\mathbb{R}^1$ applied on $\gamma_{r,\theta}(\cdot)$ yields on constant $c > 0$ independent of r, θ and k/ν such that

$$\begin{aligned}\int_{\mathbb{R}}\int_0^{2\pi} |\mathcal{M}_{\text{hel}}g_r(\theta, x_3)|^p \mathrm{d}\theta\mathrm{d}x_3 \leq& c\int_0^{2\pi}\int_{\mathbb{R}} |\gamma_{r,\theta}(x_3)|^p \mathrm{d}x_3\mathrm{d}\theta\\ =& c\int_{\mathbb{R}}^{2\pi} |g_r(\theta, x_3)|^p \mathrm{d}x_3\mathrm{d}\theta.\end{aligned}$$

Finally,

$$\|\mathcal{M}_{\text{hel}}g_r(\cdot,\cdot)\|_{L^p(\mathbb{R}^3)} \leq c\|\mathbf{g}\|_{L^p(\mathbb{R}^3)}. \tag{4.50}$$

From (4.49), (4.50) we complete the proof. □

4.1.4 Proof of Corollary 4.1

(B1): We follow the line of the proof given by Farwig, [60],[61]. Since $\nabla^2\mathbf{u}$, $\partial_3\mathbf{u} \in L^q(\mathbb{R}^3)$, $1 < q < 4$, there exists a unique matrix $B = (b_{ij})_{ij=1}^3 \in \mathbb{R}^{3\times3}$ such that

$$\nabla(\mathbf{u} - B\mathbf{x}) \in L^r(\mathbb{R}^3)^9 \quad \text{for all } r > 1,\ \frac{1}{r} \in \frac{1}{q} - \left[-\frac{1}{4}, \frac{1}{3}\right]. \tag{4.51}$$

Moreover, $\operatorname{div}\mathbf{u} \in L^q(\mathbb{R}^3)$ by assumption and $\operatorname{div}\mathbf{u} - \operatorname{tr} B \in L^r(\mathbb{R}^3)$ give us $\operatorname{tr} B = 0$. Further, analogously, since $\partial_3\mathbf{u} \in L^q(\mathbb{R}^3)$, it follows that the coefficients b_{13}, b_{23} and b_{33} vanish. Concerning $\mathbf{u}_3$ from the inequality (4.51), we get that: $-\mathbf{x}_2\partial_1\mathbf{u}_3 + \mathbf{x}_1\partial_2\mathbf{u}_3 \in L^q(\mathbb{R}^3)$ yielding

$$\frac{1}{|\mathbf{x}|}(-\mathbf{x}_2\partial_1\mathbf{u}_3 + \mathbf{x}_1\partial_2\mathbf{u}_3) \in L^q(B_1^c);$$

and

$$\nabla'\mathbf{u}_3 - (b_{31}, b_{32})^{\mathrm{T}} \in L^r(B_1^c)^2.$$

This implies

$$\frac{1}{|\mathbf{x}|}(-\mathbf{x}_2 b_{31} + \mathbf{x}_1 b_{32}) \in L^q(B_1^c) + L^r(B_1^c),$$

which is possible iff $b_{31} = b_{32} = 0$.

Summing up the previous results, we can conclude of constant the existence $\beta, \gamma, \delta \in \mathbb{R}$ satisfying

$$\mathbf{v}' := \mathbf{u}' - \beta\begin{pmatrix} -\mathbf{x}_2 \\ \mathbf{x}_1 \end{pmatrix} - \gamma\begin{pmatrix} \mathbf{x}_1 \\ -\mathbf{x}_2 \end{pmatrix} - \delta\begin{pmatrix} \mathbf{x}_2 \\ \mathbf{x}_1 \end{pmatrix}$$

and moreover $\nabla\mathbf{v}' \in L^r(\mathbb{R}^3)^6$, for all $r > 1$, $1/r = 1/q - \left[\frac{1}{4}, \frac{1}{3}\right]$. In addition,

$$\mathcal{L}'\mathbf{u}' = -\frac{1}{|x|}(-x_2\partial_1\mathbf{u}' + x_1\partial_2\mathbf{u}') + \frac{1}{|x|}(-u_2, u_1)^{\mathrm{T}} \in L^q(B_1^c)^2.$$

Let us note that

$$\mathcal{L}'\begin{pmatrix} \mathbf{x}_1 \\ -\mathbf{x}_2 \end{pmatrix} = \frac{2}{|\mathbf{x}|}\begin{pmatrix} x_2 \\ x_1 \end{pmatrix} \quad \text{and} \quad \mathcal{L}'\begin{pmatrix} \mathbf{x}_2 \\ \mathbf{x}_1 \end{pmatrix} = -\frac{2}{|\mathbf{x}|}\begin{pmatrix} \mathbf{x}_1 \\ -\mathbf{x}_2 \end{pmatrix},$$

and $\mathcal{L}'\binom{-\mathbf{x}_2}{\mathbf{x}_1} = 0$.

Case 1: $1 < q < \frac{12}{7}$.

In this case $1/r = 1/q - \frac{1}{4}$ satisfies $1 < r < 3$. Then $\nabla'\mathbf{v}' \in L^r(B_1^c)^4$, and by Galdi, see [93, Theorem II.5.1], we get the existence of a vector $\mathbf{v}'_\infty \in \mathbb{R}^2$ such that

$$\frac{\mathbf{v}' - \mathbf{v}'_\infty}{|\mathbf{x}|} \in L^r(B_1^c)^2.$$

Thus,

$$\mathcal{L}'(\mathbf{u}' - (\mathbf{v}' - \mathbf{v}'_\infty)) = \frac{2\gamma}{|\mathbf{x}|}\begin{pmatrix} \mathbf{x}_2 \\ \mathbf{x}_1 \end{pmatrix} - \frac{2\delta}{|\mathbf{x}|}\begin{pmatrix} \mathbf{x}_1 \\ -\mathbf{x}_2 \end{pmatrix} + \frac{1}{|\mathbf{x}|}\mathbf{v}'^{\perp}_\infty \in L^q(B_1^c)^2 + L^r(B_1^c)^2$$

proving $\gamma = \delta = 0$ and $v'_\infty = 0$. Consequently, $\mathbf{v} = \mathbf{u} - \widetilde{\beta}\omega \wedge \mathbf{x}$ with $\widetilde{\beta} = \beta/\omega$ satisfies $\nabla\mathbf{v} \in L^r(\mathbb{R}^3)^9$.

Case 2: $\frac{12}{7} \le q < 4$.

In case $q < 3$, define $r > 1$ by $1/r = 1/q - \frac{1}{3}$ yielding $r \ge 4$. However, if $3 \le q < 4$, let $1/r = 1/q - \frac{1}{4}$ yielding $r \ge 12$. Hence $r > 3$ and $v \in \widehat{H}^{1,r}(\mathbb{R}^3)^2 \subset C^{0,\alpha}_{\mathrm{loc}}(\mathbb{R}^3)^2$ where $\alpha = 1 - 3/r > 0$. Then by Galdi [93, Lemma II.2.6]

$$\frac{\mathbf{v}' - \mathbf{v}'_0}{|\mathbf{x}|} \in L^r(\mathbb{R}^3),$$

where $\mathbf{v}'_0 \in \mathbb{R}^2$ is well defined. As in Case 1 we get that

$$\mathcal{L}'(\mathbf{u}' - (\mathbf{v}' - \mathbf{v}'_0)) = \frac{2\gamma}{|\mathbf{x}|}\begin{pmatrix} \mathbf{x}_2 \\ \mathbf{x}_1 \end{pmatrix} - \frac{2\delta}{|\mathbf{x}|}\begin{pmatrix} \mathbf{x}_1 \\ -\mathbf{x}_2 \end{pmatrix} + \frac{1}{|\mathbf{x}|}\mathbf{v}'^{\perp}_0 \in L^q(B_1^c)^2 + L^r(B_1^c)^2.$$

Since $1/|\mathbf{x}|v_0'^{\perp} \in L^r(B_1^c)^2$, we can conclude that $\gamma = \delta = 0$. Thus $\mathbf{v} = \mathbf{u} - \widetilde{\beta} \wedge \mathbf{x}$ with $\widetilde{\beta} = \beta/\omega$ satisfy $\nabla \mathbf{v} \in L^r(\mathbb{R}^3)^9$.

(B2) From [59, Theorem 2.3] there exists $a \in \mathbb{R}^3$ such that

$$\mathbf{v} = \mathbf{u} - \boldsymbol{a} \in L^s(\mathbb{R}^3) \quad \text{for all } s > 1\ \frac{1}{s} \in \frac{1}{q} - \left[\frac{1}{2}, \frac{2}{3}\right].$$

We assume $\beta \in \mathbb{R}$ from part (B1) vanishes yielding $\nabla v \in L^r(\mathbb{R}^3)^q$ for all $r > 1$, $1/r = 1/q - [\frac{1}{4}, \frac{1}{3}]$. Let

$$\widetilde{\mathcal{L}}\mathbf{u}' := -(-\mathbf{x}_2\partial_1\mathbf{u}' + \mathbf{x}_1\partial_2\mathbf{u}') + \mathbf{u}'^{\perp} = -\partial_\theta\mathbf{u}' + \mathbf{u}'^{\perp}.$$

Since $\widetilde{\mathcal{L}}\mathbf{v}' = -\partial_\theta\mathbf{v}' + \mathbf{v}'^{\perp} = \widetilde{\mathcal{L}}\mathbf{u}' - \boldsymbol{a}'^{\perp}$, applying the integration with respect to $\theta \in [0, 2\pi]$, it implies that

$$2\pi\boldsymbol{a}'^{\perp} = \int_0^{2\pi} \widetilde{\mathcal{L}}\mathbf{u}'\mathrm{d}\theta - \int_0^{2\pi} \mathbf{v}'^{\perp}\mathrm{d}\theta \in L^q(\mathbb{R}^3)^2 + L^s(\mathbb{R}^3)^2.$$

It implies $\boldsymbol{a}' = 0$.

Lemma 4.4 *Let $r > n$ and $\mathbf{v} \in \widehat{\boldsymbol{H}}^{1,r}(\mathbb{R}^n)$. Then $\mathbf{v}_0 = v(0)$ is well defined and*

$$\int_{R^n} \frac{|\mathbf{v} - \mathbf{v}_0|^r}{|\mathbf{x}|^r} dx \leq c \int_{R^n} |\nabla\mathbf{v}|^r dx.$$

Proof of Lemma 4.4 Because $r > n$, it follows that $\mathbf{v} \in C^{1,\alpha}_{\mathrm{loc}}$, $\alpha = 1 - n/r$, and there exists a constant $C > 0$ independent of $\varepsilon > 0$ such that $|\mathbf{v}(x) - \mathbf{v}_0| \leq c|\mathbf{x}|^\alpha\|\nabla\mathbf{v}\|_{L^r(B_\varepsilon)}\ \forall x \in B_\varepsilon$. Then for all $0 < \varepsilon < R < \infty$

$$\begin{aligned}
\int_{B_c\setminus B_\varepsilon} \frac{|\mathbf{v} - \mathbf{v}_0|^r}{|\mathbf{x}|^r}\mathrm{d}x =& \frac{1}{n-r}\int_{B_r\setminus B_\varepsilon} |\mathbf{v} - \mathbf{v}_0|^r \operatorname{div}\left(\frac{\mathbf{x}}{|\mathbf{x}|^r}\right)\mathrm{d}x \\
=& -\frac{r}{n-r}\int_{B_R\setminus B_\varepsilon} (\mathbf{v} - \mathbf{v}_0)|\mathbf{v} - \mathbf{v}_0|^{r-2}\nabla v \cdot \frac{\mathbf{x}}{|\mathbf{x}|^r}\mathrm{d}x \\
&+ \frac{R^{1-r}}{n-r}\int_{\partial B_R} |\mathbf{v} - \mathbf{v}_0|^r\mathrm{d}\theta - \frac{\varepsilon^{1-r}}{n-r}\int_{\partial B_\varepsilon} |\mathbf{v} - \mathbf{v}_0|^r\mathrm{d}\sigma.
\end{aligned}$$

We omit the integral on ∂B_R since $n - r < 0$. Applying the Hölder inequality, we get

$$\begin{aligned}
\int_{B_R\setminus B_\varepsilon} \frac{|\mathbf{v} - \mathbf{v}_0|^r}{|\mathbf{x}|^r}\mathrm{d}x \leq& \frac{r}{r-n}\left(\int_{B_R\setminus B_\varepsilon} \frac{|\mathbf{v} - \mathbf{v}_0|^r}{|\mathbf{x}|^r}\mathrm{d}x\right)^{1/r'}\left(\int_{B_R\setminus B_\varepsilon} |\nabla\mathbf{v}|^r\mathrm{d}x\right)^{1/r} \\
&+ c\|\nabla\mathbf{v}\|^r_{L^r(B_\varepsilon)}.
\end{aligned}$$

Passing with $\varepsilon \to 0$, $R \to \infty$, we get the assertion with the constant $c = (r/(r - n))^r$. □

Let us prove Theorem 4.2.

Proof of Theorem 4.2 (C1) From Theorem 4.2, Corollary 4.1, it follows that a given solution $\mathbf{u} \in \widehat{\boldsymbol{H}}^{2,q}(\mathbb{R}^3)$, $\beta \in \mathbb{R}$ $\nabla'(\mathbf{u} - \beta\mathbf{u}(\mathbf{e}_3 \wedge \mathbf{x})) \in L^r(\mathbb{R}^3)^6$, $1/r = 1/q - \frac{1}{4}$ and $\partial_3(\mathbf{u} - \beta\omega(\mathbf{e}_3 \wedge \mathbf{x})) \in L^q(\mathbb{R}^3)$. Since $\mathbf{u} - \beta\omega(\mathbf{e}_3 \wedge \mathbf{x})$ solves (4.1), we can assume that $\beta = 0$. It implies for a.a. $\mathbf{x}_3 \in \mathbb{R}$ that

$$\int_{R^2} |\nabla'\mathbf{u}(\mathbf{x}', \mathbf{x}_3)|^r \mathrm{d}\mathbf{x}' < \infty, \quad \int_{\mathbb{R}^2} |\partial_3\mathbf{u}(\mathbf{x}', \mathbf{x}_3)|^q \mathrm{d}x' < \infty.$$

Then we have existence of a sequence of radii $(R_j) \subset \mathbb{R}_+$ such that

$$\left.\begin{aligned} \int_0^{2\pi} |\nabla'\mathbf{u}(R_j, \theta, \mathbf{x}_3)|^r \mathrm{d}\theta = 0(R_j^{-2}) \\ \int_0^{2\pi} |\partial_3 u(R_j, \theta, \mathbf{x}_3)|^q \mathrm{d}\theta = 0(R_j^{-2}) \end{aligned}\right\} \quad \text{as } j \to \infty. \tag{4.52}$$

Since $1 < q < 2$, $\nabla^2 u \in L^q$, by Galdi [93, Theorem II.5.1] there exists for a.a. $\mathbf{x}_3 \in \mathbb{R}$ a matrix $A(x_3) \in \mathbb{R}^{3,3}$ such that

$$\left(\int_{\mathbb{R}^2} \frac{|\nabla\mathbf{u}(\mathbf{x}', \mathbf{x}_3) - A(x_3)|^q}{|\mathbf{x}'|^q} \mathrm{d}\mathbf{x}'\right)^{1/q} \le \frac{q}{2-q} \left(\int_{\mathbb{R}^2} |\nabla'\nabla\mathbf{u}(\mathbf{x}', \mathbf{x}_3)|^q\right)^{1/q} \mathrm{d}x. \tag{4.53}$$

From [93, Lemma 5.2] it follows

$$\int_0^{2\pi} |\nabla u(R, \theta, \mathbf{x}_3) - A(\mathbf{x}_3)|^q \mathrm{d}\theta = 0(R^{q-2}) \quad \text{as } R \to \infty. \tag{4.54}$$

From (4.52), (4.54) it follows that $A(\mathbf{x}_3) = 0$. Hence, (4.53) and (4.3) yield (4.10) and (4.11) follows from (4.11) and (4.3).

(C2) If $1 < q < 3$, by Galdi [93, Theorem II.5.1] gives us estimate

$$\left(\int_{\mathbb{R}^3} \frac{|\nabla\mathbf{u}(\mathbf{x}) - A|^q}{|\mathbf{x}|^q} \mathrm{d}x\right)^{1/q} < \frac{q}{3-q} \left(\int_{\mathbb{R}^3} |\nabla^2\mathbf{u}(\mathbf{x})|^q \mathrm{d}x\right)^{1/q}$$

with a constant matrix $A \in R^{3,3}$. Moreover, by Galdi [93, Lemma 5.2]

$$\int_{|\mathbf{y}|=1} |\nabla\mathbf{u}(R\mathbf{y})|^q \mathrm{d}0(\mathbf{y}) = 0(R^{q-3})$$

as $R \to \infty$, where $\int_{|\mathbf{y}|=1}$ denotes the surface integral on the unit sphere of $\mathbb{R}^3$.

(C3) From (C2) $\mathbf{u}_3$ solves the problem

$$-\nu\nabla\mathbf{u}_3 + k\,\partial_3\mathbf{u}_3 - \omega(\mathbf{e}_3 \wedge \mathbf{x})\cdot\nabla\mathbf{u}_3 = \mathbf{f}_3.$$

Since $\omega(\mathbf{e}_3 \wedge \mathbf{x})\cdot\nabla\mathbf{u}_3 = |\omega|\partial_\theta\mathbf{u}_3$, integrating with respect to $\theta \in (0, 2\pi)$, we get θ-independent function $\mathbf{U}_3(x) = 1/2\pi\int_0^{2\pi}\mathbf{u}_3(|x'|, \theta', x_3)\mathrm{d}\theta'$ satisfying the equation

$$-\nu\Delta\mathbf{U}_3 + k\,\partial_3\mathbf{U}_3 = \frac{1}{2\pi}\int_0^{2\pi}\mathbf{f}_3\mathrm{d}\theta'.$$

Applying the Fourier transformation and the Marcinkiewicz multiplier theorem, we get that $\mathbf{U}_3$ satisfies

$$\|k\,\partial_3\mathbf{U}_3\|_q \le C\|\mathbf{f}\|_q,$$

where c is independent of k, ν. Applying Wirtinger's inequality there is a constant $c > 0$ such that for a.a. $r = |\mathbf{x}'| > 0$, $\mathbf{x}_3 \in \mathbb{R}$

$$\|\partial_3\mathbf{u}_3(r,\cdot,\mathbf{x}_3)\|_{L^q(0,2\pi)} \le c(\|\partial_\theta\partial_3\mathbf{u}_3(r,\cdot,\mathbf{x}_3)\|_{L^q(0,2\pi)} + \|\partial_3\mathbf{U}_3(r,\cdot,\mathbf{x}_3)\|_{L^q(0,2\pi)}).$$

Dividing by $1 + r$ and integrating with respect to $r > 0$ and $\mathbf{x}_3 \in \mathbb{R}^3$, we get

$$\left\|\frac{\partial_3\mathbf{u}_3}{1+|\mathbf{x}'|}\right\|_q \le c\left(\left\|\frac{\partial_\theta\partial_3\mathbf{u}_3}{1+|\mathbf{x}'|}\right\|_q + \left\|\frac{\partial_3\mathbf{U}_3}{1+|\mathbf{x}'|}\right\|_1\right).$$

Since for the first term we can apply

$$\|\partial_3\nabla'\mathbf{u}_3\|_q \le \frac{c}{\nu}\|\mathbf{f}\|_q$$

and for the second term

$$\|\partial_3\mathbf{U}_3\|_q \le \frac{c}{\nu}\|\mathbf{f}\|_q,$$

we get (4.13). □

4.1.5 *The Associated Time-Periodic Oseen Problem and Proof of Theorem 4.1*

Let also formulate the Oseen type problem in non-dimensioneless form :

$$\left\{\begin{aligned} S_{1,\tilde{\omega}}\boldsymbol{u} + \nabla p - \mathrm{Re}\,\partial_3\boldsymbol{u} &= \boldsymbol{f}, \\ \operatorname{div}\boldsymbol{u} &= 0 \end{aligned}\right\} \text{ in } \mathbb{R}^3, \tag{4.55}$$

where $\boldsymbol{u} : \mathbb{R}^3 \to \mathbb{R}^3$, $p : \mathbb{R}^3 \to \mathbb{R}$ are the velocity and pressure, respectively,

$$S_{1,\widetilde{\omega}}\boldsymbol{u} = -\,\Delta\boldsymbol{u} + \widetilde{\omega}\mathcal{R}\boldsymbol{u},$$
$$\mathcal{R}\boldsymbol{u} = (\boldsymbol{e}_3 \wedge \boldsymbol{x} \cdot \nabla\boldsymbol{u} - \boldsymbol{e}_3 \wedge \boldsymbol{u})$$

and Re, $\widetilde{\omega}$ are Reynold's and Taylor dimensionless constants.
For more details, see, e.g., [96, 98, 102].

The aim of this part is to prove elliptic L^q-estimates for solutions $(\boldsymbol{u}, p)$ to (4.55).

Let us defined the dimensionless forms of the operators

$$\mathcal{L} := S_{1,\widetilde{\omega}}\boldsymbol{u} - \text{Re}\ \partial_3\boldsymbol{u},$$
$$\mathcal{L}^* := S^*_{1,\widetilde{\omega}} + \text{Re}\ \partial_3\boldsymbol{u}$$
$$S^*_{1,\widetilde{\omega}} = -\,\Delta\boldsymbol{u} + \text{Re}\ \boldsymbol{u}.$$

For the completeness, we also formulate the existential result in a Hilbertian structure.

Theorem 4.3 *Let Re* ≥ 0, *and* $\widetilde{\omega} > 0$. *For any* $\boldsymbol{f} \in \boldsymbol{C}_0^\infty(\mathbb{R}^3)$ *there exists a solution* $(\boldsymbol{u}, p)$ *to the problem (4.55) such that*

$$\boldsymbol{u} \in \widehat{\boldsymbol{H}}^{2,2}(\mathbb{R}^3) \cap \widehat{\boldsymbol{H}}^{1,2}(\mathbb{R}^3) \cap \boldsymbol{L}^6(\mathbb{R}^3) \cap \boldsymbol{C}^\infty(\mathbb{R}^3),$$
$$p \in \widehat{H}^{1,2}(\mathbb{R}^3) \cap L^6(\mathbb{R}^3) \cap C^\infty(\mathbb{R}^3) \tag{4.56}$$

that satisfies

$$\|\nabla^2\boldsymbol{u}\|_2 + \|\nabla p\|_2 \leq C_8\|\boldsymbol{f}\|_2 \tag{4.57}$$

with C_8 *independent of Re and* $\widetilde{\omega}$. *Moreover,*

$$(\boldsymbol{u}, p) \in \bigcap_{m=1}^{\infty}\left(\widehat{\boldsymbol{H}}^{m,2}(\mathbb{R}^3) \times \widehat{H}^{m,2}(\mathbb{R}^3)\right). \tag{4.58}$$

All the statements are valid also for the adjoint problem

$$\mathcal{L}^*\boldsymbol{u} + \nabla p = \boldsymbol{f},$$
$$\text{div}\,\boldsymbol{u} = 0. \tag{4.59}$$

Theorem 4.4 *Let* $1 < q < \infty$, $\mathcal{R}_0 > 0$, $0 < Re < \mathcal{R}_0$, *and* $\widetilde{\omega} > 0$. *For any* $\boldsymbol{f} \in L^q(\mathbb{R}^3)^3$ *there exists a solution* $(\boldsymbol{u}, p) \in \widehat{\boldsymbol{H}}^{2,q}(\mathbb{R}^3) \times \widehat{H}^{1,q}(\mathbb{R}^3)$ *to (4.55) that satisfies*

$$\|\nabla^2\boldsymbol{u}\|_q + \|\nabla p\|_q \leq C_4\|\boldsymbol{f}\|_q \tag{4.60}$$

with C_4 *independent of* $\mathcal{R}_0$, *Re, and* $\widetilde{\omega}$. *Moreover,*

$$\|Re\partial_3 \boldsymbol{u}\|_q + \|\widetilde{\omega}(\boldsymbol{e}_3 \wedge \boldsymbol{x} \cdot \nabla \boldsymbol{u} - \boldsymbol{e}_3 \wedge \boldsymbol{u})\|_q \le C_5(1 + |\widetilde{\omega}|)^{-2})\|\boldsymbol{f}\|_q. \tag{4.61}$$

If $1 < q < 4$, *then*

$$\|\nabla \boldsymbol{u}\|_{\frac{4q}{4-q}} \le C_6(Re^{-\frac{1}{4}} + \widetilde{\omega}^{-\frac{1}{2}})\|\boldsymbol{f}\|_q. \tag{4.62}$$

If $1 < q < 2$, *then*

$$\|\boldsymbol{u}\|_{\frac{2q}{2-q}} \le C_7(Re^{-\frac{1}{2}} + \widetilde{\omega}^{-1})\|\boldsymbol{f}\|_q. \tag{4.63}$$

The constants C_5, C_6, C_7 *depend only on* $\mathcal{R}_0$. □

4.1.5.1 Mathematical Preliminaries

Lemma 4.5 *If the statements of Theorem 4.4 are true for all* $\boldsymbol{f} \in \boldsymbol{C}_0^\infty(\mathbb{R}^3)$, *then Theorem 4.4 is valid also for all* $\boldsymbol{f} \in \boldsymbol{L}^q(\mathbb{R}^3)$. □

Proof of Lemma 4.5 Assume that the statements Theorem 4.4 are true for all $\boldsymbol{f} \in \boldsymbol{C}_0^\infty(\mathbb{R}^3)$. Let $\boldsymbol{f} \in \boldsymbol{L}^q(\mathbb{R}^3)$ and choose $\{\boldsymbol{f}_n\}_{n=1}^\infty \subset \boldsymbol{C}_0^\infty(\mathbb{R}^3)$ with $\lim_{n\to\infty} \boldsymbol{f}_n = \boldsymbol{f}$ in $\boldsymbol{L}^q(\mathbb{R}^3)$. Let $(\boldsymbol{u}_n, p_n)$ be the solution to (4.55) with $\boldsymbol{f}_n$ as the right-hand side. Then choose $\boldsymbol{\alpha}_n, \boldsymbol{\beta}_n, \boldsymbol{\kappa}_n \in \mathbb{R}^3$ and $t_n \in \mathbb{R}$ such that

$$0 = \int_{B_1} (\partial_1 \boldsymbol{u}_n - \boldsymbol{\alpha}_n)\,\mathrm{d}\mathbf{x} = \int_{B_1} (\partial_2 \boldsymbol{u}_n - \boldsymbol{\beta}_n)\,\mathrm{d}\mathbf{x}, \tag{4.64}$$

$$0 = \int_{B_1} [\boldsymbol{u}_n - (\boldsymbol{\kappa}_n - \boldsymbol{\alpha}_n \boldsymbol{x}_1 + \boldsymbol{\beta}_n \boldsymbol{x}_2)]\,\mathrm{d}\mathbf{x}, \tag{4.65}$$

and $0 = \int_{B_1} (p_n - t_n)\,\mathrm{d}\mathbf{x}$. Put $\boldsymbol{r}_n := \boldsymbol{\kappa}_n + \boldsymbol{\alpha}_n \boldsymbol{x}_1 + \boldsymbol{\beta}_n \boldsymbol{x}_2$. From (4.60)–(4.61) we see $\{(\boldsymbol{u}_n - \boldsymbol{r}_n,\ p_n - t_n)\}_{n=1}^\infty$ is a Cauchy sequence in the Banach space X_m, where

$$X_m := \{(\boldsymbol{u}, p) \in \boldsymbol{L}^1_{\mathrm{loc}}(\mathbb{R}^3) \times L^1_{\mathrm{loc}}(\mathbb{R}^3) \mid \|(\boldsymbol{u}, p)\|_{X_m} < \infty\},$$

$$\|(\boldsymbol{u}, p)\|_{X_m} := \|\nabla^2 \boldsymbol{u}\|_q + \|\nabla p\|_q + \mathcal{R}\|\partial_3 \boldsymbol{u}\|_q + \|\boldsymbol{u}\|_{L^q(B_m)} + \|p\|_{L^q(B_m)} \qquad \forall m \in \mathbb{N},$$

where

$$B_m = \{\boldsymbol{x} \in \mathbb{R}^3 \colon |\mathbf{x}| \le m\}.$$

Then there is an element $(\boldsymbol{u}, p) \in \bigcap_{m\in\mathbb{N}} X_m$ satisfying

$$\lim_{n\to\infty} (\boldsymbol{u}_n - \boldsymbol{r}_n, p_n - t_n) = (\boldsymbol{v}, p) \quad \text{in } X_m, \quad \forall m \in \mathbb{N}.$$

It implies that

$$\lim_{n\to\infty} \mathcal{L}(\boldsymbol{u}_n - \boldsymbol{r}_n) + \nabla(p_n - t_n) = \mathcal{L}\boldsymbol{u} + \nabla p \quad \text{in } \mathcal{D}'(\mathbb{R}^3).$$

By construction we have

$$\lim_{n\to\infty} \mathcal{L}\boldsymbol{u}_n + \nabla p_n = \lim_{n\to\infty} \boldsymbol{f}_n = \boldsymbol{f} \quad \text{in } \boldsymbol{L}^q(\mathbb{R}^3).$$

Thus $\lim_{n\to\infty} \mathcal{L}\boldsymbol{r}_n = \boldsymbol{f} - (\mathcal{L}\boldsymbol{u} + \nabla p)$ in $\mathcal{D}'(\mathbb{R}^3)$. Consequently, $\boldsymbol{f} - (\mathcal{L}\boldsymbol{u} + \nabla p)$ is equal to $\mathcal{L}\boldsymbol{r}$ for some first order polynomial r. It follows that

$$(\boldsymbol{u} + \boldsymbol{r}, p) \in \widehat{\boldsymbol{H}}^{2,q}(\mathbb{R}^3) \times \widehat{H}^{1,q}(\mathbb{R}^3)$$

solves (4.55).

Moreover, $(\boldsymbol{u}_n, p_n)$ satisfies (4.60)–(4.61), and we get that $(\boldsymbol{v} + \boldsymbol{r}, p)$ satisfies (4.60)–(4.61). This proves the first part of Theorem 4.4 with respect to $\boldsymbol{f}$.

If $1 < q < 4$, we repeat the argument with the modification that we ignore (4.64) ($\boldsymbol{\alpha}_n = \boldsymbol{\beta}_n = 0$), and add the term $\|\nabla \boldsymbol{u}\|_{4q/(4-q)}$ to the X_m-norm. Then we obtain a solution to (4.55) that also satisfies (4.62). If $1 < q < 2$ we ignore (4.65), (4.64) (put $\boldsymbol{\alpha}_n = \boldsymbol{\beta}_n = \boldsymbol{\kappa}_n = 0$) and add the term $\|\boldsymbol{u}\|_{2q/(2-q)}$ to the X_n-norm. Then (4.63) holds. □

We will observe that it is enough to prove the statements of the Theorem 4.4 for either $1 < q \leq 2$ or $2 \leq q < \infty$.

Lemma 4.6 *Assume for any $\boldsymbol{\varphi} \in \boldsymbol{C}_0^\infty(\mathbb{R}^3)$ that the solution $(\boldsymbol{\psi}, \eta)$ from Lemma 4.5 to the adjoint problem*

$$\mathcal{L}^*\boldsymbol{\psi} + \nabla\eta = \boldsymbol{\varphi} \quad \text{in } \mathbb{R}^3, \tag{4.66}$$
$$\operatorname{div} \boldsymbol{\psi} = 0 \quad \text{in } \mathbb{R}^3$$

satisfies the estimates of Theorem 4.4 *for all $1 < q \leq 2$. Then the solution $(\boldsymbol{u}, p)$ to* (4.55) *from Lemma* 4.5 *satisfies the estimates of Theorem* 4.4 *for all $2 < q < \infty$. Conversely, if for all $\psi \in C_0^\infty(\mathbb{R}^3)$ the solution (ψ, η) from Lemma* 4.5 *satisfies the estimates of Theorem* 4.4 *for all $2 \leq q < \infty$, then the estimates are also true for (v, p) for all $1 < q \leq 2$.* □

Proof of Lemma 4.6 Let us assume for any $\boldsymbol{\varphi} \in \boldsymbol{C}_0^\infty(\mathbb{R}^3)$ that the solution $(\boldsymbol{\psi}, \eta)$ from Lemma 4.5 to the adjoint problem (4.66) satisfies (4.60), i.e.,

$$\|\nabla^2\boldsymbol{\psi}\|_q + \|\nabla\eta\|_q \leq c_1\|\boldsymbol{\varphi}\|_q \tag{4.67}$$

for all $q \in [2, \infty)$. Our aim is to show that the solution $(\boldsymbol{u}, p)$ from Lemma 4.5 to (4.55) satisfies (4.60) for the Hölder conjugate exponent $q' = q/(q-1)$ for any $q \in [2, \infty)$.

Let us define $\xi_N \in C_0^\infty$, $N \in \mathcal{N}$, as the radial cutoff function $\xi_N(x) := \zeta(|\mathbf{x}|/N)$, where $\zeta \in C_0^\infty$ and satisfies $\zeta = 1$ in $B_{1/2}$ and $\zeta = 0$ in $\mathbb{R}^3 \setminus B_1$. We put $\boldsymbol{u}_n := \xi_N \boldsymbol{u}$ and $\psi_N = \xi_N \psi$. Then $\lim_{N\to\infty} \boldsymbol{u}_N = \boldsymbol{u}$ in $\widehat{\boldsymbol{H}}^{1,2}(\mathbb{R}^3)) \cap \widehat{\boldsymbol{H}}^{2,2}(\mathbb{R}^3)$.

Since

$$(\boldsymbol{e}_3 \wedge \boldsymbol{x} \cdot \nabla \boldsymbol{u}_N - \boldsymbol{e}_3 \wedge \boldsymbol{u}_N) = \xi_N(\boldsymbol{e}_3 \wedge \boldsymbol{x} \cdot \nabla \boldsymbol{u} - \boldsymbol{e}_3 \wedge \boldsymbol{u}),$$

then $\lim_{N\to\infty} \mathcal{L}\boldsymbol{u}_N = \mathcal{L}\boldsymbol{u}$ in $L^2(\mathbb{R}^3)$. Similarly for ψ_M, $M \in N$

$$\begin{aligned}\int_{\mathbb{R}^3} \Delta \boldsymbol{u} \cdot \boldsymbol{\varphi}\, \mathrm{d}x &= \int_{\mathbb{R}^3} \Delta \boldsymbol{u} \cdot \mathcal{L}^* \boldsymbol{\psi}\, \mathrm{d}x = \lim_{N\to\infty} \lim_{M\to\infty} \int_{R^3} \Delta \boldsymbol{u}_N \cdot \mathcal{L}^* \boldsymbol{\psi}_M\, \mathrm{d}\boldsymbol{x} \\ &= \lim_{N\to\infty} \lim_{M\to\infty} \int_{R^3} \Delta \mathcal{L} \boldsymbol{u}_N \cdot \boldsymbol{\psi}_M\, \mathrm{d}x = \int_{\mathbb{R}^3} \boldsymbol{f} \cdot \Delta \boldsymbol{\psi}\, \mathrm{d}\boldsymbol{x}.\end{aligned}$$

We use that Δ and $\mathcal{L}$ commutes. Using (4.67) we obtain

$$\left| \int_{\mathbb{R}^3} \Delta \boldsymbol{u} \cdot \boldsymbol{\varphi}\, \mathrm{d}\boldsymbol{x} \right| \leq \|\boldsymbol{f}\|_{q'} \|\Delta \psi\|_q \leq \|\boldsymbol{f}\|_{q'} \|\boldsymbol{\varphi}\|_q.$$

Since $\boldsymbol{\varphi}$ is arbitrary, then $\|\Delta \boldsymbol{u}\|_{q'} \leq C_1 \|\boldsymbol{f}\|_{q'}$ and $\|\nabla^2 \boldsymbol{u}\|_{q'} \leq C_2 \|\boldsymbol{f}\|_{q'}$. The estimate $\|\nabla p\|_{q'} \leq C_3 \|\boldsymbol{f}\|_{q'}$ follows from the fact that $-\Delta p = \operatorname{div} \boldsymbol{f}$. Since $q \in [2, \infty)$ it implies that (4.60) is satisfied for all $q' \in (1, 2]$. □

4.1.5.2 Proof of Theorem 4.4

Let us recall the equations (see (3.13)), $\widetilde{\mathbf{u}} - \mathbf{u}_0$ being the effective oscillatory velocity part

$$\begin{aligned}\partial_t(\widetilde{\mathbf{u}} - \mathbf{u}_0) - \nu\Delta(\widetilde{\mathbf{u}} - \mathbf{u}_0) = k\partial_3(\mathbf{u} - \mathbf{u}_0) + \nabla(\widetilde{p} - p_0) = \widetilde{\mathbf{f}} - \mathcal{P}\mathbf{f} \quad &\text{in } \mathbb{R}^3 \times \mathbb{R}, \\ \operatorname{div}(\widetilde{\mathbf{u}} - \mathbf{u}_0) = 0 \quad &\text{in } \mathbb{R}^3 \times \mathbb{R}, \\ \widetilde{\mathbf{u}}(\mathbf{y}, t + 2\pi/\omega) - \mathbf{u}_0(\mathbf{y}) = \widetilde{\mathbf{u}}(\mathbf{y}, t) - \mathbf{u}_0(\mathbf{y}) \quad &\text{in } \mathbb{R}^3 \times \mathbb{R}, \\ \lim_{|\mathbf{y}|\to\infty} \widetilde{\mathbf{u}}(\mathbf{y}, \cdot) - \mathbf{u}_0(\mathbf{y}) = \mathbf{0} \quad &\text{in } \mathbb{R}.\end{aligned} \tag{4.68}$$

As in Chap. 3 we apply the group $\mathbf{G} := \mathbb{R}^3 \times \mathbb{R}/((2\pi/\omega)\mathbb{Z})$, where the $2\pi/\omega$-time periodicity is always insured, intrinsic in $\mathbf{G}$.

The transfer of topological and differentiable structures must be obtained in all spaces of functions or distributions defined in $\mathbf{G}$. So, $\mathbf{C}_0^\infty(\mathbf{G})$, $\mathbf{L}^q(\mathbf{G})$, $\mathbf{H}_0^{m,q}(\mathbf{G})$, $\widehat{\boldsymbol{H}}^{2,q}(\mathbf{G})$, and $\mathcal{S}(\mathbf{G})$ are well defined. Finally we can use the Fourier transform

$$\mathcal{F}_{\mathbf{G}}[\mathbf{w}](\boldsymbol{\xi}, l) := \frac{\omega}{2\pi}\int_0^{2\pi}\int_{\mathbb{R}^3} \mathbf{w}(\mathbf{y}, t)\mathrm{e}^{-\mathrm{i}\mathbf{y}\cdot\boldsymbol{\xi}}\mathrm{e}^{-\mathrm{i}(2\pi/\omega)lt}\mathbf{d}\mathbf{y}\mathrm{d}t$$

and the corresponding inverse transform

$$\mathcal{F}_{\mathbf{G}}^{-1}[\varphi](\mathbf{y}, t) := \sum_{l\in\mathbb{Z}}\int_{\mathbb{R}^3} \varphi(\boldsymbol{\xi}, l)\mathrm{e}^{\mathrm{i}\mathbf{y}\cdot\boldsymbol{\xi}}\mathrm{e}^{\mathrm{i}(2\pi/\omega)lt}\mathrm{d}\boldsymbol{\xi}.$$

In fact, $\mathbf{G}$ is identified with $\mathbb{R}^3 \times [0, (2\pi/\omega))$ restricting the second component to the interval of the period. Then we have the property

$$\int_{\mathbf{G}} \mathbf{w}(g)dg := \frac{\omega}{2\pi}\int_0^{2\pi}\int_{\mathbb{R}^3} \mathbf{w}(\mathbf{y}, t)\mathrm{d}\mathbf{y}\mathrm{d}t.$$

Now, let $\widetilde{\mathbf{w}}(\mathbf{y}, t) = \widetilde{\mathbf{u}}(\mathbf{y}, t) - \mathbf{u}_0(\mathbf{x})$, $\widetilde{\pi}(\mathbf{y}, t) = \widetilde{p}(\mathbf{y}, t) - p_0(\mathbf{y}, t)$, and $\widetilde{\mathbf{f}}_0 := (I - \mathcal{P})\widetilde{\mathbf{f}}$, and our system reads

$$\begin{aligned} \partial_t\widetilde{\mathbf{w}} - \nu\Delta\widetilde{\mathbf{w}} - k\partial_3\mathbf{w} + \nabla\widetilde{\pi} &= \widetilde{\mathbf{f}}_0 \quad \text{in } \mathbf{G},\\ \operatorname{div}\widetilde{\mathbf{w}} &= 0 \quad \text{in } \mathbf{G}. \end{aligned} \tag{4.69}$$

Using $\mathcal{F}_{\mathbf{G}}$, we get a solution $(\widetilde{\mathbf{w}}, \widetilde{\pi})$ in the sense of $\mathcal{S}'(\mathbf{G})$

$$\begin{aligned} \mathcal{F}_{\mathbf{G}}[\widetilde{\mathbf{f}}_0](\xi, l) &= (1 - \delta_{\mathbb{Z}}(l))\mathcal{F}_{\mathbf{G}}[\widetilde{\mathbf{f}}]\\ \left(\nu|\boldsymbol{\xi}|^2 + i(\frac{2\pi}{\omega}l - k\xi_3)\right)\mathcal{F}_{\mathbf{G}}[\widetilde{\mathbf{w}}] &= (I - \boldsymbol{\xi}\otimes\boldsymbol{\xi}/|\boldsymbol{\xi}|^2)\mathcal{F}_{\mathbf{G}}[\widetilde{\mathbf{f}}_0]\\ |\boldsymbol{\xi}|^2\mathcal{F}_{\mathbf{G}}[\widetilde{\pi}] &= i\boldsymbol{\xi}\cdot\mathcal{F}_{\mathbf{G}}[\widetilde{\mathbf{f}}_0], \end{aligned} \tag{4.70}$$

where $1 - \delta_{\mathbb{Z}}(l)$ is the symbol of $I - \mathcal{P}$, and $I - \boldsymbol{\xi}\otimes\boldsymbol{\xi}/|\boldsymbol{\xi}|^2$ is the usual symbol of the projection Π_σ from $\mathbf{L}^q(\mathbb{R}^3)$ onto $\mathbf{L}^q_\sigma(\mathbb{R}^3)$ continuous on $\mathbf{L}^q(\mathbb{R}^3)$ for any q and continuous on $\mathbf{L}^q(G)$ in this context. For the pressure, it is standard $\widetilde{\pi} = \mathcal{F}_{\mathbb{R}^3}^{-1}[i\boldsymbol{\xi}/|\boldsymbol{\xi}|^2\cdot\mathcal{F}_{\mathbb{R}^3}[\widetilde{\mathbf{f}}]]$, so we have the estimate $\|\nabla\widetilde{\pi}\|_q \leq C\|\widetilde{\mathbf{f}}_0\|_q$. Note that the quantity $\left(\nu|\boldsymbol{\xi}|^2 + i\frac{2\pi}{\omega}l\right)$ vanishes only at $(0, \mathbf{0})$, and then the Fourier multipliers M, kM, $|\boldsymbol{\xi}|M$, $|\boldsymbol{\xi}|^2M$ such that

$$M(l, \boldsymbol{\xi}) := \frac{1 - \delta_{\mathbb{Z}}(l)}{\nu|\boldsymbol{\xi}|^2 + i(\frac{2\pi}{\omega}l - k\xi_3)},$$

belong to $L^\infty(\mathbb{Z} \times \mathbb{R}^3) \cap C^\infty(\mathbb{Z} \times \mathbb{R}^3)$ and can be viewed as $L^q(\mathbb{R}^4)$-multipliers.

Multipliers theory is available in an Euclidean setting and not in the general setting of group multipliers, and therefore it is necessary to use the transfer principle. Using a transfer principle for Fourier multipliers on local compact Abelian groups gives possibility to study the properties of M via corresponding multipliers defined on $\mathbb{R}^4$. For more details see [28, 49]. From the transfer principle follows: M is an $L^q(G)$ multiplier if there is a continuous homeomorphism $\varphi\colon \hat{G} \to \mathbb{R}^4$ such that $M = m \circ \varphi$ for some $L^q(\mathbb{R}^4)$ multiplier m. Moreover, the norm of $L^q(G)$ operator corresponding to M with the norm of multiplier m in the $L^q(\mathbb{R}^4)$ corresponding to m. We introduce $\chi \in C_0^\infty(\mathbb{R}; \mathbb{R})$, $\chi(\eta) = 1$ for $|\eta| \le 1/2$, and $\chi(\eta) = 0$ for $|\eta| \ge 1$. Define $m\colon \mathbb{R}\times\mathbb{R}^3 \to C, m(\eta, \boldsymbol{\xi})\colon (1-\chi((T/2\pi)\eta))/(|\nu|\boldsymbol{\xi}|^2+i(\eta-k\xi_3))$. Further let $\varphi\colon \widehat{G} \to \mathbb{R}^{n+1}$, $\varphi(l, \boldsymbol{\xi}) = ((2\pi/T)l, \pi)$. φ is a continuous homeomorphism and $M = m \circ \varphi$. Using a standard multiplier theorem, we can get that $m \in L^q(\mathbb{R}^4)$. Since m is with nonvanishing denominator away from $(0, 0)$, then

$$(\eta, \boldsymbol{\xi}) \to \xi_1^{\epsilon_1} \cdots \xi_n^{\epsilon_n} \eta^{\epsilon_{n+1}} \partial_1^{\epsilon_1} \cdots \partial_n^{\epsilon_n} \partial_\eta^{\epsilon_{n+1}} m(\eta, \boldsymbol{\xi}) \tag{4.71}$$

stay bounded as $|(\eta, \boldsymbol{\xi})| \to \infty$. From Marcinkiewicz's multiplier theorem follows that m is $L^q(\mathbb{R}^{n+1})$ multiplier. It follows for all $\widetilde{\mathbf{f}} \in \mathbf{L}^q(\mathbf{G})$,

$$\|\mathcal{F}_{\mathbf{G}}^{-1}[M(l, \boldsymbol{\xi})\mathcal{F}_{\mathbf{G}}[\widetilde{\mathbf{f}}](l, \boldsymbol{\xi})]\|_q \le C_\omega \|\widetilde{\mathbf{f}}\|_q.$$

Since $I - \boldsymbol{\xi} \otimes \boldsymbol{\xi}/|\boldsymbol{\xi}|^2$ is symbol of the Helmholtz–Weyl projection, it means $\Pi_\sigma^q\colon \boldsymbol{L}^q(\mathbb{R}^n) \to \boldsymbol{L}_\sigma^q(\mathbb{R}^n)$, Π_σ^q is a continuous projection on $\boldsymbol{L}^q(G)$.

Now defining $(\widetilde{\mathbf{w}}, \widetilde{\pi})$ by (4.70) we can see that $(\widetilde{\mathbf{w}}, \widetilde{\pi})$ is a solution in the sense of distribution $\mathcal{S}'(G)$ and moreover

$$\|\widetilde{\mathbf{w}}\|_q = \|\mathcal{F}_{\mathbf{G}}^{-1}[M(l, \boldsymbol{\xi})\mathcal{F}_{\mathbf{G}}[\Pi_\sigma^q \widetilde{\mathbf{f}}](l, \boldsymbol{\xi})]\|_q \le C_\omega \| P_\sigma \widetilde{\mathbf{f}}_0\|_q \le C_\omega \|\widetilde{\mathbf{f}}\|_q$$

and analogously estimates for $\partial_t \widetilde{\mathbf{w}}$, $\nabla\widetilde{\mathbf{w}}$, $\nabla^2\widetilde{\mathbf{w}}$, and therefore we summarize the results.

Theorem 4.5 *Let* $\mathbf{f} \in \boldsymbol{L}^q(\mathbb{R}^3)$. *Then there exists a unique effective oscillatory velocity* $\widetilde{\mathbf{w}}$ *and its associated pressure* $\widetilde{\pi}$, *solution to problem* (4.69) *that satisfy the estimate*

$$\|\nu\nabla^2\widetilde{\mathbf{w}}\|_q + \|\widetilde{\mathbf{w}}\|_q + \|\partial_t\widetilde{\mathbf{w}}\|_q + \|\nabla\widetilde{\pi}\|_q \le cC_\omega \|\widetilde{\mathbf{f}}\|_q,$$

$$(\widetilde{\mathbf{w}}, \widetilde{\pi}) \in W_{per,\perp}^{1,2,q}(\mathbb{R} \times \mathbb{R}^3) \times D_{per,\perp}^{1,q}(\mathbb{R} \times \mathbb{R}^3).$$

As a consequence, there exists a locally integrable solution $(\mathbf{u}, p)$, $\mathbf{u} = \mathbf{u}_0 + \widetilde{\mathbf{w}}$, $p = p_0 + \widetilde{\pi}$, *to (4.1) that satisfies the inequality*

$$\|\nu\nabla^2\mathbf{u}\|_q + \|\nabla p\|_q + \|\omega\mathcal{R}\mathbf{u}\|_q \le cC_\omega \|\mathbf{f}\|_q. \tag{4.72}$$

If for some $r \in (1,\infty)$ $(\widetilde{\mathbf{w}}_1, \widetilde{\pi}_1) \in W^{1,2,4}_{per,\perp}(\mathbb{R}\times\mathbb{R}^3) \times D^{1,4}_{per,\perp}(\mathbb{R}\times\mathbb{R}^3)$ *is another solution, then* $\widetilde{\mathbf{w}}_1 = \widetilde{\mathbf{w}}$ *and* $\widetilde{\pi} = \widetilde{\pi}_1 + d(t)$ *for some* T*-periodic function* $d\colon \mathbb{R}\to\mathbb{R}$. □

Proof of Theorem 4.5 It remains to prove the uniqueness property. Assuming the existence of another solution $(\widetilde{\mathbf{w}}_1, \widetilde{\pi}_1)$ and applying the Helmholtz projection and Fourier transformation, we get

$$\left(i\frac{2\pi}{T}l + \nu|\boldsymbol{\xi}|^2 - ki\xi_3\right)\mathcal{F}_G(\widetilde{\mathbf{w}} - \widetilde{\mathbf{w}}_3) = 0.$$

The polynomial $i(2\pi/T)l + \nu|\boldsymbol{\xi}|^2 - ki\xi_3$ vanishes only in $(l,\boldsymbol{\xi}) = (0,0)$, and it implies that $\operatorname{supp}\mathcal{F}_G(\widetilde{\mathbf{w}} - \widetilde{\mathbf{w}}_1) \subset \{(0,0)\}$. Also $P_\sigma(\widetilde{\mathbf{w}} - \widetilde{\mathbf{w}}_1) = 0$ also $\delta_Z(l)\mathcal{F}_G(l,\boldsymbol{\xi}) = 0$. Then $(0,0) \notin \operatorname{supp}\mathcal{F}_G(\widetilde{\mathbf{w}} - \widetilde{\mathbf{w}}_1)$. It implies $\operatorname{supp}\mathcal{F}_G(\widetilde{\mathbf{w}} - \widetilde{\mathbf{w}}_1) = \emptyset$. So $\operatorname{supp}\mathcal{F}_G(\widetilde{\mathbf{w}} - \widetilde{\mathbf{w}}_1) = 0$ and thus $\widetilde{\mathbf{w}} = \widetilde{\mathbf{w}}_1$. □

4.2 L^q-Theory of Weak Solutions of the Oseen Type Problem in the Whole Space and in an Exterior Domain

4.2.1 *Introduction*

In this subsection we will follow the work by Kračmar, Nečasová, Penel [149, 151]. Let us recall the investigated problem.

We assume that for given $\mathbf{f}$ and g our system of equations has the following form

$$\left.\begin{aligned} -\nu\Delta\mathbf{u} + k\partial_3\mathbf{u} - (\omega(\mathbf{e}_3\wedge\mathbf{x})\cdot\nabla)\mathbf{u} + \omega\mathbf{e}_3\wedge\mathbf{u} + \nabla p = \mathbf{f}, \\ \operatorname{div}\mathbf{u} = g \end{aligned}\right\} \quad \text{in } \mathcal{D}. \tag{4.73}$$

The system of Eq. (4.73) is now complemented by the homogeneous far field condition

$$\mathbf{u} \to 0 \quad \text{as} \quad |\mathbf{x}| \to \infty \tag{4.74}$$

and the Dirichlet boundary conditions on $\partial\mathcal{D}$, either

$$\mathbf{u}|_{\partial\mathcal{D}} = 0 \tag{4.75}$$

or

$$\mathbf{u}|_{\partial\mathcal{D}} = \omega\mathbf{e}_3\wedge\mathbf{x} - e_3, \qquad \mathbf{x}\in\partial\mathcal{D}. \tag{4.76}$$

In this subsection we consider $k = \nu = 1$. We will study the boundary value problem (4.73), (4.74), (4.75). Applying the localization technique, see [144], we immediately observe that it combines both systems in the whole space and in a bounded domain. The problem in the whole space will be solved by geometrical approach using Paley–Littlewood decomposition and problem in a bounded domain case be solved as a perturbation of "classical Oseen problem."

Indeed, choose $\rho > \rho_0 > 0$ so large that $\mathcal{D} \subset B_{\rho_0} = \{\mathbf{x} \in \mathbb{R}^3 \colon |\mathbf{x}| < \rho_0\}$ and take a cutoff function $\psi \in C_0^\infty(B_\rho; [0, 1])$ such that $\psi = 1$ on B_{ρ_0} and $\mathrm{supp}(\nabla\psi) \subset \{\mathbf{x} \colon \rho_0 < |\mathbf{x}| < \rho\}$; introducing now $\mathbf{U} = (1 - \psi)\mathbf{u}$, $V = \psi\mathbf{u}$, $\pi = (1 - \psi)p$, and $\tau = \psi p$, we get

$$\mathbf{u} = \mathbf{U} + \mathbf{V}, \tag{4.77}$$

$$p = \pi + \tau \tag{4.78}$$

and two problems appear.

The whole space:

$$\begin{aligned} -\Delta\mathbf{U} + \partial_3\mathbf{U} - (\omega(\mathbf{e}_3 \wedge \mathbf{x}) \cdot \nabla)\mathbf{U} + \omega\mathbf{e}_3 \wedge \mathbf{U} + \nabla\pi &= \mathbf{F}_1(\mathbf{u}, p), \\ \mathrm{div}\,\mathbf{U} &= G_1(\mathbf{u}), \\ \mathbf{U} \to 0 \quad \text{as } |\mathbf{x}| &\to \infty, \end{aligned} \tag{4.79}$$

where

$$G_1(\mathbf{u}) = (1 - \psi)g - \nabla\psi \cdot \mathbf{u}.$$

The bounded domain $D_\rho = \mathcal{D} \cap B_\rho$:

$$\begin{aligned} -\Delta\mathbf{V} + \partial_3\mathbf{V} + \nabla\tau &= \mathbf{F}_2(\mathbf{u}, p), \\ \mathrm{div}\,\mathbf{V} &= G_2(\mathbf{u}), \\ \mathbf{V}|_{\partial\mathcal{D}_\rho} &= 0, \end{aligned} \tag{4.80}$$

where

$$G_2(\mathbf{u}) = \psi g + \nabla\psi \cdot \mathbf{u}$$

and

$$\begin{aligned} \mathbf{F}_1(\mathbf{u}, p) =& (1 - \psi)\mathbf{f} + 2(\nabla\psi \cdot \nabla)\mathbf{u} + [\Delta\psi + (\omega(\mathbf{e}_3 \wedge \mathbf{x})\nabla)\psi]\mathbf{u} \\ & - (\nabla\psi)p + (\partial_3\psi)\mathbf{u}, \\ \mathbf{F}_2(\mathbf{u}, p) =& \psi\mathbf{f} + \psi[(\omega(\mathbf{e}_3 \wedge \mathbf{x}) \cdot \nabla)\mathbf{u} - \omega\mathbf{e}_3 \wedge \mathbf{u}] - 2(\nabla\psi \cdot \nabla)\mathbf{u} \\ & - (\Delta\psi)\mathbf{u} + (\nabla\psi)p + (\partial_3\psi)\mathbf{u}. \end{aligned} \tag{4.81}$$

Let us observe that in the bounded domain D_ρ, we can equivalently write the following nonhomogeneous Stokes problem

$$\begin{aligned} -\Delta \mathbf{V} + \nabla \tau &= \mathbf{F}_2, \\ \operatorname{div} \mathbf{V} &= G_2(\mathbf{u}), \\ \mathbf{V}|_{\partial D_\rho} &= 0, \end{aligned} \tag{4.82}$$

with the modified right-hand side

$$\mathbf{F}_2 = \psi \mathbf{f} + \psi[(\omega(\mathbf{e}_3 \wedge \mathbf{x}) \cdot \nabla)\mathbf{u} - \omega \mathbf{e}_3 \wedge \mathbf{u}] - 2(\nabla\psi \cdot \nabla)\mathbf{u} - (\Delta\psi)\mathbf{u} + (\nabla\psi)p + \partial_3(\psi\mathbf{u}).$$

Let us also observe that, to prove the existence of solution in an exterior domain even if $g = 0$ in (4.73), we need to study the nonhomogeneous case (4.79) in the whole space.

In Sect. 4.2, we will give the definition of a weak solution to problem (4.73), (4.74), (4.75) and state our main results on the existence and uniqueness of its solution and the proofs of the main results. To prove it we need intermediate results for both problems (4.79) and (4.80), and Sect. 4.2.2.1 is devoted to them. Two appendices complete this subsection, the first one with the general results of Bogovskii operator by Farwig and Sohr [79], Kozono and Sohr [144] and generalization in negative Sobolev spaces by Geissert, Heck, Hieber [117], the second appendix with the technical treatment of $\partial_3\mathbf{u}$ in negative Sobolev spaces.

4.2.2 *The Main Result*

We consider the problem (4.73), (4.74), (4.75) with $g = 0$ in the exterior domain $\mathcal{D}$. Let $\mathbf{f}$ be given in $\widehat{\boldsymbol{H}}^{-1,q}(\mathcal{D})$, $1 < q < \infty$.

Definition 4.1 We call $\{\mathbf{u}, p\}$ a weak solution to $(4.73)_{g=0}$ (4.74), (4.75) if:

(1) $\{\mathbf{u}, p\} \in (\widehat{\boldsymbol{H}}^{1,q}(\mathcal{D})) \times L^q(\mathcal{D})$.
(2) $\operatorname{div} \mathbf{u} = 0$ in $L^q(\mathcal{D})$.
(3) $\partial_3\mathbf{u} - ((\omega\mathbf{e}_3 \wedge \mathbf{x}) \cdot \nabla)\mathbf{u} + \omega\mathbf{e}_3 \wedge \mathbf{u} \in \widehat{\boldsymbol{H}}^{-1,q}(\mathcal{D})$.
(4) $\langle \nabla\mathbf{u}, \nabla\boldsymbol{\varphi}\rangle + \langle \partial_3\mathbf{u} - (\omega(\mathbf{e}_3 \wedge \mathbf{x}) \cdot \nabla)\mathbf{u} + \omega\mathbf{e}_3 \wedge \mathbf{u}, \boldsymbol{\varphi}\rangle = \langle p, \operatorname{div}\boldsymbol{\varphi}\rangle + \langle \mathbf{f}, \boldsymbol{\varphi}\rangle$ for all $\boldsymbol{\varphi} \in C_0^\infty(\mathcal{D})$.

□

Our main result of Sect. 4.2 follows.

Theorem 4.6 *Let* $3/2 < q < 3$ *and suppose* $\mathbf{f}$ *is given in* $\widehat{\boldsymbol{H}}^{-1,q}(\mathcal{D})$. *Then there exists a unique weak solution up to constant* $\{\mathbf{u}, p\}$ *to* $(4.73)_{g=0}$, (4.74), (4.75), *which satisfies the estimate*

$$\|\nabla\mathbf{u}\|_{q,\mathcal{D}} + \|p\|_{q,\mathcal{D}} + \|\partial_3\mathbf{u}\|_{-1,q,\mathcal{D}} + \| -(\omega(\mathbf{e}_3 \wedge x)\cdot\nabla)\mathbf{u} + \omega\mathbf{e}_3 \wedge \mathbf{u}\|_{-1,q,\mathcal{D}} \le c_q\|\mathbf{f}\|_{-1,q,\mathcal{D}}, \tag{4.83}$$

with some constant $c_q > 0$ *depending of* q *and* ω. □

4.2.2.1 Intermediate Results

We can start with the analysis of the homogeneous problem in the whole space, then precisely of the problem (4.79) for given $\mathbf{F}_1$ and $G_1 = 0$. We denote this solution by $(\boldsymbol{U}_0, \pi_0)$ and introduce the defintion of weak solution and give statement on the existence of the weak solutions

Definition 4.2 Let $1 < q < \infty$. Given $\mathbf{F}_1 \in \widehat{\boldsymbol{H}}^{-1,q}(\mathbb{R}^3)$, we call $\{\mathbf{U}_0, \pi_0\} \in \widehat{\boldsymbol{H}}^{1,q}(\mathbb{R}^3) \times L^q(\mathbb{R}^3)$ a weak solution to (4.79) with $G_1 = 0$ if:

(1) $\operatorname{div}\mathbf{U}_0 = 0$ in $L^q(\mathbb{R}^3)$.
(2) $\partial_3\mathbf{U}_0 - ((\omega\mathbf{e}_3 \wedge \mathbf{x})\cdot\nabla)\mathbf{U}_0 + \omega\mathbf{e}_3 \wedge \mathbf{U}_0 \in \widehat{\boldsymbol{H}}^{-1,q}(\mathbb{R}^3)$.
(3) $\langle\nabla\mathbf{U}_0, \nabla\varphi\rangle + \langle\partial_3\mathbf{U}_0 - ((\omega\mathbf{e}_3\wedge\mathbf{x})\cdot\nabla)\mathbf{U}_0 + \omega\mathbf{e}_3\wedge\mathbf{U}_0, \varphi\rangle - \langle\pi_0, \operatorname{div}\varphi\rangle = \langle\mathbf{F}_1, \varphi\rangle$ for all $\varphi \in \boldsymbol{C}_0^\infty(\mathbb{R}^3)$.

□

Theorem 4.7 *Let* $1 < q < \infty$ *and let* $\mathbf{F}_1 \in \widehat{\boldsymbol{H}}^{-1,q}(\mathbb{R}^3)$ *be given; then the problem* (4.79) *with* $G_1 = 0$ *possesses a weak solution* $\{\mathbf{U}_0, \pi_0\} \in \widehat{\boldsymbol{H}}^{1,q}(\mathbb{R}^3) \times L^q(\mathbb{R}^3)$ *which satisfies*

$$\|\nabla\mathbf{U}_0\|_q + \|\pi_0\|_q + \|\partial_3\mathbf{U}_0 - ((\omega\mathbf{e}_3 \wedge \mathbf{x})\cdot\nabla)\mathbf{U}_0 + \omega\mathbf{e}_3 \wedge \mathbf{U}_0\|_{-1,q} \le c\|\mathbf{F}_1\|_{-1,q} \tag{4.84}$$

with some $c > 0$ *depending on* q. □

Theorem 4.8 *The solution* $\{\mathbf{U}_0, \pi_0\}$ *given by Theorem 4.7 is unique up to a constant multiple of* ω *for* $\mathbf{U}_0$. □

Corollary 4.2 *Let* $1 < q < 4$, $\mathbf{F}_1 \in \widehat{\boldsymbol{H}}^{-1,q}(\mathbb{R}^3)$ *and let* $\mathbf{U}_0 \in \widehat{\boldsymbol{H}}^{1,q}(\mathbb{R}^3)$ *be the unique weak solution to problem* (4.79). *Then there exists* $\alpha \in \mathbb{R}$ *such that*

$$\mathbf{U}_0 - \alpha\boldsymbol{e}_3 \in \boldsymbol{L}^s(\mathbb{R}^3) \quad \textit{for all } s > 1,\ \frac{1}{s} \in \frac{1}{q} - \left[\frac{1}{4}, \frac{1}{3}\right].$$

Moreover

$$\|\mathbf{U}_0 - \alpha \boldsymbol{e}_3\|_s \le C\|\mathbf{F}_1\|_{-1,q}$$

with a constant $C = C(\nu, k, \omega, s) > 0.$ □

Corollary 4.3 *Let* $1 < q < 3$, $\nu > 0$, $k > 0$, $\mathbf{F}_1 \in \widehat{\boldsymbol{H}}^{-1,q}(\mathbb{R}^3)$, *and let* $\mathbf{U}_0 \in \widehat{\boldsymbol{H}}^{1,q}(\mathbb{R}^3)$ *be the unique weak solution to problem* (4.79). *Then*

$$\left\|\frac{\mathbf{U}_0}{|x|}\right\|_q \le \frac{c}{\nu}\|\mathbf{F}_1\|_{-1,q}$$

with $c = c(q, \omega) > 0.$ □

Before proving Theorem 4.7 we begin with the computation of $\nabla \mathbf{U}_0$.

Computation of $\nabla \mathbf{U}_0$.

Using the fact that the space $\left\{g : g = \nabla \cdot G,\ G \in C_0^\infty\left(\mathbb{R}^3\right)^{3\times 3}\right\}$ is dense in $\widehat{W}^{-1,q}(\mathbb{R}^3)$, we can write $\mathbf{F}_1$ in the divergence form in the Oseen system (4.79).

We will work in the space of tempered distributions because we have in mind its application of the Fourier transform. We will derive the following formal expressions for $\widehat{\mathbf{U}_0}$, $\mathbf{U}_0$, and $\nabla \mathbf{U}_0$:

$$\widehat{\mathbf{U}_0}(\xi) = \int_0^\infty \mathrm{e}^{-(\nu|\boldsymbol{\xi}|^2 + \mathrm{i}k\xi_3)t} Q_{\boldsymbol{\omega}}^{\mathrm{T}}(t)\, \widehat{\mathbf{F}}_1(O_{\boldsymbol{\omega}}(t)\boldsymbol{\xi})\mathrm{d}t, \tag{4.85}$$

yielding $\mathbf{U}_0(\cdot)$ in the form

$$\mathbf{U}_0(\mathbf{x}) = \int_0^\infty K * Q_{\boldsymbol{\omega}}^{\mathrm{T}}(t)\, \mathbf{F}_1(Q_{\boldsymbol{\omega}}(t) - kt\boldsymbol{e}_3)(\mathbf{x})\mathrm{d}t, \tag{4.86}$$

where

$$K(\mathbf{x}) = \frac{1}{(4\pi\nu t)^{3/2}}\mathrm{e}^{-|\mathbf{x}|^2/4\nu t}.$$

Observing the previous integral, the solution can be rewritten as

$$\mathbf{U}_0(\mathbf{x}) = \int_{\mathbb{R}^3} \Sigma_{os}\,(\mathbf{x}, \mathbf{y})\ \nabla \cdot G\,(\mathbf{y})\, \mathrm{d}\mathbf{y},$$

where

$$\Sigma_{os}(\mathbf{x}, \mathbf{y}) = \int_0^\infty E_t\,(Q_{\boldsymbol{\omega}}(t)\mathbf{x} - \mathbf{y} - kt\boldsymbol{e}_3)\ Q_{\boldsymbol{\omega}}^{\mathrm{T}}(t)\mathrm{d}t.$$

Therefore, we can compute the gradient of $\mathbf{U}_0$,

$$\nabla \mathbf{U}_0(\mathbf{x}) = -\int_{\mathbb{R}^3} \nabla_{\mathbf{x}} \nabla_{\mathbf{y}} \Sigma_{os}(\mathbf{x}, \mathbf{y}) : G(\mathbf{y})\, \mathrm{d}\mathbf{y},$$

and come back to its Fourier transform.

Remark 4.2 To compute explicitly $\widehat{\mathbf{U}_0}$, we follow the way introduced by Farwig et al. by using geometry of the problem. It means introducing cylindrical coordinates $(r, \theta, x_3) \in (0, \infty) \times [0, 2\pi) \times \mathbb{R}$ using that $(\omega \mathbf{e}_3 \wedge \mathbf{x}) \cdot \nabla \mathbf{U}_0 = \omega(-x_2 \partial_1 \mathbf{U}_0 + x_1 \partial_2 \mathbf{U}_0)$ may be rewritten in the form

$$(\omega \mathbf{e}_3 \wedge \mathbf{x}) \cdot \nabla \mathbf{U}_0 = |\omega| \partial_\theta \mathbf{U}_0$$

using the angular derivative ∂_θ applied to $\mathbf{U}_0(r, x_3, \theta)$. Applying the Fourier transform, solving the pressure π_0, we get the first order differential equation with respect to φ for $\widehat{\mathbf{U}_0} := \widehat{\mathbf{U}_0}\left(\sqrt{\xi_1^2 + \xi_2^2},\ \varphi,\ \xi_3\right)$:

$$-\partial_\varphi \widehat{\mathbf{U}_0} + \frac{1}{\omega}(\nu|\boldsymbol{\xi}|^2 + \mathrm{i}k\xi_3)\widehat{\mathbf{U}_0} + \boldsymbol{e}_3 \wedge \widehat{\mathbf{U}_0} = \frac{1}{\omega}\widehat{\mathbf{F}}_1. \tag{4.87}$$

Solving (4.87) we get the unique $2\pi/|\boldsymbol{\omega}|$-periodic solution

$$\widehat{\mathbf{U}_0}(\boldsymbol{\xi}) = \int_0^\infty \mathrm{e}^{-(\nu|\boldsymbol{\xi}|^2 + \mathrm{i}k\xi_3)t}\, Q_{\boldsymbol{\omega}}^{\mathrm{T}}(t)\, \widehat{\mathbf{F}}_1(Q_{\boldsymbol{\omega}}(t)\boldsymbol{\xi})\mathrm{d}t. \tag{4.88}$$

$\widehat{\mathbf{U}_0}$ solves the reduced Eq. (4.87) in the Fourier space. □

- In this part we estimate the L^q-norm of each component of $TG(\mathbf{x}) := \Delta \int_{\mathbb{R}^3} \Sigma_{os}(\mathbf{x}, \mathbf{y}) : G(\mathbf{y})\mathrm{d}\mathbf{y}$, say $TG_{i,k}(\cdot)$ by the L^q-norm of $G_{i,k}(\cdot)$, and then we apply these results for L^q-estimate of $\nabla \mathbf{U}_0$.

  To this end, we follow the way used by Farwig, Hishida, Müller [70] and by Hishida [126]. By means of the Fourier transform, we have

$$\widehat{TG}(\boldsymbol{\xi}) = \frac{1}{(2\pi)^{3/2}} \int_0^\infty |\boldsymbol{\xi}|^2 \exp\left(-\nu|\boldsymbol{\xi}|^2 t\right) O_{\boldsymbol{\omega}}^{\mathrm{T}}(t)\, \widehat{G}\left((O_{\boldsymbol{\omega}}(t) \cdot -k\, t\, \boldsymbol{e}_3)\, \boldsymbol{\xi}\right) \mathrm{d}t.$$

  We can rewrite this as

$$\widehat{TG}(\boldsymbol{\xi}) = \frac{1}{\nu(2\pi)^{3/2}} \int_0^\infty |\boldsymbol{\xi}|^2 \exp\left(-|\boldsymbol{\xi}|^2 t\right) O_{\boldsymbol{\omega}/\nu}^{\mathrm{T}}(t)\, \widehat{G}\left(\left(O_{\boldsymbol{\omega}/\nu}(t) \cdot -k\frac{t}{\nu}\boldsymbol{e}_3\right)\boldsymbol{\xi}\right) \mathrm{d}t.$$

- Before applying the Littlewood–Paley decomposition, let us explain deep tool from harmonic analysis. Let us temporarily denote $TG_{i,k}(\mathbf{x})$ by $F(\mathbf{x})$ and let us define an appropriate function $\varphi(\cdot) \in C_0^\infty((0,\infty); \mathcal{S}(\mathbb{R}^3))$ such that with the square operator $S(F)(x) = \int_0^\infty |\varphi(t,\cdot) * F(x)|^2 \, dt/t$, we receive the equivalence of L^q-norms given by the theorem of E. M. Stein, [205, Chapter I, Section 8.23],

$$c_1 \|F\|_q \le \left\|\varphi(F)^{1/2}\right\|_q \le c_2 \|F\|_q .$$

The necessary properties of $\varphi(t,\cdot)$ for $t > 0$ are

$$\operatorname{supp} \widehat{\varphi}(t,\cdot) \subset \left\{ \xi \in \mathbb{R}^3 : \frac{1}{2\sqrt{t}} < |\xi| < \frac{2}{2\sqrt{t}} \right\},$$

$$\int_0^\infty \widehat{\varphi}(t,\xi)^2 \frac{\mathrm{d}t}{t} = 1, \quad \int_{\mathbb{R}^3} \varphi(t,\mathbf{x}) \, \mathrm{d}\mathbf{x} = 0.$$

We start with the Littlewood–Paley decomposition. We define $\psi \in \mathcal{S}(\mathbb{R}^3)$ by its Fourier transform

$$\widehat{\psi}(\boldsymbol{\xi}) = \frac{1}{(2\pi)^{3/2}} |\boldsymbol{\xi}|^2 \mathrm{e}^{-|\boldsymbol{\xi}|^2} \quad \text{and} \quad \widehat{\psi}_t(\boldsymbol{\xi}) = \widehat{\psi}(\sqrt{t}\boldsymbol{\xi}) \quad \text{for } t > 0 \tag{4.89}$$

and so for all $t > 0$

$$\psi_t(\mathbf{x}) = t^{-3/2} \psi\left(\frac{\mathbf{x}}{\sqrt{t}}\right), \quad \widehat{\psi}_t(\boldsymbol{\xi}) = \frac{1}{(2\pi)^{3/2}} t |\boldsymbol{\xi}|^2 \mathrm{e}^{-\nu t |\boldsymbol{\xi}|^2}. \tag{4.90}$$

Thus we get

$$\widehat{TG}(\boldsymbol{\xi}) = \frac{1}{\nu} \int_0^\infty \widehat{\psi}_t(\boldsymbol{\xi}) O^{\mathrm{T}}_{\omega/\nu}(t) \widehat{G}\left(\left(O_{\omega/\nu}(t). - k\frac{t}{\nu} \boldsymbol{e}_3\right) \boldsymbol{\xi}\right) \frac{\mathrm{d}t}{t}. \tag{4.91}$$

We define the multiplier operator Δ_j such that

$$\widehat{\Delta_j f(\boldsymbol{\xi})} := \widehat{\chi}^j(\boldsymbol{\xi}) \widehat{f}(\boldsymbol{\xi}), \tag{4.92}$$

where

$$\widehat{\chi}^j(\boldsymbol{\xi}) = \widehat{\chi}_0\left(\frac{|\boldsymbol{\xi}|}{2^{j+1}}\right) - \widehat{\chi}_0\left(\frac{|\boldsymbol{\xi}|}{2^j}\right) \tag{4.93}$$

with

$$\widehat{\chi}_0(.) \colon |\boldsymbol{\xi}| \to \mathbb{R}, \ \chi \in C^\infty, \ \widehat{\chi}_0\Big|_{\{|\boldsymbol{\xi}| \le 1/2\}} = 1, \ \widehat{\chi}_0\Big|_{\{|\boldsymbol{\xi}| \ge 1\}} = 0. \tag{4.94}$$

Note that

$$\sum_{j\in Z} \widehat{\chi}^j(\boldsymbol{\xi}) = 1\,, \tag{4.95}$$

and

$$\mathbf{f}(.) = \sum_{j\in Z} \Delta_j f(.). \tag{4.96}$$

We define χ^j for $\boldsymbol{\xi} \in \mathbb{R}^3$ and $j \in \mathbb{Z}$ by its Fourier transform

$$\widehat{\chi}^j(\boldsymbol{\xi}) = \widetilde{\chi}(2^{-j}|\boldsymbol{\xi}|), \quad \boldsymbol{\xi} \in \mathbb{R}^3,$$

yielding $\sum_{j=-\infty}^{\infty} \widehat{\chi}_j = 1$ on $\mathbb{R}^3 \setminus \{0\}$ and

$$\operatorname{supp} \widehat{\chi}^j \subset A(2^{j-1}, 2^{j+1}) := \{\xi \in \mathbb{R}^3 \colon 2^{j-1} < |\boldsymbol{\xi}| < 2^{j+1}\}. \tag{4.97}$$

Using χ^j, we put for $j \in \mathbb{Z}$

$$\psi^j = \frac{1}{(2\pi)^{n/2}} \chi_j * \psi_t, \qquad \widehat{\psi^j} = \widehat{\Delta_j \psi(.)} = \widehat{\chi}^j \cdot \widehat{\psi}_t. \tag{4.98}$$

Obviously, $\sum_{j=-\infty}^{\infty} \psi^j = \psi$ on $\mathbb{R}^3$. We start with the procedure of the Littlewood–Paley decomposition of $F = TG_{i,k}$. For $G_{i,k} \in \mathcal{S}'(\mathbb{R}^3)$ the property $TG_{i,k} \in L^p\left(\mathbb{R}^3\right)$ is equivalent to the property

$$TG_{i,k} = \sum_{j=-\infty}^{+\infty} \Delta_j TG_{i,k} \quad \text{and} \quad \left(\sum_{j=-\infty}^{+\infty} \left|\Delta_j TG_{i,k}\right|^2\right)^{1/2} \in L^p\left(\mathbb{R}^3\right),$$

where

$$\Delta^j = \mathcal{F}^{-1} \widehat{\psi}^j \left(\frac{\xi}{2^j}\right) \mathcal{F}, \quad \Delta_t^j = \mathcal{F}^{-1} \widehat{\psi}_t^j \left(\frac{\xi}{2^j}\right) \mathcal{F}.$$

We define the operator

$$\Delta \Sigma_{os} = \sum_{j\in Z} \Delta_j \Delta \Gamma \tag{4.99}$$

yielding to

$$\Delta \int_{\mathbb{R}^3} \Sigma_{os}(\mathbf{x}, \mathbf{y}) : G(\mathbf{y})\, \mathbf{dy}, \quad G \in C_0^\infty(\mathbb{R}^3)^9, \quad (G_{ik})_{1\le i\le 3, 1\le k\le 3}. \tag{4.100}$$

We introduce the linear operator

$$TG_{ik}(\mathbf{x}) = \int_{\mathbb{R}^3} \Delta\Sigma_{os}(\mathbf{x},\mathbf{y})_{ki} G_{ik}(\mathbf{y})\,\mathrm{d}\mathbf{y}. \tag{4.101}$$

Since formally $T = \sum_{j=-\infty}^{\infty} T_j$, we have to prove that this infinite series converges even in the operator norm on L^q. From general definition of Littlewood–Paley decomposition of L^q, choose $\widetilde{\varphi} \in C_0^\infty(\frac{1}{2}, 2)$ such that $0 \le \widetilde{\varphi} \le 1$ and

$$\int_0^\infty \widetilde{\varphi}(s)^2 \frac{\mathrm{d}s}{s} = \frac{1}{2}.$$

Then define $\varphi \in \mathcal{S}(\mathbb{R}^n)$ by its Fourier transform $\widehat{\varphi}(\boldsymbol{\xi}) = \widetilde{\varphi}(|\boldsymbol{\xi}|)$ yielding for every $s > 0$

$$\widehat{\varphi}_s(\boldsymbol{\xi}) = \widetilde{\varphi}(\sqrt{s}|\boldsymbol{\xi}|), \quad \operatorname{supp}\widehat{\varphi}_s \subset A\left(\frac{1}{2\sqrt{s}}, \frac{2}{\sqrt{s}}\right), \tag{4.102}$$

and the normalization

$$\int_0^\infty \widehat{\varphi}_s(\boldsymbol{\xi})^2 \,\frac{\mathrm{d}s}{s} = 1$$

for all $\boldsymbol{\xi} \in \mathbb{R}^3 \setminus \{0\}$.

Theorem 4.9 *Let $1 < q < \infty$. Then there are constants $c_1, c_2 > 0$ depending on q and φ such that for all $\mathbf{f} \in \mathbf{L}^q$*

$$c_1\|\mathbf{f}\|_q \le \left\| \left(\int_0^\infty \int_0^\infty (|\varphi_s * \mathbf{f}(.)|^2 \frac{ds}{s} \right) \right\|_{q/2} \le c_2\|\mathbf{f}\|_q,$$

where $\varphi_s \in \mathcal{S}(\mathbb{R}^n)$ is defined by (4.102). □

Proof of Theorem 4.9 See [205, Chapter I, Section 8.23]. □

We apply Theorem 4.9 to the operator $T_j G_{ik}$,

$$c_1\|T_jG_{ik}\|_q \le \left\| \int_0^\infty |(\varphi(t,.) * T_jG_{ik})(\mathbf{x})|^2 \frac{\mathrm{d}t}{t} \right\|_{q/2} \le c_2\|T_jG_{ik}\|_q. \tag{4.103}$$

Proposition 4.1 *Let $j \in \mathcal{Z}$. The linear operator T defined by* (4.101) *satisfies the estimate*

$$\|T_jG_{ik}\|_q \le c\|G_{ik}\|_q \quad \textit{for all} \quad \mathbf{G} \in \mathbf{L}^q, \; q \in (2,\infty)$$

with a constant $c = c(q, w) > 0$ independent of $\mathbf{f}$. □

Proof of Proposition 4.1 We define the sublinear operator $\mathcal{M}^j$, a modified maximal operator, by

$$\mathcal{M}^j\varphi(\mathbf{x}) = \sup_{s>0} \int_{A_s} \left(|\Delta_t^j| * |\varphi|\right)\left(Q_{\omega/\nu}(t)^{\mathrm{T}}\mathbf{x} + \frac{k}{\nu}te_3\right)\frac{\mathrm{d}t}{t}, \tag{4.104}$$

where $A_s = [s/16, 16s]$. □

First Step: We will prove the preliminary estimate

$$\|T_j G_{ik}\|_q \le c\|\Delta^j\|_1^{1/2}\|\mathcal{M}^j\|_{L^{(q/2)'}}^{1/2}\|G_{ik}\|_q, \quad j \in \mathbb{Z}. \tag{4.105}$$

To provc (4.105) we use the Littlewood–Paley decomposition of $\mathbf{L}^q$,

$$c_1^2\|\mathbf{f}\|_q^2 \le \left\|\int_0^\infty |\varphi_s * \mathbf{f}(\cdot)|^2\frac{\mathrm{d}s}{s}\right\|_{q/2} \le c_2^2\|\mathbf{f}\|_q^2. \tag{4.106}$$

By a duality argument we find some function $0 \le g \in L^{(q/2)'}$ with $\|g\|_{(q/2)'} = 1$ such that

$$\left\|\int_0^\infty |\varphi_s * T_j G_{ik}(\cdot)|^2\,\frac{\mathrm{d}s}{s}\right\|_{q/2} = \int_0^\infty\int_{\mathbb{R}^n} |\varphi_s * T_j G_{ik}(x)|^2 g(\mathbf{x})\mathrm{d}x\,\frac{\mathrm{d}s}{s}. \tag{4.107}$$

To estimate the right-hand side of (4.107), note that

$$\varphi_s * T_j G(x) = \int_0^\infty Q(t)^{\mathrm{T}}_{\omega/\nu}(t)(\varphi_s * \Delta_t^j * G_{ik})\left(Q_{\omega/\nu}(t)x - \frac{k}{\nu}te_3\right)\frac{\mathrm{d}t}{t},$$

where $\varphi_s * \Delta_t^j = 0$ unless $t \in A(s,j) := [2^{2j-4}s, 2^{2j+4}s]$. Since $\int_{t\in A(s,j)} \mathrm{d}t/t = \log 2^8$ for every $j \in \mathbb{Z}$, $s > 0$, we get by the Cauchy–Schwarz inequality and the associativity of convolutions that

$$\begin{aligned}|\varphi_s * T_j G_{ik}(\mathbf{x})|^2 \le& c\int_{A(s,j)}\left|\left(\Delta_t^j * (\varphi_s * G_{ik})\right)\left(Q_{\omega/\nu}(t)\mathbf{x} - \frac{k}{\nu}te_3\right)\right|^2\frac{\mathrm{d}t}{t}\\ \le c\|\Delta^j\|_1 &\int_{A(s,j)}\left(|\Delta_t^j| * |\varphi_s * G_{ik}|^2\right)\left(Q(t)_{\omega/\nu}\mathbf{x} + \frac{k}{\nu}te_3\right)\frac{\mathrm{d}t}{t};\end{aligned} \tag{4.108}$$

here we applied the estimate $|(\Delta_t^j * (\varphi_s * G_{ik}))(\mathbf{y})|^2 \le \|\Delta_t^j\|_1(|\Delta_t^j| * |\varphi_s * G_{ik}|^2)(\mathbf{y})$ and the identity $\|\Delta_t^j\|_1 = \|\Delta^j\|_1$, see Lemma 4.1. Thus

$$\|T_j G_{ik}\|_q^2 \tag{4.109}$$
$$\le c \int_0^\infty \int_{A(s,j)} \int_{\mathbb{R}^n} (|\Delta_t^j| * |\varphi_s * G_{ik}|^2)(\mathbf{x}) \left(Q(t)_{\omega/\nu}^{\mathrm{T}}\mathbf{x} - \frac{k}{\nu} t e_3 \right) g(\mathbf{x})\, \mathrm{d}\mathbf{x} \frac{\mathrm{d}t}{t} \frac{\mathrm{d}s}{s}$$

since Δ_t^j is radially symmetric. By the definition of $\mathcal{M}^j$ the innermost integral is bounded by $\mathcal{M}^j g(\mathbf{x})$ uniformly in $s > 0$. Hence we may proceed in (4.109) using the Hölder inequality as follows:

$$\|T_j G\|_q^2 \le c\|\Delta^j\|_1 \int_{\mathbb{R}^n} \left(\int_0^\infty |\varphi_s * G|^2(\mathbf{x}) \frac{\mathrm{d}s}{s} \right) \mathcal{M}^j g(\mathbf{x})\, \mathrm{d}\mathbf{x}. \tag{4.110}$$

Now (4.106) and the normalization $\|g\|_{(q/2)'} = 1$ complete the proof of (4.105).

Second Step: We investigate the estimate $\|\mathcal{M}^j \mathbf{g}\|_{(q/2)'}$. Since $q/2 \in (1, \infty)$ is arbitrary, we have to consider $\|\mathcal{M}^j\|_{L^p}$ for arbitrary $p \in (1, \infty)$. For this reason we introduce the classical Hardy–Littlewood maximal operator $\mathcal{M}$ on $L^p(\mathbb{R}^3)$ defined by

$$\mathcal{M}\, g(\mathbf{x}) := \sup_{s>0} \frac{1}{|B_s(\mathbf{x})|} \int_{B_s(\mathbf{x})} |g(\mathbf{x})|\, \mathrm{d}\mathbf{y}$$

and a "helical" maximal operator

$$\mathcal{M}_{\mathrm{hel}}\, g(\theta, x_3) := \sup_{s>0} \frac{1}{s} \int_{A_s} |g| \left| \theta - \frac{\omega}{\nu} t, x_3 + \frac{k}{\nu} t \right|$$

for functions g depending on (θ, x_3), which 2π periodic in θ. Since $0 \le h \in L^1(\mathbb{R}^3)$ is radially symmetric and strictly decreasing,

$$\sup_{r>0} h_r * \mathbf{U}_0(\mathbf{x}) \le c\mathcal{M}\mathbf{U}_0(\mathbf{x}).$$

Then

$$\mathcal{M}_j g(\mathbf{x}) \le c2^{-2|j|} \mathcal{M}(\mathcal{M}_{\mathrm{hel}}\, g_r(.,.)(\mathbf{x}),$$

where $g_r(\theta, x_3) = g(r, \theta, x_3) = g(\mathbf{x})$ is considered as a function of θ, x_3 and

$$\|\mathcal{M}_j g\|_p \le C2^{-2j} \|\mathcal{M}_{\mathrm{hel}}\, g_r(\cdot, \cdot)\|_{L^p(\mathbb{R}^3)}$$

due to L^p-continuity of $\mathcal{M}$.

To estimate $\mathcal{M}_{\text{hel}} g_r(\cdot,\cdot)$ in $L^p(\mathbb{R}^3)$, fix $r > 0$ and use the 2π-periodicity of g_r with respect to θ to get that

$$\begin{aligned}
&\int_{\mathbb{R}} \int_0^{2\pi} |\mathcal{M}_{\text{hel}}\, g_r(\theta, x_3)|^p \mathrm{d}\theta\, \mathrm{d}x_3 \\
&\quad \le \int_{\mathbb{R}} \int_0^{2\pi} \left| \sup_{s>0} \frac{1}{s} \int_{-16s}^{16s} |g_r| \left(\theta - \frac{\omega}{k}\left(x_3 + \frac{k}{\nu}t\right), x_3 + \frac{k}{\nu}t \right) \mathrm{d}t \right|^p \mathrm{d}\theta\, \mathrm{d}x_3 \\
&\quad = \int_{\mathbb{R}} \int_0^{2\pi} \left| \sup_{s>0} \frac{1}{s} \int_{-16s}^{16s} \gamma_{r,\theta} \left(x_3 + \frac{k}{\gamma}t \right) \mathrm{d}t \right|^p \mathrm{d}\theta\, \mathrm{d}x_3,
\end{aligned}$$

where

$$\gamma_{r,\theta}(y_3) = |g_r| \left(\theta - \frac{\omega}{k} y_3, y_3 \right).$$

Applying Hardy–Littlewood maximal operator on $\mathbb{R}^1$, we get that

$$\|\mathcal{M}_{\text{hel}}\, g_r(\cdot,\cdot)\|_{L^p(\mathbb{R}^3)} \le c\|g\|_{L^p(\mathbb{R}^3)}. \tag{4.111}$$

From Lemma 4.1 and Proposition 4.1 the operator T_j satisfies the estimates

$$\|T_j G_{ik}\|_q \le C 2^{-2|j|} \|G_{ik}\|_q, \quad j \in \mathcal{Z},\ q \in (2,\infty),\ c = c(q) > 0.$$

Then $T = \sum_{j=-\infty}^{\infty} T_j$ converges in the operator norm on $L^q(\mathbb{R}^3)^3$ and $\|TG\|_q \le c\|G\|_q$ for every $G \in S(\mathbb{R}^3)^3$.

Third Step: For $1 < q < 2$ we use the adjoint operator T^* given by

$$T^* G(\mathbf{x}) = \int_0^\infty (\Delta_t * Q_{\omega/\nu}(t) G) \left(Q^{\mathrm{T}}_{\omega/\nu}(t)\mathbf{x} + \frac{k}{\nu} t e_3 \right) \frac{\mathrm{d}t}{t} \tag{4.112}$$

with $G \in \mathcal{S}(\mathbb{R}^3)^3$, and then by the same argument we get that T^* is bounded in $L^{q/(q-1)}(\mathbb{R}^3)^9$, so T is L^q bounded for $1 < q < 2$. This implies the estimate

$$\|\nabla \mathbf{u}\|_q \le \|G\|_q. \tag{4.113}$$

Fourth Step: Now, using Farwig–Sohr Lemma 2.1 we know that there is $G \in L^q(\mathbb{R}^3)^9$ such that

$$\nabla \cdot G = \mathbf{F}_1, \quad \|G\|_{q,\mathbb{R}^3} \le C\|\mathbf{F}_1\|_{-1,q,\mathbb{R}^3}.$$

Let $G_n \in C_0^\infty(\mathbb{R}^3)^9$ be such that $\|G_n - G\|_{q,\mathbb{R}^3} \to 0$ as $n \to \infty$. Let $\mathbf{u}_n$ be solution of (4.87) with $\mathbf{F}_1 = \nabla \cdot G_n$. For each n and $m \in N$, we choose a constant vector

$\mathbf{b}_n^m \in \mathbb{R}^3$ satisfying

$$\int_{B_m} (\mathbf{u}_n(\mathbf{x}) + \mathbf{b}_n^m) d\mathbf{x} = 0$$

so that

$$\|\mathbf{u}_n + \boldsymbol{B}_n^m\|_{q,B_m} \leq C_m \|\nabla \mathbf{u}_n\|_{q,B_m} \leq C_m \|\nabla \mathbf{u}_n\|_{q,\mathbb{R}^3} \leq c_m \|G_n\|_{q,\mathbb{R}^3}$$

by the Poincaré inequality and by (4.113).

Therefore, there exist $\mathbf{u}^{(m)} \in W^{1,q}(B_m)$ and $V \in L^q(\mathbb{R}^3)^9$ such that

$$\|\mathbf{u}_n + \mathbf{b}_n^{(m)} - \mathbf{u}^{(m)}\|_{q,B_m} \to 0, \quad \|\nabla \mathbf{u}_n - V\|_{q,\mathbb{R}^3} \to 0 \quad (n \to \infty)$$

with

$$\nabla \mathbf{u}^{(m)}(\mathbf{x}) = V(\mathbf{x}) \quad \text{a.a. } \mathbf{x} \in B_m.$$

We first set

$$\tilde{\mathbf{u}} = \mathbf{u}^{(1)} \quad \text{on } B_1; \quad \mathbf{b}_n = \mathbf{b}_n^{(1)}.$$

Consider next the case $m = 2$; since $\nabla \mathbf{u}^{(2)}(\mathbf{x}) = V(\mathbf{x}) = \nabla u^{(1)}(\mathbf{x}) = \nabla \tilde{\mathbf{u}}(\mathbf{x})$ for a.a. $\mathbf{x} \in B_1 \subset B_2$, the difference $\mathbf{u}^{(2)}(\mathbf{x}) - \tilde{\mathbf{u}}(\mathbf{x}) =: \mathbf{a}$ is a constant vector and

$$\begin{aligned} |B_1|^{1/q} |\mathbf{b}_n^{(2)} - \mathbf{b}_n - \mathbf{a}| =& \|\mathbf{b}_n^{(2)} - \mathbf{b}_n - \mathbf{a}\|_{q,B_1} \\ \leq& \|\mathbf{u}_n + \mathbf{b}_n - \tilde{\mathbf{u}}\|_{q,B_1} + \|\mathbf{u}_n + \mathbf{b}_n^{(2)} - \mathbf{u}^{(2)}\|_{q,B_2} \to 0, \\ & n \to \infty. \end{aligned} \tag{4.114}$$

One extends $\tilde{\mathbf{u}}$ by

$$\tilde{\mathbf{u}} = \mathbf{u}^{(2)} - \mathbf{a} \quad \text{on } B_2.$$

Then (4.114) implies

$$\|\mathbf{u}_n + \mathbf{b}_n - \tilde{\mathbf{u}}\|_{q,B_2} \leq \|\mathbf{u}_n + \mathbf{b}_n^{(2)} - \mathbf{u}^{(2)}\|_{q,B_2} + |B_2|^{1/q} |\mathbf{b}_n^{(2)} - \mathbf{b}_n - \mathbf{a}| \to 0, \tag{4.115}$$

as $n \to \infty$. By induction there exists a function $\tilde{\mathbf{u}} \in \widehat{\boldsymbol{H}}^{1,q}(\mathbb{R}^3)$ such that

$$\|\mathbf{u}_n + \mathbf{b}_n - \tilde{\mathbf{u}}\|_{q,B_m} + |\nabla \mathbf{u}_n - \nabla \tilde{\mathbf{u}}\|_{q,\mathbb{R}^3} \to 0, \quad n \to \infty, \tag{4.116}$$

for all $m \in \mathbb{N}$.

Considering $k = \nu = 1$, then $\mathcal{L}_{\nu,\omega,k}\mathbf{u}_n = \nabla \cdot G_n$ that

$$\mathcal{L}_{\nu,\omega,k} b_n = \boldsymbol{\omega} \wedge \mathbf{b}_n = \mathcal{L}_{\nu,\omega,k}(\mathbf{u}_n + \mathbf{b}_n) - \nabla \cdot G_n \to \mathcal{L}_{\nu,\omega,k}\tilde{\mathbf{u}} - \nabla \cdot G$$
$$\text{in } \mathcal{D}'(\mathbb{R}^3)^3 \text{ as } n \to \infty$$

since there is a constant vector $\mathbf{b} \in \mathbb{R}^3$ such that

$$\boldsymbol{\omega} \wedge \mathbf{b}_n \to \boldsymbol{\omega} \wedge \mathbf{b} = \mathcal{L}\mathbf{b}$$

as $n \to \infty$. Consequently, we get

$$\mathcal{L}_{\nu,\omega,k}(\tilde{\mathbf{u}} - \mathbf{b}) = \nabla \cdot G \quad \text{in } \mathcal{D}'(\mathbb{R}^3)^3$$

and $\mathbf{U}_0 = \tilde{\mathbf{u}} - \mathbf{b}$ is the desired solution. By (4.116) we have $\|\mathbf{u}_n - \nabla \mathbf{u}\|_{q,\mathbb{R}^3} \to 0$ and, therefore, the estimate (4.84) holds.

Fifth Step: It remains to prove the uniqueness. We use the duality method. We consider the adjoint problem

$$\mathcal{L}^*\nu, \omega, k\mathbf{v} \equiv -\Delta\mathbf{v} + (\boldsymbol{\omega} \wedge \mathbf{x}) \cdot \nabla\mathbf{v} - \boldsymbol{\omega} \wedge \mathbf{v} + \frac{\partial \mathbf{U}_0}{\partial x_3} = \nabla \cdot G \tag{4.117}$$

with $G \in C_0^\infty(\mathbb{R}^3)^9$. This admits the solution

$$\widehat{\mathbf{v}}(\xi) = \int_0^\infty \mathrm{e}^{-\nu|\boldsymbol{\xi}|^2 t} Q_\omega(t)\widehat{\mathbf{f}}(Q_\omega^{\mathrm{T}}(t). - kt\boldsymbol{e}_3))(\boldsymbol{\xi})\mathrm{d}t. \tag{4.118}$$

Applying the same argument, we get

$$\|\nabla\mathbf{v}\|_{r,\mathbb{R}^3} \le C\|G\|_{r,\mathbb{R}^3} \quad \text{for all } \mathbf{v} \in \widehat{W}^{1,r}(\mathbb{R}^3), \quad r \in (1, \infty). \tag{4.119}$$

Let $\mathbf{U}_0 \in \widehat{\boldsymbol{H}}^{1,q}(\mathbb{R}^3)$ be a weak solution of $\mathcal{L}\mathbf{U}_0 = 0$ in $\widehat{\boldsymbol{H}}^{1,q}(\mathbb{R}^3)$. One can take $\mathbf{v}$ as a test function to get

$$\langle \mathcal{L}\mathbf{U}_0, v\rangle = 0.$$

Similarly, one takes $\mathbf{U}_0$ as a test function for (4.117) in $\widehat{\boldsymbol{H}}^{-1,q/(q-1)}(\mathbb{R}^3)$ to obtain

$$\langle \mathbf{U}_0, \mathcal{L}^*\mathbf{v}\rangle = \langle \mathbf{U}_0, \nabla \cdot G\rangle.$$

Therefore,

$$\langle \mathbf{U}_0, \nabla \cdot G\rangle = 0.$$

Since $G \in (C_0^\infty)^9$ is arbitrary, we obtain $\mathbf{U}_0 = 0$ in $\widehat{\boldsymbol{H}}^{1,q}(\mathbb{R}^3)$ by Theorem 4.8 $\mathbf{U}_0$ is a constant vector, but it is a constant multiple of $\boldsymbol{\omega}$ because $\boldsymbol{\omega} \wedge \mathbf{U}_0 = 0$.
To complete the proof of Theorem 4.7, we have used the following lemma.

Lemma 4.7 *Let* $\mathbf{v} \in \mathcal{S}(\mathbb{R}^3)$ *be the solution of*

$$-\Delta \mathbf{v} + \frac{\partial \mathbf{v}}{\partial x_3} - (\boldsymbol{\omega} \wedge \mathbf{x}) \cdot \nabla \mathbf{v} = \mathbf{0} \quad in\ \mathbb{R}^3.$$

Then $\operatorname{supp} \widehat{v} \subset \{0\}$. □

Proof of Lemma 4.7 See proof of Theorem 4.1, (A2). □

To complete the proof of Theorem 4.7 we use the fact that the pressure is formally obtained from the problem

$$\pi_0 = -\nabla \cdot (-\Delta)^{-1} \mathbf{F}_1.$$

Since $(-\Delta)^{-1}$ can be justified as a bounded operator from $\widehat{\boldsymbol{H}}^{-1,q}(\mathbb{R}^3)$ to $\widehat{\boldsymbol{H}}^{1,q}(\mathbb{R}^3)$, we get

$$\|\nabla \pi_0\|_q \le c \|\mathbf{F}_1\|_{1,q},$$

which implies that

$$\|\mathbf{F}_1 - \nabla \pi_0\|_{-1,q} \le c \|\mathbf{F}_1\|_{-1,q}.$$

This completes the proof of Theorem 4.7.

Remark 4.3 Without a special treatment of the third component of the gradient of $\mathbf{U}_0$, we only have:

$$\begin{aligned} &\|\partial_3 \mathbf{U}_0 - ((\omega \mathbf{e}_3 \wedge \mathbf{x}) \cdot \nabla) \mathbf{U}_0 + \omega \mathbf{e}_3 \wedge \mathbf{U}_0\|_{-1,q,\mathbb{R}^3} \\ &\quad = \|\mathbf{F}_1 + \Delta \mathbf{U}_0\|_{-1,q,\mathbb{R}^3} \le c(\|\mathbf{F}_1\|_{-1,q,\mathbb{R}^3} + \|\nabla \mathbf{U}_0\|_{q,\mathbb{R}^3}) \le c \|\mathbf{F}_1\|_{-1,q,\mathbb{R}^3} \end{aligned}$$

with some $c > 0$ depending on q. □

Proposition 4.2 *Let* $1 < q < \infty$ *and let* $\mathbf{F}_1 \in \widehat{\boldsymbol{H}}^{-1,q}(\mathbb{R}^3)$ *be given, and then the problem (4.79) with* $G_1 = 0$ *possesses a weak solution* $\{\mathbf{U}_0, \pi_0\} \in \widehat{\boldsymbol{H}}^{1,q}(\mathbb{R}^3) \times L^q(\mathbb{R}^3)$ *which satisfies (4.84) and, moreover,*

$$\|\partial_3 \mathbf{U}_0\|_{-1,q} \le c \|\mathbf{F}_1\|_{-1,q}$$

with some $c > 0$ *depending on* q. □

Corollary 4.4 *Let* $1 < q < \infty$ *and let* $\mathbf{F}_1 \in \widehat{H}^{-1,q}(\mathbb{R}^3)^3$ *be given, and then the problem (4.79) with* $G_1 = 0$ *possesses a weak solution* $\{\mathbf{U}_0, \pi_0\} \in \widehat{\boldsymbol{H}}^{1,q}(\mathbb{R}^3) \times$

$L^q(\mathbb{R}^3)$ *which satisfies:*

$$\|\nabla \mathbf{U}_0\|_q + \|\pi_0\|_q + \|\partial_3 \mathbf{U}_0\|_{-1,q} + \|-((\omega \mathbf{e}_3 \wedge \mathbf{x})\cdot\nabla)\mathbf{U}_0 + \omega \mathbf{e}_3 \wedge \mathbf{U}_0\|_{-1,q} \le c\|\mathbf{F}_1\|_{-1,q}.$$

Proof of Proposition 4.2 We need to estimate the quantity $\sup_{\|\nabla\varphi\|\le 1}\langle \partial_3 \mathbf{U}_0, \varphi\rangle$. In the work of Farwig, see [60], he got a priori bounds of $\partial_3 \mathbf{U}_0$ in L^q-norms under assumptions of bounds of $\|F_1\|_q$. Similarly we are doing by use of the multiplier theory starting from the integral representation of $\partial_3 \mathbf{U}_0$ introduced in Sect. 4.2.2.1.

We know that $\mathbf{F}_1 = \operatorname{div} H$ with $H_{\cdot k} = \mathcal{B}F_{1,k} \in L^q(\mathbb{R}^3)^3$ (see Sect. 2.2), and using the fact that the space $\{\mathbf{F}_1 \colon F_1 = \operatorname{div} H, H \in C_0^\infty(\mathbb{R}^3)^{3\times 3}\}$ is dense in $\widehat{\boldsymbol{H}}^{-1,q}(\mathbb{R}^3)^3$, we can assume first $H \in C_0^\infty(\mathbb{R}^3)^{3\times 3})$. $\mathbf{U}_0$ can be expressed in dependence on H as

$$\mathbf{U}_0(\mathbf{x}) = \int_{\mathbb{R}^3} \Sigma_{os}\,(\mathbf{x}, \mathbf{y}) \cdot \operatorname{div} H\,(\mathbf{y})\, \mathrm{d}\mathbf{y},$$

where

$$\Sigma_{os}(\mathbf{x}, \mathbf{y}) = \int_0^\infty K\,(Q_\omega(t)\mathbf{x} - \mathbf{y} - t\mathbf{e}_3)\; Q_\omega^{\mathrm{T}}(t)\mathrm{d}t$$

and

$$K = \frac{1}{(4\pi t)^{3/2}} \mathrm{e}^{-|\boldsymbol{\xi}|^2/4t}.$$

Applying the Fourier transform we get the form

$$\widehat{\mathbf{U}_{0,k}}(\boldsymbol{\xi}) = \frac{1}{D(\boldsymbol{\xi})} \int_0^{2\pi/|\omega|} \mathrm{e}^{-(|\boldsymbol{\xi}|^2 + \mathrm{i}\xi_3)t}\, O_\omega^{\mathrm{T}}(t)\, \xi_j \widehat{H}_{jk}(Q_\omega(t)\boldsymbol{\xi})\mathrm{d}t,$$

where we put

$$D(\boldsymbol{\xi}) = 1 - \mathrm{e}^{-2\pi(|\boldsymbol{\xi}|^2 + \mathrm{i}\xi_3)/|\omega|}.$$

Thus, we have

$$\widehat{\partial_3 \mathbf{U}_{0,k}}(\boldsymbol{\xi}) = \frac{\mathrm{i}\xi_3\xi_j}{D(\boldsymbol{\xi})|\omega|} \int_0^{2\pi} \mathrm{e}^{-(|\boldsymbol{\xi}|^2 + \mathrm{i}\xi_3)t/|\omega|}\, O_\omega^{\mathrm{T}}(t) \widehat{H_{jk}}(O_\omega(t)\boldsymbol{\xi})\mathrm{d}t$$

and also

$$\frac{\widehat{\partial_3 \mathbf{U}_{0,k}}(\boldsymbol{\xi})}{|\boldsymbol{\xi}|} = \int_0^{2\pi} \boldsymbol{m}_t(\boldsymbol{\xi}) \cdot Q_\omega^{\mathrm{T}}(t) \widehat{H_{\cdot k}}(Q_\omega(t). - t\mathrm{e}_3/|\omega|)(\boldsymbol{\xi})\mathrm{d}t,$$

where the multiplier $\boldsymbol{m}_t(\cdot)$ has the form

$$\boldsymbol{m}_t(\boldsymbol{\xi}) = \frac{\mathrm{i}\xi_3\boldsymbol{\xi}}{|\omega||\boldsymbol{\xi}|D(\boldsymbol{\xi})}\mathrm{e}^{-|\boldsymbol{\xi}|^2 t/|\omega|}.$$

Moreover,

$$\langle \partial_3 \mathbf{U}_{0,k}, \varphi_k \rangle = \mathrm{i}\int_{\mathbb{R}^3} \frac{\xi_3\widehat{\mathbf{U}_{0,k}}(\boldsymbol{\xi})}{|\boldsymbol{\xi}|}\cdot|\boldsymbol{\xi}|\widehat{\varphi_k}(\boldsymbol{\xi})\mathrm{d}\boldsymbol{\xi}, \tag{4.120}$$

which reads

$$\int_{\mathbb{R}^3} \widehat{V}(\boldsymbol{\xi})\cdot\widehat{\Psi}(\boldsymbol{\xi})\mathrm{d}\boldsymbol{\xi}$$

in duality L^q-$L^{q'}$. Indeed, Ψ can be taken in $L^{q'}$, and the term $\widehat{V}(\boldsymbol{\xi}) = \xi_3\widehat{\mathbf{U}_{0,k}}(\boldsymbol{\xi})/|\boldsymbol{\xi}|$ can be estimated in L^q. To this end, observing that $\|H(O_\omega(t)\cdot - t\mathbf{e}_3)\|_q = \|H\|_q$ for all $t\in(0,2\pi)$ and knowing that $\mathrm{e}^{-\mathrm{i}t\xi_3}\widehat{H}_{jk}(\cdot)$ is the Fourier transform of $H_{jk}(\cdot - t\mathbf{e}_3)$, one can show that $\widehat{\mathbf{V}}(\boldsymbol{\xi})$ satisfies the Marcinkiewicz multiplier theorem which implies the boundedness of $\mathbf{V}$ in L^q by $\mathbf{F}_1\in\widehat{W}^{-1,q}(\mathbb{R}^3)$. Therefore we obtain

$$\|\partial_3\mathbf{U}_0\|_{-1,q} \le \|H\|_q \le \|\mathbf{F}_1\|_{-1,q}.$$

For details, see Sect. 4.2.3. □

Proof of Corollary 4.2 Following [60] there exists $\alpha\in\mathbb{R}^3$ such that

$$\mathbf{v} = \mathbf{U}_0 - \boldsymbol{\alpha} \in L^s(\mathbb{R}^3), \text{ for all } s>1,\ \frac{1}{s}\in\frac{1}{q}-\left[\frac{1}{4},\frac{1}{3}\right].$$

Let

$$\tilde{\mathcal{L}}\mathbf{U}_0{}' := -\partial_\theta\mathbf{U}_0' + \mathbf{U}_0^{\perp}.$$

Since $\tilde{\mathcal{L}}\mathbf{v}' = -\partial_\theta\mathbf{v}' + \mathbf{v}^{\perp} = \tilde{\mathcal{L}}\mathbf{U}_0' - \boldsymbol{a}^{\perp}$ and applying integration with respect to θ, we get

$$2\pi\boldsymbol{a}' = \int_0^{2\pi}\tilde{\mathcal{L}}\mathbf{U}_0' d\theta - \int_0^{2\pi}\mathbf{v}' d\theta \in L^q + L^s,$$

which implies $\boldsymbol{a}' = 0$. □

Proof of Corollary 4.3 From Theorem II5.1 [95] yields the estimate

$$\left(\int_{\mathbb{R}^3}\frac{|\mathbf{U}_0(\mathbf{x})-\mathbf{u}_\infty|}{|\mathbf{x}|}d\mathbf{x}\right)^{1/q} \le \frac{q}{3-q}\left(\int_{\mathbb{R}^3}|\nabla\mathbf{U}_0(\mathbf{x})|^q d\mathbf{x}\right)^{1/q}.$$

Moreover, by Lemma 52 [95]

$$\int_{\mathbf{y}=1} |\mathbf{U}_0(R\mathbf{y})|^q do(\mathbf{y}) = o(R^{q-3})$$

as $R \to \infty$. Since $\mathbf{U}_0 \in L^s(\mathbb{R}^3)$ that $\mathbf{u}_\infty$ vanishes. □

Remark 4.4

- Let $1 < q < 3$. Then $\mathbf{U}_0 \in \boldsymbol{L}^{3q/(3-q)}(\mathbb{R}^3)$, and following Theorems 4.7 and 4.8, the solution $\{\mathbf{U}_0, \pi_0\}$ is unique.
- In the same conditions, following Corollary 4.3 from (see also [93, pages 59–62]), we have $\||\mathbf{x}|^{-1}\mathbf{U}_0\|_q \le c\|\mathbf{F}_1\|_{-1,q}$ with a positive constant $c = c(q, \omega)$.

□

Considering now the nonhomogeneous problem in the whole space, i.e., problem (4.79) for given $\mathbf{F}_1$ and G_1, we recall the following result.

Theorem 4.10 *Let $1 < q < \infty$ and let $\mathbf{F}_1 \in \widehat{\boldsymbol{H}}^{-1,q}(\mathbb{R}^3)$ be given. Suppose $G_1 \in L^q(\mathbb{R}^3)$ such that $(\boldsymbol{\omega} \wedge \mathbf{x}) \cdot G_1 \in \widehat{H}^{-1,q}(\mathbb{R}^3)$. Then the problem (4.79) possesses a weak solution $\{\mathbf{U}, \pi\} \in \widehat{\boldsymbol{H}}^{1,q}(\mathbb{R}^3) \times L^q(\mathbb{R}^3)$ which satisfies*

$$\begin{aligned} &\|\nabla \mathbf{U}\|_q + \|\pi\|_q + \|\partial_3 \mathbf{U} - ((\omega \mathbf{e}_3 \wedge \mathbf{x}) \cdot \nabla)\mathbf{U} + \omega \mathbf{e}_3 \wedge \mathbf{U}\|_{-1,q} \\ &\quad \le c(\|\mathbf{F}_1\|_{-1,q} + \|G_1\|_q + \|(\omega \mathbf{e}_3 \wedge \mathbf{x}) \cdot G_1\|_{-1,q}) \end{aligned} \tag{4.121}$$

with some $c > 0$ depending on q. □

Proof of Theorem 4.10 Let us just briefly mention the difference from the homogeneous case is the estimate of pressure.

A straightforward calculation shows that the unknown pressure p is given by the solution of the problem

$$\Delta p = \operatorname{div} \boldsymbol{f} + \nu \Delta g + \frac{\partial g}{\partial x_3} + \partial_\theta g = \operatorname{div} \boldsymbol{G} \quad \text{in } \mathbb{R}^n$$

with $\boldsymbol{G} = \boldsymbol{f} + \nu \nabla g + (\omega \mathbf{e}_3 \wedge \boldsymbol{x}) g + (\operatorname{div})^{-1} \frac{\partial g}{\partial x_3}$. Then the Hörmander–Mikhlin multiplier theorem in L^q-spaces gives us the estimate

$$\|p\|_q \le c\|\boldsymbol{G}\|_{-1,q} \le c\left(\|\boldsymbol{f}\|_{-1,q} + \left\|\frac{\partial g}{\partial x_3}\right\|_{-1,q} + \|\nu \nabla g + (\boldsymbol{\omega} \wedge \boldsymbol{x}) g\|_{-1,q}\right), \tag{4.122}$$

where $c = c(q) > 0$; in particular $\nabla p \in \boldsymbol{L}^q(\mathbb{R}^n)$. □

Remark 4.5 Due to Proposition 4.2 we are able to improve the previous estimate, and we have

$$\|\nabla \mathbf{U}\|_q + \|\pi\|_q + \|\partial_3 \mathbf{U}\|_{-1,q} + \| - ((\omega \mathbf{e}_3 \wedge \mathbf{x}) \cdot \nabla)\mathbf{U} + \omega \mathbf{e}_3 \wedge \mathbf{U}\|_{-1,q}$$
$$\leq c(\|\mathbf{F}_1\|_{-1,q} + \|G_1\|_q + \|(\omega \mathbf{e}_3 \wedge \mathbf{x}) \cdot G_1\|_{-1,q}). \tag{4.123}$$

We finally recall the well-known result, e.g., from [144], about the nonhomogeneous Stokes problem in bounded domains, solving the problem (4.82) in the domain D_ρ. So, with the used notation $\{V, \tau\}$ the theorem reads:

Theorem 4.11 *Let* $1 < q < \infty$. *Suppose that*

$$\mathbf{F}_2 \in \boldsymbol{H}^{-1,q}(\mathcal{D}_\rho), \quad G_2 \in L^q_0(\mathcal{D}_\rho),$$

so, we assume

$$\int_{\mathcal{D}_\rho} G_2(\mathbf{x})\, d\mathbf{x} = 0.$$

Then the problem (4.82) possesses a unique (up to an additive constant for τ*) weak solution* $\{\mathbf{V}, \tau\} \in \boldsymbol{H}^{1,q}_0(\mathcal{D}_\rho) \times L^q(\mathcal{D}_\rho)$, *which satisfies the estimate*

$$\|\nabla \mathbf{V}\|_{q,\mathcal{D}_\rho} + \|\tau - \bar{\tau}\|_{q,\mathcal{D}_\rho} \leq C(\|\mathbf{F}_2\|_{-1,q,\mathcal{D}_\rho} + \|G_2\|_{q,\mathcal{D}_\rho}), \tag{4.124}$$

where $\bar{\tau} = 1/|\mathcal{D}_\rho| \int_{\mathcal{D}_\rho} \tau(\mathbf{x}) dx$. □

4.2.2.2 Proof of Theorem 4.6

The proof is presented in four steps.

Step 1: Existence (Homogeneous Divergence Case)

Let $\mathbf{f} = \operatorname{div} \mathbf{F}$ with $\mathbf{F} \in C_0^\infty(\mathcal{D})^9$. In the domain $\mathcal{D}_R$ according to the support of $\mathbf{F}$, we apply the classical approach to solve

$$-\Delta \mathbf{u} + \partial_3 \mathbf{u} - ((\omega \mathbf{e}_3 \wedge \mathbf{x}) \cdot \nabla)\mathbf{u} + \omega \mathbf{e}_3 \wedge \mathbf{u} + \nabla p = \mathbf{f} = \operatorname{div} F$$

with $\operatorname{div} \mathbf{u} = 0$ and with the homogeneous Dirichlet boundary conditions on $\partial \mathcal{D}_R$.

The bilinear form $b(\mathbf{u}, \boldsymbol{\varphi})$, where

$$b(\mathbf{u}, \boldsymbol{\varphi}) = \langle \nabla \mathbf{u}, \nabla \boldsymbol{\varphi} \rangle + \langle \partial_3 \mathbf{u}, \boldsymbol{\varphi} \rangle - \langle ((\omega \mathbf{e}_3 \wedge \mathbf{x}) \cdot \nabla)\mathbf{u} - \omega \mathbf{e}_3 \wedge \mathbf{u}, \boldsymbol{\varphi} \rangle$$

is coercive on $(\boldsymbol{H}^{1,2}_\sigma(\mathcal{D}_R)) \times (H^{1,2}_\sigma(\mathcal{D}_R))^3$: $\langle \cdot, \cdot \rangle$ stands here for the L^2-inner product. One can easily verify that

$$b(\mathbf{u}_R, \mathbf{u}_R) = \|\nabla \mathbf{u}_R\|^2_{2,\mathcal{D}_R} = \langle \operatorname{div} F, \mathbf{u}_R \rangle.$$

Using the Lax–Milgram theorem, we justify the existence of a unique solution $\mathbf{u}_R \in (\boldsymbol{H}^{1,2}_\sigma(\mathcal{D}_R))$, which satisfies the estimate

$$\|\nabla \mathbf{u}_R\|_{2,\,\mathcal{D}_R} \le \|\mathbf{F}\|_{2,\mathcal{D}_R} = \|\mathbf{F}\|_{2,\mathcal{D}}.$$

We can extend $\mathbf{u}_R$ by zero in $\mathcal{D} \setminus \mathcal{D}_R$. Then we obtain $\widetilde{\mathbf{u}_R} \in (\widehat{\boldsymbol{H}}^{1,2}_\sigma(\mathcal{D}))$ satisfying the same estimate, uniform as $R \to +\infty$.

We now choose a sequence of numbers $\{R_n\}_n$ tending to infinity, so that $\widetilde{\mathbf{u}_{R_n}}$ converges weakly in $(\widehat{\boldsymbol{H}}^{1,2}_\sigma(\mathcal{D}))$. The limit $\mathbf{u}$ is unique and such that

$$\langle \nabla \mathbf{u}, \nabla \boldsymbol{\varphi}\rangle + \langle \partial_3 \mathbf{u}, \boldsymbol{\varphi}\rangle - \langle ((\omega \mathbf{e}_3 \wedge \mathbf{x}) \cdot \nabla)\mathbf{u} - \omega \mathbf{e}_3 \wedge \mathbf{u}, \boldsymbol{\varphi}\rangle - \langle \mathbf{f}, \boldsymbol{\varphi}\rangle = 0 \qquad (4.125)$$

for all $\boldsymbol{\varphi} \in \boldsymbol{C}^\infty_0(\mathcal{D})$, then for all $\boldsymbol{\varphi} \in (\widehat{\boldsymbol{H}}^{1,2}(\mathcal{D}))$.

Therefore there exists $p \in L^2_{\text{loc}}(\mathcal{D})$ (unique up to an additive constant) such that

$$-\Delta \mathbf{u} + \partial_3 \mathbf{u} - ((\omega\, \mathbf{e}_3 \wedge \mathbf{x}) \cdot \nabla)\mathbf{u} + \omega \mathbf{e}_3 \wedge \mathbf{u} - \mathbf{f} = -\nabla p.$$

Applying the localization technique in the next step, it will imply that $\mathbf{u}$ and p satisfy

$$\mathbf{u} \in \widehat{\boldsymbol{H}}^{1,q}(\mathcal{D})^3, \quad p \in L^q(\mathcal{D}) \quad \text{for } 3/2 < q \le 6.$$

Step 2: Localization Procedure

Let $\boldsymbol{\varphi} \in (\boldsymbol{C}^\infty_0(\mathbb{R}^3))$, and let ψ be always the same cutoff function as in Sect. 4.2.1. We can successively choose the following test functions φ, and our problem is decomposed into two problems:

- $\boldsymbol{\varphi} = (1-\psi)\boldsymbol{\varphi}$, we are getting the problem in the whole space by $\{\mathbf{U}, \pi\} = \{(1-\psi)\mathbf{u}, (1-\psi)p\}$ as in (4.79); the formulas describing $\mathbf{F}_1 = \mathbf{F}_1(\mathbf{u}, p)$ and $G_1 = G_1(\mathbf{u})$ are given in Sect. 4.2.1.
- $\boldsymbol{\varphi} = \psi\boldsymbol{\varphi}$, we are getting the problem in a bounded domain $\mathcal{D}_\rho$ and consider the unknown variables $\{\mathbf{V}, \tau\} = \{\psi\mathbf{u}, \psi p\}$ as in (4.80) or (4.82); see in Sect. 4.2.1 the detailed formulas for $\mathbf{F}_2 = \mathbf{F}_2(\mathbf{u}, p)$ and $G_2 = G_2(\mathbf{u})$.

Theorems 4.10 and 4.11 solve these problems under the following hypotheses

$$\mathbf{F}_1 \in (\widehat{\boldsymbol{H}}^{-1,q}(\mathbb{R}^3)), \quad G_1 \in L^q(\mathbb{R}^3), (\omega \wedge \mathbf{x}) \cdot G_1 \in \widehat{H}^{-1,q}(\mathbb{R}^3),$$

$$\mathbf{F}_2 \in (\widehat{\boldsymbol{H}}^{-1,q}(\mathcal{D}_\rho)), \quad G_2 \in L^q_0(\mathcal{D}_\rho),$$

respectively, with estimates (4.121) or (4.124).

To exploit these estimates, it remains essentially to control all the terms we have from formulas (4.81) in $\|\mathbf{F}_j(\mathbf{u}, p)\|_{-1,q,\mathbb{R}^3 or D_\rho}$ for appropriate q and $j = 1, 2$. In this way, we recall that $\nabla\psi$ and $\Delta\psi$ have a compact support in D_ρ at the most (precisely in the annulus $\{\mathbf{x} : \rho_0 < |\mathbf{x}| < \rho\}$ closed to the "obstacle" Ω), so we have:

- Either $\|\boldsymbol{\varphi}\|_{q/(q-1),\mathcal{D}_\rho} \le c(|\mathcal{D}_\rho|)\|\nabla\boldsymbol{\varphi}\|_{q/(q-1),\mathcal{D}_\rho}$ by the Friedrichs–Poincaré inequality, or $\|\boldsymbol{\varphi}\|_{q/(q-1),\mathcal{D}_\rho} \le |\ \mathcal{D}_\rho|^{1/3}\|\boldsymbol{\varphi}\|_{r,\mathcal{D}_\rho}$ by the Hölder inequality with $1/r + 1/3 = (q-1)/q$,, which implies $(q-1)/q > 1/3$ and $q > 3/2$.
  Thus, $\|\boldsymbol{\varphi}\|_{q/(q-1),\mathcal{D}_\rho} \le |\mathcal{D}_\rho|^{1/3}\|\boldsymbol{\varphi}\|_{r,\mathbb{R}^3} \le c(|\ \mathcal{D}_\rho|)\|\nabla\varphi\|_{q/(q-1),\mathbb{R}^3}$.
- $|\langle(1-\psi)\mathbf{f}, \boldsymbol{\varphi}\rangle| \le c\|\mathbf{f}\|_{-1,q,\ \mathcal{D}}\|\nabla\boldsymbol{\varphi}\|_{q/(q-1),\mathbb{R}^3}$.
- $|\langle 2(\nabla\psi\cdot\nabla)\mathbf{u} + [\Delta\psi + ((\omega\mathbf{e}_3\wedge\mathbf{x})\cdot\nabla)\psi]\mathbf{u}, \boldsymbol{\varphi}\rangle| \le |\langle(\Delta\psi)\boldsymbol{\varphi}, \nu\mathbf{u}\rangle| + |\langle 2(\nabla\psi\cdot\nabla)\boldsymbol{\varphi}, \mathbf{u}\rangle| + |\langle((\omega\mathbf{e}_3\wedge\mathbf{x})\cdot\nabla)\psi)\boldsymbol{\varphi}, \mathbf{u}\rangle| \le c\|\mathbf{u}|_{\mathcal{D}_\rho}\|_{q,\mathcal{D}_\rho}\|\nabla\boldsymbol{\varphi}\|_{q/(q-1),\mathbb{R}^3}$.
- $|\langle(\nabla\psi)p, \varphi\rangle| \le c\|p|_{\mathcal{D}_\rho}\|_{-1,q,\mathcal{D}_\rho}\|\nabla\boldsymbol{\varphi}\|_{q/(q-1),\mathbb{R}^3}$.
- $|\langle(\partial_3\psi)\mathbf{u}, \boldsymbol{\varphi}\rangle| \le c\|\mathbf{u}|_{\mathcal{D}_\rho}\|_{q,\mathcal{D}_\rho}\|\nabla\boldsymbol{\varphi}\|_{q/(q-1),\mathbb{R}^3}$.
- $\|(\omega\mathbf{e}_3\wedge\mathbf{x})(\mathbf{u}\cdot\nabla\psi)\|_{-1,q,\mathbb{R}^3} \le C\|\mathbf{u}|_{\mathcal{D}_\rho}\|_{q,\mathcal{D}_\rho}\|\nabla\boldsymbol{\varphi}\|_{q/(q-1),\mathbb{R}^3}$.

Now, since $G_1 = -\nabla\psi\cdot\mathbf{u}$ and $G_2 = \nabla\psi\cdot\mathbf{u}$, we also have

$$|\langle\nabla\psi\cdot\mathbf{u}, \boldsymbol{\varphi}\rangle| \le C\|u|_{\mathcal{D}_\rho}\|_{q,\mathcal{D}_\rho}\|\nabla\boldsymbol{\varphi}\|_{q/(q-1),\mathbb{R}^3}.$$

Then applying the estimates (4.121), (4.123), (4.124) together with the previous estimates, we get

$$\|\nabla\mathbf{U}\|_{q,\mathbb{R}^3} + \|\pi\|_{q,\mathbb{R}^3} \le c(\|\mathbf{f}\|_{-1,q} + \|\mathbf{u}|_{\mathcal{D}_\rho}\|_{q,\mathcal{D}_\rho} + \|p|_{\mathcal{D}_\rho}\|_{-1,q,\mathcal{D}_\rho}), \tag{4.126}$$

$$\begin{aligned}\|\partial_3\mathbf{U}\|_{-1,q,\mathbb{R}^3} &+ \|((\omega\mathbf{e}_3\wedge\mathbf{x})\cdot\nabla)\mathbf{U} - \omega\mathbf{e}_3\wedge\mathbf{U}\|_{-1,q,\mathbb{R}^3}\\ &\le c(\|\mathbf{f}\|_{-1,q} + \|\mathbf{u}|_{\mathcal{D}_\rho}\|_{q,\mathcal{D}_\rho} + \|p|_{\mathcal{D}_\rho}\|_{-1,q,\mathcal{D}_\rho}),\end{aligned} \tag{4.127}$$

$$\begin{aligned}&\|\nabla\mathbf{V}\|_{q,\mathcal{D}_\rho} + \|\tau\|_{q,\mathcal{D}_\rho}\\ &\le c\left(\|\mathbf{f}\|_{-1,q} + \|\mathbf{u}|_{\mathcal{D}_\rho}\|_{q,\mathcal{D}_\rho} + \|p|_{\mathcal{D}_\rho}\|_{-1,q,\mathcal{D}_\rho} + \left|\int_{\mathcal{D}_\rho}\psi(\mathbf{x})p(\mathbf{x})\mathrm{d}\mathbf{x}\right|\right),\end{aligned} \tag{4.128}$$

and we can conclude the estimate for $\|\partial_3\mathbf{u}\|_{-1,q,\mathcal{D}}$ observing that

$$\begin{aligned}\|\partial_3\mathbf{u}\|_{-1,q,\mathcal{D}} &= \|\partial_3(\mathbf{U}+\mathbf{V})\|_{-1,q,\mathcal{D}} \le \|\partial_3\mathbf{U}\|_{-1,q,\mathbb{R}^3} + \|\partial_3\mathbf{V}\|_{-1,q,\mathcal{D}_\rho}\\ &\le c(\|\partial_3\mathbf{U}\|_{-1,q,\mathbb{R}^3} + \|\partial_3\mathbf{V}\|_{q,\mathcal{D}_\rho}).\end{aligned} \tag{4.129}$$

We know that $\mathbf{u}|_{\mathcal{D}_\rho} \in \boldsymbol{H}_0^{1,2}(\mathcal{D}_\rho)$, and then by means of the embedding $\boldsymbol{H}_0^{1,2}(\mathcal{D}_\rho) \subset \mathbf{L}^q(\mathcal{D}_\rho)$, where $1/6 \le 1/q \le 2/3$, we obtain $\|\mathbf{u}|_{\mathcal{D}_\rho}\|_{q,\ \mathcal{D}_\rho} \le c$, and therefore all the inequalities make sense and $\|\mathbf{F}_j(\mathbf{u}, p)\|_{-1,q,\mathbb{R}^3 or \mathcal{D}_\rho}$ is bounded in terms of $\|\mathbf{f}\|_{-1,q}$, $\|\mathbf{u}|_{\mathcal{D}_\rho}\|_{q,\mathcal{D}_\rho}$, and $\|p|_{\mathcal{D}_\rho}\|_{-1,q,\mathcal{D}_\rho}$.

Finally $\mathbf{u} = \mathbf{U} + \mathbf{V}$, $p = \pi + \tau$,

$$\|\nabla \mathbf{u}\|_{q,\mathcal{D}} + \|p\|_{q,D} \leq c \left(\|\mathbf{f}\|_{-1,q} + \|\mathbf{u}|_{\mathcal{D}_\rho}\|_{q,\mathcal{D}_\rho} + \|p|_{\mathcal{D}_\rho}\|_{-1,q,\mathcal{D}_\rho} + \left| \int_{\mathcal{D}_\rho} \psi(\mathbf{x}) p(\mathbf{x}) \mathrm{d}x \right| \right), \tag{4.130}$$

$$\|\partial_3 \mathbf{u}\|_{-1,q,\mathcal{D}} + \|(\omega \mathbf{e}_3 \wedge \mathbf{x}) \cdot \nabla \mathbf{u} - \boldsymbol{\omega} \wedge \mathbf{u}\|_{-1,q,\mathcal{D}} \leq c \left(\|\mathbf{f}\|_{-1,q} + \|\nabla \mathbf{u}|_{\mathcal{D}_\rho}\|_q + \|p|_{\mathcal{D}_\rho}\|_q \right). \tag{4.131}$$

From the estimates (4.130), (4.131) we get immediately

$$\|\nabla \mathbf{u}\|_{q,\mathcal{D}} + \|p\|_{q,\mathcal{D}} + \|\partial_3 \mathbf{u}\|_{-1,q,\mathcal{D}} + \|(\omega \mathbf{e}_3 \wedge \mathbf{x}) \cdot \nabla \mathbf{u} - \omega \mathbf{e}_3 \wedge \mathbf{u}\|_{-1,q,\mathcal{D}} \leq c \left(\|\mathbf{f}\|_{-1,q} + \|\mathbf{u}|_{\mathcal{D}_\rho}\|_{q,\mathcal{D}_\rho} + \|p|_{\mathcal{D}_\rho}\|_{-1,q,\mathcal{D}_\rho} + \left| \int_{\mathcal{D}_\rho} \psi(\mathbf{x}) p(\mathbf{x}) \mathrm{d}\mathbf{x} \right| \right), \tag{4.132}$$

where the norms of $\mathbf{u}$ and p are computed only on the bounded set D_ρ, and then from Step 1 all norms on the right-hand side are finite. Therefore, for $\mathbf{f} = \operatorname{div} F$, $F \in (C_0^\infty(D))^{3\times 3}$, we get that the problem $(4.73)_{g=0}$, (4.74), (4.75) has a solution $\{\mathbf{u}, p\}$, $\mathbf{u} \in \widehat{\boldsymbol{H}}^{1,q}(\mathcal{D})$, $\partial_3 \mathbf{u} \in \widehat{\boldsymbol{H}}^{-1,q}(\mathcal{D})$, $(\omega \mathbf{e}_3 \wedge \mathbf{x}) \cdot \nabla \mathbf{u} - \omega \mathbf{e}_3 \wedge \mathbf{u} \in \boldsymbol{H}^{-1,q}(\mathcal{D})$, $p \in L^q(\mathcal{D})$ for all $3/2 < q \leq 6$. In this case

$$\langle \nabla \mathbf{u}, \nabla \boldsymbol{\varphi} \rangle + \langle \partial_3 \mathbf{u}, \boldsymbol{\varphi} \rangle - \langle ((\omega \mathbf{e}_3 \wedge \mathbf{x}) \cdot \nabla) \mathbf{u} - \omega \mathbf{e}_3 \wedge \mathbf{u}, \boldsymbol{\varphi} \rangle = \langle \operatorname{div} F, \boldsymbol{\varphi} \rangle, \tag{4.133}$$

where all the duality pairings have meaning for $\boldsymbol{\varphi} \in \widehat{\boldsymbol{H}}_\sigma^{1,q'}(\mathcal{D})$. If $3/2 < q < 3$, then also $\mathbf{u} \in \boldsymbol{L}^{3q/(3-q)}(\mathcal{D})$. We have got also the same results for the adjoint problem.

Step 3: Uniqueness

We shall state the following uniqueness result: Let $\{\mathbf{u}_0, p_0\}$ be a weak solution to problem $(4.73)_{\mathbf{f}=0, g=0}$, (4.74), (4.75) in the sense of Definition 4.1 such that $\mathbf{u} \in \widehat{\boldsymbol{H}}^{1,q}(\mathcal{D})$, $\partial_3 \mathbf{u} \in \widehat{\boldsymbol{H}}^{-1,q}(\mathcal{D})$, $(\omega \mathbf{e}_3 \wedge \mathbf{x}) \cdot \nabla \mathbf{u} - \omega \mathbf{e}_3 \wedge \mathbf{u} \in \boldsymbol{H}^{-1,q}(\mathcal{D})$, $p \in L^q(\mathcal{D})$ for some $3/2 < q < 3$. Then $\mathbf{u}_0 = 0$ and $p_0=0$.

Indeed, for $\{\mathbf{u}_0, p_0\}$ we have

$$\left.\begin{aligned} -\Delta \mathbf{u}_0 + \partial_3 \mathbf{u}_0 - ((\omega \mathbf{e}_3 \wedge \mathbf{x}) \cdot \nabla) \mathbf{u}_0 + \omega \mathbf{e}_3 \wedge \mathbf{u}_0 &= -\nabla p_0, \\ \operatorname{div} \mathbf{u}_0 &= 0 \end{aligned}\right\} \text{ in } \mathcal{D}. \tag{4.134}$$

The adjoint model admits a weak solution, say $\{\mathbf{u}^*, p^*\}$, so given any $F \in (C_0^\infty(\mathcal{D}))^{3\times 3}$ we have

$$\left.\begin{aligned} -\Delta \mathbf{u}^* - \partial_3 \mathbf{u}^* + ((\omega \mathbf{e}_3 \wedge \mathbf{x}) \cdot \nabla) \mathbf{u}^* - \omega \mathbf{e}_3 \wedge \mathbf{u}^* &= \operatorname{div} F - \nabla p^*, \\ \operatorname{div} \mathbf{u}^* &= 0 \end{aligned}\right\} \text{ in } \mathcal{D} \tag{4.135}$$

with $\mathbf{u}^* \in \widehat{\boldsymbol{H}}_\sigma^{1,r}(\mathcal{D})$, $p^* \in L^r(\mathcal{D})$, $3/2 < r \le 6$. Taking $\mathbf{u}^*$ as a test function in (4.134) (here we use that $q/(q-1) > 3/2$, i.e., $q < 3$) and similarly u_0 in the dual problem (4.135), we get in accordance with (4.133):

$$\begin{aligned}
&\langle \nabla \mathbf{u}_0, \nabla \mathbf{u}^* \rangle + \langle \partial_3 \mathbf{u}_0, \mathbf{u}^* \rangle - \langle (\omega \mathbf{e}_3 \wedge \mathbf{x}) \cdot \nabla \mathbf{u}_0 - \omega \mathbf{e}_3 \wedge \mathbf{u}_0, \mathbf{u}^* \rangle = 0, \\
&\langle \nabla \mathbf{u}^*, \nabla \mathbf{u}_0 \rangle - \langle \partial_3 \mathbf{u}^*, \mathbf{u}_0 \rangle + \langle (\omega \mathbf{e}_3 \wedge \mathbf{x}) \cdot \nabla \mathbf{u}^* - \omega \mathbf{e}_3 \wedge \mathbf{u}^*, \mathbf{u}_0 \rangle = \langle \operatorname{div} F, \mathbf{u}_0 \rangle.
\end{aligned}$$

From both the equalities, it follows that

$$\langle \operatorname{div} F, \mathbf{u}_0 \rangle = 0 \quad \text{for all } F \in (C_0^\infty(\mathcal{D}))^{3\times 3}.$$

Lemma 4.7 implies that $\langle \mathbf{f}, \mathbf{u}_0 \rangle = 0$ for all $\mathbf{f} \in \widehat{\boldsymbol{H}}^{-1,q/(q-1)}(\mathcal{D})$, which gives $\mathbf{u}_0 = 0$ in $\widehat{\boldsymbol{H}}^{1,q}(\mathcal{D})$ and $p_0 = 0$ in $L^q(\mathcal{D})$. This completes the proof of uniqueness.

Step 4: Proof of the estimate (4.83) **for** $\{\mathbf{u}, p\}$

Estimate (4.83) reads

$$\begin{aligned}
\|\nabla \mathbf{u}\|_{q,\mathcal{D}} + \|p\|_{q,\mathcal{D}} + \|\partial_3 \mathbf{u}\|_{-1,q,\mathcal{D}} + \| - ((\omega \mathbf{e}_3 \wedge \mathbf{x}) \cdot \nabla)\mathbf{u} + \omega \wedge \mathbf{u}\|_{-1,q,\mathcal{D}} \\
\le c_q(\| \operatorname{div} F\|_{-1,q,\mathcal{D}}).
\end{aligned}$$

Let $3/2 < q < 3$. Suppose, on the contrary, the existence of a sequence $\{\operatorname{div} F_k\}$ in $\widehat{H}^{-1,q}(\mathcal{D})^3$ tending to $f_\infty = 0$ as k tends to infinity such that for the corresponding sequence of solutions $\{\mathbf{u}_k, p_k\}$ in $\widehat{\boldsymbol{H}}_0^{1,q}(\mathcal{D})^3 \times L^q(\mathcal{D})$

$$\begin{aligned}
&\|\nabla \mathbf{u}_k\|_{q,\mathcal{D}} + \|p_k\|_{q,\mathcal{D}} + \|\partial_3 \mathbf{u}_k\|_{-1,q,\mathcal{D}} + \| - ((\omega \mathbf{e}_3 \wedge \mathbf{x}) \cdot \nabla)\mathbf{u}_k \\
&\quad + \omega \wedge \mathbf{u}_k\|_{-1,q,\mathcal{D}} = 1 \qquad (4.136)
\end{aligned}$$

for all k. We know from Step 2 that

$$\begin{aligned}
&\|\nabla \mathbf{u}_k\|_{q,\mathcal{D}} + \|p_k\|_{q,\mathcal{D}} \\
&\le c \left(\| \operatorname{div} F_k\|_{-1,q} + \|\mathbf{u}_k|_{\mathcal{D}_\rho}\|_{q,\mathcal{D}_\rho} + \|p_k|_{\mathcal{D}_\rho}\|_{-1,q,\mathcal{D}_\rho} + \left| \int_{\mathcal{D}_\rho} \psi(\mathbf{x})\, p_k(\mathbf{x}) d\mathbf{x} \right| \right), \qquad (4.137)
\end{aligned}$$

$$\begin{aligned}
&\|\partial_3 \mathbf{u}_k\|_{-1,q,\mathcal{D}} + \| - ((\omega \mathbf{e}_3 \wedge \mathbf{u}_k\|_{-1,q,\mathcal{D}} \\
&\qquad \le \| \operatorname{div} F_k\|_{-1,q} + \|\nabla \mathbf{u}_k|_{\mathcal{D}_\rho}\|_{q,\mathcal{D}_\rho} + \|p_k|_{\mathcal{D}_\rho}\|_{q,\mathcal{D}_\rho}, \qquad (4.138)
\end{aligned}$$

and from (4.132) also that

$$\begin{aligned}
&\|\nabla \mathbf{u}_k - \nabla \mathbf{u}_l\|_{q,\mathcal{D}} + \|p_k - p_l\|_{q,\mathcal{D}} + \|\partial_3 \mathbf{u}_k - \partial_3 \mathbf{u}_l\|_{-1,q,\mathcal{D}} \\
&\quad + \|(\omega \mathbf{e}_3 \wedge \mathbf{x}) \cdot \nabla(\mathbf{u}_k - \mathbf{u}_l) - \omega \mathbf{e}_3 \wedge (\mathbf{u}_k - \mathbf{u}_l)\|_{-1,q,\mathcal{D}} \\
&\quad \leq c\bigg(\|\operatorname{div}(F_k - F_l)\|_{-1,q} + \|\mathbf{u}_k - \mathbf{u}_l|_{\mathcal{D}_\rho}\|_{q,\mathcal{D}_\rho}\|_{-1,q,\mathcal{D}_\rho} \\
&\quad + \|p_k - p_l|_{\mathcal{D}_\rho} + \left|\int_{\mathcal{D}_\rho} \psi(\mathbf{x})(p_k(\mathbf{x}) - p_l(\mathbf{x}))\mathrm{d}\mathbf{x}\right|\bigg).
\end{aligned} \tag{4.139}$$

On the other hand we have (with $q < 3$)

$$\begin{aligned}
&\|\mathbf{u}_k|_{\mathcal{D}_\rho}\|_{1,q,\mathcal{D}_\rho} \leq \|\nabla \mathbf{u}_k|_{\mathcal{D}_\rho}\|_{q,\mathcal{D}_\rho} + c\|\mathbf{u}_k|_{\mathcal{D}_\rho}\|_{3q/(3-q),\mathcal{D}_\rho} \leq c\|\nabla \mathbf{u}_k\|_{q,\mathcal{D}} \leq c, \\
&\|p_k|_{\mathcal{D}_\rho}\|_{q,\mathcal{D}_\rho} \leq 1.
\end{aligned}$$

Thus we can extract subsequences $\{\mathbf{u}'_k\}$ and $\{p'_k\}$ weakly convergent in $\boldsymbol{H}^{1,q}(\mathcal{D}_\rho)$ and $L^q(\mathcal{D}_\rho)$, strongly convergent in $\boldsymbol{L}^q(\mathcal{D}_\rho)$ and $W^{-1,q}(\mathcal{D}_\rho)$ (by the Rellich theorem), to the limit, say $\{\boldsymbol{u}_\infty, p_\infty\}$. From (4.139) we can deduce that $\{\mathbf{u}'_k\}$ and $\{p'_k\}$ are Cauchy sequences in $\boldsymbol{H}^{1,q}(\mathcal{D}_\rho)$ and $L^q(\mathcal{D}_\rho)$. Then $\{\mathbf{u}_\infty, p_\infty\} \in \widehat{\boldsymbol{H}}^{1,q}(\mathcal{D}) \times L^q(\mathcal{D})$, and we obtain

$$\langle \nabla \mathbf{u}_\infty, \nabla \boldsymbol{\varphi}\rangle + \langle \partial_3 \mathbf{u}_\infty, \boldsymbol{\varphi}\rangle - \langle ((\omega \mathbf{e}_3 \wedge \mathbf{x}) \cdot \nabla)\mathbf{u}_\infty - \omega \mathbf{e}_3 \wedge \mathbf{u}_\infty, \boldsymbol{\varphi}\rangle = \langle \mathbf{f}_\infty, \boldsymbol{\varphi}\rangle = 0 \tag{4.140}$$

as in (4.133) for $\boldsymbol{\varphi} \in \widehat{\boldsymbol{H}}_\sigma^{1,q'}(\mathcal{D})^3$ with p_∞ as the associated pressure. From Step 3, it is clear that $\mathbf{u}_\infty = 0$ and $p_\infty = 0$; $\{\mathbf{u}_\infty, p_\infty\}$ is the unique weak solution to problem $(4.73)_{\mathbf{f}=0,\, g=0}$, (4.74), (4.75) with $\mathbf{f}_\infty = 0$. From the obtained strong convergence of $\{\mathbf{u}'_k\}$ and $\{p'_k\}$, it is also clear that (4.136) holds for $\{\mathbf{u}_\infty, p_\infty\}$, which leads to a contradiction.

We have completed the proof of Theorem 4.6.

4.2.2.3 Nonhomogeneous Boundary Conditions

If we replace the homogeneous Dirichlet boundary conditions by nonhomogeneous ones in the form of (4.76), we also have the following theorem:

Theorem 4.12 *Let* $3/2 < q < 3$, *and suppose* $\mathbf{f}$ *and* g *are given as previously. Then there exists a unique weak solution* $\{\mathbf{u}, p\}$ *to* $(4.73_{g=0})$, *(4.74), (4.76) (uniqueness* *up to a constant multiple of* $\boldsymbol{\omega}$ *for* $\mathbf{u}$*), which satisfies the estimate*

$$\begin{aligned}\|\nabla\mathbf{u}\|_{q,\mathcal{D}} + \|p\|_{q,\mathcal{D}} + \|\partial_3\mathbf{u}\|_{-1,q,\mathcal{D}} + \|(\omega\mathbf{e}_3\wedge\mathbf{x})\cdot\nabla\mathbf{u} - \omega\mathbf{e}_3\wedge\mathbf{u}\|_{-1,q,\mathcal{D}} \\ \le c_q(\|\mathbf{f}\|_{-1,q,\mathcal{D}} + |\omega| + |\omega|^2 + 1) \qquad (4.141)\end{aligned}$$

with some constant $c_q > 0$ *independent of* $|\omega|$. □

Proof of Theorem 4.12 The result is a corollary of Theorem 4.6. Choose a cutoff function $\xi \in C_0^\infty(\mathbb{R}^3; [0,1])$ satisfying $\xi = 1$ near the boundary $\partial\mathcal{D}$ and set

$$\mathbf{b}(\mathbf{x}) = \frac{1}{2}\operatorname{curl}\left(\xi(\mathbf{x})|\mathbf{x}|^2\omega - \frac{1}{2}e_3\wedge\nabla|\mathbf{x}|^2\right),$$

$$b|_{\partial\mathcal{D}}(\mathbf{x}) = \omega\mathbf{e}_3\wedge\mathbf{x} - e_3.$$

Let $\mathbf{v} = \mathbf{u} - \mathbf{b}$, $\operatorname{div}\mathbf{v} = 0$ since $\operatorname{div}\mathbf{u} = 0$ and $\operatorname{div}\mathbf{b} = 0$. So we obtain

$$\begin{aligned} -\Delta\mathbf{v} + \partial_3\mathbf{v} - ((\omega\mathbf{e}_3\wedge\mathbf{x})\cdot\nabla)\mathbf{v} + \omega\mathbf{e}_3\wedge\mathbf{v} + \nabla p &= \mathbf{f} + \mathbf{f}_b \quad \text{in } \mathcal{D},\\ \operatorname{div}\mathbf{v} &= 0 \quad \text{in } \mathcal{D},\\ \mathbf{v} &= 0 \quad \text{on } \partial\mathcal{D},\\ \mathbf{v} &\to 0 \quad \text{at } \infty, \qquad (4.142)\end{aligned}$$

where

$$\mathbf{f}_b = -\Delta\mathbf{b} + \partial_3\mathbf{b} - ((\omega\mathbf{e}_3\wedge\mathbf{x})\nabla)\mathbf{b} + (\omega\mathbf{e}_3\wedge\mathbf{b}).$$

Applying Theorem 4.6 we get the existence of the unique weak solution $(\mathbf{v}, p)$ with the estimate

$$\begin{aligned}\|\nabla\mathbf{v}\|_{q,\mathcal{D}} + \|p\|_{q,\mathcal{D}} + \|-((\omega\mathbf{e}_3\wedge\mathbf{x})\nabla)\mathbf{v} + \omega\mathbf{e}_3\wedge\mathbf{v}\|_{-1,q,\mathcal{D}} \\ \le c(\|\mathbf{f}\|_{-1,q,\mathcal{D}} + \|\mathbf{f}_b\|_{-1,q,\mathcal{D}}) \\ \le c(\|\mathbf{f}\|_{-1,q,\mathcal{D}} + |\omega| + |\omega|^2 + 1). \qquad (4.143)\end{aligned}$$

□

4.2.3 Details of Proof—Boundedness of $\partial_3 U_0$

We recall that $\widehat{V}(\boldsymbol{\xi})$ is proportional to the multiplier

$$\boldsymbol{m}_t(\boldsymbol{\xi}) = \frac{\mathrm{i}\xi_3\boldsymbol{\xi}}{|\omega||\boldsymbol{\xi}|D(\boldsymbol{\xi})}\mathrm{e}^{-|\boldsymbol{\xi}|^2 t/|\omega|}.$$

Let us estimate $\boldsymbol{m}_t(\boldsymbol{\xi})$ and $\xi_\alpha \partial_\alpha \boldsymbol{m}_t(\boldsymbol{\xi})$, $\alpha = 1, 2, 3$:

- In the case $|\boldsymbol{\xi}|^2 t/|\omega| \le 1$, we can neglect the exponential term, and we have to estimate $\boldsymbol{m}_0(\boldsymbol{\xi})$. So we decompose the ball $|\boldsymbol{\xi}|^2/|\omega| \le 1$ into many (but a finite number of) slices $S_n = \{\boldsymbol{\xi} \in \mathbb{R}^3\colon |\boldsymbol{\xi}|^2\ |\omega| \le 1,\ |\xi_3/|\omega| - n| \le \frac{1}{4}\}$ with appropriate values $n_* \in \mathbb{Z}$ and the remaining part S' (inside the ball, where $\operatorname{dist}(\xi_3/|\omega|, n_*) \ge 1/4$). It is easy to verify that $|D(\boldsymbol{\xi})| \ge 1$ for $\boldsymbol{\xi} \in S'$. Then $|\boldsymbol{m}_0(\boldsymbol{\xi})| \le |\boldsymbol{\xi}|/|\omega| \le 1/\sqrt{|\omega|}$ for $\boldsymbol{\xi} \in S'$.
  Now for $\boldsymbol{\xi} \in S_{n_*}$, using a Taylor expansion of $1 - \mathrm{e}^{-z}$, we get the lower bound $|D(\boldsymbol{\xi})| \ge c_0|(|\boldsymbol{\xi}|^2/|\omega| + \mathrm{i}(\xi_3/|\omega| - n_*)|$. One can deduce that $|\boldsymbol{m}_0(\boldsymbol{\xi})| \le c_1$. If $n_* \ne 0$, we precisely have

$$
\begin{aligned}
|\boldsymbol{m}_0(\boldsymbol{\xi})| \le& \frac{|\xi_3||\xi_j|}{|\omega||\boldsymbol{\xi}||D(\boldsymbol{\xi})|} \le \frac{|\xi_3||\xi_j|}{|\omega||\boldsymbol{\xi}||(|\boldsymbol{\xi}|^2/|\omega| + \mathrm{i}(\xi_3/|\omega| - n_*))|} \le \frac{1}{|\boldsymbol{\xi}|} \\
\le& \frac{1}{|\omega|(n_* - 1/4)}.
\end{aligned}
$$

- In the case $|\xi|^2 t/|\boldsymbol{\omega}| > 1$, $|D(\boldsymbol{\xi})| \ge c_2$. Then

$$
|\boldsymbol{m}_t(\boldsymbol{\xi})| \le \frac{|\xi_3|}{c_2|\boldsymbol{\xi}|^2} \le \frac{1}{c_2\sqrt{|\omega|}}.
$$

We now compute

$$
\xi_\alpha \partial_\alpha \boldsymbol{m}_t(\boldsymbol{\xi}) = \boldsymbol{m}_t(\boldsymbol{\xi}) - \frac{2\xi_\alpha^2|\boldsymbol{\xi}|^2 t}{|\omega|}\boldsymbol{m}_t(\boldsymbol{\xi}) - \xi_\alpha \frac{\partial_\alpha D(\boldsymbol{\xi})}{D(\boldsymbol{\xi})}\boldsymbol{m}_t(\xi) - \frac{\boldsymbol{\xi}_\alpha^2}{|\boldsymbol{\xi}|^2}\boldsymbol{m}_t(\boldsymbol{\xi}),
$$

with ($\delta_{3\alpha}$ denotes the Kronecker symbol)

$$
\frac{\partial_\alpha D(\boldsymbol{\xi})}{D(\boldsymbol{\xi})} = \frac{2\pi}{D(\boldsymbol{\xi})} \cdot \frac{(2\xi_\alpha^2 + \mathrm{i}\xi_\alpha\delta_{3\alpha})}{|\omega|}.
$$

All terms can be either uniformly bounded or estimated as $\boldsymbol{m}_t$. Then after some elementary calculations, for each α, we get

$$
\max_\beta \sup_{\boldsymbol{\xi}\ne 0} |\xi_\alpha^\beta \partial_\alpha^\beta \xi \boldsymbol{m}_t(\boldsymbol{\xi})| \le C(|\omega|),
$$

where β runs through the set of all multi-indices in $\{0, 1\}^3$.

We now come back to $\langle \partial_3 \mathbf{U}_0, \boldsymbol{\varphi}\rangle$: There exists a sequence of functions $H^{(l)} \in C_0^\infty$ such that $\|H^{(l)} - H\|_q \to 0$ as $l \to \infty$. Let $U_0^{(l)}$ be the solutions corresponding to $H^{(l)}$, and we apply all the previous considerations to this sequence $U_0^{(l)}$ and then to $\langle \partial_3 U_0^{(l)}, \varphi\rangle$.

Let us put ${\mathcal{H}_k}^{(l)}(t,\boldsymbol{\xi}) = O_\omega^{\mathrm{T}}(t)\,\widehat{H_{\cdot k}^{(l)}}(O_\omega(t).\ -t\mathbf{e}_3/|\omega|)(\boldsymbol{\xi})$. Marcinkiewicz multiplier theorem implies that the functions $x \to \mathcal{F}^{-1}[m_t(\boldsymbol{\xi})\cdot {\mathcal{H}_k}^{(l)}(t,\boldsymbol{\xi})](\mathbf{x})$ are bounded in L^q for all t, $0 < t < 2\pi$, by $C(|\omega|)\|H_{\cdot k}(O_\omega(t)\cdot -te_3/|\omega|)\|_q$, i.e., by $C(|\omega|)\|H_{\cdot k}\|_q$.

Then we finally get

$$\langle \partial_3 \mathbf{U}_{0,k}^{(l)}, \varphi_k\rangle = \mathrm{i}\int_{\mathbb{R}^3} \frac{\xi_3 \widehat{\mathbf{U}_{0,k}^{(l)}}(\boldsymbol{\xi})}{|\boldsymbol{\xi}|}|\boldsymbol{\xi}|\,\widehat{\varphi_k}(\boldsymbol{\xi})\,\mathrm{d}\boldsymbol{\xi} \tag{4.144}$$

in the form

$$\left\langle \int_0^{2\pi} \mathcal{F}^{-1}[m_t(\xi)\cdot {\mathcal{H}_k}^{(l)}(t,\boldsymbol{\xi})]\mathrm{d}t, \Psi\right\rangle, \tag{4.145}$$

where relation has been denoted by $\langle V^{(l)}, \Psi\rangle$. Passing to the limit as $l \to \infty$ and taking the supremum for $\|\Psi\|_{q'} \le 1$, we obtain the control of $\|\partial_3 \mathbf{U}_{0,k}\|_{-1,q}$ by $\|H_{\cdot k}\|_q$.

4.3 Further Properties of the Oseen Type Problem

The aim of this subsection is to find better regularity for the Oseen type problem in strong, weak, and very weak sense. Before we start to deal with other properties of the Oseen operator, let us summarize the results already achieved.

The whole space $\mathbb{R}^3$

- **The strong solution**

  1. Let $1 < q < \infty$, $\mathbf{f} \in L^q(\mathbb{R}^3)$, $\nu > 0$, $k \in \mathbb{R}$, and then the problem (4.1) with $g = 0$ has a solution $(\mathbf{u}, p) \in \widehat{\boldsymbol{H}}^{2,q}(\mathbb{R}^3) \times \widehat{H}^{1,q}(\mathbb{R}^3)$

$$\|\nu\nabla^2\mathbf{u}\|_q + \|\nabla p\|_q \le C\|\mathbf{f}\|_q, \tag{4.146}$$

$$\|k\partial_3\mathbf{u}\|_q + \|\omega(\mathbf{e}_3\wedge\mathbf{x})\cdot\nabla\mathbf{u} - \omega\mathbf{e}_3\wedge\mathbf{u}\|_q \le C\left(1+\frac{k^4}{\nu^2|\omega|2}\right)\|\mathbf{f}\|_q \tag{4.147}$$

  with a constant $C > 0$ independent of ν, k and ω.

2. Let $1 < q < 4$, $\mathbf{f} \in \boldsymbol{L}^q(\mathbb{R}^3)$, $g \in \widehat{H}^{1,q}(\mathbb{R}^3)$ such that $|(x_1, x_2)|g \in L^q(\mathbb{R}^3)$ and let $(\mathbf{u}, p) \in \widehat{\boldsymbol{H}}^{2,q}(\mathbb{R}^3) \times \widehat{H}^{1,q}(\mathbb{R}^3)$ be the solution of (4.1). Then there exists $\beta \in \mathbb{R}$ such that

$$\nabla'(\mathbf{u} - \beta\omega\mathbf{e}_3 \wedge \mathbf{x}) \in \boldsymbol{L}^r(\mathbb{R}^3) \quad \text{for all } r > 1, \tag{4.148}$$
$$\frac{1}{r} \in \frac{1}{q} - \left[\frac{1}{4}, \frac{1}{3}\right].$$

Moreover,

$$\|\nabla'(\mathbf{u} - \beta\omega\mathbf{e}_3 \wedge \mathbf{x})\|_r \leq C(\|\mathbf{f}\|_q + \|\nu\nabla g + (\omega\mathbf{e}_3 \wedge \mathbf{x})g - kg\boldsymbol{e}_3\|_q \tag{4.149}$$

with a constant $c = c(\nu, k, \omega, r) > 0$.
3. Let $1 < q < 2$, $\nu > 0$, $k > 0$, $\mathbf{f} \in L^q(\mathbb{R}^3)$. Then (4.1) has a solution $\mathbf{u} \in \widehat{H}^{2,q}(\mathbb{R}^3)$ satisfying the a priori estimates

$$\left\|\frac{\nabla\mathbf{u}}{|\mathbf{x}'|}\right\|_q \leq \frac{C}{\nu}\|\mathbf{f}\|_q \tag{4.150}$$

and

$$\left\|\frac{\omega((\mathbf{e}_3 \wedge \mathbf{x}) \cdot \nabla\mathbf{u} - \mathbf{e}_3 \wedge \mathbf{u})}{1 + |\mathbf{x}'|}\right\|_q \leq c\left(1 + \frac{k}{\nu}\right)\|\mathbf{f}\|_q \tag{4.151}$$

with a constant $c > 0$ independent of k, ν, ω. By $|\mathbf{x}'| = \sqrt{\mathbf{x}_1^2 + \mathbf{x}_2^2}$, $\mathbf{x}' = (\mathbf{x}_1, \mathbf{x}_2)$.
4. If $1 < q < 3$, then

$$\left\|\frac{\nabla\mathbf{u}}{|\mathbf{x}|}\right\|_q \leq \frac{1}{\nu}\|\mathbf{f}\|_q \tag{4.152}$$

and

$$\left\|\frac{\omega((\mathbf{e}_3 \wedge \mathbf{x}) \cdot \nabla\mathbf{u} - \mathbf{e}_3 \wedge \mathbf{u})}{1 + |\mathbf{x}|}\right\|_q \leq C\left(1 + \frac{k}{\nu}\right)\|\mathbf{f}\|_q \tag{4.153}$$

with a constant $c > 0$ independent of k, ν, ω.
5. For all $1 < q < \infty$ the third component $\mathbf{u}_3$ of the solution $\mathbf{u}$ satisfies the a priori estimate

$$\left\|\frac{k\partial_3\mathbf{u}_3}{1 + |\mathbf{x}'|}\right\|_q + \left\|\frac{\omega(\mathbf{e}_3 \wedge \mathbf{x}) \cdot \nabla\mathbf{u}_3}{1 + |\mathbf{x}'|}\right\|_q \leq c\left(1 + \frac{k}{\nu}\right)\|\mathbf{f}\|_q. \tag{4.154}$$

- **Weak solutions**

1. Let $1 < q < \infty$ and let $\mathbf{f} \in \widehat{\boldsymbol{H}}^{-1,q}(\mathbb{R}^3)$ be given, and then the problem (4.79) with $g = 0$ possesses a weak solution $\{\mathbf{u}, p\} \in \widehat{\boldsymbol{H}}^{1,q}(\mathbb{R}^3) \times L^q(\mathbb{R}^3)$, which satisfies

$$\|\nabla \mathbf{u}\|_q + \|p\|_q + \|\partial_3 \mathbf{u} - ((\omega \mathbf{e}_3 \wedge \mathbf{x}) \cdot \nabla)\mathbf{u} + \omega \mathbf{e}_3 \wedge \mathbf{u}\|_{-1,q} \le c\|\mathbf{f}\|_{-1,q} \tag{4.155}$$

with some $c > 0$ depending on q.

2. Let $1 < q < 4$, $\mathbf{F}_1 \in \widehat{\boldsymbol{H}}^{-1,q}(\mathbb{R}^3)$ and let $\mathbf{U}_0 \in \widehat{\boldsymbol{H}}^{1,q}(\mathbb{R}^3)$ be the unique weak solution to problem (4.79). Then there exists $\alpha \in \mathbb{R}$ such that

$$\mathbf{U}_0 - \alpha \boldsymbol{e}_3 \in \boldsymbol{L}^s(\mathbb{R}^3) \quad \text{for all } s > 1,\ \frac{1}{s} \in \frac{1}{q} - \left[\frac{1}{4}, \frac{1}{3}\right].$$

Moreover

$$\|\mathbf{U}_0 - \alpha \boldsymbol{e}_3\|_s \le C\|\mathbf{F}_1\|_{-1,q}$$

with a constant $C = C(\nu, k, \omega, s) > 0$.

3. Let $1 < q < 3$, $\nu > 0$, $k > 0$, $\mathbf{F}_1 \in \widehat{\boldsymbol{H}}^{-1,q}(\mathbb{R}^3)$, and let $\mathbf{U}_0 \in \widehat{\boldsymbol{H}}^{1,q}(\mathbb{R}^3)$ be the unique weak solution to problem (4.79). Then

$$\left\| \frac{\mathbf{U}_0}{|x|} \right\|_q \le \frac{c}{\nu}\|\mathbf{F}_1\|_{-1,q}$$

with $c = c(q, \omega) > 0$.

Exterior Domain
Let $3/2 < q < 3$ and suppose $\mathbf{f}$ is given in $\widehat{\boldsymbol{H}}^{-1,q}(\mathcal{D})$. Then there exists a unique weak solution up to constant $\{\mathbf{u}, p\}$ to (4.73), which satisfies the estimate

$$\begin{aligned}\|\nabla \mathbf{u}\|_{q,\mathcal{D}} + \|p\|_{q,\mathcal{D}} + \|\partial_3 \mathbf{u}\|_{-1,q,\mathcal{D}} + \| - (\omega(\mathbf{e}_3 \wedge x) \cdot \nabla)\mathbf{u} + \omega \mathbf{e}_3 \wedge \mathbf{u}\|_{-1,q,\mathcal{D}} \\ \le c_q\|\mathbf{f}\|_{-1,q,\mathcal{D}},\end{aligned} \tag{4.156}$$

with some constant $c_q > 0$ depending of q and ω.

4.3.1 *Further Properties of the Rotating Oseen in the Whole Space* $\mathbb{R}^3$

We follow the ideas from D. Kim, see [139].

(P1) Let X and Y be subspaces of a Banach space, and then $X + Y$ is a Banach space equipped with the norm

$$\|u\|_{X+Y} = \inf\{\|u_1\|_X + \|u_2\|_Y : u = u_1 + u_2,\ u_1 \in X, u_2 \in Y\}. \tag{4.157}$$

We introduce the anisotropic spaces (for more details see [98])

$$\begin{aligned}\widetilde{\boldsymbol{D}}^{1,q}(\mathbb{R}^3) &= \widehat{\boldsymbol{H}}^{1,q}(\mathbb{R}^3) \cap \boldsymbol{L}^{\frac{4q}{4-q}}(\mathbb{R}^3) \quad 1 < q < 4,\\ \widetilde{\boldsymbol{D}}^{2,q}(\mathbb{R}^3) &= \widehat{\boldsymbol{H}}^{2,q}(\mathbb{R}^3) \cap \widehat{\boldsymbol{H}}^{1,\frac{4q}{4-q}}(\mathbb{R}^3) \cap \boldsymbol{L}^{\frac{2q}{2-q}}(\mathbb{R}^3) \quad 1 < q < 2.\end{aligned} \tag{4.158}$$

We introduce constants:

$$\lambda = \min\left\{1, \left(\frac{|k|}{\nu}\right)^{1/4}\right\} \tag{4.159}$$

$$c_{\nu,k,\omega}(q, \mathcal{D}) = c(1 + \frac{|k|}{\nu} + \frac{|\omega|}{\nu})^M \tag{4.160}$$

$$A_{\nu,k,\omega} = 1 + \frac{k^4}{\nu^2|\omega|^2} \tag{4.161}$$

$$\alpha(q) = \begin{cases} A_{\nu,k,\omega}\lambda^{-2}, \ 2 < q \le 12/5 \\ A_{\nu,k,\omega}\lambda^{-1}, \ 12/5 < q \le 3, \\ 1, \quad 3 < q < \infty \end{cases} \tag{4.162}$$

where $c = c(q, \mathcal{D}) > 0$, $M = M(q) \ge 1$.[1] The spaces $\widetilde{\boldsymbol{D}}^{1,q}(\mathbb{R}^3)$, $\widetilde{\boldsymbol{D}}^{2,q}(\mathbb{R}^3)$ are Banach spaces equipped with the norms

$$\begin{aligned}\|\mathbf{v}\|_{\widetilde{\boldsymbol{D}}^{1,q}} &= \|\nu\nabla\mathbf{v}\|_q + \frac{\lambda}{A_{\nu,k,\omega}}\|\nu\mathbf{v}\|_{\frac{4q}{4-q}},\\ \|\mathbf{v}\|_{\widetilde{\boldsymbol{D}}^{2,q}} &= \|\nu\nabla^2\mathbf{v}\|_q + \frac{\lambda}{A_{\nu,k,\omega}}\|\nu\nabla\mathbf{v}\|_{\frac{4q}{4-q}} + \frac{\lambda}{A_{\nu,k,\omega}}\|\nu\mathbf{v}\|_{\frac{2q}{2-q}}.\end{aligned} \tag{4.163}$$

We consider the modified nonhomogeneous Oseen type problem with rotating terms in the whole space

$$\left.\begin{aligned} S_{\omega,\nu}\mathbf{u} + k\frac{\partial\mathbf{u}}{\partial x_3} + \nabla p = \mathbf{f} \\ \operatorname{div}\mathbf{u} = g \end{aligned}\right\} \quad \text{in } \mathbb{R}^3. \tag{(O)$_{\text{ws}}$}$$

Let us summarize the results from previous Sects. 4.1–4.2.

[1] M is an exponent.

Lemma 4.8 *Let* $(\mathbf{u}, p)$ *be a distributional solution of* $(O)_{ws}$ *with* $(\mathbf{f}, g) = (\mathbf{0}, 0)$. *If* $\mathbf{u}$ *is a tempered distribution, then* u_1, u_2, u_3 *and* p *are polynomials on* $\mathbb{R}^3$. *If* $\mathbf{u} \in \boldsymbol{L}^{q_1}(\mathbb{R}^3) \cap \boldsymbol{L}^{q_2}(\mathbb{R}^3)$ *for some* $1 < q_1, q_2 < \infty$ *in addition, then* $\mathbf{u} = \mathbf{0}$ *identically and* p *is constant on* $\mathbb{R}^3$. □

Lemma 4.9

(a) *Let* $1 < r < 4$. *Then for every* $\mathbf{f} \in \widehat{\boldsymbol{H}}^{-1,r}(\mathbb{R}^3)$ *and* $g = 0$ *there exists a unique solution* $(\mathbf{u}, p) \in \widetilde{\boldsymbol{D}}^{1,r}(\mathbb{R}^3) \times L^r(\mathbb{R}^3)$ *of* $(O)_{ws}$ *satisfying*

$$k\frac{\partial \mathbf{u}}{\partial x_3},\ (\boldsymbol{\omega} \wedge \mathbf{x}) \cdot \nabla \mathbf{u} - \boldsymbol{\omega} \wedge \mathbf{u} \in \widehat{\boldsymbol{H}}^{-1,r}(\mathbb{R}^3).$$

Further,

$$\|\mathbf{u}\|_{\widetilde{\boldsymbol{D}}^{1,r}} + \|p\|_r \le c\|\mathbf{f}\|_{\widehat{\boldsymbol{H}}^{-1,r}} \tag{4.164}$$

for some $c = c(r) > 0$.

(b) *Let* $1 < r < 2$. *For every* $\mathbf{f} \in \boldsymbol{L}^s(\mathbb{R}^3)$, $g \in W^{1,s}(\mathbb{R}^3)$ *with* $(\boldsymbol{\omega} \wedge \mathbf{x})g \in \boldsymbol{L}^s(\mathbb{R}^3)$, *there exists a unique solution* $(\mathbf{u}, p) \in \widetilde{\boldsymbol{D}}^{2,s}(\mathbb{R}^3) \times [\widehat{\boldsymbol{H}}^{1,s}(\mathbb{R}^3) \cap L^{s^*}(\mathbb{R}^3)]$ *of* $(O)_{ws}$, $(s^* = \frac{3s}{3-s})$ *which satisfies*

$$k\frac{\partial \mathbf{u}}{\partial x_3},\ (\boldsymbol{\omega} \wedge \mathbf{x}) \cdot \nabla \mathbf{u} - \boldsymbol{\omega} \wedge \mathbf{u} \in \boldsymbol{L}^s(\mathbb{R}^3)$$

and

$$\|\mathbf{u}\|_{\widetilde{\boldsymbol{D}}^{3,s}} + \|\nabla p\|_s \le c(\|\mathbf{f}\|_s + \|\nu \nabla g\|_s + \|kg\|_s + \|(\boldsymbol{\omega} \wedge \mathbf{x})g\|_s) \tag{4.165}$$

with $c = c(s) > 0$.

□

From Lemmas 4.9 and 4.8 using the property that sum of two Banach spaces is again a Banach space with the norm (4.163) and we receive the following lemma.

Lemma 4.10 *Let* $1 < r_1, r_2 < 4$. *Then for every* $\mathbf{f} \in \widehat{\boldsymbol{H}}^{-1,r_1}(\mathbb{R}^3) + \widehat{\boldsymbol{H}}^{-1,r_2}(\mathbb{R}^3)$ *and* $g = 0$, *there exists a unique solution* $(\mathbf{u}, p)$ *of* $(O)_{ws}$ *satisfying*

$$k\frac{\partial \mathbf{u}}{\partial x_3}, (\boldsymbol{\omega} \wedge \mathbf{x}) \cdot \nabla \mathbf{u} - \boldsymbol{\omega} \wedge \mathbf{u} \in \widehat{\boldsymbol{H}}^{-1,r_1}(\mathbb{R}^3) + \widehat{\boldsymbol{H}}^{-1,r_2}(\mathbb{R}^3).$$

Further, we have

$$\|\mathbf{u}\|_{\widetilde{\boldsymbol{D}}^{1,r_1} + \widetilde{\boldsymbol{D}}^{1,r_2}} + \|p\|_{L^{r_1} + L^{r_2}} \le c\|\mathbf{f}\|_{\widehat{\boldsymbol{H}}^{-1,r_1} + \widehat{\boldsymbol{H}}^{-1,r_2}}, \tag{4.166}$$

with

$$c = c(r_1, r_2) > 0.$$

Proof of Lemma 4.10 *Uniqueness:*

Applying Lemma 4.8 we obtain the uniqueness.

Existence:

To prove the existence of solution of the problem $(O)_{ws}$ we consider that $\mathbf{f}$ has a form $\mathbf{f} = \mathbf{f}_1 + \mathbf{f}_2$ with $\mathbf{f}_1 \in \widehat{\boldsymbol{H}}^{-1,r_1}(\mathbb{R}^3)$, $\mathbf{f}_2 \in \widehat{\boldsymbol{H}}^{-1,r_2}(\mathbb{R}^3)$. Then there exists a unique solution $(\mathbf{u}_i, p_i)$ of $(O)_{ws}$ with $(\mathbf{f}_i, 0)$ satisfying the estimate (4.164). It follows $(\mathbf{u}, p) = (\mathbf{u}_1 + \mathbf{u}_2, p_1 + p_2)$ is also a solution of $(O)_{ws}$ and

$$\begin{aligned}\|\mathbf{u}\|_{\widetilde{\boldsymbol{D}}^{1,r_1}+\widetilde{\boldsymbol{D}}^{2,r_2}} + \|p\|_{L^{r_1}+L^{r_2}} &\le \sum_{i=1}^{2}(\|\mathbf{u}_i\|_{\widetilde{\boldsymbol{D}}^{1,r_i}} + \|p_i\|_{r_i}) \\ &\le c\sum_{i=1}^{2}\|\mathbf{f}_i\|_{-1,r_i}\end{aligned} \tag{4.167}$$

for some constant $c = c(r_1, r_2)$. Since $\mathbf{f} = \mathbf{f}_1 + \mathbf{f}_2$ is arbitrary decomposition, then Lemma 4.10 is completed.

□

From Lemma 4.8–4.10 we deduce that:

Lemma 4.11 *Let* $1 < r < 4$, $1 < s < 2$, *and then for every*

$$\mathbf{f} \in \widehat{\boldsymbol{H}}^{-1,r}(\mathbb{R}^3) + \boldsymbol{L}^s(\mathbb{R}^3), \quad g \in W^{1,s}(\mathbb{R}^3)$$

with $(\boldsymbol{\omega} \wedge \mathbf{x})g \in \boldsymbol{L}^s(\mathbb{R}^3)$, *there exists a unique solution* $(\mathbf{u}, p)$ *of* $(O)_{ws}$ *satisfying*

$$k\frac{\partial \mathbf{u}}{\partial x_3}, \ (\boldsymbol{\omega} \wedge \mathbf{x})\nabla\mathbf{u} - \boldsymbol{\omega} \wedge \mathbf{u} \in \widehat{\boldsymbol{H}}^{-1,r}(\mathbb{R}^3) + \boldsymbol{L}^s(\mathbb{R}^3).$$

Further, we have

$$\begin{aligned}&\|\mathbf{u}\|_{\widetilde{\boldsymbol{D}}^{1,r}+\widetilde{\boldsymbol{D}}^{2,s}} + \|p\|_{L^r+(\widehat{H}^{1,s}\cap L^{s^*})} \\ &\le c(\|\mathbf{f}\|_{\widehat{\boldsymbol{H}}^{-1,r}+\boldsymbol{L}^s} + \|\nu\nabla g\|_s + \|kg\|_s + \|(\boldsymbol{\omega} \wedge \mathbf{x})g\|_s),\end{aligned} \tag{4.168}$$

with $c = c(r, s) > 0$. □

Corollary 4.5

(a) *Let* $1 < r < 4$, $\mathbf{f} = div\, F$, $F \in \boldsymbol{L}^r(\mathbb{R}^3)$, $g = 0$. *Then there exists a unique solution* $(\mathbf{v}, p) \in [\widehat{\boldsymbol{H}}^{1,r}(\mathbb{R}^3) \cap \boldsymbol{L}^{\frac{4r}{4-r}}(\mathbb{R}^3)] \times L^r(\mathbb{R}^3)$ *of* $(O)_{ws}$ *satisfying*

$$\frac{\lambda}{A_{\nu,k,\omega}}\|\nu\mathbf{v}\|_{\frac{4r}{4-r}} + \|\nu\nabla\mathbf{v}\|_r + \|p\|_r \le c\|F\|_r$$

for some constant $c = c(r) > 0$.

(b) *Let* $1 < s < 2$. *Then for every* $\mathbf{f} \in \boldsymbol{L}^s(\mathbb{R}^3)$, $g \in W^{1,s}(\mathbb{R}^3)$ *with* $(\boldsymbol{\omega} \wedge \mathbf{x})g \in L^s(\mathbb{R}^3)$, *there exists a unique solution*

$$(\mathbf{v}, p) \in [\widetilde{\boldsymbol{D}}^{2,s}(\mathbb{R}^3)] \times [\widehat{H}^{1,s}(\mathbb{R}^3) \cap L^{s^*}(\mathbb{R}^3)]$$

of $(O)_{ws}$ *satisfying*

$$\|\nu\nabla^2\mathbf{v}\|_s + \|\nabla p\|_s \leq c(\|\mathbf{f}\|_s + \|\nu\nabla g\|_s + \|kg\|_s + \|(\boldsymbol{\omega} \wedge \mathbf{x})g\|_s) \tag{4.169}$$

$$\begin{aligned}\lambda\|\nu\nabla\mathbf{v}\|_{\frac{4r}{4-r}} + \lambda^2\|\nu\mathbf{v}\|_{\frac{2s}{2-s}} + \left\|k\frac{\partial\mathbf{v}}{\partial x_1}\right\|_s + \|(\boldsymbol{\omega} \wedge \mathbf{x}) \cdot \nabla\mathbf{v} - \boldsymbol{\omega} \wedge \mathbf{v}\|_s \\ \leq cA_{\nu,k,\omega}(\|\mathbf{f}\|_s + \|\nu\nabla g\|_s + \|kg\|_s + \|(\boldsymbol{\omega} \wedge \mathbf{x})g\|_s),\end{aligned} \tag{4.170}$$

for some constant $c = c(s) > 0$.

□

4.3.2 Further Properties of the Rotating Oseen in the Exterior Domain

We consider the problem

$$\left.\begin{aligned} S_{\omega,\nu}\mathbf{u} + k\frac{\partial\mathbf{u}}{\partial x_3} + \nabla p = \mathbf{f} \\ \operatorname{div}\mathbf{u} = 0 \end{aligned}\right\} \quad \text{in } \mathcal{D} \tag{(O)$_{\text{ext}}$}$$

$$\mathbf{u} = \mathbf{0} \quad \text{on} \quad \partial\mathcal{D} \quad \text{or } \mathbf{u} = \mathbf{h} \text{ on } \partial\Omega$$

$$\mathbf{u}(\mathbf{x}) \to 0 \quad \text{as} \quad |\mathbf{x}| \to \infty.$$

Let $|k| + |\boldsymbol{\omega}| \leq M\nu$ for some $M \geq 1$. Then:

Lemma 4.12

(a) *Let* $2 < q < \infty$, *and assume that*

$$\mathbf{f} = \operatorname{div}\mathbf{F}, \quad \mathbf{F} \in \boldsymbol{L}^{\frac{3q}{3+q}}(\Omega), \ \mathbf{h} \in \boldsymbol{W}^{-\frac{1}{q},q}(\partial\mathcal{D}).$$

Then there exists a unique very weak solution of $(O)_{ext}$ *satisfying*

$$\mathbf{u} \in \boldsymbol{L}^q(\mathcal{D}),$$

$$\|\nu\mathbf{u}\|_{q,\mathcal{D}} \le c_{\nu,k,\omega}\alpha(q)\Big(\|\mathbf{F}\|_{\frac{3q}{3-q},\mathcal{D}} + \|\nu\mathbf{h}\|_{-1/q,q,\partial\mathcal{D}}\Big),$$

$c_{\nu,k,\omega} = c_{\nu,k,\omega}(q,\mathcal{D})$. *Moreover, if* $12/5 < q \le 3$, *then* u *satisfies the estimate*

$$\lambda\|\nu\mathbf{u}\|_{q;\mathcal{D}} \le c_M A_{\nu,k,\omega}(\|\mathbf{F}\|_{\frac{3q}{3+q};\mathcal{D}} + \|\nu\mathbf{h}\|_{-\frac{1}{q},q,(\partial\mathcal{D})}) \tag{4.171}$$

for some $c_M = c(q, M, \mathcal{D}) > 0$.

(b) *Let* $4/3 < r < 3/2$, *and assume that*

$$\mathbf{f} = \operatorname{div}\mathbf{F},\ \ \mathbf{F} \in \boldsymbol{L}^r(\mathcal{D}),\ \ \mathbf{h} \in \boldsymbol{W}^{1-\frac{1}{r},1}(\partial\mathcal{D});$$

then there exists a unique weak solution of $(O)_{ws}$ *satisfying*

$$\nabla\mathbf{u} \in \boldsymbol{L}^r(\mathcal{D}),\ \ \mathbf{u} \in \boldsymbol{L}^{\frac{4r}{4-r}}(\mathcal{D}).$$

Further, if $3/2 < r < 3$, *then* u *satisfies the estimate*

$$\lambda\|\nu\mathbf{u}\|_{\frac{4r}{4-r},\mathcal{D}} + \|\nu\nabla\mathbf{u}\|_{r;\mathcal{D}} \le c_M A_{\nu,k,\omega}(\|\mathbf{F}\|_{r;\mathcal{D}} + \|\nu\mathbf{h}\|_{1-\frac{1}{r},r(\partial\mathcal{D})}) \tag{4.172}$$

with $c_M = c(r, M, \mathcal{D}) > 0$.

□

From Lemma 4.12 it follows:

Lemma 4.13 *Assume that* $\mathbf{f} = \operatorname{div}\mathbf{F}$, $\mathbf{F} \in \boldsymbol{L}^{\frac{3}{2}}(\mathcal{D}) + \boldsymbol{L}^{\frac{12}{7}}(\mathcal{D})$, $\mathbf{h} \in \boldsymbol{W}^{-\frac{1}{3},3}(\partial\mathcal{D})$. *Then there exists a unique very weak solution of* $(O)_{ws}$. *Moreover, we have*

$$\mathbf{u} \in \boldsymbol{L}^3(\mathcal{D}) \text{ and } \lambda\|\nu\mathbf{u}\|_{3,\mathcal{D}} \le c_M A_{\nu,k,\omega}(\|F\|_{L^{\frac{3}{2}}(\mathcal{D})+L^{\frac{12}{7}}(\mathcal{D})} + \|\nu\mathbf{h}\|_{-\frac{1}{3},3,(\partial\mathcal{D})}) \tag{4.173}$$

for some constant $c_M = c(M, \mathcal{D}) > 0$. □

Proof of Lemma 4.13 *Uniqueness:* The uniqueness follows from Lemma 4.12. *Existence:* We consider $(\mathbf{F}_1, \mathbf{F}_2)$ satisfying

$$\mathbf{F} = \mathbf{F}_1 + \mathbf{F}_2 \text{ with } \mathbf{F}_1 \in \boldsymbol{L}^{\frac{3}{2}}(\mathcal{D}) \text{ and } \mathbf{F}_2 \in \boldsymbol{L}^{\frac{12}{7}}(\mathcal{D}).$$

Then applying Lemma 4.12 we know that there exist solutions $\mathbf{u}_1$ and $\mathbf{u}_2$ of $(\mathrm{O})_{\mathrm{ext}}$ with $\mathbf{f}$ replaced by $\operatorname{div}\mathbf{F}_1$ and $\operatorname{div}\mathbf{F}_2$, respectively. Further, these solutions satisfy

$$\lambda\|\nu\mathbf{u}_1\|_{3,\mathcal{D}} \le c_M\, A_{\nu,k,\omega}(\|\mathbf{F}_1\|_{\frac{3}{2},\mathcal{D}} + \|\nu\mathbf{h}\|_{-\frac{1}{3},3,\partial\mathcal{D}})$$

$$\lambda\|\nu\mathbf{u}_2\|_{3,\mathcal{D}} \le c_M\, A_{\nu,k,\omega}(\|\mathbf{F}_2\|_{\frac{12}{7},\mathcal{D}}) \tag{4.174}$$

for some constant $c_M = C(M,\mathcal{D}) > 0$. Since $\mathbf{F} = \mathbf{F}_1 + \mathbf{F}_2$ is chosen as arbitrary decomposition, it implies that $\mathbf{u} = \mathbf{u}_1 + \mathbf{u}_2 \in \boldsymbol{L}^3(\mathcal{D})$ is a very weak solution of $(\mathrm{O})_{\mathrm{ws}}$ satisfying the estimate (4.173). □

4.3.2.1 Proof of Lemma 4.12

We begin with the study of the regularity and uniqueness of very weak solution of $(\mathrm{O})_{\mathrm{ext}}$. Choosing $R > 0$ such that $\mathbb{R}^3 \setminus \mathcal{D} \subset B_{R/8}$, we define a cutoff function $\mu \in C_0^\infty(\mathbb{R}^3, [0,1])$ satisfying

$$\mu(x) = \begin{cases} 1 & \text{if } |x| \le R/4 \\ 0 & \text{if } |x| \ge 3R/4. \end{cases} \tag{4.175}$$

We denote $\mathcal{D}_r = \mathcal{D} \cap B_r$, for any $r, r \ge R/8$.

Let $\mathbf{f} = \mathbf{0}$, $1 < r < 4$, $\mathbf{f} = \operatorname{div}\mathbf{F}$, $\mathbf{F} \in \boldsymbol{L}^r(\mathcal{D})$. Defining (extension of $\mathbf{F}$ to $(\mathbb{R}^3)$) $\widetilde{\mathbf{F}}$ on $\mathbb{R}^3$ as $\widetilde{\mathbf{F}} = \mathbf{0}$ in $\mathbb{R}^3 \setminus \mathcal{D}$, $\widetilde{\mathbf{F}} = \mathbf{F}$ on $\mathcal{D}$ we have $\widetilde{\mathbf{F}} \in \boldsymbol{L}^r(\mathbb{R}^3)$ and $\|\widetilde{\mathbf{F}}\|_{r,\mathbb{R}^3} = \|\mathbf{F}\|_{r,\mathcal{D}}$. Then applying Corollary 4.5(a) we get

$$(\mathbf{v}_0, p_0) \in [\widetilde{\boldsymbol{D}}_{1,r}] \times L^r(\mathcal{D}), \tag{4.176}$$

$$\frac{\lambda}{A_{\nu,k,\omega}}\|\nu\mathbf{v}_0\|_{\frac{4q}{4-r},\mathcal{D}} + \|\nu\nabla\mathbf{v}_0\|_{r,\mathcal{D}} + \|p_0\|_{r,\mathcal{D}} \le c\|\mathbf{F}\|_{r,\mathcal{D}}, \tag{4.177}$$

with $c = (r) > 0$.

Using the trace theorem, we get $\gamma_0\mathbf{v}_0 \in W^{1-\frac{1}{r},r}(\partial\mathcal{D})$, $\|\gamma_0\mathbf{v}_0\|_{1-\frac{1}{r},r(\partial\mathcal{D})} \le c\|\mathbf{v}_0\|_{1,r,\mathcal{D}_R}$. By Sobolev imbedding

$$\|\nu\gamma_0\mathbf{v}_0\|_{1-1/r,r,\partial\mathcal{D}} \le \begin{cases} c\,\|\mathbf{F}\|_r, & 1 < r < 3 \\ c\,A_{\nu,k,\omega}\lambda^{-1}\|\mathbf{F}\|_r, & 3 \le r < 4, \end{cases} \tag{4.178}$$

$c = c(r,\mathcal{D}) > 0$.

Let $1 < s < 2$, $\mathbf{f} \in \boldsymbol{L}^s(\mathcal{D})$, again setting $\tilde{\mathbf{f}} = \mathbf{f}$ on $\mathcal{D}$, $\tilde{\mathbf{f}} = \mathbf{0}$ on $\mathbb{R}^3 \setminus \mathcal{D}$ we have $\|\tilde{\mathbf{f}}\|_{s,\mathbb{R}^3} = \|\mathbf{f}\|_{s,\mathcal{D}}$, and applying Corollary 4.5(b) there exists a solution $(\mathbf{v}_0, p_0)$ of $(\mathrm{O})_{\mathrm{ws}}$ for data $\tilde{\mathbf{f}}$ satisfying (4.169), (4.170).

Then we obtain

$$
\begin{aligned}
&(\mathbf{v}_0, p_0) \in \widetilde{\boldsymbol{D}}^{2,s}(\mathcal{D}) \times \widehat{H}^{1,s}(\mathcal{D}) \cap L^{s^*}(\mathcal{D}) \\
&\|\nu\nabla^2\mathbf{v}_0\|_{s,\mathcal{D}} + \|\nabla p_0\|_{s,\mathcal{D}} \le c\|\mathbf{f}\|_{s,\mathcal{D}}
\end{aligned}
\tag{4.179}
$$

and

$$
\begin{aligned}
&\left\| k\frac{\partial \mathbf{v}_0}{\partial x_3} \right\|_{s,\mathcal{D}} + \|(\boldsymbol{\omega}\wedge\mathbf{x})\cdot\nabla\mathbf{v}_0 - \boldsymbol{\omega}\wedge\mathbf{v}_0\|_{s,\mathcal{D}} \\
&+ \lambda\|\nu\nabla\mathbf{v}_0\|_{\frac{4s}{4-s},\mathcal{D}} + \lambda^2\|\nu\mathbf{v}_0\|_{\frac{2s}{2-s},\mathcal{D}} \le cA_{\nu,k,\omega}\|\mathbf{f}\|_{s,\mathcal{D}}
\end{aligned}
\tag{4.180}
$$

for some constant $c = c(s) > 0$.
Further, by applying the trace theorem,

$$
\gamma_0\mathbf{v}_0 \in W^{2-\frac{1}{s},s}(\partial\mathcal{D}) \text{ and } \|\gamma_0\mathbf{v}_0\|_{2-\frac{1}{s},s(\partial\mathcal{D})} \le c\|\mathbf{v}_0\|_{2,s,\mathcal{D}_R}.
$$

Using Sobolev imbedding, we get

$$
\|\nu\gamma_0\mathbf{v}_0\|_{2-\frac{1}{2},s,\partial\mathcal{D}} < c\alpha(s')\|\mathbf{f}\|_{s,\mathcal{D}}, \text{ for } c = c(s,\mathcal{D}) > 0. \tag{4.181}
$$

4.3.3 The Regularity of Very Weak Solutions

We introduce the space $\widetilde{\boldsymbol{H}}^{2,q}(\mathcal{D})$, $1 < q < 2$ such that

$$
\begin{aligned}
&\widetilde{\boldsymbol{H}}^{2,q}(\mathcal{D}) = \{\mathbf{v};\, \mathbf{v} \in \widehat{\boldsymbol{H}}^{2,q}(\mathcal{D}) \cap \widehat{\boldsymbol{H}}^{1,\frac{4q}{4-q}}(\mathcal{D}) \cap \boldsymbol{L}^{\frac{2q}{2-q}}(\mathcal{D}),\ \tfrac{\partial\mathbf{v}}{\partial x_3} \in \boldsymbol{L}^q(\mathcal{D})\} \\
&\quad 1 < q < 2,
\end{aligned}
\tag{4.182}
$$

with the norm

$$
\|\mathbf{v}\|_{\widetilde{\boldsymbol{H}}^{2,q}} = \|\nu\nabla^2\mathbf{v}\|_{q,\mathcal{D}} + \|\nu\nabla\mathbf{v}\|_{\frac{4q}{4-q},\mathcal{D}} + \|\nu\mathbf{v}\|_{\frac{2q}{2-q},\mathcal{D}} + \|k\frac{\partial\mathbf{v}}{\partial x_3}\|_{q,\mathcal{D}}.
$$

Lemma 4.14 *Let* $1 < q_1 \le q_2 < \infty$. *Let* $\mathbf{v} \in \boldsymbol{L}^{q_1}(\mathcal{D}) + \boldsymbol{L}^{q_2}(\mathcal{D})$ *and* $p \in L^1_{loc}(\mathcal{D})$ *satisfy*

$$
\begin{aligned}
&\int_{\mathcal{D}} \mathbf{v}\cdot\Big(-\nu\Delta\boldsymbol{\varphi} - k\frac{\partial\boldsymbol{\varphi}}{\partial x_3} + \nabla\psi + (\boldsymbol{\omega}\wedge\mathbf{x})\cdot\boldsymbol{\varphi} - \boldsymbol{\omega}\wedge\boldsymbol{\varphi} \\
&\qquad - p\,\mathrm{div}\,\boldsymbol{\varphi}\Big)\, d\mathbf{x} = 0 \quad \forall\, \boldsymbol{\varphi} \in C_0^\infty(\mathcal{D}),\ \psi \in C_0^\infty(\mathcal{D}).
\end{aligned}
\tag{4.183}
$$

Then:

(a) *For all* $1 < s < 2$ *we have*

$$(\mathbf{v}, p) \in \widetilde{\boldsymbol{H}}^{2,s}(B_R^c) \times \widehat{H}^{1,s}(B_R^c) \tag{4.184}$$

$$\nu\|\nabla^2\mathbf{v}\|_{s,B_R^c} + \|\nabla p\|_{s,B_R^c} \leq c_{\nu,k,\omega}\|\nu\mathbf{v}\|_{q,\mathcal{D}_{2R}} \tag{4.185}$$

$$\left\|k\frac{\partial \mathbf{v}}{\partial x_1}\right\|_{s,B_R^c} + \|(\boldsymbol{\omega}\wedge\mathbf{x})\cdot\nabla v - \boldsymbol{\omega}\wedge\mathbf{v}\|_{s;B_R^c}$$
$$+ \lambda\|\nu\nabla\mathbf{v}\|_{\frac{4s}{4-s},B_R^c} + \lambda^2\|\nu\nabla\mathbf{v}\|_{\frac{2s}{2-s},B_R^c} \leq c_{\nu,k,\omega}A_{\nu,k,\omega}\|\nu\mathbf{u}\|_{q,\mathcal{D}_{2R}} \tag{4.186}$$

for some $c_{\nu,k,\omega} = c_{\nu,k,\omega}(q, s, \mathcal{D})$.

(b) *For all* $4/3 < r < 4$ *we have*

$$(\mathbf{v}, p) \in \left(\widetilde{\boldsymbol{D}}^{1,r}(B_R^c) \cap \boldsymbol{L}^{\frac{4r}{4-r}}(B_R^c)\right) \times L^r_{loc}(B_R^c). \tag{4.187}$$

If $4/3 < r \leq 3/2$, *then there is a constant* $a \in \mathbb{R}$ *such that* $p - a \in L^{\frac{12r}{12-r}}(B_R^c)$ *and*

$$\frac{\lambda^2}{A_{\nu,k,\omega}}\|\nu\mathbf{v}\|_{\frac{4r}{4-r},B_R^c} + \frac{\lambda}{A_{\nu,k,\omega}}\|\nu\nabla\mathbf{v}\|_{r,B_R^c} + \|p-a\|_{\frac{12}{12-r},B_R^c} \leq c_{\nu,k,\omega}\|\nu\mathbf{v}\|_{q,\mathcal{D}_{2R}}. \tag{4.188}$$

If $3/2 < r < 4$, *then there is a constant* $a \in \mathbb{R}$ *such that* $p - a \in L^r(B_R^c)$ *and*

$$\frac{\lambda}{A_{\nu,k,\omega}}\|\nu\mathbf{v}\|_{\frac{4r}{4-r},B_R^c} + \|\nu\nabla\mathbf{v}\|_{r,B_R^c} + \|p-a\|_{r,B_r^c} \leq c_{\nu,k,\omega}\|\nu\mathbf{v}\|_{q;\mathcal{D}_{2R}} \tag{4.189}$$

for some $c_{\nu,k,\omega} = c_{\nu,k,\omega}(q, r, \mathcal{D})$. □

Proof of Lemma 4.14 From the regularity result of the Stokes problem, see, e.g., Galdi, [98, Section IV.4., VIII.1], we obtain

$$(\mathbf{v}, p) \in \boldsymbol{C}^\infty(\mathcal{D}) \times C^\infty(\mathcal{D}).$$

We thus set

$$\tilde{\mathbf{f}} = \nu\Delta\mathbf{v} - k\frac{\partial\mathbf{v}}{\partial x_3} + (\boldsymbol{\omega}\wedge\mathbf{x})\cdot\nabla\mathbf{v} - \boldsymbol{\omega}\wedge\mathbf{v}. \tag{4.190}$$

We obtain

$$\tilde{\mathbf{f}} \in \boldsymbol{W}^{-2,q_1}(\mathcal{D}_{2R}),\ \langle \tilde{\mathbf{f}}, \Phi \rangle = 0 \quad \forall \Phi \in C_{0,\sigma}^{\infty}(\mathcal{D}_{2R}).$$

Applying [138] (Lemma 7)) we obtain $p \in W^{-1,q_1}(\mathcal{D}_{2R})$, and there is a constant D such that

$$\|p - D\|_{-1,q_1,\mathcal{D}_{2R}} \leq c_{\nu,k,\omega} \|\nu \mathbf{v}\|_{q_1,\mathcal{D}_{2R}}, \tag{4.191}$$

where $c_{\nu,k,\omega} = c_{\nu,k,\omega}(q_1, \mathcal{D})$.
Using a bootstrap argument similar to Galdi, see [98, Section IV.4], we get

$$\|\nu \mathbf{v}\|_{1,r,B_R\setminus B_{R/4}} + \|p - D\|_{r,B_R\setminus B_{R/4}} \leq c_{\nu,k,\omega} \|\nu \mathbf{v}\|_{q_1,\mathcal{D}_{2R}}, \quad \forall\, 1 < r < \infty, \tag{4.192}$$

where $c_{\nu,k,\omega} = c_{\nu,k,\omega}\ (q, r, \mathcal{D})$. We introduce

$$\overline{\mathbf{v}} = (1 - \mu)\mathbf{v},\ \overline{p} = (1 - \mu)(p - D)$$

with $\mu \in C_0^{\infty}(\mathbb{R}^3, [0, 1])$ is a cutoff function defined in (4.175).
By simple calculation we derive

$$\overline{\mathbf{v}} \in \boldsymbol{L}^{q_1}(\mathbb{R}^3) + \boldsymbol{L}^{q_2}(\mathbb{R}^3) \tag{4.193}$$

$$\left.\begin{aligned} -\nu \Delta \overline{\mathbf{v}} + k \tfrac{\partial \overline{\mathbf{v}}}{\partial x_3} - (\boldsymbol{\omega} \wedge \mathbf{x}) \cdot \nabla \overline{\mathbf{v}} + \boldsymbol{\omega} \wedge \overline{\mathbf{v}} + \nabla \overline{p} = \overline{\mathbf{f}}, \\ \operatorname{div} \overline{\mathbf{v}} = \overline{g} \end{aligned}\right\} \text{ in } \mathbb{R}^3 \tag{4.194}$$

with

$$\begin{aligned} &\overline{\mathbf{f}} = \nu(\Delta \mu + 2\nabla \mu \cdot \nabla \overline{v}) - k\nu \tfrac{\partial \mu}{\partial x_3} + (\boldsymbol{\omega} \wedge \mathbf{x}) \cdot \nabla \mu \cdot \mathbf{v} - (p - D)\nabla \mu, \\ &\overline{q} = -\nabla \mu \cdot \mathbf{v}. \end{aligned} \tag{4.195}$$

Since $(\overline{\mathbf{f}}, \overline{g}) \in \boldsymbol{C}_0^{\infty}(\mathbb{R}^3)$ and using (4.192), we obtain

$$\begin{aligned} &\|\overline{\mathbf{f}}\|_{s,\mathbb{R}^3} + \|\nu \nabla \overline{g}\|_{s;\mathbb{R}^3} + \|k\overline{g}\|_{s,\mathbb{R}^3} + \|(\boldsymbol{\omega} \wedge \mathbf{x})\overline{g}\|_{s,\mathbb{R}^3} \\ &\leq c((\nu + |k| + |\omega|)\|\mathbf{v}\|_{1,s,B_r\setminus\overline{B_{R/4}}} + \|p - D\|_{s,B_R\setminus\overline{B_{R/4}}}) \leq c_{\nu,k,\omega} \|\nu \mathbf{v}\|_{q_1,\mathcal{D}_{2R}} \end{aligned} \tag{4.196}$$

for all $1 < s < \infty$, where $c_{\nu,k,\omega} = c_{\nu,k,\omega}(q_1, s, \mathcal{D})$.
From Corollary 4.5 (b) it implies that

$$(\overline{\mathbf{v}}, \overline{p}) \in [\widetilde{\boldsymbol{D}}^{2,q}(\mathbb{R}^3)] \times \widehat{H}^{1,s}(\mathbb{R}^3).$$

Moreover,

$$\begin{aligned}
&\|\nu\nabla^2\overline{\mathbf{v}}\|_{s,\mathbb{R}^3} + \|\nabla\overline{p}\|_{s,\mathbb{R}^3} \le c_{\nu,k,\omega}\|\nu\mathbf{v}\|_{q_1,\mathcal{D}_{2R}},\\
&\left\|k\frac{\partial\overline{\mathbf{v}}}{\partial x_3}\right\|_{s,\mathbb{R}^3} + \|(\boldsymbol{\omega}\wedge\mathbf{x})\cdot\nabla\overline{\mathbf{v}} - \boldsymbol{\omega}\wedge\overline{\mathbf{v}}\|_{s,\mathbb{R}^3}\\
&+\lambda\|\nu\nabla\overline{\mathbf{v}}\|_{\frac{4s}{4-s},\mathbb{R}^3} + \lambda^2\|\nu\overline{\mathbf{v}}\|_{\frac{2s}{2-s},\mathbb{R}^3} \le c_{\nu,k,\omega}A_{\nu,k,\omega}\|\nu\mathbf{v}\|_{q_1,\mathcal{D}_{2R}} \qquad (4.197)
\end{aligned}$$

hold for all $1 < s < 2$, where $c_{\nu,k,\omega} = c_{\nu,k,\omega}(q_1, s, \mathcal{D})$. Since $\mathbf{v} = \bar{\mathbf{v}}$, $\nabla p = \nabla\overline{p}$ in $\mathbb{R}^3 \setminus B_R$, it implies the first assertion of Lemma 4.14.
To show the second part of Lemma 4.14 we begin with **case:** $4/3 < r \le 3/2$. Setting $s = \frac{4r}{4+r}$ we have

$$1 < s < 2,\ \frac{4s}{4-s} = r,\ \frac{2s}{2-s} = \frac{4r}{4-r},\ s^* = \frac{12r}{12-r}.$$

Applying Lemma 4.13 (a) and Galdi [[98] Theorem II.6.1.] we get (4.187), (4.188).
We will focus on the **case:** $3/2 < r < 4$. Take $s = \frac{3r}{3+r}$, and there

$$1 < s < \frac{12}{7},\ s^* = r,\ \left(\frac{4s}{4-s}\right)^* = \frac{4r}{4-r}.$$

Using Lemma 4.13, we get (4.187), (4.189). □

Lemma 4.15 *Let* $(\mathbf{v}, p) \in \boldsymbol{L}^q(\mathcal{D}) \times L^1_{loc}(\mathcal{D})$ *be a very weak solution of* $(O)_{ws}$ *where*

$$\mathbf{f} = \mathbf{0},\ \mathbf{h} \in \boldsymbol{H}^{-\frac{1}{q},q}(\partial\mathcal{D}),\ 1 < q < \infty.$$

(a) *Let us assume* $\mathbf{h} \in \boldsymbol{H}^{1-1/r,r}(\partial\mathcal{D})$, $4/3 < r < 4$, $r \le q$ *and then*

$$\mathbf{v} \in \boldsymbol{H}^{1,r}(\mathcal{D}_{2R}),\ p \in L^r(\mathcal{D}_{2R}).$$

Further, we have

$$\|\nu\mathbf{v}\|_{1,r,\mathcal{D}_{2R}} + \left\|p - \frac{1}{|\mathcal{D}_{2R}|}\int_{\mathcal{D}_{2R}} p\,d\mathbf{x}\right\|_{r,\mathcal{D}_{2R}} \le c_{\nu,k,\omega}(\|\nu\mathbf{h}\|_{1-\frac{1}{r},\partial\mathcal{D}} + \|\nu\mathbf{v}\|_{q,\mathcal{D}_{2R}}),$$
$$c_{\nu,k,\omega} = c_{\nu,k,\omega}(q, r, \mathcal{D}).$$

(b) *Let us assume* $\mathbf{h} \in \boldsymbol{H}^{2-1/s,s}(\partial\mathcal{D})$, $1 < s < 2$, $4s/(4-s) \le q$. *Then*

$$\mathbf{v} \in \boldsymbol{H}^{2,s}(\mathcal{D}_{2R}),\ p \in H^{1,s}(\mathcal{D}_{2R}).$$

Moreover, we have

$$\|\nu \mathbf{v}\|_{2,s,\mathcal{D}_R} + \left\| p - \frac{1}{|\mathcal{D}_{2R}|}\int_{\mathcal{D}_{2R}} p\,dx \right\|_{1,s,\mathcal{D}_{2R}}$$
$$\leq c_{\nu,k,\omega}(\|\nu \mathbf{h}\|_{2-\frac{1}{s},s,\partial\mathcal{D}} + \|\nu \mathbf{v}\|_{q,\mathcal{D}_{2R}})$$
$$c_{\nu,k,\omega} = c_{\nu,k,\omega}(q, s, \mathcal{D}).$$

Proof of Lemma 4.15 We have to observe that $\mathbf{v} \in \boldsymbol{L}^q(\mathcal{D}_{2R})$ is a very weak solution of the Stokes problem with the right-hand side $\tilde{\mathbf{f}}$ on $\mathcal{D}_{2R}$.

$$\begin{aligned} -\nu\Delta\mathbf{v} + \nabla p &= \tilde{\mathbf{f}} \quad \text{in } \mathcal{D}_{2R} \\ \operatorname{div} \mathbf{v} &= 0 \quad \text{in } \mathcal{D}_{2R} \\ \mathbf{v} &= \mathbf{h} \quad \text{on } \partial\mathcal{D}_{2R}, \end{aligned}$$

where

$$\tilde{\mathbf{f}} = -k\frac{\partial \mathbf{v}}{\partial \mathbf{x}_3} + (\boldsymbol{\omega} \wedge \mathbf{x}) \cdot \nabla \mathbf{v} - \boldsymbol{\omega} \wedge \mathbf{v}$$

$$\mathbf{h} = \begin{cases} \nu_0 \mathbf{v} & \text{on } \partial B_{2R} \\ \mathbf{h} & \text{on } \partial\mathcal{D}. \end{cases}$$

Let us remark that

$$\tilde{\mathbf{f}} \in \boldsymbol{H}^{-1,q}(\mathcal{D}_{2R}),\ \|\tilde{\mathbf{f}}\|_{-1,q,\mathcal{D}_{2R}} \leq c(|k| + |\boldsymbol{\omega}|)\|\mathbf{v}\|_{q,\mathcal{D}_{2R}}.$$

Using Lemma 4.14 (a), the Sobolev imbedding, we obtain that $\gamma_0 \mathbf{v} \in \boldsymbol{H}^{2-\frac{1}{s},s}(\partial B_{2R})\quad \forall\, 1 < s < 2$ and

$$\begin{aligned} \|\nu\gamma_0\mathbf{v}\|_{2-\frac{1}{s},s,\partial B_R} &\leq c\Big(\|\nu\nabla^2\mathbf{v}\|_{s,B_{2R}\setminus\overline{B_R}} + \|\nu\nabla\mathbf{v}\|_{s,B_{2R\setminus\overline{B_R}}} + \|\nu\mathbf{v}\|_{s,B_{2R\setminus\overline{B_R}}}\Big) \\ &\leq c(\|\nu\nabla^2\mathbf{v}\|_{s;B_{2R\setminus\overline{B_R}}} + \|\nu\nabla\mathbf{v}\|_{s^8,B_{2R}\setminus\overline{B_R}} + \|\nu\mathbf{v}\|_{s,B_{2R}\setminus\overline{B_R}}) \\ &\leq c_{\nu,k,\omega}(\|\nu\mathbf{v}\|_{q,\mathcal{D}_2 R} + \|\nu\mathbf{v}\|_{s,\mathcal{D}_{2R}}), \end{aligned} \tag{4.198}$$

where $c_{\nu,k,\omega} = c_{\nu,k,\omega}(q, s, \mathcal{D})$.
Let us prove the first assertion of Lemma 4.15.
Suppose that $\mathbf{h} \in \boldsymbol{H}^{1-1/r,r}(\partial\mathcal{D})$, $4/3 < r < 4$, $r \leq q$ and take $s = \frac{4r}{4+r}$. Since $1 < s < 2$, $s < r = 4s/(4-s) < s^*$, it follows that $\mathbf{h} \in H^{1-1/r,r}(\partial\mathcal{D}_{2R})$ and

$$\|\nu\mathbf{h}\|_{1-\frac{1}{r},r,\partial\mathcal{D}_{2R}} \leq c(\|\nu\mathbf{h}\|_{1-\frac{1}{r},r,\partial\mathcal{D}} + \|\nu\gamma_0\mathbf{v}\|_{2-\frac{1}{s},s,\partial B_R}). \tag{4.199}$$

Using the theory of the very weak solution, see [9, 64, 112], we obtain $(\mathbf{v}, p) \in \boldsymbol{H}^{1,r}(\mathcal{D}_{2R}) \times L^r(\mathcal{D}_{2R})$ and

$$\begin{aligned}\|\nu\mathbf{v}\|_{1,r,\mathcal{D}_{2R}} &+ \left\| p - \frac{1}{|\mathcal{D}_{2R}|}\int_{\mathcal{D}_{2R}} p\mathrm{d}\mathbf{x}\right\|_{r,\mathcal{D}_{2R}} \\ &\le c(\|\nu\mathbf{h}\|_{1-\frac{1}{r},r,\partial\mathcal{D}} + \|\nu\gamma_0\mathbf{v}\|_{2-\frac{1}{s},s,\partial B_{2R}} + \|\tilde{\mathbf{f}}\|_{-1,q,\mathcal{D}_{2R}} \\ &\le c_{\nu,k,\omega}(\|\nu\mathbf{h}\|_{1-\frac{1}{r},r,\partial\mathcal{D}} + \|\nu\mathbf{v}\|_{q,\mathcal{D}_{2R}}),\end{aligned} \tag{4.200}$$

with $c_{\nu,k,\omega} = c_{\nu,k,\omega}\,(q, r, \mathcal{D})$. The last inequality can be obtained from (4.199), (4.200).

Let us consider the second assertion. Let $\mathbf{h} \in \boldsymbol{H}^{2-\frac{1}{s},s}(\partial\mathcal{D})$, $1 < s < 2, 4s/(4-s) \le q$. We set $r = 4s/(4-1)$. Since

$$W^{2-\frac{1}{s},s}(\partial\mathcal{D}) \subset W^{1-\frac{1}{r},r}(\partial\mathcal{D}),\ \frac{4}{3} < r < 4,$$

it implies that $(\mathbf{v}, p) \in \boldsymbol{H}^{1,r}(\mathcal{D}_{2R}) \times L^r(\mathcal{D}_{2R})$ and

$$\|\nu\mathbf{v}\|_{1,r,\mathcal{D}_{2R}} \le c_{\nu,k,\omega}(\|\nu\mathbf{h}\|_{1-\frac{1}{r},r,\partial\mathcal{D}} + \|\nu\mathbf{v}\|_{q,\mathcal{D}_{2R}}). \tag{4.201}$$

It yields that $\tilde{\mathbf{f}} \in \boldsymbol{L}^r(\mathcal{D}_{2R}) \subset \boldsymbol{L}^s(\mathcal{D}_{2R})$ and

$$\|\tilde{\mathbf{f}}\|_{s,\mathcal{D}_{2R}} \le c\left(\frac{|k|}{\nu} + \frac{|\omega|}{\nu}\right)\|\nu\mathbf{v}\|_{1,r,\mathcal{D}_{2R}} \le c_{\nu,k,\omega}(\|\nu\mathbf{h}\|_{2-\frac{1}{3},s,\partial\mathcal{D}} + \|\nu\mathbf{v}\|_{q;\mathcal{D}_{2R}}). \tag{4.202}$$

It implies that $(\mathbf{v}, p) \in \boldsymbol{H}^{2,s}(\mathcal{D}_{2R}) \times H^{1,s}(\mathcal{D}_{2R})$ and

$$\begin{aligned}\|\nu\mathbf{v}\|_{2,s,\mathcal{D}_{2R}} &+ \left\| p - \frac{1}{|\mathcal{D}_{2R}|}\int_{\mathcal{D}_{2R}} p\mathrm{d}\mathbf{x}\right\|_{1,s,\mathcal{D}_{2R}} \\ &\le c(\|\tilde{\mathbf{f}}\|_{s,\mathcal{D}_{2R}} + \|\nu\mathbf{h}\|_{2-\frac{1}{s},s,\partial\mathcal{D}_{2R}}) \\ &\le c(\|\tilde{\mathbf{f}}\|_{s,\mathcal{D}_{2R}} + \|\nu\mathbf{h}\|_{2-\frac{1}{s},s,\partial\mathcal{D}} + \|\nu\gamma_0\mathbf{v}\|_{2-\frac{1}{s},s,\partial B_{2R}}) \\ &\le c_{\nu,k,\omega}(\|\nu\mathbf{h}\|_{2-\frac{1}{s},s,\partial\mathcal{D}} + \|\nu\mathbf{v}\|_{q,\mathcal{D}_{2R}})\end{aligned} \tag{4.203}$$

for some constant $c_{\nu,k,\omega} = c_{\nu,k,\omega}(q, s, \mathcal{D})$. □

4.3.4 Uniqueness of the Very Weak Solution

The aim of this subsection is to prove the uniqueness of the very weak solution to $(\mathrm{O})_{\mathrm{ext}}$.

Lemma 4.16 *Let $2 < q_1 \le q_2 < \infty$. If $\mathbf{v} \in \boldsymbol{L}^{q_1}(\mathcal{D}) + \boldsymbol{L}^{q_2}(\mathcal{D})$ and $p \in L^1_{loc}(\mathcal{D})$ satisfy*

$$\int_{\mathcal{D}} \mathbf{v}\Big(-\nu\Delta\boldsymbol{\varphi} - k\frac{\partial\boldsymbol{\varphi}}{\partial x_3} + \nabla\psi + (\boldsymbol{\omega}\wedge\mathbf{x})\cdot\nabla\boldsymbol{\varphi} - \boldsymbol{\omega}\wedge\boldsymbol{\varphi} - p\,\mathrm{div}\,\boldsymbol{\varphi}\Big)\,dx = 0, \tag{4.204}$$

where $\boldsymbol{\varphi} \in \boldsymbol{C}^2_0(\overline{\mathcal{D}}) \cap \boldsymbol{C}^1_0(\mathcal{D})$, $\psi \in C^1_0(\mathcal{D})$, then $v \equiv \mathbf{0}$, p is a constant on $\mathcal{D}$. □

Proof of Lemma 4.16 From Lemma 4.14 we know that there exists $(\mathbf{v}, p)$ such that

$$\left.\begin{aligned} -\nu\Delta\mathbf{v} + k\frac{\partial\mathbf{v}}{\partial\mathbf{x}_3} - (\boldsymbol{\omega}\wedge\mathbf{x})\cdot\nabla\mathbf{v} + \boldsymbol{\omega}\wedge\mathbf{v} + \nabla p = \mathbf{0} \\ \mathrm{div}\,\mathbf{v} = 0 \end{aligned}\right\} \text{ in } \mathcal{D} \tag{4.205}$$

$$\mathbf{v} \in \widehat{\boldsymbol{H}}^{1,2}(B^c_R) \cap \boldsymbol{L}^4(B^c_R),\ \ p - a \in L^2(B^c_R),\ \ a \in \mathbb{R}. \tag{4.206}$$

Since $\mathbf{v} \in \boldsymbol{L}^{q_1}(\mathcal{D}_{2R})$ is a very weak solution of the Stokes problem in $\mathcal{D}_{2R}$ and $\bar{\gamma}_0\mathbf{v} = 0$ on $\partial\mathcal{D}$, then

$$(\mathbf{v}, p) \in \boldsymbol{H}^{1,2}(\mathcal{D}_{2R}) \times L^2(\mathcal{D}_{2R})$$

and using (4.206)

$$\mathbf{v} \in \widehat{\boldsymbol{H}}^{1,2}_0(\mathcal{D}) \cap \boldsymbol{L}^4(\mathcal{D}),\ \ p - a \in L^2(\mathcal{D}). \tag{4.207}$$

We introduce the following cutoff functions with an anisotropic decay. Let $\mu \in C^\infty([0, \infty); [0, 1])$ with $\bar{\mu}(t) = 1$ for $0 \le t \le 5/4$ and $\bar{\mu}(t) = 0$ for $t \ge 7/4$, μ a nonincreasing. For any $j \in \mathbb{N}$, $j \ge R$, we introduce

$$\mu_j(x) = \bar{\mu}\left(\sqrt{\frac{x_3^2}{j^4} + \frac{x_2^2 + x_1^2}{j^2}}\right) \qquad \forall\,\mathbf{x} = (x_1, x_2, x_3) \in \mathbb{R}^3.$$

Then $\nabla\mu_j$ is supported in the set $A_j = \{\mathbf{x} \in \mathbb{R}^3;\ 1 < x_3^2/j^4 + (x_2^2 + x_1^2)/j^2 < 2^2\}$, and there is a constant $c = c(\bar{\mu}) >$ such that

$$j^2\left\|\frac{\partial\mu_j}{\partial x_3}\right\|_{\infty,\mathbb{R}^3} + j\|\nabla\mu_j\|_{\infty,\mathbb{R}^3} \le c \tag{4.208}$$

for all $j \geq R$. Moreover, since the volume of A_j is bounded by cj^4, we obtain that

$$\left\| \frac{\partial \mu_j}{\partial x_3} \right\|_{2,\mathbb{R}^3} + \|\nabla \mu_j\|_{4,\mathbb{R}^3} \leq c \quad \forall j \geq R, \ c = c(\widetilde{\mu}) > 0.$$

We multiply (4.205) by $\mu_j \mathbf{v}$, and integrating by parts, we get

$$\nu \int_{\mathcal{D}} \mu_j |\nabla \mathbf{v}|^2 dx = -\nu \int_{A_j} (\nabla \mathbf{v} \cdot \nabla \mu_j) \cdot \mathbf{v} d\mathbf{x} + \frac{k}{2} \int_{A_j} |\mathbf{v}|^2 \frac{\partial \mu_j}{\partial x_3} d\mathbf{x} \tag{4.209}$$

$$+ \int_{\mathcal{D}} \nabla p \mu_j \mathbf{v} d\mathbf{x} + \int_{A_j} (\boldsymbol{\omega} \wedge \mathbf{x}) \cdot \nabla \mathbf{v} \cdot \mu_j \mathbf{v} d\mathbf{x} - \int_{A_j} (\boldsymbol{\omega} \wedge \mathbf{v}) \mu_j \mathbf{v} d\mathbf{x}.$$

Since

$$\int_{A_j} (\boldsymbol{\omega} \wedge \mathbf{x}) \cdot \nabla \mathbf{v} \cdot \mu_j \mathbf{v} d\mathbf{x} = \int_{A_j} (\boldsymbol{\omega} \wedge \mathbf{v}) \mu_j \mathbf{v} d\mathbf{x} = 0,$$

it implies that

$$\nu \int_{\mathcal{D}} \mu_j |\nabla \mathbf{v}|^2 d\mathbf{x} \to \nu \int_{\mathcal{D}} |\nabla \mathbf{v}|^2 d\mathbf{x}, \ j \to \infty.$$

Then from (4.207), (4.208) imply

$$\nu \int_{A_j} |(\nabla \mathbf{v} \cdot \nabla \mu_j) \cdot \mathbf{v}| d\mathbf{x} + \frac{k}{2} \int_{A_j} |\mathbf{v}|^2 |\frac{\mu_j}{\partial x_3}| d\mathbf{x} \leq \nu \|\nabla \mathbf{v}\|_{2,A_j} \|\nabla \mu_j\|_{4,\mathbb{R}^3} \|\mathbf{v}\|_{4,A_j}$$

$$+ \frac{k}{2} \|\mathbf{v}\|_{4,A_j}^2 \|\frac{\partial \mu_j}{\partial x_3}\|_{2,\mathbb{R}^3}$$

$$\leq c\nu \|\nabla \mathbf{v}\|_{2,A_j} \|\nabla \mu_j\|_{4,\mathbb{R}^3} + ck \|\mathbf{v}\|_{4,A_j}^2 \to 0 \text{ as } j \to \infty,$$

$c = c(\mu) > 0$.
Using $\nabla p = \nabla(p - a)$ we get

$$|\int_{\mathcal{D}} \nabla p \mu_j \mathbf{v} d\mathbf{x}| \leq \int_{A_j} |(p-a) \cdot \mathbf{v} \cdot \nabla \mu_j| d\mathbf{x} \leq c \|p-a\|_{2,A_j} \|\mathbf{v}\|_{4,A_j} \|\nabla \mu_j\|_{4,\mathbb{R}^3} \to 0,$$

$$j \to \infty,$$

$c = c(\mu) > 0$.

Letting $j \to \infty$ in (4.209), we get

$$\int_{\mathcal{D}} |\nabla \mathbf{v}|^2 d\mathbf{x} = 0.$$

Since $\gamma_0 \mathbf{v} = 0$ on $\partial\mathcal{D} \Rightarrow \mathbf{v} = 0$ in $\mathcal{D}$, p is a constant. □

4.3.4.1 Existence of Very Weak Solutions

We will show the existence and the regularity of the weak solution of $(\mathrm{O})_{\mathrm{ext}}$.

Lemma 4.17 *Assume that* $\mathbf{f} = \mathbf{0}$, $\mathbf{h} \in \boldsymbol{H}^{1/2,2}(\partial\mathcal{D})$. *Then there exists a unique weak solution* $(\mathbf{v}, p) \in [\widehat{\boldsymbol{H}}^{1,2}(\mathcal{D}) \cap \boldsymbol{L}^4(\mathcal{D})] \times L^2(\mathcal{D})$ *of* $(O)_{ext}$ *satisfying*

$$\|\nu \nabla \mathbf{v}\|_{2,\mathcal{D}} + \|\nu \mathbf{v}\|_{4,\mathcal{D}} \leq c_{\nu,k,\omega} \|\nu \mathbf{h}\|_{\frac{1}{2},2,\partial\mathcal{D}} \tag{4.210}$$

for some $c_{\nu,k,\omega} = c_{\nu,k,\omega}(\mathcal{D})$. □

Proof of Lemma 4.17 *Uniqueness:*
The uniqueness of solution in $\boldsymbol{L}^4(\mathcal{D})$ we can deduce from Lemma 4.16.

Existence:
To show the existence of solutions, we follow ideas from [Galdi [98], Theorem VIII.1.2]. As a first step we extend the boundary data $\mathbf{h} \in \boldsymbol{W}^{1,2,2}(\partial\mathcal{D})$ to a solenoidal function $\tilde{\mathbf{v}} \in \boldsymbol{H}^{1,2}(\mathcal{D})$ such that $g = (\boldsymbol{\omega} \wedge \mathbf{x}) \cdot \nabla\tilde{\mathbf{v}} - \boldsymbol{\omega} \wedge \tilde{\mathbf{v}}$ has a compact support,

$$\|\tilde{\mathbf{v}}\|_{1,2,\mathcal{D}} \leq c\|\mathbf{h}\|_{\frac{1}{2},\partial\mathcal{D}}, \tag{4.211}$$

$c = c(\mathcal{D}) > 0$, see [98, Theorem V.2.1], [196, Lemma 2.3]. We set $\mathbf{v} = \tilde{\mathbf{v}} + \mathbf{u}$, where $\mathbf{u}$ satisfies the problem

$$-\nu \Delta \mathbf{u} + k\frac{\partial \mathbf{u}}{\partial x_3} - (\boldsymbol{\omega} \wedge \mathbf{x}) \cdot \nabla \mathbf{u} + \boldsymbol{\omega} \wedge \mathbf{u} + \nabla p = \tilde{\mathbf{f}} \tag{4.212}$$

$$\begin{aligned} \operatorname{div} \mathbf{u} &= 0 \quad \text{in } \mathcal{D} \\ \mathbf{u} &= 0 \quad \text{on } \partial\mathcal{D} \\ \mathbf{u}(\mathbf{x}) &\to 0 \quad |\mathbf{x}| \to \infty, \end{aligned}$$

where $\tilde{\mathbf{f}} = \operatorname{div}(\nu\nabla\tilde{\mathbf{v}} + k\boldsymbol{e}_3 \otimes \tilde{\mathbf{v}}) + g \in \widehat{\boldsymbol{H}}^{-1,2}$

$$\|\tilde{\mathbf{f}}\|_{-1,2,\mathcal{D}} \leq c_{\nu,k,\omega}\|\nu\tilde{\mathbf{v}}\|_{1,2,\mathcal{D}}. \tag{4.213}$$

Multiplying $(4.212)_1$ by $\mathbf{u}$, we can get

$$\|\nu\nabla\mathbf{u}\|_{2,\mathcal{D}} \le c_{\upsilon,k,\omega}\|\nu\mathbf{h}\|_{\frac{1}{2},2,\partial\mathcal{D}}. \tag{4.214}$$

Using the Galerkin method we deduce the existence of a weak solution $\mathbf{v} \in \widehat{\boldsymbol{H}}^{1,2}(\mathcal{D}) \cap \boldsymbol{L}^6(\mathcal{D})$ of $(\mathrm{O})_{\mathrm{ext}}$ satisfying

$$\|\nu\nabla\mathbf{v}\|_{2,\mathcal{D}} + \|\nu\mathbf{v}\|_{6,\mathcal{D}} \le c_{\upsilon,k,\omega}\|\nu\mathbf{h}\|_{\frac{1}{2},2,\partial\mathcal{D}}. \tag{4.215}$$

Since $\mathbf{v} \in \boldsymbol{L}^6(\mathcal{D})$ using Lemma 4.14 (b), we get $\mathbf{v} \in \widehat{\boldsymbol{H}}^{1,\frac{12}{7}}(B_R^c) \cap \boldsymbol{L}^4(B_R^c)$,

$$\|\nu\mathbf{v}\|_{4,B_R^c} \le c\|\nu\nabla\mathbf{v}\|_{\frac{12}{7},B_R^c} \le c_{\upsilon,k,\omega}\|\nu\mathbf{v}\|_{6,\mathcal{D}_{2R}},$$

where $c_{\upsilon,k,\omega} = c_{\upsilon,k,\omega}(\mathcal{D})$. By de Rham theorem we get the existence of $p \in L^2(\mathcal{D})$. □

Proof of Lemma 4.12. The uniqueness follows from Lemma 4.16. Let $2 \le r < 4$, and let

$$\mathbf{f} = \operatorname{div}\mathbf{F},\ \ \mathbf{F} \in \boldsymbol{L}^r(\mathcal{D}),\ \ \mathbf{h} \in \boldsymbol{W}^{1-\frac{1}{r},r}(\partial B_R).$$

From the previous part there exists a weak solution $(\mathbf{v}_0, p_0)$ of $(\mathrm{O})_{\mathrm{ws}}$, $(\mathbf{f}, g) = (\operatorname{div}\widetilde{\mathbf{F}}, 0)$ satisfying (4.176), (4.178).

Since

$$\mathbf{h} - \gamma_0\mathbf{v}_0 \in \boldsymbol{W}^{1-\frac{1}{r},r}(\partial\mathcal{D}) \subset \boldsymbol{W}^{\frac{1}{2},2}(\partial\mathcal{D})$$

from Lemma 4.17 follows that there exists a weak solution $(\mathbf{v}_1, p_1) \in [\widehat{\boldsymbol{H}}^{1,2}(\mathcal{D}) \cap \boldsymbol{L}^4(\mathcal{D})] \times L^2(\mathcal{D})$ of $(\mathrm{O})_{\mathrm{ext}}$ with $(\mathbf{f}, \mathbf{h})$ replaced by $(\mathbf{0}, \mathbf{h} - \gamma_0\mathbf{v}_0)$ satisfying

$$\|\nu\mathbf{v}_1\|_{4,\mathcal{D}} \le c_{\upsilon,k,\omega}\|\nu(\mathbf{h} - \gamma_0\mathbf{v}_1)\|_{1-\frac{1}{r},r,\partial\mathcal{D}}. \tag{4.216}$$

From Lemmas 4.14 (b), and 4.15 (a) we get

$$(\mathbf{v}_1, p_1) \in [\widehat{\boldsymbol{H}}^{1,r}(\mathcal{D}) \cap \boldsymbol{L}^{\frac{4r}{4-r}}(\mathcal{D})] \times L^r(\mathcal{D})$$

satisfying the estimate

$$\frac{\lambda}{A_{\upsilon,k,\omega}}\|\nu\mathbf{v}_1\|_{\frac{4r}{4-r},\mathcal{D}} + \|\nu\nabla\mathbf{v}_1\|_{r,\mathcal{D}} \le c_{\upsilon,k,\omega}\Big(\beta(r)\|F\|_{r,\mathcal{D}} + \|\nu\mathbf{h}\|_{1-\frac{1}{r},r}\Big) \tag{4.217}$$

for some $c_{\nu,k,\omega} = c_{\nu,k,\omega}(r, \mathcal{D})$, where

$$\beta(r) = \begin{cases} 1, & \frac{3}{2} < r < 3 \\ A_{\nu,k,\omega}\lambda^{-1}, & 3 \le r < 4. \end{cases}$$

Then $(\mathbf{v}, p) = (\mathbf{v}_0 + \mathbf{v}_1, p_0 + p_1)$ is a weak solution satisfying the assertion of Lemma 4.12 in cases $2 \le r < 4$.

Case $6/5 \le s < 2$. Let $\mathbf{f} \in \boldsymbol{L}^s(\mathcal{D})$, $\mathbf{h} \in \boldsymbol{H}^{2-\frac{1}{s},s}(\partial\mathcal{D})$. From Sect. 4.3.3 it follows the existence of a strong solution $(\mathbf{v}_0, p_0)$ of $(\mathrm{O})_{\mathrm{ws}}$ with $\mathbf{f} = \tilde{\mathbf{f}}$, $g = 0$ and satisfying (4.179), (4.180), (4.183). It implies that

$$\mathbf{h} - \gamma_0\mathbf{v}_0 \in \boldsymbol{H}^{2-\frac{1}{s},s}(\partial\mathcal{D}) \hookrightarrow \boldsymbol{H}^{\frac{1}{2},2}(\partial\mathcal{D}).$$

Applying Lemma 4.17 we obtain an existence of a weak solution $(\mathbf{v}_1, p_1) \in \big(\widehat{\boldsymbol{H}}^{1,2}(\mathcal{D}) \cap \boldsymbol{L}^4(\mathcal{D})\big) \times L^2(\mathcal{D})$ of $(\mathrm{O})_{\mathrm{ws}}$ with $\mathbf{f} = \mathbf{0}$, $\mathbf{h}$ replaced by $\mathbf{h} - \gamma_0\mathbf{v}_0$, satisfying

$$\|\gamma\mathbf{v}_1\|_{4,\mathcal{D}} \le c_{\nu,k,\omega}\|\nu(\mathbf{h} - \gamma_0\mathbf{v}_0)\|_{2-\frac{r}{s},s,\partial\mathcal{D}}.$$

Further, applying Lemmas 4.14 (a), and 4.15 (b) and using the property of $\mathbf{u} \in \widehat{\boldsymbol{H}}^{1,q_\cdot}(\mathcal{D})$ there is a unique constant a such that

$$\mathbf{u} - \boldsymbol{a} \in \boldsymbol{L}^{\frac{3q}{3-q}}, \ \|\mathbf{u} - \boldsymbol{a}\|_{\frac{3q}{3-q},\mathcal{D}} \le c\|\nabla\mathbf{u}\|_{q,\mathcal{D}},$$

$c = c(q, \mathcal{D}) > 0$, we get

$$(\mathbf{v}_1, p_1) \in \widetilde{\boldsymbol{D}}^{2,s} \times [\widehat{H}^{1,s}(\mathcal{D}) \cap L^{s^*}(\mathcal{D})],$$

satisfying the estimate

$$\begin{aligned}
&\lambda\|\nu\nabla\mathbf{v}_1\|_{\frac{4q}{4-q},\mathcal{D}} + \lambda^2\|\nu\mathbf{v}_1\|_{\frac{2q}{2-q},\mathcal{D}} + \left\|k\frac{\partial\mathbf{v}_1}{\partial x_3}\right\|_{s,\mathcal{D}} \\
&\quad + \|(\boldsymbol{\omega}\wedge\mathbf{x})\cdot\nabla\mathbf{v}_1 - \boldsymbol{\omega}\wedge\mathbf{v}_1\|_{s,\mathcal{D}} \le c_{\nu,k,\omega}A_{\nu,k,\omega}\|\nu(\mathbf{h} - \gamma_0\mathbf{v}_0)\|_{2-\frac{1}{s},s,\partial\mathcal{D}} \\
&\quad \le c_{\nu,k,\omega}A_{\nu,k,\omega}(\alpha(s')\|\mathbf{f}\|_{s,\mathcal{D}} + \|\nu\mathbf{h}\|_{2-\frac{1}{s},s,\partial\mathcal{D}}), \\
&\|\nu\nabla^2\mathbf{v}_1\|_{s,\mathcal{D}} + \|\nabla p_1\|_{s,\mathcal{D}} \le c_{\nu,k,\omega}(\alpha(s')\|\mathbf{f}\|_{s,\mathcal{D}} + \|\nu\mathbf{h}\|_{2-\frac{1}{s},s,\partial\mathcal{D}})
\end{aligned}$$

for some constant $c_{\nu,k,\omega} = c_{\nu,k,\omega}(s, \mathcal{D})$. Hence $(\mathbf{v}, p) = (\mathbf{v}_0 + \mathbf{v}_1, p_0 + p_1)$ is the strong solution of $(O)_{ws}$ satisfying corresponding estimates. This gives us the existence of very weak solutions in the case $6/5 \le s < 2$.

Existence of very weak solution $2 < q \leq 6$. We use the duality argument. Let $r = \frac{3q}{3+q}$, $\mathbf{F} \in \boldsymbol{L}^r(\mathcal{D})$, $\mathbf{h} \in \boldsymbol{H}^{-\frac{1}{q},q}(\partial\mathcal{D})$. We define

$$\mathcal{H} = \{\mathbf{u} \in \boldsymbol{H}^{2,q'}(\mathcal{D}) \cap \widehat{\boldsymbol{H}}^{1,r'}_{0,\sigma}(\mathcal{D}) : (\boldsymbol{\omega} \wedge \mathbf{x}) \cdot \nabla \mathbf{u} - \boldsymbol{\omega} \wedge \mathbf{u} \in \boldsymbol{L}^{q'}\}.$$

Since $6/5 \leq q' < 2$, $r' = (q')^*$, ("$*$" means here the Sobolev exponent) from previous sections follow that $\forall\, \mathbf{w} \in \boldsymbol{L}^{q'}(\mathcal{D})$, there exists a unique pair $(\boldsymbol{\varphi}_w, \psi_w) \in \mathcal{H} \times [\widehat{H}^{1,q'}(\mathcal{D}) \cap L^{r'}(\mathcal{D})]$ such that

$$-\nu\Delta\boldsymbol{\varphi}_w - k\frac{\partial\boldsymbol{\varphi}_w}{\partial x_3} + \nabla\psi_w + (\boldsymbol{\omega} \wedge \mathbf{x}) \cdot \nabla\boldsymbol{\varphi}_w - \boldsymbol{\omega} \wedge \boldsymbol{\varphi}_w = \mathbf{w} \quad \text{in } \mathcal{D}.$$

Moreover the mapping $\mathbf{w} \mapsto (\boldsymbol{\varphi}_w, \psi_w)$ is a bounded linear operator from $\boldsymbol{L}^{q'}(\mathcal{D})$ into $\mathcal{H} \times [\widehat{H}^{1,q'}(\mathcal{D}) \times L^{r'}(\mathcal{D})]$ satisfying

$$\begin{aligned}
&\|\nu\nabla^2\boldsymbol{\varphi}_w\|_{q',\mathcal{D}} + \|\nu\nabla\boldsymbol{\varphi}_w\|_{r',\mathcal{D}} \\
&+ \|\nabla\psi_w\|_{q',\mathcal{D}} + \|\psi_w\|_{r',\mathcal{D}} \leq c_{\nu,k,\omega}\alpha(q)\|\omega\|_{q',\mathcal{D}}, \\
&c_{\nu,k,\omega} = c_{\nu,k,\omega}(q, \mathcal{D}).
\end{aligned} \tag{4.218}$$

Using (4.218), the trace theorem we obtain

$$\begin{aligned}
&|\langle \boldsymbol{f}, \boldsymbol{\varphi}_w\rangle - \nu\langle \mathbf{h}, \gamma_1\boldsymbol{\varphi}_w\rangle + \langle \mathbf{h}, (\gamma_0\psi_w)\boldsymbol{N}\rangle| \\
&\quad \leq c\left(\frac{1}{\nu}\|\mathbf{F}\|_{r,\mathcal{D}} + \|\mathbf{h}\|_{-\frac{1}{q},q,\partial\mathcal{D}}\right)(\|\nu\nabla^2\boldsymbol{\varphi}_w\|_{q',\mathcal{D}} + \|\nu\nabla\boldsymbol{\varphi}_w\|_{r',\mathcal{D}} \\
&\qquad + \|\nabla\psi_w\|_{q',\mathcal{D}} + \|\psi_w\|_{r',\mathcal{D}}) \\
&\leq c_{\nu,k,\omega}\alpha(q)\left(\frac{1}{\nu}\|\mathbf{F}\|_{r,\mathcal{D}} + \|\mathbf{h}\|_{-\frac{1}{q},q,\partial\mathcal{D}}\right)\|\mathbf{w}\|_{q',\mathcal{D}}
\end{aligned} \tag{4.219}$$

for each $\mathbf{w} \in L^{q'}(\mathcal{D})$.

It implies that the linear functional L defined as

$$L(\mathbf{w}) = \langle \boldsymbol{f}, \boldsymbol{\varphi}_w\rangle - \nu\langle \mathbf{h}, \gamma_1\boldsymbol{\varphi}_w\rangle + \langle \mathbf{h}, (\gamma_0\psi_\omega)\boldsymbol{N}\rangle \tag{4.220}$$

is bounded on $\boldsymbol{L}^{q'}(\mathcal{D})$ and

$$|||L||| \leq c_{\nu,k,\omega}\alpha(q)\left(\frac{1}{\nu}\|\mathbf{F}\|_{r,\mathcal{D}} + \|\mathbf{h}\|_{-\frac{1}{q},q,\partial\mathcal{D}}\right). \tag{4.221}$$

By the Riesz theorem, there exists a unique $\mathbf{v} \in \boldsymbol{L}^q(\mathcal{D})$ such that

$$\|\mathbf{v}\|_{q;\mathcal{D}} = |||L|||, \int_{\mathcal{D}} \mathbf{v} \cdot \mathbf{w}\, d\mathbf{x} = L(\mathbf{w}) \quad \forall \mathbf{w} \in \boldsymbol{L}^{q'}(\mathcal{D}).$$

It immediately follows that $\mathbf{v}$ is a very weak solution of $(\mathrm{O})_{\mathrm{ws}}$.

For given $\boldsymbol{\varphi} \in \mathcal{H}$, $\psi \in \widehat{H}^{1,q'}(\mathcal{D}) \cap L^{r'}(\mathcal{D})$, we define

$$\mathbf{w} = -\nu\Delta\boldsymbol{\varphi} - k\frac{\partial \boldsymbol{\varphi}}{\partial x_3} + \nabla\psi + (\boldsymbol{\omega} \wedge \mathbf{x}) \cdot \nabla\boldsymbol{\varphi} - \boldsymbol{\omega} \wedge \boldsymbol{\varphi}.$$

Applying results from Sect. 4.3.4, we obtain $(\boldsymbol{\varphi}, \psi) = (\boldsymbol{\varphi}_w, \psi_w)$ in $\mathcal{D}$. Hence, by (4.221) we obtain

$$\int_{\mathcal{D}} \mathbf{v}\left(-\nu\Delta\boldsymbol{\varphi} - k\frac{\partial \boldsymbol{\varphi}}{\partial x_3} + \nabla\psi + (\boldsymbol{\omega} \wedge \mathbf{x}) \cdot \nabla\boldsymbol{\varphi} - \boldsymbol{\omega} \wedge \boldsymbol{\varphi}\right) d\mathbf{x}$$

$$= \int_{\mathcal{D}} \mathbf{v} \cdot \mathbf{w} d\mathbf{x} = \langle \mathbf{f}, \varphi\rangle - \nu\langle \mathbf{h}, \gamma_1\boldsymbol{\varphi}\rangle + \langle \mathbf{h}, (\gamma_0\psi)\boldsymbol{N}\rangle.$$

This gives us an existence of very weak solution for $2 < q \leq 6$.

Now we need to prove the existence in case $4/3 < r < 2$.

Let $\mathbf{f} = \operatorname{div}\mathbf{F}$, $\mathbf{F} \in \boldsymbol{L}^r(\mathcal{D})$, $\mathbf{h} \in \boldsymbol{H}^{1-\frac{1}{r},r}(\partial\mathcal{D})$. From Sect. 4.3.1 it follows the existence of a weak solution $(\mathbf{u}_0, p_0)$ of $(O)_{ws}$ with $\mathbf{f} = \operatorname{div}\widetilde{\mathbf{F}}$, $g = 0$ satisfying (4.176), (4.178). The applications of the Sobolev and the trace theorems imply

$$\mathbf{h} - \gamma_0\mathbf{v}_0 \in \boldsymbol{H}^{1-\frac{1}{2},r}(\partial\mathcal{D}) \hookrightarrow \boldsymbol{H}^{-\frac{1}{r^*},r^*}(\mathcal{D})$$

$$\frac{12}{5} < r^* < 6.$$

Therefore it follows from previous sections we get that $(\mathbf{v}_1, p_1) \in \boldsymbol{L}^{r^*}(\mathcal{D}) \times L^1_{\mathrm{loc}}(\mathcal{D})$ of $(\mathrm{O})_{\mathrm{ws}}$ where we replaced $(\mathbf{f}, h)$ by $(\mathbf{0}, \mathbf{h} - \gamma_0\mathbf{v}_0)$ and

$$\|\nu\mathbf{v}_1\|_{r^*,\mathcal{D}} \leq c_{\nu,k,\omega}\alpha(r^*)\|\nu(\mathbf{h} - \gamma_0\mathbf{v}_0)\|_{1-\frac{1}{2},r,\partial\mathcal{D}}. \tag{4.222}$$

Applying the de Rham theorem we get the existence of p_1. Now, we want to prove the regularity and a priori estimates for $(\mathbf{v}_1, p_1)$.

Case $3/2 < r < 2$, then using Lemma 4.14, Lemma 4.15 (a) we obtain

$$\mathbf{v}_1 \in \widehat{\boldsymbol{H}}^{1,r}(\mathcal{D}) \cap \boldsymbol{L}^{\frac{4r}{4-r}}(\mathcal{D}) \text{ and } p_1 - a \in L^r(\mathcal{D})$$

where $a \in \mathbb{R}$. Moreover,

$$\frac{\lambda}{A_{\nu,k,\omega}}\|\nu \mathbf{v}_1\|_{\frac{4r}{4-r},\mathcal{D}} + \|\nu\nabla\mathbf{v}_1\|_{r,\mathcal{D}} \leq c_{\nu,k,\omega}\|\nu(\mathbf{h} - \gamma_0\mathbf{v}_0)\|_{1-\frac{1}{r},1,\partial\mathcal{D}}$$
$$\leq c_{\nu,k,\omega}(\|\mathbf{F}\|_{r,\mathcal{D}} + \|\nu\mathbf{h}\|_{1-\frac{1}{r},r,\partial\mathcal{D}})$$

is satisfied, for some $c_{\nu,k,\omega} = c_{\nu,k,\omega}(r, \mathcal{D})$.

Case $4/3 < r < 3/2$. From Lemmas 4.14 (b) and 4.15 (a) we get

$$\mathbf{v}_1 \in \widehat{\boldsymbol{H}}^{1,r}(\mathcal{D}) \cap \boldsymbol{L}^{\frac{4r}{4-r}}(\mathcal{D}),\ p_1 - a \in L^{\frac{12r}{12-r}}(\mathcal{D})$$

for some $a \in \mathbb{R}$. Moreover, $\alpha(r^*) = A_{\nu,k,\omega}\lambda^{-1}$, and we get from (4.222) such that

$$\lambda\|\nu\mathbf{v}_1\|_{\frac{4r}{4-r},\mathcal{D}} + \|\nu\nabla\mathbf{v}_1\|_{r,\mathcal{D}} \leq c_{\nu,k,\omega}A^2_{\nu,k,\omega}\frac{1}{\lambda^2}(\|\mathbf{F}\|_{1,\mathcal{D}} + \|\nu\mathbf{h}\|_{1-\frac{1}{r'},r,\partial\mathcal{D}}),$$
$$c_{\nu,k,\omega} = c_{\nu,k,\mathcal{D}}(r, \mathcal{D}).$$

Therefore $(\mathbf{v}, p) = (\mathbf{v}_0 + \mathbf{v}_1, p_0 + p_1 - a)$ is a weak solution of $(0)_{\text{ext}}$.

Case $1 < s < 6/5$.

Suppose that $\mathbf{f} \in \boldsymbol{L}^s(\mathcal{D})$, $h \in \boldsymbol{H}^{2-\frac{1}{s},s}(\partial\mathcal{D})$. From Sects. 4.3.3 and 4.3.4 there exists a solution $(\mathbf{v}_0, p_0)$ of $(\mathrm{O})_{\mathrm{ws}}$, with $(\mathbf{f}, g) = (\tilde{\mathbf{f}}, 0)$ satisfying (4.179), (4.180), (4.181).

For $1 < s < 6/5$, we set $r = s^*$.

We obtain

$$\mathbf{h} - \gamma_0\mathbf{v}_0 \in \boldsymbol{H}^{2-\frac{1}{s},s}(\partial\mathcal{D}) \hookrightarrow \boldsymbol{H}^{1-\frac{1}{r},r}(\partial\mathcal{D}),\ \frac{3}{2} < r < 2.$$

Moreover from previous parts, there exists a weak solution

$$(\mathbf{v}_1, p_1) \in [\widehat{\boldsymbol{H}}^{1,r}(\mathcal{D}) \cap \boldsymbol{L}^{\frac{4r}{4-r}}] \times L^r(\mathcal{D})$$

of $(\mathrm{O})_{\mathrm{ws}}$ with $(\mathbf{f}, h)$ replaced by $(\mathbf{0}, \mathbf{h} - \gamma\mathbf{u}_0)$ which satisfies

$$\|\nu\mathbf{v}_1\|_{r^*,(\mathcal{D})} \leq c\|\nu\nabla\mathbf{v}_1\|_{r,\mathcal{D}} \leq c_{\nu,k,\omega}(\|\mathbf{f}\|_{s,\mathcal{D}} + \|\nu\mathbf{h}\|_{2-\frac{1}{3},s,\partial\mathcal{D}}).$$

The proofs of regularity and corresponding estimates can be performed similarly as in the case $6/5 \leq s < 2$.

Case $6 < q < \infty$ is analogous to $2 < q \leq 6$. □

4.4 L^q-Weighted Theory of the Strong Solution to Oseen Type Problem

In this section we follow the work of Krbec, Farwig, Nečasová [72]. We consider the stationary problem—Oseen type system

$$\begin{aligned} -\nu\Delta\mathbf{u}+k\partial_3\mathbf{u}-(\omega\mathbf{e}_3\wedge\mathbf{x})\cdot\nabla\mathbf{u}+\omega\mathbf{e}_3\wedge\mathbf{u}+\nabla p =&\mathbf{f} \text{ in } \mathcal{D} \\ \operatorname{div}\mathbf{u} =&0 \text{ in } \mathcal{D} \\ \mathbf{u} \to&0 \text{ at } \infty \end{aligned} \tag{4.223}$$

together with the boundary condition $\mathbf{u}(\mathbf{x},t)=\omega\mathbf{e}_3\wedge\mathbf{x}-\mathbf{u}_\infty$ on $\partial\mathcal{D}$. Note that there is no boundary condition in the case $\mathcal{D}=\mathbb{R}^3$.

The aim of this section is to generalize the a priori estimate from previous section to weighted L^q-spaces for the whole space $\mathbb{R}^3$. Let us recall the definition of the weighted Lebesgue space

$$L^q_w(\mathbb{R}^3)=L^q_w=\left\{u\in L^1_{\mathrm{loc}}(\mathbb{R}^3)\colon \|u\|_{q,w}=\left(\int_{\mathbb{R}^n}|u(\mathbf{x})|^q w(\mathbf{x})\,\mathrm{d}\mathbf{x}\right)^{1/q}<\infty\right\},$$

where $w\in L^1_{\mathrm{loc}}$ is a nonnegative weight function and should reflect the anisotropy of the flow and the existence of a wake region in the downstream direction $x_3>0$. Our tools will include Littlewood–Paley theory, singular integral operators, multiplier operators, and maximal operators in weighted spaces so that we need weight functions satisfying Muckenhoupt type conditions.

Definition 4.3 Let $\mathcal{R}$ be a family of bounded sets R in $\mathbb{R}^n$, each of positive Lebesgue measure $|R|$. A weight function $0\le w\in L^1_{\mathrm{loc}}$ belongs to the *Muckenhoupt class* $A_q(\mathcal{R})$: $(A_q(\mathcal{R})=A_q(\mathbb{R}^n,\mathcal{R}))$, $1\le q<\infty$, if there exists a constant $C>0$ such that

$$\sup_R\left(\frac{1}{|R|}\int_R w(\mathbf{x})\,\mathrm{d}\mathbf{x}\right)\left(\frac{1}{|R|}\int_R w(\mathbf{x})^{-1/(q-1)}\,\mathrm{d}\mathbf{x}\right)^{q-1}\le C\quad\text{for any } R\in\mathcal{R}$$

if $1<q<\infty$, and

$$\sup_{R\in\mathcal{R},\,R\ni x_0}\frac{1}{|R|}\int_R w(\mathbf{x})\,\mathrm{d}\mathbf{x}\le Cw(\mathbf{x}_0)\quad\text{for a.a. } \mathbf{x}_0\in\mathbb{R}^n,$$

if $q=1$, respectively. □

Due to the anisotropic nature of our problem we shall need a variant of the classical *Muckenhoupt class* $A_q(\mathcal{C})$: $A_q(\mathcal{C})=A_q(\mathbb{R}^3,\mathcal{C})$, where $\mathcal{C}$ is the set of all cubes $Q\subset\mathbb{R}^3$ with edges parallel to the coordinate axes. Namely, $\mathcal{C}$ is replaced by $\mathcal{J}$, the set of all bounded intervals (rectangles) in $\mathbb{R}^3$, leading to the class $A_q(\mathcal{J}) = A_q(\mathbb{R}^3, \mathcal{J})$. Obviously, $A_q(\mathbb{R}^3, \mathcal{J}) \subsetneq A_q(\mathbb{R}^3, \mathcal{C})$.

Moreover, to describe the anisotropy of the wake region more precisely by weights, we have to introduce in addition to the weights on $\mathbb{R}^n$ *one-sided Muckenhoupt weights* and *one-sided maximal operators* on the real line, see Definition 4.4, Theorem 4.15 and Lemma 4.18.

Definition 4.4

(i) For every locally integrable function u on the real line, let M^+u be defined by

$$M^+u(x) = \sup_{h>0} \frac{1}{h} \int_x^{x+h} |u(t)| \mathrm{d}t.$$

Analogously,

$$M^-u(x) = \sup_{h>0} \frac{1}{h} \int_{x-h}^{x} |u(t)| \mathrm{d}t.$$

(ii) A weight function $0 \le w \in L^1_{\mathrm{loc}}(\mathbb{R})$ lies in the *weight class* A_1^- if there exists a constant $c > 0$ such that $M^+w(x) \le cw(x)$ for almost all $x \in \mathbb{R}$. Analogously, $w \in A_1^+$ if and only if $M^-w(x) \le cw(x)$ for almost all $x \in \mathbb{R}$. The smallest constant $c \ge 0$ satisfying $M^\pm w(x) \le cw(x)$ for almost all $x \in \mathbb{R}$ is called the *$A_1^\mp$-constant of w.*

(iii) A weight function $0 \le w \in L^1_{\mathrm{loc}}$ belongs to the *one-sided Muckenhoupt class* A_q^+, $1 < q < \infty$, if there exists a constant $C > 0$ such that for all $x \in \mathbb{R}$

$$\sup_{h>0} \left(\frac{1}{h} \int_{x-h}^{x} w(t) \mathrm{d}t \right) \left(\frac{1}{h} \int_x^{x+h} w(t)^{-1/(q-1)} \mathrm{d}t \right)^{q-1} \le C.$$

The smallest constant $C \ge 0$ satisfying this estimate is called the *A_q^+-constant* of w. By analogy, we define the set of weights A_q^- and the A_q^--constant of a weight in A_q^-.

□

Now we are in a position to describe the most general weights considered in this chapter. Note that these weights are independent of the angular variable θ in the cylindrical coordinate system $(r, \theta, x_3) \in [0, \infty) \times [0, 2\pi] \times \mathbb{R}$ attached to the axis of revolution $\boldsymbol{e}_3 = (0, 0, 1)^{\mathrm{T}}$. Hence we will write $w(\mathbf{x}) = w(x_1, x_2, x_3) = w_r(x_3)$ for $r = |(x_1, x_2)|$, $\mathbf{x} = (x_1, x_2, x_3)$.

Definition 4.5 For $1 \le q < \infty$ let

$$\begin{aligned}\widetilde{A}_q^- = \widetilde{A}_q^-(\mathbb{R}^3) = \{w \in A_q(\mathbb{R}^3)\colon\ & w \text{ is } \theta\text{-independent for a.a. } r > 0,\\ & w(x_1, x_2, \cdot) = w_r(\cdot) \in A_q^-(\mathbb{R})\\ & \text{with } A_q^-(\mathbb{R})\text{-constant essentially bounded in } r\}.\end{aligned} \tag{4.224}$$

□

Theorem 4.13 *Let the weight function* $0 \le w \in L^1_{loc}(\mathbb{R}^3)$ *be independent of the angular variable* θ *and satisfy the following condition depending on* $q \in (1, \infty)$*:*

$$\begin{aligned} 2 \le q < \infty\colon\ w^\tau &\in \widetilde{A}^-_{\tau q/2} \quad \textit{for some } \tau \in [1, \infty) \\ 1 < q < 2\colon\ w^\tau &\in \widetilde{A}^-_{\tau q/2} \quad \textit{for some } \tau \in \left(\frac{2}{q}, \frac{2}{2-q}\right]. \end{aligned} \tag{4.225}$$

(i) Given $\mathbf{f} \in \boldsymbol{L}^q_w(\mathbb{R}^3)$ *there exists a solution* $(\mathbf{u}, p) \in \boldsymbol{L}^1_{loc}(\mathbb{R}^3) \times L^1_{loc}(\mathbb{R}^3)$ *of* (4.223) *satisfying the estimate*

$$\|\nu\nabla^2\mathbf{u}\|_{q,w} + \|\nabla p\|_{q,w} \le c\|\mathbf{f}\|_{q,w}, \tag{4.226}$$

with a constant $c = c(q, w) > 0$ *independent of* ν*,* k *and* ω*.*

(ii) Let $\mathbf{f} \in \boldsymbol{L}^{q_1}_{w_1}(\mathbb{R}^3) \cap \boldsymbol{L}^{q_2}_{w_2}(\mathbb{R}^3)$ *such that both* (q_1, w_1) *and* (q_2, w_2) *satisfy the conditions* (4.225)*, and let* $\mathbf{u}_1, \mathbf{u}_2 \in \boldsymbol{L}^1_{loc}(\mathbb{R}^3)^3$ *together with corresponding pressure functions* $p_1, p_2 \in L^1_{loc}(\mathbb{R}^3)$ *be solutions of* (4.223) *satisfying* (4.226) *for* (q_1, w_1) *and* (q_2, w_2)*, respectively. Then there are* $\alpha, \beta \in \mathbb{R}$ *such that* $\mathbf{u}_1$ *coincides with* $\mathbf{u}_2$ *up to an affine linear field* $\alpha\mathbf{e}_3 + \beta\omega \wedge \mathbf{x}$*,* $\alpha, \beta \in \mathbb{R}$*.*

□

Corollary 4.6 *Let the weight function* $0 \le w \in L^1_{loc}(\mathbb{R}^3)$ *be independent of the angular variable* θ*. Moreover, let* w *satisfy the following condition depending on* $q \in (1, \infty)$*:*

$$\begin{aligned} 2 \le q < \infty\colon\ w^\tau &\in \widetilde{A}^-_{\tau q/2}(\mathcal{J}) \quad \textit{for some } \tau \in [1, \infty) \\ 1 < q < 2\colon\ w^\tau &\in \widetilde{A}^-_{\tau q/2}(\mathcal{J}) \quad \textit{for some } \tau \in \left(\frac{2}{q}, \frac{2}{2-q}\right], \end{aligned} \tag{4.227}$$

where the weight class $\widetilde{A}^-_\tau(\mathcal{J})$*,* $1 \le \tau < \infty$*, is defined by*

$$\widetilde{A}^-_\tau(\mathcal{J}) = \widetilde{A}^-_\tau(\mathbb{R}^3) \cap A_\tau(\mathcal{J}).$$

Given $\mathbf{f} \in \boldsymbol{L}^q_w(\mathbb{R}^3)$ *there exists a solution* $(\mathbf{u}, p) \in \boldsymbol{L}^1_{loc}(\mathbb{R}^3) \times L^1_{loc}(\mathbb{R}^3)$ *of* (4.223) *satisfying the estimate*

$$\|k\partial_3\mathbf{u}\|_{q,w} + \|(\omega\mathbf{e}_3 \wedge \mathbf{x}) \cdot \nabla\mathbf{u} - \omega\mathbf{e}_3 \wedge \mathbf{u}\|_{q,w} \le c\left(1 + \frac{k^5}{\nu^{5/2}|\omega|^{5/2}}\right)\|\mathbf{f}\|_{q,w} \tag{4.228}$$

with a constant $c = c(q, w) > 0$ *independent of* ν, k, *and* ω. □

We remark that the ω-dependent term $1+k^5/\nu^{5/2}|\omega|^{5/2}$ in (4.228) cannot be avoided in general; see [60] for an example in the space $L^2(\mathbb{R}^3)$.

As an example of anisotropic weight functions, we consider

$$w(\mathbf{x}) = \eta^\alpha_\beta(\mathbf{x}) = (1 + |\mathbf{x}|)^\alpha(1 + s(\mathbf{x}))^\beta, \quad s(\mathbf{x}) = |(x_1, x_2, x_3)| - x_3, \tag{4.229}$$

introduced in [58] to analyze the Oseen equations; see also [147],[148],[154].

Corollary 4.7 *The a priori estimate* (4.226) *holds for the anisotropic weights* $w = \eta^\alpha_\beta$, *see* (4.229), *provided that*

$$2 \le q < \infty: \ -\frac{q}{2} < \alpha < \frac{q}{2}, \quad 0 \le \beta < \frac{q}{2} \quad \textit{and} \quad \alpha + \beta > -1$$

$$1 < q < 2: \ -\frac{q}{2} < \alpha < q - 1, \quad 0 \le \beta < q - 1 \quad \textit{and} \quad \alpha + \beta > -\frac{q}{2}.$$

Note that the condition $\beta \ge 0$ will reflect the existence of a wake region in the downstream direction $x_3 > 0$ where the solution of the original nonlinear problem (4.7)–(4.7) will decay slower than in the upstream direction $x_3 < 0$.

4.4.1 Preliminaries

To prove Theorem 4.13 we need several properties of Muckenhoupt weights and of maximal operators. Recall that $\mathcal{J}$ stands for the set of all nondegenerate rectangles in $\mathbb{R}^n$ with edges parallel to the coordinate axes.

Proposition 4.3

(1) Let μ *be a nonnegative regular Borel measure such that the strong centered Hardy–Littlewood maximal operator*

$$\mathcal{M}_{\mathcal{J}}\mu(\mathbf{x}) = \sup_{R\in\mathcal{J},\, R\ni\mathbf{x}} \frac{1}{|R|}\int_R d\mu$$

is finite for almost all $\mathbf{x} \in \mathbb{R}^n$; *here* R *runs through the collection* $\mathcal{J}$ *of rectangles containing additionally the point* x, *and* $|R|$ *denotes the Lebesgue measure of* R. *Then* $(\mathcal{M}_{\mathcal{J}}\mu)^\gamma \in A_1(\mathcal{J})$ *for all* $\gamma \in [0, 1)$.

(2) For all $1 < q < \tau$ *we have* $A_1(\mathcal{J}) \subset A_q(\mathcal{J}) \subset A_\tau(\mathcal{J})$.
(3) Let $1 < q < \infty$ *and* $w \in A_q(\mathcal{J})$. *Then there are* $w_1, w_2 \in A_1(\mathcal{J})$ *such that*

$$w = \frac{w_1}{w_2^{q-1}}.$$

Conversely, given $w_1, w_2 \in A_1(\mathcal{J})$, *the weight* $w = w_1 w_2^{1-q}$ *belongs to* $A_q(\mathcal{J})$.

□

For the proofs see [115, Chapter IV, §6]. The claim (3) is a variant of Jones' factorization theorem, see [115, Chapter IV, Theorem 6.8].

For a rapidly decreasing function $u \in \mathcal{S}(\mathbb{R}^n)$ let

$$\mathcal{F}u(\boldsymbol{\xi}) = \hat{u}(\boldsymbol{\xi}) = \frac{1}{(2\pi)^{n/2}} \int_{\mathbb{R}^n} \mathrm{e}^{-\mathrm{i}\mathbf{x}\cdot\boldsymbol{\xi}} u(\mathbf{x})\, \mathrm{d}\mathbf{x}, \quad \boldsymbol{\xi} \in \mathbb{R}^n,$$

be the Fourier transform of u. Its inverse will be denoted by $\mathcal{F}^{-1}$. Moreover, we define the centered Hardy–Littlewood maximal operator

$$\mathcal{M}u(\mathbf{x}) = \sup_{Q \ni \mathbf{x}} \frac{1}{|Q|} \int_Q |u(\mathbf{y})| \mathrm{d}\mathbf{y}, \quad \mathbf{x} \in \mathbb{R}^n,$$

for $u \in L^1_{\mathrm{loc}}(\mathbb{R}^n)$, where Q runs through the set of all closed cubes centered at $\mathbf{x}$.

Theorem 4.14 *Let* $1 < q < \infty$ *and* $w \in A_q$*:*

(i) The operator $\mathcal{M}$*, defined, e.g., on* $\mathcal{S}(\mathbb{R}^n)$*, is a bounded operator from* L^q_w *to* L^q_w.
(ii) Let $m \in C^n(\mathbb{R}^n \setminus \{0\})$ *satisfy the pointwise Hörmander–Mikhlin multiplier condition*

$$|\boldsymbol{\xi}|^{|\alpha|} \, |D^\alpha m(\boldsymbol{\xi})| \leq c_\alpha \quad \textit{for all } \boldsymbol{\xi} \in \mathbb{R}^n \setminus \{0\}$$

and all multi-indices $\alpha \in \mathbb{N}_0^n$ *with* $|\alpha| \leq n_1 \in \mathbb{N}$*, where* $n_1 \geq n/2$*. Then the multiplier operator* $u \mapsto \mathcal{F}^{-1}(m\hat{u})$*,* $u \in \mathcal{S}(\mathbb{R}^n)$*, can be extended to a bounded linear operator from* L^q_w *to* L^q_w.
(iii) Let m *be of class* C^n *in each "quadrant" of* $\mathbb{R}^n$ *and let a constant* $B \geq 0$ *exist such that* $\|m\|_\infty \leq B$*,*

$$\sup_{x_{k+1},\dots,x_n} \int_{\mathcal{I}} \left| \frac{\partial^k m(\mathbf{x})}{\partial x_1 \cdots \partial x_k} \right| dx_1 \cdots dx_k \leq B$$

for any dyadic interval $\mathcal{I}$ in $\mathbb{R}^k$, $1 \le k \le n$, and also for any permutation of the variables $x_1, \dots, x_k$ within $x_1, \dots, x_n$. If $1 < p < \infty$ and $w \in A_p(\mathbb{R}^n, \mathcal{J})$, then m defines a bounded multiplier operator from $L^p_w(\mathbb{R}^n)$ to $L^p_w(\mathbb{R}^n)$.

□

Proof of Theorem 4.14 (i) See [115, Theorem IV 2.8], [173, Theorem 9]. (ii) See [115, Theorem IV 3.9] or [158, Theorem 4]. Note that the pointwise condition on m implies the integral condition in [115], [158]. For the proof of (iii) see [158]. □

Concerning *one-sided weights* and *one-sided maximal operators* on the real line, see Definition 4.4, we first recall the following duality property: $w \in A_q^+$ if and only if $w^{-q'/q} = w^{-1/(q-1)} \in A_{q'}^-$. Moreover we will need the following results:

Theorem 4.15 (Theorem 1 of [187])
Let $1 < p < \infty$ and $p' = p/(p-1)$:

(i) Let $w_1 \in A_1^+$, $w_2 \in A_1^-$. Then $w_1/w_2^{p-1} \in A_p^+$. Conversely, given $w \in A_p^+$ there exist $w_1 \in A_1^+$, $w_2 \in A_1^-$ such that $w = w_1/w_2^{p-1}$.
(ii) The operator M^+ is continuous from $L^p_w(\mathbb{R})$ to itself if and only if $w \in A_p^+$. Analogously, $M^-: L^p_w(\mathbb{R}) \to L^p_w(\mathbb{R})$ if and only if $w \in A_p^-$.

□

Obviously, $A_p \subset A_p^\pm$ where A_p denotes the usual Muckenhoupt class on the real line. Hence $|\mathbf{x}|^\alpha$, $(1 + |\mathbf{x}|)^\alpha \in A_p^\pm$ if $-1 < \alpha < p-1$, $1 < p < \infty$. However, in view of the anisotropic weight $w = \eta_\beta^\alpha$ on $\mathbb{R}^3$, see (4.229), we have to consider also *one-dimensional anisotropic weight functions* such as

$$\widetilde{w}_{\alpha,\beta}(x) = \widetilde{w}_{\alpha,\beta}(x; r) = (r^2 + x^2)^{\alpha/2} \left(\sqrt{r^2 + x^2} - x\right)^\beta, \quad x \in \mathbb{R},\ r > 0. \tag{4.230}$$

Lemma 4.18

(i) *For every $r > 0$ the univariate weight $\widetilde{w}_{\alpha,\beta}(x; r)$ lies in A_1^- if and only if $\beta \ge 0$, $\alpha \le \beta$, and $\alpha + \beta > -1$. Moreover, the A_1^--constant of $\widetilde{w}_{\alpha,\beta}$ is uniformly bounded in r.*
(ii) For every $r > 0$ the univariate weight

$$w_{\alpha,\beta}(x) = w_{\alpha,\beta}(x; r) = (1 + r^2 + x^2)^{\alpha/2} \left(1 + \sqrt{r^2 + x^2} - x\right)^\beta$$

lies in A_1^- with an A_1^--constant independent of $r > 0$ if and only if

$$\alpha \le 0 \le \beta \quad \textit{and} \quad \alpha + \beta > -1. \tag{4.231}$$

(iii) Let $1 < p < \infty$. *Then for every* $r > 0$

$$w_{\alpha,\beta}(\cdot\,; r) \in A_p^+ \quad \text{for } \alpha > -1,\ \beta \le 0,\ \alpha + \beta < p - 1 \tag{4.232}$$
$$w_{\alpha,\beta}(\cdot\,; r) \in A_p^- \quad \text{for } \alpha < p - 1,\ \beta \ge 0,\ \alpha + \beta > -1.$$

Moreover, the $A_p^{\pm}$*-constant is uniformly bounded in* $r > 0$.

□

Proof of Lemma 4.18 (i) A simple scaling argument shows that it suffices to look at the weight $\widetilde{w} = \widetilde{w}_{\alpha,\beta}$ in (4.230) for $r = 1$ only and that the A_1^--constant is independent of $r > 0$. We will consider three cases.

Case 1: $x > 0$. Then $\widetilde{w}(x) \sim (1 + |x|)^{\alpha-\beta}$, i.e., there exists a constant $c > 0$ independent of $x > 0$ such that $1/c(1 + |x|)^{\alpha-\beta} \le \widetilde{w}(x) \le c(1 + |x|)^{\alpha-\beta}$ for all $x > 0$. Hence for all $h > 0$

$$\frac{1}{h}\int_x^{x+h} \widetilde{w}(t)\mathrm{d}t \sim \frac{1}{h}\int_x^{x+h} (1+t)^{\alpha-\beta}\mathrm{d}t.$$

If $\alpha - \beta > 0$, then the term on the right-hand side is strictly increasing to $+\infty$ as $h \to \infty$. Thus we are led to the condition $\alpha \le \beta$.

Now let $\alpha \le \beta$. Then for all $h > 0$

$$\frac{1}{h}\int_x^{x+h} (1+t)^{\alpha-\beta}\mathrm{d}t \le \frac{1}{h}\int_x^{x+h} (1+x)^{\alpha-\beta}\mathrm{d}t = (1+|x|)^{\alpha-\beta} \sim \widetilde{w}(x).$$

Case 2: $x < 0$ and $0 < h < |x|$. Then $\widetilde{w}(t) \sim (1 + |t|)^{\alpha+\beta}$ for all $t \in (x, x + h)$. Assume that $\alpha + \beta = -1$ and let $h = |x|$. Then

$$\frac{1}{|x|}\int_x^0 (1+|t|)^{-1}\mathrm{d}t = \frac{\log(1+|x|)}{|x|}$$

is not bounded by $c\widetilde{w}(x) = c/|x|$ uniformly in $x < 0$ for any constant $c > 0$. Analogously, if $\alpha + \beta < -1$, then for $h = |x|$ we see that $1/|x| \int_x^0 (1 + |t|)^{\alpha+\beta}\mathrm{d}t \sim 1/|x|$ is not bounded by $c\widetilde{w}(x) = c(1 + |x|)^{\alpha+\beta}$ uniformly in $x < 0$. Hence in the following we have to assume that $\alpha + \beta > -1$. We shall consider two subcases: $h > 0$ small with respect to $|x|$ and h comparable with $|x|$. If $0 < h < \frac{1}{2}|x|$, then

$$\frac{1}{h}\int_x^{x+h} (1+|t|)^{\alpha+\beta}\mathrm{d}t \sim \frac{1}{h}\int_x^{x+h} (1+|x|)^{\alpha+\beta}\mathrm{d}t = (1+|x|)^{\alpha+\beta} \sim \widetilde{w}(x).$$

For the second subcase assume that $\frac{1}{2}|x| < h < |x|$. Then we are led to the integral

$$\frac{1}{|x|}\int_x^{x+h}(1+|t|)^{\alpha+\beta}\mathrm{d}t \le \frac{1}{|x|}\int_x^0 (1+|t|)^{\alpha+\beta}\mathrm{d}t$$
$$\sim \begin{cases} \dfrac{(1+|x|)^{\alpha+\beta+1}}{|x|}, & |x| > 1 \\ 1, & |x| < 1 \end{cases} \sim \widetilde{w}(x).$$

Case 3: $x < 0$ and $h > |x|$. In this case we have to consider the sum

$$\frac{1}{h}\int_x^0 \widetilde{w}\mathrm{d}t + \frac{1}{h}\int_0^{x+h}\widetilde{w}\mathrm{d}t \le \frac{1}{|x|}\int_x^0 \widetilde{w}\mathrm{d}t + \frac{c}{h}\int_0^{x+h}(1+t)^{\alpha-\beta}\mathrm{d}t =: I_1 + I_2,$$

where the first integral I_1 is bounded by $c\widetilde{w}(x)$ uniformly in $x < 0$, see Case 2, and where for $|x| < 1$ the second integral I_2 is bounded by $c \sim \widetilde{w}(x)$. Therefore, let $|x| > 1$ in the following. If $\alpha - \beta \le -1$, then the condition $\alpha + \beta > -1$ implies that $\beta > 0$; moreover, I_2 is easily shown to be bounded by $c\widetilde{w}(x) \sim (1 + |x|)^{\alpha+\beta}$ uniformly in $x < 0$ and $h > |x|$.

Now consider the case $\alpha - \beta > -1$. We shall investigate three possibilities of the position of h with respect to $|x|$. If $h = 2|x|$, then

$$\frac{1}{|x|}\int_0^{|x|}(1+t)^{\alpha-\beta}\mathrm{d}t = \frac{c}{|x|}((1+|x|)^{\alpha-\beta+1} - 1).$$

Since $1/|x| = o(|x|^{\alpha+\beta}) = o(\widetilde{w}(x))$ by the condition that $\alpha+\beta > -1$, the assertion $I_2 \le c\widetilde{w}(x) \sim |x|^{\alpha+\beta}$ necessarily implies that $|x|^{\alpha-\beta} \le c|x|^{\alpha+\beta}$ for $|x| > 1$. Thus β must be nonnegative.

Next, if $|x| < h < 2|x|$, then, since $\alpha - \beta \le \alpha + \beta$ and $\alpha + \beta > -1$,

$$I_2 \le \frac{c}{|x|}\int_0^{|x|}(1+t)^{\alpha-\beta}\mathrm{d}t \le c|x|^{\alpha+\beta} \sim \widetilde{w}(x).$$

Finally, if $h > 2|x| > 2$, then

$$I_2 \le \frac{c}{h}(1 + x + h)^{\alpha-\beta+1} \le ch^{\alpha-\beta} \le c|x|^{\alpha+\beta} \sim \widetilde{w}(x)$$

since $\alpha \le \beta$ (see Case 1). Summarizing the previous cases and estimates, we have proved that there exists $c > 0$ such that $M^+\widetilde{w}(x) \le c\widetilde{w}(x)$ for a.a. $x \in \mathbb{R}$ and that this result holds if and only if $\beta \ge 0$, $\alpha \le \beta$, and $\alpha + \beta > -1$.

(ii) To verify the necessity of (4.231), let $r = 1$ and $w = w_{\alpha,\beta}$. For $x > 0$ when $(1 + \sqrt{r^2 + x^2} - x)^\beta \sim 1$, we have to estimate

$$\frac{1}{h}\int_x^{x+h} w(t)\mathrm{d}t \sim \frac{1}{h}\int_x^{x+h} (1+t)^\alpha \mathrm{d}t$$

by $cw(x) \sim (1 + x)^\alpha$. As in Case 1 of Part (i) (with $\beta = 0$) we get the necessary condition $\alpha \le 0$.

Let $x < 0$. Again we shall distinguish according to the size of h with respect to $|x|$. If $0 < h < |x|$, then $w(t) \sim (1 + |t|)^{\alpha+\beta}$ for all $t \in (x, x + h)$, and

$$\frac{1}{h}\int_x^{x+h} w(t)\mathrm{d}t \sim \frac{1}{h}\int_x^{x+h} (1+|t|)^{\alpha+\beta}\mathrm{d}t$$

is bounded by $cw(x) \sim (1 + |x|)^{\alpha+\beta}$ only when $\alpha + \beta > -1$; cf. Case 2 of Part (i). Finally, when $x < 0$ and $h > |x|$, say $h = 2|x| > 2$, and when $\alpha > -1$, then

$$\frac{1}{h}\int_x^{x+h} w(t)\mathrm{d}t \sim \frac{1}{h}\int_x^{0} (1+|t|)^{\alpha+\beta}\mathrm{d}t + \frac{1}{h}\int_0^{x+h} (1+t)^\alpha \mathrm{d}t \le cw(x) + c|x|^\alpha,$$

which is bounded by $cw(x) \sim (1+|x|)^{\alpha+\beta}$ only if $\beta \ge 0$. However, if $\alpha \le -1$, then the condition $\alpha + \beta > -1$ implies that even $\beta > 0$. Hence the conditions (4.231) are necessary to prove that $w \in A_1^-$.

We shall prove that conditions (4.231) are sufficient for $w_{\alpha,\beta}(\mathbf{x}; r) \in A_1^-$ with an A_1^--constant independent of $r > 0$. Let us assume that (4.231) holds and let first $0 < r < 1$. Then

$$\begin{aligned} w(t) &\sim (1 + |t|)^\alpha \cdot \begin{cases} 1, & t > 0 \\ (1 + |t|)^\beta, & t < 0 \end{cases} \\ &\sim (1 + |t|)^{\alpha+\beta/2} \cdot \begin{cases} (1 + |t|)^{-\beta/2}, & t > 0 \\ (1 + |t|)^{\beta/2}, & t < 0 \end{cases} \sim \widetilde{w}_{\alpha',\beta'}(t; r), \end{aligned}$$

where $\alpha' = \alpha + \beta/2$, $\beta' = \beta/2$. Since the assumptions (4.231) on α, β imply that α', β' satisfy the assumptions in (i), $w \in A_1^-$ with an A_1^--constant independent of $0 < r < 1$.

Next let $r \ge 1$. An elementary calculation shows that

$$w(t) \sim \begin{cases} \widetilde{w}_{\alpha,\beta}(t; r), & t < r^2 \\ \widetilde{w}_{\alpha,0}(t; r), & t > r^2 \end{cases}.$$

Then we will consider three cases.

Case 1: $x < r^2$ and $x + h < r^2$. In this case, by Part (i),

$$\frac{1}{h}\int_x^{x+h} w(t)\mathrm{d}t \sim \frac{1}{h}\int_x^{x+h} \widetilde{w}_{\alpha,\beta}(t;r)\mathrm{d}t \le c\widetilde{w}_{\alpha,\beta}(x;r) \sim cw(x)$$

with $c > 0$ independent of $r > 1$.

Case 2: $x > r^2$ and $x + h > r^2$. Now

$$\frac{1}{h}\int_x^{x+h} w(t)\mathrm{d}t \sim \frac{1}{h}\int_x^{x+h} \widetilde{w}_{\alpha,0}(t;r)\mathrm{d}t \le c\widetilde{w}_{\alpha,0}(x;r) \sim cw(x)$$

due to Case 1 in Part (i).

Case 3: $x < r^2$ but $x + h > r^2$. Then

$$\frac{1}{h}\int_x^{x+h} w(t)\mathrm{d}t \sim \frac{1}{h}\int_x^{r^2} \widetilde{w}_{\alpha,\beta}(t;r)\mathrm{d}t + \frac{1}{h}\int_{r^2}^{x+h} \widetilde{w}_{\alpha,0}(t;r)\mathrm{d}t.$$

By Part (i), the first integral on the right-hand side is bounded by

$$\frac{r^2 - x}{h}\widetilde{w}_{\alpha,\beta}(x;r) \le \widetilde{w}_{\alpha,\beta}(x;r) \le cw(x).$$

Hence it suffices to prove that

$$\frac{1}{h}\int_{r^2}^{x+h} \widetilde{w}_{\alpha,0}(t;r)\mathrm{d}t \le cw(x).$$

If $|x| \le r^2$, then Part (i) implies that

$$\frac{1}{h}\int_{r^2}^{x+h} \widetilde{w}_{\alpha,0}(t;r)\mathrm{d}t \le \frac{x+h-r^2}{h}\widetilde{w}_{\alpha,0}(r^2;r) \le \widetilde{w}_{\alpha,0}(r^2;r) \le cr^{2\alpha},$$

where $r^{2\alpha} \le (r + |x|)^\alpha \le cw(x)$ since $\alpha \le 0 \le \beta$.

If $x < -r^2$, then $w(x) \sim |x|^{\alpha+\beta}$, and a simple scaling argument and the condition $\beta \ge 0$ allow to reduce the problem to the case $r = 1$. Actually it suffices to show the existence of $c > 0$ such that

$$J := \int_1^{x+h} t^\alpha \mathrm{d}t \le ch|x|^{\alpha+\beta} \quad \text{when} \quad x \le -1,\ x + h \ge 1.$$

If $\alpha < -1$, then J is bounded by

$$\frac{1}{|\alpha + 1|} \le c|x|^{\alpha+\beta+1} \le ch|x|^{\alpha+\beta},$$

since $\alpha + \beta > -1$ and $h > |x| > 1$. In the case $\alpha = -1$ the integral J equals

$$\log(x+h) \sim \log h + \frac{x}{h} \le c(1 + h^{\min(\beta,1)}) \le ch|x|^{\beta-1},$$

since $\beta > -1 - \alpha = 0$. Finally, for $\alpha > -1$, we may bound J by $c(x+h)^{\alpha+1}$. If $1 < |x| < h < 2|x|$, this term is bounded by $c|x| \le ch|x|^{\alpha} \le ch|x|^{\alpha+\beta}$. In the remaining case when $h > 2|x|$, we get that $(x+h)^{\alpha+1} \le ch^{\alpha+1} \le ch|x|^{\alpha+\beta}$, since $\alpha \le 0 \le \beta$.

Now (ii) is completely proved.

(iii) By Theorem 4.15 (i) and Part (ii) of this Lemma

$$w(x) = \frac{(1 + r^2 + x^2)^{\alpha_1/2}}{(1 + r^2 + x^2)^{\alpha_2(p-1)/2}(1 + \sqrt{r^2 + x^2} - x)^{\beta_2(p-1)}} \in A_p^+$$

for all $\alpha_1, \alpha_2, \beta_2$ satisfying $-1 < \alpha_1 \le 0, \alpha_2 \le 0 \le \beta_2$, and $\alpha_2 + \beta_2 > -1$. Hence, with $\alpha = \alpha_1 - \alpha_2(p-1)$, $\beta = -\beta_2(p-1)$, we get that $w = w_{\alpha,\beta}(\cdot\,; r) \in A_p^+$ for all α, β satisfying $\alpha > -1$, $\beta \le 0$, and $\alpha + \beta < p - 1$. By analogy,

$$w(x) = \frac{(1 + r^2 + x^2)^{\alpha_1/2}\left(1 + \sqrt{r^2 + x^2} - \mathbf{x}\right)^{\beta_1}}{(1 + r^2 + x^2)^{\alpha_2(p-1)/2}} \in A_p^-$$

for all $\alpha_1, \alpha_2, \beta_1$ satisfying $\alpha_1 \le 0 \le \beta_1, \alpha_1 + \beta_1 > -1, \ -1 < \alpha_2 \le 0$. Hence $w = w_{\alpha,\beta}(\cdot\,; r) \in A_p^-$ for all α, β such that $\beta \ge 0$, $\alpha < p - 1$, and $\alpha + \beta > -1$. Moreover, in both cases the $A_p^{\pm}$-constant of the weight is uniformly bounded in $r > 0$. □

Note that the univariate weights $\widetilde{w}_{\alpha,\beta}$ and $w_{\alpha,\beta}$ mainly differ for large $x > 0$. While $\widetilde{w}_{0,\beta}$ decays as $(1/x)^{\beta}$ as $x \to \infty$ for every fixed $r > 0$, the weight $w_{0,\beta}$ is bounded below by 1 as $x \to \infty$. The reason to consider the weights $w_{\alpha,\beta}$ rather than $\widetilde{w}_{\alpha,\beta}$ is based on the use of the anisotropic weights η_{β}^{α} on $\mathbb{R}^3$, see Corollary 4.6, when fixing $r = |(x_1, x_2)|$, $x_1, x_2 \in \mathbb{R}$, so that $\eta_{\beta}^{\alpha}(x_1, x_2, x_3) = w_{\alpha,\beta}(x_3; r)$.

Let us recall the geometrical approach to the problem which now will be applied in weighted L^q-spaces. We introduce cylindrical coordinates $(r, x_3, \theta) \in (0, \infty) \times \mathbb{R} \times [0, 2\pi)$ and write $u(x_1, x_2, x_3) = u(r, x_3, \theta)$. Then the term $(\mathbf{e}_3 \wedge \mathbf{x}) \cdot \nabla \mathbf{u} = -x_2\partial_1\mathbf{u} + x_1\partial_2\mathbf{u}$ may be rewritten in the form $(\mathbf{e}_3 \wedge \mathbf{x}) \cdot \nabla \mathbf{u} = \partial_\theta \mathbf{u}$ using the angular derivative ∂_θ applied to $u(r, x_3, \theta)$. Working first of all formally or in the space $\mathcal{S}'(\mathbb{R}^3)$ of tempered distributions, we apply the Fourier transform $\mathcal{F} = \widehat{\ }$ to (4.223). With the Fourier variable $\boldsymbol{\xi} = (\xi_1, \xi_2, \xi_3) \in \mathbb{R}^3$ and $s = |\boldsymbol{\xi}|$, we get from (4.223)

$$(\nu s^2 + ik\xi_3)\widehat{u} - \widetilde{\omega}(\partial_\varphi \widehat{\mathbf{u}} - \mathbf{e}_3 \wedge \widehat{\mathbf{u}}) + i\xi\widehat{p} = \widehat{\mathbf{f}}, \quad i\xi \cdot \widehat{\mathbf{u}} = 0. \tag{4.233}$$

Here $(\mathbf{e}_3 \wedge \boldsymbol{\xi}) \cdot \nabla_{\boldsymbol{\xi}} = -\xi_2\partial/\partial\xi_1 + \xi_1\partial/\partial\xi_2 = \partial_\varphi$ is the angular derivative in Fourier space when using cylindrical coordinates $(s, \xi_3, \varphi,) \in \mathbb{R}_+ \times \mathbb{R} \times [0, 2\pi)$. Since

$i\boldsymbol{\xi} \cdot \widehat{\mathbf{u}} = 0$ implies $i\boldsymbol{\xi} \cdot (\partial_\varphi \widehat{\mathbf{u}} - \omega \times \widehat{\mathbf{u}}) = 0$, the unknown pressure p is given by $-|\boldsymbol{\xi}|^2 \widehat{p} = \mathrm{i}\boldsymbol{\xi} \cdot \widehat{f}$, i.e.,

$$\widehat{\nabla p}(\boldsymbol{\xi}) = \mathrm{i}\boldsymbol{\xi} \cdot \widehat{p} = \frac{(\boldsymbol{\xi} \cdot \widehat{\mathbf{f}})\boldsymbol{\xi}}{|\boldsymbol{\xi}|^2}.$$

Then the Hörmander–Mikhlin multiplier theorem on weighted L^q-spaces (Theorem 22 (ii)) yields for every weight $w \in A_q(\mathbb{R}^3, \mathcal{C})$ the estimate

$$\|\nabla p\|_{q,w} \le c\|\mathbf{f}\|_{q,w}, \tag{4.234}$$

where $c = c(q, w) > 0$; in particular $\nabla p \in L^q_w$. Hence $\mathbf{u}$ may be considered as a (solenoidal) solution of the reduced problem

$$-\nu\Delta\mathbf{u} + k\partial_3\mathbf{u} - \widetilde{\omega}(\partial_\theta\mathbf{u} - \mathbf{e}_3 \wedge \mathbf{u}) = \mathbf{F} := \mathbf{f} - \nabla p \quad \text{in } \mathbb{R}^3, \tag{4.235}$$

or—in Fourier space—

$$(\nu s^2 + ik\xi_3)\widehat{\mathbf{u}} - \widetilde{\omega}(\partial_\varphi\widehat{\mathbf{u}} - \mathbf{e}_3 \wedge \widehat{\mathbf{u}}) = \widehat{\mathbf{F}}.$$

As shown in [61] this inhomogeneous linear differential equation of first order with respect to φ has the unique 2π-periodic solution

$$\begin{aligned}\widehat{\mathbf{u}}(\xi) =& \frac{1}{1 - \mathrm{e}^{-2\pi(\nu|\boldsymbol{\xi}|^2 + \mathrm{i}k\xi_3)/\widetilde{\omega}}} \int_0^{2\pi/\widetilde{\omega}} \mathrm{e}^{-(\nu|\boldsymbol{\xi}|^2 + \mathrm{i}k\xi_3)t}\, O_\omega^{\mathrm{T}}(t)\mathcal{F}\mathbf{F}(O_\omega(t)\boldsymbol{\xi})\mathrm{d}t,\\ =& \int_0^\infty \mathrm{e}^{-\nu|\boldsymbol{\xi}|^2 t}\, O_\omega^{\mathrm{T}}(t)(\mathcal{F}\mathbf{F}(O_\omega(t) \cdot -kt\mathbf{e}_3))(\boldsymbol{\xi})\mathrm{d}t.\end{aligned} \tag{4.236}$$

Finally note that $\mathrm{e}^{-\nu|\boldsymbol{\xi}|^2 t}$ is the Fourier transform of the heat kernel $K(\mathbf{x}) = (4\pi\nu t)^{-3/2}\mathrm{e}^{-|\mathbf{x}|^2/4\nu t}$ yielding

$$\mathbf{u}(\mathbf{x}) = \int_0^\infty K * O_\omega^{\mathrm{T}}(t)\mathbf{F}(O_\omega(t) \cdot -kt\mathbf{e}_3)(\mathbf{x})\mathrm{d}t. \tag{4.237}$$

Since $\mathbf{F} = \mathbf{f} - \nabla p$ is solenoidal, the identity $i\boldsymbol{\xi} \cdot \widehat{\mathbf{F}} = 0$ easily implies that also $\mathbf{u}$ is solenoidal.

The main ingredients of the proof of Theorem 4.13 are a weighted version of Littlewood–Paley theory and a decomposition of the integral operator

$$\begin{aligned}T\mathbf{f}(\mathbf{x}) &= \int_0^\infty \widehat{\psi}_{\nu t}(\boldsymbol{\xi})O_\omega^{\mathrm{T}}(t)\mathcal{F}\mathbf{f}(O_\omega(t) \cdot -kt\mathbf{e}_3)(\boldsymbol{\xi})\frac{\mathrm{d}t}{t}\\ &= \int_0^\infty \widehat{\psi}_t(\boldsymbol{\xi})O_{\omega/\nu}^{\mathrm{T}}(t)\mathcal{F}\mathbf{f}\left(O_{\omega/\nu}(t) \cdot -\frac{k}{\nu}t\mathbf{e}_3\right)(\boldsymbol{\xi})\frac{\mathrm{d}t}{t},\end{aligned} \tag{4.238}$$

where

$$\widehat{\psi}(\boldsymbol{\xi}) = \frac{1}{(2\pi)^{3/2}} |\boldsymbol{\xi}|^2 e^{-|\boldsymbol{\xi}|^2} \quad \text{and} \quad \widehat{\psi}_t(\boldsymbol{\xi}) = \widehat{\psi}\left(\sqrt{t}\boldsymbol{\xi}\right), \quad t > 0, \tag{4.239}$$

are the Fourier transforms of the function $\psi = -\Delta E_1 \in \mathcal{S}(\mathbb{R}^3)$ and of $\psi_t(\mathbf{x}) = t^{-3/2}\psi(\mathbf{x}/\sqrt{t})$, $t > 0$, resp. Note that due to Theorem 4.13 it suffices to find an estimate of $\|\Delta\mathbf{u}\|_{q,w}$ in order to estimate all second order derivatives $\partial_j\partial_k\mathbf{u}$ of $\mathbf{u}$.

To decompose $\widehat{\psi}_t$ choose $\widetilde{\chi} \in C_0^\infty\left(\frac{1}{2}, 2\right)$ satisfying $0 \leq \widetilde{\chi} \leq 1$ and $\sum_{j=-\infty}^{\infty} \widetilde{\chi}(2^{-j}s) = 1$ for all $s > 0$. Then define χ_j, $j \in \mathbb{Z}$, by its Fourier transform

$$\widehat{\chi}_j(\boldsymbol{\xi}) = \widetilde{\chi}(2^{-j}|\boldsymbol{\xi}|), \quad \boldsymbol{\xi} \in \mathbb{R}^n,$$

yielding $\sum_{j=-\infty}^{\infty} \widehat{\chi}_j = 1$ on $\mathbb{R}^n \setminus \{0\}$ and

$$\operatorname{supp} \widehat{\chi}_j \subset A(2^{j-1}, 2^{j+1}) := \{\boldsymbol{\xi} \in \mathbb{R}^3 : 2^{j-1} \leq |\boldsymbol{\xi}| \leq 2^{j+1}\}. \tag{4.240}$$

Using χ_j, we define for $j \in \mathbb{Z}$

$$\psi^j = \frac{1}{(2\pi)^{3/2}} \chi_j * \psi, \quad (\widehat{\psi} = \widehat{\chi}_j \cdot \widehat{\psi}). \tag{4.241}$$

Obviously, $\sum_{j=-\infty}^{\infty} \psi^j = \psi$ on $\mathbb{R}^3$. Finally, in view of (4.238), (4.241), we define the linear operators

$$\begin{aligned} T_j f(\mathbf{x}) &= \int_0^\infty \widehat{\psi}^j_{\nu t}(\boldsymbol{\xi}) O_\omega^{\mathrm{T}}(t) \mathcal{F}\mathbf{f}(O_\omega(t) \cdot - kt\mathbf{e}_3)(\xi) \frac{\mathrm{d}t}{t} \\ &= \int_0^\infty \widehat{\psi}^j_t(\boldsymbol{\xi}) O_{\omega/\nu}^{\mathrm{T}}(t) \mathcal{F}\mathbf{f}\left(O_{\omega/\nu}(t) \cdot - \frac{k}{\nu} t\mathbf{e}_3\right)(\boldsymbol{\xi}) \frac{\mathrm{d}t}{t}. \end{aligned} \tag{4.242}$$

Since formally $T = \sum_{j=-\infty}^{\infty} T_j$, we have to prove that this infinite series converges even in the operator norm on L^q_w.

To introduce a *weighted Littlewood–Paley decomposition of* L^q_w, choose $\widetilde{\varphi} \in C_0^\infty(\frac{1}{2}, 2)$ such that $0 \leq \widetilde{\varphi} \leq 1$ and $\int_0^\infty \widetilde{\varphi}(s)^2 \mathrm{d}s/s = \frac{1}{2}$. Then define $\varphi \in \mathcal{S}(\mathbb{R}^3)$ by its Fourier transform $\widehat{\varphi}(\boldsymbol{\xi}) = \widetilde{\varphi}(|\boldsymbol{\xi}|)$ yielding for every $s > 0$

$$\widehat{\varphi}_s(\boldsymbol{\xi}) = \widetilde{\varphi}(\sqrt{s}|\boldsymbol{\xi}|), \quad \operatorname{supp} \widehat{\varphi}_s \subset A\left(\frac{1}{2\sqrt{2}}, \frac{2}{\sqrt{2}}\right) \tag{4.243}$$

and the normalization $\int_0^\infty \widehat{\varphi}_s(\boldsymbol{\xi})^2 \,\mathrm{d}s/s = 1$ for all $\boldsymbol{\xi} \in \mathbb{R}^n \setminus \{0\}$.

Theorem 4.16 *Let $1 < q < \infty$ and $w \in A_q(\mathbb{R}^3)$. Then there are constants $c_1, c_2 > 0$ depending on q, w and φ such that for all $\mathbf{f} \in L^q_w$*

$$c_1\|\mathbf{f}\|_{q,w} \le \left\| \left(\int_0^\infty |\varphi_s * \mathbf{f}(\cdot)|^2 \frac{ds}{s} \right)^{1/2} \right\|_{q,w} \le c_2\|\mathbf{f}\|_{q,w}, \tag{4.244}$$

where $\varphi_s \in \mathcal{S}(\mathbb{R}^n)$ is defined by (4.243). □

Proof of Theorem 4.16 See [186, Proposition 1.9, Theorem 1.10], and also [158]. □

4.4.2 *Proofs*

As a preliminary version of Theorem 4.13 we prove the following proposition. The extension to more general weights based on complex interpolation of L^q_w-spaces will be postponed to the end of Sect. 4.4.2.

Proposition 4.4 *Let the weight $w \in L^1_{loc}(\mathbb{R}^3)$ be independent of the angle θ and define $w_r(x_3) := w(x_1, x_2, x_3)$ for fixed $r = |(x_1, x_2)| > 0$. Assume that*

$$w \in \widetilde{A}^-_{q/2} \quad \text{if } q > 2, \tag{4.245}$$

$$w \in \widetilde{A}^-_1 \text{ or } \frac{1}{w} \in \widetilde{A}^+_1 \quad \text{if } q = 2,$$

$$w^{2/(2-q)} \in \widetilde{A}^-_{q/(2-q)} \quad \text{if } 1 < q < 2.$$

Then the linear operator T defined by (4.238) *satisfies the estimate*

$$\|T\mathbf{f}\|_{q,w} \le c\|\mathbf{f}\|_{q,w} \quad \text{for all } \mathbf{f} \in \boldsymbol{L}^q_w \tag{4.246}$$

with a constant $c = c(q, w) > 0$ independent of $\mathbf{f}$. □

Proof of Proposition 4.4 *Step 1.* First we consider the case $q > 2$, $w \in \widetilde{A}^-_{q/2} \subset A_q$, and define the sublinear operator $\mathcal{M}^j$, a modified maximal operator, by

$$\mathcal{M}^j g(\mathbf{x}) = \sup_{s>0} \int_{A_s} (|\psi^j_t| * |g|) \left(O^{\mathrm{T}}_{\Omega/\nu}(t)\mathbf{x} + \frac{k}{\nu} t\mathbf{e}_3 \right) \frac{\mathrm{d}t}{t}, \tag{4.247}$$

where $A_s = [s/16, 16s]$. Then we will prove the preliminary estimate

$$\|T_j\mathbf{f}\|_{q,w} \le c\|\psi^j\|_1^{1/2} \|\mathcal{M}^j\|^{1/2}_{L^{(q/2)'}_v} \|\mathbf{f}\|_{q,w}, \quad j \in \mathbb{Z}, \tag{4.248}$$

where v denotes the θ-independent weight

$$v = w^{-(q/2)'/(q/2)} = w^{-2/(q-2)} \in \widetilde{A}^+_{(q/2)'} = \widetilde{A}^+_{q/(q-2)}. \tag{4.249}$$

To prove (4.248) we use the Littlewood–Paley decomposition of L^q_w, see (4.244), applied to $T_j\mathbf{f}$. By a duality argument, we find some function $0 \le g \in L_v^{(q/2)'} = \big(L_w^{(q/2)}\big)^*$ with $\|g\|_{(q/2)',v} = 1$ such that

$$\Big\| \int_0^\infty |\varphi_s * T_j\mathbf{f}(\cdot)|^2 \frac{\mathrm{d}s}{s} \Big\|_{q/2,w} = \int_0^\infty \int_{\mathbb{R}^3} |\varphi_s * T_j\mathbf{f}(x)|^2 g(\mathbf{x})\mathrm{d}\mathbf{x}\, \frac{\mathrm{d}s}{s}. \tag{4.250}$$

To estimate the right-hand side of (4.250) note that

$$\varphi_s * T_j\mathbf{f}(\mathbf{x}) = \int_0^\infty O^{\mathrm{T}}_{\omega/\nu}(t)(\varphi_s * \psi_t^j * \mathbf{f})\Big(O_{\omega/\nu}(t)\mathbf{x} - \frac{k}{\nu}t\mathbf{e}_3\Big)\frac{\mathrm{d}t}{t},$$

where $\varphi_s * \psi_t^j = 0$ unless $t \in A(s, j) := [2^{2j-4}s, 2^{2j+4}s]$. Since $\int_{t\in A(s,j)} \mathrm{d}t/t = \log 2^8$ for every $j \in \mathbb{Z}$, $s > 0$, we get by the inequality of Cauchy–Schwarz and the associativity of convolutions that

$$\begin{aligned}
|\varphi_s * T_j\mathbf{f}(\mathbf{x})|^2 &\le c \int_{A(s,j)} \Big|(\psi_t^j * (\varphi_s * \mathbf{f}))\Big(O_{\omega/\nu}(t)\mathbf{x} - \frac{k}{\nu}t\mathbf{e}_3\Big)\Big|^2 \frac{\mathrm{d}t}{t} \\
&\le c\|\psi^j\|_1 \int_{A(s,j)} (|\psi_t^j| * |\varphi_s * \mathbf{f}|^2)\Big(O_{\omega/\nu}(t)\mathbf{x} - \frac{k}{\nu}t\mathbf{e}_3\Big)\frac{\mathrm{d}t}{t};
\end{aligned}$$

here we used the estimate $|(\psi_t^j * (\varphi_s * \mathbf{f}))(\mathbf{y})|^2 \le \|\psi_t^j\|_1(|\psi_t^j| * |\varphi_s * \mathbf{f}|^2)(\mathbf{y})$ and the identity $\|\psi_t^j\|_1 = \|\psi^j\|_1$, see Lemma 4.1 or Lemma 3.1.

Thus

$$\begin{aligned}
&|T_j\mathbf{f}\|^2_{q,w} \\
&\le c\|\psi^j\|_1 \int_0^\infty \int_{A(s,j)} \int_{\mathbb{R}^3} (|\psi_t^j| * |\varphi_s * \mathbf{f}|^2)\Big(O_{\omega/\nu}(t)x - \frac{k}{\nu}t\mathbf{e}_3\Big) g(\mathbf{x})\mathrm{d}\mathbf{x}\frac{\mathrm{d}t}{t}\frac{\mathrm{d}s}{s} \\
&\le c\|\psi^j\|_1 \int_0^\infty \int_{A(s,j)} \int_{\mathbb{R}^3} (|\psi_t^j| * |\varphi_s * \mathbf{f}|^2)(\mathbf{x}) g\Big(O^{\mathrm{T}}_{\omega/\nu}(t)\mathbf{x} + \frac{k}{\nu}te_3\Big)\mathrm{d}x\frac{\mathrm{d}t}{t}\frac{\mathrm{d}s}{s} \\
&\le c\|\psi^j\|_1 \int_{\mathbb{R}^3}\int_0^\infty |\varphi_s * \mathbf{f}|^2(\mathbf{x}) \int_{A(s,j)} (|\psi_t^j| * g)\Big(O^{\mathrm{T}}_{\omega/\nu}(t)\mathbf{x} + \frac{k}{\nu}t\mathbf{e}_3\Big)\frac{\mathrm{d}t}{t}\frac{\mathrm{d}s}{s}\mathrm{d}\mathbf{x},
\end{aligned} \tag{4.251}$$

since ψ_t^j is radially symmetric. By definition of $\mathcal{M}^j$ the innermost integral is bounded by $\mathcal{M}^j g(\mathbf{x})$ uniformly in $s > 0$. Hence we may proceed in (4.251) using

Hölder's inequality as follows:

$$\begin{aligned}\|T_j\mathbf{f}\|_{q,w}^2 &\le c\|\psi^j\|_1 \int_{\mathbb{R}^3}\left(\int_0^\infty |\varphi_s * \mathbf{f}|^2(\mathbf{x})\,\frac{\mathrm{d}s}{s}\right)\mathcal{M}^j g(\mathbf{x})\mathrm{d}\mathbf{x}\\ &\le c\|\psi^j\|_1 \left\|\int_0^\infty |\varphi_s * \mathbf{f}|^2(\mathbf{x})\,\frac{\mathrm{d}s}{s}\right\|_{q/2,w}\|\mathcal{M}^j g\|_{(q/2)',v}.\end{aligned} \tag{4.252}$$

Now (4.244) and the normalization $\|g\|_{(q/2)',v} = 1$ complete the proof of (4.248).

Step 2. We estimate $\|\mathcal{M}^j g\|_{(q/2)',v}$. For functions γ depending on θ, x_3 only, let $\mathcal{M}_{\mathrm{hel}}$ denote the "helical" maximal operator

$$\mathcal{M}_{\mathrm{hel}}\gamma(\theta, x_3) = \sup_{s>0}\frac{1}{s}\int_{A_s}|\gamma|\left(\theta - \frac{\omega}{\nu}t, x_3 + \frac{k}{\nu}t\right)\mathrm{d}t,$$

where $A_s = [s/16, 16s]$. Then, writing $p := (q/2)'$, we claim that

$$\mathcal{M}^j g(\mathbf{x}) \le c2^{-2|j|}\mathcal{M}(\mathcal{M}_{\mathrm{hel}}g)(\mathbf{x}) \quad \text{for a.a. } \mathbf{x} \in \mathbb{R}^n, \tag{4.253}$$

$$\|\mathcal{M}^j g\|_{p,v} \le c2^{-2|j|}\|g\|_{p,v}, \tag{4.254}$$

where in (4.253) $\mathcal{M}_{\mathrm{hel}}g$ is considered as $\mathcal{M}_{\mathrm{hel}}g(r,\cdot,\cdot)$ for almost all $r > 0$.

To prove (4.253) we use the pointwise estimate $|\psi_t^j(x)| \le c2^{-2|j|}h_{t2^{-2j}}(\mathbf{x})$, see Lemma 3.1 (ii). Hence

$$\mathcal{M}^j g(x) \le c2^{-2|j|}\sup_{s>0}\int_{A_s}(h_{t2^{-2j}} * |g|)\left(O^{\mathrm{T}}_{\Omega/\nu}(t)\mathbf{x} + \frac{k}{\nu}t\mathbf{e}_3\right)\frac{\mathrm{d}t}{t}.$$

Moreover, there exists a constant $c > 0$ independent of $s > 0$, $j \in \mathbb{Z}$, such that $h_{t2^{-2j}} \le ch_{s2^{-2j}}$ for all $t \in A_s$. Consequently,

$$\begin{aligned}\mathcal{M}^j g(\mathbf{x}) &\le c2^{-2|j|}\sup_{s>0}h_{s2^{-2j}} * \int_{A_s}|g|\left(O^{\mathrm{T}}_{\Omega/\nu}(t)\mathbf{x} + \frac{k}{\nu}t\mathbf{e}_3\right)\frac{\mathrm{d}t}{t}\\ &\le c2^{-2|j|}\sup_{t>0}h_t * \mathcal{M}_{\mathrm{hel}}g(\mathbf{x}).\end{aligned}$$

Since h is nonnegative, radially decreasing, and $\|h_t\|_1 = \|h\|_1 = c_0 > 0$ for all $t > 0$, a well-known convolution estimate, see [205], II §2.1, yields the pointwise estimate (4.253).

Step 3. Note that up to now we have not yet used any specific properties of the weight $v \in A_p$. To estimate $\mathcal{M}_{\mathrm{hel}}g$ we shall work with a suitable one-sided maximal operator since our weight belongs to a Muckenhoupt class in $\mathbb{R}^3$, but a problem occurs when the weight is considered with respect to x_3 only. This naturally corresponds to the physical circumstances of the problem, where in the Oseen case the wake should appear. To estimate $\mathcal{M}_{\mathrm{hel}}g$ we write $g_r(\theta, x_3) = g(r, \theta, x_3) = g(\mathbf{x})$

and $v_r(x_3) = v(\mathbf{x})$, $r = |(x_1, x_2)| > 0$, for the θ-independent weight v. Then by the 2π-periodicity of g_r and v_r with respect to θ, we get for almost all $r > 0$

$$\begin{aligned}
&\int_{\mathbb{R}} \int_0^{2\pi} \mathcal{M}_{\mathrm{hel}} g_r(\theta, x_3)^p v_r(x_3)\, \mathrm{d}\theta\, \mathrm{d}x_3 \\
&\leq \int_{\mathbb{R}} \int_0^{2\pi} \left| \sup_{s>0} \frac{1}{s} \int_0^{16s} |g_r| \left(\theta - \frac{\omega}{k}\left(x_3 + \frac{k}{\nu} t \right), x_3 + \frac{k}{\nu} t \right) \mathrm{d}t \right|^p v_r(x_3)\, \mathrm{d}\theta\, \mathrm{d}x_3 \\
&= \int_{\mathbb{R}} \int_0^{2\pi} \left| \sup_{s>0} \frac{1}{s} \int_0^{16s} \gamma_{r,\theta} \left(x_3 + \frac{k}{\nu} t \right) \mathrm{d}t \right|^p \mathrm{d}\theta\, v_r(x_3)\, \mathrm{d}x_3 \\
&= 16 \int_0^{2\pi} \int_{\mathbb{R}} |M^+ \gamma_{r,\theta}(x_3)|^p v_r(x_3) \mathrm{d}x_3 \mathrm{d}\theta,
\end{aligned}$$

where $\gamma_{r,\theta}(y_3) = |g_r|(\theta - (\Omega/k)y_3, y_3)$ and M^+ denotes the one-sided maximal operator, see Definition 4.4. Since $w_r \in A^-_{q/2}$, by (4.249) and Theorem 4.15 (i) $v_r = w_r^{-(q/2)'/(q/2)} \in A^+_{(q/2)'} = A^+_p$ with an A^+_p-constant uniformly bounded in $r > 0$. Then Theorem 4.15 (ii) yields the estimate

$$\begin{aligned}
&\int_{\mathbb{R}} \int_0^{2\pi} \mathcal{M}_{\mathrm{hel}} g_r(\theta, x_3)^p v_r(x_3)\, \mathrm{d}\theta\, \mathrm{d}x_3 \\
&\qquad \leq c \int_0^{2\pi} \int_{\mathbb{R}} |\gamma_{r,\theta}(x_3)|^p v_r(x_3)\, \mathrm{d}x_3\, \mathrm{d}\theta = c \|g_r\|^p_{L^p(\mathbb{R}\times(0,2\pi), v_r(x_3))},
\end{aligned}$$

where $c > 0$ is independent of k, ν. Integrating with respect to $r\, dr$, $r \in (0, \infty)$, Fubini's theorem allows to consider an extension of $\mathcal{M}_{\mathrm{hel}}$ to a bounded operator from $L^p_v(\mathbb{R}^3)$ to itself with an operator norm bounded uniformly in k, ν. Moreover, $\mathcal{M}\colon L^p_v(\mathbb{R}^3) \to L^p_v(\mathbb{R}^3)$ is bounded by Theorem 4.15 (ii). Hence, (4.253) implies (4.254), and by (4.248) as well as Lemma 3.1 (ii), we get the estimate

$$\|T_j \mathbf{f}\|_{q,w} \leq c 2^{-2|j|} \|\mathbf{f}\|_{q,w}$$

for all $\mathbf{f} \in \boldsymbol{L}^q_w(\mathbb{R}^3)$ with a constant $c > 0$ independent of $j \in \mathbb{Z}$. Summarizing the previous inequalities, we proved (4.246) for $q > 2$.

Step 4. Now let $q = 2$, $w \in \widetilde{A}^-_1$. In this case the Littlewood–Paley decomposition of $T_j f$ in L^2_w implies that

$$\|T_j \mathbf{f}\|^2_{2,w} \leq c \int_0^{\infty} \int_{\mathbb{R}^n} |\varphi_s * T_j f|^2(\mathbf{x}) g(\mathbf{x}) \mathrm{d}\mathbf{x} \frac{\mathrm{d}s}{s},$$

where

$$g \in L^\infty_v, \quad v = \frac{1}{w} \quad \text{and} \quad \|g\|_{\infty,v} = \operatorname{esssup}_{\mathbb{R}^3} |gv| = 1.$$

By the same reasoning as before we arrive at (4.248), i.e.,

$$\|T_j\mathbf{f}\|_{2,w} \le c2^{-|j|}\|\mathcal{M}^j g\|_{\infty,v}^{1/2}\|\mathbf{f}\|_{2,w}, \tag{4.255}$$

and at (4.253). Concerning $\mathcal{M}_{\text{hel}}$ we use the pointwise estimate $g_r(\theta, x_3) \le w_r(x_3)$ for a.a. $\theta \in (0, 2\pi)$, $x_3 \in \mathbb{R}$, and get that

$$\mathcal{M}_{\text{hel}}g_r(\theta, x_3) \le \sup_{s>0}\frac{1}{s}\int_0^{16s} w_r\left(x_3 + \frac{k}{\nu}t\right)\mathrm{d}t \le 16M^+ w_r(x_3) \le cw_r(x_3)$$

with a constant $c > 0$ independent of $r > 0$. Since w is an $A_1(\mathbb{R}^3)$-weight, (4.253) implies that

$$\mathcal{M}^j g(\mathbf{x}) \le c2^{-2|j|}\mathcal{M}w(\mathbf{x}) \le c2^{-2|j|}w(\mathbf{x})$$

and consequently that $\|\mathcal{M}^j g\|_{\infty,v} \le c2^{-2|j|}$ with a constant $c > 0$ independent of $j \in \mathbb{Z}$. Hence $\|T_j\mathbf{f}\|_{2,w} \le c2^{-2|j|}$ proving (4.246) when $q = 2$.

Step 5. The remaining estimates are proved by duality arguments. Obviously the dual operator to T is defined by

$$T^*\mathbf{f}(\mathbf{x}) = \int_0^\infty (-\Delta)O_\Omega(t)E_t * \mathbf{f}(O_\Omega^{\mathrm{T}}(t)\mathbf{x} + kt\mathbf{e}_3)\mathrm{d}t,$$

which has the same structure as K, but with an "opposite orientation." Hence T^* is bounded on L^q_w for $q \ge 2$ and all weights $w \in \widetilde{A}^+_{q/2}$. Now let $1 < q < 2$ and $w^{2/(2-q)} \in \widetilde{A}^-_{q/(2-q)} = \widetilde{A}^-_{(q'/2)'}$. Then by simple duality arguments $w' = w^{-q'/q} \in \widetilde{A}^+_{(q'/2)}$ and

$$|\langle T\mathbf{f}, \mathbf{g}\rangle| = |\langle \mathbf{f}, T^*\mathbf{g}\rangle| \le \|\mathbf{f}\|_{q,w}\|T^*\mathbf{g}\|_{q',w'} \le c\|\mathbf{f}\|_{q,w}\|\mathbf{g}\|_{q',w'}.$$

Finally let $q = 2$ and $1/w \in \widetilde{A}^+_1$. As before,

$$|\langle T\mathbf{f}, g\rangle| \le \|\mathbf{f}\|_{2,w}\|T^*g\|_{2,1/w} \le c\|\mathbf{f}\|_{2,w}\|g\|_{2,1/w}.$$

Now Proposition 4.4 is completely proved. □

Lemma 4.19 ([14]) *Let* $1 \le p_1, p_2 < \infty$, *let* $0 < w_1, w_2$ *be weight functions,* $\delta \in (0, 1)$, *and*

$$\frac{1}{p} = \frac{1-\delta}{p_1} + \frac{\delta}{p_2}, \qquad w^{1/p} = w_1^{(1-\delta)/p_1} \cdot w_2^{\delta/p_2}.$$

Then

$$[L^{p_1}_{w_1}, L^{p_2}_{w_2}]_\delta = L^p_w$$

in the sense of complex interpolation. □

In the following we shall derive an anisotropic variant of Jones's factorization theorem tailored to our situation, when we need to work with one-sided Muckenhoupt weights with respect to x_3, satisfying the usual Muckenhoupt condition in three dimensions.

Lemma 4.20 (Anisotropic Version of Jones' Factorization Theorem) *Suppose that $w \in \widetilde{A}_q^-$. Then there exist weights $w_1 \in \widetilde{A}_1^-$ and $w_2 \in \widetilde{A}_1^+$ such that*

$$w = \frac{w_1}{w_2^{q-1}}.$$

Here $\widetilde{A}_1^+$ is defined by analogy with $\widetilde{A}_1^-$, cf. Definition 4.5, by assuming for $w_2 \in \widetilde{A}_1^+$ that $(w_2)_r \in A_1^+$ with A_1^+-constant uniformly bounded in $r > 0$. An analogous result holds for $w \in \widetilde{A}_q^+$. □

Proof of Lemma 4.20 Let $q \geq 2$. Given $w \in \widetilde{A}_q^-$, we consider the operator T defined by

$$\begin{aligned} T\mathbf{f} =& (w^{-1/q}\mathcal{M}(\mathbf{f}^{q/q'}w^{1/q}))^{q'/q} + w^{1/q}\mathcal{M}(\mathbf{w}^{-1/q}) \\ &+ (w^{-1/q}M_1^+(\mathbf{f}_r^{q/q'}w_r^{1/q}))^{q'/q} + w^{1/q}M_1^-(\mathbf{f}_r w_r^{-1/q}), \end{aligned}$$

where $r = |(x_1, x_2)|$. Then for all $\mathbf{f} \in L^q(\mathbb{R}^3)$

$$\begin{aligned} \|T\mathbf{f}\|_q^q \leq c \Big\{ & \int_{\mathbb{R}^3} w^{-q'/q}\big(\mathcal{M}(\mathbf{f}^{q/q'}w^{1/q})\big)^{q'}\mathrm{d}\mathbf{x} + \int_{\mathbb{R}^3} w(\mathcal{M}(w^{-1/q}))^q \mathrm{d}\mathbf{x} \\ &+ \int_{\mathbb{R}^2}\left(\int_{\mathbb{R}} w_r^{-q'/q}(M_1^+(\mathbf{f}_r^{q/q'}w_r^{1/q}))^{q'}\mathrm{d}x_3\right)\mathrm{d}(x_1,x_2) \\ &+ \int_{\mathbb{R}^2}\left(\int_{\mathbb{R}} w_r(M_1^+(\mathbf{f}_r w_r^{-1/q}))^q\mathrm{d}x_3\right)\mathrm{d}(x_1,x_2)\Big\} \\ \leq & A^q\|\mathbf{f}\|_q^q \end{aligned}$$

with a constant $A = A(q, w) > 0$.

Let us fix a nonnegative θ-independent function $\mathbf{f} \in L^q(\mathbb{R}^3)$ with $\|\mathbf{f}\|_q = 1$ and define

$$\eta = \sum_{k=1}^{\infty}(2A)^{-k}T^k(\mathbf{f}),$$

where $T^k(\mathbf{f}) = T(T^{k-1}(\mathbf{f}))$. Obviously $T\mathbf{f}$ and therefore also η are θ-independent. Moreover, $\eta \in L^q(\mathbb{R}^3)$ and $\|\eta\|_q \le \sum_{k=1}^{\infty} 2^{-k} = 1$. In particular, $\eta(x) < \infty$ for a.a. $\mathbf{x} \in \mathbb{R}^3$, $\eta_r(\cdot) \in L^q(\mathbb{R})$ for a.a. $(x_1, x_2) \in \mathbb{R}^2$, and $\eta_r(x_3) < \infty$ for a.a. $x_3 \in \mathbb{R}$. Since T is subadditive and positivity preserving, we get the pointwise inequality

$$T\eta \le \sum_{k=1}^{\infty}(2A)^{-k}T^{k+1}(\mathbf{f}) = \sum_{k=2}^{\infty}(2A)^{1-k}T^k(\mathbf{f}) \le (2A)\eta.$$

Now let $w_1 := w^{1/q}\eta^{q/q'}$ and $w_2 := w^{-1/q}\eta$ such that $w = w_1/w_2^{q-1}$. Then

$$\begin{aligned}
\mathcal{M}(w_1) &\le w^{1/q}(T\eta)^{q/q'} \le w^{1/q}\eta^{q/q'}(2A)^{q/q'} = (2A)^{q/q'}w_1 \\
M_1^+((w_1)_r) &\le w^{1/q}(T\eta)^{q/q'} \le w^{1/q}\eta^{q/q'}(2A)^{q/q'} = (2A)^{q/q'}(w_1)_r \\
\mathcal{M}(w_2) &\le w^{-1/q}T(\eta) \le w^{-1/q}\eta\, 2A = 2Aw_2 \\
M_1^-((w_2)_r) &\le w^{-1/q}T(\eta) \le w^{-1/q}\eta\, 2A = 2A(w_2)_r
\end{aligned}$$

proving that $w_1 \in \widetilde{A}_1^-$, $w_2 \in \widetilde{A}_1^+$.

The case $1 \le q < 2$ follows by a simple duality argument, since $w \in \widetilde{A}_q^-$ is equivalent to $w^{-q'/q} \in \widetilde{A}_{q'}^+$. □

Proof of Theorem 4.13 (i) Let $q \in (1, \infty)$ and $w \in A_q$ such that the L^q_w-estimate of ∇p holds, see (4.234). Hence it suffices to consider u defined by (4.236)–(4.237). We consider arbitrary $q_1, q_2 \in (1, \infty)$ and $\delta \in (0, 1)$ with

$$1 < q_1 < q < q_2 < \infty, \quad q_1 \le 2 \le q_2 \quad \text{and} \quad \frac{1}{q} = \frac{1-\delta}{q_1} + \frac{\delta}{q_2}, \tag{4.256}$$

and assume that $w^\tau \in \widetilde{A}^-_{\tau q/2}$ with $\tau = 2/(2 - q(1 - \delta)) \in [1, \infty)$. By Lemma 4.20 there exist weights $u \in \widetilde{A}_1^-$, $v \in \widetilde{A}_1^+$ such that

$$w^\tau = \frac{u}{v^{\tau q/2-1}} = \frac{u}{v^{q/(2-q(1-\delta))-1}}.$$

Then we define the weights w_1, w_2 by

$$w_1^{2/(2-q_1)} = \frac{u}{v^{2(q_1-1)/(2-q_1)}} \quad \text{and} \quad w_2 = \frac{u}{v^{(q_2-2)/2}}$$

yielding

$$w_1^{2/(2-q_1)} \in \widetilde{A}^-_{q_1/(2-q_1)}, \quad w_2 \in \widetilde{A}^-_{q_2/2}.$$

Since, due to an elementary calculation, $w = w_1^{q(1-\delta)/q_1} w_2^{q\delta/q_2}$, Lemma 4.20 and Proposition 4.4 prove that T is bounded on $L^q_w(\mathbb{R}^3)$. Since $u_1 \in \widetilde{A}_1^-$, $v_1 \in \widetilde{A}_1^+$ are arbitrary, we proved the boundedness of T on L^q_w for arbitrary w if

$$w^\tau \in \widetilde{A}^-_{\tau q/2}, \quad \tau = \frac{2}{2 - q(1-\delta)} \in [1, \infty).$$

Now we have to find all admissible τ subject to the restrictions given by (4.256). For this reason consider the easier term

$$s = 2\left(1 - \frac{1}{\tau}\right) = q(1-\delta) = q\frac{\frac{1}{q} - \frac{1}{q_2}}{\frac{1}{q_1} - \frac{1}{q_2}}.$$

First case: $1 < q < 2$, in which $1 < q_1 < q$ and $q_2 \geq 2$. Due to monotonicity properties of s as a function of $1/q_1$ and of $1/q_2$, it suffices to check s at the corners of the rectangle $(1/q, 1) \times (0, 1/2]$. The corresponding function values are q, 1, and $2-q$. Hence the range of s equals the interval $(2-q, q)$ yielding for $\tau = 2/(2-s)$ the condition

$$\frac{2}{q} < \tau < \frac{2}{2-q}.$$

Note that the limiting value $\tau = 2/(2-q)$ is allowed due to Proposition 4.4. Finally the condition $w^\tau \in \widetilde{A}^-_{\tau q/2}$, $2/q < \tau \leq 2/(2-q)$, easily implies that $w \in A_q$: By Lemma 4.20 there exist $u_1 \in \widetilde{A}_1^-$, $v_1 \in \widetilde{A}_1^+$ such that

$$w = \frac{u_1^{1/\tau}}{v_1^{q/2-1/\tau}}, \tag{4.257}$$

where $u_1^{1/\tau} \in \widetilde{A}_1^-$ and $q/2 - 1/\tau \leq q - 1$ yielding $v_1^{(q/2-1/\tau)/(q-1)} \in \widetilde{A}_1^+$.
Second case: $q > 2$, in which $1 < q_1 \leq 2$ and $q_2 > q$. In this case the values of s at the corners of the rectangle $[1/2, 1) \times (0, 1/q)$ in the $(1/q_1, 1/q_2)$-plane are 0, 1, and 2. Hence

$$1 < \tau < \infty,$$

and we observe that $\tau = 1$ is admissible due to Proposition 4.4. Finally note that the condition $w^\tau \in A_{\tau q/2}$ implies also $w \in \widetilde{A}_q^-$: There exist $u_1 \in \widetilde{A}_1^-$, $v_1 \in \widetilde{A}_1^+$ such that w satisfies (4.257), where again $q/2 - 1/\tau + 1 \leq q$ for all $\tau \in (1, \infty)$.
Third case: $q = 2$. In this case it suffices to interpolate between $L^2_{w_1}$ and $L^2_{w_2}$, where $w_1 \in \widetilde{A}_1^-$ and $1/w_2 \in \widetilde{A}_1^+$, see Proposition 4.4. Then T is bounded on L^2_w for all

$$w = \frac{w_1^{1-\delta}}{w_2^\delta}, \quad 0 < \delta < 1.$$

Then $w^{1/(1-\delta)} = w_1/w_2^{\delta/(1-\delta)}$, or with $\tau = 1/(1-\delta) \in (1,\infty)$,

$$w^\tau = \frac{w_1}{w_2^{\tau-1}} \in \widetilde{A}_\tau^- = \widetilde{A}_{\tau q/2}^-.$$

(ii) Note that $L^{q_i}_{w_i}(\mathbb{R}^n) \subset S'(\mathbb{R}^n)$, $i = 1,2$; indeed, $w_i \in L^1_{\mathrm{loc}}(\mathbb{R}^n)$ and $\int_{|\mathbf{x}|\geq 1} w_i(\mathbf{x})|\mathbf{x}|^{-nq_i}\,\mathrm{d}\mathbf{x} < \infty$, see [115, IV.3 (3.2)]. Since Eq. (4.223) is linear, it suffices to consider $\mathbf{f} = 0$ and a solution $u \in S'(\mathbb{R}^n)^n$ of (4.227). In the proof of [70], Theorem 1.1 (2), (3), it was shown that under these assumptions $\mathbf{u}$ is a polynomial and that $\mathbf{u}(\mathbf{x}) = \alpha\boldsymbol{\omega} + \beta\boldsymbol{\omega}\wedge\mathbf{x} + \gamma(x_1, x_2, -2x_3)^{\mathrm{T}}$, $\alpha, \beta, \gamma \in \mathbb{R}$ ($\mathbf{u}(\mathbf{x}) = \beta(-x_2, x_1)$ if $n = 2$). □

Proof of Corollary 4.6 Considering a priori estimates for $\partial u/\partial x_3$ we use the representation (4.236) of u. In order to analyze the dependence of the following estimates on the parameters k, ν and $\widetilde{\Omega}$, let

$$k' = k/\widetilde{\omega}, \quad \nu' = \nu/\widetilde{\omega} \quad \text{and} \quad D(\boldsymbol{\xi}) = 1 - \mathrm{e}^{-2\pi(\nu'|\boldsymbol{\xi}|^2 + \mathrm{i}k'\xi_3)}.$$

Then for $\mathbf{f} \in \mathcal{S}(\mathbb{R}^3)^3$ we get the identity

$$\widehat{k\partial_3\mathbf{u}(\boldsymbol{\xi})} = \frac{\mathrm{i}k'\xi_3}{D(\boldsymbol{\xi})}\int_0^{2\pi} \mathrm{e}^{-(\nu'|\boldsymbol{\xi}|^2+\mathrm{i}k'\xi_3)t}\, O^{\mathrm{T}}_{\mathbf{e}_3}(t)\widehat{F}(O_{\mathbf{e}_3}(t)\boldsymbol{\xi})\mathrm{d}t, \tag{4.258}$$

where $\mathbf{F} = \mathbf{f} - \nabla p$, see (4.235). Choose a cutoff function $\eta \in C_0^\infty(B_1(0))$ with $\eta(\boldsymbol{\xi}) = 1$ for $\boldsymbol{\xi} \in B_{1/2}(0)$ and define the multiplier functions

$$m_0(\boldsymbol{\xi}) = \frac{\mathrm{i}k'\xi_3\eta_{\nu'}(\boldsymbol{\xi})}{D(\boldsymbol{\xi})}, \quad m_1(\boldsymbol{\xi}) = \frac{k'}{\sqrt{\nu'}}\frac{1-\eta_{\nu'}(\boldsymbol{\xi})}{D(\boldsymbol{\xi})},$$

where $\eta_{\nu'}(\boldsymbol{\xi}) = \eta(\sqrt{\nu'}\boldsymbol{\xi})$, as well as

$$\mu_{0,t}(\boldsymbol{\xi}) = E^{-(\nu'|\boldsymbol{\xi}|^2+\mathrm{i}k'\xi_3)t}, \quad \mu_{1,t}(\boldsymbol{\xi}) = \mathrm{i}\xi_3\sqrt{\nu'}\mathrm{e}^{-(\nu'|\boldsymbol{\xi}|^2+\mathrm{i}k'\xi_3)t}, \quad t \in (0, 2\pi).$$

Then we get

$$\widehat{k\partial_3\mathbf{u}(\boldsymbol{\xi})} = m_0(\xi)\widehat{I_0}(\boldsymbol{\xi}) + m_1(\boldsymbol{\xi})\widehat{I_1}(\boldsymbol{\xi}),$$

where $I_0(\mathbf{x})$, $I_1(\mathbf{x})$ are defined by their Fourier transforms

$$\widehat{I_0}(\boldsymbol{\xi}) = \int_0^{2\pi} \mu_{0,t}(\boldsymbol{\xi})O^{\mathrm{T}}_{e_3}(t)\widehat{F}(O_{\mathbf{e}_3}(t)\cdot)(\boldsymbol{\xi})\mathrm{d}t,$$

$$\widehat{I_1}(\boldsymbol{\xi}) = \int_0^{2\pi} \mu_{1,t}(\boldsymbol{\xi})O^{\mathrm{T}}_{e_3}(t)\widehat{F}(O_{\mathbf{e}_3}(t)\cdot)(\boldsymbol{\xi})\mathrm{d}t.$$

Concerning the multiplier function $\mu_{0,t}$, we note that, e.g.,

$$\left|\xi_3 \frac{\partial \mu_{0,t}}{\partial \xi_3}\right| = \left|(-2\nu' t\xi_3^2 - \mathrm{i}k' t\xi_3)\mathrm{e}^{-(\nu'|\boldsymbol{\xi}|^2+\mathrm{i}k'\xi_3)t}\right|$$
$$\leq C\left(\nu' t|\boldsymbol{\xi}|^2 + \frac{k'}{\sqrt{\nu'}}\sqrt{\nu' t}|\xi_3|\right)\mathrm{e}^{-\nu'|\boldsymbol{\xi}|^2 t} \leq C\left(1+\frac{k'}{\sqrt{\nu'}}\right)$$

with a constant $C > 0$ independent of $\xi \neq 0$, $t \in (0, 2\pi)$, $k' > 0$, and $\nu' > 0$. Then it is easily seen that $\mu_{0,t}$, $\mu_{1,t}$ satisfy the pointwise multiplier estimates

$$\sup_{t\in(0,2\pi)} \max_{\alpha} \sup_{\xi\neq 0} \left(|\boldsymbol{\xi}^\alpha D_{\boldsymbol{\xi}}^\alpha \mu_{0,t}(\boldsymbol{\xi})| + \sqrt{t}|\boldsymbol{\xi}^\alpha D_{\boldsymbol{\xi}}^\alpha \mu_{1,t}(\boldsymbol{\xi})|\right) \leq C\left(1+\frac{k}{\sqrt{\nu|\omega|}}\right)$$

uniformly in $k' > 0$ and $\nu' > 0$, where $\alpha \in \mathbb{N}_0^3$ runs through the set of all multi-indices $\alpha \in \{0, 1\}^3$. Hence Theorem 4.14 (iii) and (4.234) show that

$$\|I_0\|_{q,w} \leq c\left(1+\frac{k}{\sqrt{\nu|\omega|}}\right)\int_0^{2\pi} \|F(O_{\mathbf{e}_3}(t)\cdot)\|_{q,w}\mathrm{d}t \leq c\left(1+\frac{k}{\sqrt{\nu|\omega|}}\right)\|\mathbf{f}\|_{q,w},$$
$$\|I_1\|_{q,w} \leq c\left(1+\frac{k}{\sqrt{\nu|\omega|}}\right)\int_0^{2\pi} \frac{1}{\sqrt{t}}\|F(O_{\mathbf{e}_3}(t)\cdot)\|_{q,w}\mathrm{d}t \leq c\left(1+\frac{k}{\sqrt{\nu|\omega|}}\right)\|\mathbf{f}\|_{q,w},$$

where $c > 0$ is independent of k, ω, and ν. Moreover, a lengthy, but elementary calculation proves that m_0, m_1 satisfy the pointwise estimates

$$\max_{j=0,1} \max_{\alpha} \sup_{\xi\neq 0} |\boldsymbol{\xi}^\alpha D_{\boldsymbol{\xi}}^\alpha m_j(\boldsymbol{\xi})| \leq C\left(1+\frac{k^4}{\nu^2|\omega|^2}\right)$$

with $c > 0$ independent of ν, ω, k; for details see [72]. Now another application of Theorem 4.14 (iii) yields the estimate

$$\|k\partial_3 u\|_{q,w} \leq c\left(1+\frac{k^5}{\nu^{5/2}|\omega|^{5/2}}\right)\|\mathbf{f}\|_{q,w}$$

for $\mathbf{f} \in S(\mathbb{R}^3)^3$, with a constant $c > 0$ independent of $\mathbf{f}$, k, ν and ω. Since $S(\mathbb{R}^3)$ is dense in $L^q_w(\mathbb{R}^3)$, this result extends to all $\mathbf{f} \in \mathbf{L}^q_w$; for its proof we refer to [72]. However, note that we did not estimate $\widehat{F}(O_\omega(t)\cdot\ -kt\mathbf{e}_3)(\boldsymbol{\xi})$ in $L^q(\Omega)$ as in [57]; instead we have to deal with $\widehat{F}(O_{\mathbf{e}_3}(t)\cdot)$, and the shift operator is estimated with the help of multipliers.

Now Corollary 4.6 is completely proved. □

Proof of Corollary 4.7 We have to check for which α, β the weight $w(x) = \eta^\alpha_\beta(\mathbf{x}) = (1+|\mathbf{x}|)^\alpha(1+s(\mathbf{x}))^\beta$ satisfies the conditions needed in Theorem 4.13. From [58] and [119, Theorem 52] we know that $w = \eta^\alpha_\beta \in A_p$, $1 < p < \infty$, if and only if $-1 < \beta < p-1$ and $-3 < \alpha+\beta < 3(p-1)$; moreover, by Lemma 4.18 (iii) we have to satisfy the conditions $\alpha < p-1$, $\beta \geq 0$, $\alpha+\beta > -1$ to get that $w_r(\cdot) \in A_p^-$.

Let $q > 2$. Then in view of (4.225) and (4.232) we have to analyze the convex set

$$\mathcal{C} = \Big\{(\alpha,\beta);\, \alpha < \frac{q}{2} - \frac{1}{\tau},\ \beta \geq 0,\ \alpha+\beta > -\frac{1}{\tau},\ -\frac{1}{\tau} < \beta < \frac{q}{2} - \frac{1}{\tau},$$
$$-\frac{3}{\tau} < \alpha+\beta < \frac{3q}{2} - \frac{3}{\tau} \text{ for some} \tau \in [1,\infty)\Big\}.$$

Obviously the conditions $\beta > -1/\tau$ and $-3/\tau < \alpha+\beta < 3q/2 - 3/\tau$ are redundant since $q/2 - 1/\tau$ is positive; moreover, the conditions $\alpha+\beta > -1/\tau$ and $\beta < q/2 - 1/\tau$ yield $\alpha > -q/2$. We will see that

$$\mathcal{C} = \Big\{(\alpha,\beta);\ -\frac{q}{2} < \alpha < \frac{q}{2},\ 0 \leq \beta < \frac{q}{2},\ \alpha+\beta > -1\Big\}.$$

Indeed, it suffices to consider pairs (α,β) with $\alpha < 0$. If moreover $\alpha+\beta < 0$, we find $\tau_0 > 1$ such that $\alpha+\beta = -1/\tau_0$. Then $\beta = -1/\tau_0 - \alpha < -1/\tau_0 + q/2$ and $\alpha < 0 < q/2 - 1/\tau_0$; consequently $(\alpha,\beta) \in \mathcal{C}$. If $\alpha+\beta \geq 0$, we may choose τ sufficiently large to show that $(\alpha,\beta) \in \mathcal{C}$.

Now consider the case $1 < q < 2$. As in the previous case we have to analyze the set $\mathcal{C}$ where now τ runs in the interval $(2/q, 2/(2-q)]$. Since $\tau > 2/q$, the same conditions as before are redundant; moreover, $\alpha > -q/2$. Then we will show that

$$\mathcal{C} = \Big\{(\alpha,\beta); -\frac{q}{2} < \alpha < q-1,\ 0 \leq \beta < q-1,\ \alpha+\beta > -\frac{q}{2}\Big\}.$$

Indeed, if, e.g., $\alpha < 0$ and $\alpha+\beta \leq q/2-1 < 0$, then there exists $\tau_0 \in (2/q, (2-q)/2]$ such that $\alpha+\beta = -1/\tau_0$, $\beta = -1/\tau_0 - \alpha < -1/\tau_0 + q/2$, and $\alpha < 0 < q/2 - 1/\tau_0$; however, when $\alpha+\beta > q/2-1$, we may choose $\tau = 2/(2-q)$ to see that $(\alpha,\beta) \in \mathcal{C}$. □

Let $\mathbf{K}_{\mathcal{O}}$ be the nullspace of the system (4.223), with respect to the velocity field, i.e., the set of all $\mathbf{v} \in \widehat{\mathbf{H}}^{2,q}_{w,\sigma}(\mathcal{D})$, $S_{\omega,\nu,k}\,\mathbf{v} + \nabla\pi = \mathbf{0}$, $\mathbf{v}|_{\partial\mathcal{D}} = \mathbf{0}$ for some $\pi \in \widehat{H}^{1,q}_w(\mathcal{D})$; it has the finite dimension 2 independently on q and w. $\mathcal{K}_{\mathcal{O}}$ will denote the set of $[\mathbf{v},\pi]$, $\mathbf{v} \in \mathbf{K}_{\mathcal{O}}$, π being the associate pressure to $\mathbf{v}$. We also have the property analogous to Proposition 3.7: Let $[\mathbf{u}, p]$ be some strong solution to system (4.223), ($[\mathbf{u}, p]$ not necessarily satisfying the estimates (4.226), (4.228)). Then at least one representant of $[\mathbf{u}, p] + \mathcal{K}_{\mathcal{O}}$, say $[\mathbf{u}_*, p_*]$, fulfills the estimate

$$\|\nu\nabla^2\mathbf{u}_*\|_{q,w} + \|\mathcal{R}\mathbf{u}_*\|_{q,w} + \|\partial_3\mathbf{u}_*\|_{q,w} + \|\nabla p_*\|_{q,w} \leq C\|\mathbf{f}\|_{q,w}, \quad c = c(q,w).$$

Remark 4.6 Concerning the nonhomogeneous case with $g \neq 0$, let us mention that similarly as in [73] we can get the following estimate

$$\|\nu\nabla^2\mathbf{u}, \nabla p\|_{q,w} + \|\mathcal{R}\mathbf{u}\|_{q,w} \leq c\|\mathbf{f}, \nabla g, (\boldsymbol{\omega}\wedge\mathbf{x})g, kg\,\mathbf{e}_3\|_{q,w},\ c = c(q, w). \tag{4.259}$$

Remark 4.7 Let us note that Theorem 4.13 was stated in [73] for more general type of weights which are refereed to cubes and that we need restriction to weights referred to rectangles only for the estimate (4.228). Since our aim hereafter is to introduce the notion of very weak solution for Problem (4.223), we have to consider this subclass $\widetilde{A}_q^-$ of weights. □

4.5 L^q-Weighted Theory of the Weak Solution to the Oseen Type Problem

We will follow the work by Kračmar, Krbec, Nečasová, Penel, Schumacher, see [153].

Definition 4.6 Let $1 < q < \infty$. Given $\mathbf{f} \in \widehat{H}_w^{-1,q}(\mathbb{R}^3)$, we call $\{\mathbf{u}, p\} \in \widehat{H}_w^{1,q}(\mathbb{R}^3) \times L_w^q(\mathbb{R}^3)$ weak solution to (4.223) if

$$\begin{aligned} &(1) \qquad \nabla\cdot\mathbf{u} = 0 \quad \text{in} \quad \boldsymbol{L}_w^q(\mathbb{R}^3), \\ &(2) \qquad (\omega\wedge\mathbf{x})\cdot\nabla\mathbf{u} - \omega\wedge\mathbf{u} \in \widehat{\boldsymbol{H}}_w^{-1,q}(\mathbb{R}^3), \end{aligned} \tag{4.260}$$

$\{\mathbf{u}, p\}$ satisfies $(4.223)_1$ in the sense of distributions, that is,

$$\begin{aligned} &\nu\langle\nabla\mathbf{u}, \nabla\boldsymbol{\varphi}\rangle - \langle\omega(e_3\wedge\mathbf{x})\cdot\nabla\mathbf{u} - \omega e_3\wedge\mathbf{u}, \boldsymbol{\varphi}\rangle \\ &+k\left\langle\frac{\partial\mathbf{u}}{\partial\mathbf{x}_3}, \boldsymbol{\varphi}\right\rangle - \langle p, \nabla\cdot\boldsymbol{\varphi}\rangle = \langle\mathbf{f}, \boldsymbol{\varphi}\rangle, \\ &\boldsymbol{\varphi}\in C_0^\infty(\mathbb{R}^3). \end{aligned} \tag{4.261}$$

In fact, Eq. (4.261) holds by density for all $\boldsymbol{\varphi} \in \widehat{\boldsymbol{H}}_w^{1,q/(q-1)}(\mathbb{R}^3)$.

Theorem 4.17 *Let the weight function $0 \leq w \in L_{loc}^1(\mathbb{R}^3)$ be independent of the angular variable θ and satisfy the following condition depending on $q \in (1, \infty)$:*

$$\begin{aligned} 2 \leq q < \infty : w^\tau &\in \widetilde{A}_{\tau q/2}^- \quad \textit{for some} \quad \tau\in[1,\infty) \\ 1 < q < 2 :\ w^\tau &\in \widetilde{A}_{\tau q/2}^- \quad \textit{for some} \quad \tau\in\left(\tfrac{2}{q}, \tfrac{2}{2-q}\right]. \end{aligned} \tag{4.262}$$

(i) Given $\mathbf{f} \in \boldsymbol{L}^q_w(\mathbb{R}^3)$, *there exists a solution* $(\mathbf{u}, p) \in L^1_{loc}(\mathbb{R}^3) \times L^1_{loc}(\mathbb{R}^3)$ *of*

$$\begin{aligned} -\nu\Delta\mathbf{u} + k\partial_3\mathbf{u} - (\omega\boldsymbol{e}_3 \wedge \mathbf{x}) \cdot \nabla\mathbf{u} + \omega\boldsymbol{e}_3 \wedge \mathbf{u} + \nabla p &= \mathbf{f} \ \text{ in } \ \mathbb{R}^n \\ \operatorname{div} \mathbf{u} &= 0 \ \text{ in } \ \mathbb{R}^n \end{aligned} \tag{4.263}$$

satisfying the estimate

$$\|\nu\nabla\mathbf{u}\|_{q,w} + \|p\|_{q,w} + \|\omega(\boldsymbol{e}_3 \wedge \mathbf{x}) \cdot \nabla\mathbf{u} - \omega\boldsymbol{e}_3 \wedge \mathbf{u}\|_{-1,q,w} \le c\|\mathbf{f}\|_{-1,q,w}, \tag{4.264}$$

with a constant $c = c(q, w) > 0$ *independent of* ν, k, *and* ω. □

Theorem 4.18 *The solution* $(\mathbf{u}, p)$ *given by Theorem 4.17 is unique up to a constant multiple of* $\boldsymbol{\omega}$ *for* $\mathbf{u}$. □

Theorem 4.19 *Let the assumptions of Theorem 4.18 be satisfied. Moreover let* $rot^k\mathbf{f} \in \boldsymbol{L}^q_w(\mathbb{R}^3)$ *and then*

$$\|\nu\nabla^{k+2}\mathbf{u}\|_{q,w} + \|\nabla^k p\|_{q,w} \le c\|rot^k\mathbf{f}\|_{q,w}. \tag{4.265}$$

Theorem 4.20 *Let the weight function* $0 \le w \in L^1_{loc}(\mathbb{R}^3)$ *be independent of the angular variable* θ *and satisfy the following condition depending on* $q \in (1, \infty)$*:*

$$\begin{aligned} 2 \le q < \infty : w^\tau &\in \widetilde{A}^-_{\tau q/2} \quad \text{for some} \quad \tau \in [1, \infty) \\ 1 < q < 2 : \ w^\tau &\in \widetilde{A}^-_{\tau q/2} \quad \text{for some} \quad \tau \in \left(\tfrac{2}{q}, \tfrac{2}{2-q}\right]. \end{aligned} \tag{4.266}$$

(i) Given $\mathbf{f} \in \boldsymbol{L}^q_w(\mathbb{R}^3)$ *and* $g \in H^{1,1,}_{loc}(R^n)$ *with* $\nu\nabla g + (\omega\boldsymbol{e}_3 \wedge \mathbf{x})g \in L^q_w(R^n)$, *there exists a solution* $(\mathbf{u}, p) \in \boldsymbol{L}^1_{loc}(\mathbb{R}^3) \times L^1_{loc}(\mathbb{R}^3)$ *of*

$$\begin{aligned} -\nu\Delta\mathbf{u} + k\partial_3\mathbf{u} - (\omega\boldsymbol{e}_3 \wedge \mathbf{x}) \cdot \nabla\mathbf{u} + \omega\boldsymbol{e}_3 \wedge \mathbf{u} + \nabla p &= \mathbf{f} \ \text{ in } \ \mathbb{R}^n \\ \operatorname{div} \mathbf{u} &= g \ \text{ in } \ \mathbb{R}^n \end{aligned} \tag{4.267}$$

satisfying the estimate

$$\begin{aligned} &\|\nu\nabla\mathbf{u}\|_{q,w} + \|p\|_{q,w} + \|(\omega\boldsymbol{e}_3 \wedge \mathbf{x}) \cdot \nabla\mathbf{u} - \omega\boldsymbol{e}_3 \wedge \mathbf{u}\|_{-1,q,w} \le \\ &c(\|\mathbf{f}\|_{-1,q,w} + \|\nu\nabla g + (\omega\boldsymbol{e}_3 \wedge \mathbf{x})g\|_{-1,q,w}), \end{aligned} \tag{4.268}$$

with a constant $c = c(q, w) > 0$ *independent of* ν, k, *and* ω. □

Computation of $\nabla\mathbf{u}$ and $\mathrm{rot}^k\nabla^2\mathbf{u}$

Using the fact that the space $\{\, g \mid \ g = \nabla \cdot G, \ \ G \in C_0^\infty\left(\mathbb{R}^3\right)^{3\times3}\}$ is dense in $\widehat{\boldsymbol{H}}^{-1,q}_w(\mathbb{R}^3)$, we can write either $\mathbf{f}$ in the divergence form in the Oseen system (4.223) assuming firstly $G \in C_0^\infty\left(\mathbb{R}^3\right)^{3\times3}$. We will just follow the main lines which were already proved in Sects. 4.1, 4.2, and 4.4. We will work in the space of tempered distributions because we have in mind to apply the Fourier transform. We will derive the following formal expressions of $\widehat{\mathbf{u}}$, $\mathbf{u}$, and $\nabla\mathbf{u}$:

$$\widehat{\mathbf{u}}(\boldsymbol{\xi}) = \int_0^\infty e^{-(\nu|\boldsymbol{\xi}|^2+ik\xi_3)t}\, O_\omega^T(t)\,\widehat{g}(O_\omega(t)\boldsymbol{\xi})\,dt,$$

yielding $\mathbf{u}(\cdot)$ in the form

$$\mathbf{u}(\mathbf{x}) = \int_0^\infty E_t * O_\omega^T(t)\,g(O_\omega(t).-kt\boldsymbol{e}_3)(\mathbf{x})\,dt,$$

where

$$E_t(\boldsymbol{x}) = \frac{1}{(4\pi\nu t)^{3/2}}e^{-\frac{|\mathbf{x}|^2}{4\nu t}}.$$

Observing the previous integral, the solution can be rewritten as

$$\mathbf{u}(\mathbf{x}) = \int_{\mathbb{R}^3} \Gamma\,(\mathbf{x},\mathbf{y})\ \nabla\cdot G\,(\mathbf{y})\ d\mathbf{y},$$

where

$$\Gamma(\mathbf{x},\mathbf{y}) = \int_0^\infty E_t\,(O_\omega(t)\mathbf{x}-\mathbf{y}-kt\boldsymbol{e}_3)\ O_\omega^T(t)\,dt.$$

Therefore we can compute the gradient of $\mathbf{u}$,

$$\nabla\mathbf{u}(\mathbf{x}) = -\int_{\mathbb{R}^3} \nabla_{\mathbf{x}}\nabla_{\mathbf{y}}\Gamma\,(\mathbf{x},\mathbf{y}) : G\,(\mathbf{y})\ d\mathbf{y},$$

and come back to its Fourier transform. For more details, see Sect. 4.2 or [154].

Computation of $\mathrm{rot}^{\,k}\mathbf{u}$ Due to the geometry of the problem, we introduce cylindrical coordinates $(r,\mathbf{x}_3,\theta)\in(0,\infty)\times\mathbb{R}\times[0,2\pi)$ and write $\mathbf{u}(x_1,x_2,x_3) = \mathbf{u}(r,x_3,\theta)$. Then the term $(\boldsymbol{e}_3\wedge\mathbf{x})\cdot\nabla\mathbf{u} = -x_2\partial_1\mathbf{u}+x_1\partial_2\mathbf{u}$ may be rewritten in the form $(\boldsymbol{e}_3\wedge\mathbf{x})\cdot\nabla\mathbf{u} = \partial_\theta\mathbf{u}$ using the angular derivative ∂_θ applied to $\mathbf{u}(r,x_3,\theta)$.

$$-\nu\Delta\mathbf{u}+k\partial_3\mathbf{u}-\omega(\partial_\theta\mathbf{u}-\boldsymbol{e}_3\wedge\mathbf{u}) = \boldsymbol{F} := \mathbf{f}-\nabla p \quad \text{in} \quad \mathbb{R}^3. \tag{4.269}$$

Now applying the rot^k and changing place, we get

$$\begin{aligned}-\nu\Delta rot^k\mathbf{u}+k\partial_3 rot^k\mathbf{u}-\omega(\boldsymbol{e}_3\partial_\theta rot^k\mathbf{u}-\boldsymbol{e}_3\wedge rot^k\mathbf{u}) &= rot^k\boldsymbol{F}\\ := rot^k(\mathbf{f}-\nabla p) \quad \text{in} \quad \mathbb{R}^3.&\end{aligned} \tag{4.270}$$

Setting $v = rot^k\mathbf{u}$ we can apply the same method as before.

As in previous Sects. 4.1, 4.2, and 4.4, we estimate the L^q-norm of each component of $TG(\mathbf{x}) := \Delta \int_{\mathbb{R}^3} \Gamma(\mathbf{x},\mathbf{y}) : G(\mathbf{y})\, dy$, say $TG_{i,k}(\cdot)$ by the L^q-norm of $G_{i,k}(\cdot)$, and then we apply these results for L^q-estimate of $\nabla \mathbf{u}$.

Following the same lines we solve the problem by applying the Fourier transform to get

$$\widehat{TG}(\boldsymbol{\xi}) = \frac{1}{(2\pi)^{3/2}} \int_0^\infty |\boldsymbol{\xi}|^2 \exp\left(-\nu\, |\boldsymbol{\xi}|^2\, t\right) O_\omega^T(t)\, \widehat{G}\left((O_\omega(t)\cdot -k\, t\, \boldsymbol{e}_3)\, \boldsymbol{\xi}\right) dt,$$

which we can rewrite as

$$\widehat{TG}(\boldsymbol{\xi}) = \frac{1}{\nu(2\pi)^{3/2}} \int_0^\infty |\boldsymbol{\xi}|^2 \exp\left(-\, |\boldsymbol{\xi}|^2\, t\right)$$
$$O_{\omega/\nu}^T(t)\, \widehat{G}\left(\left(O_{\omega/\nu}(t)\cdot -k\,\frac{t}{\nu}\, \boldsymbol{e}_3\right) \boldsymbol{\xi}\right) dt.$$

Let us temporarily denote $TG_{i,k}(x)$ by $F(x)$. Using the harmonic analysis we introduce an appropriate function $\varphi(\cdot) \in C_0^\infty((0,\infty);\, \mathcal{S}(\mathbb{R}^3))$ such that with the so-called square operator

$$S(F)(x) = \int_0^\infty |\varphi(t,\cdot) * F(x)|^2 \,\frac{dt}{t}$$

satisfying the equivalence of L^q-norms

$$c_1\, \|F\|_q \le \left\| S(F)^{1/2} \right\|_q \le c_2\, \|F\|_q\, .$$

The necessary properties of $\varphi(t,\cdot)$ for $t > 0$ are

$$\operatorname{supp} \widehat{\varphi}(t,\cdot) \subset \left\{ \boldsymbol{\xi} \in \mathbb{R}^n : \ \frac{1}{2\sqrt{t}} < |\boldsymbol{\xi}| < \frac{2}{2\sqrt{t}} \right\}$$
$$\int_0^\infty \widehat{\varphi}(t,\boldsymbol{\xi})^2 \frac{dt}{t} = 1$$
$$\int_{\mathbb{R}^3} \varphi(t,\mathbf{x}) d\mathbf{x} = 0$$

(see E. M. Stein Chapter I, Section 8.23 [205]).

Let us recall the Littlewood–Paley decomposition. We define $\psi \in \mathcal{S}(\mathbb{R}^3)$ by its Fourier transform

$$\widehat{\psi}(\boldsymbol{\xi}) = \frac{1}{(2\pi)^{3/2}} |\boldsymbol{\xi}|^2 e^{-|\boldsymbol{\xi}|^2} \ \text{ and } \ \widehat{\psi}_t(\boldsymbol{\xi}) = \widehat{\psi}(\sqrt{t}\boldsymbol{\xi}) \text{ for } t > 0, \qquad (4.271)$$

and so for all $t > 0$

$$\psi_t(\mathbf{x}) = t^{-3/2}\psi\left(\frac{\mathbf{x}}{\sqrt{t}}\right), \quad \widehat{\psi}_t(\boldsymbol{\xi}) = \frac{1}{(2\pi)^{3/2}}t|\boldsymbol{\xi}|^2 e^{-\nu t|\boldsymbol{\xi}|^2}. \tag{4.272}$$

It gives us

$$\widehat{TG}(\boldsymbol{\xi}) = \frac{1}{\nu}\int_0^\infty \widehat{\psi}_t(\boldsymbol{\xi})\, O^T_{\omega/\nu}(t)\, \widehat{G}\left(\left(O_{\omega/\nu}(t). - k\frac{t}{\nu}e_3\right)\boldsymbol{\xi}\right)\frac{dt}{t}. \tag{4.273}$$

We define the multiplier operator Δ_j such that

$$\widehat{\Delta_j f(\boldsymbol{\xi})} :- \widehat{\chi}^j(\boldsymbol{\xi})\widehat{f}(\boldsymbol{\xi}), \tag{4.274}$$

where

$$\widehat{\chi}^j(\boldsymbol{\xi}) = \widehat{\chi}_0\Big(\frac{\boldsymbol{\xi}}{2^{j+1}}\Big) - \widehat{\chi}_0\Big(\frac{|\boldsymbol{\xi}|}{2^j}\Big) \tag{4.275}$$

with

$$\widehat{\chi}_0(.) : |\boldsymbol{\xi}| \to \mathbb{R}, \chi \in C^\infty, \widehat{\chi}_0\Big|_{\{|\boldsymbol{\xi}|\le \frac{1}{2}\}} = 1, \widehat{\chi}_0\Big|_{\{|\boldsymbol{\xi}|\ge 1\}} = 0. \tag{4.276}$$

Note that

$$\sum_{j\in Z}\widehat{\chi}^j(\boldsymbol{\xi}) = 1. \tag{4.277}$$

and

$$f(.) = \sum_{j\in Z}\Delta_j f(.). \tag{4.278}$$

We define χ^j for $\boldsymbol{\xi} \in \mathbb{R}^3$ and $j \in \mathbb{Z}$ by its Fourier transform

$$\widehat{\chi}^j(\boldsymbol{\xi}) = \widetilde{\chi}(2^{-j}|\boldsymbol{\xi}|), \quad \boldsymbol{\xi} \in \mathbb{R}^3,$$

yielding $\sum_{j=-\infty}^{\infty}\widehat{\chi}_j = 1$ on $\mathbb{R}^3 \setminus \{0\}$ and

$$\operatorname{supp}\widehat{\chi}^j \subset A(2^{j-1}, 2^{j+1}) := \{\boldsymbol{\xi} \in \mathbb{R}^3 : 2^{j-1} < |\boldsymbol{\xi}| < 2^{j+1}\}. \tag{4.279}$$

Using χ^j we define for $j \in \mathbb{Z}$

$$\psi^j = \frac{1}{(2\pi)^{n/2}}\chi_j * \psi_t, \qquad \widehat{\psi^j} = \widehat{\Delta_j\psi(.)} = \widehat{\chi}^j \cdot \widehat{\psi}_t. \tag{4.280}$$

Obviously, $\sum_{j=-\infty}^{\infty} \psi^j = \psi$ on $\mathbb{R}^3$. We start with the procedure of the Littlewood–Paley decomposition of $F = TG_{i,k}$. For $G_{i,k} \in \mathcal{S}'(\mathbb{R}^3)$ the property $TG_{i,k} \in L^p(\mathbb{R}^3)$ is equivalent to the property

$$TG_{i,k} = \sum_{j=-\infty}^{+\infty} \Delta_j TG_{i,k} \quad \text{and} \quad \left(\sum_{j=-\infty}^{+\infty} \left|\Delta_j TG_{i,k}\right|^2 \right)^{1/2} \in L^p\left(\mathbb{R}^3\right),$$

where

$$\Delta^j = \mathcal{F}^{-1}\widehat{\psi}^j\left(\frac{\boldsymbol{\xi}}{2^j}\right)\mathcal{F},$$

$$\Delta_t^j = \mathcal{F}^{-1}\widehat{\psi}_t^j\left(\frac{\boldsymbol{\xi}}{2^j}\right)\mathcal{F}.$$

We define operator

$$\Delta\Gamma = \sum_{j\in Z} \Delta_j \Delta\Gamma \tag{4.281}$$

yielding to

$$\Delta \int_{\mathbb{R}^3} \Gamma(\mathbf{x},\mathbf{y}) : G(\mathbf{y})d\mathbf{y}, \quad G \in C_0^\infty(\mathbb{R}^3)^9, (G_{ik})_{1\le i\le 3, 1\le k\le 3}. \tag{4.282}$$

We define the linear operator

$$TG_{ik}(\mathbf{x}) = \int_{\mathbb{R}^3} \Delta\Gamma(\mathbf{x},\mathbf{y})_{ki} G_{ik}(\mathbf{y})d\mathbf{y}. \tag{4.283}$$

Since formally $T = \sum_{j=-\infty}^{\infty} T_j$, we have to prove that this infinite series converges even in the operator norm on L^q.

Let us also recall a weighted Littlewood–Paley decomposition of L^q_w and choose $\widetilde{\varphi} \in C_0^\infty(\frac{1}{2}, 2)$ such that $0 \le \widetilde{\varphi} \le 1$ and $\int_0^\infty \widetilde{\varphi}(s)^2 \frac{ds}{s} = \frac{1}{2}$. Then define $\varphi \in \mathcal{S}(\mathbb{R}^3)$ by its Fourier transform $\widehat{\varphi}(\boldsymbol{\xi}) = \widetilde{\varphi}(|\boldsymbol{\xi}|)$ yielding for every $s > 0$

$$\widehat{\varphi}_s(\boldsymbol{\xi}) = \widetilde{\varphi}(\sqrt{s}|\boldsymbol{\xi}|), \quad \operatorname{supp}\widehat{\varphi}_s \subset A\Big(\frac{1}{2\sqrt{2}}, \frac{2}{\sqrt{s}}\Big) \tag{4.284}$$

and the normalization $\int_0^\infty \widehat{\varphi}_s(\boldsymbol{\xi})^2 \frac{ds}{s} = 1$ for all $\boldsymbol{\xi} \in \mathbb{R}^n \setminus \{0\}$.

Theorem 4.21 *Let $1 < q < \infty$ and $w \in A_q(\mathbb{R}^3)$. Then there are constants $c_1, c_2 > 0$ depending on q, w and φ such that for all $f \in L^q_w$*

$$c_1\|\mathbf{f}\|_{q,w} \le \left\| \left(\int_0^\infty |\varphi_s * \mathbf{f}(\cdot)|^2 \frac{ds}{s} \right)^{1/2} \right\|_{q,w} \le c_2\|\mathbf{f}\|_{q,w}, \tag{4.285}$$

where $\varphi_s \in \mathcal{S}(\mathbb{R}^n)$ is defined by (4.284). □

For details, see [186, Proposition 1.9, Theorem 1.10], and also [158], [205]. Applying the previous theorem to the operator $T_i G_{ik}$:

$$c_1\|T_j G_{ik}\|_{q,w} \le \|(\int_0^\infty |(\varphi(t,.) * T_j G_{ik}(\mathbf{x})|^2 \frac{dt}{t})^{1/2}\|_{q,w} \le c_2\|T_j G_{ik}\|_{q,w}.$$

4.5.1 Proofs

As a preliminary version of Theorem 4.17 we prove the following proposition. The extension to more general weights based on complex interpolation of L^q_w-spaces will be described at the end of this section.

Proposition 4.5 *Let the weight $w \in L^1_{loc}(\mathbb{R}^3)$ be independent of the angle θ and define $w_r(x_3) := w(x_1, x_2, x_3)$ for fixed $r = |(x_1, x_2)| > 0$. Assume that*

$$\begin{aligned} &w \in \widetilde{A}^-_{q/2} && \text{if } q > 2,\\ &w \in \widetilde{A}^-_1 \ \text{ or } \ \tfrac{1}{w} \in \widetilde{A}^+_1 && \text{if } q = 2,\\ &w^{2/(2-q)} \in \widetilde{A}^-_{q/(2-q)} && \text{if } 1 < q < 2. \end{aligned} \tag{4.286}$$

Then the linear operator T defined by (4.283) *satisfies the estimate*

$$\|TG\|_{q,w} \le c\|G\|_{q,w} \quad \text{for all} \quad f \in L^q_w \tag{4.287}$$

with a constant $c = c(q, w) > 0$ independent of f. □

Proof of Proposition 4.5 *Step 1.* First we consider the case $q > 2$, $w \in \widetilde{A}^-_{q/2} \subset A_q$, and define the sublinear operator $\mathcal{M}^j$, a modified maximal operator, by

$$\mathcal{M}^j g(\mathbf{x}) = \sup_{s>0} \int_{A_s} \left(|\Delta^j_t| * |\varphi|\right)\left(O^T_{\Omega/\nu}(t)\mathbf{x} + \frac{k}{\nu} t e_3\right) \frac{dt}{t}, \tag{4.288}$$

where $A_s = [\frac{s}{16}, 16s]$.

First step. We will prove the preliminary estimate

$$\|T_j G_{ik}\|_{q,w} \le c\|\Delta^j\|_1^{1/2}\|\mathcal{M}^j\|^{1/2}_{L_v^{(q/2)'}}\|G_{ik}\|_{q,w}\,, \quad j \in \mathbb{Z}, \tag{4.289}$$

where v denotes the θ-independent weight

$$v = w^{-\left(\frac{q}{2}\right)'/\left(\frac{q}{2}\right)} = w^{-\frac{2}{q-2}} \in \widetilde{A}^+_{(q/2)'} = \widetilde{A}^+_{q/(q-2)}. \tag{4.290}$$

To show (4.289) we use as in previous Sect. 4.4 the Littlewood–Paley decomposition of L^q_w, see (4.285), applied to $T_j G_{ik}$. By a duality argument we find a function satisfying the properties $0 \le g \in L_v^{(q/2)'} = \left(L_w^{(q/2)}\right)^*$ with $\|g\|_{(q/2)',v} = 1$ such that

$$\left\| \int_0^\infty |\varphi_s * T_j G_{ik}(\cdot)|^2 \frac{ds}{s} \right\|_{q/2,w} = \int_0^\infty \int_{\mathbb{R}^3} |\varphi_s * T_j G_{ik}(\mathbf{x})|^2 g(\mathbf{x})\, d\mathbf{x}\, \frac{ds}{s}\,. \tag{4.291}$$

To estimate the right-hand side of (4.291) let us mention that

$$\varphi_s * T_j G_{ik}(\mathbf{x}) = \int_0^\infty O^T_{\omega/\nu}(t)(\varphi_s * \Delta^j_t * G_{ik})\Big(O_{\omega/\nu}(t)\mathbf{x} - \frac{k}{\nu} t\boldsymbol{e}_3\Big)\frac{dt}{t}\,,$$

where $\varphi_s * \Delta_t^j = 0$ unless $t \in A(s,j) := [2^{2j-4}s, 2^{2j+4}s]$. Since $\int_{t\in A(s,j)} \frac{dt}{t} = \log 2^8$ for every $j \in \mathbb{Z},\ s > 0$, we get by the inequality of Cauchy–Schwarz and the associativity of convolutions similarly as in Sect. 4.4 that

$$\begin{aligned}
|\varphi_s * T_j G_{ik}(\mathbf{x})|^2 &\le c \int_{A(s,j)} \left| \Big(\Delta_t^j * (\varphi_s * G_{ik})\Big)\Big(O_{\omega/\nu}(t)x - \frac{k}{\nu}t\boldsymbol{e}_3\Big)\right|^2 \frac{dt}{t} \\
&\le c\|\Delta^j\|_1 \int_{A(s,j)} \big(|\Delta_t^j| * |\varphi_s * G_{ik}|^2\big)\Big(O_{\omega/\nu}(t)\mathbf{x} - \frac{k}{\nu}t\boldsymbol{e}_3\Big)\frac{dt}{t}
\end{aligned}$$

(we applied the estimate $|(\Delta_t^j * (\varphi_s * G_{ik}))(\mathbf{y})|^2 \le \|\Delta_t^j\|_1 (|\Delta_t^j| * |\varphi_s * G_{ik}|^2)(\mathbf{y})$ and the identity $\|\Delta_t^j\|_1 = \|\Delta^j\|_1$, see Lemma 4.7). It implies

$$\begin{aligned}
\|T_j G_{ik}\|^2_{q,w} \le c \int_0^\infty \int_{A(s,j)} \int_{R^n} &(|\Delta_t^j| * |\varphi_s * G_{ik}|^2(\mathbf{x}) \\
&\times \int_{A(s,j)} (|\psi_t^j| * g)\Big(O^T_{\omega/\nu}(t)\mathbf{x} - \frac{k}{\nu}t\boldsymbol{e}_3\Big)\frac{dt}{t}\frac{ds}{s} d\mathbf{x}
\end{aligned} \tag{4.292}$$

(we use that Δ_t^j is radially symmetric). By the definition of $\mathcal{M}^j$ the innermost integral is estimated by $\mathcal{M}^j g(\mathbf{x})$ uniformly in $s > 0$. Hence we may proceed in (4.292) using Hölder's inequality>

$$\begin{aligned}\|T_j G_{ik}\|_{q,w}^2 &\le c\|\Delta^j\|_1 \int_{\mathbb{R}^3} \Big(\int_0^\infty |\varphi_s * f|^2(\mathbf{x}) \,\frac{ds}{s}\Big) \mathcal{M}^j g(\mathbf{x})\, d\mathbf{x} \\ &\le c\|\Delta^j\|_1 \Big\| \int_0^\infty |\varphi_s * f|^2(\mathbf{x}) \,\frac{ds}{s}\Big\|_{q/2,w} \|\mathcal{M}^j g\|_{(q/2)',v}.\end{aligned} \tag{4.293}$$

Now (4.285) and the normalization $\|g\|_{(q/2)',v} = 1$ complete the proof of (4.289).

Step 2. Let $\|\mathcal{M}^j g\|_{(q/2)',v}$. For functions γ depending on θ, x_3 only, let $\mathcal{M}_{\text{hel}}$ denote the "helical" maximal operator similarly as in Sect. 4.4

$$\mathcal{M}_{\text{hel}}\gamma(\theta, x_3) = \sup_{s>0} \frac{1}{s} \int_{A_s} |\gamma|\Big(\theta - \frac{\omega}{\nu} t, x_3 + \frac{k}{\nu} t\Big)\, dt,$$

where $A_s = \big[\frac{s}{16}, 16s\big]$. Then, writing $p := \big(\frac{q}{2}\big)'$, we claim that

$$\mathcal{M}^j g(\mathbf{x}) \le c2^{-2|j|} \mathcal{M}(\mathcal{M}_{\text{hel}} g)(\mathbf{x}) \quad \text{for a.a. } \mathbf{x} \in \mathbb{R}^n, \tag{4.294}$$

$$\|\mathcal{M}^j g\|_{p,v} \le c2^{-2|j|} \|g\|_{p,v}, \tag{4.295}$$

where in (4.294) $\mathcal{M}_{\text{hel}} g$ is considered as $\mathcal{M}_{\text{hel}} g(r, \cdot, \cdot)$ for almost all $r > 0$.

To prove (4.294) we use the pointwise estimate $|\psi_t^j(\mathbf{x})| \le c2^{-2|j|} h_{t2^{-2j}}(\mathbf{x})$, see Lemma 4.7(ii). Hence

$$\mathcal{M}^j g(\mathbf{x}) \le c2^{-2|j|} \sup_{s>0} \int_{A_s} (h_{t2^{-2j}} * |g|)\Big(O^T_{\Omega/\nu}(t)\mathbf{x} + \frac{k}{\nu} t\mathbf{e}_3\Big) \frac{dt}{t}.$$

Moreover, there exists a constant $c > 0$ independent of $s > 0$, $j \in \mathbb{Z}$, such that $h_{t2^{-2j}} \le c h_{s2^{-2j}}$ for all $t \in A_s$. Consequently,

$$\begin{aligned}\mathcal{M}^j g(\mathbf{x}) &\le c2^{-2|j|} \sup_{s>0} h_{s2^{-2j}} * \int_{A_s} |g|\Big(O^T_{\Omega/\nu}(t)\mathbf{x} + \frac{k}{\nu} t\mathbf{e}_3\Big) \frac{dt}{t} \\ &\le c2^{-2|j|} \sup_{t>0} h_t * \mathcal{M}_{\text{hel}} g(\mathbf{x}).\end{aligned}$$

From Sect. 4.2 we know that h is nonnegative, radially decreasing, and $\|h_t\|_1 = \|h\|_1 = c_0 > 0$ for all $t > 0$, and applying convolution estimate, see [205], II §2.1, it implies the pointwise estimate (4.253).

Step 3. Let us mention that up to now we have not yet used any specific properties of the weight $v \in A_p$. To estimate $\mathcal{M}_{\text{hel}} g$ we shall again work with a suitable one-sided maximal operator since our weight belongs to a Muckenhoupt class in $\mathbb{R}^3$, but a problem occurs when the weight is considered with respect to x_3 only as in previous Sects. 4.1, 4.2, and 4.4. This naturally corresponds to the physical circumstances of the problem, where in the Oseen case the wake should appear. Following same lines as before to estimate $\mathcal{M}_{\mathrm{hel}} g$ we write $g_r(\theta, x_3) = g(r, \theta, x_3) = g(\mathbf{x})$ and $v_r(x_3) = v(\mathbf{x})$, $r = |(\mathbf{x}_1, \mathbf{x}_2)| > 0$, for the θ-independent weight v. Then by the 2π-periodicity of g_r and v_r with respect to θ, we get for almost all $r > 0$

$$\begin{aligned}
&\int_{\mathbb{R}} \int_0^{2\pi} \mathcal{M}_{\mathrm{hel}} g_r(\theta, x_3)^p v_r(x_3)\, d\theta\, dx_3 \\
&\le \int_{\mathbb{R}} \int_0^{2\pi} \left| \sup_{s>0} \frac{1}{s} \int_0^{16s} |g_r|\Big(\theta - \frac{\omega}{k}\Big(x_3 + \frac{k}{\nu} t\Big), x_3 + \frac{k}{\nu} t\Big)\, dt \right|^p v_r(x_3)\, d\theta\, dx_3 \\
&= \int_{\mathbb{R}} \int_0^{2\pi} \left| \sup_{s>0} \frac{1}{s} \int_0^{16s} \gamma_{r,\theta}\Big(x_3 + \frac{k}{\nu} t\Big)\, dt \right|^p d\theta\, v_r(x_3)\, dx_3 \\
&= 16 \int_0^{2\pi} \int_{\mathbb{R}} \big|M^+ \gamma_{r,\theta}(x_3)\big|^p v_r(x_3)\, dx_3\, d\theta,
\end{aligned}$$

where $\gamma_{r,\theta}(y_3) = |g_r|(\theta - \frac{\omega}{k} y_3, y_3)$ and M^+ denotes the one-sided maximal operator, see Definition 4.4. Since $w_r \in A^-_{q/2}$, by (4.290) and Theorem 4.15 (i) $v_r = w_r^{-(q/2)'/(q/2)} \in A^+_{(q/2)'} = A^+_p$ with an A^+_p-constant uniformly bounded in $r > 0$. Then Theorem 4.15 (ii) yields the estimate

$$\begin{aligned}
&\int_{\mathbb{R}} \int_0^{2\pi} \mathcal{M}_{\mathrm{hel}} g_r(\theta, x_3)^p v_r(x_3)\, d\theta\, dx_3 \\
&\qquad \le c \int_0^{2\pi} \int_{\mathbb{R}} |\gamma_{r,\theta}(x_3)|^p v_r(x_3)\, dx_3\, d\theta = c \|g_r\|^p_{L^p(\mathbb{R} \times (0, 2\pi),\, v_r(x_3))},
\end{aligned}$$

where $c > 0$ is independent of k, ν. Integrating with respect to $r\, dr$, $r \in (0, \infty)$, Fubini's theorem allows to consider an extension of $\mathcal{M}_{\mathrm{hel}}$ to a bounded operator from $L^p_v(\mathbb{R}^3)$ to itself with an operator norm bounded uniformly in k, ν. Moreover, $\mathcal{M} : L^p_v(\mathbb{R}^3) \to L^p_v(\mathbb{R}^3)$ is bounded by Theorem 4.15 (ii). Hence, (4.294) implies (4.286), and by (4.289) as well as Lemma 3.1 (ii) we get the estimate

$$\|T_j G_{ik}\|_{q,w} \le c 2^{-2|j|} \|G_{ik}\|_{q,w}$$

for all $G_{ik} \in L^q_w(\mathbb{R}^3)$ with a constant $c > 0$ independent of $j \in \mathbb{Z}$. Summarizing the previous inequalities, we proved (4.287) for $q > 2$.

Step 4. Now let $q = 2$, $w \in \widetilde{A}^-_1$. In this case, the Littlewood–Paley decomposition of $T_j f$ in L^2_w implies that

$$\|T_j G_{ik}\|^2_{2,w} \le c \int_0^\infty \int_{\mathbb{R}^n} |\varphi_s * T_j G_{ik}|^2(\mathbf{x})\, g(\mathbf{x})\, d\mathbf{x}\, \frac{ds}{s},$$

where

$$g \in L^\infty_v, \ v = \frac{1}{w} \quad \text{and} \quad \|g\|_{\infty,v} = \operatorname*{ess\,sup}_{\mathbb{R}^3} |g\, v| = 1.$$

By the same reasoning as before we arrive at (4.289), i.e.,

$$\|T_j G_{ik}\|_{2,w} \le c 2^{-|j|} \|\mathcal{M}^j g\|^{1/2}_{\infty,v} \, \|G_{ik}\|_{2,w}, \tag{4.296}$$

and at (4.294). Concerning $\mathcal{M}_{\text{hel}}$ we use the pointwise estimate $g_r(\theta, x_3) \le w_r(x_3)$ for a.a. $\theta \in (0, 2\pi)$, $x_3 \in \mathbb{R}$, and get that

$$\mathcal{M}_{\text{hel}} g_r(\theta, x_3) \le \sup_{s>0} \frac{1}{s} \int_0^{16s} w_r\big(x_3 + \frac{k}{v} t\big)\, dt \le 16\, M^+ w_r(x_3) \le c w_r(x_3)$$

with a constant $c > 0$ independent of $r > 0$. Since w is an $A_1(\mathbb{R}^3)$-weight, (4.294) implies that

$$\mathcal{M}^j g(\mathbf{x}) \le c 2^{-2|j|} \mathcal{M} w(\mathbf{x}) \le c 2^{-2|j|} w(\mathbf{x})$$

and consequently that $\|\mathcal{M}^j g\|_{\infty,v} \le c 2^{-2|j|}$ with a constant $c > 0$ independent of $j \in \mathbb{Z}$. Hence $\|T_j G_{ik}\|_{2,w} \le c 2^{-2|j|}$ proving (4.287) when $q = 2$.

Step 5. The remaining estimates are proved by duality arguments. Obviously the dual operator to T is defined by

$$T^* G(\mathbf{x}) = \int_0^\infty (\Delta_t * O_\omega(t) G)(O^T_\omega(t)\mathbf{x} + kt\boldsymbol{e}_3)\, \frac{dt}{t},$$

which has the same structure as K, but with an "opposite orientation." Hence T^* is bounded on L^q_w for $q \ge 2$ and all weights $w \in \widetilde{A}^+_{q/2}$. Now let $1 < q < 2$ and $w^{2/(2-q)} \in \widetilde{A}^-_{q/(2-q)} = \widetilde{A}^-_{(q'/2)'}$. Then by simple duality arguments $w' = w^{-q'/q} \in \widetilde{A}^+_{(q'/2)}$ and

$$|\langle TG, g\rangle| = |\langle G, T^* g\rangle| \le \|G\|_{q,w} \|T^* g\|_{q',w'} \le c \|G\|_{q,w} \|g\|_{q',w'}.$$

Finally let $q = 2$ and $\frac{1}{w} \in \widetilde{A}^+_1$. As before,

$$|\langle TG, g\rangle| \le \|G\|_{2,w} \|T^* g\|_{2,1/w} \le c \|G\|_{2,w} \|g\|_{2,1/w}\, .$$

Now Proposition 4.5 is completely proved. □

Now, using our previous results on the Bogovskii operator, we know that there is $G \in L^q_w(\mathbb{R}^3)^9$ such that

$$\nabla \cdot G = \mathbf{f},\ \|G\|_{q,w,\mathbb{R}^3} \leq c\|\mathbf{f}\|_{-1,q,\mathbb{R}^3}.$$

Applying the density arguments Theorem 4.17 is completely proved.

Proof of Theorem 4.18 Introducing the adjoint problem

$$L^*\mathbf{v} = -\Delta\mathbf{v} + (\boldsymbol{\omega} \wedge \mathbf{x}) \cdot \nabla\mathbf{v} - \boldsymbol{\omega} \wedge \mathbf{v} + \frac{\partial \mathbf{v}}{\partial x_3} = \nabla \cdot G$$

with $G \in C_0^\infty(\mathbb{R}^3)^9$. It has the solution

$$\widehat{\mathbf{v}}(\boldsymbol{\xi}) = \int_0^\infty e^{-\nu|\mathbf{x}|^2 t} O\omega(t)(G(O_\omega^T(t). - kt\boldsymbol{e}_3))(\boldsymbol{\xi})dt.$$

Applying the duality method, we get the uniqueness up to ω. □

Proof of Theorem 4.20 Following the previous results now the pressure we can get as a solution of the following problem

$$\Delta p = div\,\mathbf{f} + \nu\Delta g + \partial_\theta g = \text{ div } G \text{ in } \mathbb{R}^n.$$

Applying the Hormander–Mikhlin multiplier and the Bogovski operator, we get the estimate

$$\begin{aligned}&\|\nu\nabla\mathbf{u}\|_{q,w} + \|p\|_{q,w} + \|\omega(\boldsymbol{e}_3 \wedge \mathbf{x}) \cdot \nabla\mathbf{u} - \omega \wedge \mathbf{u}\|_{-1,q,w}\\ &\leq c(\|\mathbf{f}\|_{-1,q,w} + \|\nu\nabla g + (\omega\boldsymbol{e}_3 \wedge \mathbf{x})g\|_{-1,q,w}).\end{aligned}$$

□

4.6 L^q-Weighted Theory of the Very Weak Solutions to the Oseen Type Problem

We will follow the work by Kračmar, Krbec, Nečasová, Penel, Schumacher, see [152].

Denoting

$$S_{\omega,\nu,k} = S_{\omega,\nu} + k\partial_3 \tag{4.297}$$

and $S'_{\omega,\nu,k} = S_{-\omega,\nu,-k}$, and denoting that the $\langle\cdot,\cdot\rangle_w$ denotes convenient dualities as in Section 3, we take the following definition:

Definition 4.7 Let $\mathbf{F} \in \widehat{\mathbf{H}}_w^{-2,q}(\mathbb{R}^3)$ and $G \in \mathcal{H}_{w,0}^{-1,q}(\mathbb{R}^3)$ be given. Then a function $\mathbf{u} \in \mathbf{L}_w^q(\mathbb{R}^3)$ is called a very weak solution to Problem (4.223), if the equation

$$\left\langle \mathbf{u}, S'_{\omega,\nu,k}\boldsymbol{\varphi} - \nabla\psi \right\rangle_w = \langle \mathbf{F}, \boldsymbol{\varphi}\rangle_w + \langle G, \psi\rangle_w \tag{4.298}$$

holds for all test functions $\boldsymbol{\varphi} \in \widehat{\mathbf{H}}_{w',\sigma}^{2,q'}(\mathbb{R}^3)$ with $\mathcal{R}\boldsymbol{\varphi} \in \mathbf{L}_{w'}^{q'}(\mathbb{R}^3)$, $\partial_3\boldsymbol{\varphi} \in \mathbf{L}_{w'}^{q'}(\mathbb{R}^3)$ and $\psi \in \widehat{H}_{w'}^{1,q'}(\mathbb{R}^3)$. □

The spaces $\mathbf{K}'_{\mathcal{O}}$, $\mathcal{K}'_{\mathcal{O}}$ being introduced as the spaces analogous to Sect. 3.2, we have the result:

Theorem 4.22 *Let $1 < q < \infty$ and w be an anisotropic q-weight function according to Theorem 4.8. Then, for every pair $[\mathbf{F}, G] \in \left(\widehat{\mathbf{H}}_w^{-2,q}(\mathcal{D}) \times \mathcal{H}_{w,0}^{-1,q}(\mathcal{D})\right) \cap {\mathcal{K}'_{\mathcal{O}}}^{\perp}$, there exists a unique very weak solution $\mathbf{u} \in \mathbf{L}_w^q(\mathbb{R}^3)$ to Problem (4.223) in the sense of Definition 4.7 and satisfying the following estimate*

$$\|\mathbf{u}\|_{q,w} \le c\left(\|\mathbf{F}\|_{\widehat{H}_w^{-2,q}} + \|G\|_{\mathcal{H}_{w,0}^{-1,q}(\mathbb{R}^3)}\right), \quad c = c(\nu, \omega, q, w). \tag{4.299}$$

Moreover, there exists an associated pressure functional $p \in \mathcal{H}_{w,0}^{-1,q}(\mathbb{R}^3)$ such that $\mathbf{u}$, p solves for all $\boldsymbol{\varphi} \in \widehat{\mathbf{H}}_{w'}^{2,q'}(\mathbb{R}^3)$ with $\mathcal{R}\boldsymbol{\varphi} \in \mathbf{L}_{w'}^{q'}(\mathbb{R}^3)$, $\partial_3\boldsymbol{\varphi} \in \mathbf{L}_{w'}^{q'}(\mathbb{R}^3)$, the equation

$$\left\langle \mathbf{u}, S'_{\omega,\nu,k}\boldsymbol{\varphi}\right\rangle_w - \langle p, \operatorname{div}\boldsymbol{\varphi}\rangle_{\mathbf{w}} = \langle \mathbf{F}, \boldsymbol{\varphi}\rangle_{\mathbf{w}}.$$

The functions $\mathbf{u}$ and p fulfill the inequality

$$\|\mathbf{u}\|_{q,w} + \|p\|_{\mathcal{H}_{w,0}^{-1,q}} \le c\left(\|\mathbf{F}\|_{\widehat{H}_w^{-2,q}} + \|G\|_{\mathcal{H}_{w,0}^{-1,q}(\mathbb{R}^3)}\right), \quad c = c(\nu, \omega, q, w). \tag{4.300}$$

Proof of Theorem 4.22 We know that w' is an anisotropic q'-weight function, which belongs to $\tilde{A}_{q'}^{+}$, see [73].

For every $\mathbf{v} \in \mathbf{L}_{w'}^{q'}(\mathbb{R}^3)$, in accordance with Theorem 4.8, there exists a couple $\{\boldsymbol{\varphi}, \psi\}$ in $\widehat{\mathbf{H}}_{w',\sigma}^{2,q'}(\mathbb{R}^3) \times \widehat{H}_{w'}^{1,q'}(\mathbb{R}^3)$, strong solution to

$$S'_{\omega,\nu,k}\boldsymbol{\varphi} - \nabla\psi = \mathbf{v}, \quad \operatorname{div}\boldsymbol{\varphi} = 0. \tag{4.301}$$

Let $\{\boldsymbol{\varphi}_v, \psi_v\}$ be a representant of $\{\boldsymbol{\varphi}, \psi\} + \mathcal{K}_{\mathcal{O}}$ which satisfies the estimate

$$\|\nu\nabla^2\boldsymbol{\varphi}_v\|_{q',w'} + \|\mathcal{R}\boldsymbol{\varphi}_v\|_{q',w'} + \|\partial_3\psi_v\|_{q',w'} + \|\nabla\psi_v\|_{q',w'} \le C\|\mathbf{v}\|_{q',w'},$$

and we now define $\mathbf{u} := \mathbf{u}(\mathbf{F}, G)$ setting

$$\langle \mathbf{u}, \mathbf{v}\rangle_w := \langle \mathbf{F}, \boldsymbol{\varphi}_v\rangle_w + \langle G, \psi_v\rangle_w,$$

given. Then one has

$$\begin{aligned}|\langle \mathbf{u}, \mathbf{v}\rangle_w| &\le |\langle \mathbf{F}, \boldsymbol{\varphi}_v\rangle_w + \langle G, \psi_v\rangle_w| \le \|\mathbf{F}\|_{\widehat{H}_w^{-2,q}} \|\nabla^2 \boldsymbol{\varphi}_v\|_{q',w'} + \|G\|_{\mathcal{H}_{w,0}^{-1,q}} \|\nabla \psi_v\|_{q',w'} \\ &\le c\left(\|\mathbf{F}\|_{\widehat{H}_w^{-2,q}} + \|G\|_{\mathcal{H}_{w,0}^{-1,q}}\right)\|\mathbf{v}\|_{q',w'}.\end{aligned}$$

Therefore $\mathbf{u} = \mathbf{u}(\mathbf{F}, G)$ belongs to $(\mathbf{L}^{q'}_{w'}(\mathbb{R}^3))' = \mathbf{L}^q_w(\mathbb{R}^3)$ and fulfills the stated estimate.

It remains to show that this function $\mathbf{u} = \mathbf{u}(\mathbf{F}, G)$ solves our problem. To see this, following the same arguments as for the modified Stokes system, let $\boldsymbol{\varphi}$, ψ be the test functions with $\boldsymbol{\varphi} \in \widehat{\mathbf{H}}^{2,q'}_{w',\sigma}(\mathbb{R}^3)$, $\mathcal{R}\boldsymbol{\varphi} \in \mathbf{L}^{q'}_{w'}(\mathbb{R}^3)$, $\partial_3\boldsymbol{\varphi} \in \mathbf{L}^{q'}_{w'}(\mathbb{R}^3)$, and $\psi \in \widehat{H}^{1,q'}_{w'}(\mathbb{R}^3)$ independently. We can choose $\mathbf{v} := S'_{\omega,\nu,k}\boldsymbol{\varphi} - \nabla\psi \in \mathbf{L}^{q'}_{w'}(\mathbb{R}^3)$, and then

$$\left\langle \mathbf{u}, S'_{\omega,\nu,k}\boldsymbol{\varphi} - \nabla\psi\right\rangle_w = \langle \mathbf{u}, \mathbf{v}\rangle_w = \langle \mathbf{F}, \boldsymbol{\varphi}\rangle_w + \langle G, \psi\rangle_w.$$

Therefore $\mathbf{u}$ is a very weak solution of our problem according to the definition.

Applying the standard de Rham theorem or Bogovski lemma, see Lemma 2.7, we get the existence of pressure which satisfies the estimate (4.300).

To prove its uniqueness, let $\mathbf{U} = \mathbf{U}(\mathbf{F}, G) \in \mathbf{L}^q_w(\mathbb{R}^3)$ be another very weak solution. We have

$$\left\langle \mathbf{U}, S'_{\omega,\nu,k}\boldsymbol{\varphi} - \nabla\psi\right\rangle_w = \langle \mathbf{F}, \boldsymbol{\varphi}\rangle_w + \langle G, \psi\rangle_w \tag{4.302}$$

for all test functions $\boldsymbol{\varphi} \in \widehat{\mathbf{H}}^{2,q'}_{w',\sigma}(\mathbb{R}^3)$ with $\mathcal{R}\boldsymbol{\varphi} \in \mathbf{L}^{q'}_{w'}(\mathbb{R}^3)$, $\partial_3\boldsymbol{\varphi} \in \mathbf{L}^{q'}_{w'}(\mathbb{R}^3)$ and $\psi \in \widehat{H}^{1,q'}_{w'}(\mathbb{R}^3)$. Thus

$$\left\langle \mathbf{U} - \mathbf{u}, S'_{\omega,\nu}\boldsymbol{\varphi} - \nabla\psi\right\rangle_w = 0,$$

where $S'_{\omega,\nu}\boldsymbol{\varphi} - \nabla\psi$ being an arbitrary element from $\mathbf{L}^{q'}_{w'}(\mathbb{R}^3)$, and it follows that $\mathbf{U} - \mathbf{u} = \mathbf{0}$ in $\mathbf{L}^q_w(\mathbb{R}^3)$. □

4.7 Notes to the Oseen Type Problem

Concluding Remarks to Previous Sections

- As we have already mentioned in Introduction of Chapter 4, Sect. 4.1 we explain two different approaches in L^q setting for the strong solution:
  - An approach introduced by Farwig et al. [60], [61], see Sect. 4.1
  - An approach introduced by Galdi, Kyed [102] for the strong solution in L^q setting, see Sect. 4.1.5
- After that we have investigated in Sect. 4.2 the weak solution in L^q setting. The existence of weak solutions in L^q framework have been shown by Kračmar, Nečasová, Penel in the whole space, see [149] and also in the exterior domain, see [147], using the parametrization by angle.
- Further properties were explained using the results by Kim [139].
- Sect. 4.4 is devoted to the theory of the strong solution in L^q setting with Muckenhoupt weights. Such theory was proven by Farwig, Krbec, Nečasová in the whole space [72].
- The results on weak solutions in L^q setting with Muckenhoupt weight w are explained in Sect. 4.5. We follow the work by Kračmar, Krbec, Nečasová, Penel, Schumacher, see [153].
- The very weak solution is studied in Sect. 4.6. We follow the work by Kračmar, Krbec, Nečasová, Penel, Schumacher, see [152].

□

Remark 4.8

- Concerning properties of Oseen type operator, see also work by Kyed, [162].
- L^q approach using the associated time-periodic Oseen problem in the case of weak solution, see [103].
- The fundamental solution of 2D Oseen type of problem was investigated recently by Silvestre [197].
- The time periodic case was investigated see Eiter, [53].
- Amrouche and Consiglieri invetstigated the generalized Oseen problem see [3].

□

Chapter 5
Representation Formula and the Asymptotic Behavior of the Steady Oseen Type Problem

The chapter is devoted to the derivation of the representation formula of the steady Oseen type problem. First a truncated exterior domain $B_R \backslash \mathcal{D}^c$ is considered. Passing with $R \to \infty$ the representation formula is derived in the case of the exterior domain. In the next parts the volume potential that arises in the problem is investigated. Moreover, the asymptotic profile is shown. Using the representation formula for the Oseen type problem also, such a formula is derived for the Navier–Stokes type problem, see Chap. 11. The asymptotic profile of the gradient of the velocity field and the second derivatives of the velocity are investigated. Finally, the Leray solution is studied.

5.1 Introduction

We consider the linear stationary problem

$$-\Delta\mathbf{u} - (\mathbf{U} + \boldsymbol{\omega} \wedge \mathbf{x})\nabla\mathbf{u} + \boldsymbol{\omega} \wedge \mathbf{u} + \nabla\pi = \mathbf{f}, \quad \operatorname{div}\mathbf{u} = 0, \text{ in } \mathcal{D}, \tag{5.1}$$

which is derived from the nonlinear stationary variant of (5.1)

$$-\Delta\mathbf{u} - (\mathbf{U} + \boldsymbol{\omega} \wedge \mathbf{x})\nabla\mathbf{u} + \boldsymbol{\omega} \wedge \mathbf{u} + (\mathbf{u} \cdot \nabla)\mathbf{u} + \nabla\pi = \mathbf{f}, \quad \operatorname{div}\mathbf{u} = 0, \text{ in } \mathcal{D}. \tag{5.2}$$

Using the transformation of variables, we suppose there is some $\tau > 0$ with $\mathbf{U} = -\tau(1, 0, 0)$; hence $\boldsymbol{\omega} = \sqrt{\varrho}(1, 0, 0)$ for some $\varrho \in \mathbb{R}\backslash\{0\}$. (For details, see (1.49).)

First, we will start with works related to this problem. In the work of Galdi and Silvestre [108] was shown that under suitable assumptions on the data, and in the case of (5.2) additionally under some smallness conditions, solutions to (5.1)

Š. Nečasová et al., *Mathematical Theory of a Fluid Flow Around a Rotating and Translating Body*, Springer Monographs in Mathematics,
https://doi.org/10.1007/978-3-031-84120-0_5

and (5.2), respectively, exist in certain Sobolev spaces. These solutions are unique in the space of functions $(\mathbf{v}, \varrho)$ satisfying the relation

$$\sup\{|\mathbf{v}(\mathbf{x})|\,|\mathbf{x}| : \mathbf{x} \in \mathbb{R}^3 \backslash B_S\} < \infty \quad \text{for some } S > 0 \text{ with } \mathcal{D}^c \subset B_S.$$

Further, Galdi and Silvestre [111] showed that under additional assumptions on the data, and after some change of variables, the solutions $(\mathbf{u}, \pi)$ constructed in [108] verify the relations

$$\sup\{|\mathbf{u}(\mathbf{x})|\,|\mathbf{x}|(1 + k(|\mathbf{x}| + x_1)) : \mathbf{x} \in \mathbb{R}^3 \backslash B_S\} < \infty, \tag{5.3}$$
$$\sup\{|\nabla\mathbf{u}(\mathbf{x})|\,|\mathbf{x}|^{3/2}(1 + k(|\mathbf{x}| + x_1))^{3/2} : \mathbf{x} \in \mathbb{R}^3 \backslash B_S\} < \infty.$$

The aim of this chapter is to get a more direct access to the decay results in (5.1). Such results are typically deduced from representation formulas.

Let us start with a heuristic explanation approach. We will explain the difficulties in applying the standard approach for deriving representation formulas (or "Green's formulas," as they are also called) for Eq. (5.1). To this end, consider the Newton potential $\mathcal{N}(\mathbf{x}) = (4\pi|\mathbf{x}|)^{-1}$, that is, the fundamental solution of the Poisson equation. Suppose that $\mathbf{u} \in C^2(\mathcal{D})$. Take $\mathbf{y} \in \mathcal{D}$, $R > 0$ with $\mathcal{D}^c \subset B_R$, and $\mathbf{y} \in B_R \backslash \mathcal{D}^c$. Let $\epsilon > 0$ with $\overline{B_\epsilon(\mathbf{y})} \subset B_R \backslash \mathcal{D}^c$. Then, following a well-known argument, we find

$$\int_{(B_R \backslash \mathcal{D}^c) \backslash B_\epsilon(\mathbf{y})} (4\pi|\mathbf{y} - \mathbf{z}|)^{-1} \Delta\mathbf{u}(\mathbf{z}) d\mathbf{z}$$
$$= \int_{\partial B_R \cup \partial\mathcal{D} \cup \partial B_\epsilon(\mathbf{y})} (4\pi)^{-1} \sum_{k=1}^{3} \Big(|\mathbf{y} - \mathbf{z}|^{-1} \partial_k \mathbf{u}(\mathbf{z}) - (y - z)_k |\mathbf{y} - \mathbf{z}|^{-3} \mathbf{u}(\mathbf{z}) \Big) n_k^{(R,\epsilon)}(\mathbf{z}) do_z,$$

where $n^{(R,\epsilon)}$ denotes the inward unit normal to $(B_R \backslash \mathcal{D}^c) \backslash B_\epsilon(\mathbf{y})$. Passing ϵ tend to zero, we obtain

$$\int_{B_R \backslash \mathcal{D}^c} (4\pi|\mathbf{y} - \mathbf{z}|)^{-1} \Delta\mathbf{u}(\mathbf{z}) d\mathbf{z}$$
$$= \int_{\partial B_R \cup \partial\mathcal{D}} (4\pi)^{-1} \sum_{k=1}^{3} \Big(|\mathbf{y} - \mathbf{z}|^{-1} \partial_k \mathbf{u}(\mathbf{z}) - (y - z)_k |\mathbf{y} - \mathbf{z}|^{-3} \mathbf{u}(\mathbf{z}) \Big)$$
$$\times\, n_k^{(R)}(z) do_z - \mathbf{u}(\mathbf{y}),$$

with $\mathbf{n}^{(R)}$ denoting the inward unit normal to $B_R\backslash\mathcal{D}^c$. If $\mathbf{u}(\mathbf{z})$ decays for $|\mathbf{z}| \to \infty$, by letting R tend to infinity we conclude that

$$\int_{\mathcal{D}} (4\pi\,|\mathbf{y}-\mathbf{z}|)^{-1}\Delta\mathbf{u}(\mathbf{z})\mathrm{d}z$$

$$= \int_{\partial\mathcal{D}} (4\pi)^{-1} \sum_{k=1}^{3} \Big(|\mathbf{y}-\mathbf{z}|^{-1}\partial_k\mathbf{u}(\mathbf{z}) - (y-z)_k|\mathbf{y}-\mathbf{z}|^{-3}\mathbf{u}(\mathbf{z})\Big)\, n_k^{\mathcal{D}}(\mathbf{z})\mathrm{d}o_z - \mathbf{u}(\mathbf{y}),$$

hence

$$\mathbf{u}(\mathbf{y}) = -\int_{\mathcal{D}} (4\pi\,|\mathbf{y}-\mathbf{z}|)^{-1}\Delta\mathbf{u}(\mathbf{z})\mathrm{d}z$$

$$+ \int_{\partial\mathcal{D}} (4\pi)^{-1} \sum_{k=1}^{3} \Big(|\mathbf{y}-\mathbf{z}|^{-1}\partial_k\mathbf{u}(\mathbf{z}) - (y-z)_k|\mathbf{y}-\mathbf{z}|^{-3}\mathbf{u}(\mathbf{z})\Big)\, n_k^{(\mathcal{D})}(z)\mathrm{d}o_z.$$

Thus the function $\mathbf{u}$ is represented as the sum of a volume integral involving $\Delta\mathbf{u}$ and of a boundary integral involving $\mathbf{u}|\partial\mathcal{D}$ and $\partial\mathbf{u}/\partial\boldsymbol{n}|\partial\mathcal{D}$.

Similarly a representation of $\mathbf{u}$ in terms of $-\Delta\mathbf{u}-(\mathbf{U}+\boldsymbol{\omega}\wedge\mathbf{z})\cdot\nabla\mathbf{u}+\omega\wedge\mathbf{u}+\nabla\pi$, $\operatorname{div}\mathbf{u}$, $\mathbf{u}|\partial\mathcal{D}$, $\partial\mathbf{u}/\partial\boldsymbol{n}|\partial\mathcal{D}$, and $\pi\,|\partial\mathcal{D}$ for $\mathbf{u} \in C^2(\mathcal{D})^3$ and $\pi \in C^1(\mathcal{D})$ is obtained. In particular, we do not need and we will not suppose that $\operatorname{div}\mathbf{u} = 0$. The role of the Newton potential will be played by the fundamental solution of (5.1) mentioned above. But that latter solution is not such a simple function as the Newton potential, so quite an effort is necessary to estimate it in a way which allows us to let ϵ tend to zero (Theorem 5.6) and R to infinity (Theorem 5.7). In particular, the transition of ϵ to zero is difficult to handle.

5.1.1 Fundamental Solution of the Stationary Problem

In this chapter we follow the work of Deuring, Kračmar, and Nečasová [35]. The aim is to rigorously derive the representation formula using the fundamental solution in the nonstationary case introduced by Gunther, Thomann, see Sect. 2.5 or [124]. Let us consider the following system of equations:

$$\begin{aligned} \partial_t\boldsymbol{u} - \nu\Delta\mathbf{u} - [(U+\boldsymbol{\omega}\wedge\mathbf{x})\cdot\nabla]\mathbf{u} + \boldsymbol{\omega}\wedge\mathbf{u} + \nabla p &= \mathbf{f} \quad \text{in } \mathcal{D}\times(0,\infty) \\ \operatorname{div}\mathbf{u} &= 0 \quad \text{in } \mathcal{D}\times(0,\infty) \\ \mathbf{u} &= \mathbf{u}_{\partial\mathcal{D}} \quad \text{on } \partial\mathcal{D}\times(0,\infty) \\ \mathbf{u}(x,t) &\to 0 \quad \text{as } |\mathbf{x}|\to\infty \end{aligned} \tag{5.4}$$

in an exterior domain $\mathcal{D}$.

We are looking for the fundamental solution which will be derived from the nonsteady case such that there is tensor $\varXi(\mathbf{y}, \mathbf{z})$ and a three-dimensional vector of distributions $\varPi(\mathbf{y}, \mathbf{z})$ such that for any vector $\boldsymbol{a} \in \mathbb{R}^3$ the distributions

$$\mathbf{v}_{\mathbf{z}}(\mathbf{y}) = \varXi(\mathbf{y}, \mathbf{z})\boldsymbol{a},$$
$$\pi_{\mathbf{z}}(\mathbf{y}) = \varPi(\mathbf{y}, \mathbf{z})\boldsymbol{a}$$

solve the system

$$\begin{aligned} \mathcal{L}\mathbf{v}_{\mathbf{z}} + \nabla \pi_{\mathbf{z}} =& \delta_{\mathbf{z}}(\mathbf{y})a \\ \operatorname{div} \mathbf{v}_{\mathbf{z}} =& 0 \end{aligned} \tag{5.5}$$

in the sense of distributions, where $\mathcal{L}$ is defined by (2.22).

The following estimates of $|\mathbf{y} - \mathbf{z}(t)|$ will play a crucial role in our argument. We would like to derive from the nonsteady fundamental solution derived by Guenther, Thomann to steady case. As a first step we derive the geometrical lemma and after that the decay property.

Lemma 5.1 (Key Geometrical Lemma) *The relation* $|e^{-t\cdot\Omega} \cdot \mathbf{v}| = |\mathbf{v}|$ *holds for* $\mathbf{v} \in \mathbb{R}^3$.
Let $R \in (0, \infty)$, $\mathbf{y}, \mathbf{z} \in B_R$ *with* $\mathbf{y} \neq \mathbf{z}$, $t \in (0, \infty)$ *with*

$$t \leq \min\{|\mathbf{y} - \mathbf{z}|/(2 \cdot |\mathbf{U}|),\ |\mathbf{y} - \mathbf{z}|/(24 \cdot |\boldsymbol{\omega}| \cdot R), \quad (\arccos(3/4))/|\boldsymbol{\omega}|\}.$$

Then $|\mathbf{y} - \mathbf{z}(t)| \geq |\mathbf{y} - \mathbf{z}|/12$. □

Proof The eigenvalues of the matrix Ω are 0, $\mathrm{i} \cdot |\boldsymbol{\omega}|$, and $-\mathrm{i} \cdot |\boldsymbol{\omega}|$. Thus there is a unitary matrix $A \in \mathbb{C}^{3\times 3}$ such that

$$\Omega = A^{-1} \cdot \begin{pmatrix} 0 & 0 & 0 \\ 0 & \mathrm{i} \cdot |\boldsymbol{\omega}| & 0 \\ 0 & 0 & -\mathrm{i} \cdot |\boldsymbol{\omega}| \end{pmatrix} \cdot A.$$

Put

$$D(t) := \begin{pmatrix} 1 & 0 & 0 \\ 0 & \mathrm{e}^{-\mathrm{i}\cdot|\boldsymbol{\omega}|\cdot t} & 0 \\ 0 & 0 & \mathrm{e}^{\mathrm{i}\cdot|\boldsymbol{\omega}|\cdot t} \end{pmatrix} \quad \text{for } t \in \mathbb{R}.$$

Then $\mathrm{e}^{-t\cdot\Omega} = A^{-1} \cdot D(t) \cdot A$. This equation implies that $|\mathrm{e}^{-t\cdot\Omega} \cdot \mathbf{v}| = |\mathbf{v}|$ for $\mathbf{v} \in \mathbb{R}^3$.

Abbreviate $\widetilde{\mathbf{y}}(t) := A \cdot (\mathbf{y} + t \cdot \boldsymbol{U})$, $\quad \widetilde{\mathbf{z}} := A \cdot \mathbf{z}$. Let us mention that in general, the vectors $\widetilde{\mathbf{y}}(t)$ and $\widetilde{\mathbf{z}}$ are complex-valued. Moreover, let us observe that

$$|\mathbf{y} - \mathbf{z}(t)| = |\widetilde{\mathbf{y}}(t) - D(t) \cdot \widetilde{\mathbf{z}}|, \quad |\mathbf{y} + t \cdot \boldsymbol{U} - \mathbf{z}| = |\widetilde{\mathbf{y}}(t) - \widetilde{\mathbf{z}}|.$$

There is an index $j \in \{1, 2, 3\}$ with $|(\widetilde{\mathbf{y}}(t) - \widetilde{\mathbf{z}})_j| \geq |\widetilde{\mathbf{y}}(t) - \widetilde{\mathbf{z}}|/3$. If $j = 1$, we obtain

$$\begin{aligned} |\mathbf{y} - \mathbf{z}(t)| =& |\widetilde{\mathbf{y}}(t) - D(t) \cdot \widetilde{\mathbf{z}}| \geq |(\widetilde{\mathbf{y}}(t) - D(t) \cdot \widetilde{\mathbf{z}})_1| = |(\widetilde{\mathbf{y}}(t) - \widetilde{\mathbf{z}})_1| \\ \geq& |\widetilde{\mathbf{y}}(t) - \widetilde{\mathbf{z}}|/3 = |\mathbf{y} + t \cdot \boldsymbol{U} - \mathbf{z}|/3 \geq (|\mathbf{y} - \mathbf{z}| - t \cdot |\boldsymbol{U}|)/3 \geq |\mathbf{y} - \mathbf{z}|/6, \end{aligned} \tag{5.6}$$

where the last inequality holds because $t \leq |\mathbf{y} - \mathbf{z}|/(2 \cdot |\boldsymbol{U}|)$. Let us assume that $j \in \{2, 3\}$. Without loss of the generality, we can take $j = 2$. Then abbreviate

$$a := \widetilde{\mathbf{y}}(t)_2, \quad b := \widetilde{\mathbf{z}}_2, \quad \gamma := |\boldsymbol{\omega}| \cdot t.$$

We find

$$\begin{aligned} |\mathbf{y} - \mathbf{z}(t)|^2 &= |\widetilde{\mathbf{y}}(t) - D(t) \cdot \widetilde{\mathbf{z}}|^2 \geq |(\widetilde{\mathbf{y}}(t) - D(t) \cdot \widetilde{\mathbf{z}})_2|^2 = |a - e^{i \cdot \gamma} \cdot b|^2 \\ &= (1 - \cos\gamma) \cdot (|a|^2 + |b|^2) + \cos\gamma \cdot |a - b|^2 \\ &\quad + 2 \cdot \sin\gamma \cdot \big(a \cdot (\mathrm{Im} b, -\mathrm{Re} b)\big) \\ &\geq \cos\gamma \cdot |a - b|^2 - 2 \cdot |\sin\gamma| \cdot |a \cdot (\mathrm{Im} b, -\mathrm{Re} b)| \\ &= \cos\gamma \cdot |a - b|^2 - 2 \cdot |\sin\gamma| \cdot |(a - b) \cdot (\mathrm{Im} b, -\mathrm{Re} b)| \\ &\geq \cos\gamma \cdot |a - b|^2 - 2 \cdot |\sin\gamma| \cdot |(a - b)| \cdot |b|, \end{aligned} \tag{5.7}$$

where the symbol $\cdot$ stands for the usual scalar product of two vectors in $\mathbb{R}^2$, and Im and Re stand for the imaginary and real parts, respectively. Next we mention that since $t \leq \big(\arccos(3/4)\big)/|\boldsymbol{\omega}|$, $\cos\gamma \geq 3/4$ holds. The assumption $t \leq |\mathbf{y} - \mathbf{z}|/(24 \cdot |\boldsymbol{\omega}| \cdot R)$ implies

$$|\sin\gamma| \leq \gamma = |\boldsymbol{\omega}| \cdot t \leq |\mathbf{y} - \mathbf{z}|/(24 \cdot R).$$

(Incidentally, since $t \leq (\arccos(3/4))/|\boldsymbol{\omega}|$, we have $\gamma < \pi/2$, hence $\sin\gamma > 0$, so we might have written $\sin\gamma$ instead of $|\sin\gamma|$.)

Moreover $|b| \leq |\widetilde{\mathbf{z}}| = |\mathbf{z}| \leq R$. Due to these observations, we may conclude from (5.7)

$$|\mathbf{y} - \mathbf{z}(t)|^2 \geq (3/4) \cdot |a - b|^2 - |\mathbf{y} - \mathbf{z}| \cdot |a - b|/12. \tag{5.8}$$

On the other hand, recalling that $j = 2$ and using the relation $t \leq |\mathbf{y} - \mathbf{z}|/(2 \cdot |U|)$, we obtain

$$|a - b| \geq |\widetilde{\mathbf{y}}(t) - \widetilde{\mathbf{z}}|/3 = |\mathbf{y} + t \cdot \boldsymbol{U} - \mathbf{z}|/3 \geq (|\mathbf{y} - \mathbf{z}| - t \cdot |\boldsymbol{U}|)/3 \geq |\mathbf{y} - \mathbf{z}|/6,$$

so that $|\mathbf{y} - \mathbf{z}| \leq 6 \cdot |a - b|$. With this relation, we conclude from (5.8) that

$$|\mathbf{y} - \mathbf{z}(t)|^2 \geq (3/4) \cdot |a - b|^2 - |a - b|^2/2 = |a - b|^2/4.$$

Since $j = 2$ and $t \leq |\mathbf{y} - \mathbf{z}|/(2 \cdot |\boldsymbol{U}|)$, we finally obtain

$$|\mathbf{y} - \mathbf{z}(t)| \geq |a - b|/2 \geq |\widetilde{\mathbf{y}}(t) - \widetilde{\mathbf{z}}|/6 \geq (|\mathbf{y} - \mathbf{z}| - t \cdot |\boldsymbol{U}|)/6 \geq |\mathbf{y} - \mathbf{z}|/12.$$

□

Theorem 5.1 *Let* $S, \delta \in (0, \infty)$, $\mu \in (1, \infty)$. *Then*

$$\int_0^\infty (|\mathbf{y} - \mathbf{z}(t)|^2 + t)^{-\mu} dt \leq C(S, \delta, \mu) \, (|\mathbf{y}| s_\tau(\mathbf{y}))^{-\mu+1/2} \tag{5.9}$$

for $\mathbf{y} \in B^c_{(1+\delta)S}$, $\mathbf{z} \in \overline{B_S}$. □

Proof Let us take $\mathbf{y} \in B^c_{(1+\delta)S}$, $\mathbf{z} \in \overline{B_S}$. We shorten $\mathbf{y}' := (y_2, y_3)$. In what follows, we will make frequent use of the equation $|\mathrm{e}^{-t\Omega} \cdot \mathbf{z}| = |\mathbf{z}|$; see Lemma 5.1.

Now, we will distinguish several cases. To begin with, we suppose that $|\mathbf{y}| \leq 8\,S$. Then, for $t \in (0, \infty)$, we get

$$|\mathbf{y} - \mathbf{z}(t)|^2 + t = |\mathbf{y} - \tau\, t e_1 - \mathrm{e}^{-t\,\Omega} \cdot \mathbf{z}|^2 + t \geq C(S)(|\mathbf{y} - \mathrm{e}^{-t\,\Omega} \cdot \mathbf{z}|^2 + t) \tag{5.10}$$

(Lemma 2.15 with $9\,S$ instead of S was applied). But since $|\mathbf{y}| \geq (1+\delta)\,S$, $|\mathbf{z}| \leq S$, the following relation holds $|\mathbf{y} - \mathrm{e}^{-t\,\Omega} \cdot \mathbf{z}| \geq |\mathbf{y}| - |\mathbf{z}| \geq \delta\, S$, so that from (5.10),

$$|\mathbf{y} - \tau\, t\, \boldsymbol{e}_1 - \mathrm{e}^{-t\,\Omega} \cdot \mathbf{z}|^2 + t \geq C(S, \delta)(1 + t) \quad \text{for } t \in (0, \infty),$$

hence

$$\begin{aligned} \int_0^\infty (|\mathbf{y} - \tau\, t\, \boldsymbol{e}_1 - \mathrm{e}^{-t\,\Omega} \cdot \mathbf{z}|^2 + t)^{-\mu} \mathrm{d}t \leq & C(S, \delta, \mu) \int_0^\infty (1 + t)^{-\mu} \mathrm{d}t \\ \leq & C(S, \delta, \mu) \leq C(S, \delta, \mu) |\mathbf{y}|^{-2\,\mu+1} \\ \leq & C(S, \delta, \mu) \, (|\mathbf{y}|\, s_\tau(\mathbf{y}))^{-\mu+1/2}\,, \end{aligned} \tag{5.11}$$

with the third inequality following from the assumption $|\mathbf{y}| \leq 8\,S$ and the last one from Lemma 2.16. In the rest of this proof, we suppose that $|\mathbf{y}| \geq 8\,S$. We note that

$$\int_0^\infty (|\mathbf{y} - \tau\, t\, \boldsymbol{e}_1 - \mathrm{e}^{-t\,\Omega} \cdot \mathbf{z}|^2 + t)^{-\mu} \mathrm{d}t = \tau^{-1} \int_0^\infty (\gamma(\mathbf{y}, \mathbf{z}, r)^2 + r/\tau)^{-\mu} \mathrm{d}r, \tag{5.12}$$

where we used the abbreviation $\gamma(\mathbf{y}, \mathbf{z}, r) := |\mathbf{y} - r\, \boldsymbol{e}_1 - \mathrm{e}^{-(r/\tau)\Omega} \cdot \mathbf{z}|$ for $r \in (0, \infty)$. In view of the assumption $|\mathbf{y}| \geq 8S$, another easy case arises if $y_1 \leq 0$. In fact, we then have

$$\gamma(\mathbf{y}, \mathbf{z}, r) \geq |\mathbf{y} - r\boldsymbol{e}_1| - |\mathbf{z}| \geq (|\mathbf{y}|^2 + r^2)^{1/2} - S \geq |\mathbf{y}|/2 + r/2 - S \geq |\mathbf{y}|/4 + r/2$$

for $r \in (0, \infty)$, so that $\gamma(\mathbf{y}, \mathbf{z}, r)^2 \geq C(|\mathbf{y}| + r)^2$, hence

$$\int_0^\infty (\gamma(\mathbf{y}, \mathbf{z}, r)^2 + r/\tau)^{-\mu} dr \leq C(\mu) \int_0^\infty (|\mathbf{y}| + r)^{-2\mu} dr \tag{5.13}$$
$$\leq C(\mu)|\mathbf{y}|^{-2\mu+1} \leq C(S, \mu)(|\mathbf{y}|s_\tau(\mathbf{y}))^{-\mu+1/2},$$

where the last inequality is a consequence of Lemma 2.16. A similar argument holds if $0 \leq y_1 \leq |\mathbf{y}|/2$. In fact, since $|\mathbf{y}| = (y_1^2 + |\mathbf{y}'|^2)^{1/2}$, we then have $|\mathbf{y}'| = (|\mathbf{y}|^2 - y_1^2)^{1/2} \geq (3|\mathbf{y}|^2/4)^{1/2} \geq |\mathbf{y}|/2$, so we get for $r \in (0, \infty)$ that

$$\gamma(\mathbf{y}, \mathbf{z}, t) \geq |\mathbf{y} - r\boldsymbol{e}_1|/2 + |\mathbf{y} - r\,\boldsymbol{e}_1|/2 - |\mathbf{z}| \geq |\mathbf{y} - r\,\boldsymbol{e}_1|/2 + |\mathbf{y}'|/2 - S$$
$$\geq |\mathbf{y} - r\,\boldsymbol{e}_1|/2 + |\mathbf{y}|/4 - S \geq |y_1 - r|/2 + |\mathbf{y}|/8,$$

where the last inequality follows from the assumption $|\mathbf{y}| \geq 8\,S$. Thus we get

$$\int_0^\infty (\gamma(\mathbf{y}, \mathbf{z}, r)^2 + r/\tau)^{-\mu} dr \leq C(\mu) \int_0^\infty (|\mathbf{y}| + |y_1 - r|)^{-2\mu} dr \tag{5.14}$$
$$\leq C(\mu) \int_{y_1}^\infty (|\mathbf{y}| + r - y_1)^{-2\mu} dr$$
$$\leq C(\mu)|\mathbf{y}|^{-2\mu+1} \leq C(S, \mu)(|\mathbf{y}|s_\tau(\mathbf{y}))^{-\mu+1/2}.$$

The last of the preceding inequalities follows from Lemma 2.16. From now we suppose that $y_1 \geq |\mathbf{y}|/2$. We thus work under the assumption that $y_1 \geq |\mathbf{y}|/2 \geq 4S$. Then we note

$$\int_0^\infty (\gamma(\mathbf{y}, \mathbf{z}, r)^2 + r/\tau)^{-\mu} dr \leq \mathfrak{A}_1 + \mathfrak{A}_2, \tag{5.15}$$

with

$$\mathfrak{A}_1 := \int_{y_1-2S}^{y_1+2S} (\gamma(\mathbf{y}, \mathbf{z}, r)^2 + r/\tau)^{-\mu} dr,$$

and with $\mathfrak{A}_2$ defined in the same way as $\mathfrak{A}_1$, but with the different domain of integration $(y_1 - 2S, y_1 + 2S)$ replaced by $(0, \infty)\backslash(y_1 - 2S, y_1 + 2S)$. We observe that for $r \in (y_1 - 2S, y_1 + 2S)$,

$$r \geq y_1 - 2S \geq |\mathbf{y}|/2 - 2S \geq |\mathbf{y}|/4,$$

because $y_1 \geq |\mathbf{y}|/2$, $|\mathbf{y}| \geq 8\,S$. Therefore

$$\mathfrak{A}_1 \leq \int_{y_1-2S}^{y_1+2S} (r/\tau)^{-\mu} dr \leq C(\mu)|\mathbf{y}|^{-\mu} \int_{y_1-2S}^{y_1+2S} dr \leq C(S, \mu)|\mathbf{y}|^{-\mu}. \tag{5.16}$$

On the other hand, for $r \in (0, \infty)\backslash(y_1 - 2S,\ y_1 + 2S)$, we have

$$\gamma(\mathbf{y}, \mathbf{z}, r) \geq |\mathbf{y} - r\boldsymbol{e}_1| - |\mathbf{z}| \geq |y_1 - r| - S \geq |y_1 - r|/2 + |y_1 - r|/2 - S \geq |y_1 - r|/2,$$

hence

$$\begin{aligned}
\mathfrak{A}_2 \leq &\int_{(0,\infty)\backslash(y_1-2S,\ y_1+2S)} ((|y_1 - r|/2)^2 + r/\tau)^{-\mu} \mathrm{d}r \\
\leq &C(\mu) \int_0^\infty (|y_1 - r| + r^{1/2})^{-2\mu} \mathrm{d}r \\
\leq &C(\mu) \left(\int_0^{y_1/2} (y_1/2)^{-2\mu} \mathrm{d}r + \int_{y_1/2}^\infty (|y_1 - r| + (y_1/2)^{1/2})^{-2\mu} \mathrm{d}r \right) \\
\leq &C(\mu) \left(y_1^{-2\mu+1} + \int_{\mathbb{R}} (|y_1 - r| + y_1^{1/2})^{-2\mu} \mathrm{d}r \right) \\
\leq &C(\mu)(y_1^{-2\mu+1} + y_1^{-\mu+1/2}) \leq C(S, \mu)|\mathbf{y}|^{-\mu+1/2},
\end{aligned} \tag{5.17}$$

with the last inequality following from the assumption $y_1 \geq |\mathbf{y}|/2 \geq 4S$. Combining (5.15)–(5.17) yields

$$\int_0^\infty (\gamma(\mathbf{y}, \mathbf{z}, r)^2 + r/\tau)^{-\mu} \mathrm{d}r \leq C(S, \mu)|\mathbf{y}|^{-\mu+1/2}.$$

Therefore, if $\tau(|\mathbf{y}| - y_1) \leq \max\{1,\ 2\tau S\}$, we have

$$\int_0^\infty (\gamma(\mathbf{y}, \mathbf{z}, r)^2 + r/\tau)^{-\mu} \mathrm{d}r \leq C(S, \mu)(|\mathbf{y}| s_\tau(\mathbf{y}))^{-\mu+1/2}. \tag{5.18}$$

Thus we are reduced to the case

$$\tau(|\mathbf{y}| - y_1) \geq \max\{1,\ 2\tau S\}, \quad y_1 \geq |\mathbf{y}|/2 \geq 4S.$$

Using the relations $\tau(|\mathbf{y}| - y_1) \geq 1,\ y_1 \geq 0$, we observe that

$$|\mathbf{y}| s_\tau(\mathbf{y}) \leq |\mathbf{y}| 2\tau(|\mathbf{y}| - y_1) = 2\tau|\mathbf{y}||\mathbf{y}'|^2/(|\mathbf{y}| + y_1) \leq 2\tau|\mathbf{y}'|^2. \tag{5.19}$$

We further mention that for $r \in (0, \infty)\backslash(y_1 - 2S,\ y_1 + 2S)$,

$$\begin{aligned}
\gamma(\mathbf{y}, \mathbf{z}, r) &\geq |\mathbf{y} - r\boldsymbol{e}_1| - |\mathbf{z}| \geq |\mathbf{y} - r\boldsymbol{e}_1|/2 + |y_1 - r|/2 - S \geq |\mathbf{y} - r\boldsymbol{e}_1|/2 \\
&\geq |y_1 - r|/4 + |\mathbf{y}'|/4,
\end{aligned}$$

so that

$$\mathfrak{A}_2 \leq C(\mu) \int_{\mathbb{R}} (|y_1 - r| + |\mathbf{y}'|)^{-2\mu} \leq C(\mu)|\mathbf{y}'|^{-2\mu+1} \leq C(\mu)(|\mathbf{y}|s_\tau(\mathbf{y}))^{-\mu+1/2}, \tag{5.20}$$

with the last inequality following from (5.19). Using (5.19) again and recalling that $\tau(|\mathbf{y}| - y_1) \geq 2\tau S$, $|\mathbf{y}| \geq 4S$, we find for $r \in (0, \infty)$ that

$$\begin{aligned} \gamma(\mathbf{y}, \mathbf{z}, r) \geq& |\mathbf{y} - r e_1| - |\mathbf{z}| \geq |\mathbf{y}'| - S \geq |\mathbf{y}'|/2 + ((2\tau)^{-1}|\mathbf{y}|s_\tau(\mathbf{y}))^{1/2}/2 - S \\ \geq& |\mathbf{y}'|/2 + (|\mathbf{y}|S)^{1/2}/2 - S \geq |\mathbf{y}'|/2 + (4S^2)^{1/2}/2 - S = |\mathbf{y}'|/2. \end{aligned}$$

It follows

$$\begin{aligned} \mathfrak{A}_1 &\leq C(\mu)|\mathbf{y}'|^{-2\mu} \int_{y_1-2S}^{y_1+2} \mathrm{d}r \leq C(S, \mu)|\mathbf{y}'|^{-2\mu} \\ &\leq C(S, \mu)(|\mathbf{y}|s_\tau(\mathbf{y}))^{-\mu} \leq C(S, \mu)(|\mathbf{y}|s_\tau(\mathbf{y}))^{-\mu+1/2}, \end{aligned} \tag{5.21}$$

where inequality (5.19) was used again. By (5.15), (5.20), and (5.21), we see that inequality (5.18) holds also in the case $\tau(|\mathbf{y}| - y_1) \geq \max\{1, 2\tau S\}$, $y_1 \geq |\mathbf{y}|/2 \geq 4S$. Inequality (5.9) follows with (5.11)–(5.14) and (5.18). As concerns estimate (5.37), it is an immediate consequence of (5.34), Lemma 2.18, and (5.9) with $\mu = -3/2 - |\alpha|/2$. □

Now, we will derive other estimate of Γ_{jk}. Let us recall Γ_{jk} which was introduced in Theorem 2.4, and it is the fundamental solution constructed by Guenther, Thomann [124] for the linearized variant of (2.20). Let us introduce the fundamental solution in a more convenient form, and we define matrices

$$\begin{aligned} G^{(1)}(\mathbf{y}, \mathbf{z}, t) :=& (\delta_{jk} - (\mathbf{y} - \mathbf{z}(t))_j \cdot (\mathbf{y} - \mathbf{z}(t))_k \cdot |\mathbf{y} - \mathbf{z}(t)|^{-2})_{1 \leq j,k \leq 3} \cdot \mathrm{e}^{-t \cdot \Omega}, \\ G^{(2)}(\mathbf{y}, \mathbf{z}, t) :=& (\delta_{jk}/3 - (\mathbf{y} - \mathbf{z}(t))_j \cdot (\mathbf{y} - \mathbf{z}(t))_k \cdot |\mathbf{y} - \mathbf{z}(t)|^{-2})_{1 \leq j,k \leq 3} \cdot \mathrm{e}^{-t \cdot \Omega} \end{aligned}$$

for $\mathbf{y}, \mathbf{z} \in \mathbb{R}^3$, $t \in (0, \infty)$ with $\mathbf{y} \neq \mathbf{z}(t)$.

The fundamental solution can be rewritten in the following form:

$$\Gamma(\mathbf{y}, \mathbf{z}, t) := K(\mathbf{y} - \mathbf{z}(t),\, t)\left[G^{(1)}(\mathbf{y}, \mathbf{z}, t) - {}_1F_1\left(1, 5/2, \frac{|\mathbf{y} - \mathbf{z}(t)|^2}{4t}\right) G^{(2)}(\mathbf{y}, \mathbf{z}, t)\right] \tag{5.22}$$

for $\mathbf{y}, \mathbf{z} \in \mathbb{R}^3$, $t \in (0, \infty)$ with $\mathbf{y} \neq \mathbf{z}(t)$, $j, k \in \{1, 2, 3\}$.

Lemma 5.2 *Let* $R \in (0, \infty)$, $\mathbf{y} \in B_R$, $\epsilon \in (0, \infty)$ *with* $\overline{B_\epsilon(\mathbf{y})} \subset B_R$, $\mathbf{z} \in B_R \backslash B_\epsilon(\mathbf{y})$. *Moreover, let* $t \in (0, \infty)$, $j, k \in \{1, 2, 3\}$, $\alpha \in \mathbb{N}_0^3$ *with* $|\alpha| \leq 1$. *Then*

$$|\partial_z^\alpha \Gamma_{jk}(\mathbf{y}, \mathbf{z}, t)| + |\partial_y^\alpha \Gamma_{jk}(\mathbf{y}, \mathbf{z}, t)| \leq \mathfrak{C}(R, \epsilon) \cdot (\chi_{(0,1]}(t) + \chi_{(1,\infty)}(t) \cdot t^{-3/2}). \tag{5.23}$$

Proof Let $\mathbf{z} \in B_R \backslash B_\epsilon(\mathbf{y})$, $t \in (0, \infty)$. Let us consider two cases. First, let

$$t \le \min\{\epsilon/(2 \cdot |\boldsymbol{U}|),\ \epsilon/(24 \cdot |\boldsymbol{\omega}| \cdot R),\ (\arccos(3/4))/|\boldsymbol{\omega}|,\ \epsilon^2\};$$

then we obtain by referring to Lemma 5.1 that $|\mathbf{y} - \mathbf{z}(t)| \ge \mathfrak{C} \cdot |\mathbf{y} - \mathbf{z}| \ge \mathfrak{C} \cdot \epsilon$, hence

$$|\mathbf{y} - \mathbf{z}(t)|^2/(4 \cdot t) \ge \mathfrak{C} \cdot \epsilon^2/t \ge \mathfrak{C} > 0.$$

Thus by Theorem 2.9,

$$|d^k/du^k[\mathrm{e}^{-\mathbf{u}} \cdot {}_1F_1(1, 5/2, u)]|_{\mathbf{u}=|\mathbf{y}-\mathbf{z}(t)|^2/(4 \cdot t)} \le C \cdot (t/\epsilon^2)^{3/2+k} \quad \text{for } k \in \{0,\ 1\},$$

and by Lemma 2.9,

$$|\partial_y^{\boldsymbol{\alpha}} K(\mathbf{y} - \mathbf{z}(t),\ t)| \le C \cdot (\epsilon^2 + t)^{-3/2-|\boldsymbol{\alpha}|/2} \quad \text{for } \boldsymbol{\alpha} \in \mathbb{N}_0^3 \text{ with } |\boldsymbol{\alpha}| \le 1.$$

Obviously,

$$|\partial_{y_l} G_{jk}^{(i)}(\mathbf{y}, \mathbf{z}, t)| \le C \cdot |\mathbf{y} - \mathbf{z}(t)|^{-1} \le C \cdot \epsilon^{-1} \quad \text{for } l \in \{1, 2, 3\},\ i \in \{1, 2\}.$$

It follows that

$$\begin{aligned}
&|\partial_{z_n} \Gamma_{jk}(\mathbf{y}, \mathbf{z}, t)| \\
&\quad \le C \cdot \sum_{l=1}^{3} (|\partial_{y_l} K(\mathbf{y} - \mathbf{z}(t),\ t)| + K(\mathbf{y} - \mathbf{z}(t),\ t) \cdot |\partial_{y_l} G_{jk}^{(1)}(\mathbf{y}, \mathbf{z}, t)| \\
&\qquad + t^{-3/2} \cdot |d/du[\mathrm{e}^{-u} \cdot {}_1F_1(1, 5/2, \mathbf{u})]|_{\mathbf{u}=|\mathbf{y}-\mathbf{z}(t)|^2/(4 \cdot t)} \cdot |\mathbf{y} - \mathbf{z}(t)|/t \\
&\qquad + t^{-3/2} \cdot |\mathrm{e}^{-\mathbf{u}} \cdot {}_1F_1(1, 5/2, \mathbf{u})|_{\mathbf{u}=|\mathbf{y}-\mathbf{z}(t)|^2/(4 \cdot t)} \cdot |\partial_{y_l} G_{jk}^{(2)}(\mathbf{y}, \mathbf{z}, t)| \\
&\quad \le C \cdot ((\epsilon^2 + t)^{-2} + (\epsilon^2 + t)^{-3/2} \cdot \epsilon^{-1} \\
&\qquad + t^{-3/2} \cdot (t/|\mathbf{y} - \mathbf{z}(t)|^2)^{5/2} \cdot |\mathbf{y} - \mathbf{z}(t)|/t + t^{-3/2} \cdot (t/\epsilon^2)^{3/2} \cdot \epsilon^{-1}) \\
&\quad \le C \cdot ((\epsilon^2 + t)^{-2} + (\epsilon^2 + t)^{-3/2} \cdot \epsilon^{-1} + \epsilon^{-4}) \le C(\epsilon).
\end{aligned} \tag{5.24}$$

Next we consider the second case, and we suppose that

$$t \ge \min\{\epsilon/(2 \cdot |U|),\ \epsilon/(24 \cdot |\omega| \cdot R),\ (\arccos(3/4))/|\omega|,\ \epsilon^2\}.$$

Then, if $t \le 1$, by Lemma 2.18 we obtain $|\partial_z^\alpha \Gamma_{jk}(\mathbf{y}, \mathbf{z}, t)| \le C(R, \epsilon)$, and if $t > 1$, it is also clear by the same reference that $|\partial_z^\alpha \Gamma_{jk}(\mathbf{y}, \mathbf{z}, t)| \le C \cdot t^{-3/2}$. Together we have found that

$$|\partial_z^\alpha \Gamma_{jk}(\mathbf{y}, \mathbf{z}, t)| \le C(R, \epsilon) \cdot (\chi_{(0,1]}(t) + \chi_{(1,\infty)}(t) \cdot t^{-3/2}).$$

The terms $|\partial_z^\alpha \Gamma_{jk}(\mathbf{y}, \mathbf{z}, t)|$ and $|\Gamma_{jk}(\mathbf{y}, \mathbf{z}, t)|$ may be handled in the same way. □

Next we give an estimate of the integral $\int_0^\infty \partial_z^\beta \Gamma_{jk}(\mathbf{y}, \mathbf{z}, t)\mathrm{d}t$ for $\beta \in \mathbb{N}_0^3$ with $|\beta| \le 1$. This estimate is interesting mainly in the case when $\mathbf{y}$ and $\mathbf{z}$ are close to each other.

Theorem 5.2 *Let $R \in (0, \infty)$. Then*

$$\int_0^\infty |\partial_{z_n} \Gamma_{jk}(\mathbf{y}, \mathbf{z}, t) - \partial_{z_n} \Lambda_{jk}(\mathbf{y} - \mathbf{z}, t)| dt \le C(R) \cdot |\mathbf{y} - \mathbf{z}|^{-3/2}, \tag{5.25}$$

$$\int_0^\infty |\Gamma_{jk}(\mathbf{y}, \mathbf{z}, t)| dt \le C(R) \cdot |\mathbf{y} - \mathbf{z}|^{-1},$$

$$\int_0^\infty |\partial_{z_n} \Gamma_{jk}(\mathbf{y}, \mathbf{z}, t)| dt \le C(R) \cdot |\mathbf{y} - \mathbf{z}|^{-2}$$

for $\mathbf{y}, \mathbf{z} \in B_R$ *with* $\mathbf{y} \ne \mathbf{z}$, $j, k, n \in \{1, 2, 3\}$. □

Proof We denote $\epsilon := \min\{C_1|y - z|, C_2\}$ with C_1, C_2 from Lemma 5.1. Further we introduce

$$\psi(\mathbf{y}, \mathbf{z}, t) := \mathrm{e}^{t\Omega} \cdot \mathbf{y} - \tau t \boldsymbol{e}_1 - \mathbf{z} \quad \text{for } t \in (0, \infty), \quad \mathfrak{F}(\mathbf{u}) := {}_1F_1(1, 5/2, \mathbf{u}) \quad \text{for } \mathbf{u} \in \mathbb{R}.$$

Recalling the choice of ϵ and referring to Lemmas 2.19 and 5.1, we find for $t \in (0, \epsilon)$, $\vartheta \in [0, 1]$ such that

$$|\psi(\mathbf{y}, \mathbf{z}, \vartheta t)| = |\mathbf{y} - \tau \vartheta t \boldsymbol{e}_1 - \mathrm{e}^{-\vartheta t \Omega} \cdot \mathbf{z}| \ge C|\mathbf{y} - \mathbf{z}| \ge C\epsilon. \tag{5.26}$$

Starting from (2.57), we split the left-hand side of (5.25) in the following way:

$$\int_0^\infty |\partial z_n \Gamma_{jk}(\mathbf{y}, \mathbf{z}, t) - \partial z_n \Lambda_{jk}(\mathbf{y} - \mathbf{z}, t)| \mathrm{d}t \tag{5.27}$$

$$\le \sum_{\nu=1}^{9} \int_0^\epsilon \mathfrak{N}_\nu(t) \mathrm{d}t + \int_\epsilon^\infty |\partial z_n \Gamma_{jk}(\mathbf{y}, \mathbf{z}, t)| \mathrm{d}t + \int_\epsilon^\infty |\partial z_n \Lambda_{jk}(\mathbf{y} - \mathbf{z}, t)| \mathrm{d}t,$$

with

$$
\begin{aligned}
\mathfrak{N}_1(t) :=& \left| \sum_{l=1}^{3} ((\mathrm{e}^{-t\Omega})_{jl} - \delta_{jl}) \partial z_n (\Lambda_{lk}(\psi(\mathbf{y}, \mathbf{z}, t),\ t)) \right|, \\
\mathfrak{N}_2(t) :=& |\partial z_n (K(\psi(\mathbf{y}, \mathbf{z}, t),\ t) - K(\mathbf{y} - \mathbf{z}, t))(\delta_{jk} - \eta_{jk}(\psi(\mathbf{y}, \mathbf{z}, t)))|, \\
\mathfrak{N}_3(t) :=& |\partial z_n (-K(\psi(\mathbf{y}, \mathbf{z}, t),\ t)\mathfrak{F}(|\psi(\mathbf{y}, \mathbf{z}, t)|^2/(4t)) \\
& + K(\mathbf{y} - \mathbf{z}, t)\mathfrak{F}(|\mathbf{y} - \mathbf{z}|^2/(4tt)))(\delta_{jk}/3 - \eta_{jk}(\psi(\mathbf{y}, \mathbf{z}, t)))|, \\
\mathfrak{N}_4(t) :=& |\partial z_n (K(\mathbf{y} - \mathbf{z}, t))(\eta_{jk}(\psi(\mathbf{y}, \mathbf{z}, t)) - \eta_{jk}(\mathbf{y} - \mathbf{z}))|, \\
\mathfrak{N}_5(t) :=& |\partial z_n (K(\mathbf{y} - \mathbf{z}, t)\mathfrak{F}(|\mathbf{y} - \mathbf{z}|^2/(4t)))(\eta_{jk}(\psi(\mathbf{y}, \mathbf{z}, t)) - \eta_{jk}(\mathbf{y} - \mathbf{z}))|.
\end{aligned}
$$

The terms $\mathfrak{N}_6(t)$ to $\mathfrak{N}_9(t)$ are defined in the same way as $\mathfrak{N}_2(t)$ to $\mathfrak{N}_5(t)$, respectively, but with the derivative ∂z_n acting on the second factor instead of the first. For example, in the definition of the term $\mathfrak{N}_6(t)$, the derivative is applied to the factor $\delta_{jk} - \eta_{jk}(\psi(\mathbf{y}, \mathbf{z}, t))$, instead of $K(\psi(\mathbf{y}, \mathbf{z}, t),\ t) - K(\mathbf{y} - \mathbf{z}, t)$ as in the definition of $\mathfrak{N}_2(t)$.

In order to estimate $\mathfrak{N}_1(t)$, we observe that the eigenvalues of the matrix Ω are 0, $\mathrm{i}|\omega|$, and $-\mathrm{i}|\omega|$. Therefore there is an invertible matrix $A \in \mathbb{C}^{3\times 3}$ such that

$$
\Omega = A \cdot \begin{pmatrix} 0 & 0 & 0 \\ 0 & \mathrm{i}|\omega| & 0 \\ 0 & 0 & -\mathrm{i}|\omega| \end{pmatrix} \cdot A^{-1},
$$

hence

$$
\mathrm{e}^{-t\Omega} = A \cdot \begin{pmatrix} 1 & 0 & 0 \\ 0 & \mathrm{e}^{-\mathrm{i}t|\omega|} & 0 \\ 0 & 0 & e^{\mathrm{i}t|\omega|} \end{pmatrix} \cdot A^{-1},
$$

so for $r, s \in \{1, 2, 3\}$,

$$
(\mathrm{e}^{-t\Omega})_{rs} - \delta_{rs}| \le C(|1 - \cos(|\omega|t)| + |\sin(|\omega|t)|) \le Ct.
$$

Therefore, with Lemma 2.18 and (5.26),

$$
\mathfrak{N}_1(t) \le Ct(|\psi(\mathbf{y}, \mathbf{z}, t)|^2 + t)^{-2} \le C|\psi(\mathbf{y}, \mathbf{z}, t)|^{-2} \le C\epsilon^{-2},
$$

hence $\int_0^\epsilon \mathfrak{N}_1(t)\mathrm{d}t \le \mathfrak{C}\epsilon^{-1}$. In view of estimating $\mathfrak{N}_2(t)$ to $\mathfrak{N}_9(t)$, we observe that

$$
|\partial^\beta \eta_{jk}(\mathbf{x})| \le C|\mathbf{x}|^{-|\beta|} \quad \text{for } \mathbf{x} \in \mathbb{R}^3\backslash\{0\},\ \beta \in \mathbb{N}_0^3 \text{ with } |\beta| \le 2; \tag{5.28}
$$

$$|\partial\vartheta(|\psi(\mathbf{y},\mathbf{z},\vartheta t)|^2)| = \left|\sum_{m=1}^{3} 2\psi(\mathbf{y},\mathbf{z},\vartheta t)_m t(\Omega\cdot e^{\vartheta t\Omega}\cdot\mathbf{y}-\tau\boldsymbol{e}_1)_m\right| \tag{5.29}$$
$$\le C|\psi(\mathbf{y},\mathbf{z},\vartheta t)|t(1+|\mathbf{y}|)$$
$$\le C(R)|\psi(\mathbf{y},\mathbf{z},\vartheta t)|t \quad \text{for } t\in(0,\epsilon),\ \vartheta\in[0,1].$$

Similarly,

$$|\partial\vartheta(\psi(\mathbf{y},\mathbf{z},\vartheta t)_s)| \le C(R)t \tag{5.30}$$

for t, ϑ as before and for $s\in\{1,2,3\}$. In order to obtain an estimate of $\mathfrak{N}_2(t)$, we apply (5.30), (5.28), and Lemma 2.9, to get

$$\mathfrak{N}_2(t) \le C\left|\int_0^1 \sum_{s=1}^{3} \partial z_s\partial z_n(K(\psi(\mathbf{y},\mathbf{z},\vartheta t),t))\partial\vartheta(\psi(\mathbf{y},\mathbf{z},\vartheta t)_s)\mathrm{d}\vartheta\right|$$
$$\le C(R)\int_0^1 (|\psi(\mathbf{y},\mathbf{z},\vartheta t)|^2+t)^{-5/2}t\,\mathrm{d}\vartheta \quad \text{for } t\in(0,\epsilon).$$

By referring to (5.26), we may conclude that $\mathfrak{N}_2(t)\le C(R)(\epsilon^2+t)^{-3/2}$ for $t\in(0,\epsilon)$, so $\int_0^\epsilon \mathfrak{N}_2(t)\mathrm{d}t \le C(R)\epsilon^{-1}$. Similar arguments yield that $\int_0^\epsilon \mathfrak{N}_6(t)\mathrm{d}t \le C(R)\epsilon^{-3/2}$. Turning to $\mathfrak{N}_3(t)$, we find that

$$\mathfrak{N}_3(t) \tag{5.31}$$
$$\le Ct^{-3/2}\left|\int_0^1 \partial\vartheta\partial z_n\left(e^{-|\psi(\mathbf{y},\mathbf{z},\vartheta t)|^2/(4t)}\mathfrak{F}(|\psi(\mathbf{y},\mathbf{z},\vartheta t)|^2/(4t))\right)\mathrm{d}\vartheta\right|$$
$$= Ct^{-3/2}\Big|\int_0^1([e^{-u}\mathfrak{F}(u)]''_{|u=|\psi(\mathbf{y},\mathbf{z},\vartheta t)|^2/(4t)}\psi(\mathbf{y},\mathbf{z},\vartheta t)_n(2t)^{-1}$$
$$\times\,\partial\vartheta(|\psi(\mathbf{y},\mathbf{z},\vartheta t)|^2)(4t)^{-1}$$
$$+\,[e^{-u}\mathfrak{F}(u)]'_{|u=|\psi(\mathbf{y},\mathbf{z},\vartheta t)|^2/(4t)}\partial\vartheta(\psi(\mathbf{y},\mathbf{z},\vartheta tt)_n)(2t)^{-1})\mathrm{d}\vartheta\Big|$$
$$\le C(R)t^{-3/2}\int_0^1 (|[e^{-u}\mathfrak{F}(u)]''|_{|u=|\psi(\mathbf{y},\mathbf{z},\vartheta t)|^2/(4t)}|\psi(\mathbf{y},\mathbf{z},\vartheta t)|^2t^{-1}$$
$$+\,|[e^{-u}\mathfrak{F}(u)]'|_{|u=|\psi(\mathbf{y},\mathbf{z},\vartheta t)|^2/(4t)})\mathrm{d}\vartheta$$
$$\le C(R)t^{-3/2}\int_0^1 (\chi_{(0,1]}(u)(u+1)+\chi_{(1,\infty)}(u)u^{-5/2})_{|u=|\psi(\mathbf{y},\mathbf{z},\vartheta t)|^2/(4t)}\mathrm{d}\vartheta$$
$$\le C(R)t^{-3/2}\int_0^1 u^{-1}{}_{|u=|\psi(\mathbf{y},\mathbf{z},\vartheta t)|^2/(4t)}\mathrm{d}\vartheta \le C(R)t^{-1/2}\int_0^1 |\psi(\mathbf{y},\mathbf{z},\vartheta t)|^{-2}\mathrm{d}\vartheta.$$

Note that we applied (5.28) in the first inequality. In the second, we used (5.29) and (5.30), whereas in the third, we applied Theorem 2.9. Concerning the next to the last inequality, we chose the upper bound u^{-1} in order to obtain suitable negative powers of t and $|\psi(\mathbf{y}, \mathbf{z}, \vartheta t)|$. Making use of (5.26), we may conclude

$$\int_0^{\epsilon} \mathfrak{N}_3(t)\mathrm{d}t \leq C(R)\epsilon^{-2} \int_0^{\epsilon} t^{-1/2}\mathrm{d}t \leq C(R)\epsilon^{-3/2}.$$

By exactly the same references and techniques, one may show that

$$\int_0^{\epsilon} \mathfrak{N}_7(t)\mathrm{d}t \leq C(R)\epsilon^{-3/2}.$$

Next we observe that by (5.28), (5.30), and (5.26),

$$\begin{aligned} &|\partial z_n(\eta_{jk}(\psi(\mathbf{y}, \mathbf{z}, t)) - \eta_{jk}(\mathbf{y} - \mathbf{z}))| \\ &= \left| \int_0^1 \sum_{s=1}^{3} \partial z_s \partial z_n(\eta_{jk}(\psi(\mathbf{y}, \mathbf{z}, \vartheta t)))\partial\vartheta(\psi(\mathbf{y}, \mathbf{z}, \vartheta t)_s)\mathrm{d}\vartheta \right| \\ &\leq C(R) \int_0^1 |\psi(\mathbf{y}, \mathbf{z}, \vartheta t)|^{-2} t\mathrm{d}\vartheta \leq C(R)\epsilon^{-2}t. \end{aligned} \tag{5.32}$$

Now we get with Lemma 2.9 that

$$\mathfrak{N}_8(t) \leq C(R)(|\mathbf{y} - \mathbf{z}|^2 + t)^{-3/2}\epsilon^{-2}t \leq C(R)\epsilon^{-2}t^{-1/2} \quad \text{for } t \in (0, \epsilon),$$

so that $\int_0^{\epsilon} \mathfrak{N}_8(t)\mathrm{d}t \leq \mathfrak{C}(R)\epsilon^{-3/2}$. A similar reasoning yields for $t \in (0, \epsilon)$ that

$$\mathfrak{N}_4(t) \leq C(R)(\epsilon^2 + t)^{-2}\epsilon^{-1}t \leq C(R)(\epsilon^2 + t)^{-3/2}\epsilon^{-1/2},$$

hence $\int_0^{\epsilon} \mathfrak{N}_4(t)\mathrm{d}t \leq \mathfrak{C}(R)\epsilon^{-3/2}$. We find with Theorem 2.9 and (5.32) that

$$\begin{aligned} \mathfrak{N}_9(t) \leq& C(R)t^{-3/2}|\mathrm{e}^{-u}\mathfrak{F}(u)|_{|u=|\mathbf{y}-\mathbf{z}|^2/(4t)}\epsilon^{-2}t \\ \leq& C(R)\epsilon^{-2}t^{-1/2}(\chi_{(0,1]}(u) + \chi_{(1,\infty)}(u)u^{-3/2})_{|u=|\mathbf{y}-\mathbf{z}|^2/(4t)} \leq C(R)\epsilon^{-2}t^{-1/2} \end{aligned}$$

for $t \in (0, \epsilon)$, hence $\int_0^{\epsilon} \mathfrak{N}_9(t)\mathrm{d}t \leq \mathfrak{C}(R)\epsilon^{-3/2}$. In the same way we get $\int_0^{\epsilon} \mathfrak{N}_5(t)\mathrm{d}t \leq \mathfrak{C}(R)\epsilon^{-3/2}$. It is an immediate consequence of Lemma 2.18 that

$$\int_{\epsilon}^{\infty} |\partial z_n \Gamma_{jk}(\mathbf{y}, \mathbf{z}, t)|\mathrm{d}t + \int_{\epsilon}^{\infty} |\partial z_n \Lambda_{jk}(\mathbf{y} - \mathbf{z}, t)|\mathrm{d}t \leq \int_{\epsilon}^{\infty} t^{-2}\mathrm{d}t \leq C\epsilon^{-1}.$$

Thus, in view of (5.27), we have shown that the left-hand side of (5.25) is bounded by $C(R)\epsilon^{-3/2}$. But since $|\mathbf{y}-\mathbf{z}| \leq 2R$ and by the choice of ϵ, we have $\epsilon \geq \mathfrak{C}(R)|\mathbf{y}-\mathbf{z}|$, so inequality (5.25) follows.

The other two inequalities stated in the theorem are readily shown. In fact, for $\alpha \in \mathbb{N}_0^3$ with $|\alpha| \leq 1$, we may refer to Lemma 2.18 and (5.26) to obtain

$$\begin{aligned}\int_0^\infty |\partial_z^\alpha \Gamma_{jk}(\mathbf{y},\mathbf{z},t)|\mathrm{d}t \leq& C \cdot \int_0^\infty (|\mathbf{y}-\mathbf{z}(t)|^2+t)^{-3/2-|\alpha|/2}\mathrm{d}t\\ \leq& C\cdot\left(\int_0^\epsilon (\epsilon^2+t)^{-3/2-|\alpha|/2}\mathrm{d}t+\int_\epsilon^\infty t^{-3/2-|\alpha|/2}\mathrm{d}t\right)\\ \leq& C\cdot\left(\int_0^\infty (\epsilon^2+t)^{-3/2-|\alpha|/2}\mathrm{d}t+\epsilon^{-1/2-|\alpha|/2}\right)\\ \leq& C\cdot\epsilon^{-1-|\alpha|}.\end{aligned}$$

□

In view of Theorem 5.2, we may define

$$\mathcal{Z}_{jk}(\mathbf{y},\mathbf{z}) := \int_0^\infty \Gamma_{jk}(\mathbf{y},\mathbf{z},t)\mathrm{d}t, \quad \mathcal{Y}_{jk}(\mathbf{x}) := \int_0^\infty \Lambda_{jk}(\mathbf{x},t)\mathrm{d}t \tag{5.33}$$

for $\mathbf{x} \in \mathbb{R}^3\backslash\{0\}$, $\mathbf{y},\mathbf{z} \in \mathbb{R}^3$ with $\mathbf{y} \neq \mathbf{z}$, $j,k \in \{1,2,3\}$. The function $(\mathcal{Z}_{jk})_{1\leq j,k\leq 3}$ is the fundamental solution of (2.20) proposed by Guenther, Thomann in [124].

Remark 5.1 Theorem 5.2 implies that $\nabla(\mathcal{Z}_{jk}-\mathcal{Y}_{jk})$ is indeed weakly singular with respect to surface integrals in $\mathbb{R}^3$. □

Lemma 5.3 *Let $j,k \in \{1,2,3\}$. Then $\mathcal{Z}_{jk} \in C^1((\mathbb{R}^3\times\mathbb{R}^3)\backslash\{(\mathbf{x},\mathbf{x})\colon \mathbf{x}\in\mathbb{R}^3\})$, $\mathcal{Y}_{jk} \in C^1(\mathbb{R}^3\backslash\{0\})$,*

$$\partial y_n \mathcal{Z}_{jk}(\mathbf{y},\mathbf{z}) = \int_0^\infty \partial y_n \Gamma_{jk}(\mathbf{y},\mathbf{z},t)dt, \tag{5.34}$$

$$\partial z_n \mathcal{Z}_{jk}(\mathbf{y},\mathbf{z}) = \int_0^\infty \partial z_n \Gamma_{jk}(\mathbf{y},\mathbf{z},t)dt,$$

$$\partial_n \mathcal{Y}_{jk}(\mathbf{x}) = \int_0^\infty \partial x_n \Lambda_{jk}(\mathbf{x},t)dt$$

for $\mathbf{y},\mathbf{z} \in \mathbb{R}^3$ with $\mathbf{y}\neq\mathbf{z}$, $\mathbf{x}\in\mathbb{R}^3\backslash\{0\}$, $n \in \{1,2,3\}$.

If $R \in (0,\infty)$, $\mathbf{y},\mathbf{z} \in B_R$ with $\mathbf{y}\neq\mathbf{z}$, $\alpha \in \mathbb{N}_0^3$ with $|\alpha|\leq 1$, we have

$$|\partial_y^\alpha \mathcal{Z}_{jk}(\mathbf{y},\mathbf{z})| + |\partial_z^\alpha \mathcal{Z}_{jk}(\mathbf{y},\mathbf{z})| \leq \mathfrak{C}(R)|\mathbf{y}-\mathbf{z}|^{-1-|\alpha|}. \tag{5.35}$$

Proof Let $U, V \subset \mathbb{R}^3$ be open and bounded, with $\overline{U} \cap \overline{V} \neq \emptyset$. Then $\epsilon :=$ dist$(U, V) > 0$, and there is $R > 0$ with $\overline{U} \cup \overline{V} \subset B_R$. Therefore inequality (5.23) holds for $\mathbf{y} \in U$, $\mathbf{z} \in V$, $t \in (0, \infty)$. Since $\int_0^\infty (\chi_{(0,1)}(t) + \chi_{(1,\infty)}(t)t^{-3/2})dt < \infty$ and in view of Corollary 2.3, the continuous differentiability of $\mathcal{Z}_{jk}$ and the first two equations in (5.34) follow by Lebesgue's theorem on dominated convergence. Estimate (5.35) is a consequence of (5.34) and Theorem 5.2. Analogous arguments hold for $\mathcal{Y}_{jk}$. □

Lemma 5.4 *Let $j, k \in \{1, 2, 3\}$. For $\boldsymbol{\alpha}, \boldsymbol{\beta} \in \mathbb{N}_0^3$ with $|\boldsymbol{\alpha} + \boldsymbol{\beta}| \leq 2$, $\mathbf{y}, \mathbf{z} \in \mathbb{R}^3$ with $\mathbf{y} \neq \mathbf{z}$, the function $(0, \infty) \ni t \mapsto \partial_y^{\alpha} \partial_z^{\beta} \Gamma_{jk}(\mathbf{y}, \mathbf{z}, t) \in \mathbb{R}$ is integrable, the derivative $\partial_y^{\alpha} \partial_z^{\beta} \mathcal{Z}_{jk}(\mathbf{y}, \mathbf{z})$ exists, and*

$$\partial_y^{\alpha} \partial_z^{\beta} \mathcal{Z}_{jk}(\mathbf{y}, \mathbf{z}) = \int_0^\infty \partial_y^{\alpha} \partial_z^{\beta} \Gamma_{jk}(\mathbf{y}, \mathbf{z}, t)dt. \tag{5.36}$$

Moreover, for $\boldsymbol{\alpha}$, $\boldsymbol{\beta}$ as before, the derivative $\partial_y^{\alpha} \partial_z^{\beta} \mathcal{Z}_{jk}(\mathbf{y}, \mathbf{z})$ is a continuous function of $\mathbf{y}, \mathbf{z} \in \mathbb{R}^3$ with $\mathbf{y} \neq \mathbf{z}$. □

Proof Let $R, \epsilon \in (0, \infty)$ with $\epsilon < R$. Let $C_1 = C_1(R, \tau, \boldsymbol{\omega})$, $C_2 = C_2(R, \tau, \boldsymbol{\omega}) > 0$ be the constants introduced in Lemma 5.1. Put $\gamma := \min\{C_2, C_1 \cdot \epsilon\}$. For $\mathbf{y}, \mathbf{z} \in B_R$ with $|\mathbf{y} - \mathbf{z}| \geq \epsilon$ and for $t \in (0, \gamma]$, we have $t \leq C_2$ and $t \leq C_1 \cdot \epsilon \leq C_1 \cdot |\mathbf{y} - \mathbf{z}|$, so that $|\mathbf{y} - \tau \cdot t \cdot \boldsymbol{e}_1 - e^{-t \cdot \Omega} \cdot \mathbf{z}| \geq |\mathbf{y} - \mathbf{z}|/12$ by Lemma 5.1. Thus, referring to Corollary 2.4, we find that for $\boldsymbol{\alpha}, \boldsymbol{\beta} \in \mathbb{N}_0^3$ with $|\boldsymbol{\alpha} + \boldsymbol{\beta}| \leq 2$, $\mathbf{y}, \mathbf{z} \in B_R$ with $|\mathbf{y} - \mathbf{z}| \geq \epsilon$,

$$\begin{aligned}
|\partial_{\mathbf{y}}^{\alpha} \partial_{\mathbf{z}}^{\beta} \Gamma_{jk}(\mathbf{y}, \mathbf{z}, t)| &\leq C \cdot (|\mathbf{y} - \tau \cdot t \cdot \boldsymbol{e}_1 - e^{-t \cdot \Omega} \cdot \mathbf{z}|^2 + t)^{-3/2 - |\alpha + \beta|/2} \\
&\leq C \cdot (\chi_{(0,\gamma]}(t) \cdot ((|\mathbf{y} - \mathbf{z}|/12)^2 + t)^{-3/2 - |\alpha + \beta|/2} \\
&\quad + \chi_{(\gamma,\infty)}(t) \cdot t^{-3/2 - |\alpha + \beta|/2}) \\
&\leq C \cdot \max\{\epsilon^{-3 - |\alpha + \beta|}, \gamma^{-|\alpha + \beta|/2}\} \\
&\quad \cdot (\chi_{(0,\gamma]}(t) + \chi_{(\gamma,\infty)}(t) \cdot t^{-3/2}).
\end{aligned}$$

But the function $(0, \infty) \ni t \mapsto \chi_{(0,\gamma]}(t) + \chi_{(\gamma,\infty)}(t) \cdot t^{-3/2} \in [0, \infty)$ is integrable. Moreover, for any $t \in (0, \infty)$, we have $\Gamma_{jk}(\cdot, \cdot, t) \in C^2(\mathbb{R}^3 \times \mathbb{R}^3)$ (Corollary 2.3). Therefore Lebesgue's theorem on dominated convergence yields that the statements of Lemma 5.4 hold for $\mathbf{y}, \mathbf{z} \in \mathbb{R}^3$ with $\epsilon < |\mathbf{y} - \mathbf{z}| < R$. Since this is true for any R, $\epsilon \in (0, \infty)$ with $\epsilon < R$, Lemma 5.4 is proved. □

Now, we are in the position to estimate $\mathcal{Z}_{jk}(\mathbf{y}, \mathbf{z})$ in the case that $|\mathbf{z}| \leq S$, $|\mathbf{y}| \geq (1 + \delta)S$ with $\delta, S > 0$ considered as given quantities. This estimate will play a crucial role in the following.

Theorem 5.3 *Let $S, \delta \in (0, \infty)$, $y \in B^c_{(1+\delta)S}$, $\mathbf{z} \in \overline{B_S}$. Then*

$$|\partial^{\boldsymbol{\alpha}}_{\mathbf{y}} \mathcal{Z}(\mathbf{y}, \mathbf{z})| + |\partial^{\boldsymbol{\alpha}}_{\mathbf{z}} \mathcal{Z}(\mathbf{y}, \mathbf{z})| \leq C(S, \delta)(|\mathbf{y}| s_\tau(\mathbf{y}))^{-1-|\alpha|/2} \tag{5.37}$$

for $j, k \in \{1, 2, 3\}$, $\alpha \in \mathbb{N}^3_0$ with $|\alpha| \leq 1$. Moreover

$$|\partial^{\boldsymbol{\alpha}}_{\mathbf{y}} \mathcal{Z}(\mathbf{y}, \mathbf{z})| + |\partial^{\boldsymbol{\alpha}}_{\mathbf{z}} \mathcal{Z}(\mathbf{y}, \mathbf{z})| \leq C(S, \delta)(|\mathbf{z}| s_\tau(\mathbf{z}))^{-1-|\alpha|/2} \tag{5.38}$$

for $\mathbf{z} \in B^c_{(1+\delta)\,S}$, $\mathbf{y} \in \overline{B_S}$, and for $j, k, \boldsymbol{\alpha}$ as in (5.37). □

Proof Inequalities (5.37) and (5.38) follow with (5.34), Lemma 2.18, and Theorem 5.1. Let us deal with (5.38). In this respect, we remark that the only property of Ω we used in the proof of Theorem 5.1 is the relation $|\mathrm{e}^{-t\Omega} \cdot \mathbf{x}| = |\mathbf{x}|$ for $\mathbf{x} \in \mathbb{R}^3$, $t \in (0, \infty)$ (Lemma 5.1). Since this relation holds, of course, for any $t \in \mathbb{R}$, and because by Lemma 5.1,

$$|\mathbf{y} - t\tau e_1 - \mathrm{e}^{-\tau\Omega} \cdot \mathbf{z}| = |-\mathbf{z} - t\tau \boldsymbol{e}_1 - \mathrm{e}^{t\Omega} \cdot (-\mathbf{y})| \quad (\mathbf{y}, \mathbf{z} \in \mathbb{R}^3,\ t \in \mathbb{R}),$$

we see that we have proved (5.9) also for $\mathbf{z} \in B^c_{(1+\delta)S}$ and $\mathbf{y} \in \overline{B_S}$, but with $\mathbf{y}$ replaced by $\mathbf{z}$ on the right-hand side. □

Theorem 5.4 *Let $S_1, S \in (0, \infty)$ with $S_1 < S$. Then*

$$|\partial^{\boldsymbol{\alpha}}_{\mathbf{y}} \partial^{\boldsymbol{\beta}}_{\mathbf{z}} \mathcal{Z}_{jk}(\mathbf{y}, \mathbf{z})| \leq C(S_1, S) \cdot (|\mathbf{y}| \cdot s_\tau(\mathbf{y}))^{-1+|\boldsymbol{\alpha}+\boldsymbol{\beta}|/2}$$

for $\mathbf{y} \in B^c_S$, $\mathbf{z} \in \overline{B_{S_1}}$, $\boldsymbol{\alpha}, \boldsymbol{\beta} \in \mathbb{N}^3_0$ with $|\boldsymbol{\alpha} + \boldsymbol{\beta}| \leq 2$, $1 \leq j, k \leq 3$. □

Proof For $\mathbf{y}, \mathbf{z}, \boldsymbol{\alpha}, \boldsymbol{\beta}, j, k$ as in the theorem, we deduce from (5.36), Corollary 2.4, and Theorem 5.1 that

$$\begin{aligned}
|\partial^{\boldsymbol{\alpha}}_{\mathbf{y}} \partial^{\boldsymbol{\beta}}_{\mathbf{z}} \mathcal{Z}_{jk}(\mathbf{y}, \mathbf{z})| \leq & \int_0^\infty |\partial^{\boldsymbol{\alpha}}_{\mathbf{y}} \partial^{\boldsymbol{\beta}}_{\mathbf{z}} \Gamma_{jk}(\mathbf{y}, \mathbf{z}, t)| \mathrm{d}t \\
\leq & C \cdot \int_0^\infty (|\mathbf{y} - \tau \cdot t \cdot \boldsymbol{e}_1 - \mathrm{e}^{-t \cdot \Omega} \cdot \mathbf{z}|^2 + t)^{-3/2-|\boldsymbol{\alpha}+\boldsymbol{\beta}|/2} \mathrm{d}t \\
\leq & C(S_1, S) \cdot (|\mathbf{y}| \cdot s_\tau(\mathbf{y}))^{-1-|\boldsymbol{\alpha}+\boldsymbol{\beta}|/2}.
\end{aligned}$$

□

5.1.2 Mathematical Preliminaries to the Representation Formula

In this subsection we will study other properties of Λ_{jk} and $\mathcal{Z}_{jk}$, which were already studied in Chap. 2 and which allow us to pass with $\epsilon \to 0$ (see the heuristic approach).

The role of the function Λ_{jk} becomes apparent from (5.25) and the ensuing lemma (see also Theorem 5.2).

Lemma 5.5 *Let $j, k, n \in \{1, 2, 3\}$, $\mathbf{x} \in \mathbb{R}^3 \backslash \{0\}$. Then*

$$\begin{aligned}\partial_n \mathcal{Y}_{jk}(\mathbf{x}) =& \int_0^\infty \partial_{z_n} \Lambda_{jk}(\mathbf{x}, t)dt \\ =& -(4 \cdot \pi)^{-1}|\mathbf{x}|^{-3} \cdot \left(\delta_{jk}x_n + \delta_{jn}x_k + \delta_{kn}x_j - 3\eta_{jk}x_n\right) \\ &+ A|\mathbf{x}|^{-3}\left(\delta_{jk}x_n/3 + \delta_{jn}x_k + \delta_{kn}x_j - 3\eta_{jk}x_n\right),\end{aligned}$$

where the constant A is independent of j, k, n, $\mathbf{x}$, and t.[1] □

Although this lemma follows from Lemma 5.1, let us indicate a simple consequence of the Hölder inequality and Theorem 5.2.

Lemma 5.6 *Let $j, k \in \{1, 2, 3\}$, $R \in (0, \infty)$, $p \in (3/2, \infty)$, $\mathbf{f} \in L^p(B_R)$, $\mathbf{y} \in B_R$. Then*

$$\int_{B_R} |\mathcal{Z}_{jk}(\mathbf{y}, \mathbf{z})|\, |\mathbf{f}(\mathbf{z})|dz \leq C(R, p)\|\mathbf{f}\|_p.$$

Proof We know by Theorem 5.2 that

$$\begin{aligned}\int_{B_R} |\mathcal{Z}_{jk}(\mathbf{y}, \mathbf{z})|\, |\mathbf{f}(\mathbf{z})|\mathrm{d}z \leq& \int_{B_R}\int_0^\infty |\Gamma_{jk}(\mathbf{y}, \mathbf{z}, t)|\, |\mathbf{f}(\mathbf{z})|\mathrm{d}t\mathrm{d}z \\ \leq& C(R)\int_{B_R} |\mathbf{y}-\mathbf{z}|^{-3/2}|\mathbf{f}(\mathbf{z})|\mathrm{d}z.\end{aligned}$$

The lemma follows from this estimate by the Hölder inequality. □

A similar argument yields:

Lemma 5.7 *Let $R \in (0, \infty)$, $p \in (3, \infty)$, $\mathbf{f} \in L^p(B_R)$, $\mathbf{y} \in B_R$. Then*

$$\int_{B_R} |\mathbf{y}-\mathbf{z}|^{-2}|\mathbf{f}(\mathbf{y})|dy \leq C(R, p)\|\mathbf{f}\|_p.$$

---

[1] As in the previous part $\eta_{jk} = \frac{x_j x_k}{|\mathbf{x}|^2}$.

Next, in Lemma 5.8 and Theorem 5.5, we prove some technical points. They constituted a major obstacle in the proof of a representation formula for smooth functions $\mathbf{u}\colon \mathcal{D} \mapsto \mathbb{R}^3$ in terms of $\mathcal{L}(\mathbf{u})+\nabla\pi$, $\mathcal{L}(\mathbf{u}) = S_{1,\omega} - k\partial_1\mathbf{u}$, $\operatorname{div}\mathbf{u}$, and $\mathbf{u}|\partial\mathcal{D}$ (Theorem 5.10). This obstacle consisted in finding a leading term in a decomposition of $\partial z_n \mathcal{Z}_{jk}(\mathbf{y}, \mathbf{z})$ such that the remainder term is weakly singular with respect to surface integrals in $\mathbb{R}^3$, see Remark 5.1. The importance of such decomposition will become apparent in the proof of Theorem 5.5. The leading term in question is in fact the function $\mathcal{Y}_{jk}(\mathbf{y}-\mathbf{z})$, which turns out to coincide with the usual fundamental solution of the Stokes system:

Lemma 5.8 *Let* $j, k \in \{1, 2, 3\}$, $\mathbf{x} \in \mathbb{R}^3\backslash\{0\}$. *Then*

$$\mathcal{Y}_{jk}(\mathbf{x}) = (8\pi|\mathbf{x}|)^{-1}(\delta_{jk} + x_j x_k|\mathbf{x}|^{-2}).$$

Proof Abbreviate $\mathfrak{F}(\mathbf{u}) := {}_1F_1(1, 5/2, \mathbf{u})$ for $\mathbf{u} \in \mathbb{R}$. Then

$$\begin{aligned}\mathcal{Y}_{jk}(\mathbf{x}) =& \big(\delta_{jk} - \eta_{jk}(\mathbf{x})\big) \int_0^\infty K(\mathbf{x}, t)\mathrm{d}t \\ &+ \big(-\delta_{jk}/3 + \eta_{jk}(\mathbf{x})\big)\,(4\pi)^{-3/2} \int_0^\infty t^{-3/2}\mathrm{e}^{-|\mathbf{x}|^2/(4t)}\mathfrak{F}\left(|\mathbf{x}|^2/(4t)\right)\mathrm{d}t \\ =&(4|\mathbf{x}|)^{-1}\pi^{-3/2}\bigg(\big(\delta_{jk} - \eta_{jk}(\mathbf{x})\big)\int_0^\infty s^{-3/2}\mathrm{e}^{-1/s}\mathrm{d}s \\ &+ \big(-\delta_{jk}/3 + \eta_{jk}(\mathbf{x})\big)\int_0^\infty s^{-3/2}\mathrm{e}^{-1/s}\mathfrak{F}(1/s)\mathrm{d}s\bigg) \\ =&(4|\mathbf{x}|)^{-1}\pi^{-3/2}\bigg(\big(\delta_{jk} - \eta_{jk}(\mathbf{x})\big)\int_0^\infty t^{-1/2}\mathrm{e}^{-t}\mathrm{d}t \\ &+ \big(-\delta_{jk}/3 + \eta_{jk}(\mathbf{x})\big)\int_0^\infty t^{-1/2}\mathrm{e}^{-t}\mathfrak{F}(t)\mathrm{d}t\bigg).\end{aligned} \tag{5.39}$$

But $\int_0^\infty t^{-1/2}\mathrm{e}^{-t}\mathrm{d}t = \pi^{1/2}$ by a result about the Gamma function. Therefore, using the abbreviation

$$A := (1/4)\pi^{-3/2}\int_0^\infty t^{-1/2}\mathrm{e}^{-t}\mathfrak{F}(t)\mathrm{d}t,$$

we conclude from (5.39) that

$$\mathcal{Y}_{jk}(\mathbf{x}) = (4\pi|\mathbf{x}|)^{-1}\left(\delta_{jk} - \eta_{jk}(\mathbf{x})\right) + A|\mathbf{x}|^{-1}\left(-\delta_{jk}/3 + \eta_{jk}(\mathbf{x})\right). \tag{5.40}$$

But $\int_0^\infty t^{-1/2}\mathrm{e}^{-t}\mathfrak{F}(t)\mathrm{d}t = 3\pi^{1/2}/2$ as follows by some standard properties of the Gamma function and by the equation $\sum_{n=1}^\infty ((2n-1)(2n+1))^{-1} = 1/2$. Therefore $A = 3(8\pi)^{-1}$, so the lemma may be deduced from (5.40). □

As a consequence of Theorem 5.2, Lemma 5.5, and the relation $\int_{\partial B_1} \eta_{jk} do = 4\pi\delta_{jk}/3$, we obtain the following theorem:

Theorem 5.5 *Let* $j, k, n \in \{1, 2, 3\}$, $\mathbf{y} \in \mathbb{R}^3$, $\epsilon_0 > 0$. *Then, for* $\mathbf{v} \in C^0\left(\overline{B_{\epsilon_0}(\mathbf{y})}\right)$,

$$\int_{\partial B_\epsilon(\mathbf{y})} \mathcal{Z}_{jk}(\mathbf{y}, \mathbf{z}) \frac{(\mathbf{y}-\mathbf{z})_n}{\epsilon} \mathbf{v}(\mathbf{z}) do_z \to 0 \quad (\epsilon \downarrow 0). \tag{5.41}$$

Let $\mathbf{w} \in C^\mu\left(\overline{B_{\epsilon_0}(\mathbf{y})}\right)$ *for some* $\mu \in (0, 1]$. *Then*

$$\int_{B_\epsilon(\mathbf{y})} \sum_{m=1}^{3} \partial_{z_m} \mathcal{Z}_{jk}(\mathbf{y}, \mathbf{z})(\mathbf{y}-\mathbf{z})_m \epsilon^{-1} \mathbf{w}(\mathbf{z}) do_z \to 2\delta_{jk}\mathbf{w}(\mathbf{y})/3 \tag{5.42}$$

for $\epsilon \downarrow 0$. □

Proof We get the first claim from Lemma 5.3 and (5.35). To prove (5.42), we choose $R > 0$ with $\overline{B_{\epsilon_0}(\mathbf{y})} \subset B_R$. For $\epsilon \in (0, \epsilon_0]$, we observe that the difference of the left- and right-hand side of (5.42) is bounded by $\sum_{\nu=1}^{3} \mathfrak{N}_\nu(\epsilon)$, with

$$\mathfrak{N}_1(\epsilon) := \int_{\partial B_\epsilon(\mathbf{y})} \sum_{m=1}^{3} |\partial z_m \mathcal{Z}_{jk}(\mathbf{y}, \mathbf{z})| \, |\mathbf{w}(\mathbf{z}) - \mathbf{w}(\mathbf{y})| \mathrm{d}o_z,$$

$$\mathfrak{N}_2(\epsilon) := |\mathbf{w}(\mathbf{y})| \sum_{m=1}^{3} \int_{\partial B_\epsilon(\mathbf{y})} |\partial z_m \mathcal{Z}_{jk}(\mathbf{y}, \mathbf{z}) - \partial z_m \mathcal{Y}_{jk}(\mathbf{y}-\mathbf{z})| \mathrm{d}o_z,$$

$$\mathfrak{N}_3(\epsilon) := \left| \mathbf{w}(\mathbf{y}) \int_{\partial B_\epsilon(\mathbf{y})} \sum_{m=1}^{3} \partial z_m \mathcal{Y}_{jk}(\mathbf{y}-\mathbf{z}) \frac{(\mathbf{y}-\mathbf{z})_m}{\epsilon} \mathrm{d}o_z - 2\delta_{jk}\mathbf{w}(\mathbf{y})/3 \right|.$$

Put

$$[\mathbf{w}]_\mu := \sup\{|\mathbf{w}(\mathbf{z}) - \mathbf{w}(\mathbf{z}')| \, |\mathbf{z}-\mathbf{z}'|^{-\mu} : \mathbf{z}, \mathbf{z}' \in \overline{B_{\epsilon_0}(\mathbf{y})}, \ \mathbf{z} \neq \mathbf{z}'\}.$$

Let $\epsilon \in (0, \epsilon_0]$. Then with (5.35), we find

$$\mathfrak{N}_1(\epsilon) \le C(R)[\mathbf{w}]_\mu \int_{\partial B_\epsilon(\mathbf{y})} |\mathbf{y}-\mathbf{z}|^{-2+\mu} \mathrm{d}o_z \le C(R)[\mathbf{w}]_\mu \epsilon^\mu.$$

Moreover, referring to (5.34) and to Theorem 5.2, we get

$$\mathfrak{N}_2(\epsilon) \le C(R)|\mathbf{w}(\mathbf{y})| \int_{\partial B_\epsilon(\mathbf{y})} |\mathbf{y}-\mathbf{z}|^{-3/2} \mathrm{d}o_z \le C(R)|\mathbf{w}(\mathbf{y})|\epsilon^{1/2}.$$

Using Lemma 5.8 and noting that $\int_{\partial B_1} \eta_{rs} \mathrm{d}o = 4\pi\delta_{rs}/3$ for $r, s \in \{1, 2, 3\}$, we find

$$\int_{\partial B_\epsilon(\mathbf{y})} \sum_{m=1}^{3} \partial z_m \mathcal{Y}_{jk}(\mathbf{y}-\mathbf{z}) \frac{(y-z)_m}{\epsilon} \mathrm{d}o_{\mathbf{z}}$$

$$=(8\pi)^{-1} \int_{\partial B_1} \sum_{m=1}^{3} \left(\delta_{jk}\eta_{mm} - \delta_{jm}\eta_{km} - \delta_{km}\eta_{jm} + 3\eta_{jk}\eta_{mm}\right) \mathrm{d}o_\eta = 2\delta_{jk}/3,$$

so that $\mathfrak{N}_3(\epsilon) = 0$. Letting ϵ tend to zero, we obtain the theorem. □

We remark that the Hölder continuity of w in the preceding theorem is needed in view of Theorem 5.2 so that the relation

$$\int_{B_\epsilon(\mathbf{y})} \sum_{m=1}^{3} \partial_{z_m} \mathcal{Z}_{jk}(\mathbf{y},\mathbf{z}) |\mathbf{w}(\mathbf{z}) - \mathbf{w}(\mathbf{y})| \mathrm{d}o_{\mathbf{z}} \to 0 \quad \text{for } \epsilon \downarrow 0$$

holds.

Also due to the relation $\int_{\partial B_1} \eta_{jk} \mathrm{d}o = (4 \cdot \pi/3)\delta_{jk}$, we get the following lemma:

Lemma 5.9 *Let* $j, k \in \{1, 2, 3\}$, $\mathbf{y} \in \mathbb{R}^3$, $\epsilon_0 > 0$, $\alpha \in (0, 1]$, $\mathbf{v} \in C^\alpha(\overline{B_{\epsilon_0}(\mathbf{y})})$. *Then*

$$\int_{\partial B_\epsilon(\mathbf{y})} (4 \cdot \pi)^{-1} (y-z)_j (y-z)_k \epsilon^{-4} \mathbf{v}(\mathbf{z}) do_z \to \delta_{jk} \mathbf{v}(\mathbf{y})/3$$

for $\epsilon \downarrow 0$. □

5.2 Derivation of the Representation Formula

Now we are in a position to state the representation formula. First we consider such a formula in a truncated exterior domain $B_R \backslash \mathcal{D}^c$.

Theorem 5.6 *Let* $R \in (0, \infty)$ *with* $\mathcal{D}^c \subset B_R$. *Let* $\mathbf{n}^{(R)}: \partial B_R \cup \partial\mathcal{D} \mapsto \mathbb{R}^3$ *denote the outward unit normal to* $B_R \backslash \mathcal{D}^c$. *Suppose that* $\mathbf{u} \in C^2(\overline{B_R} \backslash \mathcal{D}^c)$, $\pi \in C^1(\overline{B_R} \backslash \mathcal{D}^c)$, *and put* $\mathbf{f} := \mathcal{L}(\mathbf{u}) + \nabla\pi$ *with* $\mathcal{L}$ *defined as*

$$\mathcal{L}(\mathbf{u}) = -\Delta\mathbf{u} + \tau\partial_1\mathbf{u} - (\omega \wedge \mathbf{x}) \cdot \nabla\mathbf{u} + \boldsymbol{\omega} \wedge \mathbf{u}. \tag{5.43}$$

Assume that $\mathbf{y} \in B_R \backslash \mathcal{D}^c$ *and* $j \in \{1, 2, 3\}$. *Then*

$$u_j(\mathbf{y}) = \int_{B_R \backslash \mathcal{D}^c} \left(\sum_{k=1}^{3} \mathcal{Z}_{jk}(\mathbf{y}, \mathbf{z}) f_k(\mathbf{z}) + (4 \cdot \pi)^{-1} (\mathbf{y} - \mathbf{z})_j |\mathbf{y} - \mathbf{z}|^{-3} \operatorname{div} \mathbf{u}(\mathbf{z}) \right) d\mathbf{z} \tag{5.44}$$

$$+ \int_{\partial B_R \cup \partial \mathfrak{D}} \sum_{k=1}^{3} \Big[\sum_{l=1}^{3} (\mathcal{Z}_{jk}(\mathbf{y}, \mathbf{z})(\partial_l u_k(\mathbf{z}) - \delta_{kl} \pi(\mathbf{z}) + u_k(\mathbf{z})(\mathbf{U} + \boldsymbol{\omega} \wedge \mathbf{z})_l)$$

$$- \partial_{z_l} \mathcal{Z}_{jk}(\mathbf{y}, \mathbf{z}) u_k(\mathbf{z})) n_l^{(R)}(\mathbf{z}) - (4 \cdot \pi)^{-1} (\mathbf{y} - \mathbf{z})_j |\mathbf{y} - \mathbf{z}|^{-3} u_k(\mathbf{z}) n_k^{(R)}(\mathbf{z}) \Big] do_{\mathbf{z}}.$$

Proof Let $\epsilon \in (0, \infty)$ with $\overline{B_\epsilon(\mathbf{y})} \subset B_R \backslash \mathcal{D}^c$. Define $\mathbf{n}^{(R,\epsilon)}: \partial B_R \cup \partial \mathcal{D} \cup \partial B_\epsilon(\mathbf{y}) \mapsto \mathbb{R}^3$ by

$$\mathbf{n}^{(R,\epsilon)}(\mathbf{z}) := \mathbf{n}^{(R)}(\mathbf{z}) \quad \text{for } \mathbf{z} \in \partial B_R \cup \partial \mathcal{D},$$

$$\mathbf{n}^{(R,\epsilon)}(\mathbf{z}) := \epsilon^{-1}(\mathbf{y} - \mathbf{z}) \quad \text{for } \mathbf{z} \in \partial B_\epsilon(\mathbf{y}).$$

This means that $\mathbf{n}^{(R,\epsilon)}$ is the inward unit normal to $(B_R \backslash \mathcal{D}^c) \backslash B_\epsilon(\mathbf{y})$. Abbreviate

$$\mathfrak{A}_\epsilon := \int_{(B_R \backslash \mathcal{D}^c) \backslash B_\epsilon(\mathbf{y})} \sum_{k=1}^{3} \mathcal{Z}_{jk}(\mathbf{y}, \mathbf{z}) f_k(\mathbf{z}) d\mathbf{z}.$$

By Lemma 5.2, Lebesgue's theorem, and Fubini's theorem, we have

$$\mathfrak{A}_\epsilon = \lim_{S \to \infty, \delta \downarrow 0} \int_\delta^S \int_{(B_R \backslash \mathcal{D}^c) \backslash B_\epsilon(\mathbf{y})} \sum_{k=1}^{3} \Gamma_{jk}(\mathbf{y}, \mathbf{z}, t) f_k(\mathbf{z}) d\mathbf{z} dt. \tag{5.45}$$

Take $\delta, S \in (0, \infty)$ with $\delta < S$, and recall the definition of $\mathbf{f}$ in the theorem. Since $\Gamma_{jk}(\mathbf{y}, \cdot, t) \in C^\infty(\mathbb{R}^3)$ for $t \in (0, \infty)$ (Corollary 2.3), we may integrate by parts in the integral on the right-hand side of (5.45). Taking into account Theorem 2.7, we thus obtain

$$\mathfrak{A}_\epsilon = \lim_{S \to \infty, \delta \downarrow 0} \int_\delta^S \int_{(B_R \backslash \mathcal{D}^c) \backslash B_\epsilon(\mathbf{y})} \sum_{k=1}^{3} -\partial_t \Gamma_{jk}(\mathbf{y}, \mathbf{z}, t) u_k(\mathbf{z}) d\mathbf{z} dt + A_{j,\epsilon}(\mathbf{y})$$

with

$$A_{j,\epsilon}(\mathbf{y}) := \int_{\partial B_R \cup \partial \mathcal{D} \cup \partial B_\epsilon(\mathbf{y})} \sum_{k=1}^{3} \sum_{l=1}^{3} (\mathcal{Z}_{jk}(\mathbf{y}, \mathbf{z})(-\partial_l u_k(\mathbf{z}) + \delta_{kl} \pi(\mathbf{z})$$

$$- (U + \boldsymbol{\omega} \wedge z)_l u_k(\mathbf{z})) + \partial_{z_l} \mathcal{Z}_{jk}(\mathbf{y}, \mathbf{z}) u_k(\mathbf{z})) n_l^{(R,\epsilon)}(\mathbf{z}) do_{\mathbf{z}}.$$

Note that in the integral appearing in the definition of $A_{j,\epsilon}(\mathbf{y})$, we had S tending to infinity and δ to zero, and we applied Fubini's theorem. This is possible because by Lemma 5.2, we have

$$\int_0^\infty \int_{\partial B_R \cup \partial \mathcal{D} \cup \partial B_\epsilon(\mathbf{y})} (|\Gamma_{jk}(\mathbf{y}, \mathbf{z}, t)| + |\partial_{z_n} \Gamma_{jk}(\mathbf{y}, \mathbf{z}, t)|) \mathrm{d}o_{\mathbf{z}} \mathrm{d}t < \infty$$

for $1 \leq n \leq 3$. Now it follows that

$$\mathfrak{A}_\epsilon = \lim_{S \to \infty, \delta \downarrow 0} \int_{(B_R \setminus \mathcal{D}^c) \setminus B_\epsilon(\mathbf{y})} \sum_{k=1}^3 (-\Gamma_{jk}(\mathbf{y}, \mathbf{z}, S) + \Gamma_{jk}(\mathbf{y}, \mathbf{z}, \delta)) u_k(\mathbf{z}) \mathrm{d}\mathbf{z} + A_{j,\epsilon}(\mathbf{y}). \tag{5.46}$$

But by Lemma 5.2, Theorem 2.6, and Lebesgue's theorem, we have

$$\int_{(B_R \setminus \mathcal{D}^c) \setminus B_\epsilon(\mathbf{y})} \sum_{k=1}^3 \Gamma_{jk}(\mathbf{y}, \mathbf{z}, S) u_k(\mathbf{z}) \mathrm{d}\mathbf{z} \to 0 \quad \text{for } S \to \infty, \tag{5.47}$$

and

$$\int_{(B_R \setminus \mathcal{D}^c) \setminus B_\epsilon(\mathbf{y})} \sum_{k=1}^3 \Gamma_{jk}(\mathbf{y}, \mathbf{z}, \delta) u_k(\mathbf{z}) \mathrm{d}\mathbf{z} \tag{5.48}$$

$$\to -(4 \cdot \pi)^{-1} \int_{(B_R \setminus \mathcal{D}^c) \setminus B_\epsilon(\mathbf{y})} \sum_{k=1}^3 (\delta_{jk} |\mathbf{y} - \mathbf{z}|^{-3} - 3(\mathbf{y} - \mathbf{z})_j \times (\mathbf{y} - \mathbf{z})_k |\mathbf{y} - \mathbf{z}|^{-5}) u_k(\mathbf{z}) \mathrm{d}\mathbf{z} \qquad \text{for } \delta \downarrow 0.$$

Next we again perform an integration by parts, to obtain

$$\int_{(B_R \setminus \mathcal{D}^c) \setminus B_\epsilon(\mathbf{y})} - (4\pi)^{-1} \sum_{k=1}^3 (\delta_{jk} |\mathbf{y} - \mathbf{z}|^{-3} - 3(\mathbf{y} - \mathbf{z})_j (\mathbf{y} - \mathbf{z})_k |\mathbf{y} - \mathbf{z}|^{-5}) \mathbf{u}_k(\mathbf{z}) \mathrm{d}\mathbf{z} \tag{5.49}$$

$$= \int_{(B_R \setminus \mathcal{D}^c) \setminus B_\epsilon(\mathbf{y})} -(4\pi)^{-1} (\mathbf{y} - \mathbf{z})_j |\mathbf{y} - \mathbf{z}|^{-3} \operatorname{div} u(\mathbf{z}) \mathrm{d}\mathbf{z} + B_{j,\epsilon}(\mathbf{y}),$$

with

$$B_{j,\epsilon}(\mathbf{y}) := \int_{\partial B_R \cup \partial \mathcal{D} \cup \partial B_\epsilon(\mathbf{y})} (4\pi)^{-1} (\mathbf{y} - \mathbf{z})_j |\mathbf{y} - \mathbf{z}|^{-3} \sum_{k=1}^3 u_k(\mathbf{z}) n_k^{(R,\epsilon)}(\mathbf{z}) \mathrm{d}o_{\mathbf{z}}.$$

We may conclude from (5.46)–(5.49) that

$$\mathfrak{A}_\epsilon = \int_{(B_R\setminus\mathcal{D}^c)\setminus B_\epsilon(\mathbf{y})} -(4\pi)^{-1}(\mathbf{y}-\mathbf{z})_j|\mathbf{y}-\mathbf{z}|^{-3}\operatorname{div}\mathbf{u}(\mathbf{z})d\mathbf{z} + A_{j,\epsilon}(\mathbf{y}) + B_{j,\epsilon}(\mathbf{y}). \tag{5.50}$$

Let the terms $\tilde{A}_j(\mathbf{y})$ and $\tilde{B}_j(\mathbf{y})$ be defined as $A_{j,\epsilon}(\mathbf{y})$ and $B_{j,\epsilon}(\mathbf{y})$, respectively, but with the domain of integration $\partial B_R \cup \partial\mathcal{D} \cup \partial B_\epsilon$ replaced by $\partial B_R \cup \partial\mathcal{D}$. Then Theorem 5.5 and Lemma 5.9 yield

$$\begin{aligned} A_{j,\epsilon}(\mathbf{y}) + B_{j,\epsilon}(\mathbf{y}) \to & \tilde{A}_j(\mathbf{y}) + \tilde{B}_j(\mathbf{y}) + (2/3)\sum_{k=1}^{3}\delta_{jk}u_k(\mathbf{y}) + (1/3)\sum_{k=1}^{3}\delta_{jk}u_k(\mathbf{y}) \\ = & \tilde{A}_j(\mathbf{y}) + \tilde{B}_j(\mathbf{y}) + u_j(\mathbf{y}) \quad \text{for } \epsilon \downarrow 0. \end{aligned} \tag{5.51}$$

Since $\mathbf{f}$ and $\nabla\mathbf{u}$ are continuous on $\overline{B_R}\setminus\mathcal{D}^c$, Lemmas 5.6 and 5.7 imply

$$\int_{(B_R\setminus\mathcal{D}^c)\setminus B_\epsilon(\mathbf{y})} \sum_{k=1}^{3} \mathcal{Z}_{jk}(\mathbf{y},\mathbf{z}) f_k(\mathbf{y})d\mathbf{z} \to \int_{B_R\setminus\mathcal{D}^c} \sum_{k=1}^{3} \mathcal{Z}_{jk}(\mathbf{y},\mathbf{z}) f_k(\mathbf{y})d\mathbf{z} \quad \text{for } \epsilon \downarrow 0, \tag{5.52}$$

$$\begin{aligned} &\int_{(B_R\setminus\mathcal{D}^c)\setminus B_\epsilon(\mathbf{y})} (4\pi)^{-1}(\mathbf{y}-\mathbf{z})_j|\mathbf{y}-\mathbf{z}|^{-3}\operatorname{div} u(\mathbf{z})d\mathbf{z} \\ &\quad \to \int_{B_R\setminus\mathcal{D}^c} (4\pi)^{-1}(\mathbf{y}-\mathbf{z})_j|\mathbf{y}-\mathbf{z}|^{-3}\operatorname{div} u(\mathbf{z})d\mathbf{z} \quad \text{for } \epsilon \downarrow 0. \end{aligned} \tag{5.53}$$

Combining (5.50)–(5.53) yields

$$\begin{aligned} &\int_{B_R\setminus\mathcal{D}^c} \sum_{k=1}^{3} \mathcal{Z}_{jk}(\mathbf{y},\mathbf{z}) f_k(\mathbf{y})d\mathbf{z} \\ &\quad = \int_{B_R\setminus\mathcal{D}^c} -(4\pi)^{-1}(\mathbf{y}-\mathbf{z})_j|\mathbf{y}-\mathbf{z}|^{-3}\operatorname{div}\mathbf{u}(\mathbf{z})d\mathbf{z} + \tilde{A}_j(\mathbf{y}) + \tilde{B}_j(\mathbf{y}) + u_j(\mathbf{y}). \end{aligned}$$

This proves (5.44). □

By letting R tend to infinity in (5.44), the following result may be deduced from Theorem 5.6:

Theorem 5.7 *Let* $\mathbf{u} \in C^2(\overline{\mathcal{D}})$, $\pi \in C^1(\overline{\mathcal{D}})$, $\mathbf{f} \in C^0(\overline{\mathcal{D}})$ *with* $\mathbf{f} = \mathcal{L}(\mathbf{u}) + \nabla\pi$. *Suppose there is* $S > 0$ *with* $\mathcal{D}^c \subset B_S$ *such that*

$$\int_{\mathbb{R}^3\setminus B_S} |\mathbf{z}|^{-1/2}\cdot|\mathbf{f}(\mathbf{z})|dz < \infty, \qquad \int_{\mathbb{R}^3\setminus B_S} |\mathbf{z}|^{-2}\cdot|\operatorname{div}\mathbf{u}(\mathbf{z})|dz < \infty.$$

Further suppose there is a sequence (R_n) in (S, ∞) such that

$$R_n^{-1/2} \int_{\partial B_{R_n}} (|\nabla \mathbf{u}(\mathbf{z})| + |\pi(\mathbf{z})| + |\mathbf{u}(\mathbf{z})|)\, do_{\mathbf{z}} + R_n^{-2} \int_{\partial B_{R_n}} |\operatorname{div} \mathbf{u}(\mathbf{z})| do_{\mathbf{z}} \to 0$$

for $n \to \infty$. Let $j \in \{1, 2, 3\}$, $\mathbf{y} \in \mathcal{D}$. Then

$$\begin{aligned}
u_j(\mathbf{y}) = & \int_{\mathcal{D}} \left(\sum_{k=1}^{3} \mathcal{Z}_{jk}(\mathbf{y}, \mathbf{z}) \cdot f_k(\mathbf{z}) + (4 \cdot \pi)^{-1} \cdot (\mathbf{y} - \mathbf{z})_j \cdot |\mathbf{y} - \mathbf{z}|^{-3} \cdot \operatorname{div} \mathbf{u}(\mathbf{z}) \right) d\mathbf{z} \\
& - \int_{\partial \mathcal{D}} \sum_{k=1}^{3} \Bigg[\sum_{l=1}^{3} (\mathcal{Z}_{jk}(\mathbf{y}, \mathbf{z}) \cdot (\partial_l u_k(\mathbf{z}) - \delta_{kl} \cdot \pi(\mathbf{z}) + u_k(\mathbf{z}) \cdot (\mathbf{U} + \boldsymbol{\omega} \wedge \mathbf{z})_l) \\
& - \partial_{z_l} \mathcal{Z}_{jk}(\mathbf{y}, \mathbf{z}) \cdot u_k(\mathbf{z})) \cdot n_l^{(\mathcal{D})}(\mathbf{z}) \\
& - (4 \cdot \pi)^{-1} \cdot (\mathbf{y} - \mathbf{z})_j \cdot |\mathbf{y} - \mathbf{z}|^{-3} \cdot u_k(\mathbf{z}) \cdot n_k^{(\mathcal{D}})(\mathbf{z}) \Bigg] do_{\mathbf{z}}.
\end{aligned}$$

5.3 Asymptotic Behavior

5.3.1 Some Volume Potentials

In the present subsection, we will deal with the volume potentials which arise. There are two types of volume potentials, involving the kernels $\mathcal{Z}_{jk}$ and E_{4j}, respectively. We begin with the potential related to $\mathcal{Z}_{jk}$.

Lemma 5.10 *Let $p \in (1, \infty)$, $q \in (1, 2)$, $\mathbf{f} \in \boldsymbol{L}^p_{loc}(\mathbb{R}^3)$ with $\mathbf{f}|B_S^c \in \boldsymbol{L}^q(B_S^c)$ for some $S \in (0, \infty)$. Then, for $j, k \in \{1, 2, 3\}$, $\boldsymbol{\alpha} \in \mathbb{N}_0^3$ with $|\boldsymbol{\alpha}| \leq 1$, we have*

$$\int_{\mathbb{R}^3} |\partial_y^{\alpha} \mathcal{Z}_{jk}(\mathbf{y}, \mathbf{z})|\, |\mathbf{f}_k(\mathbf{z})| d\mathbf{y} < \infty \quad \textit{for a.e. } \mathbf{y} \in \mathbb{R}^3. \tag{5.54}$$

We define $\mathfrak{R}(\mathbf{f})\colon \mathbb{R}^3 \mapsto \mathbb{R}^3$ by

$$\mathfrak{R}_j(\mathbf{f})(\mathbf{y}) := \int_{\mathbb{R}^3} \sum_{k=1}^{3} \mathcal{Z}_{jk}(\mathbf{y}, \mathbf{z})\ f_k(\mathbf{z}) d\mathbf{z}$$

for $\mathbf{y} \in \mathbb{R}^3$ *such that* (5.54) *holds; else we set* $\mathfrak{R}_j(\mathbf{f})(\mathbf{y}) := 0$ $(1 \le j \le 3)$. *Then* $\mathfrak{R}(\mathbf{f}) \in \mathbf{W}^{1,1}_{loc}(\mathbb{R}^3)$ *and*

$$\partial_l \mathfrak{R}_j(\mathbf{f})(\mathbf{y}) := \int_{\mathbb{R}^3} \sum_{k=1}^{3} \partial y_l \mathcal{Z}_{jk}(\mathbf{y}, \mathbf{z}) \ f_k(\mathbf{z}) d\mathbf{z} \tag{5.55}$$

for $j, l \in \{1, 2, 3\}$ *and for a.e.* $\mathbf{y} \in \mathbb{R}^3$. *Moreover, for* $R \in (0, \infty)$ *we have*

$$\|\mathfrak{R}(\mathbf{f}|B_R), |B_R\|_p \le C(R, p) \ \|\mathbf{f}|B_R\|_p. \tag{5.56}$$

Proof Take j, k, α as in (5.54). Let $R \in (0, \infty)$. (5.35) it follows that

$$\begin{aligned}\int_{B_R} |\partial^{\boldsymbol{\alpha}}_{\mathbf{y}} \mathcal{Z}_{jk}(\mathbf{y}, \mathbf{z})| d\mathbf{z} \le& C(R) \int_{B_R} |\mathbf{y} - \mathbf{z}|^{-1-|\boldsymbol{\alpha}|} d\mathbf{z} \\ \le& C(R) \int_{B_{2R}(\mathbf{y})} |\mathbf{y} - \mathbf{z}|^{-1-|\boldsymbol{\alpha}|} d\mathbf{z} \le C(R)\end{aligned}$$

for $\mathbf{y} \in B_R$ and analogously $\int_{B_R} |\partial^{\boldsymbol{\alpha}}_{\mathbf{y}} \mathcal{Z}_{jk}(\mathbf{y}, \mathbf{z})| d\mathbf{y} \le C(R)$ for $\mathbf{z} \in B_R$. It follows by Hölder's inequality that

$$\begin{aligned}&\left(\int_{B_R} \left(\int_{B_R} |\partial^{\boldsymbol{\alpha}}_{\mathbf{y}} \mathcal{Z}_{jk}(\mathbf{y}, \mathbf{z})| \, |f_k(\mathbf{z})| d\mathbf{z}\right)^p d\mathbf{y}\right)^{1/p} \qquad (5.57)\\ &\le \left(\int_{B_R} \left(\int_{B_R} |\partial^{\boldsymbol{\alpha}}_{\mathbf{y}} \mathcal{Z}_{jk}(\mathbf{y}, \mathbf{z})| dz\right)^{p-1} \left(\int_{B_R} |\partial^{\boldsymbol{\alpha}}_{y} \mathcal{Z}_{jk}(\mathbf{y}, \mathbf{z})| \, |\mathbf{f}(z)|^p d\mathbf{z}\right) d\mathbf{y}\right)^{1/p} \\ &\le C(R, p) \left(\int_{B_R} \int_{B_R} |\partial^{\boldsymbol{\alpha}}_{\mathbf{y}} \mathcal{Z}_{jk}(\mathbf{y}, \mathbf{z})| \, |\mathbf{f}(\mathbf{z})|^p d\mathbf{z} d\mathbf{y}\right)^{1/p} \le C(R, p) \|\mathbf{f}|B_R\|_p.\end{aligned}$$

In particular it implies that the integral $\int_{B_n} |\partial^{\boldsymbol{\alpha}}_{\mathbf{y}} \mathcal{Z}_{jk}(\mathbf{y}, \mathbf{z})| \, |\mathbf{f}_k(z)| d\mathbf{z}$ is finite for a.e. $\mathbf{y} \in B_n$, $n \in \mathbb{N}$, and that inequality (5.56) is shown.

Once again take $j, k, \boldsymbol{\alpha}$ as in (5.54), and let $n \in \mathbb{N}$ with $n \ge S$. Then, using (5.38) with S replaced by $n/2$ and with $\delta = 1/2$, we find for $\mathbf{y} \in B_{n/2}$ that

$$\begin{aligned}\int_{B_n^c} |\partial^{\boldsymbol{\alpha}}_{\mathbf{y}} \mathcal{Z}_{jk}(\mathbf{y}, \mathbf{z})| \, |f_k(\mathbf{z})| d\mathbf{z} \le& C(n) \int_{B_n^c} (|\mathbf{z}| s_\tau(\mathbf{z}))^{-1-|\boldsymbol{\alpha}|/2} |\mathbf{f}(\mathbf{z})| d\mathbf{z} \qquad (5.58)\\ \le& C(n) \left(\int_{B_n^c} (|\mathbf{z}| s_\tau(\mathbf{z}))^{-q'} d\mathbf{z}\right)^{1/q'} \|\mathbf{f}|B_n^c\|_q \le C(n, q) \|\mathbf{f}|B_S^c\|_q,\end{aligned}$$

where the last inequality holds due to Lemma 2.13 and the assumption $q < 2$ (hence $q' > 2$). We thus have shown that the relation in (5.54) holds for a.e. $\mathbf{y} \in B_{n/2}$.

Since this is true for any $n \in \mathbb{N}$ with $n \geq S$, (5.54) is proved. Moreover, we deduce from (5.57) and (5.58) that

$$\int_{B_{n/2}} \int_{\mathbb{R}^3} \left| \sum_{k=1}^{3} \partial_{\mathbf{y}}^{\alpha} \mathcal{Z}_{jk}(\mathbf{y}, \mathbf{z}) f_k(\mathbf{z}) \right| \mathrm{d}\mathbf{z}\mathrm{d}\mathbf{y} \leq C(n, p, q)(\|\mathbf{f}|B_n\|_p + \|\mathbf{f}|B_S^c\|_q) \tag{5.59}$$

for $n \in \mathbb{N}$ with $n \geq S$. This means that $\mathfrak{R}_j(\mathbf{f}) \in L_{1,\mathrm{loc}}(\mathbb{R}^3)$ and that the function associating a.e. $\mathbf{y} \in \mathbb{R}^3$ with the integral $\int_{\mathbb{R}^3} \sum_{k=1}^{3} \partial y_l \mathcal{Z}_{jk}(\mathbf{y}, \mathbf{z}) f_k(\mathbf{z}) \mathrm{d}\mathbf{z}$ also belongs to $L_{1,\mathrm{loc}}(\mathbb{R}^3)$ for $1 \leq l \leq 3$. Now take $\Phi \in C_0^\infty(\mathbb{R}^3)$. Then, by (5.59) and because the support of Φ is compact,

$$\int_{\mathbb{R}^3} \partial_l \Phi(y) \mathfrak{R}_j(\mathbf{f})(\mathbf{y}) \mathrm{d}\mathbf{y} = \sum_{k=1}^{3} \lim_{\epsilon \downarrow 0} \int_{\mathbb{R}^3} \int_{\mathbb{R}^3 \setminus B_\epsilon(\mathbf{z})} \partial_l \Phi(\mathbf{y}) \mathcal{Z}_{jk}(\mathbf{y}, \mathbf{z}) \mathrm{d}y f_k(\mathbf{z}) \mathrm{d}\mathbf{z}. \tag{5.60}$$

But for any $\epsilon > 0$, we may perform a partial integration in the inner integral on the right-hand side of (5.60) (the first statement of Lemma 5.3). Due to (5.35), the term with a surface integral on $\partial B_\epsilon(\mathbf{z})$ arising in this way tends to zero for $\epsilon \downarrow 0$. (Note that for $\epsilon \in (0, 1]$, say, and for $\mathbf{y} \in \partial B_\epsilon(\mathbf{z})$, the integral with respect to $\mathbf{z}$ only extends over B_{n+1}, if $n \in \mathbb{N}$ is chosen so large that $\mathrm{supp}(\Phi) \subset B_n$.) After letting ϵ tend to zero, we obtain an equation which implies that $\mathfrak{R}_j(\mathbf{f}) \in \mathbf{W}_{\mathrm{loc}}^{1,1}(\mathbb{R}^3)$ and Eq. (5.55) holds. □

Lemma 5.11 *Take p, q,* $\mathbf{f}$ *as in Lemma* 5.10, *and suppose in addition that* $p > 3/2$.

Then the relation in (5.54) *holds for any* $\mathbf{y} \in \mathbb{R}^3$ *(without the restriction "a.e."), and the function* $\mathfrak{R}(\mathbf{f})$ *is continuous.* □

Proof We show that $\mathfrak{R}(\mathbf{f})$ is continuous. The relation in (5.54) for any $\mathbf{y} \in \mathbb{R}^3$ may be established by a similar but simpler argument.

Let $j \in \{1, 2, 3\}$, $R \in (S, \infty)$. It suffices to prove that $\mathfrak{R}_j(\mathbf{f})|B_R$ is continuous. But for $\mathbf{z} \in B_{2R}^c$, $\mathbf{y} \in B_R$, we get by (5.38) that

$$\left| \sum_{k=1}^{3} \mathcal{Z}_{jk}(\mathbf{y}, \mathbf{z}) f_k(\mathbf{z}) \right| \leq C(R) \left(|\mathbf{z}| s_\tau(\mathbf{z})\right)^{-1} |\mathbf{f}(\mathbf{z})|.$$

Since by a computation as in (5.58), the function

$$\mathbb{R}^3 \ni \mathbf{z} \mapsto \chi_{B_{2R}^c}(\mathbf{z}) \left(|\mathbf{z}| s_\tau(z)\right)^{-1} |\mathbf{f}(\mathbf{z})| \in [0, \infty)$$

is integrable, we may conclude in view of the first statement of Lemma 5.3 that the integral $\int_{B_{2R}^c} \sum_{k=1}^3 \mathcal{Z}_{jk}(\mathbf{y}, \mathbf{z}) f_k(\mathbf{z}) d\mathbf{z}$ is continuous as a function of $\mathbf{y} \in B_R$. Thus we still have to show that the function

$$I(\mathbf{y}) := \int_{B_{2R}} \sum_{k=1}^{3} \mathcal{Z}_{jk}(\mathbf{y}, \mathbf{z}) f_k(\mathbf{z}) d\mathbf{z} \quad (\mathbf{y} \in B_R)$$

is continuous as well. So take $\mathbf{y}, \mathbf{y}' \in B_R$ with $\mathbf{y} \neq \mathbf{y}'$. Then

$$|I(\mathbf{y}) - I(\mathbf{y}')| \leq \mathfrak{N}_1 + \mathfrak{N}_2, \tag{5.61}$$

with

$$\mathfrak{N}_1 := \sum_{x \in \{\mathbf{y}, \mathbf{y}'\}} \int_{B_R \cap A} \sum_{k=1}^{3} |\mathcal{Z}_{jk}(\mathbf{x}, \mathbf{z}) f_k(\mathbf{z})| d\mathbf{z},$$

$$\mathfrak{N}_2 := \int_{B_R \setminus A} \left| \int_0^1 \sum_{k,l=1}^{3} \partial x_l \mathcal{Z}_{jk}(\mathbf{x}, \mathbf{z})_{|x=y'+\vartheta(\mathbf{y}-\mathbf{y}')} (\mathbf{y} - \mathbf{y}')_l d\vartheta \right| |\mathbf{f}_k(\mathbf{z})| d\mathbf{z},$$

with $A := B_{2|\mathbf{y}-\mathbf{y}'|}(\mathbf{y})$. We get with (5.35) that

$$\begin{aligned} \mathfrak{N}_1 \leq & C(R) \sum_{\mathbf{x} \in \{\mathbf{y}, \mathbf{y}'\}} \int_{B_R \cap A} |\mathbf{x} - \mathbf{z}|^{-1} |\mathbf{f}(\mathbf{z})| d\mathbf{z} \\ \leq & C(R) \sum_{\mathbf{x} \in \{\mathbf{y}, \mathbf{y}'\}} \left(\int_{B_{3|\mathbf{y}-\mathbf{y}'|}(\mathbf{x})} |\mathbf{x} - \mathbf{z}|^{-p'} d\mathbf{z} \right)^{1/p'} \|\mathbf{f}| B_R \|_p. \end{aligned}$$

Since $p > 3/2$, hence $p' < 3$, and we may conclude that

$$\mathfrak{N}_1 \leq C(R) |\mathbf{y} - \mathbf{y}'|^{-1+3/p'} \|\mathbf{f}| B_R \|_p,$$

with $-1 + 3/p' > 0$. In order to estimate $\mathfrak{N}_2$, we note that

$$|\mathbf{y}' + \vartheta(\mathbf{y} - \mathbf{y}') - \mathbf{z}| \geq |\mathbf{y} - \mathbf{z}| - |\mathbf{y} - \mathbf{y}'| \geq |\mathbf{y} - \mathbf{z}|/2 \geq |\mathbf{y} - \mathbf{y}'|$$

for $\mathbf{z} \in \mathbb{R}^3 \setminus A$, $\vartheta \in [0, 1]$. Therefore by (5.35), if $2p' > 3$,

$$\begin{aligned} \mathfrak{N}_2 \leq & C(R) |\mathbf{y} - \mathbf{y}'| \left(\int_{B_R \setminus A} |\mathbf{y} - \mathbf{z}|^{-2p'} d\mathbf{z} \right)^{1/p'} \|\mathbf{f}| B_R \|_p \\ \leq & C(R) |\mathbf{y} - \mathbf{y}'|^{-1+3/p'} \|\mathbf{f}| B_R \|_p. \end{aligned}$$

In the case $2p' < 3$, the factor $|\mathbf{y} - \mathbf{y}'|^{-1+3/p'}$ on the right-hand side of the preceding inequality may be replaced by $|\mathbf{y} - \mathbf{y}'|$ and in the case $2p' = 3$ by $|\mathbf{y} - \mathbf{y}'| \ln\big(|\mathbf{y} - \mathbf{y}'|/(2R)\big)$. In view of (5.61), we have thus shown that $I(\mathbf{y})$ is a continuous function of $\mathbf{y} \in B_R$. This completes the proof of Lemma 5.11. □

5.3.2 Asymptotic Profile

The main and crucial idea of the proof of the next theorem consists in reducing an estimate of $\mathfrak{R}(\mathbf{f})$ to an estimate of a convolution integral involving an upper bound of an Oseen fundamental solution.

Theorem 5.8 *Let $S, S_1, \gamma \in (0, \infty)$ with $S_1 < S$, $p \in (1, \infty)$, $A \in [2, \infty)$, $B \in \mathbb{R}$, $\mathbf{f}\colon \mathbb{R}^3 \mapsto \mathbb{R}^3$ measurable with*

$$\mathbf{f}|B_{S_1} \in \boldsymbol{L}^p(B_{S_1}), \quad |\mathbf{f}(z)| \le \gamma |\mathbf{z}|^{-A} s_\tau(\mathbf{z})^{-B} \quad \textit{for } \mathbf{z} \in B_{S_1}^c, \quad A + \min\{1, B\} \ge 3.$$

Let $i, j \in \{1, 2, 3\}$, $\mathbf{y} \in B_S^c$. Then

$$|\mathfrak{R}_j(\mathbf{f})(\mathbf{y})| \le C(S, S_1, A, B)(\|\mathbf{f}|B_{S_1}\|_1 + \gamma)\,(|\mathbf{y}|s_\tau(\mathbf{y}))^{-1}\, l_{A,B}(\mathbf{y}), \tag{5.62}$$

$$\begin{aligned} |\partial_{y_i}\mathfrak{R}_j(\mathbf{f})(\mathbf{y})| \le & C(S, S_1, A, B)(\|\mathbf{f}|B_{S_1}\|_1 + \gamma) \\ & (|\mathbf{y}|s_\tau(\mathbf{y}))^{-3/2}\, s_\tau(\mathbf{y})^{\max(0,\,7/2-A-B)} l_{A,B}(\mathbf{y}), \end{aligned} \tag{5.63}$$

where $l_{A,B}(\mathbf{y}) = \begin{cases} 1 & \textit{if } A + \min\{1, B\} > 3, \\ \max(1, \ln|\mathbf{y}|) & \textit{if } A + \min\{1, B\} = 3. \end{cases}$ □

Proof By (5.37) with S, δ replaced by S_1, $S/S_1 - 1$, respectively, we find for $k \in \{1, 2, 3\}$, $\alpha \in \mathbb{N}_0^3$ with $|\boldsymbol{\alpha}| \le 1$, that

$$\int_{B_{S_1}} |\partial_{\mathbf{y}}^{\boldsymbol{\alpha}} \mathcal{Z}_{jk}(\mathbf{y}, \mathbf{z})||\mathbf{f}(\mathbf{z})|d\mathbf{z} \le C(S, S_1)\,(|\mathbf{y}|s_\tau(\mathbf{y}))^{-1-|\boldsymbol{\alpha}|/2}\, \|\mathbf{f}|B_{S_1}\|_1. \tag{5.64}$$

Recalling Lemmas 5.3, 2.18, 2.19, and 5.1, we see that

$$\begin{aligned} \mathfrak{A}_\alpha := & \int_{B_{S_1}^c} |\partial_{\mathbf{y}}^{\boldsymbol{\alpha}} \mathcal{Z}_{jk}(\mathbf{y}, \mathbf{z})|\, |\mathbf{f}(\mathbf{z})|d\mathbf{z} \\ \le & C\gamma \int_0^\infty \int_{B_{S_1}^c} (|\mathbf{y} - \tau t \boldsymbol{e}_1 - \mathrm{e}^{-t\Omega} \cdot \mathbf{z}|^2 + t)^{-3/2-|\boldsymbol{\alpha}|/2} |\mathbf{z}|^{-A} s_\tau(\mathbf{z})^{-B} d\mathbf{z}dt \end{aligned}$$

$$=C\gamma \int_0^\infty \int_{B_{S_1}^c} (|\mathbf{y} - \tau t e_1 - \mathbf{x}|^2 + t)^{-3/2-|\alpha|/2} |\mathbf{x}|^{-A} s_\tau(e^{t\Omega} \cdot \mathbf{x})^{-B} d\mathbf{x} dt$$

$$=C\gamma \int_{B_{S_1}^c} \int_0^\infty (|\mathbf{y} - \tau t e_1 - \mathbf{x}|^2 + t)^{-3/2-|\alpha|/2} dt |\mathbf{x}|^{-A} s_\tau(\mathbf{x})^{-B} d\mathbf{x},$$

where the last equation holds due to Lemmas 2.19 and 5.1. Now we apply (5.9) with $\mathbf{y}$ replaced by $\mathbf{y} - \mathbf{x}$ and with $\mathbf{z} = 0$. Moreover we use Theorem 5.2. It follows

$$\mathfrak{A}_\alpha \le C(S)\gamma \left(\int_{B_{S_1}^c \cap B_{S/2}(\mathbf{y})} |\mathbf{y} - \mathbf{x}|^{-1-|\alpha|} |\mathbf{x}|^{-A} s_\tau(\mathbf{x})^{-B} d\mathbf{x} \right. \tag{5.65}$$

$$\left. + \int_{B_{S_1}^c \setminus B_{S/2}(\mathbf{y})} (|\mathbf{y} - \mathbf{x}| s_\tau(\mathbf{y} - \mathbf{x}))^{-1-|\alpha|/2} |\mathbf{x}|^{-A} s_\tau(\mathbf{x})^{-B} d\mathbf{x} \right).$$

Next we observe that for $\mathbf{x} \in B_{S/2}(\mathbf{y})$, we have $|\mathbf{x}| \ge |\mathbf{y}| - |\mathbf{y} - \mathbf{x}| \ge |\mathbf{y}| - S/2 \ge |\mathbf{y}|/2$,

$$s_\tau(\mathbf{x})^{-1} \le C(1 + |\mathbf{y} - \mathbf{x}|) s_\tau(\mathbf{y})^{-1} \le C(S) s_\tau(\mathbf{y})^{-1}$$

(see Lemma 2.14), and similarly $s_\tau(\mathbf{y})^{-1} \le C(S) s_\tau(\mathbf{x})^{-1}$. For $\mathbf{x} \in B_{S/2}(\mathbf{y})^c$, we find

$$|\mathbf{y} - \mathbf{x}| = |\mathbf{y} - \mathbf{x}|/2 + |\mathbf{y} - \mathbf{x}|/2 \ge S/4 + |\mathbf{y} - \mathbf{x}|/2 \ge \min\{S/4, 1/2\}(1 + |\mathbf{y} - \mathbf{x}|).$$

Thus, independently of the sign of B, we may conclude from (5.65) that

$$\mathfrak{A}_\alpha \le C(S, S_1, A, B)\gamma \left(|\mathbf{y}|^{-A} s_\tau(\mathbf{y})^{-B} \int_{B_{S/2}(\mathbf{y})} |\mathbf{y} - \mathbf{x}|^{-1-|\alpha|} d\mathbf{x} \right. \tag{5.66}$$

$$\left. + \int_{B_{S_1}^c \setminus B_{S/2}(\mathbf{y})} ((1 + |\mathbf{y} - \mathbf{x}|) s_\tau(\mathbf{y} - \mathbf{x}))^{-1-|\alpha|/2} (1 + |\mathbf{x}|)^{-A} s_\tau(\mathbf{x})^{-B} d\mathbf{x} \right)$$

$$\le C(S, S_1, A, B)\gamma \left(|\mathbf{y}|^{-A} s_\tau(\mathbf{y})^{-B} \right.$$

$$\left. + \int_{\mathbb{R}^3} ((1 + |\mathbf{y} - \mathbf{x}|) s_\tau(\mathbf{y} - \mathbf{x}))^{-1-|\alpha|/2} (1 + |\mathbf{x}|)^{-A} s_\tau(\mathbf{x})^{-B} d\mathbf{x} \right).$$

In the case $\boldsymbol{\alpha} = 0$, we refer to the proof of [119, Theorem 31] and our assumptions on A and B to deduce from (5.66) that

$$\mathfrak{A}_0 \le C(S, S_1, A, B)\gamma \left(|\mathbf{y}|^{-A} s_\tau(\mathbf{y})^{-B} + (|\mathbf{y}| s_\tau(\mathbf{y}))^{-1} l_{A,B}(\mathbf{y}) \right). \tag{5.67}$$

But by Lemma 2.16 and because $A - 3/2 > 0$, $A + B \geq A + \min\{1, B\} \geq 3$, we have

$$|\mathbf{y}|^{-A} s_\tau(\mathbf{y})^{-B} \leq C(S, A)|\mathbf{y}|^{-3/2} s_\tau(\mathbf{y})^{-A+3/2-B} \leq C(S, A)|\mathbf{y}|^{-3/2} s_\tau(\mathbf{y})^{-3/2}, \tag{5.68}$$

so we may conclude from (5.67) that

$$\mathfrak{A}_0 \leq C(S, S_1, A, B)\gamma\,(|\mathbf{y}| s_\tau(\mathbf{y}),)^{-1}\, l_{A,B}(\mathbf{y}).$$

Inequality (5.62) follows from (5.64) and the preceding estimate. If $|\boldsymbol{\alpha}| = 1$, Eq. (5.66) and the proof of [119, Theorem 3.2] yield

$$\begin{aligned}\mathfrak{A}_\alpha \leq & C(S, S_1, A, B)\gamma \\ & \left(|\mathbf{y}|^{-A} s_\tau(\mathbf{y})^{-B} + (|\mathbf{y}| s_\tau(\mathbf{y}))^{-3/2}\, s_\tau(\mathbf{y})^{\max(0,7/2-A-B)} l_{A,B}(\mathbf{y})\right).\end{aligned} \tag{5.69}$$

Hence by (5.68),

$$\mathfrak{A}_\alpha \leq C(S, S_1, A, B)\gamma\,(|\mathbf{y}| s_\tau(\mathbf{y}))^{-3/2}\, s_\tau(\mathbf{y})^{\max(0,7/2-A-B)} l_{A,B}(\mathbf{y}).$$

This estimate together with (5.64) implies (5.63). □

Now we turn to volume integrals involving the kernel E_{4j}.

Lemma 5.12 *Let $p \in (1, \infty)$, $q \in (1, 3)$, $g \in L^p_{loc}(\mathbb{R}^3)$ with $g|B^c_S \in L^q(B^c_S)$ for some $S \in (0, \infty)$. Then, for $j \in \{1, 2, 3\}$,*

$$\int_{\mathbb{R}^3} |E_{4j}(\mathbf{y} - \mathbf{z})||g(\mathbf{z})| d\mathbf{y} < \infty \quad \textit{for a.e. } \mathbf{y} \in \mathbb{R}^3. \tag{5.70}$$

Thus we may define $\mathfrak{S}(g) \colon \mathbb{R}^3 \mapsto \mathbb{R}^3$ by

$$\mathfrak{S}_j(g)(\mathbf{y}) := \int_{\mathbb{R}^3} E_{4j}(\mathbf{y} - \mathbf{z}) g(z) d\mathbf{z}$$

for $\mathbf{y} \in \mathbb{R}^3$ such that (5.70) holds, and $\mathfrak{S}_j(g)(\mathbf{y}) := 0$ else, where $(1 \leq j \leq 3)$. Then $\mathfrak{S}(g) \in \mathbf{W}^{1,1}_{loc}(\mathbb{R}^3)$. For $R \in (0, \infty)$ we have

$$\|\mathfrak{S}(g|B_R)|B_R\|_p \leq C(R, p)\|g|B_R\|_p. \tag{5.71}$$

If $p > 3$, the relation in (5.70) holds for any $\mathbf{y} \in \mathbb{R}^3$ (without the restriction "a.e."), and $\mathfrak{S}(g)$ is continuous. □

Proof Lemma 5.12 may be shown by arguments analogous to those we used to prove Lemmas 5.10 and 5.11, except as concerns the claim $\mathfrak{S}(g) \in \mathbf{W}^{1,1}_{\text{loc}}(\mathbb{R}^3)$. To establish this latter point, a different reasoning based on the Calderon–Zygmund inequality is needed because the derivative $\partial_l E_{4j}$ is a singular kernel in $\mathbb{R}^3$. We refer to [93, Section IV.2] for details. □

Theorem 5.9 *Let $S, S_1, \widetilde{\gamma} \in (0, \infty)$ with $S_1 < S$, $p \in (1, \infty)$, $C \in (5/2, \infty)$, $D \in \mathbb{R}$, $g : \mathbb{R}^3 \mapsto \mathbb{R}$ measurable with*

$$g|B_{S_1} \in L^p(B_{S_1}), \quad |g(z)| \le \widetilde{\gamma}|\mathbf{z}|^{-C} s_\tau(\mathbf{z})^{-D} \quad \textit{for } \mathbf{z} \in B_{S_1}^c, \; C + \min\{1, D\} > 3.$$

Let $j \in \{1, 2, 3\}$, $\mathbf{y} \in B_S^c$. Then

$$|\mathfrak{S}_j(g)(\mathbf{y})| \le C(S, S_1, C, D)(\|g|B_{S_1}\|_1 + \widetilde{\gamma})|\mathbf{y}|^{-2}. \tag{5.72}$$

If $\operatorname{supp}(g) \subset B_{S_1}$, *we further have*

$$|\partial_n \mathfrak{S}_j(g)(\mathbf{y})| \le C(S, S_1)\|g\|_1 |\mathbf{y}|^{-3} \quad (1 \le n \le 3). \tag{5.73}$$

Proof Inequality (5.72) may be proved in the same way as Theorem 5.8, except that the reference to [119, Theorems 3.1, 3.2] is replaced by [119, Theorem 3.4] and that the argument becomes simpler due to the much simpler structure of the kernel E_{4j} compared to $\mathcal{Z}_{jk}$. As concerns (5.73), observe that $|\mathbf{y} - \mathbf{z}| \ge (1 - S_1/S)|\mathbf{y}|$ for $\mathbf{z} \in B_{S_1}$, so if $\operatorname{supp}(g) \subset B_{S_1}$, it is obvious that

$$\mathfrak{S}_j(g)|B_S^c \in C^1(B_S^c), \quad \int_{B_{S_1}} |\partial_l E_{4j}(\mathbf{y} - \mathbf{z})||g(\mathbf{z})| d\mathbf{z} < \infty,$$

$$\partial_l \mathfrak{S}_j(g)(\mathbf{y}) = \int_{B_{S_1}} \partial_l E_{4j}(\mathbf{y} - \mathbf{z}) g(\mathbf{z}) d\mathbf{z} \quad (1 \le l \le 3).$$

Inequality (5.73) now follows. □

We will use the following notational convention. If $A \subset \mathbb{R}^3$ is a measurable set and $f : A \mapsto \mathbb{R}^3$ is a measurable function, if $\widetilde{f}$ denotes the zero extension of f to $\mathbb{R}^3$, and if $\widetilde{f}$ satisfies the assumptions of Lemma 5.10, we will write $\mathfrak{R}(f)$ instead of $\mathfrak{R}(\widetilde{f})$. A similar convention is to hold with respect to $\mathfrak{S}(g)$ if $g : A \mapsto \mathbb{R}$ is a measurable function such that its zero extension to $\mathbb{R}^3$ verifies the assumptions of Lemma 5.12.

Lemma 5.13 *Let $R \in (0, \infty)$ with $\mathcal{D}^c \subset B_R$, $\mathcal{D}_R = B_r \cap \mathcal{D}$, $f \in L^1(\partial \mathcal{D}_R)$, $j, k \in \{1, 2, 3\}$, $\boldsymbol{\alpha} \in \mathbb{N}_0^3$ with $|\alpha| \le 1$. Define*

$$F(\mathbf{y}) := \int_{\partial \mathcal{D}_R} \partial_{\mathbf{z}}^{\boldsymbol{\alpha}} \mathcal{Z}_{jk}(\mathbf{y}, \mathbf{z}) f(\mathbf{z}) do_{\mathbf{z}}, \quad H(\mathbf{y}) := \int_{\partial \mathfrak{D}_R} E_{4j}(\mathbf{y} - \mathbf{z}) f(\mathbf{z}) do_{\mathbf{z}}$$

for $\mathbf{y} \in \mathfrak{D}_R$. *Then* F *and* H *are continuous. Moreover, let* $\mathbf{x} \in \mathcal{D}_R$, *and put* $\delta_{\mathbf{x}} := \operatorname{dist}(\partial\mathcal{D}_R, \mathbf{x})$. *Then*

$$|F(\mathbf{x})| + |H(\mathbf{x})| \leq C(\delta_{\mathbf{x}}, R)\|f\|_1. \tag{5.74}$$

Proof Let $U \subset \mathbb{R}^3$ be open, with $\overline{U} \subset \mathcal{D}_R$. Then $\delta_U := \operatorname{dist}(\overline{U}, \partial\mathcal{D}_R) > 0$, so we get by (5.35) that

$$|\partial_{\mathbf{z}}^{\boldsymbol{\alpha}} \mathcal{Z}_{jk}(\mathbf{y}, \mathbf{z}) f(\mathbf{z})| \leq C(R)\delta_U^{-1-|\boldsymbol{\alpha}|}|f(\mathbf{z})| \quad \text{for } \mathbf{z} \in \partial\mathcal{D}_R.$$

In view of the first statement of Lemma 5.3, we conclude that F is continuous. From (5.35), we get that $|F(\mathbf{x})| \leq \mathcal{C}(\delta_{\mathbf{x}}, R)\,\|\mathbf{f}\|_1$. Obviously $E_{4j} \in C^\infty(\mathbb{R}^3\backslash\{0\})$ and $|E_{4j}(\mathbf{x})| \leq |\mathbf{x}|^{-2}$ for $\mathbf{x} \in \mathbb{R}^3\backslash\{\mathbf{0}\}$, so the function H may be handled in the same way (and even belongs to $C^\infty(\mathcal{D}_R)$). □

Lemma 5.14 *Let* $S \in (0, \infty)$ *with* $\overline{\mathcal{D}}^c \subset B_S$. *Let* $f \in L^1(\partial\mathcal{D})$, $g \in L^1(\mathcal{D})$, $j, k \in \{1, 2, 3\}$, *and define*

$$F^{(1)}(\mathbf{y}) := \int_{\partial\mathcal{D}} \mathcal{Z}_{jk}(\mathbf{y}, \mathbf{z}) f(\mathbf{z}) do_{\mathbf{z}}, \quad F^{(2)}(\mathbf{y}) := \int_{\mathcal{D}^c} \mathcal{Z}_{jk}(\mathbf{y}, \mathbf{z}) g(\mathbf{z}) d\mathbf{z},$$

$$F^{(3)}(\mathbf{y}) := \int_{\partial\mathcal{D}} E_{4j}(\mathbf{y} - \mathbf{z}) f(\mathbf{z}) do_{\mathbf{z}}, \quad F^{(4)}(\mathbf{y}) := \int_{\mathcal{D}^c ere} \partial_k E_{4j}(\mathbf{y} - \mathbf{z}) g(\mathbf{z}) d\mathbf{z}$$

for $\mathbf{y} \in \overline{\mathfrak{D}}^c$. *Then* $F^{(i)} \in C^1(\overline{\mathcal{D}}^c)$ *for* $1 \leq i \leq 4$. *Put* $\delta := \operatorname{dist}(\overline{\mathcal{D}}, \partial B_S)$. *Then*

$$|\partial^\alpha F^{(i)}(y)| \leq C(\delta, S)\,(|\mathbf{y}|s_\tau(\mathbf{y}))^{-1-|\boldsymbol{\alpha}|/2}\,\|f\|_1, \tag{5.75}$$

$$|\partial^\alpha F^{(j)}(\mathbf{y})| \leq C(\delta, S)\,(|\mathbf{y}|s_\tau(\mathbf{y}))^{-1-|\boldsymbol{\alpha}|/2}\,\|g\|_1 \tag{5.76}$$

for $\mathbf{y} \in B_S^c$, $\alpha \in \mathbb{N}_0^3$ *with* $|\boldsymbol{\alpha}| \leq 1$, $i \in \{1, 3\}$, $j \in \{2, 4\}$. □

Proof Let $U \subset \mathbb{R}^3$ be open and bounded with $\overline{U} \subset \overline{\mathcal{D}}$. Let $R \in (0, \infty)$ with $\overline{\mathcal{D}}^c \cup \overline{U} \subset B_R$. Then an argument as in the proof of Lemma 5.13, based on (5.35) and Lemma 5.3, yields that $F^{(1)}|U \in C^1(U)$, and

$$\partial_l F^{(1)}(\mathbf{y}) = \int_{\partial\mathcal{D}} \partial y_l \mathcal{Z}_{jk}(\mathbf{y}, \mathbf{z}) f(\mathbf{z}) \mathrm{d}o_{\mathbf{z}} \quad \text{for } \mathbf{y} \in U,\ 1 \leq l \leq 3. \tag{5.77}$$

It follows that $F^{(1)} \in C^1(\overline{\mathcal{D}})$ and that Eq. (5.77) holds for $\mathbf{y} \in \overline{\mathcal{D}}$. Put $S_1 := S - \delta/2$. Then $S_1 \in (0, S)$ and $\overline{\mathcal{D}}^c \subset B_{S_1}$, so inequality (5.37), with S, δ replaced by S_1, $S/S_1 - 1$, yields

$$|\partial_{\mathbf{y}}^{\boldsymbol{\alpha}} \mathcal{Z}_{jk}(\mathbf{y}, \mathbf{z}) f(\mathbf{z})| \leq C(S, S_1)\,(|\mathbf{y}|s_\tau(\mathbf{y}))^{-1-|\boldsymbol{\alpha}|/2}\,|f(\mathbf{z})|$$

for $\mathbf{z} \in \partial\mathcal{D}$, $\mathbf{y} \in B_S^c$, $\boldsymbol{\alpha} \in \mathbb{N}_0^3$ with $|\boldsymbol{\alpha}| \le 1$. Now we get by (5.77) that

$$|\partial^{\boldsymbol{\alpha}} F^{(1)}(\mathbf{y})| \le C(\delta, S)\,(|\mathbf{y}| s_\tau(\mathbf{y}))^{-1-|\boldsymbol{\alpha}|/2}\,\|f\|_1$$

for $\mathbf{y}$, α as before. The function $F^{(2)}$ may be dealt with in a similar way. As for $F^{(3)}$ and $F^{(4)}$, we note that for $\mathbf{y} \in B_S^c$ and $\mathbf{z} \in \mathcal{D}^c$, we have $|\mathbf{y}-\mathbf{z}| \ge (1 - S_1/S)|\mathbf{y}|$. This observation and Lemma 2.16 yield the estimates of $F^{(3)}$ and $F^{(4)}$ stated in (5.75) and (5.76), respectively. □

We want to extend Lemmas 5.13 and 5.14 to the case when second order derivatives act on layer and volume potentials.

Lemma 5.15 *Let $j, k, l \in \{1, 2, 3\}$, $f \in L^1(\partial\mathcal{D})$, and put*

$$F(\mathbf{y}) := \int_{\partial\mathcal{D}} \partial z_l \mathcal{Z}_{jk}(\mathbf{y}, \mathbf{z}) \cdot f(z) do_{\mathbf{z}}, \quad F^{(1)}(\mathbf{y}) := \int_{\partial\mathcal{D}} \mathcal{Z}_{jk}(\mathbf{y}, \mathbf{z}) \cdot f(\mathbf{z}) do_{\mathbf{z}},$$

$$F^{(3)}(\mathbf{y}) := \int_{\partial\mathcal{D}} E_{4j}(\mathbf{y} - \mathbf{z}) \cdot f(\mathbf{z}) do_{\mathbf{z}} \quad \textit{for } \mathbf{y} \in \mathcal{D}.$$

Then $F \in C^1(\mathcal{D})$, $F^{(1)}$, $F^{(3)} \in C^2(\mathcal{D})$, and

$$\partial_m F(\mathbf{y}) = \int_{\partial\mathcal{D}} \partial y_m \partial z_l \mathcal{Z}_{jk}(\mathbf{y}, \mathbf{z}) \cdot f(\mathbf{z}) do_{\mathbf{z}}, \tag{5.78}$$

$$\partial^{\alpha} F^{(1)}(\mathbf{y}) = \int_{\partial\mathcal{D}} \partial_{\mathbf{y}}^{\boldsymbol{\alpha}} \mathcal{Z}_{jk}(\mathbf{y}, \mathbf{z}) \cdot f(\mathbf{z}) do_{\mathbf{z}}, \tag{5.79}$$

$$\partial^{\alpha} F^{(3)}(\mathbf{y}) = \int_{\partial\mathfrak{D}} \partial_y^{\boldsymbol{\alpha}} E_{4j}(\mathbf{y} - \mathbf{z}) \cdot f(\mathbf{z}) do_{\mathbf{z}} \tag{5.80}$$

for $1 \le m \le 3$, $\alpha \in \mathbb{N}_0^3$ with $|\boldsymbol{\alpha}| \le 2$, $\mathbf{y} \in \overline{\mathcal{D}}$.

Let $S_1, S \in (0, \infty)$ with $\mathcal{D}^c \subset B_{S_1}$, $S_1 < S$. Then

$$|\partial^{\boldsymbol{\beta}} F(\mathbf{y})| \le C(S_1, S) \cdot \|\mathbf{f}\|_1 \cdot (|\mathbf{y}| \cdot s_\tau(\mathbf{y}))^{-3/2-|\boldsymbol{\beta}|/2}, \tag{5.81}$$

$$|\partial^{\boldsymbol{\alpha}} F^{(1)}(\mathbf{y})| \le C(S_1, S) \cdot \|\mathbf{f}\|_1 \cdot (|\mathbf{y}| \cdot s_\tau(\mathbf{y}))^{-1-|\boldsymbol{\alpha}|/2}, \tag{5.82}$$

$$|\partial^{\boldsymbol{\alpha}} F^{(3)}(\mathbf{y})| \le C(S_1, S) \cdot \|\mathbf{f}\|_1 \cdot |\mathbf{y}|^{-2-|\boldsymbol{\alpha}|} \tag{5.83}$$

for $\mathbf{y} \in B_S^c$, $\boldsymbol{\alpha}, \boldsymbol{\beta} \in \mathbb{N}_0^3$ with $|\boldsymbol{\alpha}| \le 2$, $|\boldsymbol{\beta}| \le 1$. □

Proof Let $U \subset \mathbb{R}^3$ with $\overline{U} \subset \mathcal{D}$. $\mathcal{Z}_{jk}$ is a C^2-function on $(\mathbb{R}^3 \times \mathbb{R}^3)\backslash\{(x, x) : x \in \mathbb{R}^3\}$ (see Lemma 5.4) and E_{4j} is a C^∞-function on $\mathbb{R}^3\backslash\{0\}$.

Since $\operatorname{dist}(\partial\mathcal{D}, \overline{U}) > 0$, it follows from Lebesgue's theorem on dominated convergence that $F_l|U \in C^1(U)$, $F^{(1)}|U$, $F^{(3)}|U \in C^2(U)$, and Eqs. (5.78)–(5.80) hold for $\mathbf{y} \in U$. This is true for any $U \subset \mathbb{R}^3$ with $\overline{U} \subset \mathcal{D}$, so we have proved that $F \in C^1(\mathcal{D})$, $F^{(1)}$, $F^{(3)} \in C^2(\mathcal{D})$ and that Eqs. (5.78)–(5.80) hold for $y \in \mathcal{D}$.

Inequalities (5.81)–(5.83) follow from (5.78)–(5.80), Theorem 5.4, and the relations

$$|\partial^{\alpha} E_{4j}(\mathbf{x})| \leq C \cdot |\mathbf{x}|^{-2-|\boldsymbol{\alpha}|} \quad \text{for } \mathbf{x} \in \mathbb{R}^3\backslash\{\mathbf{0}\},\ \boldsymbol{\alpha} \in \mathbb{N}_0^3 \text{ with } |\boldsymbol{\alpha}| \leq 2,$$
$$|\mathbf{y}-\mathbf{z}| \geq |\mathbf{y}| - |\mathbf{z}| = (1 - S_1/S) \cdot |\mathbf{y}| + (S_1/S) \cdot |\mathbf{y}| - |\mathbf{z}| \geq (1 - S_1/S) \cdot |\mathbf{y}|. \tag{5.84}$$

□

The same type of arguments yields:

Lemma 5.16 *Let $j, k \in \{1, 2, 3\}$, $\beta \in \mathbb{N}_0^3$ with $|\beta| \leq 1$, $R > 0$, $g \in L^1(B_R)$, and put*

$$F(\mathbf{y}) := \int_{B_R} \mathcal{Z}_{jk}(\mathbf{y}, \mathbf{z}) \cdot g(\mathbf{z}) d\mathbf{z}, \quad G(\mathbf{y}) := \int_{B_R} \partial_{\mathbf{y}}^{\boldsymbol{\beta}} E_{4j}(\mathbf{y}-\mathbf{z}) \cdot g(\mathbf{z}) d\mathbf{z} \quad \textit{for } \mathbf{y} \in \overline{B_R}^c.$$

Then $F, G \in C^2(\overline{B_R}^c)$ and

$$\partial^{\alpha} F(\mathbf{y}) = \int_{B_R} \partial_{\mathbf{y}}^{\boldsymbol{\alpha}} \mathcal{Z}_{jk}(\mathbf{y}, \mathbf{z}) \cdot g(\mathbf{z}) d\mathbf{z},$$
$$\partial^{\alpha} G(\mathbf{y}) = \int_{B_R} \partial_{\mathbf{y}}^{\boldsymbol{\alpha}} \partial_{\mathbf{y}}^{\boldsymbol{\beta}} E_{4j}(\mathbf{y}-\mathbf{z}) \cdot g(\mathbf{z}) d\mathbf{z}$$

for $\boldsymbol{\alpha} \in \mathbb{N}_0^3$ with $|\boldsymbol{\alpha}| \leq 2$, $\mathbf{y} \in \overline{B_R}^c$. Let $T \in (R, \infty)$. Then

$$|\partial^{\boldsymbol{\alpha}} F(\mathbf{y})| \leq C(R, T) \cdot \|g\|_1 \cdot (|\mathbf{y}| \cdot s_{\tau}(\mathbf{y}))^{-1-|\boldsymbol{\alpha}|/2},$$
$$|\partial^{\boldsymbol{\alpha}} G(\mathbf{y})| \leq C(R, T) \cdot \|g\|_1 \cdot |\mathbf{y}|^{-2-|\boldsymbol{\alpha}+\boldsymbol{\beta}|}$$

for $\mathbf{y} \in B_T^c$, $\boldsymbol{\alpha} \in \mathbb{N}_0^3$ with $|\boldsymbol{\alpha}| \leq 2$. □

For convenience of readers we recall representation formula that was derived in the slightly different form in Theorem 5.6.

Theorem 5.10 *Let $R \in (0, \infty)$ with $\mathcal{D}^c \subset \mathbb{R}^3$, and let $\boldsymbol{n}^{(R)} : \partial B_R \cup \partial \mathcal{D} \mapsto \mathbb{R}^3$ denote the inward unit normal to $\mathcal{D}_R$. Suppose that $u \in C^2(\overline{\mathcal{D}}_R)^3$, $\pi \in C^1(\overline{\mathcal{D}_R})$, and put $F := \mathcal{L}(\mathbf{u}) + \nabla \pi$. Let $\mathbf{y} \in \mathcal{D}_R$ and $j \in \{1, 2, 3\}$. Then*

$$u_j(\mathbf{y}) = \mathfrak{R}_j(F)(\mathbf{y}) + \mathfrak{S}_j(\operatorname{div} \mathbf{u})(\mathbf{y}) + \int_{\partial \mathcal{D}_R} \mathfrak{A}_j^{(R)}(\mathbf{u}, \pi)(\mathbf{y}, \mathbf{z}) do_{\mathbf{z}}, \tag{5.85}$$

where

$$\mathfrak{A}_j^{(R)}(\mathbf{u},\pi)(\mathbf{y},\mathbf{z}) \tag{5.86}$$

$$:= \sum_{k=1}^{3}\Bigg[\sum_{l=1}^{3}(\mathcal{Z}_{jk}(\mathbf{y},\mathbf{z})(\partial_l u_k(\mathbf{z}) - \delta_{kl}\pi(\mathbf{z}) + u_k(\mathbf{z})(-\tau \boldsymbol{e}_1 + \boldsymbol{\omega}\wedge \mathbf{z})_l)$$

$$- \partial z_l \mathcal{Z}_{jk}(\mathbf{y},\mathbf{z})u_k(\mathbf{z}))n_l^{(R)}(\mathbf{z}) - E_{4j}(\mathbf{y}-\mathbf{z})u_k(\mathbf{z})n_k^{(R)}(\mathbf{z})\Bigg]$$

for $\mathbf{y} \in \mathcal{D}_R$, $\mathbf{z} \in \partial\mathcal{D}_R$. □

Outline of proof Let $\epsilon \in (0,\infty)$ with $\overline{B_\epsilon(\mathbf{y})} \subset \mathcal{D}_R$, and consider the integral

$$A_{j,\epsilon} := \int_{\mathcal{D}_R\setminus B_\epsilon(\mathbf{y})} \sum_{k=1}^{3} \mathcal{Z}_{jk}(\mathbf{y},\mathbf{z})(\mathcal{L}(\mathbf{u}) + \nabla\pi)_k(\mathbf{z})d\mathbf{z}.$$

By performing some integrations by parts, using (2.47), integrating with respect to t, and then exploiting (2.49), we obtain

$$A_{j,\epsilon} = \int_{\mathcal{D}_R\setminus B_\epsilon(\mathbf{y})} -E_{4j}(\mathbf{y}-\mathbf{z})\,\mathrm{div}\,u(\mathbf{z})d\mathbf{z} - S_{j,\epsilon}(\mathbf{y}),$$

where $S_{j,\epsilon}(\mathbf{y})$ denotes a surface integral defined in the same way as the surface integral on the right-hand side of (5.85), but with $\partial B_R \cup \partial\mathcal{D} \cup \partial B_\epsilon(\mathbf{y})$ as domain of integration instead of $\partial B_R \cup \partial\mathcal{D}$ and with $n^{(R)}$ replaced by the outward unit normal to $\mathcal{D}_R\setminus B_\epsilon(\mathbf{y})$. Equation (5.85) then follows by a passage to the limit $\epsilon \downarrow 0$, with the calculation of $\lim_{\epsilon\downarrow 0} S_{j,\epsilon}(\mathbf{y})$ based on Theorem 5.5. This reasoning requires some applications of Fubini's and Lebesgue's theorem, all of which made possible by Lemma 5.2. □

Our next aim consists in extending Eq. (5.85) to functions $\mathbf{u}$ and π which are less regular than C^2 and C^1, respectively. We begin by specifying the type of functions we will consider. From now we need that $\partial\mathcal{D}$ is of class C^2. (Theorem 5.10 also holds if $\mathcal{D}$ is only Lipschitz bounded.)

Definition 5.1 Let $p \in (1,\infty)$. Define space of regularity $\mathfrak{M}_p$ as the space of all pairs of functions such that ($\mathcal{D}_T = B_T \cap \mathcal{D}$) $\mathbf{u} \in \mathbf{W}^{2,p}_{\mathrm{loc}}(\mathcal{D})$, $\pi \in W^{1,p}_{\mathrm{loc}}(\mathcal{D})$,

$$\mathbf{u}|\mathcal{D}_T \in \mathbf{W}^{1,p}(\mathcal{D}_T), \quad \pi|\mathcal{D}_T \in L^p(\mathcal{D}_T), \quad \mathbf{u}|\partial\mathcal{D} \in \mathbf{W}^{2-1/p,p}(\partial\mathcal{D}), \tag{5.87}$$
$$\mathrm{div}\,\mathbf{u}|\mathcal{D}_T \in W^{1,p}(\mathcal{D}_T), \quad \mathcal{L}(\mathbf{u}) + \nabla\pi|\mathcal{D}_T \in L^p(\mathcal{D}_T)^3$$

for some $T \in (0,\infty)$ with $\mathcal{D}^c \subset B_T$. □

Theorem 5.11 *Let $p \in (1, \infty)$, $(\mathbf{u}, \pi) \in \mathfrak{M}_p$.*

Then $\mathbf{u}|\mathcal{D}_T \in \boldsymbol{W}^{2,p}(\mathcal{D}_T)^3$, $\pi|\mathcal{D}_T \in W^{1,p}(\mathcal{D}_T)$ for any $T \in (0, \infty)$ with $\mathcal{D}^c \subset B_T$. □

Proof The theorem follows from the regularity theory for the Stokes system. To be more specific, we first note that our assumptions imply that the relations in (5.87) hold for all $T \in (0, \infty)$ with $\mathcal{D}^c \subset B_T$. Take such a number T. Let $S \in (T, \infty)$, and choose $\zeta \in C_0^\infty(\mathbb{R}^3)$ with $\zeta|B_T = 1$, $\zeta|B_S^c = 0$. Then

$$\zeta\mathbf{u}|\mathcal{D}_S \in \boldsymbol{W}^{2,p}_{\text{loc}}(\mathcal{D}_S) \cap \boldsymbol{W}^{1,p}(\mathcal{D}_S), \quad \zeta\pi|\mathcal{D}_S \in W^{1,p}_{\text{loc}}(\mathcal{D}_S) \cap L^p(\mathcal{D}_S), \tag{5.88}$$

$$\text{div}(\zeta\mathbf{u})|\mathcal{D}_S \in W^{1,p}(\mathcal{D}_S),$$

$$\zeta\mathbf{u}|\partial\mathcal{D} = \mathbf{u}|\partial\mathcal{D} \in \boldsymbol{W}^{2-1/p,p}(\partial\mathcal{D}), \quad \text{hence } \zeta\mathbf{u}|\partial\mathcal{D}_S \in \boldsymbol{W}^{2-1/p,p}(\partial\mathcal{D}_S).$$

Moreover, since $\mathbf{u}|\mathcal{D}_S \in \boldsymbol{W}^{1,p}(\mathcal{D}_S)$, $\mathcal{L}(\mathbf{u}) + \nabla\pi|\mathcal{D}_S \in \boldsymbol{L}^p(\mathcal{D}_S)$, we have $-\Delta\mathbf{u} + \nabla\pi|\mathcal{D}_S \in \boldsymbol{L}^p(\mathcal{D}_S)$. Once more observing that $\mathbf{u}|\mathcal{D}_S \in \boldsymbol{W}^{1,p}(\mathcal{D}_S)$, $\pi|\mathcal{D}_S \in L^p(\mathcal{D}_S)$, we may conclude that

$$-\Delta(\zeta\mathbf{u}) + \nabla(\zeta\pi)|\mathcal{D}_S \in \boldsymbol{L}^p(\mathfrak{D}_S). \tag{5.89}$$

Obviously the function $\zeta\mathbf{u}$ is a weak solution of the Stokes system in $\mathcal{D}_S$ with right-hand side $-\Delta(\zeta\mathbf{u}) + \nabla(\zeta\pi)|\mathcal{D}_S$, where "weak solution" is meant in the sense of [93, (IV.1.3)]. In view of (5.88) and (5.89), it follows from [93, Lemma IV.6.1, Exercise IV.6.2] that $\zeta\mathbf{u}|\mathcal{D}_S \in \boldsymbol{W}^{2,p}(\mathcal{D}_S)$, $\zeta\pi|\mathcal{D}_S \in W^{1,p}(\mathcal{D}_S)$. This implies that $\mathbf{u}|\mathcal{D}_T \in \boldsymbol{W}^{2,p}(\mathcal{D}_T)$ and $\pi|\mathcal{D}_T \in W^{1,p}(\mathcal{D}_T)$. □

Now we are in a position to generalize Theorem 5.10 to pairs of functions $(\mathbf{u}, \pi) \in \mathfrak{M}_p$.

Theorem 5.12 *Let $p \in (1, \infty)$, $(\mathbf{u}, \pi) \in \mathfrak{M}_p$, $j \in \{1, 2, 3\}$. Put $F := \mathcal{L}(\mathbf{u}) + \nabla\pi$. Take R and $\mathbf{n}^{(R)}$ as in Theorem 5.10. Then, for a.e. $\mathbf{y} \in \mathcal{D}_R$,*

$$u_j(\mathbf{y}) = \mathfrak{R}_j(F|\mathcal{D}_R)(\mathbf{y}) + \mathfrak{S}_j(\text{div}\,\mathbf{u}|\mathcal{D}_R)(\mathbf{y}) + \int_{\partial\mathcal{D}_R} \mathfrak{A}_j^{(R)}(\mathbf{u}, \pi)(\mathbf{y}, \mathbf{z})do_{\mathbf{z}}, \tag{5.90}$$

with $\mathfrak{A}_j^{(R)}(\mathbf{u}, \pi)(\mathbf{y}, \mathbf{z})$ defined as in (5.86).

If $p > 3/2$, Eq. (5.90) holds for any $\mathbf{y} \in \mathfrak{D}_R$ (without the restriction "a.e."). □

Proof By Theorem 5.11, we have $\mathbf{u}|\mathcal{D}_R \in \boldsymbol{W}^{2,p}(\mathcal{D}_R)$ and $\pi|\mathcal{D}_R \in W^{1,p}(\mathcal{D}_R)$. Therefore (see [2, (3.18)]) there are sequences $(\mathbf{u}_n)$ in $\boldsymbol{C}^\infty(\mathbb{R}^3)$ and (π_n) in $C^\infty(\mathbb{R}^3)$ with

$$\|(\mathbf{u} - \mathbf{u}_n)|\mathcal{D}_R\|_{2,p} + \|(\pi - \pi_n)|\mathcal{D}_R\|_{1,p} \to 0. \tag{5.91}$$

By a standard trace theorem, it follows that $\mathbf{u}_k|\partial\mathcal{D}_R$, $\partial_l\mathbf{u}_k|\partial\mathcal{D}_R$, and $\pi|\partial\mathcal{D}_R$ belong to $L^1(\partial\mathcal{D}_R)$, and

$$\|(\mathbf{u}-\mathbf{u}_n)|\partial\mathcal{D}_R\|_1 + \|(\partial_l\mathbf{u}-\partial_l\mathbf{u}_n)|\partial\mathcal{D}_R\|_1 + \|(\pi-\pi_n)|\partial\mathcal{D}_R\|_1 \to 0 \tag{5.92}$$

for $n \to \infty$ $(1 \le k, l \le 3)$. Let $\mathbf{y} \in \mathcal{D}_R$. We may conclude from (5.74) and (5.92) that

$$\int_{\partial\mathcal{D}_R} \mathfrak{A}_j^{(R)}(\mathbf{u}_n, \pi_n)(\mathbf{y}, \mathbf{z})\mathrm{d}o_{\mathbf{z}} \to \int_{\partial\mathcal{D}_R} \mathfrak{A}_j^{(R)}(\mathbf{u}, \pi)(\mathbf{y}, \mathbf{z})\mathrm{d}o_{\mathbf{z}} \quad (n \to \infty), \tag{5.93}$$

where the definition of $\mathfrak{A}_j^{(R)}(\mathbf{u}_n, \pi_n)(\mathbf{y}, \mathbf{z})$ should be obvious by (5.86). For $n \in \mathbb{N}$, we set $F_n := \mathcal{L}(\mathbf{u}_n) + \nabla\pi_n$. By (5.91), we have

$$\|(F_n - F)|\mathcal{D}_R\|_p \to 0, \quad \|\operatorname{div}(\mathbf{u}-\mathbf{u}_n)|\mathcal{D}_R\|_p \to 0 \quad (n \to \infty).$$

These relations combined with (5.56) and (5.71) imply

$$\|\mathfrak{R}_j((F_n - F)|\mathcal{D}_R)|\mathcal{D}_R\|_p + \|\mathfrak{S}_j(\operatorname{div}(\mathbf{u}_n - \mathbf{u})|\mathcal{D}_R)|\mathcal{D}_R\|_p \to 0 \quad (n \to \infty).$$

Passing from L^p-convergence to pointwise convergence of subsequences and recalling (5.91), we see that there is a strictly increasing function $\sigma : \mathbb{N} \mapsto \mathbb{N}$ such that

$$\begin{aligned} &\mathfrak{R}_j(F_{\sigma(n)}|\mathcal{D}_R)(\mathbf{y}) \to \mathfrak{R}_j(F|\mathcal{D}_R)(\mathbf{y}), \\ &\mathfrak{S}_j(\operatorname{div}\mathbf{u}_{\sigma(n)}|\mathcal{D}_R)(\mathbf{y}) \to \mathfrak{S}_j(\operatorname{div}\mathbf{u}|\mathcal{D}_R)(\mathbf{y}), \quad \mathbf{u}_{\sigma(n)}(\mathbf{y}) \to \mathbf{u}(\mathbf{y}) \quad (n \to \infty) \end{aligned} \tag{5.94}$$

for a.e. $\mathbf{y} \in \mathfrak{D}_R$. On the other hand, by Theorem 5.10, Eq. (5.90) holds with $\mathbf{u}$, π replaced by $\mathbf{u}_n$, π_n, respectively, for $n \in \mathbb{N}$. Therefore we may conclude from (5.93) and (5.94) that Eq. (5.90) holds for a.e. $\mathbf{y} \in \mathcal{D}_R$.

Now suppose that $p > 3/2$. Since $(\mathbf{u}, \pi) \in \mathfrak{M}_p$ and because of the Sobolev inequality (in the case $p \le 3$), we may conclude that $\operatorname{div}\mathbf{u}|\mathcal{D}_R \in L^q(\mathcal{D}_R)$ for some $q \in (3, \infty)$. Recalling the relation $F|\mathcal{D}_R \in \mathbf{L}^p(\mathcal{D}_R)$, we thus see by Lemmas 5.11 and 5.12 with $S = R$ that $\mathfrak{R}(F|\mathcal{D}_R)$ and $\mathfrak{S}(\operatorname{div}\mathbf{u}|\mathcal{D}_R)$ are continuous. Moreover, since $p > 3/2$ and $\mathbf{u}|\mathcal{D}_R \in \mathbf{W}^{2,p}(\mathcal{D}_R)$, the Sobolev lemma implies that $\mathbf{u}$ may be considered as a continuous function on $\overline{\mathcal{D}_R}$. According to Lemma 5.13, the function associating the integral $\int_{\partial\mathcal{D}_R} \mathfrak{A}_j^{(R)}(\mathbf{y}, \mathbf{z})\mathrm{d}o_{\mathbf{z}}$ with each $\mathbf{y} \in \mathcal{D}_R$ is also continuous. Thus we may conclude that Eq. (5.90) is valid for any $\mathbf{y} \in \mathcal{D}_R$, without the restriction "a.e." □

Next we perform the transition from the representation formula on $\mathcal{D}_R$ to one on $\overline{\mathcal{D}}$. For this step, we only need the decay properties given implicitly by the relations in (5.95).

Theorem 5.13 *Let $p \in (1, \infty)$, $(\mathbf{u}, \pi) \in \mathfrak{M}_p$. Put $F := \mathcal{L}(\mathbf{u}) + \nabla\pi$, and suppose there are numbers $p_1, p_2 \in (1, 2)$, $S \in (0, \infty)$ such that $\mathcal{D}^c \subset B_S$,*

$$\begin{aligned} \mathbf{u}|B_S^c \in& \boldsymbol{L}^6(B_S^c), \quad \nabla\mathbf{u}|B_S^c \in \boldsymbol{L}^2(B_S^c)^9, \quad \pi|B_S^c \in L^2(B_S^c), \\ \mathbf{F}|B_S^c \in& \boldsymbol{L}^{p_1}(B_S^c) + \boldsymbol{L}^{p_2}(B_S^c). \end{aligned} \tag{5.95}$$

Let $j \in \{1, 2, 3\}$, and put

$$\begin{aligned} \mathfrak{B}_j(y) :=& \mathfrak{B}_j(\mathbf{u}, \pi)(\mathbf{y}) \\ :=& \int_{\partial\mathcal{D}} \sum_{k=1}^{3} \Big[\sum_{l=1}^{3} (\mathcal{Z}_{jk}(\mathbf{y}, \mathbf{z})(-\partial_l u_k(\mathbf{z}) + \delta_{kl}\pi(\mathbf{z}) + u_k(\mathbf{z})(\tau\boldsymbol{e}_1 - \boldsymbol{\omega} \wedge \mathbf{z})_l) \\ &+ \partial z_l \mathcal{Z}_{jk}(\mathbf{y}, \mathbf{z}) u_k(\mathbf{z})) n_l^{(\mathcal{D})}(\mathbf{z}) + E_{4j}(\mathbf{y} - \mathbf{z}) u_k(\mathbf{z}) n_k^{(\mathcal{D})}(\mathbf{z}) \Big] do_{\mathbf{z}} \end{aligned} \tag{5.96}$$

for $\mathbf{y} \in \mathcal{D}$. Then

$$u_j(\mathbf{y}) = \mathfrak{R}_j(F)(\mathbf{y}) + \mathfrak{S}_j(\operatorname{div}\mathbf{u})(\mathbf{y}) + \mathfrak{B}_j(\mathbf{y}) \tag{5.97}$$

for a.e. $\mathbf{y} \in \mathcal{D}$. If $p > 3/2$, Eq. (5.97) holds for any $\mathbf{y} \in \overline{\mathfrak{D}}^c$ without the restriction "a.e." □

Proof The assumptions on $\mathbf{u}$ and π yield that

$$\int_S^\infty \int_{\partial B_r} (|\mathbf{u}(\mathbf{z})|^6 + |\nabla\mathbf{u}(\mathbf{z})|^2 + |\pi(\mathbf{z})|^2) \mathrm{d}o_{\mathbf{z}} \mathrm{d}r < \infty. \tag{5.98}$$

Therefore there is an increasing sequence (R_n) in (S, ∞) with $R_n \to \infty$ and

$$\int_{\partial B_{R_n}} (|\mathbf{u}(\mathbf{z})|^6 + |\nabla\mathbf{u}(\mathbf{z})|^2 + |\pi(\mathbf{z})|^2) \mathrm{d}o_{\mathbf{z}} \le R_n^{-1} \quad \text{for } n \in \mathbb{N}. \tag{5.99}$$

Otherwise there would be a constant $C \in [S, \infty)$ such that

$$\int_{\partial B_r} (|\mathbf{u}(\mathbf{z})|^6 + |\nabla\mathbf{u}(\mathbf{z})|^2 + |\pi(\mathbf{z})|^2) \mathrm{d}o_{\mathbf{z}} \ge r^{-1} \quad \text{for } r \in [C, \infty),$$

in contradiction to (5.98). (Here we have used a standard convention from the theory of Lebesgue integration, which states that the integral of every measurable nonnegative function is defined, but may take the value ∞.) By our assumptions on F, there are functions $G^{(i)} \in L^{p_i}(B_S^c)$ for $i \in \{1,2\}$ such that $F|B_S^c = G^{(1)} + G^{(2)}$. Thus, by Lemma 5.10,

$$\int_{\mathbb{R}^3} \sum_{k=1}^{3} |\mathcal{Z}_{jk}(\mathbf{y},\mathbf{z})|(\chi_{(0,S]}(|\mathbf{z}|)|F_k(\mathbf{z})| + \chi_{(S,\infty)}(|\mathbf{z}|)(|G_k^{(1)}(\mathbf{z})| + |G_k^{(2)}(\mathbf{z})|))\mathrm{d}\mathbf{z} < \infty \tag{5.100}$$

for a.e. $\mathbf{y} \in \overline{\mathcal{D}}$. Moreover, by Lemma 5.12 with $q = 2$,

$$\int_{\mathbb{R}^3} |E_{4j}(\mathbf{y}-\mathbf{z})|\,|\operatorname{div} u(\mathbf{z})|\mathrm{d}\mathbf{z} < \infty \tag{5.101}$$

for a.e. $\mathbf{y} \in \overline{\mathcal{D}}$. Due to these observations and Theorem 5.12, we see there is a subset N of $\overline{\mathcal{D}}$ with measure zero such that the relations in (5.100) and (5.101) hold for $\mathbf{y} \in \overline{\mathcal{D}}^c \backslash N$ and such that Eq. (5.90) with R replaced by R_n holds for $n \in \mathbb{N}$ and $\mathbf{y} \in \mathcal{D}_R \backslash N$. In the case $p > 3/2$, Lemma 5.11 yields that (5.100) is valid for any $\mathbf{y} \in \overline{\mathcal{D}}$, and Theorem 5.12 implies that Eq. (5.90) with R replaced by R_n is true for $n \in \mathbb{N}$ and any $\mathbf{y} \in \overline{\mathcal{D}}$. Moreover, if $p > 3/2$, the assumption $(\mathbf{u}, \pi) \in \mathfrak{M}_p$, Lemma 5.12, and the Sobolev inequality (in the case $p \le 3$) allow to drop the restriction "a.e." in (5.101).

Take $\mathbf{y} \in \overline{\mathcal{D}}$ in the case $p > 3/2$ and $\mathbf{y} \in \overline{\mathcal{D}}^c \backslash N$ otherwise. Let $n \in \mathbb{N}$ with $R_n > |\mathbf{y}|$ (hence $y \in \mathcal{D}_{R_n}$). Then, by Eq. (5.90) with R replaced by R_n, we get

$$u_j(\mathbf{y}) = \mathfrak{R}_j(F|\mathcal{D}_{R_n})(\mathbf{y}) + \mathfrak{S}_j(\operatorname{div} \mathbf{u}|\mathfrak{D}_{R_n})(\mathbf{y}) + \mathfrak{A}_{j,n}(\mathbf{y}) + \mathfrak{B}_j(\mathbf{y}), \tag{5.102}$$

with

$$\begin{aligned}\mathfrak{A}_{j,n}(\mathbf{y}) := \int_{\partial B_{R_n}} \sum_{k=1}^{3} \Bigg[\sum_{l=1}^{3} (\mathcal{Z}_{jk}(\mathbf{y},\mathbf{z})(\partial_l u_k(\mathbf{z}) - \delta_{kl}\pi(\mathbf{z}) - \tau\delta_{1l}u_k(\mathbf{z}))\\ - \partial z_l \mathcal{Z}_{jk}(\mathbf{y},\mathbf{z})u_k(\mathbf{z}))z_l/R_n - E_{4j}(\mathbf{y}-\mathbf{z})u_k(\mathbf{z})z_k/R_n \Bigg] \mathrm{d}o_{\mathbf{z}}.\end{aligned}$$

Note that in (5.102), we used the relation $\sum_{l=1}^{3}(\omega \wedge z)_l z_l/R_n = 0$ for $\mathbf{z} \in \partial B_R$. The term $\mathfrak{B}_j(\mathbf{y})$ was defined in (5.96). Let $n \in \mathbb{N}$ with $R_n/4 \ge |\mathbf{y}|$. Observe that

$$|\mathfrak{A}_{j,n}(\mathbf{y})| \le \mathfrak{C} \sum_{\nu=1}^{4} \sum_{k=1}^{3} \mathfrak{V}_{\nu,k}(\mathbf{y}), \tag{5.103}$$

with

$$\mathfrak{V}_{1,k}(\mathbf{y}) := \left(\int_{\partial B_{R_n}} |\mathcal{Z}_{jk}(\mathbf{y},\mathbf{z})|^{6/5} do_{\mathbf{z}} \right)^{5/6} \|\mathbf{u}|\partial B_{R_n}\|_6,$$

$$\mathfrak{V}_{2,k}(\mathbf{y}) := \left(\int_{\partial B_{R_n}} |\mathcal{Z}_{jk}(\mathbf{y},\mathbf{z})|^{2} do_{\mathbf{z}} \right)^{1/2} (\|\nabla\mathbf{u}|\partial B_{R_n}\|_2 + \|\pi|\partial B_{R_n}\|_2),$$

$$\mathfrak{V}_{3,k}(\mathbf{y}) := \sum_{l=1}^{3} \left(\int_{\partial B_{R_n}} |\partial z_l \mathcal{Z}_{jk}(\mathbf{y},\mathbf{z})|^{6/5} do_{\mathbf{z}} \right)^{5/6} \|\mathbf{u}|\partial B_{R_n}\|_6,$$

$$\mathfrak{V}_{4,k}(\mathbf{y}) := \left(\int_{\partial B_{R_n}} |\mathbf{y}-\mathbf{z}|^{-12/5} do_{\mathbf{z}} \right)^{5/6} \|\mathbf{u}|\partial B_{R_n}\|_6$$

for $k \in \{1,2,3\}$. Since $|\mathbf{y}| \le R_n/4$, we may use inequality (5.38) with $S = 2|\mathbf{y}|$ in order to estimate $|\partial_z^{\boldsymbol{\alpha}} \mathcal{Z}_{jk}(\mathbf{y},\mathbf{z})|$ for $\mathbf{z} \in \partial B_{R_n}$, $\boldsymbol{\alpha} \in \mathbb{N}_0^3$ with $|\boldsymbol{\alpha}| \le 1$. We get by (5.99) and (5.38) that

$$\begin{aligned} \mathfrak{V}_{1,k}(\mathbf{y}) \le & C(|\mathbf{y}|) \left(\int_{\partial B_{R_n}} (|\mathbf{z}| s_\tau(\mathbf{z}))^{-6/5} do_{\mathbf{z}} \right)^{5/6} R_n^{-1/6} \\ \le & C(|\mathbf{y}|) \left(\int_{\partial B_{R_n}} s_\tau(\mathbf{z})^{-6/5} do_{\mathbf{z}} \right)^{5/6} R_n^{-7/6} \le C(|\mathbf{y}|) R_n^{-1/3}, \end{aligned} \tag{5.104}$$

where the last inequality follows from Lemma 2.13. The same references yield

$$|\mathfrak{V}_{2,k}(\mathbf{y})| \le C(|\mathbf{y}|) R_n^{-1}, \quad |\mathfrak{V}_{3,k}(\mathbf{y})| \le C(|\mathbf{y}|) R_n^{-5/6} \quad (1 \le k \le 3). \tag{5.105}$$

Moreover, since $|\mathbf{y}-\mathbf{z}| \ge |\mathbf{z}|/2$ for ∂B_{R_n}, we find with (5.99) that $|\mathfrak{V}_{4,k}(\mathbf{y})| \le C(|\mathbf{y}|) R_n^{-1/2}$. From (5.103)–(5.105) and the preceding inequality, we may conclude that $\mathfrak{A}_{n,j}(\mathbf{y}) \to 0$ for $n \to \infty$. Turning to $\mathfrak{R}_j(F|\mathcal{D}_{R_n})(\mathbf{y})$, we observe that by (5.100), our choice of $\mathbf{y}$, and Lebesgue's theorem on dominated convergence, we have $\mathfrak{R}_j(F|\mathcal{D}_{R_n})(\mathbf{y}) \to \mathfrak{R}_j(F)(\mathbf{y})$ for $n \to \infty$. Moreover, by (5.101) and again by the choice of y and Lebesgue's theorem, $\mathfrak{S}_j(\operatorname{div}\mathbf{u}|\mathcal{D}_{R_n})(\mathbf{y}) \to \mathfrak{S}_j(\operatorname{div}\mathbf{u})(\mathbf{y})$ for $n \to \infty$. Recalling (5.102), we thus have proved (5.97). □

In our first application of our representation formula (5.97), we state conditions on $\mathcal{L}(u) + \nabla\pi$ and $\operatorname{div}\mathbf{u}$ such that u decays as described

$$\begin{aligned} |\mathbf{u}(\mathbf{x})| =& O\left[(|\mathbf{x}|(1+\tau(|\mathbf{x}|-x_1)))^{-1}\right], \\ |\nabla\mathbf{u}(\mathbf{x})| =& O\left[(|\mathbf{x}|(1+\tau(|\mathbf{x}|-x_1)))^{-3/2}\right] \quad \text{for } |\mathbf{x}| \to \infty. \end{aligned} \tag{5.106}$$

Since in the proof of this result, we want to avoid estimates of the second derivatives of $\mathcal{Z}_{jk}$, we have to transform the integral

$\int_{\partial\mathcal{D}} \partial z_l \mathcal{Z}_{jk}(\mathbf{y}, \mathbf{z}) u_k(z) n_l^{(\mathcal{D})}(\mathbf{z}) do_{\mathbf{z}}$ appearing in the definition of $\mathfrak{B}_j(\mathbf{y})$ (see (5.96)), into a term where no differential operator acts on $\mathcal{Z}_{jk}$. This is done in the next lemma.

Lemma 5.17 *Let* $p \in (1, \infty)$, $(\mathbf{u}, \pi) \in \mathfrak{M}_p$, $j \in \{1, 2, 3\}$. *Define*

$$\mathfrak{U}_j(\mathbf{y}) := \mathfrak{U}_j(\mathbf{u})(\mathbf{y}) := \int_{\partial\mathcal{D}} \sum_{k,l=1}^{3} \partial z_l \mathcal{Z}_{jk}(\mathbf{y}, \mathbf{z}) u_k(\mathbf{z}) n_l^{(\mathcal{D})}(\mathbf{z}) do_{\mathbf{z}} \tag{5.107}$$

for $\mathbf{y} \in \overline{\mathcal{D}}$. *Let* $\mathcal{E}_p \colon \mathbf{W}^{2-1/p,p}(\partial\mathcal{D}) \mapsto \mathbf{W}^{2,p}(\mathcal{D})$ *denote a continuous extension operator ([174]). Then, for* $\mathbf{y} \in \overline{\mathcal{D}}$,

$$\begin{aligned}\mathfrak{U}_j(\mathbf{y}) = \int_{\mathcal{D}} \sum_{k=1}^{3} \Big[&\partial_k E_{4j}(\mathbf{y} - \mathbf{z}) \mathcal{E}_p(u_k)(\mathbf{z}) + \mathcal{Z}_{jk}(\mathbf{y}, \mathbf{z}) \\ &\times ((\tau \boldsymbol{e}_1 - \boldsymbol{\omega} \wedge \mathbf{z}) \nabla \mathcal{E}_p(u_k)(\mathbf{z}) + [\omega \wedge (\mathcal{E}_p(u_s)(\mathbf{z}))_{1 \le s \le 3}]_k - \Delta \mathcal{E}_p(u_k)(\mathbf{z})) \Big] dz \\ &+ \int_{\partial\mathcal{D}} \sum_{k,l=1}^{3} \mathcal{Z}_{jk}(\mathbf{y}, \mathbf{z})((-\tau \boldsymbol{e}_1 + \boldsymbol{\omega} \wedge \mathbf{z})_l u_k(\mathbf{z}) + \partial_l \mathcal{E}_p(u_k)(\mathbf{z})) n_l^{(\mathcal{D})}(\mathbf{z}) do_{\mathbf{z}}.\end{aligned} \tag{5.108}$$

Proof Let $y \in \overline{\mathcal{D}}$. Starting with (5.34), we may refer to Lemma 5.2 in order to apply Fubini's theorem and Lebesgue's theorem on dominated convergence, to obtain

$$\mathfrak{U}_j(\mathbf{y}) = \lim_{\delta \downarrow 0, T \to \infty} \int_\delta^T \int_{\partial\mathcal{D}} \sum_{k,l=1}^{3} \partial z_l \Gamma_{jk}(\mathbf{y}, \mathbf{z}, t) u_k(\mathbf{z}) n_l^{(\mathcal{D})}(z) do_{\mathbf{z}} dt.$$

Next we apply the Divergence theorem and then use (2.46). It follows

$$\begin{aligned}\mathfrak{U}_j(\mathbf{y}) = &\lim_{\delta \downarrow 0, T \to \infty} \int_\delta^T \int_{\mathcal{D}} \sum_{k=1}^{3} (\Delta_{\mathbf{z}} \Gamma_{jk}(\mathbf{y}, \mathbf{z}, t) \mathcal{E}_p(u_k)(\mathbf{z}) \\ &+ \nabla_z \Gamma(\mathbf{y}, \mathbf{z}, t) \nabla \mathcal{E}_p(u_k)(\mathbf{z})) d\mathbf{z} dt \\ = &\lim_{\delta \downarrow 0, T \to \infty} \Big[\int_\delta^T \int_{\mathcal{D}} \sum_{k=1}^{3} ((\partial_t \Gamma_{jk}(\mathbf{y}, \mathbf{z}, t) + (-\tau \boldsymbol{e}_1 + \boldsymbol{\omega} \wedge \mathbf{z}) \cdot \nabla_z \Gamma_{jk}(\mathbf{y}, \mathbf{z}, t) \\ &- [\boldsymbol{\omega} \wedge (\Gamma_{js}(\mathbf{y}, \mathbf{z}, t))_{1 \le s \le 3}]_k) \mathcal{E}_p(u_k)(\mathbf{z}) - \Gamma_{jk}(\mathbf{y}, \mathbf{z}, t) \Delta \mathcal{E}_p(u_k)(\mathbf{z})) d\mathbf{z} dt \\ &+ \int_\delta^T \int_{\partial\mathcal{D}} \sum_{k,l=1}^{3} \Gamma_{jk}(\mathbf{y}, \mathbf{z}, t) \partial_l \mathcal{E}_p(u_k)(\mathbf{z}) n_l^{(\mathcal{D})}(\mathbf{z}) do_{\mathbf{z}} dt \Big].\end{aligned} \tag{5.109}$$

As explained in the proof of Theorem 5.6, the relation in (2.49) and Lemma 5.2 yield

$$\lim_{\delta\downarrow 0, T\to\infty} \int_\delta^T \int_{\mathcal{D}} \sum_{k=1}^{3} \partial_t \Gamma_{jk}(\mathbf{y}, \mathbf{z}, t)\mathcal{E}_p(u_k)(\mathbf{z})\mathrm{d}\mathbf{z}\mathrm{d}t \tag{5.110}$$

$$= \int_{\mathcal{D}} \sum_{k=1}^{3} \partial_k E_{4j}(\mathbf{y}-\mathbf{z})\mathcal{E}_p(u_k)(\mathbf{z})\mathrm{d}\mathbf{z}.$$

For the other terms on the right-hand side of (5.109), the passage to the limit $\delta \downarrow 0$ and $T \to \infty$ presents no difficulty because due to Lemma 5.2, we may directly apply Fubini's and Lebesgue's theorem. We further use the formula $(\boldsymbol{a} \wedge \boldsymbol{b}) \cdot \boldsymbol{c} = -(a \wedge c) \cdot b$ for vectors $\boldsymbol{a}$, $\boldsymbol{b}$, $\boldsymbol{c}$ in $\mathbb{R}^3$. In this way, letting δ tend to zero and T to infinity and taking account of (5.110), we may deduce (5.108) from (5.109). □

Now we may prove a decay estimate for $\mathfrak{B}_j(\mathbf{u}, \pi)$.

Lemma 5.18 *Let* $p \in (1, \infty)$, $(\mathbf{u}, \pi) \in \mathfrak{M}_p$, $j \in \{1, 2, 3\}$. *Define* $\mathfrak{B}_j = \mathfrak{B}_j(\mathbf{u}, \pi)$ *as in* (5.96). *Then* $\mathfrak{B}_j \in C^1(\overline{\mathcal{D}})$.

Let $S \in (0, \infty)$ *with* $\overline{\mathcal{D}}^c \subset B_S$. *Put* $\delta := \mathrm{dist}(\overline{\mathcal{D}}^c, \partial B_S)$. *Let* $\boldsymbol{\alpha} \in \mathbb{N}_0^3$ *with* $|\boldsymbol{\alpha}| \le 1$, $\mathbf{y} \in B_S^c$. *Then*

$$|\partial^\alpha \mathfrak{B}_j(\mathbf{y})| \le C(S, \delta)(\|\nabla\mathbf{u}|\partial\mathcal{D}\|_1 + \|\pi|\partial\mathcal{D}\|_1 + \widetilde{C}(\mathcal{D}, p)\|\mathbf{u}|\partial\mathcal{D}\|_{2-1/p,p})(|\mathbf{y}|s_\tau(\mathbf{y}))^{-1-|\boldsymbol{\alpha}|/2}, \tag{5.111}$$

where $\widetilde{C}(\mathcal{D}, p)$ *is a constant depending only on* $\mathcal{D}$ *and* p. □

Proof We use the decomposition $\mathfrak{B}_j(\mathbf{y}) = (\mathfrak{B}_j(\mathbf{y}) - \mathfrak{U}_j(\mathbf{y})) + \mathfrak{U}_j(\mathbf{y})$ with $\mathfrak{U}_j = \mathfrak{U}_j(\mathbf{u}, \pi)$ defined in (5.107). Equation (5.108) and Lemma 5.14 yield that $\mathfrak{B}_j - \mathfrak{U}_j$ and $\mathfrak{U}_j$ belong to $C^1(\overline{\mathcal{D}})$. Therefore we have $\mathfrak{B}_j \in C^1(\overline{\mathcal{D}})$. Moreover, by (5.75), (5.76), (5.96), and (5.108),

$$|\partial^\alpha(\mathfrak{B}_j - \mathfrak{U}_j)(\mathbf{y})| + |\partial^\alpha \mathfrak{U}_j(\mathbf{y})| \le C(S, \delta)(|\mathbf{y}|s_\tau(\mathbf{y}))^{-1-|\boldsymbol{\alpha}|/2} \tag{5.112}$$

$$(\|\nabla\mathbf{u}|\partial\mathcal{D}\|_1 + \|\pi|\partial\mathcal{D}\|_1 + \|\mathbf{u}|\partial\mathcal{D}\|_1 + \sum_{k=1}^{3}(\|\mathcal{E}_p(u_k)\|_{2,1} + \|\nabla\mathcal{E}_p(u_k)|\partial\mathcal{D}\|_1)),$$

where the extension operator $\mathcal{E}_p$ was introduced in Lemma 5.17. On the other hand, by a standard trace theorem and by the choice of $\mathcal{E}_p$,

$$\begin{aligned}\|\nabla\mathcal{E}_p(\mathbf{u}_k)|\partial\mathcal{D}\|_1 \leq& C\|\nabla\mathcal{E}_p(\mathbf{u}_k)|\partial\mathcal{D}\|_p \leq C(p)\|\mathcal{E}_p(\mathbf{u}_k)\|_{2,p} \qquad (5.113)\\ \leq& C(p)\|\mathbf{u}|\partial\mathcal{D}\|_{2-1/p,p},\\ \|\mathcal{E}_p(\mathbf{u}_k)\|_{2,1} \leq& C\|\mathcal{E}_p(\mathbf{u}_k)\|_{2,p} \leq C(p)\|\mathbf{u}|_{\partial\mathcal{D}}\|_{2-1/p,p} \qquad (5.114)\end{aligned}$$

for $k \in \{1, 2, 3\}$. Inequality (5.111) is a consequence of (5.112)–(5.114). □

At this point, we are in a position to derive the decay relations (5.106) for u if $\mathcal{L}(\mathbf{u}) + \nabla\pi$ and $\operatorname{div}\mathbf{u}$ decay sufficiently fast.

Theorem 5.14 *Let $p \in (1, \infty)$, $(\mathbf{u}, \pi) \in \mathfrak{M}_p$. Put $F := \mathcal{L}(\mathbf{u}) + \nabla\pi$. Suppose there are numbers $S_1, S, \gamma \in (0, \infty)$, $A \in [2, \infty)$, $B \in \mathbb{R}$ such that $S_1 < S$, $\overline{\mathcal{D}}^c \subset B_{S_1}$,*

$$\mathbf{u}|B_S^c \in \mathbf{L}^6(B_S^c), \quad \nabla\mathbf{u}|B_S^c \in L^2(B_S^c)^9, \quad \pi|B_S^c \in L^2(B_S^c), \operatorname{supp}(\operatorname{div}\mathbf{u}) \subset B_{S_1},$$
$$A + \min\{1, B\} \geq 3, \quad |F(\mathbf{z})| \leq \gamma|\mathbf{z}|^{-A}s_\tau(\mathbf{z})^{-B} \quad \textit{for } \mathbf{z} \in B_{S_1}^c.$$

Put $\delta := \operatorname{dist}(\overline{\mathcal{D}}^c, \partial B_S)$. Let $i, j \in \{1, 2, 3\}$, $\mathbf{y} \in B_S^c$. Then

$$\begin{aligned}|u_j(\mathbf{y})| \leq& C(S, S_1, A, B, \delta)(\gamma + \|F|B_{S_1}\|_1 + \|\operatorname{div}\mathbf{u}\|_1 + \|\nabla\mathbf{u}|\partial\mathcal{D}\|_1 \qquad (5.115)\\ &+ \|\pi|\partial\mathcal{D}\|_1 + \widetilde{C}(\mathcal{D}, p)\|\mathbf{u}|\partial\mathcal{D}\|_{2-1/p,p})(|\mathbf{y}|s_\tau(\mathbf{y}))^{-1}l_{A,B}(\mathbf{y}),\\ |\partial_i u_j(\mathbf{y})| \leq& C(S, S_1, A, B, \delta)(\gamma + \|F|B_{S_1}\|_1 + \|\operatorname{div}\mathbf{u}\|_1 + \|\nabla\mathbf{u}|\partial\mathcal{D}\|_1 \qquad (5.116)\\ &+ \|\pi|\partial\mathcal{D}\|_1\\ &+ \widetilde{C}(\mathcal{D}, p)\|\mathbf{u}|\partial\mathcal{D}\|_{2-1/p,p})(|\mathbf{y}|s_\tau(\mathbf{y}))^{-3/2}s_\tau(\mathbf{y})^{\max(0,7/2-A-B)}l_{A,B}(\mathbf{y}),\end{aligned}$$

where $\widetilde{C}(\mathcal{D}, p)$ was introduced in Lemma 5.18 and function $l_{A,B}(\mathbf{y})$ in Theorem 5.8. If the assumption $\operatorname{supp}(\operatorname{div}\mathbf{u}) \subset B_{S_1}$ is replaced by the condition

$$|\operatorname{div}\mathbf{u}(\mathbf{z})| \leq \widetilde{\gamma}|\mathbf{z}|^{-C}s_\tau(\mathbf{z})^{-D} \quad \textit{for } \mathbf{z} \in B_{S_1}^c,$$

for some $\widetilde{\gamma} \in (0, \infty)$, $C \in (5/2, \infty)$, $D \in \mathbb{R}$ with $C + \min\{1, D\} > 3$, then inequality (5.115) remains valid if the term $\|\operatorname{div}\mathbf{u}\|_1$ on the right-hand side of (5.115) is replaced by $\widetilde{\gamma} + \|\operatorname{div}\mathbf{u}|B_{S_1}\|_1$. Of course, in that case the constant in (5.115) additionally depends on C and D. □

Note that if $A + \min\{1, B\} > 3$, $A + B \geq 7/2$ in Theorem 5.14, then $l_{A,B}(\mathbf{y}) = 1$ in (5.115) and $s_\tau(\mathbf{y})^{\max(0,7/2-A-B)}l_{A,B}(\mathbf{y}) = 1$ in (5.116). The preceding conditions on A and B are verified if, for example, $A = 5/2$, $B = 1$, or $B = 3/2$, and $A = 2+\epsilon$ for some $\epsilon \in (0, 1/2)$.

Proof of Theorem 5.14: By Lemma 2.13, we see that $\int_{B^c_{S_1}} |F(\mathbf{z})|^r d\mathbf{z} < \infty$ for any $r \in (1, \infty)$. Thus Theorem 5.13 yields that the representation formula (5.97) holds for a.e. $y \in \overline{\mathfrak{D}}^c$. Therefore Theorem 5.14 follows from Theorems 5.8 and 5.9 and Lemma 5.18.

In the next theorem, we present an asymptotic profile of $\mathbf{u}$ for the case that $\mathcal{L}(\mathbf{u})+\nabla\pi$ and $\operatorname{div}\mathbf{u}$ have compact support.

Theorem 5.15 *Let $p \in (1, \infty)$, $(\mathbf{u}, \pi) \in \mathfrak{M}_p$, $S, S_1 \in (0, \infty)$ with $S_1 < S$, and put $F := \mathcal{L}(\mathbf{u}) + \nabla\pi$. Suppose that*

$$\overline{\mathcal{D}} \cup \operatorname{supp}(F) \cup \operatorname{supp}(\operatorname{div}\mathbf{u}) \subset B_{S_1},$$

$$\mathbf{u}|B^c_S \in \boldsymbol{L}^6(B^c_S)^3, \quad \nabla\mathbf{u}|B^c_S \in L^2(B^c_S)^9, \quad \pi|B^c_S \in L^2(B^c_S).$$

Then there are coefficients $\beta_1, \beta_2, \beta_3 \in \mathbb{R}$ and functions $\mathfrak{F}_1, \mathfrak{F}_2, \mathfrak{F}_3 \in C^0(B^c_S)$ such that for $j \in \{1, 2, 3\}$, $\mathbf{y} \in B^c_S$,

$$u_j(\mathbf{y}) = \sum_{k=1}^{3} \beta_k \mathcal{Z}_{jk}(\mathbf{y}, 0) + \left(\int_{\partial\mathcal{D}} \mathbf{u} \cdot n^{(\mathcal{D})} do_{\mathbf{z}} + \int_{B_{S_1}} \operatorname{div}\mathbf{u}\, d\mathbf{z} \right) E_{4j}(\mathbf{y}) + \mathfrak{F}_j(\mathbf{y}), \tag{5.117}$$

and

$$\begin{aligned} |\mathfrak{F}_j(\mathbf{y})| \leq & C(S, S_1)(\|F\|_1 + \|\operatorname{div}\mathbf{u}\|_1 + \|\nabla\mathbf{u}|\partial\mathcal{D}\|_1 + \|\pi|\partial\mathcal{D}\|_1 \\ & + C(\mathcal{D}, p)\|\mathbf{u}|\partial\mathcal{D}\|_{2-1/p,p})(|\mathbf{y}| s_\tau(\mathbf{y}))^{-3/2}, \end{aligned} \tag{5.118}$$

where $C(\mathcal{D}, p) > 0$ depends only on $\mathcal{D}$ and p. (Note that $|E_{4j}(\mathbf{y})| \leq C|\mathbf{y}|^{-2}$ and $|\mathbf{y}|^{-2} \leq C(S)(|\mathbf{y}| s_\tau(\mathbf{y}))^{-1}$ for $\mathbf{y} \in B^c_S$; see Lemma 2.16.) □

Proof Take $j \in \{1, 2, 3\}$, $\mathbf{y} \in B^c_S$. Observe that

$$|\mathbf{y} - \vartheta\mathbf{z}| \geq |\mathbf{y}| - S_1 \geq (1 - S_1/S)|\mathbf{y}| > 0 \quad \text{for } \mathbf{z} \in B_{S_1},\ \vartheta \in [0, 1]. \tag{5.119}$$

In view of Lemma 5.3, we may conclude that the term $\mathcal{Z}_{jk}(\mathbf{y}, \vartheta\mathbf{z})$ is continuously differentiable with respect to $\vartheta \in [0, 1]$, for any $\mathbf{z} \in B_{S_1}$ and $k \in \{1, 2, 3\}$, with obvious derivatives. Therefore we may define

$$\begin{aligned} \mathfrak{F}_j(\mathbf{y}) := \int_{B_{S_1}} \Bigg(& \sum_{k,s=1}^{3} \int_0^1 \partial x_s \mathcal{Z}_{jk}(\mathbf{y}, \mathbf{x})_{|\mathbf{x}=\vartheta\mathbf{z}} d\vartheta\, z_s F_k(\mathbf{z}) \\ & - \sum_{s=1}^{3} \int_0^1 \partial_s E_{4j}(\mathbf{y} - \vartheta\mathbf{z}) d\vartheta\, z_s \operatorname{div}\mathbf{u}(\mathbf{z}) \Bigg) d\mathbf{z} \end{aligned}$$

$$+ \int_{\partial \mathcal{D}} \Big(\sum_{k,s=1}^{3} \int_0^1 \partial x_s \mathcal{Z}_{jk}(\mathbf{y}, \mathbf{x})_{|\mathbf{x}=\vartheta \mathbf{z}} \mathrm{d}\vartheta \, z_s$$

$$\sum_{l=1}^{3} (-\partial_l u_k(\mathbf{z}) + \delta_{kl} \pi(\mathbf{z}) + \partial_l \mathcal{E}(u_k)(\mathbf{z})) n_l^{(\mathcal{D})}(\mathbf{z})$$

$$- \sum_{s=1}^{3} \int_0^1 \partial_s E_{4j}(\mathbf{y} - \vartheta \mathbf{z}) \mathrm{d}\vartheta \, z_s u_k(\mathbf{z}) n_k^{(\mathcal{D})}(\mathbf{z}) \Big) \mathrm{d}o_{\mathbf{z}}$$

$$+ \int_{\mathcal{D}} \sum_{k=1}^{3} \Big(\partial_k E_{4j}(\mathbf{y} - \mathbf{z}) \mathcal{E}_p(u_k)(\mathbf{z})$$

$$+ \sum_{s=1}^{3} \int_0^1 \partial x_s \mathcal{Z}_{jk}(\mathbf{y}, \mathbf{x})_{|\mathbf{x}=\vartheta \mathbf{z}} \mathrm{d}\vartheta \, z_s ((\tau \boldsymbol{e}_1 - \boldsymbol{\omega} \wedge \mathbf{z}) \cdot \nabla \mathcal{E}_p(u_k)(\mathbf{z})$$

$$+ [\boldsymbol{\omega} \wedge (\mathcal{E}_p(u_s)(\mathbf{z}))_{1 \le s \le 3}]_k - \Delta \mathcal{E}_p(u_k)(\mathbf{z})) \Big) \mathrm{d}\mathbf{z},$$

where the extension operator $\mathcal{E}_p$ was introduced in Lemma 5.17. We further set

$$\beta_k := \int_{B_{S_1}} F_k(\mathbf{z}) \mathrm{d}z + \int_{\partial \mathcal{D}} \sum_{l=1}^{3} (-\partial_l u_k(\mathbf{z}) + \delta_{kl} \pi(\mathbf{z}) + \partial_l \mathcal{E}_p(u_k)(\mathbf{z})) n_l^{(\mathcal{D})}(\mathbf{z}) \mathrm{d}o_{\mathbf{z}}$$

$$+ \int_{\mathcal{D}} ((\tau \boldsymbol{e}_1 - \boldsymbol{\omega} \wedge \mathbf{z}) \cdot \nabla \mathcal{E}_p(u_k)(\mathbf{z}) + [\boldsymbol{\omega} \wedge (\mathcal{E}_p(u_s)(\mathbf{z}))_{1 \le s \le 3}]_k$$

$$- \Delta \mathcal{E}_p(u_k)(\mathbf{z})) \mathrm{d}\mathbf{z}.$$

Then, referring to (5.97), (5.96), (5.107), and (5.108), we obtain (5.117). By (5.119), the choice of $\mathcal{E}_p$ in Lemma 5.17, and (5.37), we further find

$$|\mathfrak{F}_j(\mathbf{y})| \le C(S, S_1)(|\mathbf{y}| s_\tau(\mathbf{y}))^{-3/2} \quad (5.120)$$

$$\times \Big(\|F\|_1 + \|\nabla \mathbf{u} | \partial \mathcal{D}\|_1 + \|\pi | \partial \mathcal{D}\|_1$$

$$+ \sum_{k=1}^{3} (\|\nabla \mathcal{E}_p(u_k) | \partial \mathcal{D}\|_1 + \|\mathcal{E}_p(u_k)\|_{2,1}) \Big)$$

$$+ C(S, S_1)|\mathbf{y}|^{-3} \Big(\|\operatorname{div} \mathbf{u}\|_1 + \|\mathbf{u} | \partial \mathcal{D}\|_1 + \sum_{k=1}^{3} \|\mathcal{E}_p(u_k)\|_1 \Big).$$

Inequality (5.120), Lemma 2.16, and (5.113) imply (5.117). □

5.3.3 Asymptotic Profile of the Gradient of the Velocity Field

Theorem 5.16 *Let $p \in (1, \infty)$, $(\mathbf{u}, \pi) \in \mathfrak{M}_p$, $S, S_1 \in (0, \infty)$ with $S_1 < S$. Put $\mathbf{f} := \mathcal{L}(\mathbf{u}) + \nabla \pi$, and suppose that*

$$\overline{\mathcal{D}}^c \cup \text{supp}(\mathbf{f}) \cup \text{supp}(\text{div}\, \mathbf{u}) \subset B_{S_1},$$
$$\mathbf{u}|B_S^c \in \mathcal{L}^6(B_S^c), \quad \nabla \mathbf{u}|B_S^c \in L^2(B_S^c)^9, \quad \pi|B_S^c \in L^2(B_s^c).$$

Then there are coefficients $\beta_1, \beta_2, \beta_3 \in \mathbb{R}$ and functions $\mathfrak{F}_1, \mathfrak{F}_2, \mathfrak{F}_3 \in C^1(\overline{B_{S_1}}^c)$ such that for $j \in \{1, 2, 3\}$, $\boldsymbol{\alpha} \in \mathbb{N}_0^3$ with $|\boldsymbol{\alpha}| \le 1$, $\mathbf{y} \in \overline{B_{S_1}}^c$,

$$\partial^{\boldsymbol{\alpha}} u_j(\mathbf{y}) = \sum_{k=1}^{3} \beta_k \cdot \partial_{\mathbf{y}}^{\boldsymbol{\alpha}} \mathcal{Z}_{jk}(\mathbf{y}, 0) \tag{5.121}$$
$$+ \left(\int_{\partial \mathcal{D}} \mathbf{u} \cdot n^{(\mathcal{D})} do_{\mathbf{z}} + \int_{B_{S_1}} \text{div}\, \mathbf{u} dz \right) \cdot \partial^{\alpha} E_{4j}(\mathbf{y}) + \partial^{\boldsymbol{\alpha}} \mathfrak{F}_j(\mathbf{y}),$$

and if $\mathbf{y} \in B_S^c$,

$$|\partial^{\boldsymbol{\alpha}} \mathfrak{F}_j(\mathbf{y})| \le C(S, S_1) \cdot (\|\mathbf{f}\|_1 + \|\,\text{div}\, \mathbf{u}\|_1 + \|\nabla \mathbf{u}|\partial \mathcal{D}\|_1 \tag{5.122}$$
$$+ \|\pi|\partial \mathcal{D}\|_1 + \|\mathbf{u}|\partial \mathcal{D}\|_1)(|\mathbf{y}| \cdot s_\tau(\mathbf{y}))^{-3/2-|\boldsymbol{\alpha}|/2}.$$

Remark 5.2 The new feature presented by Theorem 5.16 is that the case $|\boldsymbol{\alpha}| = 1$ is admitted. □

Proof of Theorem 5.16 Take $j \in \{1, 2, 3\}$, $\mathbf{y} \in \overline{B_S}^c$. Observe that

$$|\mathbf{y} - \vartheta \cdot \mathbf{z}| \ge |\mathbf{y}| - S_1 \ge (1 - S_1/S) \cdot |\mathbf{y}| > 0 \quad \text{for } \mathbf{z} \in B_{S_1},\ \vartheta \in [0, 1]; \tag{5.123}$$

compare (5.84). Note that (5.123) holds in particular if $\mathbf{z} \in \mathcal{D}^c$. We put

$$\beta_k := \int_{B_{S_1}} f_k(\mathbf{z}) \text{d}\mathbf{z}$$
$$+ \int_{\partial \mathcal{D}} \sum_{l=1}^{3} (-\partial_l u_k(\mathbf{z}) + \delta_{kl} \cdot \pi(\mathbf{z}) + u_k(\mathbf{z}) \cdot (\tau \cdot \boldsymbol{e}_1 - \boldsymbol{\omega} \wedge z)_l) \cdot n_l^{(\mathcal{D})}(\mathbf{z}) \text{d}o_{\mathbf{z}}$$

for $1 \le k \le 3$,

$$\mathfrak{F}_j(y) := \int_{B_{S_1}} \left(\sum_{k=1}^{3} [(\mathcal{Z}_{jk}(\mathbf{y}, \mathbf{z}) - \mathcal{Z}_{jk}(\mathbf{y}, 0)) \cdot f_k(\mathbf{z})] \right. \tag{5.124}$$

$$+ (E_{4j}(\mathbf{y}-\mathbf{z}) - E_{4j}(\mathbf{y})) \cdot \operatorname{div} \mathbf{u}(\mathbf{z}) \Big) d\mathbf{z}$$

$$+ \int_{\partial\mathcal{D}} \sum_{k=1}^{3} ((\mathcal{Z}_{jk}(\mathbf{y}, \mathbf{z}) - \mathcal{Z}_{jk}(\mathbf{y}, 0))$$

$$\times \sum_{l=1}^{3} (-\partial_l u_k(\mathbf{z}) + \delta_{kl} \cdot \pi(\mathbf{z}) + u_k(\mathbf{z}) \cdot (\tau \cdot \boldsymbol{e}_1 - \boldsymbol{\omega} \wedge \mathbf{z})_l) \cdot n_l^{(\mathcal{D})}(\mathbf{z})$$

$$+ (E_{4j}(\mathbf{y}-\mathbf{z}) - E_{4j}(\mathbf{y})) \cdot u_k(\mathbf{z}) \cdot n_k^{(\mathcal{D})}(\mathbf{z})) do_{\mathbf{z}}$$

$$+ \int_{\partial\mathcal{D}} \sum_{k,l=1}^{3} \partial z_l \mathcal{Z}_{jk}(\mathbf{y}, \mathbf{z}) \cdot u_k(\mathbf{z}) \cdot n_l^{(\mathcal{D})}(\mathbf{z}) d\mathbf{z}$$

for $\mathbf{y} \in \overline{B_{S_1}}^c$, $1 \le j \le 3$. Then, by Lemmas 5.15 and 5.16, we may conclude that $\mathfrak{F}_j \in C^1(\overline{B_{S_1}}^c)$, and the derivative $\partial y_m \mathfrak{F}_j(\mathbf{y})$ equals the right-hand side of (5.124), but with the differential operator ∂y_m acting on $\mathcal{Z}_{jk}(\mathbf{y}, \mathbf{z}) - \mathcal{Z}_{jk}(\mathbf{y}, 0)$, $E_{4j}(\mathbf{y} - \mathbf{z}) - E_{4j}(\mathbf{y})$, and $\partial z_l \mathcal{Z}_{jk}(\mathbf{y}, \mathbf{z})$ ($\mathbf{z} \in B_{S_1}$ or $\mathbf{z} \in \partial\mathcal{D}$, $\mathbf{y} \in \overline{B_{S_1}}^c$, $1 \le k, l, m \le 3$). Obviously Eq. (5.121) holds. Now we recall that $\mathcal{Z}_{jk} \in C^2((\mathbb{R}^3 \times \mathbb{R}^3) \backslash \{(\mathbf{x}, \mathbf{x}) : \mathbf{x} \in \mathbb{R}^3\})$ (Lemma 5.4). Thus, in view of (5.123), we find for $\mathbf{y} \in B_S^c$, $z \in B_{S_1}$, $\boldsymbol{\alpha} \in \mathbb{N}_0^3$ with $|\boldsymbol{\alpha}| \le 1$, $1 \le k \le 3$ that

$$\begin{aligned} |\partial_{\mathbf{y}}^{\boldsymbol{\alpha}} \mathcal{Z}_{jk}(\mathbf{y}, \mathbf{z}) - \partial_{\mathbf{y}}^{\boldsymbol{\alpha}} \mathcal{Z}_{jk}(\mathbf{y}, 0)| &= \left| \sum_{s=1}^{3} \int_0^1 \partial_{\mathbf{y}}^{\boldsymbol{\alpha}} \partial x_s \mathcal{Z}_{jk}(\mathbf{y}, \mathbf{x})_{|\mathbf{x}=\vartheta \cdot \mathbf{z}} \cdot z_s d\vartheta \right| \\ &\le C(S_1, S) \cdot (|\mathbf{y}| \cdot s_\tau(\mathbf{y}))^{-3/2-|\boldsymbol{\alpha}|/2}, \end{aligned} \tag{5.125}$$

where the last inequality follows from Theorem 5.4; note that $\vartheta \cdot \mathbf{z} \in B_{S_1}$ for $\mathbf{z} \in B_{S_1}$, $\vartheta \in [0, 1]$. Similarly, since

$$|\partial^{\boldsymbol{\beta}} E_{4j}(\mathbf{x})| \le C \cdot |\mathbf{x}|^{-2-|\boldsymbol{\beta}|} \quad \text{for } \mathbf{x} \in \mathbb{R}^3 \backslash \{0\}, \ \boldsymbol{\beta} \in \mathbb{N}_0^3 \quad \text{with } |\boldsymbol{\beta}| \le 2$$

and because of (5.123), we have

$$\begin{aligned} |\partial_{\mathbf{y}}^{\boldsymbol{\alpha}} E_{4j}(\mathbf{y}-\mathbf{z}) - \partial_{\mathbf{y}}^{\boldsymbol{\alpha}} E_{4j}(\mathbf{y})| &= \left| \sum_{s=1}^{3} \int_0^1 \partial_x^{\boldsymbol{\alpha}+e_s} E_{4j}(\mathbf{x})_{|\mathbf{x}=\mathbf{y}-\vartheta \cdot \mathbf{z}} \cdot z_s d\vartheta \right| \\ &\le C \cdot \int_0^1 |\mathbf{y} - \vartheta \cdot \mathbf{z}|^{-3-|\boldsymbol{\alpha}|} d\vartheta \le C(S_1, S) \cdot |\mathbf{y}|^{-3-|\boldsymbol{\alpha}|} \\ &\le C(S_1, S) \cdot (|\mathbf{y}| \cdot s_\tau(\mathbf{y}))^{-3/2-|\boldsymbol{\alpha}|/2} \end{aligned} \tag{5.126}$$

for $\mathbf{z} \in B_{S_1}$, $\alpha \in \mathbb{N}_0^3$ with $|\alpha| \leq 1$, where the last inequality of (5.126) follows from Lemma 2.16. Further note that by Theorem 5.4,

$$|\partial_y^\alpha \partial z_l \mathcal{Z}_{jk}(\mathbf{y}, \mathbf{z})| \leq C(S_1, S) \cdot (|\mathbf{y}| \cdot s_\tau(\mathbf{y}))^{-3/2-|\boldsymbol{\alpha}|/2} \tag{5.127}$$

for $\mathbf{z} \in \partial\mathcal{D}$, $\boldsymbol{\alpha} \in \mathbb{N}_0^3$ with $|\boldsymbol{\alpha}| \leq 1$, $1 \leq k, l \leq 3$. Inequalities (5.125)–(5.127) together with (5.124) yield (5.122). □

5.3.4 Decay Estimates of the Second Derivatives of the Velocity

Due to the integral $\int_{\partial\mathcal{D}} \partial z_l \mathcal{Z}_{jk}(\mathbf{y}, \mathbf{z}) \cdot u_k(\mathbf{z}) \cdot n_l^{(\mathcal{D})}(\mathbf{z}) d\mathbf{z}$ appearing in the definition of $\mathfrak{B}_j(\mathbf{y})$, the second derivatives of $\mathfrak{B}_j(y)$ cannot be evaluated directly because we do not have estimate of the third-order derivatives of $\mathcal{Z}_{jk}$. But the differential operator ∂z_l acting on $\mathcal{Z}_{jk}$ in the above integral may be moved away by a partial integration. We recall the corresponding result as from previous part.

Lemma 5.19 (Lemma 5.17) *Let $p \in (1, \infty)$, $(u, \pi) \in \mathfrak{M}_p$, $j \in \{1, 2, 3\}$. Define*

$$\mathfrak{U}_j(\mathbf{y}) := \mathfrak{U}_j(\mathbf{u})(\mathbf{y}) := \int_{\partial\mathcal{D}} \sum_{k,l=1}^{3} \partial z_l \mathcal{Z}_{jk}(\mathbf{y}, \mathbf{z}) \cdot u_k(\mathbf{z}) \cdot n_l^{(\mathcal{D})}(\mathbf{z}) do_{\mathbf{z}}$$

for $\mathbf{y} \in \overline{\mathcal{D}}$. Let $\mathcal{E}_p\colon \mathbf{W}^{2-1/p,p}(\partial\mathfrak{D}) \mapsto \mathbf{W}^{2,p}(\mathcal{D}^c)$ denote a continuous extension operator (see [174]*). Then, for $\mathbf{y} \in \overline{\mathcal{D}}$,*

$$\begin{aligned}\mathfrak{U}_j(\mathbf{y}) = &\int_{\mathcal{D}}^{c} \sum_{k=1}^{3} [\partial_k E_{4j}(\mathbf{y} - \mathbf{z}) \cdot \mathcal{E}_p(u_k)(\mathbf{z}) + \mathcal{Z}_{jk}(\mathbf{y}, \mathbf{z}) \cdot ((\tau \cdot \boldsymbol{e}_1 - \boldsymbol{\omega} \wedge \mathbf{z}) \\ &\cdot \nabla \mathcal{E}_p(u_k)(\mathbf{z}) + [\boldsymbol{\omega} \wedge (\mathcal{E}_p(u_s)(\mathbf{z}))_{1 \leq s \leq 3}]_k - \Delta \mathcal{E}_p(u_k)(\mathbf{z}))] d\mathbf{z} \\ &+ \int_{\partial\mathcal{D}} \sum_{k,l=1}^{3} \mathcal{Z}_{jk}(\mathbf{y}, \mathbf{z}) \cdot ((-\tau \cdot \boldsymbol{e}_1 + \boldsymbol{\omega} \wedge \mathbf{z})_l \cdot u_k(\mathbf{z}) + \partial_l \mathcal{E}_p(u_k)(\mathbf{z})) \\ &\cdot n_l^{(\mathcal{D})}(\mathbf{z}) do_{\mathbf{z}}.\end{aligned}$$

Writing $\mathfrak{B}_j(\mathbf{y}) = (\mathfrak{B}_j(\mathbf{y}) - \mathfrak{U}_j(\mathbf{y})) + \mathfrak{U}_j(\mathbf{y})$, we see that we have transformed $\mathfrak{B}_j(\mathbf{y})$ in such a way that no derivative acts on $\mathcal{Z}_{jk}$ any more. But on the other hand, the term $\|\mathbf{u}|\partial\mathcal{D}\|_{2-1/p,p}$ appears in the upper bound of the terms involving $\mathcal{E}_p(\mathbf{u})$. In fact, we obtain the following:

Lemma 5.20 *Let $p \in (1, \infty)$, $(\mathbf{u}, \pi) \in \mathfrak{M}_p$, $S_1, S \in (0, \infty)$ with $\overline{\mathcal{D}}^c \subset B_{S_1}$, and $S_1 < S$. Then*

$$|\partial^\alpha \mathfrak{B}_j(\mathbf{y})| \leq C(S_1, S) \cdot (\|\nabla \mathbf{u}|\partial\mathcal{D}\|_1 + \|\pi|\partial\mathcal{D}\|_1 + C_p \cdot \|\mathbf{u}|\partial\mathcal{D}\|_{2-1/p,p}) \cdot (|\mathbf{y}| \cdot s_\tau(\mathbf{y}))^{-1-|\alpha|/2}$$

for $\mathbf{y} \in B_S^c$, $1 \leq j \leq 3$, $\alpha \in \mathbb{N}_0^3$ with $|\alpha| \leq 2$, where C_p is a constant with $\|\mathcal{E}_p(\mathbf{v})\|_{2,p} \leq C_p \cdot \|\mathbf{v}|\partial\mathcal{D}\|_{2-1/p,p}$ for $\mathbf{v} \in W^{2-1/p,p}(\partial\Omega)$. This constant may, of course, be chosen in such a way that it depends only on $\mathcal{D}$ and p. □

Proof Use Lemmas 5.15 and 5.16 (with r, T replaced by S_1, S, respectively), and note that $|\mathbf{y}|^{-1} \leq C(S) \cdot s_\tau(\mathbf{y})^{-1}$ for $\mathbf{y} \in B_S^c$ by Lemma 2.16. □

Exploiting the representation formula (5.90), we may now estimate the second derivatives of $\mathbf{u}$, under the assumption that $\mathbf{f}$ and $\operatorname{div} \mathbf{u}$ have compact support:

Theorem 5.17 *Let $p \in (1, \infty)$, $(\mathbf{u}, \pi) \in \mathfrak{M}_p$. Put $\mathbf{f} := \mathcal{L}(\mathbf{u}) + \nabla\pi$. Suppose there are numbers $S_1, S \in (0, \infty)$, with $S_1 < S$, $\overline{\mathcal{D}}^c \cup \operatorname{supp}(\mathbf{f}) \cup \operatorname{supp}(\operatorname{div} \mathbf{u}) \subset B_{S_1}$,*

$$\mathbf{u}|B_S^c \in \mathbf{L}^6(B_S^c), \quad \nabla\mathbf{u}|B_S^c \in L^2(B_S^c)^9, \quad \pi|B_S^c \in L^2(B_S^c).$$

Let $j \in \{1, 2, 3\}$, $\mathbf{y} \in B_S^c$, $\alpha \in \mathbb{N}_0^3$ with $|\alpha| \leq 2$. Then

$$|\partial^\alpha u_j(\mathbf{y})| \leq C(S, S_1) \cdot (\|\mathbf{f}|B_{S_1}\|_1 + \|\operatorname{div} \mathbf{u}\|_1 + \|\nabla\mathbf{u}|\partial\mathcal{D}\|_1 + \|\pi|\partial\mathcal{D}\|_1 + C_p \cdot \|\mathbf{u}|\partial\mathcal{D}\|_{2-1/p,p}) \cdot (|\mathbf{y}| \cdot s_\tau(\mathbf{y}))^{-1-|\alpha|/2},$$

where C_p was introduced in Lemma 5.20. □

Proof Starting from (5.90), we estimate the boundary potential $\partial^\alpha \mathfrak{B}_j(\mathbf{y})$ by applying Lemma 5.20 and the volume potentials $\partial^\alpha \mathfrak{R}_j(\mathbf{f})(\mathbf{y})$ and $\partial^\alpha \mathfrak{S}_j(\operatorname{div} \mathbf{u})(\mathbf{y})$ by referring to Lemma 5.16 (with R, T replaced by S_1, S, respectively). In the estimate of $\partial^\alpha \mathfrak{S}_j(\operatorname{div} \mathbf{u})(\mathbf{y})$, we further take account of the inequality $|\mathbf{y}|^{-1} \leq C(S) \cdot s_\tau(\mathbf{y})^{-1}$ (Lemma 2.16). □

5.4 Leray Solution

5.4.1 Introduction

We are interested in a linearized version of (5.2) that reads as follows:

$$\mathcal{L}(\mathbf{u}) + \nabla\pi = f, \quad \operatorname{div} \mathbf{u} = 0 \quad \text{in } \mathcal{D}, \tag{5.128}$$

It is well known ([98]) that for data of arbitrary size, both problems (5.2) and (5.128) admit a "Leray solution" characterized by the relations

$$\mathbf{u} \in \boldsymbol{L}^6(\mathcal{D}), \quad \nabla \mathbf{u} \in L^2(\mathcal{D})^9, \quad \pi \in L^2_{\text{loc}}\mathcal{D}). \tag{5.129}$$

Galdi, Kyed in [100] showed that if the right-hand side $\mathbf{f}$ in the nonlinear problem (5.2) has bounded support, then the velocity part $\mathbf{u}$ of such a solution $(\mathbf{u}, \pi)$ decays as follows:

$$|\partial^\alpha \mathbf{u}(\mathbf{x})| = O[(|\mathbf{x}| \cdot s_\tau(\mathbf{x}))^{-1-|\alpha|/2}] \quad (|\mathbf{x}| \to \infty), \tag{5.130}$$

where $\boldsymbol{\alpha} \in \mathbb{N}_0^3$ with $|\boldsymbol{\alpha}| := \alpha_1 + \alpha_2 + \alpha_3 \leq 1$ (decay of $\mathbf{u}$ and $\nabla \mathbf{u}$). The term $s_\tau(\mathbf{x})$ in (5.130) is defined by

$$s_\tau(\mathbf{x}) := 1 + \tau(|\mathbf{x}| - x_1) \quad (\mathbf{x} \in \mathbb{R}^3). \tag{5.131}$$

In the previous part of this chapter (see also [35]–[37]), we considered a different kind of solution, which we called "weak solution," and which satisfies the relations

$$\mathbf{u}_{|B_S^c} \in \boldsymbol{L}^6(B_S^c), \; \nabla \mathbf{u}_{|B_S^c} \in L^2(B_S^c)^9, \; \pi|_{B_S^c} \in L^2(B_S^c) \tag{5.132}$$

for some $S > 0$ with $\mathcal{D}^c \subset B_S$, as well as some additional regularity assumptions, which require in particular that $\pi \in W^{1,p}_{\text{loc}}(\mathcal{D})$ and $\pi_{|B_T \backslash \overline{\mathcal{D}^c}} \in L^p(B_T \backslash \overline{\mathcal{D}^c})$ for some $p \in (1, \infty)$ and for any $T \in (0, \infty)$ with $\overline{\mathcal{D}^c} \subset B_T$. Although these additional regularity assumptions and (5.132) do not imply (5.129) if $p < 6/5$, it is clear that weak solutions should be considered as less general than Leray ones, in particular in view of the requirement on π in (5.129) and (5.132), respectively.

In this part, we revisit the theory in [35]–[37], showing that it remains valid when the condition $\pi|_{B_S^c} \in L^2(B_S^c)$ in (5.132) is dropped.

5.4.2 Auxiliary Results

The key auxiliary result of this section is as follows:

Theorem 5.18 *Let* $q \in (1, 2)$, $p \in (1, \infty)$, $S > 0$, $\mathbf{f} \in \boldsymbol{L}^q(\overline{B_S}^c) + \boldsymbol{L}^{3/2}(\overline{B_S}^c)$, $\mathbf{u} \in \boldsymbol{L}^6(\overline{B_S}^c) \cap \boldsymbol{W}^{2,p}_{loc}(\overline{B_S}^c)$, $\pi \in W^{1,p}_{loc}(\overline{B_S}^c)$ *with* $\mathcal{L}(\mathbf{u}) + \nabla \pi = \mathbf{f}$, $\operatorname{div} \mathbf{u} = 0$.

Then there is $c \in \mathbb{R}$ *with* $\pi + c_{|B_{2S}^c} \in L^{3q/(3-q)}(B_{2S}^c) + L^3(B_{2S}^c)$. □

Proof We use the approach from the proof of [101, Theorem 44]. By the cutoff procedure from that proof and since $W^{1,q}_{loc}(\mathbb{R}^3) \subset L^{3/2}_{loc}(\mathbb{R}^3)$, we see there are functions $\mathbf{F} \in \boldsymbol{L}^q(\mathbb{R}^3) + \boldsymbol{L}^{3/2}(\mathbb{R}^3)$, $\mathbf{U} \in \boldsymbol{L}^6(\mathbb{R}^3) \cap \boldsymbol{W}^{2,p}_{loc}(\mathbb{R}^3)$, $\Pi \in W^{1,p}_{loc}(\mathbb{R}^3)$ such that

$$\mathcal{L}(\mathbf{U}) + \nabla\Pi = \mathbf{F}, \quad \operatorname{div}\mathbf{U} = 0,$$

$$\mathbf{U}(\mathbf{x}) = \mathbf{u}(\mathbf{x}) + \beta|\mathbf{x}|^{-3}\mathbf{x} \text{ for } \mathbf{x} \in B^c_{2S}, \quad \Pi_{|B^c_{2S}} = \pi_{|B^c_{2S}}$$

with some constant $\beta \in \mathbb{R}$. Note that the argument from that proof strongly simplifies in the present situation because we consider the linear problem (5.128) instead of the nonlinear one (5.2). By the assumptions on $\mathbf{F}$, there are $\mathbf{F}^{(1)} \in \boldsymbol{L}^q(\mathbb{R}^3)$, $\boldsymbol{F}^{(2)} \in \boldsymbol{L}^{3/2}(\mathbb{R}^3)$ such that $\mathbf{F} = \mathbf{F}^{(1)} + \mathbf{F}^{(2)}$. But according to [57] and Section 4.1, there are pairs of functions $(\mathbf{U}^{(1)}, \Pi^{(1)}) \in X_q$, $(\mathbf{U}^{(2)}, \Pi^{(2)}) \in X_{3/2}$ such that

$$\mathcal{L}(\mathbf{U}^{(\kappa)}) + \nabla\Pi^{(\kappa)} = \mathbf{F}^{(\kappa)}, \quad \operatorname{div}\mathbf{U}^{(\kappa)} = 0 \quad \text{for } \kappa \in \{1, 2\}.$$

This means in particular that $\mathbf{U}^{(1)} \in \boldsymbol{W}^{2,q}_{loc}(\mathbb{R}^3) \cap \boldsymbol{L}^{2q/(2-q)}(\mathbb{R}^3)$, $\Pi^{(1)} \in W^{1,q}_{loc}(\mathbb{R}^3)$, $\mathbf{U}^{(2)} \in \boldsymbol{L}^6(\mathbb{R}^3)$ (hence $\mathbf{U} - \mathbf{U}^{(2)} \in \boldsymbol{L}^6(\mathbb{R}^3)$), and $\mathbf{U}^{(2)} \in \boldsymbol{W}^{2,3/2}_{loc}(\mathbb{R}^3)$, $\Pi^{(2)} \in W^{1,3/2}_{loc}(\mathbb{R}^3)$. We may thus apply [101, Lemma 4.1] with $s = \min\{p, q, 3/2\}$, $q_1 = 2q/(2-q)$, $q_2 = 6$, $\mathbf{f} = \mathbf{F}^{(1)}$, $(\mathbf{v}_1, p_1) = (\mathbf{U}^{(1)}, \Pi^{(1)})$, $(\mathbf{v}_2, p_2) = (\mathbf{U} - \mathbf{U}^{(2)}, \Pi - \Pi^{(2)})$. It follows that $\mathbf{U}^{(1)} = \mathbf{U} - \mathbf{U}^{(2)}$, $\Pi^{(1)} = \Pi - \Pi^{(2)} + c$ for some $c \in \mathbb{R}$. Hence $\Pi + c \in L^{3q/(3-q)}(\mathbb{R}^3) + L^3(\mathbb{R}^3)$, so that $\pi + c_{|B^c_{2S}} \in L^{3q/(3-q)}(B^c_{2S}) + L^3(B^c_{2S})$. □

Theorem 5.19 *Let $S \in (0, \infty)$. Then $|\mathcal{Z}(\mathbf{y}, \mathbf{z})| \le C(S)\ (|\mathbf{z}|s_\tau(z))^{-1}$ for $\mathbf{y} \in \overline{B_S}$, $\mathbf{z} \in B^c_{2S}$.* □

Proof This theorem is a special case of Theorem 5.3. □

5.4.3 Representation Formula for the Leray Solution

Theorem 5.20 $\sum_{k=1}^3 \partial z_k \mathcal{Z}_{jk}(\mathbf{y}, \mathbf{z}) = 0$ *for* $1 \le j \le 3$, $\mathbf{y}, \mathbf{z} \in \mathbb{R}^3$ *with* $\mathbf{y} \ne \mathbf{z}$. □

Proof It follows from Lemma 5.3. □

Theorem 5.21 *Let $p \in (1, \infty)$, $(\mathbf{u}, \pi) \in \mathfrak{M}_p$. Put $\mathbf{F} := \mathcal{L}(\mathbf{u}) + \nabla\pi$, and suppose there are numbers $q \in (1, 2)$, $S \in (0, \infty)$ such that*

$$\mathcal{D}^c \cup \operatorname{supp}(\operatorname{div} u) \subset B_S,$$

$$\mathbf{u}_{|B^c_S} \in \mathcal{L}^6(B^c_S), \quad \nabla\mathbf{u}_{|B^c_S} \in L^2(B^c_S)^9, \quad \mathbf{F}_{|B^c_S} \in \boldsymbol{L}^q(B^c_S) + \boldsymbol{L}^{3/2}(B^c_S).$$

Let $j \in \{1, 2, 3\}$, and put

$$\mathfrak{B}_j(\mathbf{y}) := \mathfrak{B}_j(\mathbf{u}, \pi)(\mathbf{y}) \tag{5.133}$$
$$:= \int_{\partial\mathcal{D}} \sum_{k=1}^{3} \Big[\sum_{l=1}^{3} (\mathcal{Z}_{jk}(\mathbf{y}, \mathbf{z})(-\partial_l u_k(\mathbf{z}) + \delta_{kl}\pi(\mathbf{z}) + u_k(\mathbf{z})(\tau \boldsymbol{e}_1 - \boldsymbol{\omega} \wedge \mathbf{z})_l)$$
$$+ \partial z_l \mathcal{Z}_{jk}(\mathbf{y}, \mathbf{z}) u_k(\mathbf{z})) n_l^{(\mathcal{D})}(\mathbf{z}) + E_j(\mathbf{y} - \mathbf{z}) u_k(\mathbf{z}) n_k^{(\mathcal{D})}(\mathbf{z}) \Big] do_{\mathbf{z}}$$

for $\mathbf{y} \in \mathcal{D}$. Then

$$u_j(\mathbf{y}) = \mathfrak{R}_j(\mathbf{F})(\mathbf{y}) + \mathfrak{S}_j(\operatorname{div} \mathbf{u})(\mathbf{y}) + \mathfrak{B}_j(\mathbf{y}) \tag{5.134}$$

for a.e. $\mathbf{y} \in \mathcal{D}$. If $p > 3/2$, (5.134) holds for any $\mathbf{y} \in \mathcal{D}$, without the restriction "a.e." □

Proof By Theorem 5.18, there is $c \in \mathbb{R}$, $\pi_1 \in L^{3q/(3-q)}(B_{2S}^c)$, $\pi_2 \in L^3(B_{2S}^c)$ such that $\pi_{|B_{2S}^c} = \pi_1 + \pi_2 + c$. From this fact and our assumptions on u, it follows that

$$\int_{2S}^{\infty} \int_{\partial B_r} (|\mathbf{u}(\mathbf{z})|^6 + |\nabla \mathbf{u}(\mathbf{z})|^2 + |\pi_1(\mathbf{z})|^{3q/(3-q)} + |\pi_2(\mathbf{z})|^3) \mathrm{d}o_{\mathbf{z}} \mathrm{d}r < \infty.$$

Thus there is an increasing sequence (R_n) in $(2S, \infty)$ with $R_n \to \infty$ and

$$\int_{\partial B_{R_n}} (|\mathbf{u}(\mathbf{z})|^6 + |\nabla \mathbf{u}(\mathbf{z})|^2 + |\pi_1(\mathbf{z})|^{3q/(3-q)} + |\pi_2(\mathbf{z})|^3) \mathrm{d}o_{\mathbf{z}} \le R_n^{-1} \quad \text{for } n \in \mathbb{N}. \tag{5.135}$$

By our assumptions on $\mathbf{F}$, there are functions $\mathbf{G}^{(1)} \in \boldsymbol{L}^q(B_S^c)$, $\boldsymbol{G}^{(2)} \in \boldsymbol{L}^{3/2}(B_S^c)$ such that $\mathbf{F}|_{B_S^c} = \mathbf{G}^{(1)} + \mathbf{G}^{(2)}$. Thus by Lemmas 5.10 and 5.11

$$\int_{\overline{\mathfrak{D}}^c} \sum_{k=1}^{3} |\mathcal{Z}_{jk}(\mathbf{y}, \mathbf{z})| (\chi_{(0,S)}(|\mathbf{z}|) |F_k(\mathbf{z})| + \chi_{(S,\infty)}(|\mathbf{z}|)(|\mathbf{G}_k^{(1)}(\mathbf{z})| + |\mathbf{G}_k^{(2)}(\mathbf{z})|)) \mathrm{d}\mathbf{z} < \infty \tag{5.136}$$

for a.e. $\mathbf{y} \in \mathcal{D}$. Moreover, by Lemma 5.12 with $p = q = 2$,

$$\int_{\mathcal{D}} |E_j(\mathbf{y} - \mathbf{z})| \, |\operatorname{div} \mathbf{u}(\mathbf{z})| \mathrm{d}\mathbf{z} < \infty \tag{5.137}$$

for a.e. $\mathbf{y} \in \overline{\mathcal{D}}$. Due to these observations and Theorem 5.12, we see there is a subset N of $\mathcal{D}$ with measure zero such that the relations in (5.136), (5.137) hold for $\mathbf{y} \in \mathcal{D} \setminus N$ and such that (5.90) with R replaced by R_n holds for $n \in \mathbb{N}$ and $y \in \mathcal{D}_{R_n} \setminus N$.

In the case $p > 3/2$, Lemmas 5.10 and 5.11 yield that (5.90) is valid for any $\mathbf{y} \in \mathcal{D}$, and Theorem 5.12 states that (5.90) with R replaced by R_n is true for $n \in \mathbb{N}$ and for any $\mathbf{y} \in \mathcal{D}$. Moreover, since $(\mathbf{u}, \pi) \in \mathfrak{M}_p$ and in view of Theorem 5.11, we obtain $\operatorname{div} \mathbf{u}_{|\mathcal{D}_T} \in W^{1,p}(\mathcal{D}_T)$ for $T \in (0, \infty)$ with $\mathcal{D}^c \subset B_T$. But $\operatorname{supp}(\operatorname{div} \mathbf{u}) \subset B_S$, so $\operatorname{div} \mathbf{u} \in W^{1,p}(\mathcal{D})$. Thus, if $p > 3/2$, the Sobolev inequality implies there is $s \in (3, \infty]$ with $\operatorname{div} \mathbf{u} \in L^s(\mathcal{D})$, so Lemma 5.11 yields that the restriction "a.e." may be dropped in (5.137) as well.

Take $\mathbf{y} \in \mathcal{D}$ in the case $p > 3/2$ and $\mathbf{y} \in \mathcal{D} \setminus N$ otherwise. Let $n \in \mathbb{N}$ with $R_n > |\mathbf{y}|$ (hence $\mathbf{y} \in \mathcal{D}_{R_n}$). Then, by (5.90) with R replaced by R_n and π by $\pi - c$, we get

$$u_j(\mathbf{y}) = \mathfrak{R}_j(F_{|\mathcal{D}_{R_n}})(\mathbf{y}) + \mathfrak{S}_j(\operatorname{div} u_{|\mathcal{D}_{R_n}})(\mathbf{y}) + U_{in,j}(\mathbf{y}) + \mathfrak{B}_j(\mathbf{y}) \tag{5.138}$$

with

$$U_{in,j}(\mathbf{y}) := \int_{\partial B_{R_n}} \sum_{k=1}^{3} \Big[\sum_{l=1}^{3} (\mathcal{Z}_{jk}(\mathbf{y}, \mathbf{z})(\partial_l u_k(\mathbf{z}) - \delta_{kl}(\pi - c)(\mathbf{z}) - \tau\delta_{1l} u_k(\mathbf{z}))$$
$$- \partial z_l \mathcal{Z}_{jk}(\mathbf{y}, \mathbf{z})(u_k(\mathbf{z})))\frac{z_l}{R_n} - E_j(\mathbf{y} - \mathbf{z}) u_k(\mathbf{z}) \frac{z_k}{R_n} \Big] do_{\mathbf{z}},$$

where we used the relation $\sum_{l=1}^{3} (\boldsymbol{\omega} \wedge \mathbf{z})_l z_l = 0$ for $\mathbf{z} \in \partial B_{R_n}$. Concerning $\mathfrak{B}_j(\mathbf{y})$, we note that

$$\int_{\partial \mathcal{D}} \sum_{l=1}^{3} \mathcal{Z}_{jk}(\mathbf{y}, \mathbf{z}) n_k^{(\mathcal{D})}(\mathbf{z}) do_{\mathbf{z}} = 0 \quad \text{for } \mathbf{y} \in \mathcal{D}_R,\ 1 \le j \le 3;$$

see Theorems 5.3 and 5.20, Lemma 2.17, and Corollary 2.3. Thus the definition of $\mathfrak{B}_j(\mathbf{y})$ need not be modified even though we replace π by $\pi - c$.

Let $n \in \mathbb{N}$ with $R_n/4 \ge |\mathbf{y}|$. Observe that

$$|U_{in,j}(\mathbf{y})| \le \mathfrak{C} \sum_{v=1}^{5} \sum_{k=1}^{3} \mathcal{B}_{v,k}(\mathbf{y})$$

with

$$\mathcal{B}_{1,k}(\mathbf{y}) := \left(\int_{\partial B_{R_n}} |\mathcal{Z}_{jk}(\mathbf{y}, \mathbf{z})|^{6/5} do_{\mathbf{z}} \right)^{5/6} \|u_{|\partial B_{R_n}}\|_6,$$

$$\mathcal{B}_{2,k}(\mathbf{y}) = \left(\int_{\partial B_{R_n}} |\mathcal{Z}_{jk}(\mathbf{y}, \mathbf{z})|^2 do_{\mathbf{z}} \right)^{1/2} \|\nabla \mathbf{u}_{|\partial B_{R_n}}\|_2,$$

$$\mathcal{B}_{3,k}(\mathbf{y}) = \left(\int_{\partial B_{R_n}} |\mathcal{Z}_{jk}(\mathbf{y},\mathbf{z})|^{3q/(4q-3)} \mathrm{d}o_{\mathbf{z}} \right)^{(4q-3)/3q} \|\pi_{1|\partial B_{R_n}}\|_{3q/(3-q)},$$

$$\mathcal{B}_{4,k}(\mathbf{y}) = \left(\int_{\partial B_{R_n}} |\mathcal{Z}_{jk}(\mathbf{y},\mathbf{z})|^{3/2} \mathrm{d}o_{\mathbf{z}} \right)^{2/3} \|\pi_{2|\partial B_{R_n}}\|_{3},$$

$$\mathcal{B}_{5,k}(\mathbf{y}) = \sum_{l=1}^{3} \left(\int_{\partial B_{\mathbb{R}_n}} |\partial z_l \mathcal{Z}_{jk}(\mathbf{y},\mathbf{z})|^{6/5} \mathrm{d}o_{\mathbf{z}} \right)^{5/6} \|u_{|\partial B_{R_n}}\|_{6},$$

$$\mathcal{B}_{6,k}(\mathbf{y}) = \left(\int_{\partial B_{R_n}} |\mathbf{y}-\mathbf{z}|^{-12/5} \mathrm{d}o_{\mathbf{z}} \right)^{5/6} \|u_{|\partial B_{R_n}}\|_{6}$$

for $k \in \{1, 2, 3\}$. As in the proof of Theorem 5.13, we get

$$|\mathcal{B}_{1,k}(k)| \le C(|\mathbf{y}|) R_n^{-1/3}, \quad |\mathcal{B}_{2,k}(\mathbf{y})| \le C(|\mathbf{y}|) R_n^{-1}, \quad |\mathcal{B}_{5,k}(\mathbf{y})| \le C(|\mathbf{y}|) R_n^{-5/6},$$
$$|\mathcal{B}_{6,k}(\mathbf{y})| \le C(|\mathbf{y}|) R_n^{-1/2}.$$

Now applying Theorem 5.19, Lemma 2.13, and (5.135), we obtain

$$\begin{aligned}
\mathcal{B}_{3,k}(\mathbf{y}) \le & C(|\mathbf{y}|) \left(\int_{\partial B_{R_n}} (|\mathbf{z}| s_\tau(\mathbf{z}))^{-3q/(4q-3)} \mathrm{d}o_{\mathbf{z}} \right)^{(4q-3)/3q} \|\pi_1|\partial B_{R_n}\|_{3q/(3-q)} \\
\le & C(|\mathbf{y}|) R_n^{-1} \left(\int_{\partial B_{R_n}} |s_\tau(\mathbf{z})|^{-3q/(4q-3)} \mathrm{d}o_{\mathbf{z}} \right)^{(4q-3)/3q} R_n^{-(3-q)/3q} \\
\le & C(|\mathbf{y}|) R_n^{-1} R_n^{(4q-3)/3q} R_n^{-(3-q)/3q} \le C(|\mathbf{y}|) R^{(2q-6)/3q}, \\
\mathcal{B}_{4,k}(\mathbf{y}) \le & C(|\mathbf{y}|) \left(\int_{\partial B_{R_n}} (|\mathbf{z}| s_\tau(z))^{-3/2} \mathrm{d}o_{\mathbf{z}} \right)^{2/3} \|\pi_2|\partial B_{R_n}\|_{3} \\
\le & C(|\mathbf{y}|) R_n^{-1} \left(\int_{\partial B_{R_n}} |s_\tau(\mathbf{z})|^{-3/2} \mathrm{d}o_{\mathbf{z}} \right)^{2/3} R_n^{-1/3} \le C(|\mathbf{y}|) R^{-2/3}.
\end{aligned}$$

Thus we conclude that $U_{in,j}(\mathbf{y}) \to 0$ for $n \to \infty$. Turning to $\mathfrak{R}_j(F_{|\mathcal{D}_{R_n}})(\mathbf{y})$ and $\mathfrak{S}_j(\operatorname{div} u_{|\mathcal{D}_{R_n}})(\mathbf{y})$ and applying (5.136), (5.137), and Lebesgue's theorem, it follows that

$$\mathfrak{R}_j(F_{|\mathcal{D}_{R_n}})(\mathbf{y}) \to \mathfrak{R}_j(\mathbf{F})|(\mathbf{y}), \quad \mathfrak{S}_j(\operatorname{div} \mathbf{u}_{|\mathcal{D}_{R_n}})(\mathbf{y}) \to \mathfrak{S}_j(\operatorname{div} \mathbf{u})(\mathbf{y})$$

for $n \to \infty$. Theorem 5.21 now follows with (5.138). □

5.4.4 Asymptotic Profile

We begin by considering the decay of the velocity and its gradient in the linear case.

Theorem 5.22 *Let $p \in (1, \infty)$, $(\mathbf{u}, \pi) \in \mathfrak{M}_p$. Put $\mathbf{F} := \mathcal{L}(\mathbf{u}) + \nabla\pi$. Suppose there are numbers $S_1, S, \gamma \in (0, \infty)$, $A \in [2, \infty)$, $B \in \mathbb{R}$ such that $S_1 < S$,*

$$\mathcal{D}^c \cup \operatorname{supp}(\operatorname{div} u) \subset B_{S_1}, \quad \mathbf{u}_{|B_S^c} \in \boldsymbol{L}^6(B_S^c), \quad \nabla\mathbf{u}_{|B_S^c} \in L^2(B_S^c)^9,$$

$$A + \min\{1, B\} \geq 3, \quad |\mathbf{F}(\mathbf{z})| \leq \gamma|\mathbf{z}|^{-A} s_\tau(\mathbf{z})^{-B} \quad \text{for } \mathbf{z} \in B_{S_1}^c.$$

Let $i, j \in \{1, 2, 3\}$, $\mathbf{y} \in B_S^c$. Then

$$\begin{aligned}|\mathbf{u}_j(\mathbf{y})| \leq & C(S, S_1, A, B)(\gamma + \|F_{|\mathcal{D}_{S_1}}\|_1 + \|\operatorname{div}\mathbf{u}\|_1 \\ & + \|\mathbf{u}_{|\partial\mathcal{D}}\|_1 + \|\nabla\mathbf{u}_{|\partial\mathcal{D}}\|_1 + \|\pi_{|\partial\mathcal{D}}\|_1)(|\mathbf{y}| s_\tau(\mathbf{y}))^{-1} l_{A,B}(\mathbf{y}),\end{aligned} \tag{5.139}$$

$$\begin{aligned}|\partial_i u_j(\mathbf{y})| \leq & C(S, S_1, A, B)(\gamma + \|\mathbf{F}_{|\mathcal{D}_{S_1}}\|_1 \\ & + \|\mathbf{u}_{|\partial\mathcal{D}}\|_1 + \|\operatorname{div}\mathbf{u}\|_1 + \|\nabla\mathbf{u}_{|\partial\mathcal{D}}\|_1 + \|\pi_{|\partial\mathcal{D}}\|_1) \\ & (|\mathbf{y}| s_\tau(\mathbf{y}))^{-3/2} s_\tau(\mathbf{y})^{\max(0, 7/2 - A - B)} l_{A,B}(\mathbf{y}),\end{aligned} \tag{5.140}$$

where $l_{A,B}(\mathbf{y}) := 1$ if $A + \min\{1, B\} > 3$ and $l_{A,B}(\mathbf{y}) := \max(1, \ln|\mathbf{y}|)$ if $A + \min\{1, B\} = 3$. □

Proof Theorem 5.22 may be deduced from Theorems 5.21 and 5.14. □

Next we turn to the decay of derivatives of the velocity up to order 2.

Theorem 5.23 *Let $p \in (1, \infty)$, $(\mathbf{u}, \pi) \in \mathfrak{M}_p$. Put $\mathbf{F} := \mathcal{L}(\mathbf{u}) + \nabla\pi$. Suppose there are numbers $S_1, S \in (0, \infty)$ with $S_1 < S$,*

$$\mathcal{D}^c \cup \operatorname{supp}(F) \cup \operatorname{supp}(\operatorname{div} u) \subset B_{S_1}, \quad \mathbf{u}_{|B_S^c} \in \boldsymbol{L}^6(B_S^c), \quad \nabla\mathbf{u}_{|B_S^c} \in L^2(B_S^c)^9.$$

Let $\mathcal{E}_p\colon \boldsymbol{W}^{2-1/p,p}(\partial\mathcal{D}) \mapsto \boldsymbol{W}^{2,p}(\mathcal{D})$ be an extension operator with $\|\mathcal{E}_p(\mathbf{v})\|_{2,p} \leq C_p\|\mathbf{v}\|_{2-1/2,p}$ for $\mathbf{v} \in \boldsymbol{W}^{2-1/p,p}(\partial\mathcal{D})$ for some $C_p > 0$.

Let $j \in \{1, 2, 3\}$, $\mathbf{y} \in B_S^c$, $\boldsymbol{\alpha} \in \mathbb{N}_0^3$ with $|\boldsymbol{\alpha}| \leq 2$. Then

$$\begin{aligned}|\partial^\alpha u_j(\mathbf{y})| \leq & C(S, S_1) \cdot (\|\mathbf{F}_{|\mathcal{D}_{S_1}}\|_1 + \|\operatorname{div}\mathbf{u}\|_1 + \|\nabla\mathbf{u}_{|\partial\mathcal{D}}\|_1 \\ & + \|\pi_{|\partial\mathcal{D}}\|_1 + C_p \cdot \|\mathbf{u}_{|\partial\mathcal{D}}\|_{2-1/p,p}) \cdot (|\mathbf{y}| \cdot s_\tau(\mathbf{y}))^{-1-|\alpha|/2}.\end{aligned} \tag{5.141}$$

□

Proof Theorem 5.23 follows from Theorems 5.21 and 5.14. □

We shall study an asymptotic profile of the velocity and its gradient of the linear case:

Theorem 5.24 *Let $p \in (1,\infty)$, $(\mathbf{u},\pi) \in \mathfrak{M}_p$, $S, S_1 \in (0,\infty)$ with $S_1 < S$. Put $F := \mathcal{L}(\mathbf{u}) + \nabla\pi$. Suppose that*

$$\mathcal{D}^c \cup \operatorname{supp}(F) \cup \operatorname{supp}(\operatorname{div}\mathbf{u}) \subset B_{S_1}, \quad \mathbf{u}|_{B_S^c} \in \boldsymbol{L}^6(B_S^c), \quad \nabla\mathbf{u}_{|B_S^c} \in L^2(B_S^c)^9.$$

Then there are coefficients $\beta_1, \beta_2, \beta_3 \in \mathbb{R}$ and functions $\mathfrak{F}_1, \mathfrak{F}_2, \mathfrak{F}_3 \in C^1(\overline{B_{S_1}}^c)$ such that for $j \in \{1,2,3\}$, $\boldsymbol{\alpha} \in \mathbb{N}_0^3$ with $|\boldsymbol{\alpha}| \le 1$,

$$\partial^{\boldsymbol{\alpha}} u_j(\mathbf{y}) = \sum_{k=1}^{3} \beta_k \partial^{\alpha_y} \mathcal{Z}_{jk}(\mathbf{y},0) + \left(\int_{\partial\mathcal{D}} u \cdot n^{(\mathcal{D})} do_{\mathbf{z}} + \int_{\mathcal{D}_{S_1}} \operatorname{div}\mathbf{u} d\mathbf{z}\right) \partial^{\alpha} E_j(\mathbf{y})$$

$$+ \partial^{\alpha}\mathfrak{F}_j(\mathbf{y}) \textit{ for } \mathbf{y} \in \overline{B_{S_1}}^c,$$

and

$$|\partial^{\boldsymbol{\alpha}}\mathfrak{F}_j(\mathbf{y})| \le C(S,S_1)(\|F\|_1 + \|\operatorname{div}\mathbf{u}\|_1 + \|\nabla\mathbf{u}_{|\partial\mathcal{D}}\|_1$$
$$+ \|\pi_{|\partial\mathcal{D}}\|_1 + \|\mathbf{u}_{|\partial\mathcal{D}}\|_1)(|\mathbf{y}|s_\tau(\mathbf{y}))^{-3/2-|\boldsymbol{\alpha}|/2} \quad \textit{for } \mathbf{y} \in B_S^c.$$

□

Proof Theorem 5.24 may be deduced from Theorem 5.21 in the same way as [38, Theorem 1.1] from [36, Theorem 4.6]. □

Remark 5.3 An explicit formula for β_i, $\mathfrak{F}_i$, $i = 1,\ldots,3$,, is given in proof of Theorem 5.15. □

We may use Theorems 5.22 and 5.23 in order to derive a decay estimate as in (5.2) for Leray solutions of the linear problem (5.128). This may be done by considering the restriction of such a solution to $B_{S_0}^c$ for some $S_0 > 0$ sufficiently large. The idea is to apply Theorems 5.22 and 5.23 with $\mathcal{D}^c$ replaced by B_{S_0}. In this way, the behavior of the solution in question near the boundary of its domain and the regularity of that boundary do not matter. A crucial technical result in this respect is the ensuing observation on the interior regularity of generalized solutions to (5.128).

Theorem 5.25 *Let $U \subset \mathbb{R}^3$ be open and bounded, $p \in (1,\infty)$, $\mathbf{f} \in \boldsymbol{L}^p_{loc}(\overline{U}^c)$, $\mathbf{u} \in \boldsymbol{W}^{1,2}_{loc}(\overline{U}^c)$, $\pi \in L^p_{loc}(\overline{U}^c)$,*

$$\int_{\overline{U}^c} (\nabla\mathbf{u}\cdot\nabla\boldsymbol{\varphi} + (\tau\partial_1\mathbf{u} - (\boldsymbol{\omega}\wedge\mathbf{x})\cdot\nabla\mathbf{u} + \boldsymbol{\omega}\wedge\mathbf{u})\varphi - \pi\operatorname{div}\boldsymbol{\varphi} - \mathbf{f}\boldsymbol{\varphi})d\mathbf{x} = 0 \tag{5.142}$$

$$\operatorname{div}\mathbf{u} = 0,$$

for $\boldsymbol{\varphi} \in \boldsymbol{C}_0^\infty(\overline{U}^c)$.

Then $\mathbf{u} \in \mathbf{W}^{2,\min\{2,p\}}_{loc}(\overline{U}^c)$, $\pi \in W^{1,\min\{2,p\}}_{loc}(\overline{U}^c)$, *and* $\mathcal{L}(\mathbf{u}) + \nabla\pi = \mathbf{f}$. □

Proof This theorem follows from the interior regularity of Stokes flows, as stated in [93, Theorem IV.4.1]. □

Now we may establish (5.139) for Leray solutions of (5.128).

Corollary 5.1 *Let* $U \subset \mathbb{R}^3$ *be open and bounded. Let* $p \in (1, \infty)$, $\mathbf{f} \in \mathbf{L}^p_{loc}(\overline{U}^c)$, $\gamma, S_1 \in (0, \infty)$ *with* $\overline{U} \subset B_{S_1}$, $A \in [2, \infty)$, $B \in \mathbb{R}$ *with* $A + \min\{1, B\} > 3$, $|\mathbf{f}(\mathbf{z})| \le \gamma \cdot |\mathbf{z}|^{-A} s_\tau(\mathbf{z})^{-B}$ *for* $\mathbf{z} \in B^c_{S_1}$.

Let $\mathbf{u} \in \mathbf{W}^{1,1}_{loc}(\overline{U}^c)$ *with* $\mathbf{u} \in \mathbf{L}^6(\overline{U}^c)$ *and* $\nabla\mathbf{u} \in L^2(\overline{U}^c)^9$. *Let* $\pi \in L^2_{loc}(\overline{U}^c)$, *and suppose that (5.142) holds.*

Choose some $S_0 \in (0, S_1)$ *with* $\overline{U} \subset B_{S_0}$, *and let* $S \in (S_1, \infty)$. *Then, for* $\mathbf{z} \in B^c_S$, $1 \le i, j \le 3$, *inequalities (5.139) and (5.140) are valid, but with* $\mathcal{D}^c$ *replaced by* B_{S_0}.

Now, suppose that $\text{supp}(\mathbf{f}) \subset B_{S_1}$, *put* $s := s(p) := \min\{2, p\}$, *and let* $\mathcal{E}_s \colon \mathbf{W}^{2-1/s,s}(\partial B_{S_0}) \mapsto \mathbf{W}^{2,s}(B_{S_0})$ *be a continuous extension operator. Let* $C_p > 0$ *be a constant with*

$$\|\mathcal{E}_s(\mathbf{v})\|_{2,s} \le C_p \cdot \|\mathbf{v}\|_{2-1/s,s} \quad \textit{for } \mathbf{v} \in W^{2-1/s,s}(\partial B_{S_0}).$$

Then, for $1 \le j \le 3$, $\boldsymbol{\alpha} \in \mathbb{N}^3_0$ *with* $|\boldsymbol{\alpha}| \le 2$, $\mathbf{x} \in B^c_S$, *inequality (5.141) holds with* $\mathcal{D}^c$ *replaced by* B_{S_0} *and with the norm* $\|\cdot\|_{2-1/s,s}$ *in the place of* $\|\cdot\|_{2-1/p,p}$. □

Proof Theorem 5.25 yields that $\mathbf{u} \in \mathbf{W}^{2,\min\{2,p\}}_{\text{loc}}(\overline{U}^c)$, $\pi \in W^{1,\min\{2,p\}}_{\text{loc}}(\overline{U}^c)$, and $\mathcal{L}(\mathbf{u}) + \nabla\pi = \mathbf{f}$. We may conclude in particular that the pair $(\mathbf{u}_{|B^c_{S_0}}, \pi_{|B^c_{S_0}})$ belongs to $\text{div}\,\mathfrak{M}_{\min\{2,p\}}$, with $\overline{B_{S_0}}^c$ in the place of $\mathcal{D}$ as the domain of reference in the definition of $\mathfrak{M}_{\min\{2,p\}}$. Therefore the corollary follows from Theorems 5.22 and 5.23. □

5.5 Notes on Chap. 5

Chapter 5 is based on works by Deuring, Kračmar, and Nečasová:

- Derivation of the representation formula [35]
- Asymptotic behavior of Oseen type problem [36, 37], and [38]
- Leray solution [40]

Let us mention that by the different approach the Oseen type of problem was studied in book by G.P. Galdi, Chapter VIII [98]. Moreover, Galdi and Silvestre, see [108], have shown that under suitable assumptions on the data the solutions exist in certain Sobolev spaces and in [111] showed that such solutions satisfied the relations (5.3).

The study on the asymptotic properties of Oseen type problem can also be found in the work by M. Kyed [160, 161], and [162]. Further, time-periodic Oseen problem was investigated by M.Kyed [164] and Galdi and Kyed [104] and recently by Eiter and Kyed [51, 52].

Chapter 6
Artificial Boundary Conditions for the Oseen Type Problem

The chapter is devoted to the study of the linearized stationary incompressible flow (Oseen type flow) around rotating and translating body in the exterior domain $\mathcal{D}$. One of the goals is to obtain the pointwise estimate for the pressure. Moreover, we consider the linearized problem in a truncation domain $\mathcal{D}_R := B_R \cap \overline{\mathcal{D}}$ under certain artificial boundary conditions on the truncating boundary ∂B_R and then compare this solution with the solution in the exterior domain $\overline{\mathcal{D}}$ to get the truncation error estimate.

6.1 Introduction

We consider the following system of equations

$$\begin{aligned} & -\Delta u(\mathbf{z}) + (\tau \boldsymbol{e}_1 - \sqrt{\varrho}\boldsymbol{e}_1 \wedge \mathbf{z})\nabla \mathbf{u}(\mathbf{z}) + \sqrt{\varrho}\boldsymbol{e}_1 \wedge \mathbf{u}(\mathbf{z}) + \nabla \pi(\mathbf{z}) = \boldsymbol{F}(\mathbf{z}) \\ & \operatorname{div} \mathbf{u}(\mathbf{z}) = 0 \quad \text{for } \mathbf{z} \in \overline{\mathcal{D}}, \end{aligned} \tag{6.1}$$

where $\mathcal{D} \subset \mathbb{R}^3$ is open and bounded, with Lipschitz boundary.

The goal of this chapter is twofolds:

- We derive the pointwise estimates for the pressure.[1]
- We solve the system (6.1) in a truncation $\mathcal{D}_R := B_R \cap \overline{\mathcal{D}}$ of the exterior domain $\overline{\mathcal{D}}$ under certain artificial boundary conditions on the truncating boundary ∂B_R. Further, we compare this solution with the solution of (6.1) in the exterior domain, i.e., to get some type of error estimates of the method of an artificial

[1] Let us mention that the decay of pressure was also investigated in the work of Galdi, Kyed [101] and in case of pure rotation, see [63].

Š. Nečasová et al., *Mathematical Theory of a Fluid Flow Around a Rotating and Translating Body*, Springer Monographs in Mathematics,
https://doi.org/10.1007/978-3-031-84120-0_6

boundary condition. For this goal we use pointwise estimates of the velocity derived in Chap. 5 and the pointwise estimate of the pressure derived in this chapter.

Let us mention that mathematical analysis of the problem of the Navier–Stokes equations with artificial boundary condition was performed by many authors but without considering the rotation of body, see, e.g., [19, 20, 33, 34]. The rotation of body was considering just by Deuring, Kračmar and Nečasová in [43], and we will follow this approach.

6.2 Preliminaries

First let us recall notations which we will use in this chapter:

- $s(\mathbf{y}) := 1 + |\mathbf{y}| - \mathbf{y}_1$ for $\mathbf{y} \in \mathbb{R}^3$.
- $\mathcal{D}_R$ is the truncation of the exterior domain $\overline{\mathcal{D}}$ by the ball B_R. The boundary $\partial\mathcal{D}_R$ consists of parts $\partial\mathcal{D}$ and ∂B_R, the latter we call the truncating boundary.
- Fix $\tau \in (0, \infty)$, $\rho \in \mathbb{R} \setminus \{0\}$, and put

$$\boldsymbol{e}_1 := (1, 0, 0), \quad \Omega := \rho \begin{pmatrix} 0 & 0 & 0 \\ 0 & 0 & -1 \\ 0 & 1 & 0 \end{pmatrix},$$

  so that $\Omega \cdot \mathbf{z} = \rho \boldsymbol{e}_1 \wedge \mathbf{z}$ for $z \in \mathbb{R}^3$.
- For $U \subset \mathbb{R}^3$ open, $\mathbf{u} \in \boldsymbol{W}^{2,1}_{\text{loc}}(U)^3$, $\mathbf{z} \in U$, put

$$\begin{aligned} \mathcal{L}(\mathbf{u})(\mathbf{z}) &:= -\,\Delta\mathbf{u}(\mathbf{z}) + \tau\partial_1\mathbf{u}(\mathbf{z}) - (\sqrt{\rho}\boldsymbol{e}_1 \wedge \mathbf{z})\nabla\mathbf{u}(\mathbf{z}) + \sqrt{\rho}\boldsymbol{e}_1 \wedge \mathbf{u}(\mathbf{z}), \\ \mathcal{L}^*(\mathbf{u})(\mathbf{z}) &:= -\,\Delta\mathbf{u}(\mathbf{z}) - \tau\partial_1\mathbf{u}(\mathbf{z}) + (\sqrt{\rho}\boldsymbol{e}_1 \wedge \mathbf{z})\nabla\mathbf{u}(\mathbf{z}) - \sqrt{\rho}\boldsymbol{e}_1 \wedge \mathbf{u}(\mathbf{z}). \end{aligned}$$

-

$$\begin{aligned} K(\mathbf{z}, t) :=& (4\pi t)^{-3/2} \mathrm{e}^{-|\mathbf{z}|^2/(4t)} \quad (\mathbf{z} \in \mathbb{R}^3,\ t \in (0, \infty)), \\ \Lambda(\mathbf{z}, t) :=& \left(K(\mathbf{z}, t)\delta_{jk} + \partial z_j \partial z_k \left(\int_{\mathbb{R}^3} (4\pi|\mathbf{z} - \mathbf{y}|)^{-1} K(\mathbf{y}, t) \mathrm{d}\mathbf{y} \right) \right)_{1 \le j,k \le 3} \\ & (\mathbf{z} \in \mathbb{R}^3,\ t > 0), \\ \Gamma(\mathbf{x}, \mathbf{y}, t) :=& \Lambda(\mathbf{x} - \tau t \boldsymbol{e}_1 - \mathrm{e}^{-t\Omega}\mathbf{y}, t) \cdot \mathrm{e}^{-t\Omega}, \\ \widetilde{\Gamma}(\mathbf{x}, \mathbf{y}, t) :=& \Lambda(\mathbf{x} + \tau t \boldsymbol{e}_1 - \mathrm{e}^{t\Omega}\mathbf{y}, t) \cdot \mathrm{e}^{t\Omega} \quad (\mathbf{x}, \mathbf{y} \in \mathbb{R}^3, t > 0), \\ \mathcal{Z}(\mathbf{x}, \mathbf{y}) :=& \int_0^\infty \Gamma(\mathbf{x}, \mathbf{y}, t) \mathrm{d}t, \quad \widetilde{\mathcal{Z}}(\mathbf{x}, \mathbf{y}) := \int_0^\infty \widetilde{\Gamma}(\mathbf{x}, \mathbf{y}, t) \mathrm{d}t, \\ & (\mathbf{x}, \mathbf{y} \in \mathbb{R}^3,\ \mathbf{x} \neq \mathbf{y}). \end{aligned}$$

- For $q \in (1, 2)$, $\boldsymbol{F} \in \boldsymbol{L}^q(\mathbb{R}^3)$, put

$$\mathfrak{R}(f)(\mathbf{x}) := \int_{\mathbb{R}^3} \mathcal{Z}(\mathbf{x}, \mathbf{y})\boldsymbol{F}(\mathbf{y})d\mathbf{y} \quad (\mathbf{x} \in \mathbb{R}^3),$$

  for details see Lemma 5.10.
- We will use the space

$$\widehat{\boldsymbol{H}}_0^{1,2}(\overline{\mathcal{D}}) := \{\mathbf{v} \in \boldsymbol{L}^6(\overline{\mathcal{D}}) \cap \boldsymbol{W}^{1,2}_{\text{loc}}(\overline{\mathcal{D}}) : \nabla\mathbf{v} \in L^2(\overline{\mathcal{D}}), \quad \mathbf{v}|\partial\mathcal{D} = 0\}$$

  equipped with the norm $\|\nabla\mathbf{u}\|_2$, where $\mathbf{v}|\partial\mathcal{D}$ means the trace of $\mathbf{v}$ on $\partial\mathcal{D}$.
- For $p \in (1, \infty)$, define $\mathfrak{M}_p$ as the space of all pairs of functions $(\mathbf{u}, \pi)$ such that $\mathbf{u} \in \boldsymbol{W}^{2,p}_{\text{loc}}(\overline{\mathcal{D}})$, $\pi \in W^{1,p}_{\text{loc}}(\overline{\mathcal{D}})$,

$$\mathbf{u}|\mathcal{D}_R \in \boldsymbol{W}^{1,p}(\mathcal{D}_R), \quad \pi|\mathcal{D}_R \in L^p(\mathcal{D}_R), \quad \mathbf{u}|\partial\mathcal{D} \in \boldsymbol{W}^{2-1/p,p}(\partial\mathcal{D}),$$

$$\operatorname{div}\mathbf{u}|\mathcal{D}_R \in W^{1,p}(\mathcal{D}_R), \quad \mathcal{L}(\mathbf{u}) + \nabla\pi|\mathcal{D}_R \in \boldsymbol{L}^p(\mathcal{D}_R)$$

  for some $R \in (0, \infty)$ with $\overline{\mathcal{D}} \subset B_R$, see Definition 5.1.

  We write C for generic constants. It should be clear from context which are the parameters these constants depend on. In order to lift possible ambiguities, we sometimes use the notation $C(\gamma_1, \ldots, \gamma_n)$ in order to indicate that the constant in question depends in particular on $\gamma_1, \ldots, \gamma_n \in (0, \infty)$ for some $n \in \mathbb{N}$. But the relevant constant may depend on other parameters as well.

6.3 Auxiliary Results to Asymptotic Behavior of the Pressure

Lemma 6.1 (Weyl's Lemma) *Let $n \in \mathbb{N}$, $U \subset \mathbb{R}^n$ open, $u \in L^1_{\text{loc}}(U)$ with $\int_U u\Delta l d\mathbf{x} = 0$ for $l \in C_0^\infty(U)$. Then $u \in C^\infty(U)$ and $\Delta u = 0$.* □

Proof An elementary proof is given in [198, Appendix]. □

For $q \in (1, 3/2)$, $h \in L^q(\mathbb{R}^3)$, put

$$\mathcal{N}(h)(\mathbf{x}) := \int_{\mathbb{R}^3} -(4\pi|\mathbf{x} - \mathbf{y}|)^{-1}h(\mathbf{y})d\mathbf{y} \quad (\mathbf{x} \in \mathbb{R}^3).$$

For $q \in (1, 3)$, $h \in L^q(\mathbb{R}^3)$, put

$$\mathfrak{S}(h)(\mathbf{x}) := \left(\int_{\mathbb{R}^3} (4\pi|\mathbf{x} - \mathbf{y}|^3)^{-1}(\mathbf{x} - \mathbf{y})_j h(\mathbf{y})d\mathbf{y}\right)_{1\le j\le 3} \quad (\mathbf{x} \in \mathbb{R}^3).$$

For $q \in (1,3)$, $\mathbf{h} \in \boldsymbol{L}^q(\mathbb{R}^3)$, put

$$\mathcal{P}(\mathbf{h})(\mathbf{x}) := \int_{\mathbb{R}^3} (4\pi |\mathbf{x}-\mathbf{y}|^3)^{-1}((\mathbf{x}-\mathbf{y})\mathbf{h}(\mathbf{y}))d\mathbf{y} \quad (\mathbf{x} \in \mathbb{R}^3).$$

Remark 6.1 Note that $\mathfrak{S}(h)$ is a vector-valued function with h being scalar, whereas $\mathcal{P}(\mathbf{h})$ is a scalar function with $\mathbf{h}$ being vector-valued. □

Lemma 6.2 *Let $q \in (1,3/2)$, $h \in L^q(\mathbb{R}^3)$. Then $\mathcal{N}(h) \in W^{2,q}_{\text{loc}}(\mathbb{R}^3) \cap L^{(1/q-2/3)^{-1}}(\mathbb{R}^3)$, $\Delta\mathcal{N}(h) = h$. If $h \in W^{1,q}(\mathbb{R}^3)$, then $\partial_l \mathcal{N}(h) = \mathcal{N}(\partial_l h)$ $(1 \le l \le 3)$.*

Let $q \in (1,3)$, $h \in L^q(\mathbb{R}^3)$. Then $\mathfrak{S}(h) \in \mathbf{W}^{1,q}_{\text{loc}}(\mathbb{R}^3)$, $\operatorname{div}\mathfrak{S}(h) = h$. If $q \in (1,3/2)$, then $\nabla\mathcal{N}(h) = \mathfrak{S}(h)$. If $h \in W^{1,q}(\mathbb{R}^3)$, then $\mathfrak{S}(h) \in \mathbf{W}^{2,q}_{\text{loc}}(\mathbb{R}^3)$.

Let $q \in (1,3)$, $\mathbf{h} \in \boldsymbol{L}^q(\mathbb{R}^3)$. Then

$$\mathcal{P}(\mathbf{h}) \in W^{1,q}_{\text{loc}}(\mathbb{R}^3) \cap L^{(1/q-1/3)^{-1}}(\mathbb{R}^3),$$

$$\left(\int_{\mathbb{R}^3}\left(\int_{\mathbb{R}^3} |\mathbf{x}-\mathbf{y}|^{-2}|h(\mathbf{y})|d\mathbf{y}\right)^{(1/q-1/3)^{-1}} d\mathbf{x}\right)^{1/q-1/3} \le C\|\mathbf{h}\|_q.$$

□

Proof The assertion of Lemma 6.2 follows from the well-known Hardy–Littlewood–Sobolev inequality, Calderon–Zygmund inequality, and density arguments. □

Lemma 6.3 ([119, Lemma 2.2]) *Let $B \in \mathbb{R}$, $S \in (0,\infty)$. Then*

$$\int_{\partial B_R} s(\mathbf{x})^{-B} do_{\mathbf{x}} \le C(S,B)R^{2-\min\{1,B\}} l_{A,B}(R) \tag{6.2}$$

for $R \in [S,\infty)$, with $l_{A,B}(R) := 1$ if $B \ne 1$, and $\sigma(R) = \ln(1+R)$ if $B = 1$. □

6.4 Decay Estimate of the Pressure

To begin with we recall some known results from Chap. 5, see also [36] and [38] about the decay of the velocity part of the solution of the system (6.1), and in order to get the full decay characterization of the solution, we derive the decay of the pressure part of solution of (6.1).

Our starting point is a decay result from Sect. 6.3 [see also [38]] for the velocity part u of a solution to (6.1). For convenience of reader we recall this theorem:

Theorem 6.1 (Theorem 5.8) *Suppose that $\mathcal{D}^c$ is C^2-bounded. Let $p \in (1,\infty)$, $(\mathbf{u},\pi) \in \mathfrak{M}_p$. Let us set $\mathbf{F} = \mathcal{L}(\mathbf{u}) + \nabla\pi$. Suppose there are numbers $S_1, S, \gamma \in$*

$(0, \infty)$, $A \in [2, \infty)$, $B \in \mathbb{R}$ *such that* $S_1 < S$,

$$\overline{\mathcal{D}}^c \cup \operatorname{supp}(\operatorname{div} \mathbf{u}) \subset B_{S_1}, \quad \mathbf{u}|B_S^c \in \boldsymbol{L}^6(B_S^c), \quad \nabla\mathbf{u}|B_S^c \in L^2(B_S^c)^9,$$

$$A + \min\{1, B\} \geq 3, \quad |\mathbf{F}(\mathbf{z})| \leq \gamma |\mathbf{z}|^{-A} s(\mathbf{z})^{-B} \quad \textit{for } \mathbf{z} \in B_{S_1}^c.$$

Then

$$|\mathbf{u}(\mathbf{y})| \leq C(|\mathbf{y}|s(\mathbf{y}))^{-1} l_{A,B}(\mathbf{y}), \tag{6.3}$$

$$|\nabla\mathbf{u}(\mathbf{y})| \leq C(|\mathbf{y}|s(\mathbf{y}))^{-3/2} s(\mathbf{y})^{\max(0,7/2-A-B)} l_{A,B}(|\mathbf{y}|) \tag{6.4}$$

for $\mathbf{y} \in B_S^c$, *where function* $l_{A,B}$ *is given by*

$$l_{A,B}(\mathbf{x}) = \begin{cases} 1 & \textit{if } A + \min\{1, B\} > 3 \\ \max(1, \ln(|y)|) & \textit{if } A + \min\{1, B\} = 3. \end{cases}$$

The assumptions $\mathbf{u}|B_S^c \in \boldsymbol{L}^6(B_S^c)$, $\nabla\mathbf{u}|B_S^c \in L^2(B_S^c)^9$ should be interpreted as decay conditions on $\mathbf{u}$.

It may be deduced from Theorem 6.1 that inequalities (6.3) and (6.4) hold under assumptions weaker than those stated in that theorem. We specify this more general situation in the ensuing corollary, which in addition indicates some properties of $\mathbf{F}$ that will be useful in the following.

Corollary 6.1 *Let* $p \in (1, \infty)$, $\gamma, S_1, S \in (0, \infty)$ *with* $\mathcal{D}^c \subset B_{S_1}$, $S_1 < S$, $A \in [2, \infty)$, $B \in \mathbb{R}$ *with* $A + \min\{1, B\} \geq 3$. *Let* $F : \overline{\mathcal{D}} \mapsto \mathbb{R}$ *be measurable with* $\mathbf{F}|\mathcal{D}_{S_1} \in \boldsymbol{L}^p(\mathcal{D}_{S_1})^3$ *and* $|F(\mathbf{z})| \leq \gamma |\mathbf{z}|^{-A} s(\mathbf{z})^{-B}$ *for* $\mathbf{z} \in B_{S_1}^c$. *Let* $\mathbf{u} \in \boldsymbol{W}^{1,p}_{\mathrm{loc}}(\overline{\mathcal{D}})$ *with* $\mathbf{u}|B_S^c \in \boldsymbol{L}^6(B_S^c)$, $\nabla\mathbf{u}|B_S^c \in L^2(B_S^c)^9$, $\operatorname{supp}(\operatorname{div} \mathbf{u}) \subset B_{S_1}$,

$$\int_{\overline{\mathcal{D}}} \big[\nabla\mathbf{u}\nabla\boldsymbol{\varphi} + (\tau\partial_1\mathbf{u} - (\sqrt{\varrho}\boldsymbol{e}_1 \wedge \mathbf{z})\nabla\mathbf{u} + (\sqrt{\varrho}\boldsymbol{e}_1 \wedge \mathbf{u}) - F)\boldsymbol{\varphi}\big] d\mathbf{z}$$

$$= 0 \quad \textit{for } \boldsymbol{\varphi} \in \boldsymbol{C}_0^\infty(\overline{\mathcal{D}}) \quad \textit{with } \operatorname{div} \boldsymbol{\varphi} = 0. \tag{6.5}$$

Then inequalities (6.3) *and* (4.23) *hold for* $\mathbf{y} \in B_S^c$.

Moreover $\mathbf{F} \in \boldsymbol{L}^q(\overline{\mathcal{D}})$ *for* $q \in (1, p]$. *If* $p \geq 6/5$, *the function* $\mathbf{F}$ *may be considered as a bounded linear functional on* $\widehat{\boldsymbol{H}}_0^{1,2}(\overline{\mathcal{D}})$, *in the usual sense.*

Let $\pi \in L^p_{\mathrm{loc}}(\overline{\mathcal{D}})$ *with*

$$\int_{\overline{\mathcal{D}}} [\nabla\mathbf{u}\nabla\boldsymbol{\varphi} + (\tau\partial_1\mathbf{u} - (\sqrt{\varrho}\boldsymbol{e}_1 \wedge \mathbf{z})\nabla\mathbf{u} + (\sqrt{\varrho}\boldsymbol{e}_1 \wedge \mathbf{u}) - \mathbf{F})\boldsymbol{\varphi} - \pi \operatorname{div} \boldsymbol{\varphi}] d\mathbf{z} = 0 \tag{6.6}$$

for $\varphi \in C_0^\infty(\overline{\mathcal{D}}^c)$. *Fix some number* $S_0 \in (0, S_1)$ *with* $\overline{\mathcal{D}}^c \cup \operatorname{supp}(\operatorname{div}\mathbf{u}) \subset B_{S_0}$. *Then the relations* $\mathbf{u}|\overline{B_{S_0}}^c \in \boldsymbol{W}^{2,p}_{\text{loc}}(\overline{B_{S_0}}^c)$, $\pi \in W^{1,p}_{\text{loc}}(\overline{B_{S_0}}^c)$, *and* $\mathcal{L}(\mathbf{u}|\overline{B_{S_0}}^c) + \nabla\pi = \mathbf{F}|\overline{B_{S_0}}^c$ *hold.* □

Proof For $\mathbf{z} \in B^c_{S_1}$, we have

$$\begin{aligned}|\mathbf{F}(\mathbf{z})| &\le \gamma C(S_1, A)|\mathbf{z}|^{-2}s(\mathbf{z})^{-A+2-B} \le \gamma C(S_1, A)|\mathbf{z}|^{-2}s(\mathbf{z})^{-A+2-\min\{1,B\}}\\ &\le \gamma C(S_1, A)|\mathbf{z}|^{-2}s(\mathbf{z})^{-1}.\end{aligned}$$

Thus for $q \in (1, \infty)$, with Lemma 6.3,

$$\int_{B^c_{S_1}} |\mathbf{F}(\mathbf{z})|^q \mathrm{d}\mathbf{z} \le C \int_{S_1}^{\infty} r^{-2q} \int_{\partial B_r} s(\mathbf{z})^{-q}\mathrm{d}o_{\mathbf{z}}\mathrm{d}r \le C \int_{S_1}^{\infty} r^{-2q+1}\mathrm{d}r < \infty.$$

It follows that $\mathbf{F} \in \boldsymbol{L}^q(\overline{\mathcal{D}})$ for $q \in (1, p]$. From [98, Theorem II.6.1], the inequality $\|\mathbf{v}\|_6 \le C\|\nabla\mathbf{v}\|_2$ holds for $\mathbf{v} \in \widehat{\boldsymbol{H}}_0^{1,2}(\overline{\mathcal{D}})$. Thus, if $p \ge 6/5$, hence $\mathbf{F} \in \boldsymbol{L}^{6/5}(\overline{\mathcal{D}})$, this function $\mathbf{F}$ may be considered as a linear bounded functional on $\widehat{\boldsymbol{H}}_0^{1,2}(\overline{\mathcal{D}})$. The L^p-integrability of $\mathbf{F}$ and the assumptions on $\mathbf{u}$ imply that the function

$$\boldsymbol{G}(\mathbf{z}) := \boldsymbol{F}(\mathbf{z}) - (\tau\boldsymbol{e}_1 - (\sqrt{\varrho}\boldsymbol{e}_1 \wedge \mathbf{z})) \cdot \nabla\mathbf{u}(\mathbf{z}) - (\sqrt{\varrho}\,\boldsymbol{e}_1 \wedge \mathbf{u}(\mathbf{z}))\mathbf{z} \in \overline{\mathcal{D}}, \tag{6.7}$$

belongs to $\boldsymbol{L}^p_{\text{loc}}(\overline{D})$. The choice of S_0 (see at the end of Corollary 6.1) means in particular that $\operatorname{div}(\mathbf{u}|\overline{B_{S_0}}^c) = 0$. This equation, (6.6), the relation $\mathbf{G} \in \boldsymbol{L}^p_{loc}(\overline{\mathcal{D}})$, and interior regularity of solutions to the Stokes system (see [98, Theorem IV.4.1] for example) imply the statements in the last sentence of Corollary 6.1.

Put $S_0' := (S_0 + S_1)/2$, □

$A_{S_0',R} := B_R\backslash B_{S_0'}$ for $R \in (S_0', \infty)$. Then $\mathbf{u}|A_{S_0',R} \in \boldsymbol{W}^{2,p}(A_{S_0',R})$ and $\pi|A_{S_0',R} \in W^{1,p}(A_{S_0',R})$ for $R \in (S_0', \infty)$, so $(\mathbf{u}|A_{S_0',R}, \pi|A_{S_0',R}) \in \mathfrak{M}_p$, with $B_{S_0'}$ in the role of $\mathcal{D}$. Note that $S_0 < S_o' < S_1 < S$. Thus the assumptions of Theorem 6.1 are satisfied with $\mathcal{D}^c$ replaced by $B_{S_0'}$. As a consequence inequalities (6.3) and (4.23) hold. □

Remark 6.2 Solutions as considered in Corollary 6.1 exist if, for example, Dirichlet boundary conditions are prescribed on $\partial\mathcal{D}$. In fact, as stated in [98, Theorem VIII.1.2], if $\mathbf{F}$ is a bounded linear functional on the space $\widehat{\boldsymbol{H}}_0^{1,2}(\overline{\mathcal{D}})$, and if $\mathbf{b} \in \boldsymbol{H}^{1/2}(\partial\overline{\mathcal{D}})$, then there is a function $\mathbf{u} \in \boldsymbol{L}^6(\overline{\mathcal{D}}) \cap \boldsymbol{W}^{1,1}_{\text{loc}}(\overline{\mathcal{D}})$ such that $\nabla\mathbf{u} \in L^2(\overline{\mathcal{D}})^9$ and $\mathbf{u}$ satisfies Eqs. (6.5) and $\operatorname{div}\mathbf{u} = 0$ (weak form of (4.16)), as well as the boundary conditions $\mathbf{u}|\partial\mathcal{D} = \mathbf{b}$. Existence of a pressure $\pi \in L^p_{\text{loc}}(\overline{\mathcal{D}})$ with (6.6) holds according to [98, Lemma VIII.1.1]. □

Let us state the main result of this section.

Theorem 6.2 *Let p, γ, S_1, S, A, B, F, $\mathbf{u}$ be given as in Corollary* 6.1, *but with the stronger assumptions $A = 5/2$, $B \in (1/2, \infty)$ on A and B. Let $\pi \in L^p_{\text{loc}}(\overline{\mathcal{D}})$ such that* (6.6) *holds. Then there is $c_0 \in \mathbb{R}$ such that*

$$|\pi(\mathbf{x}) + c_0| \le C|\mathbf{x}|^{-2} \quad \textit{for } \mathbf{x} \in B_S^c. \tag{6.8}$$

Proof By Corollary 6.1 we have $\mathbf{F} \in \boldsymbol{L}^q(\overline{\mathcal{D}})$ for $q \in (1, p]$. Fix some number $S_0 \in (0, S_1)$ with $\overline{\mathcal{D}}^c \cup \text{supp}(\text{div}\,\mathbf{u}) \subset B_{S_0}$. Then again by Corollary 6.1, the relations $\mathbf{u}|\overline{B_{S_0}}^c \in \boldsymbol{W}^{2,p}_{\text{loc}}(\overline{B_{S_0}})$, $\pi|\overline{B_{S_0}}^c \in W^{1,p}_{\text{loc}}(\overline{B_{S_0}}^c)$, and $\mathcal{L}(\mathbf{u}|\overline{B_{S_0}}^c) + \nabla(\pi|\overline{B_{S_0}}^c) = \mathbf{F}|\overline{B_{S_0}}^c$ hold. Note that $S_0 < S_1 < S$. Take $\varphi \in C^\infty(\mathbb{R}^3)$ with

$$\varphi|B_{S_1+1/4(S-S_1)} = 0, \quad \varphi|B^c_{S_1+3/4(S-S_1)} = 1,$$

and put $\widetilde{\mathbf{u}} := \varphi \cdot \mathbf{u}$, $\widetilde{\pi} := \varphi \cdot \pi$, with $\widetilde{\mathbf{u}}, \widetilde{\pi}$ to be considered as functions in $\mathbb{R}^3$. By the choice of φ and the properties of $\mathbf{u}$ and π, we get $\widetilde{\mathbf{u}} \in \boldsymbol{W}^{2,q}_{\text{loc}}(\mathbb{R}^3)$, $\widetilde{\pi} \in W^{1,q}_{\text{loc}}(\mathbb{R}^3)$ for $q \in [1, p]$, $\widetilde{\mathbf{u}}|B_S^c = \mathbf{u}|B_S^c \in \boldsymbol{L}^6(\mathbb{R}^3)$, $\nabla\widetilde{\mathbf{u}}|B_S^c = \nabla\mathbf{u}|B_S^c \in L^2(\mathbb{R}^3)^9$. Put

$$\begin{aligned} g_l(\mathbf{z}) := &-\sum_{k=1}^{3} \partial_k\varphi(\mathbf{z})\partial_k u_l(\mathbf{z}) - \Delta\varphi(\mathbf{z})u_l(\mathbf{z}) + \tau\partial_1\varphi(\mathbf{z})u_l(\mathbf{z}) \\ &-\sum_{k=1}^{3}(\tau \boldsymbol{e}_1 \wedge \mathbf{z})_k \cdot \partial_k\varphi(\mathbf{z}) \cdot u_l(\mathbf{z}) + \partial_l\varphi(\mathbf{z})\pi(\mathbf{z}) \end{aligned}$$

for $\mathbf{z} \in \mathbb{R}^3$, $1 \le l \le 3$, and set $\gamma := \text{div}\,\widetilde{\mathbf{u}}$. Then

$$\begin{aligned} &\text{supp}(\mathbf{g}) \subset \overline{B_{S_1+3(S-S_1)/4}} \setminus B_{S_1+(S-S_1)/4}, \quad \mathbf{g} \in L^q(\mathbb{R}^3) \quad \text{for } q \in [1, p], \\ &\mathcal{L}\widetilde{\mathbf{u}} + \nabla\widetilde{\pi} = \mathbf{g} + \varphi \cdot \mathbf{F}, \quad \gamma = \nabla\varphi \cdot \mathbf{u}, \end{aligned} \tag{6.9}$$

in particular $\text{supp}(\gamma) \subset \overline{B_{S_1+3(S-S_1)/4}} \setminus B_{S_1+(S-S_1)/4}$, $\gamma \in W^{2,q}(\mathbb{R}^3)$, $\mathbf{g} + \varphi\mathbf{F} \in \boldsymbol{L}^q(\mathbb{R}^3)$ for $q \in (1, p]$. Let $\mathbf{x} \in \mathbb{R}^3$, $\varepsilon > 0$ with $\overline{B_\varepsilon(x)} \subset B_{S_1}$, where $B_\varepsilon(\mathbf{x}) = \{\mathbf{y} \in \mathbb{R}^3;\ |\mathbf{y} - \mathbf{x}| < \varepsilon\}$. Since $\widetilde{\mathbf{u}}|B_{S_1} = 0$, $\widetilde{\pi}|B_{S_1} = 0$, it follows from Theorem 5.21 with $\mathcal{D}^c$ replaced by $B_\varepsilon(\mathbf{x})$ that

$$\widetilde{\mathbf{u}}(\mathbf{y}) = \mathfrak{R}(\mathbf{g} + \varphi\mathbf{F})(\mathbf{y}) + \mathfrak{S}(\gamma)(\mathbf{y}) \quad \text{for } \mathbf{y} \in \overline{B_\varepsilon(\mathbf{x})}^c.$$

Since this is true for any $\mathbf{x} \in \mathbb{R}^3$, $\varepsilon > 0$ with $\overline{B_\varepsilon(\mathbf{x})} \subset B_{S_1}$, it follows that

$$\widetilde{\mathbf{u}} = \mathfrak{R}(\mathbf{g} + \varphi\boldsymbol{F}) + \mathfrak{S}(\gamma) \quad \text{in } \mathbb{R}^3. \tag{6.10}$$

Since $\mathfrak{S}(\gamma) \in \boldsymbol{W}^{2,q}_{\text{loc}}(\mathbb{R}^3)$ for $q \in [1, \min\{3, p\})$ by Lemma 6.2, and from (6.10)

$$\mathfrak{R}(\mathbf{g} + \varphi\mathbf{F}) \in \boldsymbol{W}^{2,q}_{\text{loc}}(\mathbb{R}^3) \quad \text{for } q \in [1, \min\{3, p\}).$$

This relation, (5.64), and the inequality following (5.68) imply

$$\sup_{\mathbf{z}\in\mathbb{R}^3}\int_{\mathbb{R}^3}|\mathcal{Z}(\mathbf{z},\mathbf{y})(\mathbf{g}+\varphi\mathbf{F})(\mathbf{y})|\mathrm{d}\mathbf{y}<\infty. \tag{6.11}$$

Let $\boldsymbol{\psi}\in\boldsymbol{C}_0^\infty(\mathbb{R}^3)$. Due to (6.11), we may apply Fubini's theorem, to obtain

$$\begin{aligned}A:=&\int_{\mathbb{R}^3}\boldsymbol{\psi}(\mathbf{x})(\mathcal{L}\mathfrak{R}(\mathbf{g}+\varphi\mathbf{F}))(\mathbf{x})\mathrm{d}\mathbf{x}=\int_{\mathbb{R}^3}(\mathcal{L}^*\boldsymbol{\psi})(\mathbf{x})\mathfrak{R}(\mathbf{g}+\varphi\mathbf{F})(\mathbf{x})\mathrm{d}\mathbf{x}\\ =&\int_{\mathbb{R}^3}\int_{\mathbb{R}^3}[(\mathcal{L}^*\boldsymbol{\psi})(\mathbf{x})]^\top\mathcal{Z}(\mathbf{x},\mathbf{y})(\mathbf{g}+\varphi\mathbf{F})(\mathbf{y})\mathrm{d}\mathbf{y}\mathrm{d}\mathbf{x}\\ =&\int_{\mathbb{R}^3}\int_{\mathbb{R}^3}[(\mathcal{L}^*\boldsymbol{\psi})(\mathbf{z})]^\top\mathcal{Z}(\mathbf{x},\mathbf{y})(\mathbf{g}+\varphi\mathbf{F})(\mathbf{y})\mathrm{d}\mathbf{x}\mathrm{d}\mathbf{y}.\end{aligned} \tag{6.12}$$

But for $\mathbf{a},\boldsymbol{b},\mathbf{x},\mathbf{y}\in\mathbb{R}^3$ with $\mathbf{x}\neq\mathbf{y}$,

$$\boldsymbol{a}^\top\mathcal{Z}(\mathbf{x},\mathbf{y})\boldsymbol{b}=\int_0^\infty\boldsymbol{a}^\top\Gamma(\mathbf{x},\mathbf{y},t)\boldsymbol{b}\mathrm{d}t,$$

hence with Lemma 2.20,

$$\begin{aligned}\boldsymbol{a}^\top\cdot\mathcal{Z}(\mathbf{x},\mathbf{y})\cdot\boldsymbol{b}&=\int_0^\infty\boldsymbol{a}^\top\mathrm{e}^{-t\Omega}\Lambda(\mathrm{e}^{t\Omega}\mathbf{x}-\tau t\boldsymbol{e}_1-\mathbf{y},t)\cdot\boldsymbol{b}\,\mathrm{d}t\\ &=\int_0^\infty\boldsymbol{b}^\top[\mathrm{e}^{-t\Omega}\Lambda(\mathrm{e}^{t\Omega}\mathbf{x}-\tau t\boldsymbol{e}_1-\mathbf{y},t)]^\top\boldsymbol{a}\,\mathrm{d}t\\ &=\int_0^\infty\boldsymbol{b}^\top\cdot\Lambda(\mathrm{e}^{t\Omega}\mathbf{x}-\tau t\boldsymbol{e}_1-\mathbf{y},t)\mathrm{e}^{t\Omega}\cdot\boldsymbol{a}\,\mathrm{d}t\\ &=\int_0^\infty\boldsymbol{b}^\top\Lambda(\mathbf{y}+\tau t\boldsymbol{e}_1-\mathrm{e}^{t\Omega}\mathbf{x},t)\mathrm{e}^{t\Omega}\boldsymbol{a}\,\mathrm{d}t\\ &=\int_0^\infty\boldsymbol{b}^\top\widetilde{\Gamma}(\mathbf{y},\mathbf{x},t)\cdot\boldsymbol{a}\,\mathrm{d}t=\boldsymbol{b}^\top\widetilde{\mathcal{Z}}(\mathbf{y},\mathbf{x})\cdot\boldsymbol{a}.\end{aligned}$$

Thus from (6.12)

$$A=\int_{\mathbb{R}^3}(\mathbf{g}+\varphi\mathbf{F})(\mathbf{y})^\top\cdot\int_{\mathbb{R}^3}\widetilde{\mathcal{Z}}(\mathbf{y},\mathbf{x})(\mathcal{L}^*\boldsymbol{\psi})(\mathbf{x})\,\mathrm{d}\mathbf{x}\,\mathrm{d}\mathbf{y}. \tag{6.13}$$

Since $\boldsymbol{\psi}\in\boldsymbol{C}_0^\infty(\mathbb{R}^3)$, we may choose $\mathbf{x}_0\in\mathbb{R}^3$, $\varepsilon>0$ such that

$$\overline{B_\varepsilon(\mathbf{x}_0)}\subset\mathbb{R}^3\setminus\operatorname{supp}(\boldsymbol{\psi}).$$

Therefore we get from Theorem 5.7 with $\mathcal{D}$, U, ω, $\mathbf{u}$, $\mathcal{L}$ replaced by $B_\varepsilon(\mathbf{x}_0)$, $\tau \boldsymbol{e_1}$, $-\rho \boldsymbol{e_1}$, ψ, $\mathcal{L}^*$, respectively, and with $\pi = 0$, that

$$\int_{\mathbb{R}^3} \widetilde{\mathcal{Z}}(\mathbf{y}, \mathbf{x})(\mathcal{L}^* \boldsymbol{\psi})(\mathbf{x}) \mathrm{d}x = \boldsymbol{\psi}(\mathbf{y}) - \mathcal{S}(\operatorname{div} \boldsymbol{\psi})(\mathbf{y})$$

for $\mathbf{y} \in \mathbb{R}^3 \setminus \overline{B_\varepsilon(\mathbf{x}_0)}$. Since this is true for any $\mathbf{x}_0 \in \mathbb{R}^3$, $\varepsilon > 0$ with $\overline{B_\varepsilon(\mathbf{x}_0)} \subset \mathbb{R}^3 \setminus \operatorname{supp}(\boldsymbol{\psi})$, the preceding equation holds for any $\mathbf{y} \in \mathbb{R}^3$. It follows from (6.13)

$$\int_{\mathbb{R}^3} \boldsymbol{\psi}(\mathbf{x})(\mathcal{L}\mathfrak{R}(\mathbf{g} + \varphi \mathbf{F}))(\mathbf{x}) \mathrm{d}\mathbf{x} = \int_{\mathbb{R}^3} (\mathbf{g} + \varphi \mathbf{F})(\mathbf{y})(\boldsymbol{\psi}(\mathbf{y}) - \mathfrak{S}(\operatorname{div} \boldsymbol{\psi})(\mathbf{y})) \mathrm{d}\mathbf{y}. \tag{6.14}$$

Recalling that $\mathbf{g} + \varphi \mathbf{F} \in \boldsymbol{L}^q(\mathbb{R}^3)$ for $q \in (1, p]$, we get with Lemma 6.2 that

$$\int_{\mathbb{R}^3} \boldsymbol{\psi}(\mathbf{x}) \nabla \mathcal{P}(\mathbf{g} + \varphi \mathbf{F})(\mathbf{x}) \mathrm{d}\mathbf{x} = \int_{\mathbb{R}^3} -\operatorname{div} \boldsymbol{\psi}(\mathbf{x}) \mathcal{P}(\mathbf{g} + \varphi \mathbf{F})(\mathbf{x}) \mathrm{d}\mathbf{x}. \tag{6.15}$$

Put $q_0 := \min\{6/5, p\}$, and let us note that $q_0 \in (1, 3/2)$, $q_0 \leq p$.
Thus, by Hölder's inequality and Lemma 6.2,

$$\begin{aligned}
&\int_{\mathbb{R}^3} |\operatorname{div} \boldsymbol{\psi}(\mathbf{x}) \mathcal{P}(\mathbf{g} + \varphi \mathbf{F})(\mathbf{x})| \mathrm{d}\mathbf{x} \\
&\leq \int_{\mathbb{R}^3} \int_{\mathbb{R}^3} |\operatorname{div} \boldsymbol{\psi}(\mathbf{x})(4\pi |\mathbf{x} - \mathbf{y}|^3)^{-1} (\mathbf{x} - \mathbf{y}) \cdot (\mathbf{g} + \varphi \mathbf{F})(\mathbf{y})| \mathrm{d}\mathbf{y} \mathrm{d}\mathbf{x} \\
&\leq \| \operatorname{div} \boldsymbol{\psi} \|_{(4/3 - 1/q_0)^{-1}} \\
&\quad \left(\int_{\mathbb{R}^3} \left(\int_{\mathbb{R}^3} (4\pi |\mathbf{x} - \mathbf{y}|^2)^{-1} |(\mathbf{g} + \varphi \mathbf{F})(\mathbf{y})| \mathrm{d}\mathbf{y} \right)^{(1/q_0 - 1/3)^{-1}} \mathrm{d}x \right)^{1/q_0 - 1/3} \\
&\leq C \| \operatorname{div} \boldsymbol{\psi} \|_{(4/3 - 1/q_0)^{-1}} \| \mathbf{g} + \varphi \mathbf{F} \|_{q_0} < \infty.
\end{aligned}$$

As a consequence, we may apply Fubini's theorem to deduce from (6.15) that

$$\begin{aligned}
&\int_{\mathbb{R}^3} \psi(\mathbf{x}) \nabla \mathcal{P}(\mathbf{g} + \varphi \mathbf{F})(\mathbf{x}) \mathrm{d}\mathbf{x} \\
&= - \int_{\mathbb{R}^3} \int_{\mathbb{R}^3} (\operatorname{div} \boldsymbol{\psi})(\mathbf{x})(4\pi |\mathbf{x} - \mathbf{y}|^3)^{-1} (\mathbf{x} - \mathbf{y})(\mathbf{g} + \varphi \mathbf{F})(\mathbf{y}) \mathrm{d}\mathbf{x} \mathrm{d}\mathbf{y} \\
&= \int_{\mathbb{R}^3} (\mathbf{g} + \varphi \mathbf{F})(\mathbf{y}) \mathfrak{S}(\operatorname{div} \boldsymbol{\psi})(\mathbf{y}) \mathrm{d}\mathbf{y}.
\end{aligned} \tag{6.16}$$

From (6.14) and (6.16) it implies

$$\int_{\mathbb{R}^3} \boldsymbol{\psi}(\mathbf{x})((\mathcal{L}\mathfrak{R}(\mathbf{g}+\varphi\mathbf{F}))(\mathbf{x}) + \nabla\mathcal{P}(\mathbf{g}+\varphi\mathbf{F})(\mathbf{x}))\mathrm{d}\mathbf{x} = \int_{\mathbb{R}^3} \boldsymbol{\psi}(\mathbf{x})(\mathbf{g}+\varphi\mathbf{F})(\mathbf{x})\mathrm{d}\mathbf{x}.$$

Since this is true for any $\boldsymbol{\psi} \in \boldsymbol{C}_0^\infty(\mathbb{R}^3)$, we have found that

$$\mathcal{L}\mathfrak{R}(\mathbf{g}+\varphi\mathbf{F}) + \nabla\mathcal{P}(\mathbf{g}+\varphi\mathbf{F}) = \mathbf{g}+\varphi\mathbf{F}. \tag{6.17}$$

On the other hand, by (6.10) and (6.9)

$$\mathcal{L}\mathfrak{R}(\mathbf{g}+\varphi\mathbf{F}) + \mathcal{L}\mathfrak{S}(\gamma) + \nabla\widetilde{\pi} = L\widetilde{\mathbf{u}} + \nabla\widetilde{\pi} = \mathbf{g}+\varphi\mathbf{F}.$$

By subtracting this equation from (6.17), we get

$$\nabla\mathcal{P}(\mathbf{g}+\varphi\mathbf{F}) - \mathcal{L}\mathfrak{S}(\gamma) - \nabla\widetilde{\pi} = 0. \tag{6.18}$$

Next we study the term $\operatorname{div}(\mathcal{L}\mathfrak{S}(\gamma))$. Recalling that $q_0 < 3/2$, $q_0 \le p$ and $\gamma \in W^{2,q}(\mathbb{R}^3)$ for $q \in [1, p]$ (see (6.9)), applying Lemma 6.2, we obtain

$$\mathfrak{S}(\gamma) \in \boldsymbol{W}^{2,q_0}_{\mathrm{loc}}(\mathbb{R}^3), \quad \operatorname{div}\mathfrak{S}(\gamma) = \gamma, \quad \mathcal{N}(\gamma) \in W^{2,q_0}_{\mathrm{loc}}(\mathbb{R}^3), \quad \nabla\mathcal{N}(\gamma) = \mathfrak{S}(\gamma). \tag{6.19}$$

Since $\boldsymbol{e}_1 \wedge \mathfrak{S}(\gamma) = (0, -\mathfrak{S}_3(\gamma), \mathfrak{S}_2(\gamma))$ and because of the equation $\nabla\mathcal{N}(\gamma) = \mathfrak{S}(\gamma)$ in (6.19), we may conclude that

$$\operatorname{div}(\boldsymbol{e}_1 \wedge \mathfrak{S}(\gamma)) = -\partial_2\mathfrak{S}_3(\gamma) + \partial_3\mathfrak{S}_2(\gamma) = -\partial_2\partial_3\mathcal{N}(\gamma) + \partial_3\partial_2\mathcal{N}(\gamma) = 0. \tag{6.20}$$

Moreover, for $\mathbf{z} \in \mathbb{R}^3$, $1 \le j \le 3$,

$$(\boldsymbol{e}_1 \wedge \mathbf{z})\nabla\mathfrak{S}_j(\gamma)(\mathbf{z}) = -z_3\partial_2\mathfrak{S}_j(\gamma)(\mathbf{z}) + z_2\partial_3\mathfrak{S}_j(\gamma)(\mathbf{z}). \tag{6.21}$$

Put $\varphi(\mathbf{z}) := -z_3\partial_2\gamma(\mathbf{z}) + z_2\partial_3\gamma(\mathbf{z})$ for $\mathbf{z} \in \mathbb{R}^3$. Then with (6.21), the equation $\operatorname{div}\mathcal{S}(\gamma) = \gamma$ in (6.19), and the second and third equations in (6.20),

$$\begin{aligned}
\operatorname{div}_z((\boldsymbol{e}_1 \wedge \mathbf{z})\nabla\mathfrak{S}(\gamma)(\mathbf{z})) =& \sum_{j=1}^3 \partial z_j((\boldsymbol{e}_1 \wedge \mathbf{z})\nabla\mathfrak{S}_j(\gamma)(\mathbf{z}))\\
=& - z_3\partial_2\operatorname{div}_{\mathbf{z}}\mathfrak{S}(\gamma)(\mathbf{z}) + z_2\partial_3\operatorname{div}_{\mathbf{z}}\mathfrak{S}(\gamma)(\mathbf{z})\\
&+ \sum_{j=1}^3 (\partial z_j(-z_3)\partial_2\mathfrak{S}_j(\gamma)(\mathbf{z}) + \partial z_j(z_2)\partial_3\mathfrak{S}_j(\gamma)(\mathbf{z}))\\
=& \varphi(\mathbf{z}) - \partial_2\mathfrak{S}_3(\gamma)(\mathbf{z}) + \partial_3\mathfrak{S}_2(\gamma)(\mathbf{z}) = \varphi(\mathbf{z}).
\end{aligned}$$

Let $\psi \in C_0^\infty(\mathbb{R}^3)$. Then it follows that

$$\begin{aligned}\int_{\mathbb{R}^3} \nabla\psi(\rho \boldsymbol{e}_1 \wedge \mathfrak{S}(\gamma))\mathrm{d}\mathbf{x} =&0,\\ \int_{\mathbb{R}^3} \nabla\psi(\mathbf{z})[(\rho \boldsymbol{e}_1 \wedge \mathbf{z})\nabla\mathfrak{S}(\gamma)(\mathbf{z})]\mathrm{d}\mathbf{z} =& \int_{\mathbb{R}^3} \psi(\mathbf{z})(-\varphi(\mathbf{z}))\mathrm{d}\mathbf{z}.\end{aligned}$$

Again with (6.19),

$$\int_{\mathbb{R}^3} \nabla\psi \Delta\mathfrak{S}(\gamma)\mathrm{d}\mathbf{x} = \int_{\mathbb{R}^3} \nabla\Delta\psi \cdot \mathfrak{S}(\gamma)\mathrm{d}\mathbf{x} = -\int_{\mathbb{R}^3} \Delta\psi \cdot \gamma\mathrm{d}\mathbf{x} = \int_{\mathbb{R}^3} \psi \cdot (-\Delta\gamma)\mathrm{d}\mathbf{x},$$

and similarly,

$$\int_{\mathbb{R}^3} \nabla\psi(\tau\partial_1\mathfrak{S}(\gamma))\mathrm{d}\mathbf{x} = \int_{\mathbb{R}^3} \psi(-\tau\partial_1\gamma)\mathrm{d}\mathbf{x}.$$

Combining these equations, we get

$$\int_{\mathbb{R}^3} \nabla\psi \cdot \mathcal{L}\mathfrak{S}(\gamma)\mathrm{d}\mathbf{x} = \int_{\mathbb{R}^3} \psi(\varphi + \Delta\gamma - \tau\partial_1\gamma)\mathrm{d}\mathbf{x}.$$

Further from (6.18)

$$\int_{\mathbb{R}^3} \nabla\psi[\nabla\mathcal{P}(g + \varphi\mathbf{F}) - \nabla(\varphi\pi)]\mathrm{d}\mathbf{x} = \int_{\mathbb{R}^3} \psi(\varphi + \Delta\gamma - \tau\partial_1\gamma)\mathrm{d}\mathbf{x}. \tag{6.22}$$

Since $\gamma \in W^{2,q}(\mathbb{R}^3)$ for $q \in [1, p]$ and $\mathrm{supp}(\gamma) \subset B_S \backslash B_{S_1}$ due to (6.9)), it follows that $\varphi + \Delta\gamma - \tau\partial_1\gamma \in L^q(\mathbb{R}^3)$ for $q \in [1, p]$, so we may consider $\mathcal{N}(\varphi + \Delta\gamma - \tau\partial_1\gamma)$. Lemma 6.2 yields

$$\begin{aligned}\mathcal{N}(\varphi + \Delta\gamma - \tau\partial_1\gamma) \in& W^{2,q_0}_{\mathrm{loc}}(\mathbb{R}^3),\\ \Delta\mathcal{N}(\varphi + \Delta\gamma - \tau\partial_1\gamma) =& \varphi + \Delta\gamma - \tau\partial_1\gamma.\end{aligned}$$

Therefore from (6.22)

$$\int_{\mathbb{R}^3} \nabla\psi[\nabla\mathcal{P}(\mathbf{g} + \varphi F) - \nabla\mathcal{N}(\varphi + \Delta\varphi - \tau\partial_1\gamma) - \nabla(\varphi\pi)]\mathrm{d}\mathbf{x} = 0.$$

Lemma 6.1 now yields

$$Q := \mathcal{P}(\mathbf{g} + \varphi\mathbf{F}) - \mathcal{N}(\varphi + \Delta\gamma - \tau\partial_1\gamma) - \varphi\pi \in C^\infty(\mathbb{R}^3), \quad \Delta Q = 0. \tag{6.23}$$

Now we again apply Lemma 6.2. Since $\mathbf{g} + \varphi \cdot \mathbf{F} \in \boldsymbol{L}^{q_0}(\mathbb{R}^3)$, we have

$$\mathcal{P}(\mathbf{g} + \varphi\mathbf{F}) \in L^{(1/q_0-1/3)^{-1}}(\mathbb{R}^3).$$

Moreover $\varphi + \Delta\gamma - \tau\partial_1\gamma \in L^{q_0}(\mathbb{R}^3)$, so

$$\mathcal{N}(\varphi + \Delta\gamma - \tau\partial_1\gamma) \in L^{(1/q_0-2/3)^{-1}}(\mathbb{R}^3).$$

Since $q_0 \leq p$, and in view of our remarks at the beginning of this proof, we know that $\mathbf{u}|\overline{B_{S_0}}^c \in \boldsymbol{W}^{2,q_0}_{\text{loc}}(\overline{B_{S_0}}^c)$, $\pi|\overline{B_{S_0}}^c \in W^{1,q_0}_{\text{loc}}(\overline{B_{S_0}}^c)$, $\mathcal{L}(\mathbf{u}|\overline{B_{S_0}}^c) + \nabla(\pi|\overline{B_{S_0}}^c) = \mathbf{F}|\overline{B_{S_0}}^c$, $\operatorname{div}(\mathbf{u}|\overline{B_{S_0}}^c) = 0$, and $\mathbf{F} \in \boldsymbol{L}^{q_0}(\mathbb{R}^3)$. By the choice of $\mathbf{u}$ in Corollary 6.1, we have $\mathbf{u}|B_S^c \in L^6(B_S^c)^3$. Therefore, Theorem 5.18 yields there is $c_0 \in \mathbb{R}$ such that

$$\pi + c_0 \mid B_{2S}^c \in L^{3q_0/(3-q_0)}(B_{2S}^c) + L^3(B_{2S}^c).$$

But by (6.23),

$$Q - c_0 = \mathcal{P}(\mathbf{g} + \varphi\mathbf{F}) - \mathcal{N}(\varphi + \Delta\gamma - \tau\partial_1\gamma) - \varphi(\pi + c_0) + (\varphi - 1)c_0,$$

where $\operatorname{supp}(\varphi - 1) \subset B_S$ and $\operatorname{supp}(\varphi) \subset B_{S_1}^c$. We may conclude that

$$\begin{aligned} Q - c_0 \in & L^{(1/q_0-1/3)^{-1}}(\mathbb{R}^3) + L^{(1/q_0-2/3)^{-1}}(\mathbb{R}^3) + L^{q_0}(\mathbb{R}^3) \\ & + L^{3q_0/(3-q_0)}(\mathbb{R}^3) + L^3(\mathbb{R}^3). \end{aligned} \tag{6.24}$$

Let $\varepsilon \in (0, \infty)$, and let $(Q - c_0)_\varepsilon$ be the usual Friedrich's mollifier of $Q - c_0$ associated with ε.

Due to (6.23), (6.24) and by standard properties Friedrich's mollifier, the function $(Q - c_0)_\varepsilon$ is bounded and $\Delta(Q - c_0)_\varepsilon = 0$. Now Liouville's theorem yields $(Q - c_0)_\varepsilon = 0$. Since this is true for any $\varepsilon > 0$ and because $Q \in C^\infty(\mathbb{R}^3)$, we may conclude that $Q - c_0 = 0$, that is,

$$\varphi(\pi + c_0) = \mathcal{P}(g + \varphi F) - \mathcal{N}(\varphi + \Delta\gamma - \tau\partial_1\gamma) + (\varphi - 1)\, c_0,$$

hence

$$\pi + c_0 \mid B_S^c = \mathcal{P}(g + \varphi F) - \mathcal{N}(\varphi + \Delta\gamma - \tau\partial_1\gamma)|B_S^c, \tag{6.25}$$

where we used that $\operatorname{supp}(\varphi - 1) \subset B_S$ and $\varphi|B_S^c = 1$. Since $\operatorname{supp}(\mathbf{g}) \subset \overline{B_{S_1+3(S-S_1)/4}}$, we have

$$|\mathcal{P}(\mathbf{g})(\mathbf{x})| \leq c \cdot |\mathbf{x}|^{-2} \quad \text{for } \mathbf{x} \in B_S^c. \tag{6.26}$$

Due to the assumptions $A = 5/2$, $B \in (1/2, \infty)$ and because

$$\varphi \mathbf{F}|B_{S_1+(S-S_1)/4} = 0 \quad \text{and} \quad S_1 < S_1 + (S - S_1)/4 < S,$$

we get by [58, Theorem 3.2] or [119, Theorem 3.4] that

$$|\mathcal{P}(\varphi \mathbf{F})(\mathbf{x})| \le c|\mathbf{x}|^{-2} \quad \text{for } \mathbf{x} \in B_S^c. \tag{6.27}$$

Recalling that according to [119, Theorem 3.4 (iii)], a logarithmic factor should be added on the right-hand side of (6.27) in the case $A = 5/2$, $B = 1$. But this factor is superfluous. In fact, if the relation $|\mathbf{F}(\mathbf{z})| \le \gamma|\mathbf{z}|^{-A} s(\mathbf{z})^{-B}$ ($\mathbf{z} \in B_{S_1}^c$) is valid with $A = 5/2$, $B = 1$, it holds in the case $A = 5/2$, $B = 3/4$, too. But then [119, Theorem 3.4 (i), (iii)] yields that (6.27) holds as it is, without additional factor.

Define $\zeta(\mathbf{x}) = -x_3\gamma(\mathbf{x})$, $\tilde{\zeta}(\mathbf{x}) := x_2\gamma(\mathbf{x})$ for $\mathbf{x} \in \mathbb{R}^3$. Then $\text{supp}(\zeta) \cup \text{supp}(\tilde{\zeta}) \subset B_S \setminus \overline{B_{S_1}}$,

$$\begin{aligned} \zeta, \tilde{\zeta} &\in W^{2,q}(\mathbb{R}^3) \quad \text{for } q \in [1, p], \\ \varphi &= \partial_2\zeta + \partial_3\tilde{\zeta}. \end{aligned}$$

It follows with Lemma 6.2 that

$$\mathcal{N}(\varphi) = \partial_2\mathcal{N}(\zeta) + \partial_3\mathcal{N}(\tilde{\zeta}) = \mathfrak{S}_2(\zeta) + \mathfrak{S}_3(\tilde{\zeta}).$$

Similarly, since $\text{supp}(\gamma) \subset B_S \setminus \overline{B_{S_1}}$, $\gamma \in W^{2,q}(\mathbb{R}^3)$ for $q \in [1, p]$,

$$\mathcal{N}(\Delta\gamma - \tau\partial_1\gamma) = \sum_{k=1}^{3} \mathfrak{S}_k(\partial_k\gamma) - \tau\mathfrak{S}_1(\gamma).$$

Together

$$\mathcal{N}(\varphi + \Delta\gamma - \tau\partial_1\gamma) = \mathfrak{S}_2(\zeta) + \mathfrak{S}_3(\tilde{\zeta}) + \sum_{k=1}^{3} \mathfrak{S}_k(\partial_k\gamma) - \tau\mathfrak{S}_1(\gamma).$$

Since $\text{supp}(\zeta) \cup \text{supp}(\tilde{\zeta}) \cup \text{supp}(\gamma) \subset B_{S_1+3(S-S_1)/4}$, we may conclude that

$$|\mathcal{N}(\varphi + \Delta\gamma - \tau\partial_1\gamma)(\mathbf{x})| \le C|\mathbf{x}|^{-2} \quad \text{for } |\mathbf{x}| \in B_S^c. \tag{6.28}$$

Inequality (6.8) follows from (6.25)–(6.28). □

We remark that Theorem 6.2 remains valid if the assumptions on A and B are replaced by the conditions $A \ge 5/2$, $A + \min\{1, B\} > 1$, which are weaker than those in Corollary 6.1. This observation is made precise by the ensuing corollary. Its proof is obvious, but this modified version of Theorem 6.2 still is interesting because its requirements on A and B are closer to the ones in Corollary 6.1 than those stated in Theorem 6.2.

Corollary 6.2 *Let p, γ, S_1, S, A, B, F, u be given as in Corollary* 6.1, *but with the stronger assumptions $A \geq 5/2$, $A + \min\{1, B\} > 3$ on A and B. Let $\pi \in L^p_{\text{loc}}(\overline{\mathcal{D}})$ such that* (6.6) *holds. Then there is $c_0 \in \mathbb{R}$ such that inequality* (6.8) *is valid.* □

Proof Put $B' := A - 5/2 + \min\{1, B\}$. Since $A + \min\{1, B\} > 3$, we have $B' \in (1/2, \infty)$. Moreover, since $A \geq 5/2$, we find for $\mathbf{z} \in B^c_{S_1}$ that

$$|\mathbf{F}(\mathbf{z})| \leq \gamma C(S_1, A)|\mathbf{z}|^{-5/2} s(\mathbf{z})^{-A+5/2-B} \leq \gamma C(S_1, A)|\mathbf{z}|^{-5/2} s(z)^{-B'}.$$

Thus the assumptions of Theorem 6.2 are satisfied with B replaced by B' and with a modified parameter γ. This implies the conclusion of Theorem 6.2. □

6.5 Formulation of the Problem with Artificial Boundary Conditions

Let us recall that we defined $\mathcal{D}_R = B_R \cap \overline{\mathcal{D}}$. We introduce the subspace $\boldsymbol{W}_R$ of $\boldsymbol{H}^1(\mathcal{D}_R)$ denoting

$$\boldsymbol{W}_R := \{\mathbf{v} \in \boldsymbol{H}^1(\mathcal{D}_R) \colon \mathbf{v}|\partial\mathcal{D} = 0\},$$

where $\mathbf{v}|\partial\mathcal{D}$ means the trace of $\mathbf{v}$ on $\partial\mathcal{D}$.

Lemma 6.4 ([30, Lemma 4.1]) *The estimate*

$$\|\mathbf{u}\|_2 \leq C\,(R\|\nabla\mathbf{u}\|_2 + R^{1/2}\,\|\mathbf{u} \mid \partial B_R\|_2)$$

holds for $R \in (0, \infty)$ with $\overline{\mathcal{D}}^c \subset B_R$ and for $\mathbf{u} \in \boldsymbol{W}_R$. □

We introduce an inner product $(\cdot, \cdot)^{(R)}$ in $\boldsymbol{W}_R$ by defining

$$(\mathbf{v}, \mathbf{w})^{(R)} = \int_{\mathcal{D}_R} \nabla\mathbf{v}\nabla\mathbf{w}\, d\mathbf{x} + \int_{\partial B_R} (\tau/2)\mathbf{v}\mathbf{w} d o_{\mathbf{x}} \quad \text{for } \mathbf{v}, \mathbf{w} \in \boldsymbol{W}_R.$$

The space $\boldsymbol{W}_R$ equipped with this inner product is a Hilbert space. The norm generated by this scalar product $(\cdot, \cdot)^{(R)}$ is denoted by $|\cdot|^{(R)}$, that is,

$$|\mathbf{v}|^{(R)} := (\|\nabla\mathbf{v}\|_2^2 + (\tau/2)\|\mathbf{v} \mid \partial B_R\|_2^2)^{1/2} \quad \text{for } \mathbf{v} \in \boldsymbol{W}_R.$$

We define the bilinear forms

$$
\begin{aligned}
a_R &: \boldsymbol{H}^1(\mathcal{D}_R) \times \boldsymbol{H}^1(\mathcal{D}_R) \to \mathbb{R},\\
\beta_R &: \boldsymbol{H}^1(\mathcal{D}_R) \times L^2(\mathcal{D}_R) \to \mathbb{R},\\
\delta_R &: \boldsymbol{H}^1(\mathcal{D}_R) \times \boldsymbol{H}^1(\mathcal{D}_R)^3 \to \mathbb{R},
\end{aligned}
$$

$$
a_R(\mathbf{u}, \mathbf{w}) := \int_{\mathcal{D}_R} [\nabla \mathbf{u} \nabla \mathbf{w} + \tau \partial_1 \mathbf{u}\mathbf{w}] \mathrm{d}\mathbf{x} + \frac{\tau}{2} \int_{\partial B_R} (\mathbf{u}(\mathbf{x})\mathbf{w}(\mathbf{x})) \left(1 - \frac{x_1}{R}\right) \mathrm{d}o_{\mathbf{x}},
$$

$$
\beta_R(\mathbf{w}, \sigma) := - \int_{\mathcal{D}_R} (\operatorname{div} \mathbf{w}) \sigma \, \mathrm{d}\mathbf{x},
$$

$$
\delta_R(\mathbf{u}, \mathbf{w}) := \int_{\mathcal{D}_R} [-((\sqrt{\varrho}\boldsymbol{e}_1 \wedge \mathbf{x})\nabla)\mathbf{u}(\mathbf{x}) + (\sqrt{\varrho}\boldsymbol{e}_1 \wedge \mathbf{u}(\mathbf{x}))] \mathbf{w}(\mathbf{x}) \mathrm{d}\mathbf{x}
$$

for $\mathbf{u}, \mathbf{w} \in \boldsymbol{H}^1(\mathcal{D}_R)$, $\sigma \in L^2(\mathcal{D}_R)$, $R \in (0, \infty)$ with $\overline{\mathcal{D}^c} \subset B_R$.

Lemma 6.5 *Let $R \in (0, \infty)$ with $\overline{\mathcal{D}}^c \subset B_R$. Then*

$$
|a_R(\mathbf{u}, \mathbf{w}) + \delta_R(\mathbf{u}, \mathbf{w})| \leq C(R) |\mathbf{u}|^{(R)} |\mathbf{w}|^{(R)}
$$

for $\mathbf{u}, \mathbf{w} \in \boldsymbol{H}^1(\mathcal{D}_R)$. □

Proof The proof of Lemma 6.5 follows directly Lemma 6.4. □

The key observation in this section is stated in the following lemma, which is the basis of the theory presented in this section.

Lemma 6.6 *Let $R \in (0, \infty)$ with $\overline{\mathcal{D}}^c \subset B_R$, and let $\mathbf{w} \in \boldsymbol{W}_R$. Then the equation $(|\mathbf{w}|^{(R)})^2 = a_R(\mathbf{w}, \mathbf{w}) + \delta_R(\mathbf{w}, \mathbf{w})$ holds.* □

Proof Using the definitions $a_R(\cdot, \cdot)$, $\delta_R(\cdot, \cdot)$, we get

$$
\begin{aligned}
a_R(\mathbf{w}, \mathbf{w}) + \delta_R(\mathbf{w}, \mathbf{w}) &= \int_{\mathcal{D}_R} \left[|\nabla \mathbf{w}|^2 + \tau \partial_1 \left(\frac{|\mathbf{w}|^2}{2} \right) - (\sqrt{\varrho}\boldsymbol{e}_1 \wedge \mathbf{x}) \nabla \left(\frac{|\mathbf{w}|^2}{2} \right) \right] \mathrm{d}\mathbf{x} \\
&\quad + \frac{\tau}{2} \int_{\partial B_R} |\mathbf{w}(\mathbf{x})|^2 \left(1 - \frac{x_1}{R}\right) \mathrm{d}o_{\mathbf{x}} \\
&= \int_{\mathcal{D}_R} |\nabla \mathbf{w}|^2 \mathrm{d}\mathbf{x} \\
&\quad + \int_{\partial B_R} \left(\frac{\tau}{2} |\mathbf{w}(\mathbf{x})|^2 \frac{x_1}{R} - \frac{1}{2} (\sqrt{\varrho}\boldsymbol{e}_1 \wedge \mathbf{x}) \frac{\mathbf{x}}{R} |\mathbf{w}(\mathbf{x})|^2 \right) \mathrm{d}o_{\mathbf{x}} \\
&\quad + \frac{\tau}{2} \int_{\partial B_R} |\mathbf{w}(\mathbf{x})|^2 \left(1 - \frac{x_1}{R}\right) \mathrm{d}o_{\mathbf{x}} \\
&= \int_{\mathcal{D}_R} |\nabla \mathbf{w}|^2 \mathrm{d}\mathbf{x} + \frac{\tau}{2} \int_{\partial B_R} |\mathbf{w}(\mathbf{x})|^2 \mathrm{d}O_{\mathbf{x}} = (|\mathbf{w}|^{(R)})^2.
\end{aligned}
$$

We applied that

$$(\boldsymbol{\omega} \wedge \mathbf{x}) \cdot \mathbf{x} = 0 \quad \text{for } \mathbf{x}, \boldsymbol{\omega} \in \mathbb{R}^3.$$

As in [33], we obtain that the bilinear form β_R is stable: □

Theorem 6.3 ([33, Corollary 4.3]) *Let $R > 0$ with $\overline{\mathcal{D}}^c \subset B_R$. Then*

$$\inf_{\substack{\rho \in L^2(\mathcal{D}_R) \\ \rho \neq 0}} \sup_{\substack{\mathbf{v} \in \boldsymbol{W}_R \\ \mathbf{v} \neq 0}} \frac{\beta_R(\mathbf{v}, \rho)}{|\mathbf{v}|^{(R)} \|\rho\|_2} \geq C(R).$$

We note that functions from $W^{1,1}_{\text{loc}}(\overline{\mathcal{D}})$ with L^2-integrable gradient are L^2-integrable on truncated exterior domains:

Lemma 6.7 ([98], Lemma II.6.1) *Let $\mathbf{w} \in \boldsymbol{W}^{1,1}_{\text{loc}}(\overline{\mathcal{D}})$ with $\nabla \mathbf{w} \in \boldsymbol{L}^2(\overline{\mathcal{D}})$, and let $R \in (0, \infty)$ with $\overline{\mathcal{D}}^c \subset B_R$. Then $\mathbf{w}_{|\mathcal{D}_R} \in \boldsymbol{L}^2(\mathcal{D}_R)$. In particular the trace of $\mathbf{w}$ on $\partial\mathcal{D}$ is well defined.* □

The preceding lemma is implicitly used in the ensuing theorem, where we introduce an extension operator $\mathfrak{E}\colon \boldsymbol{H}^{1/2}(\partial\mathcal{D})^3 \mapsto \boldsymbol{W}^{1,1}_{\text{loc}}(\overline{\mathcal{D}})^3$ such that $\operatorname{div} \mathfrak{E}(\boldsymbol{b}) = 0$.

Theorem 6.4 ([98], Exercise III.3.8) *There is an operator $\mathfrak{E}$ from $\boldsymbol{H}^{1/2}(\partial\mathcal{D})$ into $\boldsymbol{W}^{1,1}_{\text{loc}}(\overline{\mathcal{D}}^c)$ satisfying the relations $\nabla\mathfrak{E}(\boldsymbol{b}) \in \boldsymbol{L}^2(\overline{\mathcal{D}})^9$, $\mathfrak{E}(\boldsymbol{b}) \mid \partial\mathcal{D} = \boldsymbol{b}$, and $\operatorname{div} \mathfrak{E}(\boldsymbol{b}) = 0$ for $\boldsymbol{b} \in \boldsymbol{H}^{1/2}(\partial\mathcal{D})$.* □

In view of Lemmas 6.5 and 6.6, Theorems 6.4 and 6.3, the theory of mixed variational problems yields:

Theorem 6.5 *Let $S > 0$ with $\overline{\mathcal{D}}^c \subset B_S$, $R \in [2S, \infty)$, $\mathbf{F} \in \boldsymbol{L}^{6/5}(\mathcal{D}_R)$, $\mathbf{b} \in \boldsymbol{H}^{1/2}(\partial\mathcal{D})$. Then there is a uniquely determined pair of functions $(\widetilde{\boldsymbol{V}}, P) = (\widetilde{\boldsymbol{V}}(R, \mathbf{F}, \boldsymbol{b}), P(R, \mathbf{F}, \boldsymbol{b})) \in \boldsymbol{W}_R \times L^2(\mathcal{D}_R)$ such that*

$$\begin{aligned} a_R(\widetilde{\boldsymbol{V}}, g) + \delta_R(\widetilde{\boldsymbol{V}}, g) + \beta_R(\mathbf{g}, P) = \int_{\mathcal{D}_R} \mathbf{F} \cdot \mathbf{g} \, d\mathbf{x} - a_R(\mathfrak{E}(\mathbf{b})|\mathcal{D}_R, \mathbf{g}) \\ - \delta_R(\mathfrak{E}(\mathbf{b})|\mathcal{D}_R, \mathbf{g}) \quad \text{for } \mathbf{g} \in \boldsymbol{W}_R, \end{aligned} \tag{6.29}$$

$$\beta_R(\widetilde{\boldsymbol{V}}, \sigma) = 0 \quad \text{for } \sigma \in L^2(\mathcal{D}_R), \tag{6.30}$$

where the operator $\mathfrak{E}$ was introduced in Theorem 6.4. □

Let us interpret variational problem (6.29), (6.30) as a boundary value problem. Define the expression used in the boundary condition on the artificial boundary ∂B_R:

$$\mathcal{L}_R(\mathbf{u},\pi)(\mathbf{x}) := \left(\sum_{j=1}^{3} \partial_j u_k(\mathbf{x}) \frac{x_j}{R} - \pi(\mathbf{x}) \frac{x_k}{R} + \frac{\tau}{2}\left(1 - \frac{x_1}{R}\right) u_k(\mathbf{x}) \right)_{1 \le k \le 3}$$

for $\mathbf{x} \in \partial B_R$, $R \in (0,\infty)$ with $\overline{\mathcal{D}}^c \subset B_R$, $\mathbf{u} \in W^{2,6/5}(\mathcal{D}_R)$, $\pi \in W^{1,6/5}(\mathcal{D}_R)$.

Lemma 6.8 *Assume that $\mathcal{D}^c$ is C^2-bounded. Let $S \in (0,\infty)$ with $\overline{\mathcal{D}}^c \subset B_S$, $R \in [2S,\infty)$, $\mathbf{F} \in \boldsymbol{L}^{6/5}(\mathcal{D}_R)$, and $\mathbf{b} \in \boldsymbol{W}^{7/6,6/5}(\partial\mathcal{D})$. Put $V := \widetilde{V}(R,\mathbf{F},\boldsymbol{b}) + \mathfrak{E}(\boldsymbol{b})|\mathcal{D}_R$, with $V(R,\mathbf{F},\boldsymbol{b})$ from Theorem* 6.5 *and $\mathfrak{E}(\boldsymbol{b})$ from Theorem* 6.4. *Suppose that $\mathbf{V} \in \boldsymbol{W}^{2,6/5}(\mathcal{D}_R)$ and $P = P(R,\mathbf{F},\boldsymbol{b}) \in W^{1,6/5}(\mathcal{D}_R)$, with $P(R,\mathbf{F},\boldsymbol{b})$ also introduced in Theorem* 6.5. *Then*

$$\begin{aligned} -\Delta\mathbf{V}(\mathbf{z}) + (\tau\boldsymbol{e}_1 - \sqrt{\varrho}\boldsymbol{e}_1 \wedge \mathbf{z})\nabla\mathbf{V}(\mathbf{z}) + \sqrt{\varrho}\boldsymbol{e}_1 \wedge \mathbf{V}(\mathbf{z}) + \nabla P(\mathbf{z}) &= \mathbf{F}(\mathbf{z}), \\ \operatorname{div}\mathbf{V}(\mathbf{z}) &= 0 \end{aligned} \qquad (6.31)$$

for $\mathbf{z} \in \mathcal{D}_R$, and $\mathbf{V}|\partial\mathcal{D} = \mathbf{b}$, $\mathcal{L}_R(\mathbf{V},P) = 0$. □

The proof of Lemma 6.8 is standard.

This lemma means that a solution of variational problem (6.29), (6.30) may be considered as a weak solution of the modified Oseen system with rotation in $\mathcal{D}_R$, under the Dirichlet boundary condition on $\partial\mathcal{D}$ and under the artificial boundary condition $\mathcal{L}_R(\mathbf{V},P) = 0$ on ∂B_R. The solution of (6.29), (6.30) will be now compared to the exterior modified Oseen flow introduced in Corollary 6.1:

Theorem 6.6 *Suppose that $\mathcal{D}^c$ is C^2-bounded. Let $\gamma, S_1 \in (0,\infty)$ with $\overline{\mathcal{D}}^c \subset B_{S_1}$, $A \in [5/2,\infty)$, $B \in \mathbb{R}$ with $A + \min\{1,B\} > 3$. Let $\mathbf{F}\colon \overline{\mathcal{D}} \mapsto \mathbb{R}^3$ be measurable with $\mathbf{F}|\mathcal{D}_{S_1} \in \boldsymbol{L}^{6/5}(\mathcal{D}_{S_1})$ and $|\mathbf{F}(z)| \le \gamma|\mathbf{z}|^{-A}s(\mathbf{z})^{-B}$ for $\mathbf{z} \in B^c_{S_1}$.*

Let $\mathbf{b} \in \boldsymbol{W}^{7/6,6/5}(\partial\mathcal{D})$, $\mathbf{u} \in \boldsymbol{W}^{1,1}_{\text{loc}}(\overline{\mathcal{D}}) \cap \boldsymbol{L}^6(\overline{\mathcal{D}})$ such that $\nabla\mathbf{u} \in L^2(\overline{\mathcal{D}})^9$, $\operatorname{div}\mathbf{u} = 0$, $\mathbf{u} \mid \partial\mathcal{D} = \mathbf{b}$, and Eq. (6.5) *is satisfied.*

For $R \in [2S_1,\infty)$, put $V_R := \widetilde{V}(R,\mathbf{F},\mathbf{b}) + \mathfrak{E}(\mathbf{b})$, with $\mathfrak{E}(\mathbf{b})$ from Theorem 6.4, *and $\widetilde{V}(R,F,b)$ from Theorem* 6.5. *Then*

$$|\mathbf{u}|_{\mathcal{D}_R} - V_R|^{(R)} \le CR^{-1} \quad \textit{for } R \in [2S,\infty).$$

We note that since $W^{2,6/5}(\mathcal{D}) \subset H^1(\mathcal{D})$ by a Sobolev inequality, we have $W^{7/6,6/5}(\partial\mathcal{D}) \subset H^{1/2}(\partial\mathcal{D})$, as follows with the usual lifting and trace properties. As a consequence, $\boldsymbol{b} \in \boldsymbol{H}^{1/2}(\partial\mathcal{D})$, so the term $\mathfrak{E}(b)$ is well defined. We further remark that by Corollary 6.1 with $p = 6/5$, the function $\mathbf{F}$ may be considered as a bounded linear functional on $\widehat{\boldsymbol{H}}_0^{1,2}(\overline{\mathcal{D}})$. Therefore, as explained in Remark 6.2, a function $\mathbf{u}$ with properties as stated in Theorem 6.6 does in fact exist.

Proof of Theorem 6.6 All conditions in Corollary 6.1 are verified if γ, S_1, A, B, $\mathbf{F}$, $\mathbf{u}$ are given as in Theorem 6.6, and if $p = 6/5$ and $S = 2S_1$. Note in this respect that the conditions on $\mathbf{u}$ in Theorem 6.6 obviously imply $\mathbf{u} \in \boldsymbol{W}^{1,6/5}_{\text{loc}}(\overline{\mathcal{D}})$. Corollary 6.1 now yields that $\mathbf{F} \in \boldsymbol{L}^{6/5}(\overline{\mathcal{D}})$ and that the function $\mathbf{u}$ satisfies inequalities (6.3) and (4.23) with $S = 2S_1$.

On the other hand, since $\mathbf{u} \in \boldsymbol{W}^{1,6/5}_{\text{loc}}(\overline{\mathcal{D}})$, the function $\boldsymbol{G}$ already considered in the proof of Corollary 6.1 (see (6.7)) belongs to $\boldsymbol{L}^{6/5}_{\text{loc}}(\overline{\mathcal{D}})$. Therefore, by interior regularity of solutions to the Stokes system (see [98, Theorem IV.4.1]), we may deduce from Eqs. (6.5) and $\operatorname{div} \mathbf{u} = 0$ that $\mathbf{u} \in \boldsymbol{W}^{2,6/5}_{\text{loc}}(\overline{\mathcal{D}})$ and that there is $\pi \in W^{1,6/5}_{\text{loc}}(\overline{\mathcal{D}})$ with $\mathcal{L}(\mathbf{u}) + \nabla\pi = \mathbf{F}$. In particular the pair $(\mathbf{u}, \pi)$ verifies (6.6). In view of our assumptions on A and B, we thus see that the requirements in Corollary 6.2 are fulfilled for γ, S_1, A, B, $\mathbf{F}$, $\mathbf{u}$ as in Theorem 6.6 and for $p = 6/5$ and $S = 2S_1$. As a consequence, Corollary 6.2 yields that there is $c_0 \in \mathbb{R}$ such that (6.8) holds with $S = 2S_1$.

Take $R \in [2S_1, \infty)$. Since $\mathbf{u} \in \boldsymbol{W}^{2,6/5}_{\text{loc}}(\overline{\mathcal{D}})$, we have $\mathbf{u}|\partial B_R \in \boldsymbol{W}^{7/5,6/5}(\partial B_R)$. Combining this relation with the assumption $\mathbf{b} \in \boldsymbol{W}^{7/5,6/5}(\partial\mathcal{D})$ and the boundary condition $\mathbf{u}|\partial\mathcal{D} = \mathbf{b}$, we get $\mathbf{u}|\partial\mathcal{D}_R \in \boldsymbol{W}^{7/5,6/5}(\partial\mathcal{D}_R)$. Moreover, our requirements on $\mathbf{u}$ yield that $\mathbf{u}|\mathcal{D}_R \in \boldsymbol{W}^{1,6/5}(\mathcal{D}_R)$. Since $\mathbf{F} \in \boldsymbol{L}^{6/5}(\overline{\mathcal{D}})$, as already mentioned, we get $\mathbf{G}|\mathcal{D}_R \in \boldsymbol{L}^{6/5}(\mathcal{D}_R)$, with G from (6.7). Recalling that $\mathcal{D}$ is supposed to be C^2-bounded, we may now apply the result in [98, Lemma IV.6.1] on boundary regularity of solutions to the Stokes system. This reference yields that $\mathbf{u}|\mathcal{D}_R \in \boldsymbol{W}^{2,6/5}(\mathcal{D}_R)$, $\pi|\mathcal{D}_R \in W^{1,6/5}(\mathcal{D}_R)$ and that the pair $(\mathbf{u}, \pi)$ solves (4.16).

Let $P_R := P(R, \mathbf{F}, b)$ be given as in Theorem 6.5, and put $\mathbf{w} := \mathbf{u} - \mathbf{V}_R$, $\kappa := \pi - P_R$, and let $\mathbf{g} \in \boldsymbol{W}_R$. Let us note that by Theorem 6.5, we have

$$a_R(\mathbf{V}_R, \mathbf{g}) + \delta_R(\mathbf{V}_R, \mathbf{g}) + \beta_R(\mathbf{g}, P_R) = \int_{\mathcal{D}_R} F \cdot \mathbf{g}\, d\mathbf{x}.$$

Thus

$$\begin{aligned}
&a_R(\mathbf{w}, \mathbf{g}) + \delta_R(\mathbf{w}, \mathbf{g}) + \beta_R(\mathbf{g}, \kappa) \\
&= a_R(\mathbf{u}|_{\mathcal{D}_R}, \mathbf{g}) + \delta_R(\mathbf{u}|_{\mathcal{D}_R}, \mathbf{g}) + \beta_R(\mathbf{g}, \pi|_{\mathcal{D}_R}) \\
&- \underbrace{(a_R(\mathbf{V}_R, \mathbf{g}) + \delta_R(\mathbf{V}_R, \mathbf{g}) + \beta_R(\mathbf{g}, P_R))}_{=\int_{\mathcal{D}_R} \mathbf{F}\cdot\mathbf{g} d\mathbf{x}} \\
&= \int_{\mathcal{D}_R} (\nabla\mathbf{u} \cdot \nabla\mathbf{g} + \tau\partial_1\mathbf{u} \\
&\cdot \mathbf{g} - (\sqrt{\varrho}\boldsymbol{e}_1 \wedge \mathbf{x}) \cdot \nabla\mathbf{u} \cdot \mathbf{g} + (\sqrt{\varrho}\boldsymbol{e}_1 \wedge \mathbf{u}) \cdot \mathbf{g} - \pi \operatorname{div} \mathbf{g} - \mathbf{F} \cdot \mathbf{g}) d\mathbf{x} \\
&+ \frac{\tau}{2} \int_{\partial B_R} \mathbf{u}(\mathbf{x}) \cdot \mathbf{g}(\mathbf{x}) \left(1 - \frac{x_1}{R}\right) d o_{\mathbf{x}}
\end{aligned}$$

$$= \int_{\mathcal{D}_R} [-\Delta \mathbf{u} \cdot \mathbf{g} + \tau \partial_1 \mathbf{u} \cdot \mathbf{g} - (\sqrt{\varrho} \boldsymbol{e}_1 \wedge \mathbf{x}) \cdot \nabla \mathbf{u} \cdot \mathbf{g} + (\sqrt{\varrho} \boldsymbol{e}_1 \wedge \mathbf{u}) \cdot \mathbf{g}$$

$$+ \nabla \pi \mathbf{g} - F \cdot \mathbf{g}] \mathrm{d}\mathbf{x}$$

$$\underbrace{+ \int_{\partial B_R} \left(\sum_{j,k=1}^{3} \left[\partial_j u_k(\mathbf{x}) g_k(\mathbf{x}) \frac{x_j}{R} - \pi(\mathbf{x}) \delta_{jk} g_k(\mathbf{x}) \frac{x_j}{R} \right] + \frac{\tau}{2} \mathbf{u}(\mathbf{x}) \cdot \mathbf{g}(\mathbf{x}) \left(1 - \frac{x_1}{R}\right) \right) \mathrm{d}o_{\mathbf{x}}.}_{= \int_{\partial B_R} \mathcal{L}_R(\mathbf{u}, \pi) \mathbf{g} \mathrm{d}o}$$

Since the pair $(\mathbf{u}, \pi)$ solves (4.16), we now get

$$a_R(\mathbf{w}, \mathbf{g}) + \delta_R(\mathbf{w}, \mathbf{g}) + \beta_R(\mathbf{g}, \kappa) = \int_{\partial B_R} \mathcal{L}_R(\mathbf{u}, \pi)(\mathbf{x}) \mathbf{g}(\mathbf{x}) \mathrm{d}o_{\mathbf{x}}. \tag{6.32}$$

Let $c \in \mathbb{R}$ be an arbitrary constant. For $\mathbf{g} := \mathbf{w}$ we get with Lemma 6.6 that

$$(|\mathbf{w}|^{(R)})^2 = a_R(\mathbf{w}, \mathbf{w}) + \delta_R(\mathbf{w}, \mathbf{w}) + \beta_R(\mathbf{w}, \kappa) = \int_{\partial B_R} \mathcal{L}_R(\mathbf{u}, \pi + c)(\mathbf{x}) \cdot \mathbf{w}(\mathbf{x}) \mathrm{d}o_{\mathbf{x}}, \tag{6.33}$$

because by the assumptions on $\mathbf{u}$ and Theorems 6.4 and 6.5,

$$\int_{\partial B_R} \left[\sum_{j,k=1}^{3} c \delta_{jk} w_k(\mathbf{x}) \frac{x_j}{R} \right] \mathrm{d}o_{\mathbf{x}} = \int_{\partial \mathcal{D}} c \mathbf{w} \boldsymbol{n} \, \mathrm{d}o_{\mathbf{x}} + \int_{\mathcal{D}_R} c \operatorname{div} \mathbf{w} \, \mathrm{d}\mathbf{x} = 0,$$

where $\boldsymbol{n}$ denotes the outward unit normal to $\mathcal{D}^c$. Let c_0 be the constant introduced above as part of estimate (6.8). Because

$$\begin{aligned} \int_{\partial B_R} \mathcal{L}_R(\mathbf{u}, \pi + c_0)(\mathbf{x}) \cdot \mathbf{w}(\mathbf{x}) \mathrm{d}o_{\mathbf{x}} &\leq \|\mathcal{L}_R(\mathbf{u}, \pi + c_0)\|_2 \|\mathbf{w}|_{\partial B_R}\|_2 \\ &\leq C \|\mathcal{L}_R(\mathbf{u}, \pi + c_0)\|_2 \cdot |\mathbf{w}|^{(R)}, \end{aligned}$$

we get from (6.33)

$$|\mathbf{w}|^{(R)} \leq C \|\mathcal{L}_R(\mathbf{u}, \pi + c_0)\|_2.$$

The last step is estimation: $\|\mathcal{L}_R(\mathbf{u}, \pi + c_0)\|_2 \leq C \, R^{-1}$. We start by observing that

$$\begin{aligned} &\|\mathcal{L}_R(\mathbf{u}, \pi + c_0)\|_2 \\ &\leq C \left[\|\nabla \mathbf{u}|_{\partial B_R}\|_2 + \|[\pi(\mathbf{x}) + c_0]|_{\partial B_R}\|_2 + \left(\int_{\partial B_R} \left(1 - \frac{x_1}{R}\right)^2 |\mathbf{u}(\mathbf{x})|^2 \mathrm{d}o_{\mathbf{x}} \right)^{1/2} \right]. \end{aligned}$$

As explained above, inequalities (6.3), (4.23), and (6.8) are valid with $S = 2S_1$. According to (6.3) and (6.8), we have $|\mathbf{u}(\mathbf{x})| \le C(|\mathbf{x}|s(\mathbf{x}))^{-1}$, and $|\pi(\mathbf{x}) + c_0| \le C|\mathbf{x}|^{-2}$ for $\mathbf{x} \in B^c_{2S_1}$. Inequality (4.23) yields $|\nabla\mathbf{u}(\mathbf{x})| \le C|\mathbf{x}|^{-3/2}s(\mathbf{x})^{-B'}$ for $\mathbf{x}$ as before, with $B' := 3/2 - \max\{0, 7/2 - A - B\}$. If $B \ge 1$, we recall that $A \ge 5/2$, getting $A + B \ge 7/2$, hence $B' = 3/2$. On the other hand, if $B < 1$, then $\min\{1, B\} = B$, so that the assumption $A + \min\{1, B\} > 3$ becomes $A + B > 3$, hence $B' > 1$. Thus we get in any case that $B' > 1 > 1/2$. In view of these observations, and with Lemma 6.3, we obtain

$$\begin{aligned}
\|\mathcal{L}_R(\mathbf{u}, \pi + c_0)\|_2 \le& C\Bigg[\Big(\int_{\partial B_R} |\mathbf{x}|^{-3}s(\mathbf{x})^{-2B'}\mathrm{d}o_{\mathbf{x}}\Big)^{1/2} \\
&+ \Big(\int_{\partial B_R} |\pi(\mathbf{x}) + c_0|^2\mathrm{d}o_{\mathbf{x}}\Big)^{1/2} \\
&+ \Big(\frac{1}{R^2}\int_{\partial B_R} (|\mathbf{x}| - x_1)^2|\mathbf{u}(\mathbf{x})|^2\mathrm{d}o_{\mathbf{x}}\Big)^{1/2}\Bigg] \\
\le& C\Bigg[\Big(\frac{1}{R^3}\int_{\partial B_R} s(\mathbf{x})^{-2B'}\mathrm{d}o_{\mathbf{x}}\Big)^{1/2} + \Big(\frac{1}{R^4}\int_{\partial B_R} 1\mathrm{d}o_{\mathbf{x}}\Big)^{1/2} \\
&+ \Big(\frac{1}{R^2}\int_{\partial B_R} s(\mathbf{x})^2(|\mathbf{x}|s(\mathbf{x}))^{-2}\mathrm{d}o_{\mathbf{x}}\Big)^{1/2}\Bigg] \\
\le& C\Bigg[\Big(\frac{1}{R^2}\Big)^{1/2} + \Big(\frac{1}{R^2}\Big)^{1/2} + \Big(\frac{1}{R^4}\int_{\partial B_R} 1\mathrm{d}o_{\mathbf{x}}\Big)^{1/2}\Bigg] \le CR^{-1}.
\end{aligned}$$

This completes the proof of Theorem 6.6. □

Chapter 7
The Oseen-Type Problem: A Weak Solution in Anisotropic L^2-Spaces

7.1 Introduction

In this chapter we investigate again the Oseen-type problem in an exterior domain $\mathcal{D} \subset \mathbb{R}^3$ or the whole space $\mathbb{R}^3$

$$S_{\omega,\nu}\mathbf{u} + k\partial_3\mathbf{u} + \nabla p = \mathbf{f} \quad \text{in } \mathcal{D} \tag{7.1}$$

$$\operatorname{div}\mathbf{u} = g \quad \text{in } \mathcal{D} \tag{7.2}$$

$$\mathbf{u} \to \mathbf{0} \quad \text{as } |\mathbf{x}| \to \infty \tag{7.3}$$

with the Dirichlet boundary condition, if $\mathcal{D} \neq \mathbb{R}^3$

$$\mathbf{u} = \omega(\mathbf{e}_3 \wedge \mathbf{x}) - k\mathbf{e}_3 \quad \text{on } \partial\mathcal{D} \tag{7.4}$$

by means of a pure variational approach. Let us recall that $S_{\omega,\nu}\mathbf{u} = -\nu\Delta\mathbf{u} - \boldsymbol{\omega} \wedge \mathbf{x} \cdot \nabla\mathbf{u} + \boldsymbol{\omega} \wedge \mathbf{u}$. The condition (7.3) is assumed to be satisfied if $\mathbf{u}$ belongs to an appropriate weighted Sobolev space.

The fundamental tensor (similarly as the fundamental tensor of the classical Oseen problem) exhibits the anisotropic behavior at infinity. To reflect this property and to test the solutions to (7.1), (7.2), we introduce the following weight functions:

$$\eta_\beta^\alpha(\mathbf{x}) = \eta_{\beta,\varepsilon}^{\alpha,\delta}(\mathbf{x}) = (1+\delta r)^\alpha (1+\varepsilon s)^\beta,$$

with $r = r(\mathbf{x}) = |\mathbf{x}|$, $s = s(\mathbf{x}) = r - x_3$, α, β, ε, δ being from real numbers (with $\varepsilon, \delta > 0$). Let $\mathbf{L}^2_{\alpha,\beta}(\mathcal{D})$ denote the corresponding weighted space $\mathbf{L}^2_{\eta_\beta^\alpha}(\mathcal{D}) = L^2_{\eta_\beta^\alpha}(\mathcal{D})^3$ with the norm $\|\cdot\|_{2,\alpha,\beta}$. The weight functions η_β^α belong to the Muckenhoupt class if $-1 < \beta < 1$ and $-3 < \alpha + \beta < 3$.

Š. Nečasová et al., *Mathematical Theory of a Fluid Flow Around a Rotating and Translating Body*, Springer Monographs in Mathematics,
https://doi.org/10.1007/978-3-031-84120-0_7

We will use the following notations: $\mathbf{H}^1(\mathcal{D}; \eta_{\beta_0}^{\alpha_0}, \eta_{\beta_1}^{\alpha_1})$ as the set of $\mathbf{u} \in \mathbf{L}^2_{\alpha_0,\beta_0}(\mathcal{D})$ with all the weak derivatives $\partial_i \mathbf{u} \in \mathbf{L}^2_{\alpha_1,\beta_1}(\mathcal{D})$, the corresponding norm is given by

$$\|\mathbf{u}\|_{\mathbf{H}^1(\mathcal{D};\eta_{\beta_0}^{\alpha_0},\eta_{\beta_1}^{\alpha_1})} = \left(\int_{\mathcal{D}} |\mathbf{u}|^2 \eta_{\beta_0}^{\alpha_0} \mathrm{d}\mathbf{x} + \int_{\mathcal{D}} |\nabla \mathbf{u}|^2 \eta_{\beta_1}^{\alpha_1} \mathrm{d}\mathbf{x}\right)^{1/2},$$

and the closure of $\mathbf{C}_0^\infty(\mathcal{D})$ in this space is denoted by $\overset{\circ}{\mathbf{H}}{}^1(\mathcal{D}; \eta_{\beta_0}^{\alpha_0}, \eta_{\beta_1}^{\alpha_1})$.

We now introduce two Hilbert spaces for $\alpha \geq 0$, $\beta > 0$, $\alpha + \beta < 3$,

$$\mathbf{H}_{\alpha,\beta}(\mathcal{D}) := \overset{\circ}{\mathbf{H}}{}^1(\mathcal{D}; \eta_{\beta-1}^{\alpha-1}, \eta_\beta^\alpha),$$

$$\mathbf{H}_{\alpha,\beta}(\overline{\mathcal{D}}) := \overset{\circ}{\mathbf{H}}{}^1(\overline{\mathcal{D}}; \eta_{\beta-1}^{\alpha-1}, \eta_\beta^\alpha),$$

$$\mathbf{V}_{\alpha,\beta}(\mathcal{D}) := \overset{\circ}{\mathbf{H}}{}^1(\mathcal{D}; \eta_{\beta}^{\alpha-1}, \eta_\beta^\alpha),$$

$$\mathbf{V}_{\alpha,\beta}(\overline{\mathcal{D}}) := \overset{\circ}{\mathbf{H}}{}^1(\overline{\mathcal{D}}; \eta_{\beta}^{\alpha-1}, \eta_\beta^\alpha).$$

Omitting the domain will indicate that $\mathcal{D} = \mathbb{R}^3$, i.e., $\mathbf{H}^1 = \mathbf{H}^1(\mathbb{R}^3) = \mathbf{H}^1(\mathbb{R}^3; 1, 1)$.

We will also use the notation of sets $B_R = \{\mathbf{x} \in \mathbb{R}^3; |\mathbf{x}| \leq R\}$, $B^R = \{\mathbf{x} \in \mathbb{R}^3; |\mathbf{x}| \geq R\}$, $\mathcal{D}_R = \mathcal{D} \cap B_R$, $\mathcal{D}^R = \mathcal{D} \cap B^R$, $B_{R_2}^{R_1} = B^{R_1} \cap B_{R_2}$, $\mathcal{D}_{R_2}^{R_1} = |mathcal D \cap B_{R_2}^{R_1}$ for positive numbers R, R_1, R_2.

Remark 7.1 Let us mention that for $\eta_{\beta,\epsilon}^{\alpha,\delta}$ and for any $\delta_1, \delta_2, \epsilon_1, \epsilon_2 > 0$, we have

$$c_{min} \eta_{\beta,\epsilon_2}^{\alpha,\delta_2} \leq \eta_{\beta,\epsilon_1}^{\alpha,\delta_1} \leq c_{max} \eta_{\beta,\epsilon_2}^{\alpha,\delta_2}, \tag{7.5}$$

$c_{min} = \min(1, (\delta_1/\delta_2)^\alpha \cdot \min(1, (\epsilon_1/\epsilon_2)^\beta)$, $c_{max} = \max(1, (\delta_1/\delta_2)^\alpha) \cdot \max(1, (\epsilon_1/\epsilon_2)^\beta)$. The parameters δ and ϵ are used in a way to rescale the isotropic and anisotropic parts of the weight function η_β^α □

The next three theorems represent the main results of this chapter.

Theorem 7.1 *Let $\beta > 0$. There are positive constants R_0, c_0, c_1 depending on α, β, δ, ε (explicit expressions of these constants are given by Lemma 7.1, essentially $c_0 = O(\varepsilon^{-2} + \delta^{-2})$ and $c_1 = O(\varepsilon^{-1}\delta^{-1})$ for δ and ε tending to zero) such that for all $\mathbf{v} \in \mathbf{H}_{\alpha,\beta}$*

$$\|\mathbf{v}\|_{2,\alpha-1,\beta-1}^2 \leq c_0 \int_{B_{R_0}} |\nabla \mathbf{v}|^2 \eta_\beta^\alpha \mathrm{d}\mathbf{x} + c_1 \int_{B^{R_0}} |\nabla \mathbf{v}|^2 \eta_\beta^\alpha \mathrm{d}\mathbf{x}. \tag{7.6}$$

Theorem 7.2 (Existence and Uniqueness) *Let* $0 < \beta \leq 1$, $0 \leq \alpha < y_1\beta$, $\mathbf{f} \in \mathbf{L}^2_{\alpha+1,\beta}$, $g \in H^1_{\text{loc}}$ *such that* $\nu\nabla g - kg\mathbf{e}_3 + \omega g(\mathbf{e}_3 \wedge \mathbf{x}) \in \mathbf{L}^2_{\alpha+1,\beta}$; y_1 *will be given in Lemma* 7.4. *Then there exists a unique weak solution* $\{\mathbf{u}, p\}$ *of the problem* (7.1)–(7.3) *such that* $\mathbf{u} \in \mathbf{V}_{\alpha,\beta}$, $p \in L^2_{\alpha,\beta-1}$, $\nabla p \in \mathbf{L}^2_{\alpha+1,\beta}$, *and*

$$\begin{aligned}\|\mathbf{u}\|_{2,\alpha-1,\beta} + \|\nabla\mathbf{u}\|_{2,\alpha,\beta} + \|p\|_{2,\alpha,\beta-1} + \|\nabla p\|_{2,\alpha+1,\beta}\\ \leq C(\|\mathbf{f}\|_{2,\alpha+1,\beta} + \|\nu\nabla g - kg\mathbf{e}_3 + \omega g(\mathbf{e}_3 \wedge \mathbf{x})\|_{2,\alpha+1,\beta}).\end{aligned}$$

In the second part of the chapter we extend the results of the first part onto exterior domains.

Theorem 7.3 *Let* $\mathcal{D} \subset \mathbb{R}^3$ *be an exterior domain and* $0 < \beta \leq 1$, $0 \leq \alpha < y_1 \cdot \beta$; y_1 *is given in Lemma* 7.4, $\mathbf{f} \in \mathbf{L}^2_{\alpha+1,\beta}(\mathcal{D})$, $g \in \overset{\circ}{H}{}^1(\mathcal{D})$ *with* $\operatorname{supp} g = K \subset\subset \mathcal{D}$ *and* $\int_{\mathcal{D}} g\mathrm{d}\mathbf{x} = 0$. *Then there exists a weak solution* $\{\mathbf{u}, p\}$ *of the problem* (7.1)–(7.4) *such that* $\mathbf{u} \in \mathbf{V}_{\alpha,\beta}(\overline{\mathcal{D}})$, $p \in L^2_{\alpha,\beta-1}(\mathcal{D})$, $\nabla p \in \mathbf{L}^2_{\alpha+1,\beta}(\mathcal{D})$, *and*

$$\begin{aligned}\|\mathbf{u}\|_{2,\alpha-1,\beta} + \|\nabla\mathbf{u}\|_{2,\alpha,\beta} + \|p\|_{2,\alpha,\beta-1} + \|\nabla p\|_{2,\alpha+1,\beta}\\ \leq C(\|\mathbf{f}\|_{2,\alpha+1,\beta} + \|g\|_{1,2} + \omega^2 + \omega + k^2 + k).\end{aligned}$$

Remark 7.2 Concerning $\partial_3\mathbf{u}$ and $\nabla\mathbf{u}$, our analysis did not catch any difference in the dependence of the parameters α and β. The reason appears inside the proofs of Theorems 7.2 and 7.3 when we ask for the coercivity of the bilinear form $\widetilde{Q}(\cdot,\cdot)$, testing Eq. (7.20) by $\mathbf{u}\eta^\alpha_\beta$. On the other hand, we have no heuristic argument for not expecting better decay behavior of $\partial_1\mathbf{u}$ like ∇p as in Farwig's result, see [58]. □

Remark 7.3 The important feature of the Friedrichs–Poincaré type inequality is that we are able to evaluate its coefficients, precisely expressed in Lemma 7.1 separately near the obstacle and far from the obstacle. □

Remark 7.4 For $\alpha > 0$, using these coefficients, negative values of the function $F_{\alpha,\beta}(\cdot,\cdot)$ defined by the formula (7.12) can be compensated by the viscous Dirichlet integral; this analysis was not required in [58] because $F_{\alpha,\beta}(\cdot,\cdot)$ is positive when $\alpha < 0$. □

Remark 7.5 The previous compensation cannot be associated with a large interval of positive values for α: So, we receive the technical condition $\alpha/\beta < y_1$. Using other type of weight functions characterized by some parameters, one can get another technical condition on these parameters. □

7.2 Friedrichs–Poincaré Inequality

In this subsection we derive an inequality of the Friedrichs–Poincaré type in appropriate weighted Sobolev spaces. We also recall some necessary technical assertions; for more details see Kračmar and Penel [156].

Proposition 7.1 *For arbitrary $\alpha, \beta \geq 0$ and $\mathbf{x} \in \mathbb{R}^3$, $\mathbf{x} \neq \mathbf{0}$:*

$$\Delta\eta_\beta^\alpha(\mathbf{x}) \geq 2\beta \min(1, \beta)\varepsilon\delta\eta_{\beta-1}^{\alpha-1}(\mathbf{x}).$$

Proof We introduce $\beta^* = \min(\beta, 1)$ in an explicit expression of $\Delta\eta_\beta^\alpha$:

$$\begin{aligned}\Delta\eta_\beta^\alpha = &\left\{\left(\alpha^2\delta^2\frac{1+\varepsilon s}{1+\delta r} - \alpha\delta^2\frac{1+\varepsilon s}{1+\delta r}\right) + 2\alpha\beta\delta\varepsilon\frac{s}{r}\right.\\ &+ 2\beta\,(\beta-1)\,\frac{\varepsilon}{r}\,(1+\delta r)\,\frac{\varepsilon s}{1+\varepsilon s}\\ &\left.+2\alpha\delta^2(1+\varepsilon s)\frac{1}{\delta r} + (1-\beta^*+\beta^*)2\beta\frac{\varepsilon}{r}(1+\delta r)\right\}\eta_{\beta-1}^{\alpha-1},\end{aligned}$$

for $r > 0$. We denote the five terms in { } by $T_1, T_2, \ldots, T_5$ and overwrite the previous relation as $\Delta\eta_\beta^\alpha = \{[T_1 + T_4] + T_2 + [T_3 + (1-\beta^*)T_5] + \beta^* T_5\}\eta_{\beta-1}^{\alpha-1}$. Observing that $T_5 \geq 2\beta\varepsilon\delta$, the proposition is trivial. □

Proposition 7.2 *Let $\alpha \geq 0$, $\beta \geq 0$, $\delta > 0$, $\varepsilon > 0$, and $\kappa > 1$. Then for $\mathbf{x} \in \mathbb{R}^3$, $|\mathbf{x}| \geq |\delta^{-1} - (2\varepsilon)^{-1}|(\kappa-1)^{-1}$:*

$$|\nabla\eta_\beta^\alpha(\mathbf{x})|^2 \leq 2\kappa\delta\varepsilon(\alpha+\beta)^2(\eta_{\beta-1/2}^{\alpha-1/2}(\mathbf{x}))^2. \tag{7.7}$$

Let $\alpha \geq 0$, $\beta \geq 0$, $\delta > 0$, $\varepsilon > 0$, and $(\beta-\alpha)(2\varepsilon-\delta) \geq 0$. Then for $\mathbf{x} \in \mathbb{R}^3$, $\mathbf{x} \neq 0$:

$$|\nabla\eta_\beta^\alpha(\mathbf{x})|^2 \leq (\alpha\delta + 2\beta\varepsilon)^2(\eta_{\beta-1/2}^{\alpha-1/2}(\mathbf{x}))^2. \tag{7.8}$$

Proof If $\beta = 0$ and $\alpha = 0$, then both inequalities (7.7) and (7.8) are valid. Let us concentrate on the nontrivial cases.

For $r > 0$, $s \in [0, 2r]$, we have that $\partial g/\partial s > 0$, where g is a function defined by relations:

$$\begin{aligned}|\nabla\eta_\beta^\alpha(\mathbf{x})|^2 =& g(s(\mathbf{x}), r(\mathbf{x}))\left(\eta_{\beta-1/2}^{\alpha-1/2}(\mathbf{x})\right)^2,\\ g(s,r) \equiv& \alpha^2\delta^2\left(\frac{1+\varepsilon s}{1+\delta r}\right) + 2\alpha\beta\delta\varepsilon\frac{s}{r} + 2\beta^2\varepsilon^2\left(\frac{1+\delta r}{1+\varepsilon s}\right)\frac{s}{r}.\end{aligned}$$

So, $g(s, r)$ is increasing as a function of s and

$$
\begin{aligned}
G(r) \equiv& \max_{s\in[0,2r]} g(s,r) = g(2r,r)\\
=&\alpha^2\delta^2\frac{1+2\varepsilon r}{1+\delta r} + 4\alpha\beta\delta\varepsilon + 4\beta^2\varepsilon^2\frac{1+\delta r}{1+2\varepsilon r} \le 2\kappa(\alpha+\beta)^2\delta\varepsilon \qquad (7.9)
\end{aligned}
$$

for $\kappa > 1$ and $r \ge |\delta^{-1} - (2\varepsilon)^{-1}|(\kappa-1)^{-1}$. So, inequality (7.7) is proved.

To justify the second inequality (7.8), we observe that for the given values of α, β, δ, ε and for $r > 0$, $G(r) \le G(0)$. □

Next we derive an inequality of the Friedrichs–Poincaré type in the space $\mathbf{H}_{\alpha,\beta}$. It is necessary for our aim to get expressions of constants in this inequality. It follows from Proposition 7.1.

Lemma 7.1 *Let $\alpha \ge 0$, $\beta > 0$, $\alpha+\beta < 3$, $\kappa > 1$. Let δ and ε be arbitrary positive constants, such that $(\beta-\alpha)(2\varepsilon-\delta) \ge 0$. Then for all $\mathbf{u} \in \mathbf{H}_{\alpha,\beta}$,*

$$
\|\mathbf{u}\|^2_{2,\alpha-1,\beta-1} \le c_0\|\nabla\mathbf{u}|B_{R_0}\|^2_{2,\alpha,\beta} + c_1\|\nabla\mathbf{u}|B^{R_0}\|^2_{2,\alpha,\beta}, \qquad (7.10)
$$

where $c_0 = \left[(\alpha\delta+2\beta\varepsilon)/(\beta\beta^*\delta\varepsilon)\right]^2$, $c_1 = [(2\kappa)/(\delta\varepsilon)]\cdot\left[(\alpha+\beta)/(\beta\beta^*)\right]^2$, and $R_0 \ge |\delta^{-1} - (2\varepsilon)^{-1}|(\kappa-1)^{-1}$. □

Remark 7.6 Let us observe that if additionally $\delta < 2\varepsilon$ and $1 < \kappa \le 2\varepsilon/\delta + \delta/(2\varepsilon) - 1$, then $c_0 \ge c_1$. □

Proof of Lemma 7.1 Due to the density of $\mathbf{C}_0^\infty$ in $\mathbf{H}_{\alpha,\beta}$, it is sufficient to prove the inequality for all $\mathbf{u} \in \mathbf{C}_0^\infty$. From Proposition 7.1, it follows that for $\mathbf{v} \in \mathbf{C}_0^\infty$

$$
\begin{aligned}
2\beta\beta^*\delta\varepsilon \int_{\mathbb{R}^3\setminus B_\rho} \mathbf{v}^2\eta^{\alpha-1}_{\beta-1}d\mathbf{x} \le& \int_{\mathbb{R}^3\setminus B_\rho} \mathbf{v}^2\Delta\eta^\alpha_\beta d\mathbf{x}\\
=&-2\int_{\mathbb{R}^3\setminus B_\rho}(\mathbf{v}\cdot\nabla)\mathbf{v}\cdot\nabla\eta^\alpha_\beta d\mathbf{x} + \int_{\partial B_\rho}\mathbf{v}^2\nabla\eta^\alpha_\beta\cdot\mathbf{n}dS\\
\le&\beta\beta^*\delta\varepsilon\int_{\mathbb{R}^3\setminus B_\rho}\mathbf{v}^2\eta^{\alpha-1}_{\beta-1}d\mathbf{x} + \frac{1}{\beta\beta^*\delta\varepsilon}\int_{\mathbb{R}^3\setminus B_\rho}|\nabla\mathbf{v}|^2|\nabla\eta^\alpha_\beta|^2\eta^{-\alpha+1}_{-\beta+1}d\mathbf{x}\\
&+\int_{\partial B_\rho}\mathbf{v}^2\nabla\eta^\alpha_\beta\cdot\mathbf{n}dS.
\end{aligned}
$$

Hence, because the surface integral is a value of the order $O(\rho^2)$, we have

$$\beta\beta^*\delta\varepsilon \int_{\mathbb{R}^3} \mathbf{v}^2\eta_{\beta-1}^{\alpha-1}\mathbf{dx} \le \frac{1}{\beta\beta^*\delta\varepsilon}\int_{\mathbb{R}^3} |\nabla\mathbf{v}|^2|\nabla\eta_\beta^\alpha|^2\eta_{-\beta+1}^{-\alpha+1}\mathbf{dx}. \tag{7.11}$$

By means of the Cauchy–Schwarz inequality and from Proposition 7.2 with $R_0 \ge |\delta^{-1} - (2\varepsilon)^{-1}|/(\kappa - 1)$, we finally get (7.10). □

To begin with we shall introduce some technical lemmas. Let us define $F_{\alpha,\beta}(s,r)$ by the relation:

$$F_{\alpha,\beta}(s,r)\cdot\eta_{\beta-1}^{\alpha-1} \equiv -\nu|\nabla\eta_\beta^\alpha|^2/\eta_\beta^\alpha - k\partial_3\eta_\beta^\alpha. \tag{7.12}$$

The following lemma gives the evaluation of $F_{\alpha,\beta}(s,r)$ from below:

Lemma 7.2 *Let* $0 \le \alpha < \beta$, $\kappa > 1$, $0 < \varepsilon \le (1/(2\kappa))\cdot(k/\nu)\cdot((\beta-\alpha)/\beta^2)$, *and* δ, ν, k. *Then*

$$F_{\alpha,\beta}(s,r) - (1-\kappa^{-1})k\delta\varepsilon(\beta-\alpha)s \ge -\alpha\delta k(1+\nu k^{-1}\alpha\delta) \tag{7.13}$$

for all $r > 0$ *and* $s \in [0, 2r]$. □

Proof Expressing the function $F_{\alpha,\beta}(s,r)$ explicitly, we get

$$\begin{aligned} F_{\alpha,\beta}(s,r) = &-\nu\alpha^2\delta^2\left(\frac{1+\varepsilon s}{1+\delta r}\right) - 2\nu\alpha\beta\delta\varepsilon\frac{s}{r} - 2\nu\beta^2\varepsilon^2\left(\frac{1+\delta r}{1+\varepsilon s}\right)\frac{s}{r} \\ &- k\alpha\delta(1+\varepsilon s)\frac{r-s}{r} + k\beta\varepsilon(1+\delta r)\frac{s}{r}. \end{aligned}$$

For convenient use we subtract $(1-\kappa^{-1})k\delta\varepsilon(\beta-\alpha)s$ from $F_{\alpha,\beta}(s,r)$. We observe (see Sect. 7.5.2 Appendix A) that, for the given α, β, ε, κ for all δ, ν, $k > 0$ and for $r > 0$, $F_{\alpha,\beta}(s,r) - (1-\kappa^{-1})k\delta\varepsilon(\beta-\alpha)s \ge F_{\alpha,\beta}(0,r)$, which immediately gives inequality (7.13). □

7.3 Pressure in $\mathbb{R}^3$ and Uniqueness Theorems

In this section, we will start with the question about the existence of the pressure in $\mathbb{R}^3$. Then we formulate and prove two theorems about the uniqueness of the weak solution of the problem (7.1)–(7.2) in $\mathbb{R}^3$ and in an exterior domain $D \ne \mathbb{R}^3$. The presented approach will be used in the next subsection, in the proof of existence of a solution verifying solenoidality of the constructed function $\mathbf{u}$.

7.3.1 The Pressure in $\mathbb{R}^3$

If there exist distributions $\mathbf{u}p$ satisfying (7.1)–(7.2) in $\mathcal{D} = \mathbb{R}^3$, then pressure p satisfies the equation

$$\Delta p = \operatorname{div} \mathbf{F}, \quad \text{where } \mathbf{F} = \mathbf{f} + \nu \nabla g - kg\mathbf{e}_3 + \omega g(\mathbf{e}_3 \wedge \mathbf{x}), \tag{7.14}$$

because $\operatorname{div}(\omega \mathcal{R}\mathbf{u}) = \omega(\mathbf{e}_3 \wedge \mathbf{x}) \cdot \nabla \operatorname{div} \mathbf{u} = \operatorname{div}[\omega g(\mathbf{e}_3 \wedge \mathbf{x})]$.

Let $\mathcal{E}$ be the fundamental solution of the Laplace equation, i.e., $\mathcal{E} = -1/(4\pi r)$. Assuming firstly $\mathbf{F} \in \mathbf{C}_0^\infty$, we have $p = \mathcal{E} \star \operatorname{div} \mathbf{F}$ and $\nabla p = \nabla \mathcal{E} \star \operatorname{div} \mathbf{F}$, and so, $p = \nabla \mathcal{E} \star \mathbf{F}$ and $\nabla p = \nabla^2 \mathcal{E} \star \mathbf{F}$. It is well known that both formulas can be extended for $\mathbf{F} \in \mathbf{L}^2_{\alpha+1,\beta}$ with $0 < \beta < 1$ and $-2 < \alpha + \beta < 2$ (the last convolution $\nabla p = \nabla^2 \mathcal{E} \star \mathbf{F}$ due to the fact that $\nabla^2 \mathcal{E}$ is the singular kernel of the Calderon–Zygmund type and that $\eta_\beta^{\alpha+1}$ belongs to the Muckenhoupt class of weights A_2), see [58, Theorem 3.2, Theorem 5.5], [155, Theorem 4.4, Theorem 5.4], where the theorems are formulated for the pressure part $\mathcal{P}$ of the fundamental solution of the classical Oseen problem, so $\mathcal{P} = \nabla \mathcal{E}$ and $\nabla \mathcal{P} = \nabla^2 \mathcal{E}$. For $\mathbf{F} \in \mathbf{L}^2_{\alpha+1,\beta}$, we get $p \in L^2_{\alpha,\beta-1}$ and $\nabla p \in \mathbf{L}^2_{\alpha+1,\beta}$, and there are positive constants C_1, C_2 such that the following estimates are satisfied:

$$\|p\|_{2,\alpha,\beta-1} \le C_1 \|\mathbf{F}\|_{2,\alpha+1,\beta}, \quad \|\nabla p\|_{2,\alpha+1,\beta} \le C_2 \|\mathbf{F}\|_{2,\alpha+1,\beta}. \tag{7.15}$$

Remark 7.7 Another possibility of construction of the pressure is the use of Hörmander–Michlin multiplier theorem. Both techniques can be used in L^2- and L^q-framework to get an estimate of ∇p. □

7.3.2 Uniqueness in $\mathbb{R}^3$

Theorem 7.4 (Uniqueness in $\mathbb{R}^3$) *Let $\{\mathbf{u}, p\}$ be a distributional solution of the problem (7.1)–(7.2) with $\mathbf{f} = 0$, $g = 0$ such that $\mathbf{u} \in \widehat{\mathbf{H}}_0^{1,2}$ and $p \in L^2_{\text{loc}}$. Then $\mathbf{u} = 0$ and $p = \text{const}$.* □

Proof From the condition $\mathbf{u} \in \widehat{\mathbf{H}}_0^{1,2}$, we get $\nabla \mathbf{u} \in \mathbf{L}^2$, $\mathbf{u} \in \mathbf{L}^6$, $\mathbf{u} \in \mathcal{S}'$. Because $\operatorname{div}(\omega \mathcal{R}\mathbf{u}) = \omega(\mathbf{e}_3 \wedge \mathbf{x}) \cdot \nabla \operatorname{div} \mathbf{u} = 0$, we have $\Delta p = 0$. Hence, applying Laplacian and the Fourier transform, we get

$$\begin{aligned} \Delta(-\nu \Delta \mathbf{u} + k\partial_3 \mathbf{u} - \omega(\mathbf{e}_3 \wedge \mathbf{x}) \cdot \nabla \mathbf{u} + \omega \mathbf{e}_3 \wedge \mathbf{u}) &= \mathbf{0}, \\ |\boldsymbol{\xi}|^2(\nu|\boldsymbol{\xi}|^2 \widehat{\mathbf{u}} + ik\xi_3 \widehat{\mathbf{u}} - (\boldsymbol{\omega} \wedge \boldsymbol{\xi}) \cdot \nabla_\xi \widehat{\mathbf{u}} + \boldsymbol{\omega} \wedge \widehat{\mathbf{u}}) &= \mathbf{0} \quad \text{in } \mathcal{S}'. \end{aligned}$$

Assuming the equation in cylindrical coordinates (ρ, φ, ξ_3) and denoting $T(\varphi)\widehat{\mathbf{v}} = \widehat{\mathbf{u}}(\rho, \varphi, \xi_3)$, where

$$T(\varphi) = \begin{bmatrix} \cos(\varphi), & -\sin(\varphi), & 0 \\ \sin(\varphi), & \cos(\varphi), & 0 \\ 0, & 0, & 1 \end{bmatrix},$$

we get

$$|\boldsymbol{\xi}|^2\{-\partial_\varphi \widehat{\mathbf{v}} + [(\nu/\widetilde{\omega})|\boldsymbol{\xi}|^2 + \mathrm{i}(k/\widetilde{\omega})\xi_3]\widehat{\mathbf{v}}\} = \mathbf{0} \quad \text{in } \mathcal{S}'. \tag{7.16}$$

We will show that from this equation follows that $\operatorname{supp}\widehat{\mathbf{v}} \subset \{0\}$, and due to the definition of $\widehat{\mathbf{v}}$, we will have also $\operatorname{supp}\widehat{\mathbf{u}} \subset \{\mathbf{0}\}$. This means that $\mathbf{u}$ is a polynomial of x_1, x_2, x_3. Because $\mathbf{u} \in \mathbf{L}^6$, we get $\mathbf{u} = 0$. Substituting into (7.1), we get $\nabla p = 0$ and $p = \text{const}$.

So, we have to prove that for an arbitrary real vector function $\Psi \in \mathbf{C}_0^\infty(\mathbb{R}^3 \setminus \{\mathbf{0}\})$ defined for $[\xi_1, \xi_2, \xi_3] \in \mathbb{R}^3$, we have $\langle \widehat{\mathbf{v}}, \Psi \rangle = 0$. If for each $\Psi \in \mathbf{C}_0^\infty(\mathbb{R}^3 \setminus \{\mathbf{0}\})$ there is a function $\Phi \in \mathbf{C}_0^\infty(\mathbb{R}^3 \setminus \{\mathbf{0}\})$ such that

$$\partial_\varphi(|\boldsymbol{\xi}|^2 \Phi) + [(\nu/\widetilde{\omega})|\boldsymbol{\xi}|^2 + \mathrm{i}(k/\widetilde{\omega})\xi_3](|\boldsymbol{\xi}|^2 \Phi) = \Psi, \tag{7.17}$$

then from (7.16) follows

$$\begin{aligned} 0 &= \left\langle |\boldsymbol{\xi}|^2 \left\{ -\partial_\varphi \widehat{\mathbf{v}} + \left[(\nu/\widetilde{\omega})|\boldsymbol{\xi}|^2 + \mathrm{i}(k/\widetilde{\omega})\xi_3 \right] \widehat{\mathbf{v}} \right\}, \Phi \right\rangle \\ &= \left\langle \widehat{\mathbf{v}}, \partial_\varphi \left(|\boldsymbol{\xi}|^2 \Phi \right) + \left[(\nu/\widetilde{\omega})|\boldsymbol{\xi}|^2 + \mathrm{i}(k/\widetilde{\omega})\xi_3 \right] \left(|\boldsymbol{\xi}|^2 \Phi \right) \right\rangle = \langle \widehat{\mathbf{v}}, \Psi \rangle. \end{aligned}$$

Hence, the proof of $\operatorname{supp}\widehat{\mathbf{v}} \subset \{0\}$ is reduced to the solvability of (7.17). First we note that it is sufficient to solve the equation

$$\partial_\varphi \zeta + \left((\nu/\widetilde{\omega})|\boldsymbol{\xi}|^2 + \mathrm{i}(k/\widetilde{\omega})\xi_3 \right) \zeta = \Psi \tag{7.18}$$

because the division on the expression $|\boldsymbol{\xi}|^2$ defines the one-to-one correspondence of the space $\mathbf{C}_0^\infty(\mathbb{R}^3 \setminus \{\mathbf{0}\})$ onto $\mathbb{C}_0^\infty(\mathbb{R}^3 \setminus \{\mathbf{0}\})$.

Let us analyze Eq. (7.18) in cylindrical coordinates $[\rho, \varphi, \xi_3]$, where $\rho = (\xi_1^2 + \xi_2^2)^{1/2}$. For an arbitrary real vector function $\Psi \in \mathbf{C}_0^\infty(\mathbb{R}^3 \setminus \{\mathbf{0}\})$ defined for $[\xi_1, \xi_2, \xi_3] \in \mathbb{R}^3$, we define $f(t) := \Psi(\rho \cos t \rho \sin t, \xi_3)$, $a := (\nu/\widetilde{\omega})|\boldsymbol{\xi}|^2 + \mathrm{i}(k/\widetilde{\omega})\xi_3$, assuming $\widetilde{\omega} > 0$.

Now, we will use the following technical proposition about the existence of a solution of an ordinary differential equation in a space of periodical functions (and later also in the proof of existence of a solution of the problem for checking solenoidality of a constructed solution, see the proof of Theorem 7.6): □

Proposition 7.3 *Let $a \in \mathbb{C}$, $\operatorname{Re} a > 0$. Let $f \in C^\infty(\mathbb{R})$ be a 2π-periodical complex function. Then there is a unique 2π-periodical solution $g \in C^\infty(\mathbb{R})$ of the equation*

$$g' + ag = f,$$

and the solution g can be expressed in the following form:

$$g(\varphi) = (\mathrm{e}^{2\pi a} - 1)^{-1} \int_0^{2\pi} \mathrm{e}^{at} f(\varphi + t)\mathrm{d}t = \mathrm{e}^{-a\varphi} \int_{-\infty}^{\varphi} \mathrm{e}^{at} f(t)\mathrm{d}t.$$

Proof of the proposition follows from standard computations.

Using Proposition 7.3 we get the solution of (7.18) in the form

$$\zeta(\rho, \varphi, \xi_3) = \left\{ \exp\left[2\pi \left(\frac{\nu}{\widetilde{\omega}} |\boldsymbol{\xi}|^2 + \mathrm{i}\frac{k}{\widetilde{\omega}} \right) \right] - 1 \right\}^{-1}$$
$$\times \int_0^{2\pi} \exp\left[\left(\frac{\nu}{\widetilde{\omega}} |\boldsymbol{\xi}|^2 + \mathrm{i}\frac{k}{\widetilde{\omega}} \xi_3 \right) t \right] \Psi\left(\rho\cos(t + \varphi), \rho\sin(t + \varphi), \xi_3\right) \mathrm{d}t.$$

It is easy to see that function ζ as the function of $[\xi_1, \xi_2, \xi_3]$ is infinitely differentiable with respect to these variables and $\zeta \in \mathbf{C}_0^\infty(\mathbb{R}^3 \setminus \{\mathbf{0}\})$. Finally we put $\Phi = \zeta / |\boldsymbol{\xi}|^2$.

7.3.3 Uniqueness in an Exterior Domain $\mathcal{D} \subset \mathbb{R}^3$

The last two sections are devoted to the problem in an exterior domain. We start with the question of uniqueness. The uniqueness theorem proved in this section together with the uniqueness theorem in $\mathbb{R}^3$ from the previous subsection will be used in the next section in the proof of the existence of a solution in an exterior domain, in the localization procedure. The homogenous Dirichlet boundary condition on $\partial\mathcal{D}$ for $\mathbf{u}$ in the next theorem follows from the assumption $\mathbf{u} \in \mathbf{V}_{0,0}(\mathcal{D})$.

Theorem 7.5 *Let $\{\mathbf{u}, p\}$ be a distributional solution of the problem* (7.1)–(7.3) *with $\mathbf{f} = \mathbf{0}$ and $g = 0$ such that $\mathbf{u} \in \mathbf{V}_{0,0}(\mathcal{D})$ and $p \in L^2_{-1,0}(\mathcal{D})$. Then $\mathbf{u} = \mathbf{0}$ and $p = 0$.* □

Proof Let $\Phi = \Phi(z) \in C_0^\infty(\langle 0, +\infty))$ be a nonincreasing cutoff function such that $\Phi(z) \equiv 1$ for $z < 1/2$ and $\Phi(z) \equiv 0$ for $z > 1$. Let $|\Phi'| \leq 3$. Let $\Phi_R \equiv \Phi_R(\mathbf{x}) \equiv \Phi(|\mathbf{x}|/R)$. We have $|\nabla\Phi_R| \leq 3/R$ and $|\partial_3\Phi_R| \leq 3/R$ for $\mathbf{x} \in \mathbb{R}^3$, $R/2 \leq |\mathbf{x}| \leq R$. Let $\{R_j\} \in \mathbb{R}$ be an increasing sequence of radii with the limit $+\infty$. So we have that $\mathbf{u}_j \equiv \mathbf{u} \cdot \Phi_{R_j} \in \overset{\circ}{\mathbf{H}}{}^1(\mathcal{D})$, and $\{\mathbf{u}_j\}$ is a sequence of functions with limit $\mathbf{u}$ in the space $\mathbf{V}_{0,0}(\mathcal{D})$. Using the (non-solenoidal) test functions $\varphi = \mathbf{u}\Phi^2_{R_j} = \mathbf{u}_j\Phi_{R_j} \in \overset{\circ}{\mathbf{H}}{}^1(\mathcal{D})$ for Eq. (7.1), we get

$$\nu \int_{\mathcal{D}} \nabla \mathbf{u} \colon \nabla(\mathbf{u}\Phi^2_{R_j})\mathbf{dx} + k \int_{\mathcal{D}} \partial_3 \mathbf{u} \cdot \mathbf{u}\Phi^2_{R_j}\mathbf{dx}$$
$$+ \omega \int_{\mathcal{D}} (\mathbf{e}_3 \wedge \boldsymbol{x}) \cdot \nabla \mathbf{u} \cdot \mathbf{u}\Phi^2_{R_j}\mathbf{dx} + \int_{\mathcal{D}} \nabla p \cdot \mathbf{u}\Phi^2_{R_j}\mathbf{dx} = 0. \qquad (7.19)$$

Using in (7.19) relation $\nabla \mathbf{u} : \nabla(\mathbf{u}\Phi^2_{R_j}) = |\nabla \mathbf{u}_j|^2 - \nabla\Phi_{R_j} \cdot \nabla\Phi_{R_j}|\mathbf{u}|^2$, integrating by parts, we get after some evident rearrangements

$$\nu \int_{|\mathcal{D}} |\nabla \mathbf{u}_j|^2 \mathbf{dx} - \frac{\omega}{2} \int_{\mathcal{D}} \operatorname{div}(\mathbf{e}_3 \wedge \boldsymbol{x}) \left|\mathbf{u}_j\right|^2 d\mathbf{x}$$
$$- \frac{k}{2} \int_{\mathcal{D}} |\mathbf{u}|^2 \partial_3 \Phi^2_{R_j}\mathbf{dx} - \frac{\omega}{2} \int_{\mathcal{D}} |\mathbf{u}|^2 (\mathbf{e}_3 \wedge \boldsymbol{x}) \cdot \nabla\Phi^2_{R_j}\mathbf{dx}$$
$$- \nu \int_{\mathcal{D}} |\nabla\Phi_{R_j}|^2 |\mathbf{u}|^2 \mathbf{dx} - \int_{\mathcal{D}} p\mathbf{u} \cdot \nabla(\Phi^2_{R_j})\mathbf{dx} = 0.$$

$$\nu \int_{\mathcal{D}} |\nabla \mathbf{u}_j|^2 \mathbf{dx} \le C \left(\int_{\mathcal{D}^{R_j/2}_{R_j}} |\mathbf{u}|^2 r^{-1} \mathbf{dx} + \int_{\mathcal{D}^{R_j/2}_{R_j}} |p|\,|\mathbf{u}| r^{-1} \mathbf{dx} \right).$$

$\mathbf{u} \in \mathbf{L}^2_{-1,0}(\mathcal{D})$, $p \in L^2_{-1,0}(\mathcal{D})$, $p\mathbf{u} \in \mathbf{L}^1_{-1,0}(\mathcal{D})$. So, for $j \to \infty$ we get $\int_{\mathcal{D}} |\nabla \mathbf{u}|^2 dx \le 0$. Hence, the function $\nabla \mathbf{u} = 0$ a.e. in $\mathcal{D}$, and this means $\mathbf{u}$ is a constant a.e. in $\mathcal{D}$. From $\mathbf{u} \in \mathbf{L}^2_{-1,0}(\mathcal{D})$ follows that $\mathbf{u} = \mathbf{0}$ a.e. in D. Using now an arbitrary test function φ for Eq. (7.1), we get $\int_{\mathcal{D}} \nabla p\varphi \mathbf{dx} = 0$. So, the function $\nabla p = 0$ a.e. in $\mathcal{D}$, and this means p is a constant a.e. in $\mathcal{D}$. From $p \in L^2_{-1,0}(\mathcal{D})$ follows that $p = 0$ a.e. in $\mathcal{D}$, and the uniqueness is proved. □

7.4 Existence of a Solution in the Whole Space

7.4.1 *The Problem in B_R*

In this subsection we will study the existence of a weak solution in a bounded domain B_R, with the right-hand side $\tilde{f}$, where $\tilde{f} = .f - \nabla p \in \boldsymbol{L}^2 - \alpha + 1, \beta$ (the pressure is assumed to be known):

$$S_{\nu,\omega}\mathbf{u} + k\partial_3 \mathbf{u} = \tilde{\mathbf{f}} \quad \text{in } B_R \qquad (7.20)$$
$$\mathbf{u} = 0 \quad \text{on } \partial B_R. \qquad (7.21)$$

We show the existence of a weak solution $\mathbf{u}_R \in \overset{\circ}{\mathbf{H}}{}^1(B_R)$. Let us introduce a continuous bilinear form $\widetilde{Q}(\cdot,\cdot)$ on $\overset{\circ}{\mathbf{H}}{}^1(B_R) \times \overset{\circ}{\mathbf{H}}{}^1(B_R)$:

$$\begin{aligned}\widetilde{Q}(\mathbf{u},\mathbf{v}) =& \nu \int_{B_R} \nabla\mathbf{u} \colon \nabla(\mathbf{v}\eta^0_{\beta_0})\mathbf{dx} + k\int_{B_R} \partial_3\mathbf{u}\cdot(\mathbf{v}\eta^0_{\beta_0})\mathbf{dx}\\ &+\omega\int_{B_R}(\mathbf{e}_3\wedge\mathbf{x})\cdot\nabla\mathbf{u}(\mathbf{v}\eta^0_{\beta_0})\mathbf{dx} + \omega\int_{B_R}(\mathbf{e}_3\wedge\mathbf{u})\cdot(\mathbf{v}\eta^0_{\beta_0})\mathbf{dx},\\ \widetilde{Q}(\mathbf{v},\mathbf{v}) \geq& 2^{-1}\nu\int_{B_R}|\nabla\mathbf{v}|^2\eta^0_{\beta_0}\mathbf{dx} + 2^{-1}\int_{B_R}\mathbf{v}^2 F_{0,\beta_0}(s,r;\nu)\eta^{-1}_{\beta_0-1}\mathbf{dx}.\end{aligned} \tag{7.22}$$

Lemma 7.3 *Let* $0<\beta_0\leq 1$. *Then, for all* $\tilde{\mathbf{f}} \in \mathbf{L}^2_{1,\beta_0}(B_R)$, $\varepsilon_0 < \frac{1}{2}(k/\nu)\cdot(1/\beta_0)$, $\eta^\alpha_{\beta_0} \equiv \eta^{\alpha,\varepsilon_0}_{\beta_0,\varepsilon_0}$, *there exists unique* $\mathbf{u}_R \in \overset{\circ}{\mathbf{H}}{}^1(B_R)$ *such that for all* $\mathbf{v} \in \overset{\circ}{\mathbf{H}}{}^1(B_R)$

$$\widetilde{Q}(\mathbf{u}_R,\mathbf{v}) = \int_{B_R}\widetilde{\mathbf{f}}\cdot\mathbf{v}\eta^0_{\beta_0}\mathbf{dx}. \tag{7.23}$$

Proof A bilinear form $\widetilde{Q}$ is coercive, i.e., there exists a constant $C_R > 0$ such that $\widetilde{Q}(\mathbf{v},\mathbf{v}) \geq C_R\,\|\mathbf{v}\|^2$, where $\|\cdot\|$ is the norm in the space $\overset{\circ}{\mathbf{H}}{}^1(B_R)$. Indeed, we get

$$\widetilde{Q}(\mathbf{v},\mathbf{v}) \geq \frac{\nu}{2}\int_{B_R}|\nabla\mathbf{v}|^2\eta^0_{\beta_0}\mathbf{dx} + \frac{1}{2}\int_{B_R}\mathbf{v}^2F_{0,\beta_0}(s,r)\eta^{-1}_{\beta_0-1}\mathbf{dx}.$$

Because $\varepsilon_0 < \frac{1}{2}(k/\nu)\cdot(1/\beta_0)$, there is a constant κ satisfying all previous conditions and additionally $\varepsilon_0 \leq (1/2\kappa)\cdot(k/\nu)\cdot(1/\beta_0)$. Because $\alpha = 0$, we get from Lemma 7.2

$$\begin{aligned}&\int_{B_R}\mathbf{v}^2F_{0,\beta_0}(s,r)\eta^{-1}_{\beta_0-1}\mathbf{dx} \geq (1-\kappa^{-1})k\varepsilon_0^2\beta_0\int_{B_R}\mathbf{v}^2\eta^{-1}_{\beta_0-1}s\mathbf{dx},\\ &\widetilde{Q}(\mathbf{v},\mathbf{v}) \geq \frac{\nu}{2}\int_{B_R}|\nabla\mathbf{v}|^2\eta^0_{\beta_0}d\mathbf{x} + \frac{1}{2}\left(1-\frac{1}{\kappa}\right)k\varepsilon_0\beta_0\int_{B_R}\mathbf{v}^2\eta^{-1}_{\beta_0-1}(\varepsilon_0 s)\mathbf{dx}.\end{aligned}$$

Using Lemma 7.1 and Remark 7.6, we derive

$$\begin{aligned}\widetilde{Q}(\mathbf{v},\mathbf{v}) \geq& \frac{\nu}{4}\int_{B_R}|\nabla\mathbf{v}|^2\eta^0_{\beta_0}\mathbf{dx} + \frac{\nu}{16}\varepsilon_0^2\beta_0^2\int_{B_R}\mathbf{v}^2\eta^{-1}_{\beta_0-1}\mathbf{dx}\\ &+\frac{1}{2}\left(1-\frac{1}{\kappa}\right)k\varepsilon_0\beta_0\int_{B_R}\mathbf{v}^2\eta^{-1}_{\beta_0-1}(\varepsilon_0 s)\mathbf{dx}\\ \geq&\left(1-\frac{1}{\kappa}\right)\frac{\nu}{4}\min\left\{1,\frac{1}{4}\varepsilon_0^2\beta_0^2, 2\frac{k}{\nu}\beta_0\varepsilon_0\right\}\end{aligned}$$

$$\left(\int_{B_R} |\nabla v|^2 \eta^0_{\beta_0} \mathrm{d}\mathbf{x} + \int_{B_R} \mathbf{v}^2 \eta^{-1}_{\beta_0} \mathrm{d}\mathbf{x}\right) \tag{7.24}$$

$$\widetilde{Q}(\mathbf{v}, \mathbf{v}) \geq C_R \left(\int_{B_R} |\nabla \mathbf{v}|^2 \mathrm{d}\mathbf{x} + \int_{B_R} \mathbf{v}^2 \mathrm{d}\mathbf{x}\right) = C_R \|\mathbf{v}\|^2, \tag{7.25}$$

where $C_R = (\nu/4)\cdot(1-\kappa^{-1}) \min\{1, \varepsilon_0^2\beta_0^2/4, 2(k/\nu)\beta\varepsilon_0\}\cdot(1+\varepsilon_0 R)$. Using the Lax–Milgram theorem, we get that there is $\mathbf{u}_R \in \overset{\circ}{\mathbf{H}}{}^1(B_R)$ such that (7.23) is satisfied.

□

Remark 7.8 An arbitrary function $\boldsymbol{\Phi} \in \overset{\circ}{\mathbf{H}}{}^1(B_R)$ can be expressed in the form $\boldsymbol{\Phi} = \mathbf{v}\eta^0_{\beta_0}$, where $\mathbf{v} \in \overset{\circ}{\mathbf{H}}{}^1(B_R)$. Therefore, for all $\boldsymbol{\Phi} \in \overset{\circ}{\mathbf{H}}{}^1(B_R)$,

$$Q(\mathbf{u}_R, \boldsymbol{\Phi}) = \int_{B_R} \tilde{\mathbf{f}} \cdot \Phi \mathrm{d}\mathbf{x}, \tag{7.26}$$

where by the definition $Q(\mathbf{u}_R, \boldsymbol{\Phi}) \equiv Q(\mathbf{u}_R, \mathbf{v}\eta^0_{\beta_0}) \equiv \widetilde{Q}(\mathbf{u}_R, \mathbf{v})$. □

7.4.1.1 Uniform Estimates of $\mathbf{u}_R$ in $\mathbb{R}^3$

Our next aim is to prove that the weak solutions $\mathbf{u}_R$ of (7.23) are uniformly bounded in $\mathbf{V}_{\alpha,\beta}$ as $R \to +\infty$.

Let y_1 be the unique real solution of the algebraic equation $4y^3 + 8y^2 + 5y - 1 = 0$. It is easy to verify that $y_1 \in (0, 1)$. We will explain later why the control of α/β by y_1 is necessary.

Lemma 7.4 *Let* $0 < \beta \leq 1$, $0 \leq \alpha < y_1\beta$, $\tilde{\mathbf{f}} \in \mathbf{L}^2_{\alpha+1,\beta}$. *Then, as* $R \to +\infty$, *the weak solutions* $\mathbf{u}_R$ *of (7.23) given by Lemma 7.3 are uniformly bounded in* $\mathbf{V}_{\alpha,\beta}$. *There is a constant* $c > 0$, *which does not depend on* R *such that*

$$\int_{\mathbb{R}^3} \widetilde{\mathbf{u}}^2_R \eta^{\alpha-1}_\beta \mathrm{d}\mathbf{x} + \int_{\mathbb{R}^3} |\nabla \widetilde{\mathbf{u}}_R|^2 \eta^\alpha_\beta \mathrm{d}\mathbf{x} \leq c \int_{\mathbb{R}^3} |\tilde{\mathbf{f}}|^2 \eta^{\alpha+1}_\beta \mathrm{d}\mathbf{x} \tag{7.27}$$

for all R *greater than some* $R_0 > 0$, $\widetilde{\mathbf{u}}_R$ *being extension by zero of* $\mathbf{u}_R$ *on* $\mathbb{R}^3 \setminus B_R$.

□

Proof First, we derive an estimate of $\mathbf{u}_R$ on a bounded subdomain $B_{R_0} \subset B_R$. The choice of R_0 will be given in the next part of the proof. Our aim is to get an estimate with a constant not depending on R. Let us substitute $\mathbf{v} = \mathbf{u}_R$ into (7.23). Hence, we get from (7.24)

$$\widetilde{Q}(\mathbf{u}_R, \mathbf{u}_R) = \int_{B_R} \tilde{\mathbf{f}}\mathbf{u}_R \eta^0_{\beta_0} \mathrm{d}\mathbf{x} \geq C_1 \left(\int_{B_R} |\nabla \mathbf{u}_R|^2 \eta^0_{\beta_0} \mathrm{d}\mathbf{x} + \int_{B_R} \mathbf{u}^2_R \eta^{-1}_{\beta_0} \mathrm{d}\mathbf{x}\right),$$

with the constant $C_1 > 0$ stated in (7.24). Let R_0 be some fixed positive number such that $0 < R_0 < R$. We get

$$\int_{B_{R_0}} |\nabla \mathbf{u}_R|^2 \eta_\beta^\alpha \mathbf{dx} + \int_{B_{R_0}} \mathbf{u}_R^2 \eta_\beta^{\alpha-1} \mathbf{dx} \leq C_2 \int_{B_R} |\tilde{\mathbf{f}}|\, |\mathbf{u}_R| \eta_\beta^\alpha \mathbf{dx}, \tag{7.28}$$

where the constant $C_2 = C_1^{-1}(1+\varepsilon_0 R_0)^\alpha (1+\varepsilon_0 2R_0)^{|\beta-\beta_0|}$ depends on k, ν, α, β, β_0, ε_0, R_0, κ but does not depend on R.

Now, we are going to derive an estimate of $\mathbf{u}_R$ on domain B_R. Using the test function $\boldsymbol{\Phi} = \mathbf{u}_R \eta_\beta^\alpha = \mathbf{u}_R (1+\delta r)^\alpha (1+\varepsilon s)^\beta \in \overset{\circ}{\mathbf{H}}{}^1(B_R)$ in (7.26), we get after integration by parts

$$\nu \int_{B_R} |\nabla \mathbf{u}_R|^2 \eta_\beta^\alpha \mathbf{dx} + \nu \int_{B_R} (\mathbf{u}_R \cdot \nabla \mathbf{u}_R) \cdot \nabla \eta_\beta^\alpha \mathbf{dx} - \frac{k}{2} \int_{B_R} \mathbf{u}_R^2 \partial_3 \eta_\beta^\alpha \mathbf{dx}$$
$$= \int_{B_R} \tilde{\mathbf{f}} \mathbf{u}_R \eta_\beta^\alpha \mathbf{dx}.$$

So, we get for some $\kappa > 1$

$$\frac{\nu}{2} \int_{B_R} |\nabla \mathbf{u}_R|^2 \eta_\beta^\alpha \mathbf{dx} + \frac{1}{2} \int_{B_R} \mathbf{u}_R^2 F_{\alpha,\beta}(s,r) \eta_{\beta-1}^{\alpha-1} \mathbf{dx} \leq \int_{B_R} |\tilde{\mathbf{f}}|\, |\mathbf{u}_R| \eta_\beta^\alpha \mathbf{dx}.$$

Let $R_0 \geq |\delta^{-1} - (2\varepsilon)^{-1}|(\kappa-1)^{-1}$. Using Lemma 7.2 (with $0 \leq \alpha < \beta$, $\varepsilon \leq (1/(2\kappa))(k/\nu)((\beta-\alpha)/\beta^2)$) and Lemma 7.1 (with $\delta < 2\varepsilon$), the second term in the previous estimate can be evaluated from below:

$$\int_{B_R} \mathbf{u}_R^2\, F_{\alpha,\beta}(s,r)) \eta_{\beta-1}^{\alpha-1} \mathbf{dx}$$
$$\geq -\alpha\delta k \left(1 + \frac{\nu\kappa}{k}\alpha\delta\right) \frac{2\kappa}{\delta\varepsilon} \left(\frac{\alpha+\beta}{\beta\beta^*}\right)^2 \int_{B_R^{R_0}} |\nabla \mathbf{u}_R|^2 \eta_\beta^\alpha \mathbf{dx}$$
$$+ (1-\kappa^{-1}) k\delta\varepsilon(\beta-\alpha) \int_{B_R^{R_0}} \mathbf{u}_R^2 \eta_{\beta-1}^{\alpha-1} s \mathbf{dx} - 2C_4 \int_{B_{R_0}} |\nabla \mathbf{u}_R|^2 \eta_\beta^\alpha \mathbf{dx}.$$

Denote $C_5 = \alpha\delta k(1+\nu\kappa\alpha\delta/k)(\kappa/(\delta\varepsilon))((\alpha+\beta)/(\beta\beta^*))^2$. It is clear that $C_5 \leq \nu/(2\kappa^2) < \nu/(2\kappa)$ if $1 + \nu\kappa\alpha\delta/k \leq \kappa$ (i.e., $\delta \leq (k/\nu)((\kappa-1))/(\kappa\beta))$ and $\alpha \leq (1/(2\kappa^4))(\nu/k)((\beta\beta^*)/(\alpha+\beta))^2\varepsilon$. We have

$$\frac{\nu}{2\kappa} \int_{B_R} |\nabla \mathbf{u}_R|^2 \eta_\beta^\alpha \mathbf{dx} + \frac{1}{2}\left(1 - \frac{1}{\kappa}\right) k\delta\varepsilon\,(\beta-\alpha) \int_{B_R} \mathbf{u}_R^2 \eta_{\beta-1}^{\alpha-1} s \mathbf{dx}$$
$$- C_6 \int_{B_{R_0}} \mathbf{u}_R^2 \eta_{\beta-1}^{\alpha-1} \mathbf{dx} - C_7 \int_{B_{R_0}} |\nabla \mathbf{u}_R|^2 \eta_\beta^\alpha \mathbf{dx} \leq \int_{B_R} |\tilde{\mathbf{f}}|\, |\mathbf{u}_R| \eta_\beta^\alpha \mathbf{dx}.$$

We use now relation (7.28) in order to estimate the integrals computed on the domain B_{R_0}. Before using the mentioned inequality, we will rescale it with respect to new values ε, δ, see Remark 7.1. We denote the new constant in (7.28) after rescaling as C_2'.

$$\frac{\nu}{\kappa}\int_{B_R}|\nabla\mathbf{u}_R|^2\eta_\beta^\alpha\mathbf{dx}+k\delta\varepsilon(\beta-\alpha)\int_{B_R}\mathbf{u}_R^2\eta_{\beta-1}^{\alpha-1}s\mathbf{dx}\le C_8\int_{B_R}|\tilde{\mathbf{f}}|\,|\mathbf{u}_R|\eta_\beta^\alpha\mathbf{dx},$$

where $C_8=\{1+C_2'\max(C_6,C_7)\}\cdot 2(1-\kappa^{-1})^{-1}$. We use Lemma 7.1 and Remark 7.6. So, if $\delta<2\varepsilon$ and $1<\kappa\le 2\varepsilon/\delta+\delta/(2\varepsilon)-1$, we get

$$\frac{\nu}{2\kappa}\left(\frac{\beta\beta^*\delta\varepsilon}{\alpha\delta+2\beta\varepsilon}\right)^2\int_{B_R}\mathbf{u}_R^2\eta_{\beta-1}^{\alpha-1}\mathbf{dx}\le\frac{\nu}{2\kappa}\int_{B_R}|\nabla\mathbf{u}_R|^2\eta_\beta^\alpha\mathbf{dx},$$

$$\frac{\nu}{2\kappa}\int_{B_R}|\nabla\mathbf{u}_R|^2\eta_\beta^\alpha\mathrm{dx}+\frac{\nu}{2\kappa}\left(\frac{\beta\beta^*\delta\varepsilon}{\alpha\delta+2\beta\varepsilon}\right)^2\int_{B_R}\mathbf{u}_R^2\eta_{\beta-1}^{\alpha-1}\mathbf{dx}$$
$$+k\delta\varepsilon(\beta-\alpha)\int_{B_R}\mathbf{u}_R^2\eta_{\beta-1}^{\alpha-1}s\mathbf{dx}\le C_8\int_{B_R}|\tilde{\mathbf{f}}|\,|\mathbf{u}_R|\eta_\beta^\alpha\mathbf{dx}.$$

So we get

$$\int_{B_R}|\nabla\mathbf{u}_R|^2\eta_\beta^\alpha\mathbf{dx}+2\int_{B_R}\mathbf{u}_R^2\eta_{\beta-1}^{\alpha-1}\mathbf{dx}+2\varepsilon\int_{B_R}\mathbf{u}_R^2\eta_{\beta-1}^{\alpha-1}s\mathbf{dx}$$
$$=\int_{B_R}|\nabla\mathbf{u}_R|^2\eta_\beta^\alpha\mathbf{dx}+2\int_{B_R}\mathbf{u}_R^2\eta_\beta^{\alpha-1}\mathbf{dx}\le C_{10}\int_{B_R}|\tilde{\mathbf{f}}|\,|\mathbf{u}_R|\eta_\beta^\alpha\mathbf{dx},$$

$C_9=\min(\nu/(2\kappa),(\nu/(2\kappa))(\beta\beta^*\delta\varepsilon/\alpha\delta+2\beta\varepsilon))^2,k\delta(\beta-\alpha)/2$ and $C_{10}=C_8/C_9$. We have also

$$\int_{B_R}|\tilde{\mathbf{f}}|\,|\mathbf{u}_R|\,\eta_\beta^\alpha\mathrm{dx}\le\frac{t}{2}\int_{B_R}\mathbf{u}_R^2\eta_\beta^{\alpha-1}\mathbf{dx}+\frac{1}{2t}\int_{B_R}|\tilde{\mathbf{f}}|^2\eta_\beta^{\alpha+1}\mathrm{dx}.$$

So, if we choose $t=2\cdot C_{10}^{-1}$, then we get

$$\int_{B_R}|\nabla\mathbf{u}_R|^2\eta_\beta^\alpha\mathbf{dx}+\int_{B_R}\mathbf{u}_R^2\eta_\beta^{\alpha-1}\mathbf{dx}\le c\int_{\mathbb{R}^3}|\tilde{\mathbf{f}}|^2\eta_\beta^{\alpha+1}\mathbf{dx}.$$

It can be easily shown that all conditions on α, β, δ, ε, κ used in the proof are compatible if $0\le\alpha<y_1\beta$, see Sect. 7.5.3 Appendix B. □

7.4.1.2 The Problem in $\mathbb{R}^3$

Let y_1 be the same as in Lemma 7.4.

Theorem 7.6 (Existence and Uniqueness in $\mathbb{R}^3$) *Let $0 < \beta \leq 1$, $0 \leq \alpha < y_1\beta$, $\mathbf{f} \in \mathbf{L}^2_{\alpha+1,\beta}$, $g \in H^1_{loc}$ such that $\nu\nabla g - kg\mathbf{e}_1 + \omega g(\mathbf{e}_3 \wedge \mathbf{x}) \in L^2_{\alpha+1,\beta}$. Then there exists a unique weak solution $\{\mathbf{u}, p\}$ of the problem*

$$-\nu\,\Delta\mathbf{u} + k\partial_3\mathbf{u} - (\boldsymbol{\omega} \wedge x)\cdot\nabla\mathbf{u} + \boldsymbol{\omega} \wedge u + \nabla p = \mathbf{f} \quad \text{in } \mathbb{R}^3, \tag{7.29}$$

$$\operatorname{div}\mathbf{u} = g \quad \text{in } \mathbb{R}^3 \tag{7.30}$$

such that $\mathbf{u} \in \mathbf{V}_{\alpha,\beta}$, $p \in L^2_{\alpha,\beta-1}$, $\nabla p \in \mathbf{L}^2_{\alpha+1,\beta}$, and

$$\begin{aligned}&\|\mathbf{u}\|_{2,\alpha-1,\beta} + \|\nabla\mathbf{u}\|_{2,\alpha,\beta} + \|p\|_{2,\alpha,\beta-1} + \|\nabla p\|_{2,\alpha+1,\beta}\\ &\quad\leq C\left(\|\mathbf{f}\|_{2,\alpha+1,\beta} + \|\nu\nabla g - kg\mathbf{e}_3 + \omega g(\mathbf{e}_3 \wedge \mathbf{x})\|_{2,\alpha+1,\beta}\right).\end{aligned} \tag{7.31}$$

Proof The uniqueness of the solution follows from Theorem 7.4, and we now justify the existence. Let p be the same as in Sect. 7.3.1. Let $\{R_n\}$ be a sequence of positive real numbers, converging to $+\infty$. Let $\mathbf{u}_{R_n}$ be the weak solution of (7.20), (7.21) on B_{R_n}. Extending $\mathbf{u}_{R_n}$ by zero on $\mathbb{R}^3 \setminus B_{R_n}$ to a function $\tilde{\mathbf{u}}_n \in \mathbf{V}_{\alpha,\beta}$, we get a bounded sequence $\{\tilde{\mathbf{u}}_n\}$ in $\mathbf{V}_{\alpha,\beta}$. Thus, there is a subsequence $\tilde{\mathbf{u}}_{n_k}$ of $\tilde{\mathbf{u}}_n$ with a weak limit $\mathbf{u}$ in $\mathbf{V}_{\alpha,\beta}$. Obviously, $\mathbf{u}$ is a weak solution of (7.29) and

$$\begin{aligned}\|\mathbf{u}\|^2_{2,\alpha-1,\beta} + \|\nabla\mathbf{u}\|^2_{2,\alpha,\beta} &\leq \liminf_{k\in\mathbb{N}}\left(\int_{\mathbb{R}^3}\tilde{\mathbf{u}}^2_{n_k}\eta_\beta^{\alpha-1}\mathbf{dx} + \int_{\mathbb{R}^3}|\nabla\tilde{\mathbf{u}}_{n_k}|^2\eta_\beta^{\alpha}\mathbf{dx}\right)\\ &\leq c|\tilde{\mathbf{f}}|^2\eta_\beta^{\alpha+1}\mathbf{dx} = c\int_{\mathbb{R}^3}|\mathbf{f} - \nabla p|^2\eta_\beta^{\alpha+1}\mathbf{dx}.\end{aligned}$$

Taking into account also relation (7.15), we get (7.31).

Let us also check that for $\mathbf{u}$ Eq. (7.30) is satisfied. Let us mention that $\mathbf{u} \in \mathbf{H}^2_{loc}$ because $\mathbf{f} - \nabla p \in \mathbf{L}^2_{\alpha+1,\,\beta}$. So, computing the divergence of (7.29), we get

$$-\nu\,\Delta(\operatorname{div}\mathbf{u}) + k\partial_3(\operatorname{div}\mathbf{u}) - \omega(\mathbf{e}_3 \wedge \mathbf{x})\cdot\nabla(\operatorname{div}\mathbf{u}) = \operatorname{div}\mathbf{f} - \Delta p \tag{7.32}$$

in the distributional sense. From (7.14) we have

$$-\nu\Delta\gamma + k\partial_3\gamma - (\boldsymbol{\omega} \wedge \mathbf{x})\cdot\nabla\gamma = 0$$

for $\gamma = \operatorname{div}\mathbf{u} - g \in L^2_{\alpha,\beta} \subset L^2$. Using Fourier transform we get

$$(\nu|\xi|^2 + \mathrm{i}k\xi_3)\widehat{\gamma} - (\boldsymbol{\omega} \wedge \boldsymbol{\xi})\cdot\nabla_\xi\widehat{\gamma} = 0 \quad \text{in } \mathcal{S}'.$$

Assuming $\widehat{\gamma}$ in cylindrical coordinates $[\rho, \varphi, \xi_3]$, $\rho = (\xi_1^2 + \xi_2^2)^{1/2}$, we can overwrite the equation in the form:

$$-\partial_\varphi \widehat{\gamma} + [(\nu/\widetilde{\omega})\, |\xi|^2 + \mathrm{i}(k/\widetilde{\omega})\xi_3]\widehat{\gamma} = 0.$$

Using the same approach as in the proof of the uniqueness Theorem 7.4, we prove that $\operatorname{supp} \widehat{\boldsymbol{\gamma}} \subset \{0\}$. The proof of this fact is reduced to the solvability of Eq. (7.18) which was proved for arbitrary $\Psi \in C_0^\infty(\mathbb{R}^3 \setminus \{0\})$ in the proof of Theorem 7.4. So, by the same procedure we derive that γ is a polynomial in $\mathbb{R}^3$, and because $\gamma \in L^2$ we get $\gamma \equiv 0$, i.e., (7.30). □

7.5 Existence of Solution in Exterior Domains

In this section we assume problem (7.1)–(7.4) in an exterior domain $\mathcal{D}$. First we assume the case of the homogenous Dirichlet boundary condition on $\partial\mathcal{D}$.

7.5.1 Homogenous Dirichlet Boundary Conditions

Function g is assumed to be zero, and $\mathbf{f} = \operatorname{div} \mathbf{F}$ with $\mathbf{F} \in C_0^\infty(\mathcal{D})^9$. We will prove that the problem has a weak solution $\{\mathbf{u}, p\} \in \widehat{\mathbf{H}}_0^1(\mathcal{D}) \times L^2_{\mathrm{loc}}(\mathcal{D})$. So we assume the following sequence of problems on domains $\mathcal{D}_R = B_R \cap \mathcal{D}$:

$$-\nu\, \Delta \mathbf{u}_R + k\partial_3 \mathbf{u}_R + (\boldsymbol{\omega} \wedge \mathbf{x}) \cdot \nabla \mathbf{u}_R - \boldsymbol{\omega} \wedge \mathbf{u}_R + \nabla p_R = \operatorname{div} \mathbf{F} \quad \text{in } \mathcal{D}_R \tag{7.33}$$

$$\operatorname{div} \mathbf{u}_R = 0 \quad \text{in } \mathcal{D}_R \tag{7.34}$$

$$\mathbf{u}_R = 0 \quad \text{on } \partial\mathcal{D}_R. \tag{7.35}$$

Following Girault–Raviart [121], we formulate each problem in the following mixed variational form: To find $\{\mathbf{u}_R, p_R\} \in \mathbf{W}_R \times \Pi_R$, such that for all $\mathbf{v} \in \mathbf{W}_R$, $\pi \in \Pi_R$:

$$a(\mathbf{u}_R, \mathbf{v}) + b(\mathbf{v}, p_R) = \langle \operatorname{div} \mathbf{F}, \mathbf{v} \rangle \tag{7.36}$$

$$b(\mathbf{u}_R, \pi) = 0, \tag{7.37}$$

where $\mathbf{W}_R = \widehat{\mathbf{H}}_0^1(\mathcal{D}_R)$, $\Pi_R = \{\pi \in L^2(\mathcal{D}_R); \int_{\mathcal{D}_R} \pi \,\mathrm{d}\mathbf{x} = 0\}$ with usual norms $\|\boldsymbol{\varphi}\|_{W_R} = \|\nabla\boldsymbol{\varphi}\|_2$, $\|\pi\|_{\Pi_R} = \|\pi\|_2$, and

$$a(\boldsymbol{\varphi}, \boldsymbol{\psi}) = \nu \int_{\mathcal{D}_R} \nabla\boldsymbol{\varphi} : \nabla\boldsymbol{\psi} \,\mathrm{d}\mathbf{x} + k \int_{\mathcal{D}_R} \partial_3\boldsymbol{\varphi} \cdot \boldsymbol{\psi} \,\mathrm{d}\mathbf{x}$$

$$+\int_{D_R} [(\omega \wedge \mathbf{x})\nabla\boldsymbol{\varphi} - \boldsymbol{\omega} \wedge \boldsymbol{\varphi}] \cdot \boldsymbol{\psi} d\mathbf{x}$$

$$b(\boldsymbol{\varphi}, \pi) = -\int_{D_R} \pi \operatorname{div} \boldsymbol{\varphi} d\mathbf{x}.$$

These bilinear forms are continuous on $\mathbf{W}_R \times \mathbf{W}_R$ and $\mathbf{W}_R \times \Pi_R$, respectively. It is easy to see that $a(\boldsymbol{\varphi}, \boldsymbol{\varphi}) \geq \nu \|\boldsymbol{\varphi}\|^2_{\mathbf{W}_R}$, and it is known that

$$\sup_{\mathbf{v} \in \mathbf{W}_R} \frac{(\pi, \operatorname{div} \mathbf{v})}{|\mathbf{v}|_{\mathbf{W}_R}} \geq C_0 \|\pi\|_{\Pi_R}$$

for some $C_0 = C_0(R) > 0$. Hence, there exists a weak solution $\{\mathbf{u}_R, p_R\}$ of the problem and $\|\mathbf{u}_R\|_{\mathbf{W}_R} + \|p_R\|_{\Pi_R} \leq C_1 \| \operatorname{div} \mathbf{F}\|_{-1}$ for some $C_1 = C_1(R) > 0$. Testing now (7.36) by $\mathbf{v} = \mathbf{u}_R$, we get

$$\nu \int_{\mathcal{D}_R} |\nabla \mathbf{u}_R|^2 d\mathbf{x} = \int_{\mathcal{D}_R} (\operatorname{div} \mathbf{F}) \cdot \mathbf{u}_R d\mathbf{x} = \int_{\mathcal{D}_R} \mathbf{F} \colon \nabla \mathbf{u}_R d\mathbf{x} \leq \|\mathbf{F}\|_2 \|\nabla \mathbf{u}_R\|_2$$

$$\|\nabla \mathbf{u}_R\|_2 \leq \nu^{-1} \|\mathbf{F}\|_2. \tag{7.38}$$

Since the a priori estimate (7.38) is available, where $\mathbf{u}_R$ is understood as its extension by setting zero in $\mathcal{D} \setminus \mathcal{D}_R$, there exists $\mathbf{u} \in \widehat{\mathbf{H}}^1_0(\mathcal{D})$ and a sequence $\{R_n\} \to \infty$ so that $\mathbf{u}_{R_n} \rightharpoonup \mathbf{u}$ weakly in $\widehat{\mathbf{H}}^1_0(\mathcal{D})$ as $n \to \infty$.

Let us show that $\operatorname{div} \mathbf{u} = 0$ in $L^2(\mathcal{D})$. From the same inequality follows the weak convergence of $\operatorname{div} \mathbf{u}_{R_n}$ in $L^2(\mathcal{D})$. From (7.37) we get $\operatorname{div} \mathbf{u}_{R_n} \equiv C_n$ on D_{R_n} for some real constant C_n depending on n. In spite of (7.38) we get that the weak limit of $\operatorname{div} \mathbf{u}_{R_n}$ is zero in $L^2(\mathcal{D})$.

Finally, for all $\boldsymbol{\varphi} \in \mathbf{C}^\infty_0(\mathcal{D})$ with $\operatorname{div} \boldsymbol{\varphi} = 0$, we have from (7.36) after $R_n \to \infty$

$$\langle L\mathbf{u} - \operatorname{div} \mathbf{F}\boldsymbol{\varphi}\rangle = 0,$$

$$L\mathbf{u} \equiv -\nu \Delta \mathbf{u} + k\partial_3 \mathbf{u} + (\boldsymbol{\omega} \wedge \mathbf{x}) \cdot \nabla \mathbf{u} - \boldsymbol{\omega} \wedge \mathbf{u}.$$

By a result of de Rham, there is a distribution p such that $-\nabla p = L\mathbf{u} - \operatorname{div} \mathbf{F}$ in $\mathcal{C}^{\infty'}_0(\mathcal{D})$. Because the right-hand side belongs to $H^{-1}(\mathcal{D}_R)$ for every sufficiently large $R > 0$, we have that $p \in L^2(\mathcal{D}_R)$, and so, $p \in L^2_{\text{loc}}(\mathcal{D})$.

Now we use the following:

Lemma 7.5 (Kozono and Sohr [144, Lemma 2.2, Corollary 2.3] [Lemma2.3]) *Let $\Omega \subset \mathbb{R}^n$ $(n \geq 2)$ be any domain, and let $1 < q < \infty$. For all $g \in \widehat{W}^{-1,q}(\Omega)$, there is $\boldsymbol{G} \in \boldsymbol{L}^q(\Omega)$ such that*

$$\operatorname{div} \boldsymbol{G} = g, \quad \|\boldsymbol{G}\|_{q,\Omega} \leq C \|g\|_{-1,q,\Omega}$$

with some $C > 0$. *As a result, the space* $\{\operatorname{div} \mathbf{G};\ \mathbf{G} \in \mathbf{C}_0^\infty(\Omega)\}$ *is dense in* $\widehat{\mathbf{W}}^{-1,q}(\Omega)$. □

Hence, we get the existence of solution $\{\mathbf{u}, p\} \in \widehat{\mathbf{H}}_0^1(\mathcal{D}) \times L^2_{\text{loc}}(\mathcal{D})$ for an arbitrary function $\widetilde{\mathbf{f}} \in \widehat{\mathbf{H}}^{-1}(\mathcal{D})$.

For the extension of Theorem 7.6 to the case of an exterior domain, we use the localization procedure, see [144]. Let now $\mathbf{f} \in \mathbf{L}^2_{\alpha+1,\beta}(\mathcal{D})$. We define for an arbitrary $R > 0$

$$\mathbf{f}_R = \begin{cases} \mathbf{f}, & \mathbf{x} \in \mathcal{D}_R, \\ \mathbf{0}, & \mathbf{x} \in \mathcal{D} \setminus \mathcal{D}_R. \end{cases}$$

It can be shown that $\mathbf{f}_R$ belongs to $\widehat{\mathbf{H}}^{-1}(\mathcal{D}) \cap \mathbf{L}^2_{\alpha+1,\beta}(\mathcal{D})$. By use of cutoff function Ψ we decompose the solution $\{\mathbf{u}, p\}$ of the problem (7.1)–(7.4) (with the homogenous Dirichlet boundary condition) on the solution of a problem in $\mathbb{R}^3$ and the solution of a Stokes problem in a bounded domain:

$$\begin{aligned} \mathbf{u} &= \mathbf{U} + \mathbf{V}, \quad \text{where } \mathbf{U} = (1 - \Psi)\mathbf{u},\ \mathbf{V} = \Psi \mathbf{u} \\ p &= \sigma + \tau, \quad \text{where } \sigma = (1 - \Psi)p,\ \tau = \Psi p, \end{aligned}$$

where $\Psi \in C_0^\infty$, $\operatorname{supp} \Psi \subset\subset B_{\rho_1}$ such that $\Psi \equiv 1$ on B_{ρ_0}, $0 < \rho_0 < \rho_1 < \rho$ so that $\mathbb{R}^3 \setminus \mathcal{D} \subset B_{\rho_0}$. We get that $\{\mathbf{U}, \sigma\}$ is a weak solution of the modified Oseen problem in $\mathbb{R}^3$

$$-\nu \Delta \mathbf{U} + k\partial_3 \mathbf{U} - (\boldsymbol{\omega} \wedge \mathbf{x})\nabla \mathbf{U} + \boldsymbol{\omega} \wedge \mathbf{U} + \nabla \sigma = \mathbf{Z}_1 \tag{7.39}$$

$$\operatorname{div} \mathbf{U} = -\nabla \Psi \mathbf{u}, \tag{7.40}$$

and $\{\mathbf{V}, \boldsymbol{\tau}\}$ is a weak solution of the Stokes problem in a bounded domain $\mathcal{D}_\rho$

$$-\nu \Delta \mathbf{V} + \nabla \tau = \mathbf{Z}_2 \quad \text{in } \mathcal{D}_\rho \tag{7.41}$$

$$\operatorname{div} \mathbf{V} = \nabla \Psi \mathbf{u} \quad \text{in } \mathcal{D}_\rho \tag{7.42}$$

$$\mathbf{V}|_{\partial \mathcal{D}_\rho} = 0, \tag{7.43}$$

where the right-hand sides are given by $\mathbf{Z}_1$ and $\mathbf{Z}_2$.

$$\begin{aligned} \mathbb{Z}_1 &= 2\nabla \Psi \nabla \mathbf{u} + \mathbf{u}\Delta \Psi - k\partial_3 \Psi \mathbf{u} + (\nabla \Psi(\boldsymbol{\omega} \times \mathbf{x}))\mathbf{u} - \nabla \Psi p + (1 - \Psi)\mathbf{f}_R, \\ \mathbb{Z}_2 &= -2\nabla \Psi \nabla \mathbf{u} - \mathbf{u}\Delta \Psi + k\partial_3 \Psi \mathbf{u} + \Psi[(\boldsymbol{\omega} \times \mathbf{x})\nabla \mathbf{u} - \boldsymbol{\omega} \times \mathbf{u}] + \nabla \Psi p + \Psi \mathbf{f}_R. \end{aligned}$$

Let us mention that $\mathbf{Z}_1 \in \mathbf{L}^2_{\alpha+1,\beta}(\mathcal{D})$. To solve the Stokes problem on the bounded domain, we use the following lemma, see [144]:

Lemma 7.6 (The Stokes Problem on a Bounded Domain) *Let Ω be a bounded domain of $\mathbb{R}^n$, $n \geq 2$, of class C^{m+2}, $m \geq 0$. For any*

$$\mathbf{f} \in \mathbf{W}^{m,q}(\Omega), \quad g \in W^{m+1,q}(\Omega), \quad \mathbf{v}_* \in \mathbf{W}^{m+2-1/q,q}(\partial\Omega),$$

$1 < q < \infty$, *with*

$$\int_{\partial\Omega} \mathbf{v}_* \cdot \mathbf{n} dS = \int_{\Omega} g d\mathbf{x}, \tag{7.44}$$

there exists one and only one solution $\{\mathbf{V}, \tau\}$ *to the Stokes system*

$$\begin{aligned} -\Delta\mathbf{V} + \nabla\tau &= \mathbf{f} \quad \text{in } \Omega, \\ \operatorname{div}, \mathbf{V} &= g \quad \text{in } \Omega \\ \mathbf{V} &= \mathbf{v}_* \quad \text{on } \partial\Omega \end{aligned}$$

such that $\mathbf{V} \in \mathbf{W}^{m+2,q}(\Omega)$, $\tau \in W^{m+1,q}(\Omega)$, *and*

$$\|\mathbf{V}\|_{m+2,q} + \|\tau - \bar{\tau}\|_{m+1,q} \leq c(\|\mathbf{f}\|_{m,q} + \|\mathbf{v}_*\|_{m+2-1/q,q} + \|g\|_{m+1,q}), \tag{7.45}$$

where $\bar{\tau} = |\Omega|^{-1} \int_{\Omega} \tau d\mathbf{x}$ *and* $c = c(m, n, q, \Omega)$.

Furthermore, for Ω of class C^2, for every

$$\mathbf{f} \in \mathbf{W}_0^{-1,q}(\Omega), \quad g \in L^q(\Omega), \quad \mathbf{v}_* \in \mathbf{W}^{1-1/q,q}(\partial\Omega),$$

$1 < q < \infty$, *with* (7.44) *there exists one and only one q-generalized solution* $\{\mathbf{V}, \tau\}$ *to the Stokes system such that* $\mathbf{V} \in \mathbf{W}^{1,q}(\Omega)$, $\tau \in L^q(\Omega)$, *and the estimate* (7.45) *is valid with* $m = -1$. □

From the results about the existence and uniqueness of solutions of the Oseen problem in $\mathbb{R}^3$ (7.39), (7.40), i.e., from Theorem 7.6 and Theorem 7.4 follows that a solution $\{\mathbf{U}, \sigma\}$ is subject of the estimate (7.31), with $\mathbf{f}$ and g replaced by $\mathbf{Z}_1$ and $-\nabla\Psi \cdot \mathbf{u}$, respectively. Using also the respective results in a bounded domain for (7.41)–(7.43), see Lemma 7.6 with $m = 0$ and bounded domain D_ρ, we get the following lemma for an exterior domain:

Lemma 7.7 *Let $\mathcal{D} \subset \mathbb{R}^3$ be an exterior domain and $0 < \beta \leq 1$, $0 \leq \alpha < y_1 \cdot \beta$; y_1 is given in Lemma 7.4. Then there exists a weak solution $\{\mathbf{u}, p\}$ of the problem (7.1)–(7.3) with the homogenous Dirichlet boundary condition, $\mathbf{f} := \mathbf{f}_R$, and $g = 0$, such that $\mathbf{u} \in \mathbf{V}_{\alpha,\beta}(\mathcal{D})$, $p \in L^2_{\alpha,\beta-1}(\mathcal{D})$, $\nabla p \in \mathbf{L}^2_{\alpha+1,\beta}(\mathcal{D})$, and*

$$\begin{aligned} &\|\mathbf{u}\|_{2,\alpha-1,\beta} + \|\nabla\mathbf{u}\|_{2,\alpha,\beta} + \|p\|_{2,\alpha,\beta-1} + \|\nabla p\|_{2,\alpha+1,\beta} \\ &\qquad \leq C_1 \left(\|\mathbf{f}_R\|_{2,\alpha+1,\beta} + \|\mathbf{u}\|_{1,2;A_\rho} + \|p\|_{0,2;\mathcal{D}_\rho} \right), \end{aligned} \tag{7.46}$$

where $A_\rho := B_\rho \setminus B_{\rho/2}$, and constant C_1 does not depend on R. □

Now, we would like to show that the preceding estimate is valid (with another constant) also if we add to the left-hand side the L^2-norm of second gradient of $\mathbf{u}$ on some compact subset of $\mathcal{D}$. Taking into account the assertion of Lemma 7.6 for $m = 0$, we get that $\mathbf{u} \in \mathbf{W}^{2,2}_{\mathrm{loc}}(\mathcal{D})$, $p \in W^{1,2}_{\mathrm{loc}}(\mathcal{D})$. Multiplying the relation (7.1)–(7.4) in an exterior domain $\mathcal{D}$ (with $g = 0$ and the homogenous Dirichlet boundary condition on $\partial\mathcal{D}$) by $\Delta\mathbf{u}$ and integrating over the compact set K_1 with $A_\rho \subset K_1 \subset \mathcal{D}$, we get

$$\|\Delta\mathbf{u}\|_{2;K_1} \le C_2\left(\|\mathbf{u}\|_{2;K_1} + \|\nabla\mathbf{u}\|_{2;K_1} + \|p\|_{2;K_1} + \|\nabla p\|_{2;K_1} + \|\mathbf{f}_R\|_{2;K_1}\right). \tag{7.47}$$

Using (7.46), (7.47), and the known relation

$$\|\nabla^2\mathbf{u}\|_{2;K} \le c\left(\|\Delta\mathbf{u}\|_{2;K_1} + \|\nabla\mathbf{u}\|_{2;K_1}\right)$$

with $A_\rho \subset K \subset K_1$, we get the following corollary:

Corollary 7.1 *In conditions of Lemma 7.7 the following estimate is valid, and constant C does not depend on R:*

$$\begin{aligned}\|\mathbf{u}\|_{2,\alpha-1,\beta} + \|\nabla\mathbf{u}\|_{2,\alpha,\beta} + \|\nabla^2\mathbf{u}\|_{2;\,A_\rho} + \|p\|_{2,\alpha,\beta-1} + \|\nabla p\|_{2,\alpha+1,\beta}\\ \le C\left(\|\mathbf{f}_R\|_{2,\alpha+1,\beta} + \|\mathbf{u}\|_{1,2;A_\rho} + \|p\|_{0,2;D_\rho}\right).\end{aligned} \tag{7.48}$$

Now, we will prove that the estimate (7.48) is valid without the right-hand side terms containing $\mathbf{u}$ and p with constant c which does not depend on R, i.e., we will prove

$$\|\mathbf{u}\|_{2,\alpha-1,\beta} + \|\nabla\mathbf{u}\|_{2,\alpha,\beta} + \|\nabla^2\mathbf{u}\|_{2;A_\rho} + \|p\|_{2,\alpha,\beta-1} + \|\nabla p\|_{2,\alpha+1,\beta} \le c\|\mathbf{f}_R\|_{2,\alpha+1,\beta}. \tag{7.49}$$

Let us define the norms:

$$\begin{aligned}\|(\mathbf{v},q)\|_{(1)} &:= \|\mathbf{v}\|_{1,2;A_\rho} + \|q\|_{0,2;D_\rho}\\ \|(\mathbf{v},q)\|_{(2)} &:= \|\mathbf{v}\|_{2,\alpha-1,\beta} + \|\nabla\mathbf{v}\|_{2,\alpha,\beta} + \|\nabla^2\mathbf{v}\|_{2;A_\rho}\\ &\quad + \|q\|_{2,\alpha,\beta-1} + \|\nabla q\|_{2,\alpha+1,\beta}.\end{aligned}$$

For the corresponding Hilbert spaces H_1 and H_2, we have $H_2 \hookrightarrow\hookrightarrow H_1$. Let us assume that the estimate (7.49) is not true. This means that there is a sequence of functions $\{\mathbf{f}^{(k)}_{R_k}\}_{k=1}^\infty$ with $R_k \to +\infty$, a sequence of corresponding solutions

$\{(\mathbf{u}_k, p_k)\}_{k=1}^{\infty}$, and a sequence of constants $\{c_k\}_{k=1}^{\infty} \to \infty$ such that

$$\begin{aligned} 1 \equiv &\|\mathbf{u}_k\|_{2,\alpha-1,\beta} + \|\nabla \mathbf{u}_k\|_{2,\alpha,\beta} + \|\nabla^2 \mathbf{u}_k\|_{2;A_\rho} + \|p_k\|_{2,\alpha,\beta-1} + \|\nabla p_k\|_{2,\alpha+1,\beta} \\ &\equiv \|(\mathbf{u}_k, p_k)\|_{(2)} \geq c_k \left\|\mathbf{f}_{R_k}^{(k)}\right\|_{2,\alpha+1,\beta}. \end{aligned}$$

So we get $\left\{\|\mathbf{f}_{R_k}^{(k)}\|_{2,\alpha+1,\beta}\right\}_k \to 0$. The sequence $\{(\mathbf{u}_k, p_k)\}_{k=1}^{\infty}$ is bounded in the norm $\|\cdot\|_{(2)}$, so there is a subsequence of this sequence (we will denote this subsequence using the same notation) with the weak limit $(\mathbf{u}, p)$ in the corresponding Hilbert space H_2. So, $(\mathbf{u}, p)$ is a solution of the problem with the zero right-hand side. Due to uniqueness given by Theorem 7.5, we conclude that $\|(\mathbf{u}, p)\|_{(2)} = 0$. Because $H_2 \hookrightarrow\hookrightarrow H_1$, we have $\|(\mathbf{u} - \mathbf{u}_k, p - p_k)\|_{(1)} \to 0$. From Corollary 7.1, we also get

$$\|(\mathbf{u} - \mathbf{u}_k, p - p_k)\|_{(2)} \to 0,$$

i.e., $\{(\mathbf{u}_k, p_k)\}_{k=1}^{\infty}$ converges strongly in H_2. Because $\|(\mathbf{u}_k, p_k)\|_{(2)} = 1$ for $k \in \mathbb{N}$, we also get $\|(\mathbf{u}, p)\|_{(2)} = 1$. This is the contradiction. So, we proved the following:

Theorem 7.7 *Let $\mathcal{D} \subset \mathbb{R}^3$ be an exterior domain and $0 < \beta \leq 1$, $0 \leq \alpha < y_1\beta$; y_1 is given in Lemma 7.4, $\mathbf{f} \in \mathbf{L}^2_{\alpha+1,\beta}(\mathcal{D})$. Then there exists a weak solution $\{\mathbf{u}, p\}$ of the problem (7.1)–(7.3) with the homogenous Dirichlet boundary condition on $\partial\mathcal{D}$, $g = 0$, such that $\mathbf{u} \in \mathbf{V}_{\alpha,\beta}(\mathcal{D})$, $p \in L^2_{\alpha,\beta-1}(\mathcal{D})$, $\nabla p \in \mathbf{L}^2_{\alpha+1,\beta}(\mathcal{D})$, and*

$$\|\mathbf{u}\|_{2,\alpha-1,\beta} + \|\nabla \mathbf{u}\|_{2,\alpha,\beta} + \|p\|_{2,\alpha,\beta-1} + \|\nabla p\|_{2,\alpha+1,\beta} \leq C\|\mathbf{f}\|_{2,\alpha+1,\beta}.$$

Remark 7.9 The used contradiction argument is based on a subtle choice of the sequence $\{\mathbf{f}_{R_k}^{(k)}\}_k$ with $R_k \to +\infty$. We cannot construct a contradiction separately for $\mathbf{f}_R$ with fixed R because then the constant c in (7.49) may depend on R. □

7.5.1.1 Nonhomogenous Cases

In this subsection we take into account the nonhomogenous Dirichlet boundary condition and the nonhomogenous continuity equation.

We can prove the following extension of Theorem 7.7 for the case $g \neq 0$:

Corollary 7.2 *Let $\mathcal{D} \subset \mathbb{R}^3$ be an exterior domain and $0 < \beta \leq 1$, $0 \leq \alpha < y_1 \cdot \beta$; y_1 is given in Lemma 7.4, $\mathbf{f} \in \mathbf{L}^2_{\alpha+1,\beta}(\mathcal{D})$, $g \in W_0^{1,2}(\mathcal{D})$, with $\operatorname{supp} g = K \subset\subset \mathcal{D}$ and $\int_D g\mathrm{d}\mathbf{x} = 0$. Then there exists a weak solution $\{\mathbf{u}, p\}$ of the problem (7.1)–(7.3) with the homogenous boundary condition on ∂D such that $\mathbf{u} \in \mathbf{V}_{\alpha,\beta}(\mathcal{D})$, $p \in L^2_{\alpha,\beta-1}(\mathcal{D})$, $\nabla p \in \mathbf{L}^2_{\alpha+1,\beta}(\mathcal{D})$, and*

$$\|\mathbf{u}\|_{2,\alpha-1,\beta} + \|\nabla \mathbf{u}\|_{2,\alpha,\beta} + \|p\|_{2,\alpha,\beta-1} + \|\nabla p\|_{2,\alpha+1,\beta} \leq C(\|\mathbf{f}\|_{2,\alpha+1,\beta} + \|g\|_{1,2}).$$

First of all let us recall the lemma which will be used for the extension of our results to the case with nonzero divergence:

Lemma 7.8 ((M. E. Bogovskii, G. P. Galdi, and H. Sohr), Lemma 2.2) *Let $\Omega \subseteq \mathbb{R}^n$, $n \geq 2$, be a bounded Lipschitz domain, and $1 < q < \infty$, $n \in \mathbb{N}$. Then for each $g \in W_0^{k,q}(\Omega)$ with $\int_\Omega g\mathrm{d}x = 0$, there exists $\mathbf{G} \in \left(W_0^{k+1,q}(\Omega)\right)^n$ satisfying*

$$\operatorname{div}\mathbf{G} = g, \quad \|\mathbf{G}\|_{(W_0^{k+1,q}(\Omega))^n} \leq C\, \|g\|_{W_0^{k,q}(\Omega)}$$

with some constant $C = C(q, k, \Omega) > 0$. □

For the proof and further references, see, e.g., [201, Lemma 2.3.1].

Proof of Corollary 7.2 Using Lemma 7.8 we find $\mathbf{G} \in \mathbf{W}_0^{2,2}(\mathcal{D})$, $\operatorname{supp}\mathbf{G} \subset \mathcal{K}$, where $\mathcal{K}$ is a bounded Lipschitz domain being contained in ε-neighborhood $\mathcal{K}_\varepsilon$ of compact set K for an arbitrary $\varepsilon > 0$, $\operatorname{div}\mathbf{G} = g$, $\|\mathbf{G}\|_{2,2} \leq C\|g\|_{1,2}$. We choose ε such that $\mathcal{K}_\varepsilon \subset \mathcal{D}$. Let us assume the following problem:

$$\begin{aligned} -\nu\Delta\mathbf{U} + k\partial_3\mathbf{U} - (\boldsymbol{\omega}\wedge\mathbf{x})\cdot\nabla\mathbf{U} + \boldsymbol{\omega}\wedge\mathbf{U} + \nabla p =& \mathbf{F} \quad \text{in } \mathcal{D} \\ \operatorname{div}\mathbf{U} =& 0 \quad \text{in } \mathcal{D} \end{aligned}$$

with the homogenous Dirichlet boundary condition for $\mathbf{U}$, where $\mathbf{U} = \mathbf{u} - \mathbf{G}$, $\mathbf{F} = \mathbf{f} + \nu\Delta\mathbf{G} - k\partial_1\mathbf{G} + (\boldsymbol{\omega}\wedge\mathbf{x})\cdot\nabla\mathbf{G} - \boldsymbol{\omega}\wedge\mathbf{G}$. The assertion of Corollary 7.2 follows from Theorem 7.7. □

Now we justify our third main theorem.

Theorem 7.8 *Let $\mathcal{D} \subset \mathbb{R}^3$ be an exterior domain and $0 < \beta \leq 1$, $0 \leq \alpha < y_1 \cdot \beta$; y_1 is given in Lemma 7.4, $\mathbf{f} \in \mathbf{L}^2_{\alpha+1,\beta}(\mathcal{D})$, $g \in W_0^{1,2}(\mathcal{D})$, with $\operatorname{supp} g = K \subset\subset \mathcal{D}$ and $\int_\mathcal{D} g\mathrm{d}\mathbf{x} = 0$. Then there exists a weak solution $\{\mathbf{u}, p\}$ of the problem (7.1)–(7.4) such that $\mathbf{u} \in \mathbf{V}_{\alpha,\beta}(\overline{\mathcal{D}})$, $p \in L^2_{\alpha,\beta-1}(\mathcal{D})$, $\nabla p \in \mathbf{L}^2_{\alpha+1,\beta}(\mathcal{D})$, and*

$$\begin{aligned} &\|\mathbf{u}\|_{2,\alpha-1,\beta} + \|\nabla\mathbf{u}\|_{2,\alpha,\beta} + \|p\|_{2,\alpha,\beta-1} + \|\nabla p\|_{2,\alpha+1,\beta} \\ &\qquad \leq C(\|\mathbf{f}\|_{2,\alpha+1,\beta} + \|g\|_{1,2} + \omega^2 + \omega + k^2 + k). \end{aligned}$$

Proof Let $\rho > 0$ be such that $\mathbb{R}^3 \setminus B_{\rho/2} \subset \mathcal{D}$. Let $\Phi = \Phi(z) \in C_0^\infty(\langle 0, +\infty))$ be a nonincreasing cutoff function such that $\Phi(z) \equiv 1$ for $z < 1/2$ and $\Phi(z) \equiv 0$ for $z > 1$. Let $|\Phi'| \leq 3$. Let $\Phi_\rho \equiv \Phi_\rho(\mathbf{x}) \equiv \Phi(|\mathbf{x}|/\rho)$. We have $|\nabla\Phi_\rho| \leq 3/\rho$ and $|\partial_3\Phi_\rho| \leq 3/\rho$ for $\mathbf{x} \in \mathbb{R}^3$, $\rho/2 \leq |\mathbf{x}| \leq \rho$. Let us define $\widetilde{\mathbf{u}} = \mathbf{u} - [(\omega\wedge\mathbf{x}) - k\mathbf{e}_1]\cdot \Phi_\rho(\mathbf{x})$. Then function $(\widetilde{\mathbf{u}}, p)$ satisfies to (7.1)–(7.3) with the homogenous Dirichlet boundary condition, where $\mathbf{f} \in \mathbf{L}^2_{\alpha+1,\beta}(\mathcal{D})$ is replaced by some other function $\widetilde{\mathbf{f}} \in \mathbf{L}^2_{\alpha+1,\beta}(\mathcal{D})$ and g by another function $\widetilde{g} \in C_0^\infty(\mathcal{D})$ with $\operatorname{supp} \widetilde{g} = K \cup A_\rho$, $A_\rho := B_\rho \setminus \overline{B_{\rho/2}} \subset\subset \mathcal{D}$ and

$$\int_{\mathcal{D}} \widetilde{g} \mathrm{d}\mathbf{x} = 0.$$

So, using now Corollary 7.2, we get the assertion of Theorem 7.8. □

7.5.2 Appendix A

Relation (7.13) follows from an estimate of the derivative of F_1:

$$\begin{aligned}
\frac{\partial}{\partial s} F_1(s,r) \equiv & \frac{\partial}{\partial s}\{F_{\alpha,\beta}(s,r) - (1-\kappa^{-1})k\delta\varepsilon(\beta-\alpha)s\} \\
= & -\nu\alpha^2\delta^2\varepsilon\frac{1}{1+\delta r} - 2\nu\alpha\beta\delta\varepsilon\frac{1}{r} - 2\nu\beta^2\varepsilon^2\frac{1+\delta r}{r}\frac{1}{(1+\varepsilon s)^2} \\
& - k\alpha\delta\varepsilon + k\alpha\delta\frac{1}{r}(1+2\varepsilon s) + k\beta\varepsilon(1+\delta r)\frac{1}{r} \\
& - (1-\kappa^{-1})k\delta\varepsilon(\beta-\alpha) \\
\geq & \delta\varepsilon\{r^{-1}[k(\alpha/\varepsilon + \beta/\delta) - \nu\alpha^2 - 2\nu\alpha\beta - 2\nu\beta^2\varepsilon/\delta] \\
& + [-2\nu\beta^2\varepsilon + k(\beta-\alpha)/\kappa]\} \geq 0.
\end{aligned}$$

The last inequality follows from the fact that we have $k\alpha/\varepsilon \geq \nu\alpha^2 + 2\nu\alpha\beta$, $k\beta/\delta \geq 2\nu\beta^2\varepsilon/\delta$, $k(\beta-\alpha)/\kappa \geq 2\nu\beta^2\varepsilon$ if $\varepsilon \leq (1/(2\kappa))(k/\nu)((\beta-\alpha)/\beta^2)$. Hence, if the last inequality (which is included in the conditions of Lemma 7.2) is satisfied, then $(\partial/\partial s)F_1(s,r) \geq 0$. So, we get immediately

$$F_1(s,r) \geq F_1(0,r) \equiv -k\alpha\delta - \nu\alpha^2\delta^2(1+\delta r)^{-1} \geq -\alpha\delta k(1+\nu k^{-1}\alpha\delta).$$

7.5.3 Appendix B

Let us show that all conditions on α, β, δ, ε, κ used in the proof of Lemma 7.4 are compatible if $0 < \beta \leq 1$, $0 \leq \alpha < y_1\beta$. Let us collect these assumptions: $0 < \delta < 2\varepsilon$, $1 < \kappa \leq 2\varepsilon/\delta + \delta/(2\varepsilon) - 1$, $0 \leq \alpha < \beta$, $\varepsilon \leq (1/(2\kappa^2)) \cdot (k/\nu) \cdot ((\beta-\alpha)/\beta^2)$, $\delta \leq (k/\nu) \cdot (\kappa-1)/(\kappa\beta)$, $\alpha \leq (1/(2\kappa^4)) \cdot (k/\nu) \cdot (\beta\beta^*/(\alpha+\beta))^2\varepsilon$.

From $\alpha \leq (1/(2\kappa^4)) \cdot (k/\nu) \cdot (\beta\beta^*/(\alpha+\beta))^2\varepsilon$, and $\varepsilon \leq (1/(2\kappa^2)) \cdot (k/\nu) \cdot ((\beta-\alpha)/\beta^2)$, we get $\alpha \leq (1/(4\kappa^6)) \cdot (\beta^*)^2(\beta-\alpha)/(\alpha+\beta)^2$. So we get ($\kappa > 1$,

$\beta \le 1$): $\alpha/\beta \le (1/(4\kappa^6))(1-\alpha/\beta)/(1+\alpha/\beta)^2$. By substitution $y = \alpha/\beta$, we get the inequality

$$4y^3 + 8y^2 + 4y + \kappa^{-6} \cdot (y-1) \le 0. \tag{7.50}$$

Taking into account the condition $0 \le \alpha < \beta$, we seek for solutions from $[0, 1)$. It is clear that the equation $4y^3 + 8y^2 + y + \kappa^{-6}(y-1) = 0$ has a unique real solution $y_\kappa \in (0, 1)$ for $\kappa > 1$. It is also clear that arbitrary $y \in [0, y_\kappa)$ solves (7.50). The value y_κ as a function of κ is decreasing. For $\kappa \to 1$ we get the inequality $4y^3 + 8y^2 + 5y - 1 \le 0$. This respective equation has a unique solution $y_1 = \left(\sqrt{13}/\left(6\sqrt{6}\right) + 53/216\right)^{1/3} + (1/30)\left(\sqrt{13}/\left(6\sqrt{6}\right) + 53/216\right)^{-1/3}$. Approximately, with an error less than 10^{-8}, we have $y_1 \doteq 0.1582981$, $(y_1 > 1/7)$. If $0 \le \alpha < y_1\beta$, then there is $\kappa > 1$ sufficiently close to number 1, such that $0 \le \alpha \le y_\kappa\beta$, so the relation $\alpha \le \left(1/\left(4\kappa^6\right)\right) \cdot (\beta^*)^2 (\beta-\alpha)/(\alpha+\beta)^2$ is satisfied. Then we can define $\varepsilon = 1/\left(2\kappa^2\right) \cdot (k/\nu) \cdot \left((\beta-\alpha)/\left(\beta^2\right)\right)$. The relation $\varepsilon \le (1/(2\kappa)) \cdot (k/\nu) \cdot (1/\beta)$ is satisfied. Then we take sufficiently small $\delta > 0$ such that $0 < \delta < 2\varepsilon$ and $1 < \kappa \le 2\varepsilon/\delta + \delta/(2\varepsilon) - 1$. Hence, all conditions which we assume in the proof of Lemma 7.4 are satisfied.

7.6 Notes on Chap. 7

This chapter is the extension of the previous results by Kračmar and Penel on the generalized Oseen problem, see [156, 157].

Chapter 8
Stokes and Oseen Type Operators: Spectral Theory and Generated Semigroups

The aim of Chap. 8 is to investigate the spectral theory of the Stokes-type and Oseen-type operators arising from the problems 3.1 and 4.1 in L^2 and L^q spaces in the whole space $\mathbb{R}^3$ and in the exterior domain $\mathcal{D}$. The content of this chapter is essentially based on the series of five papers [62, 74–77] of the first and third author of this book with R. Farwig.

8.1 Associated Stokes and Oseen Type Operators and Their Basic Properties

Recall that $\mathcal{D}$ denotes an exterior domain in $\mathbb{R}^3$. We assume that the boundary $\partial\mathcal{D}$ is of the class $C^{1,1}$ in this chapter.

We use the notation:

- $\boldsymbol{C}^{\infty}_{0,\sigma}(\mathcal{D})$ is the space of infinitely differentiable divergence-free vector functions in $\mathcal{D}$. For $1 < q < \infty$, $\boldsymbol{L}^q_\sigma(\mathcal{D})$ denotes the closure of $\boldsymbol{C}^{\infty}_{0,\sigma}(\mathcal{D})$ in $\boldsymbol{L}^q(\mathcal{D})$. The space $\boldsymbol{L}^q_\sigma(\mathcal{D})$ can be characterized as a space of functions from $\boldsymbol{L}^q(\mathcal{D})$ that are divergence-free in the sense of distributions, and their normal component on $\partial\mathcal{D}$ is equal to zero in the sense of traces. (See [98, Section III.2].)
- The space $\boldsymbol{L}^q(\mathcal{D})$ can be expressed as a direct sum $\boldsymbol{L}^q_\sigma(\mathcal{D}) \oplus \boldsymbol{G}^q(\mathcal{D})$, where $\boldsymbol{G}^q(\mathcal{D}) := \{\nabla p \in \boldsymbol{L}^q(\mathcal{D});\ p \in W^{1,q}_{loc}(\mathcal{D})\}$. This is the so-called *Helmholtz decomposition.* Note that if $q = 2$, then the Helmholtz decomposition of $\boldsymbol{L}^2(\mathcal{D})$ holds for any domain $\mathcal{D} \subset \mathbb{R}^3$. The existence of the Helmholtz decomposition of $\boldsymbol{L}^q(\mathcal{D})$ for general $q \in (1,\infty)$ is a subtle problem; however, the assumption that $\mathcal{D}$ is an exterior domain with the boundary of the class $C^{1,1}$ guarantees that the Helmholtz decomposition of $\boldsymbol{L}^q(\mathcal{D})$ exists, see [199, Theorem 1.4]. We denote

Š. Nečasová et al., *Mathematical Theory of a Fluid Flow Around a Rotating and Translating Body*, Springer Monographs in Mathematics,
https://doi.org/10.1007/978-3-031-84120-0_8

by Π_σ^q the projection of $\boldsymbol{L}^q(\mathcal{D})$ onto $\boldsymbol{L}_\sigma^q(\mathcal{D})$, associated with the Helmholtz decomposition. The operator Π_σ^q is called the *Helmholtz projection.*

- $\boldsymbol{W}_{0,\sigma}^{1,q}(\mathcal{D}) := \boldsymbol{W}_0^{1,2}(\mathcal{D}) \cap \boldsymbol{L}_\sigma^q(\mathcal{D})$.
- Recall the notation $\boldsymbol{\omega} := \omega \mathbf{e}_3$ and $S_{\nu,\omega}\mathbf{u} := -\nu\Delta\mathbf{u} - \omega(\mathbf{e}_3 \wedge \mathbf{x}) \cdot \nabla\mathbf{u} + \boldsymbol{\omega} \wedge \mathbf{u}$. Assuming that $\omega \in \mathbb{R}$, $k \in \mathbb{R}$, and $\mathbf{u} \in D(\mathscr{A}_k^\omega)$, where

$$D(\mathscr{A}_k^\omega) := \{\mathbf{u} \in \boldsymbol{W}^{2,q}(\mathcal{D}) \cap \boldsymbol{W}_{0,\sigma}^{1,q}(\mathcal{D});\ (\boldsymbol{\omega} \wedge \mathbf{x}) \cdot \nabla\mathbf{u} \in \boldsymbol{L}^q(\mathcal{D})\},$$

we denote

$$\begin{aligned} \mathscr{A}_k^\omega \mathbf{u} &:= -\Pi_\sigma^q S_{\nu,\omega}\mathbf{u} + k\Pi_\sigma^q \partial_3\mathbf{u} \\ &= \nu\Pi_\sigma^q \Delta\mathbf{u} + \Pi_\sigma^q[(\boldsymbol{\omega} \wedge \mathbf{x}) \cdot \nabla\mathbf{u} - \boldsymbol{\omega} \wedge \mathbf{u}] + k\Pi_\sigma^q \partial_3\mathbf{u}. \end{aligned} \tag{8.1}$$

If $\omega = 0$ and $k = 0$, then $\mathscr{A}_k^\omega \equiv \mathscr{A}_0^0$ is called the *Stokes operator*. If $\omega = 0$ and $k \neq 0$, then $\mathscr{A}_k^\omega \equiv \mathscr{A}_k^0$ is said to be the *Oseen operator*. If $\omega \neq 0$, then we just add "–type" to the name of the operator and call $\mathscr{A}_0^\omega$ the *Stokes type operator* and $\mathscr{A}_k^\omega$ (for $k \neq 0$) the *Oseen type operator.*
- For $R > 0$, we denote $\mathcal{D}_R := \mathcal{D} \cap B_R(\mathbf{0})$.
- $R_0 := \inf\{R > 0;\ \mathbb{R}^3 \setminus \mathcal{D}_R \subset \mathcal{D}\}$.

The next lemma comes from Sect. 3.1, Theorem 3.1, Sect. 4.1, Theorem 4.1, Sect. 4.1.5, Theorem 4.4 and from [77].

Lemma 8.1 *There exists a constant $c_1 > 0$, such that if $\mathbf{u} \in D(\mathscr{A}_k^\omega)$ and $\mathbf{f} \in \boldsymbol{L}_\sigma^q(\mathcal{D})$ satisfy the equation $\mathscr{A}_k^\omega\mathbf{u} = \mathbf{f}$, then*

$$\|\mathbf{u}\|_{2,q} + \|(\boldsymbol{\omega} \wedge \mathbf{x}) \cdot \nabla\mathbf{u}\|_q \ \le\ c_1\big(\|\mathbf{f}\|_q + \|\mathbf{u}\|_q\big). \tag{8.2}$$

Proof Note that if $\mathcal{D} = \mathbb{R}^3$, then (8.2) follows from Theorems 3.1, 4.1, 4.4, Theorem [57, Theorem 11] and an interpolation argument. See directly from the estimates (3.21) and (3.22)) in the case $k = 0$ and $k \neq 0$ (4.3). For more details see Sects. 3.1, 4.1, 4.1.5.

Now, consider an exterior domain $\mathcal{D} \subset \mathbb{R}^3$ with the boundary of the class $C^{1,1}$. Let $\mathbf{u} \in D(\mathscr{A}_k^\omega)$ and $\mathbf{f} \in \boldsymbol{L}_\sigma^q(\mathcal{D})$ satisfy the equation $\mathscr{A}_k^\omega\mathbf{u} = \mathbf{f}$. Then there exists a scalar function p such that

$$\nabla p \ =\ \nu\Delta\mathbf{u} + (\boldsymbol{\omega} \wedge \mathbf{x}) \cdot \nabla\mathbf{u} - \boldsymbol{\omega} \wedge \mathbf{u} - k\partial_3\mathbf{u} - \mathbf{f} \ \in \boldsymbol{L}^q(\mathcal{D}).$$

Since p is determined uniquely up to a real additive constant, we can assume without loss of generality that $\int_{\mathcal{D}_R} p\,\mathrm{d}\mathbf{x} = 0$, where $R > 0$ is chosen so that $\mathbb{R}^3 \setminus B_{R-1}(\mathbf{0}) \subset \mathcal{D}$. Let $\eta \in C_0^\infty(\mathbb{R}^3)$ be a cutoff function with values in $[0, 1]$, such that

$$\eta(\mathbf{x}) = \begin{cases} 0 \text{ for } |\mathbf{x}| \le R - 1, \\ 1 \text{ for } |\mathbf{x}| \ge R. \end{cases}$$

Then $(\eta\mathbf{u}, \eta p)$ can be considered to be a solution of the whole space problem

$$\nu\Delta\mathbf{v} + (\boldsymbol{\omega}\wedge\mathbf{x})\cdot\nabla\mathbf{v} - \boldsymbol{\omega}\wedge\mathbf{v} - k\partial_3\mathbf{v} - \nabla\tilde{p} = \mathbf{f}_1, \quad \nabla\cdot\mathbf{v} = g_1 \quad \text{in } \mathbb{R}^3,$$

where

$$\begin{aligned}\mathbf{f}_1 &= \eta\mathbf{f} + 2\nu\nabla\eta\cdot\nabla\mathbf{u} + \nu\mathbf{u}\Delta\eta + \big((\boldsymbol{\omega}\wedge\mathbf{x})\cdot\nabla\eta\big)\mathbf{u} - k(\partial_3\eta)\mathbf{u} - p\nabla\eta,\\ g_1 &= \mathbf{u}\cdot\nabla\eta.\end{aligned}$$

Note that $\mathbf{f}_1$ coincides with $\eta\mathbf{f}$ up to perturbation terms, supported in $\operatorname{supp}(\nabla\eta) \subset \overline{\mathcal{D}}_R$. Similarly, $\operatorname{supp} g_1 \subset \overline{\mathcal{D}}_R$. By Theorems 3.1, 4.1, 4.4, [57, Theorem 11], there exists a solution $(\mathbf{v}, \tilde{p})$ satisfying the estimate

$$\begin{aligned}\|\nabla^2\mathbf{v}\|_{q;\,\mathbb{R}^3} + \|(\boldsymbol{\omega}\times\mathbf{x})\cdot\nabla\mathbf{v} - \boldsymbol{\omega}\wedge\mathbf{v}\|_{q;\,\mathbb{R}^3} + \|\nabla\tilde{p}\|_{q;\,\mathbb{R}^3}\\ \le c\left(\|\mathbf{f}\|_{q;\,\mathbb{R}^3} + \|\mathbf{v}\|_{1,q;\,\mathcal{D}_R} + \|p\|_{q;\,\mathcal{D}_R}\right).\end{aligned}$$

Moreover, as the solution is unique (see Theorem 3.1, Theorem 4.1, Theorem 4.4, [57, Theorem 1.1]), $(\eta\mathbf{u}, \eta p)$ satisfies the same estimate. Hence we get the inequality

$$\begin{aligned}\|\nabla^2(\eta\mathbf{u})\|_{q;\,\mathbb{R}^3} + \|(\boldsymbol{\omega}\wedge\mathbf{x})\cdot\nabla(\eta\mathbf{u})\|_{q;\,\mathbb{R}^3} + \|\nabla(\eta p)\|_{q;\,\mathbb{R}^3}\\ \le c\left(\|\mathbf{f}\|_q + \|\mathbf{u}\|_q + \|\mathbf{v}\|_{1,q;\,\mathcal{D}_R} + \|p\|_{q;\,\mathcal{D}_R}\right).\end{aligned} \tag{8.3}$$

Next we treat $((1-\eta)\mathbf{u}, (1-\eta)p)$ as a solution of the Stokes problem

$$\nu\Delta\mathbf{v} - \nabla\tilde{p} = \mathbf{f}_2, \quad \operatorname{div}\mathbf{v} = g_2 \quad \text{in } \mathcal{D}_R, \qquad \mathbf{v} = \mathbf{0} \quad \text{on } \partial\mathcal{D}_R,$$

where $g_2 = -\mathbf{u}\cdot\nabla\eta$ and

$$\mathbf{f}_2 = (1-\eta)\big[\mathbf{f} - (\boldsymbol{\omega}\wedge\mathbf{x})\cdot\nabla\mathbf{u} + \boldsymbol{\omega}\wedge\mathbf{v} + k\partial_3\mathbf{u}\big] - 2\nu\nabla\eta\cdot\nabla\mathbf{u} - \nu\,\mathbf{u}\,\Delta\eta + p\,\nabla\eta.$$

Classical estimates for the Stokes system in the bounded domain $\mathcal{D}_R$ and the precise form of $\mathbf{f}_2$ imply that

$$\|\nabla^2((1-\eta)\mathbf{u})\|_{q;\,\mathcal{D}_R} + \|\nabla(1-\eta)p\|_{q;\,\mathcal{D}_R} \le c\left(\|\mathbf{f}\|_q + \|\mathbf{u}\|_{1,q;\,\mathcal{D}_R} + \|p\|_{q;\,\mathcal{D}_R}\right). \tag{8.4}$$

Summing (8.3) and (8.4) and using an interpolation estimate for $\nabla\mathbf{u}$, we obtain

$$\|\nabla^2\mathbf{u}\|_q + \|(\boldsymbol{\omega}\wedge\mathbf{x})\cdot\nabla\mathbf{u}\|_q + \|\nabla p\|_q \le c\left(\|\mathbf{f}\|_q + \|\mathbf{u}\|_q + \|p\|_{q;\,\mathcal{D}_R}\right), \tag{8.5}$$

where c is independent of $\mathbf{f}$, $\mathbf{u}$, and p.

Now we will prove (8.2) by contradiction, using (8.5). Suppose that to every $n \in \mathbb{N}$ there exist $\mathbf{u}_n \in D(\mathscr{A}_k^{(\omega)})$ and p_n with $\nabla p_n \in \boldsymbol{L}^q(\mathcal{D})$, $\int_{\mathcal{D}_R} p_n \, \mathrm{d}\mathbf{x} = 0$, such that

$$1 = \|\nabla^2 \mathbf{u}_n\|_q + \|(\boldsymbol{\omega} \wedge \mathbf{x}) \cdot \nabla \mathbf{u}_n\|_q + \|\nabla p_n\|_q \geq n \left(\|\mathbf{f}_n\|_q + \|\mathbf{u}_n\|_q\right),$$

where $\mathbf{f}_n = \mathscr{A}_k^{\omega} \mathbf{u}_n$. Then we find a subsequence of $\{\mathbf{u}_n, p_n\}$, again denoted by $\{\mathbf{u}_n, p_n\}$, and $\mathbf{u} \in D(\mathscr{A}_k^{(\omega)})$ and p with $\nabla p \in \boldsymbol{L}^q(\mathcal{D})$ such that in the weak sense

$$\nabla^2 \mathbf{u}_n \rightharpoonup \nabla^2 \mathbf{u}, \quad (\boldsymbol{\omega} \wedge \mathbf{x}) \cdot \nabla \mathbf{u}_n \rightharpoonup (\boldsymbol{\omega} \wedge \mathbf{x}) \cdot \nabla \mathbf{u}, \quad \nabla p_n \rightharpoonup \nabla p \tag{8.6}$$

in the L^q-norm over $\mathcal{D}$, and in the strong sense

$$\mathbf{f}_n \to \mathbf{0}, \quad \mathbf{v}_n \to \mathbf{0} \quad \text{in } \boldsymbol{L}^q(\mathcal{D}), \quad p_n \to p \quad \text{in } L^q(\mathcal{D}_R), \quad \int_{\mathcal{D}_R} p \, \mathrm{d}\mathbf{x} = 0, \tag{8.7}$$

as $n \to \infty$, and

$$\begin{aligned} \nu \Delta \mathbf{u} + (\boldsymbol{\omega} \wedge \mathbf{x}) \cdot \nabla \mathbf{u} - \boldsymbol{\omega} \wedge \mathbf{u} - k \partial_3 \mathbf{u} - \nabla p &= 0 && \text{in } \mathcal{D}, \\ \operatorname{div} \mathbf{u} &= 0 && \text{in } \mathcal{D}, \\ \mathbf{u} &= \mathbf{0} && \text{on } \partial \mathcal{D}. \end{aligned}$$

By (8.6), (8.7), we conclude that $\mathbf{u} = 0$ and $\nabla p = \mathbf{0}$ in $\mathcal{D}$. Since the limit function p was chosen so that its mean value $\mathcal{D}_R$ equals zero, we have $p = 0$ in $\mathcal{D}$. Using now (8.5) with $\mathbf{u}_n,\ p_n,\ \mathbf{f}_n$ instead of $\mathbf{u},\ p,\ \mathbf{f}$, we observe that the left-hand side equals one, while the right -hand side tends to zero as $n \to \infty$. This is the contradiction with the assumption that (8.2) was wrong. □

Theorem 8.1 *The operator $\mathscr{A}_k^{\omega}$ is densely defined and closed in $\boldsymbol{L}^q_\sigma(\mathcal{D})$. The adjoint operator (as an operator in $\boldsymbol{L}^{q'}_\sigma(\mathcal{D})$, where q' is the conjugate exponent to q) has the form*

$$(\mathscr{A}_k^{(\omega)})^* \mathbf{u} = \nu \Pi_\sigma^{q'} \Delta \mathbf{u} - \Pi_\sigma^{q'} [(\boldsymbol{\omega} \wedge \mathbf{x}) \cdot \nabla \mathbf{u} - \boldsymbol{\omega} \wedge \mathbf{u}] - k \Pi_\sigma^{q'} \partial_3 \mathbf{u} \tag{8.8}$$

with $D((\mathscr{A}_k^{(\omega)})^*) = \{\mathbf{u} \in \boldsymbol{W}^{2,q'}(\mathcal{D}) \cap \boldsymbol{W}^{1,q'}_{0,\sigma}(\mathcal{D});\ (\boldsymbol{\omega} \wedge \mathbf{x}) \cdot \nabla \mathbf{u} \in \boldsymbol{L}^{q'}(\mathcal{D})\}$. □

Proof Obviously, since $\boldsymbol{C}^\infty_{0,\sigma}(\mathcal{D})$ is dense in $\boldsymbol{L}^q_\sigma(\mathcal{D})$ and $\boldsymbol{C}^\infty_{0,\sigma}(\mathcal{D}) \subset D(\mathscr{A}_k^{(\omega)})$, $D(\mathscr{A}_k^{(\omega)})$ is dense in $\boldsymbol{L}^q_\sigma(\mathcal{D})$.

Further, we need to show that the graph of $\mathscr{A}_k^{\omega}$ is closed in $\boldsymbol{L}^q_\sigma(\mathcal{D}) \times \boldsymbol{L}^q_\sigma(\mathcal{D})$. Thus, let $\mathbf{u}_n \in D(\mathscr{A}_k^{(\omega)})$, $\mathbf{u}_n \to \mathbf{u}$ in $\boldsymbol{L}^q_\sigma(\mathcal{D})$, and $\mathbf{f}_n := \mathscr{A}_k^{\omega} \mathbf{u}_n \to \mathbf{f}$ in $\boldsymbol{L}^q_\sigma(\mathcal{D})$. Then $\mathscr{A}_k^{(\omega)}(\mathbf{u}_n - \mathbf{u}_m) = \mathbf{f}_n - \mathbf{f}_m$, and due to (8.2), we have

$$\|\mathbf{u}_n - \mathbf{u}_m\|_{2,q} + \|(\boldsymbol{\omega} \wedge \mathbf{x}) \cdot \nabla(\mathbf{u}_n - \mathbf{u}_m)\|_q \leq c_1 \left(\|\mathbf{f}_n - \mathbf{f}_m\|_q + \|\mathbf{u}_n - \mathbf{u}_m\|_q\right).$$

Thus, $\mathbf{u}_n \to \mathbf{u}$ in $\boldsymbol{W}^{2,q}(\mathcal{D}) \cap \boldsymbol{W}_0^{1,q}(\mathcal{D})$, and the sequence $\{(\boldsymbol{\omega} \wedge \mathbf{x}) \cdot \nabla \mathbf{u}_n\}$ converges to some function $\mathbf{h}$ in $\boldsymbol{L}^q(\mathcal{D})$. Since $(\boldsymbol{\omega} \wedge \mathbf{x}) \cdot \nabla \mathbf{u}_n \to (\boldsymbol{\omega} \wedge \mathbf{x}) \cdot \nabla \mathbf{u}$ in $\boldsymbol{L}^q(\mathcal{D}_R)$ for any $R > R_0$ arbitrarily large, we deduce that $\mathbf{h} = (\boldsymbol{\omega} \wedge \mathbf{x}) \cdot \nabla \mathbf{u}$. This implies that $\mathbf{u} \in D(\mathscr{A}_k^{\omega})$ and $\mathscr{A}_k^{\omega}\mathbf{u} = \mathbf{f}$. Thus, the operator $\mathscr{A}_k^{\omega}$ is closed.

It follows, e.g., from [116, Proposition 4.3] that $R(\zeta I - \mathscr{A}_k^{\omega}) = \boldsymbol{L}^q_{\sigma}(\mathcal{D})$ for $\zeta > 0$ sufficiently large. Let us denote by $\mathscr{T}_k^{\omega}$ the operator on the right-hand side of (8.8) with the domain $D(\mathscr{T}_k^{\omega}) := \{\mathbf{u} \in \boldsymbol{W}^{2,q'}(\mathcal{D}) \cap \boldsymbol{W}^{1,q'}_{0,\sigma}(\mathcal{D});\ \omega(\mathbf{e}_3 \wedge \mathbf{x}) \cdot \nabla \mathbf{u} \in \boldsymbol{L}^{q'}(\mathcal{D})\}$. By analogy with $\mathscr{A}_k^{\omega}$, $\mathscr{T}_k^{\omega}$ is closed in $\boldsymbol{L}^{q'}_{\sigma}(\mathcal{D})$ and $R(\zeta I - \mathscr{T}_k^{\omega}) = \boldsymbol{L}^{q'}_{\sigma}(\mathcal{D})$ if $\zeta > 0$ is sufficiently large. It is easy to verify that operators $\mathscr{A}_k^{\omega}$ and $\mathscr{T}_k^{\omega}$ are adjoint to each other in the sense of Kato [137, page 167]; hence, $\mathscr{T}_k^{\omega} \subset (\mathscr{A}_k^{\omega})^*$. In order to show that $\mathscr{T}_k^{\omega} = (\mathscr{A}_k^{\omega})^*$, we need to verify that $\mathscr{T}_k^{\omega}$ is the maximal operator adjoint to $\mathscr{A}_k^{\omega}$. Suppose that $\mathbf{v} \in D((\mathscr{A}_k^{\omega})^*)$ and $\mathbf{f} = (\zeta I - (\mathscr{A}_k^{\omega})^*)\mathbf{v}$. Since $\mathbf{f} \in R(\zeta I - \mathscr{T}_k^{\omega})$, it can also be expressed in the form $\mathbf{f} = (\zeta I - \mathscr{T}_k^{\omega})\mathbf{w}$ for some $\mathbf{w} \in D(\mathscr{T}_k^{\omega})$. Hence $(\zeta I - (\mathscr{A}_k^{\omega})^*)\mathbf{v} = (\zeta I - \mathscr{T}_k^{\omega})\mathbf{w}$, which is the equality in $\boldsymbol{L}^{q'}_{\sigma}(\mathcal{D})$. If $\mathbf{u} \in D(\mathscr{A}_k^{\omega})$ $(\subset \boldsymbol{L}^q_{\sigma}(\mathcal{D}))$, then

$$\int_{\mathcal{D}} (\zeta I - (\mathscr{A}_k^{\omega})^*)\mathbf{v} \cdot \mathbf{u}\, \mathrm{d}\mathbf{x} = \int_{\mathcal{D}} (\zeta I - \mathscr{T}_k^{\omega})\mathbf{w} \cdot \mathbf{u}\, \mathrm{d}\mathbf{x},$$

$$\int_{\mathcal{D}} \mathbf{v} \cdot (\zeta I - \mathscr{A}_k^{\omega})\mathbf{u}\, \mathrm{d}\mathbf{x} = \int_{\mathcal{D}} \mathbf{w} \cdot (\zeta I - \mathscr{A}_k^{\omega})\mathbf{u}\, \mathrm{d}\mathbf{x}.$$

As this holds for all $\mathbf{u} \in D(\mathscr{A}_k^{\omega})$, we get $\mathbf{v} = \mathbf{w} \in D((\mathscr{A}_k^{\omega})^*)$. Consequently, $(\mathscr{A}_k^{\omega})^* = \mathscr{T}_k^{\omega}$. □

The next lemma follows from Sect. 3.1, Theorem 3.1, Sect. 4.1, Theorem 4.1, Sect. 4.1.5, [108] and [62] (the case $k = 0$) or from [76] (the case $q = 2$).

Lemma 8.2 *If* $\mathbf{u} \in D(\mathscr{A}_k^{\omega})$, *then both the terms* $(\boldsymbol{\omega} \wedge \mathbf{x}) \cdot \nabla \mathbf{u} - \boldsymbol{\omega} \wedge \mathbf{u}$ *and* $k\,\partial_3 \mathbf{u}$ *belong to* $\boldsymbol{L}^q_{\sigma}(\mathcal{D})$. □

Proof Let us begin with the term $(\boldsymbol{\omega} \wedge \mathbf{x}) \cdot \nabla \mathbf{u} - \boldsymbol{\omega} \wedge \mathbf{u}$. It is sufficient to show that the integral (in $\mathcal{D}$) of the product of $(\boldsymbol{\omega} \wedge \mathbf{x}) \cdot \nabla \mathbf{u} - \boldsymbol{\omega} \wedge \mathbf{u}$ with any function $\nabla \psi$ in $\boldsymbol{L}^{q'}(\mathcal{D})$ (for $\psi \in W^{1,q'}_{\mathrm{loc}}(\mathcal{D})$) is zero. (See [98, Lemma III.2.1].) Since $\boldsymbol{C}^{\infty}_{0,\sigma}(\mathcal{D})$ is dense in $D(\mathscr{A}_k^{\omega})$ in the topology of $\boldsymbol{W}^{1,q}(\mathcal{D})$, it is sufficient to consider only $\mathbf{u} \in \boldsymbol{C}^{\infty}_{0,\sigma}(\mathcal{D})$. Then

$$\int_{\mathcal{D}} [(\boldsymbol{\omega} \wedge \mathbf{x}) \cdot \nabla \mathbf{u} - \boldsymbol{\omega} \wedge \mathbf{u}] \cdot \nabla \psi\, \mathrm{d}\mathbf{x} = -\int_{\mathcal{D}} \operatorname{div}[(\boldsymbol{\omega} \wedge \mathbf{x}) \cdot \nabla \mathbf{u} - \boldsymbol{\omega} \wedge \mathbf{u}]\, \psi\, \mathrm{d}\mathbf{x}.$$

The last integral equals zero, because $\operatorname{div}[(\boldsymbol{\omega} \wedge \mathbf{x}) \cdot \nabla \mathbf{u} - \boldsymbol{\omega} \wedge \mathbf{u}] = 0$ in $\mathcal{D}$.

The inclusion $k\,\partial_3 \mathbf{u} \in \boldsymbol{L}^q_{\sigma}(\mathcal{D})$ can be proven in the same way. □

Lemma 8.2 enables us to omit the second and the third projections Π_σ^q in the definition of $\mathscr{A}_k^{\omega}$ and to write

$$\mathscr{A}_k^{\omega}\mathbf{u} = \mathscr{A}_0^{0}\mathbf{u} + (\boldsymbol{\omega} \wedge \mathbf{x}) \cdot \nabla\mathbf{u} - \boldsymbol{\omega} \wedge \mathbf{u} + k\,\partial_3\mathbf{u}. \tag{8.9}$$

The adjoint operator $(\mathscr{A}_k^{\omega})^*$ can also be simplified in the same way.

The form of the operator $\mathscr{A}_k^{\omega}$ can also be further simplified if one uses the cylindrical coordinates: We denote by r, φ, and x_3 the cylindrical coordinate system whose axis is the x_3-axis. The corresponding cylindrical components of vector functions are denoted by the indices r, φ, and 3. In order to distinguish between the Cartesian and the cylindrical components of vectors, we write the Cartesian components in parentheses and the cylindrical components in brackets. In this sense, we have $(u_1, u_2, u_3) \triangleq [u_r, u_\varphi, u_3]$. Using the transformations

$$u_r = u_1 \cos\varphi + u_2 \sin\varphi, \qquad u_1 = u_r \cos\varphi - u_\varphi \sin\varphi,$$
$$u_\varphi = -u_1 \sin\varphi + u_2 \cos\varphi, \qquad u_2 = u_r \sin\varphi + u_\varphi \cos\varphi,$$

one can calculate that

$$\begin{aligned}
(\boldsymbol{\omega} \wedge \mathbf{x}) \cdot \nabla\mathbf{u} - \boldsymbol{\omega} \wedge \mathbf{u} &= \omega\,\partial_\varphi\mathbf{u} - (\boldsymbol{\omega} \wedge \mathbf{u}) = \omega\,\partial_\varphi(u_1, u_2, u_3) - \omega\,(-u_2, u_1, 0) \\
&= \omega\,\partial_\varphi \begin{pmatrix} u_r \cos\varphi - u_\varphi \sin\varphi \\ u_r \sin\varphi + u_\varphi \cos\varphi \\ u_3 \end{pmatrix}^T - \omega \begin{pmatrix} -u_r \sin\varphi - u_\varphi \cos\varphi \\ u_r \cos\varphi - u_\varphi \sin\varphi \\ 0 \end{pmatrix}^T \\
&= \omega \begin{pmatrix} (\partial_\varphi u_r) \cos\varphi - (\partial_\varphi u_\varphi) \sin\varphi \\ (\partial_\varphi u_r) \sin\varphi + (\partial_\varphi u_\varphi) \cos\varphi \\ \partial_\varphi u_3 \end{pmatrix}^T \triangleq \omega \begin{bmatrix} \partial_\varphi u_r \\ \partial_\varphi u_\varphi \\ \partial_\varphi u_3 \end{bmatrix}^T \\
&= \omega\,\partial_\varphi\big[u_r, u_\varphi, u_3\big].
\end{aligned}$$

Thus, if $\mathbf{u}$ is identified with $[u_r, u_\varphi, u_3]$, then (8.9) can be written in the form

$$\mathscr{A}_k^{\omega}\mathbf{u} = \mathscr{A}_0^{0}\mathbf{u} + \omega\,\partial_\varphi\mathbf{u} + k\,\partial_3\mathbf{u}, \tag{8.10}$$

where $\mathscr{A}_0^0$ naturally denotes the operator acting in the cylindrical coordinates.

Let us further denote by $D_0(\mathscr{A}_k^{\omega})$ the subspace of $D(\mathscr{A}_k^{\omega})$ which contains only functions that have a compact support in $\overline{\mathcal{D}}$.

Theorem 8.2 *$D_0(\mathscr{A}_k^{\omega})$ is a core of the operator $\mathscr{A}_k^{\omega}$, i.e., the graph of the restriction of the operator $\mathscr{A}_k^{\omega}$ to $D_0(\mathscr{A}_k^{\omega})$ is dense in the graph of $\mathscr{A}_k^{\omega}$ in the norm of $\mathbf{L}_\sigma^q(\mathcal{D}) \times \mathbf{L}_\sigma^q(\mathcal{D})$.* □

Proof Let $\mathbf{v} \in D(\mathscr{A}_k^{\omega})$ and $\mathbf{f} = \mathscr{A}_k^{\omega}\mathbf{v}$. We will show that $[\mathbf{v}, \mathscr{A}_k^{\omega}\mathbf{v}]$ can be approximated by a sequence of elements $[\mathbf{v}^n, \mathscr{A}_k^{\omega}\mathbf{v}^n]$ where $\mathbf{v}^n \in D_0(\mathscr{A}_k^{\omega})$, $n \in \mathbb{N}$.

Let $\eta \in C_0^\infty(\mathbb{R}^3)$ be radially symmetric, with values in [0, 1], such that

$$\eta(\mathbf{x}) = \begin{cases} 1 \text{ for } |\mathbf{x}| \le 1, \\ 0 \text{ for } 2 \le |\mathbf{x}|. \end{cases}$$

Denote $K_1 := \{\mathbf{x} \in \mathbb{R}^3;\ 1 < |\mathbf{x}| < 2\}$ and, more generally, for $r > 0$, let $K_r := \{\mathbf{x} \in \mathbb{R}^3;\ r < |\mathbf{x}| < 2r\}$. It follows from Sect. 2.2 that there exists a bounded linear operator $\mathfrak{B} : L^q(K_1) \to \boldsymbol{W}_0^{1,q}(K_1)$ such that $\operatorname{div} \mathfrak{B} f = f$ for all $f \in L^q(K_1)$ satisfying $\int_{K_1} f \, \mathrm{d}\mathbf{x} = 0$. The operator $\mathfrak{B}$ is bounded from $W_0^{1,q}(K_1)$ to $\boldsymbol{W}_0^{2,q}(K_1)$ as well (See also [15]).

Let $n \in \mathbb{N}$ be so large that $\mathbb{R}^3 \setminus B_n(\mathbf{0}) \subset \mathcal{D}$. Put $\mathbf{v}^n(\mathbf{x}) := \eta(\mathbf{x}/n)\,\mathbf{v}(\mathbf{x}) - \boldsymbol{V}^n(\mathbf{x})$ with the correction term $\boldsymbol{V}^n(\mathbf{x})$ being equal to $\boldsymbol{U}^n(\mathbf{x}/n)$, where

$$\boldsymbol{U}^n(\mathbf{y}) = \begin{cases} \mathfrak{B}\big[\nabla\eta(\mathbf{y}) \cdot \mathbf{v}(n\mathbf{y})\big] & \text{for } \mathbf{y} \in K_1, \\ \mathbf{0} & \text{for } \mathbf{y} \in \mathbb{R}^3 \setminus K_1. \end{cases}$$

The function $\mathbf{v}^n$ is divergence-free, it coincides with $\mathbf{v}$ in $\mathcal{D} \cap B_n(\mathbf{0})$, and its support is a subset of the closure of $\mathcal{D} \cap B_{2n}(\mathbf{0})$. Due to the continuity of the operator $\mathfrak{B}$ from $L^q(K_1)$ to $\boldsymbol{W}_0^{1,q}(K_1)$ and from $W_0^{1,q}(K_1)$ to $\boldsymbol{W}_0^{2,q}(K_1)$, the function $\boldsymbol{U}^n$ satisfies the estimates

$$\|\nabla \boldsymbol{U}^n\|_{q;\,K_1} \le c\,\big\|\nabla\eta \cdot \mathbf{v}(n\,\cdot)\big\|_{q;\,K_1} \le c\,\big\|\mathbf{v}(n\,\cdot)\big\|_{q;K_1} = c\,n^{-3/q}\,\|\mathbf{v}\|_{q;\,K_n}\,,$$

$$\begin{aligned} \|\nabla^2 \boldsymbol{U}^n\|_{q;K_1} &\le c\,\big\|\nabla\big(\nabla\eta \cdot \mathbf{v}(n\,\cdot)\big)\big\|_{q;\,K_1} \le c\,\big\|\mathbf{v}(n\,\cdot)\big\|_{q;\,K_1} + c\,\big\|\nabla\mathbf{v}(n\,\cdot)\big\|_{q;\,K_1} \\ &= c\,n^{-3/q}\,\|\mathbf{v}\|_{q;\,K_n} + c\,n^{1-3/q}\,\|\nabla\mathbf{v}\|_{q;\,K_n}\,. \end{aligned}$$

This means that

$$\|\nabla \boldsymbol{V}^n\|_{q;\,K_n} = n^{-1+3/q}\,\|\nabla \boldsymbol{U}^n\|_{q;\,K_1} \le \frac{c}{n}\,\|\mathbf{v}\|_{q;\,K_n},$$

$$\|\nabla^2 \boldsymbol{V}^n\|_{q;\,K_n} = n^{-2+3/q}\,\|\nabla^2 \boldsymbol{U}^n\|_{q;\,K_1} \le \frac{c}{n^2}\,\|\mathbf{v}\|_{q;\,K_n} + \frac{c}{n}\,\|\nabla\mathbf{v}\|_{q;\,K_n}\,.$$

Using also the fact that $\boldsymbol{V}^n(\mathbf{x}) = \mathbf{0}$ for $|\mathbf{x}| = n$, we derive that

$$\|\boldsymbol{V}^n\|_{q;\,K_n} \le cn\,\|\nabla \boldsymbol{V}^n\|_{q;\,K_n} \le c\,\|\mathbf{v}\|_{q;\,K_n}.$$

All generic constants c are independent of n. Thus, $\|\boldsymbol{V}^n\|_{2,q} \to 0$ as $n \to \infty$. The same result also holds on the $\|\,.\,\|_{2,q}$-norm of the difference $\eta(\mathbf{x}/n)\,\mathbf{v}(\mathbf{x}) - \mathbf{v}(\mathbf{x})$. Consequently,

$$\mathbf{v}^n \longrightarrow \mathbf{v} \qquad \text{in } \boldsymbol{W}^{2,q}(\mathcal{D}) \quad \text{for } n \to \infty. \tag{8.11}$$

Furthermore, since $(\boldsymbol{\omega} \wedge \mathbf{x}) \cdot \nabla\eta(\mathbf{x}/n) = 0$, we have

$$
\begin{aligned}
&\|\mathscr{A}_k^{\omega}\mathbf{v} - \mathscr{A}_k^{\omega}\mathbf{v}^n\|_{q;\,\mathcal{D}} \\
&\quad \leq c\left(\|\mathbf{v} - \mathbf{v}_n\|_{2,q;\,\mathcal{D}} + \left\|(\boldsymbol{\omega} \wedge \mathbf{x}) \cdot \nabla\mathbf{v} - (\boldsymbol{\omega} \wedge \mathbf{x}) \cdot \nabla\mathbf{v}^n\right\|_{q;\,\mathcal{D}}\right) \\
&\quad \leq c\left(\|\mathbf{v} - \mathbf{v}_n\|_{2,q;\,\mathcal{D}} + \left\|(\boldsymbol{\omega} \wedge \mathbf{x}) \cdot \left(1 - \eta(\mathbf{x}/n)\right) \nabla\mathbf{v}\right\|_{q;\,\mathcal{D}}\right. \\
&\qquad \left. + \left\|(\boldsymbol{\omega} \wedge \mathbf{x}) \cdot \nabla\boldsymbol{V}^n\right\|_{q;\,\mathcal{D}}\right) \\
&\quad \leq c\left(\|\mathbf{v} - \mathbf{v}_n\|_{2,q;\,\mathcal{D}} + \|(\boldsymbol{\omega} \wedge \mathbf{x}) \cdot \nabla\mathbf{v}\|_{q;\,\mathcal{D}\setminus B_n(0)} + n\,\|\nabla\boldsymbol{V}^n\|_{q;\,\mathcal{D}}\right) \\
&\quad \longrightarrow 0 \quad \text{for} \quad n \to \infty. \qquad (8.12)
\end{aligned}
$$

We can now observe from (8.11) and (8.12) that

$$
[\mathbf{v}^n, \mathscr{A}_k^{\omega}\mathbf{v}^n] \longrightarrow [\mathbf{v}, \mathscr{A}_k^{\omega}\mathbf{v}] \qquad \text{in } \boldsymbol{L}^q_{\sigma}(\mathcal{D}) \times \boldsymbol{L}^q_{\sigma}(\mathcal{D})
$$

as $n \to \infty$. The proof is completed. □

If $\boldsymbol{U}$ is a steady solution to the problem (1.35)–(1.38) and $\mathbf{v}$ is another solution, then $\mathbf{v}$ can be written in the form $\mathbf{v} = \boldsymbol{U} + \mathbf{u}$, where the perturbation $\mathbf{u}$ satisfies the operator equation

$$
\frac{\mathrm{d}\mathbf{u}}{\mathrm{d}t} = \mathscr{A}_k^{\omega}\mathbf{u} + \mathscr{B}\mathbf{u} + \mathscr{N}\mathbf{u}
$$

in space $\boldsymbol{L}^q_{\sigma}(\mathcal{D})$, where formally $\mathscr{B}\mathbf{u} = \Pi^q_{\sigma}[\boldsymbol{U}\cdot\nabla\mathbf{u} + \mathbf{u}\cdot\nabla\boldsymbol{U}]$ and $\mathscr{N}\mathbf{u} = \Pi^q_{\sigma}[\mathbf{u}\cdot\nabla\mathbf{u}]$. The linear operators $\mathscr{A}_k^{\omega}$ and $\mathscr{B}$ play a fundamental role in studies of properties of solutions to the nonlinear problem (1.23)–(1.24), (1.30)–(1.31), like, e.g., regularity, bifurcations, and short- and long-time behavior. (See, e.g., [105] and [106].) We further consider the operator $\mathscr{B}$ on $D(\mathscr{A}_k^{\omega})$ in a more general form:

$$
\mathscr{B}\mathbf{u} := \Pi^q_{\sigma} B(\mathbf{x}) \cdot \mathbf{u} + \Pi^q_{\sigma}\mathbf{b}(\mathbf{x}) \cdot \nabla\mathbf{u}. \qquad (8.13)
$$

We suppose that there exists $\iota > 0$, such that the 3×3 matrix-valued function B and the vector-valued function $\mathbf{b}$ satisfy

$$
B \in L^{\frac{3}{2}+\iota}(\mathcal{D})^{3\times 3} \quad \text{if } 1 < q < \tfrac{3}{2}, \qquad B \in L^{q+\iota}(\mathcal{D})^{3\times 3} \quad \text{if } q \geq \tfrac{3}{2}, \qquad (8.14)
$$

$$
\mathbf{b} \in \boldsymbol{L}^{3+\iota}(\mathcal{D}) \quad \text{if } 1 < q < 3, \qquad \mathbf{b} \in \boldsymbol{L}^{q+\iota}(\mathcal{D}) \quad \text{if } q \geq 3. \qquad (8.15)
$$

These conditions guarantee, except others, that the products $B(\mathbf{x}) \cdot \mathbf{u}$ and $\mathbf{b}(\mathbf{x}) \cdot \nabla\mathbf{u}$ are in $\boldsymbol{L}^q(\mathcal{D})$, provided that $\mathbf{u} \in D(\mathscr{A}_k^{\omega})$.

The next lemma is needed in the discussion on the spectrum of the operator $\mathscr{L}_k^{\omega}$, where

$$\mathscr{L}_k^{\omega}\mathbf{u} := \mathscr{A}_k^{\omega}\mathbf{u} + \mathscr{B}\mathbf{u}. \tag{8.16}$$

Lemma 8.3 *The operator $\mathscr{B}$ is $\mathscr{A}_k^{\omega}$-compact in $\boldsymbol{L}^q_{\sigma}(\mathcal{D})$.* □

Proof We show the proof only for $1 < q < \frac{3}{2}$. Using appropriately the conditions (8.14) and (8.15), the proof for other q can be done in a similar way.

Let $\{\mathbf{u}_n\}$ be a bounded sequence in $\boldsymbol{L}^q_{\sigma}(\mathcal{D})$, such that the sequence $\{\mathscr{A}_k^{\omega}\mathbf{u}_n\}$ is also bounded in $\boldsymbol{L}^q_{\sigma}(\mathcal{D})$. Then, due to estimate (8.2), the sequence $\{\mathbf{u}_n\}$ is bounded in $\boldsymbol{W}^{2,q}(\mathcal{D})$. Applying the continuous imbeddings $\boldsymbol{W}^{2,q}(\mathcal{D}) \hookrightarrow \boldsymbol{W}^{1,\frac{3q}{3-q}}(\mathcal{D})$ and $\boldsymbol{W}^{2,q}(\mathcal{D}) \hookrightarrow \boldsymbol{L}^{\frac{3q}{3-2q}}(\mathcal{D})$, we deduce that the sequence $\mathbf{u}_n$ is also bounded in $\boldsymbol{W}^{1,\frac{3q}{3-q}}(\mathcal{D})$ and in $\boldsymbol{L}^{\frac{3q}{3-2q}}(\mathcal{D})$. As the space $\boldsymbol{W}^{2,q}(\mathcal{D})$ is reflexive, there exists a subsequence $\{\mathbf{u}_n^{R_0}\}$ of $\{\mathbf{u}_n\}$ that converges weakly in $\boldsymbol{W}^{2,q}(\mathcal{D})$ to a limit function $\mathbf{u}$. Since $\boldsymbol{W}^{2,q}(\mathcal{D}_{R_0})$ is compactly imbedded to $\boldsymbol{W}^{1,r}(\mathcal{D}_{R_0})$ for each $r \in (1, \frac{3q}{3-q})$, $\mathbf{u}_n^{R_0} \to \mathbf{u}$ strongly in $\boldsymbol{W}^{1,r}(\mathcal{D}_{R_0})$. By analogy, there exists a subsequence $\{\mathbf{u}_n^{R_0+1}\}$ of $\{\mathbf{u}_n^{R_0}\}$, such that $\mathbf{u}_n^{R_0+1} \to \mathbf{u}$ strongly in $\boldsymbol{W}^{1,r}(\mathcal{D}_{R_0+1})$. Proceeding in this way, we get a subsequence $\{\mathbf{u}_n^{R_0+2}\}$ of $\{\mathbf{u}_n^{R_0+1}\}$, such that $\mathbf{u}_n^{R_0+2} \to \mathbf{u}$ strongly in $\boldsymbol{W}^{1,r}(\mathcal{D}_{R_0+2})$, etc. If we put $\mathbf{v}_n = \mathbf{u}_n^{R_0+n}$, we obtain a subsequence of $\{\mathbf{u}_n\}$ that converges strongly in $\boldsymbol{W}^{1,r}(\mathcal{D}_R)$ (for every $R \geq R_0$) to function $\mathbf{u}$. From this, one can also deduce that $\mathbf{v}_n \to \mathbf{u}$ strongly in $\boldsymbol{L}^{\frac{3r}{3-r}}(\mathcal{D}_R)$, i.e., also in $\boldsymbol{L}^s(\mathcal{D}_R)$ for any $s \in (1, \frac{3q}{3-2q})$. (This holds for each $R \geq R_0$.) We claim that $\mathscr{B}\mathbf{v}_n \to \mathscr{B}\mathbf{u}$ in $\boldsymbol{L}^q_{\sigma}(\mathcal{D})$. Let r and s be chosen so that

$$r = \frac{(3+\iota)q}{(3+\iota)-q} \qquad \text{and} \qquad s = \frac{(3+2\iota)q}{(3+2\iota)-2q}.$$

(Recall that ι is the number from (8.14) and (8.15).) Then r and s also satisfy

$$q < r < \frac{3q}{3-q} \qquad \text{and} \qquad q < s < \frac{3q}{3-2q}.$$

The choice of r implies that $\frac{qr}{r-q} = 3 + \iota$, which means that $\mathbf{b} \in L^{\frac{qr}{r-q}}(D)$. The choice of s guarantees that $\frac{qs}{s-q} = \frac{3}{2} + \iota$, which means that $B \in L^{\frac{qs}{s-q}}(\mathcal{D})$. Thus, for every $n \in \mathbb{N}$ and $R > R_0$, we have

$$\|\mathscr{B}\mathbf{v}_n - \mathscr{B}\mathbf{u}\|_q^q \leq c(q) \int_{\mathcal{D}} \Big(\big|B \cdot (\mathbf{v}_n - \mathbf{u})\big|^q + \big|\mathbf{b} \cdot \nabla(\mathbf{v}_n - \mathbf{u})\big|^q \Big)\, \mathrm{d}\mathbf{x}$$

$$= c(q) \left(\int_{\mathcal{D}_R} \cdots + \int_{\mathcal{D}-\mathcal{D}_R} \cdots \right) \leq c(q,r,s)\,(\kappa_1 + \kappa_2 + \kappa_3 + \kappa_4),$$

where

$$\kappa_1 = \left(\int_{\mathcal{D}_R} |B|^{\frac{qs}{s-q}}\, \mathrm{d}\mathbf{x}\right)^{\frac{s-q}{s}} \left(\int_{\mathcal{D}_R} |\mathbf{v}_n - \mathbf{u}|^s\, \mathrm{d}\mathbf{x}\right)^{\frac{q}{s}},$$

$$\kappa_2 = \left(\int_{\mathcal{D}\setminus\mathcal{D}_R} |B|^{\frac{qs}{s-q}}\, \mathrm{d}\mathbf{x}\right)^{\frac{s-q}{s}} \left(\int_{\mathcal{D}\setminus\mathcal{D}_R} |\mathbf{v}_n - \mathbf{u}|^s\, \mathrm{d}\mathbf{x}\right)^{\frac{q}{s}},$$

$$\kappa_3 = \left(\int_{\mathcal{D}_R} |\mathbf{b}|^{\frac{qr}{r-q}}\, \mathrm{d}\mathbf{x}\right)^{\frac{r-q}{r}} \left(\int_{\mathcal{D}_R} |\nabla(\mathbf{v}_n - \mathbf{u})|^r\, \mathrm{d}\mathbf{x}\right)^{\frac{q}{r}},$$

$$\kappa_4 = \left(\int_{\mathcal{D}\setminus\mathcal{D}_R} |\mathbf{b}|^{\frac{qr}{r-q}}\, \mathrm{d}\mathbf{x}\right)^{\frac{r-q}{r}} \left(\int_{\mathcal{D}\setminus\mathcal{D}_R} |\nabla(\mathbf{v}_n - \mathbf{u})|^r\, \mathrm{d}\mathbf{x}\right)^{\frac{q}{r}}.$$

Here, κ_2, κ_4 can be made arbitrarily small by choosing R sufficiently large. Afterward, κ_1, κ_3 can be made arbitrarily small by choosing n sufficiently large. This confirms that $\mathscr{B}\mathbf{v}_n \to \mathscr{B}\mathbf{u}$ in $\boldsymbol{L}^q_\sigma(\mathcal{D})$ for $1 < q < \frac{3}{2}$, indeed. □

Theorem 8.3 *Suppose that $q = 2$, i.e., $\boldsymbol{L}^q_\sigma(\mathcal{D}) \equiv \boldsymbol{L}^2_\sigma(\mathcal{D})$ is a Hilbert space. Then:*

(i) The Stokes type operator $\mathscr{A}_0^\omega$ is normal if and only if domain $\mathcal{D}$ is axially symmetric about the x_3-axis.
(ii) The Oseen type operator $\mathscr{A}_k^\omega$ is normal (where $k \neq 0$) if and only if $\mathcal{B} = \emptyset$ (which means that $\mathcal{D} = \mathbb{R}^3$). □

Proof Recall that the operator $\mathscr{A}_k^\omega$ is normal if and only if it commutes with its adjoint, which means that $\mathscr{A}_k^\omega(\mathscr{A}_k^\omega)^* = (\mathscr{A}_k^\omega)^*\mathscr{A}_k^\omega$, see Sect. 8.2. This notion has a sense if $\mathscr{A}_k^\omega$ and $(\mathscr{A}_k^\omega)^*$ are operators in the same space, which is true in the case $q = 2$, when both $\mathscr{A}_k^\omega$ and $(\mathscr{A}_k^\omega)^*$ are operators in $\boldsymbol{L}^2_\sigma(\mathcal{D})$:

(1) Let us at first assume that $\mathcal{D}$ is axially symmetric and show that $\mathscr{A}_0^\omega$ and $(\mathscr{A}_0^\omega)^*$ commute. Equality (8.10) implies that $(\mathscr{A}_0^\omega)^*\mathbf{u} = \mathscr{A}_0^0\mathbf{u} - \omega\,\partial_\varphi\mathbf{u} = \nu\Pi_\sigma\Delta\mathbf{u} - \omega\partial_\varphi\mathbf{u}$.

Suppose that $\mathbf{u} \in D(\mathscr{A}_0^\omega(\mathscr{A}_0^\omega)^*)$, i.e., $\mathbf{u} \in D((\mathscr{A}_0^\omega)^*)$ and $(\mathscr{A}_0^\omega)^*\mathbf{u} \in D(\mathscr{A}_0^\omega) = D((\mathscr{A}_0^\omega)^*)$. The latter means that $\mathscr{A}_0^0\mathbf{u} - \omega\partial_\varphi\mathbf{u} \in D(\mathscr{A}_0^0)$ and $\partial_\varphi(\mathscr{A}_0^0\mathbf{u} - \omega\partial_\varphi\mathbf{u}) \in \boldsymbol{L}^2(\mathcal{D})$, implying that $\mathbf{u} \in D((\mathscr{A}_0^\omega)^{*2})$. Put $\mathbf{w} = \mathscr{A}_0^\omega(\mathscr{A}_0^\omega)^*\mathbf{u}$.

In order to show that $\mathbf{u} \in D((\mathscr{A}_0^\omega)^*\mathscr{A}_0^\omega)$, we treat the scalar product $(\mathscr{A}_0^\omega\mathbf{u}, \mathscr{A}_0^\omega\mathbf{v})_2$ for $\mathbf{v} \in D(\mathscr{A}_0^\omega)$ as follows:

$$\begin{aligned}(\mathscr{A}_0^\omega\mathbf{u}, \mathscr{A}_0^\omega\mathbf{v})_2 &= \big((\mathscr{A}_0^\omega)^*\mathbf{u}, \mathscr{A}_0^\omega\mathbf{v}\big)_2 + 2(\omega\partial_\varphi\mathbf{u}, \mathscr{A}_0^\omega\mathbf{v})_2 \\ &= \big((\mathscr{A}_0^\omega)^*\mathbf{u}, \mathscr{A}_0^\omega\mathbf{v}\big)_2 + 2(\omega\partial_\varphi\mathbf{u}, \mathscr{A}_0^0\mathbf{v})_2 + 2(\omega\partial_\varphi\mathbf{u}, \omega\partial_\varphi\mathbf{v})_2. \qquad (8.17)\end{aligned}$$

Assume at first that $\mathbf{v}$ has a compact support in $\overline{\mathcal{D}}$. Then

$$\begin{aligned}(\partial_\varphi \mathbf{u}, \mathscr{A}_0^0 \mathbf{v})_2 &= \nu \int_{\mathcal{D}} \partial_\varphi \mathbf{u} \cdot \Delta \mathbf{v}\, \mathrm{d}\mathbf{x} = -\nu \int_{\mathcal{D}} \nabla \partial_\varphi \mathbf{u} \cdot \nabla \mathbf{v}\, \mathrm{d}\mathbf{x} \\ &= \nu \int_{\mathcal{D}} \nabla \mathbf{u} \cdot \partial_\varphi \nabla \mathbf{v}\, \mathrm{d}\mathbf{x} = -\nu \int_{\mathcal{D}} \Delta \mathbf{u} \cdot \partial_\varphi \mathbf{v}\, \mathrm{d}\mathbf{x} = -(\mathscr{A}_0^0 \mathbf{u}, \partial_\varphi \mathbf{v})_2.\end{aligned}$$

Substituting this identity into (8.17), we obtain

$$\begin{aligned}(\mathscr{A}_0^\omega \mathbf{u}, \mathscr{A}_0^\omega \mathbf{v})_2 &= \big((\mathscr{A}_0^\omega)^* \mathbf{u}, \mathscr{A}_0^\omega \mathbf{v}\big)_2 - 2(\mathscr{A}_0^0 \mathbf{u}, \omega \partial_\varphi \mathbf{v})_2 \\ &\quad + 2(\omega \partial_\varphi \mathbf{u}, \omega \partial_\varphi \mathbf{v})_2 \\ &= \big((\mathscr{A}_0^\omega)^* \mathbf{u}, \mathscr{A}_0^\omega \mathbf{v}\big)_2 - 2\big((\mathscr{A}_0^\omega)^* \mathbf{u}, \omega \partial_\varphi \mathbf{v}\big)_2 \\ &= \big((\mathscr{A}_0^\omega)^* \mathbf{u}, (\mathscr{A}_0^\omega)^* \mathbf{v}\big)_2 = \big(\mathscr{A}_0^\omega (\mathscr{A}_0^\omega)^* \mathbf{u}, \mathbf{v}\big)_2 = (\mathbf{w}, \mathbf{v})_2. \end{aligned} \tag{8.18}$$

In fact, (8.18) holds for all $\mathbf{v} \in D(\mathscr{A}_0^\omega)$ because the set $\{\mathbf{v} \in D(\mathscr{A}_0^\omega);\, \mathbf{v}$ has a compact support in $\overline{\mathcal{D}}\}$ is a core of $\mathscr{A}_0^\omega$. (See Theorem 8.2.) Now, (8.18) shows that for fixed $\mathbf{u}$, $(\mathscr{A}_0^\omega \mathbf{u}, \mathscr{A}_0^\omega \mathbf{v})_2$ can be extended to a continuous linear functional of $\mathbf{v} \in \boldsymbol{L}^2_\sigma(\mathcal{D})$. Thus, $\mathbf{u} \in D((\mathscr{A}_0^\omega)^* \mathscr{A}_0^\omega)$.

We have proved the inclusion $D(\mathscr{A}_0^\omega (\mathscr{A}_0^\omega)^*) \subset D((\mathscr{A}_0^\omega)^* \mathscr{A}_0^\omega)$. The opposite inclusion can be proved in the same way. Moreover, (8.18) implies that

$$\big((\mathscr{A}_0^\omega)^* \mathscr{A}_0^\omega \mathbf{u}, \mathbf{v}\big)_2 = (\mathscr{A}_0^\omega (\mathscr{A}_0^\omega)^* \mathbf{u}, \mathbf{v})_2$$

for all $\mathbf{v} \in D(\mathscr{A}_0^\omega)$ and even for all $\mathbf{v} \in \boldsymbol{L}^2_\sigma(\mathcal{D})$, which confirms that the operators $\mathscr{A}_0^\omega$ and $(\mathscr{A}_0^\omega)^*$ commute.

(2) Now, we assume that $\mathcal{D}$ is not axially symmetric about the x_3-axis and show that the operator $\mathscr{A}_0^\omega$ is not normal in $\boldsymbol{L}^2_\sigma(\mathcal{D})$. We will prove that $D((\mathscr{A}_0^\omega)^* \mathscr{A}_0^\omega) \neq D(\mathscr{A}_0^\omega (\mathscr{A}_0^\omega)^*)$, which implies that $(\mathscr{A}_0^\omega)^* \mathscr{A}_0^\omega \neq \mathscr{A}_0^\omega (\mathscr{A}_0^\omega)^*$. Let $R > R_0$ and let $(\mathscr{A}_0^0)_{\mathcal{D}_R}$ be the Stokes operator in the space $\boldsymbol{L}^2_\sigma(\mathcal{D}_R)$ with the dense domain $D((\mathscr{A}_0^0)_{\mathcal{D}_R}) = \boldsymbol{W}^{2,2}(\mathcal{D}_R) \cap \boldsymbol{W}^{1,2}_{0,\sigma}(\mathcal{D}_R)$. The spectrum of $(\mathscr{A}_0^0)_{\mathcal{D}_R}$ (as well as the spectrum of $(\mathscr{A}_0^0)_{\mathcal{D}_R} + \omega \partial_\varphi)$ consists of a countable number of isolated eigenvalues with finite multiplicities and negative real parts. Choose an eigenvalue ζ of $(\mathscr{A}_0^0)_{\mathcal{D}_R}$ and denote by $\mathbf{v}$ an associated eigenfunction. The equation

$$(\mathscr{A}_0^0)_{\mathcal{D}_R} \mathbf{u} + \omega \partial_\varphi \mathbf{u} \;=\; \mathbf{v} \tag{8.19}$$

has a unique solution $\mathbf{u} \in D((\mathscr{A}_0^0)_{\mathcal{D}_R})$, $\mathbf{u} \not\equiv \mathbf{0}$. Let us show, by contradiction, that $\partial_\varphi \mathbf{u} \not\equiv \mathbf{0}$ on $\partial\mathcal{D}$. Assume the opposite, i.e., $\partial_\varphi \mathbf{u} \equiv \mathbf{0}$ on $\partial\mathcal{D}$. Then

$\partial_\varphi \mathbf{u} \in \boldsymbol{V}_{\mathcal{D}_R}$, where $\boldsymbol{V}_{\mathcal{D}_R} := \boldsymbol{W}_0^{1,2}(\mathcal{D}_R) \cap \boldsymbol{L}^2_\sigma(\mathcal{D}_R)$. The operator $((\mathscr{A}_0^0)_{\mathcal{D}_R}$ can be extended to the one-to-one continuous linear operator mapping $\boldsymbol{V}_{\mathcal{D}_R}$ onto the dual space $\boldsymbol{V}'_{\mathcal{D}_R}$. Moreover, ∂_φ maps $\boldsymbol{V}_{\mathcal{D}_R}$ into $\boldsymbol{L}^2_\sigma(\mathcal{D})$ and $(\mathscr{A}_0^{(\omega)})_{\mathcal{D}_R} + \omega\partial_\varphi$ is an injection from $\boldsymbol{V}_{\mathcal{D}_R}$ into $\boldsymbol{V}'_{\mathcal{D}_R}$, because 0 is not an eigenvalue of $(\mathscr{A}_0^0)_{\mathcal{D}_R} + \omega\partial_\varphi$. Equation (8.19) shows that $(\mathscr{A}_0^0)_{\mathcal{D}_R}\mathbf{u}$ also belongs to $\boldsymbol{V}_{\mathcal{D}_R}$. Now, $(\mathscr{A}_0^0)_{\mathcal{D}_R}\partial_\varphi\mathbf{u} \in \boldsymbol{V}'_{\mathcal{D}_R}$, and it can be simply shown that it equals $\partial_\varphi(\mathscr{A}_0^0)_{\mathcal{D}_R}\mathbf{u}$, which is in $\boldsymbol{L}^2_\sigma(\mathcal{D})$. Indeed, if $\boldsymbol{\varphi} \in C^\infty_{0,\sigma}(\mathcal{D}_R)$, the duality between the spaces $\boldsymbol{V}'_{\mathcal{D}_R}$ and $\boldsymbol{V}_{\mathcal{D}_R}$ yields

$$
\begin{aligned}
\langle(\mathscr{A}_0^0)_{\mathcal{D}_R}\partial_\varphi\mathbf{u}, \boldsymbol{\varphi}\rangle &= -\nu\int_{\mathcal{D}_R}\nabla\partial_\varphi\mathbf{u}\cdot\nabla\boldsymbol{\varphi}\,\mathrm{d}\mathbf{x} = \nu\int_{\mathcal{D}_R}\partial_\varphi\mathbf{u}\cdot\Delta\boldsymbol{\varphi}\,\mathrm{d}\mathbf{x}\\
&= -\nu\int_{\mathcal{D}_R}\mathbf{u}\cdot\partial_\varphi\Delta\boldsymbol{\varphi}\,\mathrm{d}\mathbf{x} = -\int_{\mathcal{D}_R}\mathbf{u}\cdot\Delta\partial_\varphi\boldsymbol{\varphi}\,\mathrm{d}\mathbf{x}\\
&= -\int_{\mathcal{D}_R}\mathbf{u}\cdot(\mathscr{A}_0^0)_{\mathcal{D}_R}\partial_\varphi\boldsymbol{\varphi}\,\mathrm{d}\mathbf{x} = -\int_{\mathcal{D}_R}(\mathscr{A}_0^0)_{\mathcal{D}_R}\mathbf{u}\cdot\partial_\varphi\boldsymbol{\varphi}\,\mathrm{d}\mathbf{x}\\
&= \int_{\mathcal{D}_R}\partial_\varphi(\mathscr{A}_0^0)_{\mathcal{D}_R}\mathbf{u}\cdot\boldsymbol{\varphi}\,\mathrm{d}\mathbf{x}.
\end{aligned}
$$

Hence we get

$$
\begin{aligned}
\mathbf{0} &= ((\mathscr{A}_0^0)_{\mathcal{D}_R} - \zeta I)\mathbf{v} = ((\mathscr{A}_0^0)_{\mathcal{D}_R} - \zeta I)((\mathscr{A}_0^0)_{\mathcal{D}_R} + \omega\partial_\varphi)\mathbf{u}\\
&= ((\mathscr{A}_0^0)_{\mathcal{D}_R})^2\mathbf{u} + \omega(\mathscr{A}_0^0)_{\mathcal{D}_R}\partial_\varphi\mathbf{u} - \zeta(\mathscr{A}_0^0)_{\mathcal{D}_R}\mathbf{u} - \zeta\omega\partial_\varphi\mathbf{u}\\
&= ((\mathscr{A}_0^0)_{\mathcal{D}_R})^2\mathbf{u} + \omega\partial_\varphi(\mathscr{A}_0^0)_{\mathcal{D}_R}\mathbf{u} - (\mathscr{A}_0^0)_{\mathcal{D}_R}\zeta\mathbf{u} - \omega\partial_\varphi\zeta\mathbf{u}\\
&= ((\mathscr{A}_0^0)_{\mathcal{D}_R} + \omega\partial_\varphi)((\mathscr{A}_0^0)_{\mathcal{D}_R} - \zeta I)\mathbf{u}.
\end{aligned}
$$

This implies that $((\mathscr{A}_0^0)_{\mathcal{D}_R} - \zeta I)\mathbf{u} = \mathbf{0}$, which means that $\mathbf{u}$ is an eigenfunction of $(\mathscr{A}_0^0)_{\mathcal{D}_R}$ associated with the eigenvalue ζ, too. Since the space generated by such eigenfunctions is finite dimensional, $\mathbf{v}$ can be chosen so that $\mu\mathbf{u} = \mathbf{v}$ for some $\mu \in \mathbb{C}$. Then Eq. (8.19) implies

$$
\omega\,\partial_\varphi\mathbf{u} = (\mu - \zeta)\mathbf{u} \tag{8.20}
$$

in $\mathcal{D}_R$. Since $\mathcal{D}_R$ is not axially symmetric, there exists $\mathbf{x}_0 \in \partial\mathcal{D}$ such that in a neighborhood $U(\mathbf{x}_0)$, intersected with $\partial\mathcal{D}$, ∂_φ is not the tangential derivative. Consider (8.20) as a first order linear differential equation in φ with initial values related to points in $U(\mathbf{x}_0)$. The boundary condition $\mathbf{u} = \mathbf{0}$ on $\partial\mathcal{D}$ enables us to conclude that $\mathbf{u}$ vanishes identically in an open subset of $\mathcal{D}_R$. Now the unique continuation principle for elliptic equations (see [167]) applied to $\boldsymbol{\omega} = \mathbf{curl}\,\mathbf{u}$ shows that $\boldsymbol{\omega} \equiv \mathbf{0}$ and consequently that also $\mathbf{u} \equiv \mathbf{0}$ in $\mathcal{D}_R$ which is impossible because $\mathbf{u} \neq \mathbf{0}$ in $\mathcal{D}_R$. The assumption $\partial_\varphi\mathbf{u} \equiv \mathbf{0}$

on $\partial\mathcal{D}$ thus leads to the contradiction, hence $\partial_\varphi \mathbf{u} \not\equiv \mathbf{0}$ on $\partial\mathcal{D}$. Using an appropriate cutoff function procedure, analogous to the one used in the proof of Lemma 8.7, we can construct a function $\mathbf{z}$ in $D((\mathscr{A}_0^\omega)^*\mathscr{A}_0^\omega)$ that coincides with the function $\mathbf{u}$ in the neighborhood of $\partial\mathcal{D}$ and equals $\mathbf{0}$ outside $\mathcal{D}_R$. Hence $\partial_\varphi \mathbf{z} \not\equiv \mathbf{0}$ on $\partial\mathcal{D}$. However, then $\mathbf{z}$ cannot belong to $D(\mathscr{A}_0^\omega(\mathscr{A}_0^\omega)^*)$ because all functions $\mathbf{z} \in D((\mathscr{A}_0^\omega)^*\mathscr{A}_0^\omega) \cap D(\mathscr{A}_0^\omega(\mathscr{A}_0^\omega)^*)$ satisfy $\mathbf{z} = \mathscr{A}_0^0\mathbf{z} + \omega\partial_\varphi\mathbf{z} = \mathscr{A}_0^0\mathbf{z} - \omega\partial_\varphi\mathbf{z} = \mathbf{0}$ on $\partial\mathcal{D}$, which implies that $\partial_\varphi\mathbf{z} \equiv \mathbf{0}$ on $\partial\mathcal{D}$. This confirms that $D((\mathscr{A}_0^\omega)^*\mathscr{A}_0^\omega) \neq D(\mathscr{A}_0^\omega(\mathscr{A}_0^\omega)^*)$.

(3) Now, assume that $k \neq 0$ and $\mathcal{D} = \mathbb{R}^3$. In this case, one can almost copy the procedure from part 1) and show that the operator $\mathscr{A}_k^\omega$ is normal. The presence of the term $k\,\partial_3\mathbf{u}$ requires only small technical modifications.

(4) Finally, let $k \neq 0$ and $\mathcal{D} \neq \mathbb{R}^3$. We will show that $\mathscr{A}_k^\omega$ is not normal in $\boldsymbol{L}^2_\sigma(\mathcal{D})$. Since the value of ω plays no role in this part, we assume for simplicity that $\omega = 0$ and $\mathscr{A}_k^\omega$ reduces to the "classical" Oseen operator $\mathscr{A}_k^0$. We will construct a function $\mathbf{z}$ in the domain of $D((\mathscr{A}_k^0)^*\mathscr{A}_k^0)$, that is not in $D(\mathscr{A}_k^0(\mathscr{A}_k^0)^*)$. This will imply that $D((\mathscr{A}_k^0)^*\mathscr{A}_k^0) \neq D(\mathscr{A}_k^0(\mathscr{A}_k^0)^*)$, and consequently, $(\mathscr{A}_k^0)^*\mathscr{A}_k^0 \neq \mathscr{A}_k^0(\mathscr{A}_k^0)^*$. Let $R > 0$ be so large that the body $\mathcal{B}$ is contained in the cube $(-R, R)^3$. Define the set

$$\begin{aligned}\mathcal{D}_{\text{per}} := \big\{\mathbf{x} = (x_1, x_2, x_3) \in \mathbb{R}^3;\ \exists l_1, l_2, l_3 \in \mathbb{Z} :\\ \big(x_1 - l_1R,\ x_2 - l_2R,\ x_3 - l_3R\big) \in [-R, R]^3 - \mathcal{B}\big\}.\end{aligned}$$

Note that $\mathcal{D}_{\text{per}}$ is a domain in $\mathbb{R}^3$ which consists of infinitely many copies of the set $[-R, R]^3 - \mathcal{B}$, periodically repeated in directions parallel with the x_1-, x_2-, and x_3-axes. We shall use the following function spaces:

- $\boldsymbol{C}^\infty_{0,\text{per}}$ is the space of infinitely differentiable vector functions $\boldsymbol{\varphi}$ in $\mathcal{D}_{\text{per}}$, R-periodic in the directions of all Cartesian axes and such that the distance between supp $\boldsymbol{\varphi}$ and $\mathcal{B}$ is positive.
- $\boldsymbol{L}^2_{\text{per}}$ is the completion of $\boldsymbol{C}^\infty_{0,\text{per}}$ in the norm identical with the L^2-norm on $(-R, R)^3 \setminus \mathcal{B}$. The spaces $\boldsymbol{W}^{1,2}_{0,\text{per}}$ and $\boldsymbol{W}^{2,2}_{\text{per}}$ are defined by analogy.
- $\boldsymbol{L}^2_{\sigma,\text{per}}$ is the closure of the space of divergence-free functions from $\boldsymbol{C}^\infty_{0,\text{per}}$ in $\boldsymbol{L}^2_{\text{per}}$.

Let $\mathscr{A}^0_{0,\text{per}}$ denote the Stokes operator in the space $\boldsymbol{L}^2_{\sigma,\text{per}}$ with the dense domain $D(\mathscr{A}^0_{0,\text{per}}) = \boldsymbol{W}^{2,2}_{\text{per}} \cap \boldsymbol{W}^{1,2}_{0,\text{per}} \cap \boldsymbol{L}^2_{\sigma,\text{per}}$. Then $\mathscr{A}^0_{0,\text{per}}$ has a compact resolvent, and its spectrum, as well as the spectrum of the operator $\mathscr{A}^0_{0,\text{per}} - k\,\partial_3$ (with the same domain), consists of a countable number of isolated eigenvalues with finite multiplicities and negative real parts. Let ζ be an eigenvalue of $\mathscr{A}^0_{0,\text{per}} - k\,\partial_3$ and $\mathbf{u}$ be an associated eigenfunction. Then the equation

$$\mathscr{A}^0_{0,\text{per}}\mathbf{u} - k\,\partial_3\mathbf{u} - \zeta\mathbf{u} = \mathbf{0} \tag{8.21}$$

holds in $\mathcal{D}_{\rm per}$. Let us show, by contradiction, that the eigenfunction $\mathbf{u}$ can be chosen so that $\partial_3\mathbf{u} \not\equiv \mathbf{0}$ on $\partial\mathcal{D}_{\rm per}$. Assume the opposite, i.e., that all eigenfunctions $\mathbf{v}$ of the operator $\mathscr{A}^0_{0,\rm per} - k\,\partial_3$, corresponding to the eigenvalue ζ, satisfy $\partial_3\mathbf{v} \equiv \mathbf{0}$ on $\partial\mathcal{D}_{\rm per}$. Then, for each eigenfunction $\mathbf{v}$, there are two possibilities: Either $\partial_3\mathbf{v} \equiv \mathbf{0}$ in $\mathcal{D}_{\rm per}$ (i.e., $\partial_3\mathbf{v} = \mathbf{0}$ a.e. on $\mathcal{D}_{\rm per}$), which can be easily excluded, or $\partial_3\mathbf{v}$ is also an eigenfunction of $\mathscr{A}^0_{0,\rm per} - k\,\partial_3$ corresponding to the same eigenvalue ζ. Since the eigenspace of $\mathscr{A}^0_{0,\rm per} - k\,\partial_3$, generated by all such eigenfunctions, is finite dimensional, we can choose an eigenfunction $\mathbf{u}$ so that

$$k\,\partial_3\mathbf{u} = \mu\,\mathbf{u} \tag{8.22}$$

with an appropriate constant μ. Since $\mathbf{u} = \mathbf{0}$ on $\partial\mathcal{D}_{\rm per}$ and Eq. (8.22) is satisfied in $\mathcal{D}_{\rm per}$, the integration of (8.22) on line segments parallel with the x_3-axis and starting from the boundary of $\mathcal{D}_{\rm per}$ yields that $\mathbf{u} = \mathbf{0}$ on all such line segments. Thus, $\mathbf{u}$ vanishes identically in an open subset of $\mathcal{D}_{\rm per}$. Now the unique continuation principle, see, e.g., R. Leis [167], applied to $\mathbf{w} = \mathbf{curl}\,\mathbf{u}$, shows that $\mathbf{w} \equiv \mathbf{0}$ in $\mathcal{D}_{\rm per}$. Consequently, $\mathbf{u} \equiv \mathbf{0}$ in $\mathcal{D}_{\rm per}$, which is impossible because $\mathbf{u}$ is an eigenfunction. Since the assumption that $\partial_3\mathbf{u} \equiv \mathbf{0}$ on $\partial\mathcal{D}_{\rm per}$ leads to a contradiction, we have $\partial_3\mathbf{u} \not\equiv \mathbf{0}$ on $\partial\mathcal{D}_{\rm per}$. Note that $\partial\mathcal{D} \subset \partial\mathcal{D}_{\rm per}$ and $\partial\mathcal{D}_{\rm per}$ consists of infinitely many copies of $\partial\mathcal{D}$ repeated periodically with the period R in the direction of each Cartesian coordinate. Now, we multiply the function $\mathbf{u}$ by an infinitely differentiable cutoff function η_R that equals one in the neighborhood of $\partial\mathcal{D}$ and whose support is contained in $(-R, R)^3 \setminus \mathcal{B}$, and correct the product $\eta_R\mathbf{u}$ by an appropriate function $\boldsymbol{U}_R$ that guarantees the identity $\operatorname{div}(\eta_R\mathbf{u} - \boldsymbol{U}_R) = 0$. In this way, we can obtain a function $\mathbf{z}$ in $D((\mathscr{A}^0_k)^*\mathscr{A}^0_k)$ that coincides with the function $\mathbf{u}$ constructed above in the neighborhood of $\mathcal{D}$ and equals zero outside $(-R, R)^3$. The function $\mathbf{z}$ satisfies $\partial_3\mathbf{z} \not\equiv \mathbf{0}$ on $\partial\mathcal{D}$. Then $\mathbf{z}$ cannot belong to $D(\mathscr{A}^0_k(\mathscr{A}^0_k)^*)$ because all functions from $D((\mathscr{A}^0_k)^*\mathscr{A}^0_k) \cap D(\mathscr{A}^0_k(\mathscr{A}^0_k)^*)$ satisfy on $\partial\mathcal{D}$ the conditions $\mathbf{z} = \mathscr{A}^0_0\mathbf{z} + k\,\partial_3\mathbf{z} = \mathscr{A}^0_0\mathbf{z} - k\,\partial_3\mathbf{z} = \mathbf{0}$, which implies that $\partial_3\mathbf{z} = \mathbf{0}$ on $\partial\mathcal{D}$. □

For completeness of the content of this section, note that the operators $\mathscr{A}^0_0$ and $\mathscr{A}^0_k$ generate analytic semigroups and the operators $\mathscr{A}^\omega_0$ and $\mathscr{A}^\omega_k$ generate C_0-semigroups in $\boldsymbol{L}^q_\sigma(\mathcal{D})$. More details are explained in Sect. 8.5. Note that the closedness of the operator $\mathscr{A}^\omega_k$, stated in Theorem 8.1, now directly follows from the fact that $\mathscr{A}^\omega_k$ generates a C_0-semigroup.

8.2 A Brief Survey of Basic Notions from Spectral Theory of Operators in a Banach Space

Since we further pay a lot of attention to the shape and structure of spectra of the operators $\mathscr{A}^\omega_0$ and $\mathscr{A}^\omega_k$ in this chapter, we recall some general notions from the spectral theory of linear operators in a Banach or Hilbert space in this section. Thus, let X be a Banach space with the norm $\|\,.\,\|$ and $\mathscr{T}$ be a closed linear operator in X with a dense domain $D(\mathscr{T})$. Recall that the operator $\mathscr{T}$ is said to be *closed* if its graph is closed in $X \times X$. The density of $D(\mathscr{T})$ in X guarantees that the adjoint operator $\mathscr{T}^*$ exists, see Kato [137, p. 167]. We denote by $N(\mathscr{T})$ the null space of $\mathscr{T}$ and by $R(\mathscr{T})$ the range of $\mathscr{T}$. Furthermore, we use this notation:

- $\mathrm{nul}(\mathscr{T})$ (the *nullity* of $\mathscr{T}$) $:= \dim N(\mathscr{T})$.
- $\mathrm{def}(\mathscr{T})$ (the *deficiency* of $\mathscr{T}$) $:= \dim X/R(\mathscr{T})$.
- $\mathrm{ind}(\mathscr{T})$ (the *index* of $\mathscr{T}$) $:= \mathrm{nul}(\mathscr{T}) - \mathrm{def}(\mathscr{T})$.
- $\mathrm{nul}'(\mathscr{T})$ (the *approximate nullity* of $\mathscr{T}$) $:=$ the greatest number $m \in \mathbb{N} \cup \{\infty\}$ such that to any $\epsilon > 0$ there exists an m-dimensional closed linear manifold $M_\epsilon \subset D(\mathscr{T})$ with the property that $\|\mathscr{T}u\| \le \epsilon\,\|u\|$ for all $u \in M_\epsilon$.
- $\mathrm{def}'(\mathscr{T})$ (the *approximate deficiency* of $\mathscr{T}$) $:= \mathrm{nul}'(\mathscr{T}^*)$.

These numbers satisfy the inequalities $\mathrm{nul}'(\mathscr{T}) \ge \mathrm{nul}(\mathscr{T})$ and $\mathrm{def}'(\mathscr{T}) \ge \mathrm{def}(\mathscr{T})$, see Kato [137, pages 230–233]. Moreover, if $R(\mathscr{T})$ is closed, then $\mathrm{nul}'(\mathscr{T}) = \mathrm{nul}(\mathscr{T})$ and $\mathrm{def}'(\mathscr{T}) = \mathrm{def}(\mathscr{T})$. On the other hand, if $R(\mathscr{T})$ is not closed, then $\mathrm{nul}'(\mathscr{T}) = \mathrm{def}'(\mathscr{T}) = \infty$. The identity $\mathrm{nul}'(\mathscr{T}) = \infty$ is equivalent to the existence of a non-compact sequence $\{u_n\}$ on the unit sphere in X such that $\mathscr{T}u_n \to 0$ (= the zero element in X)for $n \to \infty$ (see [137, p. 233]).

We say that $\mathscr{T}$ is a *Fredholm operator* if its range $R(\mathscr{T})$ is closed in X and both the numbers $\mathrm{nul}(\mathscr{T})$ and $\mathrm{def}(\mathscr{T})$ are finite. The operator $\mathscr{T}$ is called *semi-Fredholm* if the range $R(\mathscr{T})$ is closed in X and at least one of the numbers $\mathrm{nul}(\mathscr{T})$ and $\mathrm{def}(\mathscr{T})$ is finite. Consequently, $\mathscr{T}$ is semi-Fredholm if and only if at least one of the numbers $\mathrm{nul}'(\mathscr{T})$ and $\mathrm{def}'(\mathscr{T})$ is finite.

- $\rho(\mathscr{T})$ (the *resolvent set* of $\mathscr{T}$) is the set of all $\lambda \in \mathbb{C}$ such that $\mathscr{T} - \lambda I$ has a bounded inverse operator, defined on the whole space X.

The set $\rho(\mathscr{T})$ can also be characterized as the set of $\lambda \in \mathbb{C}$ such that $\mathrm{nul}(\mathscr{T} - \lambda I) = \mathrm{def}(\mathscr{T} - \lambda I) = \mathrm{nul}'(\mathscr{T} - \lambda I) = \mathrm{def}'(\mathscr{T} - \lambda I) = 0$. It is well known that $\rho(\mathscr{T})$ is open in $\mathbb{C}$.

- $\mathrm{Sp}(\mathscr{T})$ (the *spectrum* of $\mathscr{T}$) is defined to be the complement of $\rho(\mathscr{T})$ in $\mathbb{C}$. $\mathrm{Sp}(\mathscr{T})$ is a closed subset of $\mathbb{C}$.
- $\mathrm{Sp}_{\mathrm{p}}(\mathscr{T})$ (the *point spectrum* of $\mathscr{T}$) consists of all eigenvalues of $\mathscr{T}$. It is the set of $\lambda \in \mathbb{C}$ such that $\mathrm{nul}(\mathscr{T} - \lambda I) > 0$.

$\mathrm{Sp}_{\mathrm{p}}(\mathscr{T})$ can also be defined as the set of all $\lambda \in \mathbb{C}$ such that the operator $\mathscr{T} - \lambda I$ is not injective. The *geometric multiplicity* (or just the *multiplicity*) of an eigenvalue λ is defined to be the dimension of $N(\mathscr{T} - \lambda I)$. Each nonzero element of $N(\mathscr{T} - \lambda I)$ is said to be an *eigenvector* (or *eigenfunction*, if X is the space of functions) of $\mathscr{T}$, corresponding to the eigenvalue λ.

If λ is an isolated eigenvalue (in the sense that there exists a neighborhood $U(\lambda)$ in $\mathbb{C}$, such that $U(\lambda) \setminus \{\lambda\} \subset \rho(\mathscr{T})$), then the so-called *algebraic multiplicity* of λ denotes the dimension of the range of the associated eigenprojection $P(\lambda)$ (also called the *Riesz projection*), which is defined by the formula

$$P_\lambda := -\frac{1}{2\pi \mathrm{i}} \int_C (\mathscr{T} - \zeta I)^{-1} \, \mathrm{d}\zeta.$$

Here, C is a simple closed curve in $U(\lambda) \setminus \{\lambda\}$, oriented counter-clockwise, which encloses λ. If x is an eigenvector of $\mathscr{T}$, corresponding to the eigenvalue λ, then

$$\begin{aligned} P_\lambda x &= -\frac{1}{2\pi \mathrm{i}} \int_C \frac{(\mathscr{T} - \zeta I)^{-1}}{\zeta - \lambda} (\zeta - \lambda) x \, \mathrm{d}\zeta \\ &= \frac{1}{2\pi \mathrm{i}} \int_C \frac{(\mathscr{T} - \zeta I)^{-1}}{\zeta - \lambda} (\mathscr{T} - \zeta I) x \, \mathrm{d}\zeta = 2\pi \mathrm{i} \int_C \frac{x}{\zeta - \lambda} \, \mathrm{d}\zeta = x. \end{aligned}$$

Thus, $x \in R(P_\lambda)$. From this, one can observe that the geometric multiplicity of λ is less than or equal to the algebraic multiplicity.

If $x \neq 0$ and $x \in N(\mathscr{T} - \lambda I)^n$ for some $n \in \mathbb{N}$, $n > 1$, then x is called a *generalized eigenvector* (or a *generalized eigenfunction*) of $\mathscr{T}$, corresponding to the eigenvalue λ. By analogy with the eigenvector, it is not difficult to show that if x is a generalized eigenvector of $\mathscr{T}$, corresponding to eigenvalue λ, then $P_\lambda x = x$. Consequently, $x \in R(P_\lambda)$. The opposite implication, i.e., $x \neq 0$, $x \in R(P_\lambda)$ $\Longrightarrow x$ is a generalized eigenvector of $\mathscr{T}$ (corresponding to eigenvalue λ), holds if the algebraic multiplicity of λ is finite. If the algebraic multiplicity is infinite, then $R(P_\lambda)$ need not generally coincide with $\bigcup_{n\in\mathbb{N}} N((\mathscr{T} - \lambda I)^n)$. However, independently of whether the algebraic multiplicity of λ is finite or infinite, $R(P_\lambda)$ satisfies

$$R(P_\lambda) = \left\{x \in X;\ \lim_{n\to\infty} \|(\mathscr{T} - \lambda I)^n x\|^{1/n} = 0\right\},$$

see, e.g., Heuser [133, p. 206].

- $\mathrm{Sp_c}(\mathscr{T})$ (the *continuous spectrum* of $\mathscr{T}$) is the set of such $\lambda \in \mathbb{C}$ that $\mathrm{nul}(\mathscr{T} - \lambda I) = 0$, $R(\mathscr{T} - \lambda I)$ is dense in X, but $R(\mathscr{T} - \lambda I) \neq X$.

In this case, $\mathrm{nul}'(\mathscr{T} - \lambda I) = \mathrm{def}(\mathscr{T} - \lambda I) = \mathrm{def}'(\mathscr{T} - \lambda I) = \infty$. (Note that the equality $\mathrm{def}(\mathscr{T} - \lambda I) = \infty$ follows, e.g., from the non-closedness of $R(\mathscr{T} - \lambda I)$ and from Theorem IV.5.2 and Problem IV.5.7 in [137].)

- $\mathrm{Sp_r}(\mathscr{T})$ (the *residual spectrum* of $\mathscr{T}$) is the set of such $\lambda \in \mathbb{C}$ that $\mathrm{nul}(\mathscr{T} - \lambda I) = 0$ and $R(\mathscr{T} - \lambda I)$ is not dense in X.

In this case, $\mathrm{def}(\mathscr{T} - \lambda I) > 0$. The residual spectrum can be characterized as the set of $\lambda \in \mathbb{C}$ such that $\bar{\lambda} \in \mathrm{Sp_p}(\mathscr{T}^*)$ and $\lambda \notin \mathrm{Sp_p}(\mathscr{T})$.

Obviously, $\mathrm{Sp_p}(\mathscr{T})$, $\mathrm{Sp_c}(\mathscr{T})$, and $\mathrm{Sp_r}(\mathscr{T})$ are pairwise disjoint and $\mathrm{Sp}(\mathscr{T}) = \mathrm{Sp_p}(\mathscr{T}) \cup \mathrm{Sp_c}(\mathscr{T}) \cup \mathrm{Sp_r}(\mathscr{T})$.

- $\mathrm{Sp_{ess}}(\mathscr{T})$ (the *essential spectrum* of $\mathscr{T}$) is the set of all $\lambda \in \mathbb{C}$ such that $\mathscr{T} - \lambda I$ is not semi-Fredholm, which means that $\mathrm{nul}'(\mathscr{T} - \lambda I) = \mathrm{def}'(\mathscr{T} - \lambda I) = \infty$.

Obviously, $\mathrm{Sp_{ess}}(\mathscr{T}) \supset \mathrm{Sp_c}(\mathscr{T})$. The essential spectrum is a closed subset of $\mathbb{C}$, and $\mathrm{ind}(\mathscr{T} - \lambda I)$ is constant in each component G of $\mathbb{C} \setminus \mathrm{Sp_{ess}}(\mathscr{T})$. Moreover, $\mathrm{nul}(\mathscr{T} - \lambda I)$ and $\mathrm{def}(\mathscr{T} - \lambda I)$ are constant in G with a possible exception of an at most countable set of isolated eigenvalues of finite algebraic multiplicities which can cluster only at points of $\mathrm{Sp_{ess}}(\mathscr{T})$. (See [137, p. 243].) There are no generally valid relations between $\mathrm{Sp_p}(\mathscr{T})$, $\mathrm{Sp_r}(\mathscr{T})$ on one hand and $\mathrm{Sp_{ess}}(\mathscr{T})$ on the other hand. However, any point on the boundary of $\mathrm{Sp}(\mathscr{T})$ belongs to $\mathrm{Sp_{ess}}(\mathscr{T})$ unless it is an isolated point of $\mathrm{Sp}(\mathscr{T})$. (See [137, p. 244].)

Note that one can also find a different definition of the essential spectrum in the literature, according to which λ belongs to the essential spectrum of $\mathscr{T}$ if the operator $\mathscr{T} - \lambda I$ is not Fredholm. (See, e.g., the footnote on p. 243 in [188].) In order to distinguish between the two definitions, we denote by $\mathrm{Sp^*_{ess}}(\mathscr{T})$ the essential spectrum of operator $\mathscr{T}$, satisfying the second definition. (We shall also denote by star some other quantities, related to the second definition of the essential spectrum.) Obviously, $\mathrm{Sp_{ess}}(\mathscr{T}) \subset \mathrm{Sp^*_{ess}}(\mathscr{T})$. Recall that while $\mathrm{Sp_{ess}}(\mathscr{T})$ is preserved under relatively compact additive perturbations of $\mathscr{T}$, see [?, Theorem IV.5.35], $\mathrm{Sp^*_{ess}}(\mathscr{T})$ is preserved under compact additive perturbations of $\mathscr{T}$, see [56, p. 248] or [188, Corollary 2.2].

- $\mathrm{Sp_c^{Gl}}(\mathscr{T})$ denotes the set of those $\lambda \in \mathbb{C}$ for which there exists a non-compact sequence $\{u_n\}$ in the unit sphere in X such that $(\mathscr{T} - \lambda I)u_n \to 0$ for $n \to \infty$. It is equivalent to the equality $\mathrm{nul}'(\mathscr{T} - \lambda I) = \infty$. (See [137, Theorem IV.5.11].)

The definition of $\mathrm{Sp_c^{Gl}}(\mathscr{T})$ is due to Glazman, see [122]. It is proven in [122, page 20] that $\mathrm{Sp_c^{Gl}}(\mathscr{T})$ is a closed set in $\mathbb{C}$. One can deduce from the previous definitions that $\mathrm{Sp_c}(\mathscr{T}) \subset \mathrm{Sp_{ess}}(\mathscr{T}) \subset \mathrm{Sp_c^{Gl}}(\mathscr{T}) \subset \mathrm{Sp}(\mathscr{T})$.

- $\mathrm{Sp_{ap}}(\mathscr{T})$ (the *approximate point spectrum* of $\mathscr{T}$) consists of all points $\lambda \in \mathbb{C}$ with the property that there exists a sequence $\{u_n\}$ in $D(\mathscr{T})$ such that $\|u_n\| = 1$ and $(\mathscr{T} - \lambda I)\mathbf{u}_n \to 0$ as $n \to \infty$.

The inequality $\mathrm{nul}'(\mathscr{T} - \lambda I) > 0$ (for $\lambda \in \mathrm{Sp_{ap}}(\mathscr{T})$) implies that $\lambda \in \mathrm{Sp}(\mathscr{T})$, which means that $\mathrm{Sp_{ap}}(\mathscr{T}) \subset \mathrm{Sp}(\mathscr{T})$. The inclusions $\mathrm{Sp_p}(\mathscr{T}) \subset \mathrm{Sp_{ap}}(\mathscr{T})$ and $\mathrm{Sp_c^{Gl}}(\mathscr{T}) \subset \mathrm{Sp_{ap}}(\mathscr{T})$ directly follow from the definitions of $\mathrm{Sp_p}(\mathscr{T})$, $\mathrm{Sp_c^{Gl}}(\mathscr{T})$, and $\mathrm{Sp_{ap}}(\mathscr{T})$. Hence $\mathrm{Sp_p}(\mathscr{T}) \cup \mathrm{Sp_c^{Gl}}(\mathscr{T}) \subset \mathrm{Sp_{ap}}(\mathscr{T}) \subset \mathrm{Sp}(\mathscr{T})$. As $\mathrm{Sp}(\mathscr{T}) = \mathrm{Sp_p}(\mathscr{T}) \cup \mathrm{Sp_c}(\mathscr{T}) \cup \mathrm{Sp_r}(\mathscr{T}) \subset \mathrm{Sp_p}(\mathscr{T}) \cup \mathrm{Sp_c^{Gl}}(\mathscr{T}) \cup \mathrm{Sp_r}(\mathscr{T}) \subset \mathrm{Sp_{ap}}(\mathscr{T}) \cup \mathrm{Sp_r}(\mathscr{T})$, we also have the equality $\mathrm{Sp_{ap}}(\mathscr{T}) \cup \mathrm{Sp_r}(\mathscr{T}) = \mathrm{Sp}(\mathscr{T})$. Note that the approximate point spectrum has been mainly introduced for the sake of completeness of this brief survey and we use it only in Sect. 8.5.2.1.

If X is a Hilbert space, then its dual can be identified with X, and it makes therefore sense to consider the adjoint operator $\mathscr{T}^*$ to be an operator in X. In this case, operator $\mathscr{T}$ is said to be *normal* if $\mathscr{T}^*\mathscr{T} = \mathscr{T}\mathscr{T}^*$. If $\mathscr{T}$ is normal, then $\mathscr{T}$ and $\mathscr{T}^*$ have the same null space, see [137, page 277]. Moreover, if operator $\mathscr{T}$ is normal, then $\mathrm{Sp_r}(\mathscr{T}) = \emptyset$ and $\mathrm{Sp_{ess}}(\mathscr{T}) = \mathrm{Sp_c^{Gl}}(\mathscr{T})$. (The first equality easily follows from the identities $R(\mathscr{T} - \lambda I)^\perp = N(\mathscr{T}^* - \bar{\lambda} I) = N(\mathscr{T} - \lambda I)$, and the second equality is proven, e.g., in [62].)

In this chapter, we further study the spectra of the operators $\mathscr{A}_0^\omega$ (the Stokes type operator) and $\mathscr{L}_0^\omega \equiv \mathscr{A}_k^\omega + \mathscr{B}$ (the perturbed Stokes type operator), $\mathscr{A}_k^\omega$ (the Oseen type operator) and $\mathscr{L}_k^\omega \equiv \mathscr{A}_k^\omega + \mathscr{B}$ (the perturbed Oseen type operator). We always begin with the L^2-theory, i.e., with the case $q = 2$, when $\mathscr{A}_0^\omega$ and $\mathscr{A}_k^\omega$ are operators in $\boldsymbol{L}^2_\sigma(\mathcal{D})$ (Sects. 8.3.1 and 8.4.1), and afterward we present the L^q-theory for general $1 < q < \infty$ (Sects. 8.3.2 and 8.4.2). Note that the advantage of the case $q = 2$, when $\boldsymbol{L}^2_\sigma(\mathcal{D})$ is a Hilbert space, is that it enables one to apply and clearly illustrate possibilities of the Fourier expansion in the cylindrical variable φ. On the other hand, the main tools, used in the general case $1 < q < \infty$, are the Fourier transform in the whole space case $\mathcal{D} = \mathbb{R}^3$ and the transfer of main results to the case of an exterior domain $\mathcal{D}$.

8.3 Spectral Theory of the Stokes Type Operator

8.3.1 L^2-Theory

The content of this subsection is mainly based on the paper [62] by Farwig and Neustupa. We consider $q = 2$. In this case, $\boldsymbol{G}^2(\mathcal{D})$ is an orthogonal complement of $\boldsymbol{L}^2_\sigma(\mathcal{D})$ in $\boldsymbol{L}^2(\mathcal{D})$.

If $\omega = 0$, then $\mathscr{A}_0^\omega$ is the Stokes operator, which is well known to be self-adjoint in $\boldsymbol{L}^2_\sigma(\mathcal{D})$, see, e.g., [201, page 128]. Its spectra satisfy $\mathrm{Sp_p}(\mathscr{A}_0^0) = \mathrm{Sp_r}(\mathscr{A}_0^0) = \emptyset$ and $\mathrm{Sp}(\mathscr{A}_0^0) = \mathrm{Sp_c}(\mathscr{A}_0^0) = (-\infty, 0]$. Let us sketch the proof of these identities: $\mathrm{Sp_r}(\mathscr{A}_0^0 = \emptyset$, because $\mathscr{A}_0^0$ is normal. The identity $\mathrm{Sp}(\mathscr{A}_0^0) = (-\infty, 0]$ is mentioned, e.g., in [165], and it can also be deduced from [122]. In order to show that $\mathrm{Sp_p}(\mathscr{A}_0^0) = \emptyset$, assume, by contradiction, that $\lambda \in \mathrm{Sp}(\mathscr{A}_0^0)$ is an eigenvalue of $\mathscr{A}_0^0$ and $\mathbf{u}$ is an associated eigenfunction. Multiplying the equation $\mathscr{A}_0^0\mathbf{u} = \lambda\mathbf{u}$ by $\bar{\mathbf{u}}$, we can show that $\lambda < 0$. The vector function $\mathbf{w} = \mathbf{curl}\,\mathbf{u}$ satisfies $\Delta\mathbf{w} - \lambda\mathbf{w} = \mathbf{0}$ in $\mathcal{D}$. Then there exists $R_1 > 0$ such that $\mathbf{w} = \mathbf{0}$ for all $\mathbf{x}$, satisfying $|\mathbf{x}| > R_1$. (This follows from [136, Theorem 1].) Due to the unique continuation principle, see Leis [167], we have $\mathbf{w} = \mathbf{0}$ in $\mathcal{D}$. Since domain $\mathcal{D}$ is simply connected, $\mathbf{u}$ has the form $\nabla\varphi$, where φ is an appropriate scalar function in $\mathcal{D}$. Using the equation $\mathrm{div}\,\mathbf{v} = 0$ and the boundary condition $\mathbf{u} = \mathbf{0}$ on $\partial\mathcal{D}$, we derive that $\mathbf{u} = \mathbf{0}$ in $\mathcal{D}$. Thus, $\mathbf{u}$ cannot be an eigenfunction and λ cannot be an eigenvalue.

8.3.1.1 The Case of an Axially Symmetric Domain $\mathcal{D}$

We assume that the domain $\mathcal{D} \subset \mathbb{R}^3$ is axially symmetric with respect to the x_3-axis, and we consistently identify $\mathbf{u}$ with $[u_r, u_\varphi, u_3]$ in this paragraph. Thus, since we now assume that $k = 0$, formula (8.10) reduces to the form

$$\mathscr{A}_0^\omega \mathbf{u} = \mathscr{A}_0^0 \mathbf{u} + \omega\, \partial_\varphi \mathbf{u}. \tag{8.23}$$

Let l be an integer. We introduce the following spaces and operators:

- $\boldsymbol{L}^2(\mathcal{D})_l := \{\mathbf{v} \in \boldsymbol{L}^2(\mathcal{D});\ \mathbf{v} \text{ has the form } \mathbf{v} = \boldsymbol{V}(r, x_3)\, \mathrm{e}^{\mathrm{i}l\varphi}\}$.
- $\boldsymbol{C}^\infty_{0,\sigma}(\mathcal{D})_l := \boldsymbol{C}^\infty_{0,\sigma}(\mathcal{D}) \cap \boldsymbol{L}^2(\mathcal{D})_l$.
- $\boldsymbol{L}^2_\sigma(\mathcal{D})_l$—the closure of $\boldsymbol{C}^\infty_{0,\sigma}(\mathcal{D})_l$ in $\boldsymbol{L}^2(\mathcal{D})_l$.
- P_l—the orthogonal projection of $\boldsymbol{L}^2(\mathcal{D})$ onto $\boldsymbol{L}^2(\mathcal{D})_l$.
- $(\mathscr{A}_0^0)_l$—the restriction of the operator $\mathscr{A}_0^0$ to $\boldsymbol{L}^2_\sigma(\mathcal{D})_l$.
- $(\mathscr{A}_0^\omega)_l$—the restriction of the operator $\mathscr{A}_0^\omega$ to $\boldsymbol{L}^2_\sigma(\mathcal{D})_l$.

Obviously, $\boldsymbol{L}^2(\mathcal{D})_l$ is a closed subspace of $\boldsymbol{L}^2(\mathcal{D})$, and $\boldsymbol{L}^2_\sigma(\mathcal{D})_l$ is a closed subspace of $\boldsymbol{L}^2_\sigma(\mathcal{D})$. The domain of $(\mathscr{A}_0^\omega)_l$ equals $D(\mathscr{A}_0^0) \cap \boldsymbol{L}^2_\sigma(\mathcal{D})_l$.

Each function from $\boldsymbol{L}^2(\mathcal{D})$ can be uniquely written in the form of a convergent Fourier series (with respect to the variable φ) of terms from $\boldsymbol{L}^2(\mathcal{D})_l$, $l \in \mathbb{Z}$:

$$\mathbf{v}(r, \varphi, x_3) = \sum_{l=-\infty}^{\infty} \boldsymbol{V}^k(r, x_3)\, \mathrm{e}^{\mathrm{i}l\varphi}, \tag{8.24}$$

where

$$\boldsymbol{V}^l(r, x_3) = \frac{1}{2\pi} \int_0^{2\pi} \mathbf{v}(r, \varphi, x_3) \mathrm{e}^{-\mathrm{i}l\varphi}\, \mathrm{d}\varphi.$$

Thus, we have the decomposition $\boldsymbol{L}^2(\mathcal{D}) = \cdots \oplus \boldsymbol{L}^2(\mathcal{D})_{-2} \oplus \boldsymbol{L}^2(\mathcal{D})_{-1} \oplus \boldsymbol{L}^2(\mathcal{D})_0 \oplus \boldsymbol{L}^2(\mathcal{D})_1 \oplus \boldsymbol{L}^2(\mathcal{D})_2 \oplus \ldots$.

Lemma 8.4 *Let $l \in \mathbb{Z}$. Then $\Pi^2_\sigma \boldsymbol{L}^2(\mathcal{D})_l = \boldsymbol{L}^2_\sigma(\mathcal{D}) \cap \boldsymbol{L}^2(\mathcal{D})_l = \boldsymbol{L}^2_\sigma(\mathcal{D})_l = P_l \boldsymbol{L}^2_\sigma(\mathcal{D})$.* □

Proof Suppose that $\mathbf{v} \in \boldsymbol{W}^{1,2}(\mathcal{D}) \cap \boldsymbol{L}^2(\mathcal{D})_l$. The solution of the Neumann problem

$$\Delta\varphi = \operatorname{div} \mathbf{v} \quad \text{in } \mathcal{D}, \qquad \frac{\partial \varphi}{\partial \mathbf{n}} = \mathbf{v} \cdot \mathbf{n} \quad \text{on } \partial\mathcal{D}$$

can be found in the form $\varphi = \Phi(r, x_3)\, \mathrm{e}^{\mathrm{i}l\varphi}$. Then $\Pi_\sigma \mathbf{v} = \mathbf{v} - \nabla\varphi \in \boldsymbol{L}^2(\mathcal{D})_l$. Using the density of $\boldsymbol{W}^{1,2}(\mathcal{D}) \cap \boldsymbol{L}^2(\mathcal{D})_l$ in $\boldsymbol{L}^2(\mathcal{D})_l$, we can show that this is true for all $\mathbf{v} \in \boldsymbol{L}^2(\mathcal{D})_l$, i.e., $\Pi_\sigma \boldsymbol{L}^2(\mathcal{D})_l \subset \boldsymbol{L}^2_\sigma(\mathcal{D})_l \cap \boldsymbol{L}^2(\mathcal{D})_l$. The opposite inclusion is obvious.

Since $\boldsymbol{C}^{\infty}_{0,\sigma}(\mathcal{D})_l$ is dense in $\boldsymbol{L}^2_{\sigma}(\mathcal{D})_l \cap \boldsymbol{L}^2(\mathcal{D})_l$, $\boldsymbol{L}^2_{\sigma}(\mathcal{D})_l$ (as a closure of $\boldsymbol{C}^{\infty}_{0,\sigma}(\mathcal{D})_l$) equals $\boldsymbol{L}^2_{\sigma}(\mathcal{D})_l \cap \boldsymbol{L}^2(\mathcal{D})_l$.

In order to prove the last equality, we consider $\mathbf{v} \in \boldsymbol{C}^{\infty}_{0,\sigma}(\mathcal{D})$ with the Fourier expansion (8.24). Denote by V^l_r, V^l_{φ}, V^l_3 the components of V^l. Since

$$0 = \sum_{m=-\infty}^{\infty} \operatorname{div}\Big[\boldsymbol{V}^l(r,x_3)\mathrm{e}^{\mathrm{i}l\varphi}\Big] = \sum_{m=-\infty}^{\infty} \Big[\frac{1}{r}\partial_r(rV^l_r) + \frac{1}{r}\,\mathrm{i}lV^l_{\varphi} + (\partial_3 V^l_3)\Big]\mathrm{e}^{\mathrm{i}l\varphi},$$

we deduce that $\operatorname{div}[\boldsymbol{V}^l(r,x_3)\,\mathrm{e}^{\mathrm{i}l\varphi}] = 0$ for each $l \in \mathbb{Z}$. Hence $P_l\mathbf{v} = \boldsymbol{V}^l(r,x_3)\mathrm{e}^{\mathrm{i}l\varphi} \in \boldsymbol{C}^{\infty}_{0,\sigma}(\mathcal{D})_l$, which is a subset of $\boldsymbol{L}^2_{\sigma}(\mathcal{D})_l \cap \boldsymbol{L}^2(\mathcal{D})_l$. Since $\boldsymbol{C}^{\infty}_{0,\sigma}(\mathcal{D})$ is dense in $\boldsymbol{L}^2_{\sigma}(\mathcal{D})_l$, we obtain $P_l\boldsymbol{L}^2_{\sigma}(\mathcal{D})_l \subset \boldsymbol{L}^2_{\sigma}(\mathcal{D})_l \cap \boldsymbol{L}^2(\mathcal{D})_l$. On the other hand, if $\mathbf{v} \in \boldsymbol{L}^2_{\sigma}(\mathcal{D})_l \cap \boldsymbol{L}^2(\mathcal{D})_l$, then $P_l\mathbf{v} = \mathbf{v}$; hence it also belongs to $P_l\boldsymbol{L}^2_{\sigma}(\mathcal{D})_l$. Thus, the opposite inclusion $\boldsymbol{L}^2_{\sigma}(\mathcal{D})_l \cap \boldsymbol{L}^2(\mathcal{D})_l \subset P_l\boldsymbol{L}^2_{\sigma}(\mathcal{D})_l$ is also true. □

Lemma 8.5 *$(\mathscr{A}^0_0)_l$ is a self-adjoint operator in the space $\boldsymbol{L}^2_{\sigma}(\mathcal{D})_l$ with the domain $D((\mathscr{A}^0_0)_l) = P_l[D(\mathscr{A}^0_0)]$.* □

Proof Let $\mathbf{v} \in D(\mathscr{A}^0_0) \equiv \boldsymbol{W}^{2,2}(\mathcal{D}) \cap \boldsymbol{W}^{1,2}_{0,\sigma}(\mathcal{D})$ and let (8.24) be its Fourier expansion in the variable φ. Then $\boldsymbol{V}^l(r,x_3)\,\mathrm{e}^{\mathrm{i}l\varphi} \equiv P_l\mathbf{v} \in \boldsymbol{W}^{2,2}(\mathcal{D})$, and, due to the axial symmetry of $\mathcal{D}$ and the boundary condition $\mathbf{v} = \mathbf{0}$ on $\partial\mathcal{D}$, $\boldsymbol{V}^l(r,x_3)\,\mathrm{e}^{\mathrm{i}l\varphi} \in \boldsymbol{W}^{1,2}_0(\mathcal{D})$, too. We have already seen in the proof of Lemma 8.4 that $\boldsymbol{V}^l(r,x_3)\,\mathrm{e}^{\mathrm{i}l\varphi} \in \boldsymbol{L}^2_{\sigma}(\mathcal{D})_l$. Hence $P_l[D(\mathscr{A}^0_0)] \subset D((\mathscr{A}^0_0)_l)$.

On the other hand, if $\mathbf{v} \in D((\mathscr{A}^0_0)_l)$, then $\mathbf{v} \in D(\mathscr{A}^0_0)$, and since $P_l\mathbf{v} = \mathbf{v}$, $\mathbf{v} \in \boldsymbol{L}^2(\mathcal{D})_l$. Hence $\mathbf{v} \in D(\mathscr{A}^0_0) \cap \boldsymbol{L}^2(\mathcal{D})_l = D(\mathscr{A}^0_0) \cap \boldsymbol{L}^2_{\sigma}(\mathcal{D})_l = P_l[D(\mathscr{A}^0_0)]$. We have proven the identity $D((\mathscr{A}^0_0)_l) = P_l[D(\mathscr{A}^0_0)]$.

If $\mathbf{v} \in D((\mathscr{A}^0_0)_l)$, then $\Delta\mathbf{v} \in \boldsymbol{L}^2(\mathcal{D})_l$, and due to Lemma 8.4, $(\mathscr{A}^0_0)_l\mathbf{v} \equiv \mathscr{A}^0_0\mathbf{v} = \nu\Pi_{\sigma}\Delta\mathbf{v} \in \boldsymbol{L}^2_{\sigma}(\mathcal{D})_l$. Hence $(\mathscr{A}^0_0)_l$ is an operator in $\boldsymbol{L}^2_{\sigma}(\mathcal{D})_l$.

The domain $D((\mathscr{A}^0_0)_l)$ is dense in $\boldsymbol{L}^2_{\sigma}(\mathcal{D})_l$ because it contains $\boldsymbol{C}^{\infty}_{0,\sigma}(\mathcal{D})_l$. Moreover, the operator $(\mathscr{A}^0_0)_l$ is symmetric because it is the part of the symmetric operator $\mathscr{A}^0_0$ in $L^2_{\sigma}(\mathcal{D})_l$, and $(\mathscr{A}^0_0)_l$ is closed because it is a restriction of the closed operator $\mathscr{A}^0_0$ to a closed subspace of $\boldsymbol{L}^2_{\sigma}(\mathcal{D})$. Thus, in order to show that $(\mathscr{A}^0_0)_l$ is self-adjoint, it is sufficient to show that $\rho((\mathscr{A}^0_0)_l)$ contains at least one real number. (See [137, page 271].) Indeed, if $\zeta \in \mathbb{R}$, $\zeta > 0$, and $\mathbf{f} \in \boldsymbol{L}^2_{\sigma}(\mathcal{D})_l$, then it can be verified that $\mathbf{u} := (\mathscr{A}^0_0 - \zeta I)^{-1}\mathbf{f}$ represents the unique solution of the equation $((\mathscr{A}^0_0)_l - \zeta I)\mathbf{u} = \mathbf{f}$ in $\boldsymbol{L}^2_{\sigma}(\mathcal{D})_l$. Thus, $\zeta \in \rho((\mathscr{A}^0_0)_l)$. □

Lemma 8.6 *$Sp((\mathscr{A}^0_0)_l) = Sp_C((\mathscr{A}^0_0)_l) = (-\infty, 0]$.* □

Proof Since the operator $(\mathscr{A}^0_0)_l$ is self-adjoint, $\mathrm{Sp}_{\mathrm{r}}((\mathscr{A}^0_0)_l) = \emptyset$ and $\mathrm{Sp}^{\mathrm{Gl}}_{\mathrm{c}}((\mathscr{A}^0_0)_l) = \mathrm{Sp}_{\mathrm{ess}}((\mathscr{A}^0_0)_l)$. Moreover, since $(\mathscr{A}^0_0)_l$ is a part of $\mathscr{A}^0_0$, we get $\mathrm{Sp}^{\mathrm{Gl}}_{\mathrm{c}}((\mathscr{A}^0_0)_l) \subset \mathrm{Sp}^{\mathrm{Gl}}_{\mathrm{c}}(\mathscr{A}^0_0) = (-\infty, 0]$.

Let us at first prove that $\mathrm{Sp}_{\mathrm{c}}^{\mathrm{Gl}}((\mathscr{A}_0^0)_l)$ covers the whole interval $(-\infty, 0]$. Since $\mathrm{Sp}_{\mathrm{c}}^{\mathrm{Gl}}((\mathscr{A}_0^0)_l)$ is a closed set, it is sufficient to show that it is also open in $(-\infty, 0]$ and non-empty. The last property is, however, clear because $(\mathscr{A}_0^0)_l$ is self-adjoint: The equality $\mathrm{Sp}_{\mathrm{c}}^{\mathrm{Gl}}((\mathscr{A}_0^0)_l) = \emptyset$ would imply $\mathrm{Sp}((\mathscr{A}_0^0)_l) = \mathrm{Sp}_{\mathrm{p}}((\mathscr{A}_0^0)_l)$. The last set is empty, because $\mathrm{Sp}_{\mathrm{p}}((\mathscr{A}_0^0)_l) \subset \mathrm{Sp}_{\mathrm{p}}(\mathscr{A}_0^0) = \emptyset$. Hence $\mathrm{Sp}_{\mathrm{c}}^{\mathrm{Gl}}((\mathscr{A}_0^0)_l) \neq \emptyset$. Let us show that $\mathrm{Sp}_{\mathrm{c}}^{\mathrm{Gl}}((\mathscr{A}_0^0)_l)$ is open in $(-\infty, 0]$. Suppose that $\lambda \in \mathrm{Sp}_{\mathrm{c}}^{\mathrm{Gl}}((\mathscr{A}_0^0)_l)$. Then $\mathrm{nul}'((\mathscr{A}_0^0)_l - \lambda I) = \infty$. This enables us to construct, by mathematical induction, an orthonormal sequence $\{\mathbf{v}_n\}$ in $\boldsymbol{L}^2_\sigma(\mathcal{D})_l$ such that

$$((\mathscr{A}_0^0)_l - \lambda I)\,\mathbf{v}_n \;\equiv \epsilon_n \to 0 \quad \text{in } \boldsymbol{L}^2_\sigma(\mathcal{D})_l \quad \text{for } n \to \infty. \tag{8.25}$$

Indeed, suppose that we have already constructed $\mathbf{v}_1, \ldots, \mathbf{v}_n$ so that $\|((\mathscr{A}_0^0)_l - \lambda I)\mathbf{v}_j\|_2 \leq 1/j$ for $j = 1, \ldots, n$. Denote by $\mathbf{N}_n$ the linear hull of $\mathbf{v}_1, \ldots, \mathbf{v}_n$. To $\epsilon_{n+1} = 1/(n+1)$ there exists an infinite-dimensional linear manifold $\mathbf{M}_{n+1}$ in $\boldsymbol{L}^2_\sigma(\mathcal{D})_l$ such that $\|((\mathscr{A}_0^0)_l - \lambda I)\mathbf{u}\|_2 \leq 1/(n+1)$ for all $\mathbf{u} \in \mathbf{M}_{n+1}$. Due to [137, Lemma IV.2.3], there exists $\mathbf{v}_{n+1} \in \mathbf{M}_{n+1}$ such that $\|\mathbf{v}_{n+1}\|_2 = 1$ and the distance between $\mathbf{v}_{n+1}$ and $\mathbf{N}_n$ also equals 1. It can be simply shown that $\mathbf{v}_{n+1}$ is orthogonal to N_n.

In order to complete the proof, we shall use the next lemma. □

Lemma 8.7 *Let $\{\mathbf{v}_n\} \subset D((\mathscr{A}_0^0)_l)$ be an orthonormal sequence in $\boldsymbol{L}^2_\sigma(\mathcal{D})_l$ satisfying (8.25). Then there exists $R > R_0$ and a non-compact sequence $\{\mathbf{u}_n\}$ in $D((\mathscr{A}_0^0)_l)$ such that $\|\mathbf{u}_n\|_2 = 1$, $\mathbf{u}_n = \mathbf{0}$ in $\mathcal{D}_R$, and*

$$((\mathscr{A}_0^0)_l - \lambda I)\mathbf{u}_n \to 0 \quad \textit{in } \boldsymbol{L}^2_\sigma(\mathcal{D})_l \quad \textit{for } n \to \infty. \tag{8.26}$$

Proof Obviously, $\{\mathbf{v}_n\}$ converges to the zero function weakly in $\boldsymbol{L}^2_\sigma(\mathcal{D})_l$. The estimate

$$\|\nabla \mathbf{v}\|_2 + \|\nabla^2 \mathbf{v}\|_2 \leq c\left(\|\mathscr{A}_0^0 \mathbf{v}\|_2 + \|\mathbf{v}\|_2\right) \tag{8.27}$$

(see [107, pages 205, 279]) shows that the sequence $\{\mathbf{v}_n\}$ is bounded in $\boldsymbol{W}_0^{1,2}(\mathcal{D}) \cap \boldsymbol{W}^{2,2}(\mathcal{D})$. Then there exists a subsequence, again denoted by $\{\mathbf{v}_n\}$, which is weakly convergent to $\mathbf{0}$ in $\boldsymbol{W}_0^{1,2}(\mathcal{D}) \cap \boldsymbol{W}^{2,2}(\mathcal{D})$. Suppose that $R \geq R_0 + 3$ is a fixed number. The compact imbedding $\boldsymbol{W}^{2,2}(\mathcal{D}_R) \hookrightarrow\hookrightarrow \boldsymbol{W}^{1,2}(\mathcal{D}_R)$ yields

$$\mathbf{v}_n \to \mathbf{0} \quad \text{strongly in } \boldsymbol{W}^{1,2}(\mathcal{D}_R). \tag{8.28}$$

The first part of (8.25) can be written in the form

$$\nu \Delta \mathbf{v}_n - \lambda \mathbf{v}_n + \nabla q_n = \boldsymbol{\epsilon}_n, \tag{8.29}$$

where q_n is an appropriate scalar function. It follows from (8.29) that $\nabla q_n \to \mathbf{0}$ weakly in $\boldsymbol{L}^2(\mathcal{D})$. Thus, functions q_n (which are given uniquely up to an additive constant by (8.29)) can be chosen so that $q_n \to q \equiv$ const. strongly in $L^2(\mathcal{D}_R)$. The constant can be chosen so that $q = 0$.

Denote by η an infinitely differentiable cutoff function in $\mathcal{D}$ such that

$$\eta(\mathbf{x}) = \begin{cases} 0 \text{ if } |\mathbf{x}| < R - 2, \\ 1 \text{ if } |\mathbf{x}| > R - 1, \end{cases}$$

$0 \leq \eta(\mathbf{x}) \leq 1$ if $R - 2 \leq |\mathbf{x}| \leq R - 1$, and that η is independent of φ. Put $\mathbf{u}_n = \eta \mathbf{v}_n - \boldsymbol{V}_n$, where $\operatorname{div} \boldsymbol{V}_n = \nabla\eta \cdot \mathbf{v}_n$. Although $\boldsymbol{V}_n$ is not given uniquely, the results on solutions of the equation $\operatorname{div} \boldsymbol{V} = \mathbf{f}$ (see, e.g., [98]) show that the function $\boldsymbol{V}_n$ can be chosen such that $\operatorname{supp} \boldsymbol{V}_n \subset \{\mathbf{x} \in \mathcal{D};\ R - 3 < |\mathbf{x}| < R\}$ and

$$\|\boldsymbol{V}_n\|_{2,2} \leq c\, \|\nabla\eta \cdot \mathbf{v}_n\|_{1,2} \leq c. \tag{8.30}$$

(The generic constant c is independent of n.) Moreover, since $\nabla\eta$ is independent of φ and $\mathbf{v}_n \in \boldsymbol{L}^2_\sigma(\mathcal{D})_l$, the function $\boldsymbol{V}_n$ can be constructed so that it belongs to $\boldsymbol{L}^2(\mathcal{D})_l$.

The function $\mathbf{u}_n$ is divergence-free, equals $\mathbf{0}$ in $\mathcal{D}_{R-3}$, equals $\mathbf{v}_n$ in $\mathcal{D} - \mathcal{D}_R$, and belongs to $\boldsymbol{L}^2(\mathcal{D})_l$. Due to the properties of the functions η and $\boldsymbol{V}_n$, we get $\mathbf{u}_n \in D((\mathscr{A}_0^0)_l)$. Obviously $\mathbf{u}_n$ satisfies

$$\begin{aligned} \nu\Delta\mathbf{u}_n - \lambda\mathbf{u}_n + \nabla(\eta q_n) &= \eta[\nu\Delta\mathbf{v}_n - \lambda\mathbf{v}_n] + 2\nu\nabla\eta\cdot\nabla\mathbf{v}_n \\ &\quad + \nu(\Delta\eta)\mathbf{v}_n - \nu\Delta\boldsymbol{V}_n + \lambda\boldsymbol{V}_n + \nabla(\eta q_n) \\ &= \eta\boldsymbol{\epsilon}_n + 2\nu\nabla\eta\cdot\nabla\mathbf{v}_n + \nu(\Delta\eta)\mathbf{v}_n - \nu\Delta\boldsymbol{V}_n + \lambda\boldsymbol{V}_n + (\nabla\eta)q_n, \end{aligned} \tag{8.31}$$

where $\eta\boldsymbol{\epsilon}_n \to \mathbf{0}$ in $\boldsymbol{L}^2(\mathcal{D})$ due to (8.25), and $\nu[2\nabla\eta \cdot \nabla\mathbf{v}_n + (\Delta\eta)\mathbf{v}_n] \to \mathbf{0}$ in $\boldsymbol{L}^2(\mathcal{D})$ because $\nabla\eta$ and $\Delta\eta$ are supported in $\mathcal{D}_R$ and due to (8.28). Furthermore, $(\nu\Delta\boldsymbol{V}_n - \lambda\boldsymbol{V}_n) \to \mathbf{0}$ in $\boldsymbol{L}^2(\mathcal{D})$ due to (8.30), (8.28). Finally, $(\nabla\eta)q_n \to \mathbf{0}$ in $\boldsymbol{L}^2(\mathcal{D})$ because $q_n \to 0$ in $L^2(\mathcal{D}_R)$ and $\nabla\eta$ is supported in $\mathcal{D}_R$. Thus,

$$\nu\Delta\mathbf{u}_n - \lambda\mathbf{u}_n + \nabla(\eta q_n) \to \mathbf{0} \quad \text{in } \boldsymbol{L}^2(\mathcal{D}) \quad \text{for } n \to \infty,$$

and therefore $\{\mathbf{u}_n\}$ satisfies (8.26). We have

$$\|\mathbf{u}_n\|_2^2 \geq \int_{|x|>R} |\mathbf{u}_n(\mathbf{x})|^2\, \mathrm{d}\mathbf{x} = \int_{|x|>R} |\mathbf{v}_n(\mathbf{x})|^2\, \mathrm{d}\mathbf{x} \to 1 \quad \text{for } n \to \infty$$

because $\|\mathbf{v}_n\|_2 = 1$ and due to (8.28). If we divide each of the functions $\mathbf{u}_n$ by its norm $\|\mathbf{u}_n\|_2$ and denote the new function again by $\mathbf{u}_n$, we obtain the sequence $\{\mathbf{u}_n\}$ with all the properties stated in the lemma. Finally, the orthonormality of $\{\mathbf{v}_n\}$ and (8.28) imply the non-compactness of the sequence $\{\mathbf{u}_n\}$. □

Completion of the Proof of Lemma 8.6 Consider the sequence $\{\mathbf{u}_n\}$, provided by Lemma 8.7. There exists $0 < \zeta_0 < 1$ such that for any $\zeta > \zeta_0$ the functions

$$\mathbf{u}_n^\zeta(\mathbf{x}) = \begin{cases} \dfrac{1}{\zeta^{3/2}}\mathbf{u}_n\left(\dfrac{\mathbf{x}}{\zeta}\right) & \text{for } \mathbf{x}/\zeta \in \mathcal{D}, \\ 0 & \text{for } \mathbf{x}/\zeta \notin \mathcal{D} \end{cases}$$

have supports outside $\overline{\mathcal{D}}_{R_0}$. Thus $\mathbf{u}_n^\zeta \in D((\mathscr{A}_0^0)_l)$, $\{\mathbf{u}_n^\zeta\}$ is a non-compact sequence in $\boldsymbol{L}^2_\sigma(\mathcal{D})_l$ and

$$\|\mathbf{u}_n^\zeta\|_2^2 = \int_{\mathcal{D}} |\mathbf{u}_n^\zeta(\mathbf{x})|^2\,\mathrm{d}\mathbf{x} = \frac{1}{\zeta^3}\int_{x/\zeta\in\mathcal{D}} \left|\mathbf{u}_n\left(\frac{\mathbf{x}}{\zeta}\right)\right|^2 \mathrm{d}\mathbf{x} = \int_{\mathcal{D}} |\mathbf{u}_n(\mathbf{y})|^2\,\mathrm{d}\mathbf{y} = 1.$$

Since $\nu\Delta_x\mathbf{u}_n^\zeta(\mathbf{x}) - \lambda/\zeta^2\mathbf{u}_n^\zeta(\mathbf{x}) = 1/\zeta^{7/2}\left(\Delta_y\mathbf{u}_n(\mathbf{y}) - \lambda\mathbf{u}_n(\mathbf{y})\right)$ for all $\mathbf{x}$ and $\mathbf{y}$ from $\mathcal{D}$ such that $\mathbf{y} = \mathbf{x}/\zeta$,

$$\left((\mathscr{A}_0^0)_l\mathbf{u}_n^\zeta - \frac{\lambda}{\zeta^2}I\right)\mathbf{u}_n^\zeta \to \mathbf{0} \quad \text{for } n \to \infty. \tag{8.32}$$

This shows that $\lambda/\zeta^2 \in \mathrm{Sp}_\mathrm{c}^{\mathrm{Gl}}((\mathscr{A}_0^0)_l)$. Since ζ can be chosen arbitrarily in the interval (ζ_0, ∞), some neighborhood of λ in $(-\infty, 0]$ is contained in $\mathrm{Sp}_\mathrm{c}^{\mathrm{Gl}}((\mathscr{A}_0^0)_l)$. Thus, $\mathrm{Sp}_\mathrm{c}^{\mathrm{Gl}}((\mathscr{A}_0^0)_l)$ is open in $(-\infty, 0]$.

We have proved that $\mathrm{Sp}_\mathrm{c}^{\mathrm{Gl}}((\mathscr{A}_0^0)_l) = (-\infty, 0]$. Since $\mathrm{Sp}_\mathrm{c}^{\mathrm{Gl}}((\mathscr{A}_0^0)_l) = \mathrm{Sp}_{\mathrm{ess}}((\mathscr{A}_0^0)_l)$, both sets are also equal to $\mathrm{Sp}_\mathrm{c}((\mathscr{A}_0^0)_l)$ because $\mathrm{Sp}_\mathrm{p}((\mathscr{A}_0^0)_l) = \mathrm{Sp}_\mathrm{r}((\mathscr{A}_0^0)_l) = \emptyset$. □

Let us further deal with the operator $(\mathscr{A}_0^\omega)_l$. Its domain is the same as the domain of $(\mathscr{A}_0^0)_l$, i.e.,

$$D((\mathscr{A}_0^\omega)_l) = D((\mathscr{A}_0^0)_l) \equiv \boldsymbol{W}^{2,2}(\mathcal{D}) \cap \boldsymbol{W}^{1,2}(\mathcal{D}) \cap \boldsymbol{L}^2_\sigma(\mathcal{D})_l.$$

If $\mathbf{u} \in \boldsymbol{L}^2_\sigma(\mathcal{D})_l$, then it has the form $\boldsymbol{U}(r, x_3)\mathrm{e}^{\mathrm{i}l\varphi}$ and $\partial_\varphi\mathbf{u} = \mathrm{i}l\boldsymbol{U}\mathrm{e}^{\mathrm{i}l\varphi} = \mathrm{i}l\mathbf{u}$. Therefore, $(\mathscr{A}_0^\omega)_l$ can be written as

$$(\mathscr{A}_0^\omega)_l\mathbf{u} = (\mathscr{A}_0^0)_l\mathbf{u} + \omega\partial_\varphi\mathbf{u} = (\mathscr{A}_0^0)_l\mathbf{u} + \mathrm{i}l\omega\mathbf{u}. \tag{8.33}$$

Lemma 8.8 *$(\mathscr{A}_0^\omega)_l$ is a normal operator in $\boldsymbol{L}^2_\sigma(\mathcal{D})_l$ and*

$$Sp((\mathscr{A}_0^\omega)_l) = Sp_c((\mathscr{A}_0^\omega)_l) = \{\lambda = \alpha + il\omega;\ \alpha \le 0\}.$$

Proof Since $(\mathscr{A}_0^0)_l$ is reduced to $\boldsymbol{L}^2_\sigma(\mathcal{D})_l$, the operator $(\mathscr{A}_0^\omega)_l$ is an operator in $\boldsymbol{L}^2_\sigma(\mathcal{D})_l$ due to (8.33). Moreover, as $(\mathscr{A}_0^0)_l$ is self-adjoint, the operator $(\mathscr{A}_0^\omega)_l$ is densely defined and closed. The adjoint operator to $(\mathscr{A}_0^\omega)_l$ has the form

$$((\mathscr{A}_0^\omega)_l)^*\mathbf{u} = (\mathscr{A}_0^0)_l\mathbf{u} - \omega\partial_\varphi\mathbf{u} = (\mathscr{A}_0^0)_l\mathbf{u} - \mathrm{i}l\omega\mathbf{u}. \tag{8.34}$$

This operator commutes with $(\mathscr{A}_0^\omega)_l$, and hence $(\mathscr{A}_0^\omega)_l$ is normal. The characterization of $\mathrm{Sp}((\mathscr{A}_0^\omega)_l)$ follows from the representation (8.33) of $(\mathscr{A}_0^\omega)_l$ and from Lemma 8.6. □

Lemma 8.9 $Sp_P(\mathscr{A}_0^\omega) = Sp_r(\mathscr{A}_0^\omega) = \emptyset$. □

Proof Note that $\mathrm{Sp_r}(\mathscr{A}_0^\omega) = \emptyset$ because $\mathscr{A}_0^\omega$ is normal. Suppose that λ is an eigenvalue of $\mathscr{A}_0^\omega$ and $\mathbf{u}$ is a corresponding eigenfunction. The equation $\mathscr{A}_0^\omega\mathbf{u} - \lambda\mathbf{u} = \mathbf{0}$ means that there exists a scalar function p such that $\nu\Delta\mathbf{u} + \omega\partial_\varphi\mathbf{u} + \nabla p - \lambda\mathbf{u} = \mathbf{0}$ in $\mathcal{D}$. Multiplying this equation by $\bar{\mathbf{u}}$ and integrating on $\mathcal{D}$, we can verify that $\mathrm{Re}\,\lambda < 0$. Furthermore, expanding $\mathbf{u}$ and p to the Fourier series in variable φ and denoting the coefficients by $\boldsymbol{U}^l(r, x_3)$ and $P^l(r, x_3)$ (for $l \in \mathbb{Z}$), we can deduce that

$$\nu\Delta(\boldsymbol{U}^l\mathrm{e}^{\mathrm{i}l\varphi}) + \omega\mathrm{i}l\boldsymbol{U}^l\mathrm{e}^{\mathrm{i}l\varphi} + \nabla(P^l\mathrm{e}^{\mathrm{i}l\varphi}) - \lambda\boldsymbol{U}^l\mathrm{e}^{\mathrm{i}l\varphi} = \mathbf{0} \tag{8.35}$$

in $\mathcal{D}$. Moreover, $\mathrm{div}(\boldsymbol{U}^l\mathrm{e}^{\mathrm{i}l\varphi}) = 0$ and $\boldsymbol{U}^l = \mathbf{0}$ on $\partial\mathcal{D}$. This implies that $[\mathscr{A}_0^0 + (\omega\mathrm{i}l - \lambda)](\boldsymbol{U}^l\mathrm{e}^{\mathrm{i}l\varphi}) = \mathbf{0}$. Since the Stokes operator $\mathscr{A}_0^0$ has no eigenvalues, we obtain $\boldsymbol{U}^l = \mathbf{0}$. This identity holds for all $l \in \mathbb{Z}$, hence $\mathbf{u} = \mathbf{0}$. This is a contradiction with the assumption that $\mathbf{u}$ is the eigenfunction. □

Further, we denote

$$\Lambda_0^\omega := \bigcup_{l\in\mathbb{Z}} \{\lambda = \alpha + \mathrm{i}l\omega;\ \alpha \leq 0\}. \tag{8.36}$$

Lemma 8.10 $Sp_C(\mathscr{A}_0^\omega) = \Lambda_0^\omega$. □

Proof Lemma 8.9 and the inclusion $\mathrm{Sp_c}(\mathscr{A}_0^\omega) \subset \mathrm{Sp_c^{Gl}}(\mathscr{A}_0^\omega)$ imply that $\mathrm{Sp_c}(\mathscr{A}_0^\omega) = \mathrm{Sp_c^{Gl}}(\mathscr{A}_0^\omega)$.

Suppose that $\lambda = \alpha + \mathrm{i}\beta \in \mathrm{Sp_c^{Gl}}(\mathscr{A}_0^\omega)$. Then there exists a non-compact sequence $\{\mathbf{u}_n\}$ in the unit sphere in $\boldsymbol{L}^2_\sigma(\mathcal{D})$ such that

$$(\mathscr{A}_0^\omega - \lambda I)\mathbf{u}_n = \boldsymbol{\epsilon}_n \to 0 \quad \text{in } \boldsymbol{L}^2_\sigma(\mathcal{D}) \quad \text{as } n \to \infty. \tag{8.37}$$

Let us write $\mathbf{u}_n$ in the form $\mathbf{u}_n^{-\infty,K_1} + \mathbf{u}_n^{K_1,K_2} + \mathbf{u}_n^{K_2,\infty}$, where $K_1, K_2 \in \mathbb{Z}$, $K_1 \leq K_2$,

$$\mathbf{u}_n^{-\infty,K_1}(r, x_3, \varphi) = \sum_{m=-\infty}^{K_1-1} \boldsymbol{U}_n^l(r, x_3)\mathrm{e}^{\mathrm{i}l\varphi}$$

and $\mathbf{u}_n^{K_1,K_2}$, $\mathbf{u}_n^{K_2,\infty}$ are defined by similar sums, where m runs from K_1 to K_2 or from K_2+1 to ∞. Obviously, $\boldsymbol{U}_n^l \mathrm{e}^{\mathrm{i}l\varphi} = P_l \mathbf{u}_n$. Since $\mathbf{u}_n \in D(\mathscr{A}_0^{\omega}) \subset D(\mathscr{A}_0^0)$, part a) of Lemma 8.5 implies that $\boldsymbol{U}_n^l \mathrm{e}^{\mathrm{i}l\varphi} \in D((\mathscr{A}_0^0)_l) \equiv D((\mathscr{A}_0^{\omega})_l)$. The identity

$$1 = \|\mathbf{u}_n\|_2^2 = \|\mathbf{u}_n^{-\infty,K_1}\|_2^2 + \|\mathbf{u}_n^{-K_1,K_2}\|_2^2 + \|\mathbf{u}_n^{K_2,\infty}\|_2^2 \tag{8.38}$$

implies that there exists a subsequence of $\{\mathbf{u}_n\}$ (we shall preserve the same notation for the subsequence) such that at least one of the following three statements is true:

(A) There exists an increasing sequence $\{K_2^n\}$ of integer numbers which tends to ∞ as $n \to \infty$ and $\|\mathbf{u}_n^{K_2^n,\infty}\|_2 > 1/\sqrt{3}$ for all $n \in \mathbb{N}$.
(B) There exists a decreasing sequence $\{K_1^n\}$ of integer numbers which tends to $-\infty$ as $n \to \infty$ and $\|\mathbf{u}_n^{-\infty,K_1^n}\|_2 > 1/\sqrt{3}$ for all $n \in \mathbb{N}$.
(C) There exist fixed $K_1, K_2 \in \mathbb{Z}$, $K_1 < K_2$, such that $\|\mathbf{u}_n^{K_1,K_2}\|_2 > 1/\sqrt{3}$ for all $n \in \mathbb{N}$.

Suppose that statement (A) is true. Let us multiply (8.37) by $\overline{\boldsymbol{U}_n^l}\,\mathrm{e}^{-\mathrm{i}l\varphi}$, integrate on $\mathcal{D}$, and sum over m from K_2^n to ∞. We obtain

$$-\|\nabla\mathbf{u}_n^{K_2^n,\infty}\|_2^2 - \lambda\|\mathbf{u}_n^{K_2^n,\infty}\|_2^2 + \mathrm{i}\omega \sum_{m=K_2^n}^{\infty} m\|\boldsymbol{U}_n^l \mathrm{e}^{\mathrm{i}l\varphi}\|_2^2 = \left(\boldsymbol{\epsilon}_n, \mathbf{u}_n^{K_2^n,\infty}\right)_2.$$

Note that the right-hand side tends to zero as $n \to \infty$. However, the imaginary part of the left-hand side equals

$$\begin{aligned}
&-\beta \left\|\mathbf{u}_n^{K_2^n,\infty}\right\|_2^2 + \omega \sum_{m=K_2^n}^{\infty} m \left\|\boldsymbol{U}_n^l \mathrm{e}^{\mathrm{i}l\varphi}\right\|_2^2 \\
&\qquad > -\beta \left\|\mathbf{u}_n^{K_2^n,\infty}\right\|_2^2 + \omega K_2^n \sum_{m=K_2^n}^{\infty} \left\|\boldsymbol{U}_n^l \mathrm{e}^{\mathrm{i}l\varphi}\right\|_2^2 \\
&\qquad = (-\beta + \omega K_2^n) \left\|\mathbf{u}_n^{K_2^n,\infty}\right\|_2^2 \geq \frac{1}{3}(-\beta + \omega K_2^n),
\end{aligned}$$

which tends to ∞. This is a contradiction. If statement (B) is true, then we derive a similar contradiction.

Let us finally assume that (C) is true. Then there exists an integer m in the interval $[K_1, K_2]$, a $\delta > 0$, and a subsequence of $\{\mathbf{u}_n\}$ (again denoted by $\{\mathbf{u}_n\}$) such that $\|\mathbf{u}_n^{m,m}\|_2 \equiv \|\boldsymbol{U}_n^l\, \mathrm{e}^{\mathrm{i}l\varphi}\|_2 > \delta$. The first part of (8.37) can be written in the form

$$\sum_{m=-\infty}^{\infty} (\mathscr{A}_0^{\omega} - \lambda I)[\boldsymbol{U}_n^l \mathrm{e}^{\mathrm{i}l\varphi}] = \sum_{m=-\infty}^{\infty} (\mathscr{A}_0^0 - \lambda I + \mathrm{i}l\omega I)[\boldsymbol{U}_n^l \mathrm{e}^{\mathrm{i}l\varphi}]$$

$$= \sum_{m=-\infty}^{\infty} \mathcal{E}_n^l \mathrm{e}^{\mathrm{i}l\varphi},$$

where $\mathcal{E}_n^l$ are the coefficients in the Fourier expansion of $\boldsymbol{\epsilon}_n$ in the variable φ. Hence

$$(\mathscr{A}_0^0 - \lambda I + \mathrm{i}l\omega I)\,[\boldsymbol{U}_n^l\,\mathrm{e}^{\mathrm{i}l\varphi}] = \mathcal{E}_n^l \mathrm{e}^{\mathrm{i}l\varphi} \tag{8.39}$$

for all $n \in \mathbb{N}$. Multiplying (8.39) by $\overline{\boldsymbol{U}_n^l}\,\mathrm{e}^{-\mathrm{i}l\varphi}$ and integrating on $\mathcal{D}$, we obtain

$$-\,\|\nabla(\boldsymbol{U}_n^l \mathrm{e}^{\mathrm{i}l\varphi})\|_2^2 - (\lambda - \mathrm{i}l\omega)\|\boldsymbol{U}_n^l \mathrm{e}^{\mathrm{i}l\varphi}\|_2^2 = \left(\mathcal{E}_n^l \mathrm{e}^{\mathrm{i}l\varphi}, \boldsymbol{U}_n^l \mathrm{e}^{\mathrm{i}l\varphi}\right)_2. \tag{8.40}$$

The imaginary part of the left-hand side is $-(\beta - m\omega)\,\|\boldsymbol{U}_n^l\,\mathrm{e}^{\mathrm{i}l\varphi}\|_2^2$, whereas the right-hand side tends to zero for $n \to \infty$, due to (8.37). Hence $\beta = m\omega$. Finally, (8.39) shows that $\lambda - \mathrm{i}\,m\omega = \alpha \in \mathrm{Sp}(\mathscr{A}_0^0)$. However, since $\mathrm{Sp}(\mathscr{A}_0^0) = (-\infty, 0]$, α is nonpositive.

We have proved the inclusion $\mathrm{Sp}_\mathrm{C}^{\mathrm{Gl}}(\mathscr{A}_0^\omega) \subset \{z = \alpha + \mathrm{i}l\omega;\, l \in \mathbb{Z}, \alpha \leq 0\}$. The opposite inclusion follows from the fact that each of the operators $(\mathscr{A}_0^\omega)_l$ is a part of $\mathscr{A}_0^\omega$ and so $\mathrm{Sp}_\mathrm{C}^{\mathrm{Gl}}((\mathscr{A}_0^\omega)_l) \equiv \{z = \alpha + \mathrm{i}l\omega;\ \alpha \leq 0\} \subset \mathrm{Sp}_\mathrm{C}^{\mathrm{Gl}}(\mathscr{A}_0^\omega)$ for all $l \in \mathbb{Z}$. □

Combining Lemmas 8.9 and 8.10, we obtain the theorem:

Theorem 8.4 *If domain $\mathcal{D}$ is axially symmetric, then the spectra of the operator $\mathscr{A}_0^\omega$ (as an operator in $\boldsymbol{L}_\sigma^2(\mathcal{D})$) satisfy $Sp_P(\mathscr{A}_0^\omega) = Sp_r(\mathscr{A}_0^\omega) = \emptyset$ and $Sp_C(\mathscr{A}_0^\omega) = Sp_{ess}(\mathscr{A}_0^\omega) = \Lambda_0^\omega$.* □

(The theorem includes the case $\mathcal{D} = \mathbb{R}^3$.)

8.3.1.2 The Case of a General Exterior Domain $\mathcal{D}$

We assume that $\mathcal{D}$ is a general exterior domain in $\mathbb{R}^3$ with the boundary of the class $C^{1,1}$ in this paragraph.

We denote by $(\mathscr{A}_0^\omega)_{\mathbb{R}^3}$ the operator, which is defined in the same way as $\mathscr{A}_0^\omega$, however on the whole space $\mathbb{R}^3$ instead of the exterior domain $\mathcal{D} \subset \mathbb{R}^3$. The operator $(\mathscr{A}_0^\omega)_{\mathbb{R}^3}$ has all the properties derived in paragraph 8.3.1.1.

Lemma 8.11 *$Sp_C^{Gl}(\mathscr{A}_0^\omega) = Sp_C^{Gl}((\mathscr{A}_0^\omega)_{\mathbb{R}^3})$.* □

Proof Suppose that $\lambda \in \mathrm{Sp}_\mathrm{C}^{\mathrm{Gl}}(\mathscr{A}_0^\omega)$. Then there exists an orthonormal sequence $\{\mathbf{v}_n\} \subset D(\mathscr{A}_0^\omega)$ in $L_\sigma^2(\mathcal{D})$ such that $\|\mathbf{v}_n\|_2 = 1$ and $\{\mathbf{v}_n\}$ satisfies

$$(\mathscr{A}_0^\omega - \lambda I)\mathbf{v}_n \to 0 \quad \text{in } \boldsymbol{L}_\sigma^2(\mathcal{D}) \quad \text{for } n \to \infty. \tag{8.41}$$

Using the same cutoff procedure as in the proof of Lemma 8.7, we can prove that there exists a non-compact sequence $\{\mathbf{u}_n\}$ in $D(\mathscr{A}_0^{\omega})$ such that $\|\mathbf{u}_n\|_2 = 1$, $\mathbf{u}_n = \mathbf{0}$ in $\mathcal{D}_R$ (where $R \geq R_0$) and

$$(\mathscr{A}_0^{\omega} - \lambda I)\mathbf{u}_n \to 0 \quad \text{in } \boldsymbol{L}^2_{\sigma}(\mathcal{D}) \quad \text{for } n \to \infty. \tag{8.42}$$

All functions $\mathbf{u}_n$, extended by zero from $\mathcal{D}$ to the whole $\mathbb{R}^3$, belong to the domain of operator $(\mathscr{A}_0^{\omega})_{\mathbb{R}^3}$. Thus, (8.42) shows that $\lambda \in \mathrm{Sp}_{\mathrm{C}}^{\mathrm{Gl}}((\mathscr{A}_0^{\omega})_{\mathbb{R}^3})$.

On the other hand, if $\lambda \in \mathrm{Sp}_{\mathrm{C}}^{\mathrm{Gl}}((\mathscr{A}_0^{\omega})_{\mathbb{R}^3})$, then we can use analogous arguments and prove that λ also belongs to $\mathrm{Sp}_{\mathrm{C}}^{\mathrm{Gl}}(\mathscr{A}_0^{\omega})$. □

Note that $\mathrm{Sp}_{\mathrm{C}}^{\mathrm{Gl}}((\mathscr{A}_0^{\omega})_{\mathbb{R}^3})$ equals $\mathrm{Sp}_{\mathrm{C}}((\mathscr{A}_0^{\omega})_{\mathbb{R}^3})$ and is described by Lemma 8.10. Since $\mathrm{Sp}_{\mathrm{ess}}(\mathscr{A}_0^{\omega})$ is closed and a subset of $\mathrm{Sp}_{\mathrm{C}}^{\mathrm{Gl}}(\mathscr{A}_0^{\omega})$, the open set $G := \mathbb{C} \setminus \mathrm{Sp}_{\mathrm{ess}}(\mathscr{A}_0^{\omega})$ has just one component and $\mathrm{ind}(\mathscr{A}_0^{\omega} - \lambda I)$ is constant in G. Using the inclusion $\rho(\mathscr{A}_0^{\omega}) \subset G$, we get $\mathrm{ind}(\mathscr{A}_0^{\omega} - \lambda I) = 0$ in G. This shows that $G \cap \mathrm{Sp}_{\mathrm{C}}^{\mathrm{Gl}}(\mathscr{A}_0^{\omega}) = \emptyset$. We have proven the theorem:

Theorem 8.5 *The essential spectrum of the operator $\mathscr{A}_0^{\omega}$ (as an operator in $\boldsymbol{L}^2_{\sigma}(\mathcal{D})$) satisfies $\mathrm{Sp}_{ess}(\mathscr{A}_0^{\omega}) = \mathrm{Sp}_C^{Gl}(\mathscr{A}_0^{\omega}) = \Lambda_0^{\omega}$.* □

Since, in the case of a general exterior domain $\mathcal{D}$ in $\mathbb{R}^3$, the results on the whole spectrum $\mathrm{Sp}(\mathscr{A}_0^{\omega})$ of operator $\mathscr{A}_0^{\omega}$ and the spectrum $\mathrm{Sp}(\mathscr{L}_0^{\omega})$ of operator $\mathscr{L}_0^{\omega}$ in the case $q = 2$ do not differ from the results in the general case $q \in (1, \infty)$, we refer to Theorems 8.9 and 8.11 for the information on $\mathrm{Sp}(\mathscr{A}_0^{\omega})$ and $\mathrm{Sp}(\mathscr{L}_0^{\omega})$.

8.3.2 *L^q-Theory*

The content of this subsection is based on the papers [74] and [75] by Farwig, Nečasová and Neustupa. We assume that $1 < q < \infty$. The text is split to the two parts: the case $\mathcal{D} = \mathbb{R}^3$ and the case of an exterior domain $\mathcal{D}$ in $\mathbb{R}^3$.

8.3.2.1 The Case $\mathcal{D} = \mathbb{R}^3$

If $\mathcal{D} = \mathbb{R}^3$ and $\mathbf{u} \in D(\mathscr{A}_0^{\omega})$, then $\Delta\mathbf{u}$ belongs to $\boldsymbol{L}^q_{\sigma}(\mathbb{R}^3)$. Thus, the projection Π^q_{σ} in the definition of $\mathscr{A}_0^{\omega}$ can be omitted in front of $\Delta\mathbf{u}$, and we can write

$$\mathscr{A}_0^{\omega}\mathbf{u} = \nu\Delta\mathbf{u} + (\boldsymbol{\omega} \wedge \mathbf{x}) \cdot \nabla\mathbf{u} - \boldsymbol{\omega} \wedge \mathbf{u}.$$

Theorem 8.6 $\mathbb{C} \setminus \Lambda_0^{\omega} \subset \rho(\mathscr{A}_0^{\omega})$. □

Proof It is sufficient to show that to each $\lambda \in \mathbb{C} \setminus \Lambda_0^\omega$ there exists $c > 0$ such that if $\mathbf{f} \in \boldsymbol{L}^q_\sigma(\mathbb{R}^3)$ then the resolvent equation

$$\mathscr{A}_0^\omega \mathbf{u} - \lambda \mathbf{u} = \mathbf{f} \tag{8.43}$$

has a unique solution $\mathbf{u} \in D(\mathscr{A}_0^\omega)$, satisfying the estimate

$$\|\mathbf{u}\|_q \le c\, \|\mathbf{f}\|_q. \tag{8.44}$$

Equation (8.43) means that

$$\nu\Delta\mathbf{u} + (\boldsymbol{\omega} \wedge \mathbf{x}) \cdot \nabla\mathbf{u} - \boldsymbol{\omega} \wedge \mathbf{u} - \lambda\mathbf{u} = \mathbf{f}, \quad \operatorname{div}\mathbf{u} = 0. \tag{8.45}$$

Due to the geometry of the problem, it is again reasonable to use cylindrical coordinates (r, φ, x_3) in $\mathbb{R}^3$. Then $\mathscr{A}_0^\omega \mathbf{u} = \mathscr{A}_0^0 \mathbf{u} + \omega\partial_\varphi \mathbf{u}$, see (8.10).

Suppose at first that $\mathbf{f}$ belongs to Schwartz' space $\boldsymbol{S}(\mathbb{R}^3)$ of the so-called rapidly decreasing functions (in our case vectorial functions), see, e.g., [123, p. 105]. We denote by $\mathcal{F}$ the Fourier transform, by $\mathcal{F}_{-1}$ its inverse, by $\widehat{\ }$ Fourier images of functions, by $\boldsymbol{\xi} = (\xi_1, \xi_2, \xi_3)$ their Cartesian variables, and we put $s = |\boldsymbol{\xi}|$. Then we look for a solution $\mathbf{u}$ of (8.43) in the form

$$\mathbf{u}(\mathbf{x}) = \mathcal{F}_{-1}[\widehat{\mathbf{u}}](\mathbf{x}) = \frac{1}{(2\pi)^{3/2}} \int_{\mathbb{R}^3} e^{i\mathbf{x}\cdot\boldsymbol{\xi}} \widehat{\mathbf{u}}(\boldsymbol{\xi})\, d\boldsymbol{\xi}. \tag{8.46}$$

Substituting this to (8.45), we obtain

$$(\lambda + \nu s^2)\widehat{\mathbf{u}} - \partial_\varphi \widehat{\mathbf{u}} + \boldsymbol{\omega} \wedge \widehat{\mathbf{u}} = -\widehat{\mathbf{f}}. \tag{8.47}$$

Here $\partial_\varphi \widehat{\mathbf{u}}$ denotes the angular derivative

$$\partial_\varphi \widehat{\mathbf{u}} = (\boldsymbol{\omega} \wedge \boldsymbol{\xi}) \cdot \nabla\widehat{\mathbf{u}} \equiv -\xi_2 \frac{\partial \widehat{\mathbf{u}}}{\partial \xi_1} + \xi_1 \frac{\partial \widehat{\mathbf{u}}}{\partial \xi_2}$$

when using cylindrical coordinates (r, φ, ξ_3) in the space of Fourier variables. The equation $\operatorname{div}\mathbf{u} = 0$ leads to the condition $i\boldsymbol{\xi} \cdot \widehat{\mathbf{u}} = 0$. Now $\widehat{\mathbf{u}}$ can be considered to be a solution of the first order ordinary differential equation (8.47) with respect to the angular variable φ. Writing $\widehat{\mathbf{u}}$ in the form

$$\widehat{\mathbf{u}}(r, \varphi, \xi_3) = Q(\varphi)\widehat{\mathbf{v}}(r, \varphi, \xi_3),$$

where $Q(\varphi)$ is defined in (1.17), one verifies that $\partial_\varphi \widehat{\mathbf{u}} = Q(\varphi)\, \partial_\varphi \widehat{\mathbf{v}} + \boldsymbol{\omega} \wedge [Q(\varphi)\widehat{\mathbf{v}}]$ and that (8.47) is equivalent to the equation

$$-\partial_\varphi \widehat{\mathbf{v}} + (\lambda + \nu s^2)\widehat{\mathbf{v}} = -Q(\varphi)^T \widehat{\mathbf{f}}.$$

Its solution $\widehat{\mathbf{v}}$ satisfies

$$\begin{aligned}\widehat{\mathbf{v}}(r,\varphi+2\pi,\xi_3) &= \mathrm{e}^{2\pi(\lambda+\nu s^2)}\widehat{\mathbf{v}}(r,\varphi,\xi_3)\\ &\quad+\int_{\varphi}^{\varphi+2\pi}\mathrm{e}^{(\lambda+\nu s^2)(\varphi+2\pi-t)}\,Q(t)^T\widehat{\mathbf{f}}(r,t,\xi_3)\,\mathrm{d}t\\ &= \mathrm{e}^{2\pi(\lambda+\nu s^2)}\widehat{\mathbf{v}}(r,\varphi,\xi_3)\\ &\quad+\int_0^{2\pi}\mathrm{e}^{(\lambda+\nu s^2)(2\pi-t)}\,Q(t+\varphi)^T\widehat{\mathbf{f}}(r,t+\varphi,\xi_3)\,\mathrm{d}t.\end{aligned}$$

Since $\widehat{\mathbf{v}}$ is a 2π-periodic function in the variable φ, we have

$$\widehat{\mathbf{v}}(r,\varphi,\xi_3) = \frac{1}{1-\mathrm{e}^{2\pi(\lambda+\nu s^2)}}\int_0^{2\pi}\mathrm{e}^{(\lambda+\nu s^2)(2\pi-t)}\,Q(t+\varphi)^T\widehat{\mathbf{f}}(r,t+\varphi,\xi_3)\,\mathrm{d}t,$$

$$\widehat{\mathbf{u}}(r,\varphi,\xi_3) = \int_0^{2\pi}\frac{\mathrm{e}^{(\lambda+\nu s^2)(2\pi-t)}}{1-\mathrm{e}^{2\pi(\lambda+\nu s^2)}}\,Q(t)^T\widehat{\mathbf{f}}(r,t+\varphi,\xi_3)\,\mathrm{d}t. \tag{8.48}$$

Note that $s^2=|\boldsymbol{\xi}|^2=r^2+\xi_3^2$. Returning to the Cartesian variables $\boldsymbol{\xi}=(\xi_1,\xi_2,\xi_3)$ in (8.48), we obtain

$$\begin{aligned}\widehat{\mathbf{u}}(\boldsymbol{\xi}) &= \int_0^{2\pi}\Psi(\lambda,\boldsymbol{\xi},t)\,Q(t)^T\widehat{\mathbf{f}}(Q(t)\boldsymbol{\xi})\,\mathrm{d}t\\ &= \int_0^{2\pi}\Psi(\lambda,\boldsymbol{\xi},t)\,Q(t)^T\,\mathcal{F}[\mathbf{f}(Q(t)\cdot)](\boldsymbol{\xi})\,\mathrm{d}t,\end{aligned} \tag{8.49}$$

where

$$\Psi(\lambda,\boldsymbol{\xi},t) = \frac{\mathrm{e}^{(\lambda+\nu s^2)(2\pi-t)}}{1-\mathrm{e}^{2\pi(\lambda+\nu s^2)}}.$$

Denote $\alpha := \operatorname{Re}\lambda$ and $\beta := \operatorname{Im}\lambda$. Function Ψ satisfies the estimate

$$\begin{aligned}|\Psi(\lambda,\boldsymbol{\xi},t)| &= \mathrm{e}^{-\alpha t}\left|\frac{\mathrm{e}^{-\nu s^2 t}}{\mathrm{e}^{-2\pi(\alpha+\nu s^2)}-\mathrm{e}^{2\pi\mathrm{i}\beta}}\right|\\ &\le \begin{cases}\dfrac{\mathrm{e}^{-2\pi\alpha}}{\operatorname{dist}(\mathrm{e}^{2\pi\mathrm{i}\beta};\mathbb{R}_+)} & \text{if } \alpha\le 0,\ \beta\notin\mathbb{Z},\\[2ex] \dfrac{1}{1-\mathrm{e}^{-2\pi\alpha}} & \text{if } \alpha>0.\end{cases}\end{aligned}$$

(Here, $\mathbb{R}_+$ denotes the positive part of the real axis.) Thus, for fixed $\lambda = \alpha + \mathrm{i}\beta$ in $\mathbb{C} \setminus \Lambda_0^\omega$, $|\Psi(\lambda, \cdot\, t)|$ is bounded in $\mathbb{R}^3$ with an upper bound independent of t and $\boldsymbol{\xi}$. Further, if $i \in \{1; 2; 3\}$, then

$$\frac{\partial \Psi}{\partial \xi_i} = 2\nu\xi_i \frac{(2\pi - t)[\mathrm{e}^{(\lambda+\nu s^2)(4\pi - t)} - \mathrm{e}^{(\lambda+\nu s^2)(2\pi - t)}] - 2\pi \mathrm{e}^{(\lambda+\nu s^2)(4\pi - t)}}{\left[1 - \mathrm{e}^{2\pi(\lambda+\nu s^2)}\right]^2}$$

$$= -2\nu\xi_i \frac{(2\pi - t)\mathrm{e}^{-(\lambda+\nu s^2)(2\pi + t)} + t\mathrm{e}^{-(\lambda+\nu s^2)t}}{\left[\mathrm{e}^{-2\pi(\lambda+\nu s^2)} - 1\right]^2},$$

and $|\xi_i|\, |\partial\Psi/\partial\xi_i|$ can be estimated as follows:

$$|\xi_i| \left|\frac{\partial \Psi}{\partial \xi_i}\right| \le 2\nu s^2 \frac{(2\pi - t)\mathrm{e}^{-(\alpha+\nu s^2)(2\pi + t)} + t\mathrm{e}^{-(\alpha+\nu s^2)t}}{\left|\mathrm{e}^{-2\pi(\lambda+\nu s^2)} - 1\right|^2}$$

$$\le 2\nu \mathrm{e}^{-\alpha t} \frac{(2\pi - t)s^2 \mathrm{e}^{-2\pi\nu s^2} \mathrm{e}^{-2\pi\alpha} + s^2 t \mathrm{e}^{-\nu s^2 t}}{|\mathrm{e}^{-2\pi\mathrm{i}\beta}|^2 \left|\mathrm{e}^{-2\pi(\alpha+\nu s^2)} - \mathrm{e}^{2\pi\mathrm{i}\beta}\right|^2}$$

$$\le \begin{cases} 2\nu \mathrm{e}^{-2\pi\alpha} \dfrac{2\pi c\mathrm{e}^{-2\pi\alpha} + c}{[\mathrm{dist}(\mathrm{e}^{2\pi\mathrm{i}\beta}; \mathbb{R}_+)]^2} & \text{if } \alpha \le 0,\ \beta \notin \mathbb{Z}, \\ 2\nu \dfrac{2\pi c + c}{[1 - \mathrm{e}^{-2\pi\alpha}]^2} & \text{if } \alpha > 0. \end{cases}$$

The upper bound of $|\xi_i|\, |\partial\Psi/\partial\xi_i|$ does not depend on t and on $\boldsymbol{\xi}$. We can similarly estimate all other terms of the form

$$|\xi_1|^{\kappa_1} |\xi_2|^{\kappa_2} |\xi_3|^{\kappa_3} \left|\frac{\partial^{\kappa_1+\kappa_2+\kappa_3}\Psi}{\partial\xi_1^{\kappa_1}\partial\xi_2^{\kappa_2}\partial\xi_3^{\kappa_3}}\right|,$$

where $\kappa_1, \kappa_2, \kappa_3$ are equal to zero or one.

Let us now define the function $\mathbf{u}$ by the formula

$$\mathbf{u}(\mathbf{x}) = \int_0^{2\pi} Q(t)^T \mathcal{F}_{-1}\left[\Psi(\lambda, \boldsymbol{\xi}, t)\, \mathcal{F}\big[\mathbf{f}(Q(t)\cdot)\big](\boldsymbol{\xi})\right](\mathbf{x})\, \mathrm{d}t. \tag{8.50}$$

Applying Lizorkin's multiplier theorem (see, e.g., [93], page 375) and using the estimates of Ψ and its derivatives discussed above, we derive the inequality

$$\left\|\mathcal{F}_{-1}\left[\Psi(\lambda, \boldsymbol{\xi}, t)\mathcal{F}[\mathbf{f}(Q(t)\cdot)](\boldsymbol{\xi})\right]\right\|_q \le c\, \|\mathbf{f}\|_q, \tag{8.51}$$

where $c = c(\lambda, q)$. Then (8.50) and (8.51) imply that $\mathbf{u} \in \boldsymbol{L}^q_\sigma(\mathbb{R}^3)$ and that it satisfies (8.44). Moreover, by construction, $\mathbf{u}$ is a solution of (8.45) in the sense of distributions. In order to show that $\mathbf{u}$ is a strong solution, we still need to verify that it belongs to the domain of the operator $\mathscr{A}_0^{\omega}$. Consider the auxiliary problem

$$\nu\Delta\mathbf{v} + (\boldsymbol{\omega}\wedge\mathbf{x})\cdot\nabla\mathbf{v} - \boldsymbol{\omega}\wedge\mathbf{v} - \nabla p = \mathbf{f} + \lambda\mathbf{u} \qquad \text{in } \mathbb{R}^3, \tag{8.52}$$

$$\operatorname{div}\mathbf{v} = 0 \qquad \text{in } \mathbb{R}^3$$

for the unknown function $\mathbf{v}$. Theorem 3.1 (Theorem 1.1 in [70]) says that the problem (8.52) has a locally integrable solution $(\mathbf{v}, p)$ which satisfies

$$\|\nu\nabla^2\mathbf{v}\|_q + \|\partial_\varphi\mathbf{v} - \boldsymbol{\omega}\wedge\mathbf{v}\|_q + \|\nabla p\|_q \le c\,\big(\|\mathbf{f}\|_q + |\lambda|\|\mathbf{u}\|_q\big), \tag{8.53}$$

where $c = c(q)$. Moreover, the set of all solutions of (8.52) has the structure $(\mathbf{v} + \mathbf{w}, p)$ where

$$\mathbf{w} = a\,\boldsymbol{\omega} + b\,\boldsymbol{\omega}\wedge\mathbf{x} + c(x_1, x_2, -2x_3); \quad a, b, c \in \mathbb{R}.$$

Therefore, $\mathbf{u}$ and $\mathbf{v}$ differ at most in the additive linear vector field $a\,\boldsymbol{\omega} + b\,\boldsymbol{\omega}\wedge\mathbf{x} + c(x_1, x_2, -2x_3)$ which, however, does not affect the validity of (8.53) because $\nabla^2\mathbf{w} = \mathbf{0}$ and $\partial_\varphi\mathbf{w} - \boldsymbol{\omega}\wedge\mathbf{w} = \mathbf{0}$. Hence $\mathbf{u}$ satisfies (8.53), too, and this inequality and (8.44) yield

$$\nu\,\|\nabla^2\mathbf{u}\|_q + \|\partial_\varphi\mathbf{u} - \boldsymbol{\omega}\wedge\mathbf{u}\|_q + \|\nabla p\|_q \le c\,(1 + |\lambda|)\|\mathbf{f}\|_q. \tag{8.54}$$

In particular, $\mathbf{u}$ defined by (8.50) lies in $D(\mathscr{A}_0^{\omega})$ and is a strong solution of (8.43).

The extension of this result from $\mathbf{f} \in \boldsymbol{S}(\mathbb{R}^3)$ to all $\mathbf{f} \in \boldsymbol{L}^q(\mathbb{R}^3)$ follows from the density of $\boldsymbol{S}(\mathbb{R}^3)$ in $\boldsymbol{L}^q(\mathbb{R}^3)$. □

Theorem 8.7 $Sp(\mathscr{A}_0^{\omega}) = Sp_{ess}(\mathscr{A}_0^{\omega}) = \Lambda_0^{\omega}$. □

Proof Transforming the problem (8.45) to the cylindrical coordinates (r, φ, x_3) and denoting by $\mathbf{u}$ (respectively, $\mathbf{f}$) the triplet (u_r, u_φ, u_3) (respectively (f_r, f_φ, f_3)), we obtain that

$$\nu\Delta\mathbf{u} + \partial_\varphi\mathbf{u} - \lambda\mathbf{u} = \mathbf{f}, \quad \partial_r(ru_r) + \partial_\varphi u_\varphi + \partial_3(ru_3) = 0. \tag{8.55}$$

Each $\lambda \in \Lambda_0^{\omega}$ has the form $\lambda = \alpha + il$ where $\alpha \le 0$ and $l \in \mathbb{Z}$. Assume that $\alpha < 0$ and $l \in \mathbb{Z}$. (The assumption $l \neq 0$ is used just for simplicity; the case $l = 0$ could be treated similarly.) $l \neq 0$. Let $\eta \in C^\infty([0, 1])$ be equal to zero on $[0, \frac{1}{4}]$, increasing on $[\frac{1}{4}, \frac{3}{4}]$ and equal to one on $[\frac{3}{4}, 1]$. We put

$$v_r^n(r) = \begin{cases} 0 & \text{for } 0 \le r \le n, \\ \eta\left(\frac{r-n}{n}\right) & \text{for } n \le r \le 2n, \\ \cos\left[\zeta(r - 2n)\right] & \text{for } 2n \le r \le 2n + n^2T, \\ \eta\left(\frac{3n+n^2T-r}{n}\right) & \text{for } 2n + n^2T \le r \le 3n + n^2T, \\ 0 & \text{for } 3n + n^2T \le r, \end{cases},$$

where $\zeta = \sqrt{-\alpha/\nu}$ and $T = 2\pi/\zeta$ is the period of the function $\cos[\zeta(r-2n)]$, as a function of the variable r. Moreover, let $\varphi \in C_0^\infty(-\infty,\infty)$ be equal to 1 on $[-\frac{1}{2},\frac{1}{2}]$ and to 0 on $(-\infty,-1]\cup[1,\infty)$. Then we define $\mathbf{u}^n = \big(u_r^n,\, u_\varphi^n,\, u_3^n\big)$ by the formulas

$$
\begin{aligned}
u_r^n(r,\varphi,x_3) &:= \gamma_n v_r^n(r)\varphi\left(\frac{x_3}{n}\right)\mathrm{e}^{\mathrm{i}l\varphi},\\
u_\varphi^n(r,\varphi,x_3) &:= -\frac{1}{\mathrm{i}l}\,\partial_r\left[r u_r^n(r,\varphi,x_3)\right]\\
&= -\frac{\gamma_n}{\mathrm{i}l}\left[v_r^n(r)+r\frac{\mathrm{d}v_r^n(r)}{\mathrm{d}r}\right]\varphi\left(\frac{x_3}{n}\right)\mathrm{e}^{\mathrm{i}l\varphi},\\
u_3^n(r,\varphi,x_3) &:= 0,
\end{aligned}
$$

where the constant γ_n is chosen so that $\|\mathbf{u}^n\|_q = 1$. Note that the function $\mathbf{u}^n$ satisfies the second equation in (8.55), which is the equation of continuity. If we calculate γ_n, then the decisive term for large n comes from $u_\varphi^n(r,\varphi,x_3)$ for $2n \le r \le 2n+n^2T$. (To be more precise, it comes from $-\,(\gamma_n/\mathrm{i}l)\,r\,(\mathrm{d}v_r^n/\mathrm{d}r)\varphi\mathrm{e}^{\mathrm{i}l\varphi}$.) All other terms are of lower order in powers of n. Now the L^q-norm of the mentioned term can be estimated from below as follows:

$$
\begin{aligned}
\left\|\frac{\gamma_n}{\mathrm{i}l}\,r\,\frac{\mathrm{d}v_r^n(r)}{\mathrm{d}r}\varphi\left(\frac{x_3}{n}\right)\mathrm{e}^{\mathrm{i}l\varphi}\right\|_q &= \frac{\gamma_n}{|l|}\left(\int_{2n}^{2n+n^2T} r^q\left|\frac{\mathrm{d}}{\mathrm{d}r}\cos[\zeta(r-2n)]\right|^q r\,\mathrm{d}r\right.\\
&\qquad\left.\cdot\int_0^{2\pi}|\mathrm{e}^{\mathrm{i}l\varphi}|^q\,\mathrm{d}\varphi\int_{-\infty}^{\infty}\left|\varphi\left(\frac{x_3}{n}\right)\right|^q\mathrm{d}x_3\right)^{1/q}\\
&= c\gamma_n\left(\int_{2n}^{2n+n^2T} r^{q+1}\,|\sin[\zeta(r-2n)]|^q\,\mathrm{d}r\right)^{1/q}\left(n\int_{-1}^{1}|\varphi(s)|^q\,\mathrm{d}s\right)^{1/q}\\
&= c\gamma_n\left(\sum_{i=1}^{n^2}\int_{2n+(i-1)T}^{2n+iT} r^{q+1}\,|\sin[\zeta(r-2n)]|^q\,\mathrm{d}r\right)^{1/q} n^{1/q}\\
&\ge c\gamma_n\left(\sum_{i=1}^{n^2}[2n+(i-1)T]^{q+1}\int_{2n+(i-1)T}^{2n+iT}|\sin[\zeta(r-2n)]|^q\,\mathrm{d}r\right)^{1/q} n^{1/q}\\
&\ge c\gamma_n\left(\sum_{i=1}^{n^2}(i-1)^{q+1}\right)^{1/q} n^{1/q} \ge c\gamma_n n^{2+5/q}.
\end{aligned}
$$

The generic constant c is independent of n. It is even easier to derive the upper estimate

$$\left\| \frac{\gamma_n}{\mathrm{i}l}\, r\, \frac{\mathrm{d}v_r^n(r)}{\mathrm{d}r}\, \varphi\left(\frac{x_3}{n}\right) \mathrm{e}^{\mathrm{i}l\varphi} \right\|_q \leq c\gamma_n n^{2+5/q}.$$

From these estimates and from the condition $\|\mathbf{u}^n\|_q = 1$, we obtain the inequalities

$$\frac{c_2}{n^{2+5/q}} \leq \gamma_n \leq \frac{c_3}{n^{2+5/q}} \quad \text{for all } n \in \mathbb{N}, \tag{8.56}$$

where $c_2 = c_2(\alpha, l, q)$ and $c_3 = c_3(\alpha, l, q)$ are appropriate positive constants.

Substituting $\mathbf{u}^n$ to the left-hand side in the first equation of (8.55), we calculate that the corresponding right-hand side is $\mathbf{f}^n = \mathbf{g}^n\, \mathrm{e}^{\mathrm{i}l\varphi}$, where $\mathbf{g}^n \equiv (g_r^n, g_\varphi^n, g_3^n)$ depends only on the variables r, x_3. To be more precise,

$$g_r^n(r, x_3) = 0 \quad \text{for } 0 \leq r \leq n, \tag{8.57}$$

$$\begin{aligned} g_r^n(r, x_3) = {} & \frac{\gamma_n\nu}{n^2}\eta''\left(\frac{r-n}{n}\right)\varphi\left(\frac{x_3}{n}\right) + \frac{\gamma_n\nu}{rn}\eta'\left(\frac{r-n}{n}\right)\varphi\left(\frac{x_3}{n}\right) \\ & - \frac{\gamma_n\nu l^2}{r^2}\eta\left(\frac{r-n}{n}\right)\varphi\left(\frac{x_3}{n}\right) + \frac{\gamma_n\nu}{n^2}\eta\left(\frac{r-n}{n}\right)\varphi''\left(\frac{x_3}{n}\right) \\ & - \gamma_n\alpha\eta\left(\frac{r-n}{n}\right)\varphi\left(\frac{x_3}{n}\right) \qquad \text{for } n \leq r \leq 2n, \end{aligned} \tag{8.58}$$

$$\begin{aligned} g_r^n(r, x_3) = {} & \frac{\gamma_n\nu}{n^2}\cos\big[\zeta(r-2n)\big]\varphi''\left(\frac{x_3}{n}\right) - \frac{\gamma_n\nu}{r}\zeta\sin\big[\zeta(r-2n)\big]\varphi\left(\frac{x_3}{n}\right) \\ & - \frac{\gamma_n l^2\nu}{r^2}\cos\big[\zeta(r-2n)\big]\varphi\left(\frac{x_3}{n}\right) \quad \text{for } 2n \leq r \leq 2n + n^2T \end{aligned} \tag{8.59}$$

and

$$\begin{aligned} g_r^n(r, x_3) = {} & \frac{\gamma_n\nu}{n^2}\eta''\left(\frac{3n+n^2T-r}{n}\right)\varphi\left(\frac{x_3}{n}\right) - \frac{\gamma_n\nu}{rn}\eta'\left(\frac{3n+n^2T-r}{n}\right)\varphi\left(\frac{x_3}{n}\right) \\ & - \frac{\gamma_n\nu l^2}{r^2}\eta\left(\frac{3n+n^2T-r}{n}\right)\varphi\left(\frac{x_3}{n}\right) + \frac{\gamma_n\nu}{n^2}\eta\left(\frac{3n+n^2T-r}{n}\right)\varphi''\left(\frac{x_3}{n}\right) \\ & - \gamma_n\alpha\eta\left(\frac{3n+n^2T-r}{n}\right)\varphi\left(\frac{x_3}{n}\right) \quad \text{for } 2n+n^2T \leq r \leq 3n+n^2T, \end{aligned} \tag{8.60}$$

$$g_r^n(r, x_3) = 0 \quad \text{for } 3n + n^2T \leq r. \tag{8.61}$$

By analogy with the relation between u_r^n and u_φ^n, the function g_φ^n equals $-\partial_r\big[r\, g_r^n(r, x_3)\big]/\mathrm{i}l$. Furthermore, $g_3^n(r, x_3) = 0$ for all $r \geq 0$ and $x_3 \in (-\infty, \infty)$.

Now we need to estimate the norm of $\mathbf{f}^n$ in $L^q(\mathbb{R}^3)$ and to show that

$$\|\mathbf{f}^n\|_q \to 0 \quad \text{for } n \to \infty. \tag{8.62}$$

Let us begin with the first term on the right-hand side of (8.58), multiplied by $\mathrm{e}^{\mathrm{i}l\varphi}$. This term is defined for $n \le r \le 2n$, $0 \le \varphi \le 2\pi$, and $-\infty < x_3 < \infty$. Thus, its contribution to the upper bound of $\|\mathbf{f}^n\|_q$ is

$$\begin{aligned}
&\left(\int_n^{2n}\int_0^{2\pi}\int_{-\infty}^{\infty}\left|\frac{\gamma_n\nu}{n^2}\eta''\left(\frac{r-n}{n}\right)\varphi\left(\frac{x_3}{n}\right)\mathrm{e}^{\mathrm{i}l\varphi}\right|^q r\,\mathrm{d}x_3\,\mathrm{d}\varphi\,\mathrm{d}r\right)^{1/q}\\
&\qquad \le c\frac{\gamma_n}{n^2}\left(\int_n^{2n}\left|\eta''\left(\frac{r-n}{n}\right)\right|^q r\,\mathrm{d}r\int_{-\infty}^{\infty}\left|\varphi\left(\frac{x_3}{n}\right)\right|^q\mathrm{d}x_3\right)^{1/q} \le c\frac{\gamma_n}{n^2}n^{3/q},
\end{aligned}$$

where c is independent of n. The right-hand side tends to zero as $n \to \infty$ due to (8.56). We arrive at the same conclusion when dealing with other terms on the right- hand sides of (8.57)–(8.61) and also with all corresponding terms in the expression of g_φ^n. (Naturally, all terms must be multiplied by $\mathrm{e}^{\mathrm{i}l\varphi}$.) Let us consider one more example: The first term on the right hand of (8.59) is defined for $2n \le r \le 2n + n^2T$, $0 \le \varphi \le 2\pi$, and $-\infty < x_3 < \infty$. Its contribution to an upper bound of $\|\mathbf{f}^n\|_q$ is

$$\begin{aligned}
&\left(\int_{2n}^{2n+n^2T}\int_0^{2\pi}\int_{-\infty}^{\infty}\left|\frac{\gamma_n\nu}{n^2}\cos\left[\zeta(r-2n)\right]\varphi''\left(\frac{x_3}{n}\right)\mathrm{e}^{\mathrm{i}l\varphi}\right|^q r\,\mathrm{d}x_3\,\mathrm{d}\varphi\,\mathrm{d}r\right)^{1/q}\\
&\qquad \le c\frac{\gamma_n}{n^2}\left(\int_{2n}^{2n+n^2T} r\,\mathrm{d}r\int_{-\infty}^{\infty}\left|\varphi''\left(\frac{x_3}{n}\right)\right|^q\mathrm{d}x_3\right)^{1/q} \le c\,\frac{\gamma_n}{n^2}\,n^{5/q}.
\end{aligned}$$

Due to (8.56), the right-hand side again tends to zero as $n \to \infty$. In this way we successively verify (8.62).

Thus we have constructed a sequence $\{\mathbf{u}^n\}$ in the unit sphere in $L^q_\sigma(\mathbb{R}^3)$ such that $\|(\mathscr{A}_0^\omega - \lambda I)\mathbf{u}^n\|_q \to \mathbf{0}$ as $n \to \infty$. The sequence $\{\mathbf{u}^n\}$ is not compact in $L^q_\sigma(\mathbb{R}^3)$ because the intersection of supports of any infinite family of functions chosen from this sequence is empty. This proves that the chosen number $\lambda = \alpha + \mathrm{i}l$ ($\alpha < 0$, $l \in \mathbb{Z}$) is in the essential spectrum of the operator $\mathscr{A}_0^\omega$. Due to the closedness of the essential spectrum, each λ of this form belongs to the essential spectrum of $\mathscr{A}_0^\omega$ even if $\alpha \le 0$. □

Theorem 8.7 says that the spectrum of $\mathscr{A}_0^\omega$ coincides with the essential spectrum and equals the set Λ_0^ω. However, it does not specify whether $\lambda \in \Lambda_0^\omega$ belongs to the point spectrum, or the continuous spectrum, or the residual spectrum. This question is studied in detail in paper [75]. The corresponding result from [75], which we cite without proof, says:

(i) If $1 < q < \frac{3}{2}$, then $(\Lambda_0^{\omega})^{\circ} = \mathrm{Sp_r}(\mathscr{A}_0^{\omega})$, and the codimension of $\overline{R(\lambda I - \mathscr{A}_0^{\omega})}$ equals infinity for each $\lambda \in (\Lambda_0^{\omega})^{\circ}$.
(ii) If $\frac{3}{2} \le q \le 3$, then $\Lambda_0^{\omega} = \mathrm{Sp_c}(\mathscr{A}_0^{\omega})$.
(iii) If $3 < q < \infty$, then $(\Lambda_0^{\omega})^{\circ} = \mathrm{Sp_p}(\mathscr{A}_0^{\omega})$ and the geometric multiplicity of each eigenvalue $\lambda \in (\Lambda_0^{\omega})^{\circ}$ is infinite.

(Here, $(\Lambda_0^{\omega})^{\circ}$ denotes the set $\bigcup_{l\in\mathbb{Z}} \{\lambda = \alpha + \mathrm{i}l\omega;\, \alpha < 0\}$.)

8.3.2.2 The Case of a General Exterior Domain $\mathcal{D}$

We assume that $\mathcal{D}$ is an exterior domain in $\mathbb{R}^3$ with the boundary of the class $C^{1,1}$ in this paragraph.

Theorem 8.8 *The essential spectrum of the operator $\mathscr{A}_0^{\omega}$ (as an operator in $\boldsymbol{L}^q_{\sigma}(\mathcal{D})$) has the same form as in the case $\mathcal{D} = \mathbb{R}^3$, i.e., $Sp_{ess}(\mathscr{A}_0^{\omega}) = \Lambda_0^{\omega}$.* □

Proof Let us at first show that $\Lambda_0^{\omega} \subset \mathrm{Sp_{ess}}(\mathscr{A}_0^{\omega})$. Let $\lambda = \alpha + \mathrm{i}l$, where $\alpha < 0$ and $l \in \mathbb{Z}$. The supports of the functions $\mathbf{u}^n$, constructed in the proof of Theorem 8.7, are subsets of $\{\mathbf{x} = (x_1, x_2, x_3) \in \mathbb{R}^3;\ \sqrt{x_1^2 + x_2^2} > n\}$ and consequently also subsets of $\mathcal{D}$ for $n \ge n_0$, n_0 sufficiently large. It means that the sequence $\{\mathbf{u}^n\}$ (starting from $n = n_0$) is a non-compact sequence in the unit sphere in $L^q_{\sigma}(\mathcal{D})$ satisfying

$$\|(\mathscr{A}_0^{\omega} - \lambda I)\mathbf{u}^n\|_q \to 0 \quad \text{as } n \to \infty. \tag{8.63}$$

Consequently, $\lambda \in \mathrm{Sp_{ess}}(\mathscr{A}_0^{\omega})$. The inclusion $\Lambda_0^{\omega} \subset \mathrm{Sp_{ess}}(\mathscr{A}_0^{\omega})$ now follows from the fact that $\mathrm{Sp_{ess}}(\mathscr{A}_0^{\omega})$ is closed.

Let us now verify the opposite inclusion, i.e., $\mathrm{Sp_{ess}}(\mathscr{A}_0^{\omega}) \subset \Lambda_0^{\omega}$. By contradiction: suppose that there exists $\lambda \in \mathrm{Sp_{ess}}(\mathscr{A}_0^{\omega})$ such that $\lambda \notin \Lambda_0^{\omega}$. Then $\mathrm{nul}'(\mathscr{A}_0^{\omega} - \lambda I) = \infty$. Using this information, we will construct by mathematical induction a sequence $\{\mathbf{u}^n\}$ in $\boldsymbol{L}^q_{\sigma}(\mathcal{D})$ satisfying $\|\mathbf{u}^n\|_q = 1$, (8.63) and

$$\mathrm{dist}\left(\mathbf{u}^{n+1}; \mathcal{L}_n\right) = 1 \tag{8.64}$$

for all $n \in \mathbb{N}$, where $\mathcal{L}_n$ denotes the linear hull of the functions $\mathbf{u}^1, \dots, \mathbf{u}^n$. Suppose that we have already constructed $\mathbf{u}^1, \dots, \mathbf{u}^m$ satisfying $\|(\mathscr{A}_0^{\omega} - \lambda I)\mathbf{u}^j\|_q \le 1/j$ for $j = 1, \dots, m$ and (8.64) for all $n = 1, \dots, m-1$. To $\epsilon_{m+1} = 1/(m+1)$ there exists an infinite-dimensional linear manifold M_{m+1} in $\boldsymbol{L}^q_{\sigma}(\mathcal{D})$ such that $\|(\mathscr{A}_0^{\omega} - \lambda I)\mathbf{u}\|_q \le \epsilon_{m+1}$ for all $\mathbf{u} \in M_{m+1}$. Then due to Lemma IV.2.3 in [137], we find $\mathbf{u}^{m+1} \in M_{m+1}$ such that $\|\mathbf{u}^{m+1}\|_q = 1$ and $\mathrm{dist}\left(\mathbf{u}^{m+1}; \mathcal{L}_m\right) = 1$.

The sequence $\{\mathbf{u}^n\}$ obviously satisfies

$$\|(\mathscr{A}_0^{\omega} - \lambda I)\mathbf{u}^n\|_q \le \frac{1}{n} \quad \text{for all } n \in \mathbb{N}. \tag{8.65}$$

Denote $\mathbf{f}^n := (\mathscr{A}_0^{\omega} - \lambda I)\mathbf{u}^n$. Then there exists $\nabla p^n \in \boldsymbol{L}^q(\mathcal{D})$ such that

$$\nu\Delta\mathbf{u}^n + (\boldsymbol{\omega}\wedge\mathbf{x})\cdot\nabla\mathbf{u}^n - \boldsymbol{\omega}\wedge\mathbf{u}^n - \lambda\mathbf{u}^n - \nabla p^n = \mathbf{f}^n, \quad \operatorname{div}\mathbf{u}^n = 0 \tag{8.66}$$

in $\mathcal{D}$. Using estimate (8.2), we deduce that

$$\|\nu\nabla^2\mathbf{u}^n\|_q + \|(\boldsymbol{\omega}\wedge\mathbf{x})\cdot\nabla\mathbf{u}^n - \boldsymbol{\omega}\wedge\mathbf{u}^n\|_q + \|\nabla p^n\|_q \le c\left(\|\mathbf{f}^n\|_q + |\lambda|\right) \le c, \tag{8.67}$$

where the generic constant c does not depend on n. Due to (8.67) and the identity $\|\mathbf{u}^n\|_q = 1$, the sequence $\{\mathbf{u}^n\}$ is bounded in the space $D(\mathscr{A}_0^{\omega})$. (Here, we consider $D(\mathscr{A}_0^{\omega})$ to be a Banach space with the norm $\|\mathbf{u}\|_{D(\mathscr{A}_0^{\omega})} := \|\mathbf{u}\|_{2,q} + \|(\boldsymbol{\omega}\wedge\mathbf{x})\cdot\nabla\mathbf{u}\|_q$.) Hence there exists a subsequence, weakly convergent in $D(\mathscr{A}_0^{\omega})$. We denote the subsequence again by $\{\mathbf{u}^n\}$. The subsequence naturally preserves the property (8.64).

Put $\mathbf{v}^n := (\mathbf{u}^{n+1} - \mathbf{u}^n)/\delta_n$, where $\delta_n = \|\mathbf{u}^{n+1} - \mathbf{u}^n\|_q \ge 1$. Then $\{\mathbf{v}^n\}$ is a sequence in the unit sphere in $\boldsymbol{L}^q_\sigma(\mathcal{D})$, which converges weakly to zero in $D(\mathscr{A}_0^{\omega})$. Hence $\{\mathbf{v}^n\}$ converges strongly to zero in $\boldsymbol{W}^{1,q}(\mathcal{D}_R)$ for each $R > 0$. Note that the function $\mathbf{v}^n$ satisfies the equation

$$\nu\Delta\mathbf{v}^n + (\boldsymbol{\omega}\wedge\mathbf{x})\cdot\nabla\mathbf{v}^n - \boldsymbol{\omega}\wedge\mathbf{v}^n - \lambda\mathbf{v}^n - \frac{1}{\delta_n}\nabla(p^{n+1} - p^n) = \frac{1}{\delta_n}\left(\mathbf{f}^{n+1} - \mathbf{f}^n\right) \tag{8.68}$$

in $\mathcal{D}$. This equation, together with the information on the weak convergence of $\{\mathbf{v}^n\}$ to zero in $D(\mathscr{A}_0^{\omega})$, implies that the sequence $\{\nabla(p^{n+1} - p^n)\}$ weakly converges to zero in $\boldsymbol{L}^q(\mathcal{D})$. Thus, the functions p^n, which are given uniquely up to an additive constant, can be chosen so that $p^{n+1} - p^n \to 0$ strongly in $L^q(\mathcal{D}_R)$ for each $R > R_0$.

The sequence $\{\mathbf{v}^n\}$ does not contain any subsequence, convergent in $\boldsymbol{L}^q_\sigma(\mathcal{D})$. Let us prove this statement by contradiction: Assume that $\{\mathbf{v}^{m_n}\}$ is a convergent subsequence of $\{\mathbf{v}^n\}$ in $\boldsymbol{L}^q_\sigma(\mathcal{D})$. This subsequence has the same weak limit as $\{\mathbf{v}^n\}$, hence $\mathbf{v}^{m_n} \rightharpoonup \mathbf{0}$ in $\boldsymbol{L}^q_\sigma(\mathcal{D})$ as $n \to \infty$. The strong limit in $\boldsymbol{L}^q_\sigma(\mathcal{D})$ cannot differ from the weak limit, hence $\mathbf{v}^{m_n} \to \mathbf{0}$ in $\boldsymbol{L}^q_\sigma(\mathcal{D})$ as $n \to \infty$. However, this is impossible because $\|\mathbf{v}^{m_n}\|_q = \|\mathbf{u}^{m_n+1} - \mathbf{u}^{m_n}\|_q \ge 1$.

Let ψ be an infinitely differentiable cutoff function defined in $\mathbb{R}^3$ such that

$$\psi(\mathbf{x}) = \begin{cases} 0 \text{ for } 0 \le |\mathbf{x}| \le \frac{5}{4}R, \\ 1 \text{ for } \frac{7}{4}R \le |\mathbf{x}|. \end{cases}$$

Put $\mathbf{w}^n(\mathbf{x}) := \psi(\mathbf{x})\mathbf{v}^n(\mathbf{x}) - \boldsymbol{V}^n(\mathbf{x})$ where $\boldsymbol{V}^n$ is a function that corrects the divergence of $\psi\mathbf{v}^n$ so that $\mathbf{w}^n$ is divergence-free, i.e., $\operatorname{div}\boldsymbol{V}^n = \nabla\psi\cdot\mathbf{v}^n$. Since the support of

$\nabla\psi\cdot\mathbf{v}^n$ is contained in $\{\mathbf{x}\in\mathbb{R};\ \frac{5}{4}R\le|\mathbf{x}|\le\frac{7}{4}R\}$, it follows, e.g., from [18] that one can construct such a function $\boldsymbol{V}^n$ so that it is supported in $\{\mathbf{x}\in\mathbb{R};\ R\le|\mathbf{x}|\le 2R\}$ and satisfies

$$\|\boldsymbol{V}^n\|_{2,q}\le c\,\|\nabla\psi\cdot\mathbf{v}^n\|_{1,q}\le c\quad\text{for all } n\in\mathbb{N}, \tag{8.69}$$

where the generic constant c is independent of n and $\boldsymbol{V}^n$. Using Eq. (8.68), we derive that $\mathbf{w}^n$ satisfies

$$\begin{aligned}
&\nu\Delta\mathbf{w}^n+(\boldsymbol{\omega}\wedge\mathbf{x})\cdot\nabla\mathbf{w}^n-\boldsymbol{\omega}\wedge\mathbf{w}^n-\lambda\mathbf{w}^n-\frac{1}{\delta_n}\nabla\Big[\psi(p^{n+1}-p^n)\Big]\\
&=\frac{\psi}{\delta_n}(\mathbf{f}^{n+1}-\mathbf{f}^n)-\frac{1}{\delta_n}\nabla\psi(p^{n+1}-p^n)+\nu(\Delta\psi)\mathbf{v}^n+2\nu\nabla\psi\cdot\nabla\mathbf{v}^n\\
&\quad-\nu\Delta\boldsymbol{V}^n+(\boldsymbol{\omega}\wedge\mathbf{x})\cdot(\nabla\psi\otimes\mathbf{v}^n)-(\boldsymbol{\omega}\wedge\mathbf{x})\cdot\nabla\boldsymbol{V}^n+\boldsymbol{\omega}\wedge\boldsymbol{V}^n+\lambda\boldsymbol{V}^n.
\end{aligned} \tag{8.70}$$

The right-hand side converges strongly to zero in $\boldsymbol{L}^q(\mathcal{D})$ as $n\to\infty$. (This follows from the strong convergence of $\{\mathbf{v}^n\}$ to zero in $\boldsymbol{W}^{1,2}(\mathcal{D}_R)$, the strong convergence of $\{p^{n+1}-p^n\}$ to zero in $L^q(\mathcal{D}_R)$, from (8.69) and from the information on the support of $\nabla\psi$ and $\Delta\psi$.) Hence

$$\|(\mathscr{A}_0^\omega-\lambda I)\mathbf{w}^n\|_q\to 0\quad\text{as } n\to\infty. \tag{8.71}$$

In order to estimate the norm of $\mathbf{w}_n$ for large n, let us choose $\epsilon>0$ arbitrarily small. Then for all n sufficiently large

$$\begin{aligned}
\|\mathbf{w}^n\|_q&\le\bigg(\int_{|x|<2R}|\psi\mathbf{v}^n-\boldsymbol{V}^n|^q\,\mathrm{d}\mathbf{x}\bigg)^{1/q}+\bigg(\int_{2R<|x|}|\mathbf{v}^n|^q\,\mathrm{d}\mathbf{x}\bigg)^{1/q}\le\epsilon+1,\\
\|\mathbf{w}^n\|_q&\ge\bigg(\int_{2R<|x|}|\mathbf{v}^n|^q\,\mathrm{d}\mathbf{x}\bigg)^{1/q}\\
&\ge\bigg(\int_{\mathcal{D}}|\mathbf{v}^n|^q\,\mathrm{d}\mathbf{x}\bigg)^{1/q}-\bigg(\int_{|x|<2R}|\mathbf{v}^n|^q\,\mathrm{d}\mathbf{x}\bigg)^{1/q}\ge 1-\epsilon.
\end{aligned}$$

Let us now normalize the sequence $\{\mathbf{w}^n\}$ by dividing each of the functions $\mathbf{w}^n$ by its norm in $\boldsymbol{L}^q_\sigma(\mathcal{D})$. In order to preserve a simple notation, we denote the normalized functions again by $\mathbf{w}^n$. If we finally put $\mathbf{w}^n(\mathbf{x})=\mathbf{0}$ for $\mathbf{x}\in\mathbb{R}^3-\mathcal{D}$, we obtain a sequence in the unit sphere in $\boldsymbol{L}^q_\sigma(\mathbb{R}^3)$, which does not contain a subsequence converging in $\boldsymbol{L}^q_\sigma(\mathbb{R}^3)$ and satisfying (8.71) with $\|\cdot\|_q$ representing the norm in $\boldsymbol{L}^q(\mathbb{R}^3)$. The existence of the sequence $\{\mathbf{w}^n\}$ with all these properties implies that λ belongs to the essential spectrum of the operator $\mathscr{A}_0^\omega$, considered as an operator in $\boldsymbol{L}^q_\sigma(\mathbb{R}^3)$. However, we know from Theorem 8.7 that the essential spectrum coincides with the set Λ_0^ω. Hence $\lambda\in\Lambda_0^\omega$, which is the contradiction. The proof is completed.

□

In contrast to Theorem 8.7, which provides the description of the full spectrum, Theorem 8.8 only describes the essential spectrum of operator $\mathscr{A}_0^{\omega}$. The reason is that we can construct a solution of the resolvent Eq. (8.43) in the case $\mathcal{D} = \mathbb{R}^3$ by means of the Fourier transform, but the same method cannot be used if $\mathcal{D}$ is general exterior domain in $\mathbb{R}^3$. The next theorem concerns the set $\mathbb{C} \setminus \Lambda_0^{\omega}$, and it completes the information, given by Theorem 8.7.

Theorem 8.9 *$\mathbb{C} \setminus \Lambda_0^{\omega}$ is a subset of $\rho(\mathscr{A}_0^{\omega})$, with a possible exception of at most a countable set of isolated eigenvalues of $\mathscr{A}_0^{\omega}$ with finite algebraic multiplicities.*

Let $\lambda \in \mathbb{C} \setminus \Lambda_0^{\omega}$. Then:

(i) If $Re\,\lambda \geq 0$, then $\lambda \in \rho(\mathscr{A}_0^{\omega})$.
(ii) If $Re\,\lambda < 0$, then either λ is an eigenvalue of $\mathscr{A}_0^{\omega}$ for all $q \in (1, \infty)$, whose both algebraic and geometric multiplicities are finite and independent of q, or $\lambda \in \rho(\mathscr{A}_0^{\omega})$ for all $q \in (1, \infty)$. Each associated eigenfunction and each associated generalized eigenfunction belong to $D(\mathscr{A}_0^{\omega})$, independently of $q \in (1, \infty)$.

Proof Since $\Lambda_0^{\omega} = \mathrm{Sp}_{\mathrm{ess}}(\mathscr{A}_0^{\omega})$, the set $\mathbb{C} \setminus \Lambda_0^{\omega}$ is a subset of $\rho(\mathscr{A}_0^{\omega})$, up to an at most countable set of isolated eigenvalues of $\mathscr{A}_0^{\omega}$, which have finite algebraic multiplicities, see [137, paragraph IV.5.6].

Thus, assume that $\mathbf{w}$ is an eigenfunction of $\mathscr{A}_0^{\omega}$, corresponding to an eigenvalue $\lambda \notin \Lambda_0^{\omega}$. The fact that then λ is an eigenvalue of $\mathscr{A}_0^{\omega}$ for all $1 < q < \infty$ is proven in [77]. The idea of the proof is as follows: Let $R > 0$ be so large that $\mathbb{R}^3 \setminus B_{R-1}(\mathbf{0}) \subset \mathcal{D}$. We split $\mathbf{w}$ by means of an appropriate cutoff function procedure, and the Bogovskij operator to the sum $\mathbf{w}_1 + \mathbf{w}_2$, where both $\mathbf{w}_1$ and $\mathbf{w}_2$ belong to $D(\mathscr{A}_0^{\omega})$, $\mathbf{w}_1$ is supported in $\mathbb{R}^3 \setminus B_{R-1}(\mathbf{0})$ and $\mathbf{w}_2$ is supported in $\mathcal{D} \cap B_R(\mathbf{0})$. Then $(\lambda I - \mathscr{A}_0^{\omega})\mathbf{w}_1 = \mathbf{f}_1$, where $\mathbf{f}_1$ can be explicitly calculated and is supported in $B_R(\mathbf{0}) \setminus B_{R-1}(\mathbf{0})$. Since $\mathbf{w} \in D(\mathscr{A}_0^{\omega})$, elliptic regularity theory and Sobolev's embedding theorem prove that $\mathbf{f}_1 \in \boldsymbol{L}^s_{\sigma}(\mathcal{D})$ for all $1 < s < q^*$, where $q^* = nq/(n-q)$ if $q \leq 3$ and $q^* = \infty$ if $q > 3$. Extending $\mathbf{f}_1$ by zero to $\mathbb{R}^3 \setminus \mathcal{D}$, we get a function from $\boldsymbol{L}^s_{\sigma}(\mathbb{R}^3)$. Applying Theorem 8.6, we deduce that $\mathbf{w}_1$, extended by zero to $\mathbb{R}^3 \setminus \mathcal{D}$, belongs to $\boldsymbol{L}^s_{\sigma}(\mathbb{R}^3)$, $s < q'$, as well. Applying further estimate (8.2), we obtain that $\mathbf{w}_1 \in D((\mathscr{A}_0^{\omega})_{\mathbb{R}^3})$. Furthermore, since $\mathbf{w}_2$ is supported in a bounded subdomain of $\mathcal{D}$, we verify that $\mathbf{w}_2 \in D(\mathscr{A}_0^{\omega})$ (where $\mathscr{A}_0^{\omega}$ is considered to be an operator in $\boldsymbol{L}^s_{\sigma}(\mathcal{D})$) for all $1 < s < q'$. Repeating this step finitely many times, if necessary, we see that $\mathbf{w} = \mathbf{w}_1 + \mathbf{w}_2 \in D(\mathscr{A}_0^{\omega})$ for all $1 < s < \infty$. Therefore, $\mathbf{w}$ is an eigenfunction of $\mathscr{A}_0^{\omega}$ associated with the eigenvalue λ for all $1 < s < \infty$.

Since the geometric multiplicity of λ is the maximum number of linearly independent associated eigenfunctions $\mathbf{w}$, and these eigenfunctions are independent of s, the geometric multiplicity of λ is also independent of s.

The algebraic multiplicity of λ, since it is finite, equals the sum of the lengths of all linearly independent chains of the so-called generalized eigenfunctions, associated with the eigenvalue λ. If $\mathbf{w}^1, \dots, \mathbf{w}^l$ is such a chain, then $(\lambda I - \mathscr{A}_0^{\omega})\mathbf{w}^1 = \mathbf{0}$ and $(\lambda I - \mathscr{A}_0^{\omega})\mathbf{w}^m = \mathbf{w}^{m-1}$ for $m = 2, \dots, l$. By analogy with the eigenfunction $\mathbf{w}$ discussed above, one can successively show that all the functions $\mathbf{w}^1, \dots, \mathbf{w}^l$ also belong to $D(\mathscr{A}_0^\omega)$ for all $1 < s < \infty$. Consequently, the algebraic multiplicity of λ, as an eigenfunction of $\mathscr{A}_0^\omega$, is independent of s as well.

Finally, if $\lambda \in \mathbb{C} \setminus \Lambda_0^\omega$, $\operatorname{Re}\lambda \geq 0$, then one can prove that the operator $\lambda I - \mathscr{A}_0^\omega$ (where $\mathscr{A}_0^\omega$ is considered to be an operator in $\boldsymbol{L}^2_\sigma(\mathcal{D})$) has a bounded inverse in $\boldsymbol{L}^2_\sigma(\mathcal{D})$, just multiplying the resolvent equation $(\lambda I - \mathscr{A}_0^\omega)\mathbf{u} = \mathbf{f}$ by $\mathbf{u}$ and integrating on $\mathcal{D}$. Hence $\lambda \in \rho(\mathscr{A}_0^\omega)$ for $q = 2$. Due to the explanation given above, λ is not an eigenvalue of $\mathscr{A}_0^\omega$ for any $q \in (1,\infty)$ and λ also cannot belong to $\mathrm{Sp}_\mathrm{r}(\mathscr{A}_0^\omega)$. Hence $\lambda \in \rho(\mathscr{A}_0^\omega)$, independently of q. □

The next theorem shows that the set of eigenvalues of $\mathscr{A}_0^\omega$ outside Λ_0^ω is empty if domain $\mathcal{D}$ is axially symmetric.

Theorem 8.10 *Let domain $\mathcal{D}$ be axially symmetric about the x_3-axis and $\lambda \in \mathbb{C} \setminus \Lambda_0^\omega$. Then $\lambda \in \rho(\mathscr{A}_0^\omega)$.* □

Proof We have shown in Sect. 8.3.1 (in the case $q = 2$) that if $\mathcal{D}$ is axially symmetric about the x_3-axis then $\mathrm{Sp}(\mathscr{A}_0^\omega) = \Lambda_0^\omega$. Hence all $\lambda \in \mathbb{C} \setminus \Lambda_0^\omega$ belong to $\rho(\mathscr{A}_0^\omega)$. Due to Theorem 8.9, these λ belong to $\rho(\mathscr{A}_0^\omega)$ for all $q \in (1,\infty)$ as well. □

Now, it follows from Lemma 8.3, Theorem 8.8, and [137, Theorem IV535] (which says that the essential spectrum is preserved under relatively compact perturbations) that $\mathrm{Sp}_\mathrm{ess}(\mathscr{L}_0^\omega) = \mathrm{Sp}_\mathrm{ess}(\mathscr{A}_0^\omega)$. Moreover, the index of $(\mathscr{L}_0^\omega - \lambda I)$ equals zero for $\lambda \in \mathbb{C} \setminus \mathrm{Sp}_\mathrm{ess}(\mathscr{L}_0^\omega)$. (The index is constant in $\mathbb{C} \setminus \mathrm{Sp}_\mathrm{ess}(\mathscr{L}_0^\omega)$, see [137, pages 242–243]. Since $\operatorname{ind}(\mathscr{L}_0^\omega - \lambda I) = 0$ for λ with "sufficiently large" positive real part, we also have $\operatorname{ind}(\mathscr{L}_0^\omega) - \lambda I) = 0$ for $\lambda \in \mathbb{C} \setminus \mathrm{Sp}_\mathrm{ess}(\mathscr{L}_0^\omega)$.) The set of $\lambda \in \mathbb{C} \setminus \mathrm{Sp}_\mathrm{ess}(\mathscr{L}_0^\omega)$, where $\operatorname{nul}(\mathscr{L}_0^\omega - \lambda I) = \operatorname{def}(\mathscr{L}_0^\omega - \lambda I) > 0$, consists of isolated points, which can possibly cluster only on the boundary of $\mathrm{Sp}_\mathrm{ess}(\mathscr{L}_0^\omega)$. These λ are the eigenvalues of $\mathscr{L}_0^\omega$. The algebraic multiplicities of these eigenvalues are finite, see [137, page 243].) The next theorem summarizes these findings:

Theorem 8.11 *The essential spectrum of operator $\mathscr{L}_0^\omega$ (as an operator in $\boldsymbol{L}^q_\sigma(\mathcal{D})$) is the same as the essential spectrum of $\mathscr{A}_0^\omega$ and equals the set Λ_0^ω. Moreover, $\mathrm{Sp}(\mathscr{L}_0^\omega) = \Lambda_0^\omega \cup \Gamma$, where set Γ consists of at most countable number of isolated eigenvalues of $\mathscr{L}_0^\omega$, which can possibly cluster only on the boundary of $\mathrm{Sp}_{ess}(\mathscr{L}_0^\omega)$, and they all have finite algebraic multiplicities.* □

8.4 Spectral Theory of the Oseen Type Operator

Recall that $\mathcal{D}$ denotes an exterior domain in $\mathbb{R}^3$ with the boundary of the class $C^{1,1}$. We often use the estimate (8.2) of a solution of the equation $\mathscr{A}_k^\omega\mathbf{u} = f$ in $\boldsymbol{L}^q_\sigma(\mathcal{D})$, where $1 < q < \infty$ and $\mathscr{A}_k^\omega$ is the Oseen type operator, defined by (8.27). (See also the simplified form (8.9) of $\mathscr{A}_k^\omega$.) Recall that $\mathscr{A}_k^\omega$ is a densely defined and closed operator in $\boldsymbol{L}^q_\sigma(\mathcal{D})$, see Theorem 8.1.

We consider the case $k \neq 0$ in this section. We may assume, without loss of generality, that $k > 0$.

8.4.1 L^2-Theory

The content of this subsection comes from the paper [76] by Farwig and Neustupa.

The spectrum of the "classical" Oseen operator $\mathscr{A}_k^0$ was studied by Babenko in [12]. Using a different approach, the same description of $\mathrm{Sp}(\mathscr{A}_k^0)$, as in [12], was also obtained in [76]. The corresponding result says:

Theorem 8.12 *The spectrum of the Oseen operator* $\mathscr{A}_k^0$ *in* $\boldsymbol{L}^2_\sigma(\mathcal{D})$ *satisfies* $Sp(\mathscr{A}_k^0) = Sp_{ess}(\mathscr{A}_k^0) = \Lambda_k^0 := \{\lambda = \alpha + i\beta \in \mathbb{C};\ \alpha, \beta \in \mathbb{R},\ \alpha \le -\nu\beta^2/k^2\}$. □

Proof We present the proof from [76]:

(1) Let us at first prove the inclusion $\mathbb{C} \setminus \Lambda_k^0 \subset \rho(\mathscr{A}_k^0)$. Suppose, by contradiction, that $\lambda = \alpha + i\beta \in (\mathbb{C} \setminus \Lambda_k^0) \cap \mathrm{Sp}(\mathscr{A}_k^0)$. Assume at first that $\lambda \in \mathrm{Sp_p}(\mathscr{A}_k^0) \cup \mathrm{Sp_c}(\mathscr{A}_k^0)$. Then there exists a sequence $\mathbf{v}^n$ of generally complex-valued functions in the unit sphere in $\boldsymbol{L}^2_\sigma(\mathcal{D})$ such that

$$(\mathscr{A}_k^0 - \lambda I)\mathbf{v}^n = \boldsymbol{\epsilon}^n \longrightarrow \mathbf{0} \qquad \text{in } \boldsymbol{L}^2_\sigma(\mathcal{D}) \quad \text{as } n \to \infty. \tag{8.72}$$

Testing (8.72) by $\mathbf{v}^n$ in L^2-sense of complex-valued functions, we get

$$-\nu\,\|\nabla\mathbf{v}^n\|_2^2 - \lambda\,\|\mathbf{v}^n\|_2^2 - k\,(\partial_3\mathbf{v}^n, \mathbf{v}^n)_2 = (\boldsymbol{\epsilon}^n, \mathbf{v}^n)_2. \tag{8.73}$$

Note that $\mathrm{Re}\,(\partial_3\mathbf{v}^n, \mathbf{v}^n)_2 = 0$. Considering the real part of (8.73), we obtain

$$\nu\,\|\nabla\mathbf{v}^n\|_2^2 = -\alpha\,\|\mathbf{v}^n\|_2^2 - \mathrm{Re}\,(\boldsymbol{\epsilon}^n, \mathbf{v}^n)_2 \le -\alpha + \|\boldsymbol{\epsilon}^n\|_2. \tag{8.74}$$

Since $\|\boldsymbol{\epsilon}^n\|_2 \to 0$ for $n \to \infty$, we get $\alpha \le 0$. From (8.74), we obtain

$$\|\nabla\mathbf{v}^n\|_2 \le \Big(-\frac{\alpha}{\nu} + \frac{1}{\nu}\,\|\boldsymbol{\epsilon}^n\|_2\Big)^{1/2}. \tag{8.75}$$

Considering the imaginary part in (8.73), we get

$$\beta = \beta\,\|\mathbf{v}^n\|_2^2 = -k\,\mathrm{Im}\,(\partial_3\mathbf{v}^n, \mathbf{v}^n)_2 - \mathrm{Im}\,(\boldsymbol{\epsilon}^n, \mathbf{v}^n)_2.$$

Hence $|\beta| \le k\,\|\nabla\mathbf{v}^n\|_2 + \|\boldsymbol{\epsilon}^n\|_2$. Estimating $\|\nabla\mathbf{v}^n\|_2$ by means of (8.75), we obtain the inequality

$$|\beta| \le k\Big(-\frac{\alpha}{\nu} + \frac{1}{\nu}\,\|\boldsymbol{\epsilon}^n\|_2\Big)^{1/2} + \|\boldsymbol{\epsilon}^n\|_2\,.$$

The limit transition $n \to \infty$ implies that $|\beta| \le k\sqrt{-\alpha/\nu}$, i.e., $\alpha \le \nu\beta^2/k^2$. This inequality is in contradiction with the assumption $\lambda \in \mathbb{C} \setminus \Lambda_k^0$. Hence $\lambda \in \mathbb{C} \setminus \Lambda_k^0$ cannot belong to $\mathrm{Sp_p}(\mathscr{A}_k^0) \cup \mathrm{Sp_c}(\mathscr{A}_k^0)$. If $\lambda \in \mathrm{Sp_r}(\mathscr{A}_k^0)$, then $\overline{\lambda}$ belongs to the point spectrum of the adjoint operator $(\mathscr{A}_k^0)^*$, which leads to the same contradiction as if $\lambda \in \mathrm{Sp_p}(\mathscr{A}_k^0)$. Thus, $\lambda \in \rho(\mathscr{A}_k^0)$, which yields $\mathbb{C} \setminus \Lambda_k^0 \subset \rho(\mathscr{A}_k^0)$.

(2) We will now prove that $\Lambda_k^0 \subset \mathrm{Sp_c^{Gl}}(\mathscr{A}_k^0)$. Let $\lambda = \alpha + \mathrm{i}\beta \in (\Lambda_k^0)^\circ$ be given. (We denote by $(\Lambda_k^0)^\circ$ the interior of Λ_k^0, i.e., the set of $\alpha + \mathrm{i}\beta \in \mathbb{C}$ such that $\alpha < -\nu\beta^2/k^2$.) Then α can be written in the form $\alpha = \alpha_1 + \alpha_2$ where $\alpha_1 = -\nu\beta^2/k^2$ and $\alpha_2 < 0$. We shall explicitly define functions $\mathbf{v}^n \in \boldsymbol{L}^2_\sigma(\mathcal{D})$ such that $\|\mathbf{v}^n\|_2 = 1$, $(\mathscr{A}_k^0 - \lambda I)\mathbf{v}^n \to \mathbf{0}$ in $\boldsymbol{L}^2_\sigma(\mathcal{D})$ for $n \to \infty$ and the sequence $\{\mathbf{v}^n\}$ does not contain any subsequence, convergent in $\boldsymbol{L}^2_\sigma(\mathcal{D})$. Let us denote by v_r^n, v_φ^n, and v_3^n the cylindrical components of $\mathbf{v}^n$. Put

$$\begin{aligned} v_r^n(r,\varphi,x_3) &:= \kappa_n\, U^n(x_3)\, V^n(r)\, \mathrm{e}^{\mathrm{i}l\varphi}, \\ v_\varphi^n(r,\varphi,x_3) &:= -\frac{1}{\mathrm{i}l}\, \partial_r\, [r\, v_r^n(x_3,r,\varphi)] \\ &= -\frac{1}{\mathrm{i}l}\, \kappa_n\, U^n(x_3) \Big[V^n(r) + r\, \frac{\mathrm{d}V^n(r)}{\mathrm{d}r} \Big]\, \mathrm{e}^{\mathrm{i}l\varphi}, \\ v_3^n(r,\varphi,x_3) &:= 0, \end{aligned}$$

where l is an arbitrary, but fixed nonzero integer. The function $\mathbf{v}^n$ is divergence-free:

$$\operatorname{div} \mathbf{v}^n \;\equiv\; \frac{1}{r}\, \partial_r(r v_r^n) + \frac{1}{r}\, \partial_\varphi v_\varphi^n + \partial_3 v_3^n \;=\; 0.$$

Suppose that U^n has the form

$$U^n(x_3) \;:=\; \eta_1^n(x_3)\, Y(x_3), \tag{8.76}$$

where η_1^n is an infinitely differentiable function on $(-\infty,\infty)$ with values in the interval $[0,1]$,

$$\eta_1^n(x_3) \;=\; \begin{cases} 0 \;\text{ for } x_3 \le -n-n^2 \;\text{ and }\; n+n^2 \le x_3, \\ 1 \;\text{ for } -n^2 \le x_3 \le n^2, \end{cases}$$

and $Y(x_3) = \mathrm{e}^{\mathrm{i}ax_3}$. The identity $\alpha_1 = -\nu\beta^2/k^2$ guarantees that the characteristic equation $\nu\zeta^2 - k\zeta - (\alpha_1 + \mathrm{i}\beta) = 0$, corresponding to the differential equation (8.77) below, has the root $\zeta_1 = ia$ where $a = -\beta/k$. Thus, the function Y is a bounded nontrivial solution of the ordinary differential equation

$$\nu\, Y''(x_3) - k\, Y'(x_3) - (\alpha_1 + \mathrm{i}\beta)\, Y(x_3) \;=\; 0 \tag{8.77}$$

in the interval $(-\infty, \infty)$. The function V^n is supposed to have the form

$$V^n(r) \;:=\; \eta_2^n(r)\, \mathrm{e}^{\mathrm{i}br}; \qquad b = \sqrt{-\frac{\alpha_2}{\nu}}, \tag{8.78}$$

where η_2^n is an infinitely differentiable function on $[0, \infty)$ such that $0 \le \eta_2^n \le 1$ and

$$\eta_2^n(r) \;=\; \begin{cases} 0 & \text{for } 0 \le r \le n \text{ and } 3n + n^2 \le r, \\ 1 & \text{for } 2n \le r \le 2n + n^2. \end{cases}$$

Both the functions η_1 and η_2 can be chosen so that their derivatives are of the order $1/n$. If follows from the definition of V^n that

$$\nu\, \frac{\mathrm{d}^2}{\mathrm{d}r^2}\, V^n(r) - \alpha_2\, V^n(r) \;=\; 0 \tag{8.79}$$

for $2n < r < 2n + n^2$. Finally, the constant κ_n is chosen so that $\|\mathbf{v}^n\|_2 = 1$. The support of $\mathbf{v}^n$ is a subset of

$$\begin{aligned} S^n := \big\{ &\mathbf{x} = [r, \varphi, x_3] \in \mathbb{R}^3; \\ &n \le r \le 3n + n^2,\ \ 0 \le \varphi < 2\pi,\ \ -n - n^2 \le x_3 \le n + n^2 \big\}. \end{aligned} \tag{8.80}$$

We can observe that for large n the decisive contribution to $\|\mathbf{v}^n\|_2$ comes from the integral of $|v_\varphi^n|^2$, namely of its part $\big|(-1/\mathrm{i}l)\, \kappa_n\, U^n\, r\, (\mathrm{d}V^n/\mathrm{d}r)\, \mathrm{e}^{\mathrm{i}l\varphi}\big|^2$, on the region

$$\begin{aligned} D^n := \big\{ &\mathbf{x} = [r, \varphi, x_3] \in \mathbb{R}^3; \\ &2n < r < 2n + n^2,\ \ 0 < \varphi < 2\pi,\ \ -n^2 < x_3 < n^2 \big\}. \end{aligned} \tag{8.81}$$

The integrals of all other parts on other regions are of a lower order in n. Calculating the integral of $\big|(-1/\mathrm{i}l)\, \kappa_n\, U^n\, r\, (\mathrm{d}V^n/\mathrm{d}r)\, \mathrm{e}^{\mathrm{i}l\varphi}\big|^2$ on the domain D^n, we obtain

$$\begin{aligned} &\int_{-n^2}^{n^2} \int_{2n}^{2n+n^2} \int_0^{2\pi} \left| \frac{\kappa_n}{\mathrm{i}l}\, U^n(x_3)\, r\, \frac{\mathrm{d}V^n(r)}{\mathrm{d}r} \right|^2 r\, \mathrm{d}\varphi\, \mathrm{d}r\, \mathrm{d}x_3 \\ &\quad = 2\pi\, \frac{\kappa_n^2}{l^2} \int_{-n^2}^{n^2} |U^n(x_3)|^2\, \mathrm{d}x_3 \int_{2n}^{2n+n^2} r^3 \left| \frac{\mathrm{d}V^n(r)}{\mathrm{d}r} \right|^2 \mathrm{d}r \end{aligned}$$

$$= 2\pi \, \frac{\kappa_n^2}{l^2} \, 2n^2 \, \frac{b^2}{4} \, \Big((2n + n^2)^4 - (2n)^4 \Big).$$

We have used the equalities $\eta_1^n(x_3) = \eta_2^n(r) = 1$, which imply that $|U^n(x_3)| = |V^n(r)| = 1$ for $(r, \varphi, x_3) \in D^n$. Thus, there exist $n_0 \in \mathbb{N}$ and positive constants c_4 and c_5 (independent of n) such that

$$\forall\, n \in \mathbb{N},\ n \geq n_0 \ : \quad \frac{c_4}{n^5} \leq \kappa_n \leq \frac{c_5}{n^5}\,. \tag{8.82}$$

Looking at $(\mathscr{A}_k^0 - \lambda I)\mathbf{v}^n$, one can omit the projection Π_σ^2 in front of the Laplace operator in $\mathscr{A}_k^0 \mathbf{v}^n$ because $\Delta \mathbf{v}^n$ is divergence-free and has a compact support in $\mathcal{D}$. Thus, $(\mathscr{A}_k^0 - \lambda I)\mathbf{v}^n = \nu \Delta \mathbf{v}^n - k\,\partial_3 \mathbf{v}^n - \lambda \mathbf{v}^n$. Calculating the norm of this expression in $\boldsymbol{L}_\sigma^2(\mathcal{D})$, we observe that the contributions coming from $\mathcal{D} \setminus D^n$ tend to zero as $n \to \infty$ because they represent square roots of integrals of functions bounded by $c\kappa_n^2 r^2$ on $S^n \setminus D^n$. Due to (8.82), these contributions are of the order $n^{-1/2}$. Concerning the integral on D^n, the decisive part again comes from $(\nu\Delta - k\,\partial_3 - \lambda I)v_\varphi^n$, namely from $(\nu\Delta - k\,\partial_3 - \lambda I)$ applied to the term $(-1/\mathrm{i}l)\,\kappa_n\, U^n\, r\, (\mathrm{d}V^n/\mathrm{d}r)\, \mathrm{e}^{\mathrm{i}l\varphi}$ because of the factor r inside this term. Note that due to (8.77) and (8.79)

$$\begin{aligned}
&(\nu\Delta - k\,\partial_3 - \lambda I)\Big(\frac{\kappa_n}{\mathrm{i}l}\, U^n(x_3)\, r\, \frac{\mathrm{d}V^n(r)}{\mathrm{d}r}\, \mathrm{e}^{\mathrm{i}l\varphi} \Big) \\
&\quad = \Big(\nu\partial_r^2 + \frac{\nu}{r}\,\partial_r + \frac{\nu}{r^2}\,\partial_\varphi^2 + \nu\partial_3^2 - k\,\partial_3 - \lambda I \Big)\Big(\frac{\kappa_n}{\mathrm{i}l}\, U^n(x_3)\, r\, \frac{\mathrm{d}V^n(r)}{\mathrm{d}r}\, \mathrm{e}^{\mathrm{i}l\varphi} \Big) \\
&\quad = \frac{\kappa_n}{\mathrm{i}l}\, U^n(x_3) \Big(\nu\, \frac{\mathrm{d}^2}{\mathrm{d}r^2} + \frac{\nu}{r}\, \frac{\mathrm{d}}{\mathrm{d}r} - \alpha_2 I \Big) \Big[r\, \frac{\mathrm{d}V^n(r)}{\mathrm{d}r} \Big] \mathrm{e}^{\mathrm{i}l\varphi} \\
&\qquad + \frac{\kappa_n}{\mathrm{i}l} \Big(\nu Y''(x_3) - kY'(x_3) - [\alpha_1 + \mathrm{i}\beta]\, Y(x_3) \Big) \Big[r\, \frac{\mathrm{d}V^n(r)}{\mathrm{d}r} \Big] \mathrm{e}^{\mathrm{i}l\varphi} \\
&\qquad - \frac{\kappa_n}{\mathrm{i}l}\, \frac{\nu l^2}{r^2}\, U^n(x_3)\, r\, \frac{\mathrm{d}V^n(r)}{\mathrm{d}r}\, \mathrm{e}^{\mathrm{i}l\varphi} \\
&\quad = \bigg\{ \frac{\kappa_n}{\mathrm{i}l}\, U^n(x_3)\, r\, \frac{\mathrm{d}}{\mathrm{d}r} \Big[\nu\, \frac{\mathrm{d}^2 V^n(r)}{\mathrm{d}r^2} - \alpha_2\, V^n(r) \Big] + \frac{\kappa_n}{\mathrm{i}l}\, U^n(x_3)\, 2\nu\, \frac{\mathrm{d}^2 V^n(r)}{\mathrm{d}r^2} \\
&\qquad + \frac{\kappa_n}{\mathrm{i}l}\, U^n(x_3)\, \frac{\nu}{r}\, \frac{\mathrm{d}}{\mathrm{d}r} \Big[r\, \frac{\mathrm{d}V^n(r)}{\mathrm{d}r} \Big] - \frac{\kappa_n}{\mathrm{i}l}\, U^n(x_3)\, \frac{\nu l^2}{r}\, \frac{\mathrm{d}V^n(r)}{\mathrm{d}r} \bigg\} \mathrm{e}^{\mathrm{i}l\varphi} \\
&\quad = \frac{\nu\kappa_n}{\mathrm{i}l} \Big(-3b^2 + \frac{\mathrm{i}b}{r} - \frac{l^2\,\mathrm{i}b}{r} \Big) \mathrm{e}^{\mathrm{i}(ax_3 + br)}\, \mathrm{e}^{\mathrm{i}l\varphi},
\end{aligned}$$

where we used the simple forms of the functions U^n and V^n on D^n, i.e., $U^n(x_3) = \mathrm{e}^{\mathrm{i}ax_3}$ and $V^n(r) = \mathrm{e}^{\mathrm{i}br}$, in the last step. Hence

$$\int_{-n^2}^{n^2}\int_{2n}^{2n+n^2}\int_0^{2\pi}\left|\left(\nu\Delta - k\,\partial_3 - \lambda I\right)\left(\frac{\kappa_n}{\mathrm{i}l}\,U^n(x_3)\,r\,\frac{\mathrm{d}V^n(r)}{\mathrm{d}r}\,\mathrm{e}^{\mathrm{i}l\varphi}\right)\right|^2 r\,\mathrm{d}\varphi\,\mathrm{d}r\,\mathrm{d}x_3$$

$$\leq c(\nu,l,b)\,\kappa_n^2\int_{-n^2}^{n^2}\int_{2n}^{2n+n^2} r\,\mathrm{d}r\,\mathrm{d}x_3$$

$$= c(\nu,l,b)\,\kappa_n^2\,n^2\left[(2n+n^2)^2-(2n)^2\right].$$

The last term tends to zero as $n \to \infty$ due to (8.82). Hence $\|(\mathscr{A}_k^0 - \lambda I)\mathbf{v}^n\|_2 \to 0$ as $n \to \infty$. The sequence $\{\mathbf{v}^n\}$ does not contain any convergent subsequence because the intersection of supports of any infinite family of functions, chosen from $\{\mathbf{v}^n\}$, is empty. Since λ was an arbitrarily chosen number from $(\Lambda_k^0)^\circ$, we have obtained the inclusion $(\Lambda_k^0)^\circ \subset \mathrm{Sp}_{\mathrm{C}}^{\mathrm{Gl}}(\mathscr{A}_k^0)$. It means that $\mathrm{nul}'(\mathscr{A}_k^0 - \lambda I) = \infty$ for $\lambda \in (\Lambda_k^0)^\circ$. Since the operators $\mathscr{A}_k^0$ and $(\mathscr{A}_k^0)^*$ differ only in the sign in front of $k\,\partial_3$, we can prove in the same way that $\mathrm{nul}'((\mathscr{A}_k^0)^* - \overline{\lambda} I) = \infty$. It means that $\mathrm{def}'(\mathscr{A}_k^0 - \lambda I) = \infty$, and consequently, $\lambda \in \mathrm{Sp}_{\mathrm{ess}}(\mathscr{A}_k^0)$. The essential spectrum is a closed set, hence $\Lambda_k^0 \subset \mathrm{Sp}_{\mathrm{ess}}(\mathscr{A}_k^0)$. □

8.4.1.1 The Case of an Axially Symmetric Domain $\mathcal{D}$

We assume that the domain $\mathcal{D} \subset \mathbb{R}^3$ is axially symmetric with respect to the x_3-axis in this paragraph.

Let l be an integer. We will use the spaces $\boldsymbol{C}_{0,\sigma}^\infty(\mathcal{D})_l$ and $\boldsymbol{L}_\sigma(\mathcal{D})_l$ and the projection P_l, defined in paragraph 8.3.1.1. By analogy with $(\mathscr{A}_0^0)_l$, we denote by $(\mathscr{A}_k^0)_l$ the restriction of the operator $\mathscr{A}_k^0$ to the space $\boldsymbol{L}_\sigma^2(\mathcal{D})_l$. Obviously, the domain of $(\mathscr{A}_k^0)_l$ equals $D(\mathscr{A}_k^0) \cap \boldsymbol{L}_\sigma^2(\mathcal{D})_l$. Recall that $\Pi_\sigma^2\boldsymbol{L}^2(\mathcal{D})_l = \boldsymbol{L}_\sigma^2(\mathcal{D}) \cap \boldsymbol{L}^2(\mathcal{D})_l = \boldsymbol{L}_\sigma^2(\mathcal{D})_l = P_l\boldsymbol{L}_\sigma^2(\mathcal{D})$ due to Lemma 8.4.

The next lemma generalizes some results from [62].

Lemma 8.12 *Let $l \in \mathbb{Z}$. Then $(\mathscr{A}_k^0)_l$ is a closed operator in $\boldsymbol{L}_\sigma^2(\mathcal{D})_l$ with the dense domain $D((\mathscr{A}_k^0)_l)$. Moreover, $D((\mathscr{A}_k^0)_l) = P_l[D(\mathscr{A}_k^0)]$ and $R((\mathscr{A}_k^0)_l) \subset \boldsymbol{L}_\sigma^2(\mathcal{D})_l$.* □

Proof The operator $(\mathscr{A}_k^0)_l$ is closed because it is a restriction of the closed operator $\mathscr{A}_k^0$ onto a closed subspace of $\boldsymbol{L}_\sigma^2(\mathcal{D})$. The domain of $(\mathscr{A}_k^0)_l$ is the set of functions from $\boldsymbol{L}_\sigma^2(\mathcal{D})_l$ that belong to $\boldsymbol{W}^{2,2}(\mathcal{D}) \cap \boldsymbol{W}_0^{1,2}(\mathcal{D})$. This set contains $\boldsymbol{C}_{0,\sigma}^\infty(\mathcal{D})$, and hence it is dense in $\boldsymbol{L}_\sigma^2(\mathcal{D})_l$.

Let $\mathbf{v} \in D(\mathscr{A}_k^0) \equiv \boldsymbol{W}^{2,2}(\mathcal{D}) \cap \boldsymbol{W}_0^{1,2}(\mathcal{D}) \cap \boldsymbol{L}_\sigma^2(\mathcal{D})$ and let (8.24) be its Fourier expansion in the variable φ. Then $\boldsymbol{V}^l(x_1, r)\,\mathrm{e}^{\mathrm{i}l\varphi} \equiv P_l\mathbf{v} \in \boldsymbol{W}^{2,2}(\mathcal{D})$, and, due to the axial symmetry of $\mathcal{D}$ and the boundary condition satisfied by $\mathbf{v}$ on $\partial\mathcal{D}$, $\boldsymbol{V}^l(r, x_3)\,\mathrm{e}^{\mathrm{i}l\varphi}$ also belongs to $\boldsymbol{W}_0^{1,2}(\mathcal{D})$. Since $\operatorname{div}\mathbf{v} = 0$ and the functions $\operatorname{div}\big[\boldsymbol{V}^l(r, x_3)\,\mathrm{e}^{\mathrm{i}l\varphi}\big]$ are orthogonal in $\boldsymbol{L}^2(\mathcal{D})$ (for different l), we obtain $\operatorname{div}\big[\boldsymbol{V}^l(r, x_3)\,\mathrm{e}^{\mathrm{i}l\varphi}\big] = 0$. Hence $\boldsymbol{V}^l(r, x_3)\,\mathrm{e}^{\mathrm{i}l\varphi} \in \boldsymbol{L}_\sigma^2(\mathcal{D})_l$ and consequently, $P_l[D(\mathscr{A}_k^0)] \subset D((\mathscr{A}_k^0)_l)$.

On the other hand, if $\mathbf{v} \in D((\mathscr{A}_k^0)_l)$, then it belongs to $D(\mathscr{A}_k^0)$, and since $P_l\mathbf{v} = \mathbf{v}$, it also belongs to $\boldsymbol{L}^2(\mathcal{D})_l$. Hence $\mathbf{v} \in D(\mathscr{A}_k^0) \cap \boldsymbol{L}^2(\mathcal{D})_l = D(\mathscr{A}_k^0) \cap \boldsymbol{L}_\sigma^2(\mathcal{D})_l = P_l[D(\mathscr{A}_k^0)]$.

If $\mathbf{v} \in D((\mathscr{A}_k^0)_l)$, then $\Delta\mathbf{v}$, $\partial_3\mathbf{v} \in \boldsymbol{L}^2(\mathcal{D})_l$, and due to Lemma 8.4, $\mathscr{A}_k^0\mathbf{v} = \nu\Pi_\sigma\Delta\mathbf{v} - k\,\partial_3\mathbf{v} \in \boldsymbol{L}_\sigma^2(\mathcal{D})_l$. Hence $\mathscr{A}_k^0\mathbf{v} \in \boldsymbol{L}_\sigma^2(\mathcal{D})_l$, which yields $R((\mathscr{A}_k^0)_l) \subset \boldsymbol{L}_\sigma^2(\mathcal{D})_l$. □

Lemma 8.13 *Let $l \in \mathbb{Z}$. Then $Sp((\mathscr{A}_k^0)_l) = Sp_{ess}((\mathscr{A}_k^0)_l) = \Lambda_k^0$, where Λ_k^0 is the parabolic region in $\mathbb{C}$, described in Theorem 8.12: $\Lambda_k^0 = \big\{\lambda = \alpha + i\beta \in \mathbb{C};\ \alpha, \beta \in \mathbb{R},\ \alpha \le -\nu\beta^2/k^2\big\}$.* □

Proof The operator $(\mathscr{A}_k^0)_l$ is a part of $\mathscr{A}_k^0$, and hence $\mathrm{Sp}((\mathscr{A}_k^0)_l) \subset \mathrm{Sp}(\mathscr{A}_k^0) = \Lambda_k^0$.

On the other hand, for $\lambda \in (\Lambda_k^0)^\circ$, we have shown the existence of a non-compact sequence $\mathbf{v}^n \in \boldsymbol{L}_\sigma^2(\mathcal{D})$ such that $\|\mathbf{v}^n\|_2 = 1$ and $(\mathscr{A}_k^0 - \lambda I)\mathbf{v}^n \to \mathbf{0}$ in $\boldsymbol{L}_\sigma^2(\mathcal{D})$ for $n \to \infty$ in the proof of Theorem 8.12, part II. The construction of $\mathbf{v}^n$ involved the choice of an arbitrary nonzero integer l. An easy examination shows that the functions $\mathbf{v}^n$ actually belong not only to $\boldsymbol{L}_\sigma^2(\mathcal{D})$, but to $\boldsymbol{L}_\sigma^2(\mathcal{D})_l$. Thus, we obtain that $(\Lambda_k^0)^\circ \subset \mathrm{Sp}_{\mathrm{C}}^{\mathrm{Gl}}((\mathscr{A}_k^0)_l)$ for $l \neq 0$. Using the same arguments as at the end of the proof of Theorem 8.12, we deduce that $\Lambda_k^0 \subset \mathrm{Sp}_{\mathrm{ess}}((\mathscr{A}_k^0)_l)$ for $l \neq 0$. It completes the proof in the case when $l \neq 0$.

The case $l = 0$ must be treated separately. Suppose that $\lambda = \alpha + \mathrm{i}\beta \in (\Lambda_k^0)^\circ$. Let us construct a non-compact sequence $\{\mathbf{v}^n\}$ in the unit sphere in $\boldsymbol{L}_\sigma^2(\mathcal{D})_0$ such that $(\mathscr{A}_k^0 - \lambda I)\mathbf{v}^n \to \mathbf{0}$ as $n \to \infty$. The inclusion $\mathbf{v}^n \in \boldsymbol{L}_\sigma^2(\mathcal{D})_0$ means that $\mathbf{v}^n \equiv [v_r^n, v_\varphi^n, v_3^n]$ does not depend on φ. Then the condition $\operatorname{div}\mathbf{v}^n = 0$ says that $\partial_r(rv_r^n) + \partial_3(rv_3^n) = 0$. This equation is automatically satisfied if $\mathbf{v}^n$ has the cylindrical components

$$v_r^n(r, x_3) = -\frac{1}{r}\,\partial_3\psi^n(r, x_3), \qquad v_\varphi^n(r, x_3) = 0, \qquad v_3^n(r, x_3) = \frac{1}{r}\,\partial_r\psi^n(r, x_3).$$

Put $\psi^n(r, x_3) = \delta_n\, U^n(x_3)\, V^n(r)$ where U^n and V^n are the same functions as in the proof of Theorem 8.12, i.e., the functions given by (8.76) and (8.78), and the factor δ_n is be chosen so that $\|\mathbf{v}^n\|_2 = 1$. Calculating the norm of $\|\mathbf{v}^n\|_2$, we can observe that for large n the decisive contribution comes from the integral on D^n, see (8.81). The contribution of the integral over $\mathcal{D} \setminus D^n$ is of a lower order in powers of n. The cutoff functions η_1^n and η_2^n are both equal to 1 on D^n. Hence $U^n(x_3) = \mathrm{e}^{\mathrm{i}ax_3}$, where $a = -\beta/k$, and $V^n(r) = \mathrm{e}^{\mathrm{i}br}$, where $b = \sqrt{-\alpha_2/\nu}$, see the proof of Theorem 8.12.

Thus, $\psi(r, x_3) = \delta_n \, \mathrm{e}^{\mathrm{i}(ax_3+br)}$ on D^n and

$$\begin{aligned}
\int_{D^n} |\mathbf{v}^n|^2 \, \mathrm{d}\mathbf{x} &= \int_{-n^2}^{n^2} \int_{2n}^{2n+n^2} \int_0^{2\pi} \big(|v_3^n|^2 + |v_r^n|^2\big) \, \mathrm{d}\varphi \, r \, \mathrm{d}r \, \mathrm{d}x_3 \\
&= 2\pi \int_{-n^2}^{n^2} \int_{2n}^{2n+n^2} \frac{1}{r} \big(|\partial_r \psi^n|^2 + |\partial_3 \psi^n|^2\big) \, \mathrm{d}r \, \mathrm{d}x_3 \\
&= 2\pi \, \delta_n^2 \int_{-n^2}^{n^2} \int_{2n}^{2n+n^2} \frac{1}{r} (a^2+b^2) \, \mathrm{d}r \, \mathrm{d}x_3 \\
&= 2\pi \, \delta_n^2 \, (a^2+b^2) \, 2n^2 \Big(\ln(2n+n^2) - \ln(2n)\Big).
\end{aligned}$$

The condition that this tends to 1 as $n \to \infty$ leads to the existence of constants $c_6, \ c_7 > 0$ and $n_0 \in \mathbb{N}$ such that

$$\forall n \in \mathbb{N}, \ n \geq n_0 : \qquad \frac{c_6}{n\sqrt{\ln n}} \leq \delta_n \leq \frac{c_7}{n\sqrt{\ln n}}. \tag{8.83}$$

Suppose that n is so large that the support of $\mathbf{v}^n$ is a subset of $\mathcal{D}$. Then, since $(\mathscr{A}_k^0 - \lambda I)\mathbf{v}^n = \nu \Delta \mathbf{v}^n - k \, \partial_3 \mathbf{v}^n - \lambda \mathbf{v}^n$, we get

$$\begin{aligned}
\|(\mathscr{A}_k^0 - \lambda I)\mathbf{v}^n\|_2^2 = \delta_n^2 & \int_{S^n - D^n} \big|\nu \Delta \mathbf{v}^n - k \, \partial_3 \mathbf{v}^n - \lambda \mathbf{v}^n\big|^2 \, \mathrm{d}\mathbf{x} \\
& + \delta_n^2 \int_{D^n} \big|\nu \Delta \mathbf{v}^n - k \, \partial_3 \mathbf{v}^n - \lambda \mathbf{v}^n\big|^2 \, \mathrm{d}\mathbf{x},
\end{aligned} \tag{8.84}$$

where S^n is defined by (8.80). The integrands are less than or equal to c/r^2. Thus, the first term on the right-hand side of (8.84) can be estimated from above by

$$\begin{aligned}
& c\delta_n^2 \int_{-n-n^2}^{n+n^2} \mathrm{d}x_3 \left(\int_n^{2n} + \int_{2n+n^2}^{3n+n^2} \right) \frac{r \, \mathrm{d}r}{r^2} \\
& \quad + c\delta_n^2 \left(\int_{-n-n^2}^{-n^2} + \int_{n^2}^{n+n^2} \right) \mathrm{d}x_3 \int_n^{3n+n^2} \frac{r \, \mathrm{d}r}{r^2} \\
& \leq c\delta_n^2 \, (2n + 2n^2) \left(\ln \frac{2n}{n} + \ln \frac{3n+n^2}{2n+n^2} \right) + c\delta_n^2 \left(2n \, \ln \frac{3n+n^2}{n} \right).
\end{aligned}$$

Using (8.83), we verify that the right-hand side tends to zero as $n \to \infty$. In the second term on the right- hand side of (8.84), we use the identity $\psi^n(r, x_3) = \delta_n \, \mathrm{e}^{\mathrm{i}(ax_3+br)}$, which holds on D^n. Note that the function $Y(x_3) = \mathrm{e}^{\mathrm{i}ax_3}$ satisfies the differential equation (8.77) and that the function $\mathrm{e}^{\mathrm{i}br}$ satisfies the differential equation (8.79). Then, calculating the expression $\nu \Delta \mathbf{v}^n - k \, \partial_3 \mathbf{v}^n - \lambda \mathbf{v}^n$, we find that

$|\nu\Delta\mathbf{v}^n - k\,\partial_3\mathbf{v}^n - \lambda\mathbf{v}^n| \leq c\delta_n/r^2$. Consequently, the second term on the right-hand side of (8.84) is less than or equal to

$$c\delta_n^2 \int_{-n^2}^{n^2} \int_{2n}^{2n+n^2} \int_0^{2\pi} \frac{1}{r^4}\,\mathrm{d}\varphi\, r\,\mathrm{d}r\,\mathrm{d}x_3 \;=\; c\delta_n^2\, n^2 \int_{2n}^{2n+n^2} \frac{\mathrm{d}r}{r^3} \;\leq\; c\delta_n^2\,.$$

Due to (8.83), the right-hand side tends to zero as $n \to \infty$. Hence $\|(\mathscr{A}_k^0 - \lambda I)\mathbf{v}^n\|_2 \to 0$ as $n \to \infty$. The sequence $\{\mathbf{v}^n\}$ is non-compact because the intersection of the supports of any infinite family of functions chosen from $\{\mathbf{v}^n\}$ is empty. Since all the functions $\mathbf{v}^n$ belong to $\boldsymbol{L}^2_\sigma(\mathcal{D})_0$, we obtain $\lambda \in \mathrm{Sp}_{\mathrm{c}}^{\mathrm{Gl}}((\mathscr{A}_k^0)_0)$. Applying once again the same arguments as at the end of the proof of Theorem 8.12, we observe that $\lambda \in \mathrm{Sp}_{\mathrm{ess}}((\mathscr{A}_k^0)_0)$, and due to the closedness of $\mathrm{Sp}_{\mathrm{ess}}((\mathscr{A}_k^0)_0)$, we obtain the inclusion $\Lambda_k^0 \subset \mathrm{Sp}_{\mathrm{ess}}((\mathscr{A}_k^0)_0)$. This completes the proof in the case $l = 0$. □

Since $-(\mathscr{A}_k^0)_l$ is, by definition, the reduction of the sectorial operator $-\mathscr{A}_k^0$ to the space $\boldsymbol{L}^2_\sigma(\mathcal{D})_l$, $-(\mathscr{A}_k^0)_l$ is a sectorial operator in $\boldsymbol{L}^2_\sigma(\mathcal{D})_l$.

We deal with the decomposition of the operator $\mathscr{A}_k^\omega$ further on in this paragraph.

Let $l \in \mathbb{Z}$. We denote by $(\mathscr{A}_k^\omega)_l$ the restriction of $\mathscr{A}_k^\omega$ to $\boldsymbol{L}^2_\sigma(\mathcal{D})_l$. The domain of $(\mathscr{A}_k^\omega)_l$ is the same as the domain of $(\mathscr{A}_k^0)_l$, i.e.,

$$D((\mathscr{A}_k^\omega)_l) = D((\mathscr{A}_k^0)_l) = D(\mathscr{A}_k^0) \cap \boldsymbol{L}^2_\sigma(\mathcal{D})_l = \boldsymbol{W}^{2,2}(\mathcal{D}) \cap \boldsymbol{W}_0^{1,2}(\mathcal{D}) \cap \boldsymbol{L}^2_\sigma(\mathcal{D})_l.$$

If $\mathbf{u} \in \boldsymbol{L}^2_\sigma(\mathcal{D})_l$, then $\mathbf{u}$ has the form $\mathbf{u}(r,\varphi,x_3) = \boldsymbol{U}(r,x_3)\,\mathrm{e}^{\mathrm{i}l\varphi}$ and $\partial_\varphi\mathbf{u} = \mathrm{i}\,l\,\boldsymbol{U}(r,x_3)\,\mathrm{e}^{\mathrm{i}l\varphi} = \mathrm{i}\,l\,\mathbf{u}$. Hence

$$(\mathscr{A}_k^\omega)_l\mathbf{u} \;=\; (\mathscr{A}_k^0)_l\mathbf{u} + \omega\,\partial_\varphi\mathbf{u} \;=\; (\mathscr{A}_k^0)_l\mathbf{u} + \mathrm{i}k\omega\,\mathbf{u}. \tag{8.85}$$

This shows that $(\mathscr{A}_k^\omega)_l$ is a closed and densely defined operator in $\boldsymbol{L}^2_\sigma(\mathcal{D})_l$. The identity (8.85) and Lemma 8.13 imply that

$$\mathrm{Sp}((\mathscr{A}_k^\omega)_l) \;=\; \mathrm{Sp}_{\mathrm{ess}}((\mathscr{A}_k^\omega)_l) \;=\; \{\lambda \in \mathbb{C};\ \lambda - \mathrm{i}k\omega \in \Lambda_k^0\}. \tag{8.86}$$

The next theorem provides an information on the spectrum of the full operator $\mathscr{A}_k^\omega$.

Theorem 8.13 *Let domain $\mathcal{D}$ be axially symmetric about the x_3-axis. Then the spectrum of the Oseen type operator $\mathscr{A}_k^\omega$ in $\boldsymbol{L}^2_\sigma(\mathcal{D})$ satisfies $Sp(\mathscr{A}_k^\omega) = Sp_{ess}(\mathscr{A}_k^\omega) = \Lambda_k^\omega$, where*

$$\Lambda_k^\omega := \bigcup_{l\in\mathbb{Z}} \Big\{\lambda = \alpha + i\beta + il\omega \in \mathbb{C};\ \alpha,\beta \in \mathbb{R},\ \alpha \leq -\frac{\nu\beta^2}{\gamma^2}\Big\}. \tag{8.87}$$

Proof Each operator $(\mathscr{A}_k^{\omega})_l$, for $l \in \mathbb{Z}$, is a part of $\mathscr{A}_k^{\omega}$, hence $\mathrm{Sp_{ess}}((\mathscr{A}_k^{\omega})_l) \subset \mathrm{Sp_{ess}}(\mathscr{A}_k^{\omega})$. Thus, $\cup_{l\in\mathbb{Z}} \mathrm{Sp_{ess}}((\mathscr{A}_k^{\omega})_l) = \Lambda_k^{\omega} \subset \mathrm{Sp_{ess}}(\mathscr{A}_k^{\omega}) \subset \mathrm{Sp}(\mathscr{A}_k^{\omega})$.

It remains to prove the opposite inclusion, i.e., that $\mathrm{Sp}(\mathscr{A}_k^{\omega}) \subset \Lambda_k^{\omega}$ or, equivalently, that $(\mathbb{C} \setminus \Lambda_k^{\omega}) \subset \rho(\mathscr{A}_k^{\omega})$. Suppose that $\lambda \equiv \alpha + \mathrm{i}\beta \in \mathbb{C} \setminus \Lambda_k^{\omega}$. We will show that the operator $\mathscr{A}_k^{\omega} - \lambda I$ has a bounded inverse in $\boldsymbol{L}^2_{\sigma}(\mathcal{D})$. Let

$$\mathbf{f}(r,\varphi,x_3) = \sum_{l=-\infty}^{\infty} \mathbf{f}_l(r,x_3)\,\mathrm{e}^{\mathrm{i}l\varphi} \in \boldsymbol{L}^2_{\sigma}(\mathcal{D}),$$

where $\mathbf{f}_l\,\mathrm{e}^{\mathrm{i}l\varphi} \in \boldsymbol{L}^2_{\sigma}(\mathcal{D})_l$, be given. Let us at first deal with the equation $((\mathscr{A}_k^{\omega})_l - \lambda I)\mathbf{w}_l = \mathbf{f}_l\,\mathrm{e}^{\mathrm{i}l\varphi}$ in $\boldsymbol{L}^2_{\sigma}(\mathcal{D})_l$ for the unknown $\mathbf{w}_l$. Put $\mathbf{w}_l = \mathbf{u}_l\,\mathrm{e}^{\mathrm{i}l\varphi}$, where $\mathbf{u}_l$ is a new unknown function. Using (8.85), we observe that this equation is equivalent to

$$(A_k^0)_l(\mathbf{u}_l\,\mathrm{e}^{\mathrm{i}l\varphi}) - (\alpha + \mathrm{i}\beta - \mathrm{i}k\omega)(\mathbf{u}_l\,\mathrm{e}^{\mathrm{i}l\varphi}) = \mathbf{f}_l\,\mathrm{e}^{\mathrm{i}l\varphi}. \tag{8.88}$$

Due to Lemma 8.13 we have $\alpha + \mathrm{i}\beta - \mathrm{i}k\omega \in \rho((\mathscr{A}_k^0)_l)$ for all $l \in \mathbb{Z}$. Since the operator $-(\mathscr{A}_k^0)_l$ is sectorial, we deduce from resolvent estimates for sectorial operators, see, e.g., D. Henry [132], p. 23, that there exists a constant $M > 0$, independent of l, such that

$$\|\mathbf{u}_l\|_2 = \|\mathbf{u}_l\,\mathrm{e}^{\mathrm{i}l\varphi}\|_2 \le \frac{M}{1+|l|}\,\|\mathbf{f}_l\,\mathrm{e}^{\mathrm{i}l\varphi}\|_2 = \frac{M}{1+|l|}\,\|\mathbf{f}_l\|_2\,. \tag{8.89}$$

Then the series $\sum_{l=-\infty}^{\infty} \mathbf{u}_l\ \mathrm{e}^{\mathrm{i}l\varphi}$ converges in $\boldsymbol{L}^2_{\sigma}(\mathcal{D})$ and its sum $\mathbf{u}$ satisfies

$$\|\mathbf{u}\|_2^2 = \sum_{l=-\infty}^{\infty} \|\mathbf{u}_l\|_2^2 \le \sum_{l=-\infty}^{\infty} \frac{M^2}{(1+|l|)^2}\,\|\mathbf{f}_l\|_2^2 \le M^2\,\|\mathbf{f}\|_2^2\,. \tag{8.90}$$

Using Eq. (8.88) and the estimate (8.89), we get

$$\big\|(\mathscr{A}_k^{\omega})_l(\mathbf{u}_l\,\mathrm{e}^{\mathrm{i}l\varphi})\big\|_2 \le \|\mathbf{f}_l\,\mathrm{e}^{\mathrm{i}l\varphi}\|_2 + |\alpha+\mathrm{i}\beta|\,\|\mathbf{u}_l\,\mathrm{e}^{\mathrm{i}l\varphi}\|_2 \le c\,\|\mathbf{f}_l\,\mathrm{e}^{\mathrm{i}l\varphi}\|_2,$$

where c is independent of l. Using these inequalities and the closedness of the operator $\mathscr{A}_k^{\omega}$, we deduce that $\mathbf{u} \in D(\mathscr{A}_k^{\omega})$ and $(\mathscr{A}_k^{\omega} - \lambda I)\mathbf{u} = \mathbf{f}$. This, together with (8.90), completes the proof. □

Note that the set Λ_k^{ω} is a union of infinitely many equally shifted filled parabolas in the left half-plane of $\mathbb{C}$.

8.4.1.2 The Case of a General Exterior Domain $\mathcal{D}$

Let $\mathcal{B} \neq \emptyset$ (which means that $\mathcal{D} \equiv \mathbb{R}^3 \setminus \mathcal{B} \neq \mathbb{R}^3$). Recall that $R_0 = \max\{|\mathbf{x}|;\ \mathbf{x} \in \mathcal{B}\}$ and $\mathcal{D}_R = \mathcal{D} \cap B_R(\mathbf{0})$.

Lemma 8.14 *Let $\lambda \in Sp_C^{Gl}(\mathscr{A}_k^{\omega})$. Then there exists $R > R_0$ and a non-compact sequence $\{\mathbf{u}^n\}$ in $D(\mathscr{A}_k^{\omega})$ such that $\|\mathbf{u}^n\|_2 = 1$, $\mathbf{u}^n = \mathbf{0}$ in $\mathcal{D}_R$ and*

$$(\mathscr{A}_k^{\omega} - \lambda I)\,\mathbf{u}^n \longrightarrow 0 \quad \text{in } \boldsymbol{L}^2_{\sigma}(\mathcal{D}) \quad \text{for } n \to \infty. \tag{8.91}$$

Proof The condition $\lambda \in \mathrm{Sp}_{\mathrm{C}}^{\mathrm{Gl}}(\mathscr{A}_k^{\omega})$ means that $\mathrm{nul}'(\mathscr{A}_k^{\omega} - \lambda I) = \infty$. Then there exists an orthonormal sequence $\{\mathbf{v}^n\}$ in $\boldsymbol{L}^2_{\sigma}(\mathcal{D})$ such that

$$(\mathscr{A}_k^{\omega} - \lambda I)\,\mathbf{v}^n = \boldsymbol{\epsilon}^n \longrightarrow 0 \quad \text{in } \boldsymbol{L}^2_{\sigma}(\mathcal{D}) \quad \text{for } n \to \infty. \tag{8.92}$$

The construction of the sequence $\{\mathbf{v}^n\}$ is based on the same arguments as in the proof of Lemma 8.6. Obviously, $\mathbf{v}^n \rightharpoonup \mathbf{0}$ in $\boldsymbol{L}^2_{\sigma}(\mathcal{D})$. Using (8.92) and the estimate (8.2), we deduce that the sequence $\{\mathbf{v}^n\}$ is bounded in $\boldsymbol{W}_0^{1,2}(\mathcal{D}) \cap \boldsymbol{W}^{2,2}(\mathcal{D})$ and the sequence $\{\omega(\mathbf{e}_3 \wedge \mathbf{x}) \cdot \nabla \mathbf{v}^n\}$ is bounded in $\boldsymbol{L}^2_{\sigma}(\mathcal{D})$. Hence there exists a subsequence, which we again denote by $\{\mathbf{v}^n\}$, such that $\mathbf{v}^n \rightharpoonup \mathbf{0}$ in $\boldsymbol{W}_0^{1,2}(\mathcal{D}) \cap \boldsymbol{W}^{2,2}(\mathcal{D})$ and $\omega(\mathbf{e}_3 \wedge \mathbf{x}) \cdot \nabla \mathbf{v}^n \rightharpoonup \mathbf{0}$ in $\boldsymbol{L}^2_{\sigma}(\mathcal{D})$. Suppose that $R \geq R_0 + 3$ is a fixed number. The compact imbedding of $\boldsymbol{W}^{2,2}(\mathcal{D}_R)$ into $\boldsymbol{W}^{1,2}(\mathcal{D}_R)$ yields

$$\mathbf{v}^n \longrightarrow \mathbf{0} \qquad \text{strongly in } \boldsymbol{W}^{1,2}(\mathcal{D}_R). \tag{8.93}$$

The identity $(\mathscr{A}_k^{\omega} - \lambda I)\mathbf{v}^n = \boldsymbol{\epsilon}^n$ means that

$$\nu \Delta \mathbf{v}^n + \omega(\mathbf{e}_3 \wedge \mathbf{x}) \cdot \nabla \mathbf{v}^n - \omega \mathbf{e}_3 \wedge \mathbf{v}^n - k\,\partial_3 \mathbf{v}^n - \lambda \mathbf{v}^n + \nabla q^n = \boldsymbol{\epsilon}^n, \tag{8.94}$$

where q^n is an appropriate scalar function. It follows from (8.92), (8.94) and the aforementioned weak convergence of $\{\mathbf{v}^n\}$ and $\{\omega(\mathbf{e}_3 \wedge \mathbf{x}) \cdot \nabla \mathbf{v}^n\}$ that $\nabla q^n \to \mathbf{0}$ weakly in $\boldsymbol{L}^2(\mathcal{D})$. Thus, the functions q^n, which are given uniquely up to an additive constant by (8.94), can be chosen so that $q^n \to q \equiv const.$ strongly in $L^2(\mathcal{D}_R)$. Denote by η an infinitely differentiable cutoff function in $\mathcal{D}$ such that

$$\eta(\mathbf{x}) = \begin{cases} 0 & \text{if} \quad |\mathbf{x}| < R - 2, \\ 1 & \text{if} \quad |\mathbf{x}| > R - 1, \end{cases}$$

and $0 \leq \eta(\mathbf{x}) \leq 1$ if $R - 2 \leq |\mathbf{x}| \leq R - 1$. Put $\mathbf{u}^n = \eta \mathbf{v}^n - \boldsymbol{V}_n$, where $\operatorname{div} \boldsymbol{V}_n = \nabla \eta \cdot \mathbf{v}^n$. Although $\boldsymbol{V}_n$ is not given uniquely, the results on solutions of the equation $\operatorname{div} \boldsymbol{V} = \mathbf{f}$, see, e.g., [18], show that the function $\boldsymbol{V}_n$ can be chosen so that $\operatorname{supp} \boldsymbol{V}_n \subset \{\mathbf{x} \in \mathcal{D};\ R - 3 < |\mathbf{x}| < R\}$ and there exist $c > 0$ such that

$$\|\boldsymbol{V}_n\|_{2,2} = \|\boldsymbol{V}_n\|_{2,2;\,\mathcal{D}_R} \leq c\,\|\nabla \eta \cdot \mathbf{v}^n\|_{1,2;\,\mathcal{D}_R} \longrightarrow 0 \tag{8.95}$$

as $n \to \infty$. The function $\mathbf{u}^n$ is divergence-free, equals $\mathbf{0}$ in $\mathcal{D}_{R-3}$, equals $\mathbf{v}^n$ in $\mathcal{D} \setminus \mathcal{D}_R$, and belongs to $\boldsymbol{L}^2(\mathcal{D})$. Due to the properties of the functions η and $\boldsymbol{V}_n$ we get $\mathbf{u}^n \in D(\mathscr{A}_k^{(\omega)})$. Obviously $\mathbf{u}^n$ satisfies

$$
\begin{aligned}
&\nu\Delta\mathbf{u}^n + \omega(\mathbf{e}_3 \wedge \mathbf{x}) \cdot \nabla\mathbf{u}^n - \omega\mathbf{e}_3 \wedge \mathbf{u}^n - k\,\partial_3\mathbf{u}^n - \lambda\mathbf{u}^n + \nabla(\eta q^n) \\
&= \eta\left[\nu\Delta\mathbf{v}^n + \omega(\mathbf{e}_3 \wedge \mathbf{x}) \cdot \nabla\mathbf{v}^n - \omega\mathbf{e}_3 \wedge \mathbf{v}^n - k\,\partial_3\mathbf{u}_v - \lambda\mathbf{v}^n\right] + 2\nu\nabla\eta \cdot \nabla\mathbf{v}^n \\
&\quad + \nu(\Delta\eta)\mathbf{v}^n - \nu\Delta\boldsymbol{V}_n + \left[\omega(\mathbf{e}_3 \wedge \mathbf{x}) \cdot \nabla\eta\right]\mathbf{v}^n - k\,(\partial_3\eta)\,\mathbf{v}^n - \omega(\mathbf{e}_3 \wedge \mathbf{x}) \cdot \nabla\boldsymbol{V}_n \\
&\quad + \omega\mathbf{e}_3 \wedge \boldsymbol{V}_n + k\,\partial_3\boldsymbol{V}_n + \lambda\boldsymbol{V}_n + \nabla(\eta q^n) \\
&= \eta\boldsymbol{\epsilon}^n + 2\nu\nabla\eta \cdot \nabla\mathbf{v}^n + \nu(\Delta\eta)\mathbf{v}^n - \nu\Delta\boldsymbol{V}_n + \left[\omega(\mathbf{e}_3 \wedge \mathbf{x}) \cdot \nabla\eta\right]\mathbf{v}^n - k\,(\partial_3\eta)\,\mathbf{v}^n \\
&\quad - \omega(\mathbf{e}_3 \wedge \mathbf{x}) \cdot \nabla\boldsymbol{V}_n + \omega\mathbf{e}_3 \wedge \boldsymbol{V}_n + k\,\partial_3\boldsymbol{V}_n + \lambda\boldsymbol{V}_n + (\nabla\eta)q^n,
\end{aligned}
\tag{8.96}
$$

where $\eta\boldsymbol{\epsilon}^n \to \mathbf{0}$ in $\boldsymbol{L}^2(\mathcal{D})$ due to (8.92), and $\nu[2\nabla\eta \cdot \nabla\mathbf{v}^n + (\Delta\eta)\mathbf{v}^n] \to \mathbf{0}$ in $\boldsymbol{L}^2(\mathcal{D})$ because $\nabla\eta$ and $\Delta\eta$ are supported in $\mathcal{D}_R$ and due to (8.93). The terms $\left[\omega(\mathbf{e}_3 \wedge \mathbf{x}) \cdot \nabla\eta\right]\mathbf{v}^n$ and $k\,(\partial_3\eta)\,\mathbf{v}^n$ also tend to zero in $\boldsymbol{L}^2(\mathcal{D})$ for the same reasons. Furthermore, all terms involving $\boldsymbol{V}_n$ tend to $\mathbf{0}$ in $\boldsymbol{L}^2(\mathcal{D})$ due to (8.95). Finally, $(\nabla\eta)q^n \to \mathbf{0}$ in $\boldsymbol{L}^2(\mathcal{D})$ because $q^n \to 0$ in $L^2(\mathcal{D}_R)$ and $\nabla\eta$ is supported in $\mathcal{D}_R$. Thus,

$$
\nu\Delta\mathbf{u}^n + \omega(\mathbf{e}_3 \wedge \mathbf{x}) \cdot \nabla\mathbf{u}^n - \omega\mathbf{e}_3 \wedge \mathbf{u}^n - k\,\partial_3\mathbf{u}^n - \lambda\mathbf{u}^n + \nabla(\eta q^n) \longrightarrow \mathbf{0}
$$

in $\boldsymbol{L}^2(\mathcal{D})$ for $n \to \infty$. Hence $\{\mathbf{u}^n\}$ satisfies (8.91). Moreover, we have

$$
\|\mathbf{u}^n\|_2^2 \geq \int_{|x|>R} |\mathbf{u}^n(\mathbf{x})|^2\,\mathrm{d}\mathbf{x} = \int_{|x|>R} |\mathbf{v}^n(\mathbf{x})|^2\,\mathrm{d}\mathbf{x} \longrightarrow 1 \quad \text{for } n \to \infty
$$

because $\|\mathbf{v}^n\|_2 = 1$ and due to (8.93). If we divide each unction $\mathbf{u}^n$ by its norm in $\boldsymbol{L}^2_\sigma(\mathcal{D})$ and denote the new function again by $\mathbf{u}^n$, we obtain the sequence $\{\mathbf{u}^n\}$ with all the properties stated in Lemma 8.14. Finally, the orthonormality of $\{\mathbf{v}^n\}$ and (8.93) imply the non-compactness of the sequence $\{\mathbf{u}^n\}$. □

Denote by $\widehat{\mathscr{A}_k^{\omega}}$ the operator which is defined in the same way as $\mathscr{A}_k^{(\omega)}$, however acting on the whole space $\mathbb{R}^3$ instead of the exterior domain $\mathcal{D} \subset \mathbb{R}^3$. Obviously, the operator $\widehat{\mathscr{A}_k^{\omega}}$ has all the properties derived in Sect. 8.4.1.1. In particular, $\mathrm{Sp}(\widehat{\mathscr{A}_k^{(\omega)}}) = \mathrm{Sp}_{\mathrm{ess}}(\widehat{\mathscr{A}_k^{\omega}}) = \Lambda_k^{\omega}$.

Lemma 8.15 $Sp_C^{Gl}(\widehat{\mathscr{A}_k^{\omega}}) = Sp_C^{Gl}(\mathscr{A}_k^{(\omega)}) = \Lambda_k^{\omega}$. □

Proof Suppose that $\lambda \in \mathrm{Sp}_{\mathrm{C}}^{\mathrm{Gl}}(\mathscr{A}_k^{(\omega)})$. Let $R > 0$ and $\{\mathbf{u}^n\}$ be a number and a sequence, respectively, with the properties named in Lemma 8.14. All functions $\mathbf{u}^n$, extended by zero from $\mathcal{D}$ to the whole space $\mathbb{R}^3$, belong to the domain of operator $\widehat{\mathscr{A}_k^{\omega}}$. Thus, (8.91) shows that $\lambda \in \mathrm{Sp}_{\mathrm{C}}^{\mathrm{Gl}}(\widehat{\mathscr{A}_k^{\omega}})$. On the other hand, if $\lambda \in \mathrm{Sp}_{\mathrm{C}}^{\mathrm{Gl}}(\widehat{\mathscr{A}_k^{\omega}})$, then we can use analogous arguments and prove that λ also belongs to $\mathrm{Sp}_{\mathrm{C}}^{\mathrm{Gl}}(\mathscr{A}_k^{(\omega)})$. □

Theorem 8.14 *Let $\mathcal{D}$ be a general exterior domain in $\mathbb{R}^3$ with a $C^{1,1}$-boundary $\partial\mathcal{D}$. Then:*

(i) $Sp_{ess}(\mathscr{A}_k^{\omega}) = \Lambda_k^{\omega}$.
(ii) $\mathbb{C} \setminus \Lambda_k^{\omega}$ *is a subset of* $\rho(\mathscr{A}_k^{\omega})$, *with a possible exception of at most a countable set of isolated eigenvalues of* $\mathscr{A}_k^{\omega}$ *with finite algebraic multiplicities, which can possibly cluster only at points on the boundary of* Λ_k^{ω}. *Each such eigenvalue has a negative real part.*

□

Proof Let us at first show that $\mathrm{Sp}_\mathrm{c}^\mathrm{Gl}(\mathscr{A}_k^{\omega}) \subset \mathrm{Sp}_\mathrm{ess}(\mathscr{A}_k^{\omega})$. For $\lambda \in \mathrm{Sp}_\mathrm{c}^\mathrm{Gl}(\mathscr{A}_k^{\omega})$ we have that $\mathrm{nul}'(\mathscr{A}_k^{\omega} - \lambda I) = \infty$. Moreover, $\overline{\lambda} \in \mathrm{Sp}_\mathrm{c}^\mathrm{Gl}((\mathscr{A}_k^{\omega})^*) = \mathrm{Sp}_\mathrm{c}^\mathrm{Gl}(A_{-\gamma}^{-\omega})$, so that $\mathrm{nul}'((\mathscr{A}_k^{\omega})^* - \overline{\lambda} I) = \infty$. Hence $\mathrm{def}'(\mathscr{A}_k^{\omega} - \lambda I) = \infty$, which shows that $\lambda \in \mathrm{Sp}_\mathrm{ess}(\mathscr{A}_k^{\omega})$, see [137], p. 234. Since the opposite inclusion $\mathrm{Sp}_\mathrm{ess}(\mathscr{A}_k^{\omega}) \subset \mathrm{Sp}_\mathrm{c}^\mathrm{Gl}(\mathscr{A}_k^{\omega})$ is trivial, see Sect. 8.2, we obtain the equality $\mathrm{Sp}_\mathrm{c}^\mathrm{Gl}(\mathscr{A}_k^{\omega}) = \mathrm{Sp}_\mathrm{ess}(\widehat{\mathscr{A}_k^{\omega}})$. Due to Lemma 8.15, we get $\mathrm{Sp}_\mathrm{ess}(\mathscr{A}_k^{\omega}) = \Lambda_k^{\omega}$.

Since $\mathrm{ind}(\mathscr{A}_k^{\omega} - \lambda I) = 0$ in $\mathbb{C} \setminus \mathrm{Sp}_\mathrm{ess}(\mathscr{A}_k^{\omega})$ and due to [137, paragraph IV.5.6], $\mathbb{C} \setminus \mathrm{Sp}_\mathrm{ess}(\mathscr{A}_k^{\omega})$ contains at most countably many eigenvalues λ of $\mathscr{A}_k^{\omega}$, each of whose has a finite algebraic multiplicity and which can cluster only on the boundary of $\mathbb{C} \setminus \mathrm{Sp}_\mathrm{ess}(\mathscr{A}_k^{\omega})$. Moreover, $0 < \mathrm{nul}(\mathscr{A}_k^{\omega} - \lambda I) = \mathrm{def}(\mathscr{A}_k^{\omega} - \lambda I) < \infty$ holds at each of them.

If $\lambda = \alpha + \mathrm{i}\beta$ is an eigenvalue of $\mathscr{A}_k^{\omega}$ and $\mathbf{v}$ is a corresponding eigenfunction, then, multiplying the equation $\mathscr{A}_k^{\omega}\mathbf{v} = \lambda\,\mathbf{v}$ by $\overline{\mathbf{v}}$ and integrating on $\mathcal{D}$, we obtain the identity $-\nu\,\|\nabla\mathbf{v}\|_2^2 = \alpha\,\|\mathbf{v}\|_2^2$, by analogy with the part of the proof of Theorem 8.12, which leads to (8.75). This shows that $\mathrm{Re}\,\lambda < 0$. □

Theorem IV.5.35 in [137] and Lemma 8.3 imply that the essential spectrum of the full operator $\mathscr{L}_k^{\omega}$ is the same as $\mathrm{Sp}_\mathrm{ess}(\mathscr{A}_k^{\omega})$. (Recall that $\mathscr{L}_k^{\omega}$ is defined by formula (8.16).) Moreover, since $\mathrm{ind}(\mathscr{L}_k^{\omega} - \lambda I) = 0$ in $\mathbb{C} \setminus \mathrm{Sp}_\mathrm{ess}(\mathscr{L}_k^{\omega})$, we can apply [137, Theorem IV.5.31] and use the same arguments as above, so that we obtain the next theorem:

Theorem 8.15 *Let $\mathcal{D}$ be a general exterior domain in $\mathbb{R}^3$ with a $C^{1,1}$-boundary $\partial\mathcal{D}$. Then the spectrum of the operator $\mathscr{L}_k^{\omega}$ in $\mathbf{L}_\sigma^2(\mathcal{D})$ satisfies:*

(i) $Sp_{ess}(\mathscr{L}_k^{\omega}) = Sp_{ess}(\mathscr{A}_k^{\omega}) = \Lambda_k^{\omega}$.
(ii) $Sp(\mathscr{L}_k^{\omega}) = Sp_{ess}(\mathscr{L}_k^{\omega}) \cup \Gamma$, *where* Γ *consists of an at most countable set of isolated eigenvalues of* $\mathscr{L}_k^{\omega}$, *which can possibly cluster only at points of* $Sp(\mathscr{L}_k^{\omega})$. *Each of them has a finite algebraic multiplicity.*

□

8.4.2 L^q-Theory

The content of this subsection is based on the paper [77] by Farwig and Neustupa.

Let $1 < q < \infty$. By analogy with $D(\mathscr{A}_0^{(\omega)})$, we treat the domain $D(\mathscr{A}_k^{(\omega)})$ as a Banach space with the norm $\|\mathbf{v}\|_{D(\mathscr{A}_k^{(\omega)})} := \|\mathbf{v}\|_{2,q} + \|(\boldsymbol{\omega}\wedge\mathbf{x})\cdot\nabla\mathbf{v}\|_q$. Our approach is based on the Fourier transform in the whole space and on the transfer of results to the exterior domain. The same idea has already been used in Sect. 8.3.2 in the case $k = 0$.

8.4.2.1 The Case $\mathcal{D} = \mathbb{R}^3$

The next lemma provides an information on solutions of the resolvent equation

$$\mathscr{A}_k^{(\omega)}\mathbf{v} - \lambda\mathbf{v} = \mathbf{f} \tag{8.97}$$

for $\lambda \in \mathbb{C}\setminus\Lambda_k^{\omega}$ and $\mathbf{f}\in \boldsymbol{L}^q_\sigma(\mathbb{R}^3)$. Recall that the set Λ_k^{ω} is defined in (8.87).

Lemma 8.16 *Suppose that* $\lambda\in\mathbb{C}\setminus\Lambda_k^{\omega}$. *There exists a constant* $c = c(\lambda, q) > 0$ *such that if* $\mathbf{f}\in\boldsymbol{L}^q_\sigma(\mathbb{R}^3)$ *and* $\mathbf{v}\in D(\mathscr{A}_k^{(\omega)})$ *satisfy the resolvent Eq. (8.97), then*

$$\|\mathbf{v}\|_q \le c\,\|\mathbf{f}\|_q. \tag{8.98}$$

Proof The linear space $D_0(\mathscr{A}_k^{(\omega)})$ is a core of $\mathscr{A}_k^{\omega} - \lambda I$, see Theorem 8.2. Thus, it is sufficient to prove (8.98) only for $\mathbf{v}\in D_0(\mathscr{A}_k^{(\omega)})$.

Due to Lemma 8.2, Eq. (8.97) can be written in the form

$$\nu\Delta\mathbf{v} + (\boldsymbol{\omega}\wedge\mathbf{x})\cdot\nabla\mathbf{v} - \boldsymbol{\omega}\wedge\mathbf{v} - k\,\partial_3\mathbf{v} - \lambda\mathbf{v} = \mathbf{f}. \tag{8.99}$$

Due to the geometry of the problem, it is reasonable to use the cylindrical coordinates (r, φ, x_3) in $\mathbb{R}^3$; then $r^2 = x_1^2 + x_2^2$. The term $(\boldsymbol{\omega}\wedge\mathbf{x})\cdot\nabla\mathbf{v}$ in (8.99), which equals $\omega(-x_2\,\partial_1\mathbf{v} + x_1\,\partial_2\mathbf{v})$, can be simplified to

$$(\boldsymbol{\omega}\wedge\mathbf{x})\cdot\nabla\mathbf{v} = \omega\,\partial_\varphi\mathbf{v}. \tag{8.100}$$

We use the same notation as in Sect. 8.3.2, i.e., $\mathcal{F}$ is the Fourier transform , $\mathcal{F}_{-1}$ is the inverse Fourier transform, $\widehat{\ }$ denotes Fourier images of functions, $\boldsymbol{\xi} = (\xi_1, \xi_2, \xi_3)$ denotes their Cartesian variables, and $s = |\boldsymbol{\xi}|$. Applying $\mathcal{F}$ to (8.99), we obtain

$$(\lambda + \mathrm{i}k\xi_3 + \nu s^2)\widehat{\mathbf{v}} - \omega\,\partial_\varphi\widehat{\mathbf{v}} + \boldsymbol{\omega}\wedge\widehat{\mathbf{v}} = -\widehat{\mathbf{f}}. \tag{8.101}$$

Here, $\partial_\varphi\widehat{\mathbf{v}}$ denotes the angular derivative

$$\partial_\varphi\widehat{\mathbf{v}} = (\mathbf{e}_3\wedge\boldsymbol{\xi})\cdot\nabla\widehat{\mathbf{v}} \equiv -\xi_2\,\frac{\partial\widehat{\mathbf{v}}}{\partial\xi_1} + \xi_1\,\frac{\partial\widehat{\mathbf{v}}}{\partial\xi_2}$$

when using cylindrical coordinates (r, φ, ξ_3) in the space of the Fourier variables. The equation $\operatorname{div} \mathbf{v} = 0$ (following from the fact that $\mathbf{v} \in D(\mathscr{A}_k^{(\omega)})$) leads to the condition $\mathrm{i}\boldsymbol{\xi} \cdot \widehat{\mathbf{v}} = 0$. The function $\widehat{\mathbf{v}}$ can be considered to be a solution of the first order ordinary differential equation (8.101) with respect to the angular variable φ. Writing $\widehat{\mathbf{v}}$ in the form

$$\widehat{\mathbf{v}}(r, \varphi, \xi_3) \;=\; Q(-\varphi)\, \widehat{\mathbf{w}}(r, \varphi, \xi_3),$$

where Q is the matrix from (1.17), one verifies that

$$\omega\, \partial_\varphi \widehat{\mathbf{v}} \;=\; Q(-\varphi)\, \omega\, \partial_\varphi \widehat{\mathbf{w}} + \boldsymbol{\omega} \wedge [Q(-\varphi)\widehat{\mathbf{w}}].$$

Hence (8.101) is equivalent to the equation

$$-\,\omega\, \partial_\varphi \widehat{\mathbf{w}} + (\lambda + \mathrm{i}k\xi_3 + \nu s^2)\widehat{\mathbf{w}} \;=\; -Q(\varphi)\, \widehat{\mathbf{f}}, \tag{8.102}$$

or using the definition

$$a(\boldsymbol{\xi}) = \lambda + \mathrm{i}k\xi_3 + \nu s^2, \quad s = |\boldsymbol{\xi}|,$$

to the inhomogeneous ordinary first order linear differential equation with respect to φ

$$\omega\, \partial_\varphi \widehat{\mathbf{w}} - a(\boldsymbol{\xi})\widehat{\mathbf{w}} = Q(\varphi)\widehat{\mathbf{f}}.$$

Its solution $\widehat{\mathbf{w}}$ satisfies

$$\widehat{\mathbf{w}}(r, \varphi + 2\pi, \xi_3)$$
$$= \mathrm{e}^{2\pi\, a(\boldsymbol{\xi})/\omega}\, \widehat{\mathbf{w}}(r, \varphi, \xi_3)] + \frac{1}{\omega} \int_0^{2\pi} \mathrm{e}^{(2\pi - t)a(\boldsymbol{\xi})/\omega}\, Q(t + \varphi)\, \widehat{\mathbf{f}}(r, t + \varphi, \xi_3)\, \mathrm{d}t.$$

Since $\widehat{\mathbf{w}}$ is 2π-periodic in the variable φ, we have

$$\widehat{\mathbf{w}}(r, \varphi, \xi_3) \;=\; \frac{1}{\omega} \int_0^{2\pi} \frac{\mathrm{e}^{(2\pi - t)a(\boldsymbol{\xi})/\omega}}{1 - \mathrm{e}^{2\pi\, a(\boldsymbol{\xi})/\omega}}\, Q(t + \varphi)\, \widehat{\mathbf{f}}(r, t + \varphi, \xi_3)\, \mathrm{d}t$$

and consequently

$$\widehat{\mathbf{v}}(r, \varphi, \xi_3) \;=\; \frac{1}{\omega} \int_0^{2\pi} \frac{\mathrm{e}^{-ta(\boldsymbol{\xi})/\omega}}{\mathrm{e}^{-2\pi\, a(\boldsymbol{\xi})/\omega} - 1}\, Q(t)\, \widehat{\mathbf{f}}(r, t + \varphi, \xi_3)\, \mathrm{d}t. \tag{8.103}$$

Returning to the Cartesian variables $\boldsymbol{\xi} = (\xi_1, \xi_2, \xi_3)$ in (8.103), we obtain

$$\begin{aligned}\widehat{\mathbf{v}}(\boldsymbol{\xi}) &= \frac{1}{\omega} \int_0^{2\pi} \Psi(\lambda, \boldsymbol{\xi}, t)\; Q(t)\widehat{\mathbf{f}}\big(Q(-t)\boldsymbol{\xi}\big)\, \mathrm{d}t \\ &= \frac{1}{\omega} \int_0^{2\pi} \Psi(\lambda, \boldsymbol{\xi}, t)\; Q(t)\, \mathcal{F}[\mathbf{f}(Q(-t)\cdot)](\boldsymbol{\xi})\, \mathrm{d}t, \end{aligned} \tag{8.104}$$

where

$$\Psi(\lambda, \boldsymbol{\xi}, t) \;=\; \frac{\mathrm{e}^{-ta(\boldsymbol{\xi})/\omega}}{\mathrm{e}^{-2\pi\, a(\boldsymbol{\xi})/\omega} - 1}.$$

In order to complete the proof, we need the next lemma. □

Lemma 8.17 *If $\lambda \notin \Lambda_k^{\omega}$, then there exists $c > 0$, depending only on k, ω and on the position of λ in $\mathbb{C} \setminus \Lambda_k^{\omega}$, such that*

$$\left| e^{-2\pi\,(\lambda + ik\xi_3 + \nu s^2)/\omega} - 1 \right| \;\geq\; c \qquad \textit{for all } \boldsymbol{\xi} \in \mathbb{R}^3.$$

Proof The modulus of $\mathrm{e}^{-2\pi\, a(\boldsymbol{\xi})/\omega} - 1$ is bounded below by a positive constant depending only on λ, k and ω (i.e., independent of $\boldsymbol{\xi}$) if the distance $d_\lambda(\boldsymbol{\xi}, l)$ of the complex number $-2\pi\, a(\boldsymbol{\xi})/\omega$ from $2\pi\, \mathrm{i}l$ (for $l \in \mathbb{Z}$) is bounded below by another positive constant, independent of $\boldsymbol{\xi}$ and l. Let $\lambda = \alpha + \mathrm{i}\beta$ and $r^2 = \xi_1^2 + \xi_2^2$ so that

$$\begin{aligned} d_\lambda^2(\boldsymbol{\xi}, l) &= \frac{4\pi^2}{\omega^2}\, \big|\alpha + \mathrm{i}\beta + \mathrm{i}k\xi_3 + \nu\xi_3^2 + \nu r^2 + \mathrm{i}l\omega\big|^2 \\ &= \frac{4\pi^2}{\omega^2}\, \big(\alpha + \nu\xi_3^2 + \nu r^2\big)^2 + \frac{4\pi^2}{\omega^2}\, \big(\beta + k\xi_3 + l\omega\big)^2. \end{aligned}$$

Assume that $\min_{\xi \in \mathbb{R}^3} d_\lambda^2(\boldsymbol{\xi}, l) = 0$ for some fixed $l \in \mathbb{Z}$; the minimum exists because $\lim_{|\xi| \to \infty} d_\lambda^2(\boldsymbol{\xi}, k) = \infty$. Then $\xi_3 = -(\beta + l\omega)/k$ and

$$\alpha \;=\; -\nu r^2 - \frac{\nu}{k^2}\, (\beta + l\omega)^2 \;\leq\; -\frac{\nu}{k^2}\, (\beta + l\omega)^2,$$

but this inequality contradicts with $\lambda = \alpha + \mathrm{i}\beta \notin \Lambda_k^{\omega}$. Hence

$$D_\lambda(l) \;:=\; \min_{\xi \in \mathbb{R}^3} d_\lambda(\boldsymbol{\xi}, l) \;>\; 0$$

for every $l \in \mathbb{Z}$. Moreover, the sequence $\{D_\lambda(l)\}_{l \in \mathbb{Z}}$ does not converge to zero as $l \to \infty$ or $l \to -\infty$, because this would imply the existence of $\boldsymbol{\xi}(l) \in \mathbb{R}^3$ such that $d_\lambda\big(\boldsymbol{\xi}(l), l\big) \to 0$ as $l \to \infty$ or $l \to -\infty$. Consequently $\beta + k\xi_3(k) + l\omega \to 0$,

$|\xi_3(l)| \to \infty$ and $\alpha + \nu\xi_3(k)^2 + \nu r(k)^2 \to \infty$, which is in contradiction with $d_\lambda\big(\boldsymbol{\xi}(l), l\big) \to 0$. Thus Lemma 8.17 is proved. □

Continuation of the Proof of Lemma 8.16 Lemma 8.17 implies that, for fixed $\lambda \in \mathbb{C} \setminus \Lambda_k^\omega$, the modulus of $\Psi(\lambda, \boldsymbol{\xi}, t)$ is bounded uniformly with respect to $t \in (0, 2\pi)$ and $\boldsymbol{\xi} \in \mathbb{R}^3$. Further, if $i \in \{1, 2\}$ and $j \in \{1, 2, 3\}$, then

$$\frac{\partial\Psi}{\partial\xi_i} = \frac{2\nu\xi_i}{\omega}\, \frac{-t\,\mathrm{e}^{-ta(\boldsymbol{\xi})/\omega}\big(\mathrm{e}^{-2\pi a(\boldsymbol{\xi})/\omega} - 1\big) + 2\pi\,\mathrm{e}^{-(2\pi+t)a(\boldsymbol{\xi})/\omega}}{\big[\mathrm{e}^{-2\pi\,a(\boldsymbol{\xi})/\omega} - 1\big]^2}$$

and $|\xi_i|\,\big|\partial\Psi/\partial\xi_i\big|$ can be estimated as follows:

$$|\xi_i| \left|\frac{\partial\Psi}{\partial\xi_i}\right| \le \left|\frac{2\nu s^2}{\omega}\, \frac{t\,\mathrm{e}^{-ta(\boldsymbol{\xi})/\omega} + (2\pi - t)\,\mathrm{e}^{-(2\pi+t)a(\boldsymbol{\xi})/\omega}}{\big[\mathrm{e}^{-2\pi\,a(\boldsymbol{\xi})/\omega} - 1\big]^2}\right|$$

$$\le c(\lambda)\left[s^2\,\frac{t}{\omega}\,\mathrm{e}^{-\nu s^2 t/\omega} + s^2\,\frac{2\pi + t}{\omega}\,\mathrm{e}^{-\nu s^2\,(2\pi+t)/\omega}\right].$$

We observe that the right-hand side is less than or equal to a constant independent of $\boldsymbol{\xi}$ and t. We can similarly estimate $|\xi_j|\,\big|\partial\Psi/\partial\xi_3\big|$ and all other terms of the form

$$|\xi_1|^{\kappa_1}\,|\xi_2|^{\kappa_2}\,|\xi_3|^{\kappa_3}\left|\frac{\partial^{\kappa_1+\kappa_2+\kappa_3}\Psi}{\partial\xi_1^{\kappa_1}\,\partial\xi_2^{\kappa_2}\,\partial\xi_3^{\kappa_3}}\right|, \qquad \kappa_1,\, \kappa_2,\, \kappa_3 \in \{0, 1\}.$$

Applying the inverse Fourier transform to (8.104), we obtain the formula

$$\mathbf{v}(\mathbf{x}) := \frac{1}{\omega}\int_0^{2\pi} Q(t)\,\mathcal{F}_{-1}\Big[\Psi(\lambda, \boldsymbol{\xi}, t)\,\mathcal{F}[\mathbf{f}(Q(-t)\,\cdot\,)](\boldsymbol{\xi})\Big](\mathbf{x})\,\mathrm{d}t.$$

Using Lizorkin's multiplier theorem (see, e.g., [95, p. 375]) and the estimates of Ψ and its derivatives discussed above, we derive the inequality

$$\Big\|\mathcal{F}_{-1}\Big[\Psi(\lambda, \boldsymbol{\xi}, t)\,\mathcal{F}[\mathbf{f}(Q(-t)\,\cdot\,)](\boldsymbol{\xi})\Big]\Big\|_q \le c\,\|\mathbf{f}\|_q\,, \quad t \in [0, 2\pi]\,, \tag{8.105}$$

where $c = c(\lambda, q)$. The formula (8.104) and the inequality (8.105) now imply the estimate (8.98). □

Lemma 8.18 *Suppose that $\lambda \in \mathbb{C} \setminus \Lambda_k^\omega$. Then $\lambda \in \rho(\mathscr{A}_k^\omega)$.* □

Proof The estimates (8.2) and (8.98) in Lemmas 8.1 and 8.16 imply that the range of $\mathscr{A}_k^\omega - \lambda I$ is closed in $\boldsymbol{L}^q_\sigma(\mathbb{R}^3)$ and that the operator $\mathscr{A}_k^\omega - \lambda I$ is injective. The same statements also hold on the adjoin operator $(\mathscr{A}_k^\omega)^* - \overline{\lambda} I$ in the space $\boldsymbol{L}^{q'}_\sigma(\mathbb{R}^3)$. Since $(\mathscr{A}_k^\omega)^* - \overline{\lambda} I$ is injective, we deduce that $R(\mathscr{A}_k^\omega - \lambda I) = \boldsymbol{L}^q_\sigma(\mathbb{R}^3)$. Hence $\lambda \in \rho(\mathscr{A}_k^\omega)$. □

Lemma 8.19 $Sp_P(\mathscr{A}_k^{\omega}) = \emptyset$. □

Proof Let $\lambda = \alpha + \mathrm{i}\beta \in \Lambda_k^{\omega}$, $\alpha \leq 0$, $\beta \in \mathbb{R}$, and let $\mathbf{v} \in D(\mathscr{A}_k^{\omega})$ satisfy the equation $\mathscr{A}_k^{\omega}\mathbf{v} - \lambda\mathbf{v} = \mathbf{0}$. Applying the Fourier transform, we derive the identity

$$\omega\,\partial_{\varphi}\widehat{\mathbf{v}} - (\lambda + \mathrm{i}k\xi_3 + \nu|\boldsymbol{\xi}|^2)\widehat{\mathbf{v}} - \boldsymbol{\omega}\wedge\widehat{\mathbf{v}} \;=\; \mathbf{0}.$$

(We use the same notation as in the proof of Lemma 8.16.) First we consider the simpler case $1 < q \leq 2$, when $\widehat{\mathbf{v}}$ is a function from $\boldsymbol{L}^{q'}(\mathbb{R}^3)$. Denoting $\widehat{\mathbf{w}}(r,\varphi,\xi_3) = Q(-\varphi)^T\,\widehat{\mathbf{v}}(r,\varphi,\xi_3)$, we derive, by analogy with (8.102), the equation

$$\omega\,\partial_{\varphi}\widehat{\mathbf{w}} - (\lambda + \mathrm{i}k\xi_3 + \nu s^2)\widehat{\mathbf{w}} \;=\; \mathbf{0}, \qquad s = |\boldsymbol{\xi}|\,. \tag{8.106}$$

Solving explicitly this ordinary differential equation, we obtain

$$\widehat{\mathbf{w}}(r,\varphi+2\pi,\xi_3) \;=\; \widehat{\mathbf{w}}(r,\varphi,\xi_3)\,\mathrm{e}^{2\pi\,(\lambda+\mathrm{i}k\xi_3+\nu s^2)}.$$

Since $\widehat{\mathbf{w}}$ is 2π-periodic in the variable φ, we get either $\widehat{\mathbf{w}} = \mathbf{0}$, or $\lambda + \mathrm{i}k\xi_3 + \nu s^2 = 0$, i.e., $\alpha + \nu s^2 = 0$ and $\beta + k\xi_3 = 0$ for all $\boldsymbol{\xi} \in \mathbb{R}^3$. Since the latter is impossible, $\widehat{\mathbf{w}} = \mathbf{0}$ a.e. in $\mathbb{R}^3$, which implies that $\widehat{\mathbf{v}} = \mathbf{0}$ a.e. in $\mathbb{R}^3$ and also $\mathbf{v} = \mathbf{0}$. Thus, λ cannot be an eigenvalue of $\mathscr{A}_k^{\omega}$ and $\mathbf{v}$ cannot be an associated eigenfunction.

In the general case $q > 2$, let us fix $\lambda \in \Lambda_k^{\omega}$, $\lambda = \alpha + \mathrm{i}\beta + \mathrm{i}l\omega$, where $\alpha \leq -\nu\beta^2/k^2$ and $l \in \mathbb{Z}$. Consider $\mathbf{v} \in D(\mathscr{A}_k^{\omega})$ and $\widehat{\mathbf{w}}(r,\varphi,\xi_3) = Q(\varphi)\widehat{\mathbf{v}}(r,\varphi,\xi_3)$ as above. Since the coefficients of $Q(\varphi)$ are either constant or $\cos\varphi = \xi_2/r$ or $\sin\varphi = \xi_3/r$, the function $\mathbf{w}$ (the inverse Fourier transform of $\widehat{\mathbf{w}}$) is defined by an application of the 2D-Riesz transform to $\mathbf{v}$. Hence $\mathbf{w} \in \boldsymbol{L}^q(\mathbb{R}^3)$.

We at first determine the support of tempered distributions $\widehat{\mathbf{w}}$ that solve Eq. (8.106). □

Assertion 1 *Let* $\widehat{\mathbf{w}} \in \boldsymbol{\mathcal{S}}(\mathbb{R}^3)$ *be a solution of* (8.106). *Then*

$$\operatorname{supp}\widehat{\mathbf{w}} \subset D := \{\boldsymbol{\xi} \in \mathbb{R}^3;\, \alpha + \nu s^2 = 0,\ \beta + k\xi_3 \in \omega\mathbb{Z}\}.$$

Proof of Assertion 1 Given $\boldsymbol{\psi} \in \boldsymbol{C}_0^{\infty}(\mathbb{R}^3 \setminus D)$, we solve the equation

$$-\,\omega\,\partial_{\varphi}\boldsymbol{\Psi} - a(\boldsymbol{\xi})\boldsymbol{\Psi} \;=\; \boldsymbol{\psi}, \tag{8.107}$$

where $a(\boldsymbol{\xi}) = \lambda + \mathrm{i}k\xi_3 + \nu s^2$, $s = |\boldsymbol{\xi}|$. The solution of (8.107) is

$$\boldsymbol{\Psi}(\varphi) \;=\; \mathrm{e}^{-a(\boldsymbol{\xi})\varphi/\omega}\Big(\boldsymbol{\Psi}_0 - \frac{1}{\omega}\int_0^{\varphi} \mathrm{e}^{a(\boldsymbol{\xi})\varphi'/\omega}\boldsymbol{\psi}(\varphi')\,\mathrm{d}\varphi'\Big).$$

(We omit writing the variables r, ξ_3 in $\boldsymbol{\Psi}$ and $\boldsymbol{\psi}$.) Since $\boldsymbol{\Psi}$ is 2π-periodic in φ, the initial value $\boldsymbol{\Psi}_0$ must satisfy the condition

$$\left(\mathrm{e}^{-2\pi a(\boldsymbol{\xi})/\omega} - 1\right)\boldsymbol{\Psi}_0 \;=\; \frac{1}{\omega}\,\mathrm{e}^{-2\pi a(\boldsymbol{\xi})/\omega} \int_0^{2\pi} \mathrm{e}^{a(\boldsymbol{\xi})\varphi'/\omega}\boldsymbol{\psi}(\varphi')\,\mathrm{d}\varphi'.$$

This equation is uniquely solvable for the unknown $\boldsymbol{\Psi}_0$ if

$$2\pi a(\boldsymbol{\xi})/\omega \notin 2\pi\mathrm{i}\mathbb{Z} \iff \lambda + \mathrm{i}k\xi_3 + \nu s^2 \neq k\mathrm{i}\omega \;\text{ for all } k \in \mathbb{Z} \iff \boldsymbol{\xi} \notin D.$$

For $\boldsymbol{\xi} \notin D$, the unique solution of (8.107) is

$$\boldsymbol{\Psi}(\varphi) \;=\; \frac{1}{\omega}\,\frac{1}{1-\mathrm{e}^{2\pi a(\boldsymbol{\xi})/\omega}} \int_0^{2\pi} \mathrm{e}^{a(\boldsymbol{\xi})\varphi'/\omega}\boldsymbol{\psi}(\varphi'+\varphi)\,\mathrm{d}\varphi',$$

in view of (8.103) with ω replaced by $-\omega$. Since $\operatorname{supp}\boldsymbol{\psi} \subset \mathbb{R}^3 \setminus D$, we obtain that $\boldsymbol{\Psi} \in \boldsymbol{C}_0^\infty(\mathbb{R}^3 \setminus D)$.

Using now (8.106), we get

$$\langle \widehat{\mathbf{w}}, \boldsymbol{\psi}\rangle = \langle \widehat{\mathbf{w}}, -\omega\partial_\varphi\boldsymbol{\Psi} - a(\boldsymbol{\xi})\boldsymbol{\Psi}\rangle = \langle \omega\partial_\varphi\widehat{\mathbf{w}} - (\lambda + \mathrm{i}\xi_3 + \nu s^2)\widehat{\mathbf{w}}\,,\boldsymbol{\Psi}\rangle = 0$$

for all $\boldsymbol{\psi} \in \boldsymbol{C}_0^\infty(\mathbb{R}^3 \setminus D)$. This identity proves that $\operatorname{supp}\widehat{\mathbf{w}} \subset D$. □

Continuation of the Proof of Lemma 8.19 1 The set D can be written as the union of finitely many disjoint sets

$$D_l := \{\boldsymbol{\xi} \in \mathbb{R}^3;\; \alpha + \nu s^2 = 0,\; \beta + l\xi_3 = \omega k\}, \quad l \in \mathbb{Z}.$$

Each non-void set D_l defines a circle in $\mathbb{R}^3$ parallel to the $\xi_1\xi_2$-plane with the center $(0, 0, \xi_3^{(l)})$, where $\xi_3^{(l)} = (\omega l - \beta)/k$. At least the set D_0 is non-void since $\lambda \in \Lambda_k^\omega$. Using a suitable partition of unity with respect to the variable ξ_3, we may write $\widehat{\mathbf{w}} \in \boldsymbol{S}'(\mathbb{R}^3)$ (where $\boldsymbol{S}'(\mathbb{R}^3)$ denotes the dual to $\boldsymbol{S}(\mathbb{R}^3)$; its elements are called *tempered distributions*), a solution of (8.106), as a finite linear combination of tempered distributions $\widehat{\mathbf{w}}_l$ with $\operatorname{supp}\widehat{\mathbf{w}}_l \subset D_l$. Since $\mathbf{w}$ is in $\boldsymbol{L}^q(\mathbb{R}^3)$, $\mathbf{w}_l$ is also in $\boldsymbol{L}^q(\mathbb{R}^3)$ for each k. Moreover, since $\mathbf{w}$ is in $\boldsymbol{C}^\infty(\mathbb{R}^3)$ by the elliptic regularity theory, $\mathbf{w}_l$ is also in $\boldsymbol{C}^\infty(\mathbb{R}^3)$. We further need the assertion:

Assertion 2 *Let $h \in S'(\mathbb{R}^3) \cap C^\infty(\mathbb{R}^3)$ satisfy $\operatorname{supp}\widehat{h} \subset \{0\} \times \mathbb{R}^2$. Then h is a polynomial with respect to the variable x_3.*

Proof of Assertion 2 Given $h \in S'(\mathbb{R}^3)$, we find $N \in \mathbb{N}$ and a constant $c > 0$ such that

$$|\langle \widehat{h}, \psi\rangle| \;\le\; c\,\Big\|(1+|\,.\,|)^N \sum_{0\le|\boldsymbol{\mu}|\le N} |\partial^{\boldsymbol{\mu}}\psi|\Big\|_{\infty;\,\mathbb{R}^3}$$

for all test functions $\psi \in \mathcal{S}(\mathbb{R}^3)$; here $\boldsymbol{\mu} \in \mathbb{N}^3$ denotes a multi-index. Let $\eta \in C_0^\infty(\mathbb{R})$ be a cutoff function such that $\eta(\xi_3) = 1$ for ξ_3 in a neighborhood of 0, and let $\eta_\epsilon(\xi_3) = \eta(\xi_3/\epsilon)$ for $\epsilon > 0$. Note that $\|\partial_3^m \eta_\epsilon\|_\infty \leq c_m \epsilon^{-m}$ for all $m \in \mathbb{N}$. Since $\operatorname{supp} \widehat{h} \subset \{0\} \times \mathbb{R}^2$, we get that $\langle \widehat{h}, \psi\rangle = \langle \widehat{h}, \eta_\epsilon \psi\rangle$ for all $\epsilon > 0$. Hence

$$\left|\langle \xi_3^{N+1}\widehat{h},\, \psi\rangle\right| = \left|\langle \widehat{h},\, \xi_3^{N+1}\eta_\epsilon \psi\rangle\right| \leq c\,\epsilon \left\|(1+|\,.\,|)^N \sum_{0\leq|\boldsymbol{\mu}|\leq N} |\partial^{\boldsymbol{\mu}}\psi|\right\|_\infty,$$

where $c > 0$ is independent of $\epsilon \in (0, 1)$. Consequently, $\xi_3^{N+1}\widehat{h} = 0$. Now we conclude that $\partial_3^{N+1} h = 0$ and that $h \in C^\infty(\mathbb{R}^3)$ is a polynomial with respect to the variable x_3 of order at most N. □

Completion of the Proof of Lemma 8.19 2 To complete the proof of Lemma 8.19, let $\mathbf{v} \in D(\mathscr{A}_k^{\omega})$ satisfy the equation $\mathscr{A}_k^{\omega}\mathbf{v} - \lambda\mathbf{v} = \mathbf{0}$. By the above arguments, it suffices to show that each $\mathbf{w}_l$ in the partition of $\mathbf{w}$ vanishes. Here, $\mathbf{w}_l \in \boldsymbol{L}^q(\mathbb{R}^3) \cap \boldsymbol{C}^\infty(\mathbb{R}^3)$ has the property $\operatorname{supp} \widehat{\mathbf{w}}_l \subset D_l$, i.e., $\operatorname{supp} \widehat{\mathbf{w}}_l \subset \{\xi_3^{(l)}\} \times \mathbb{R}^2$ with $\xi_3^{(l)} = (\omega l - \beta)/k$. Then the function $\mathbf{x} \mapsto \mathrm{e}^{\mathrm{i}x_3\xi_3^{(l)}}\, \mathbf{w}_l(\mathbf{x})$, which satisfies the assumptions of Assertion 2, is a polynomial in the variable x_3. Since this function lies in $\boldsymbol{L}^q(\mathbb{R}^3)$, it must vanish identically. We proved that $\mathbf{w} = \mathbf{0}$ and also $\mathbf{v} = \mathbf{0}$.

Lemma 8.20 $Sp_r(\mathscr{A}_k^{\omega}) = \emptyset$. □

The equality $\mathrm{Sp_r}(\mathscr{A}_k^{\omega}) = \emptyset$ follows from Lemma 8.19 and duality arguments.

Lemma 8.21 $Sp(\mathscr{A}_k^{\omega}) = Sp_c(\mathscr{A}_k^{\omega}) = Sp_{ess}(\mathscr{A}_k^{\omega}) = \Lambda_k^{\omega}$. □

Proof Lemmas 8.18–8.20 imply that $\mathrm{Sp}(\mathscr{A}_k^{\omega}) = \mathrm{Sp_c}(\mathscr{A}_k^{\omega}) \subset \Lambda_k^{\omega}$. Thus, we need to show that $\Lambda_k^{\omega} \subset \mathrm{Sp_c}(\mathscr{A}_k^{\omega})$.

Suppose at first that $\lambda \in (\Lambda_k^{\omega})^\circ$, the interior of Λ_k^{ω}. It means that there exist $\alpha, \beta \in \mathbb{R}$ and $l \in \mathbb{Z}$ such that $\lambda = \alpha + \mathrm{i}\beta + \mathrm{i}l\omega$ and $\alpha < -\nu\beta^2/k^2$. The number α can be written in the form $\alpha = \alpha_1 + \alpha_2$ where $\alpha_1 = -\nu\beta^2/k^2$ and $\alpha_2 < 0$. Assume that $l \neq 0$. The procedure in the case $l = 0$ would be analogous.

We shall explicitly define a sequence of functions $\{\mathbf{v}^n\}$ in $D(\mathscr{A}_k^{\omega})$ such that $\|\mathbf{v}^n\|_q = 1$ and $\|(\mathscr{A}_k^{\omega} - \lambda I)\mathbf{v}^n\|_q \to 0$ as $n \to \infty$. It will imply that $\mathrm{nul}'(\mathscr{A}_k^{\omega} - \lambda I) > 0$.

Let us denote by v_1^n, v_r^n and v_φ^n the cylindrical components of $\mathbf{v}^n$. Put

$$\begin{aligned} v_1^n(r,\varphi,x_3) &:= 0, \\ v_r^n(r,\varphi,x_3) &:= \kappa_n\, U^n(x_3)\, V^n(r)\, \mathrm{e}^{\mathrm{i}l\varphi}, \\ v_\varphi^n(r,\varphi,x_3) &:= -\frac{1}{\mathrm{i}l}\, \partial_r\, [r\, v_r^n(r,\varphi,x_3)] \\ &= -\frac{1}{\mathrm{i}l}\, \kappa_n\, U^n(x_3) \left[V^n(r) + r\, \frac{\mathrm{d}V^n(r)}{\mathrm{d}r}\right] \mathrm{e}^{\mathrm{i}l\varphi}. \end{aligned}$$

The function U^n has the form

$$U^n(x_3) := \eta_1^n(x_3)\, Y(x_3), \quad \text{where} \quad Y(x_3) = \mathrm{e}^{\mathrm{i}ax_3},\ a = -\frac{\beta}{k},$$

and where η_1^n is an infinitely differentiable function on $(-\infty, +\infty)$ such that $0 \leq \eta_1^n \leq 1$,

$$\eta_1^n(x_3) = \begin{cases} 0 & \text{for } x_3 \leq -n - n^2 \text{ and } n + n^2 \leq x_3, \\ 1 & \text{for } -n^2 \leq x_3 \leq n^2. \end{cases}$$

The identity $\alpha_1 = -\nu\beta^2/k^2$ guarantees that the characteristic equation $\nu\zeta^2 - k\zeta - (\alpha_1 + \mathrm{i}\beta) = 0$, corresponding to Eq. (8.108) below, has the root $\zeta_1 = ia$. Thus, the function Y is a bounded nontrivial solution of the ordinary differential equation

$$\nu\, Y''(x_3) - k\, Y'(x_3) - (\alpha_1 + \mathrm{i}\beta)\, Y(x_3) = 0 \tag{8.108}$$

in the interval $(-\infty, \infty)$. The function V^n is supposed to have the form

$$V^n(r) := \eta_2^n(r)\, \mathrm{e}^{\mathrm{i}br}, \quad b = \sqrt{-\frac{\alpha_2}{\nu}},$$

where η_2^n is an infinitely differentiable function on $[0, \infty)$ such that $0 \leq \eta_2^n \leq 1$ and

$$\eta_2^n(r) = \begin{cases} 0 & \text{for } 0 \leq r \leq n \text{ and } 3n + n^2 \leq r, \\ 1 & \text{for } 2n \leq r \leq 2n + n^2. \end{cases}$$

Both the functions η_1^n and η_2^n can be chosen so that their derivatives are of the order $1/n$. The definition of V^n guarantees that it satisfies

$$\nu\, \frac{\mathrm{d}^2}{\mathrm{d}r^2} V^n(r) - \alpha_2\, V^n(r) = 0 \tag{8.109}$$

for $2n < r < 2n + n^2$. Finally, the constant κ_n is chosen so that $\|\mathbf{v}^n\|_q = 1$.

One can easily check that $\mathbf{v}^n$ satisfies the condition of incompressibility

$$\nabla \cdot \mathbf{v}^n \equiv \partial_3 v_1^n + \frac{1}{r}\, \partial_r (r v_r^n) + \frac{1}{r}\, \partial_\varphi v_\varphi^n = 0.$$

The support of $\mathbf{v}^n$ is a subset of

$$S^n := \big\{ \mathbf{x} = [r, \varphi, x_3] \in \mathbb{R}^3;\ |x_3| \leq n + n^2,\ n \leq r \leq 3n + n^2, \\ 0 \leq \varphi < 2\pi \big\}. \tag{8.110}$$

Considering the norm of $\mathbf{v}^n$, we observe that for large n the decisive contribution comes from the integral of $|v_\varphi^n|^q$, namely of its part $\big|(-1/\mathrm{i}l)\,\kappa_n\, U^n\, r\, (\mathrm{d}V^n/\mathrm{d}r)\,\mathrm{e}^{\mathrm{i}l\varphi}\big|^q$, on the region

$$S_0^n := \big\{\mathbf{x} = [r, \varphi, x_3] \in \mathbb{R}^3;\ |x_3| < n^2,\ 2n < r < 2n + n^2,\ 0 < \varphi < 2\pi\big\}.$$

The integrals of all other parts on other regions are of a lower order in n. Calculating the integral of $\big|(-1/\mathrm{i}l)\,\kappa_n\, U^n\, r\, (\mathrm{d}V^n/\mathrm{d}r)\,\mathrm{e}^{\mathrm{i}l\varphi}\big|^q$ on the domain S_0^n, we obtain

$$\begin{aligned}
&\int_{-n^2}^{n^2}\int_{2n}^{2n+n^2}\int_0^{2\pi}\left|\frac{\kappa_n}{\mathrm{i}l}\, U^n(x_3)\, r\, \frac{\mathrm{d}V^n(r)}{\mathrm{d}r}\right|^q 2r\ \mathrm{d}\varphi\, \mathrm{d}r\, \mathrm{d}x_3\\
&= 2\pi\, \frac{\kappa_n^q}{|l|^q}\int_{-n^2}^{n^2}|U^n(x_3)|^q\, \mathrm{d}x_3\int_{2n}^{2n+n^2} r^{q+1}\left|\frac{\mathrm{d}V^n(r)}{\mathrm{d}r}\right|^q \mathrm{d}r\\
&= 2\pi\, \frac{\kappa_n^q}{|l|^q}\, 2n^2\, \frac{b^q}{q+2}\Big((2n+n^2)^{q+2} - (2n)^{q+2}\Big).
\end{aligned}$$

Here we have used the equalities $\eta_1^n(x_3) = \eta_2^n(r) = 1$, hence $|U^n(x_3)| = |V^n(r)| = 1$ for $(r, \varphi, x_3) \in S_0^n$. Thus, there exist $n_0 \in \mathbb{N}$ and positive constants c_8 and c_9 (independent of n) such that

$$\frac{c_8}{n^{2+6/q}} \le \kappa_n \le \frac{c_9}{n^{2+6/q}} \qquad \text{for all } n \in \mathbb{N} \text{ with } n \ge n_0. \tag{8.111}$$

We can naturally use the form (8.9) of the operator $\mathscr{A}_k^\omega$. Moreover, if we identify $\mathbf{v}$ with the triplet of its cylindrical coordinates, i.e., $\mathbf{v} \triangleq [v_r, v_\varphi, v_3]^T$, then we can write $\mathscr{A}_k^\omega\mathbf{v}$ in the form $\mathscr{A}_k^\omega\mathbf{v} = \nu\Delta\mathbf{v} + \omega\,\partial_\varphi\mathbf{v} - k\,\partial_3\mathbf{v}$, see (8.10). Hence

$$\begin{aligned}
&(\mathscr{A}_k^\omega - \lambda I)\mathbf{v}^n\\
&\quad = (\nu\Delta + \omega\,\partial_\varphi - k\,\partial_3 - \lambda I)\begin{bmatrix} \kappa_n\, U^n(x_3)\, V^n(r)\, \mathrm{e}^{\mathrm{i}k\varphi}\\ -\dfrac{1}{\mathrm{i}k}\,\kappa_n\, U^n(x_3)\Big[V^n(r) + r\,\dfrac{\mathrm{d}V^n(r)}{\mathrm{d}r}\Big]\mathrm{e}^{\mathrm{i}k\varphi}\\ 0\end{bmatrix}.
\end{aligned}$$

Calculating now the norm of $(\mathscr{A}_k^\omega - \lambda I)\mathbf{v}^n$ in $\boldsymbol{L}_\sigma^q(\mathbb{R}^3)$, we observe that the contributions coming from $\mathbb{R}^3 \setminus S_0^n$ tend to zero as $n \to \infty$, because they represent q-roots of integrals of functions bounded by $c\kappa_n^q r^q$ on $S^n \setminus S_0^n$. Due to (8.111), this contribution is of the order $n^{-1/q}$. Concerning the integral on S_0^n, the decisive part again comes from $(\nu\Delta + \omega\partial_\varphi - k\,\partial_3 - \lambda I)v_\varphi^n$, namely from $(\nu\Delta + \omega\partial_\varphi - k\,\partial_3 - \lambda I)$

applied to the term $(-1/\mathrm{i}l)\,\kappa_n\,U^n\,r\,(\mathrm{d}V^n/\mathrm{d}r)\,\mathrm{e}^{\mathrm{i}l\varphi}$ because of the factor r inside this term. By the equality $\lambda = \alpha_1 + \alpha_2 + \mathrm{i}\beta + \mathrm{i}l\omega$ and (8.108) and (8.109), we have

$$
\begin{aligned}
&\big(\nu\Delta + \omega\partial_\varphi - k\,\partial_3 - \lambda I\big)\bigg(\frac{\kappa_n}{\mathrm{i}l}\,U^n(x_3)\,r\,\frac{\mathrm{d}V^n(r)}{\mathrm{d}r}\,\mathrm{e}^{\mathrm{i}l\varphi}\bigg)\\
&= \frac{\kappa_n}{\mathrm{i}l}\Big(\nu\partial_3^2 + \nu\partial_r^2 + \frac{\nu}{r}\,\partial_r + \frac{\nu}{r^2}\,\partial_\varphi^2 + \omega\,\partial_\varphi - k\,\partial_3 - \lambda I\Big)\\
&\qquad \bigg(U^n(x_3)\,r\,\frac{\mathrm{d}V^n(r)}{\mathrm{d}r}\,\mathrm{e}^{\mathrm{i}l\varphi}\bigg)\\
&= \frac{\kappa_n}{\mathrm{i}l}\,U^n(x_3)\,\bigg(\nu\,\frac{\mathrm{d}^2}{\mathrm{d}r^2} + \frac{\nu}{r}\,\frac{\mathrm{d}}{\mathrm{d}r} - \alpha_2 I\bigg)\bigg[r\,\frac{\mathrm{d}V^n(r)}{\mathrm{d}r}\bigg]\mathrm{e}^{\mathrm{i}l\varphi}\\
&\quad + \frac{\kappa_n}{\mathrm{i}l}\Big(\nu Y''(x_3) - kY'(x_3) - [\alpha_1 + \mathrm{i}\beta]\,Y(x_3)\Big)\bigg[r\,\frac{\mathrm{d}V^n(r)}{\mathrm{d}r}\bigg]\mathrm{e}^{\mathrm{i}l\varphi}\\
&\quad - \frac{\kappa_n}{\mathrm{i}l}\,\frac{\nu k^2}{r^2}\,U^n(x_3)\,r\,\frac{\mathrm{d}V^n(r)}{\mathrm{d}r}\,\mathrm{e}^{\mathrm{i}l\varphi}\\
&= \frac{\kappa_n}{\mathrm{i}l}\bigg\{U^n(x_3)\,r\,\frac{\mathrm{d}}{\mathrm{d}r}\bigg[\nu\,\frac{\mathrm{d}^2V^n(r)}{\mathrm{d}r^2} - \alpha_2\,V^n(r)\bigg] + U^n(x_3)\,2\nu\,\frac{\mathrm{d}^2V^n(r)}{\mathrm{d}r^2}\\
&\quad + U^n(x_3)\,\frac{\nu}{r}\,\frac{\mathrm{d}}{\mathrm{d}r}\bigg[r\,\frac{\mathrm{d}V^n(r)}{\mathrm{d}r}\bigg] - \frac{\kappa_n}{\mathrm{i}l}\,U^n(x_3)\,\frac{\nu l^2}{r}\,\frac{\mathrm{d}V^n(r)}{\mathrm{d}r}\bigg\}\,\mathrm{e}^{\mathrm{i}l\varphi}\\
&= \frac{\nu\kappa_n}{\mathrm{i}l}\Big(-3b^2 + \frac{\mathrm{i}b}{r} - \frac{l^2\,\mathrm{i}b}{r}\Big)\,\mathrm{e}^{\mathrm{i}(ax_3+br)}\,\mathrm{e}^{\mathrm{i}l\varphi},
\end{aligned}
$$

where, in the last step, we used the simple forms of the functions U^n and V^n on S_0^n, i.e., $U^n(x_3) = \mathrm{e}^{\mathrm{i}ax_3}$ and $V^n(r) = \mathrm{e}^{\mathrm{i}br}$. Hence

$$
\begin{aligned}
&\int_{-n^2}^{n^2}\int_{2n}^{2n+n^2}\int_0^{2\pi}\bigg|\Big(\nu\Delta + \omega\partial_\varphi - k\,\partial_3 - \lambda I\Big)\bigg(\frac{\kappa_n}{\mathrm{i}l}\,U^n(x_3)\,r\,\frac{\mathrm{d}V^n(r)}{\mathrm{d}r}\,\mathrm{e}^{\mathrm{i}l\varphi}\bigg)\bigg|^q\\
&\qquad\qquad r\,\mathrm{d}\varphi\,\mathrm{d}r\,\mathrm{d}x_3\\
&\qquad\qquad \le\; c(\nu,l,b)\,\kappa_n^q\int_{-n^2}^{n^2}\int_{2n}^{2n+n^2} r\;\mathrm{d}r\,\mathrm{d}x_3\\
&\qquad\qquad =\; c(\nu,l,b)\,\kappa_n^q\,\big\{2n^2\big[(2n+n^2)^2 - (2n)^2\big]\big\}.
\end{aligned}
$$

The last term tends to zero as $n\to\infty$ by (8.111). In this way, we show that $\|(\mathscr{A}_k^\omega - \lambda I)\mathbf{v}^n\|_q \to 0$ as $n\to\infty$.

We have proved that $\mathrm{nul}'(\mathscr{A}_k^\omega - \lambda I) > 0$. Hence $\mathrm{nul}'(\mathscr{A}_k^\omega - \lambda I)$ differs from $\mathrm{nul}(\mathscr{A}_k^\omega - \lambda I)$, which equals zero due to Lemma 8.19. It means that the range

$R(\mathscr{A}_k^\omega - \lambda I)$ is not closed in $\boldsymbol{L}^q_\sigma(\mathbb{R}^3)$. Consequently, $\text{nul}'(\mathscr{A}_k^\omega - \lambda I) = \text{def}'(\mathscr{A}_k^\omega - \lambda I) = \infty$ and $\lambda \in \text{Sp}_{\text{ess}}(\mathscr{A}_k^\omega)$. Since $\text{Sp}_{\text{ess}}(\mathscr{A}_k^\omega)$ is closed, Λ_k^ω is a subset of $\text{Sp}_{\text{ess}}(\mathscr{A}_k^\omega)$. By Lemmas 8.19 and 8.20, we obtain the inclusion $\Lambda_k^\omega \subset \text{Sp}_{\text{c}}(\mathscr{A}_k^\omega)$. □

The next theorem resumes the results of Lemmas 8.18–8.21.

Theorem 8.16 *Let* $1 < q < \infty$, $k \neq 0$, *and* $\mathcal{D} = \mathbb{R}^3$. *Then the spectra of the Oseen type operator* $\mathscr{A}_k^\omega$ *in* $\boldsymbol{L}^q_\sigma(\mathbb{R}^3)$ *satisfy* $Sp_p(\mathscr{A}_k^\omega) = Sp_r(\mathscr{A}_k^\omega) = \emptyset$ *and* $Sp(\mathscr{A}_k^\omega) = Sp_c(\mathscr{A}_k^\omega) = \Lambda_k^\omega$, *where set* Λ_k^ω *is defined by (8.87).* □

8.4.2.2 The Case of an Exterior Domain $\mathcal{D}$

We assume that $\mathcal{D} \subset \mathbb{R}^3$, $\mathcal{D} \neq \mathbb{R}^3$ is an exterior domain with the boundary of class $C^{1,1}$ in this paragraph.

Lemma 8.22 $\Lambda_k^\omega \subset Sp_{ess}(\mathscr{A}_k^\omega)$. □

Proof We prove in almost the same way as in the proof of Lemma 8.21 that if $\lambda \in (\Lambda_k^\omega)^\circ$ then $\text{nul}'(\mathscr{A}_k^\omega - \lambda I) > 0$. Unfortunately, since an analogue to Lemma 8.19 is not available, we cannot deduce directly from this inequality that $\lambda \in \text{Sp}_{\text{ess}}(\mathscr{A}_k^\omega)$ as in the proof of Lemma 8.21. However, as $\text{supp}\, \mathbf{v}^n \subset S^n$ (see (8.110)), there exists $n_0 \in \mathbb{N}$ so large that $\mathbf{v}^n$ belongs to the domain of $\mathscr{A}_k^\omega$ for $n \geq n_0$. We observe that any infinite subsequence $\{S^{k_n}\}$ of $\{S^n\}$ has the intersection property $\bigcap_{n=1}^\infty S^{k_n} = \emptyset$. It implies that the sequence $\{\mathbf{v}^n\}_{n \geq n_0}$ is not compact in $\boldsymbol{L}^q_\sigma(\mathcal{D})$. Consequently, $\text{nul}'(\mathscr{A}_k^\omega - \lambda I) = \infty$. We can prove in the same way that $\text{nul}'\big((\mathscr{A}_k^\omega)^* - \overline{\lambda} I\big) = \infty$. Hence $\text{def}'(\mathscr{A}_k^\omega - \lambda I) = \infty$, and the operator $\mathscr{A}_k^\omega - \lambda I$ is not semi-Fredholm. Thus, $\lambda \in \text{Sp}_{\text{ess}}(\mathscr{A}_k^\omega)$. The inclusion $\Lambda_k^\omega \subset \text{Sp}_{\text{ess}}(\mathscr{A}_k^\omega)$ now follows from the closedness of $\text{Sp}_{\text{ess}}(\mathscr{A}_k^\omega)$. □

Lemma 8.23 $Sp_{ess}(\mathscr{A}_k^\omega) \subset \Lambda_k^\omega$. □

Proof Let $\lambda \in \text{Sp}_{\text{ess}}(\mathscr{A}_k^\omega)$. Then $\text{nul}'(\mathscr{A}_k^\omega - \lambda I) = \infty$. This information enables us to construct, by mathematical induction, a sequence $\{\mathbf{u}^n\}$ in $D(\mathscr{A}_k^\omega)$ satisfying $\|\mathbf{u}^n\|_q = 1$, $\|(\mathscr{A}_k^\omega - \lambda I)\mathbf{u}^n\|_q \to 0$ as $n \to \infty$ and

$$\text{dist}\big(\mathbf{u}^n;\ \mathcal{L}_{n-1}\big) = 1 \tag{8.112}$$

for all $n \in \mathbb{N}$, where $\mathcal{L}_{n-1}$ denotes the linear hull of the functions $\mathbf{u}^1, \ldots, \mathbf{u}^{n-1}$: Suppose that we have already constructed $\mathbf{u}^1, \ldots, \mathbf{u}^l$ satisfying $\|(\mathscr{A}_k^\omega - \lambda I)\mathbf{u}^j\|_q \leq 1/j$ for $j = 1, \ldots, l$ and (8.112) for all $n = 1, \ldots, l$. To $\epsilon_{l+1} = 1/(l+1)$ there exists an infinite-dimensional linear manifold M_{l+1} in $D(\mathscr{A}_k^\omega)$ such that $\|(\mathscr{A}_k^\omega - \lambda I)\mathbf{u}\|_q \leq \epsilon_{l+1}$ for all $\mathbf{u} \in M_{l+1}$. Then due to [137, Lemma IV.2.3], we find $\mathbf{u}^{l+1} \in M_{l+1}$ such that $\|\mathbf{u}^{l+1}\|_q = 1$ and $\text{dist}\big(\mathbf{u}^{l+1};\ \mathcal{L}_l\big) = 1$. The sequence $\{\mathbf{u}^n\}$ satisfies

$$\|(\mathscr{A}_k^\omega - \lambda I)\mathbf{u}^n\|_q \leq \frac{1}{n} \qquad \text{for all } n \in \mathbb{N}. \tag{8.113}$$

Denote $\mathbf{f}^n := (\mathscr{A}_k^{\omega} - \lambda I)\mathbf{u}^n$. Due to Lemma 8.1, we have

$$\|\mathbf{u}^n\|_{2,q} + \|(\boldsymbol{\omega} \wedge \mathbf{x}) \cdot \nabla \mathbf{u}^n\|_q \le c\,\|\mathbf{f}^n\|_q + c\,\big(1 + |\lambda|\big)\,\|\mathbf{u}^n\|_q$$
$$\le c + c\,\big(1 + |\lambda|\big) \;=\; c, \tag{8.114}$$

where the generic constant c is independent of $n \in \mathbb{N}$. Furthermore, there exists $\nabla p^n \in \boldsymbol{L}^q(\mathcal{D})$ such that

$$\nu\Delta\mathbf{u}^n + (\boldsymbol{\omega} \wedge \mathbf{x}) \cdot \nabla\mathbf{u}^n - \boldsymbol{\omega} \wedge \mathbf{u}^n - k\,\partial_3\mathbf{v}^n - \lambda\mathbf{u}^n - \nabla p^n = \mathbf{f}^n$$

in $\mathcal{D}$ and, by (8.114), $\|\nabla p^n\|_q \le c$. The sequence $\{\mathbf{u}^n\}$ is bounded in the space $D(\mathscr{A}_k^{\omega})$. Hence there exists a subsequence, denoted again by $\{\mathbf{u}^n\}$, which is weakly convergent in $D(\mathscr{A}_k^{\omega})$. This subsequence naturally preserves the property (8.113).

Put $\mathbf{v}^n := (\mathbf{u}^{n+1} - \mathbf{u}^n)/\delta_n$ where $\delta_n = \|\mathbf{u}^{n+1} - \mathbf{u}^n\|_q$. Then $\{\mathbf{v}^n\}$ is a sequence in the unit sphere in $\boldsymbol{L}^q_\sigma(\mathcal{D})$. It converges weakly to zero in $D(\mathscr{A}_k^{\omega})$ because $(\mathbf{u}^{n+1} - \mathbf{u}^n) \rightharpoonup \mathbf{0}$ in $\boldsymbol{L}^q_\sigma(\mathcal{D})$ as $n \to \infty$ and by (8.112) $\delta_n \ge 1$. Hence $\{\mathbf{v}^n\}$ converges strongly to $\mathbf{0}$ in $\boldsymbol{W}^{1,q}(\mathcal{D} \cap B_R(\mathbf{0}))$ for each $R > 0$. Note that the function $\mathbf{v}^n$ satisfies the equation

$$\nu\Delta\mathbf{v}^n + (\boldsymbol{\omega} \wedge \mathbf{x}) \cdot \nabla\mathbf{v}^n - \boldsymbol{\omega} \wedge \mathbf{v}^n - k\,\partial_3\mathbf{v}^n - \lambda\mathbf{v}^n - \frac{1}{\delta_n}\,\nabla(p^{n+1} - p^n)$$
$$= \frac{1}{\delta_n}\,\big(\mathbf{f}^{n+1} - \mathbf{f}^n\big) \tag{8.115}$$

in $\mathcal{D}$. This equation, together with the information on the weak convergence of $\{\mathbf{v}^n\}$ to zero in $D(\mathscr{A}_k^{\omega})$, implies that the sequence $\{\nabla(p^{n+1} - p^n)\}$ weakly converges to zero in $\boldsymbol{L}^q(\mathcal{D})$. Thus, the functions p^n, which are given uniquely up to an additive constant, can be chosen so that $p^{n+1} - p^n \to 0$ strongly in $L^q(\mathcal{D} \cap B_R(\mathbf{0}))$ for each $R > 0$.

The sequence $\{\mathbf{v}^n\}$ does not contain any subsequence, convergent in $\boldsymbol{L}^q_\sigma(\mathcal{D})$. Indeed, assume that $\{\mathbf{v}^{k_n}\}$ is a convergent subsequence of $\{\mathbf{v}^n\}$ in $\boldsymbol{L}^q_\sigma(\mathcal{D})$. This subsequence has the same weak limit as $\{\mathbf{v}^n\}$, hence $\mathbf{v}^{k_n} \rightharpoonup \mathbf{0}$ in $\boldsymbol{L}^q_\sigma(\mathcal{D})$ as $n \to \infty$. Since the subsequence is convergent, we even have $\mathbf{v}^{k_n} \to \mathbf{0}$ in $\boldsymbol{L}^q_\sigma(\mathcal{D})$ as $n \to \infty$. However, this is impossible because $\|\mathbf{v}^{k_n}\|_q = 1$.

Suppose that $R > 0$ is so large that the domain $\{\mathbf{x} \in \mathbb{R}^3;\ |\mathbf{x}| > R\}$ is a subset of $\mathcal{D}$. Let η be an infinitely differentiable cutoff function in $\mathcal{D}$, with values in $[0, 1]$, such that

$$\eta(\mathbf{x}) = \begin{cases} 0 \text{ for } |\mathbf{x}| \le R, \\ 1 \text{ for } R + 1 \le |\mathbf{x}|. \end{cases}$$

Put $K_R := \{\mathbf{x} \in \mathbb{R}^3;\ R < |\mathbf{x}| < R + 1\}$ and let $\mathfrak{B} : W_0^{1,q}(K_R) \mapsto \boldsymbol{W}_0^{2,q}(K_R)$ be the Bogovskij operator (see the proof of Theorem 8.2). Then $V^n := \mathfrak{B}\big(\nabla\eta \cdot \mathbf{v}^n\big)$

belongs to $W_0^{2,q}(K_R)$. If we extend it by zero to $\mathcal{D} \setminus K_R$, it can be considered to be an element of $W_0^{2,q}(\mathcal{D})$. Due to the continuity of the operator $\mathfrak{B}$ and the strong convergence of $\{\mathbf{v}^n\}$ to $\mathbf{0}$ in $W^{1,q}\big(\mathcal{D} \cap B_{R+1}(\mathbf{0})\big)$, we get that $V^n \to \mathbf{0}$ in $W_0^{2,q}(\mathcal{D})$.

Now we define $\mathbf{w}^n(\mathbf{x}) := \eta(\mathbf{x})\,\mathbf{v}^n(\mathbf{x}) - V^n(\mathbf{x})$. The function $\mathbf{w}^n$ belongs to $D(\mathscr{A}_k^{\omega})$ and, due to (8.115), satisfies the equation

$$\begin{aligned}
&\nu\Delta\mathbf{w}^n + (\boldsymbol{\omega}\wedge\mathbf{x})\cdot\nabla\mathbf{w}^n - \boldsymbol{\omega}\wedge\mathbf{w}^n - k\,\partial_3\mathbf{w}^n - \lambda\mathbf{w}^n - \frac{1}{\delta_n}\,\nabla\big[\eta(p^{n+1}-p^n)\big] \\
&\quad = \frac{\eta}{\delta_n}\,(\mathbf{f}^{n+1}-\mathbf{f}^n) - \frac{1}{\delta_n}\,\nabla\eta\,(p^{n+1}-p^n) + \nu\,(\Delta\eta)\,\mathbf{v}^n + 2\nu\,\nabla\eta\cdot\nabla\mathbf{v}^n \\
&\qquad - \nu\Delta V^n + \lambda V^n + (\boldsymbol{\omega}\wedge\mathbf{x})\cdot(\nabla\eta\otimes\mathbf{v}^n) - (\boldsymbol{\omega}\wedge\mathbf{x})\cdot\nabla V^n \\
&\qquad + \boldsymbol{\omega}\wedge V^n - k\,(\partial_3\eta)\mathbf{v}^n + k\,\partial_3 V^n.
\end{aligned} \tag{8.116}$$

The right-hand side converges strongly to zero in $L^q(\mathcal{D})$ as $n \to \infty$; this follows from the strong convergence of $\{\mathbf{v}^n\}$ to zero in $W^{1,q}(\mathcal{D} \cap B_{R+1}(\mathbf{0}))$, the strong convergence of $\{p^{n+1} - p^n\}$ to zero in $L^q(\mathcal{D} \cap B_{R+1}(\mathbf{0}))$, from the information on the support of $\nabla\eta$ and from the strong convergence of $\{V^n\}$ to zero in $W^{2,q}(\mathcal{D})$. Hence

$$\|(\mathscr{A}_k^{\omega} - \lambda I)\mathbf{w}^n\|_q \longrightarrow 0 \qquad \text{as } n \to \infty. \tag{8.117}$$

Moreover, given $\epsilon > 0$, there exists $n_0 \in \mathbb{N}$ such that if $n \in \mathbb{N}$, $n \geq n_0$, then

$$\|\mathbf{w}^n\|_q \leq \Big(\int_{|\mathbf{x}|<R+1} |\eta\mathbf{v}^n - V^n|^q\,\mathrm{d}\mathbf{x}\Big)^{\frac{1}{q}} + \Big(\int_{R+1<|\mathbf{x}|} |\mathbf{v}^n|^q\,\mathrm{d}\mathbf{x}\Big)^{\frac{1}{q}} \leq \epsilon + 1,$$

$$\begin{aligned}
\|\mathbf{w}^n\|_q &\geq \Big(\int_{R+1<|\mathbf{x}|} |\mathbf{v}^n|^q\,\mathrm{d}\mathbf{x}\Big)^{\frac{1}{q}} \geq \Big(\int_{\mathcal{D}} |\mathbf{v}^n|^q\,\mathrm{d}\mathbf{x}\Big)^{\frac{1}{q}} - \Big(\int_{|\mathbf{x}|<R+1} |\mathbf{v}^n|^q\,\mathrm{d}\mathbf{x}\Big)^{\frac{1}{q}} \\
&\geq 1 - \epsilon.
\end{aligned}$$

Let us now normalize the sequence $\{\mathbf{w}^n\}$ by dividing each of the functions $\mathbf{w}^n$ by its norm in $L^q_\sigma(\mathcal{D})$. In order to preserve a simple notation, we denote the normalized functions again by $\mathbf{w}^n$. If we finally put $\mathbf{w}^n(\mathbf{x}) = \mathbf{0}$ for $\mathbf{x} \in \mathbb{R}^3 \setminus \mathcal{D}$, we obtain a non-compact sequence in the unit sphere in $L^q_\sigma(\mathbb{R}^3)$, satisfying (8.117) with $\|\,.\,\|_q$ being the norm in $L^q_\sigma(\mathbb{R}^3)$. Let us denote, for a while, by $(\mathscr{A}_k^{\omega})_{\mathbb{R}^3}$ the operator $\mathscr{A}_k^{\omega}$, considered in $L^q_\sigma(\mathbb{R}^3)$. The existence of the sequence $\{\mathbf{w}^n\}$ with the above properties implies that $\mathrm{nul}'\big((\mathscr{A}_k^{\omega})_{\mathbb{R}^3} - \lambda I\big) = \infty$. Hence $\lambda \notin \rho\big((\mathscr{A}_k^{\omega})_{\mathbb{R}^3}\big)$, which, due to Theorem 8.16, yields $\lambda \in \Lambda_k^{\omega}$. □

Lemma 8.24 *Let $\lambda \in \mathbb{C} \setminus \Lambda_k^{\omega}$. Then either $\lambda \in Sp_p(\mathscr{A}_k^{\omega})$ for all $1 < q < \infty$ or $\lambda \in \rho(\mathscr{A}_k^{\omega})$ for all $1 < q < \infty$. Moreover, $\lambda \in \rho(\mathscr{A}_k^{\omega})$ when $Re\,\lambda \geq 0$.* □

Proof Assume that $1 < q < \infty$ and that $\lambda \in \mathbb{C} \setminus \Lambda_k^\omega$ is an eigenvalue of $\mathscr{A}_k^\omega$ in $\boldsymbol{L}_\sigma^q(\mathcal{D})$ with a corresponding eigenfunction $\mathbf{v}$. Let p with $\nabla p \in \boldsymbol{L}^q(\mathcal{D})$ be a corresponding pressure function. Using the cutoff function η, the set K_R, and Bogovskij's operator $\mathfrak{B} : W_0^{1,q}(K_R) \mapsto \boldsymbol{W}_0^{2,q}(K_R)$ from the proof of Theorem 8.2, we get that $\mathbf{w}_1 := \eta\mathbf{v} - \boldsymbol{V}$ (where $\boldsymbol{V} = \mathfrak{B}(\nabla\eta\cdot\mathbf{v})$ in K_R, $\boldsymbol{V} = \mathbf{0}$ in $\mathbb{R}^3 \setminus K_R$) solves the equation

$$\begin{aligned}\nu\Delta\mathbf{w}_1 + (\boldsymbol{\omega}\wedge\mathbf{x})\cdot\nabla\mathbf{w}_1 - \boldsymbol{\omega}\wedge\mathbf{w}_1 - k\,\partial_3\mathbf{w}_1 - \lambda\mathbf{w}_1 - \nabla(\eta p) = \mathbf{f}\\ := -\nabla\eta\, p + \nu\,(\Delta\eta)\,\mathbf{v} + 2\nu\,\nabla\eta\cdot\nabla\mathbf{v} - \nu\Delta\boldsymbol{V} + \lambda\boldsymbol{V} + (\boldsymbol{\omega}\wedge\mathbf{x})\cdot(\nabla\eta\otimes\mathbf{v})\\ - (\boldsymbol{\omega}\wedge\mathbf{x})\cdot\nabla\boldsymbol{V} + \boldsymbol{\omega}\wedge\boldsymbol{V} - k\,(\partial_3\eta)\mathbf{v} + k\,\partial_3\boldsymbol{V}\end{aligned} \tag{8.118}$$

in $\mathbb{R}^3$ (cf. (8.116)).

Assume at first that $q \geq 3$. Due to the boundedness of the operator $\mathfrak{B}$ from $W_0^{2,q}(K_R)$ to $\boldsymbol{W}_0^{3,q}(K_R)$ (see, e.g., [95, p. 130]), the restriction of the vector field $\boldsymbol{V}$ to K_R belongs to $\boldsymbol{W}_0^{3,q}(K_R)$. From this, we deduce that $\mathbf{f} \in W^{1,q}(K_R)$. Hence, by Sobolev's embedding theorem, $\mathbf{f} \in \boldsymbol{L}^s(K_R)$ for $1 < s < \infty$. Since $\mathbf{f}$ is supported in $\overline{K}_R$, we have that $\mathbf{f} \in \boldsymbol{L}^s(\mathbb{R}^3)$ for $1 < s < \infty$. Now, we may apply Theorem 8.16 (with s instead of q) to the whole space problem (8.118) and conclude that $\mathbf{w}_1 \in D\big((\mathscr{A}_k^\omega)_{\mathbb{R}^3,s}\big) \subset W^{2,s}(\mathbb{R}^3)$, where $(\mathscr{A}_k^\omega)_{\mathbb{R}^3,s}$ denotes the operator $\mathscr{A}_k^\omega$ in $\boldsymbol{L}_\sigma^s(\mathbb{R}^3)$. Moreover, $\nabla(\eta p) \in \boldsymbol{L}^s(\mathbb{R}^3)$ because $\lambda \in \mathbb{C} \setminus \Lambda_k^\omega$ belongs to the resolvent set of $(\mathscr{A}_k^\omega)_{\mathbb{R}^3,s}$. Similarly, we derive that $\mathbf{w}_2 := (1-\eta)\mathbf{v} + \boldsymbol{V}$, the solution of a problem analogous to (8.118) in the bounded domain $\mathcal{D}_{R+1} := \mathcal{D} \cap B_{R+1}(\mathbf{0})$, satisfies $\mathbf{w}_2 \in W^{2,s}(\mathcal{D}_{R+1})$. Consequently, $\mathbf{v} \equiv \mathbf{w}_1 + \mathbf{w}_2 \in D(\mathscr{A}_k^\omega)$ and λ is an eigenvalue of $\mathscr{A}_k^\omega$, as an operator in $\boldsymbol{L}_\sigma^s(\mathcal{D})$.

If $1 < q < 3$, then we obtain the same result for $1 < s < 3q/(3-q)$. However, repeating finitely many times the same argument, we can extend the result to all $1 < s < \infty$.

Finally, when $s = 2$, a variational argument implies that $\operatorname{Re}\lambda < 0$ for all $\lambda \in \operatorname{Sp}(\mathscr{A}_k^\omega)$ (cf. [77, Theorem 1.1]). □

The results of Lemmas 8.22–8.24 are summarized in the next theorem:

Theorem 8.17 *Let $1 < q < \infty$, $k \neq 0$, and $\mathcal{D}$ be a general exterior domain in $\mathbb{R}^3$ with the $C^{1,1}$-boundary. Then:*

(i) $Sp_{ess}(\mathscr{A}_k^\omega) = \Lambda_k^\omega$.
(ii) If $Re\,\lambda \geq 0$, then $\lambda \in \rho(\mathscr{A}_k^\omega)$.
(iii) If $\lambda \in \mathbb{C} \setminus \Lambda_k^\omega$, $Re\,\lambda < 0$, then either $\lambda \in \rho(\mathscr{A}_k^\omega)$ for all $1 < q < \infty$ or $\lambda \in Sp_p(\mathscr{A}_k^\omega)$ for all $1 < q < \infty$.

Note that the same theorem as Theorem 8.15 also holds (for the same reasons) on the operator $\mathscr{L}_k^\omega$, as an operator in $\boldsymbol{L}_\sigma^q(\mathcal{D})$.

8.5 Semigroups Generated by the Stokes and Oseen Type Operators

The fact that the Stokes operator $\mathscr{A}_0^0$ generates an analytic semigroup in $\boldsymbol{L}^q_\sigma(\mathcal{D})$ was proved by Giga [120]. The same statement on the Oseen operator $\mathscr{A}_k^0$ comes from Miyakawa [172]. These results imply that the operator $-\mathscr{A}_k^0$ and particularly also $-\mathscr{A}_0^0$ are sectorial, see Henry [132], p. 20–21. However, if $\omega \neq 0$ (which is our subject of interest in this book) then the operator $\mathscr{A}_k^\omega$ generates only a C_0-semigroup , see [125] or [116] for $k = 0$ and [194] for $k \in \mathbb{R}$. Let us further explain these and also some other related results in greater detail. Our survey respects the chronological order, in which the corresponding papers were published.

8.5.1 *Basic Information on the Semigroups That Involve the Rotational Effect*

The first result regarding the semigroups in the case $\omega \neq 0$ was published by T. Hishida, see [125] (1999). Here, the author considers the Stokes type operator $\mathscr{A}_0^\omega$ in a smooth exterior domain $\mathcal{D}$ and proves that $\mathscr{A}_0^\omega$ generates a C_0-semigroup $e^{\mathscr{A}_0^\omega t}$ of contraction in the space $\boldsymbol{L}^2_\sigma(\mathcal{D})$. The key step in the proof is the analysis of the equation

$$\mathbf{u} - \mathscr{A}_0^\omega \mathbf{u} = \mathbf{f} \tag{8.119}$$

for $\mathbf{f} \in \boldsymbol{L}^2_\sigma(\mathcal{D})$. The author proves that the equation has a unique solution in $D(\mathscr{A}_0^\omega) \equiv \{\mathbf{u} \in W^{2,2}(\mathcal{D}) \cap \boldsymbol{W}^{1,2}_{0,\sigma}(\mathcal{D});\ (\boldsymbol{\omega} \wedge \mathbf{x}) \cdot \nabla \mathbf{u} \in L^2(\mathcal{D})\}$, which satisfies the estimate

$$\|\Delta \mathbf{u}\|_2 + \|(\boldsymbol{\omega} \wedge \mathbf{x}) \cdot \nabla \mathbf{u}\|_2 \leq c\, \|\mathbf{f}\|_2,$$

where the constant c is independent of $\mathbf{f}$ and $\mathbf{u}$. Then, applying the Lumer–Phillips theorem, one can deduce that $\mathscr{A}_0^\omega$ generates a C_0.. semigroup of contraction in $\boldsymbol{L}^2_\sigma(\mathcal{D})$. The mentioned properties of a solution of Eq. (8.119) are derived in [125] in this way: The author at first considers the case $\mathcal{D} = \mathbb{R}^3$ and derived the results by means of the Fourier transform. Then, applying an appropriate cutoff function procedure splits the problem in domain $\mathcal{D}$ into the problem in $\mathbb{R}^3$ and the problem in a bounded domain. Since $(\boldsymbol{\omega} \wedge \mathbf{x}) \cdot \nabla \mathbf{u}$ can be treated as a lower-order perturbation of the Stokes operator $\mathscr{A}_0^0 \equiv \Pi^2_\sigma \Delta$ in the bounded part of $\mathcal{D}$, the author obtains the desired estimates from the theory of the Stokes operator in a bounded domain.

Note that the information on the unique solution of Eq. (8.119) and the estimate (8.5.1) also directly follow from Lemma 8.1 and Theorem 8.9. This lemma and theorem are, however, of later date than Hishida's paper [125].

Hishida's results from [125] were generalized to the L^q-framework (where $1 < q < \infty$) by M. Geissert, H. Heck and M. Hieber [116] (2006). The authors considered the operator

$$\mathscr{A}_{\mathcal{D}} := \Pi_\sigma^q(\Delta\mathbf{u} + M\mathbf{x}\cdot\nabla\mathbf{u} - M\mathbf{u})$$

and the perturbed operator

$$\mathscr{A}_{\mathcal{D},\mathbf{b}} := \mathscr{A}_{\mathcal{D}}\mathbf{u} + \Pi_\sigma^q(\mathbf{b}\cdot\nabla\mathbf{u} + \mathbf{u}\cdot\nabla\mathbf{b})$$

in $\boldsymbol{L}^q_\sigma(\mathcal{D})$, where $1 < q < \infty$, $\mathcal{D}$ is an exterior domain in $\mathbb{R}^n$ with a $C^{1,1}$ boundary, M is an $n \times n$ real matrix, and $\mathbf{b}$ is an infinitely differentiable vector function in $\overline{\mathcal{D}}$ with a compact support. They at first considered the case $\mathcal{D} = \mathbb{R}^n$ and showed that the operator $\mathscr{A}_{\mathbb{R}^n}$ generates a C_0-semigroup $\mathrm{e}^{\mathscr{A}_{\mathbb{R}^n}t}$ in $\boldsymbol{L}^q_\sigma(\mathbb{R}^n)$. Moreover, if $1 < q \le r < \infty$, then there exist $c > 0$ and $\alpha > 0$ such that the semigroup satisfies the estimates

$$\|\mathrm{e}^{\mathscr{A}_{\mathbb{R}^n}t}\mathbf{f}\|_{r;\,\mathbb{R}^n} \le c\,t^{-\frac{n}{2}\left(\frac{1}{q}-\frac{1}{r}\right)}\,\mathrm{e}^{\alpha t}\,\|\mathbf{f}\|_{q;\,\mathbb{R}^n}, \tag{8.120}$$

$$\|\nabla\mathrm{e}^{\mathscr{A}_{\mathbb{R}^n}t}\mathbf{f}\|_{q;\,\mathbb{R}^n} \le c\,t^{-\frac{1}{2}}\,\mathrm{e}^{\alpha t}\,\|\mathbf{f}\|_{q;\,\mathbb{R}^n} \tag{8.121}$$

for $t > 0$ and $\mathbf{f} \in \boldsymbol{L}^p_\sigma(\mathbb{R}^n)$. Here, the important used tool is the theory of Ornstein–Uhlenbeck operators . The next step is the derivation of analogous results in a bounded domain $\Omega \subset \mathbb{R}^n$ with a $C^{1,1}$-boundary. In this part, the authors consider the operator

$$\mathscr{A}_{\Omega,\mathbf{b}}\mathbf{u} := \mathscr{A}_{\Omega}\mathbf{u} + \Pi_\sigma^q(\mathbf{b}\cdot\nabla\mathbf{u} + \mathbf{u}\cdot\nabla\mathbf{b}),$$

where the operator $\mathscr{A}_\Omega$ is defined by analogy with $\mathscr{A}_{\mathcal{D}}$ and the function $\mathbf{b}$ is supposed to be in $\boldsymbol{C}^\infty_c(\overline{\Omega})$. Treating the terms $\Pi_\sigma^q(M\mathbf{x}\cdot\nabla\mathbf{u} - M\mathbf{u})$ and $\Pi_\sigma^q(\mathbf{b}\cdot\nabla\mathbf{u} + \mathbf{u}\cdot\nabla\mathbf{b})$ as lower-order perturbation of the Stokes operator $\Pi_\sigma^q\Delta$ and applying the theory of perturbations of analytic semigroups, the authors show that $\mathscr{A}_{\Omega,\mathbf{b}}$ generates an analytic semigroup $\mathrm{e}^{\mathscr{A}_{\Omega,\mathbf{b}}t}$ on $\boldsymbol{L}^q_\sigma(\Omega)$. Moreover, there exist $c > 0$ and $\alpha > 0$ such that

$$\|\mathrm{e}^{\mathscr{A}_{\Omega,\mathbf{b}}t}\mathbf{f}\|_{r;\,\Omega} \le c\,t^{-\frac{n}{2}\left(\frac{1}{q}-\frac{1}{r}\right)}\,\mathrm{e}^{\alpha t}\,\|\mathbf{f}\|_{q;\,\Omega}, \tag{8.122}$$

$$\|\nabla\mathrm{e}^{\mathscr{A}_{\Omega,\mathbf{b}}t}\mathbf{f}\|_{q;\,\Omega} \le c\,t^{-\frac{1}{2}}\,\mathrm{e}^{\alpha t}\,\|\mathbf{f}\|_{q;\,\Omega}, \tag{8.123}$$

$$\|\nabla^2\mathrm{e}^{\mathscr{A}_{\Omega,\mathbf{b}}t}\mathbf{f}\|_{q;\,\Omega} \le c\,t^{-1}\,\mathrm{e}^{\alpha t}\,\|\mathbf{f}\|_{q;\,\Omega} \tag{8.124}$$

for $t > 0$ and $\mathbf{f} \in \boldsymbol{L}^p_\sigma(\mathbb{R}^n)$. Then the authors combine the results from $\mathbb{R}^n$ and the bounded domain Ω and prove that $\mathscr{A}_{\mathcal{D},\mathbf{b}}$ generates a C_0-semigroup $\mathrm{e}^{\mathscr{A}_{\mathcal{D},\mathbf{b}}t}$ in

$\boldsymbol{L}^q_\sigma(\Omega)$. The semigroup is constructed by the technique of iterated convolutions and satisfies the same estimates as (8.122) and (8.123).

Analogous results on the semigroup, generated by the Stokes type operator $\mathscr{A}_0^\omega$, were also obtained by T. Hishida and Y. Shibata [129] (2009). In this chapter, in addition to the estimates of $\mathrm{e}^{\mathscr{A}_0^\omega t}\mathbf{f}$ and $\nabla \mathrm{e}^{\mathscr{A}_0^\omega t}\mathbf{f}$ in L^q-norms, the authors also derived estimates in the norms of the Lorentz spaces $\boldsymbol{L}^{r,\kappa}(\mathcal{D})$ and $L^{r,\kappa}(\mathcal{D})^{2\times 2}$, where $1 \leq \kappa < \infty$.

The semigroup, generated by the Oseen type operator $\mathscr{A}_k^\omega$ (for $k \neq 0$), was studied by Y. Shibata [194] (2007) and [195] (2010). The author assumes that $\mathcal{D}$ is an exterior domain in $\mathbb{R}^3$ with the $C^{1,1}$-boundary. The main results of [194] say that $\mathscr{A}_k^\omega$ generates a C_0-semigroup $\mathrm{e}^{\mathscr{A}_k^\omega t}$ in $\boldsymbol{L}^q_\sigma(\mathcal{D})$, where $1 < q < \infty$. The semigroup satisfies the estimates

$$\|\mathrm{e}^{\mathscr{A}_k^\omega t}\mathbf{f}\|_r \leq c\, t^{-\frac{3}{2}\left(\frac{1}{q}-\frac{1}{r}\right)}\,\|\mathbf{f}\|_q \qquad \text{for } 1 < q \leq r \leq \infty,\ q \neq \infty, \tag{8.125}$$

$$\|\nabla \mathrm{e}^{\mathscr{A}_k^\omega t}\mathbf{f}\|_r \leq c\, t^{-\frac{1}{2}-\frac{3}{2}\left(\frac{1}{q}-\frac{1}{r}\right)}\,\|\mathbf{f}\|_q \quad \text{for } 1 < q \leq r \leq 3 \tag{8.126}$$

for $t > 0$ and $\mathbf{f} \in \boldsymbol{L}^q_\sigma(\mathcal{D})$. (See [194, Theorem 1.3].) The generic constant c on the right-hand sides depends on $\mathcal{D}$, ω, k, q, and r. However, if $0 < |\omega| < \omega_0$ and $0 < |k| < k_0$, then c depends only on $\mathcal{D}$, ω_0, k_0, q, and r. (Recall that ω and $\boldsymbol{\omega}$ are related through the formula $\boldsymbol{\omega} := \omega\mathbf{e}_3$.) The author again at first makes the corresponding analysis in the whole space $\mathbb{R}^3$, applying especially the Fourier transform, and then transforms the results to the exterior domain $\mathcal{D}$, using an appropriate cutoff function technique. In addition to the estimates (8.125) and (8.126), one can also deduce from [195, Theorem 11] that there exist $\alpha > 0$ and $c > 0$ such that the estimate

$$\|\nabla^2 \mathrm{e}^{\mathscr{A}_k^\omega t}\mathbf{f}\|_q \leq \frac{c}{t}\,\mathrm{e}^{\alpha t}\,\|\mathbf{f}\|_q \tag{8.127}$$

holds for all $\mathbf{f} \in \mathrm{d}(\mathscr{A}_k^\omega)$ and $t > 0$.

8.5.2 On the Uniform Growth Bound of the Semigroups

The question whether the so-called uniform growth bound of a C_0-semigroup equals the "spectral bound" of the generator belongs to classical problems in the theory of the C_0-semigroups. In this subsection, we will discuss this question in greater detail in the particular cases of semigroups, generated by the Oseen and perturbed Oseen type operators. Note that the content of this subsection essentially comes from the paper [91] by G. P. Galdi and J. Neustupa.

8.5.2.1 The Uniform Growth Bound of a General C_0-Semigroup

In this subsection, we suppose that $\mathrm{e}^{\mathscr{T}t}$ is a general C_0-semigroup in a Banach space X (with the norm denoted by $\| \,.\, \|$), generated by a linear operator $\mathscr{T}$. The *uniform growth bound* $\gamma(\mathrm{e}^{\mathscr{T}t})$ of the semigroup $\mathrm{e}^{\mathscr{T}t}$ is defined by the formula

$$\gamma(\mathrm{e}^{\mathscr{T}t}) := \inf \big\{\mu \in \mathbb{R};\ \exists M_\mu > 0\ \forall t > 0\ \forall x \in X\ :\ \|\mathrm{e}^{\mathscr{T}t}x\| \leq M_\mu\, \mathrm{e}^{\mu t}\, \|x\|\big\}.$$

The often studied problem in the theory of C_0-semigroups is the question under which conditions the uniform growth bound $\gamma(\mathrm{e}^{\mathscr{T}t})$ equals the *spectral bound* of the operator $\mathscr{T}$, which is the number

$$s(\mathscr{T})\ :=\ \sup\ \big\{\mathrm{Re}\,\lambda;\ \lambda \in \mathrm{Sp}(\mathscr{T})\big\}.$$

Generally, one has $s(\mathscr{T}) \leq \gamma(\mathrm{e}^{\mathscr{T}t})$. The equality holds under various additional assumptions, see, e.g., the books [56] (by K. J. Engel and R. Nagel) or [179] (by J. van Neerven). It is known that $s(\mathscr{T}) = \gamma(\mathrm{e}^{\mathscr{T}t})$ if the semigroup $\mathrm{e}^{\mathscr{T}t}$ satisfies the spectral mapping theorem, i.e., if the equality $\mathrm{Sp}(\mathrm{e}^{\mathscr{T}t}) \smallsetminus \{0\} = \exp\big(t\,\mathrm{Sp}(\mathscr{T})\big)$ holds for some $t > 0$. However, while the identities $\mathrm{Sp}_{\mathrm{p}}(\mathrm{e}^{\mathscr{T}t}) \smallsetminus \{0\} = \exp\big(t\,\mathrm{Sp}_{\mathrm{p}}(\mathscr{T})\big)$ and $\mathrm{Sp}_{\mathrm{r}}(\mathrm{e}^{\mathscr{T}t}) \smallsetminus \{0\} = \exp\big(t\,\mathrm{Sp}_{\mathrm{r}}(\mathscr{T})\big)$ hold for the point spectrum and the residual spectrum, the approximate point spectrum generally satisfies only the inclusion $\exp\big(t\,\mathrm{Sp}_{\mathrm{ap}}(\mathscr{T})\big) \subset \mathrm{Sp}_{\mathrm{ap}}(\mathrm{e}^{\mathscr{T}t}) \smallsetminus \{0\}$. The identity $\exp\big(t\,\mathrm{Sp}_{\mathrm{ap}}(\mathscr{T})\big) = \mathrm{Sp}_{\mathrm{ap}}(\mathrm{e}^{\mathscr{T}t}) \smallsetminus \{0\}$ (which implies the validity of the spectral mapping theorem) is satisfied, e.g., if the semigroup $\mathrm{e}^{\mathscr{T}t}$ is eventually uniformly continuous and, in particular, if it is analytic. (See [56, Theorem IV.3.10] or [179, Theorem 2.3.2].)

The following notions and statements play an important role in studies of the C_0-semigroups: The *essential spectral radius* of the semigroup $\mathrm{e}^{\mathscr{T}t}$ at time t is the number

$$r^*_{\mathrm{ess}}(\mathrm{e}^{\mathscr{T}t})\ :=\ \sup\ \big\{|\lambda|;\ \lambda \in \mathrm{Sp}^*_{\mathrm{ess}}(\mathrm{e}^{\mathscr{T}t})\big\}.$$

(Recall that the set $\mathrm{Sp}^*_{\mathrm{ess}}(\,.\,)$ is defined in Sect. 8.2.) The essential spectral radius of $\mathrm{e}^{\mathscr{T}t}$ can be characterized as the infimum of the set of $\rho > 0$ such that the implication "$\zeta \in \mathrm{Sp}(\mathrm{e}^{\mathscr{T}t}),\ |\zeta| > \rho \implies \zeta$ is an eigenvalue of $\mathrm{e}^{\mathscr{T}t}$ with a finite algebraic multiplicity" holds. The *essential growth bound* of the semigroup $\mathrm{e}^{\mathscr{T}t}$ is defined to be

$$\gamma^*_{\mathrm{ess}}(\mathrm{e}^{\mathscr{T}t})\ :=\ \frac{1}{t_0}\ \ln r^*_{\mathrm{ess}}(\mathrm{e}^{\mathscr{T}t_0}) \qquad \text{(for any } t_0 > 0).$$

(One can verify that the right-hand side is independent of t_0 for $t_0 > 0$.) The uniform growth bound of the semigroup $\mathrm{e}^{\mathscr{T}t}$ satisfies the formula $\gamma(\mathrm{e}^{\mathscr{T}t}) = \max\{\gamma^*_{\mathrm{ess}}(\mathrm{e}^{\mathscr{T}t});\ s(\mathscr{T})\}$. (See, e.g., [56, pp. 249–258].)

8.5.2.2 The Uniform Growth Bounds of the Semigroups Generated by the Oseen and Perturbed Oseen Type Operators

Recall that the Oseen type operator $\mathscr{A}_k^\omega$ (i.e., the Oseen operator with the rotational effect) is defined in Sect. 8.1. We will also deal with the perturbed Oseen type operator $\mathscr{L}_k^\omega := \mathscr{A}_k^\omega + \mathscr{B}$, see (8.16), where the operator $\mathscr{B}$ is defined in (8.13).

Concerning the semigroup $\mathrm{e}^{\mathscr{A}_k^\omega t}$, it follows immediately from Theorem 8.17 and from the shape of the set Λ_k^ω that $s(\mathscr{A}_k^\omega) = 0$. The inequality (8.125), used with $r = q$, which implies that $\gamma(\mathrm{e}^{\mathscr{A}_k^\omega t}) \leq 0$, and the inequality $\gamma(\mathrm{e}^{\mathscr{A}_k^\omega t}) \geq s(\mathscr{A}_k^\omega)$ yield $\gamma(\mathrm{e}^{\mathscr{A}_k^\omega t}) = 0$. Thus, we obtain the identity

$$\gamma(\mathrm{e}^{\mathscr{A}_k^\omega t}) = s(\mathscr{A}_k^\omega) = 0. \tag{8.128}$$

The essential spectral radius $r_{\mathrm{ess}}^*(\mathrm{e}^{\mathscr{A}_k^\omega t})$ is greater than or equal to one for each $t > 0$, because $r_{\mathrm{ess}}^*(\mathrm{e}^{\mathscr{A}_k^\omega t})$ is the infimum of the set of $\rho > 0$ such that $\{\zeta \in \mathrm{Sp}(\mathrm{e}^{\mathscr{A}_k^\omega t});\ |\zeta| > \rho\}$ consists of isolated eigenvalues of $\mathrm{e}^{\mathscr{A}_k^\omega t}$ with finite algebraic multiplicities, and each set $\{\zeta \in \mathrm{Sp}(\mathrm{e}^{\mathscr{A}_k^\omega t});\ |\zeta| > \rho\}$ (for $0 < \rho < 1$) contains the set $\{\mathrm{e}^{\lambda t};\ \lambda \in \mathrm{Sp}(\mathscr{A}_k^\omega),\ \ln\rho < \Re\,\lambda\}$ (due to the spectral inclusion theorem, see [56, p. 276]), which is not isolated. On the other hand, since $\gamma_{\mathrm{ess}}^*(\mathrm{e}^{\mathscr{A}_k^\omega t}) \leq \gamma(\mathrm{e}^{\mathscr{A}_k^\omega t}) = 0$, we obtain $\gamma_{\mathrm{ess}}^*(\mathrm{e}^{\mathscr{A}_k^\omega t}) = 0$. Thus, in addition to (8.128), we have proven that $\gamma_{\mathrm{ess}}^*(\mathrm{e}^{\mathscr{A}_k^\omega t}) = 0$.

Let us now focus on the perturbed operator $\mathscr{L}_k^\omega$ and the semigroup generated by this operator. Recall that if $k = 0$ and the perturbation has the special form $\mathscr{B}\mathbf{u} = \Pi_\sigma^q(\mathbf{b}\cdot\nabla\mathbf{u} + \mathbf{u}\cdot\nabla\mathbf{b})$, where $\mathbf{b}$ is an infinitely differentiable vector function with a compact support, then the perturbed Stokes type operator $\mathscr{A}_0^\omega + \mathscr{B}$ generates a C_0-semigroup in $\boldsymbol{L}_\sigma^q(\mathcal{D})$. (See [116] and Sect. 8.5.1.) However, to our knowledge, if $\mathscr{B}$ has the form (8.13), where B and $\mathbf{b}$ satisfy (8.14) and (8.15), then there is no analogous result in the literature. We will partially fill in this gap in this subsection. However, we further deal only with the case $q = 2$, where one can use the dissipativity of the operator $\mathscr{L}_k^\omega - \xi I$ for some $\xi > 0$ and apply the Lumer–Phillips theorem.

Lemma 8.25 *The operator $\mathscr{L}_k^\omega$ generates a C_0-semigroup $e^{\mathscr{L}_k^\omega t}$ in $\boldsymbol{L}_\sigma^2(\mathcal{D})$.* □

Proof Let us at first show that $\mathscr{L}_k^\omega - \xi I$ is dissipative in $\boldsymbol{L}_\sigma^2(\mathcal{D})$ for large positive ξ. If $\boldsymbol{\varphi} \in D(\mathscr{L}_k^\omega) \equiv D(\mathscr{A}_k^\omega)$, then

$$\begin{aligned}
&\int_{\mathcal{D}} \mathscr{L}_k^\omega\boldsymbol{\varphi}\cdot\boldsymbol{\varphi}\,\mathrm{d}\mathbf{x}\\
&\quad= \int_{\mathcal{D}} [\nu\Delta\boldsymbol{\varphi} + (\boldsymbol{\omega}\wedge\mathbf{x})\cdot\nabla\boldsymbol{\varphi} - \boldsymbol{\omega}\wedge\boldsymbol{\varphi} + k\,\partial_3\boldsymbol{\varphi} + B\cdot\boldsymbol{\varphi} + \mathbf{b}\cdot\nabla\boldsymbol{\varphi}]\cdot\boldsymbol{\varphi}\,\mathrm{d}\mathbf{x}\\
&\quad= -\nu\,\|\nabla\boldsymbol{\varphi}\|_2^2 + \int_{\mathcal{D}} [B\cdot\boldsymbol{\varphi} + \mathbf{b}\cdot\nabla\boldsymbol{\varphi}]\cdot\boldsymbol{\varphi}\,\mathrm{d}\mathbf{x},
\end{aligned} \tag{8.129}$$

where the right-hand side can be estimated by means of the Young and Sobolev inequalities (see, e.g., [98, p. 42 and p. 54]):

$$\left|\int_{\mathcal{D}} B\boldsymbol{\varphi}\cdot\boldsymbol{\varphi}\,\mathrm{d}\mathbf{x}\right| \le \|B\|_{2+\iota}\,\|\boldsymbol{\varphi}\|_2^{\frac{1+2\iota}{2(2+\iota)}}\,\|\boldsymbol{\varphi}\|_6^{\frac{3}{2(2+\iota)}} \le \|B\|_{2+\iota}\,\|\boldsymbol{\varphi}\|_2^{\frac{1+2\iota}{2(2+\iota)}}\,c\,\|\nabla\boldsymbol{\varphi}\|_2^{\frac{3}{2(2+\iota)}}$$

$$\le \|B\|_{2+\iota}\left(\epsilon\,\|\nabla\boldsymbol{\varphi}\|_2^2 + c(\epsilon)\,\|\boldsymbol{\varphi}\|_2^2\right),$$

$$\left|\int_{\mathcal{D}} \mathbf{b}\cdot\nabla\boldsymbol{\varphi}\cdot\boldsymbol{\varphi}\,\mathrm{d}\mathbf{x}\right| \le \|\mathbf{b}\|_{3+\iota}\,\|\nabla\boldsymbol{\varphi}\|_2\,\|\boldsymbol{\varphi}\|_{\frac{6+2\iota}{1+\iota}}$$

$$\le \|\mathbf{b}\|_{3+\iota}\,\|\nabla\boldsymbol{\varphi}\|_2\,\|\boldsymbol{\varphi}\|_2^{\frac{\iota}{3+\iota}}\,\|\boldsymbol{\varphi}\|_6^{\frac{3}{3+\iota}} \le \|\mathbf{b}\|_{3+\iota}\,\|\nabla\boldsymbol{\varphi}\|_2\,\|\boldsymbol{\varphi}\|_2^{\frac{\iota}{3+\iota}}\,c\,\|\nabla\boldsymbol{\varphi}\|_2^{\frac{3}{3+\iota}}$$

$$\le \|\mathbf{b}\|_{3+\iota}\left(\delta\,\|\nabla\boldsymbol{\varphi}\|_2^2 + c(\delta)\,\|\boldsymbol{\varphi}\|_2^2\right).$$

If $\epsilon > 0$ and $\delta > 0$ are chosen sufficiently small, then these inequalities and (8.129) imply that

$$\int_{\mathcal{D}} (\mathscr{L}_k^{\omega}\boldsymbol{\varphi} - \xi\boldsymbol{\varphi})\cdot\boldsymbol{\varphi}\,\mathrm{d}\mathbf{x} \le 0$$

for some $\xi > 0$ sufficiently large. Thus, the operator $\mathscr{L}_k^{\omega} - \xi I$ is dissipative in $\boldsymbol{L}^2_{\sigma}(\mathcal{D})$. By Lumer–Phillips theorem (see, e.g., [56, p. 83]), $\mathscr{L}_k^{\omega} - \xi I$ generates a C_0-semigroup of contraction. Thus, $\mathscr{L}_k^{\omega}$ itself generates a C_0-semigroup. □

Our next objective is to show that $\gamma(\mathrm{e}^{\mathscr{L}_k^{\omega} t}) = s(\mathscr{L}_k^{\omega})$.

This equality holds if $\omega = 0$, because then the semigroup $\mathrm{e}^{\mathscr{L}_k^{\omega} t} \equiv \mathrm{e}^{\mathscr{L}_k^{0} t}$ is analytic. The analyticity of $\mathrm{e}^{\mathscr{L}_k^{0} t}$ follows from the analyticity of $\mathrm{e}^{\mathscr{A}_k^{0} t}$ (see [172]), Lemma 8.3 (which implies that the perturbation $\mathscr{B}$ is $\mathscr{A}_k^0$–compact), Lemma 2.16 [56, p. 179] (which implies that $\mathscr{B}$ is $\mathscr{A}_k^0$-bounded with the relative $\mathscr{A}_k^0$-bound equal to zero) and from Theorem 2.10 [56, p. 176] (on perturbation of a generator of an analytic semigroup by a relatively bounded operator with the zero relative bound).

The question of validity of the equality $\gamma(\mathrm{e}^{\mathscr{L}_k^{\omega} t}) = s(\mathscr{L}_k^{\omega})$ in the case $\omega \neq 0$ is, however, a subtler problem. In order to study this question, we will use the variation of parameters formula

$$\mathrm{e}^{\mathscr{L}_k^{\omega} t} = \mathrm{e}^{\mathscr{A}_k^{\omega} t} + \int_0^t \mathrm{e}^{\mathscr{L}_k^{\omega}\tau}\,\mathscr{B}\,\mathrm{e}^{\mathscr{A}_k^{\omega}(t-\tau)}\,\mathrm{d}\tau \tag{8.130}$$

for $t \ge 0$, see e.g. [56, p. 161]. (The formula is in fact derived in [56] under the assumption that operator $\mathscr{B}$ is bounded, but using the "smoothing" properties of $\mathrm{e}^{\mathscr{A}_k^{\omega} t}$ following from (8.123) and (8.127), and applying the same arguments as in [56], one also obtains (8.130) for the unbounded operator $\mathscr{B}$ we deal with.) Let us verify that the integral on the right-hand side of (8.130) converges in the topology of the space of bounded linear operators in $\boldsymbol{L}^2_\sigma(\mathcal{D})$. The integrand satisfies the estimates

$$\begin{aligned}
\big\| e^{\mathscr{L}_k^\omega \tau}\, \mathscr{B}\, e^{\mathscr{A}_k^\omega (t-\tau)} \boldsymbol{\varphi} \big\|_2 \, &\leq \, c\, M_\mu\, e^{\mu\tau}\, \| \mathscr{B}\, e^{\mathscr{A}_k^\omega (t-\tau)} \boldsymbol{\varphi} \|_2 \\
&\leq \, c\, M_\mu\, e^{\mu\tau}\, t^{-\frac{1}{4}} \big(\| B \cdot e^{\mathscr{A}_k^\omega (t-\tau)} \boldsymbol{\varphi} \|_2 + \| \mathbf{b} \cdot \nabla e^{\mathscr{A}_k^\omega (t-\tau)} \boldsymbol{\varphi} \|_2 \big) \\
&\leq \, c\, M_\mu\, e^{\mu\tau}\, t^{-\frac{1}{4}} \big(\| B \|_{2+\iota}\, \| e^{\mathscr{A}_k^\omega (t-\tau)} \boldsymbol{\varphi} \|_s + \| \mathbf{b} \|_{3+\iota}\, \| \nabla e^{\mathscr{A}_k^\omega (t-\tau)} \boldsymbol{\varphi} \|_r \big) \\
&\leq \, c\, M_\mu\, e^{\mu\tau}\, t^{-\frac{1}{4}} \big(\| B \|_{2+\iota}\, (t-\tau)^{-\frac{3}{2}(\frac{1}{2}-\frac{1}{s})} \\
&\qquad + \| \mathbf{b} \|_{3+\iota}\, (t-\tau)^{-\frac{1}{2}-\frac{3}{2}(\frac{1}{2}-\frac{1}{r})} \big)\, \| \boldsymbol{\varphi} \|_2 \qquad (8.131)
\end{aligned}$$

for all $\boldsymbol{\varphi} \in \boldsymbol{L}^2_\sigma(\mathcal{D})$ and $\mu > \gamma(e^{\mathscr{L}_k^\omega t})$. where

$$s = \frac{2(2+\iota)}{\iota} \qquad \text{and} \qquad r = \frac{3(3+\iota)}{3+2\iota}. \qquad (8.132)$$

One can verify that $-\frac{3}{2}(\frac{1}{2}-\frac{1}{s}) = -\frac{3}{2}/(2+\iota) > -1$ and $-\frac{1}{2}-\frac{3}{2}(\frac{1}{2}-\frac{1}{r}) = -\frac{1}{2}-\frac{3}{2}/(3+\iota) > -1$. Hence the right-hand side of (8.131) is integrable, as a function of τ, on the interval $(0, t)$.

Lemma 8.26 *The operator $\mathscr{B}\, e^{\mathscr{A}_k^\omega (t-\tau)}$ is compact in $\boldsymbol{L}^2_\sigma(\mathcal{D})$ for each $t > 0$ and $0 \leq \tau < t$.* □

Proof Let $t > 0$ and $0 \leq \tau < t$ be fixed. It follows from the inequalities (8.125)–(8.127) that the operator $e^{\mathscr{A}_k^\omega (t-\tau)}$, restricted to $D(\mathscr{A}_k^\omega)$, is bounded as an operator from $\boldsymbol{L}^2_\sigma(\mathcal{D})$ to $\boldsymbol{W}^{2,2}(\mathcal{D})$.

Let us further show that $\mathscr{B}$ is a compact operator from $\boldsymbol{W}^{2,2}(\mathcal{D})$ to $\boldsymbol{L}^2_\sigma(\mathcal{D})$. Let $\{\boldsymbol{\varphi}_n\}$ be a bounded sequence in $\boldsymbol{W}^{2,2}(\mathcal{D})$ and $R > 0$. Then

$$\begin{aligned}
\| \mathscr{B}\boldsymbol{\varphi}_n - \mathscr{B}\boldsymbol{\varphi}_m \|_2 \, &= \, \big\| \mathbf{b} \cdot \nabla(\boldsymbol{\varphi}_n - \boldsymbol{\varphi}_m) + B \cdot (\boldsymbol{\varphi}_n - \boldsymbol{\varphi}_m) \big\|_2 \\
&\leq \, \big\| \mathbf{b} \cdot \nabla(\boldsymbol{\varphi}_n - \boldsymbol{\varphi}_m) + B \cdot (\boldsymbol{\varphi}_n - \boldsymbol{\varphi}_m) \big\|_{2;\, \mathcal{D}_R} \\
&\quad + \big\| \mathbf{b} \cdot \nabla(\boldsymbol{\varphi}_n - \boldsymbol{\varphi}_m) + B \cdot (\boldsymbol{\varphi}_n - \boldsymbol{\varphi}_m) \big\|_{2;\, \mathcal{D} \smallsetminus \mathcal{D}_R} \\
&\leq \, \| \mathbf{b} \|_{3+\iota;\, \mathcal{D}_R}\, \| \nabla(\boldsymbol{\varphi}_n - \boldsymbol{\varphi}_m) \|_{r;\, \mathcal{D}_R} \\
&\quad + \| B \|_{2+\iota;\, \mathcal{D}_R}\, \| \boldsymbol{\varphi}_n - \boldsymbol{\varphi}_m \|_{s;\, \mathcal{D}_R} \\
&\quad + \| \mathbf{b} \|_{3+\iota;\, \mathcal{D} \smallsetminus \mathcal{D}_R}\, \| \nabla(\boldsymbol{\varphi}_n - \boldsymbol{\varphi}_m) \|_{r;\, \mathcal{D} \smallsetminus \mathcal{D}_R} \\
&\quad + \| B \|_{2+\iota;\, \mathcal{D} \smallsetminus \mathcal{D}_R}\, \| \boldsymbol{\varphi}_n - \boldsymbol{\varphi}_m \|_{s;\, \mathcal{D} \smallsetminus \mathcal{D}_R} \qquad (8.133)
\end{aligned}$$

for $m, n \in \mathbb{N}$, where $\mathcal{D}_R = \mathcal{D} \cap B_R(\mathbf{0})$. (Here and further on in this section, r and s are the numbers, given by (8.132).) Notice that the inclusion $\{\boldsymbol{\varphi}_n\} \subset \boldsymbol{W}^{2,2}(\mathcal{D})$ and the imbedding inequalities imply that the sequences $\{\nabla(\boldsymbol{\varphi}_n - \boldsymbol{\varphi}_m)\}$ and $\{\boldsymbol{\varphi}_n - \boldsymbol{\varphi}_m\}$

are bounded in $L^r(\mathcal{D})^{3\times3}$ and $\boldsymbol{L}^s(\mathcal{D})$, respectively. Let $\epsilon > 0$ be given. Due to the assumptions (8.14) and (8.15) on the functions B and $\mathbf{b}$, there exists $R > 0$ so large that the terms containing the norms on $\mathcal{D} \setminus \mathcal{D}_R$ are altogether less than or equal to $\epsilon/2$, independently of m and n. Applying the compact imbedding $\boldsymbol{W}^{2,2}(\mathcal{D}_R) \hookrightarrow\hookrightarrow \boldsymbol{W}^{1,r}(\mathcal{D}_R)$ and $\boldsymbol{W}^{2,2}(\mathcal{D}_R) \hookrightarrow\hookrightarrow \boldsymbol{L}^s(\mathcal{D}_R)$, one can show that there exists a subsequence of $\{\boldsymbol{\varphi}_n\}$ (which we denote again by $\{\boldsymbol{\varphi}_n\}$), such that the sequence $\{\nabla\boldsymbol{\varphi}_n\}$ converges in $L^r(\mathcal{D}_R)^{3\times3}$ and the sequence $\{\boldsymbol{\varphi}_b\}$ converges in $\boldsymbol{L}^s(\mathcal{D}_R)$. Hence the terms containing the norms on $\mathcal{D}_R$ are altogether also less than or equal to $\epsilon/2$ for m and n sufficiently large. This shows that $\{\mathscr{B}\boldsymbol{\varphi}_n\}$ is a Cauchy sequence in $\boldsymbol{L}^2_\sigma(\mathcal{D})$. The proof of the compactness of $\mathscr{B}$, as an operator from $\boldsymbol{W}^{2,2}(\mathcal{D})$ to $\boldsymbol{L}^2_\sigma(\mathcal{D})$, is completed.

Now, we observe that the composite operator $\mathscr{B}\,\mathrm{e}^{\mathscr{A}^\omega_k(t-\tau)}$ is compact in $\boldsymbol{L}^2_\sigma(\mathcal{D})$, because $\mathrm{e}^{\mathscr{A}^\omega_k(t-\tau)}$ is bounded from $\boldsymbol{L}^2_\sigma(\mathcal{D})$ to $\boldsymbol{W}^{2,2}(\mathcal{D})$ and $\mathscr{B}$ is compact from $\boldsymbol{W}^{2,2}(\mathcal{D})$ to $\boldsymbol{L}^2_\sigma(\mathcal{D})$. □

Lemma 8.27 *Let $t > 0$. Then the operator $\int_0^t \mathrm{e}^{\mathscr{L}^\omega_k\tau}\,\mathscr{B}\,\mathrm{e}^{\mathscr{A}^\omega_k(t-\tau)}\,d\tau$ is compact in $\boldsymbol{L}^2_\sigma(\mathcal{D})$.* □

Proof Let us at first show that for each given $\boldsymbol{\varphi} \in \boldsymbol{L}^2_\sigma(\mathcal{D})$, the function $\mathrm{e}^{\mathscr{L}^\omega_k\tau}\,\mathscr{B}\,\mathrm{e}^{\mathscr{A}^\omega_k(t-\tau)}\boldsymbol{\varphi}$ is continuous (in the norm of $\boldsymbol{L}^2_\sigma(\mathcal{D})$) in dependence on τ in the interval $[0,t)$. Thus, let $\tau \in [0,t)$ be fixed and let δ satisfy $-\tau < \delta < t-\tau$. We have

$$
\begin{aligned}
&\big\|\mathrm{e}^{\mathscr{L}^\omega_k(\tau+\delta)}\,\mathscr{B}\,\mathrm{e}^{\mathscr{A}^\omega_k(t-\tau-\delta)}\boldsymbol{\varphi} - \mathrm{e}^{\mathscr{L}^\omega_k\tau}\,\mathscr{B}\,\mathrm{e}^{\mathscr{A}^\omega_k(t-\tau)}\boldsymbol{\varphi}\big\|_2 \\
&\qquad \leq \big\|\mathrm{e}^{\mathscr{L}^\omega_k(\tau+\delta)}\,\mathscr{B}\,(\mathrm{e}^{\mathscr{A}^\omega_k(t-\tau-\delta)} - \mathrm{e}^{\mathscr{A}^\omega_k(t-\tau)})\boldsymbol{\varphi}\big\|_2 \\
&\qquad\quad + \big\|(\mathrm{e}^{\mathscr{L}^\omega_k(\tau+\delta)} - \mathrm{e}^{\mathscr{L}^\omega_k\tau})\,\mathscr{B}\,\mathrm{e}^{\mathscr{A}^\omega_k(t-\tau)}\boldsymbol{\varphi}\big\|_2.
\end{aligned}
$$

The second term on the right-hand side tends to zero for $\delta \to 0$ due to the strong continuity of the semigroup $\mathrm{e}^{\mathscr{L}^\omega_k t}$. The first term on the right-hand side is less than or equal to

$$
\begin{aligned}
&M_\mu\,\mathrm{e}^{\mu(\tau+\delta)}\,\big\|\mathscr{B}\,(\mathrm{e}^{\mathscr{A}^\omega_k(t-\tau-\delta)} - \mathrm{e}^{\mathscr{A}^\omega_k(t-\tau)})\boldsymbol{\varphi}\big\|_2 \\
&\quad \leq M_\mu\,\mathrm{e}^{\mu(\tau+\delta)}\,\big(\|B\cdot(\mathrm{e}^{\mathscr{A}^\omega_k(t-\tau-\delta)} - \mathrm{e}^{\mathscr{A}^\omega_k(t-\tau)})\boldsymbol{\varphi}\|_2 \\
&\qquad + \|\mathbf{b}\cdot\nabla(\mathrm{e}^{\mathscr{A}^\omega_k(t-\tau-\delta)} - \mathrm{e}^{\mathscr{A}^\omega_k(t-\tau)})\boldsymbol{\varphi}\|_2\big) \\
&\quad \leq M_\mu\,\mathrm{e}^{\mu(\tau+\delta)}\,\big(\|B\|_{2+\iota}\,\|(\mathrm{e}^{\mathscr{A}^\omega_k(t-\tau-\delta)} - \mathrm{e}^{\mathscr{A}^\omega_k(t-\tau)})\boldsymbol{\varphi}\|_s \\
&\qquad + \|\mathbf{b}\|_{3+\iota}\,\|\nabla(\mathrm{e}^{\mathscr{A}^\omega_k(t-\tau-\delta)} - \mathrm{e}^{\mathscr{A}^\omega_k(t-\tau)})\boldsymbol{\varphi}\|_r\big) \\
&\quad \leq M_\mu\,\mathrm{e}^{\mu(\tau+\delta)}\,\big(\|B\|_{2+\iota}\,\|\mathrm{e}^{\mathscr{A}^\omega_k(t-\tau)/2}\,(\mathrm{e}^{\mathscr{A}^\omega_k((t-\tau)/2-\delta)} - \mathrm{e}^{\mathscr{A}^\omega_k(t-\tau)/2})\boldsymbol{\varphi}\|_s \\
&\qquad + \|\mathbf{b}\|_{3+\iota}\,\|\nabla(\,\mathrm{e}^{\mathscr{A}^\omega_k(t-\tau)/2}\,(\mathrm{e}^{\mathscr{A}^\omega_k((t-\tau)/2-\delta)} - \mathrm{e}^{\mathscr{A}^\omega_k(t-\tau)/2})\boldsymbol{\varphi}\|_r\big)
\end{aligned}
$$

$$\leq M_\mu \, e^{\mu(\tau+\delta)} \Big(\|B\|_{2+\iota}\, c\,(t-\tau)^{-\frac{3}{4}(\frac{1}{2}-\frac{1}{s})} \, \|e^{\mathscr{A}_k^\omega((t-\tau)/2-\delta)} - e^{\mathscr{A}_k^\omega(t-\tau)/2})\boldsymbol{\varphi}\|_2$$
$$+ \|\mathbf{b}\|_{3+\iota}\, c\,(t-\tau)^{-\frac{1}{4}-\frac{3}{4}(\frac{1}{2}-\frac{1}{r})} \, \|e^{\mathscr{A}_k^\omega((t-\tau)/2-\delta)} - e^{\mathscr{A}_k^\omega(t-\tau)/2})\boldsymbol{\varphi}\|_2 \Big),$$

which tends to zero for $\delta \to 0$ due to the strong continuity of the semigroup $e^{\mathscr{A}_k^\omega t}$. The continuity of $e^{\mathscr{L}_k^\omega \tau}\, \mathscr{B}\, e^{\mathscr{A}_k^\omega(t-\tau)} \boldsymbol{\varphi}$ in dependence on τ is proven.

Thus, $e^{\mathscr{L}_k^\omega \tau}\, \mathscr{B}\, e^{\mathscr{A}_k^\omega(t-\tau)}$ is a family of compact linear operators in $\boldsymbol{L}^2_\sigma(\mathcal{D})$, strongly continuous in dependence on τ for $\tau \in [0,\xi]$ for every $0 < \xi < t$. This information enables us to apply Theorem C.7 from [56, p. 525] and conclude that $\int_0^{t-\xi} e^{\mathscr{L}_k^\omega \tau}\, \mathscr{B}\, e^{\mathscr{A}_k^\omega(t-\tau)}\, d\tau$ is a compact operator in $\boldsymbol{L}_\sigma$ for every $\xi \in (0,t)$. Since

$$\int_0^t e^{\mathscr{L}_k^\omega \tau}\, \mathscr{B}\, e^{\mathscr{A}_k^\omega(t-\tau)}\, d\tau = \lim_{\xi\to 0+} \int_0^{t-\xi} e^{\mathscr{L}_k^\omega \tau}\, \mathscr{B}\, e^{\mathscr{A}_k^\omega(t-\tau)}\, d\tau$$

in the topology of $B(\text{Ł}^2(\mathcal{D}))$ (the space of bounded linear operators from $\boldsymbol{L}^2_\sigma(\mathcal{D})$ to itself) and the subspace of compact linear operators in $\boldsymbol{L}^2_\sigma(\mathcal{D})$ is closed in $B(\boldsymbol{L}^2_\sigma(\mathcal{D}))$, we observe that the operator $\int_0^t e^{\mathscr{L}_k^\omega \tau}\, \mathscr{B}\, e^{\mathscr{A}_k^\omega(t-\tau)}\, d\tau$ is compact, too.

□

The formula (8.130) and Lemma 8.27 show that, for any $t > 0$, the operators $e^{\mathscr{L}_k^\omega t}$ and $e^{\mathscr{A}_k^\omega t}$ differ just by an additive compact operator. Thus, $\mathrm{Sp}^*_{\mathrm{ess}}(e^{\mathscr{L}_k^\omega t}) = \mathrm{Sp}^*_{\mathrm{ess}}(e^{\mathscr{A}_k^\omega t})$. Consequently, $r^*_{\mathrm{ess}}(e^{\mathscr{L}_k^\omega t}) = r^*_{\mathrm{ess}}(e^{\mathscr{A}_k^\omega t})$ for each $t > 0$ and $\gamma^*_{\mathrm{ess}}(e^{\mathscr{L}_k^\omega t}) = \gamma^*_{\mathrm{ess}}(e^{\mathscr{A}_k^\omega t}) = 0$. Since $\gamma(e^{\mathscr{L}_k^\omega t}) = \max\{\gamma^*_{\mathrm{ess}}(e^{\mathscr{L}_k^\omega t});\ s(\mathscr{L}_k^\omega)\} = \max\{0;\ s(\mathscr{L}_k^\omega)\}$, we obtain the equality

$$\gamma(e^{\mathscr{L}_k^\omega t}) = s(\mathscr{L}_k^\omega).$$

The next theorem summarizes our knowledge of the semigroup, generated by the perturbed Oseen type operator $\mathscr{L}_k^\omega$.

Theorem 8.18 *The operator $\mathscr{L}_k^\omega$ generates a C_0-semigroup $e^{\mathscr{L}_k^\omega t}$ in $\boldsymbol{L}^2_\sigma(\mathcal{D})$. The uniform growth bound $\gamma(e^{\mathscr{L}_k^\omega t})$ of the semigroup $e^{\mathscr{L}_k^\omega t}$ and the spectral bound $s(\mathscr{L}_k^\omega)$ of the operator $\mathscr{L}_k^\omega$ are equal.* □

Part III
Navier–Stokes Type Equations

Chapter 9
A Brief Introduction to the Existence Theory for the Stationary Navier–Stokes Type Problem Around Translating and Rotating Body

The studies of stationary Navier–Stokes type equations, describing flow of a viscous incompressible fluid around a rotating body, were mainly initiated by the series of pioneering works of G. P. Galdi and his coworkers, see e.g., the papers [92, 96, 97, 100–106, 108–111], where the existence of solutions was obtained in Sobolev spaces with the Hilbert space structure. Important existential results in the general L^p-framework, including results in Lorentz spaces, were also obtained by R. Farwig and T. Hishida [66] in the purely rotating case, i.e., the case when the considered body is only rotating, but not translating. Both rotation and translation of the body were considered by H. Heck, H. Kim, and H. Kozono in paper [131], where the authors obtained results in the generalized L^p-framework.

The purpose of this chapter is to provide a brief introduction and survey of the existential theory for the stationary Navier–Stokes type problems, related to flows around a rotating and translating body. Note that the problems are stationary in the coordinate system, attached to the body—the existing solutions are, however, time periodic from the point of view of an external observer.

9.1 L^2 Theory

9.1.1 Navier–Stokes Type Problem Without the Oseen Term

We are interested in the following problem in the exterior domain $\mathcal{D}$:

$$\left.\begin{aligned} \mathrm{Re}(\mathbf{v}\cdot\nabla\mathbf{v} - \boldsymbol{\mu}\wedge\mathbf{x}\cdot\nabla\mathbf{v} + \boldsymbol{\mu}\wedge\mathbf{v}) &= \Delta\mathbf{v} - \nabla p \\ \operatorname{div}\mathbf{v} &= 0 \\ \lim_{|\mathbf{x}|\to\infty}\mathbf{v}(\mathbf{x}) &= 0 \end{aligned}\right\} \text{ in } \mathcal{D} \tag{9.1}$$

Š. Nečasová et al., *Mathematical Theory of a Fluid Flow Around a Rotating and Translating Body*, Springer Monographs in Mathematics,
https://doi.org/10.1007/978-3-031-84120-0_9

$$\mathbf{v}(\mathbf{x}) = \boldsymbol{\mu} \wedge \mathbf{x}, \ \mathbf{x} \in \partial\mathcal{D}, \tag{9.2}$$

see also (1.47).

G. P. Galdi [97] showed that if the number Re is not too Large, then there exists a unique steady solution, such that the velocity $\mathbf{v}(\mathbf{x})$ and its gradient $\nabla\mathbf{v}(\mathbf{x})$ decay like $|\mathbf{x}|^{-1}$ and $|\mathbf{x}|^{-2}$, respectively. Moreover, the pressure field $p(\mathbf{x})$ and its gradient $\nabla p(\mathbf{x})$ decay like $|\mathbf{x}|^{-2}$ and $|\mathbf{x}|^{-3}$, respectively. These solutions are "physically reasonable" in the sense of Finn, see [82].

Let us introduce the following norms:

$$[\|\mathbf{f}\|]_k = \text{ess sup}_{\mathbf{x}\in\mathcal{D}}((|x|^k + 1)|\mathbf{f}(\mathbf{x})|),$$

$$[\|g\|]_k = \text{ess sup}_{\mathbf{x}\in\mathcal{D}}((|x|^k + 1)|g(\mathbf{x})|).$$

The main result from the work [97] reads as follows:

We denote $\mathcal{D}_R = \mathcal{D} \cap B_R$, $\mathcal{D}^R = \mathcal{D} \cap B_R^c$, where B_R is a ball with radius R.

Theorem 9.1 (The Navier–Stokes Type Problem in an Exterior Domain) *Let $\mathcal{D} \equiv \mathbb{R}^3 \setminus \mathcal{B}$ be an exterior domain of class C^2, $R > d(\mathcal{B})$, $q > 1$, where $d(\mathcal{B})$ is the diameter of $\mathcal{B}$. Then there exists a constant* $\text{Re}_0 > 0$ *depending on $\mathcal{D}$, R and q, such that if* $0 < \text{Re} < \text{Re}_0$, *then the problem* (9.1) *admits a unique solution* $(\mathbf{v}, p)$, *satisfying*

$$\|\mathbf{v}\|_{2,q,\mathcal{D}_R} + \|\nabla^2\mathbf{v}\|_2 + [\|\mathbf{v}\|]_1 + \|\nabla\mathbf{v}\|_2 < \infty \tag{9.3}$$

$$[\|p\|]_2 + [\|\nabla p\|]_{3,\mathcal{D}^R} < \infty. \tag{9.4}$$

Moreover, $\mathbf{v}, p \in C^\infty(\mathcal{D})$. □

To prove this result, the author began with the linear theory in the whole space and in the exterior domain and then applied an appropriate fixed point argument in order to extend the result to the nonlinear problem. Concretely, as intermediate steps, he proved the next two theorems, which concern the linear problems:

Theorem 9.2 (The Stokes Type Problem in the Whole Space) *Assume that:*

(A) *$F = \{F_{ij}\}$ is a second order tensor field in $\mathbb{R}^3$ such that*

$$[\|F\|]_2 + [\|\partial_i F_{ij}\boldsymbol{e}_j\|]_3 + [\|\partial_j F_{ij}\boldsymbol{e}_i\|]_4 < \infty.$$

(B) $\mathbf{f} \in W_0^{1,q}(B_\varrho)$ *for some* $\varrho > 0$ *and* $q > 3$. □

Then the problem

$$\left.\begin{array}{c}\Delta\mathbf{u} + \text{Re}(\boldsymbol{\mu} \wedge \mathbf{x}\cdot\nabla\mathbf{u} - \boldsymbol{\mu} \wedge \mathbf{u}) = \nabla\varphi + \partial_i F_{ij}\boldsymbol{e}_j + \mathbf{f} \\ \operatorname{div}\mathbf{u} = 0\end{array}\right\} \text{ in } \mathbb{R}^3 \tag{9.5}$$

has at least one solution $(\mathbf{u}, \varphi)$ *such that*

$$\begin{aligned}&\mathbf{u} \in \boldsymbol{W}^{2,2}_{\text{loc}}(\mathbb{R}^3) \cap \widehat{\boldsymbol{H}}^{2,2}(\mathbb{R}^3) \cap \widehat{\boldsymbol{H}}^{1,2}(\mathbb{R}^3) \cap \boldsymbol{L}^6(\mathbb{R}^3) \cap \boldsymbol{L}^\infty(\mathbb{R}^3), \\ &[\|\mathbf{u}\|]_1 + [\|\nabla\mathbf{u}\|]_2 < \infty, \\ &\varphi \in W^{1,r}(\mathbb{R}^3) \cap \widehat{H}^{2,s}(\mathbb{R}^3), \quad \textit{for all } q \ge s > 1,\ r > \frac{3}{2}, \\ &[\|\varphi\|]_2 + [\|\nabla\varphi\|]_3 < \infty.\end{aligned} \tag{9.6}$$

Moreover, the following estimate holds

$$\begin{aligned}&\|\nabla^2\mathbf{u}\|_2 + [\|\mathbf{u}\|]_1 + [\|\nabla\,\mathbf{u}\|]_2 + [\|\varphi\|]_2 + [\|\nabla\,\varphi\|]_3 + \|D^2\varphi\|_s \\ &\quad \le c(\||F|\|_2 + [\|\partial_i F_{ij} e_j\|]_3 + [\|\partial_j\partial_i F_{ij}\|]_4 + \|\mathbf{f}\|_{q,B_\varrho} + \|\operatorname{div}\mathbf{f}\|_{q,B_\varrho}).\end{aligned} \tag{9.7}$$

Finally, if $(\mathbf{u}_1, \varphi_1)$ *is another solution to (9.5) with*

$$\begin{aligned}\mathbf{u}_1 &\in \boldsymbol{W}^{2,2}_{\text{loc}}(\mathbb{R}^3) \cap \widehat{\boldsymbol{H}}^{1,2}(\mathbb{R}^3) \cap \boldsymbol{L}^6(\mathbb{R}^3), \\ \varphi_1 &\in W^{1,2}_{\text{loc}}(\mathbb{R}^3),\end{aligned}$$

then $\mathbf{u} \equiv \mathbf{u}_1$, $\varphi \equiv \varphi_1 + \text{const}$.

The next theorem concerns the problem in the exterior domain $\mathcal{D}$:

$$\begin{aligned}\Delta\mathbf{v} + \operatorname{Re}(\boldsymbol{\mu} \wedge \mathbf{x} \cdot \nabla\mathbf{v} - \boldsymbol{\mu} \wedge \mathbf{v}) &= \nabla p + \partial_i \mathcal{F}_{ij} \boldsymbol{e}_j \\ \operatorname{div}\mathbf{v} &= 0 \\ \mathbf{v} &= \mathbf{v}_*, \quad \mathbf{x} \in \partial\mathcal{D}.\end{aligned} \tag{9.8}$$

Theorem 9.3 (The Stokes Type Problem in the Exterior Domain) *Let* $\mathcal{D}$ *be an exterior domain of class* C^2 *and let* $\mathcal{F} = \{\mathcal{F}_{ij}\}$ *be a second order tensor field in* $\mathcal{D}$ *such that*

$$[\|\mathcal{F}\|]_2 + [\|\partial_i \mathcal{F}_{ij} e_j\|]_3 + [\|\partial_j\partial_i \mathcal{F}_{ij}\|]_4 < \infty.$$

Let $\mathbf{v}_* \in W^{2k-1/q,q}(\partial\mathcal{D})$, $q > 1$ *satisfy* $\int_{\partial\mathcal{D}} \mathbf{v}_* \cdot \boldsymbol{n} = 0$. *Then the problem* (9.8) *has a unique solution* (v, p), *satisfying*

$$\begin{aligned}&\mathbf{v} \in \boldsymbol{W}^{2,q}_{\text{loc}}(\overline{\mathcal{D}}) \cap \widehat{\boldsymbol{H}}^{2,2}(\mathcal{D}), \quad \textit{for all } q \ge 1, \\ &[\|\mathbf{v}\|]_1 + [\|\nabla\mathbf{v}\|]_2 < \infty \\ &p \in W^{1,q}_{loc}(\overline{\mathcal{D}}), \quad \textit{for all } q \ge 1,\ [\|\,p\,\|]_2 + [\|\,\nabla p\,\|]_{3,\mathcal{D}^R}, \\ &\textit{for all } R > d(\mathcal{B}).\end{aligned} \tag{9.9}$$

Moreover. the estimate

$$\|\mathbf{v}\|_{2,q,\mathcal{D}_R} + \|\nabla^2\mathbf{v}\|_2 + [\|\mathbf{v}\|]_1 + [\|\nabla\mathbf{v}\|]_2 + [\|p\|]_2 + [\|\nabla p\|]_{3,\mathcal{D}^R}$$
$$\leq c([\|\mathcal{F}\|]_2 + [\|\partial_i\mathcal{F}_{ij}e_j\|]_3 + [\|\partial_j\partial_i\mathcal{F}_{ij}\|]_4 + \|\mathbf{v}_*\|_{2-1/q,q,\partial\mathcal{D}}), \tag{9.10}$$

holds with c being a constant depending on $\mathcal{D}$, q, R, and B (whenever Re $\in [0, B]$*), and $\mathcal{D}_R = \mathcal{D}\cap B_R$, $\mathcal{D}^R = \mathcal{D}\cap B_R^c$.* □

Proof of Theorem 9.1 The proof of Theorem 9.1 is based on Theorem 9.3 and on an appropriate fixed point theorem. We sketch the idea, used in [97].

Let us fix $R > d(\mathcal{B})$, $q > 1$. We introduce the space

$$K_{R,q} = \{\boldsymbol{\varphi} \in W^{1,2}_{\text{loc}}(\mathcal{D})\colon\ \operatorname{div}\boldsymbol{\varphi} = 0 \text{ in } \Omega,\ \|\boldsymbol{\varphi}\|_{2,q,\mathcal{D}_R}$$
$$+ \|D^2\boldsymbol{\varphi}\|_2 + [\|\boldsymbol{\varphi}\|]_1 + [\|\nabla\boldsymbol{\varphi}\|]_2 < \infty\},$$

where $K_{R,q}$ is the Banach space with the norm

$$\|\boldsymbol{\varphi}\|_{K_{R,q}} \equiv \|\boldsymbol{\varphi}\|_{2,q,\mathcal{D}_R} + \|D^2\boldsymbol{\varphi}\|_2 + \||\boldsymbol{\varphi}|\|_1 + \||\nabla\boldsymbol{\varphi}|\|_2.$$

Consider the mapping

$$\mathcal{H}\colon \boldsymbol{\varphi} \in K_{R,q} \mapsto \mathbf{v},$$

where $\mathbf{v}$ is a solution of the problem

$$\begin{aligned} \Delta\mathbf{v} + \text{Re}(\boldsymbol{\mu}\wedge\mathbf{x}\cdot\nabla\mathbf{v} - \boldsymbol{\mu}\wedge\mathbf{v}) &= \nabla p + \partial_i\mathcal{F}_{ij}\boldsymbol{e}_j, \\ \operatorname{div} v &= 0 \\ \mathbf{v} &= \boldsymbol{\mu}\wedge\mathbf{x},\ \mathbf{x}\in\partial\mathcal{D}, \end{aligned} \tag{9.11}$$

where $\mathcal{F}_{ij} = \text{Re}\,\varphi_i\varphi_j$. Applying the operator div to Eq. (9.11), we obtain

$$\partial_j\mathcal{F}_{ij}\boldsymbol{e}_j = \text{Re}\,\boldsymbol{\varphi}\cdot\nabla\boldsymbol{\varphi},\quad \partial_i\partial_j\mathcal{F}_{ij} = \text{Re}\,\nabla\boldsymbol{\varphi}\cdot(\nabla\boldsymbol{\varphi})^T.$$

Since $\boldsymbol{\varphi} \in K_{R,q}$, we have

$$[\|\mathcal{F}\|]_2 + [\|\partial_i\mathcal{F}_{ij}e_j\|]_3 + [\|\partial_i\partial_j\mathcal{F}_{ij}\|]_4 \leq c\ \text{Re}\ \|\varphi\|^2_{K_{R,q}}.$$

Due to Theorem 9.3, the solution $\mathbf{v}$ of (9.11) is unique and

$$\|\mathbf{v}\|_{K_{R,q}} + [\||p|\|]_2 + [\|\nabla p\|]_{3,\mathcal{D}^R} \leq C\,(\text{Re}\|\varphi\|^2_{K_{R,q}} + 1). \tag{9.12}$$

Put $\mathbf{v}_1 = \mathcal{H}(\boldsymbol{\varphi}_1)$, $\mathbf{v}_2 = \mathcal{H}(\boldsymbol{\varphi}_2)$, $\boldsymbol{\varphi}_1, \boldsymbol{\varphi}_2 \in K_{R,q}$. Denoting $\mathbf{v} = \mathbf{v}_1 - \mathbf{v}_2$, $\boldsymbol{\varphi} = \boldsymbol{\varphi}_1 - \boldsymbol{\varphi}_2$ we obtain

$$\left.\begin{aligned} \Delta\mathbf{v} + \mathrm{Re}\,(\boldsymbol{\mu} \wedge \mathbf{x} \cdot \nabla\mathbf{v} - \boldsymbol{\mu} \wedge \mathbf{v}) = \nabla p + \mathrm{Re}\,\partial_i(\varphi_i\varphi_{1j} + \varphi_{2i}\varphi_j)\boldsymbol{e}_j \\ \operatorname{div}\mathbf{v} = 0 \end{aligned}\right\} \text{in } \mathcal{D} \tag{9.13}$$

$$\mathbf{v} = 0 \qquad \text{at} \qquad \partial\mathcal{D} \tag{9.14}$$

for some p. Then, applying Theorem 9.3, we get

$$\|\mathbf{v}_1 - \mathbf{v}_2\|_{K_{R,q}} \le c\,\mathrm{Re}\,(\|\boldsymbol{\varphi}_1\|_{K_{R,q}} + \|\boldsymbol{\varphi}_2\|_{K_{R,q}})\,\|\varphi_1 - \varphi_2\|_{K_{R,q}}. \tag{9.15}$$

One can deduce from (9.13), (9.15) that the mapping $\mathcal{H}$ has a fixed point in $K_{R,q}$ for Re sufficiently small. It follows from (9.13) that the pressure satisfies the estimate from Theorem 9.3. Using the boot-strap argument and the regularity of solutions of the Stokes type problem, we deduce that (v, p) is of class $C^\infty(\mathcal{D})$.

The last step is to show the uniqueness of the solution $\mathbf{u}$ of (9.1).

Let $\mathbf{u} = \mathbf{v}_1 - \mathbf{v}_2$, $\varphi = p_2 - p_1$. with $\{\mathbf{v}_i, p_i\}$ two smooth solutions to (9.1). Then

$$\mathrm{Re}(\mathbf{u} \cdot \nabla\mathbf{u} + \mathbf{u} \cdot \nabla\mathbf{v}_1 + \mathbf{v}_1 \cdot \nabla\mathbf{u} - \mathbf{u} \wedge \mathbf{x} \cdot \nabla\mathbf{u} + \boldsymbol{\mu} \wedge \mathbf{u}) = \Delta\mathbf{u} - \nabla\varphi \tag{9.16}$$

$$\operatorname{div}\mathbf{u} = 0$$

$$\lim_{|x|\to\infty} \mathbf{u} = 0$$

$$\mathbf{u}(x) = 0 \quad \text{on } \partial\mathcal{D}.$$

Multiplying (9.16) by $\mathbf{u}$, integrating by parts over $\mathcal{D}_R$, we get

$$\int_{\mathcal{D}_R} |\nabla\mathbf{u}|^2 = \int_{\partial B_R} \left\{\frac{\partial\mathbf{u}}{\partial\mathbf{n}} \cdot \mathbf{u} - \mathrm{Re}\left[\frac{1}{2}|\mathbf{u}|^2(\mathbf{u} + \mathbf{v}_1) \cdot N - \varphi N\right]\right\} + \mathrm{Re}\int_{\mathcal{D}_R} \mathbf{u}\nabla\mathbf{v}_1 \cdot \mathbf{u}, \tag{9.17}$$

where

$$\mathbf{n} = \frac{\mathbf{x}}{|\mathbf{x}|}.$$

Passing $R \to \infty$, we get

$$\int_{\mathcal{D}} |\nabla\mathbf{u}|^2 = \mathrm{Re}\int_{\mathcal{D}} \mathbf{u} \cdot \nabla\mathbf{v}_1 \cdot \mathbf{u}.$$

Since $\|\nabla \mathbf{v}_1\|_2 \le c$, the above identity implies

$$\int_{\mathcal{D}} |\nabla \mathbf{u}|^2 \le c_1 \mathrm{Re} \int \frac{|\mathbf{u}|^2}{|\mathbf{x}|^2}. \tag{9.18}$$

It follows from [93] that

$$\int_{\mathcal{D}} \frac{|\mathbf{u}|^2}{|\mathbf{x}|^2} \le 4 \int_{\mathcal{D}} |\nabla \mathbf{u}|^2.$$

Now, it follows from (9.18) and $(9.16)_4$ that if Re is sufficiently small, then $\mathbf{u} = \mathbf{0}$.

□

9.1.2 The Navier–Stokes Type Problem with the Oseen Term

The question of existence of a solution in the case when the velocity at infinity approaches some nonzero constant vector field was studied by G. P. Galdi and A. L. Silvestre in papers [109, 110]. They considered the problem

$$\left.\begin{aligned} \Delta \mathbf{u} = \tau \left(\mathbf{u} \cdot \nabla \mathbf{u} - \frac{\partial \mathbf{u}}{\partial x_3} \right) &+ \sqrt{\rho}\,(\mathbf{e}_3 \wedge \mathbf{u} - \mathbf{e}_3 \wedge \mathbf{x} \cdot \nabla \mathbf{u}) \\ +\nabla p + \tau \mathbf{f} & \\ \nabla \cdot \mathbf{u} = 0 & \end{aligned}\right\} \text{ in } \mathcal{D} \tag{9.19}$$

$$\begin{aligned} &\mathbf{u} = \mathbf{u}_* \text{ at } \partial D \\ &\lim_{|\mathbf{x}| \to \infty} \mathbf{u}(\mathbf{x}) = 0, \end{aligned} \tag{9.20}$$

compare with (1.50). The main result of [110] reads as follows.

Theorem 9.4 *Let $\mathcal{D}$ be an exterior domain of class C^2 and let $\tau \in [0, B_1]$, $\rho \in [0, B_2]$ for given $B_1, B_2 > 0$. Moreover, let $\mathbf{f} = \nabla \cdot \mathcal{F} \in L^2(\mathcal{D})$ with $[\|\mathcal{F}\|]_2 < \infty$ and $\mathbf{u}_* \in W^{3/2,2}(\partial\mathcal{D})$. Then:*

(i) Existence:
There exists a positive constant $K_1 = K_1(\mathcal{D}, B_1, B_2)$, such that if

$$\tau(\|\mathbf{u}_*\|_{3/2,2} + \|\nabla \cdot \mathcal{F}\|_2 + [\|\mathcal{F}\|]_2) < K_1, \tag{9.21}$$

then the problem (9.19)–(9.20) has a solution $(\mathbf{u}, p)$, satisfying

$$\begin{aligned} &\mathbf{u} \in W^{2,2}_{\mathrm{loc}}(\overline{\mathcal{D}}) \cap \widehat{\boldsymbol{H}}^{2,2}(\mathcal{D}) \cap \widehat{\boldsymbol{H}}^{1,2}(\mathcal{D}), \\ &[|\mathbf{u}|]_1 < \infty, \\ &p \in W^{1,2}(\mathcal{D}). \end{aligned}$$

Moreover, the solution satisfies

$$|\mathbf{u}|_{2,2} + |\mathbf{u}|_{1,2} + [\|\mathbf{u}\|]_1 + |p|_{1,2} + \|p\|_2 \le C(\|\nabla \cdot \mathcal{F}\|_2 + [\|\mathcal{F}\|]_2 + \|\mathbf{u}_*\|_{3/2,2})$$

with $C = C(\mathcal{D}, B_1, B_2) > 0$, *and the energy equality*

$$2\|D(\mathbf{u})\|_2^2 - \int_{\partial\mathcal{D}} \mathbf{u}_* \cdot T(\mathbf{u}, p) \cdot n + \frac{\tau}{2}\int_{\partial\mathcal{D}} |\mathbf{u}_*|^2 \mathbf{u}_* \cdot \mathbf{n}$$
$$- \frac{1}{2}\int_{\partial\mathcal{D}} |\mathbf{u}_*|^2 (\tau\,\mathbf{e}_1 + \sqrt{\rho}\,\mathbf{e}_1 \wedge \mathbf{x}) \cdot \mathbf{n} + \tau \int_{\mathcal{D}} \mathbf{f} \cdot \mathbf{n} = 0,$$

where $T(\mathbf{u}, p) = 2D(\mathbf{u}) - pI$ *(the Cauchy stress tensor),* I *is the identity matrix,* $2D(\mathbf{u}) = \nabla\mathbf{u} + (\nabla\mathbf{u})^T$.

(ii) Uniqueness:
Let $(\mathbf{u}_1, p_1)$ *be another solution to (9.19) with*

$$\mathbf{u}_1 \in W^{2,2}_{\text{loc}}(\overline{\mathcal{D}}) \cap \widehat{H}^{1,2}(\mathcal{D}),\ [|\mathbf{u}|]_1 < \infty, \quad p_1 \in W^{1,2}_{\text{loc}}(\overline{\mathcal{D}}) \cap L^2(\mathcal{D}).$$

If condition (9.21) holds, then $\mathbf{u} = \mathbf{u}_1$, $p = p_1$.

□

To show the existence of a solution, the authors at first successively study the Oseen problem in the whole space and in the exterior domain. For completeness we also cite these results and then briefly describe the proof of Theorem 9.4.

The next auxiliary theorem concerns the L^2-theory for the Oseen type problem in the whole space:

$$\Delta\mathbf{u} = -\,\tau\frac{\partial\mathbf{u}}{\partial\mathbf{x}_3} + \sqrt{\rho}\,(\mathbf{e}_3 \wedge \mathbf{u} - \mathbf{e}_3 \wedge \mathbf{x} \cdot \nabla\mathbf{u}) + \nabla p + \nabla \cdot \mathcal{F}, \tag{9.22}$$
$$\operatorname{div}\mathbf{u} = 0.$$

Existence of a strong solution to these equations in the L^2-framework was investigated in [96] and [109]. The corresponding theorem says:

Theorem 9.5 *Let* $\nabla \cdot \mathcal{F} \in L^2(\mathbb{R}^3)$ *with* $[|\mathcal{F}|]_2 < \infty$ *and* $\tau \in [0, B_1]$ *for some* $B_1 > 0$. *Then the problem* (9.22) *has at least one solution such that:*

(i) $\mathbf{u} \in W^{2,2}_{\text{loc}}(\mathbb{R}^3) \cap \widehat{H}^{2,2}(\mathbb{R}^3) \cap \widehat{H}^{1,2}(\mathbb{R}^3)$, $[|\mathbf{u}|]_1 < \infty$.
(ii) $p \in W^{1,2}_{\text{loc}}(\mathbb{R}^3) \cap \widehat{H}^{1,2}(\mathbb{R}^3) \cap L^s(\mathbb{R}^3)$, $s > 3/2$.

Moreover, the solution satisfies the estimate

$$|\mathbf{u}|_{2,2} + |\mathbf{u}|_{1,2} + [|\mathbf{u}|]_1 + |p|_{1,2} + \|p\|_s \le C(\|\nabla \cdot F\|_2 + [|\mathcal{F}|]_2),$$

where $C = C(s, B_1) > 0$. *Finally, if* $(\mathbf{u}_1, p_1)$ *is another solution to* (9.22) *with* $\mathbf{u}_1 \in \boldsymbol{W}^{2,2}_{loc}(\mathbb{R}^3) \cap \widehat{\boldsymbol{H}}^{1,2}(\mathbb{R}^3) \cap \boldsymbol{L}^6(\mathbb{R}^3)$, $p_1 \in W^{1,2}_{loc}(\mathbb{R}^3)$, *then* $\mathbf{u} = \mathbf{u}_1$ *and* $p = p_1$.

□

The other auxiliary theorem deals with this Oseen type problem in the exterior domain $\mathcal{D}$:

$$\Delta \mathbf{u} + \tau \frac{\partial \mathbf{u}}{\partial \mathbf{x}_3} + \sqrt{\rho}\,(\mathbf{e}_3 \wedge \mathbf{x} \cdot \nabla \mathbf{u} - \mathbf{e}_3 \wedge \mathbf{u}) = \nabla p + \tau\, \nabla \cdot \mathcal{F}, \tag{9.23}$$

$$\operatorname{div} \mathbf{u} = 0,$$

$$\mathbf{u} = \mathbf{u}_* \quad \text{on } \partial\mathcal{D},$$

$$\lim_{|\mathbf{x}| \to \infty} \mathbf{u}(\mathbf{x}) = 0.$$

Theorem 9.6 *Let* $\mathcal{D}$ *be an exterior domain of class* C^2, *and let* $\tau \in [0, B_1]$, $\rho \in [0, B_2]$ *for some* $B_1, B_2 > 0$. *Assume that* $\mathcal{F}$ *is a second order tensor field on* $\mathcal{D}$ *such that* $\nabla \cdot \mathcal{F} \in L^2(\mathcal{D})$, $[|\mathcal{F}|]_2 < \infty$ *and* u_* *is an element of* $W^{3/2,2}(\partial\mathcal{D})$. *Then:*

(i) Existence:
The problem (9.23) *has at least one solution* $(\mathbf{u}, p)$, *satisfying*

$$\mathbf{u} \in \boldsymbol{W}^{2,2}_{loc}(\overline{\mathcal{D}}) \cap \widehat{\boldsymbol{H}}^{1,2}(\mathcal{D}) \cap \widehat{\boldsymbol{H}}^{2,2}(\mathcal{D}), \quad [\|\mathbf{u}\|]_1 < \infty, \quad p \in W^{1,2}(\mathcal{D})$$

along with the estimate

$$|\mathbf{u}|_{2,2} + |\mathbf{u}|_{1,2} + [\|\mathbf{u}\|]_1 + \|p\|_{1,2} \le C(\|\nabla \cdot \mathcal{F}\|_{2,\mathcal{D}} + [\|\mathcal{F}\|]_{2,\mathcal{D}} + \|\mathbf{u}_*\|_{3/2,2,\partial\mathcal{D}})$$

with $C = C(\mathcal{D}, B_1, B_2) > 0$.

(ii) Uniqueness:
If $(\mathbf{u}_1, p_1)$ *is another solution to* (9.23) *with*

$$\mathbf{u}_1 \in \boldsymbol{W}^{2,2}_{loc}(\overline{\mathcal{D}}), \quad [|\mathbf{u}_1|]_1 < \infty, \quad p_1 \in W^{1,2}_{loc}(\overline{\mathcal{D}}) \cap L^2(\mathcal{D}), \tag{9.24}$$

then $(\mathbf{u}, p) = (\mathbf{u}_1, p_1)$.

□

Before we sketch the proof of Theorem 9.4, let us introduce a special cutoff function with an anisotropic decay, which will be used in the proof.

Lemma 9.1 *Let* $\mathcal{D} := \mathbb{R}^3 \setminus \mathcal{B}$ *be an exterior domain and let* $d(\mathcal{B})$ *be diameter of* $\mathcal{B}$. *For* $\alpha > 1$ *and for each* $R > d(\mathcal{B})$, *there exists a cutoff function* $\psi_{\alpha,R} \in C^\infty_0(\overline{\mathcal{D}})$ *such that* $0 \le \psi_{\alpha,R}(\mathbf{x}) \le 1$ *for each* $\mathbf{x} \in \mathcal{D}$, *and* $\lim_{R\to\infty} \psi_{\alpha,R}(\mathbf{x}) = 1$ *uniformly with respect to* $\mathbf{x}$ *for all* $\alpha > 1$, *such that*

$$\left| \frac{\partial \psi_{\alpha,R}}{\partial x_3} |\mathbf{x}| \right| \le \frac{C_1}{R^\alpha}, \quad \left| \frac{\partial \psi_{\alpha,R}}{\partial x_i}(\mathbf{x}) \right| < \frac{C_1}{R}, \quad i = 1, 2$$

$$|\Delta\psi_{\alpha,R}(\mathbf{x})| \le \frac{C_2}{R_2},$$

$$\nabla\psi_{\alpha,R}(x)(\mathbf{e}_3 \wedge \mathbf{x}) = 0,$$

where C_1, C_2 are positive constants independent of $\mathbf{x}$ *and R. Moreover, the support of $\nabla\psi_{\alpha,R}$ is contained in the cylindrical shell*

$$S = \left\{\mathbf{x} \in \mathbb{R}^3 : \frac{R}{\sqrt{2}} < r < \sqrt{2}R, \frac{R^\alpha}{\sqrt{2}} < |x_3| < \sqrt{2}R^\alpha \quad \text{or} \quad -\frac{R^\alpha}{\sqrt{2}} \le x_3 \le \frac{R^\alpha}{\sqrt{2}}\right\},$$

where $r = \sqrt{x_1^2 + x_2^2}$. Finally,

$$\lim_{R\to\infty} \int_{\mathcal{D}} \left|\frac{\partial\psi_{\alpha,R}}{\partial x_3}\right| \frac{1}{|\mathbf{x}|^2} = 0.$$

□

See [109] for the proof.

Let us introduce the set

$$X := \{\boldsymbol{\psi} \in \mathbf{W}^{2,2}_{\text{loc}}(\overline{\mathcal{D}}) : \nabla\cdot\boldsymbol{\psi} = 0 \text{ in } \mathcal{D},\ \|\nabla^2\boldsymbol{\psi}\|_2 + \|\nabla\boldsymbol{\psi}\|_2 + [\|\boldsymbol{\psi}\|]_1 < \infty\}.$$

X is a Banach space with the norm

$$\|\boldsymbol{\psi}\|_X := \|\nabla^2\boldsymbol{\psi}\|_2 + \|\nabla\boldsymbol{\psi}\|_2 + [\|\boldsymbol{\psi}\|]_1.$$

Idea of the Proof of Theorem 9.4 Define

$$D := \|\mathbf{u}_*\|_{3/2,2} + \|\nabla\cdot\mathcal{F}\|_2 + [\|\mathcal{F}\|]_2, \quad K = \max\{1, B_1\}\,C$$

$$X' := \{\boldsymbol{\psi} \in X : \|\boldsymbol{\psi}\|_X \le 2\,K\,D\},$$

where C is the constant from Theorem 9.6, for details see [109].

Let $\mathcal{M}$ be the map

$$\mathcal{M} : X' \to X',$$

such that for each $\boldsymbol{\psi} \in X$, the pair $(\boldsymbol{\varphi}, p) := (\mathcal{M}(\boldsymbol{\psi}), p)$ is a solution to the problem

$$\left.\begin{aligned} \Delta\boldsymbol{\varphi} + \tau\frac{\partial\boldsymbol{\varphi}}{\partial x_3} + \sqrt{\rho}\,(\mathbf{e}_3\wedge\mathbf{x}\cdot\nabla\boldsymbol{\varphi} - \mathbf{e}_3\wedge\boldsymbol{\varphi}) \\ = \nabla p + \tau\,\nabla\cdot(\mathcal{F} + \boldsymbol{\psi}\otimes\boldsymbol{\psi}) \end{aligned}\right\} \text{ in } \mathcal{D}$$

$$\nabla\cdot\varphi = 0$$

$$\boldsymbol{\varphi} = \mathbf{u}_* \text{ on } \partial\mathcal{D}$$
$$\lim_{|x|\to\infty} \boldsymbol{\varphi}(\mathbf{x}) = 0.$$

Since

$$\nabla \cdot (\mathcal{F} + \boldsymbol{\psi} \otimes \boldsymbol{\psi}) \in L^2(\mathcal{D}), \textit{and}\, [\|\mathcal{F} + \boldsymbol{\psi} \otimes \boldsymbol{\psi}\|]_2 \leq [\|\mathcal{F}\|]_2 + \|\boldsymbol{\psi}\|_X^2$$

and under assumption

$$4\tau\, CKD < 1, \tag{9.25}$$

it follows that $\mathcal{M}(\boldsymbol{\psi})$ is well defined. Under the condition (9.25), using Theorem 9.6, we can find $(\boldsymbol{\varphi}, p)$, satisfying

$$\|\boldsymbol{\varphi}\|_X + \|\nabla p\|_2 + \|p\|_2 \leq C(\tau\|\nabla \cdot \mathcal{F}\|_2 + \tau[\|\mathcal{F}\|]_2 + \|\mathbf{u}_*\|_{3/2,2}$$
$$+\tau\|\psi\|_X^2) \leq KD(1 + 4\tau\, C\, K\, D) \leq 2\, KD.$$

Further, let $\psi_1, \psi_2 \in X'$, with

$$\boldsymbol{\varphi}_i = M(\psi_i),\ i = 1, 2.$$

Applying Theorem 9.6, we obtain

$$\|\boldsymbol{\varphi}_1 - \boldsymbol{\varphi}_2\|_X \leq C\tau(\|\boldsymbol{\varphi}_1\|_X + \|\boldsymbol{\varphi}_2\|_X)\|\boldsymbol{\psi}_1 - \boldsymbol{\psi}_2\|_X.$$

The assumption (9.25) guarantees that $\mathcal{M}$ is a contraction. It implies that $\mathcal{M}$ has a fixed point $\mathbf{u} \in X'$, satisfying the estimate

$$\|\mathbf{u}\|_X \leq 2K(\|\mathbf{u}_*\|_{3/2,2} + \|\nabla \cdot \mathcal{F}\|_2 + [\|\mathcal{F}\|]_2). \tag{9.26}$$

As a last step, let us investigate the uniqueness: Let $(\mathbf{u}_1, p_1)$ be another solution to (9.19) with

$$\mathbf{u}_1 \in \boldsymbol{W}^{2,2}_{\text{loc}}(\bar{\mathcal{D}}) \cap \widehat{\boldsymbol{H}}^{1,2}(\mathcal{D})$$
$$[\|\mathbf{u}\|]_1 < \infty,\ p_1 \in W^{1,2}_{\text{loc}}(\bar{\mathcal{D}}) \cap L^2(\mathcal{D}).$$

Set $(\mathbf{U}, P) := (\mathbf{u} - \mathbf{u}_1, p - p_1)$. The pair $(\mathbf{U}, P)$ satisfies

$$\int_{\mathcal{D}} \psi_{R,\alpha} |\nabla \mathbf{U}|^2 = \frac{1}{2} \int_{\mathcal{D}} \Delta \psi_{R,\alpha} |\mathbf{U}|^2 + \int_{\mathcal{D}} P \nabla \psi_{R,\alpha} \cdot \mathbf{U}$$
$$+ \frac{\tau}{2} \int_{\mathcal{D}} \nabla \psi_{R,\alpha} \cdot \mathbf{u} |\mathbf{U}|^2 + \frac{\tau}{2} \left(\int_{\mathcal{D}} (\nabla \psi_{R,2} \cdot \mathbf{U}(|\mathbf{U}|^2 + 2\mathbf{U} \cdot \mathbf{u}) - \int_{\mathcal{D}} \frac{\partial \psi_{R,\alpha}}{\partial x_3} |\mathbf{U}|^2 \right)$$
$$+ \tau \int_{\mathcal{D}} \psi_{R,\alpha} \mathbf{U} \cdot \nabla \mathbf{U} \cdot \mathbf{u},$$

where $\psi_{R,\alpha}$ is the cutoff function introduced in Lemma 9.1. Let $R \to \infty$. Using the property of uniqueness from Theorem 9.6 and the rate of regularity of $\mathbf{U}$, P, we get

$$\|\nabla \mathbf{U}\|_2^2 = \tau \int_{\mathcal{D}} \mathbf{U} \cdot \nabla \mathbf{U} \cdot \mathbf{u}. \tag{9.27}$$

From (9.26), we at first obtain that

$$|\mathbf{u}(\mathbf{x})| \leq \frac{2K}{|\mathbf{x}|} (\|\mathbf{u}_*\|_{3/2,2} + \|\nabla \cdot \mathcal{F}\|_2 + [\|\mathcal{F}\|]_2) = \frac{2\,KD}{|\mathbf{x}|}, \quad \mathbf{x} \in \mathcal{D}.$$

Using the inequality $\|\mathbf{U}/\mathbf{x}\|_2 \leq 2\|\nabla \mathbf{U}\|_2$ and (9.27), we get

$$\|\nabla \mathbf{U}\|_2^2 \leq 2\tau\, KD \|\nabla \mathbf{U}\|_2 \|\mathbf{U}/|\mathbf{x}|\|_2 \leq 4\tau\, K\, D \|\nabla \mathbf{U}\|_2^2.$$

Since $4\tau K\, D < 1$, the above inequality implies $\mathbf{U} = 0$, $P = 0$.

The last step is to show that the energy equality holds. Multiplying $(9.19)_1$ by $\psi_{R,\alpha}\mathbf{u}$, integrating over $\mathcal{D}$, and applying the integration by parts, we obtain

$$\int_{\partial\mathcal{D}} \mathbf{u}_* \cdot T(\mathbf{u}, p)\mathbf{n} - 2 \int_{\mathcal{D}} \psi_{R,\alpha} |D(\mathbf{u})|^2 - \int_{\mathcal{D}} |\nabla \psi_{R,\alpha} \otimes \mathbf{u} : T(\mathbf{u}, p)$$
$$= \frac{\tau}{2} \int_{\partial\mathcal{D}} |\mathbf{u}_*|^2 \mathbf{u}_* \cdot \mathbf{n} - \frac{1}{2} \int_{\mathcal{D}} \nabla \psi_{R,\alpha} \cdot \mathbf{u} |\mathbf{u}|^2$$
$$- \frac{1}{2} \int_{\partial\mathcal{D}} |\mathbf{u}_*|^2 (\tau\, \mathbf{e}_3 + \sqrt{\rho} \mathbf{e}_3 \wedge \mathbf{x}) \cdot \mathbf{n}$$
$$- \tau \int_{\mathcal{D}} \psi_{R,\alpha} \mathbf{u} \cdot \mathbf{f}.$$

Passing with R to ∞, using the properties of $\psi_{R,\alpha}$, $\mathbf{u}$ and p, we get the energy inequality. □

The functional analytic approach to the existence theory introduced by G. P. Galdi can be found in [113] and also in the survey paper by G. P. Galdi and J. Neustupa, see [114].

9.2 L^p-Theory

The existence of a weak solution in the exterior domain $\mathcal{D}$ was investigated by R. Farwig and T. Hishida [66]. The authors showed the existence and uniqueness of the solution $(\mathbf{u},\ p)$ in the class $L_{3/2,\infty}$, provided that the external body force has the form $\mathbf{f} = \operatorname{div} \mathbf{F}$, where $\mathbf{F} \in L_{3/2,\infty}$ and $\boldsymbol{\omega}$ is small enough. (Recall that $L_{3/2,\infty}$ is the Lorentz space.) More concretely, the problem, investigated by R. Farwig and T. Hishida [66], consists of the equations[1]

$$\left.\begin{aligned} \mathbf{u}\cdot\nabla\mathbf{u} &= \Delta\mathbf{u} + (\boldsymbol{\omega}\wedge\mathbf{x})\cdot\mathbf{u} - \boldsymbol{\omega}\wedge\mathbf{u} - \nabla p + \mathbf{f}, \\ \operatorname{div}\mathbf{u} &= 0, \end{aligned}\right\}\quad \text{in } \mathcal{D} \tag{9.28}$$

and the conditions

$$\begin{aligned} \mathbf{u} &= \boldsymbol{\omega}\wedge\mathbf{x}, \\ \mathbf{u} \to 0 &\ \text{ as } |x| \mapsto \infty. \end{aligned} \tag{9.29}$$

Before we state the theorem on existence, let us recall that the Lorentz space $L_{q,r}(\mathcal{D})$ with the norm $\|\cdot\|_{q,r,\mathcal{D}}$ for $1 < q < \infty$, $1 < r \le \infty$ is defined by means of the real interpolation

$$L_{q,r}(\mathcal{D}) = (L_1(\mathcal{D}), L_\infty(\mathcal{D}))_{1-1/q,r},$$

see, e.g., Bergh, Löfström, [14] for details. By analogy, the homogeneous Sobolev–Lorentz spaces are defined in this way: for $1 < q_0 < q < q_1 < \infty$, $1/q = (1-\theta)/q_0 + \theta/q_1$, $1 \le r \le \infty$, we define

$$\widehat{H}^1_{q,r}(\mathcal{D}) = (\widehat{H}^1_{q_0}(\mathcal{D}), \widehat{H}^1_{q_1}(\mathcal{D}))_{\theta,r},$$

which is independent of the choose of $(q_0, q-1)$ with the norm $\|\nabla(\cdot)\|_{q,r,\mathcal{D}}$. See [14] for more details.

Theorem 9.7 *There is a constant* $\eta = \eta(\mathcal{D}) > 0$ *such that for*

$$\mathbf{f} \in \widehat{\boldsymbol{H}}^{-1}_{3/2,\infty}(\mathcal{D}) \tag{9.30}$$

with

$$|\boldsymbol{\omega}| + \|\mathbf{f}\|_{\widehat{\boldsymbol{H}}^{-1}_{3/2,\infty}(\mathcal{D})} \le \eta, \tag{9.31}$$

[1] Let us mention that system (9.28) and (9.29) have different non-dimensional form. Here the notations of Farwig and Hishida is used.

the problem (9.28–9.29) possesses a unique weak solution $(\mathbf{u}, p)$ *with*

$$\nabla \mathbf{u}, p \in L_{3/2,\infty}(\mathcal{D}), \tag{9.32}$$

subject to the estimate

$$\|\nabla \mathbf{u}\|_{3/2,\infty} + \|\mathbf{u}\|_{1,\infty} + \|p\|_{3/2,\infty} \leq C(|\boldsymbol{\omega}| + \|\mathbf{f}\|_{\widehat{\boldsymbol{H}}^{-1}_{3/2,\infty}(\mathcal{D})}) \tag{9.33}$$

with some C *independent of* $|\boldsymbol{\omega}|$ *and* $\mathbf{f}$.

□

The proof in [66] is based on the linear theory in Lorentz spaces. The authors at first use the previous results of T. Hishida [127] concerning the existence of a weak solution in the Stokes type case in L^q-setting, generalized by means of the real interpolation to the Lorentz spaces. At second, R. Farwig and T. Hishida generalized the former Cattabriga's result (see [22]) in the bounded domain to the case of Lorentz spaces. As a third step, they investigated the exterior Stokes type problem in Lorentz spaces, applying the technique of localization, see, e.g., [144]. Finally, using the fixed point argument they proved Theorem 9.7.

The existence theory in the case, when the Oseen term is considered in the full nonlinear problem in the L^p-setting, can be found in the paper [131] by H. Heck, H. Kim, H. Kozono, where the used procedure is based on the linear theory of the Oseen type problem in L^q-setting, developed by S. Kračmar, Š Nečasová and P. Penel [149–151]. See also Sect. 4.2. Further, the authors generalize the results to the Lorentz spaces and apply the fixed point argument as in the previous case. In this way, they prove the existence of a weak solution of the Navier–Stokes type problem.

Chapter 10
Pointwise Decay for the Stationary Navier–Stokes Type Problem

10.1 Introduction

We will consider a stationary viscous incompressible flow around a translating and rotating body. The goal of this section is to derive the optimal rates of decay for the velocity and its gradient, on the basis of a representation formula involving a fundamental solution constructed by R. B. Guenther and E. A. Thomann [124] for a linearized system and developed work by Deuring et al. [35–38, 40], see also Chap. 5.

We consider the stationary incompressible Navier–Stokes type system:

$$-\Delta \mathbf{u} + \tau \partial_1 \mathbf{u} + \tau (\mathbf{u} \cdot \nabla)\mathbf{u} - (\boldsymbol{\omega} \wedge \mathbf{x}) \cdot \nabla \mathbf{u} + \boldsymbol{\omega} \wedge \mathbf{u} + \nabla \pi = \mathbf{f}, \quad \operatorname{div} \mathbf{u} = 0 \tag{10.1}$$

for $\mathbf{x} \in \mathcal{D}$ with zero flow at infinity:

$$\mathbf{u}(x) \to 0 \quad \text{for } |\mathbf{x}| \to \infty. \tag{10.2}$$

Problem (10.1), (10.2)[1] together with some boundary condition on $\partial\mathcal{D}$ constitutes a mathematical model describing stationary flow of a viscous incompressible fluid around a rigid body which moves at a constant velocity and rotates at a constant angular velocity.

The open, bounded set $\mathcal{D}^c \subset \mathbb{R}^3$ describes the rigid body, the given function $\mathbf{f}\colon \overline{\mathcal{D}} \mapsto \mathbb{R}^3$ stands for an exterior force, and the unknowns $\mathbf{u}\colon \overline{\mathcal{D}} \mapsto \mathbb{R}^3$ and $\pi : \overline{\mathcal{D}} \mapsto \mathbb{R}$ correspond, respectively, to the normalized velocity and pressure field of the fluid. In this model, the translational velocity of the rigid body is given by the vector $-\tau \cdot (1, 0, 0)$ and its angular velocity by $\boldsymbol{\omega} := \sqrt{\varrho} \cdot (1, 0, 0)$,

[1] In Chaps. 10, 11 we are using non-dimensional form introduced in papers [35–38, 40].

Š. Nečasová et al., *Mathematical Theory of a Fluid Flow Around a Rotating and Translating Body*, Springer Monographs in Mathematics,
https://doi.org/10.1007/978-3-031-84120-0_10

for some $\tau \in (0, \infty)$ (Reynolds number) and $\varrho \in \mathbb{R}\backslash\{0\}$ (Taylor number) (see (1.50). The velocity $\mathbf{u}$ was shifted in such a way that the limit velocity at infinity is zero (see (10.2)). For more extensive background information on (10.1), we refer to [96, 98, 109].

We will consider that $\mathbf{f} \in \boldsymbol{L}^{p_0}(\mathbb{R}^3)$ for some $p_0 \in (1, \infty)$ and $\mathbf{f}$ has compact support. Further suppose there is a pair of functions $(\mathbf{u}, \pi)$ with $\mathbf{u} \in \boldsymbol{L}^6(\overline{\mathcal{D}})$, $\nabla\mathbf{u} \in L^2(\overline{\mathcal{D}})^9$, and $\pi \in L^2_{\mathrm{loc}}(\overline{\mathcal{D}})$ satisfying (10.1) in the distributional sense ("Leray solution").[2]

Remark 10.1 Note that the condition $\mathbf{u} \in \boldsymbol{L}^6(\overline{\mathcal{D}}^c)^3$, $\nabla\mathbf{u} \in L^2(\overline{\mathcal{D}}^c)^9$ means in particular that (10.2) holds in a weak sense; compare [95, Theorem II.5.1]. □

Under this case, Galdi and Kyed [101] showed that

$$|\partial^\alpha \mathbf{u}(\mathbf{x})| = O[(|\mathbf{x}|s_1(\mathbf{x}))^{-1-|\alpha|/2}] \quad (|\mathbf{x}| \to \infty), \tag{10.3}$$

where $\alpha \in \mathbb{N}_0^3$ with $|\alpha| := \alpha_1 + \alpha_2 + \alpha_3 \leq 1$ (decay of $\mathbf{u}$ and $\nabla\mathbf{u}$). The term $s_1(\mathbf{x})$ in (10.3) is defined by

$$s_1(\mathbf{x}) := 1 + |\mathbf{x}| - \mathbf{x}_1 \quad (\mathbf{x} \in \mathbb{R}^3). \tag{10.4}$$

Its presence in (10.3) may be considered as a mathematical manifestation of the wake extending downstream behind the rigid body. Even in the linear nonrotational case, that is, in the case of solutions to the Oseen system

$$-\Delta\mathbf{u} + \tau\partial_1\mathbf{u} + \nabla\pi = \mathbf{f}, \quad \operatorname{div}\mathbf{u} = 0, \tag{10.5}$$

the velocity cannot be expected to decay more rapidly than $(|\mathbf{x}|s_1(\mathbf{x}))^{-1}$ for $|\mathbf{x}| \to \infty$ or its gradient more rapidly than $(|\mathbf{x}|s_1(\mathbf{x}))^{-3/2}$ (see [155]). Therefore, the decay rate in (10.3) should be best possible in the present case, too.

Let us state the main result of this chapter:

Theorem 10.1 *Let $\tau \in (0, \infty)$, $\varrho \in \mathbb{R}\backslash\{0\}$, $\mathcal{D}^c \subset \mathbb{R}^3$ be open and bounded.*

Take $\gamma, S_1 \in (0, \infty)$, $p_0 \in (1, \infty)$, $A \in (2, \infty)$, $B \in [0, 3/2]$, $\mathbf{f}\colon \mathbb{R}^3 \mapsto \mathbb{R}^3$ measurable with $\overline{\mathcal{D}}^c \subset B_{S_1}$, $A + \min\{B, 1\} > 3$, $A + B \geq 7/2$, $\mathbf{f}|B_{S_1} \in \boldsymbol{L}^{p_0}(B_{S_1})$,

$$|\mathbf{f}(\mathbf{y})| \leq \gamma \cdot |\mathbf{y}|^{-A} \cdot s_1(\mathbf{y})^{-B} \quad \text{for } \mathbf{y} \in B_{S_1}^c. \tag{10.6}$$

Let $\mathbf{u} \in \boldsymbol{L}^6(\overline{\mathcal{D}}) \cap \boldsymbol{W}^{1,1}_{\mathrm{loc}}(\overline{\mathcal{D}})$, $\pi \in L^2_{\mathrm{loc}}(\overline{\mathcal{D}})$ with $\nabla\mathbf{u} \in L^2(\overline{\mathcal{D}})^9$, $\operatorname{div}\mathbf{u} = 0$, and

$$\begin{aligned}\int_{\overline{\mathcal{D}}} [\nabla\mathbf{u} \cdot \nabla\varphi + (\tau \cdot \partial_1\mathbf{u} + \tau \cdot (\mathbf{u} \cdot \nabla)\mathbf{u} \\ - (\boldsymbol{\omega} \wedge \mathbf{x}) \cdot \nabla\mathbf{u} + \boldsymbol{\omega} \wedge \mathbf{u}) \cdot \boldsymbol{\varphi} - \pi \cdot \operatorname{div}\boldsymbol{\varphi} - \mathbf{f} \cdot \boldsymbol{\varphi}]\,\mathrm{d}x = 0,\end{aligned}$$

[2] Such a solution exists under suitable assumptions on $\partial\mathcal{D}$, $\mathbf{u}|\partial\mathcal{D}$, and p_0 [98, Theorem XI.3.1].

for $\varphi \in C_0^\infty(\overline{\mathcal{D}})$. *Let* $S \in (S_1, \infty)$. *Then*

$$|\partial^\alpha \mathbf{u}(\mathbf{x})| \le D \cdot (|\mathbf{x}| \cdot s_1(\mathbf{x}))^{-1-|\alpha|/2} \quad \textit{for } \mathbf{x} \in B_S^c,\ \alpha \in \mathbb{N}_0^3 \textit{ with } |\alpha| \le 1, \tag{10.7}$$

with the constant D *depending on* γ, S_1, p_0, A, B, $\|\mathbf{f}|B_{S_1}\|_1$, $\mathbf{u}$, π, S, *and an arbitrary but fixed number* $S_0 \in (0, S_1)$ *with* $\overline{\mathcal{D}}^c \subset B_{S_0}$. □

Remark 10.2

- Note that we do not impose any conditions on the regularity of the boundary of $\mathcal{D}$.
- We need not prescribe any boundary conditions for u on $\partial\mathcal{D}$.
- With an improvement with respect to the theory in [101], we do not require $\mathbf{f}$ to have compact support. Instead we only consider that $\mathbf{f}$ decays as stated in (10.6).
- In the case when $\mathbf{f}$ decays less fast, the inequality (10.7) cannot be expected to hold. This may be seen by the very detailed study presented in [155] on the asymptotic behavior of solutions to the Oseen system (10.5) in the whole space $\mathbb{R}^3$. □

This chapter is not motivated by the more general assumptions on $\mathbf{f}$ we admit compared to the theory in [101]. Instead, our main purpose consists in introducing a new direction to obtain (10.7). In [101], this relation was reduced to estimates of solutions of the time-dependent Oseen system in the whole space $\mathbb{R}^3$. Here we will deduce (10.7) from a representation formula for u stated in Theorem 10.12 below and established in [35] and [40]. This formula involves a fundamental solution introduced by Guenther and Thomann in [124] for the linearized system

$$-\Delta\mathbf{u} + \tau \cdot \partial_1\mathbf{u} - (\boldsymbol{\omega} \wedge \mathbf{z}) \cdot \nabla\mathbf{u} + \boldsymbol{\omega} \wedge \mathbf{u} + \nabla\pi = \mathbf{f}, \quad \operatorname{div}\mathbf{u} = 0. \tag{10.8}$$

Deriving (10.7) from a representation formula is interesting because if $\omega = 0$, that is, in the case of a rigid body moving steadily but without rotation, the asymptotic behavior of the flow field is usually deduced from such formulas (see [94, Section IX.6] for example). Our way is made difficult by the structure of the Guenther–Thomann fundamental solution [124]. In fact, as was already pointed out in Chap. 3 for the case $\tau = 0$ (see also [70]), a fundamental solution $\mathcal{Z}(\mathbf{x}, \mathbf{y})$ to (10.8) cannot be bounded by $c \cdot |\mathbf{x} - \mathbf{y}|^{-1}$ uniformly in $\mathbf{x}, \mathbf{y} \in \mathbb{R}^3$ with $|\mathbf{x}|$ and $|\mathbf{y}|$ large, contrary to what may be expected in view of the situation in the Stokes and Oseen case. Actually it seems that no uniform bound $c \cdot |\mathbf{x} - \mathbf{y}|^{-\epsilon}$ exists, for whatever $\epsilon \in (0, \infty)$. But due to Lemma 10.17 and Theorem 10.11 below, we are able to circumvent this difficulty. Both of these references pertain to convolutions of the Guenther–Thomann fundamental solution [124]: The first provides a key element for obtaining pointwise bounds of these convolutions, whereas the second presents L^p-bounds. As a by-product of the latter result, we obtain the ensuing existence theorem for solutions to the linear system (10.8) in the whole space $\mathbb{R}^3$:

Theorem 10.2 *Let $p \in (1, 2)$. Then there is $\mathbf{u} \in \boldsymbol{W}^{2,p}_{\mathrm{loc}}(\mathbb{R}^3) \cap \boldsymbol{L}^{(1/p-1/2)^{-1}}(\mathbb{R}^3)$, $\pi \in W^{1,p}_{\mathrm{loc}}(\mathbb{R}^3) \cap L^{(1/p-1/3)^{-1}}(\mathbb{R}^3)$ such that $\partial_l \mathbf{u} \in \boldsymbol{L}^{(1/p-1/4)^{-1}}(\mathbb{R}^3)$, $\partial_l \partial_m \mathbf{u}$, $\partial_1 \mathbf{u} \in \boldsymbol{L}^p(\mathbb{R}^3)$, $\partial_l \pi \in L^p(\mathbb{R}^3)$ for $1 \le l, m \le 3$, and such that $(\mathbf{u}, \pi)$ solves (10.8) in $\mathbb{R}^3$.* □

This theorem was proved by Farwig, Hishida, and Müller, see [57, 70], and we gave complete proofs in Chaps. 3 and 4. These authors established it by adapting the proof of Lizorkin's multiplier theorem [170, Theorem 8] to problem (10.8), a procedure involving the harmonic analysis of Littlewood–Paley theory. Following this approach, the weighted analogy was investigated, see Chap. 3 and Sect. 4.4.

We are able to deduce L^p-estimates of convolutions of the Guenther–Thomann fundamental solution—and thus Theorem 10.2—directly from Lizorkin's multiplier theorem. This access to Theorem 10.2, although still rather technical, is considerable but maybe less involved than the one in [70].

In previous chapters we proved a representation formula, a decay estimates as in (10.7), and asymptotic expansions for weak solutions of the linear problem (10.8), as well as a representation formula for weak solutions of (10.1). In the context of these results, a weak solution $(\mathbf{u}, \pi)$ of (10.8) or (10.1) is characterized by the assumptions that $\mathbf{u}$ is $\boldsymbol{L}^6$-integrable outside a ball containing $\overline{\mathcal{D}}^c$, and $\nabla \mathbf{u}$ and π are L^2-integrable outside such a ball. The essential difference between this type of solution and a Leray solution as specified in Theorem 10.1 consists in the fact that *the pressure part of a Leray solution needs to be L^2-integrable only locally, but not globally, in the complement of a ball around $\overline{\mathcal{D}}^c$*. It should be insisted that in the present context, the notions of weak solution and Leray solution do not refer to the behavior of $\mathbf{u}$ or π near $\partial \mathcal{D}$. In Chap. 5, Sect. 5.4 we extended the results from Chaps. 2 and 5 and Sects. 5.1–5.3 to Leray solutions of (10.8) and (10.1), respectively.

Remark 10.3 There is another type of solutions which is of interest in relation to (10.1) and (10.8), that is, physically reasonable (p.r.) solutions. They are characterized by the assumption that the velocity decays as $O(|\mathbf{x}|^{-1})$ for $|\mathbf{x}| \to \infty$. References [108–112] treat existence, uniqueness, and the validity of (10.7) for these solutions. Note that the theory in [101] and in the work at hand means in particular that Leray solutions are physically reasonable. □

Remark 10.4 Concerning further articles related to the work at hand, we mention Farwig and Hishida [67, 68], who considered the linear equation (10.8) and the nonlinear one (10.1) without the Oseen term $\tau \cdot \partial_1 \mathbf{u}$ (flow around a body that rotates but does not perform a translation). It turned out that a fundamental solution of (10.8) with $\tau = 0$ may be constructed in two steps: First a suitable rotational term is introduced into the fundamental solution of the time-dependent Stokes system, and then the function obtained in this way is integrated with respect to time. With this fundamental solution as starting point, Farwig and Hishida succeeded in exhibiting detailed profiles of the flow in question, both in the linear [68] and in the nonlinear case [67]. The profiles we obtained in Chap. 5, and see also [36, 38] for the case

$\tau \neq 0$ are less elaborated. This is due to the markedly more complicated structure of the Guenther–Thomann fundamental solution compared to the function constructed by Farwig and Hishida.

Stability estimates in the L^2-setting are proved in [111] and in the $L^{3,\infty}$-setting in [129]. □

Let us briefly indicate how we will proceed in the following. In Sect. 10.2, we will present various auxiliary results, most of them proved elsewhere, with the notable exception of Lemma 10.17. As indicated above, this lemma is an intermediate but crucial step in view of obtaining pointwise bounds for convolutions of the Guenther–Thomann fundamental solution. Section 10.3 deals with L^p-bounds of such convolutions. Establishing such bounds constitutes the main difficulty of our theory. We recall that as a by-product, these L^p-bounds yield existence of solutions to the linear system (10.8) in the whole space (Theorem 10.2). This point will also be discussed in Sect. 10.3.

At the beginning of Sect. 10.4, we will introduce a solution $(\mathbf{u}, \pi)$ of problem (10.1), (10.2) with properties as in Theorem 10.1. This solution will be kept fixed throughout the rest of this work. Then, in Theorem 10.12, we will apply [40, Theorem 5], expressing $\mathbf{u}$ as a sum of certain integrals. This is the representation formula mentioned above; it will be used frequently. For this formula, additional assumptions on regularity of $\mathbf{u}$ and π near the boundary of the domain under consideration in that formula are needed. Thus, in order to avoid additional regularity assumptions near $\partial\mathcal{D}$ for $\mathbf{u}$ and π, we will apply the representation formula in question not on $\overline{\mathcal{D}}^c$, but on $B_{S_1}^c$, for some ball B_{S_1} containing $\overline{\mathcal{D}}$. In this way, we will only need results on interior regularity of $\mathbf{u}$ and π and results that follow by the L^p-theory of the Stokes system and thus need not be required as assumptions. In Sect. 10.4, we will exploit Theorem 10.12 in order to show that $\mathbf{u}$ belongs to $L^{12/5}(\overline{\mathcal{D}}^c)^3$ and $\nabla\mathbf{u}$ to $L^{12/7}(\overline{\mathcal{D}}^c)^9$ (see Theorem 10.15 and Corollary 10.11). Besides Theorem 10.12, the proof of this result involves L^p-estimates from Sect. 10.3 and Banach's fixed point theorem. Being rather lengthy, this proof will be split into several steps, beginning with Lemma 10.20 and ending with Corollary 10.11. Less effort will be necessary to establish that $\nabla\mathbf{u}$ belongs to $\boldsymbol{L}^4$ outside a ball around $\overline{\mathcal{D}}$ (see Lemma 10.21).

Section 10.5 deals with pointwise estimates of $\mathbf{u}$. A key role in this respect will be played by a function $\varphi(S)$ defined as the sup of $|\mathbf{u}(\mathbf{x})|$ with respect to $\mathbf{x}$ from outside B_S, for suitable $S > 0$. Applying results from Sect. 10.2, in particular Lemma 10.17, as well as Corollary 10.11, we will show that $\varphi(S) \to 0$ for $S \to \infty$, and $\varphi(S) \leq \mathfrak{C} \cdot (S^{-1} + \varphi(S/2)^{7/6})$ if S is sufficiently large (see Theorem 10.16). At this point, a result by Babenko [11] implies that $|\mathbf{u}(\mathbf{x})|$ decays as $|\mathbf{x}|^{-1}$ for $|\mathbf{x}| \to \infty$ (see Theorem 10.17). Then, by an argument based on [155, Theorem 3.2], we will obtain an estimate of $\mathbf{u}$ corresponding to (10.7) with $\alpha = 0$ (see Theorem 10.18).

The last section (Sect. 10.6) treats the decay of $\nabla\mathbf{u}$. In a first step, we will exploit the preceding estimate of $\mathbf{u}$ and L^p-estimates of $\nabla\mathbf{u}$ from Sect. 10.4 in order to show that $\nabla\mathbf{u}$ is bounded outside a ball around $\overline{\mathcal{D}}$ (see Theorem 10.19). Once this result is available, the rest is canonical: We will again refer to [155, Theorem 3.2], applying an iteration argument that starts with Theorem 10.19 in order to establish (10.7) for $\alpha \in \mathbb{N}^3$ with $|\alpha| = 1$ (see Theorem 10.20).

10.2 Auxiliary Results

Lemma 10.1 *Let* $k \in \mathbb{N}$, $\mathbf{z} \in \mathbb{R}^3$, $s, r \in [0, \zeta]$. *Then*

$$|\mathbf{z} - \tau \cdot (k \cdot \zeta + s) \cdot \boldsymbol{e}_1|^2 + k \cdot \zeta + s \geq C(\tau, \varrho) \cdot (|\mathbf{z} - \tau \cdot (k \cdot \zeta + r) \cdot \boldsymbol{e}_1|^2 + k \cdot \zeta + r). \tag{10.9}$$

Proof First consider the case $|\mathbf{z} - \tau \cdot k \cdot \zeta \cdot \boldsymbol{e}_1| \leq 2 \cdot \tau \cdot \zeta$. Then

$$\begin{aligned} 3 \cdot \tau \cdot \zeta \geq & |\mathbf{z} - \tau \cdot k \cdot \zeta \cdot \boldsymbol{e}_1| + \tau \cdot \zeta \geq |\mathbf{z} - \tau \cdot k \cdot \zeta \cdot \boldsymbol{e}_1| + |\tau \cdot r \cdot \boldsymbol{e}_1| \\ \geq & |\mathbf{z} - \tau \cdot (k \cdot \zeta + r) \cdot \boldsymbol{e}_1|. \end{aligned}$$

Since on the other hand the left-hand side of (10.9) is bounded from below by $k \cdot \zeta$ and

$$k \cdot \zeta \geq \zeta/2 + k \cdot \zeta/2 = (3 \cdot \tau \cdot \zeta)^2/(18 \cdot \tau^2 \cdot \zeta) + 2 \cdot k \cdot \zeta/4,$$

with $2 \cdot k \cdot \zeta \geq k \cdot \zeta + r$, inequality (10.9) follows in the case under consideration. Now assume that $|\mathbf{z} - \tau \cdot k \cdot \zeta \cdot \boldsymbol{e}_1| \geq 2 \cdot \tau \cdot \zeta$. Then we find

$$|\mathbf{z} - \tau \cdot (k \cdot \zeta + s) \cdot \boldsymbol{e}_1| \geq |\mathbf{z} - \tau \cdot k \cdot \zeta \cdot \boldsymbol{e}_1| - \tau \cdot s \geq |\mathbf{z} - \tau \cdot k \cdot \zeta \cdot \boldsymbol{e}_1|/2,$$

and

$$\begin{aligned} |\mathbf{z} - \tau \cdot k \cdot \zeta \cdot \boldsymbol{e}_1| \geq & |\mathbf{z} - \tau \cdot k \cdot \zeta \cdot \boldsymbol{e}_1|/2 + \tau \cdot \zeta \\ \geq & |\mathbf{z} - \tau \cdot (k \cdot \zeta + r) \cdot \boldsymbol{e}_1|/2 - \tau \cdot r/2 + \tau \cdot \zeta \\ \geq & |\mathbf{z} - \tau \cdot (k \cdot \zeta + r) \cdot \boldsymbol{e}_1|/2. \end{aligned}$$

Observing that $k \cdot \zeta + s \geq k \cdot \zeta \geq k \cdot \zeta/2 + r/2$, we thus see that inequality (10.9) holds in the case $|\mathbf{z} - \tau \cdot k \cdot \zeta \cdot \boldsymbol{e}_1| \geq 2 \cdot \tau \cdot \zeta$ as well. □

We will recall the following Sobolev inequality:

Theorem 10.3 ([95, Theorem I.5.1]) *Let* $R > 0$, $v \in L^6(B_R^c) \cap W_{\mathrm{loc}}^{1,1}(B_R^c)$ *with* $\nabla v \in \boldsymbol{L}^2(B_R^c) \cap \boldsymbol{L}^{12/5}(B_R^c)$. *Then* $v \in L^{10:12}(B_R^c)$. □

In order to estimate the Navier–Stokes nonlinearity, we will frequently refer to the following application of Hölder's inequality.

Lemma 10.2 *Let $p, q \in (1, \infty)$ with $r := p \cdot q/(p+q) \geq 1$, $A \subset \mathbb{R}^3$ measurable. Then there is $C_0(p,q) > 0$ with $\|\mathbf{v}^T \cdot W\|_r \leq C_0(p,q) \cdot \|\mathbf{v}\|_p \cdot \|W\|_q$ for $\mathbf{v} \in \boldsymbol{L}^p(A)^3$, $W \in L^q(A)^{3\times 3}$.*

In particular, $\mathbf{v}^T \cdot W \in \boldsymbol{L}^{3/2}(\mathbb{R}^3)$ if $p = 6$, $q = 2$. □

Proof Obviously $q/(p+q) < 1$, so $r < p$. Similarly we see that $r < q$. Thus the lemma follows from Hölder's inequality. □

Lemma 10.3 *The function $\mathrm{e}^{t \cdot \Omega}$ $(t \in \mathbb{R})$ is periodic, with period ζ. Moreover, $|\mathrm{e}^{t \cdot \Omega} \cdot \mathbf{z}| = |\mathbf{z}|$ for $\mathbf{z} \in \mathbb{R}^3$, $t \in \mathbb{R}$, for definition of Ω see Lemma 5.1.* □

Proof The eigenvalues of the matrix Ω are 0, $i \cdot |\varrho|$, and $-i \cdot |\varrho|$. □

Lemma 10.4 ([58, Lemma 4.3]) *Let $b \in (1, \infty)$. Then $\int_{\partial B_r} s_1(\mathbf{x})^{-b} \mathrm{d}o_{\mathbf{x}} \leq C(b) \cdot r$ for $r \in (0, \infty)$.* □

Corollary 10.1 *Let $a \in (2, \infty)$, $b \in (1, \infty)$, $R \in (0, \infty)$. Then $\int_{B_R^c} |\mathbf{x}|^{-a} \cdot s_1(\mathbf{x})^{-b} \mathrm{d}\mathbf{x} \leq C(a, b) \cdot R^{-a+2}$.* □

Lemma 10.5 *Let $R \in (0, \infty)$. Then $|\mathbf{x}| \geq C(R) \cdot s_1(\mathbf{x})$ for $\mathbf{x} \in B_R^c$.* □

Proof Proof is obvious. □

Lemma 10.6 ([32, Lemma 2]) *Let $M \in (0, \infty)$. Then*

$$|\mathbf{x} - \tau \cdot t \cdot \boldsymbol{e}_1|^2 + t \geq C(\tau, M) \cdot (|\mathbf{x}|^2 + t) \quad \text{for } \mathbf{x} \in B_M,\ t \in (0, \infty),$$

$$|\mathbf{x} - \tau \cdot t \cdot \boldsymbol{e}_1|^2 + t \geq C(\tau, M) \cdot (|\mathbf{x}| \cdot s_1(\mathbf{x}) + t) \quad \text{for } \mathbf{x} \in B_M^c,\ t \in (0, \infty).$$

Theorem 10.4 ([155, Theorem 3.2]) *Let $a \in (1/2, \infty)$, $b \in [0, 3/2]$. Then*

$$\int_{\mathbb{R}^3} [(1 + |\mathbf{y} - \mathbf{z}|) \cdot \nu(\mathbf{y}-\mathbf{z})]^{-3/2} \cdot (1 + |\mathbf{z}|)^{-a} \cdot \nu(\mathbf{z})^{-b} \mathrm{d}\mathbf{z}$$
$$\leq C(a, b) \cdot (1 + |\mathbf{y}|)^{-c} \cdot s(\mathbf{y})^{-d} \quad \text{for } \mathbf{y} \in \mathbb{R}^3,$$

where $c := a - 1/2$, $d := b$ if $a \leq 2$, $a < b + 1$, $a + \min\{1, b\} \neq 3$, and $a \neq 1$ or $b \neq 0$; $c := (a + b)/2$, $d := (a + b)/2 - 1/2$ if $b + 1 < a$ and $1 < a + b < 3$; $c := d := 3/2$ if $a + b \geq 7/2$ and $a + \min\{1, b\} > 3$. □

Theorem 10.5 *Let $R, \delta \in (0, \infty)$, $a \in (1, \infty)$. Then*

$$\int_0^\infty (|\mathbf{y} - \tau \cdot t \cdot \boldsymbol{e}_1 - \mathrm{e}^{-t \cdot \Omega} \cdot \mathbf{z}|^2 + t)^{-a} \mathrm{d}t \leq C(\tau, R, \delta, a) \cdot (|\mathbf{y}| \cdot s(\mathbf{y}))^{-a+1/2}$$

for $\mathbf{y} \in B^c_{(1+\delta)\cdot R}$, $\mathbf{z} \in \overline{B_R}$, *and*

$$\int_0^\infty (|\mathbf{y} - \tau \cdot t \cdot \boldsymbol{e}_1 - \mathrm{e}^{-t\cdot\Omega} \cdot \mathbf{z}|^2 + t)^{-a} \mathrm{d}t \le C(\tau, R, \delta, a) \cdot (|\mathbf{z}| \cdot \nu(\mathbf{z}))^{-a+1/2}$$

for $\mathbf{z} \in B^c_{(1+\delta)\cdot R}$, $\mathbf{y} \in \overline{B_R}$. □

Proof For the first inequality, see Theorem 5.1 and Theorem 5.3. The second inequality holds according to the last paragraph of the proof of Theorem 5.1. This last paragraph further explains why the constants in the two inequalities in Theorem 10.5 do not depend on ϱ. □

Theorem 10.6 (Theorem 5.2) *Let* $k \in \{0, 1\}$, $R \in (0, \infty)$, $\mathbf{y}, \mathbf{z} \in B_R$ *with* $\mathbf{y} \neq \mathbf{z}$. *Then*

$$\int_0^\infty (|\mathbf{y} - \tau \cdot t \cdot \boldsymbol{e}_1 - \mathrm{e}^{-t\cdot\Omega} \cdot \mathbf{z}|^2 + t)^{-3/2-k/2} \mathrm{d}t \le C(\tau, \varrho, R) \cdot |\mathbf{y} - \mathbf{z}|^{-1-k}.$$

Let us recall some notations from Chaps. 2, 5, and 6. Put:

- $\mathfrak{E}(\mathbf{x}) := (4 \cdot \pi \cdot |\mathbf{x}|)^{-1}$ for $\mathbf{x} \in \mathbb{R}^3\backslash\{0\}$ (the fundamental solution of the Poisson equation, the kernel of the "Newton potential").
- $E(\mathbf{x}) := \mathbf{x}/(4 \cdot \pi \cdot |\mathbf{x}|^3)$ for $\mathbf{x} \in \mathbb{R}^3\backslash\{0\}$.
- $K(\mathbf{x}, t) := (4 \cdot \pi \cdot t)^{-3/2} \cdot e^{-|\mathbf{x}|^2/(4\cdot t)}$ for $\mathbf{x} \in \mathbb{R}^3$, $t \in (0, \infty)$, (the fundamental solution of the heat equation).
- ${}_1F_1(1, c, u) := \sum_{n=0}^\infty (\Gamma(c)/\Gamma(n + c)) \cdot u^n$ for $u \in \mathbb{R}$, $c \in (0, \infty)$ ("Kummer function"), where the letter Γ denotes the usual Gamma function .
- $N_{jk}(\mathbf{x}) := x_j \cdot x_k \cdot |\mathbf{x}|^{-2}$.
- $\Lambda_{jk}(x, t) := K(\mathbf{x}, t)\cdot(\delta_{jk} - N_{jk}(\mathbf{x}) - {}_1F_1(1, 5/2, |x|^2/(4\cdot t))\cdot(\delta_{jk}/3 - N_{jk}(\mathbf{x})))$ for $\mathbf{x} \in \mathbb{R}^3\backslash\{0\}$, $t \in (0, \infty)$, $j, k \in \{1, 2, 3\}$.
- $$\Gamma(\mathbf{y}, \mathbf{z}, t) := \Lambda(\mathbf{y} - \tau \cdot t \cdot \boldsymbol{e}_1 - \mathrm{e}^{-t\cdot\Omega} \cdot \mathbf{z}, t) \cdot \mathrm{e}^{-t\cdot\Omega} \tag{10.10}$$

  for $\mathbf{y}, \mathbf{z} \in \mathbb{R}^3$, $t \in (0, \infty)$ with $\mathbf{y} - \tau \cdot t \cdot \boldsymbol{e}_1 - \mathrm{e}^{-t\cdot\Omega} \cdot \mathbf{z} \neq 0$.
- $\mathfrak{K}(\mathbf{y}, \mathbf{z}, t) := K(\mathbf{y} - \tau \cdot t \cdot \boldsymbol{e}_1 - \mathrm{e}^{-t\cdot\Omega} \cdot \mathbf{z}, t)$ for $\mathbf{y}, \mathbf{z} \in \mathbb{R}^3$, $t \in (0, \infty)$;
  $\Psi(\mathbf{y}, \mathbf{z}, t) := (4 \cdot \pi)^{-1} \cdot \int_{\mathbb{R}^3} |\mathbf{y} - \mathbf{x}|^{-1} \cdot \mathfrak{K}(\mathbf{x}, \mathbf{z}, t)\, \mathrm{d}\mathbf{x}$ for $\mathbf{y}, \mathbf{z}, t$ as before;
  $\mathcal{Z}(\mathbf{y}, \mathbf{z}) := \int_0^\infty \Gamma(\mathbf{y}, \mathbf{z}, t)\mathrm{d}t$ for $\mathbf{y}, \mathbf{z} \in \mathbb{R}^3$ with $\mathbf{y} \neq \mathbf{z}$. [3]

[3] Note that in the following, the letter Γ will not denote the standard Gamma function, but the function defined in (10.10), which constitutes the velocity part of the fundamental solution introduced by Guenther, Thomann [124] for the time-dependent variant of (10.8). The related pressure part is given by the function E introduced above. We further indicate that $\mathcal{Z}$ is the velocity part of the Guenther–Thomann fundamental solution of (10.8) mentioned in previous chapters and E its pressure part.

Theorem 10.7 *For $h \in C_0^\infty(\mathbb{R}^3)$, $s \in \{1, 2\}$, $\mathbf{x} \in \mathbb{R}^3$, we have $\int_{\mathbb{R}^3} |\mathbf{x} - \mathbf{y}|^{-s} \cdot |h(\mathbf{y})| d\mathbf{y} < \infty$, so we may define $\mathcal{N}(h)(\mathbf{x}) := \int_{\mathbb{R}^3} \mathfrak{E}(\mathbf{x} - \mathbf{y}) \cdot h(\mathbf{y})\, d\mathbf{y}$. Let $g \in C_0^\infty(\mathbb{R}^3)$. Then $\mathcal{N}(g) \in C^\infty(\mathbb{R}^3)$, $\partial^\alpha \mathcal{N}(g) = \mathcal{N}(\partial^\alpha g)$ for $\alpha \in \mathbb{N}_0^3$, and*

$$\partial_l \mathcal{N}(g)(\mathbf{x}) = \int_{\mathbb{R}^3} \partial_l \mathfrak{E}(\mathbf{x} - \mathbf{y}) \cdot g(\mathbf{y}) d\mathbf{y} \quad (1 \le l \le 3,\ \mathbf{x} \in \mathbb{R}^3).$$

If $p \in (1, 3)$, the inequality $\|\partial_l N(g)\|_{(1/p-1/3)^{-1}} \le C(p) \cdot \|g\|_p$ holds for $1 \le l \le 3$. In addition, $\Delta \mathcal{N}(g) = -g$ and $\|\partial^\alpha \mathcal{N}(g)\|_p \le C(p) \cdot \|g\|_p$ for $\alpha \in \mathbb{N}_0^3$ with $|\alpha| = 2$, $p \in (1, \infty)$. □

We will give a sketch of the proof.

Proof On writing $\mathcal{N}(g)(\mathbf{x}) = \int_{\mathbb{R}^3} \mathfrak{E}(\mathbf{z}) \cdot g(\mathbf{x} - \mathbf{z}) d\mathbf{z}$ and observing that $\mathfrak{E} \in L^1_{loc}(\mathbb{R}^3 \backslash \{0\})$, we apply Lebesgue's theorem to obtain that $\mathcal{N}(g) \in C^\infty(\mathbb{R}^3)$ and $\partial^\alpha \mathcal{N}(g) = \mathcal{N}(\partial^\alpha g)$ for $\alpha \in \mathbb{N}_0^3$. Starting from the equation $\partial_l \mathcal{N}(g) = \mathcal{N}(\partial_l g)$, then performing an integration by parts on $\mathbb{R}^3 \backslash B_\epsilon(\mathbf{x})$ for $\epsilon > 0$, and after that letting ϵ tend to zero, we obtain the formula for $\partial_l \mathcal{N}(g)(\mathbf{x})$ stated in the theorem. This formula and the Hardy–Littlewood–Sobolev inequality (see [204, page 119, Theorem 1]) yield the estimate $\|\partial_l \mathcal{N}(g)\|_{(1/p-1/3)^{-1}} \le C(p) \cdot \|g\|_p$ for $p \in (1, 3)$, $1 \le l \le 3$. The last claim of the theorem may be obtained by twice integrating by parts in the expression $\mathcal{N}(\partial_l \partial_m g)(\mathbf{x})$ $(1 \le l, m \le 3,\ \mathbf{x} \in \mathbb{R}^3)$ and by applying Calderon–Zygmund's inequality (see [204, page 39, Theorem 3]). □

Let us state the following variant of the Helmholtz decomposition.

Theorem 10.8 *Let $p, q \in (1, \infty)$, $\mathbf{f} \in \boldsymbol{L}^p(\mathbb{R}^3) \cap \boldsymbol{L}^q(\mathbb{R}^3)$. Then there are sequences $(\boldsymbol{\varphi}_n)$, (φ_n), $(\widetilde{\varphi}_n)$ with $\boldsymbol{\varphi}_n \in \boldsymbol{C}_0^\infty(\mathbb{R}^3)$, $\operatorname{div} \boldsymbol{\varphi}_n = 0$, $\varphi_n, \widetilde{\varphi}_n \in C^\infty(\mathbb{R}^3)$ for $n \in \mathbb{N}$, and $\|\boldsymbol{\varphi}_n + \nabla \varphi_n - \mathbf{f}\|_p \to 0$, $\|\varphi_n + \nabla \widetilde{\varphi}_n - \mathbf{f}\|_q \to 0$, $(n \to \infty)$.* □

Proof Let us choose a sequence $(\mathbf{g}_n)$ in $\boldsymbol{C}_0^\infty(\mathbb{R}^3)$ with $\|\mathbf{f} - \mathbf{g}_n\|_s \to 0$, $(n \to \infty)$ for $s = p$ and $s = q$. Let $n \in \mathbb{N}$, and abbreviate $\boldsymbol{U}_n := \nabla \mathcal{N}(\operatorname{div} \mathbf{g}_n)$, with $\mathcal{N}(\operatorname{div} \mathbf{g}_n)$ introduced in Theorem 10.7. By that theorem, we have $\boldsymbol{U}_n \in \boldsymbol{C}^\infty(\mathbb{R}^3)$ and $\operatorname{div}(\mathbf{g}_n + \boldsymbol{U}_n) = 0$. Moreover, again by Theorem 10.7,

$$U_{n,j} = \sum_{k=1}^{3} \partial_j \partial_k \mathcal{N}(g_k), \quad \partial_l U_{n,j} = \sum_{k=1}^{3} \partial_j \partial_k N(\partial_l g_k) \quad \text{for } 1 \le j, l \le 3.$$

Hence, once more by Theorem 10.7, $\boldsymbol{U}_n \in \boldsymbol{W}^{1,p}(\mathbb{R}^3) \cap \boldsymbol{W}^{1,q}(\mathbb{R}^3)$. Now we may conclude by Galdi [95, Theorem III.6.1] there is a sequence $(\boldsymbol{\varphi}_m^{(n)})_{m \ge 1}$ in $\boldsymbol{C}_0^\infty(\mathbb{R}^3)$ with $\operatorname{div} \boldsymbol{\varphi}_m^{(n)} = 0$ for $m \in \mathbb{N}$, and $\|\mathbf{g}_n + \boldsymbol{U}_n - \boldsymbol{\varphi}_m^{(n)}\|_s \to 0$ $(m \to \infty)$ for $s = p$ and $s = q$. If $p > 3$, we further have

$$\mathcal{N}(\operatorname{div} \mathbf{g}) = \sum_{k=1}^{3} \partial_k \mathcal{N}(g_k) \in L^{(1/p-1/3)^{-1}}(\mathbb{R}^3)$$

by Theorem 10.7. Thus we may refer to [95, Theorem II.6.2, II.5.1] to obtain a sequence $(\varphi_m^{(n)})_{m\geq 1}$ in $C_0^\infty(\mathbb{R}^3)$ with $\|\nabla\varphi_m^{(n)} + \boldsymbol{U}_n\|_p \to 0$ $(m \to \infty)$. By analogous argument yields a sequence $(\widetilde{\varphi}_m^{(n)})_{m\geq 1}$ with $\|\nabla\widetilde{\varphi}_m^{(n)} + \boldsymbol{U}_n\|_q \to 0$ $(m \to \infty)$. These observations imply Theorem 10.8. □

Theorem 10.9 *Let* $p \in (1, 3)$, $\mathbf{g} \in \boldsymbol{L}^p(\mathbb{R}^3)$. *Then, for a.e.* $\mathbf{x} \in \mathbb{R}^3$, *the function* $\mathbf{y} \mapsto |\mathbf{x} - \mathbf{y}|^{-2} \cdot |\mathbf{g}(\mathbf{y})|$ *is integrable on* $\mathbb{R}^3$. *We may thus define*

$$Q(\mathbf{g})(\mathbf{x}) := \int_{\mathbb{R}^3} E(\mathbf{x}-\mathbf{y}) \cdot \mathbf{g}(\mathbf{y})d\mathbf{y} \quad \textit{for } \mathbf{x} \in \mathbb{R}^3.$$

Then $\|Q(\mathbf{g})\|_{(1/p-1/3)^{-1}} \leq C(p) \cdot \|\mathbf{g}\|_p$. *If* $\mathbf{g} \in \boldsymbol{C}_0^\infty(\mathbb{R}^3)$, *we have* $Q(\mathbf{g}) \in C^\infty(\mathbb{R}^3)$ *and* $\partial^\alpha Q(\mathbf{g}) = Q(\partial^\alpha \mathbf{g})$ *for* $\alpha \in \mathbb{N}_0^3$.

Moreover $Q(\mathbf{g}) \in W^{1,1}_{\text{loc}}(\mathbb{R}^3)$, *and if in addition* $\mathbf{g} \in \boldsymbol{L}^s(\mathbb{R}^3)$ *for some* $s \in (1, \infty)$, *the inequality* $\|\nabla Q(\mathbf{g})\|_s \leq C(s) \cdot \|\mathbf{g}\|_s$ *holds.* □

Proof The first part of Theorem 10.9, up to and including the estimate of $Q(\mathbf{g})$ in the norm of $L^{(1/p-1/3)^{-1}}(\mathbb{R}^3)$, follows from the Hardy-Littlewood–Sobolev inequality ([204, page 119, Theorem 1]). The claims pertaining to the case $\mathbf{g} \in \boldsymbol{C}_0^\infty(\mathbb{R}^3)$ hold by Theorem 10.7. The last part of Theorem 10.9 follows from the same reference by a density argument. □

As a next step we turn to the fundamental solution Γ of the time-dependent variant of (10.8).

Lemma 10.7 *The functions* $\mathfrak{K}$, Ψ, *and* Γ_{jk} *belong to* $C^\infty(\mathbb{R}^3 \times \mathbb{R}^3 \times (0, \infty))$ *for* $1 \leq j, k \leq 3$. *Moreover*

$$\Gamma(\mathbf{y}, \mathbf{z}, t) = (\mathfrak{K}(\mathbf{y}, \mathbf{z}, t) \cdot \delta_{rs} + \partial y_r \partial y_s \Psi(\mathbf{y}, \mathbf{z}, t))_{1\leq r,s\leq 3} \cdot \mathrm{e}^{-t\cdot\Omega} \tag{10.11}$$

for $\mathbf{y}, \mathbf{z} \in \mathbb{R}^3$, $t > 0$. □

Proof Obviously $\mathfrak{K}$ is a C^∞-function on $\mathbb{R}^3 \times \mathbb{R}^3 \times (0, \infty)$. In view of [124, (3.11)], the same is true of Ψ. Equation (10.11) holds according to [124, (3.9)]. The relation $\Gamma \in C^\infty(\mathbb{R}^3 \times \mathbb{R}^3 \times (0, \infty))^{3\times 3}$ now follows from (10.11). Note that in Chap. 2, we proved this relation in a different way; see Lemma 2.17 and Corollary 2.3. □

Lemma 10.8 $\sum_{k=1}^3 \partial z_k \Gamma_{jk}(\mathbf{y}, \mathbf{z}, t) = 0$ *for* $\mathbf{y}, \mathbf{z} \in \mathbb{R}^3$, $t \in (0, \infty)$, $j \in \{1, 2, 3\}$. □

Proof See [124, Theorem 1.3], or verify by a direct computation. □

Lemma 10.9 $|\partial_{\mathbf{x}}^\alpha K(\mathbf{x}, t)| \leq C \cdot (|\mathbf{x}|^2 + t)^{-3/2-|\alpha|/2}$ *for* $\mathbf{x} \in \mathbb{R}^3$, $t \in (0, \infty)$, $\alpha \in \mathbb{N}_0^3$ *with* $|\alpha| \leq 1$. □

Proof The reader may either refer to [202] or prove the lemma directly by distinguishing the cases $|\mathbf{x}|^2 \leq t$ and $|\mathbf{x}|^2 > t$. □

Lemma 10.10 (Lemma 2.18) *For* $\mathbf{y}, \mathbf{z} \in \mathbb{R}^3$, $t \in (0, \infty)$, $\alpha \in \mathbb{N}_0^3$, $\alpha| \leq 1$, *we have*

$$|\partial_y^\alpha \Gamma(\mathbf{y}, \mathbf{z}, t)| + |\partial_z^\alpha \Gamma(\mathbf{y}, \mathbf{z}, t)| \leq C \cdot (|\mathbf{y} - \tau \cdot t \cdot \boldsymbol{e}_1 - \mathrm{e}^{-t \cdot \Omega} \cdot \mathbf{z}|^2 + t)^{-3/2-|\alpha|/2}.$$

Corollary 10.2 *For* $\mathbf{y}, \mathbf{z} \in \mathbb{R}^3$ *with* $\mathbf{y} \neq \mathbf{z}$, $\alpha \in \mathbb{N}_0^3$ *with* $|\alpha| \leq 1$, *we have*

$$\int_0^\infty (|\partial_y^\alpha \mathfrak{K}(\mathbf{y}, \mathbf{z}, t)| + |\partial_z^\alpha \mathfrak{K}(\mathbf{y}, \mathbf{z}, t)|)\mathrm{d}t < \infty.$$

Proof It follows from Lemma 10.9 and Theorem 10.6. □

Corollary 10.3 $\int_0^\infty (|\partial_y^\alpha \Gamma(\mathbf{y}, \mathbf{z}, t)| + |\partial_z^\alpha \Gamma(\mathbf{y}, \mathbf{z}, t)|)\mathrm{d}t < \infty$ *for* $\mathbf{y}, \mathbf{z} \in \mathbb{R}^3$ *with* $\mathbf{y} \neq \mathbf{z}$, $\alpha \in \mathbb{N}_0^3$ *with* $|\alpha| \leq 1$. *This means in particular that the function* $\mathcal{Z}$ *is well defined.* □

Proof It follows from Lemma 10.10 and Theorem 10.6. □

Corollary 10.4 *If* $\mathbf{y}, \mathbf{z} \in \mathbb{R}^3$ *with* $\mathbf{y} \neq \mathbf{z}$, $\alpha \in \mathbb{N}_0^3$ *with* $|\alpha| \leq 1$, $j, k \in \{1, 2, 3\}$, *we have*

$$\int_0^\infty (|\partial_y^\alpha \partial y_j \partial y_k \Psi(\mathbf{y}, \mathbf{z}, t)| + |\partial_z^\alpha \partial y_j \partial y_k \Psi(\mathbf{y}, \mathbf{z}, t)|)\mathrm{d}t < \infty.$$

Proof It follows from Eq. (10.11) and Corollaries 10.2 and 10.3. □

Lemma 10.11 $\mathcal{Z} \in C^1((\mathbb{R}^3 \times \mathbb{R}^3)\backslash\mathrm{diag}(\mathbb{R}^3 \times \mathbb{R}^3))^{3\times 3}$. *If* $\mathbf{y}, \mathbf{z} \in \mathbb{R}^3$ *with* $\mathbf{y} \neq \mathbf{z}$, $\alpha \in \mathbb{N}_0^3$ *with* $|\alpha| \leq 1$, *then*

$$\partial_y^\alpha \mathcal{Z}(\mathbf{y}, \mathbf{z}) = \int_0^\infty \partial_y^\alpha \Gamma(\mathbf{y}, \mathbf{z}, t)\mathrm{d}t, \quad \partial_z^\alpha \mathcal{Z}(\mathbf{y}, \mathbf{z}) = \int_0^\infty \partial_z^\alpha \Gamma(\mathbf{y}, \mathbf{z}, t)\mathrm{d}t, \tag{10.12}$$

$$|\partial_y^\alpha \mathcal{Z}(\mathbf{y}, \mathbf{z})| + |\partial_z^\alpha \mathcal{Z}(\mathbf{y}, \mathbf{z})| \leq C \cdot \int_0^\infty (|\mathbf{y} - \tau \cdot t \cdot \boldsymbol{e}_1 - \mathrm{e}^{-t \cdot \Omega} \cdot \mathbf{z}|^2 + t)^{-3/2-|\alpha|/2}\mathrm{d}t. \tag{10.13}$$

Proof Using Lemma 5.3 for the first statement of Lemma 10.11 and for (10.12) inequality (10.13) follows from (10.12) and Lemma 10.10. □

Corollary 10.5 *Let* $R \in (0, \infty)$.
Then $|\partial_y^\alpha \mathcal{Z}(\mathbf{y}, \mathbf{z})| + |\partial_z^\alpha \mathcal{Z}(\mathbf{y}, \mathbf{z})| \leq C(\tau, \varrho, R) \cdot |\mathbf{y} - \mathbf{z}|^{-1-|\alpha|/2}$ *for* $\mathbf{y}, \mathbf{z} \in B_R$ *with* $\mathbf{x} \neq \mathbf{y}$, $\alpha \in \mathbb{N}_0^3$ *with* $|\alpha| \leq 1$. □

Proof It follows from (10.13) and Theorem 10.6. □

Corollary 10.6 *Let* $R, \delta \in (0, \infty)$. *Then*

$$|\partial_y^\alpha \mathcal{Z}(\mathbf{y}, \mathbf{z})| + |\partial_z^\alpha \mathcal{Z}(\mathbf{y}, \mathbf{z})| \leq C(\tau, R, \delta) \cdot (|\mathbf{z}| \cdot s_1(\mathbf{z}))^{-1-|\alpha|/2}$$

for $\mathbf{z} \in B^c_{(1+\delta)\cdot R}$, $\mathbf{y} \in \overline{B_R}$, $\alpha \in \mathbb{N}_0^3$ *with* $|\alpha| \le 1$. *If instead* $\mathbf{y} \in B^c_{(1+\delta)\cdot R}$, $\mathbf{z} \in \overline{B_R}$ *and* α *is given in the same way, then*

$$|\partial_y^\alpha \mathcal{Z}(\mathbf{y},\mathbf{z})| + |\partial_z^\alpha \mathcal{Z}(\mathbf{y},\mathbf{z})| \le C(\tau, R, \delta) \cdot (|\mathbf{y}| \cdot s(\mathbf{y}))^{-1-|\alpha|/2}.$$

Proof It follows from (10.13) and Theorem 10.5. □

Lemma 10.12 *Let* $S \in (0,\infty)$, $g \in L^1(\partial B_S)$, $y \in (\partial B_S)^c$, $\alpha \in \mathbb{N}_0^3$ *with* $|\alpha| \le 1$. *Then*

$$\int_{\partial B_S} (|\partial_z^\alpha \mathcal{Z}(\mathbf{y},\mathbf{z})| + |E(\mathbf{y}-\mathbf{z})|) \cdot |g(\mathbf{z})| do_z < \infty.$$

Proof It follows from the first statement of Lemma 10.11. □

Let us recall already obtained results from previous chapters. As before we introduce a volume potential with $\mathcal{Z}$ as kernel. Due to its role in the representation formula in Theorem 10.12 below, this potential is the fundamental mathematical object of this investigation.

Lemma 10.13 (Lemma 5.10) *Let* $A \subset \mathbb{R}^3$ *be measurable,* $p \in (1,2)$, $\mathbf{f} \in \boldsymbol{L}^p(A)$. *Let* $\widetilde{\mathbf{f}}$ *denote the zero extension of* $\mathbf{f}$ *to* $\mathbb{R}^3$. *Then*

$$\int_{\mathbb{R}^3} (|\partial_y^\alpha \mathcal{Z}(\mathbf{y},\mathbf{z})| + |\partial_z^\alpha \mathcal{Z}(\mathbf{y},\mathbf{z})|) \cdot |\widetilde{\mathbf{f}}(\mathbf{z})| \, d\mathbf{z} < \infty$$

for a.e. $\mathbf{y} \in \mathbb{R}^3$, $\alpha \in \mathbb{N}_0^3$ *with* $|\alpha| \le 1$. *Define*

$$\mathfrak{R}(\mathbf{f})(\mathbf{y}) := \int_{\mathbb{R}^3} \mathcal{Z}(\mathbf{y},\mathbf{z}) \cdot \widetilde{\mathbf{f}}(\mathbf{z}) \, d\mathbf{z}$$

for $\mathbf{y} \in \mathbb{R}^3$. *Then* $\mathfrak{R}(\mathbf{f}) \in W^{1,1}_{\text{loc}}(\mathbb{R}^3)^3$ *and*

$$\partial_y^\alpha \mathfrak{R}(\mathbf{f})(\mathbf{y}) = \int_{\mathbb{R}^3} \partial_y^\alpha \mathcal{Z}(\mathbf{y},\mathbf{z}) \cdot \widetilde{\mathbf{f}}(\mathbf{z}) \, d\mathbf{z}$$

for $\mathbf{y}$, α *as above.* □

Let us list some properties of $\mathfrak{R}(\mathbf{f})$.

Lemma 10.14 (Lemma 5.11) *Let* $p \in (1,2)$, $R \in (0,\infty)$, $\mathbf{f} \in \boldsymbol{L}^p(\mathbb{R}^3)$, $\alpha \in \mathbb{N}_0^3$ *with* $|\alpha| \le 1$. *Then* $\|\partial^\alpha \mathfrak{R}(\mathbf{f})|B_R\|_p \le C(\tau, \varrho, R, p) \cdot \|\mathbf{f}\|_p$. □

Corollary 10.7 *Let* $p \in (1,2)$, $\mathbf{f} \in \boldsymbol{L}^p(\mathbb{R}^3)$. *Take sequences* $(\boldsymbol{\varphi}_n)$, (φ_n) *as in Theorem 10.8. Then there is a subsequence* $(\mathbf{g}_n)$ *of* $(\boldsymbol{\varphi}_n + \nabla\varphi_n)$ *such that* $\partial^\alpha \mathfrak{R}(\mathbf{g}_n)(\mathbf{x}) \to \partial^\alpha \mathfrak{R}(\mathbf{f})(\mathbf{x})$ $(n \to \infty)$ *for a.e.* $\mathbf{x} \in \mathbb{R}^3$ *and for* $\alpha \in \mathbb{N}_0^3$ *with* $|\alpha| \le 1$. □

Proof Theorem 10.8 and Lemma 10.14 imply that $\|(\partial^\alpha \mathfrak{R}(\mathbf{f}) - \partial^\alpha \mathfrak{R}(\boldsymbol{\varphi}_n + \nabla \varphi_n)) \mid B_R\|_p \to 0$ when n tends to ∞, for any $R > 0$ and for α as above. □

Lemma 10.15 *Let* $\alpha \in \mathbb{N}_0^3$ *with* $|\alpha| \le 1$, $\boldsymbol{\varphi} \in \boldsymbol{C}_0^\infty(\mathbb{R}^3)$ *with* $\operatorname{div} \boldsymbol{\varphi} = 0$, $\varphi \in C_0^\infty(\mathbb{R}^3)$, $\mathbf{y} \in \mathbb{R}^3$. *Then*

$$\int_0^\infty \int_{\mathbb{R}^3} |\partial_y^\alpha \mathfrak{K}(\mathbf{y}, \mathbf{z}, t) \cdot (\mathrm{e}^{-t \cdot \Omega} \cdot \boldsymbol{\varphi})(\mathbf{z})| \, \mathrm{d}\mathbf{z} \, \mathrm{d}t < \infty$$

and

$$\partial^\alpha \mathfrak{R}(\boldsymbol{\varphi} + \nabla \varphi)(\mathbf{y}) = \int_0^\infty \int_{\mathbb{R}^3} \partial_y^\alpha \mathfrak{K}(\mathbf{y}, \mathbf{z}, t) \cdot (\mathrm{e}^{-t \cdot \Omega} \cdot \boldsymbol{\varphi})(\mathbf{z}) \, \mathrm{d}\mathbf{z} \, \mathrm{d}t.$$

Proof There is $R > 0$ with $\operatorname{supp}(\boldsymbol{\varphi} + \nabla \varphi) \subset B_R$. Therefore, by (10.13) and Theorem 10.6, we see that $|\partial_y^\alpha \Gamma(\mathbf{y}, \mathbf{z}, t)| \cdot |(\boldsymbol{\varphi} + \nabla \varphi)(\mathbf{z})|$ as a function of $\mathbf{z} \in \mathbb{R}^3$, $t \in (0, \infty)$ is integrable. It follows by Lemmas 10.13 and 10.11 that

$$\partial^\alpha \mathfrak{R}(\boldsymbol{\varphi} + \nabla \varphi)(\mathbf{y}) = \int_0^\infty \int_{\mathbb{R}^3} \partial_y^\alpha \Gamma(\mathbf{y}, \mathbf{z}, t) \cdot (\varphi + \nabla \varphi)(\mathbf{z}) \, \mathrm{d}\mathbf{z} \, \mathrm{d}t. \tag{10.14}$$

In view of the first claim in Lemma 10.7 and by Lemma 10.8, we obtain by an integration by parts that

$$\int_{\mathbb{R}^3} \partial_y^\alpha \Gamma(\mathbf{y}, \mathbf{z}, t) \cdot \nabla \varphi(\mathbf{z}) \, \mathrm{d}\mathbf{z} = 0$$

for $t \in (0, \infty)$. Moreover, we use the splitting of $\partial_y^\alpha \Gamma(\mathbf{y}, \mathbf{z}, t)$ in (10.11), thus arriving at the equation

$$\partial^\alpha \mathfrak{R}(\boldsymbol{\varphi} + \nabla \varphi)(\mathbf{y}) = \mathfrak{A}_1 + \mathfrak{A}_2,$$

with

$$\mathfrak{A}_m := \int_0^\infty \int_{\mathbb{R}^3} H_m(\mathbf{y}, \mathbf{z}, t) \cdot (\mathrm{e}^{-t \cdot \Omega} \cdot \varphi)(\mathbf{z}) \mathrm{d}\mathbf{z} \mathrm{d}t$$

for $m \in \{1, 2\}$, where

$$H_1(\mathbf{y}, \mathbf{z}, t) := \mathfrak{K}(\mathbf{y}, \mathbf{z}, t) \quad \text{and} \quad H_2(\mathbf{y}, \mathbf{z}, t) := (\partial_y^\alpha \partial y_j \partial y_k \Psi(\mathbf{y}, \mathbf{z}, t))_{1 \le j,k \le 3}.$$

Once more using the first claim in Lemma 10.7 in order to perform an integration by parts, taking into account the equation

$$\nabla_y \Psi(\mathbf{y}, \mathbf{z}, t) = -\nabla_z \Psi(\mathbf{y}, \mathbf{z}, t) \cdot \mathrm{e}^{t \cdot \Omega},$$

and recalling the assumption $\operatorname{div}\varphi = 0$, we obtain $A_2 = 0$. Therefore Lemma 10.15 follows from (10.14). □

Lemma 10.16 *Let $a \in (1/2, \infty)$, $b \in [0, 3/2]$, $R, S \in (0, \infty)$ with $R < S$, $\mathbf{y} \in B_S^c$, and $l \in \{1, 2, 3\}$. Then*

$$\int_{B_R^c} (|\partial y_l \mathcal{Z}(\mathbf{y}, \mathbf{z})| + |\partial z_l \mathcal{Z}(\mathbf{y}, \mathbf{z})|) \cdot |\mathbf{z}|^{-a} \cdot s_1(\mathbf{z})^{-b} \, d\mathbf{z}$$

$$\leq C(\tau, \varrho, R, S, a, b) \cdot \Bigg(|\mathbf{y}|^{-a} \cdot s(\mathbf{y})^{-b} + \int_{\mathbb{R}^3} [(1 + |\mathbf{y} - \mathbf{z}|) \cdot s(\mathbf{y} - \mathbf{z})]^{-3/2} \cdot (1 + |\mathbf{z}|)^{-a} \cdot s(\mathbf{z})^{-b} \, d\mathbf{z} \Bigg).$$

Proof Replicate the proof of Theorem 5.3 with S_1, A, B replaced by R, a, b, respectively, and with $\alpha = e_l$. □

Lemma 10.17 *Let $A \subset \mathbb{R}^3$ be measurable, with $\Omega \cdot A = A$. Let $g\colon A \mapsto \mathbb{R}$ be a measurable function, and let $\mathbf{y} \in \mathbb{R}^3$, $l \in \{1, 2, 3\}$. Then*

$$\int_A (|\partial y_l \mathcal{Z}(\mathbf{y}, \mathbf{z})| + |\partial z_l \mathcal{Z}(\mathbf{y}, \mathbf{z})|) \cdot |g(\mathbf{z})| \, d\mathbf{z}$$

$$\leq C(\tau, \varrho) \cdot \int_0^\zeta \int_A |\mathbf{g}(e^{t \cdot \Omega} \cdot \mathbf{z})| \cdot W(\mathbf{y} - \mathbf{z}, t) \, d\mathbf{z} \, dt,$$

where $W(\mathbf{x}, t)$ for $\mathbf{x} \in \mathbb{R}^3$, $t \in (0, \zeta)$, is defined by

$$W(\mathbf{x}, t) := (|\mathbf{x} - \tau \cdot t \cdot \boldsymbol{e}_1|^2 + t)^{-2} + \int_\zeta^\infty (|\mathbf{x} - \tau \cdot s \cdot \boldsymbol{e}_1|^2 + s)^{-2} ds,$$

and where the quantity ζ was introduced at the beginning of this section. □

Proof Let the left-hand side of the estimate in Lemma 10.17 be denoted by $\mathfrak{V}$. By (10.13),

$$\mathfrak{V} \leq C(\tau, \varrho) \cdot \int_0^\infty \int_A (|\mathbf{y} - \tau \cdot t \cdot \boldsymbol{e}_1 - e^{-t \cdot \Omega} \cdot \mathbf{z}|^2 + t)^{-2} \cdot |\mathbf{g}(\mathbf{z})| \, d\mathbf{z} \, dt.$$

On performing a change of variables and taking account of the fact that the components of the matrix $e^{t \cdot \Omega}$ are continuous functions of t with period ζ (see Lemma 10.3), and in particular are bounded, we may conclude that $\mathfrak{V} \leq C(\tau, \varrho) \cdot (\mathfrak{V}_1 + \mathfrak{V}_2)$, with

$$\mathfrak{V}_1 := \int_0^\zeta \int_A \sum_{k=1}^\infty (|\mathbf{y} - \mathbf{x} - \tau \cdot (k \cdot \zeta + t) \cdot \boldsymbol{e}_1|^2 + k \cdot \zeta + t)^{-2} \cdot |\mathbf{g}(e^{t \cdot \Omega} \cdot \mathbf{x})| \, d\mathbf{x} \, dt, \tag{10.15}$$

and with $\mathfrak{V}_2$ defined in the same way as $\mathfrak{V}_1$, except that the sum with respect to k is replaced by the term $(|\mathbf{y}-\mathbf{x}-\tau\cdot t\cdot \boldsymbol{e}_1|^2+t)^{-2}$. In other words, $\mathfrak{V}_2$ is related to the index $k=0$ in the sum appearing in (10.15). By Lemma 10.1 with $s=t$, $r=0$, we obtain

$$\mathfrak{V}_1 \le C(\tau,\varrho)\cdot \int_A G(\mathbf{x})\cdot \sum_{k=1}^{\infty}(|\mathbf{y}-\mathbf{x}-\tau\cdot k\cdot\zeta\cdot \boldsymbol{e}_1|^2+k\cdot\zeta)^{-2}\,d\mathbf{x},$$

where we used the abbreviation

$$G(\mathbf{x}) := \int_0^{\zeta} |\mathbf{g}(\mathrm{e}^{t\cdot\Omega}\cdot\mathbf{x})|\,dt,$$

for $\mathbf{x}\in A$. Applying the identity $a=\zeta^{-1}\cdot\int_0^{\zeta} a\,ds$, valid for any $a\in\mathbb{R}$, and exploiting Lemma 10.1 again, this time with $s=0$, we arrive at the inequality

$$\mathfrak{V}_1 \le C(\tau,\varrho)\cdot \int_A G(\mathbf{x})\cdot \sum_{k=1}^{\infty}\int_0^{\zeta} (|\mathbf{y}-\mathbf{x}-\tau\cdot(k\cdot\zeta+r)\cdot \boldsymbol{e}_1|^2+k\cdot\zeta+r)^{-2}\,dr\,d\mathbf{x}.$$

But the sum in the preceding estimate equals

$$\int_{\zeta}^{\infty} (|\mathbf{y}-\mathbf{x}-\tau\cdot r\cdot \boldsymbol{e}_1|^2+r)^{-2}dr.$$

Lemma 10.17 follows with the inequality $\mathfrak{V}\le C(\tau,\varrho)\cdot(\mathfrak{V}_1+\mathfrak{V}_2)$ derived above. □

In the rest of this section, we show that the potential $\mathfrak{R}(\mathbf{f})$ (Lemma 10.13) is the velocity part, and $Q(\mathbf{f})$ (see Theorem 10.9) the pressure part, of a solution to (10.8) in the whole space $\mathbb{R}^3$, under the assumption that $\mathbf{f}$ is smooth. We start with a representation formula for the "velocity part" of C^{∞}-solutions in the whole space $\mathbb{R}^3$ to the system adjoint to the momentum equation of the linear system (10.8), with vanishing "pressure." Such a "velocity" need not be solenoidal.

Theorem 10.10 *Let $\widetilde{\Gamma}$ be defined in the same way as Γ, but with τ and ϱ replaced by $-\tau$ and $-\varrho$, respectively. Then*

$$\int_0^{\infty} |\widetilde{\Gamma}(\mathbf{y},\mathbf{z},t)|dt \le C(\tau,\varrho,R)\cdot|\mathbf{y}-\mathbf{z}|^{-1}$$

for $\mathbf{y},\mathbf{z}\in B_R$, $\mathbf{y}\ne\mathbf{z}$, $R\in(0,\infty)$. Thus, for $\boldsymbol{\varphi}\in \mathbf{C}_0^{\infty}(\mathbb{R}^3)$, we may define $\widetilde{\mathfrak{R}}(\boldsymbol{\varphi})$ in the same manner as $\mathfrak{R}(\boldsymbol{\varphi})$ (see Lemma 10.11), but with Γ replaced by $\widetilde{\Gamma}$.

For $\varphi \in C_0^\infty(\mathbb{R}^3)$*, the equation*

$$\varphi = \widetilde{\mathfrak{R}}(\widetilde{L}(\varphi)) - \nabla \mathcal{N}(\operatorname{div} \varphi)$$

holds, with $\mathcal{N}(\operatorname{div} \varphi)$ *introduced in Theorem 10.7.* □

Proof We refer to Theorem 5.2 with $\boldsymbol{U} = \tau \cdot \boldsymbol{e}_1$ and with $-\boldsymbol{\omega}$ instead of $\boldsymbol{\omega}$, and to Lemma 5.7, with $\mathcal{D} = \emptyset$, $\pi = 0$, $\mathbf{u} = \varphi$, $\boldsymbol{U} = \tau \cdot \boldsymbol{e}_1$, and again with $-\boldsymbol{\omega}$ in the place of $\boldsymbol{\omega}$. Note that the term

$$\int_{\mathbb{R}^3} (4 \cdot \pi \cdot |\mathbf{y} - \mathbf{z}|^3)^{-1} \cdot (y - z)_j \cdot \operatorname{div} \varphi(\mathbf{z}) \, d\mathbf{z}$$

arising due to that latter reference equals $-\partial_j N(\operatorname{div} \varphi)(\mathbf{x})$ for $\mathbf{y} \in \mathbb{R}^3$, $1 \le j \le 3$ (see Theorem 10.7). □

Lemma 10.18 *Let* $\mathbf{g} \in C_0^\infty(\mathbb{R}^3)$*. Then* $\mathfrak{R}(\mathbf{g}) \in C^\infty(\mathbb{R}^3)$*.* □

Proof Obviously, for $t \in (0, \infty)$, the function $\mathbf{z} \mapsto \mathbf{g}(e^{t \cdot \Omega} \cdot \mathbf{z})$ is C^∞ on $\mathbb{R}^3$. Moreover, we see by Lemma 10.3 that for $\alpha \in \mathbb{N}_0^3$, there is $c_\alpha > 0$ with $|\partial_z^\alpha \mathbf{g}(e^{t \cdot \Omega} \cdot \mathbf{z})| \le c_\alpha$ for $\mathbf{z} \in \mathbb{R}^3$, $t > 0$. Fix $R > 0$ with $\operatorname{supp}(\mathbf{g}) \subset B_R$, and let $S > 0$. By Lemma 10.3, we have $|e^{t \cdot \Omega} \cdot (\mathbf{y} - \mathbf{z})| \ge R$ for $\mathbf{y} \in B_S$, $\mathbf{z} \in B_{R+S}^c$, $t > 0$, so that $|\partial_y^\alpha \mathbf{g}(e^{t \cdot \Omega} \cdot (\mathbf{y} - \mathbf{z}))| \le C(\alpha, c_\alpha) \cdot \chi_{B_{R+S}}(\mathbf{z})$ for $\mathbf{y}$, $\mathbf{z}$, t as before and for $\alpha \in \mathbb{N}_0^3$.

On the other hand, $|\Lambda(\mathbf{z}, t)|$ is bounded by $C \cdot (|\mathbf{z}|^2 + t)^{-3/2}$ for $\mathbf{z} \in \mathbb{R}^3$, $t > 0$ (see Lemma 2.18), so with Lemma 10.6,

$$|\Lambda(\mathbf{z} - \tau \cdot t \cdot \boldsymbol{e}_1, t)| \le C(\tau, R, S) \cdot (|\mathbf{z}|^2 + t)^{-3/2}$$

for $\mathbf{z} \in B_{R+S}$, $t > 0$. Therefore, the term $|\Lambda(\mathbf{z} - \tau \cdot t \cdot \boldsymbol{e}_1, t)| \cdot \chi_{B_{R+S}}(\mathbf{z})$ is integrable with respect to $(\mathbf{z}, t) \in \mathbb{R}^3 \times (0, \infty)$.

Now it follows by Lebesgue's and Fubini's theorem that $\mathfrak{R}(\mathbf{g})|B_S \in C^\infty(B_S)^3$. This proves the lemma. □

Corollary 10.8 *Let* $\mathbf{g} \in C_0^\infty(\mathbb{R}^3)$*. Then* $L(\mathfrak{R}\mathbf{g}) + \nabla Q(\mathbf{g}) = \mathbf{g}$*, with* $Q(\mathbf{g})$ *defined in Theorem* 10.9. □

Proof Let $\varphi \in C_0^\infty(\mathbb{R}^3)$. Then

$$\int_{\mathbb{R}^3} \varphi \cdot [L(\mathfrak{R}(\mathbf{g})) + \nabla Q(\mathbf{g})] \, d\mathbf{z} = \int_{\mathbb{R}^3} (\widetilde{L}(\varphi) \cdot \mathfrak{R}(\mathbf{g}) - \operatorname{div} \varphi \cdot Q(\mathbf{g})) \, d\mathbf{z}. \qquad (10.16)$$

But $\Gamma(\mathbf{x}, \mathbf{y}, t) = e^{-t \cdot \Omega} \cdot \Lambda(\mathbf{y} + \tau \cdot t \cdot \boldsymbol{e}_1 - e^{t \cdot \Omega} \cdot \mathbf{x}, t)$ for $\mathbf{x}, \mathbf{y} \in \mathbb{R}^3$, $t \in (0, \infty)$, and $(e^{-t \cdot \Omega})^\top = e^{t \cdot \Omega}$, $\Lambda^T = \Lambda$. Thus, noting that $\widetilde{L}(\varphi) \in C_0^\infty(\mathbb{R}^3)$, we have

$$\int_{\mathbb{R}^3} \widetilde{L}(\varphi) \cdot \mathfrak{R}(\mathbf{g}) \, d\mathbf{z} = \int_{\mathbb{R}^3} \widetilde{\mathfrak{R}}(\widetilde{L}(\varphi)) \cdot \mathbf{g} \, d\mathbf{z}.$$

Moreover, we have

$$\int_{\mathbb{R}^3} \operatorname{div} \boldsymbol{\varphi} \cdot Q(\mathbf{g})\, d\mathbf{z} = \int_{\mathbb{R}^3} \nabla \mathcal{N}(\operatorname{div} \boldsymbol{\varphi}) \cdot \mathbf{g}\, d\mathbf{z}$$

by Theorem 10.7. Thus the left-hand side of (10.16) equals

$$\int_{\mathbb{R}^3} [\widetilde{\mathfrak{R}}(\widetilde{L}(\boldsymbol{\varphi})) - \nabla \mathcal{N}(\operatorname{div} \boldsymbol{\varphi})] \cdot \mathbf{g}\, d\mathbf{x}.$$

Now Theorem 10.10 implies that the left-hand side of (10.16) coincides with $\int_{\mathbb{R}^3} \boldsymbol{\varphi} \cdot \mathbf{g} dx$. This proves the corollary. □

10.3 L^p-Estimates of the Volume Potential $\mathfrak{R}(f)$

The ensuing theorem, which constitutes the main difficulty of our argument, will be proved by reduction to a multiplier theorem by Lizorkin [170].

Theorem 10.11 *Let* $p \in (1, 2)$, $q \in (1, 4)$, $r \in (1, \infty)$. *Then there are constants* $C_1(p) = C_1(p, \varrho, \tau)$, $C_2(q) = C_2(q, \varrho, \tau)$, $C_3(r) = C_3(r, \varrho, \tau) > 0$ *such that*

$$\begin{aligned}
\|\mathfrak{R}(\boldsymbol{\Phi} + \nabla\varphi)\|_{(1/p-1/2)^{-1}} &\le C_1(p) \cdot \|\boldsymbol{\Phi} + \nabla\varphi\|_p, \\
\|\nabla\mathfrak{R}(\boldsymbol{\Phi} + \nabla\varphi)\|_{(1/q-1/4)^{-1}} &\le C_2(q) \cdot \|\boldsymbol{\Phi} + \nabla\varphi\|_q, \\
\|\partial_1\mathfrak{R}(\boldsymbol{\Phi} + \nabla\varphi)\|_r + \|\nabla^2\mathfrak{R}(\boldsymbol{\Phi} + \nabla\varphi)\|_r &\le C_3(r) \cdot \|\boldsymbol{\Phi} + \nabla\varphi\|_r,
\end{aligned}$$

for $\boldsymbol{\Phi} \in \boldsymbol{C}_0^\infty(\mathbb{R}^3)$ *with* $\operatorname{div} \boldsymbol{\Phi} = 0$, $\varphi \in C_0^\infty(\mathbb{R}^3)$. □

Proof We will establish the second inequality. The other ones may be shown by a similar reasoning. There is a minor additional difficulty related to the estimate of the second derivatives of $\mathfrak{R}(\boldsymbol{\Phi} + \nabla\varphi)$. We will give an indication in this respect at the end of this proof.

For a rapidly decreasing function $\varphi\colon \mathbb{R}^3 \mapsto \mathbb{R}$, we define its Fourier transform $\widehat{\gamma}$ by $\widehat{\gamma}(\boldsymbol{\xi}) := (2\cdot\pi)^{-3/2} \cdot \int_{\mathbb{R}^3} e^{-i\cdot\boldsymbol{\xi}\cdot\mathbf{x}} \cdot \gamma(\mathbf{x})\, d\mathbf{x}$, $(\boldsymbol{\xi} \in \mathbb{R}^3)$, and consequently its inverse Fourier transform $\check{\gamma}$ by $\check{\gamma}(\boldsymbol{\xi}) := (2\cdot\pi)^{-3/2} \cdot \int_{\mathbb{R}^3} e^{i\cdot\boldsymbol{\xi}\cdot\mathbf{x}} \cdot \gamma(\mathbf{x})\, d\mathbf{x}$, $(\boldsymbol{\xi} \in \mathbb{R}^3)$. It is well known that $[K(\cdot, t)]^\wedge(\boldsymbol{\xi}) = (2\cdot\pi)^{-3/2} \cdot e^{-|\boldsymbol{\xi}|^2\cdot t}$ for $t > 0$, $\boldsymbol{\xi} \in \mathbb{R}^3$.

Let $l \in \{1, 2, 3\}$, $\boldsymbol{\Phi} \in \boldsymbol{C}_0^\infty(\mathbb{R}^3)$ with $\operatorname{div} \boldsymbol{\Phi} = 0$, $\varphi \in C_0^\infty(\mathbb{R}^3)$, $\mathbf{y} \in \mathbb{R}^3$. Setting $h(\mathbf{x}, t) := e^{-t\cdot\Omega} \cdot \boldsymbol{\Phi}(e^{t\cdot\Omega} \cdot \mathbf{x})$ for $\mathbf{x} \in \mathbb{R}^3$, $t \in (0, \infty)$, we get with Lemmas 10.15 and 10.3 and a change of variables that

$$\partial_l\mathfrak{R}(\boldsymbol{\Phi} + \nabla\varphi)(\mathbf{y}) = \int_0^\infty \int_{\mathbb{R}^3} \partial y_l K(\mathbf{y} - \tau \cdot t \cdot \boldsymbol{e}_1 - \mathbf{x}, t) \cdot h(\mathbf{x}, t) d\mathbf{x} dt.$$

By Plancherel's theorem and the above formula for $[K(\cdot,t)]^\wedge$, we may conclude that

$$\partial_l \mathfrak{R}(\boldsymbol{\Phi} + \nabla\varphi)(\mathbf{y}) = \int_0^\infty \int_{\mathbb{R}^3} \mathrm{e}^{-\mathrm{i}\cdot\boldsymbol{\xi}\cdot\mathbf{y}} \cdot \mathfrak{S}(\boldsymbol{\xi},t) \cdot [h(\cdot,t)]^\wedge(\boldsymbol{\xi})\mathrm{d}\boldsymbol{\xi}\mathrm{d}t,$$

where $\mathfrak{S}(\boldsymbol{\xi},t) := \mathfrak{S}_l(\boldsymbol{\xi},t) := -(2\cdot\pi)^{-3/2}\cdot \mathrm{i}\cdot\xi_l\cdot \mathrm{e}^{t\cdot(\mathrm{i}\cdot\tau\cdot\xi_1-|\boldsymbol{\xi}|^2)}$ for $\boldsymbol{\xi}\in\mathbb{R}^3$, $t>0$. By Lemma 10.3, we have

$$[h(\cdot,t)]^\wedge(\boldsymbol{\xi}) = [h(\cdot,t+\zeta)]^\wedge(\boldsymbol{\xi}) \quad \text{for } \boldsymbol{\xi}\in\mathbb{R}^3,\ t>0, \tag{10.17}$$

so we arrive at the following equation:

$$\partial_l \mathfrak{R}(\boldsymbol{\Phi}+\nabla\varphi)(\mathbf{y}) = \sum_{k=0}^{\infty}\int_0^\zeta \int_{\mathbb{R}^3} \mathrm{e}^{-\mathrm{i}\cdot\boldsymbol{\xi}\cdot\mathbf{y}}\cdot\mathfrak{S}(\boldsymbol{\xi},t+k\cdot\zeta)\cdot[h(\cdot,t)]^\wedge(\boldsymbol{\xi})\mathrm{d}\boldsymbol{\xi}\mathrm{d}t. \tag{10.18}$$

Observe that $|\mathfrak{S}(\boldsymbol{\xi},t)| \le |\boldsymbol{\xi}|\cdot \mathrm{e}^{-|\boldsymbol{\xi}|^2\cdot t}$ and $|[h(\cdot,t)]^\wedge(\boldsymbol{\xi})| \le |\widehat{\boldsymbol{\Phi}}(\mathrm{e}^{t\cdot\Omega}\cdot\boldsymbol{\xi})|$ ($\boldsymbol{\xi}\in\mathbb{R}^3$, $t>0$), where the last inequality follows with Lemma 10.3. Therefore

$$\begin{aligned}&\sum_{k=0}^{\infty}\int_0^\zeta \int_{\mathbb{R}^3} \big|\mathrm{e}^{-\mathrm{i}\cdot\boldsymbol{\xi}\cdot\mathbf{y}}\cdot\mathfrak{S}(\boldsymbol{\xi},t+k\cdot\zeta)\cdot[h(\cdot,t)]^\wedge(\boldsymbol{\xi})\big|\,\mathrm{d}\boldsymbol{\xi}\mathrm{d}t\\ &\qquad \le C\cdot\int_0^\infty\int_{\mathbb{R}^3}|\boldsymbol{\eta}|\cdot\mathrm{e}^{-|\boldsymbol{\eta}|^2\cdot t}\cdot|\widehat{\boldsymbol{\Phi}}(\boldsymbol{\eta})|\,\mathrm{d}\boldsymbol{\eta}\,\mathrm{d}t \le C\cdot\int_{\mathbb{R}^3}|\boldsymbol{\eta}|^{-1}\cdot|\widehat{\boldsymbol{\Phi}}(\boldsymbol{\eta})|\,\mathrm{d}\boldsymbol{\eta},\end{aligned} \tag{10.19}$$

where we used Lemma 10.3 again. But since $\boldsymbol{\Phi}\in C_0^\infty(\mathbb{R}^3)$, the function $\widehat{\boldsymbol{\Phi}}$ is rapidly decreasing, so $\int_{\mathbb{R}^3}|\boldsymbol{\eta}|^{-1}\cdot|\widehat{\boldsymbol{\Phi}}(\boldsymbol{\eta})|\mathrm{d}\boldsymbol{\eta} < \infty$. Hence on the right-hand side of (10.18), we may reorder integration and summation, thus arriving at the equation

$$\partial_l \mathfrak{R}(\boldsymbol{\Phi}+\nabla\varphi)(\mathbf{y}) = \int_0^\zeta \int_{\mathbb{R}^3} \mathrm{e}^{-\mathrm{i}\cdot\boldsymbol{\xi}\cdot\mathbf{y}}\cdot\sum_{k=0}^{\infty}\mathfrak{S}(\boldsymbol{\xi},t+k\cdot\zeta)\cdot[h(\cdot,t)]^\wedge(\boldsymbol{\xi})\mathrm{d}\boldsymbol{\xi}\mathrm{d}t. \tag{10.20}$$

The idea now is to apply the multiplier theorem [170, Theorem 8] to the inner integral on the right-hand side of (10.20). To this end, we have to derive a suitable estimate of $\sum_{k=0}^{\infty}\mathfrak{S}(\boldsymbol{\xi},t+k\cdot\zeta)$. So let $t\in(0,\zeta]$, and define

$$\begin{aligned}h_k^{(\kappa)}(\boldsymbol{\xi}) :=&\, \mathrm{i}\cdot\xi_l\cdot(-\tau\cdot\mathrm{i}+2\cdot\xi_1)^{\kappa_1}\cdot(2\cdot\xi_2)^{\kappa_2}\cdot(2\cdot\xi_3)^{\kappa_3}\cdot(-t-k\cdot\zeta)^{|\kappa|},\\ \widetilde{h}_k^{(\kappa)}(\boldsymbol{\xi}) :=&\, \delta_{1,\kappa_l}\cdot\mathrm{i}\cdot(-\tau\cdot\mathrm{i}+2\cdot\xi_1)^{\kappa_1-\delta_{1l}}\cdot(2\cdot\xi_2)^{\kappa_2-\delta_{2l}}\cdot(2\cdot\xi_3)^{\kappa_3-\delta_{3l}}\\ &\times(-t-k\cdot\zeta)^{|\kappa|-1}\end{aligned}$$

for $\boldsymbol{\xi} \in \mathbb{R}^3$, $k \in \mathbb{N}_0$, and $\kappa \in \{0, 1\}^3$. Note that for such κ, we have $\kappa_j = 0$ or $\kappa_j = 1$ for $j \in \{1, 2, 3\}$, as well as $|\kappa| \le 3$. Since

$$|(h_k^{(\kappa)}(\boldsymbol{\xi}) + \widetilde{h}_k^{(\kappa)}(\boldsymbol{\xi})) \cdot \mathrm{e}^{(t+k\cdot\zeta)\cdot(\tau\cdot\mathrm{i}\cdot\xi_1 - |\boldsymbol{\xi}|^2)}| \le C(\tau, \varrho) \cdot R^4 \cdot (1+k)^3 \cdot \mathrm{e}^{-k\cdot\zeta\cdot\delta^2}$$

for $\kappa \in \{0, 1\}^3$, $k \in \mathbb{N}_0$, $\delta \in (0, 1)$, $R \in (1, \infty)$, $\boldsymbol{\xi} \in B_R \backslash B_\delta$, and because $\sum_{k=0}^{\infty}(1 + k)^3 \cdot \mathrm{e}^{-k\cdot\zeta\cdot\delta^2} < \infty$ for $\delta > 0$, we may conclude that the derivative $\partial_\xi^\kappa \left(\sum_{k=0}^\infty \mathfrak{S}(\boldsymbol{\xi}, t + k\cdot\zeta)\right)$ exists, is continuous, and

$$\partial_\xi^\kappa \left(\sum_{k=0}^{\infty} \mathfrak{S}(\boldsymbol{\xi}, t + k \cdot \zeta)\right) = -(2\cdot\pi)^{-3/2} \cdot \sum_{k=0}^{\infty}(h_k^{(\kappa)}(\boldsymbol{\xi}) + \widetilde{h}_k^{(\kappa)}(\boldsymbol{\xi})) \times \mathrm{e}^{(t+k\cdot\zeta)\cdot(\tau\cdot\mathrm{i}\cdot\xi_1 - |\boldsymbol{\xi}|^2)} \tag{10.21}$$

for $\kappa \in \{0, 1\}^3$ and $\boldsymbol{\xi} \in \mathbb{R}^3 \backslash \{0\}$. Moreover, we may choose $\epsilon_0 = \epsilon_0(\tau, \varrho) \in (0, 1]$ such that for $\boldsymbol{\xi} \in B_{\epsilon_0}$

$$\begin{aligned} |\sin(\tau\cdot\zeta\cdot\xi_1)| &\ge \tau\cdot\zeta\cdot|\xi_1|/2, \quad 1 - \mathrm{e}^{-\zeta\cdot|\boldsymbol{\xi}|^2} \ge \zeta\cdot|\boldsymbol{\xi}|^2/2, \\ \mathrm{e}^{-\zeta\cdot|\boldsymbol{\xi}|^2} &\ge 1/2, \quad 1 - \cos(\tau\cdot\zeta\cdot\xi_1) \ge \tau^2\cdot\zeta^2\cdot\xi_1^2/4. \end{aligned} \tag{10.22}$$

Let $\kappa \in \{0, 1\}^3$ and $\boldsymbol{\xi} \in \mathbb{R}^3 \backslash \{0\}$. In view of estimating the right-hand side of (10.21), we distinguish the cases $|\boldsymbol{\xi}| \le \epsilon_0$ and $|\boldsymbol{\xi}| > \epsilon_0$. First suppose that $|\boldsymbol{\xi}| \le \epsilon_0$. Then, observing that $|-\tau\cdot\mathrm{i} + 2\cdot\xi_1| \le \tau + \epsilon_0$ and $|\mathrm{e}^{t\cdot(\tau\cdot\mathrm{i}\cdot\xi_1 - |\boldsymbol{\xi}|^2)}| \le 1$, we get from (10.21)

$$\begin{aligned} &\left|\partial_\xi^\kappa \left(\sum_{k=0}^{\infty} \mathfrak{S}(\boldsymbol{\xi}, t + k\cdot\zeta)\right)\right| \\ &\le C(\tau, \varrho) \cdot \left(|\boldsymbol{\xi}|^{1+\kappa_2+\kappa_3} \cdot \left|\sum_{k=0}^{\infty}(-t - k\cdot\zeta)^{|\kappa|} \cdot \mathrm{e}^{\zeta\cdot(\tau\cdot\mathrm{i}\cdot\xi_1 - |\boldsymbol{\xi}|^2)\cdot k}\right| \right. \\ &\left. + \delta_{1,\kappa_l} \cdot |\boldsymbol{\xi}|^{\kappa_2+\kappa_3-\delta_{2l}-\delta_{3l}} \cdot \left|\sum_{k=0}^{\infty}(-t - k\cdot\zeta)^{|\kappa|-1} \cdot \mathrm{e}^{\zeta\cdot(\tau\cdot\mathrm{i}\cdot\xi_1 - |\boldsymbol{\xi}|^2)\cdot k}\right| \right). \end{aligned} \tag{10.23}$$

Next we note that for $k \in \mathbb{N}_0$, we have

$$(-t - k\cdot\zeta)^{|\kappa|} = \sum_{m=0}^{|\kappa|} a_{m,|\kappa|} \cdot \prod_{j=0}^{m-1}(k - j), \tag{10.24}$$

with coefficients $a_{0,|\kappa|}, \ldots, a_{|\kappa|,|\kappa|}$ that are polynomials in t and ζ and do not depend on k, which may thus be estimated by some constant $C(\tau, \varrho) > 0$. In fact, if $|\kappa| = 0$, we take $a_{0,|\kappa|} = 1$. In the case $|\kappa| = 1$, we choose $a_{0,|\kappa|} = -t$ and $a_{1,|\kappa|} = -\zeta$, and if $|\kappa| = 2$, we may set $a_{0,|\kappa|} = t^2$, $a_{1,|\kappa|} = 2 \cdot t \cdot \zeta + \zeta^2$, $a_{2,|\kappa|} = \zeta^2$. We leave it to the reader to determine $a_{0,|\kappa|}, \ldots, a_{3,|\kappa|}$ if $|\kappa| = 3$. In the case $\kappa_l = 1$, an equation analogous to (10.24) holds for $(-t - k \cdot \zeta)^{|\kappa|-1}$. Thus we get from (10.23) that

$$\begin{aligned}
&\left|\partial_\xi^\kappa \left(\sum_{k=0}^{\infty} \mathfrak{S}(\boldsymbol{\xi}, t + k \cdot \zeta)\right)\right| \\
\leq & C(\tau, \varrho) \cdot \left(|\boldsymbol{\xi}|^{1+\kappa_2+\kappa_3} \cdot \sum_{m=0}^{|\kappa|} \left| \sum_{k=0}^{\infty} \left(\prod_{j=0}^{m-1} (k-j) \right) \cdot e^{\zeta \cdot (\tau \cdot i \cdot \xi_1 - |\boldsymbol{\xi}|^2) \cdot k} \right| \right. \\
&\left. + \delta_{1,\kappa_l} \cdot |\boldsymbol{\xi}|^{\kappa_2+\kappa_3-\delta_{2l}-\delta_{3l}} \cdot \sum_{m=0}^{|\kappa|-1} \left| \sum_{k=0}^{\infty} \left(\prod_{j=0}^{m-1} (k-j) \right) \cdot e^{\zeta \cdot (\tau \cdot i \cdot \xi_1 - |\boldsymbol{\xi}|^2) \cdot k} \right| \right).
\end{aligned}$$

Note that for $k, m \in \mathbb{N}_0$ with $k < m$, the product $\prod_{j=0}^{m-1} (k-j)$ vanishes, so we may start summation with respect to k only with $k = m$ instead of $k = 0$. On applying the formula $\sum_{k=m}^{\infty} \left(\prod_{j=0}^{m-1} (k-j) \right) \cdot a^{k-m} = m! \cdot (1-a)^{-m-1}$, which is valid for $a \in \mathbb{C}$ with $|a| < 1$ and for $m \in \mathbb{N}_0$, we may thus deduce from the preceding inequality that

$$\begin{aligned}
&\left|\partial_\xi^\kappa \left(\sum_{k=0}^{\infty} \mathfrak{S}(\boldsymbol{\xi}, t + k \cdot \zeta)\right)\right| \\
\leq & C(\tau, \varrho) \cdot \left(|\boldsymbol{\xi}|^{1+\kappa_2+\kappa_3} \cdot \sum_{m=0}^{|\kappa|} \left| m! \cdot (1 - e^{\zeta \cdot (\tau \cdot i \cdot \xi_1 - |\boldsymbol{\xi}|^2)})^{-m-1} \right| \right. \\
&\left. + \delta_{1,\kappa_l} \cdot |\boldsymbol{\xi}|^{\kappa_2+\kappa_3-\delta_{2l}-\delta_{3l}} \cdot \sum_{m=0}^{|\kappa|-1} \left| m! \cdot (1 - e^{\zeta \cdot (\tau \cdot i \cdot \xi_1 - |\boldsymbol{\xi}|^2)})^{-m-1} \right| \right). \qquad (10.25)
\end{aligned}$$

On the other hand, recalling the choice of ϵ_0 in (10.22) and observing that $(a^2 + b^2)^{1/2} \geq (a+b) \cdot 2^{-1/2}$ for $a, b \in [0, \infty)$, we find

$$\begin{aligned}
&|1 - e^{\zeta \cdot (\tau \cdot i \cdot \xi_1 - |\boldsymbol{\xi}|^2)}| \\
&\geq \left[|\sin(\zeta \cdot \tau \cdot \xi_1)| \cdot e^{-\zeta \cdot |\boldsymbol{\xi}|^2} + \left(1 - \cos(\zeta \cdot \tau \cdot \xi_1) \cdot e^{-\zeta \cdot |\boldsymbol{\xi}|^2} \right) \right] \cdot 2^{-1/2} \\
&\geq C(\tau, \varrho) \cdot (|\xi_1| + |\boldsymbol{\xi}|^2 + \xi_1^2) \geq C(\tau, \varrho) \cdot (|\xi_1| + |\boldsymbol{\xi}|^2).
\end{aligned}$$

Moreover, because of the assumption $|\xi| \le \epsilon_0$, we find for $m \in \{0, \ldots, |\kappa|-1\}$ that the inequality $(|\xi_1| + |\boldsymbol{\xi}|^2)^{-m-1} \le C(\tau, \varrho) \cdot (|\xi_1| + |\boldsymbol{\xi}|^2)^{-a}$ holds with $a = |\kappa| + 1$ and $a = |\kappa|$. Thus, from (10.25)

$$\begin{aligned}\left|\partial_\xi^\kappa \left(\sum_{k=0}^{\infty} \mathfrak{S}(\boldsymbol{\xi}, t + k \cdot \zeta)\right)\right| \le & C(\tau, \varrho) \cdot [(|\xi_1| + |\boldsymbol{\xi}|^2)^{-|\kappa|-1/2+(\kappa_2+\kappa_3)/2} \\ & + \delta_{1,\kappa_l} \cdot (|\xi_1| + |\boldsymbol{\xi}|^2)^{-|\kappa|+(\kappa_2+\kappa_3)/2-(\delta_{2l}+\delta_{3l})/2}] \\ \le & C(\tau, \varrho) \cdot (|\xi_1| + |\boldsymbol{\xi}|^2)^{-|\kappa|-1/2+(\kappa_2+\kappa_3)/2},\end{aligned}$$

where in the case $l = 1$, $\kappa_l = 1$, the last inequality follows with the assumption $|\boldsymbol{\xi}| \le \epsilon_0$. Therefore

$$\begin{aligned}|\xi_1|^{\kappa_1+1/4} \cdot & \left|\partial_\xi^\kappa \left(\sum_{k=0}^{\infty} \mathfrak{S}(\boldsymbol{\xi}, t + k \cdot \zeta)\right)\right| \\ \le & C(\tau, \varrho) \cdot (|\xi_1| + |\boldsymbol{\xi}|^2)^{-|\kappa|-1/4+\kappa_1+(\kappa_2+\kappa_3)/2} \\ \le & C(\tau, \varrho) \cdot |\boldsymbol{\xi}|^{-2\cdot|\kappa|-1/2+2\cdot\kappa_1+\kappa_2+\kappa_3} = C(\tau, \varrho) \cdot |\boldsymbol{\xi}|^{-\kappa_2-\kappa_3-1/2}.\end{aligned}$$

Hence

$$\prod_{j=1}^{3} |\xi_j|^{\kappa_j+1/4} \cdot \left|\partial_\xi^\kappa \left(\sum_{k=0}^{\infty} \mathfrak{S}(\boldsymbol{\xi}, t + k \cdot \zeta)\right)\right| \le C(\tau, \varrho). \tag{10.26}$$

Now suppose that $|\boldsymbol{\xi}| > \epsilon_0$. Then we have $|-\tau \cdot \mathrm{i} + 2 \cdot \xi_1| \le C(\tau, \varrho) \cdot |\boldsymbol{\xi}|$, so we may deduce from (10.21) that

$$\begin{aligned}& \left|\partial_\xi^\kappa \left(\sum_{k=0}^{\infty} \mathfrak{S}(\boldsymbol{\xi}, t + k \cdot \zeta)\right)\right| \\ \le & C(\tau, \varrho) \cdot \sum_{k=0}^{\infty} \left(|\boldsymbol{\xi}|^{1+|\kappa|} \cdot (t + k \cdot \zeta)^{|\kappa|} + \delta_{1,\kappa_l} \cdot |\boldsymbol{\xi}|^{|\kappa|-1} \cdot (t + k \cdot \zeta)^{|\kappa|-1}\right) \\ & \times \mathrm{e}^{-(t+k\cdot\zeta)\cdot|\boldsymbol{\xi}|^2}.\end{aligned}$$

Thus, when we multiply the left-hand side of the preceding inequality by $\prod_{j=1}^{3} |\xi_j|^{\kappa_j+1/4}$, we get an upper bound of the form of a series with respect to $k \in \mathbb{N}_0$ of the products

$$(|\boldsymbol{\xi}|^2 \cdot (t + k \cdot \zeta))^{|\kappa|-1/8} \cdot (|\boldsymbol{\xi}|^2 \cdot (t + k \cdot \zeta) + \delta_{1,\kappa_l}) \cdot (t + k \cdot \zeta)^{-7/8}.$$

But for $m \in \{0, 1\}$, the term $((t + k \cdot \zeta) \cdot |\xi|^2)^{|\kappa|-1/8+m} \cdot e^{-(t+k\cdot\zeta)\cdot|\xi|^2/2}$ is bounded by a constant only depending on τ and ϱ. Thus we obtain

$$\begin{aligned}\prod_{j=1}^{3} |\xi_j|^{\kappa_j+1/4} \cdot \left|\partial_\xi^\kappa \left(\sum_{k=0}^{\infty} \mathfrak{S}(\boldsymbol{\xi}, t + k \cdot \zeta)\right)\right| &\leq C(\tau, \varrho) \cdot \sum_{k=0}^{\infty} (t + k \cdot \zeta)^{-7/8} \cdot e^{-(t+k\cdot\zeta)\cdot|\boldsymbol{\xi}|^2/2} \\ &\leq C(\tau, \varrho) \cdot t^{-7/8} \cdot \sum_{k=0}^{\infty} e^{-k\cdot\zeta\cdot\epsilon_0^2/2} \leq C(\tau, \varrho) \cdot t^{-7/8}, \end{aligned} \tag{10.27}$$

where in the second inequality we again used the assumption $|\boldsymbol{\xi}| \geq \epsilon_0$. In view of (10.26), we may thus conclude that the preceding estimate holds for any $\boldsymbol{\xi} \in \mathbb{R}^3\backslash\{0\}$. Recall that t was chosen arbitrarily in $(0, \zeta]$. Now, for such t, define $\boldsymbol{N}(t) \colon \mathbb{R}^3 \mapsto \mathbb{R}^3$ by

$$\boldsymbol{N}(t)(\boldsymbol{y}) := \int_{\mathbb{R}^3} e^{-i\cdot\boldsymbol{\xi}\cdot\boldsymbol{y}} \cdot \sum_{k=0}^{\infty} \mathfrak{S}(\boldsymbol{\xi}, t + k \cdot \zeta) \cdot [h(\cdot, t)]^\wedge(\boldsymbol{\xi}) d\boldsymbol{\xi}$$

for $\mathbf{y} \in \mathbb{R}^3$. Then inequality (10.27) and [170, Theorem 8] imply that $\|\boldsymbol{N}(t)\|_{(1/q-1/4)^{-1}} \leq C(\tau, \varrho) \cdot t^{-7/8} \cdot \|h(\cdot, t)\|_q$ for $t \in (0, \zeta]$. Therefore with (10.20) and Minkowski's inequality,

$$\|\partial_l \mathfrak{R}(\boldsymbol{\Phi} + \nabla\varphi)\|_{(1/q-1/4)^{-1}} \leq C(\tau, \varrho) \cdot \int_0^\zeta t^{-7/8} \cdot \|h(\cdot, t)\|_q dt \tag{10.28}$$

for t as before. But $\|h(\cdot, t)\|_q = \|\boldsymbol{\Phi}\|_q$ by Lemma 10.3 and a change of variables. Since in addition $\|\boldsymbol{\Phi}\|_q \leq C(q) \cdot \|\boldsymbol{\Phi} + \nabla\varphi\|_q$ by [95, Theorem III.1.2], the second inequality in Theorem 10.11 follows from (10.28).

Concerning an estimate of the second derivatives of $\mathfrak{R}(\boldsymbol{\Phi} + \nabla\varphi)$, there is an additional problem which consists in establishing an analog to (10.18) for these derivatives although no analog of the formula at the end of Lemma 10.13 is available for them. In this respect, we indicate that

$$|\partial y_m(e^{i\cdot\boldsymbol{\xi}\cdot\mathbf{y}}) \cdot \mathfrak{S}(\boldsymbol{\xi}, t + k \cdot \zeta) \cdot [h(\cdot, t)]^\wedge(\boldsymbol{\xi})| \leq C \cdot |\boldsymbol{\xi}|^2 \cdot e^{-|\boldsymbol{\xi}|^2\cdot(t+k\cdot\zeta)} \cdot |\widehat{\boldsymbol{\Phi}}(e^{t\cdot\Omega} \cdot \boldsymbol{\xi})|$$

for $\mathbf{y}, \boldsymbol{\xi} \in \mathbb{R}^3$, $t \in (0, \zeta]$, $1 \leq l, m \leq 3$, where the function $\mathfrak{S} = \mathfrak{S}_l$ is defined as above. But the term $\sum_{k=0}^{\infty} \int_0^\zeta \int_{\mathbb{R}^3} |\boldsymbol{\xi}|^2 \cdot e^{-|\boldsymbol{\xi}|^2\cdot(t+k\cdot\zeta)} \cdot |\widehat{\boldsymbol{\Phi}}(e^{t\cdot\Omega} \cdot \boldsymbol{\xi})| d\boldsymbol{\xi} dt$ is finite, as follows by the same reasoning as used in (10.19). In view of (10.18), this means that $\partial_m \partial_l \mathfrak{R}(\boldsymbol{\Phi} + \nabla\varphi)(\mathbf{y})$ equals the right-hand side of (10.18), but with the term $e^{i\cdot\boldsymbol{\xi}\cdot\mathbf{y}}$ replaced by $\partial y_m(e^{i\cdot\boldsymbol{\xi}\cdot\mathbf{y}})$. Now we may estimate $\|\partial_m \partial_l \mathfrak{R}(\boldsymbol{\Phi} + \nabla\varphi)\|_r$ by the same techniques as used above in order to find an upper bound for $\|\partial_l \mathfrak{R}(\boldsymbol{\Phi} + \nabla\varphi)\|_{(1/q-1/4)^{-1}}$. For more details concerning the proof of the third inequality, see [39]. □

Corollary 10.9 *Let $p \in (1, 2)$ and $\mathbf{f} \in \boldsymbol{L}^p(\mathbb{R}^3)$. Then $\mathfrak{R}(\mathbf{f}) \in \boldsymbol{W}^{2,p}_{\mathrm{loc}}(\mathbb{R}^3)$, and the inequality $\|\mathfrak{R}(\mathbf{f})\|_{(1/p-1/2)^{-1}} \le C_1(p) \cdot \|\mathbf{f}\|_p$ holds.*

If $q \in (1, 4)$ and $\mathbf{f} \in \boldsymbol{L}^p(\mathbb{R}^3) \cap \boldsymbol{L}^q(\mathbb{R}^3)$, we have $\|\nabla\mathfrak{R}(\mathbf{f})\|_{(1/q-1/4)^{-1}} \le C_2(q) \cdot \|\mathbf{f}\|_q$.

If $r \in (1, \infty)$ and $\mathbf{f} \in \boldsymbol{L}^p(\mathbb{R}^3) \cap \boldsymbol{L}^r(\mathbb{R}^3)$, we further have $\|\partial_1\mathfrak{R}(\mathbf{f})\|_r + \|\nabla^2\mathfrak{R}(\mathbf{f})\|_r \le C_3(r) \cdot \|\mathbf{f}\|_r$. □

Proof By Theorem 10.8, we may choose sequences $(\boldsymbol{\Phi}_n)$, (φ_n) with $\boldsymbol{\Phi}_n \in \boldsymbol{C}_0^\infty(\mathbb{R}^3)$, $\varphi_n \in C_0^\infty(\mathbb{R}^3)$, div, $\boldsymbol{\Phi}_n = 0$ for $n \in \mathbb{N}$, and $\|\mathbf{f} - \boldsymbol{\Phi}_n - \nabla\varphi_n\|_p \to 0$ $(n \to \infty)$. Therefore, by Corollary 10.7, there is a subsequence $(\mathbf{g}_n)$ of $(\boldsymbol{\Phi}_n + \nabla\varphi_n)$ such that $\mathfrak{R}(\mathbf{g}_n)(x) \to \mathfrak{R}(\mathbf{f})(\mathbf{x})$ $(n \to \infty)$ for a.e. $\mathbf{x} \in \mathbb{R}^3$. At this point we may conclude from Theorem 10.11 that $\mathfrak{R}(\mathbf{f}) \in \boldsymbol{W}^{2,p}_{\mathrm{loc}}(\mathbb{R}^3)$ and $\|\partial^\alpha\mathfrak{R}(\mathbf{f}) - \partial^\alpha\mathfrak{R}(\boldsymbol{\Phi}_n + \nabla\varphi_n)\|_s \to 0$ $(n \to \infty)$, $\|\partial^\alpha\mathfrak{R}(\mathbf{f})\|_s \le c(\alpha, p) \cdot \|\mathbf{f}\|_p$ for $\alpha \in \mathbb{N}_0^3$ with $|\alpha| \le 2$, where $s = (1/p - 1/2)^{-1}$, $c(\alpha, p) = C_1(p)$ if $\alpha = 0$, $s = (1/p - 1/4)^{-1}$, $c(\alpha, p) = C_2(p)$ if $|\alpha| = 1$, and $s = p$, $c(\alpha, p) = C_3(p)$ in the case $|\alpha| = 2$.

Now take $q \in (1, 4)$, and suppose in addition that $\mathbf{f} \in \boldsymbol{L}^q(\mathbb{R}^3)$. Again referring to Theorem 10.8, we choose sequences $(\boldsymbol{\Phi}_n)$, (φ_n), $(\widetilde{\varphi}_n)$, with $(\boldsymbol{\Phi}_n)$, (φ_n) having properties as above, and such that $\widetilde{\varphi}_n \in C_0^\infty(\mathbb{R}^3)$ for $n \in \mathbb{N}$ and $\|\mathbf{f} - \boldsymbol{\Phi}_n - \nabla\widetilde{\varphi}_n\|_q \to 0$ $(n \to \infty)$. Let $l \in \{1, 2, 3\}$. As shown above, $\|\partial_l\mathfrak{R}(\mathbf{f}) - \partial_l\mathfrak{R}(\boldsymbol{\Phi}_n + \nabla\varphi_n)\|_{(1/p-1/4)^{-1}} \to 0$ $(n \to \infty)$. In addition we deduce from Theorem 10.11 there is a function $\boldsymbol{G} \in \boldsymbol{L}^{(1/q-1/4)^{-1}}(\mathbb{R}^3)$ such that $\|\boldsymbol{G} - \partial_l\mathfrak{R}(\boldsymbol{\Phi}_n + \nabla\widetilde{\varphi}_n)\|_{(1/q-1/4)^{-1}} \to 0$ $(n \to \infty)$. In particular, there exists a strictly increasing function $\sigma : \mathbb{N} \mapsto \mathbb{N}$ such that $\partial_l\mathfrak{R}(\boldsymbol{\Phi}_{\sigma(n)} + \nabla\varphi_{\sigma(n)})(\mathbf{x}) \to \partial_l\mathfrak{R}(\mathbf{f})(\mathbf{x})$ and $\partial_l\mathfrak{R}(\boldsymbol{\Phi}_{\sigma(n)} + \nabla\widetilde{\varphi}_{\sigma(n)})(\mathbf{x}) \to G(\mathbf{x})$ $(n \to \infty)$ for a.e. $\mathbf{x} \in \mathbb{R}^3$. But we know from Lemma 10.15 that $\mathfrak{R}(\boldsymbol{\Phi}_n + \nabla\varphi_n) = \mathfrak{R}(\boldsymbol{\Phi}_n + \nabla\widetilde{\varphi}_n)$ for $n \in \mathbb{N}$. Thus we obtain $\partial_l\mathfrak{R}(\mathbf{f}) = G$, and we may now conclude with Theorem 10.11 and the relation $\|\mathbf{f} - \boldsymbol{\Phi}_n - \nabla\widetilde{\varphi}_n\|_q \to 0$ that $\|\partial_l\mathfrak{R}(\mathbf{f})\|_{(1/q-1/4)^{-1}} \le C_2(q) \cdot \|\mathbf{f}\|_q$. The last statement of the corollary may be shown in an analogous way. □

The preceding results immediately imply Theorem 10.2. In fact, we may give the following:

Proof of Theorem 10.2 Combine Corollary 10.9, Theorem 10.9, and Corollary 10.8. □

10.4 L^p-Estimates of the Velocity and Its Gradient

Take a set $\mathcal{D}$, parameters γ, S_1, p_0, A, B, and functions $\mathbf{f}$, $\mathbf{u}$, and π as in Theorem 10.1. The set $\mathcal{D}$ and the preceding parameters and functions (in particular, $\mathbf{u}$ and π) are to be fixed for the rest of this section. Since $\bar{\mathcal{D}}^c \subset B_{S_1}$, we may choose

$S_0 \in (0, S_1)$ with $\bar{\mathcal{D}}^c \subset B_{S_0}$. In the following, we will use the letter $\mathfrak{C}$ for constants that may depend on $\tau, \varrho, \gamma, S_0, S_1, p_0, A, B, \|\mathbf{f}|B_{S_1}\|_1, u$, and π. If such a constant additionally depends on $\gamma_1, \ldots, \gamma_n \in (0, \infty)$ for some $n \in \mathbb{N}$, we will denote it by $\mathfrak{C}(\gamma_1, \ldots, \gamma_n)$. We recall that by a previous convention, we write C for numerical constants and $C(\gamma_1, \ldots, \gamma_n)$ for constants only depending on $\gamma_1, \ldots, \gamma_n$.

The assumptions on $\mathbf{f}$ in Theorem 10.1 mean in particular that for any $p > 1$, the function $\mathbf{f}$ is L^p-integrable outside B_{S_1}:

Lemma 10.19 $\mathbf{f}|B_{S_1}^c \in L^p(S_{S_1}^c)$ *for* $p \in (1, \infty]$. □

Proof Put $B^* := \min\{1, B\}$, $\epsilon := \min\{(A-2)/2,\ (A+B^*-3)/2\}$. By our choice of A and B in Theorem 10.1, we have $\epsilon > 0$. For $y \in B_{S_1}^c$, we find as in the proof of [29, Lemma 2.12] that

$$|\mathbf{y}|^{-A} \cdot s(\mathbf{y})^{-B} \le C(A) \cdot |\mathbf{y}|^{-2-\epsilon} \cdot s(\mathbf{y})^{-(A+B^*-3)/2-1}.$$

Let $p \in (1, \infty)$. Since $A + B^* - 3 > 0$ by our assumptions on A and B, the relation $[-(A+B^*-3)/2-1] \cdot p < -1$ holds. Hence $\int_{B_{S_1}^c} (|\mathbf{y}|^{-A} \cdot s(\mathbf{y})^{-B})^p d\mathbf{y} \le C(A, B, S_1, p)$ by Corollary 10.1. Therefore Lemma 10.19 follows with (10.6). □

The preceding lemma means in particular that the potential $\mathfrak{R}(\mathbf{f})$ is well defined and belongs to $W^{1,1}_{loc}(\mathbb{R}^3)$ (see Lemma 10.13). Since $(\mathbf{u} \cdot \nabla)\mathbf{u} \in L^{3/2}(\bar{\mathcal{D}})$ by Lemma 10.2, the same is true of $\mathfrak{R}((\mathbf{u} \cdot \nabla)\mathbf{u})$.

The starting point of our estimates is the representation formula from Chap. 5 already recalled in Sects. 10.1 and 10.2. This formula refers to the velocity part of Leray solutions to the nonlinear system (10.1) in exterior domains and may thus be applied to $\mathbf{u}$.

Theorem 10.12 *Put* $\widetilde{p} := \min\{3/2, p_0\}$, *with* p_0 *from the assumptions on* f. *Then* $\mathbf{u} \in W^{2,\widetilde{p}}_{loc}(\bar{\mathcal{D}})$, $\pi \in W^{1,\widetilde{p}}_{loc}(\bar{\mathcal{D}})$. *Define*[4]

$$\begin{aligned}\mathfrak{B}_j(\mathbf{u}, \pi)(\mathbf{y}) := \int_{\partial B_{S_0}} \sum_{k=1}^{3} \Big[\sum_{l=1}^{3} &(\mathcal{Z}_{jk}(\mathbf{y}, \mathbf{z}) \cdot (-\partial_l u_k(\mathbf{z}) + \delta_{kl} \cdot \pi(\mathbf{z}) + u_k(\mathbf{z}) \\ &\times (\tau \cdot \boldsymbol{e}_1 - \boldsymbol{\omega} \wedge \mathbf{z})_l) + \partial z_l \mathcal{Z}_{jk}(\mathbf{y}, \mathbf{z}) \cdot u_k(\mathbf{z})) \cdot (z_l/S_0) \\ &+ E_j(\mathbf{y} - \mathbf{z}) \cdot u_k(\mathbf{z}) \cdot (z_k/S_0) \Big] do_z\end{aligned}$$

for $\mathbf{y} \in \overline{B_{S_0}}^c$, $1 \le j \le 3$. *Then* $\mathbf{u}(\mathbf{y}) = \mathfrak{R}(\mathbf{f} - \tau \cdot (\mathbf{u} \cdot \nabla)\mathbf{u}|B_{S_0}^c)(\mathbf{y}) + \mathfrak{B}(\mathbf{u}, \pi)(\mathbf{y})$, *for* $\mathbf{y}$ *as before.* □

Proof We derive L_p estimates of the velocity from the integral representation formulas of $\mathbf{u}$, of $\mathfrak{B}(\mathbf{u}, \nu)$ and of the potential $\mathfrak{R}(\mathbf{f} - (\mathbf{u} \cdot \nabla)\mathbf{u})$, and from respective

[4] Note that the letter $\mathfrak{B}$ does not denote the Bogovskii operator in this chapter.

different integrability properties to establish. The first statement of Theorem 10.12 follows from the interior regularity theory for the Stokes system. For more details, we refer to the proof of Theorem 11.7. The assumptions on **f** in that reference hold according to Lemma 10.19. From the first claim in Theorem 10.12, we may conclude that $\mathbf{u}|\partial B_{S_0} \in \boldsymbol{W}^{2-1/\tilde{p},\tilde{p}}(\partial B_{S_0})$. Due to this observation and Lemma 10.12, the function $\mathfrak{B}(u, \pi)$ is well defined. The main claim of Theorem 10.12, that is, the representation formula in the last sentence of this theorem, may now be deduced from Theorem 5.20 and Theorem 5.21 with $\mathcal{D}^c$ replaced by B_{S_0} and from Lemma 10.19. □

As a consequence, we have the following theorem:

Theorem 10.13 *Let $S \in (S_1, \infty)$. Then*

$$|\partial^\alpha \mathbf{u}(\mathbf{x}) - \partial^\alpha \mathfrak{R}(\tau \cdot (\mathbf{u} \cdot \nabla)\mathbf{u}|B^c_{S_0})(\mathbf{x})|$$
$$\leq C(S_0, S_1, S, A, B) \cdot \mathfrak{M} \cdot (|\mathbf{x}| \cdot s_1(\mathbf{x}))^{-1-|\alpha|/2} \tag{10.29}$$

for $\mathbf{x} \in B^c_S$, $\alpha \in \mathbb{N}_0^3$ with $|\alpha| \leq 1$, where we used the abbreviation

$$\mathfrak{M} := \gamma + \|\mathbf{f}|B_{S_1}\|_1 + \|u|\partial B_{S_0}\|_1 + \|\nabla \mathbf{u}|\partial B_{S_0}\|_1 + \|\pi|\partial B_{S_0}\|_1.$$

In particular,

$$\mathbf{u} - \mathfrak{R}(\tau \cdot (\mathbf{u} \cdot \nabla)\mathbf{u}|B^c_{S_0})|B^c_S \in \boldsymbol{L}^q(B^c_S) \quad \text{for } q \in (2, \infty), \tag{10.30}$$

$$\nabla \mathbf{u} - \nabla \mathfrak{R}(\tau \cdot (\mathbf{u} \cdot \nabla)\mathbf{u}|B^c_{S_0})|B^c_S \in L^q(B^c_S)^9 \quad \text{for } q \in (4/3, \infty). \tag{10.31}$$

Proof Consider the function $\mathfrak{B}(\mathbf{u}, \pi)$ from Theorem 10.12. By Lemma 5.20 with $\mathcal{D}^c$ replaced by B_{S_0}, we see that $|\partial^\alpha \mathfrak{B}(\mathbf{u}, \pi)(\mathbf{x})|$ bounded by

$$C(S_0, S_1, S) \cdot (\|\mathbf{u} \mid \partial B_{S_0}\|_1 + \|\nabla \mathbf{u} \mid \partial B_{S_0}\|_1 + \|\pi \mid \partial B_{S_0}\|_1) \cdot (|\mathbf{x}| \cdot s_1(\mathbf{x}))^{-1-|\alpha|/2}$$

for $\mathbf{x}$, α as in (10.29). On the other hand, according to Lemma 5.10 (see also [36, Theorem 3.3]), we have

$$|\partial^\alpha \mathfrak{R}(\mathbf{f} \mid B^c_{S_0})(\mathbf{x})| \leq C(S_1, S, A, B) \cdot (\gamma + \|\mathbf{f} \mid B_{S_1}\|_1) \cdot (|\mathbf{x}| \cdot s_1(\mathbf{x}))^{-1-|\alpha|/2},$$

again for $\mathbf{x}$, α as in (10.29). Now (10.29) follows from Theorem 10.12. The remaining statements of Theorem 10.13 follow from (10.29) and Corollary 10.1. □

Lemma 10.20 $\nabla \mathbf{u} \mid B^c_S \in L^{12/5}(B^c_S)^9$ *for $S \in (S_1, \infty)$.* □

Proof Since $(\mathbf{u} \cdot \nabla)\mathbf{u} \in \boldsymbol{L}^{3/2}(\bar{\mathcal{D}})$ (see Lemma 10.2), Corollary 10.9 with $q = 3/2$ yields $\nabla \mathfrak{R}((\mathbf{u} \cdot \nabla)\mathbf{u} \mid B^c_{S_0}) \in L^{12/5}(\mathbb{R}^3)^3$. The lemma follows from this observation and (10.31). □

Lemma 10.20 and Theorem 10.3 imply:

Corollary 10.10 $\mathbf{u} \mid B_S^c \in \boldsymbol{L}^{10:12}(B_S^c)$ *for* $S \in (S_1, \infty)$. □

Theorem 10.14 $\mathbf{u} \in \boldsymbol{L}^4(\bar{\mathcal{D}})$. □

Proof Due to the assumption $\nabla\mathbf{u} \in L^2(\bar{\mathcal{D}})^9$, we may choose $r \in (S_1, \infty)$ such that

$$\|\nabla\mathbf{u} \mid B_r^c\|_2 \leq \min\{(2\cdot\tau\cdot C_0(4,2)\cdot C_1(4/3))^{-1},\ (2\cdot\tau\cdot C_0(6,2)\cdot C_1(3/2))^{-1}\}, \tag{10.32}$$

with $C_0(4,2)$, $C_0(6,2)$ from Lemma 10.2 and $C_1(4/3)$, $C_1(3/2)$ from Theorem 10.11. For $\varphi \in L^4(B_r^c)^3$, we define $\mathfrak{A}(\varphi)(x) := u(x)$ for $\mathbf{x} \in B_r \backslash B_{S_0}$, $\mathfrak{A}(\varphi)(\mathbf{x}) := \varphi(\mathbf{x})$ for $\mathbf{x} \in B_r^c$.

Since $\mathbf{u} \in \boldsymbol{L}^6(\bar{\mathcal{D}})$, we have $\mathfrak{A}(\boldsymbol{\varphi}) \in \boldsymbol{L}^4(B_{S_0}^c)$. Therefore, Lemma 10.2 with $p = 4, q = 2$ yields $(\mathfrak{A}(\boldsymbol{\varphi})\cdot\nabla)\mathbf{u} \in \boldsymbol{L}^{4/3}(B_{S_0}^c)$ if $\boldsymbol{\varphi} \in \boldsymbol{L}^4(B_r^c)$, so by Corollary 10.9 with $p = 4/3$ and (10.30), we may define a mapping $\mathfrak{T}\colon L^4(B_r^c)^3 \mapsto L^4(B_r^c)^3$ by setting

$$\mathfrak{T}(\varphi) := (\mathbf{u} - \tau\cdot\mathfrak{R}((\mathbf{u}\cdot\nabla)\mathbf{u} \mid B_{S_0}^c) + \tau\cdot\mathfrak{R}([\mathfrak{A}(\boldsymbol{\varphi})\cdot\nabla]\mathbf{u})) \mid B_r^c \quad \text{for } \boldsymbol{\varphi} \in \boldsymbol{L}^4(B_r^c).$$

By the same references, we find for $\mathbf{w}_1, \mathbf{w}_2 \in \boldsymbol{L}^4(B_r^c)$ that

$$\begin{aligned}\|\mathfrak{T}(\mathbf{w}_1) - \mathfrak{T}(\mathbf{w}_2)\|_4 \leq& \tau\cdot\|\mathfrak{R}([(\mathbf{w}_1 - \mathbf{w}_2)\cdot\nabla]\mathbf{u} \mid B_r^c)\|_4\\ =& \tau\cdot C_1(4/3)\cdot\|[(\mathbf{w}_1 - \mathbf{w}_2)\cdot\nabla]\mathbf{u} \mid B_r^c\|_{4/3}\\ \leq& \tau\cdot C_1(4/3)\cdot C_0(4,2)\cdot\|\mathbf{w}_1 - \mathbf{w}_2\|_4\cdot\|\nabla\mathbf{u} \mid B_r^c\|_2,\end{aligned}$$

so that by (10.32) $\|\mathfrak{T}(\mathbf{w}_1) - \mathfrak{T}(\mathbf{w}_2)\|_4 \leq \|\mathbf{w}_1 - \mathbf{w}_2\|_4/2$. Thus there is $\mathbf{v} \in \boldsymbol{L}^4(B_r^c)$ with $\mathfrak{T}(\mathbf{v}) = \mathbf{v}$, that is,

$$\mathbf{v} = (\mathbf{u} - \tau\cdot\mathfrak{R}((\mathbf{u}\cdot\nabla)\mathbf{u} \mid B_{S_0}^c) + \tau\cdot\mathfrak{R}([\mathfrak{A}(\mathbf{v})\cdot\nabla]\mathbf{u})) \mid B_r^c. \tag{10.33}$$

Lemma 10.2 with $p = 4, q = 12/5$ and $p = 6, q = 2$, respectively, yields

$$\|(\mathfrak{A}(\mathbf{v})\cdot\nabla)\mathbf{u}\|_{3/2} \leq C_0(4, 12/5)\cdot\|\mathbf{v}\|_4\cdot\|\nabla\mathbf{u}|B_r^c\|_{12/5} + C_0(6,2)\cdot\|\mathbf{u}\|_6\cdot\|\nabla\mathbf{u}\|_2,$$

so $(\mathfrak{A}(\mathbf{v})\cdot\nabla)\mathbf{u} \in \boldsymbol{L}^{3/2}(B_{S_0}^c)$ in view of Lemma 10.20. It follows with Corollary 10.9 with $p = 3/2$ that $\mathfrak{R}([\mathfrak{A}(\mathbf{v})\cdot\nabla]\mathbf{u}) \in \boldsymbol{L}^6(\mathbb{R}^3)$. We may conclude by (10.30) and (10.33) that $\mathbf{v} \in \boldsymbol{L}^6(B_r^c)$, so $\mathfrak{A}(\mathbf{v}) \in \boldsymbol{L}^6(B_{S_0}^c)$. Now we return to Lemma 10.2 and Corollary 10.9, applying the former one with $p = 6, q = 2$ and the latter one with $p = 3/2$, to deduce from (10.33) and (10.32) that

$$\begin{aligned}\|\mathbf{v} - \mathbf{u}|B_r^c\|_6 \leq& \tau\cdot\|\mathfrak{R}([(\mathbf{v} - \mathbf{u})\cdot\nabla]\mathbf{u} \mid B_r^c)\|_6\\ \leq& \tau\cdot C_1(3/2)\cdot C_0(6,2)\cdot\|\mathbf{v} - \mathbf{u}|B_r^c\|_6\cdot\|\nabla\mathbf{u}|B_r^c\|_2\\ \leq& \|\mathbf{v} - \mathbf{u}|B_r^c\|_6/2.\end{aligned}$$

This means that $\mathbf{v} - \mathbf{u} \mid B_r^c = 0$, so $\mathbf{u} \mid B_r^c = \mathbf{v} \in \boldsymbol{L}^4(B_r^c)$. Since $\mathbf{u} \in \boldsymbol{L}^6(\bar{\mathcal{D}})$, we thus have proved that $\mathbf{u} \in \boldsymbol{L}^4(\bar{\mathcal{D}})$. □

Theorem 10.15 $\mathbf{u} \in L^3(\bar{\mathcal{D}})$ *and* $\nabla\mathbf{u} \in L^{12/7}(\bar{\mathcal{D}})^9$. □

Proof Since $\mathbf{u} \in \boldsymbol{L}^6(\bar{\mathcal{D}})$ by assumption and $\mathbf{u} \in \boldsymbol{L}^4(\bar{\mathcal{D}})$ by Theorem 10.14, we may choose $r \in (S_1, \infty)$ with

$$\|\mathbf{u}|B_r^c\|_6 + \|\mathbf{u}|B_r^c\|_4 \le \min\{(a_j \cdot \tau \cdot A_j \cdot B_j)^{-1} \colon j \in \{1, \dots, 4\}\}, \tag{10.34}$$

with $A_1 := C_1(6/5)$, $A_2 := C_2(6/5)$, $B_1 := B_2 := C_0(4, 12/7)$, $a_1 := a_2 := 4$, $A_3 := C_1(3/2)$, $B_3 := C_0(6, 2)$, $a_3 := 1$, $A_4 := C_2(4/3)$, $B_4 := C_0(4, 2)$, $a_4 := 2$. (The preceding constants were introduced in Lemma 10.2 and Theorem 10.11, respectively.) Put

$$\mathfrak{W} := \{\boldsymbol{\varphi} \in \boldsymbol{L}^3(\bar{B}_r^c) \cap \boldsymbol{W}^{1,1}_{\mathrm{loc}}(\bar{B}_r^c) \colon \nabla\boldsymbol{\varphi} \in L^{12/7}(\bar{B}_r^c)^9\},$$
$$\|\boldsymbol{\varphi}\|_{\mathfrak{W}} := \|\boldsymbol{\varphi}\|_3 + \|\nabla\boldsymbol{\varphi}\|_{12/7}$$

for $\boldsymbol{\varphi} \in \mathfrak{W}$. The mapping $\| \ \|_{\mathfrak{W}}$ is a norm on $\mathfrak{W}$, and $\mathfrak{W}$ equipped with this norm is a Banach space. For $\boldsymbol{\varphi} \in \mathfrak{W}$, define $\mathfrak{A}(\boldsymbol{\varphi})(\mathbf{x}) := (\partial_k u_l(\mathbf{x}))_{1\le k,l\le 3}$ for $\mathbf{x} \in B_r \backslash B_{S_0}$, $\mathfrak{A}(\boldsymbol{\varphi})(\mathbf{x}) := (\partial_k \varphi_l(\mathbf{x}))_{1\le k,l\le 3}$ for $\mathbf{x} \in B_r^c$. Since $\nabla\mathbf{u} \in L^2(\bar{\mathcal{D}})^9$ and by the definition of $\mathfrak{W}$, we have $\mathfrak{A}(\boldsymbol{\varphi}) \in L^{12/7}(B_{S_0}^c)^9$ $(\boldsymbol{\varphi} \in \mathfrak{W})$. Therefore by Lemma 10.2 with $p = 4, q = 12/7$ and by Theorem 10.14, we get $\mathbf{u}^T \cdot \mathfrak{A}(\boldsymbol{\varphi}) \in \boldsymbol{L}^{6/5}(B_{S_0}^c)$ for $\boldsymbol{\varphi} \in \mathfrak{W}$. Thus Corollary 10.9 with $p = q = 6/5$ implies $\mathfrak{R}(\mathbf{u}^T \cdot \mathfrak{A}(\boldsymbol{\varphi})) \in \boldsymbol{L}^3(\mathbb{R}^3)$ and $\nabla\mathfrak{R}(\mathbf{u}^T \cdot \mathfrak{A}(\boldsymbol{\varphi})) \in L^{12/7}(\mathbb{R}^3)^9$. In view of (10.30) and (10.31), we may thus define $\mathfrak{T} \colon \mathfrak{W} \mapsto \mathfrak{W}$ by setting

$$\mathfrak{T}(\boldsymbol{\varphi}) := [\mathbf{u} - \tau \cdot \mathfrak{R}((\mathbf{u} \cdot \nabla)\mathbf{u} \mid B_{S_0}^c) + \tau \cdot \mathfrak{R}(\mathbf{u}^T \cdot \mathfrak{A}(\boldsymbol{\varphi}))] \mid B_r^c \quad (\boldsymbol{\varphi} \in \mathfrak{W}).$$

For $\mathbf{w}_1, \mathbf{w}_2 \in \mathfrak{W}$, we obtain with Lemma 10.2 ($p = 4$, $q = 12/7$), Corollary 10.9 ($p = q = 6/5$), and (10.34) that

$$\begin{aligned}
\|\mathfrak{T}(\mathbf{w}_1) - \mathfrak{T}(\mathbf{w}_2)\|_3 \le& \tau \cdot \|\mathfrak{R}((\mathbf{u} \mid B_r^c \cdot \nabla)(\mathbf{w}_1 - \mathbf{w}_2))\|_3 \\
\le& \tau \cdot C_1(6/5) \cdot C_0(4, 12/7) \cdot \|\mathbf{u} \mid B_r^c\|_4 \\
& \times \|\nabla(\mathbf{w}_1 - \mathbf{w}_2)\|_{12/7} \\
\le& \|\mathbf{w}_1 - \mathbf{w}_2\|_{\mathfrak{W}}/4, \\
\|\nabla(\mathfrak{T}(\mathbf{w}_1) - \mathfrak{T}(\mathbf{w}_2))\|_{12/7} \le& \tau \cdot \|\nabla\mathfrak{R}((\mathbf{u} \mid B_r^c \cdot \nabla)(\mathbf{w}_1 - \mathbf{w}_2))\|_{12/7} \\
\le& \tau \cdot C_2(6/5) \cdot C_0(4, 12/7) \\
& \times \|\mathbf{u} \mid B_r^c\|_4 \cdot \|\nabla(\mathbf{w}_1 - \mathbf{w}_2)\|_{12/7} \\
\le& \|\mathbf{w}_1 - \mathbf{w}_2\|_{\mathfrak{W}}/4.
\end{aligned}$$

Therefore $\|\mathfrak{T}(\mathbf{w}_1) - \mathfrak{T}(\mathbf{w}_2)\|_{\mathfrak{W}} \le \|\mathbf{w}_1 - \mathbf{w}_2\|_{\mathfrak{W}}/2$ for $\mathbf{w}_1, \mathbf{w}_2 \in \mathfrak{W}$, so there is $\mathbf{v} \in \mathfrak{W}$ with $\mathbf{v} = \mathfrak{T}(\mathbf{v})$, that is,

$$\mathbf{v} = [\mathbf{u} - \tau \cdot \mathfrak{R}((\mathbf{u} \cdot \nabla)\mathbf{u} \mid B^c_{S_0}) + \tau \cdot \mathfrak{R}(\mathbf{u}^T \cdot \mathfrak{A}(\mathbf{v}))] \mid B^c_r. \tag{10.35}$$

As mentioned above, we have $\mathfrak{A}(\boldsymbol{\varphi}) \in \boldsymbol{L}^{12/7}(B^c_{S_0})$ for $\boldsymbol{\varphi} \in \mathfrak{W}$. Therefore Lemma 10.2 with $p = 6$, $q = 12/7$ implies $\mathbf{u}^T \cdot \mathfrak{A}(\mathbf{v}) \in \boldsymbol{L}^{4/3}(B^c_{S_0})$, so by Corollary 10.9 with $q = 4/3$, the relation $\nabla\mathfrak{R}(\mathbf{u}^T \cdot \mathfrak{A}(\mathbf{v})) \in L^2(\mathbb{R}^3)^9$ holds. It follows with (10.31) and (10.35) that $\nabla\mathbf{v} \in L^2(B^c_r)^9$. But then we may conclude that $\mathfrak{A}(\mathbf{v}) \in L^2(B^c_{S_0})^9$, so Lemma 10.2 with $p = 6$, $q = 2$ yields $u^T \cdot \mathfrak{A}(\mathbf{v}) \in \boldsymbol{L}^{3/2}(B^c_{S_0})$. Hence $\mathfrak{R}(\mathbf{u}^T \cdot \mathfrak{A}(\mathbf{v})) \in \boldsymbol{L}^6(\mathbb{R}^3)$ due to Corollary 10.9 with $p = 3/2$. Referring to (10.30) and (10.35), we may conclude that $\mathbf{v} \in \boldsymbol{L}^6(B^c_r)$. We may now apply (10.35) and then Corollary 10.9 with $p = 3/2$, $q = 4/3$, Lemma 10.2 with $p = 6$, $q = 2$ and $p = 4$, $q = 2$, respectively, as well as (10.34), to obtain

$$\begin{aligned}
\|\mathbf{v} - \mathbf{u}|B^c_r\|_6 &\le \tau \cdot \|\mathfrak{R}((\mathbf{u} \cdot \nabla)(\mathbf{v} - \mathbf{u} \mid B^c_r))\|_6 \\
&\le \tau \cdot C_1(3/2) \cdot C_0(6,2) \cdot \|\mathbf{u} \mid B^c_r\|_6 \cdot \|\nabla(\mathbf{v} - \mathbf{u}|B^c_r)\|_2 \\
&\le \|\nabla(\mathbf{v} - \mathbf{u}|B^c_r)\|_2, \\
\|\nabla(\mathbf{v} - \mathbf{u}|B^c_r)\|_2 &\le \tau \cdot \|\nabla\mathfrak{R}((\mathbf{u} \cdot \nabla)(\mathbf{v} - \mathbf{u} \mid B^c_r))\|_2 \\
&\le \tau \cdot C_2(4/3) \cdot C_0(4,2) \cdot \|\mathbf{u} \mid B^c_r\|_4 \cdot \|\nabla(\mathbf{v} - \mathbf{u}|B^c_r)\|_2 \\
&\le \|\nabla(\mathbf{v} - \mathbf{u} \mid B^c_r)\|_2/2.
\end{aligned}$$

The second estimate implies $\|\nabla(\mathbf{v} - \mathbf{u} \mid B^c_r)\|_2 = 0$, so the first yields $\|\mathbf{v} - \mathbf{u} \mid B^c_r\|_6 = 0$. Therefore $\mathbf{u} \mid B^c_r = v \in \boldsymbol{L}^3(B^c_r)$, and $\nabla(\mathbf{u} \mid B^c_r) = \nabla\mathbf{v} \in L^{12/7}(B^c_r)^9$. Since $\mathbf{u} \in \boldsymbol{L}^6(\bar{\mathfrak{D}})$ and $\nabla\mathbf{u} \in L^2(\bar{\mathfrak{D}})^9$, Theorem 10.15 follows. □

Corollary 10.11 $\mathbf{u} \in \boldsymbol{L}^p(\bar{\mathfrak{D}})$ *for* $p \in [12/5, 6]$, *in particular for* $p = 8/3$ *and* $p = 5/2$. □

Proof Theorem 10.15 and Lemma 10.2 with $p = 3$, $q = 12/7$ imply $(\mathbf{u} \cdot \nabla)\mathbf{u} \in \boldsymbol{L}^{12/11}(\bar{\mathfrak{D}})$, so Corollary 10.9 with $p = 12/11$ yields $\mathfrak{R}((\mathbf{u} \cdot \nabla)\mathbf{u}) \in \boldsymbol{L}^{12/5}(\mathbb{R}^3)$. Now the corollary follows from (10.30). □

Lemma 10.21 *Let* $S \in (S_1, \infty)$. *Then* $\nabla\mathbf{u} \mid B^c_S \in L^4(B^c_S)^9$. □

Proof Put $S_2 := S_1 + (S - S_1)/3$, $S_3 := S_1 + 2 \cdot (S - S_1)/3$. Let $l \in \{1, 2, 3\}$. By Lemma 10.13,

$$\partial_l \mathfrak{R}((\mathbf{u} \cdot \nabla)\mathbf{u}|B^c_{S_0}) = \mathfrak{A} + \partial_l \mathfrak{R}((\mathbf{u} \cdot \nabla)\mathbf{u}|B^c_{S_2}), \tag{10.36}$$

with

$$\mathfrak{A}(\mathbf{y}) := \int_{B_{S_2} \setminus B_{S_0}} \partial y_l \mathcal{Z}(\mathbf{y}, \mathbf{z}) \cdot ((\mathbf{u} \cdot \nabla)\mathbf{u})(\mathbf{z}) d\mathbf{z} \quad \text{for } \mathbf{y} \in \mathbb{R}^3.$$

For $\mathbf{y} \in B^c_{S_3}$, we get by Corollary 10.6 with R, δ replaced by S_2, $S_3/S_2 - 1$, respectively, that

$$|\mathfrak{A}(\mathbf{y})| \leq C(\tau, S_1, S) \cdot \|(\mathbf{u} \cdot \nabla)\mathbf{u} \mid B_{S_2} \backslash B_{S_0}\|_1 \cdot (|\mathbf{y}| \cdot s(\mathbf{y}))^{-3/2}.$$

But $\|(\mathbf{u} \cdot \nabla)\mathbf{u} \mid B_{S_2} \backslash B_{S_0}\|_1 \leq C(S_1, S) \cdot \|(\mathbf{u} \cdot \nabla)\mathbf{u}\|_{3/2} \leq C(S_1, S) \cdot \|\mathbf{u}\|_6 \cdot \|\nabla\mathbf{u}\|_2 \leq \mathfrak{C}(S)$, so $|\mathfrak{A}(\mathbf{y})| \leq \mathfrak{C}(S) \cdot (|\mathbf{y}| \cdot s(\mathbf{y}))^{-3/2}$ for $\mathbf{y} \in B^c_{S_3}$. Corollary 10.1 now implies $\mathfrak{A}|B^c_{S_3} \in \boldsymbol{L}^3(B^c_{S_3})$. Since $\nabla\mathbf{u} \mid B^c_{S_2} \in L^{12/5}(B^c_{S_2})^9$ (see Lemma 10.20) and $\mathbf{u} \in \boldsymbol{L}^6(\bar{\mathcal{D}})$, we may apply Lemma 10.2 with $p = 6$, $q = 12/5$ and Corollary 10.9 with $q = 12/7$ and $\mathbf{g} = (\mathbf{u} \cdot \nabla)\mathbf{u} \mid B^c_{S_2}$, to obtain $\partial_l\mathfrak{R}((\mathbf{u} \cdot \nabla)\mathbf{u} \mid B^c_{S_2}) \in \boldsymbol{L}^3(\mathbb{R}^3)$. On recalling (10.31) and (10.36), we thus see that $\partial_l\mathbf{u}|B^c_{S_3} \in \boldsymbol{L}^3(B^c_{S_3})$. Again take $l \in \{1, 2, 3\}$. Then

$$\partial_l\mathfrak{R}((\mathbf{u} \cdot \nabla)\mathbf{u}|B^c_{S_0}) = \widetilde{\mathfrak{A}} + \partial_l\mathfrak{R}((\mathbf{u} \cdot \nabla)\mathbf{u} \mid B^c_{S_3}), \tag{10.37}$$

with $\widetilde{\mathfrak{A}}(\mathbf{y})$ defined in the same way as $\mathfrak{A}(\mathbf{y})$, except that the domain of integration $B_{S_2} \backslash B_{S_0}$ is replaced by $B_{S_3} \backslash B_{S_0}$. Applying Corollary 10.6 again, but this time with R, δ replaced by S_3, $S/S_3 - 1$, respectively, we get that $|\widetilde{\mathfrak{A}}(\mathbf{y})| \leq \mathfrak{C}(S) \cdot (|\mathbf{y}| \cdot s(\mathbf{y}))^{-3/2}$ for $\mathbf{y} \in B^c_S$, in particular $\widetilde{\mathfrak{A}} \mid B^c_S \in \boldsymbol{L}^4(B^c_S)$ by Corollary 10.1. Moreover, referring to Lemma 10.2 with $p = 6$, $q = 3$ and to Corollary 10.9 with $q = 2$ and g replaced by $(\mathbf{u} \cdot \nabla)\mathbf{u} \mid B^c_{S_3}$, we may conclude that $\partial_l\mathfrak{R}((\mathbf{u} \cdot \nabla)\mathbf{u} \mid B^c_{S_3}) \in \boldsymbol{L}^4(\mathbb{R}^3)$. Lemma 10.21 now follows with (10.31) and (10.37). □

10.5 Pointwise Estimates of the Velocity

In a first step, we transform $\mathfrak{R}_j((\mathbf{u} \cdot \nabla)\mathbf{u} \mid B^c_{S_0})$ by an integration by parts:

Lemma 10.22 *Let* $j \in \{1, 2, 3\}$, $\mathbf{y} \in \bar{B}^c_{S_0}$. *Then* $\mathfrak{R}_j((\mathbf{u} \cdot \nabla)\mathbf{u} \mid B^c_{S_0})(\mathbf{y}) = -\mathfrak{F}_j(\mathbf{y}) - \widetilde{\mathfrak{B}}_j(\mathbf{y})$ *with*

$$\mathfrak{F}_j(\mathbf{y}) := \sum_{k,l=1}^{3} \int_{B^c_{S_0}} \partial z_l \mathcal{Z}_{jk}(\mathbf{y}, \mathbf{z}) \cdot (u_l \cdot u_k)(\mathbf{z}) d\mathbf{z},$$

$$\widetilde{\mathfrak{B}}_j(\mathbf{y}) := \sum_{k,l=1}^{3} \int_{\partial B_{S_0}} \mathcal{Z}_{jk}(\mathbf{y}, \mathbf{z}) \cdot (u_l \cdot u_k)(\mathbf{z}) \cdot z_l/S_0 do_z.$$

Proof Since $(\mathbf{u}\cdot\nabla)\mathbf{u}$ and $u_k\cdot\mathbf{u}$ belong to $\boldsymbol{L}^{3/2}(\bar{\mathcal{D}})$ (see Lemma 10.2 and Theorem 10.15), Lemma 10.13 yields that

$$\int_{B_{S_0}} |\partial_z^\alpha \mathcal{Z}_{jk}(\mathbf{y},\mathbf{z})|\cdot|g(\mathbf{z})|d\mathbf{z}<\infty$$

for $\alpha=0$, $g=u_l\cdot\partial_l u_k$, and for $\alpha=e_l$, $\mathbf{g}=u_l\cdot u_k$, with $1\le k,l\le 3$.

Let $\psi\in C_0^\infty(\mathbb{R}^3)$ with $\psi\mid B_1=0$, $\psi|B_2^c=1$, $0\le\psi\le 1$. For $\epsilon>0$, put $\psi_\epsilon(z):=\psi(\epsilon^{-1}\cdot\mathbf{z})$ for $\mathbf{z}\in\mathbb{R}^3$. We may conclude with the first sentence of this proof and with Lebesgue's theorem that

$$\mathfrak{R}((\mathbf{u}\cdot\nabla)\mathbf{u}\mid B_{S_0}^c)(\mathbf{y})=\lim_{R\to\infty}\lim_{\epsilon\downarrow 0}\int_{B_R\backslash B_{S_0}}\psi_\epsilon(\mathbf{y}-\mathbf{z})\cdot\mathcal{Z}(\mathbf{y},\mathbf{z})\cdot(\mathbf{u}(\mathbf{z})\cdot\nabla)\mathbf{u}(\mathbf{z})d\mathbf{z}. \tag{10.38}$$

Put $\delta:=\min\{(|\mathbf{y}|-S_0)/2,\ S_0/2\}$, and note that for $R\in(S_0+|\mathbf{y}|,\infty)$, $\epsilon\in(0,\delta)$, we have $\bar{B}_{S_0}\subset B_R$ and $B_{2\cdot\epsilon}(\mathbf{y})\subset B_R\backslash B_{S_0}$. Moreover, $\psi_\epsilon(\mathbf{y}-\cdot)\cdot\mathcal{Z}_{jk}(\mathbf{y}-\cdot)\in C^1(\mathbb{R}^3)$ for $1\le j,k\le 3$, $\epsilon>0$ by Lemma 10.11. Thus we may deduce from (10.38) by an integration by parts that

$$\mathfrak{R}_j((\mathbf{u}\cdot\nabla)\mathbf{u}\mid B_{S_0}^c)(\mathbf{y})=\lim_{R\to\infty,\ R>|\mathbf{y}|+S_0}\ \lim_{\epsilon\downarrow 0,\ \epsilon<\delta}\left(-\widetilde{\mathfrak{B}}_j(\mathbf{y})+\mathfrak{F}_{1,\epsilon}+\mathfrak{F}_{2,R}-\mathfrak{F}_{3,R,\epsilon}\right), \tag{10.39}$$

where

$$\mathfrak{F}_{1,\epsilon}:=\sum_{k,l=1}^{3}\int_{B_{2\cdot\epsilon}(\mathbf{y})\backslash B_\epsilon(\mathbf{y})}(\partial_l\psi_\epsilon)(\mathbf{y}-\mathbf{z})\cdot\mathcal{Z}_{jk}(\mathbf{y},\mathbf{z})\cdot(u_l\cdot u_k)(\mathbf{z})d\mathbf{z},$$

$$\mathfrak{F}_{2,R}:=\sum_{k,l=1}^{3}\int_{\partial B_R}\mathcal{Z}_{jk}(\mathbf{y},\mathbf{z})\cdot(u_l\cdot u_k)(\mathbf{z})\cdot z_l/R\,do_z,$$

and where $\mathfrak{F}_{3,R,\epsilon}$ is defined as the term $\mathfrak{F}_j(\mathbf{y})$ in the lemma, but with the additional factor $\psi_\epsilon(\mathbf{y}-\mathbf{z})$ under the integral and with the domain of integration $B_{S_0}^c$ replaced by $B_R\backslash B_{S_0}$. For $\epsilon\in(0,\delta)$, $\mathbf{z}\in B_{2\cdot\epsilon}(\mathbf{y})\backslash B_\epsilon(\mathbf{y})$, we have $\epsilon\le|\mathbf{y}-\mathbf{z}|$ and $\mathbf{y},\mathbf{z}\in B_R$ (see above), so Corollary 10.5 yields $|\mathcal{Z}(\mathbf{y},\mathbf{z})|\le C(\tau,\varrho,R)\cdot|\mathbf{y}-\mathbf{z}|^{-1}\le C(\tau,\varrho,R)\cdot\epsilon^{-1}$. Since by Hölder's inequality $\int_{B_{2\cdot\epsilon}(\mathbf{y})\backslash B_\epsilon(\mathbf{y})}|\mathbf{u}(\mathbf{z})|^2d\mathbf{z}\le C\cdot\epsilon^2\cdot\|\mathbf{u}|B_{2\cdot\epsilon}(\mathbf{y})\|_6^2$ for $\epsilon\in(0,\delta)$, and because $|\nabla\psi(\mathbf{x})|\le C\cdot\epsilon^{-1}$ for $\mathbf{x}\in\mathbb{R}^3$, $\epsilon>0$, we may conclude that $|\mathfrak{F}_{1,\epsilon}|\le C(\tau,\varrho,R)\cdot\|\mathbf{u}|B_{2\cdot\epsilon}(\mathbf{y})\|_6^2$, so $\mathfrak{F}_{1,\epsilon}\to 0$ for $\epsilon\downarrow 0$. Moreover, Lebesgue's theorem and the first sentence of this proof yield $\mathfrak{F}_{3,R,\epsilon}\to\mathfrak{F}_{3,R}$ $(\epsilon\downarrow 0)$, with $\mathfrak{F}_{3,R}$

defined in the same way as $\mathfrak{F}_j(\mathbf{y})$, but with the domain of integration $B^c_{S_0}$ replaced by $B_R \backslash B_{S_0}$ $(R \in (|\mathbf{y}| + S_0, \infty)$. Therefore, from (10.39),

$$\mathfrak{R}_j((\mathbf{u} \cdot \nabla)\mathbf{u} \mid B^c_{S_0})(\mathbf{y}) = \lim_{R \to \infty,\, R > |\mathbf{y}| + S_0} (-\widetilde{\mathfrak{B}}_j(\mathbf{y}) + \mathfrak{F}_{2,R} - \mathfrak{F}_{3,R}). \tag{10.40}$$

Next we observe that

$$\int_{B^c_R} |\mathbf{u}(\mathbf{z})|^6 d\mathbf{z} = \int_R^\infty \int_{\partial B_r} |\mathbf{u}(\mathbf{z})|^6 do_z dr \quad \text{for } R \in (S_0, \infty).$$

Therefore the assumption that there is $R_0 \in (S_0, \infty)$ with $\int_{\partial B_r} |\mathbf{u}(\mathbf{z})|^6 do_z \geq 1/r$ for any $r \in [R_0, \infty)$ immediately leads to a contradiction to the relation $\mathbf{u} \in \boldsymbol{L}^6(\bar{\mathcal{D}})$. Thus there is a sequence (R_n) in $[2 \cdot (S_0 + |\mathbf{y}|), \infty)$ with $R_n \to \infty$ and $\int_{\partial B_{R_n}} |\mathbf{u}(\mathbf{z})|^6 do_z \leq 1/R_n$ for $n \in \mathbb{N}$. On the other hand, by Corollary 10.6 with $\delta = 1$, $R = S_0 + |\mathbf{y}|$, we have

$$|\mathcal{Z}(\mathbf{y}, \mathbf{z})| \leq C(\tau, S_0 + |\mathbf{y}|) \cdot (R_n \cdot s(\mathbf{z}))^{-1} \quad \text{for } \mathbf{z} \in B_{R_n},\ n \in \mathbb{N}.$$

As a consequence $|\mathfrak{F}_{2,R_n}| \leq C(\tau, S_0 + |\mathbf{y}|) \cdot R_n^{-1} \cdot \int_{\partial B_{R_n}} s(\mathbf{z})^{-1} \cdot |\mathbf{u}(\mathbf{z})|^2 do_z$ for $n \in \mathbb{N}$. Therefore by Hölder's inequality and Corollary 10.1 with $a = 0$, $b = 3/2$, we get $|\mathfrak{F}_{2,R_n}| \leq C(\tau, S_0 + |y|) \cdot R_n^{-1/3} \cdot \|\mathbf{u} \mid \partial B_{R_n}\|_6^2$, so $\mathfrak{F}_{2,R_n} \to 0$ $(n \to \infty)$ by the choice of the sequence (R_n). Finally, Lebesgue's theorem and the first sentence of this proof imply $\mathfrak{F}_{3,R_n} \to \mathfrak{F}_j(\mathbf{y})$ $(n \to \infty)$. Lemma 10.22 thus follows from (10.40). □

Corollary 10.12 *Let $S \in (S_1, \infty)$. Then*

$$|\mathbf{u}(\mathbf{x}) - \tau \cdot \mathfrak{F}(\mathbf{x})| \leq C(S_0, S_1, S, A, B) \cdot (\mathfrak{M} + \|\mathbf{u}|\partial B_{S_0}\|_2^2) \cdot (|\mathbf{x}| \cdot s_1(\mathbf{x}))^{-1} \ \textit{for}\ \mathbf{x} \in B^c_S,$$

with $\mathfrak{M}$ from Theorem 10.13. □

Proof Combine (10.29) with $\alpha = 0$, Lemma 10.22, and Corollary 10.6 with $R = S_0$, $\delta = S/S_0 - 1$. □

Theorem 10.16 *Put $\varphi(S) := \sup\{|\mathbf{u}(\mathbf{y})| : \mathbf{y} \in B^c_S\}$ for $S \in (S_1, \infty)$. Then $\varphi(S) < \infty$ for $S \in (S_1, \infty)$, $\varphi(S) \to 0$ $(S \to \infty)$, and $\varphi(S) \leq \mathfrak{C} \cdot (S^{-1} + \varphi(S/2)^{7/6})$ for $S \in (2 \cdot S_1, \infty)$.* □

Proof We note that $e^{t \cdot \Omega} \cdot B^c_R = B^c_R$ for $t \in \mathbb{R}$, $R > 0$ (see Lemma 10.3).

Let $S \in (S_1, \infty)$, and put $S_2 := (S + S_1)/2$, $S_3 := \min\{S_1/2, (S - S_1)/2\}$. Let $\mathbf{y} \in B^c_S$. Then Lemma 10.17 yields $|\mathfrak{F}(\mathbf{y})| \leq C(\tau, \varrho) \cdot \sum_{m=1}^3 \mathfrak{K}_m(\mathbf{y})$, with $\mathfrak{F}(\mathbf{y})$ defined in Lemma 10.22, and

$$\mathfrak{K}_m(\mathbf{y}) := \int_0^\zeta \int_{A_m} W(\mathbf{y} - \mathbf{z}, t) \cdot |\mathbf{u}(e^{t \cdot \Omega} \cdot \mathbf{z})|^2 d\mathbf{z}$$

for $m \in \{1, 2, 3\}$, with $A_1 := B_{S_2} \backslash B_{S_0}$, $A_2 := B^c_{S_2} \cap B_{S_1}(\mathbf{y})$, $A_3 := B^c_{S_2} \backslash B_{S_1}(\mathbf{y})$. The function W was introduced in Lemma 10.17.

For $\mathbf{z} \in B_{S_2}$, we have $|\mathbf{y} - \mathbf{z}| \geq S - S_2 = (S - S_1)/2 \geq S_3$. Thus we may apply Lemma 10.6 with $M = S_3$ as well as Theorem 10.5 with $z = 0$, $a = 2$, $R = S_3/2$, $\delta = 1$, and with $\mathbf{y}$ replaced by $\mathbf{y} - \mathbf{z}$, to obtain for $\mathbf{z} \in B_{S_2}$, $t \in (0, \zeta)$ that

$$\begin{aligned} W(\mathbf{y}-\mathbf{z}, t) \leq & C(\tau, S_3) \cdot [(|\mathbf{y}-\mathbf{z}| \cdot s(\mathbf{y}-\mathbf{z}) + t)^{-2} + (|\mathbf{y}-\mathbf{z}| \cdot s(\mathbf{y}-\mathbf{z}))^{-3/2}] \\ \leq & C(\tau, S_3) \cdot (|\mathbf{y}-\mathbf{z}| \cdot s(\mathbf{y}-\mathbf{z}))^{-3/2}. \end{aligned}$$

Therefore, according to Hölder's inequality, the term $\mathfrak{K}_1(\mathbf{y})$ is bounded by

$$\begin{aligned} C(\tau, S_3) \cdot \int_0^\zeta & \left(\int_{B_{S_2} \backslash B_{S_0}} |\mathbf{u}(e^{t \cdot \Omega} \cdot \mathbf{z})|^{8/3} d\mathbf{z} \right)^{3/4} \\ & \times \left(\int_{B_{S_2}} (|\mathbf{y}-\mathbf{z}| \cdot s(\mathbf{y}-\mathbf{z}))^{-6} d\mathbf{z} \right)^{1/4} dt. \end{aligned}$$

But the first of the preceding integrals with respect to $\mathbf{z}$ admits the upper bound $\|\mathbf{u}\|^2_{8/3}$, as follows by a change of variables and Lemma 10.3. Since $|\mathbf{y} - \mathbf{z}| \geq (S - S_1)/2$ for $\mathbf{z} \in B_{S_2}$, as observed above, so that $B_{S_2} \subset B_{(S-S_1)/2}(\mathbf{y})^c$, we may use Corollary 10.1 with $a = b = 6$, obtaining that the second integral with respect to $\mathbf{z}$ is bounded by $C \cdot (S - S_1)^{-4}$. Recalling Corollary 10.11, we thus get

$$\mathfrak{K}_1(\mathbf{y}) \leq C(\tau, S_3) \cdot \|\mathbf{u}\|^2_{8/3} \cdot (S - S_1)^{-1} \leq \mathfrak{C}(S_3) \cdot (S - S_1)^{-1}. \tag{10.41}$$

For $\mathbf{z} \in B_{S_1}(\mathbf{y})$ with $\mathbf{z} \neq \mathbf{y}$, $t \in (0, \zeta)$, Lemma 10.6 with $M = S_1$ yields

$$\begin{aligned} W(\mathbf{y}-\mathbf{z}, t) \leq & C(\tau, S_1) \cdot \left((|\mathbf{y}-\mathbf{z}|^2 + t)^{-2} + \int_0^\infty (|\mathbf{y}-\mathbf{z}|^2 + s)^{-2} ds \right) \\ \leq & C(\tau, S_1) \cdot ((|\mathbf{y}-\mathbf{z}| + t^{1/2})^{-4} + |\mathbf{y}-\mathbf{z}|^{-2}). \end{aligned}$$

Therefore, by Hölder's inequality,

$$\begin{aligned} \mathfrak{K}_2(\mathbf{y}) \leq & C(\tau, S_1) \cdot \int_0^\zeta \left(\int_{B_{S_1}(\mathbf{y})} ((|\mathbf{y}-\mathbf{z}| + t^{1/2})^{-4} + |\mathbf{y}-\mathbf{z}|^{-2})^{6/5} d\mathbf{z} \right)^{5/6} \\ & \times \left(\int_{B^c_{S_2}} |\mathbf{u}(e^{t \cdot \Omega} \cdot z)|^{10:12} d\mathbf{z} \right)^{1/6} dt. \end{aligned}$$

The preceding integral over $B^c_{S_2}$ equals $\|\mathbf{u}|B^c_{S_2}\|^2_{10:12}$. Concerning the integral over $B_{S_1}(\mathbf{y})$, it is bounded by $C(S_1)\cdot(t^{-9/10}+1)$. Thus, on integrating with respect to t, we get $\mathfrak{K}_2(\mathbf{y}) \le C(\tau, S_1)\cdot\|\mathbf{u}|B_{S_2}\|^2_{10:12}$.

Lemma 10.6 with $M = S_1$ and Theorem 10.5 with $\mathbf{z} = 0$, $R = S_1/2$, $\delta = 1$, $a = 2$, and $\mathbf{y}$ replaced by $\mathbf{y}-\mathbf{z}$ imply the following estimate, for $t \in (0, \zeta)$, $\mathbf{z} \in B_{S_1}(\mathbf{y})^c$:

$$\begin{aligned} W(\mathbf{y}-\mathbf{z}, t) \le& C(\tau, S_1)\cdot[(|\mathbf{y}-\mathbf{z}|\cdot s(\mathbf{y}-\mathbf{z})+t)^{-2} + (|\mathbf{y}-\mathbf{z}|\cdot s(\mathbf{y}-\mathbf{z}))^{-3/2}] \\ \le& C(\tau, S_1)\cdot(|\mathbf{y}-\mathbf{z}|\cdot s(\mathbf{y}-\mathbf{z}))^{-3/2}. \end{aligned}$$

We may conclude with Hölder's inequality that

$$\begin{aligned} \mathfrak{K}_3(\mathbf{y}) \le& C(\tau, S_1)\cdot \int_0^{\zeta}\left(\int_{B_{S_1}(\mathbf{y})^c}(|\mathbf{y}-\mathbf{z}|\cdot s(\mathbf{y}-\mathbf{z}))^{-9/4}\mathrm{d}\mathbf{z}\right)^{2/3} \\ &\times\left(\int_{B^c_{S_2}}|\mathbf{u}(\mathrm{e}^{t\cdot\Omega}\cdot\mathbf{z})|^6\mathrm{d}\mathbf{z}\right)^{1/3}\mathrm{d}t. \end{aligned}$$

Corollary 10.1 with $a = b = 9/4$ yields that the preceding integral over $B^c_{S_1}(\mathbf{y})$ is bounded by a constant $C(S_1)$. Moreover, the integral over $B^c_{S_2}$ equals $\|\mathbf{u}|B^c_{S_2}\|^6_6$. Thus we have found that $\mathfrak{K}_3(\mathbf{y}) \le C(\tau, S_1)\cdot\|\mathbf{u}|B^c_{S_2}\|^2_6$.

Combining the preceding estimates of $\mathfrak{K}_1(\mathbf{y})$, $\mathfrak{K}_2(\mathbf{y})$, and $\mathfrak{K}_3(\mathbf{y})$, we arrive at the inequality

$$|\mathfrak{F}(\mathbf{y})| \le \mathfrak{C}(S_3)\cdot((S-S_1)^{-1} + \|\mathbf{u}|B^c_{S_2}\|^2_{10:12} + \|\mathbf{u}|B^c_{S_2}\|^2_6), \tag{10.42}$$

where $\mathbf{y}$ was arbitrary from $B_S(\mathbf{y})^c$ and S arbitrary from (S_1,∞). Abbreviating $A(S) := \sup\{|\mathfrak{F}(\mathbf{y})| : \mathbf{y} \in B^c_S\}$ for $S \in (S_1,\infty)$, we may conclude from (10.42) and Corollary 10.10 that $A(S) < \infty$ for $S \in (S_1,\infty)$. We further observe that the parameter S_2 in (10.42) satisfies the relation $S_2 > S/2$, and the quantity S_3 coincides with $S_1/2$ if $S \in (2\cdot S_1, \infty)$. Thus we may deduce from (10.42) that

$$|A(S)| \le \mathfrak{C}\cdot(S^{-1} + \|\mathbf{u}|B^c_{S/2}\|^2_{10:12} + \|\mathbf{u}|B^c_{S/2}\|^2_6) \quad \text{for } S \in (2\cdot S_1, \infty). \tag{10.43}$$

Again referring to Corollary 10.10, we see that (10.43) implies $A(S) \to 0$ for $S \to \infty$. On the other hand, put $B(S) := \sup\{|\mathbf{u}(\mathbf{y}) - \tau\cdot\mathfrak{F}(\mathbf{y})| : \mathbf{y} \in B^c_S\}$ for $S \in (S_1,\infty)$. Then Corollary 10.12 implies that $B(S) < \infty$ for $S \in (S_1,\infty)$. In addition, the same reference yields $|\mathbf{u}(\mathbf{y}) - \tau\cdot\mathfrak{F}(\mathbf{y})| \le \mathfrak{C}\cdot|\mathbf{y}|^{-1}$ for $\mathbf{y} \in B^c_{2\cdot S_1}$. Hence $B(S) \to 0$ for $S \to \infty$. Thus we have shown that $\varphi(S) < \infty$ for $S \in (S_1,\infty)$ and $\varphi(S) \to 0$ for $S \to \infty$.

Now let $S \in (2\cdot S_1,\infty)$ and $\mathbf{y} \in B^c_S$. Lemma 10.17 and Lemma 10.3 yield

$$|\mathfrak{F}(\mathbf{y})| \le C(\tau,\varrho)\cdot[\mathfrak{K}_1(\mathbf{y}) + \varphi(S_2)^{7/6}\cdot(\widetilde{\mathfrak{K}}_2(\mathbf{y}) + \widetilde{\mathfrak{K}}_3(\mathbf{y}))],$$

where $\widetilde{\mathfrak{K}}_2(\mathbf{y})$, $\widetilde{\mathfrak{K}}_3(\mathbf{y})$ are defined as $\mathfrak{K}_2(\mathbf{y})$, $\mathfrak{K}_3(\mathbf{y})$, respectively, but with the term $|\mathbf{u}(e^{\tau\cdot\Omega}\cdot z)|^2$ replaced by $|\mathbf{u}(e^{\tau\cdot\Omega}\cdot \mathbf{z})|^{5/6}$. The quantity S_2 is defined at the beginning of this proof. Since $S \geq 2\cdot S_1$, inequality (10.41) yields $\mathfrak{K}_1(\mathbf{y}) \leq \mathfrak{C}\cdot S^{-1}$. The terms $\widetilde{\mathfrak{K}}_2(\mathbf{y})$ and $\widetilde{\mathfrak{K}}_3(\mathbf{y})$ may be estimated in the same way as $\mathfrak{K}_2(\mathbf{y})$ and $\mathfrak{K}_3(\mathbf{y})$, respectively. We obtain

$$\widetilde{\mathfrak{K}}_2(\mathbf{y}) + \widetilde{\mathfrak{K}}_3(\mathbf{y}) \leq C(\tau, S_1)\cdot(\|\mathbf{u} \mid B^c_{S_2}\|_5^{5/6} + \|\mathbf{u}|B^c_{S_2}\|_{5/2}^{5/6}).$$

Observing that $S_2 \geq S/2$, hence $\varphi(S_2) \leq \varphi(S/2)$, and recalling Corollary 10.11, we thus get

$$|\mathfrak{F}(\mathbf{y})| \leq \mathfrak{C}\cdot(S^{-1} + \varphi(S/2)^{7/6}\cdot(\|\mathbf{u}\|_5^{5/6} + \|\mathbf{u}\|_{5/2}^{5/6})) \leq \mathfrak{C}\cdot(S^{-1} + \varphi(S/2)^{7/6}).$$

On the other hand, Corollary 10.12 with S replaced by $2\cdot S_1$ yields $|\mathbf{u}(\mathbf{y}) - \tau\cdot\mathfrak{F}(\mathbf{y})| \leq \mathfrak{C}\cdot|\mathbf{y}|^{-1}$. Altogether we have $|\mathbf{u}(\mathbf{y})| \leq \mathfrak{C}\cdot(S^{-1} + \varphi(S/2)^{7/6})$. Therefore $\varphi(S) \leq \mathfrak{C}\cdot(S^{-1} + \varphi(S/2)^{7/6})$. □

Theorem 10.17 *Let $S \in (S_1, \infty)$. Then $|\mathbf{u}(\mathbf{x})| \leq \mathfrak{C}(S)\cdot|\mathbf{x}|^{-1}$ for $\mathbf{x} \in B^c_S$.* □

Proof The theorem follows from Theorem 10.16 by an argument due to Babenko [11]. For details, we refer to the proof of [31, Theorem 4.1]. □

Now we are in a position to establish the looked-for decay estimate of $\mathbf{u}$:

Theorem 10.18 *Let $S \in (S_1, \infty)$. Then $|\mathbf{u}(\mathbf{y})| \leq \mathfrak{C}(S)\cdot(|\mathbf{y}|\cdot s(\mathbf{y}))^{-1}$ for $\mathbf{y} \in B^c_S$.* □

Proof By Theorem 10.16, we have $|\mathbf{u}(\mathbf{y})| \leq \mathfrak{C}(S)$ for $\mathbf{y} \in B^c_S$, so

$$|\mathbf{u}(\mathbf{y})| \leq \mathfrak{C}(S)\cdot(|\mathbf{y}|\cdot s(\mathbf{y}))^{-1} \quad \text{for } \mathbf{y} \in B_{2\cdot S}\backslash B_S. \tag{10.44}$$

Let $\mathbf{y} \in B^c_{2\cdot S}$. Then by Corollary 10.12,

$$|\mathbf{u}(\mathbf{y})| \leq \mathfrak{C}(S)\cdot(|\mathbf{y}|\cdot s(\mathbf{y}))^{-1} + \mathfrak{A}_1(\mathbf{y}) + \mathfrak{A}_2(\mathbf{y}), \tag{10.45}$$

with $\mathfrak{A}_1(\mathbf{y}) := \sum_{j,k,l=1}^3 \int_{B_S\backslash B_{S_0}} |\partial \mathbf{z}_l \mathcal{Z}_{jk}(\mathbf{y},\mathbf{z})\cdot u_l(\mathbf{z})\cdot u_k(\mathbf{z})|\,d\mathbf{z}$, and with $\mathfrak{A}_2(\mathbf{y})$ defined in the same way as $\mathfrak{A}_1(\mathbf{y})$, but with the domain of integration $B_S\backslash B_{S_0}$ replaced by B^c_S. Corollary 10.6 with $R = S$, $\delta = 1$, $\alpha = e_l$ for $1 \leq l \leq 3$ yields

$$\mathfrak{A}_1(\mathbf{y}) \leq C(\tau, S)\cdot(|\mathbf{y}|\cdot s(\mathbf{y}))^{-1}\cdot\|\mathbf{u}|B_S\backslash B_{S_0}\|_2^2 \leq \mathfrak{C}(S)\cdot(|\mathbf{y}|\cdot s(\mathbf{y}))^{-1}. \tag{10.46}$$

Note that $\|\mathbf{u} \mid B_S\backslash B_{S_0}\|_2 \leq C(S)\cdot\|\mathbf{u}\|_6$. Moreover, by Theorem 10.17,

$$\mathfrak{A}_2(\mathbf{y}) \leq \mathfrak{C}(S)\cdot\sum_{j,k,l=1}^{3}\int_{B^c_S} |\partial z_l \mathcal{Z}_{jk}(\mathbf{y},\mathbf{z})|\cdot|\mathbf{z}|^{-2}\,d\mathbf{z}. \tag{10.47}$$

Hence with Lemma 10.16 and Theorem 10.4 with $a = 2, b = 0$,

$$\mathfrak{A}_2(\mathbf{y}) \leq \mathfrak{C}(S) \cdot (|\mathbf{y}|^{-2} + |\mathbf{y}|^{-1} \cdot s(\mathbf{y})^{-1/2}) \leq \mathfrak{C}(S) \cdot |\mathbf{y}|^{-1} \cdot s(\mathbf{y})^{-1/2},$$

where the last inequality holds by Lemma 10.5. This estimate, (10.45), and (10.46) yield $|\mathbf{u}(\mathbf{y})| \leq \mathfrak{C}(S) \cdot |\mathbf{y}|^{-1} \cdot s(\mathbf{y})^{-1/2}$ for $\mathbf{y} \in B^c_{2 \cdot S}$. Due to (10.44), the preceding inequality even holds for $\mathbf{y} \in B^c_S$. With this result at hand, we return to the term $\mathfrak{A}_2(y)$, which may now be estimated as in (10.47), but with an additional factor $s(\mathbf{z})^{-1}$ under the integral. Observing that $s(\mathbf{z})^{-1} \leq s(\mathbf{z})^{-1/2}$ ($\mathbf{z} \in \mathbb{R}^3$), we again apply Lemma 10.16, and then Theorem 10.4, this time with $a = 2, b = 1/2$, to obtain $\mathfrak{A}_2(\mathbf{y}) \leq \mathfrak{C}(S) \cdot (|\mathbf{y}|^{-2} \cdot s(\mathbf{y})^{-1/2} + |\mathbf{y}|^{-5/4} \cdot s(\mathbf{y})^{-3/4})$ for $\mathbf{y} \in B^c_{2 \cdot S}$, and thus $\mathfrak{A}_2(\mathbf{y}) \leq \mathfrak{C}(S) \cdot (|\mathbf{y}| \cdot s(\mathbf{y}))^{-1}$ due to Lemma 10.5. Recalling (10.45) and (10.46), we see that $|\mathbf{u}(\mathbf{y})| \leq \mathfrak{C}(S) \cdot (|\mathbf{y}| \cdot s(\mathbf{y}))^{-1}$ for $\mathbf{y} \in B^c_{2 \cdot S}$. Theorem 10.18 now follows with (10.44). □

10.6 Pointwise Estimates of the Gradient of the Velocity

Let us first show that $\nabla \mathbf{u}$ is bounded outside any ball B_S with $S > S_1$.

Theorem 10.19 *Let $S \in (S_1, \infty)$. Then $|\nabla \mathbf{u}(\mathbf{y})| \leq \mathfrak{C}(S)$ for $\mathbf{y} \in B^c_S$.* □

Proof Put $S_2 := (S + S_1)/2$. Let $l \in \{1, 2, 3\}$, $\mathbf{y} \in B^c_S$. By (10.29) and Lemma 10.13,

$$|\partial_l \mathbf{u}(\mathbf{y})| \leq \mathfrak{C}(S) \cdot (|\mathbf{y}| \cdot s(\mathbf{y}))^{-3/2} + \mathfrak{A}_1(\mathbf{y}) + \mathfrak{A}_2(\mathbf{y}), \tag{10.48}$$

with $\mathfrak{A}_1(\mathbf{y}) := \int_{B_{S_2} \setminus B_{S_0}} |\partial y_l \mathcal{Z}(\mathbf{y}, \mathbf{z}) \cdot (\mathbf{u}(\mathbf{z}) \cdot \nabla)\mathbf{u}(\mathbf{z})| d\mathbf{z}$, and with $\mathfrak{A}_2(\mathbf{y})$ defined in the same way as $\mathfrak{A}_1(\mathbf{y})$, but with the domain of integration $B_{S_2} \setminus B_{S_0}$ replaced by $B^c_{S_2}$. Corollary 10.6 with $R = S_2$, $\delta = S/S_2 - 1$, $\alpha = e_l$ yields

$$\begin{aligned}
\mathfrak{A}_1(\mathbf{y}) \leq & C(\tau, S_1, S) \cdot (|\mathbf{y}| \cdot s(\mathbf{y}))^{-3/2} \cdot \|(\mathbf{u} \cdot \nabla)\mathbf{u} \mid B_{S_2} \setminus B_{S_0}\|_1 \\
\leq & C(\tau, S_0, S_1, S) \cdot (|\mathbf{y}| \cdot s(\mathbf{y}))^{-3/2} \cdot \|\mathbf{u}\|_6 \cdot \|\nabla \mathbf{u}\|_2 \\
\leq & \mathfrak{C}(S) \cdot (|\mathbf{y}| \cdot s(\mathbf{y}))^{-3/2}.
\end{aligned} \tag{10.49}$$

Moreover, we deduce from Lemmas 10.3 and 10.17 with $A = B^c_{S_2}$, $\mathbf{g}(\mathbf{z}) := \tau \cdot |(\mathbf{u}(\mathbf{z}) \cdot \nabla)\mathbf{u}(\mathbf{z})|$ for $\mathbf{z} \in B^c_{S_2}$ that

$$\mathfrak{A}_2(\mathbf{y}) \leq C(\tau, \varrho) \cdot (\mathfrak{K}_1(\mathbf{y}) + \mathfrak{K}_2(\mathbf{y})), \tag{10.50}$$

where $\mathfrak{K}_1(\mathbf{y}) := \int_0^\zeta \int_{B^c_{S_2} \cap B_{S_1}(\mathbf{y})} W(\mathbf{y} - \mathbf{z}, t) \cdot \mathbf{g}(e^{t \cdot \Omega} \cdot \mathbf{z}) dz dt$, and where $\mathfrak{K}_2(\mathbf{y})$ is defined in the same way as $\mathfrak{K}_1(\mathbf{y})$, but with the domain of integration $B^c_{S_2} \cap B_{S_1}(\mathbf{y})$

replaced by $B^c_{S_2} \backslash B_{S_1}(\mathbf{y})$. The function W was introduced in Lemma 10.17. For $\mathbf{z} \in B_{S_1}(\mathbf{y})$, $\mathbf{z} \neq \mathbf{y}$, $t \in (0, \zeta)$, we deduce from Lemma 10.6 with $M = S_1$ that

$$W(\mathbf{y}-\mathbf{z}, t) \le C(S_1) \cdot ((|\mathbf{y}-\mathbf{z}|^2 + t)^{-2} + |\mathbf{y}-\mathbf{z}|^{-2}).$$

Therefore by Hölder's inequality,

$$\begin{aligned} \mathfrak{R}_1(\mathbf{y}) \le & C(\tau, S_1) \cdot \int_0^\zeta \left(\int_{B^c_{S_2}} g(e^{t\cdot\Omega} \cdot \mathbf{z})^4 \, \mathrm{d}\mathbf{z} \right)^{1/4} \\ & \times \left(\int_{B_{S_1}(\mathbf{y})} ((|\mathbf{y}-\mathbf{z}|^2 + t)^{-2} + |\mathbf{y}-\mathbf{z}|^{-2})^{4/3} \, \mathrm{d}z \right)^{3/4} \mathrm{d}t. \end{aligned} \tag{10.51}$$

By a change of variables and Lemma 10.3, we see that the preceding integral over $B^c_{S_2}$ is bounded by $\|(\mathbf{u} \cdot \nabla)\mathbf{u} \mid B^c_{S_2}\|_4$. But $|\mathbf{u}(\mathbf{z})| \le \mathfrak{C}(S)$ for $\mathbf{z} \in B^c_{S_2}$ (see Theorem 10.17), so with Lemma 10.21 $\|(\mathbf{u} \cdot \nabla)\mathbf{u} \mid B^c_{S_2}\|_4 \le \mathfrak{C}(S) \cdot \|\nabla\mathbf{u}|B^c_{S_2}\|_4 \le \mathfrak{C}(S)$. We further observe that

$$\int_{B_{S_1}(\mathbf{y})} (|\mathbf{y}-\mathbf{z}| + t^{1/2})^{-16/3} \mathrm{d}z \le C(S_1) \cdot t^{-7/6}$$

and

$$\int_{B_{S_1}(\mathbf{y})} |\mathbf{y}-\mathbf{z}|^{-8/3} \mathrm{d}z \le C(S_1).$$

Thus we may deduce from (10.51) that $\mathfrak{R}_1(\mathbf{y}) \le \mathfrak{C}(S) \cdot \int_0^\zeta (t^{-7/8} + 1)\mathrm{d}t \le \mathfrak{C}(S)$. Turning to $\mathfrak{R}_2(\mathbf{y})$, we remark that for $\mathbf{z} \in B^c_{S_1}(\mathbf{y})$, we have

$$\int_\zeta^\infty (|\mathbf{y}-\mathbf{z}-\tau \cdot t \cdot \boldsymbol{e}_1|^2 + t)^{-2} \mathrm{d}t \le C(\tau, S_1) \cdot (|\mathbf{y}-\mathbf{z}| \cdot s(\mathbf{y}-\mathbf{z}))^{-3/2},$$

as follows from Theorem 10.5 with $R = S_1/2$, $\delta = 1$, $z = 0$, $a = 2$ and with $\mathbf{y}-\mathbf{z}$ in the place of $\mathbf{y}$. Therefore by Lemma 10.6 with $M = S_1$, we see that for $\mathbf{z} \in B^c_{S_1}(\mathbf{y})$, the term $W(\mathbf{y}-\mathbf{z}, t)$ is bounded by $C(\tau, S_1) \cdot \sum_{\sigma \in \{3/2, 2\}} (|\mathbf{y}-\mathbf{z}| \cdot s(\mathbf{y}-\mathbf{z}))^{-\sigma}$ and hence by $C(\tau, S_1) \cdot (|\mathbf{y}-\mathbf{z}| \cdot s(\mathbf{y}-\mathbf{z}))^{-3/2}$. It follows with Hölder's inequality

$$\begin{aligned} \mathfrak{R}_2(\mathbf{y}) \le & C(\tau, S_1) \cdot \int_0^\zeta \left(\int_{B^c_{S_2}} \mathbf{g}(e^{t\cdot\Omega} \cdot \mathbf{z})^3 \mathrm{d}\mathbf{z} \right)^{1/3} \\ & \times \left(\int_{B_{S_1}(\mathbf{y})^c} (|\mathbf{y}-\mathbf{z}| \cdot s(\mathbf{y}-\mathbf{z}))^{-9/4} \mathrm{d}\mathbf{z} \right)^{2/3} \mathrm{d}t. \end{aligned}$$

By the same reasoning as used above for $\mathfrak{K}_1(\mathbf{y})$, we see that the preceding integral over $B^c_{S_2}$ is bounded by $\mathfrak{C}(S) \cdot \|\nabla\mathbf{u}|B^c_{S_2}\|_3$ and hence by $\mathfrak{C}(S) \cdot (\|\nabla\mathbf{u}|B^c_{S_2}\|_2 + \|\nabla\mathbf{u}|B^c_{S_2}\|_4) \le \mathfrak{C}(S)$. Concerning the integral over $B^c_{S_1}(\mathbf{y})$, it is bounded by a constant $C(S_1)$, according to Corollary 10.1. Therefore $\mathfrak{K}_2(\mathbf{y}) \le \mathfrak{C}(S)$. Theorem 10.19 follows from (10.48)–(10.50) and the preceding estimate of $\mathfrak{K}_1(\mathbf{y})$ and $\mathfrak{K}_2(\mathbf{y})$. □

Now we may derive the looked-for estimate of the gradient of $\mathbf{u}$. Together with Theorem 10.18, it implies Theorem 10.1.

Theorem 10.20 *Let $S \in (S_1, \infty)$. Then $|\nabla\mathbf{u}(\mathbf{y})| \le \mathfrak{C}(S)\cdot(|\mathbf{y}|\cdot s(\mathbf{y}))^{-3/2}$ for $\mathbf{y} \in B^c_S$.* □

Proof Recalling that $|\nabla\mathbf{u}(\mathbf{y})| \le \mathfrak{C}(S)$ for $\mathbf{y} \in B^c_S$ by the preceding theorem, we see that

$$|\nabla\mathbf{u}(\mathbf{y})| \le \mathfrak{C}(S) \cdot (|\mathbf{y}| \cdot s(\mathbf{y}))^{-3/2} \quad \text{for } \mathbf{y} \in B_{2\cdot S}\backslash B_S. \tag{10.52}$$

Inequality (10.29) and Lemma 10.13 yield

$$|\nabla\mathbf{u}(\mathbf{y})| \le \mathfrak{C}(S) \cdot (|\mathbf{y}| \cdot s(\mathbf{y}))^{-3/2} + \mathfrak{A}_1(\mathbf{y}) + \mathfrak{A}_2(\mathbf{y}) \quad \text{for } \mathbf{y} \in B^c_S, \tag{10.53}$$

with

$$\mathfrak{A}_1(\mathbf{y}) := \sum_{l=1}^{3} \int_{B_S\backslash B_{S_0}} |\partial y_l \mathcal{Z}(\mathbf{y}, \mathbf{z}) \cdot (\mathbf{u}(\mathbf{z}) \cdot \nabla)\mathbf{u}(\mathbf{z})| \, d\mathbf{z},$$

and with $\mathfrak{A}_2(\mathbf{y})$ defined in the same way as $\mathfrak{A}_1(\mathbf{y})$, but with the domain of integration $B_S\backslash B_{S_0}$ replaced by B^c_S. By Corollary 10.6 with $R = S$, $\delta = 1$, $\alpha = e_l$ for $1 \le l \le 3$, we find

$$\begin{aligned}\mathfrak{A}_1(\mathbf{y}) \le& C(\tau, S) \cdot (|\mathbf{y}| \cdot s(\mathbf{y}))^{-3/2} \cdot \|(\mathbf{u} \cdot \nabla)\mathbf{u} \mid B_S\backslash B_{S_0}\|_1 \\ \le& C(\tau, S) \cdot (|\mathbf{y}| \cdot s(\mathbf{y}))^{-3/2} \cdot \|\mathbf{u}\|_6 \cdot \|\nabla\mathbf{u}\|_2 \le \mathfrak{C}(S) \cdot (|\mathbf{y}| \cdot s(\mathbf{y}))^{-3/2} \\ &\text{for } \mathbf{y} \in B^c_{2\cdot S},\end{aligned} \tag{10.54}$$

where we used that $\|\mathbf{u}|B_S\backslash B_{S_0}\|_2 \le C(S) \cdot \|u\|_6 \cdot \|\nabla u\|_2$. Let $\mathbf{y} \in B^c_{2\cdot S}$. By Theorems 10.18 and 10.19, we get

$$\mathfrak{A}_2(\mathbf{y}) \le \mathfrak{C}(S) \cdot \sum_{l=1}^{3} \int_{B^c_S} |\partial y_l \mathcal{Z}(\mathbf{y}, \mathbf{z})| \cdot (|\mathbf{z}| \cdot s(\mathbf{z}))^{-1} \, d\mathbf{z}.$$

It follows by Lemma 10.16 and Theorem 10.4 with $a = b = 1$ that

$$\mathfrak{A}_2(\mathbf{y}) \le \mathfrak{C}(S) \cdot [(|\mathbf{y}| \cdot s(\mathbf{y}))^{-1} + |\mathbf{y}|^{-1/2} \cdot s(\mathbf{y})^{-1}] \le \mathfrak{C}(S) \cdot |\mathbf{y}|^{-1/2} \cdot s(\mathbf{y})^{-1}.$$

This inequality and (10.52)–(10.54) imply that

$$|\nabla \mathbf{u}(\mathbf{y})| \le \mathfrak{C}(S) \cdot |\mathbf{y}|^{-1/2} \cdot s(\mathbf{y})^{-1} \le \mathfrak{C}(S) \cdot (|\mathbf{y}| \cdot s(\mathbf{y}))^{-1/2} \quad \text{for } \mathbf{y} \in B_S^c.$$

On using this inequality and again Theorem 10.18, we find

$$\mathfrak{A}_2(\mathbf{y}) \le \mathfrak{C}(S) \cdot \sum_{l=1}^{3} \int_{B_S^c} |\partial y_l \mathcal{Z}(\mathbf{y}, \mathbf{z})| \cdot (|\mathbf{z}| \cdot s(\mathbf{z}))^{-3/2} \, d\mathbf{z} \quad \text{for } \mathbf{y} \in B_{2\cdot S}^c.$$

Hence by Lemma 10.16 and Theorem 10.4 with $a = b = 3/2$,

$$\begin{aligned} \mathfrak{A}_2(\mathbf{y}) \le & \mathfrak{C}(S) \cdot [(|\mathbf{y}| \cdot s(\mathbf{y}))^{-3/2} + |\mathbf{y}|^{-1} \cdot s(\mathbf{y})^{-3/2}] \\ \le & \mathfrak{C}(S) \cdot |y|^{-1} \cdot s(\mathbf{y})^{-3/2} \quad (\mathbf{y} \in B_{2\cdot S}^c). \end{aligned}$$

Again referring to (10.52)–(10.54), we conclude $|\nabla \mathbf{u}(\mathbf{y})| \le \mathfrak{C}(S) \cdot |\mathbf{y}|^{-1} \cdot s(\mathbf{y})^{-3/2}$ for $\mathbf{y} \in B_S^c$. Returning to $\mathfrak{A}_2(\mathbf{y})$ a third time, we deduce from the preceding estimate and Theorem 10.18 that

$$\mathfrak{A}_2(\mathbf{y}) \le \mathfrak{C}(S) \cdot \sum_{l=1}^{3} \int_{B_S^c} |\partial y_l \mathcal{Z}(\mathbf{y}, \mathbf{z})| \cdot |\mathbf{z}|^{-2} \cdot s(\mathbf{z})^{-3/4} d\mathbf{z} \quad (\mathbf{y} \in B_{2\cdot S}^c).$$

Hence with Lemma 10.16 and Theorem 10.4 with $a = 2$, $b = 3/4$,

$$\begin{aligned} \mathfrak{A}_2(\mathbf{y}) \le & \mathfrak{C}(S) \cdot (|\mathbf{y}|^{-2} \cdot s(\mathbf{y})^{-3/4} + |\mathbf{y}|^{-11/8} \cdot s(\mathbf{y})^{-7/8}) \\ \le & \mathfrak{C}(S) \cdot |\mathbf{y}|^{-11/8} \cdot s(\mathbf{y})^{-7/8} \quad (\mathbf{y} \in B_{2\cdot S}^c), \end{aligned}$$

where we used Lemma 10.5 in the last inequality. Now (10.52)–(10.54) yield

$$|\nabla \mathbf{u}(\mathbf{y})| \le \mathfrak{C}(S) \cdot |\mathbf{y}|^{-11/8} \cdot s(\mathbf{y})^{-7/8} \le \mathfrak{C}(S) \cdot |\mathbf{y}|^{-11/8} \cdot s(\mathbf{y})^{-1/2} \quad \text{for } \mathbf{y} \in B_S^c.$$

Finally we use the same sequence of references again, in particular Theorem 10.4, this time with $a = 19/8$ and $b = 3/2$, obtaining

$$\begin{aligned} \mathfrak{A}_2(\mathbf{y}) \le & \mathfrak{C}(S) \cdot [|\mathbf{y}|^{-19/8} \cdot s(\mathbf{y})^{-3/2} + (|\mathbf{y}| \cdot s(\mathbf{y}))^{-3/2}] \\ \le & \mathfrak{C}(S) \cdot (|\mathbf{y}| \cdot s(\mathbf{y}))^{-3/2} \quad (\mathbf{y} \in B_{2\cdot S}^c). \end{aligned}$$

Theorem 10.20 follows with (10.52)–(10.54). □

Chapter 11
Asymptotic Behavior of the Solutions of the Navier–Stokes Type Problem

This chapter is devoted to the problem of the motion of incompressible flow around a rotating and translating rigid body $\mathcal{B}$ in an exterior domain $\mathcal{D}$. $\mathcal{B}$ is assumed open and bounded, with Lipschitz boundary. We focus on the asymptotic behavior of the problem. It is shown that the velocity may be split into a constant times the first column of the fundamental solution of the Oseen system, plus a remainder term decaying pointwise near infinity at a rate that is higher than the decay rate of the Oseen tensor. Moreover the optimal rates of the velocity, gradient of the velocity, and the pressure are derived.

11.1 Introduction

Consider a rigid body moving with constant velocity and rotating with constant angular velocity in a viscous incompressible fluid. Suppose the flow around the body is steady ("developed"). Then the usual mathematical model of this flow, with respect to a reference frame in which the body is at rest, is given by the Navier–Stokes system with an Oseen term and rotational terms. Written in normalized form (see, e.g., [98, 109] for details), this system takes the form

$$\begin{aligned}
&-\Delta \mathbf{u}(\mathbf{x}) + \tau\partial_1\mathbf{u}(\mathbf{x}) + \tau\,(\mathbf{u}(\mathbf{x})\cdot\nabla)\mathbf{u}(\mathbf{x}) - (\sqrt{\varrho}\boldsymbol{e}_1\wedge\mathbf{x})\cdot\nabla\mathbf{u}(\mathbf{x}) + \sqrt{\varrho}\boldsymbol{e}_1\wedge\mathbf{u}(\mathbf{x})\\
&+\nabla\pi(\mathbf{x}) = \mathbf{f}(\mathbf{x}),\\
&\operatorname{div}\mathbf{u}(\mathbf{x}) = 0 \quad \text{for } \mathbf{x}\in\overline{\mathcal{D}}
\end{aligned}\tag{11.1}$$

and is supplemented by homogeneous Dirichlet boundary conditions at infinity

$$|\mathbf{u}(\mathbf{x})| \to 0 \quad \text{for } |\mathbf{x}|\to\infty. \tag{11.2}$$

Š. Nečasová et al., *Mathematical Theory of a Fluid Flow Around a Rotating and Translating Body*, Springer Monographs in Mathematics,
https://doi.org/10.1007/978-3-031-84120-0_11

In this model, the open bounded set $\mathcal{B} = \mathcal{D}^c \subset \mathbb{R}^3$ corresponds to the rigid body, the unknown functions $\mathbf{u}$ and π describe, respectively, the velocity and the pressure field of the flow, whereas the given function $\mathbf{f}$ represents a prescribed volume force acting on the fluid. The parameter $\tau \in (0, \infty)$ is the Reynolds number and $\varrho \in \mathbb{R}\backslash\{0\}$ the Taylor number. We assume the Reynolds number and the Taylor number to be fixed throughout the chapter. The symbol $\boldsymbol{e}_1$ stands for the unit vector $(1, 0, 0)$. The constant velocity of the body, as seen from an observer at rest, is given by $-\tau\, \boldsymbol{e}_1$ and its constant angular velocity by $\sqrt{\varrho}\, \boldsymbol{e}_1$. In particular, the direction of translation and the axis of rotation of the rigid body are parallel, as should be expected in the case of a steady-state flow around that body.

11.2 Discussion and Main Results

First the aim of this section is to find the leading term of problem (11.1). The leading term to solve the problem is relevant in many mathematical applications; Numerical computation of exterior flows often involves bounded computational domains, with an artificial boundary condition on the outer boundary of such domains. The choice of such boundary conditions, and related error estimates, is determined by the asymptotic structure of the flow field and thus by a leading term, see [4, 10, 86].

- A leading term L with all relevant features could be exhibited, for some types of flows. Let us mention the case $\varrho = 0$ and the Oseen term in (11.1) is dropped (case of a body which neither rotates nor translates), which is rotationally symmetric and rotates but does not translate. In these situations, Korolev and Šverák [143] showed that L is given by the so-called Landau solution of the stationary Navier–Stokes system

$$- \Delta \mathbf{w} + \tau\, (\mathbf{w} \cdot \nabla)\mathbf{w} + \nabla q = 0, \quad \operatorname{div} \mathbf{w} = 0 \quad \text{in } \mathbb{R}^3\backslash\{0\},$$

  under the assumption that the velocity is small in a suitable sense.
- In the case where only the rotation is considered, Farwig and Hishida [67–69] and Farwig, Galdi, and Kyed [63] developed a satisfactory theory. It refers to the case that the Oseen term $\tau\, \partial_1 \mathbf{u}$ is not present in (11.1), but the parameter ϱ does not vanish (flow around a body that rotates but does not translate; "purely rotational case"). In this situation, if ϱ is small, it turned out that a leading term is again given by the Landau solution.
- On the other hand, if the Oseen term is present, but ϱ vanishes (case of a body that translates but does not rotate; "purely translational case"), it was shown first for solutions that are "strong" in a suitable sense—the work by Finn (see [83, 87])—and then for Leray solution by Babenko and Galdi (see [11, 98], and the references in [98]) that a leading term may be given by an appropriate linear combination of the columns of the (matrix-valued) velocity part of a fundamental

solution E—defined in Sect. 10.3 to the stationary Oseen system

$$-\,\Delta\mathbf{v}+\tau\,\partial_1\mathbf{v}+\nabla\gamma=\mathbf{g},\quad \operatorname{div}\mathbf{v}=0. \tag{11.3}$$

We focus on the asymptotic behavior of the solutions to the problem. It is shown that the velocity may be split in to a constant times the first column of the fundamental solution of the Oseen system (see Sect. 10.3 in the previous chapter), plus a remainder term already defined in Chap. 5 (see Sect. 5.3, Theorem 5.13) this remainder term decaying pointwise near infinity at a rate that is higher than the decay rate of the Oseen tensor. Moreover, it is derived from the optimal rates of the velocity, gradient of the velocity, and of the pressure.

We are interested in *Leray solutions* to (11.1), (11.2), that is, solutions $(\mathbf{u},\pi)$ such that $\mathbf{u}\in L^6(\overline{\mathcal{D}})$, $\nabla\mathbf{u}\in L^2(\overline{\mathcal{D}})$, and $\pi\in L^2_{loc}(\overline{\mathcal{D}})$.

Let us mention that from [98, Theorem XI.3.1] it follows that this type of solution is known to exist even for large data, under suitable regularity assumptions on $\partial\mathcal{D}$ and $\mathbf{f}$. Note that the conditions $\mathbf{u}\in L^6(\overline{\mathcal{D}})$, $\nabla\mathbf{u}\in L^2(\overline{\mathcal{D}})$ mean in particular that (11.2) holds in a weak sense (see below 11.4). We state this property in the form of [Theorem 10.1].

Theorem 11.1 *Take* $S_1,\ S\in(0,\infty)$, $p\in(1,\infty)$, $\mathbf{f}\in L^p(\mathbb{R}^3)$ *with* $S_1<S$, *and* $\overline{\mathcal{D}^c}\cup supp(\mathbf{f})\subset B_{S_1}$. *Let* $\mathbf{u}\in L^6(\overline{\mathcal{D}})\cap W^{1,1}_{loc}(\overline{\mathcal{D}})$, $\pi\in L^2_{loc}(\overline{\mathfrak{D}}^c)$ *with* $\nabla\mathbf{u}\in L^2(\overline{\mathfrak{D}})^9$, $div\,\mathbf{u}=0$, *and*

$$\int_{\overline{\mathcal{D}}}\Big[\nabla\mathbf{u}\cdot\nabla\boldsymbol{\varphi}+\big(\tau\,\partial_1\mathbf{u}+\tau\,(\mathbf{u}\cdot\nabla)\mathbf{u}-(\sqrt{\varrho}\,\boldsymbol{e}_1\wedge\mathbf{z})\cdot\nabla\mathbf{u}+\sqrt{\varrho}\,\boldsymbol{e}_1\wedge\mathbf{u}\big)\cdot\boldsymbol{\varphi}$$

$$-\pi\,div\,\boldsymbol{\varphi}-\mathbf{f}\cdot\boldsymbol{\varphi}\Big]\,d\mathbf{z}=0\quad for\ \ \boldsymbol{\varphi}\in C_0^\infty(\overline{\mathcal{D}}). \tag{11.4}$$

Then

$$|\partial^\alpha\mathbf{u}(\mathbf{x})|\le\mathfrak{C}\left(|\mathbf{x}|\,s(\mathbf{x})\right)^{-1-|\alpha|/2}]\quad for\ \ \mathbf{x}\in\mathbb{R}^3\ with\ |\mathbf{x}|\ge S, \tag{11.5}$$

with $\alpha\in\mathbb{N}_0^3$, $|\alpha|\le 1$, $|\alpha|:=\alpha_1+\alpha_2+\alpha_3$ *and*

$$s(\mathbf{x}):=1+\tau\,(|\mathbf{x}|-x_1)\quad(\mathbf{x}\in\mathbb{R}^3). \tag{11.6}$$

□

Note that no boundary conditions on $\mathbf{u}$ or π are imposed in Theorem 11.1. Moreover it is only for simplicity that we suppose that $\mathbf{f}$ has bounded support. Actually in Theorem 11.1 this assumption is replaced by a decay condition on $\mathbf{f}$. Further note that in the case $\alpha=0$, the relation in (11.5) describes the decay of $\mathbf{u}$, whereas the asymptotic behavior of ∇u is specified by (11.5) if $\alpha\in\mathbb{N}_0^3$ with $|\alpha|=1$. Since the vector $-\,\tau\,\boldsymbol{e}_1$ corresponds to the velocity of the body as seen from an observer at rest, the factor $s(\mathbf{x})^{-1-|\alpha|/2}$ in (11.5) should be considered as **a**

mathematical manifestation of the wake extending behind a body moving in a viscous incompressible fluid.

In view of (11.5), we can look for a vector-valued function L ("leading term") with the property that $L(\mathbf{x})$ decays with exactly the rate given by the right-hand side of (11.5) with $\alpha = 0$, whereas the term $R(\mathbf{x}) := (\mathbf{u} - L)(\mathbf{x})$ ("remainder") decays pointwise at a rate which is higher than the one in (11.5) with $\alpha = 0$. In addition, an analogous statement should hold for $\partial^\alpha L(\mathbf{x})$ if $|\alpha| = 1$. Identifying such a function L is highly interesting from a physical point of view because it gives a good idea of the structure ("profile") of the flow as observed at some distance from the rigid body. This interpretation of L explains why a leading term should be expressed, if possible, as an explicit combination of elementary functions. Actually, beyond the physical meaning of such a term, it might even be argued that in any case, the notion of leading term refers to functions that are given in a more or less explicit way.

The starting point of the chapter at hand is references [163] by Kyed and [41] by Deuring, Kračmar, Nečasová. These articles presented partial results on leading terms for the flow considered here, that is, the flow around a body performing a rotation *and* a translation. Kyed [163] showed that a leading term is given by the product of a coefficient γ times the column $(E_{j1})_{1\le j\le 3}$ of the fundamental solution E—mentioned above—of the Oseen system (11.3) (see Sect. 10.3). The term $E_{j1}(x)$ may be expressed explicitly in terms of elementary functions and does not depend on the quantity ϱ associated with the rotation of the fluid. The coefficient γ is given explicitly, too, its definition involving the Cauchy stress tensor. In particular the equations

$$\left.\begin{aligned} u_j(\mathbf{x}) &= \gamma\, E_{j1}(\mathbf{x}) + R_j(\mathbf{x}), \\ \partial_l u_j(\mathbf{x}) &= \beta\, \partial_l E_{j1}(\mathbf{x}) + S_{jl}(\mathbf{x}) \quad (\mathbf{x} \in \overline{\mathcal{D}},\ 1 \le j, l \le 3) \end{aligned}\right\}, \tag{11.7}$$

where the remainder terms $R_j(\mathbf{x})$ and $S_{jl}(\mathbf{x})$ in (11.7) are characterized by the relations $R \in L^q(\overline{\mathcal{D}})$ for $q \in (4/3, \infty)$ and $S \in L^q(\overline{\mathcal{D}})^9$ for $q \in (1, \infty)$. It is known from [94, Section VII.3] that $E_{j1}|B_r^c \notin L^q(B_r^c)$ for $q \in [1, 2]$ and $\partial_l E_{j1}|B_r^c \notin L^q(B_r^c)$ for $q \in [1, 4/3]$, where $j, l \in \{1, 2, 3\}$, $r > 0$, and B_r^c denotes the complement of the open ball B_r in $\mathbb{R}^3$. Therefore the function R_j decays faster than E_{j1} and S_{jl} faster than $\partial_l E_{j1}$, in the sense of L^q-integrability. Thus the equations in (11.7) may in fact be considered as asymptotic expansions of u and ∇u, respectively.

However, the theory in [163] is valid only under the assumptions that f vanishes and u verifies no slip boundary conditions, which here take the form

$$\mathbf{u}(\mathbf{x}) = -\tau\, \boldsymbol{e}_1 + \sqrt{\varrho}\, (\boldsymbol{e}_1 \wedge \mathbf{x}) \quad \text{for } \mathbf{x} \in \partial\mathcal{D}. \tag{11.8}$$

This indicates but does not prove that the remainder $R(\mathbf{x})$ *might decay faster for* $|\mathbf{x}| \to \infty$ *than does* $\big(|\mathbf{x}|\, s(\mathbf{x})\big)^{-1}$, *that is, faster than the rate given by (11.5) for* $\mathbf{u}(\mathbf{x})$, *and similarly* $\nabla R(\mathbf{x})$ *might coincide with* $S(\mathbf{x})$ *and exhibit a more rapid pointwise decay than the one given by (11.5) for* $\nabla \mathbf{u}(\mathbf{x})$. The question remained open whether this is actually the case. In fact, reference [163] does not deal with pointwise decay of R and S. However, in [161] Kyed shows that $|R(\mathbf{x})|$ behaves as $O(|\mathbf{x}|^{-4/3+\epsilon})$ if $|\mathbf{x}| \to \infty$, for some arbitrary but fixed $\epsilon > 0$.

Concerning the argument adopted in [163] in order to derive these results, it is based on certain estimates of solutions to the time-periodic Oseen system in the whole space $\mathbb{R}^3$. These estimates, in turn, are obtained via Fourier expansions of these solutions, an approach that ultimately reduces to a study of solutions to the standard Oseen system (11.3) in the whole space $\mathbb{R}^3$.

On the other hand, reference [41]—the second reference constituting the starting point of this work—states in Theorem 11.1 that for $\mathbf{x} \in \mathbb{R}^3$ with $|\mathbf{x}| > S_1$, $1 \leq j \leq 3$,

$$u_j(\mathbf{x}) = \sum_{k=1}^{3} \beta_k \, \mathcal{Z}_{jk}(\mathbf{x}, 0) + \Big(\int_{\partial \mathcal{B}} \mathbf{u} \cdot \mathbf{n} \, do_{\mathbf{x}} \Big) \, x_j \, (4\pi \, |\mathbf{x}|^3)^{-1} + \mathfrak{F}_j(\mathbf{x}). \tag{11.9}$$

Here S_1 is a sufficiently large positive real number, n stands for the outward unit normal to $\mathcal{D}$, $(\mathcal{Z}_{jk})_{1 \leq j,k \leq 3}$ is the velocity part of the fundamental solution constructed by Guenther, Thomann [124], see details in Chaps. 2 and 5 (also see [65]) for the linearization

$$-\Delta \mathbf{v} + \tau \, \partial_1 \mathbf{v} - \sqrt{\varrho} \, (\boldsymbol{e}_1 \wedge \mathbf{x}) \cdot \nabla \mathbf{v} + \sqrt{\varrho} \, \boldsymbol{e}_1 \wedge \mathbf{v} + \nabla \sigma = \mathbf{g}, \quad \operatorname{div} \mathbf{v} = 0 \tag{11.10}$$

of (11.1) (see Sect. 10.3), β_1, β_2, and β_3 are coefficients defined in terms of $\mathbf{u}$, π, and f (see Theorem 11.2 below), and $\mathfrak{F}$ is a function from $\boldsymbol{C}^1(\mathbb{R}^3 \backslash \overline{B_{S_1}})$ given explicitly in terms of $\mathcal{Z}$, u, and π (again see Theorem 11.2 below). As is shown in [41] or in the proof of Theorem 11.2, this function $\mathfrak{F}$ decays pointwise, in the sense that

$$|\partial^\alpha \mathfrak{F}(\mathbf{x})| = O\big((|\mathbf{x}| \, s(\mathbf{x}))^{-3/2-|\alpha|/2} \ln(2 + |\mathbf{x}|) \big) \;\text{ for } |\mathbf{x}| \to \infty, \tag{11.11}$$

where $\alpha \in \mathbb{N}_0^3$ with $|\alpha| \leq 1$.

No boundary conditions are imposed in [41], and the function $\mathbf{f}$ need not vanish; it is only required to have bounded support. One might even replace this assumption by a decay condition on $\mathbf{f}$, but this point is not elaborated in [41].

It is known from [36, Theorem 2.19]—and restated below in Corollary 11.2—that

$$|\partial^\alpha \mathcal{Z}(\mathbf{x}, 0)| = O\big((|\mathbf{x}| \, s(\mathbf{x}))^{-1-|\alpha|/2} \big) \quad \text{for } |\mathbf{x}| \to \infty, \tag{11.12}$$

where α is the same as in (11.11).

So, if the decay rate in (11.12) is sharp, Eq. (11.9) may be considered as an asymptotic expansion in the usual sense: The remainder exhibits a faster pointwise decay than the leading term (see (11.11)). However, since the definition of the term

$\mathcal{Z}(\mathbf{x}, 0)$ involves an integral over $(0, \infty)$, the leading term $\sum_{k=1}^{3} \beta_k\, \mathcal{Z}_{jk}(\mathbf{x}, 0)$ in (11.9) is not as explicit as one would like it to be.

At this point it is natural to ask whether the rather explicit leading term from [163] may be kept, while, as in [41], boundary conditions are dropped, a nonvanishing function **f** is admitted, and pointwise decay estimates of the remainder are established.

The main point we want to highlight in this chapter is that reference [42] gives a positive answer to this question. In fact, by setting

$$\mathfrak{G}_j(\mathbf{x}) := \sum_{k=2}^{3} \beta_k\, \mathcal{Z}_{jk}(\mathbf{x}, 0) + \mathfrak{F}_j(\mathbf{x}) \quad (\mathbf{x} \in \mathbb{R}^3 \tag{11.13}$$

with $|\mathbf{x}| > S_1,\ 1 \leq j \leq 3$,
it may be deduced from (11.9) and (11.11) that

$$u_j(\mathbf{x}) = \beta_1\, E_{j1}(\mathbf{x}) + \Big(\int_{\partial \mathcal{B}} u \cdot n\, do_x \Big)\, x_j\, (4\pi\, |\mathbf{x}|^3)^{-1} + \mathfrak{G}_j(\mathbf{x}) \tag{11.14}$$

for $\mathbf{x} \in \mathbb{R}^3$ with $|\mathbf{x}| > S_1,\ 1 \leq j \leq 3$, and

$$|\partial^\alpha \mathfrak{G}(\mathbf{x})| = O\big((|\mathbf{x}|\, s(\mathbf{x}))^{-3/2-|\alpha|/2}\, \ln(2 + |\mathbf{x}|) \big) \quad \text{for } |\mathbf{x}| \to \infty \tag{11.15}$$

($\alpha \in \mathbb{N}_0^3$ with $|\alpha| \leq 1$); see Theorem 11.3 and Corollary 11.1. The decay rate of $\partial^\alpha R(\mathbf{x})$ stated in (11.15) is optimal in the sense that derivatives of E_{j1} of order $|\alpha|+1$ exhibit the same rate (see [119]), except for the logarithmic factor $\ln(2+|\mathbf{x}|)$. As in [41], no boundary conditions are imposed in [42], and **f** need not vanish. Instead it is again assumed that $supp(\mathbf{f})$ is compact, and once more this assumption may be replaced by suitable decay conditions on **f**. The theory in [41] and [42] will be stated in more detail in Theorems 11.2 and 11.3 and Corollary 11.1 below, and their proof will be discussed in Sects. 11.4–11.8.

If we compare how the coefficient γ from (11.7) is defined in [163] and the coefficient β_1 from (11.14) introduced by P. Deuring et al. [41] (see Theorem 11.3 below), we see that γ and β_1 coincide, provided that **u** satisfies boundary condition (11.8) imposed in [163]. Moreover the function $\boldsymbol{R}$ from (11.13) belongs to $\boldsymbol{L}^p(B_{S_1}^c)$ for $p > 4/3$ and $\partial_j \boldsymbol{R}$ to $\boldsymbol{L}^p(B_{S_1}^c)$ for $p > 1,\ 1 \leq j \leq 3$ (Lemma 11.20). **Thus the theory of (11.7) covers the one in [163] as a special case.**

Let us state the main theorems of the chapter. The first goal is to find the leading term of the velocity and its gradient for the Navier–Stokes problem with rotational terms. Let us recall that the quantities $\tau,\ \boldsymbol{\omega}$ will be recalled in Sect. 11.3. We study the case **f** has a compact support in $\overline{\mathcal{D}}^c$.

Theorem 11.2 *Let $S, S_1 \in (0, \infty)$ with $\overline{\mathcal{D}^c} \subset B_{S_1},\ p \in (1, \infty),\ \mathbf{f} \in \boldsymbol{L}^p(\overline{\mathcal{D}})$ with $supp(\mathbf{f}) \subset B_{S_1},\ \mathbf{u} \in \boldsymbol{L}^6(\overline{\mathcal{D}}) \cap \mathbf{W}^{1,1}_{loc}(\overline{\mathcal{D}})$ with $\nabla \mathbf{u} \in L^2(\overline{\mathcal{D}})$ and $\mathbf{u}|\partial\mathcal{D} \in \mathbf{W}^{2-1/p,\,p}(\partial\mathcal{D}),\ \pi \in L^2_{loc}(\overline{\mathcal{D}})$ with $\pi|\mathcal{D}_{S_1} \in L^p(\mathcal{D}_{S_1})$. Suppose that the pair $(\mathbf{u}, \pi)$*

is a weak solution of the Navier–Stokes system with Oseen and rotational terms and with right-hand side $\mathbf{f}$ *in the sense of (11.4). Suppose in addition that*

$$\mathcal{D}^c \text{ is } C^2\text{-bounded}, \quad \mathbf{u}|\partial\mathcal{D} \in \mathbf{W}^{2-1/p,\,p}(\partial\mathcal{D}), \tag{11.16}$$

$$\pi\,|B_{S_1}\backslash\overline{\mathcal{D}^c} \in L^p(B_{S_1}\backslash\overline{\mathcal{D}^c}). \tag{11.17}$$

Let $\boldsymbol{n}$ *denote the outward unit normal to* $\mathcal{B}$. *Then there are coefficients* $\beta_1, \beta_2, \beta_3 \in \mathbb{R}$ *and functions* $\mathfrak{F}_1, \mathfrak{F}_2, \mathfrak{F}_3 \in C^1(\overline{B_{S_1}}^c)$ *such that for* $j \in \{1,2,3\}$, $\alpha \in \mathbb{N}_0^3$ *with* $|\alpha| \le 1$, $\mathbf{x} \in \overline{B_{S_1}}^c$,

$$\partial^\alpha u_j(\mathbf{x}) = \partial^\alpha L_j(\mathbf{x}) + \partial^\alpha \mathfrak{F}_j(\mathbf{x}), \tag{11.18}$$

where

$$L_j(\mathbf{x}) = \left\{ \sum_{k=1}^{3} \beta_k\, \mathcal{Z}_{jk}(\mathbf{x}, 0) + \left(\int_{\partial\mathcal{D}} \mathbf{u} \cdot \boldsymbol{n}^{(\mathcal{D})}\, do_{\mathbf{z}} \right) E_{4j}(\mathbf{x}) \right\}, \tag{11.19}$$

$$\begin{aligned}
\beta_k :=& \int_{\mathcal{D}_{S_1}} f_k(\mathbf{y})\, d\mathbf{y} \\
&+ \int_{\partial\mathcal{B}} \sum_{l=1}^{3} \Big(-\partial_l u_k(\mathbf{y}) + \delta_{kl}\, \pi(\mathbf{y}) + (\tau\, \boldsymbol{e}_1 - \sqrt{\varrho}\, \boldsymbol{e}_1 \wedge \mathbf{y})_l\, u_k(\mathbf{y}) \\
&+ \tau\, (u_l\, u_k)(\mathbf{y}) \Big)\, n_l(\mathbf{y})\, do_{\mathbf{y}}
\end{aligned}$$

for $1 \le k \le 3$,

$$\begin{aligned}
\mathfrak{F}_j(\mathbf{x}) :=& \int_{\overline{\mathcal{D}}} \Big[\sum_{k=1}^{3} \big(\mathcal{Z}_{jk}(\mathbf{x}, \mathbf{y}) - \mathcal{Z}_{jk}(\mathbf{x}, 0) \big)\, f_k(\mathbf{y}) - \tau \cdot \sum_{k,l=1}^{3} \mathcal{Z}_{jk}(\mathbf{x}, \mathbf{y})\, (u_l\, \partial_l u_k)(\mathbf{y}) \Big]\, d\mathbf{y} \\
&+ \int_{\partial\mathcal{D}} \sum_{k=1}^{3} \Big[\big(\mathcal{Z}_{jk}(\mathbf{x}, \mathbf{y}) - \mathcal{Z}_{jk}(\mathbf{x}, 0) \big) \sum_{l=1}^{3} \Big(-\partial_l u_k(\mathbf{y}) + \delta_{kl}\, \pi(\mathbf{y}) \\
&+ (\tau\, \boldsymbol{e}_1 - \sqrt{\varrho}\, \boldsymbol{e}_1 \wedge \mathbf{y})_l\, u_k(\mathbf{y}) \Big)\, n_l(\mathbf{y}) + \big(E_{4j}(\mathbf{x} - \mathbf{y}) - E_{4j}(\mathbf{x}) \big)\, u_k(\mathbf{y})\, n_k(\mathbf{y}) \\
&+ \sum_{l=1}^{3} \Big(\partial y_l \mathcal{Z}_{jk}(\mathbf{x}, \mathbf{y})\, (u_k\, n_l)(\mathbf{y}) - \tau\, \mathcal{Z}_{jk}(\mathbf{x}, 0)\, (u_l\, u_k\, n_l)(\mathbf{y}) \Big) \Big]\, do_{\mathbf{y}}
\end{aligned}$$

□

for $\mathbf{x} \in \overline{B_{S_1}}^c$, $1 \le j \le 3$. *The preceding integrals are absolutely convergent. Moreover, if* $S \in (S_1, \infty)$, $\mathbf{x} \in B_S^c$,

$$|\partial^\alpha \mathfrak{F}(\mathbf{x})| \le \mathfrak{C}\, (|\mathbf{x}|\, s_\tau(\mathbf{x}))^{-3/2-|\alpha|/2} \ln(2 + |\mathbf{x}|), \tag{11.20}$$

where $\mathfrak{C}$ depends on τ, ω, p, S_1, S, certain norms of $\mathbf{u}$, π, and f, and the constant D from (11.36).

Remark 11.1 The rate of decay stated in Theorem 11.2 is optimal in the following sense:

- Firstly, $|\mathfrak{G}_j(\mathbf{x})|$ will be seen to behave as $O\big[\,\big(|\mathbf{x}|\,s_\tau(\mathbf{x})\big)^{-3/2}\ln(2+|\mathbf{x}|)\,\big]$ for $|\mathbf{x}|\to\infty$ and thus in the same way as $|\nabla\mathfrak{O}(\mathbf{x})|$ except for the logarithmic factor $\ln(2+|\mathbf{x}|)$, where $\mathfrak{O}$ is the fundamental solution of the scalar Oseen problem; see (11.27) for the expression, and also see [119].
- Secondly we will show that $|\nabla\mathfrak{G}_j(\mathbf{x})|$ decays as $O\big[\,\big(|\mathbf{x}|\,s_\tau(\mathbf{x})\big)^{-2}\ln(2+|\mathbf{x}|)\,\big]$ for $|\mathbf{x}|\to\infty$, that is, with the same rate as $|D^2\mathfrak{O}(\mathbf{x})|$, once more up to the logarithmic factor $\ln(2+|\mathbf{x}|)$; again see [119]. □

Remark 11.2 In the case of [161, 163], the term $\big(\int_{\partial\mathcal{D}}\mathbf{u}\cdot\mathbf{n}^{(\mathcal{D})}\,do_{\mathbf{z}}\big)\,E_{4j}(\mathbf{x})$ is zero because of the boundary conditions. □

In other words, the decay rates of the remainder and its gradient are optimal in the sense that they correspond to those of an additional derivative of $\mathfrak{O}$ apart from a logarithmic factor. A comparison with $\mathfrak{O}$ seems to be reasonable in view of the fact that we want to arrive at a leading term of the form coefficient times a column of $\mathfrak{O}$.

The second main theorem deals with the point that the velocity may be split into a constant times the first column of the fundamental solution of the Oseen system, plus a remainder term decaying pointwise near infinity at a rate which is higher than the decay rate of the Oseen tensor.

Theorem 11.3 *Let $\mathcal{B}$, p, f, S, S_1, $\mathbf{u}$, π satisfy the assumptions of Theorem 11.2, including (11.4). Let β_1, β_2, β_3, and $\mathfrak{F}$ be defined as in Theorem 11.2. Define the function $\mathfrak{G}$*

$$\mathfrak{G}_j(\mathbf{x}) := \sum_{k=2}^{3}\beta_k\,\mathcal{Z}_{jk}(\mathbf{x},0)+\mathfrak{F}_j(\mathbf{x})\quad (\mathbf{x}\in\overline{B_{S_1}}^{\,c},\ 1\le j\le 3). \tag{11.21}$$

Then $\mathfrak{G}\in C^1(\overline{B_{S_1}}^{\,c})$, the equation

$$u_j(\mathbf{x})=\beta_1\,E_{j1}(\mathbf{x})+\Big(\int_{\partial\mathcal{B}}\mathbf{u}\cdot\mathbf{n}\,do_{\mathbf{x}}\Big)\,x_j\,(4\pi\,|\mathbf{x}|^3)^{-1}+\mathfrak{G}_j(\mathbf{x}) \tag{11.22}$$

holds for $(\mathbf{x}\in\overline{B_{S_1}}^{\,c},\ 1\le j\le 3)$, and for any $S\in(S_1,\infty)$, there is a constant $C>0$ which depends on τ, ϱ, S_1, S, $\mathbf{f}$, $\mathbf{u}$, and π and which is such that

$$|\partial^\alpha\mathfrak{G}(\mathbf{x})|\le C\,\big(|\mathbf{x}|\,s(\mathbf{x})\big)^{-3/2-|\alpha|/2}\ln(2+|\mathbf{x}|)\quad \textit{for}\ \ \mathbf{x}\in\overline{B_S}^{\,c},\ \alpha\in\mathbb{N}_0^3$$

with $|\alpha|\le 1$. □

We recall that the asymptotic behaviour of the function E appearing in the leading term in (11.22) is described in Theorem 11.5. As explained above, we may drop the assumptions in (11.16) if we replace $(\mathbf{u}, \pi)$ by $(\mathbf{u}|\overline{B_{S_0}}^c, \pi|\overline{B_{S_0}}^c)$, with some suitably chosen number S_0. Here are the details.

Corollary 11.1 *Take $\mathcal{B}$, p, $\mathbf{f}$, S, S_1, $\mathbf{u}$, π as in Theorem 11.2, but without requiring (11.16). (This means that $(\mathbf{u}, \pi)$ is only assumed to be a Leray solution of (11.1) and (11.2).) Put $\widetilde{p} := \min\{3/2, p\}$.*

Then $\mathbf{u} \in \mathbf{W}^{2,\widetilde{p}}_{loc}(\overline{\mathcal{D}})$ and $\pi \in W^{1,\widetilde{p}}_{loc}(\overline{\mathcal{D}})$.

Fix some number $S_0 \in (0, S_1)$ with $\overline{\mathcal{B}} \cup supp(\mathbf{f}) \subset B_{S_0}$, and define β_1, β_2, β_3, and $\mathfrak{F}$ as in Theorem 11.2, but with $\mathcal{B}$ replaced by B_{S_0} and $\mathbf{n}(\mathbf{x})$ by $S_0^{-1}\,\mathbf{x}$, for $\mathbf{x} \in \partial B_{S_0}$. Moreover, define $\mathfrak{G}$ as in (11.21).

Then all the conclusions of Theorem 11.3 are valid. □

Finally, we derive the pointwise estimates for the pressure.

Theorem 11.4 *Let $\tau \in (0, \infty)$, $\boldsymbol{\omega} \in \mathbb{R}^3\backslash\{\mathbf{0}\}$, $\mathcal{D}^c \subset \mathbb{R}^3$ be open and bounded. Take γ, $S_1 \in (0, \infty)$, $p_0 \in (1, \infty)$, $A \in (2, \infty)$, $B \in [0, 3/2]$, $\mathbf{f} : \mathbb{R}^3 \mapsto \mathbb{R}^3$ measurable with $\overline{\mathcal{D}} \subset B_{S_1}$, $A + \min\{B, 1\} > 3$, $A + B \geq 7/2$, $\mathbf{f}|B_{S_1} \in L^{p_0}(B_{S_1})^3$,*

$$|\mathbf{f}(\mathbf{y})| \leq \gamma\, |\mathbf{y}|^{-A}\, s_\tau(\mathbf{y})^{-B} \quad \textit{for } \mathbf{y} \in B^c_{S_1}. \tag{11.23}$$

Let $\mathbf{u} \in \mathbf{L}^6(\overline{\mathcal{D}}) \cap \mathbf{W}^{1,1}_{loc}(\overline{\mathcal{D}})$, $\pi \in L^2_{loc}(\overline{\mathcal{D}})$, $\nabla\mathbf{u} \in L^2(\overline{\mathcal{D}})$, $div\,\mathbf{u} = 0$, and

$$\int_{\overline{\mathcal{D}}} \Big[\, \nabla\mathbf{u} \cdot \nabla\boldsymbol{\varphi} + \big(\tau\, \partial_1\mathbf{u} + \tau\, (\mathbf{u}\cdot\nabla)\mathbf{u} \tag{11.24}$$

$$-(\boldsymbol{\omega} \wedge \mathbf{z}) \cdot \nabla\mathbf{u} + \boldsymbol{\omega} \wedge \mathbf{u}\big) \cdot \boldsymbol{\varphi} - \pi\, div\, \boldsymbol{\varphi} - \mathbf{f} \cdot \boldsymbol{\varphi} \,\Big]\, d\mathbf{z} = 0, \tag{11.25}$$

for $\boldsymbol{\varphi} \in C^\infty_0(\overline{\mathcal{D}})$. Let $S \in (S_1, \infty)$. Suppose in addition that $A \geq 5/2$. Then there is $c_0 \in \mathbb{R}$ such that inequality

$$|\pi(\mathbf{x}) + c_0| \leq C\, |\mathbf{x}|^{-2} \quad \textit{for } \mathbf{x} \in B^c_S \tag{11.26}$$

holds. □

11.3 Historical Remarks

The fundamental contributions to studies of asymptotic behavior of solutions of the Navier–Stokes and Navier–Stokes type problems in infinity come from R. Finn, K. I. Babenko, and G. P. Galdi. The first two authors considered just the flow around a rigid body without the effect of rotation. The rotation of the rigid body was taken into account by G. P. Galdi.

R. Finn [82–85] proved that if $(\mathbf{v}, p)$ is a sufficiently smooth solution with $\mathbf{f}$ having a bounded support, such that $|\mathbf{v}(\mathbf{x})| \le c|\mathbf{x}|^{-\alpha}$ for some $\alpha > 1/2$ and all "large" $|\mathbf{x}|$, then the pointwise asymptotic structure of $(\mathbf{v}, p)$ can be sharply evaluated. He showed that the solution exhibits a paraboloidal "wake region" $\mathcal{R}$ with the property that the velocity field $\mathbf{v}$ inside $\mathcal{R}$ decays pointwise slower than outside $\mathcal{R}$. R. Finn left open the question of the asymptotic behavior of second derivatives of $\mathbf{v}$ outside $\mathcal{R}$, which was solved by P. Deuring [46]. R. Finn called solutions with these properties "physically reasonable." Later, D. Clark [24] proved that the vorticity field of any "physically reasonable" solution decays at infinity exponentially fast outside of $\mathcal{R}$ and sufficiently far from the body.

K. I. Babenko investigated the relation between weak and physically reasonable solutions. Using Lizorkin's multiplier theory with anisotropic Sobolev-like inequalities and representation formula for the solution invented by R. Finn, he was able to show that in the case when the external force $\mathbf{f}$ has a bounded support, every weak solution is physically reasonable in the sense of R. Finn. More details can be found in the papers [99] by G. P. Galdi and [80] by R. Farwig and H. Sohr.

G. P. Galdi introduced a completely different and very general approach to the problem of asymptotic behavior of solutions at infinity. He considered a sequence of appropriate time-dependent solutions $\mathbf{v}_n$ to the linear Oseen type problem and showed that the weak solution $\mathbf{v}$ to the Navier–Stokes type problem can be viewed as a limit (for time growing to infinity and $n \to \infty$) of $\mathbf{v}_n(t)$. For more details, see [97, 98, 101, 108, 110, 111].

Moreover, P. Deuring, S. Kračmar, and Š. Nečasová [35–44] used the fundamental solution, constructed by Thomann and Guenther for the nonsteady Oseen problem with the rotational and translational terms. From that, the authors derived the form of the fundamental solution for the steady Oseen type problem. Moreover the representation formula for a solution of the nonlinear steady Navier–Stokes problem with "rotating" terms is obtained. Using the representation formula, the new information on the asymptotic behavior of a solution is shown. The obtained results are very general and independent of the boundary conditions.

11.4 Notation and Auxiliary Results

Let us denote $s(\mathbf{x}) := s_1(\mathbf{x}) = 1 + (|\mathbf{x}| - x_1)$. We recall that the function s_τ was defined in Chap. 10. We begin by introducing the fundamental solutions used in what follows.

$$\mathfrak{O}(\mathbf{x}) := (4\pi\,|\mathbf{x}|)^{-1}\, e^{-\tau\,(|\mathbf{x}|-x_1)/2} \quad \text{for } \mathbf{x} \in \mathbb{R}^3 \backslash \{0\} \tag{11.27}$$

(the fundamental solution of the scalar Oseen equation $-\Delta v + \tau\,\partial_1 v = g$ in $\mathbb{R}^3$)

$$\psi(r) := \int_0^r (1 - e^{-t})\, t^{-1}\, dt \quad (r \in \mathbb{R}), \qquad \Phi(\mathbf{x}) := (4\pi\,\tau)^{-1}\, \psi\big(\tau\,(|\mathbf{x}| - x_1)/2\big)$$

$$(\mathbf{x} \in \mathbb{R}^3),$$
$$E_{jk}(\mathbf{x}) := (\delta_{jk}\,\Delta - \partial_j\partial_k)\Phi(\mathbf{x}), \quad E_{4k}(\mathbf{x}) := x_k\,(4\,\pi\,|\mathbf{x}|^3)^{-1}$$
$$(\mathbf{x} \in \mathbb{R}^3\backslash\{0\},\ 1 \le j,k \le 3)$$

(the fundamental solution of the Oseen system)

$$-\,\Delta\mathbf{v} + \tau\,\partial_1\mathbf{v} + \nabla\sigma = \mathbf{g}, \quad \operatorname{div}\mathbf{v} = 0 \quad \text{in}\ \mathbb{R}^3. \tag{11.28}$$

We further define

$$F^{(\lambda)}(\boldsymbol{\xi}) := (2\,\pi)^{-3/2}\,(\lambda + |\boldsymbol{\xi}|^2 + i\,\tau\,\xi_1)^{-1} \quad \text{for}\ \boldsymbol{\xi} \in \mathbb{R}^3,\ \lambda \in (0,\infty),$$

the Fourier transform of $\mathfrak{O}^{(\lambda)}$; see Theorem 11.9, and see also [42, Theorem 4.1].

We write $\mathfrak{C}(\beta_1, \beta_2, .., \beta_n)$ for positive constants that additionally depend on parameters $\beta_1, \beta_2, ..., \beta_n \in \mathbb{R}$, for some $n \in \mathbb{N}$. $C(\gamma_1, \gamma_2, ..., \gamma_n)$ means a positive constant only depending on $\gamma_1, \gamma_2, ..., \gamma_n$.

We recall some important lemmas and theorems which will be used in proofs of main theorems.

Lemma 11.1 ([58], [Lemma 2.13]) *Let* $\beta \in (1,\infty)$. *Then* $\int_{\partial B_r} s_\tau(\mathbf{x})^{-\beta} do_{\mathbf{x}} \le \mathfrak{C}(\beta)$ *for* $r \in (0,\infty)$.

Lemma 11.2 (Corollary 2.18) *The function* Γ *may be continuously extended to a function from* $C^\infty(\mathbb{R}^3 \times \mathbb{R}^3 \times (0,\infty))$.

According to [35, Theorem 3.1], we have the following lemma:

Lemma 11.3

$$\int_0^\infty |\Gamma_{jk}(\mathbf{y},\mathbf{z},t)|\mathrm{d}t < \infty$$

for $\mathbf{y}, \mathbf{z} \in \mathbb{R}^3$ *with* $\mathbf{y} \ne \mathbf{z}$, $1 \le j,k \le 3$.

Thus we may define

$$\mathcal{Z}_{jk}(\mathbf{y},\mathbf{z}) := \int_0^\infty \Gamma_{jk}(\mathbf{y},\mathbf{z},t)\mathrm{d}t$$

for $\mathbf{y}, \mathbf{z} \in \mathbb{R}^3$ with $\mathbf{y} \ne \mathbf{z}$, $1 \le j,k \le 3$.
We will use the following technical lemmas:

Lemma 11.4 *Let* $\delta > 0$. *Assuming* $\mathbf{z} \in B_\delta(\mathbf{x})$, *we have*

$$|\mathbf{z}| \ge |\mathbf{x}|/2, \quad \text{for}\ |\mathbf{x}| \ge 2\delta, \tag{11.29}$$
$$s_\tau(\mathbf{z})^{-1} \le \mathfrak{C}(1 + |\mathbf{x}-\mathbf{z}|)\,s_\tau(\mathbf{x})^{-1} \le \mathfrak{C}(\delta)s_\tau(\mathbf{x})^{-1}. \tag{11.30}$$

Proof For $|\mathbf{x}| \geq 2\delta$ we have $|\mathbf{z}| \geq |\mathbf{x}| - |\mathbf{x}-\mathbf{z}| \geq |\mathbf{x}| - \delta \geq |\mathbf{x}|/2$, i.e., the relation (11.29) is satisfied. For the proof of (11.30), see Lemma 2.14 or [34, Lemma 4.8]. □

Lemma 11.5 (Corollary 2.4) *Let $j, k \in \{1, 2, 3\}$, $\alpha, \beta \in \mathbb{N}_0^3$ with $|\alpha + \beta| \leq 2$, $\mathbf{y}, \mathbf{z} \in \mathbb{R}^3$, $t \in (0, \infty)$. Then*

$$|\partial_y^\alpha \partial_z^\beta \Gamma_{jk}(\mathbf{y}, \mathbf{z}, t)| \leq \mathfrak{C}(|\mathbf{y} - \tau t \boldsymbol{e}_1 - \mathrm{e}^{-t\cdot\Omega} \cdot \mathbf{z}|^2 + t)^{-3/2-|\alpha+\beta|/2}.$$

Lemma 11.6 ([36, Theorem 2.19], Theorem 5.1) *Let $S_1, S \in (0, \infty)$ with $S_1 < S$, $\nu \in (1, \infty)$. Then*

$$\int_0^\infty (|\mathbf{y} - \tau t \boldsymbol{e}_1 - \mathrm{e}^{-t\cdot\Omega} \cdot \mathbf{z}|^2 + t)^{-\nu} \mathrm{d}t \leq \mathfrak{C}(S_1, S, \nu)(|\mathbf{y}| \cdot s_\tau(\mathbf{y}))^{-\nu+1/2} \tag{11.31}$$

for $\mathbf{y} \in B_S^c$, $\mathbf{z} \in \overline{B_{S_1}}$.

Lemma 11.7 ([38, Lemma 3.2], Lemma 5.4) *Let $j, k \in \{1, 2, 3\}$. For $\alpha, \beta \in \mathbb{N}_0^3$ with $|\alpha + \beta| \leq 2$, $\mathbf{y}, \mathbf{z} \in \mathbb{R}^3$ with $\mathbf{y} \neq \mathbf{z}$, the function $(0, \infty) \ni t \mapsto \partial_y^\alpha \partial_z^\beta \Gamma_{jk}(\mathbf{y}, \mathbf{z}, t) \in \mathbb{R}$ is integrable, the derivative $\partial_y^\alpha \partial_z^\beta \mathcal{Z}_{jk}(\mathbf{y}, \mathbf{z})$ exists, and*

$$\partial_y^\alpha \partial_z^\beta \mathcal{Z}_{jk}(\mathbf{y}, \mathbf{z}) = \int_0^\infty \partial_y^\alpha \partial_z^\beta \Gamma_{jk}(\mathbf{y}, \mathbf{z}, t) \mathrm{d}t. \tag{11.32}$$

Moreover, for α, β as before, the derivative $\partial_y^\alpha \partial_z^\beta \mathcal{Z}_{jk}(\mathbf{y}, \mathbf{z})$ is a continuous function of $\mathbf{y}, \mathbf{z} \in \mathbb{R}^3$ with $\mathbf{y} \neq \mathbf{z}$.

Lemma 11.8 (Theorems 5.3 and 5.4) *Let $S_1, S \in (0, \infty)$ with $S_1 < S$, $\alpha, \beta \in \mathbb{N}_0^3$ with $|\alpha + \beta| \leq 2$, $1 \leq j, k \leq 3$. Then*

$$\begin{aligned}
|\partial_y^\alpha \partial_z^\beta \mathcal{Z}_{jk}(\mathbf{y}, \mathbf{z})| &\leq \mathfrak{C}(S_1, S)(|\mathbf{y}| \cdot s_\tau(\mathbf{y}))^{-1-|\alpha+\beta|/2} \quad \textit{for } \mathbf{y} \in B_S^c, \mathbf{z} \in \overline{B_{S_1}}, \\
|\partial_y^\alpha \partial_z^\beta \mathcal{Z}_{jk}(\mathbf{y}, \mathbf{z})| &\leq \mathfrak{C}(S_1, S)(|\mathbf{z}| \cdot s_\tau(\mathbf{z}))^{-1-|\alpha+\beta|/2} \quad \textit{for } \mathbf{z} \in B_S^c,\ \mathbf{y} \in \overline{B_{S_1}}.
\end{aligned}$$

Proof: See Lemmas 11.6–11.7.

Lemma 11.9 ([35, Theorem 3.1], Theorems 5.2 and 10.6) *Let $k \in \{0, 1\}$, $R \in (0, \infty)$, $\mathbf{y}, \mathbf{z} \in B_R$ with $\mathbf{y} \neq \mathbf{z}$. Then*

$$\int_0^\infty (|\mathbf{y} - \tau t \boldsymbol{e}_1 - \mathrm{e}^{-t\Omega} \cdot \mathbf{z}|^2 + t)^{-3/2-k/2} \mathrm{d}t \leq \mathfrak{C}(R)|\mathbf{y} - \mathbf{z}|^{-1-k}.$$

Due to Lemma 11.5, this means for $\mathbf{y}, \mathbf{z}$ as above, and for $j, k \in \{1, 2, 3\}$, $\alpha \in \mathbb{N}_0^3$ with $|\alpha| \leq 1$ that

$$|\partial_y^\alpha \mathcal{Z}(\mathbf{y}, \mathbf{z})| + |\partial_z^\alpha \mathcal{Z}(\mathbf{y}, \mathbf{z})| \leq \mathfrak{C}(R)|\mathbf{y} - \mathbf{z}|^{-1-|\alpha|}.$$

Lemma 11.10 ([[38, Lemma 4.1], Lemma 5.14]) *Let* $j, k \in \{1, 2, 3\}$, $\mathbf{g} \in L^1(\partial\mathcal{D})$, *and put*

$$\mathbf{F}(\mathbf{y}) := \int_{\partial\mathcal{D}} \mathcal{Z}_{jk}(\mathbf{y}, \mathbf{z})\mathbf{g}(\mathbf{z})\mathrm{d}o_{\mathbf{z}} \quad \text{for } \mathbf{y} \in \overline{\mathcal{D}}.$$

Then $\mathbf{F} \in \boldsymbol{C}^1(\overline{\mathcal{D}})$ *and*

$$\partial_m \mathbf{F}(\mathbf{y}) = \int_{\partial\mathcal{D}} \partial y_m \mathcal{Z}_{jk}(\mathbf{y}, \mathbf{z}) g(\mathbf{z})\mathrm{d}o_{\mathbf{z}} \quad \text{for } 1 \le m \le 3,\ \mathbf{y} \in \overline{\mathcal{D}}. \tag{11.33}$$

The next lemma is well known. It was already used in [69], for example. For the convenience of the reader, we give a proof.

Lemma 11.11 *Let* $t \in \mathbb{R}$. *Then*

$$\mathrm{e}^{t\Omega} = \begin{pmatrix} 1 & 0 & 0 \\ 0 & \cos(t\sqrt{\varrho}) & -\sin(t\sqrt{\varrho}) \\ 0 & \sin(t\sqrt{\varrho}) & \cos(t\sqrt{\varrho}) \end{pmatrix}.$$

Proof Put

$$\Omega' := \begin{pmatrix} 0 & -1 \\ 1 & 0 \end{pmatrix}, \quad T := \begin{pmatrix} 1 & 1 \\ -\mathrm{i} & -\mathrm{i} \end{pmatrix}, \quad A := \begin{pmatrix} i & 0 \\ 0 & -\mathrm{i} \end{pmatrix}.$$

Obviously

$$\mathrm{e}^{tA} = \begin{pmatrix} \mathrm{e}^{t\mathrm{i}} & 0 \\ 0 & \mathrm{e}^{-t\mathrm{i}} \end{pmatrix}.$$

But $T \cdot A \cdot T^{-1} = \Omega'$, so $T \cdot \mathrm{e}^{tA} \cdot T^{-1} = \mathrm{e}^{t\Omega'}$. On computing the elements of the matrix on the left-hand side of the preceding equation, we obtain the lemma. □

Lemma 11.12 ([38, Lemma 4.2], Lemma 5.13) *Let* $j, k, l \in \{1, 2, 3\}$, $R > 0$, $g \in L^1(B_R)$, *and put*

$$\mathbf{F}(\mathbf{y}) := \int_{B_R} \partial_{z_l} \mathcal{Z}_{jk}(\mathbf{y}, \mathbf{z})\mathbf{g}(\mathbf{z})\mathrm{d}\mathbf{z} \quad \text{for } \mathbf{y} \in \overline{B_R}^c.$$

Then $\mathbf{F} \in C^1(\overline{B_R}^c)$ *and*

$$\partial_m \mathbf{F}(\mathbf{y}) = \int_{B_R} \partial_{y_m} \partial_{z_l} \mathcal{Z}_{jk}(\mathbf{y}, \mathbf{z})\mathbf{g}(\mathbf{z})\mathrm{d}\mathbf{z} \quad \text{for } \mathbf{y} \in \overline{B_R}^c,\ 1 \le m \le 3.$$

Lemma 11.13 *Let $f \in \mathcal{S}(\mathbb{R}^3)$, and put $\mathcal{N}(f)(\mathbf{x}) := \int_{\mathbb{R}^3} \mathfrak{E}(\mathbf{x}-\mathbf{y}) f(\mathbf{y})\,d\mathbf{y}$ for $\mathbf{x} \in \mathbb{R}^3$. Then $\mathcal{N}(f) \in C^\infty(\mathbb{R}^3)$ and $\partial^\alpha \mathcal{N}(f)(x) = \int_{\mathbb{R}^3} \mathfrak{E}(\mathbf{x}-\mathbf{y}) \partial^\alpha f(\mathbf{y})\,d\mathbf{y}$ for $\mathbf{x} \in \mathbb{R}^3$, $\alpha \in \mathbb{N}_0^3$.*

Lemma 11.14 ([50]) *Let $K \in C^\infty(\mathbb{R}^3 \times (0,\infty))$ and*

$$|\partial_t^l \partial_x^\alpha K(\mathbf{x},t)| \le \mathcal{C}(\alpha,l)(|\mathbf{x}|^2+t)^{-3/2-|\alpha|/2-l} e^{-|\mathbf{x}|^2/(8t)}$$

for $\mathbf{x} \in \mathbb{R}^3$, $t \in (0,\infty)$, $\alpha \in \mathbb{N}_0^3$, $l \in \mathbb{N}_0$. In particular $K(\cdot,t) \in L^1(\mathbb{R}^3) \cap \mathcal{S}(\mathbb{R}^3)$ for $t > 0$.

Theorem 11.5 ([119]) *$E_{jk} \in C^\infty(\mathbb{R}^3\backslash\{0\})$ and*

$$|\partial^\alpha E_{jk}(\mathbf{x})| \le \mathcal{C}(|\mathbf{x}|s(\mathbf{x}))^{-1-|\alpha|/2} \max\{1, |\mathbf{x}|^{-|\alpha|/2}\}$$

for $\mathbf{x} \in \mathbb{R}^3\backslash\{0\}$, $1 \le j,k \le 3$, $\alpha \in \mathbb{N}_0^3$ with $|\alpha| \le 1$. □

As a consequence of Theorem 11.5, we have $E_{jk} \in L^1_{loc}(\mathbb{R}^3)$ and $E_{jk}|B_1^c$ bounded ($1 \le j,k \le 3$). The same properties are obvious for $\mathfrak{E}$, $\mathfrak{O}$, and $\mathfrak{O}^{(\lambda)}$. Moreover, $|\Phi(\mathbf{x})| \le \mathcal{C}(1+|\mathbf{x}|)$ ($x \in \mathbb{R}^3$). In view of these observations, the Fourier transforms of these functions will be considered as tempered distributions (which, of course, will turn out to be represented by functions). Following Solonnikov [203, (40)], we use Lemmas 11.13 and 11.14 to introduce the velocity part $(\Lambda_{jk})_{1\le j,k\le 3}$ of a fundamental solution of the time-dependent Stokes system, setting

$$\Lambda_{jk}(\mathbf{x},t) := \delta_{jk} K(\mathbf{x},t) + \partial_j \partial_k \left(\int_{\mathbb{R}^3} \mathfrak{E}(\mathbf{x}-\mathbf{y})\, K(\mathbf{y},t) d\mathbf{y} \right)$$

($\mathbf{x} \in \mathbb{R}^3,\ t > 0,\ 1 \le j,k \le 3$).

Lemma 11.15 ([203, Lemma 13], [202]) *$\Lambda_{jk} \in C^\infty(\mathbb{R}^3 \times (0,\infty))$ and*

$$|\partial_t^l \partial_x^\alpha \Lambda_{jk}(\mathbf{x},t)| \le \mathcal{C}(\alpha, l, (|\mathbf{x}|^2+t)^{-3/2-|\alpha|/2-l}$$

for $\mathbf{x} \in \mathbb{R}^3$, $t \in (0,\infty)$, $1 \le j,k \le 3$, $\alpha \in \mathbb{N}_0^3$, $l \in \mathbb{N}_0$. □

Lemma 11.15 yields that $\Lambda_{jk}(\ \cdot\ ;\ ,t) \in L^2(\mathbb{R}^3)$ but does not imply $\Lambda_{jk}(\cdot,t) \in L^1(\mathbb{R}^3)$ ($t > 0$). So the Fourier transform of this function should be understood either as a transform of an L^2-function or as a tempered distribution.

Corollary 11.2 *The function $\mathcal{Z}(\cdot,0)$ belongs to $C^1(\mathbb{R}^3\backslash\{0\})^{3\times 3}$.*

Let $S \in (0,\infty)$. Then $|\partial_x^\alpha \mathcal{Z}(\mathbf{x},0)| \le \mathcal{C}(S)(|\mathbf{x}|s(\mathbf{x}))^{-1-|\alpha|/2}$ for $\mathbf{x} \in B_S^c$, $\alpha \in \mathbb{N}_0^3$ with $|\alpha| \le 1$.

Moreover, $|\mathcal{Z}(\mathbf{x},0)| \le \mathcal{C}|\mathbf{x}|^{-1}$ for $\mathbf{x} \in B_1\backslash\{0\}$. □

Proof The first part of the corollary follows from Lemmas 5.1 and 2.18. The last estimate is a consequence of Corollary 10.11 and Theorem 5.1. □

Finally, we need some pointwise estimates of convolutions of the type $\eta_{-b}^{-a} * \eta_{-d}^{-c} \leq c\eta_{-f}^{-e}$, where $\eta_{\beta}^{\alpha}(\mathbf{x}) := (1 + |\mathbf{x}|)^{\alpha} s(\mathbf{x})^{\beta}$. The logarithmic factor on the right-hand side of these estimates is also admitted. For the systematic study for all real values of parameters a, b, c, d of convolutions in $\mathbb{R}^3$, we refer to [58, 119]. We need these types of estimates in two concrete cases.

Lemma 11.16 *Let $\gamma \in (1/4, \infty)$. Then there is a constant $C(\gamma) > 0$ such that for all $\mathbf{x} \in \mathbb{R}^3$*

$$\int_{\mathbb{R}^3} [(1 + |\mathbf{x} - \mathbf{y}|), s(\mathbf{x} - \mathbf{y})]^{-3/2} [(1 + |\mathbf{y}|) s(\mathbf{y})]^{-\gamma} \mathrm{d}\mathbf{y}$$
$$\leq C(\gamma)(1 + |\mathbf{x}|)^{-c} s(\mathbf{x})^{-d} \ln^k (2 + |\mathbf{x}|),$$

where

$$c := \begin{cases} \gamma - 1/2 & \text{if } \gamma \in (1/4,\ 2] \\ 3/2 & \text{if } \gamma \in (2,\ +\infty) \end{cases} \qquad d := \begin{cases} \gamma & \text{if } \gamma \in (1/4,\ 3/2] \\ 3/2 & \text{if } \gamma \in (3/2,\ +\infty) \end{cases}$$

$$k := \begin{cases} 0 & \text{if } \gamma \neq 2 \\ 1 & \text{if } \gamma = 2. \end{cases}$$

Lemma 11.17 (Theorem 3.2, [119]) *There exists a constant $C > 0$ such that for all $\mathbf{x} \in \mathbb{R}^3$*

$$\int_{\mathbb{R}^3} [(1 + |\mathbf{x} - \mathbf{y}|) s(\mathbf{x} - \mathbf{y})]^{-2} [(1 + |\mathbf{y}|) s(\mathbf{y})]^{-2} \mathrm{d}\mathbf{y}$$
$$\leq C[(1 + |\mathbf{x}|) s(\mathbf{x})]^{-2} \ln(2 + |\mathbf{x}|).$$

The starting point of our considerations will be the following theorem about the integrability and pointwise decays of the velocity and its gradient, where the velocity is a solution of the rotational Navier–Stokes equations:

Theorem 11.6 ([39, Theorem 1.1], Theorem 10.1) *Let $\tau \in (0, \infty)$, $\boldsymbol{\omega} \in \mathbb{R}^3 \backslash \{\mathbf{0}\}$, $\mathcal{D}^c \subset \mathbb{R}^3$ be open and bounded. Take $\gamma, S_1 \in (0, \infty)$, $p_0 \in (1, \infty)$, $A \in (2, \infty)$, $B \in [0, 3/2]$, $\mathbf{f} \mathbb{R}^3 \mapsto \mathbb{R}^3$ measurable with $\overline{\mathcal{D}}^c \subset B_{S_1}$, $A + \min\{B, 1\} > 3$, $A + B \geq 7/2$, $\mathbf{f}|B_{S_1} \in \boldsymbol{L}^{p_0}(B_{S_1})$,*

$$|\mathbf{f}(\mathbf{y})| \leq \gamma |\mathbf{y}|^{-A} s_{\tau}(\mathbf{y})^{-B} \quad \text{for } \mathbf{y} \in B_{S_1}^c. \tag{11.34}$$

Let $\mathbf{u} \in \boldsymbol{L}^6(\overline{\mathcal{D}}) \cap \boldsymbol{W}^{1,1}_{\mathrm{loc}}(\overline{\mathcal{D}})$, $\pi \in L^2_{\mathrm{loc}}(\overline{\mathcal{D}})$, $\nabla\mathbf{u} \in L^2(\overline{\mathcal{D}})^9$, *and*

$$\int_{\overline{\mathcal{D}}} [\nabla\mathbf{u}\cdot\nabla\boldsymbol{\varphi} + (\tau\partial_1\mathbf{u} + \tau(\mathbf{u}\cdot\nabla)\mathbf{u} \\ - \boldsymbol{\omega}\wedge\mathbf{z})\cdot\nabla\mathbf{u} + \boldsymbol{\omega}\wedge\mathbf{u})\cdot\boldsymbol{\varphi} - \pi\,\mathrm{div}\,\boldsymbol{\varphi} - \mathbf{f}\cdot\boldsymbol{\varphi}]\,d\mathbf{z} = 0, \quad \mathrm{div}\,\mathbf{u} = 0 \tag{11.35}$$

for $\boldsymbol{\varphi} \in \boldsymbol{C}^\infty_0(\overline{\mathcal{D}})$. *Let* $S \in (S_1, \infty)$. *Then*

$$|\partial^\alpha\mathbf{u}(\mathbf{y})| \le D(|\mathbf{y}|s_\tau(\mathbf{y}))^{-1-|\alpha|/2} \quad \textit{for } \mathbf{x} \in B^c_S,\ \alpha \in \mathbb{N}^3_0,\ |\alpha| \le 1, \tag{11.36}$$

with the constant D depending on τ, ρ, γ, S_1, p_0, A, B, $\|\mathbf{f}|B_{S_1}\|_1$, $\mathbf{u}$, π, S, *and an arbitrary but fixed number* $S_0 \in (0, S_1)$ *with* $\overline{\mathcal{D}} \subset B_{S_0}$. □

Let $p \in (1,\infty)$, $q \in (1,2)$, $\mathbf{f} \in \boldsymbol{L}^p_{\mathrm{loc}}(\mathbb{R}^3)$ with $\mathbf{f}|^c_S \in \boldsymbol{L}^q(B^c_S)$ for some $S \in (0,\infty)$.

For $\mathbf{y} \in \mathbb{R}^3$, $j \in \{1,2,3\}$, we set

$$\mathfrak{R}_j(\mathbf{f})(\mathbf{y}) := \int_{\mathbb{R}^3} \sum_{k=1}^{3} \mathcal{Z}_{jk}(\mathbf{y},\mathbf{z}) f_k(\mathbf{z})d\mathbf{z}.$$

According to [36, Lemma 3.1], the integral appearing in the definition of $\mathfrak{R}_j(\mathbf{f})$ is well defined at least for almost every $\mathbf{y} \in \mathbb{R}^3$. If $\mathbf{f}$ is a function on $\overline{\mathcal{D}}$, the function $\mathbf{f}$ in the previous definition is to be replaced by the extension of f by zero to $\mathbb{R}^3$.

11.5 Representation Formula for the Navier–Stokes System

Theorem 11.7 *Let* $\mathbf{u} \in \boldsymbol{W}^{1,1}_{\mathrm{loc}}(\overline{\mathcal{D}}) \cap \boldsymbol{L}^6(\overline{\mathcal{D}})$, $\mathbf{u}_{|\partial\mathcal{D}} \in \boldsymbol{W}^{2-1/p,p}(\partial\mathcal{D})$ *with* $\nabla\mathbf{u} \in L^2(\overline{\mathcal{D}})^9$. *Let* $\pi \in L^2(\overline{\mathcal{D}})$, $p \in (1,\infty)$, $q \in (1,2)$, $f : \overline{\mathcal{D}} \mapsto \mathbb{R}^3$ *be a function with* $\mathbf{f}|\mathcal{D}_T \in L^p(\mathcal{D}_T)^3$ *for* $T \in (0,\infty)$ *with* $\overline{\mathcal{D}}^c \subset B_T$, $\mathbf{f}|B^c_S \in L^q(B^c_S)^3$ *for some* $S \in (0,\infty)$ *with* $\overline{\mathcal{D}}^c \subset B_S$.

Suppose that the pair $(\mathbf{u}, \pi)$ *is a weak solution of the Navier–Stokes system with Oseen and rotational terms and with the right-hand side* f, *that is,*

$$\int_{\overline{\mathcal{D}}} \big(\nabla\mathbf{u}\cdot\nabla\boldsymbol{\varphi} + \big(\tau(\mathbf{u}\cdot\nabla)\mathbf{u} + \tau\partial_1\mathbf{u} - (\boldsymbol{\omega}\wedge\mathbf{z})\cdot\nabla\mathbf{u} + \boldsymbol{\omega}\wedge\mathbf{u}\big)\cdot\boldsymbol{\varphi} + \pi\,\mathrm{div}\,\boldsymbol{\varphi}\big)\,d\mathbf{z} \\ = \int_{\overline{\mathcal{D}}} \mathbf{f}\cdot\boldsymbol{\varphi}\,d\mathbf{z} \quad \textit{for } \boldsymbol{\varphi} \in \boldsymbol{C}^\infty_0(\overline{\mathcal{D}}), \quad \mathrm{div}\,\mathbf{u} = 0.$$

Then

$$u_j(\mathbf{y}) = \mathfrak{R}_j(\mathbf{f} - \tau(\mathbf{u}\cdot\nabla)\mathbf{u})(\mathbf{y}) + \mathfrak{B}_j(\mathbf{u},\pi)(\mathbf{y}) \tag{11.37}$$

for $j \in \{1,2,3\}$, *a.e.* $\mathbf{y} \in \overline{\mathcal{D}}$, *where* $\mathfrak{B}_j(\mathbf{u},\pi)$ *was defined in* (5.96). □

Proof Since $\mathbf{u} \in \boldsymbol{L}^6(\overline{\mathcal{D}})$ and $\nabla\mathbf{u} \in L^2(\overline{\mathcal{D}})^9$, Hölder's inequality yields $\tau(\mathbf{u}\cdot\nabla)\mathbf{u} \in \boldsymbol{L}^{3/2}(\overline{\mathcal{D}})$. It further follows that the term $\tau\partial_1\mathbf{u}(\mathbf{z}) - (\boldsymbol{\omega}\wedge\mathbf{z})\cdot\nabla\mathbf{u}(\mathbf{z}) + \boldsymbol{\omega}\wedge\mathbf{u}(\mathbf{z})$, considered as a function of $\mathbf{z} \in \mathcal{D}_T$, belongs to $\boldsymbol{L}^2(\mathcal{D}_T)$ for any $T \in (0,\infty)$ with $\mathcal{D}^c \subset B_T$. Therefore, putting

$$\mathbf{F}(\mathbf{z}) := \mathbf{f}(\mathbf{z}) - \tau(\mathbf{u}(\mathbf{z})\cdot\nabla)\mathbf{u}(\mathbf{z}) - \tau\partial_1\mathbf{u}(\mathbf{z}) + (\boldsymbol{\omega}\wedge\mathbf{z})\cdot\nabla\mathbf{u}(\mathbf{z}) - \boldsymbol{\omega}\wedge\mathbf{u}(\mathbf{z})$$

for $\mathbf{z} \in \overline{\mathcal{D}}$, we see that $\mathbf{F}|\mathcal{D}_T \in \boldsymbol{L}^{\min\{p,3/2\}}(\mathcal{D}_T)$ for T as above. Thus, considering the pair $(\mathbf{u},\pi)$ as a weak solution (in the sense of [95, (IV.1.3)]) of the Stokes system with the right-hand side F, we may refer to [95, Theorem IV.4.1] (interior regularity for the Stokes system), to obtain that

$$\mathbf{u}|\mathcal{D}_T \in \boldsymbol{W}^{2,\min\{p,3/2\}}_{\text{loc}}(\mathcal{D}_T), \quad \pi|\mathcal{D}_T \in W^{1,\min\{p,3/2\}}_{\text{loc}}(\mathcal{D}_T) \quad (T \text{ as above}),$$
$$-\Delta\mathbf{u} + \nabla\pi = \mathbf{F}, \quad \text{hence} \quad \mathcal{L}(\mathbf{u}) + \nabla\pi = \mathbf{f} - \tau(\mathbf{u}\cdot\nabla)\mathbf{u}.$$

As $\tau(\mathbf{u}\cdot\nabla)\mathbf{u} \in \boldsymbol{L}^{3/2}(\overline{\mathcal{D}})$, we now conclude that $\mathcal{L}(\mathbf{u}) + \nabla\pi|\mathcal{D}_T \in \boldsymbol{L}^{\min\{p,3/2\}}(\mathcal{D}_T)$ for T as above, so $(\mathbf{u},\pi) \in \mathfrak{M}_{\min\{p,3/2\}}$, see Definition 5.1. The preceding observations mean that the assumptions of Theorem 5.13 are satisfied with p and p_1 replaced by $\min\{p,3/2\}$ and q, respectively, and with $p_2 = 3/2$. Thus Eq. (11.37) follows from Theorem 5.13. □

Theorem 11.8 *Let* $\mathbf{u} \in \boldsymbol{W}^{1,1}_{\text{loc}}(\overline{\mathcal{D}}) \cap \boldsymbol{L}^6(\overline{\mathcal{D}})$ *with* $\nabla\mathbf{u} \in \boldsymbol{L}^2(\overline{\mathcal{D}})^9$. *Let* $p \in (1,\infty)$, $q \in (1,2)$, $f : \overline{\mathcal{D}} \mapsto \mathbb{R}^3$ *a function with* $\mathbf{f}|\mathcal{D}_T \in \boldsymbol{L}^p(\mathcal{D}_T)$ *for* $T \in (0,\infty)$ *with* $\overline{\mathcal{D}}^c \subset B_T$, $\mathbf{f} \mid B_S^c \in \boldsymbol{L}^q(B_S^c)$ *for some* $S \in (0,\infty)$ *with* $\overline{\mathcal{D}} \subset B_S$. *Further assume that* $\mathbf{u}|\partial\mathcal{D} \in W^{2-1/p,p}(\partial\mathcal{D})$ *and* $\pi : \overline{\mathcal{D}} \mapsto \mathbb{R}$ *is a function with* $\pi|\mathcal{D}_T \in L^p(\mathcal{D}_T)$ *for* T *as above.*

Suppose that the pair $(\mathbf{u},\pi)$ *is a weak solution of the Navier–Stokes system with Oseen and rotational terms and with right-hand side* $\mathbf{f}$ *in the sense of (11.35). Then* $\mathbf{u} \in W^{2,\min\{p,3/2\}}(\mathcal{D}_T)^3$, $\pi \in W^{1,\min\{p,3/2\}}(\mathcal{D}_T)$ *for any* $T \in (0,\infty)$ *with* $\overline{\mathcal{D}}^c \subset B_T$,

$$u_j(\mathbf{y}) = \mathfrak{R}_j(\mathbf{f} - \tau(\mathbf{u}\cdot\nabla)\mathbf{u})(\mathbf{y}) + \mathfrak{B}_j(\mathbf{u},\pi)(\mathbf{y}) \tag{11.38}$$

for $j \in \{1, 2, 3\}$, *a.e.* $\mathbf{y} \in \overline{\mathcal{D}}$, *where* $\mathfrak{B}_j(\mathbf{u}, \pi)$ *is defined by*

$$\begin{aligned}
&\mathfrak{B}_j(\mathbf{u}, \pi)(\mathbf{y}) \\
&:= \int_{\partial\mathcal{D}} \sum_{k=1}^{3} \Bigg[\sum_{l=1}^{3} (\mathcal{Z}_{jk}(\mathbf{y}, \mathbf{z})(-\partial_l u_k(\mathbf{z}) + \delta_{kl}\pi(\mathbf{z}) + u_k(\mathbf{z})(\tau \boldsymbol{e}_1 - \boldsymbol{\omega} \wedge \mathbf{z})_l) \\
&\quad + \partial z_l \mathcal{Z}_{jk}(\mathbf{y}, \mathbf{z}) u_k(\mathbf{z})) n_l^{(\mathcal{D})}(\mathbf{z}) + E_{4j}(\mathbf{y} - \mathbf{z}) u_k(\mathbf{z}) n_k^{(\mathcal{D})}(\mathbf{z}) \Bigg] do_{\mathbf{z}} \qquad (11.39)
\end{aligned}$$

for $\mathbf{y} \in \overline{\mathcal{D}}$. □

Proof Since $\mathbf{u} \in \boldsymbol{L}^6(\overline{\mathcal{D}})$ and $\nabla\mathbf{u} \in L^2(\overline{\mathcal{D}})$, we have $(\mathbf{u} \cdot \nabla)\mathbf{u} \in \boldsymbol{L}^{3/2}(\overline{\mathcal{D}})$; hence $\mathbf{f} - \tau \cdot \mathbf{u} \cdot \nabla\mathbf{u}|_{\mathcal{D}_T} \in \boldsymbol{L}^{\min\{p,3/2\}}(\mathcal{D}_T)$ for T as in the theorem. Using theory of the interior regularity of the Stokes problem, we thus get $\mathbf{u} \in \boldsymbol{W}^{2,\min\{p,3/2\}}_{\mathrm{loc}}(\overline{\mathcal{D}})$, $\pi \in W^{1,\min\{p,3/2\}}_{\mathrm{loc}}(\overline{\mathcal{D}})$, so we may conclude that $(\mathbf{u}, \pi) \in \mathfrak{M}_{\min\{p,3/2\}}$. Moreover $\mathbf{f} - \tau \cdot \mathbf{u} \cdot \nabla\mathbf{u}|_{B_S^c} \in \boldsymbol{L}^q(B_S^c) + \boldsymbol{L}^{3/2}(B_S^c)$. Thus (11.38) follows from Theorem 5.21. □

In comparison with the linear case we will need some additional lemma:

Lemma 11.18 *Let* $\varphi \in W^{1,1}_{\mathrm{loc}}(U)$ *for some open set* $U \subset \mathbb{R}^3$ *and* $A \in \mathbb{R}^{3\times3}$ *such that* $A^{-1} = A^T$. *Then,*

$$A\nabla_z(\varphi(A\mathbf{z})) = \nabla\varphi(A\mathbf{z}).$$

Proof Indeed,

$$\frac{\partial}{\partial z_l}(\varphi(A\mathbf{z})) = \sum_{k=1}^{3} \partial_k\varphi(A\mathbf{z}) \frac{\partial(A\mathbf{z})_k}{\partial z_l} = \sum_{k=1}^{3} \partial_k\varphi(A\mathbf{z}) A_{kl} = \sum_{k=1}^{3} A^T_{lk} \partial_k\varphi(A\mathbf{z}),$$

i.e., $\nabla_z(\varphi(A\mathbf{z})) = A^T\nabla\varphi(A\mathbf{z})$, which gives the mentioned formula. □

Corollary 11.3 *In the situation of Theorem 11.6, we get for* $\mathbf{z} \in \overline{B_{S_1}}^c$ *that*

$$\begin{aligned}
\sum_{l=1}^{3} (u_l \partial_l \mathbf{u})(\mathrm{e}^{t\Omega}\mathbf{z}) &= \sum_{l=1}^{3} \partial_l (u_l \mathbf{u})(\mathrm{e}^{t\Omega}\mathbf{z}) \\
&= \sum_{l=1}^{3}\sum_{k=1}^{3} (\mathrm{e}^{t\Omega})_{lk} \frac{\partial}{\partial z_k}[(u_l\mathbf{u})(\mathrm{e}^{t\Omega}\mathbf{z})] = \sum_{l=1}^{3} (\mathrm{e}^{t\Omega}\nabla_z)_l [(u_l\mathbf{u})(\mathrm{e}^{t\Omega}\mathbf{z})].
\end{aligned}$$

Lemma 11.19 *In the situation of Theorem 11.6, we have*

$$\int_{\overline{\mathcal{D}}} |\partial_x^\alpha \mathcal{Z}(\mathbf{x}, \mathbf{y})[(\mathbf{u} \cdot \nabla)\mathbf{u}](\mathbf{y})| \, d\mathbf{y} < \infty \qquad (11.40)$$

for $\mathbf{x} \in \overline{B_{S_1}}^c$, $\alpha \in \mathbb{N}_0^3$ *with* $|\alpha| \leq 1$. *Moreover, the function* $\mathfrak{V}(\mathbf{x}) := \int_{\overline{\mathcal{D}}} \mathcal{Z}(\mathbf{x}, \mathbf{y})(\mathbf{u} \cdot \nabla)\mathbf{u}](\mathbf{y})dy$ $(\mathbf{x} \in \overline{B_{S_1}}^c)$ *belongs to* $\boldsymbol{C}^1(\overline{B_{S_1}}^c)$ *with*

$$\partial^\alpha \mathfrak{V}(bfx) = \int_{\overline{\mathcal{D}}} \partial_x^\alpha \mathcal{Z}(\mathbf{x}, \mathbf{y})[(\mathbf{u} \cdot \nabla)\mathbf{u}(\mathbf{y})|\,\mathrm{d}\mathbf{y} \tag{11.41}$$

for $\mathbf{x}$, α *as in* (11.40). □

Proof Let $U \subset \mathbb{R}^3$ be open and bounded with $\overline{U} \subset \overline{B_{S_1}}^c$. It is enough to show that (11.40) holds for $x \in U$, $\mathfrak{V}|U \in \boldsymbol{C}^1(U)$, and (11.41) is valid for $\mathbf{x} \in U$.

Due to our assumptions on U, we may choose $R, S \in (S_1, \infty)$ such that $\overline{B_S} \cap \overline{U} = \emptyset$ and $\overline{U} \subset B_R$. In particular we have $\operatorname{dist}(B_S, U) > 0$ and $\operatorname{dist}(U, B_R^c) > 0$. This observation and Lemma 11.16 give us $|\partial_{\mathbf{x}}^\alpha \mathcal{Z}(\mathbf{x}, \mathbf{y})| \leq C_0$ for $\mathbf{x} \in U$, $\mathbf{y} \in B_S \backslash \overline{\mathcal{D}}^c$, $\alpha \in \mathbb{N}_0^3$ with $|\alpha| \leq 1$, where C_0 is independent of $\mathbf{x}$ and $\mathbf{y}$. We further observe that $(\mathbf{u} \cdot \nabla)\mathbf{u} \in \boldsymbol{L}^{3/2}(\overline{\mathcal{D}})$, hence $(\mathbf{u} \cdot \nabla)\mathbf{u}|B_S \backslash \overline{\mathcal{D}}^c \in \boldsymbol{L}^1(B_S \backslash \overline{\mathcal{D}}^c)$. Lemma 11.8 and (11.36) yield that $|\partial_x^\alpha \mathcal{Z}(\mathbf{x}, \mathbf{y})[(\mathbf{u} \cdot \nabla)\mathbf{u}](\mathbf{y})| \leq C_1 \cdot |\mathbf{y}|^{-7/2-|\alpha|/2}$ for $\mathbf{x} \in U$, $\mathbf{y} \in B_R^c$ with C_1 again being independent of $\mathbf{x}$ and $\mathbf{y}$. In view of the last statement of Lemma 11.7, we may thus conclude by Lebesgue's theorem that the function

$$y \mapsto \partial_x^\alpha \mathcal{Z}(\mathbf{x}, \mathbf{y})[(\mathbf{u} \cdot \nabla)\mathbf{u}](\mathbf{y}), \quad \mathbf{y} \in A := (B_S \backslash \overline{\mathcal{D}}^c) \cup B_R^c$$

is integrable for $\mathbf{x} \in U$, $\alpha \in \mathbb{N}_0^3$ with $|\alpha| \leq 1$, that the function

$$\mathfrak{V}^{(I)}(\mathbf{x}) := \int_A \mathcal{Z}(\mathbf{x}, \mathbf{y})[(\mathbf{u} \cdot \nabla)\mathbf{u}](\mathbf{y})\,\mathrm{d}\mathbf{y}, \quad \mathbf{x} \in U,$$

belongs to $\boldsymbol{C}^1(U)$, and that $\partial^\alpha \mathfrak{V}^{(I)}(\mathbf{x}) = \int_A \partial_x^\alpha \mathcal{Z}(\mathbf{x}, \mathbf{y})[(\mathbf{u} \cdot \nabla)\mathbf{u}](\mathbf{y})\,\mathrm{d}\mathbf{y}$ for $\mathbf{x}$, α as before.

Let $\varphi \in C_0^\infty(\mathbb{R}^3)$ with $0 \leq \varphi \leq 1$, $\varphi|B_{1/2} = 0$, $\varphi|B_1^c = 1$, and define $\varphi_\delta(\mathbf{x}) := \varphi(\delta^{-1}\mathbf{x})$ for $\mathbf{x} \in \mathbb{R}^3$, $\delta > 0$. Then $\varphi_\delta \in C_0^\infty(\mathbb{R}^3)$, $0 \leq \varphi_\delta \leq 1$, $\varphi_\delta|B_{\delta/2} = 0$, $\varphi_\delta|B_\delta^c = 1$, and $|\nabla\varphi_\delta(x)| \leq C\delta^{-1}$ for $\mathbf{x} \in \mathbb{R}^3$, $\delta > 0$.

Using Lemma 11.9 and Theorem 11.6, we see there are constants C_2, C_3 with

$$\begin{aligned} &|\partial_x^\alpha(\mathcal{Z}(\mathbf{x}, \mathbf{y})\varphi_\delta(\mathbf{x} - \mathbf{y}))[(\mathbf{u} \cdot \nabla)\mathbf{u}](\mathbf{y})| \\ &\leq C_2(|\mathbf{x} - \mathbf{y}|^{-2} + \delta^{-1}|\mathbf{x} - \mathbf{y}|^{-1})\chi_{(\delta/2,\infty)}(|\mathbf{x} - \mathbf{y}|) \leq C_3\delta^{-2} \end{aligned} \tag{11.42}$$

for $\mathbf{x} \in U$, $\mathbf{y} \in B_R \backslash B_S$, $\alpha \in \mathbb{N}_0^3$ with $|\alpha| \leq 1$. In addition, if $\mathbf{y} \in B_R \backslash B_S$, the function $\mathbf{x} \mapsto \mathcal{Z}(\mathbf{x}, \mathbf{y})\varphi_\delta(\mathbf{x} - \mathbf{y})[(\mathbf{u} \cdot \nabla)\mathbf{u}](\mathbf{y})$, $\mathbf{x} \in U$, is continuously differentiable, as follows from Lemma 2.6. Now we may conclude from Lebesgue's theorem that the function $\mathbf{y} \mapsto \partial_{\mathbf{x}}^\alpha(\mathcal{Z}(\mathbf{x}, \mathbf{y})\varphi_\delta(\mathbf{x} - \mathbf{y}))[(\mathbf{u} \cdot \nabla)\mathbf{u}](\mathbf{y})$, $\mathbf{y} \in B_R \backslash B_S$, is integrable for any $\delta > 0$, $\mathbf{x} \in U$, $\alpha \in \mathbb{N}_0^3$ with $|\alpha| \leq 1$, the function

$$\mathfrak{B}_\delta(\mathbf{x}) := \int_{B_R \backslash B_S} \mathcal{Z}(\mathbf{x}, \mathbf{y})\varphi_\delta(\mathbf{x} - \mathbf{y})[(\mathbf{u} \cdot \nabla)\mathbf{u}](\mathbf{y})\,\mathrm{d}\mathbf{y}, \quad \mathbf{x} \in U,$$

belongs to $C^1(U)^3$ for any $\delta > 0$, and

$$\partial^\alpha \mathfrak{B}_\delta(\mathbf{x}) = \int_{B_R \setminus B_S} \partial_x^\alpha (\mathcal{Z}(\mathbf{x}, \mathbf{y})\varphi_\delta(\mathbf{x} - \mathbf{y}))[(\mathbf{u} \cdot \nabla)\mathbf{u}](\mathbf{y})\, \mathrm{d}\mathbf{y}$$

for δ, $\mathbf{x}$, α as before. Proceeding as in (11.42), we further obtain

$$\begin{aligned}\int_{B_R \setminus B_S} &|\partial_x^\alpha (\mathcal{Z}(\mathbf{x}, \mathbf{y})\varphi_\delta(\mathbf{x} - \mathbf{y}) - \mathcal{Z}(\mathbf{x}, \mathbf{y}))[(\mathbf{u} \cdot \nabla)\mathbf{u}](\mathbf{y})|\, \mathrm{d}\mathbf{y} \\ &\le C_4 \int_{B_R \setminus B_S} \chi_{B_\delta}(\mathbf{x} - \mathbf{y})(|\mathbf{x} - \mathbf{y}|^{-2} + \delta^{-1}|\mathbf{x} - \mathbf{y}|^{-1})\, \mathrm{d}\mathbf{y} \le C_5\delta\end{aligned}$$

for $\mathbf{x} \in U$, $\alpha \in \mathbb{N}_0^3$ with $|\alpha| \le 1$ with the C_4, C_5 denoting constants independent of δ and $\mathbf{x}$. Therefore, by an argument involving uniform convergence of $\mathfrak{B}_\delta$ and $\nabla\mathfrak{B}_\delta$ for $\delta \downarrow 0$, we may conclude that the function

$$\mathfrak{V}^{(II)}(x) := \int_{B_R \setminus B_S} \mathcal{Z}(\mathbf{x}, \mathbf{y})[(\mathbf{u} \cdot \nabla)\mathbf{u}](\mathbf{y})\, \mathrm{d}\mathbf{y}, \quad \mathbf{x} \in U,$$

belongs to $\boldsymbol{C}^1(U)$, and

$$\partial^\alpha \mathfrak{V}^{(II)}(\mathbf{x}) = \int_{B_R \setminus B_S} \partial_x^\alpha \mathcal{Z}(\mathbf{x}, \mathbf{y})[(\mathbf{u} \cdot \nabla)\mathbf{u}](\mathbf{y})\, \mathrm{d}\mathbf{y}$$

for $\mathbf{x} \in U$, $\alpha \in \mathbb{N}_0^3$ with $|\alpha| \le 1$. Since $\mathfrak{V}(\mathbf{x}) = \mathfrak{V}^{(I)}(\mathbf{x}) + \mathfrak{V}^{(II)}(\mathbf{x})$ for $\mathbf{x} \in U$, the proof of the lemma is complete. □

11.6 Proof of Theorem 11.2

The aim of this part is to show the leading term of the velocity and its gradient for the Navier–Stokes problem with rotation.

Remark 11.3 In Theorem 11.2, the estimate presented in Theorem 5.24 for the linear case is extended to the nonlinear one. Note that by [124, (3.9)], the function $\mathcal{Z}(\mathbf{x}, 0)$ in the leading term on the right-hand side of (11.18) corresponds to the time integral of a fundamental solution of the evolutionary Oseen system multiplied by a rotation depending on time. □

Proof of Theorem 11.2 We will denote the term of (11.18) contained $L_j(\boldsymbol{x})$ in braces $\{\dots\}$ "the leading term," and $\mathfrak{F}$ "the remainder." From Theorem 11.7 we have

$$u_j(\mathbf{x}) = \mathfrak{R}_j(\mathbf{f} - \tau(\mathbf{u} \cdot \nabla)\mathbf{u})(\mathbf{x}) + \mathfrak{B}_j(\mathbf{u}, \pi)(\mathbf{x}), \quad j \in \{1, 2, 3\} \text{ for a.e. } x \in \overline{\mathcal{D}}, \tag{11.43}$$

where $\mathfrak{B}_j(\mathbf{u}, \pi)$ was defined in (11.39).

We put

$$\begin{aligned}
\beta_k :=& \beta_k^{(I)} - \tau \beta_k^{(II)} \\
\beta_k^{(I)} :=& \int_{B_{S_1}} f_k(\mathbf{y})\, \mathrm{d}\mathbf{y} \\
&+ \int_{\partial\mathcal{D}} \sum_{l=1}^{3} (-\partial_l u_k(\mathbf{y}) + \delta_{kl}\pi(\mathbf{y}) + u_k(\mathbf{y})(\tau \boldsymbol{e}_1 - \boldsymbol{\omega} \wedge \mathbf{y})_l) n_l^{(\mathcal{D})}(\mathbf{y}) \mathrm{d}o_{\mathbf{y}} \\
\beta_k^{(II)} :=& \int_{\partial\overline{\mathcal{D}}} \sum_{m=1}^{3} (n_m u_m u_k)(\mathbf{y}) \mathrm{d}o_{\mathbf{y}}
\end{aligned}$$

for $1 \le k \le 3$. By the definition of β_k the leading term in formula (11.18) is determined. Because (11.18) is in fact rearrangement of formula (11.43), we now define the value $\mathfrak{F}_j$ as the difference of the right-hand side of the representation formula (11.43) minus the leading term. We will distinguish $\mathfrak{F}^{(I)}$ coming from the linear terms and $\mathfrak{F}^{(II)}$ arising from the nonlinear part, i.e., from $\mathfrak{R}_j((\mathbf{u} \cdot \nabla)\mathbf{u})$:

$$\begin{aligned}
\mathfrak{F}_j(\mathbf{x}) :=& \mathfrak{F}_j^{(I)}(\mathbf{x}) - \tau \mathfrak{F}_j^{(II)}(\mathbf{x}), \\
\mathfrak{F}_j^{(I)}(\mathbf{x}) :=& \int_{B_{S_1}} \left(\sum_{k=1}^{3} [(\mathcal{Z}_{jk}(\mathbf{x}, \mathbf{y}) - \mathcal{Z}_{jk}(\mathbf{x}, 0)) f_k(\mathbf{y})] \right) \mathrm{d}\mathbf{y} \\
&+ \int_{\partial\mathcal{D}} \sum_{k=1}^{3} ((\mathcal{Z}_{jk}(\mathbf{x}, \mathbf{y}) - \mathcal{Z}_{jk}(\mathbf{x}, 0)) \\
&\sum_{l=1}^{3} (-\partial_l u_k(\mathbf{y}) + \delta_{kl}\pi(\mathbf{y}) + u_k(\mathbf{y})(\tau \boldsymbol{e}_1 - \boldsymbol{\omega} \wedge \mathbf{y})_l) n_l^{(\mathcal{D})}(\mathbf{y}) \\
&+ (E_{4j}(\mathbf{x} - \mathbf{y}) - E_{4j}(\mathbf{x})) u_k(\mathbf{y}) n_k^{(\mathcal{D})}(\mathbf{y})) \mathrm{d}o_{\mathbf{y}} \\
&+ \int_{\partial\mathcal{D}} \sum_{k,l=1}^{3} \partial y_l\, \mathcal{Z}_{jk}(\mathbf{x}, \mathbf{y}) u_k(\mathbf{y}) n_l^{(\mathcal{D})}(\mathbf{y}) \mathrm{d}o_{\mathbf{y}},
\end{aligned} \tag{11.44}$$

$$\begin{aligned}
\mathfrak{F}_j^{(II)}(\mathbf{x}) :=& \int_{\overline{\mathcal{D}}^c} \sum_{k,l=1}^{3} \mathcal{Z}_{jk}(\mathbf{x}, \mathbf{y})(u_l \partial_l u_k)(\mathbf{y}) \mathrm{d}\mathbf{y} \\
&- \int_{\partial\mathcal{D}} \sum_{k,l=1}^{3} \mathcal{Z}_{jk}(\mathbf{x}, 0)(n_l u_l\, u_k)(\mathbf{y}) \mathrm{d}o_{\mathbf{y}}
\end{aligned} \tag{11.45}$$

for $\mathbf{x} \in \overline{B_{S_1}}^c$, $1 \le j \le 3$. Then by (11.43) we get (11.18).

The assertion of the theorem will be proved in five steps:

1. Estimates and continuity of $\partial^\alpha \mathfrak{F}^{(I)}$, where $|\alpha| = 0$ or $|\alpha| = 1$

By exactly the same proof as it was given in the proof of [34, Theorem 1.1], we obtain that $\mathfrak{F}^{(I)} \in C^1(\overline{B_{S_1}}^c)^3$ and

$$|\partial^\alpha \mathfrak{F}^{(I)}(\mathbf{x})| \leq \mathfrak{C}(\|f\|_1 + \|\nabla u|\partial\mathcal{D}\|_1 + \|\pi|\partial\mathcal{D}\|_1 + \|\mathbf{u}|\partial\mathcal{D}\|_1)(|\mathbf{x}|s_\tau(\mathbf{x}))^{-3/2-|\alpha|/2}$$

for $S \in (S_1, \infty)$, $\mathbf{x} \in \overline{B_S}^c$, $\alpha \in \mathbb{N}_0^3$ with $|\alpha| \leq 1$, with $\mathfrak{C}$ depending on τ, ω, p, S_1, and S.

2. C^1-continuity of $\mathfrak{F}^{II}$

By Lemma 11.19, the function $\mathfrak{F}^{(II)} \in C^1(\overline{B_{S_1}}^c)^3$ and the first order derivatives may be moved into the volume integral appearing on the right-hand side of (11.45).

3. Estimates of $\partial^\alpha \mathfrak{F}^{(II)}$: first steps

Let $\mathbf{x} \in B_S^c$. Recalling that

$$\mathfrak{F}^{(II)}(\mathbf{x}) = \int_{\overline{\mathcal{D}}} \sum_{l=1}^{3} \mathcal{Z}(\mathbf{x}, \mathbf{y})(u_l \partial_l \mathbf{u})(\mathbf{y})\, d\mathbf{y} \quad - \int_{\partial\mathcal{D}} \sum_{l=1}^{3} \mathcal{Z}(\mathbf{x}, 0)(n_l u_l u)(\mathbf{y}) do_{\mathbf{y}},$$

we apply firstly the integration by parts and then split the resulting volume integral in an integral $\mathfrak{B}_R$ on the bounded domain $B_R \setminus \overline{\mathcal{D}}^c$ and integral $\mathcal{E}_R$ on the exterior domain $(B_R)^c$, where $R = (S_1 + S)/2$. Thus $\mathfrak{F}^{II}(\mathbf{x})$ has a form

$$\begin{aligned}
&\int_{\partial\mathcal{D}} \sum_{l=1}^{3} [\mathcal{Z}(\mathbf{x}, \mathbf{y}) - \mathcal{Z}(\mathbf{x}, 0)](n_l u_l \mathbf{u})(\mathbf{y}) do_{\mathbf{y}} \\
&\qquad - \left\{ \int_{B_R \setminus \overline{\mathcal{D}}^c} + \int_{(B_R)^c} \right\} \sum_{l=1}^{3} \partial_{y_l} \mathcal{Z}(\mathbf{x}, \mathbf{y})(u_l \mathbf{u})(\mathbf{y})\, d\mathbf{y} \\
&\qquad = \mathcal{S}_{\partial\mathcal{D}} - \mathcal{B}_R - \mathcal{E}_R. \qquad (11.46)
\end{aligned}$$

Let us mention that, here and in similar situations in the following, a partial integration has to be performed first on a bounded domain, where $B_T \setminus (\overline{B_R} \cup B_\epsilon(\mathbf{x}))$ with $T > \max\{2R, 2|\mathbf{x}|\}$, $0 < \epsilon$, is a good choice for such a domain. In the next step we let ϵ tend to zero. This passage to the limit may be handled by referring to Lemma 11.9 and (11.36). Finally we let T tend to infinity. The surface integral on ∂B_T which came up in the partial integration then vanishes, as follows from Lemma 11.8 and (11.36). The same references imply that all the volume integrals involved tend to integrals on B_R^c when $T \to \infty$.

In volume integral $\mathcal{E}_R$ over the exterior domain $(B_R)^c$, we use firstly the definition of $\mathcal{Z}$ (11.32) and Fubini's theorem and then the domain invariant transformation $y = e^{t\Omega}\mathbf{z}$ for fixed $t > 0$. The reason why we use the mentioned transformation is that we would like to avoid a periodic term in the right-hand side of (11.51):

$$\begin{aligned}\mathcal{E}_R &= \int_{(B_R)^c} \sum_{l=1}^{3} \partial_{y_l} \mathcal{Z}(\mathbf{x}, \mathbf{y})(u_l \mathbf{u})(\mathbf{y})\, \mathrm{d}\mathbf{y} \\ &= \int_0^\infty \int_{(B_R)^c} \sum_{l=1}^{3} \partial_{y_l} \Gamma(\mathbf{x}, \mathbf{y}, t)[u_l \mathbf{u}](\mathbf{y})\, \mathrm{d}\mathbf{y}\mathrm{d}t \\ &= \int_0^\infty \int_{(B_R)^c} \sum_{l=1}^{3} \partial_{y_l} \Gamma(\mathbf{x}, \mathbf{y}, t)_{|\mathbf{y}=e^{t\Omega}\mathbf{z}}[u_l \mathbf{u}](e^{t\Omega}\mathbf{z})\, \mathrm{d}\mathbf{z}\mathrm{d}t.\end{aligned}$$

Finally, we split the exterior domain of integration B_R^c on two domains: Let δ be a sufficiently small positive number comparing to 1, R, and $S - S_1$, f.e. $\delta := \min\{1, (S - S_1)/2, R/2\} = \min\{1, S - R, R/2\}$. Note that $\overline{B_\delta(x)} \subset B_R^c$. We obtain

$$\begin{aligned}\mathcal{E}_R =&\left\{\int_0^\infty \int_{B_\delta(x)} + \int_0^\infty \int_{(B_R)^c \setminus B_\delta(\mathbf{x})}\right\} \\ &\sum_{l=1}^{3} \partial_{y_l} \Gamma(\mathbf{x}, \mathbf{y}, t)_{|\mathbf{y}=e^{t\Omega}z}[(u_l \mathbf{u})(e^{t\Omega}\mathbf{z})\, \mathrm{d}\mathbf{z}\mathrm{d}t \\ =&\mathcal{V}_\delta + \mathcal{V}_{R,\delta}.\end{aligned}$$

Substituting the expression of $\mathcal{E}_R$ into (11.46), we get

$$\mathfrak{F}^{II} = \mathcal{S}_{\partial\mathcal{D}} - \mathcal{B}_R - \mathcal{V}_\delta - \mathcal{V}_{R,\delta}; \tag{11.47}$$

$\partial_x^\alpha \mathcal{B}_R$, $\partial_x^\alpha \mathcal{S}_{\partial\mathcal{D}}$: Estimating the behavior of the first two terms and their derivatives ∂_x^α for $|\alpha| = 0, 1$, we get the following estimate:

$$|\partial_x^\alpha \mathcal{B}_R| + |\partial_x^\alpha \mathcal{S}_{\partial\mathcal{D}}| \leq \mathfrak{C}(S_1, S)(|x| s_\tau(\mathbf{x}))^{-3/2-|\alpha|/2}, \quad \mathbf{x} \in B_S^c. \tag{11.48}$$

Indeed, from Lemma 11.8 for $\mathbf{y} \in B_R$, $\mathbf{x} \in B_S^c$:

$$|\partial_x^\alpha \partial_{y_l} \mathcal{Z}(\mathbf{x}, \mathbf{y})| \leq \mathfrak{C}(S_1, S)(|\mathbf{x}| s_\tau(\mathbf{x}))^{-3/2-|\alpha|/2}, \tag{11.49}$$

$$\left|\partial_x^\alpha (\mathcal{Z}(\mathbf{x}, \mathbf{y}) - \mathcal{Z}(\mathbf{x}, 0)\right| = \left|\sum_{k=1}^{3} \partial_x^\alpha \partial_{y_k} \mathcal{Z}(\mathbf{x}, \theta\mathbf{y}) y_k\right| \leq \mathfrak{C}(S_1, S)(|\mathbf{x}| s_\tau(\mathbf{x}))^{-3/2-|\alpha|/2} \tag{11.50}$$

for some $0 \le \theta \le 1$. So, with Lemma 11.12 and (11.49)

$$\begin{aligned}|\partial_x^\alpha \mathcal{B}_R| \le & \left| \int_{B_R \setminus \overline{\mathcal{D}}^c} \sum_{l=1}^3 \partial_x^\alpha \partial_{y_l} \mathcal{Z}(\mathbf{x}, \mathbf{y})(u_j \mathbf{u})(\mathbf{y})\, \mathrm{d}\mathbf{y} \right| \\ \le & \mathfrak{C}(S_1, S)(|\mathbf{x}| s_\tau(\mathbf{x}))^{-3/2-|\alpha|/2} \int_{B_R \setminus \overline{\mathcal{D}}} \sum_{l=1}^3 |(u_l\, \mathbf{u})(\mathbf{y})|\, \mathrm{d}\mathbf{y} \\ \le & \mathfrak{C}(S_1, S)(|\mathbf{x}| s_\tau(\mathbf{x}))^{-3/2-|\alpha|/2},\end{aligned}$$

because $|\mathbf{u}|^2$ is L^1-integrable on bounded domain $B_R \setminus \overline{\mathcal{D}}^c$. Similarly, we have with (11.33) and (11.50)

$$\begin{aligned}|\partial_x^\alpha \mathcal{S}_{\partial\mathcal{D}}| \le & \mathfrak{C}(S_1, R)(|\mathbf{x}| s_\tau(\mathbf{x}))^{-3/2-|\alpha|/2} \int_{\partial\mathcal{D}} \sum_{l=1}^3 |(n_l u_l\, \mathbf{u})(\mathbf{y})| \mathrm{d}o_{\mathbf{y}} \\ \le & \mathfrak{C}(S_1, R)(|\mathbf{x}| s_\tau(\mathbf{x}))^{-3/2-|\alpha|/2}.\end{aligned}$$

4. Estimates of $\partial^\alpha \mathfrak{F}^{(II)}$ for $\alpha = 0$

$\mathcal{V}_\delta$: For the estimation of this term we use Lemma 11.5 for the first order derivatives of Γ: We have (for $\mathbf{x} \ne \mathrm{e}^{t\Omega}\mathbf{z}$)

$$|\partial_{y_j} \Gamma(\mathbf{x}, \mathbf{y}, t)_{|\mathbf{y}=\mathrm{e}^{t\Omega}\mathbf{z}}| \le \mathfrak{C}(|\mathbf{x} - \tau t \mathrm{e}_1 - \mathbf{z}|^2 + t)^{-2}. \tag{11.51}$$

From Theorem 11.6 we have for $\mathbf{y} \in B_R^c$

$$|\mathbf{u}(\mathbf{y})|^2 \le \mathfrak{C}(R)(|\mathbf{y}| s_\tau(\mathbf{y}))^{-2}.$$

If $\mathbf{z} \in (B_R)^c$, then $\mathrm{e}^{t\Omega}\mathbf{z} \in (B_R)^c$, and we get

$$|\mathbf{u}(\mathrm{e}^{t\Omega}\mathbf{z})|^2 \le \mathfrak{C}(R)(|\mathrm{e}^{t\Omega}\mathbf{z}| s_\tau(\mathrm{e}^{t\Omega}\mathbf{z}))^{-2} = \mathfrak{C}(R)(|\mathbf{z}|\, s_\tau(\mathbf{z}))^{-2}. \tag{11.52}$$

Since $\overline{B_\delta(\mathbf{x})} \subset B_R^c$, we thus get due to (11.29) and (11.30)

$$|\mathbf{u}(\mathrm{e}^{t\Omega}\mathbf{z})| \le \mathfrak{C}(R, \delta)(|\mathbf{x}| s_\tau(\mathbf{x}))^{-2} \quad \text{for } \mathbf{z} \in \overline{B_\delta(\mathbf{x})}. \tag{11.53}$$

So, we have

$$\begin{aligned}|\mathcal{V}_\delta| &= \left|\int_0^\infty \int_{B_\delta(\mathbf{x})} \sum_{j=1}^3 \partial_{y_j} \Gamma(\mathbf{x}, \mathbf{y}, t)_{|\mathbf{y}=\mathrm{e}^{t\Omega}\mathbf{z}} [(u_j \mathbf{u})(\mathrm{e}^{t\Omega}\mathbf{z})]\,\mathrm{d}\mathbf{z}\,\mathrm{d}t\right| \\ &\le \mathfrak{C}(R) \int_{B_\delta(\mathbf{x})} \int_0^\infty (|\mathbf{x} - \tau t \mathrm{e}_1 - \mathbf{z}|^2 + t)^{-2} (|\mathbf{x}|\, s_\tau(\mathbf{x}))^{-2}\,\mathrm{d}t\,\mathrm{d}\mathbf{z} \\ &\le \mathfrak{C}(R, \delta)(|\mathbf{x}| s_\tau(\mathbf{x}))^{-2} \int_{B_\delta(\mathbf{x})} |\mathbf{x} - \mathbf{z}|^{-2} \mathrm{d}\mathbf{z} \le \mathfrak{C}(R, \delta)(|\mathbf{x}| s_\tau(\mathbf{x}))^{-2},\end{aligned}$$

where the integral with respect to variable t is estimated using Lemma 11.9, choosing in its application $\mathbf{y} := \mathbf{x} - \mathbf{z}$, $\mathbf{z} := 0$.

$\mathcal{V}_{R,\delta}$: Similarly as in the previous case using (11.51) and (11.52),

$$\begin{aligned}|\mathcal{V}_{R,\delta}| \le& \left|\int_0^\infty \int_{B_R^c \setminus B_\delta(\mathbf{x})} \sum_{j=1}^3 \partial_{y_j} \Gamma(\mathbf{x}, \mathbf{y}, t)_{|\mathbf{y}=\mathrm{e}^{t\Omega}\mathbf{z}} [(u_j \mathbf{u})(\mathrm{e}^{t\Omega}\mathbf{z})]\,\mathrm{d}\mathbf{z}\,\mathrm{d}t\right| \\ \le& \mathfrak{C}(R) \int_0^\infty \int_{B_R^c \setminus B_\delta(\mathbf{x})} (|\mathbf{x} - \tau t \mathrm{e}_1 - \mathbf{z}|^2 + t)^{-2} (|\mathbf{z}| s_\tau(\mathbf{z}))^{-2}\,\mathrm{d}\mathbf{z}\,\mathrm{d}t.\end{aligned}$$

Now, the integral with respect to t can be estimated using Lemma 11.6, $\mathbf{y} := \mathbf{x} - \mathbf{z}$, $\mathbf{z} := 0$:

$$\int_0^\infty (|\mathbf{x} - \tau t \mathrm{e}_1 - \mathbf{z}|^2 + t)^{-2} \mathrm{d}t \le \mathfrak{C}(S_1, S)(|\mathbf{x} - \mathbf{z}| s_\tau(\mathbf{x} - \mathbf{z}))^{-3/2}, \quad \mathbf{z} \in B_R^c \setminus B_\delta(\mathbf{x})$$

$$\begin{aligned}|\mathcal{V}_{R,\delta}| \le& \mathfrak{C}(S_1, S) \int_{B_R^c \setminus B_\delta(\mathbf{x})} (|\mathbf{x} - \mathbf{z}| s_\tau(\mathbf{x} - \mathbf{z}))^{-3/2} (|\mathbf{z}|\, |s_\tau(\mathbf{z})|)^{-2} \mathrm{d}\mathbf{z} \\ \le& \mathfrak{C}(S_1, S) \ln(2 + |\mathbf{x}|)(|\mathbf{x}||s_\tau(\mathbf{x})|)^{-3/2}.\end{aligned}$$

The last inequality follows from Lemma 11.16 ($\gamma = 2$).

5. Estimates of $\partial^\alpha \mathfrak{F}^{(II)}$ for $|\alpha| = 1$

Let us mention that S, S_1, R, δ are the same as in the previous section, so $\overline{B_\delta(\mathbf{x})} \subset B_R^c$. The aim of this part is to find the leading term of the gradient of velocity for the Navier–Stokes problem with rotation: The difference with the previous case is that we cannot apply the integration by parts over the whole domain $\overline{\mathcal{D}}$ because we have to protect the neighborhood $B_\delta(\mathbf{x})$ due to singularities of the second order derivatives of $\mathcal{Z}$. On the other hand, to avoid some technical difficulties, we are able to handle the integrals with respect to t only in domains invariant with respect to the transformation $\mathbf{y} = \mathrm{e}^{t\Omega}\mathbf{z}$, $t > 0$. These facts cause some additional computations. So, we use Lemma 11.19 and split the domain of integration into the bounded part $B_R \setminus \overline{\mathcal{D}}$ and the exterior domain $(B_R)^c$, and we apply the integration by parts first only on the bounded domain:

$$\begin{aligned}
\partial_x^\alpha \mathfrak{F}_j^{(II)}(\mathbf{x}) &= \int_{\overline{\mathcal{D}}} \sum_{k,l=1}^{3} \partial_{\mathbf{x}}^\alpha \mathcal{Z}_{jk}(\mathbf{x},\mathbf{y})(u_l \partial_l u_k)(\mathbf{y})\mathrm{d}\mathbf{y} \\
&\quad - \int_{\partial\mathcal{D}} \sum_{k,l=1}^{3} \partial_x^\alpha \mathcal{Z}_{jk}(\mathbf{x},0)(n_l u_l u_k)(\mathbf{y})\mathrm{d}o_{\mathbf{y}} \\
&= \left\{ \int_{B_R\setminus\overline{\mathcal{D}}^c} + \int_{(B_R)^c} \right\} \sum_{k,l=1}^{3} \partial_x^\alpha \mathcal{Z}_{jk}(\mathbf{x},\mathbf{y})(u_l \partial_l u_k)(\mathbf{y})\mathrm{d}\mathbf{y} \\
&\quad - \int_{\partial\mathcal{D}} \sum_{k,l=1}^{3} \partial_x^\alpha \mathcal{Z}_{jk}(\mathbf{x},0)(n_l u_l u_k)(\mathbf{y})\mathrm{d}o_{\mathbf{y}} \\
&= \int_{\partial\overline{\mathcal{D}}} \sum_{k,l=1}^{3} [\partial_x^\alpha \mathcal{Z}_{jk}(\mathbf{x},\mathbf{y}) - \partial_x^\alpha \mathcal{Z}_{jk}(\mathbf{x},0)](n_l u_l u_k)(\mathbf{y})\mathrm{d}o_{\mathbf{y}} \\
&\quad - \int_{B_R\setminus\overline{\mathcal{D}}^c} \sum_{k,l=1}^{3} \partial y_l \partial_x^\alpha \mathcal{Z}_{jk}(\mathbf{x},\mathbf{y})(u_l u_k)(\mathbf{y})\mathrm{d}\mathbf{y} \\
&\quad + \int_{\partial B_R} \sum_{k,l=1}^{3} \partial_x^\alpha \mathcal{Z}_{jk}(\mathbf{x},\mathbf{y})(u_l\, u_k)(\mathbf{y})\frac{y_l}{R}\mathrm{d}o_{\mathbf{y}} \\
&\quad + \int_{(B_R)^c} \sum_{k,l=1}^{3} \partial_x^\alpha \mathcal{Z}_{jk}(\mathbf{x},\mathbf{y})(u_l\, \partial_l u_k)(\mathbf{y})\mathrm{d}\mathbf{y}.
\end{aligned}$$

So we get

$$\partial_x^\alpha \mathfrak{F}^{(II)}(\mathbf{x}) = \partial^\alpha \mathcal{S}_{\partial\mathcal{D}} - \partial^\alpha \mathcal{B}_R + \mathcal{S}_R' + \mathcal{E}_R'. \tag{11.54}$$

Evaluation of the last term in (11.54) with (11.32) is

$$\begin{aligned}
\mathcal{E}_R'(\mathbf{x})_j &= \int_{(B_R)^c} \sum_{k,l=1}^{3} \partial_x^\alpha \mathcal{Z}_{jk}(\mathbf{x},\mathbf{y})(u_l \partial_l u_k)(\mathbf{y})\mathrm{d}\mathbf{y} \\
&= \int_{(B_R)^c} \int_0^\infty \sum_{k,l=1}^{3} \partial_x^\alpha \Gamma_{jk}(\mathbf{x},\mathbf{y},t)(u_l \partial_l u_k)(\mathbf{y})\mathrm{d}\mathbf{y}\mathrm{d}t.
\end{aligned}$$

The domain of integration of $\mathcal{E}_R'$ is $(B_R)^c$. This exterior domain is invariant with respect to the transformation $\mathbf{y} = \mathrm{e}^{t\Omega}\mathbf{z}$, $t > 0$. We use the same transformation to avoid periodic terms as in the case $|\alpha| = 0$:

$$\mathcal{E}'_R(x)_j = \int_0^{+\infty} \int_{(B_R)^c} \sum_{k,l=1}^{3} \partial_x^\alpha \Gamma_{jk}(\mathbf{x}, \mathrm{e}^{\tau\Omega}\mathbf{z}, t)(u_l \partial_l u_k)(\mathrm{e}^{\tau\Omega}\mathbf{z})\, \mathrm{d}\mathbf{z}\, \mathrm{d}t.$$

Unlike the case $|\alpha| = 0$, the mentioned transformation is used *before* the integration by parts. We split the domain of integration into two domains $B_\delta(\mathbf{x})$ and $(B_R)^c \setminus B_\delta(\mathbf{x})$. In the integral over the unbounded domain, we apply the identity from Corollary 11.3 and integrate by parts:

$$\begin{aligned}
\mathcal{E}'_R(\mathbf{x})_j =& \int_0^{\infty} \int_{B_\delta(\mathbf{x})} \sum_{k,l=1}^{3} \partial_{\mathbf{x}}^\alpha \Gamma_{jk}(\mathbf{x}, \mathrm{e}^{\tau\Omega}\mathbf{z}, t)(u_l\, \partial_l u_k)(\mathrm{e}^{\tau\Omega}\mathbf{z})\, \mathrm{d}\mathbf{z}\, \mathrm{d}t \\
&+ \int_0^{\infty} \int_{\partial B_\delta(\mathbf{x})} \sum_{k,l=1}^{3} \partial_{\mathbf{x}}^\alpha \Gamma_{jk}(\mathbf{x}, \mathrm{e}^{t\Omega}\mathbf{z}, t)[(u_l u_k)(\mathrm{e}^{t\Omega}\mathbf{z})] \left(\mathrm{e}^{t\Omega}\frac{\mathbf{x}-\mathbf{z}}{\delta}\right)_l \mathrm{d}o_{\mathbf{z}}\mathrm{d}t \\
&+ \int_0^{\infty} \int_{\partial B_R} \sum_{k,l=1}^{3} \partial_{\mathbf{x}}^\alpha \Gamma_{jk}(\mathbf{x}, \mathrm{e}^{t\Omega}\mathbf{z}, t)[(u_l u_k)(\mathrm{e}^{t\Omega}\mathbf{z})] \left(\mathrm{e}^{t\Omega}\frac{-\mathbf{z}}{R}\right)_l \mathrm{d}o_{\mathbf{z}}\mathrm{d}t \\
&- \int_0^{\infty} \int_{B_R^c \setminus B_\delta(\mathbf{x})} \sum_{k,l=1}^{3} (\mathrm{e}^{t\Omega}\nabla_{\mathbf{z}})_l \partial_{\mathbf{x}}^\alpha \Gamma_{jk}(\mathbf{x}, \mathrm{e}^{t\Omega}\mathbf{z}, t)[(u_l u_k)(\mathrm{e}^{t\Omega}\mathbf{z})]\, \mathrm{d}\mathbf{z}\, \mathrm{d}t \\
=&(\mathcal{U}_\delta)_j + (\mathcal{S}'_\delta)_j + (-\mathcal{S}'_R)_j + (\mathcal{U}_{R,\delta})_j.
\end{aligned}$$

Substituting the expression of $\mathcal{E}'_R(\mathbf{x})$ into (11.54) and using (11.32), we get finally

$$\partial_x^\alpha \mathfrak{F}^{(II)}(x) = \partial^\alpha \mathcal{S}_{\partial\mathcal{D}} - \partial^\alpha \mathcal{B}_R + \mathcal{U}_\delta + \mathcal{S}'_\delta + \mathcal{U}_{R,\delta}. \tag{11.55}$$

Now we will estimate all terms of (11.55) for $|\alpha| = 1$:

$\partial^\alpha \mathcal{S}_{\partial\mathcal{D}}$, $\partial^\alpha \mathcal{B}_{B_R}$: From (11.48) we know that $|\partial_x^\alpha \mathcal{S}_{\partial\mathcal{D}}| + |\partial_x^\alpha \mathcal{B}_R| \le \mathfrak{C}_1(S_1, S)$ $(|\mathbf{x}|\, s_\tau(\mathbf{x}))^{-2}$.

$\mathcal{U}_\delta$: Estimates of this term are completely analogous to the evaluation of $\mathcal{V}_\delta$ in the case $|\alpha| = 0$. Only difference is that from Theorem 10.1 $|\mathbf{u}(\mathbf{y})||\nabla\mathbf{u}(y)| \le \mathfrak{C}(S)(|\mathbf{y}|s_\tau(\mathbf{y}))^{-5/2}$ for $\mathbf{y} \in (B_R)^c$. We get

$$|\mathcal{U}_\delta| \le \mathfrak{C}(S_1, S)(|\mathbf{x}||s_\tau(\mathbf{x})|)^{-5/2}.$$

$\mathcal{S}'_\delta$: From Lemma 11.5 for the first order derivatives of Γ, $\mathbf{x} \neq e^{t\Omega}\mathbf{z}$:

$$|\partial_x^\alpha \Gamma(\mathbf{x}, e^{t\Omega}\mathbf{z}, t)| \leq \mathfrak{C}(|\mathbf{x} - \tau t\mathbf{e}_1 - \mathbf{z}|^2 + t)^{-2}.$$

By (11.53)

$$|\mathbf{u}(e^{t\Omega}\mathbf{z})|^2 \leq \mathfrak{C}(R, \delta)(|\mathbf{x}|s_\tau(\mathbf{x}))^{-2} \quad \text{for } \mathbf{z} \in \partial B_\delta(\mathbf{x}).$$

It is also clear that $|e^{t\Omega}(\mathbf{x} - \mathbf{z})/\delta| = 1$ for $\mathbf{z} \in B_\delta(\mathbf{x})$.
So, we have

$$\begin{aligned}|\mathcal{S}'_\delta| &\leq \mathfrak{C}(R) \int_{\partial B_\delta(\mathbf{x})} \int_0^\infty (|\mathbf{x} - \tau t\mathbf{e}_1 - \mathbf{z}|^2 + t)^{-2}(|\mathbf{x}|s_\tau(\mathbf{x}))^{-2} \mathrm{d}t\, \mathrm{d}o_\mathbf{z} \\ &\leq \mathfrak{C}(R, \delta)(|\mathbf{x}|s_\tau(\mathbf{x}))^{-2} \int_{\partial B_\delta(\mathbf{x})} |\mathbf{x} - \mathbf{z}|^{-2} \mathrm{d}o_z \leq \mathfrak{C}(R, \delta)(|\mathbf{x}|s_\tau(\mathbf{x}))^{-2},\end{aligned}$$

where the integral with respect to variable t is estimated using Lemma 11.9, $\mathbf{y} := \mathbf{x} - \mathbf{z}$, $\mathbf{z} := \mathbf{0}$. So, the integral $\mathcal{S}'_\delta$ belongs to the remainder.

$\mathcal{U}_{R,\delta}$: We shall use Lemma 11.5, for the evaluation of the second order derivatives of the function Γ:

$$|(e^{t\Omega}\nabla_\mathbf{z})_j \partial_\mathbf{x}^\alpha \Gamma(\mathbf{x}, e^{t\Omega}\mathbf{z}, t)| \leq \mathfrak{C}(|\mathbf{x} - \tau t\mathbf{e}_1 - \mathbf{z}|^2 + t)^{-5/2}.$$

The integral with respect to t of the right-hand side can be estimated using Lemma 11.6 choosing $\mathbf{y}, \mathbf{z}$ from the lemma by the following way: $\mathbf{y} := \mathbf{x} - \mathbf{z}$, $\mathbf{z} := 0$. Hence,

$$\int_0^\infty (|\mathbf{y} - \tau t\mathbf{e}_1 - \mathbf{z}|^2 + t)^{-5/2} \mathrm{d}t \leq \mathfrak{C}(S_1, S)(|\mathbf{x} - \mathbf{z}|s_\tau(\mathbf{x} - \mathbf{z}))^{-2}.$$

Using (11.52), we find

$$\begin{aligned}|\mathcal{U}_{R,\delta}| &\leq \left| \int_0^\infty \int_{B_R^c \setminus B_\delta(\mathbf{x})} \sum_{j=1}^3 (e^{t\Omega}\nabla_\mathbf{z})_j \partial \mathbf{x}_m \Gamma(\mathbf{x}, e^{t\Omega}\mathbf{z}, t)[(u_j\mathbf{u})(e^{t\Omega}\mathbf{z})]\, \mathrm{d}\mathbf{z}\, \mathrm{d}t \right| \\ &\leq \mathfrak{C}(S_1, S) \int_{B_R^c \setminus B_\delta(\mathbf{x})} (|\mathbf{x} - \mathbf{z}|s_\tau(\mathbf{x} - \mathbf{z}))^{-2}(|\mathbf{z}||s_\tau(\mathbf{z})|)^{-2} \mathrm{d}\mathbf{z} \\ &\leq \mathfrak{C}(S_1, S)(|\mathbf{x}|s_\tau(\mathbf{x}))^{-2} \ln(2 + |\mathbf{x}|).\end{aligned}$$

This is the last inequality we get by Lemma 11.17. □

Remark 11.4 So, finally we get that the leading term of $\partial^\alpha u_j$ is expressed as in the linear case

$$\sum_{k=1}^{3} \beta_k \partial_x^\alpha \mathcal{Z}_{jk}(\mathbf{x}, 0),$$

where $\beta = (\beta_1, \beta_2, \beta_3)$ contains additionally the term $\int_{\partial\mathcal{D}} \sum_{j=1}^{3} (n_j u_j \mathbf{u})(\mathbf{y}) do_{\mathbf{y}}$. □

Corollary 11.4 *Let $U \subset \mathbb{R}^3$ be open and bounded, $S_1 \in (0, \infty)$ with $\overline{U} \subset B_{S_1}$, $p \in (1, \infty)$, $\boldsymbol{f} \in L^p(U)$ with* $\operatorname{supp}(\mathbf{f}) \subset B_{S_1}$. *Let $\mathbf{u} \in \boldsymbol{W}^{1,1}_{\text{loc}}(\overline{U}^c)$ with $\mathbf{u} \in \boldsymbol{L}^6(\overline{U}^c)$, $\nabla\mathbf{u} \in L^2(\overline{U}^c)^9$, $\pi \in L^2_{\text{loc}}(\overline{U}^c)$, and suppose that $\mathbf{u}$, π satisfy (11.35) (a weak form of rotational Navier–Stokes system, as in Theorem* (10.1)*) with U in the place of $\mathcal{D}$. Let $S_0 \in (0, S_1)$ with $\overline{U} \subset B_{S_0}$. Then the conclusions of Theorem 11.2 hold, with $\mathcal{D}$ replaced by B_{S_0}.* □

Proof Obviously $(\mathbf{u} \cdot \nabla)\mathbf{u} \in \boldsymbol{L}^{3/2}(\overline{U}^c)$ so that $\mathbf{f} - (\mathbf{u} \cdot \nabla)\mathbf{u} \in \boldsymbol{L}^{\min\{p,3/2\}}_{\text{loc}}(\overline{U}^c)$. Therefore, by interior regularity of the Stokes system, as stated in [95, Theorem IV.4.1], we have $\mathbf{u} \in \boldsymbol{W}^{2,\min\{p,3/2\}}_{\text{loc}}(\overline{U}^c)$ and $\pi \in W^{1,\min\{p,3/2\}}_{\text{loc}}(\overline{U}^c)$; also see the proof of [36, Theorem 5.5]. Now the corollary follows from Theorem 11.2, with $p_0 = \min(p, 3/2)$ and B_{S_0} in the place of $\mathcal{D}^c$. □

11.7 Preliminaries to Fourier Transforms

In this section we show that $Z_{j1}(\cdot, 0) = E_{j1}$. To this end, we prove that the Fourier transforms of these two functions coincide. To begin with, we recall some well known facts about the Fourier transforms of some of the fundamental solutions introduced in Sect. 10.2. Other intermediate results in this section may also be well known (e.g., Corollary 11.6), but since their proofs are very short, we present them for completeness.

Theorem 11.9 *For $\boldsymbol{\xi} \in \mathbb{R}^3 \backslash \{0\}$, we have $\widehat{\mathfrak{E}}(\boldsymbol{\xi}) = (2\pi)^{-3/2} |\boldsymbol{\xi}|^{-2}$. If $f \in \mathcal{S}(\mathbb{R}^3)$ and $\mathcal{N}(f)(\mathbf{x}) := \int_{\mathbb{R}^3} \mathfrak{E}(\mathbf{x} - \mathbf{y}) f(\mathbf{y})\, d\mathbf{y}$ for $\mathbf{x} \in \mathbb{R}^3$, then $\widehat{\mathcal{N}}(f)(\boldsymbol{\xi}) = |\boldsymbol{\xi}|^{-2} \widehat{f}(\boldsymbol{\xi})$ for ξ as above.*

Moreover, $[K(\cdot, t)]^\wedge(\boldsymbol{\xi}) = (2\pi)^{-3/2} e^{-t|\boldsymbol{\xi}|^2}$ *and* $\widehat{\mathfrak{O}^{(\lambda)}}(\boldsymbol{\xi}) = F^{(\lambda)}(\boldsymbol{\xi})$ *for $t \in (0, \infty)$, $\boldsymbol{\xi} \in \mathbb{R}^3$, and $\lambda \in (0, \infty)$.* □

Proof For the first formula, the reader may consult [178, Proposition 2.1.1] and its proof. The second equation follows from the first by a well known formula for the Fourier transform of a convolution. As a direct reference we mention [205, Lemma V.1.1]. The third equation is well known, and as concerns the fourth, we refer to [45, Theorem 2.1]. □

Corollary 11.5 *$\widehat{\mathfrak{O}}(\mathbf{z}) = (2\pi)^{-3/2}(\mathrm{i}\tau z_1 + |\mathbf{z}|^2)^{-1}$ and $\int_0^\infty K(\mathbf{z}-\tau t\boldsymbol{e}_1, t)\mathrm{d}t = \mathfrak{O}(\mathbf{z})$ for $\mathbf{z} \in \mathbb{R}^3\backslash\{0\}$.* □

Proof Let $\varphi \in \mathcal{S}(\mathbb{R}^3)$. For $n \in \mathbb{N}$, $\boldsymbol{\xi} \in \mathbb{R}^3$, we have $|F^{(1/n)}(\xi)\varphi(\boldsymbol{\xi})| \leq C|\boldsymbol{\xi}|^{-2}|\varphi(\boldsymbol{\xi})|$. But $\int_{\mathbb{R}^3} |\boldsymbol{\xi}|^{-2}|\varphi(\boldsymbol{\xi})|\mathrm{d}\boldsymbol{\xi} < \infty$, because φ is rapidly decreasing. Thus we get from Lebesgue's theorem

$$\mathfrak{A} := (2\pi)^{-3/2} \int_{\mathbb{R}^3} (\mathrm{i}\tau\xi_1 + |\boldsymbol{\xi}|^2)^{-1}\varphi(\boldsymbol{\xi})\mathrm{d}\xi = \lim_{n\to\infty} \int_{\mathbb{R}^3} F^{(1/n)}(\boldsymbol{\xi})\varphi(\boldsymbol{\xi})\, \mathrm{d}\boldsymbol{\xi}.$$

Due to the last equation in Theorem 11.9, we may conclude

$$\mathfrak{A} = \lim_{n\to\infty} \int_{\mathbb{R}^3} \mathfrak{O}^{(1/n)}(\mathbf{x})\widehat{\varphi}(\mathbf{x})\, \mathrm{d}\mathbf{x}. \tag{11.56}$$

But $|\mathfrak{O}^{(1/n)}(\mathbf{x})\widehat{\varphi}(\mathbf{x})| \leq C|\mathbf{x}|^{-1}|\widehat{\varphi}(\mathbf{x})|$ for $n \in \mathbb{N}$, $\mathbf{x} \in \mathbb{R}^3\backslash\{0\}$ with

$$\int_{\mathbb{R}^3} |\mathbf{x}|^{-1}|\widehat{\varphi}(\mathbf{x})|\, \mathrm{d}\mathbf{x} < \infty$$

because φ, hence $\widehat{\varphi}$, is rapidly decreasing. Thus Eq. (11.56) and Lebesgue's theorem yield $\mathfrak{A} = \int_{\mathbb{R}^3} \mathfrak{O}(\mathbf{x})\widehat{\varphi}(\mathbf{x})\, \mathrm{d}\mathbf{x}$. Since this is true for any $\varphi \in \mathcal{S}(\mathbb{R}^3)$, the first equation in the corollary follows. The second is a consequence of the first and the formula for $[K(\,\cdot\,, t)]^\wedge$ in Theorem 11.9. □

Corollary 11.6 *Let $t \in (0, \infty)$, $j, k \in \{1, 2, 3\}$. Then*

$$[\Lambda_{jk}(\,\cdot\,, t)]^\wedge(\boldsymbol{\xi}) = (2\pi)^{-3/2}(\delta_{jk} - \xi_j\xi_k|\boldsymbol{\xi}|^{-2})\mathrm{e}^{-t|\boldsymbol{\xi}|^2} \quad \textit{for } \boldsymbol{\xi} \in \mathbb{R}^3\backslash\{0\}.$$

Proof We have $K(\,\cdot\,, t) \in \mathcal{S}(\mathbb{R}^3)$ (Lemma 11.14). Therefore by Lemma 11.13, $\Lambda_{jk}(\mathbf{x}, t) = \delta_{jk}K(\mathbf{x}, t) + \int_{\mathbb{R}^3} \mathfrak{E}(\mathbf{x} - \mathbf{y})\partial_j\partial_k K(\mathbf{y}, t)\mathrm{d}y$ ($\mathbf{x} \in \mathbb{R}^3$). Since $K(\,\cdot\,, t)$ belongs to $\mathfrak{S}(\mathbb{R}^3)$, hence $\partial_j\partial_k K(\,\cdot\,, t)$ does, too, Corollary 11.6 follows from Theorem 11.9. □

Corollary 11.7 *Let $j \in \{1, 2, 3\}$, $t \in (0, \infty)$. Then*

$$[\Gamma_{j1}(\,\cdot\,, 0, t)]^\wedge(\boldsymbol{\xi}) = (2\pi)^{-3/2}(\delta_{j1} - \xi_j\xi_1|\boldsymbol{\xi}|^{-2})\mathrm{e}^{-t(\mathrm{i}\tau\xi_1+|\boldsymbol{\xi}|^2)}$$

for $\boldsymbol{\xi} \in \mathbb{R}^3\backslash\{0\}$. □

Proof By Lemma 11.11, we have $\Gamma_{j1}(\mathbf{x}, 0, t) = (\Lambda(\mathbf{x}-\tau t\boldsymbol{e}_1, t)\mathrm{e}^{-t\Omega})_{j1} = \Lambda_{j1}(\mathbf{x}-\tau t\boldsymbol{e}_1, t)$, so Corollary 11.7 follows from Corollary 11.6. □

Corollary 11.8 *Let $j \in \{1, 2, 3\}$, $t \in (0, \infty)$. Then*

$$[Z_{j1}(\cdot, 0)]^\wedge(\boldsymbol{\xi}) = (2\pi)^{-3/2}(\delta_{j1} - \xi_j\xi_1\,|\boldsymbol{\xi}|^{-2})(i\tau\xi_1 + |\boldsymbol{\xi}|^2)^{-1}$$

for $\boldsymbol{\xi} \in \mathbb{R}^3\backslash\{0\}$. □

Proof Let $\varphi \in \mathcal{S}(\mathbb{R}^3)$. With Corollary 11.5, we get

$$A := \int_{\mathbb{R}^3}\int_0^\infty |\Gamma_{j1}(\mathbf{x}, 0, t)\widehat{\varphi}(\mathbf{x})|\,dt\,d\mathbf{x} \le \mathcal{C}\int_{\mathbb{R}^3}\int_0^\infty (|\mathbf{x}-\tau t\boldsymbol{e}_1|^2+t)^{-3/2}|\widehat{\varphi}(\mathbf{x})|\,dt\,d\mathbf{x}.$$

By Lemmas 2.14 and 2.15 and because $\hat{\varphi}$ belongs to $\mathcal{S}(\mathbb{R}^3)$, we get that A is bounded by

$$\mathcal{C}\Big(\int_{\mathbb{R}^3}\int_1^\infty t^{-3/2}|\widehat{\varphi}(\mathbf{x})|\,dt\,d\mathbf{x} \\ + \int_{B_1}\int_0^1 |\mathbf{x}|^{-3/2}t^{-3/4}|\widehat{\varphi}(\mathbf{x})|\,dt\,d\mathbf{x} + \int_{B_1^c}\int_0^1 |\widehat{\varphi}(\mathbf{x})|\,dt\,d\mathbf{x}\Big),$$

and hence $A < \infty$. Therefore, we may apply Fubini's theorem, to obtain

$$\begin{aligned}\int_{\mathbb{R}^3} Z_{j1}(\mathbf{x}, 0)\widehat{\varphi}(\mathbf{x})\,d\mathbf{x} &= \int_0^\infty\int_{\mathbb{R}^3} \Gamma_{j1}(\mathbf{x}, 0, t)\widehat{\varphi}(\mathbf{x})\,d\mathbf{x}\,dt, \\ &= \int_0^\infty\int_{\mathbb{R}^3} (2\pi)^{-3/2}(\delta_{j1} - \xi_j\xi_1|\boldsymbol{\xi}|^{-2})e^{-t(i\tau\xi_1+|\boldsymbol{\xi}|^2)}\varphi(\boldsymbol{\xi})\,d\boldsymbol{\xi}\,dt,\end{aligned}$$

where the last equation follows from Corollary 11.7. But

$$\begin{aligned}&\int_0^\infty\int_{\mathbb{R}^3} |(\delta_{j1} - \xi_j\xi_1|\boldsymbol{\xi}|^{-2})e^{-t(i\tau\xi_1+|\boldsymbol{\xi}|^2)}\varphi(\boldsymbol{\xi})|\,d\boldsymbol{\xi}\,dt \\ &\qquad \le \mathcal{C}\int_0^\infty\int_{\mathbb{R}^3} e^{-t|\boldsymbol{\xi}|^2}|\varphi(\boldsymbol{\xi})|\,d\boldsymbol{\xi}\,dt < \infty.\end{aligned}$$

Thus, we may use Fubini's theorem again, arriving at the equation

$$\int_{\mathbb{R}^3} Z_{j1}(\mathbf{x}, 0)\widehat{\varphi}(\mathbf{x})d\mathbf{x} = \int_{\mathbb{R}^3} (2\pi)^{-3/2}(\delta_{j1} - \xi_j\xi_1|\boldsymbol{\xi}|^{-2})(i\tau\xi_1 + |\boldsymbol{\xi}|^2)^{-1}\varphi(\boldsymbol{\xi})\,d\boldsymbol{\xi}.$$

This proves Corollary 11.8. □

Theorem 11.10 *Let $j, k \in \{1, 2, 3\}$. Then for $\boldsymbol{\xi} \in \mathbb{R}^3\backslash\{0\}$,*

$$\widehat{E}_{jk}(\boldsymbol{\xi}) = (2\pi)^{-3/2}(\delta_{jk} - \xi_j\xi_j|\boldsymbol{\xi}|^{-2})(i\tau\xi_1 + |\boldsymbol{\xi}|^2)^{-1}.$$

Proof For $\mathbf{x} \in \mathbb{R}^3\backslash\{0\}$, we find

$$\begin{aligned}\partial_1 \Phi(\mathbf{x}) =&(4\pi\tau)^{-1}\psi'(\tau(|\mathbf{x}|-x_1)/2)\tau(x_1/|\mathbf{x}|-1)/2\\ =&(4\pi\tau|\mathbf{x}|)^{-1}(\mathrm{e}^{-\tau(|\mathbf{x}|-x_1)/2}-1)=\tau^{-1}(\mathfrak{O}(\mathbf{x})-\mathfrak{E}(\mathbf{x})).\end{aligned}$$

Hence with Corollary 11.5 and Theorem 11.9 for $\boldsymbol{\xi} \in \mathbb{R}^3\backslash\{0\}$,

$$\begin{aligned}\mathrm{i}\xi_1\widehat{\Phi}(\boldsymbol{\xi}) =&\widehat{\partial_1\Phi}(\boldsymbol{\xi})=\tau^{-1}(2\pi)^{-3/2}((\mathrm{i}\tau\xi_1+|\boldsymbol{\xi}|^2)^{-1}-|\boldsymbol{\xi}|^{-2})\\ =&-\mathrm{i}(2\pi)^{-3/2}\xi_1((\mathrm{i}\tau\xi_1+|\boldsymbol{\xi}|^2)|\boldsymbol{\xi}|^2)^{-1}.\end{aligned}$$

As a consequence $\widehat{\Phi}(\boldsymbol{\xi})=-(2\pi)^{-3/2}((\mathrm{i}\tau\xi_1+|\boldsymbol{\xi}|^2)|\boldsymbol{\xi}|^2)^{-1}$, so the theorem follows by the definition of E_{jk}. □

Theorem 11.10 may be deduced also from the results in [93, Chapter VII]. In fact, it is shown in [93, Section VII.3] that the convolution $\mathfrak{O} * f$, for $\mathbf{f} \in C_0^\infty(\mathbb{R}^3)$, belongs to $C^\infty(\mathbb{R}^3)$ and is the velocity part of a solution to the Oseen system (11.28) in $\mathbb{R}^3$. On the other hand, by [93, Section VII.4], the inverse Fourier transform of the function $(2\pi)^{-3/2}(\mathrm{i}\tau\xi_1+|\boldsymbol{\xi}|^2)^{-1}\widehat{f}(\boldsymbol{\xi})(\delta_{jk}-\xi_j\xi_k\,|\boldsymbol{\xi}|^{-2})_{1\le j,k\le 3}$ also solves (11.28) in $\mathbb{R}^3$ and belongs to certain Sobolev spaces. A uniqueness result would yield that the two solutions coincide, implying Theorem 11.10. However, we prefer to carry out a direct proof of this theorem, instead of relying on the rather lengthy theory in [93, Chapter VII], which in fact yields much stronger results, not needed here, than Theorem 11.10.

Combining Theorem 11.10 and Corollary 11.8, we arrive at the main result of this section.

Corollary 11.9 $Z_{j1}(\,\cdot\,,0)=E_{j1}$ *for* $1\le j\le 3$. □

11.8 Proof of Theorem 11.3 and Corollary 11.1

We first show that in the case $k \in \{2,3\}$, the function $\partial^\alpha_{jk} Z(\,\cdot\,,0)$ decays faster for $|\mathbf{x}|\to\infty$ than indicated by Corollary 11.2.

Theorem 11.11 *Let* $S \in [2\tau\pi/|\sqrt{\varrho}|,\infty)$. *Then*

$$|\partial^\alpha_x Z_{jk}(\mathbf{x},0)| \le \mathcal{C}(S)(|\mathbf{x}|s(\mathbf{x}))^{-3/2-|\alpha|/2}$$

for $\mathbf{x} \in B^c_{S+\tau\pi/|\sqrt{\varrho}|}$, $\alpha \in \mathbb{N}_0^3$ *with* $|\alpha|\le 1$, $j \in \{1,2,3\}$, $k \in \{2,3\}$. □

Proof Take $\mathbf{x}, \alpha, j, k$ as in the theorem. We get with Lemma 10.11 (see also Chap. 2) that

$$\partial_x^\alpha Z_{jk}(\mathbf{x},0) = \int_0^\infty \partial_x^\alpha \Gamma_{jk}(\mathbf{x},0,t)\mathrm{d}t = \int_0^\infty [\partial_x^\alpha \Lambda(\mathbf{x}-\tau t\boldsymbol{e}_1,t)\cdot \mathrm{e}^{-t\Omega}]_{jk}\mathrm{d}t,$$

so with Lemma 11.11 in the case $k = 2$,

$$\partial_x^\alpha Z_{jk}(\mathbf{x},0) = \int_0^\infty (\partial_x^\alpha \Lambda_{j2}(\mathbf{x}-\tau t\boldsymbol{e}_1,t)\cos(\sqrt{\varrho}t) - \partial_x^\alpha \Lambda_{j3}(\mathbf{x}-\tau t\boldsymbol{e}_1,t) \times \sin(\sqrt{\varrho}t)r)\mathrm{d}t, \tag{11.57}$$

with a similar formula in the case $k = 3$. Let $\varphi\colon \mathbb{R} \mapsto \mathbb{R}$ be defined by either $\varphi(t) := \cos(\sqrt{\varrho}t)$ for $t \in \mathbb{R}$ or $\varphi(t) := \sin(\sqrt{\varrho}t)$ for $t \in \mathbb{R}$. Let $m \in \{1,2,3\}$. Then, since $\varphi(t+\pi/|\sqrt{\varrho}|) = -\varphi(t)$ for $t \in \mathbb{R}$,

$$\begin{aligned}
&\int_0^\infty \partial_x^\alpha \Lambda_{jm}(\mathbf{x}-\tau t\boldsymbol{e}_1,t)\varphi(t)\mathrm{d}t = \sum_{n=0}^\infty \int_{2n\pi/|\sqrt{\varrho}|}^{2(n+1)\pi/|\sqrt{\varrho}|} \partial_x^\alpha \Lambda_{jm}(\mathbf{x}-\tau t\boldsymbol{e}_1,t)\varphi(t)\mathrm{d}t\\
&= \sum_{n=0}^\infty \int_{2n\pi/|\sqrt{\varrho}|}^{(2n+1)\pi/|\sqrt{\varrho}|} (\partial_x^\alpha \Lambda_{jm}(\mathbf{x}-\tau t\boldsymbol{e}_1,t)(-\partial_x^\alpha \Lambda_{jm}(\mathbf{x}-\tau(t+\pi/|\sqrt{\varrho}|)\boldsymbol{e}_1,t\\
&\quad +\pi/|\sqrt{\varrho}|))\varphi(t)\mathrm{d}t\\
&= \sum_{n=0}^\infty \int_{2n\pi/|\sqrt{\varrho}|}^{(2n+1)\pi/|\sqrt{\varrho}|} \int_0^1 (-\tau\partial_x^{\alpha+e_1} + \partial_x^\alpha\partial_4)\\
&\quad \Lambda_{jm}(x-\tau(t+\vartheta\pi/|\sqrt{\varrho}|)\boldsymbol{e}_1, t+\vartheta\pi/|\sqrt{\varrho}|)\\
&\quad \times(-\pi/|\sqrt{\varrho}|)\varphi(t)\mathrm{d}\vartheta\,\mathrm{d}t.
\end{aligned}$$

Therefore by Lemma 11.15,

$$\begin{aligned}
A :=& \left|\int_0^\infty \partial_x^\alpha \Lambda_{jm}(\mathbf{x}-\tau t\boldsymbol{e}_1,t)\varphi(t)\mathrm{d}t\right|\\
\le& C\sum_{n=0}^\infty\sum_{m=1}^2 \int_{2n\pi/|\sqrt{\varrho}|}^{(2n+1)\pi/|\sqrt{\varrho}|}\\
&\int_0^1 (|\mathbf{x}-\tau(t+\vartheta\pi/|\sqrt{\varrho}|)\boldsymbol{e}_1|^2 + t+\vartheta\pi/|\sqrt{\varrho}|)^{-3/2-|\alpha|/2-m/2}\mathrm{d}\vartheta\,\mathrm{d}t\\
\le& C\sum_{m=1}^2 \int_0^1\int_0^\infty \left(|\mathbf{x}-(\tau\vartheta\pi/|\sqrt{\varrho}|)\boldsymbol{e}_1-\tau t\boldsymbol{e}_1|^2+t\right)^{-3/2-|\alpha|/2-m/2}\mathrm{d}t\,\mathrm{d}\vartheta.
\end{aligned}$$

Since $\mathbf{x} \in B^c_{S+\tau\pi/|\sqrt{\varrho}|}$, we have $|\mathbf{x} - (\tau\vartheta\pi/|\sqrt{\varrho}|)\boldsymbol{e}_1| \geq S$ for $\vartheta \in [0, 1]$, so we may apply Lemma 11.6 with $z = 0$, $R_2 = S$, $R_1 = S/2$, $y = x - (\tau\vartheta\pi/|\sqrt{\varrho}|)e_1$, $\nu = 3/2 + |\alpha|/2 + l/2$, to obtain

$$A \leq \mathcal{C}(S) \sum_{m=1}^{2} \int_0^1 [|\mathbf{x} - (\tau\vartheta\pi/|\sqrt{\varrho}|)\boldsymbol{e}_1| s(\mathbf{x} - (\tau\vartheta\pi/|\sqrt{\varrho}|)\boldsymbol{e}_1)]^{-1-|\alpha|/2-m/2} d\vartheta. \tag{11.58}$$

But for $\vartheta \in [0, 1]$, we have $|\mathbf{x} - (\tau\vartheta\pi/|\sqrt{\varrho}|)\boldsymbol{e}_1| \geq |\mathbf{x}|/2 + S/2 - \tau\vartheta\pi/|\sqrt{\varrho}| \geq |\mathbf{x}|/2$, where the last inequality holds because $S \geq 2\tau\pi/|\sqrt{\varrho}|$. Moreover we get from Lemma 11.4 that $s(\mathbf{x} - (\tau\vartheta\pi/|\sqrt{\varrho}|)\boldsymbol{e}_1)^{-1} \leq \mathcal{C}s(\mathbf{x})^{-1}$ for $\vartheta \in [0, 1]$. Therefore from (11.58),

$$A \leq \mathcal{C}(S) \sum_{m=1}^{2} (|\mathbf{x}|s(\mathbf{x}))^{-1-|\alpha|/2-m/2} \leq \mathcal{C}(S)(|\mathbf{x}|s(\mathbf{x}))^{-3/2-|\alpha|/2}.$$

Theorem 11.11 follows with Eq. (11.57) and its analog for $k = 3$. □

Corollary 11.10 *Let $S \in (0, \infty)$. Then $|\partial^\alpha_x Z_{jk}(\mathbf{x}, 0)| \leq \mathcal{C}(S)(|\mathbf{x}|s(\mathbf{x}))^{-3/2-|\alpha|/2}$ for $\mathbf{x} \in B^c_S$ and for α, j, k as in Theorem 11.11.* □

Proof Let $\mathbf{x} \in B^c_S$, and take α, j, k as in Theorem 11.11. By Corollary 11.2, we have $|\partial^\alpha_x Z_{jk}(\mathbf{x}, 0)| \leq \mathcal{C}(S)(|\mathbf{x}|s(\mathbf{x}))^{-1-|\alpha|/2}$.

Suppose that $S \geq 2\tau\pi/|\sqrt{\varrho}|$. Then we distinguish the cases $\mathbf{x} \in B^c_{S+\tau\pi/|\sqrt{\varrho}|}$ and $\mathbf{x} \in B_{S+\tau\pi/|\sqrt{\varrho}|} \backslash B_S$. If $\mathbf{x} \in B^c_{S+\tau\pi/|\sqrt{\varrho}|}$, the inequality stated in the corollary follows from Theorem 11.11. In the second case, we observe that $1 \leq (S+\tau\pi/|\sqrt{\varrho}|)|\mathbf{x}|^{-1} \leq \mathcal{C}(S)(|\mathbf{x}|s(\mathbf{x}))^{-1/2}$, so the inequality claimed in Corollary 11.10 may be deduced from the estimate stated at the beginning of this proof.

Now suppose that $S < 2\tau\pi/|\sqrt{\varrho}|$. Then we use that either $\mathbf{x} \in B^c_{3\tau\pi/|\sqrt{\varrho}|}$ or $\mathbf{x} \in B_{3\tau\pi/|\sqrt{\varrho}|} \backslash B_S$. If $\mathbf{x} \in B^c_{3\tau\pi/|\sqrt{\varrho}|}$, the inequality we want to show follows from Theorem 11.11 with $2\tau\pi/|\sqrt{\varrho}|$ in the place of S. In the case $\mathbf{x} \in B_{3\tau\pi/|\sqrt{\varrho}|} \backslash B_S$, we use the relation $1 \leq (3\tau\pi/|\sqrt{\varrho}|)|\mathbf{x}|^{-1} \leq \mathcal{C}(S)(|\mathbf{x}|s(\mathbf{x}))^{-1/2}$ and again the estimate from the beginning of the proof, once more obtaining an upper bound $\mathcal{C}(S)(|\mathbf{x}|s(\mathbf{x}))^{-3/2-|\alpha|/2}$ for $|\partial^\alpha_x Z_{jk}(\mathbf{x}, 0)|$, as stated in Corollary 11.10. □

The proofs of Theorem 11.3 and Corollary 11.1 are now obvious.

Proof of Theorem 11.3: Combine Theorem 11.2 and Corollaries 11.9 and 11.10. □

Proof of Corollary 11.1: From interior regularity of solutions to the Stokes system ([93, Theorem IV.4.1]) and the assumption $\mathbf{f} \in \boldsymbol{L}^p(\mathbb{R}^3)$, we may conclude that $\boldsymbol{u} \in \mathbf{W}^{2,\widetilde{p}}_{loc}(\overline{\mathcal{B}}^c)$ and $\pi \in W^{1,\widetilde{p}}_{loc}(\overline{\mathcal{B}}^c)$, with $\widetilde{p}$ from Corollary 11.1. More details about this conclusion may be found in the proof of [36, Theorem 5.5]. It follows that $\mathbf{u}|\partial B_{S_0} \in W^{2-1/\widetilde{p},\widetilde{p}}(\partial B_{S_0})$ and $\pi|B_R\backslash\overline{B_{S_0}} \in L^{\widetilde{p}}(B_R\backslash\overline{B_{S_0}})$ for any $R \in (S_0, \infty)$. Now we may apply Theorem 11.3 with $\mathcal{B}$, $\mathbf{f}$, $\mathbf{u}$, and π replaced by B_{S_0}, $\mathbf{f}|\overline{B_{S_0}}^c$, $\mathbf{u}|\overline{B_{S_0}}^c$, and $\pi|\overline{B_{S_0}}^c$, respectively. Corollary 11.1 then follows from Theorem 11.3. □

We add a lemma that shows that the pointwise decay properties of our remainder imply L^p-integrability as derived by Kyed [163] and restated in Sects. 11.1–11.2.

Lemma 11.20 $\mathfrak{G}|B^c_{S_1} \in L^p(B^c_{S_1})$ *for* $p \in (4/3, \infty]$ *and* $\partial_j\mathfrak{G}|B^c_{S_1} \in L^p(B^c_{S_1})$ *for* $p \in (1, \infty]$, $1 \le j \le 3$. □

Proof Take $p \in (1, \infty)$ and $j \in \{1, 2, 3\}$. We show that $\partial_j\mathfrak{G}|B^c_{S_1} \in L^p(B^c_{S_1})$. The same type of argument yields $\mathfrak{G}|B^c_{S_1} \in L^p(B^c_{S_1})$ if $p > 4/3$. Since $2 - 2/p > 0$, we may fix some $\epsilon \in (0, 2 - 2/p)$. Take $\epsilon := (2 - 2/p)/2$ in order to specify how this parameter depends on p. The relation $\epsilon > 0$ implies the term $|\mathbf{x}|^{-\epsilon}\ln(2 + |\mathbf{x}|)$ is bounded uniformly in $\mathbf{x} \in B^c_{S_1}$. Therefore with Theorem 11.3,

$$\begin{aligned}\int_{B^c_{S_1}} |\partial_i G(\mathbf{x})|^p d\mathbf{x} \le& C^p \int_{B^c_{S_1}} ((|\mathbf{x}|s(\mathbf{x}))^{-2}\ln(2 + |\mathbf{x}|))^p d\mathbf{x}\\ \le& C^p C(p) \int_{B^c_{S_1}} (|\mathbf{x}|s(\mathbf{x}))^{-2p+\epsilon p} d\mathbf{x},\end{aligned}$$

where the constant C was introduced in Theorem 11.3. Since $\epsilon < 2 - 2/p$, we have $-2p + \epsilon p < -2$, so we get with Lemma 10.4

$$\begin{aligned}\int_{B^c_{S_1}} |\partial_i G(\mathbf{x})|^p d\mathbf{x} \le& C^p C(p) \int_{S_1}^{\infty} r^{-2p+\epsilon p} \int_{\partial B_r} s(\mathbf{x})^{-2p+\epsilon p} do_{\mathbf{x}} dr\\ \le& C^p C(p) \int_{S_1}^{\infty} r^{-2p+\epsilon p+1} dr \le C^p C(p).\end{aligned}$$

□

11.9 Proof of Theorem 11.4

Our starting point is a result on the decay of the pressure in the linearized problem (11.59) below which was stated in Theorem 6.2 and proved in Chap. 6. Let us recall this result.

Theorem 11.12 *Let* $p \in (1, \infty)$, γ_0, S_1, $S \in (0, \infty)$, $B \in (1/2, \infty)$ *such that* $\overline{\mathcal{D}} \subset B_{S_1}$, $S_1 < S$, $5/2 + \min\{1, B\} \ge 3$.

Take $\mathbf{F} : \overline{\mathcal{D}}^C \mapsto \mathbb{R}^3$ *measurable with* $\mathbf{F}|\mathcal{D}_{S_1} \in \boldsymbol{L}^p(\mathcal{D}_{S_1})$ *and* $|\mathbf{F}(\mathbf{z})| \le \gamma_0\, |\mathbf{z}|^{-5/2}\, s(\mathbf{z})^{-B}$ *for* $\mathbf{z} \in B^c_{S_1}$.

Let $\mathbf{u} \in \boldsymbol{W}^{1,p}_{loc}(\overline{\mathcal{D}})$, $\pi \in L^p_{loc}(\overline{\mathcal{D}})$ *with* $\mathbf{u}|B^c_S \in \boldsymbol{L}^6(B^c_S)$, $\nabla\mathbf{u}|B^c_S \in L^2(B^c_S)^9$, $div\, \mathbf{u} = 0$, *and*

$$\int_{\mathcal{D}} \Big[\nabla\mathbf{u}\cdot\nabla\varphi + \big(\tau\,\partial_1\mathbf{u} - (\sqrt{\varrho}\,\boldsymbol{e}_1 \wedge \mathbf{z})\cdot\nabla\mathbf{u} + \sqrt{\varrho}\,\boldsymbol{e}_1 \wedge \mathbf{u}\big)\cdot\varphi \tag{11.59}$$

$$-\pi\, div\,\varphi - \boldsymbol{F}\cdot\varphi \Big]\, d\mathbf{z} = 0 \quad for\ \ \varphi \in \boldsymbol{C}^\infty_0(\overline{\mathcal{D}}).$$

(Note that (11.59) is a weak form of (11.10).) Then there is $c_0 \in \mathbb{R}$ *such that* $|\pi(\mathbf{x}) + c_0| \le \mathfrak{C}\, |x|^{-2}$ *for* $x \in B^c_S$. □

Proof of Theorem 11.4 By the assumptions of Theorem 11.4, we have $\mathbf{u} \in \boldsymbol{L}^6(\overline{\mathcal{D}}) \cap \boldsymbol{W}^{1,1}_{loc}(\overline{\mathcal{D}})$, $\nabla\mathbf{u} \in L^2(\overline{\mathcal{D}})^9$, and $\pi \in L^2_{loc}(\overline{\mathcal{D}})$, so by Hölder's inequality $-\tau\,(\mathbf{u}\cdot\nabla)\mathbf{u} \in \boldsymbol{L}^{3/2}(\overline{\mathcal{D}})$. Put $\widetilde{p} := \min\{3/2,\ p\}$ and $\widetilde{\mathbf{F}} := \mathbf{f} - \tau\,(\mathbf{u}\cdot\nabla)\mathbf{u}$. Since $\mathbf{f} \in \boldsymbol{L}^p(\mathbb{R}^3)$ and $supp(\mathbf{f}) \subset B_{S_1}$, we see that $\mathbf{F}|\mathcal{D}_{(S_1+S)/2} \in L^{\widetilde{p}}(\mathcal{D}_{(S_1+S)/2})^3$. Moreover $\widetilde{p} < 2$, so we may further conclude that $\mathbf{u} \in \boldsymbol{W}^{1,\widetilde{p}}_{loc}(\overline{\mathcal{D}})$ and $\pi \in L^{\widetilde{p}}_{loc}(\overline{\mathcal{D}})$. Obviously Eq. (11.59) follows from (11.35) and the definition of $\mathbf{F}$. In addition, from Theorem 11.6 with $(S_1 + S)/2$ in the place of S, it follows that there is $\gamma_0 \in (0, \infty)$ such that $|\tau\,(\mathbf{u}(\mathbf{z})\cdot\nabla)\mathbf{u}(\mathbf{z})| \le \gamma_0\,\big(|\mathbf{z}|\, s(\mathbf{z})\big)^{-5/2}$ for $\mathbf{z} \in B^c_{S_1+S)/2}$. Since $supp(\mathbf{f}) \subset B_{S_1}$, we may conclude that $|\mathbf{F}(\mathbf{z})| \le \gamma_0\,\big(|\mathbf{z}|\, s(\mathbf{z})\big)^{-5/2}$ for such $\mathbf{z}$. Altogether we may apply Theorem 11.12 with $\widetilde{p}$ and $(S_1 + S)/2$ in the place of, respectively, p and S_1 and with $B = 5/2$. Then Theorem 11.4 follows from Theorem 11.12. □

11.10 Notes on Chap. 11

Chapter 10 follows from the works by Deuring et al., see [41, 42].

- Navier–Stokes type system—pure rotation case
  - L^2 theory on the existence of the Navier–Stokes type system can be found, e.g., in the works of Galdi [97, 98] and Galdi and Silvestre [109–111].
  - For the L^q theory we refer to see write work of Farwig and Hsihida [66].
  - For the asymptotic behavior of the velocity field, we refer, e.g., to works of Galdi [100, 101], Farwig and Hishida [67, 69], and for the behavior of the pressure, see [63]. Generalization see [125]. For generalization of anisotropic Oseen semigroup see [209].

- Navier–Stokes case with rotation and translation
  - L^2 theory on the existence of the Navier–Stokes type system can be found, e.g., in the work of Galdi [98].
  - For the existence of Navier–Stokes in L^q setting with the Oseen term, we refer to the work of Heck et al. [131].
  - For the asymptotic behavior, see, e.g., works of Kyed [161, 163].
  - See also the recent work by Kim [140] concerning very weak solutions.
- Let us mention also the recent work by Galdi concerning time-periodic case, see [92].

Chapter 12
A Brief Introduction to the Existence Theory for the Nonstationary Navier–Stokes Type Problem Around Translating and Rotating Body

The qualitative mathematical analysis of the problem (1.23), (1.24), (1.27) (1.30) was started by W. Borchers in his habilitation thesis [17]. Here, the author proved the global in time existence of a Leray–Hopf type weak solution, provided that the initial velocity $\mathbf{v}_0$ lies in $\boldsymbol{L}^2_\sigma(\mathcal{D})$ and the external body force is zero. The weak solution satisfies the inclusion $\mathbf{v} \in L^\infty(0, T;\ \boldsymbol{L}^2_\sigma(\mathcal{D})) \cap L^2(0, T;\ \boldsymbol{W}^{1,2}(\mathcal{D}))$ (expressed by means of our notation) for any $T > 0$. An important role in the proof is played by the fact that the "rotational terms" $\boldsymbol{\omega} \wedge \mathbf{v}$, $(\boldsymbol{\omega} \wedge \mathbf{x}) \cdot \nabla\mathbf{v}$, as well as the "translational term" $k\mathbf{e}_3 \cdot \nabla\mathbf{v}$, are formally orthogonal to $\mathbf{v}$ in $\boldsymbol{L}^2(\mathcal{D})$, which enables one to obtain an analogous energy-type inequality as in the case of the "pure" Navier–Stokes equations without the "rotational terms." It is worth mentioning that W. Borchers also obtained the same result in the case when the vectors $\boldsymbol{\omega}$ and $\dot{\mathbf{a}}$ depend (in a certain appropriate sense) on time. (See Chap. 1 for the meaning of $\dot{\mathbf{a}}$.)

Further results on this theme were obtained by T. Hishida [125], M. Hieber et al. [116], and G. P. Galdi et al. [96]. In [125], T. Hishida proved the existence and uniqueness of a strong solution locally in time provided that the initial velocity belongs to the class $H^{1/2}$. This regularity assumption coincides with that in the famous paper of Fujita, Kato [90]. An essential part of the proof consists in the deduction of a certain smoothing property together with estimates near $t = 0$ of the semigroup, generated by the corresponding linearized operator, although the semigroup is only of the class C_0 and is not analytic.

T. Hishida noted that the time derivative of the solution is interconnected through the momentum equation with the nonlinear convective term, with the gradient of the pressure and also with the terms, arising from the rotation and the translation. It is therefore difficult to specify precisely the regularity of each of these terms.

The generalization of Hishida's results to the L^p framework was done by M. Hieber, H. Heck, and M. Geissert in [116], where the authors proved the existence of a unique local mild solution to the Navier–Stokes type problem.

Š. Nečasová et al., *Mathematical Theory of a Fluid Flow Around a Rotating and Translating Body*, Springer Monographs in Mathematics,
https://doi.org/10.1007/978-3-031-84120-0_12

In this chapter, we further provide a brief overview of some results on the existence of strong solutions of the nonsteady Navier–Stokes problem around the rotating and translating rigid body in L^2 frame. In Sect. 12.1, we deal with local in time existential results for strong solutions in three spatial dimensions and global in time results in two spatial dimensions. The corresponding theory is based on the local transformation introduced by A. Inoue and M. Wakimoto [135]. The presented results are due to P. Cumsille and M. Tucsnak, see [26]. Section 12.2 is devoted to the global in time existence of strong solutions. Here, we mainly come from the paper [108] by G. P. Galdi and A. L. Silvestre. The used approach is based on the global transformation, described in Chap. 1 (Introduction). The presented theory and results also involve the question of convergence to the steady solution (steady in the coordinate system attached to the body).

12.1 Local in Time Existence of a Strong Solution

The contents of this subsection are based on the results of P. Cumsille and M. Tucsnak, see [26].

12.1.1 Introduction

Following [26], we consider not only the case when the compact rigid body rotates in $\mathbb{R}^3$ (we put $n = 3$) but also the two-dimensional case, when the body is planar and rotates in $\mathbb{R}^2$ (we consider $n = 2$). Recall that $\mathcal{B}(t)$ denotes the subset of $\mathbb{R}^n$ occupied by the body, and $\mathcal{D}(t) = \mathbb{R}^n \setminus \mathcal{B}(t)$ is the exterior domain occupied by the fluid at time t.

The system of equations modeling the motion of the viscous incompressible fluid around the rotating obstacle is described by the system

$$\frac{\partial \mathbf{v}}{\partial t} - \nu \Delta \mathbf{v} + (\mathbf{v} \cdot \nabla)\mathbf{v} + \nabla q = \mathbf{f}, \ \mathbf{x} \in \mathcal{D}(t), \ t \in (0, T), \tag{12.1}$$

$$\operatorname{div} \mathbf{v} = 0, \ \mathbf{x} \in \mathcal{D}(t), \ t \in (0, T), \tag{12.2}$$

$$\mathbf{v}(\mathbf{x}, t) = \boldsymbol{\omega} \wedge \mathbf{x}, \ \mathbf{x} \in \partial\mathcal{B}(t), \ t \in (0, T), \tag{12.3}$$

$$\mathbf{v}(\mathbf{x}, 0) = \mathbf{v}_0(\mathbf{x}), \ \mathbf{x} \in \mathcal{D}(0). \tag{12.4}$$

Note that if $n = 2$, then $\boldsymbol{\omega}$ is just a scalar ω and $\boldsymbol{\omega} \wedge \mathbf{x} = \omega \mathbf{x}^{\perp} = \omega(-x_2, x_1)$.

The aim of this part is to present results on the local in time existence and uniqueness of a strong solution of (12.1)–(12.4) and to sketch main ideas of the proofs. The details can be found in [26].

Let us at first introduce some notation that will be used later in this subsection. Recall that $\mathcal{D} = \mathcal{D}(0)$ and $\mathcal{B} = \mathcal{B}(0)$. Furthermore, we denote

$$\mathcal{H}^m(t) = \boldsymbol{H}(\mathcal{D}(t)) = [H^m(\mathcal{D}(t))]^3, \qquad \mathcal{H}^m = \mathcal{H}^m(0),$$

$$\mathcal{L}^p(t) = \boldsymbol{L}^p(\mathcal{D}(t)) = [L^p(\mathcal{D}(t))]^3, \qquad \mathcal{L}^p = \mathcal{L}^p(0).$$

Suppose that $\boldsymbol{\psi} : \mathbb{R}^3 \times [0, \infty) \to \mathbb{R}^3$ is a C^∞-diffeomorphism, transforming for each $t \geq 0$ $\mathcal{D}$ to $\mathcal{D}(t)$, such that the mapping $(\mathbf{y}, t) \mapsto D_t D^\alpha \boldsymbol{\psi}(\mathbf{y}, t)$, $\boldsymbol{\alpha} \in \mathbb{N}^3$, being a multi-index, is continuous and of bounded support in $\mathcal{D}$. (The change of variables with these properties is constructed in Sect. 12.1.3.1.) For $\mathbf{v}(\,.\,, t) : \mathcal{D}(t) \to \mathbb{R}^n$, we denote $\mathbf{u}(\mathbf{y}, t) := \mathbf{v}(\boldsymbol{\psi}(\mathbf{y}, t), t)$ for all $t \geq 0$ and $\mathbf{y} \in \mathcal{D}$. We introduce the following function spaces:

$$L^2(0, T;\ \mathcal{H}^2(t)) \equiv L^2(0, T;\ \boldsymbol{H}^2(\mathcal{D}(t))) := \{\mathbf{v};\ \mathbf{u} \in L^2(0, T;\ \boldsymbol{H}^2(\mathcal{D}))\},$$

$$H^1(0, T;\ \mathcal{L}^2(t)) \equiv H^1(0, T;\ \boldsymbol{L}^2(\mathcal{D}(t))) := \{\mathbf{v};\ \mathbf{u} \in H^1(0, T;\ \boldsymbol{L}^2(\mathcal{D}))\},$$

$$C([0, T];\ \mathcal{H}^1(t)) \equiv C([0, T];\ \boldsymbol{H}^1(\mathcal{D}(t))) := \{\mathbf{v};\ \mathbf{u} \in C([0, T];\ \boldsymbol{H}^1(\mathcal{D}))\},$$

$$L^2(0, T; \mathcal{H}^1(t)) \equiv L^2(0, T; \boldsymbol{H}^1(\mathcal{D}(t))) := \{\mathbf{v};\ \mathbf{u} \in L^2(0, T;\ \boldsymbol{H}^1(\mathcal{D}))\},$$

$$\mathcal{U}(0, T; \mathcal{D}(t)) = L^2(0, T; \mathcal{H}^2(t)) \cap C([0, T], \mathcal{H}^1(t)) \cap H^1(0, T; \mathcal{L}^2(t)). \tag{12.5}$$

Theorem 12.1 *Suppose that $\partial\mathcal{B}$ is a $C^{2+\mu}$-boundary, with $\mu \in (0, 1)$. Let $\mathbf{f} \in L^2_{\text{loc}}(0, \infty; [\boldsymbol{W}^{1,\infty}(\mathbb{R}^n)])$ and $\mathbf{v}_0 \in \mathcal{H}^1$ be such that*

$$\operatorname{div} \mathbf{v}_0 = 0 \quad \text{in } \mathcal{D},$$

$$\mathbf{v}_0(x) = \boldsymbol{\omega} \wedge \mathbf{x}, \qquad \mathbf{x} \in \partial\mathcal{B}.$$

Let us consider a constant $C_0 > 0$ such that $\|\mathbf{v}_0\|_{\mathcal{H}^1} \leq C_0$. Then there exists a time T_0 depending just on C_0 such that the system (12.1)–(12.4) admits a unique strong solution

$$(\mathbf{u}, q) \in \mathcal{U}(0, T, \mathcal{D}(t)) \times L^2(0, T; \widehat{H}^1(\mathcal{D}(t)))$$

for any $T \in (0, T_0)$.

Moreover, we can choose T_0 such that one of the following possibilities holds true:

(i) $T_0 = \infty$.
(ii) The function $t \to \|\mathbf{v}(t)\|_{\mathcal{H}^1(t)}$ is not bounded in $[0, T_0)$. □

Theorem 12.2 *Assume that the hypothesis in Theorem* 12.1 *holds true, and suppose that $n = 2$. Then, the strong solution given by Theorem* 12.1 *is global in time, i.e., the alternative (i) in Theorem* 12.1 *holds true.* □

12.1.2 Preliminaries

To sketch the proof of Theorem 12.1, we at first reduce the system (12.1)–(12.4) to the problem in the cylindrical domain $\mathcal{D} \times (0, T)$. Let us introduce a change of variables which coincides with the simple rotation in the neighborhood of the rotating body, but it is equal to identity far from the rotating body. (For more details regarding such a change of variables, see Inoue and Wakimoto [135].)

We obtain a system equivalent to (12.1)–(12.4), which has the form

$$\frac{\partial \mathbf{u}}{\partial t} - \nu(\boldsymbol{L}\mathbf{u}) + (\boldsymbol{M}\mathbf{u}) + (\boldsymbol{N}\mathbf{u}) + (\boldsymbol{G}p) = 0 \quad \text{in } \mathcal{D} \times (0, T),$$

$$\operatorname{div} \mathbf{u} = 0, \ \mathcal{D} \times (0, T),$$

$$\mathbf{u}(\mathbf{y}, t) = \boldsymbol{\omega} \wedge \mathbf{y}, \ \partial\mathcal{B} \times (0, T),$$

$$\mathbf{u}(\mathbf{y}, 0) = \mathbf{v}_0(\mathbf{y}), \ \mathbf{y} \in \mathcal{D}.$$

The unknowns are $\mathbf{u}(\mathbf{y}, t)$ and $p(\mathbf{y}, t)$. The symbols $(\boldsymbol{L}\mathbf{u})$, $(\boldsymbol{M}\mathbf{u})$, $(\boldsymbol{N}\mathbf{u})$, and $(\boldsymbol{G}p)$, which successively denote the transformations of the terms in (12.1), are precisely defined in Sect. 12.1.4. Here, we only note that $(\boldsymbol{L}\mathbf{u})$ is the transformation of $\Delta\mathbf{v}$, $\boldsymbol{M}$ denotes a certain linear operator, $(\boldsymbol{N}\mathbf{u})$ corresponds to $(\mathbf{v} \cdot \nabla)\mathbf{v}$, and $(\boldsymbol{G}p)$ is the transformation of ∇p. The coefficients which are containing in $(\boldsymbol{L}\mathbf{u})$, $(\boldsymbol{M}\mathbf{u})$, $(\boldsymbol{N}\mathbf{u})$, and $(\boldsymbol{G}p)$ depend smoothly on the time t so that $(\boldsymbol{L}\mathbf{u})$ is close to $\Delta\mathbf{u}$, $(\boldsymbol{G}p)$ is close to ∇p, etc., for close to 0.

We also construct a smooth and compactly supported function $\Lambda : \mathbb{R}^n \to \mathbb{R}^n$ satisfying

$$\operatorname{div} \Lambda = 0 \quad \text{in } \mathbb{R}^n,$$

$$\Lambda(\mathbf{y}) = \boldsymbol{\omega} \wedge \mathbf{y} \quad \text{in a ball containing } \mathcal{B},$$

see Sect. 12.1.3.1. Further, we put

$$\boldsymbol{U}(\mathbf{y}, t) = \mathbf{u}(\mathbf{y}, t) - \Lambda(\mathbf{y}),$$

$$P(\mathbf{y}, t) = p(\mathbf{y}, t),$$

where $(\mathbf{u}, p)$ is a solution of the problem. Then $(\boldsymbol{U}, P)$ is a solution of the equivalent problem:

$$\text{(P)} \qquad \begin{cases} \dfrac{\partial \boldsymbol{U}}{\partial t} - \nu(\boldsymbol{L}\boldsymbol{U}) + (\boldsymbol{M}\boldsymbol{U}) + (\boldsymbol{N}\boldsymbol{U}) + (\boldsymbol{B}\boldsymbol{U}) + (\boldsymbol{G}P) = \boldsymbol{F}, \\ \operatorname{div} \boldsymbol{U} = 0, \\ \boldsymbol{U}(\mathbf{y}, t) = 0, \\ \boldsymbol{U}(\mathbf{y}, t) = \boldsymbol{U}_0(\mathbf{y}), \end{cases}$$

where $(\boldsymbol{B}\boldsymbol{U})$ depends linearly on $\boldsymbol{U}$ and $\nabla\boldsymbol{U}$. (As one can see below, $(\boldsymbol{B}\boldsymbol{U})$ also depends on Λ and $\nabla\Lambda$.) The functions $\boldsymbol{F}$ and $\boldsymbol{U}_0$ are given by the formulas

$$\begin{aligned}\boldsymbol{F} &= \nu(\boldsymbol{L}\Lambda) - (\boldsymbol{M}\Lambda) - (\boldsymbol{N}\Lambda),\\ \boldsymbol{U}_0 &= \mathbf{v}_0 - \Lambda.\end{aligned}$$

The solution $(\boldsymbol{U}, P)$ of problem (P) can be found as a fixed point of the mapping

$$\mathcal{N} : (\boldsymbol{W}, Q) \to (\boldsymbol{U}, P),$$

where $(\boldsymbol{U}, P)$ is a solution of the Stokes problem

$$\begin{aligned}\frac{\partial \boldsymbol{U}}{\partial t} - \nu\Delta\boldsymbol{U} + \nabla p = \boldsymbol{F} \quad &\text{in } \mathcal{D} \times (0, T),\\ \operatorname{div}\boldsymbol{U} = 0 \quad &\text{in } \mathcal{D} \times (0, T),\\ \boldsymbol{U}(\mathbf{y}, t) = 0 \quad &\text{on } \partial\mathcal{B} \times (0, T),\\ \boldsymbol{U}(\mathbf{y}, 0) = \boldsymbol{U}_0(\mathbf{y}),\ \ &\mathbf{y} \in \mathcal{D}\end{aligned}$$

with

$$\boldsymbol{F} = \nu(\boldsymbol{L}\boldsymbol{\Lambda}) - (\boldsymbol{M}\boldsymbol{\Lambda}) - (\boldsymbol{N}\boldsymbol{\Lambda}) + \nu((\boldsymbol{L}-\Delta)\boldsymbol{W}) - (\boldsymbol{M}\boldsymbol{W}) - (\boldsymbol{N}\boldsymbol{W}) - (\boldsymbol{B}\boldsymbol{W}) + ((\nabla - \mathcal{I})Q).$$

For T_0 small enough, we will show the existence of a closed ball $\mathcal{K}$ such that $\mathcal{N}$ maps $\mathcal{K}$ into itself and the restriction of $\mathcal{N}$ to this ball is a contraction for all $T \in (0, T_0)$. This will complete the proof of the local in time existence and uniqueness of the strong solution.

12.1.3 *The Local Transformation*

12.1.3.1 The Change of Variables

Recall that $d(\mathcal{B})$ is the diameter of the set $\mathcal{B}$. Let $r > d(\mathcal{B})$. We denote, for brevity, by B_r the open ball centered at the origin and of radius r and by $\mathcal{B}(t)$ the position of $\mathcal{B} \equiv \mathcal{B}(0)$ at time t, caused by the rotation. Obviously, $\mathcal{B}(t) \subset B_r$ for all $t \geq 0$.

Let $\boldsymbol{\theta} \in C^\infty(\mathbb{R}^n, \mathbb{R})$, and let be a cutoff function with support constrained in B_{2r}, $\theta \equiv 1$ in $\overline{B}_r$.

Now, we define the function $\boldsymbol{\Lambda} : \mathbb{R}^n \to \mathbb{R}^n$ by the formula

$$\boldsymbol{\Lambda}(\mathbf{x}) = -\frac{1}{2}\operatorname{curl}\,(\boldsymbol{\theta}|\mathbf{x}|^2\boldsymbol{\omega}) = \begin{cases} \begin{pmatrix} -\dfrac{1}{2}\dfrac{\partial\boldsymbol{\theta}}{\partial x_2}|\mathbf{x}|^2 - \boldsymbol{\theta}x_2 \\ \dfrac{1}{2}\dfrac{\partial\boldsymbol{\theta}}{\partial x_1}|\mathbf{x}|^2 + \boldsymbol{\theta}x_1 \end{pmatrix} & \text{for } n = 2 \\ \begin{pmatrix} -\dfrac{1}{2}\dfrac{\partial\boldsymbol{\theta}}{\partial x_2}|\mathbf{x}|^2 - \boldsymbol{\theta}x_2 \\ \dfrac{1}{2}\dfrac{\partial\boldsymbol{\theta}}{\partial x_2}|\mathbf{x}|^2 + \theta x_1 \\ 0 \end{pmatrix} & \text{for } n = 3. \end{cases} \tag{12.6}$$

$\boldsymbol{\Lambda}$ is a $\boldsymbol{C}^\infty$ function with these properties:

Lemma 12.1 *The mapping $\boldsymbol{\Lambda}$ defined by (12.6) satisfies:*

(i) $\boldsymbol{\Lambda} = \mathbf{0}$ *outside of* B_{2r}
(ii) $\operatorname{div}\boldsymbol{\Lambda} = \mathbf{0}$ *in* $\mathbb{R}^n$
(iii) $\boldsymbol{\Lambda}(\mathbf{x}) = \boldsymbol{\omega} \wedge \mathbf{x}$ *if* $\mathbf{x} \in \mathcal{B}(t)$, $t \geq 0$

Let $\boldsymbol{X}(\cdot, t)$ denote the time-dependent vector field satisfying

$$\left.\begin{aligned} \frac{\partial \boldsymbol{X}}{\partial t}(\mathbf{y}, t) = \boldsymbol{\Lambda}(\boldsymbol{X}(\mathbf{y}, t)),\ t > 0 \\ \boldsymbol{X}(\mathbf{y}, 0) = \mathbf{y} \in \mathbb{R}^n \end{aligned}\right\}. \tag{12.7}$$

Lemma 12.2 *For all* $\mathbf{y} \in \mathbb{R}^n$*, the initial-valued problem (12.7) admits a unique solution* $\boldsymbol{X}(\mathbf{y}, \cdot) : [0, \infty) \to \mathbb{R}^n$ *which is a* $\boldsymbol{C}^\infty$*-function in* $\mathbb{R}^n \times [0, \infty)$*. Moreover, for all* $t \geq 0$*, we have that the mapping* $\mathbf{y} \to \boldsymbol{X}(\mathbf{y}, t)$ *is a* C^∞*-diffeomorphism from* $\mathbb{R}^n$ *onto itself and from* $\mathcal{D}$ *onto* $\mathcal{D}(t)$*.*
Furthermore, its inverse denoted by $\boldsymbol{Y}(\mathbf{x}, t)$ *is also a* $\boldsymbol{C}^\infty$*-function is* $\mathbb{R}^n \times [0, \infty[$*.* □

The proofs of Lemmas 12.1 and 12.2 can be found in [26].

Remark 12.1 Let X be as in Lemma 12.2. Denote for each $t \geq 0$ by

$$J_X = \left(\frac{\partial X_i}{\partial y_j}\right)_{i,j}, \quad i, j \in \{1, \ldots, n\}$$

the Jacobi matrix of the transformation $\mathbf{y} \to \boldsymbol{X}(\mathbf{y}, t)$. Then

$$\det J_X(\mathbf{y}, t) = 1 \quad \forall\, \mathbf{y} \in \mathbb{R}^n,\ \forall\, t \geq 0.$$

12.1.4 The System in the Cylindrical Domain

Let us set

$$\mathbf{u}(\mathbf{y}, t) = J_Y(\boldsymbol{X}(\mathbf{y}, t), t)\mathbf{v}(\boldsymbol{X}(\mathbf{y}, t), t) \quad \forall\, \mathbf{y} \in \mathcal{D}, \ \ \forall\, t \geq 0. \tag{12.8}$$

Hence

$$u_i(\mathbf{y}, t) = \sum_{j=1}^{n} \frac{\partial Y_i}{\partial x_j}(\boldsymbol{X}(\mathbf{y}, t), t)\, v_j(\boldsymbol{X}(\mathbf{y}, t), t) \tag{12.9}$$

for $i \in \{1, \ldots, n\}$ and

$$p(\mathbf{y}, t) = q(\boldsymbol{X}(\mathbf{y}, t), t) \quad \forall\, \mathbf{y} \in \mathcal{D}, \ \ \forall\, t \geq 0, \tag{12.10}$$

where $\boldsymbol{X}$ and $\boldsymbol{Y}$ were introduced in Lemma 12.2.

Since $J_Y(\mathbf{y}, t) J_Y(\boldsymbol{X}(\mathbf{y}, t), t) = \mathrm{Id} \quad \forall\, (\mathbf{y}, t) \in \mathcal{D} \times [0, \infty[$, we get the following lemma:

Lemma 12.3 *Suppose that* $\boldsymbol{X}$ *and* $\mathbf{u}$ *are defined as before. Then*

$$\operatorname{div} \mathbf{u}(\mathbf{y}, t) = \operatorname{div} \mathbf{v}(\boldsymbol{X}(\mathbf{y}, t), t) \quad \forall\, (\mathbf{y}, t) \in \mathcal{D} \times [0, \infty[.$$

Now, we define $(\boldsymbol{L}\mathbf{u})$, $(\boldsymbol{N}\mathbf{u})$, $(\boldsymbol{M}\mathbf{u})$, $(\boldsymbol{G}p)$ by the formulas

$$\begin{aligned}
(\boldsymbol{L}u)_i &= \sum_{j,k=1}^{n} \frac{\partial}{\partial y_j}\left(g^{jk}\frac{\partial u_i}{\partial y_k}\right) + 2\sum_{j,k,l=1}^{n} g^{kl}\Gamma^i_{jk}\frac{\partial u_j}{\partial y_l} \\
&\quad + \sum_{j,k,l=1}^{n}\left\{\frac{\partial}{\partial y_k}(g^{kl}\Gamma^i_{jl}) + \sum_{m=1}^{n} g^{kl}\Gamma^m_{jl}\Gamma^i_{km}\right\} u_j \\
(\boldsymbol{N}u)_i &= \sum_{j=1}^{n} u_j \frac{\partial u_i}{\partial y_j} + \sum_{j,k=1}^{n} \Gamma^i_{jk} u_j u_k \\
(\boldsymbol{M}u)_i &= \sum_{j=1}^{n} \frac{\partial \boldsymbol{Y}_j}{\partial t}\frac{\partial u_i}{\partial y_j} + \sum_{j,k=1}^{n}\left\{\Gamma^i_{jk}\frac{\partial \boldsymbol{Y}_k}{\partial t} + \frac{\partial \boldsymbol{Y}_i}{\partial x_k}\frac{\partial^2 \boldsymbol{X}_k}{\partial t \partial y_j}\right\} u_j \\
(\boldsymbol{G}p)_i &= \sum_{j=1}^{n} g^{ij}\frac{\partial p}{\partial y_j},
\end{aligned}$$

where, for each $i, j, k \in \{1, \dots, n\}$,

$$g^{ij} = \sum_{k=1}^{n} \frac{\partial Y_i}{\partial x_k}(X(\mathbf{y}, t), t)\frac{\partial Y_j}{\partial x_k}(X(\mathbf{y}, t), t)$$

(the metric contravariant tensor),

$$g_{ij} = \sum_{k=1}^{n} \frac{\partial X_i}{\partial y_k}(\mathbf{y}, t)\frac{\partial X_j}{\partial y_k}(\mathbf{y}, t)$$

(the metric covariant tensor), and

$$\Gamma_{ij}^{k} = \frac{1}{2}\sum_{i=1}^{n} g^{kl}\left\{\frac{\partial g_{il}}{\partial y_j} + \frac{\partial g_{jl}}{\partial y_i} - \frac{\partial g_{ij}}{\partial y_l}\right)$$

(Christoffel's symbol).

Proposition 12.1 *The pair* $(\mathbf{v}, q)$ *satisfies*

$$(\mathbf{v}, q) \in \mathcal{U}(0, T; \mathcal{D}(t)) \times L^2(0, T; \widehat{H}^1(\mathcal{D}(t)))$$

together with (12.1)–(12.4) *iff the pair* $(\mathbf{u}, p)$ *defined by* (12.8)–(12.10) *satisfies the conditions*

$$(\mathbf{u}, p) \in \mathcal{U}(0, T; \mathcal{D}) \times L^2(0, T; \widehat{H}^1(\mathcal{D}))$$

together with

$$\frac{\partial \mathbf{u}}{\partial t} - \nu(\boldsymbol{L}\mathbf{u}) + (\boldsymbol{M}\mathbf{u}) + (\boldsymbol{N}\mathbf{u}) + (\boldsymbol{G}p) = \mathbf{0} \quad \text{in } \mathcal{D} \times (0, T) \tag{12.11}$$

$$\operatorname{div} \mathbf{u} = 0 \quad \text{in } \mathcal{D} \times (0, T) \tag{12.12}$$

$$\mathbf{u}(\mathbf{y}, t) = \boldsymbol{\omega} \wedge \mathbf{y} \quad \text{on } \partial\mathcal{B} \times (0, T), \tag{12.13}$$

$$\mathbf{u}(\mathbf{y}, 0) = \mathbf{v}_0(\mathbf{y}), \ \mathbf{y} \in \mathcal{D}. \tag{12.14}$$

For the proof we refer to [135].

12.1.5 *Proof of Theorem 12.1*

12.1.5.1 The Linear System

We introduce the Banach space

$$\widetilde{\mathcal{U}} \equiv \widetilde{\mathcal{U}}(0, T; \mathcal{D}) := L^2(0, T; \mathcal{H}^2) \cap C([0, T], \mathcal{H}^1) \cap H^1(0, T, \mathcal{L}^2) \tag{12.15}$$

with the norm

$$\|\boldsymbol{U}\|_{\widetilde{\mathcal{U}}} = \|\boldsymbol{U}\|_{L^2(0,T;\mathcal{H}^2)} + \|\boldsymbol{U}\|_{L^\infty(0,T;\mathcal{H}^1)} + \|\boldsymbol{U}\|_{H^1(0,T;\mathcal{L}^2)}.$$

Let $(\mathbf{u}, p)$ be a solution of (12.11)–(12.14), and let $\boldsymbol{U}(\mathbf{y}, t) = \mathbf{u}(\mathbf{y}, t) - \boldsymbol{\Lambda}(\mathbf{y})$, $P(\mathbf{y}, t) = p(\mathbf{y}, t)$, where $\boldsymbol{\Lambda}$ is given by (12.6).

$(\boldsymbol{U}, P)$ satisfies the system

$$\frac{\partial \boldsymbol{U}}{\partial t} - \nu(\boldsymbol{L}\boldsymbol{U}) + (\boldsymbol{M}\boldsymbol{U}) + (\boldsymbol{N}\boldsymbol{U}) + (\boldsymbol{B}\boldsymbol{U}) + (\boldsymbol{G}p) = \boldsymbol{F} \quad \text{in } \mathcal{D} \times (0, T), \tag{12.16}$$

$$\operatorname{div} \boldsymbol{U} = 0 \quad \text{in } \mathcal{D} \times (0, T), \tag{12.17}$$

$$\boldsymbol{U}(\mathbf{y}, t) = 0 \quad \text{on } \partial\mathcal{B} \times (0, T), \tag{12.18}$$

$$\boldsymbol{U}(\mathbf{y}, 0) = \boldsymbol{U}_0(\mathbf{y}), \ \mathbf{y} \in \mathcal{D}, \tag{12.19}$$

where, for each $i \in \{1, \ldots, n\}$, we have set

$$(\boldsymbol{B}\boldsymbol{U})_i = \sum_{j=1}^{n} \left(\boldsymbol{U}_j \frac{\partial \boldsymbol{\Lambda}_i}{\partial x_j} + \boldsymbol{\Lambda}_j \frac{\partial \boldsymbol{U}_i}{\partial x_j} \right) + 2 \sum_{j,k=1}^{n} \Gamma^i_{j,k} \boldsymbol{U}_j \boldsymbol{\Lambda}_k \tag{12.20}$$

and

$$\left. \begin{array}{c} \boldsymbol{F} = \nu(\boldsymbol{L}\boldsymbol{\Lambda}) - (\boldsymbol{M}\boldsymbol{\Lambda}) - (\boldsymbol{N}\boldsymbol{\Lambda}) \\ \boldsymbol{U}_0 = \mathbf{v}_0 - \boldsymbol{\Lambda} \end{array} \right\}. \tag{12.21}$$

It implies that problem (12.11)–(12.14) is equivalent to system (12.16)–(12.21), which has homogeneous boundary conditions. The solution of the problem (12.16)–(12.21) can be solved as a fixed point of the mapping $\mathcal{N} : (\boldsymbol{W}, Q) \to (\boldsymbol{U}, P)$ defined from

$$\widetilde{\mathcal{U}} \times L^2(0, T; \widehat{H}^1(\mathcal{D}))$$

onto itself, where $(\boldsymbol{U}, P)$ satisfies the classical Stokes system

$$\frac{\partial \boldsymbol{U}}{\partial t} - \nu \Delta \boldsymbol{U} + \nabla p = \boldsymbol{F} \quad \text{in } \mathcal{D} \times (0, T), \tag{12.22}$$

$$\operatorname{div} \boldsymbol{U} = 0 \quad \text{in } \mathcal{D} \times (0, T), \tag{12.23}$$

$$\boldsymbol{U}(\mathbf{y}, t) = \mathbf{0} \quad \text{on } \partial\mathcal{B} \times (0, T), \tag{12.24}$$

$$\boldsymbol{U}(\mathbf{y}, 0) = \boldsymbol{U}_0(\mathbf{y}), \ \mathbf{y} \in \mathcal{D} \tag{12.25}$$

with

$$\begin{aligned} \boldsymbol{F} = {} & \nu(\boldsymbol{L}\boldsymbol{\Lambda}) - (\boldsymbol{M}\boldsymbol{\Lambda}) - (\boldsymbol{N}\boldsymbol{\Lambda}) + \nu((\boldsymbol{L} - \Delta)\boldsymbol{W}) - (\boldsymbol{M}\boldsymbol{W}) - (\boldsymbol{N}\boldsymbol{W}) \\ & - (\boldsymbol{B}\boldsymbol{W}) + ((\nabla - G)Q), \end{aligned} \tag{12.26}$$

where the operators $\boldsymbol{L}$, $\boldsymbol{M}$, $\boldsymbol{N}$, $\boldsymbol{G}$ are introduced in (12.11)–(12.17) and $\boldsymbol{B}$ is defined by (12.20).

Proposition 12.2 *Let $T > 0$, $\boldsymbol{F} \in L^2(0, T; \mathcal{L}^2)$, $\boldsymbol{U}_0 \in \mathcal{H}^1$ be such that*

$$\operatorname{div} \boldsymbol{U}_0 = 0 \quad \textit{in } \mathcal{D},$$
$$\boldsymbol{U}_0 = \mathbf{0} \quad \textit{on } \partial\mathcal{B}.$$

Then the linearized problem (12.22)–(12.25) admits a unique strong solution

$$\boldsymbol{U} \in \widetilde{\boldsymbol{\mathcal{U}}}, \quad P \in L^2(0, T; \widehat{H}^1(\mathcal{D})).$$

Moreover, $(\boldsymbol{U}, P)$ satisfies

$$\begin{aligned} \|\boldsymbol{U}\|_{L^2(0,T;\boldsymbol{\mathcal{H}}^2)} + \|\boldsymbol{U}\|_{L^\infty(0,T;\boldsymbol{\mathcal{H}}^1)} + \|\boldsymbol{U}\|_{H^1(0,T;\boldsymbol{\mathcal{L}}^2)} + \|\nabla P\|_{L^2(0,T;\mathcal{L}^2)} \\ \le C\exp(T)(\|\boldsymbol{U}_0\|_{\boldsymbol{\mathcal{L}}^2} + \|\boldsymbol{F}\|_{L^2(0,T;\boldsymbol{\mathcal{L}}^2)}) \end{aligned}$$

for some constant $C > 0$ depending only on $\boldsymbol{\Lambda}$ and ν. □

To prove this result, we apply the theory of semigroups.
In accordance with the notation, introduced in Sect. 8, let

$$\boldsymbol{C}^\infty_{0,\sigma}(\mathcal{D}) := \{\varphi \in \boldsymbol{C}^\infty_0(\mathcal{D}), \operatorname{div} \varphi = 0\},$$

$$\boldsymbol{L}^2_\sigma(\mathcal{D}) := \text{the closure of } \boldsymbol{C}^\infty_{0,\sigma} \text{ in } \boldsymbol{\mathcal{L}}^2,$$

$$\boldsymbol{H}^1_{0,\sigma}(\mathcal{D}) := \text{the closure of } \boldsymbol{C}^\infty_{0,\sigma}(\mathcal{D}) \text{ in } \boldsymbol{\mathcal{H}}^1,$$

where the spaces denoted by boldface letters are spaces of vector functions with values in $\mathbb{R}^n$. The spaces $\boldsymbol{L}^2_\sigma(\mathcal{D})$ and $\boldsymbol{H}^1_{0,\sigma}(\mathcal{D})$ can be characterized as follows:

$$\boldsymbol{L}^2_\sigma(\mathcal{D}) = \{\mathbf{v} \in \boldsymbol{L}^2(\mathcal{D});\ \operatorname{div}\mathbf{v} = 0 \text{ in } \mathcal{D}'(\mathcal{D}),\ \mathbf{v}\cdot\mathbf{n} = 0 \text{ in } H^{-1/2}(\partial\mathcal{D})\},$$
$$\boldsymbol{H}^1_{0,\sigma}(\mathcal{D}) = \{\mathbf{v} \in \boldsymbol{H}^1_0(\mathcal{D});\ \operatorname{div}\mathbf{v} = 0 \text{ in } L^2(\mathcal{D})\}.$$

We will further use the Stokes operator in $\boldsymbol{L}^2_\sigma(\mathcal{D})$. This operator, denoted by $\mathscr{A}^0_0$, has already been studied in Chap. 8. (The subscript 0 corresponds to zero translation, and the superscript 0 expresses the zero rotation.) Recall that $D(\mathscr{A}^0_0) = \boldsymbol{W}^{2,2}(\mathcal{D}) \cap \boldsymbol{W}^{1,2}_{0,\sigma}(\mathcal{D})$ and $\mathscr{A}^0_0$ is defined by the formula $\mathscr{A}^0_0 := -\Pi^2_\sigma \nu \Delta$. In this section, just to simplify the notation, we write only $\mathscr{A}$ instead of $\mathscr{A}^0_0$.

Proposition 12.3 *The operator $\mathscr{A}$ is self-adjoint and positive. Moreover, for any $\boldsymbol{U} \in D(\mathscr{A})$, we have that*

$$\|\boldsymbol{U}\|_{\mathcal{H}^2} \le C\|(I + A)U\|_{\mathcal{L}^2} \tag{12.27}$$

for some constant $C > 0$ depending only on $\mathcal{D}$ and ν. □

For the proof we refer to [81]. The fact that the operator $\mathscr{A}$ is self-adjoint also follows from [201] or Theorem 8.1. The positivity also follows from the form of the spectrum of $\mathscr{A}^0_0$, which is explained at the beginning of Sect. 8.3.1.
Before the proof of Proposition 12.2, we present an observation.
Observation 1: $D(\mathscr{A}^{1/2}) = \boldsymbol{W}^{1,2}_{0,\sigma}(\mathcal{D})$.
To prove Observation 1, it is sufficient to take into account that

$$\|\boldsymbol{U}\|^2_{D(\mathscr{A}^{1/2})} = (\boldsymbol{U}, \boldsymbol{U})_{\mathcal{L}^2} + (\mathscr{A}\boldsymbol{U}, \boldsymbol{U})_{\mathcal{L}^2} = \|\boldsymbol{U}\|^2_{\mathcal{L}^2} + \nu\|\nabla\boldsymbol{U}\|^2_{[\boldsymbol{L}^2(\mathcal{D})]}$$

for all $\boldsymbol{U} \in D(\mathscr{A})$, which implies that this norm is equivalent to the $\mathcal{H}^1$-norm. Since $\boldsymbol{C}^\infty_{0,\sigma}(\mathscr{A})$, we deduce that $D(\mathscr{A}^{1/2}) = \boldsymbol{W}^{1,2}_{0,\sigma}(\mathcal{D})$.

Proof of Proposition 12.2 It follows from Observation 1 that $\boldsymbol{U}_0 = D(\mathscr{A}^{1/2})$. Due to Lemma 3.3, Theorem 3.1 [13], and Proposition 12.3, we obtain that the initial-valued problem

$$\boldsymbol{U}' + \mathscr{A}\boldsymbol{U} = \Pi^2_\sigma \boldsymbol{F}, \quad U(0) = \boldsymbol{U}_0 \tag{12.28}$$

admits a unique solution

$$\boldsymbol{U} \in L^2(0, T; D(\mathscr{A})) \cap C([0, T], D(\mathscr{A}^{1/2})) \cap H^1(0, T; \boldsymbol{L}^2_\sigma(\mathcal{D})).$$

Further, there exists $k > 0$, depending on $\mathcal{D}$, ν, T, nondecreasing with respect to T such that

$$\|\boldsymbol{U}\|_{L^2(0,T;D(\mathscr{A}))} + \|\boldsymbol{U}\|_{C([0,T],D(\mathscr{A}^{1/2}))} + \|\boldsymbol{U}\|_{H^1(0,T;\boldsymbol{L}^2_\sigma(\mathcal{D}))} \qquad (12.29)$$
$$\leq k(\|\boldsymbol{U}_0\|_{D(\mathscr{A}^{1/2})} + \|{\Pi_\sigma}^2\boldsymbol{F}\|_{L^2(0,T;\boldsymbol{L}^2_\sigma(\mathcal{D}))}).$$

In our case, we multiply (12.28) by $\boldsymbol{U}$ and $\mathscr{A}\boldsymbol{U}$, and we get that $k = C\exp(T)$, for some $C > 0$ depending only on $\mathcal{D}$, ν. Moreover, multiplying (12.28) by $\boldsymbol{\varphi} \in \mathcal{V}(\mathcal{D})$, we get

$$\int_{\mathcal{D}} [\boldsymbol{U}'(t) - \nu\Delta\boldsymbol{U}(t) - \boldsymbol{F}(t)] \cdot \boldsymbol{\varphi} d\mathbf{y} = 0 \text{ a.e. in } (0, T).$$

Using Propositions 1.1 and 1.2 from [210], we get the existence of pressure $P \in L^2(0, T; \widehat{H}^1(\mathcal{D}))$ such that

$$\boldsymbol{U}'(t) - \nu\Delta\boldsymbol{U}(t) + \nabla P(t) = \boldsymbol{F}(t) \quad \text{in } \mathcal{D} \times (0, T). \qquad (12.30)$$

Since $\boldsymbol{U}(t) \in \boldsymbol{W}^{1,2}_{0,\sigma}(\mathcal{D})$, we have

$$\begin{aligned} \operatorname{div} \boldsymbol{U} &= 0 \quad \text{in } \mathcal{D} \times (0, T), \\ \boldsymbol{U} &= 0 \quad \text{on } \partial\mathcal{D} \times (0, T). \end{aligned}$$

This gives us that from (12.30) $(\boldsymbol{U}, P)$ is a strong solution to the problem (12.22)–(12.29).

To show the uniqueness, it follows from the fact that all solutions of (12.22)–(12.25) are also solutions of (12.28), whose solutions are unique. Finally, (12.26) follows from (12.29) and (12.27).

Let us mention the estimates on the coefficients. The proof can be found in [26]. □

Observation 2. From Definition (12.6) of $\boldsymbol{\Lambda}$, we get the existence of a constant $K = K(r) > 0$ such that

$$\|D^\alpha \boldsymbol{\Lambda}\|_{\boldsymbol{L}^\infty(\mathbb{R}^n)} \leq K, \quad \forall \alpha \in \mathbb{N}^n, \ |\alpha| \leq 3. \qquad (12.31)$$

Lemma 12.4 *Let $T > 0$. Then, there exists a constant $K = K(r) > 0$ such that the function $\boldsymbol{X}$ introduced in (12.7) satisfies the following estimates:*

$$\left\| \frac{\partial \boldsymbol{X}_i}{\partial y_i} \right\|_{L^\infty(\mathbb{R}^n \times (0,T))} \leq \exp(kT),$$

$$\left\| \frac{\partial^2 \boldsymbol{X}_i}{\partial y_i \partial y_k} \right\|_{L^\infty(\mathbb{R}^n \times (0,T))} \leq KT\exp(2kT),$$

$$\left\| \frac{\partial^3 \boldsymbol{X}_i}{\partial y_j \partial y_k \partial y_l} \right\| \leq KT\exp(2KT)[1 + 2kT\exp(kT)],$$

for all $i, j, k, l \in \{1, \ldots, n\}$.

Moreover, the same estimates are valid if we replace $\partial \boldsymbol{X}_i/\partial y$, by $\partial \boldsymbol{Y}_i/\partial x_j$, $\partial^2 \boldsymbol{X}_i/\partial y_j \partial y_k$ by $\partial^2 \boldsymbol{Y}_i/\partial x_j \partial x_k$, and $\partial^3 \boldsymbol{X}_i/\partial y_j \partial y_k \partial y_l$ by $\partial^3 \boldsymbol{Y}_i/\partial x_j \partial x_k \partial x_l$, respectively. □

For proof see [26].

Corollary 12.1 *There exists a constant $K = K(r) > 0$ such that for all $m, l \in \{1, \ldots, n\}$*

$$\left\| \frac{\partial \boldsymbol{X}_m}{\partial y_l} - \delta_{ml} \right\|_{L^\infty(\mathbb{R}^n \times (0,T))} \leq nKT\exp(KT) \tag{12.32}$$

$$\|g_{ml} - \delta_{ml}\|_{L^\infty(\mathbb{R}^n \times (0,T))} \leq nkT(\exp(KT))[1 + n\exp(kT)]. \tag{12.33}$$

For proof see [26].

12.1.6 Proof of the Local Existence and Uniqueness Result

From Proposition 12.1 we have to show that the mapping of $\mathcal{N}$ is well defined. For $T, R > 0$,

$$\mathcal{K} = \{(\boldsymbol{W}, Q) \in \widetilde{\boldsymbol{\mathcal{U}}} \times L^2(0, T; \widehat{H}^1(\mathcal{D})) | \|\boldsymbol{W}\|_{\mathcal{U}} + \|\nabla Q\|_{L^2(0,T;\mathcal{L}^2)} \leq R\}. \tag{12.34}$$

We set by K a constant satisfying the following:

(*) K is positive and depends only on r ($r \geq d(\mathcal{B})$) and n.

The goal of this part is to show that for T_0 small enough, and R large enough, $\mathcal{N}(\mathcal{K}) \subset \mathcal{K}$ and $\mathcal{N} \cdot \mathcal{K} \to \mathcal{K}$ is a contraction for any $T \in (0, T_0)$.

Lemma 12.5 *Suppose that* $(\boldsymbol{W}, Q) \in \mathcal{K}$, $\boldsymbol{L}, \boldsymbol{M}, \boldsymbol{G}, \boldsymbol{B}$ *defined by (12.11), (12.13), (12.14), and (12.20). Then there exist constants* K, K_1 *satisfying* $(*)$ *such that:*

(i) $\|\nu((\boldsymbol{L} - \Delta)\boldsymbol{W})\|_{L^2(0,T;\mathcal{L}^2)} \leq KTR\exp(K_1T)$
(ii) $\|(\boldsymbol{M}\boldsymbol{W})\|_{L^2(0,T,\mathcal{L}^2)} \leq KT^{1/2}R\exp(K_1T)$
(iii) $\|(\boldsymbol{B}\boldsymbol{W})\|_{L^2(0,T,\mathcal{L}^2)} \leq KT^{1/2}R\exp(K_1T)$
(iv) $\|(\nabla - \boldsymbol{G})\varphi\|_{L^2(0,T,\mathcal{L}^2)} \leq KTR\exp(K_1T)$ □

Proof See [26]. □

Lemma 12.6 *Let* $\mathcal{U}$ *be the space defined by (12.15). Then for* $\boldsymbol{V}, \boldsymbol{W} \in \widetilde{\mathcal{U}}$ *we have:*

(i) $(\boldsymbol{W} \cdot \nabla)\boldsymbol{V} \in L^{5/2}(0, T, \mathcal{L}^2)$
(ii) $\boldsymbol{W}_i\boldsymbol{V}_j \in L^\infty(0, T; L^2(\mathcal{D}))$ $\forall i, j \in \{1, \ldots, n\}$

Moreover, there exists a constant $C > 0$ *depending on* $\mathcal{D}$ *such that*

$$\|(\boldsymbol{W}\nabla)\boldsymbol{V}\|_{L^{5/2}(0,T;\mathcal{L}^2)} \leq c\|\boldsymbol{W}\|_{L^\infty(0,T,\mathcal{H}^1)}\|\boldsymbol{V}\|^{1/5}_{L^\infty(0,T,\mathcal{H})}\|\boldsymbol{V}\|^{4/5}_{L^2(0,T,\mathcal{H}^2)} \tag{12.35}$$

$$\|\boldsymbol{W}_i\boldsymbol{V}_j\|_{L^\infty(0,T;L^2(\mathcal{D}))} \leq C\|\boldsymbol{W}\|_{L^\infty(0,T,\mathcal{H}^1)} \cdot \|\boldsymbol{V}\|_{L^\infty(0,T,\mathcal{H}^1)}. \tag{12.36}$$

See [26] for the proof.

Corollary 12.2 *There exists constant* K_1 *satisfying* $(*)$ *and* $c = c(n, \mathcal{D}) > 0$ *such that for all* $(\boldsymbol{W}, Q) \in \mathcal{K}$

$$\|(\boldsymbol{N}\boldsymbol{W})\|_{L^2(0,T;\mathcal{L}^2)} \leq CT^{1/10}R^2\exp(K_1T).$$

Proof See [26]. □

Corollary 12.3 *For* R *large enough and* T_0 *small enough, we have* $\mathcal{N}(\mathcal{K}) \subset \mathcal{K}$ *for any* $T \in (0, T_0)$. □

Proof See [26]. □

In order to apply the Banach fixed point theorem, we have to show that for R and T as in previous Corollary 12.3 the mapping $\mathcal{N} : \mathcal{K} \to \mathcal{K}$ is a contraction for any $T \in (0, T_0)$.

Let us consider

$$(W^1, Q^1), (W^2, Q^2) \in \mathcal{K}.$$

By F^i we set the function defined by (12.26) corresponding to (W^i, Q^i), $i \in \{1, 2\}$, and by $\mathcal{N}(\boldsymbol{W}^i, Q^i) = (\boldsymbol{U}^i, \boldsymbol{P}^i)$, $i \in \{1, 2\}$. We set

$$\boldsymbol{W} = \boldsymbol{W}^1 - \boldsymbol{W}^2$$

$$Q = Q^1 - Q^2$$

$$U = U^1 - U^2$$
$$P = P^1 - P^2$$
$$F = F^1 - F^2.$$

Then (U, P) is the solution of the system (the Stokes problem)

$$\frac{\partial U}{\partial t} - \nu \Delta U = \nabla P = F \quad \text{in } \mathcal{D} \times (0, T)$$
$$\operatorname{div} U = 0 \quad \text{in} \mathcal{D} \times (0, T)$$
$$U(\mathbf{y}, t) = 0 \quad \text{on } \partial\mathcal{B} \times (0, T)$$
$$U(\mathbf{y}, 0) = 0, \ \mathbf{y} \in \mathcal{D},$$

where

$$F = \nu((L-\Delta)W)-(MW)-(NW^1)+(NW^2)-(BW)+((\nabla-G)Q), \qquad (12.37)$$

and the estimates of terms in F are the following:

Lemma 12.7 *There exist constants K, K_1 satisfying $(*)$ such that for all $(W, Q) \in$ $|boldsymbol\widetilde{\mathcal{U}} \times L^2(0, T, \widehat{H}^1(\mathcal{D}))$:*

(i) $\|\nu((L - \Delta)W)\|_{L^2(0,T,\mathcal{L}^2)} \leq KT\|W\|_{\mathcal{U}}\exp(K_1T)$
(ii) $\|(MW)\|_{L^2(0,T;\mathcal{L}^2)} \leq KT^{1/2}\|W\|_{\mathcal{U}}\exp(K_1, T)$
(iii) $\|(BW)\|L_{2(0,T;\mathcal{L}^2)} \leq KT^{1/2}\|W\|_{\mathcal{U}}\exp(K_1T)$
(iv) $\|((\nabla - G)Q\|L_{L^2(0,T;\mathcal{L}^2)} \leq KT\|\nabla Q\|_{L^2(0,T;\mathcal{L}^2)}\exp(K_1T)$ □

Proof See [26]. □

Lemma 12.8 *There exists a constant K_1 satisfying $(*)$ and $c = c(n, \mathcal{D})$ such that for all $(W^1(Q^1))$, $(W^1, W^2) \in \mathcal{K}$*

$$\|(NW^1) - (NW^2)\|_{L^2(0,T;\mathcal{L}^2)} \leq CRT^{1/10}\|W\|_{\mathcal{U}}\exp(K_1T).$$

Proof See [26]. □

Proof of Theorem 12.1 From Lemmas 12.7 and 12.9 and using (12.37) and (12.24), we get

$$\begin{aligned} \|U\|_{\widetilde{\mathcal{U}}} + \|\nabla P\|_{L^2(0,T;\mathcal{L}^2)} \leq & C\exp(T)\{KT\|W\|_{\mathcal{U}} \exp(K_1T) \\ & + 2KT^{1/2}\|W\|_{\mathcal{U}} \exp(K_1T) \\ & + KT\|\nabla Q\|L_{L^2(0,T;\mathcal{L}^2)} \exp(K_1T) \\ & + c_1RT^{1/10}\|W\|_{\mathcal{U}} \exp(K_1T)\}. \end{aligned} \qquad (12.38)$$

Therefore, for $R > 0$ and $T_0 > 0$ from Corollary 12.3, we get from (12.38) that

$$\begin{aligned}\|U\|_{\widetilde{\mathcal{U}}} + \|\nabla P\|_{L^2(0,T;\mathcal{L}^2)} \leq& \widetilde{C}\{(\widetilde{C}_2/2)T_0\|W\|_{\mathcal{U}} + \widetilde{C}_2 T_0^{1/2}\|W\|_{\mathcal{U}} \\ &+ (\widetilde{C}_2/2)T_0\|\nabla Q\|_{L^2(0,T;\mathcal{L}^2)} + \widetilde{C}_1 R T_0^{1/10}\|W\|_{\mathcal{U}}\} \\ \leq& \widetilde{C}\widetilde{C}_3 T_0^{1/10} R\{\|W\|_{\mathcal{U}} + \|\nabla Q\|_{L^2(0,T,\mathcal{L}^2)}\},\end{aligned}$$

where $\widetilde{C}$, $\widetilde{C}_1$, $\widetilde{C}_2$ follow from the proof of Corollary 12.2. Then, for R and T_0 as in Corollary 12.3, respectively, we obtain from previous inequality $\mathcal{N}; \mathcal{K} \to \mathcal{K}$ is a contraction for all $T \in (0, T_0)$. This gives us that $\mathcal{N}$ has a unique fixed point, and it implies that problem (12.16)–(12.21) admits a unique strong solution

$$(U, P) \in \widetilde{\mathcal{U}} \wedge L^2(0, T; \widehat{H}^1(\Omega))$$

for any $T \in (0, T_0)$.

Further, we have to show that we can choose the time T_0 such that one of the alternatives (i) or (ii) holds true. The result follows in a standard manner from the fact that the local existence time T_0 obtained in Corollary 12.2 is uniform with respect to $\mathbf{v}_0$, provided that $\|\mathbf{v}_0\|_{\mathcal{H}^1} \leq C_0$. We have thus proved our local existence and uniqueness result. □

12.1.7 Proof the Global Existence and Uniqueness in 2D

From Theorem 12.1, we would like to show that in H^1 norm of v in $2D$ does not blow up in finite time. Therefore, the aim is to show

$$t \to \|v(t)\|_{\mathcal{H}^1(t)}$$

is bounded on $[0, T)$ for all $T \in (0, T_0)$, where T_0 is the maximal time existence of the solution $\mathbf{v}$. We define a new change of variables

$$\mathbf{w}(\mathbf{y}, t) = \mathbf{v}(\mathbf{y}, t) - \boldsymbol{\Lambda}(\mathbf{y}), \tag{12.39}$$

where $(\mathbf{v}, q)$ is a solution to problem (12.1)–(12.4) given by Theorem 12.1 and $\boldsymbol{\Lambda}$ is defined by (12.6). Clearly $(\mathbf{w}, q)$ satisfies

$$\frac{\partial \mathbf{w}}{\partial t} - \nu\Delta\mathbf{w} + (\mathbf{w}\cdot\nabla)\mathbf{w} + \nabla q = \boldsymbol{F} - (\mathbf{w}\cdot\nabla)\boldsymbol{\Lambda} - (\boldsymbol{\Lambda}\cdot\nabla)\mathbf{w} \tag{12.40}$$

$$\operatorname{div}\mathbf{w} = 0 \tag{12.41}$$

$$\mathbf{w}(\mathbf{x}, t) = \mathbf{0}\ \partial\mathcal{B}(t) \tag{12.42}$$

$$\mathbf{w}(\mathbf{x}, 0) = \mathbf{w}_0(\mathbf{x}) \quad \text{in } \mathcal{D}, \tag{12.43}$$

where

$$F = \nu \Delta \boldsymbol{\Lambda} - (\boldsymbol{\Lambda} \cdot \nabla)\boldsymbol{\Lambda} \tag{12.44}$$

$$\mathbf{w}_0 = \mathbf{v}_0 - \boldsymbol{\Lambda}. \tag{12.45}$$

Lemma 12.9 *Let* $(\mathbf{v}, q)$ *be the strong solution to problem (12.1)–(12.4) given by Theorem 12.1. It implies that* $\mathbf{v}$ *is bounded in space* $\mathbf{v} \in L^\infty(0,T;\mathcal{L}^2(t)) \cap L^2(0,T,\mathcal{H}^1(\mathcal{D}))$*, i.e., for any* $T \in (0,T_0)$ *there exists a constant* $C_T > 0$ *such that*

$$\|\mathbf{v}\|_{L^\infty(0,T;\mathcal{L}^2(t))} + \|\mathbf{v}\|_{L^2(0,T;\mathcal{H}^1(t))} \leq C_T.$$

Proof: See [26].

Lemma 12.10 *Let* $\mathbf{w} \in L^2(0,T;\mathcal{H}^2(t)) \cap C([0,T];\mathcal{H}^1(t) \cap H^1(0,T;\mathcal{L}^2(t))$ *be such that* $\mathbf{w}|_{\partial\mathcal{B}(t)} = 0$ *for all* $t \in [0,T]$*. Define*

$$\frac{d^\Lambda \mathbf{w}}{dt} = \frac{\partial \mathbf{w}}{\partial t} + (\boldsymbol{\Lambda} \cdot \nabla)\mathbf{w}.$$

Then we have that for a.e. $t \in (0,T)$

$$\frac{1}{2}\frac{d}{dt}\int_{\mathcal{D}(t)} |\nabla \mathbf{w}|^2 dx = \int_{D(t)} \Delta \mathbf{w} \cdot \frac{d^\Lambda \mathbf{w}}{dt} dx - \int_{D(t)} \nabla \mathbf{w} : [\nabla \mathbf{w} \cdot \nabla \Lambda] dx. \tag{12.46}$$

Proof See [26]. □

Proposition 12.4 *Let* $(\mathbf{v}, q)$ *be the strong solution to problem (12.1)–(12.4) as in Theorem 12.1, and let* $n = 2$*. Then, the mapping* $t \to \|\mathbf{v}(t)\|_{\mathcal{H}^1(t)}$ *is bounded in* $[0,T)$ *for all* $T \in (0,T_0)$*, i.e., there exists a constant* $C_T > 0$ *such that*

$$\|\mathbf{v}(t)\|_{\mathcal{H}^1(t)} \leq C_T, \quad \forall t \in [0,T).$$

Proof It is enough to show that

$$\|\mathbf{w}(t)\|_{\mathcal{H}^1(t)} \leq C_T \quad \forall t \in [0,T),$$

where $\mathbf{w}$ is defined by (12.39). Let us study the function $\boldsymbol{\varphi}$ defined by $\boldsymbol{\varphi} = (d^\Lambda \mathbf{w}/dt) - (\mathbf{w} \cdot \nabla)\boldsymbol{\Lambda}$. From the previous lemma, it follows that:

(i) $\boldsymbol{\varphi} \in L^2(0,T;\mathcal{L}^2(t))$ for any $T \in (0,T_0)$.
(ii) $\operatorname{div} \boldsymbol{\varphi} = 0$ in $\mathcal{D}'(\mathcal{D}(t))$ a.e. in $(0,T)$.
(iii) $\boldsymbol{\varphi} \cdot n = 0$ in $H^{-1/2}(\partial B(t))$ a.e. in $(0,T)$.

Moreover, from the definition of $\boldsymbol{\varphi}$ and Eq. (12.40), we can rewrite

$$\boldsymbol{\varphi} - \nu\Delta\mathbf{w} + (\mathbf{w}\cdot\nabla)\mathbf{w} + \nabla q = \boldsymbol{F} - 2(\mathbf{w}\cdot\nabla)\boldsymbol{\Lambda} \text{ in } \mathcal{D}(t),\ t \in (0,T).$$

We take inner product $\mathcal{L}^2(t)$ with φ

$$\begin{aligned}\|\boldsymbol{\varphi}\|^2_{\mathcal{L}^2(t)} &- \nu(\Delta\mathbf{w},\boldsymbol{\varphi})_{\mathcal{L}^2(t)} + ((\mathbf{w}\cdot\nabla)\mathbf{w},\boldsymbol{\varphi})_{\mathcal{L}^2(t)} \\ &= (F,\boldsymbol{\varphi})_{\mathcal{L}^2(t)} - 2((\mathbf{w}\cdot\nabla)\boldsymbol{\Lambda}),\boldsymbol{\varphi})_{\mathcal{L}^2(t)}.\end{aligned} \tag{12.47}$$

Integrating by parts and applying Lemma 12.10,

$$\begin{aligned}-\nu\int_{\mathcal{D}(t)}\Delta\mathbf{w}\cdot\boldsymbol{\varphi}d\mathbf{x} =& \frac{\nu}{2}\frac{d}{dt}\int_{\mathcal{D}(t)}|\nabla\mathbf{w}|^2 d\mathbf{x} \\ &+\nu\int_{\mathcal{D}(t)}\nabla\mathbf{w} : [\nabla u\cdot\nabla\boldsymbol{\Lambda}]dx - \nu\int_{\mathcal{D}(t)}\nabla\mathbf{w} : \nabla[(\mathbf{w}\cdot\nabla)\boldsymbol{\Lambda})d\mathbf{x}.\end{aligned} \tag{12.48}$$

Then from (12.47) using (12.48), we obtain

$$\begin{aligned}&\|\boldsymbol{\varphi}\|^2_{\mathcal{L}^2(t)} + \frac{\nu}{2}\frac{d}{dt}\|\nabla\mathbf{w}(t)\|^2_{[L^2(\mathcal{D}(t))]^4} \\ &\quad=\nu\int_{\mathcal{D}(t)}\nabla\mathbf{w} : \nabla[(\mathbf{w}\cdot\nabla)\boldsymbol{\Lambda}]dx - \nu\int_{\mathcal{D}(t)}\nabla\mathbf{w} : [\nabla\mathbf{w}\cdot\nabla\boldsymbol{\Lambda}]dx \\ &\qquad - ((\mathbf{w}\cdot\nabla)\mathbf{w},\boldsymbol{\varphi})_{\mathcal{L}^2(t)} + (F,\boldsymbol{\varphi})_{\mathcal{L}^2(t)} - 2((\mathbf{w}\cdot\nabla)\boldsymbol{\Lambda},\boldsymbol{\varphi})_{\mathcal{L}^2(t)} \text{ a.e. in } (0,T).\end{aligned}$$

Applying the Hölder inequality, Young's inequality, and (12.31), we get

$$\begin{aligned}&\|\boldsymbol{\varphi}^2\|_{\mathcal{L}^2(t)} = \frac{\nu}{2}\frac{d}{dt}\|\nabla\mathbf{w}(t)\|^2_{[L^2(\mathcal{D}(t))]^4} \\ &\quad\le 2\nu K\|\nabla\mathbf{w}\|^2_{[L^2(\mathcal{D}(t))]^4} + \nu K\|\nabla\mathbf{w}\|_{[L^2(\mathcal{D}(t))]} \\ &\qquad + (\mathbf{w}\cdot\nabla)\mathbf{w}\|_{L(t)}\|\varphi\|_{\mathcal{L}^2(t)} + C(\|\varphi\|_{\mathcal{L}^2(t)} + 2K\|\mathbf{w}\|_{\mathcal{L}^2(t)}\|\varphi\|_{\mathcal{L}^2(T)} \text{ a.e. in } (0,T).\end{aligned} \tag{12.49}$$

Using the inequality

$$(\varepsilon a)\cdot\left(\frac{1}{\varepsilon}b\right) \le \frac{1}{2}\varepsilon^2 a^2 + \frac{1}{2\varepsilon^2}b^2 \quad \forall a,b,\varepsilon > 0 \tag{12.50}$$

and integrating (12.49) with respect to t,

$$\begin{aligned}
&\frac{1}{4}\int_0^t \|\boldsymbol{\varphi}\|^2_{\mathcal{L}^2(t)}dx + \frac{\nu}{2}\|\nabla \mathbf{w}(t)\|^2_{[L^2(\mathcal{D}(t))]^4} \\
&\le \frac{\nu}{2}\|\nabla \mathbf{w}_0\|^2_{[L^2(\Omega)]^4} + 2\nu k\int_0^t \|\nabla \mathbf{w}\|^2_{[L^2(\Omega(s))]}ds \\
&+ \nu K\int_0^t \|\nabla \mathbf{w}\|_{[L^2(\Omega)(s))]^4}\|\mathbf{w}\|_{\mathcal{L}^2(s)} \\
&+ \int_0^t \|(\mathbf{w}\cdot\nabla)\mathbf{w}\|L^2_{\mathcal{L}^2(s)} + c^2T \\
&+ 4K^2\int_0^t \|\mathbf{w}\|^2_{\mathcal{L}^2(s)}ds \quad \forall t\in[0,T].
\end{aligned}$$

Using Hölder inequality and Lemma 12.9, it implies

$$\begin{aligned}
&\frac{1}{4}\int_0^t \|\boldsymbol{\varphi}\|^2_{\mathcal{L}^2(s)} + \frac{\nu}{2}\|\nabla \mathbf{w}(t)\|^2_{L^2(\mathcal{D}(t))^4} \\
&\le M + \int_0^t \|(\mathbf{w}\cdot\nabla)\mathbf{w}\|^2_{\mathcal{L}^2(s)}ds \quad \forall t\in[0,T),
\end{aligned} \tag{12.51}$$

where

$$\begin{aligned}
M =&(\nu/2)\|\nabla \mathbf{w}_0\|^2_{[L^2(\mathcal{D}(t))]^4} + 2\nu K\widetilde{C}^2 \\
&+ \nu K\widetilde{C}^2T^{1/2} + C^2T + 4K^2\widetilde{C}^2T, \\
\widetilde{C} =&\widetilde{C}(\nu, r, \|\mathbf{v}_0\|_{\mathcal{L}^2}, T) > 0
\end{aligned}$$

$$\|\mathbf{w}\|_{L^\infty(0,T;\mathcal{L}^2(t))} + \|\mathbf{w}\|_{L^2(0,T;\mathcal{H}^1(t))} \le \widetilde{C}. \tag{12.52}$$

We need to get the estimation of $\mathbf{w}\cdot\nabla\mathbf{w}$. Using the change of variable $\mathbf{X} = Q(t)\mathbf{y}$, we study $\mathbf{z}(\mathbf{y}, t) = Q^t(t)\mathbf{w}(Q(t)\mathbf{y}, t)$, $p(\mathbf{y}, t) = q(Q(t)\mathbf{y}, t)$ with $(\mathbf{w}, q)$ solutions of (12.40)–(12.47). Then we obtain

$$\int_{\mathcal{D}(t)} |(\mathbf{w}\cdot\nabla_x)\mathbf{w}|^2dx = \int_{\mathcal{D}(t)} |(\mathbf{z}\cdot\nabla_y)\mathbf{z}|^2d\mathbf{y} \tag{12.53}$$

$$\begin{aligned}
\int_\Omega |(\mathbf{z}\cdot\nabla_y)\mathbf{z})|^2d\mathbf{y} \le&\|\mathbf{z}\|^2_{[L^4(\mathcal{D}))]^2}\|\nabla\mathbf{z}\|^2_{[L^4(\mathcal{D}))]^4} \\
\le&C_1\|\mathbf{z}\|_{L^2}\|\nabla\mathbf{z}\|_{[L^2(\Omega)]^4}\|\nabla\mathbf{z}\|_{[L^2(\Omega))^4]}\|z\|_{\mathcal{H}^2}, \\
C_1 =& C_1(\mathcal{D}) > 0.
\end{aligned}$$

Moreover, $\mathbf{z} \in \mathcal{D}(A)$, and from Proposition 12.3

$$\|\mathbf{z}\|_{\mathcal{H}^2} \leq C_2(\|\mathbf{z}\|_{\mathcal{L}^2} + \|A\mathbf{z}\|_{\mathcal{L}^2}) \tag{12.54}$$
$$C_2 = C_2(\nu, \mathcal{D}) > 0, \text{ positive constant.}$$

Now, we investigate $(\mathbf{w}, q)$ as the solution of the homogeneous resolvent Stokes problem at some fixed time $t > 0$:

$$\begin{aligned} \mathbf{w} - \nu\Delta\mathbf{w} + \nabla q &= \mathbf{f} + \mathbf{w} \\ \operatorname{div}\mathbf{w} &= 0 \\ \mathbf{w} &= \mathbf{0} \end{aligned}$$
$$\mathbf{f} = \boldsymbol{F} - 2(\mathbf{w}\cdot\nabla)\boldsymbol{\Lambda} - \boldsymbol{\varphi} - (\mathbf{w}\cdot\nabla)\mathbf{w}. \tag{12.55}$$

It follows from an easy observation that $(\mathbf{z}, p)$ satisfies the similar homogeneous resolvent Stokes problem

$$\begin{aligned} \mathbf{z} - \nu\Delta\mathbf{z} + \nabla p &= \tilde{\mathbf{f}} + \mathbf{z} \quad \text{in } \mathcal{D}, \\ \operatorname{div}\mathbf{z} &= 0 \quad \text{in } \mathcal{D}, \\ \mathbf{z} &= \mathbf{0} \quad \text{on } \partial\mathcal{D} \end{aligned}$$

with $\tilde{\mathbf{f}}(\mathbf{y}, t) = Q^T(t)\mathbf{f}(Q(t)\mathbf{y}, t)$. Then from previous estimates, we get

$$\|\mathbf{z}\|_{\mathcal{L}^2} + \|A\mathbf{z}\|_{\mathcal{L}^2} \leq C_3(\|\mathbf{f}\|_{\mathcal{L}^2(t)} + \|\mathbf{z}\|_{\mathcal{L}^2}),$$
$$C_3 = C_3(\nu, \mathcal{D}) > 0.$$

From (12.53)–(12.54), we deduce that

$$\begin{aligned} &\int_{\mathcal{D}(t)} t(\mathbf{w}\cdot\nabla_x)\mathbf{w}|^2 dx \\ &\leq C_4\|\mathbf{w}\|_{\mathcal{L}^2(t)}\|\nabla\mathbf{w}\|^2_{[L^2(\mathcal{D}(t))]^4}(\|\mathbf{f}\|_{\mathcal{L}^2(t)} + \|\mathbf{w}\|_{\mathcal{L}^2(t)}), \\ &C_4 = C_4(\nu, \Omega) > 0. \end{aligned} \tag{12.56}$$

From (12.55) and (12.31),

$$\|\mathbf{f}\|_{\mathcal{L}^2(t)} \leq \|\boldsymbol{F}\|_{\mathcal{L}^2(t)} + 2K\|\mathbf{w}\|_{\mathcal{L}^2(t)} + \|\boldsymbol{\varphi}\|_{\mathcal{L}^2(t)} + \|(\mathbf{w}\cdot\nabla)\mathbf{w}\|_{L^2(t)}. \tag{*}$$

Now, using (∗) in (12.56), it follows that

$$\begin{aligned}\|(\mathbf{w}\cdot\nabla)\mathbf{w}\|^2_{\mathcal{L}^2(t)} \leq & C_4\|\mathbf{w}\|^2_{\mathcal{L}^2(t)}\|\nabla w\|^2_{[L^2(\mathcal{D}(t))]^4} \\ & + C_4\|\mathbf{w}\|_{L^2(t)}\|\nabla \mathbf{w}\|^2_{[L^2(\mathcal{D}(t))]^4}(\|\boldsymbol{F}\|_{\mathcal{L}^2(t)} + 2K\|\mathbf{w}\|_{L^2(t)}) \\ & + \|\boldsymbol{\varphi}\|_{\mathcal{L}^2(t)} + \|(\mathbf{w}\cdot\nabla)\mathbf{w}\|_{\mathcal{L}^2(t)}.\end{aligned} \tag{12.57}$$

Integrating (12.57) with respect to t and using Lemma 12.9, we get

$$\begin{aligned}\int_0^t \|(\mathbf{w}\cdot\nabla)\mathbf{w}\|^2_{\mathcal{L}^2(s)}ds \leq & L + C_4\widetilde{C}\int_0^t \|\nabla\mathbf{w}\|^2_{(L^2(\mathcal{D}(s))^2} \\ & + C_4\widetilde{C}\int_0^t \|\nabla\mathbf{w}\|^2_{[L^2(\mathcal{D}(t)]^4}\|(\mathbf{w}\cdot\nabla)\mathbf{w}\|_{\mathcal{L}^2(s)}\end{aligned} \tag{12.58}$$

with $L = C_4\widetilde{C}^4 + C_4\widetilde{C}^3C + 2K\widetilde{C}^4$, $\widetilde{C}$, C constant. Applying (12.50), from (12.58), we obtain

$$\begin{aligned}\int_0^t \|(\mathbf{w}\cdot\nabla)\mathbf{w}\|^2_{\mathcal{L}^2(s)}ds \leq L + & \frac{C_4^2\widetilde{C}^2}{2\varepsilon^2}\int_0^t \|\nabla w\|^4_{[L^2(\mathcal{D}(s))]^4} \\ & + \varepsilon^2\int_0^t (\|\boldsymbol{\varphi}\|^2\mathcal{L}^2(s) + \|\mathbf{w}\cdot\nabla)\mathbf{w}\|^2_{\mathcal{L}^2(s)}.\end{aligned}$$

Let us choose ε small enough, and using previous estimates, we obtain

$$\begin{aligned}\frac{1}{5}\int_0^t & \|\boldsymbol{\varphi}\|^2_{\mathcal{L}^2(s)}ds + \frac{2\nu}{5}\|\nabla\mathbf{w}(t)\|^2_{[L^2(\mathcal{D}(t))]^4} \\ & \leq \frac{4M}{5} + L + \frac{5C_4^2\widetilde{C}^2}{2}\int_0^t \|\nabla\mathbf{w}\|^4_{[L^2(\mathcal{D}(s))]^4}ds \\ & + \frac{1}{5}\int_0^t \|\boldsymbol{\varphi}\|_{\mathcal{L}^2(s)}ds, \quad \forall t\in[0,T].\end{aligned}$$

Hence, we obtain

$$\|\nabla\mathbf{w}\|^2_{[L^2(\mathcal{D}(t))^4]} \leq \frac{2M}{\nu} + \frac{5L}{2\nu} + \frac{25C_4^2\widetilde{C}^2}{4\nu}\int_0^t \|\nabla u\|^4_{[L^2(\mathcal{D}(t))]^4}.$$

Applying Gronwall's lemma, we obtain

$$\|\nabla w\|^2_{[L^2(\mathcal{D}(t)]^2} \leq \frac{4M+3L}{2\nu}\exp\frac{25C_2^2\widetilde{C}^2}{4\nu}\int_0^T \|\nabla\mathbf{w}\|^2_{[L^2(\mathcal{D}(s))]^4}ds,$$

which together with Lemma 12.9 implies that the mapping

$$t \to \|\nabla \mathbf{w}(t)\|_{[L^2(\mathcal{D}(t))]^4}$$

is bounded in $[0, T)$ which implies the existence and uniqueness in 2D case. □

Remark 12.2 The generalization in L^p case and with several rigid bodies we refer to [47]. □

12.2 Global in Time Existence of a Strong Solution

This section is based on the results by Galdi and Silvestre, see [108] where the global in time existence of a strong solution was proven. We just give the sketch of their proofs, and the details can be found in [108]. We consider the system

$$\begin{aligned}
\partial_t \mathbf{u} + \mathbf{u} \cdot \nabla \mathbf{u} &= \nu \Delta \mathbf{u} - \nabla p + \boldsymbol{\omega} \wedge \mathbf{x} \cdot \nabla \mathbf{u} - \boldsymbol{\omega} \wedge \mathbf{u}, \ \mathcal{D} \times (0, \infty), \\
\nabla \cdot \mathbf{u} &= 0, \ \mathcal{D} \times (0, \infty), \\
\mathbf{u} &= \boldsymbol{\omega} \wedge \mathbf{x}, \quad \text{on } \partial\mathcal{D} \times (0, \infty), \\
\lim_{|x| \to \infty} \mathbf{u}(x, t) &= 0, \ t \in (0, \infty), \\
\mathbf{u}(x, 0) &= \mathbf{u}_0(x), \ x \in \mathcal{D}
\end{aligned} \tag{12.59}$$

and the corresponding steady problem for the fields $\mathbf{u}^s$, p^s:

$$\begin{aligned}
\nu \Delta \mathbf{u}^s - \nabla p^s &= \mathbf{u}^s \cdot \nabla \mathbf{u}^s + \boldsymbol{\omega} \wedge \mathbf{u}^s - \boldsymbol{\omega} \wedge \mathbf{x} \cdot \nabla \mathbf{u}^s \\
\nabla \cdot \mathbf{u}^s &= 0 \\
\mathbf{u}^s &= \boldsymbol{\omega} \wedge \mathbf{x} \\
\lim_{|\mathbf{x}| \to \infty} \mathbf{u}^s(\mathbf{x}) &= 0.
\end{aligned} \tag{12.60}$$

The aim of this subsection is to show the existence of strong solutions to (12.59) in the sense that $\mathbf{u}$, p have the following properties:

$$\begin{aligned}
&(\mathbf{u} - \mathbf{b}), \ \partial_t \mathbf{u} \in L^\infty(0, T; \boldsymbol{L}^2(\mathcal{D})); \ \nabla \mathbf{u}, \nabla^2 \mathbf{u} \in L^2(0, T; \boldsymbol{L}^2(\mathcal{D})); \\
&\nabla p \in L^2(0, T; L^2(\mathcal{D}))
\end{aligned} \tag{12.61}$$

for some $T > 0$, where $\mathbf{b}$ is a suitable extension of the boundary data $\boldsymbol{\omega} \wedge \mathbf{x}|_{\partial\mathcal{D}}$ to $\mathcal{D}$.

First we will focus on the proof of the *local existence:*
Let $\mathbf{u}_0 \in H^2(\mathcal{D})$ satisfy

$$\nabla \cdot \mathbf{u}_0 = 0 \ \text{ in } \mathcal{D}, \ \mathbf{u}|_{\partial\mathcal{D}} = \boldsymbol{\omega} \wedge \mathbf{x}, \tag{12.62}$$

and assume

$$\boldsymbol{\omega} \wedge \mathbf{x} \cdot \nabla \mathbf{u}_0 \in L^2(\mathcal{D});$$

then, there exists $T' > 0$ such that (12.59) has strong solutions in $[0, T')$ satisfying (12.61) with $T = T'$ and $\mathbf{b}$ smooth and of bounded support in $\mathcal{D}$.
Moreover,

$$\mathbf{u} \in C([0, T'); \boldsymbol{H}^1(\mathcal{D})) \cap C((0, T'); \boldsymbol{H}^2(\mathcal{D})),$$
$$\lim_{t \to 0^+} \|\mathbf{u}(t) - \mathbf{u}_0\|_{\boldsymbol{H}^1(\mathcal{D})} = 0.$$

12.2.1 Global Existence and Convergence to Steady State

Let $(\mathbf{u}_0 - \mathbf{u}^s) \in \boldsymbol{H}^2(\mathcal{D})$ with $\mathbf{u}_0$ satisfying (12.62) and $(\mathbf{u}^s, p^s)$ a solution to (12.60) satisfying

$$\begin{aligned} &\|\nabla^2 \mathbf{u}^s\|_2 + \sup_{x \in \mathcal{D}}((1 + |\mathbf{x}|^2)|\nabla \mathbf{u}^s(\mathbf{x})|) + \sup_{\mathbf{x} \in \mathcal{D}}((1 + |\mathbf{x}|)|\mathbf{u}^s(\mathbf{x})|) \le C|\boldsymbol{\omega}| \\ &\|\nabla p^s\|_2 + \sup_{\mathbf{x} \in \mathcal{D}}((1 + |\mathbf{x}|^2)|p^s(\mathbf{x})|) + \sup_{\mathbf{x} \in \mathcal{D}^R}((1 + |\mathbf{x}|^3)|\nabla p^s(\mathbf{x})|) \le C|\boldsymbol{\omega}|, \end{aligned} \tag{12.63}$$

where $C = C(\mathcal{D}, \nu) > 0$ and $\mathcal{D}^R = \{\mathbf{x} \in \mathcal{D}; |\mathbf{x}| > R\}$.
Moreover,

$$\boldsymbol{\omega} \wedge \mathbf{x} \cdot \nabla(\mathbf{u}_0 - \mathbf{u}^s) \in \boldsymbol{L}^2(\mathcal{D}).$$

Then, there is a constant $C = C(\mathcal{D}, \nu) > C$ such that if

$$\|\mathbf{u}_0 - \mathbf{u}^s\|_{H^1(\mathcal{D})} + |\boldsymbol{\omega}| < C,$$

problem (12.59) has a global strong solution in $[0, \infty)$, with $\mathbf{u}$ satisfying $(12.63)_1$ with $T = \infty$, $\mathbf{b} = \mathbf{u}^s$.

Moreover, for all $T > 0$,

$$
\begin{aligned}
(\mathbf{u} - \mathbf{u}^s) &\in C([0, T];\ \boldsymbol{L}^2(\mathcal{D})),\\
\nabla \mathbf{u} &\in C([0, T];\ \boldsymbol{L}^2(\mathcal{D})),\\
\nabla^2 \mathbf{u} &\in C((0, T];\ L^2(\mathcal{D})),\\
\nabla p &\in L^2(0, T;\ L^2(\mathcal{D})),\\
\text{and } \lim_{t\to 0^+} &\|\mathbf{u}(t) - \mathbf{u}_0\|_{H^1(\mathcal{D})} = 0.
\end{aligned}
$$

Finally, $\mathbf{u}(t)$ converges to $\mathbf{u}^s$, as $t \to \infty$ in the sense

$$\lim_{t\to\infty} \|\nabla(\mathbf{u} - \mathbf{u}^s)\|_{L^2(\mathcal{D})} = 0.$$

First we rewrite our system (12.59) in the form $(\mathbf{u}, p) = (\mathbf{v} + \mathbf{b}, p)$, where $\mathbf{b}$ is a suitable extension of the boundary data $\boldsymbol{\omega} \wedge \mathbf{x}|_{\partial\mathcal{D}}$ and $(\mathbf{u}, p)$ solves the following problem:

$$
\begin{aligned}
&\partial_t \mathbf{v} + \mathbf{v}\cdot\nabla\mathbf{v} = \nu\Delta\mathbf{v} - \nabla p - \mathbf{v}\cdot\nabla\mathbf{b} - \mathbf{b}\nabla\mathbf{v} + \boldsymbol{\omega}\wedge\mathbf{x}\cdot\nabla\mathbf{v} - \boldsymbol{\omega}\wedge\mathbf{x} + \mathbf{f},\ \mathcal{D}\times(0,\infty),\\
&\qquad \nabla\cdot\mathbf{v} = 0,\ \mathcal{D}\times(0,\infty),\\
&\qquad \mathbf{v} = 0 \text{ at } \partial\mathcal{D}\times(0,\infty),\\
&\lim_{|\mathbf{x}|\to\infty} \mathbf{v}(\mathbf{x}, t) = 0,\ t\in(0,\infty),\\
&\qquad \mathbf{v}(\mathbf{x}, 0) = \mathbf{u}_0 - \mathbf{b}(\mathbf{x}) \equiv \mathbf{a}(\mathbf{x}),\ \mathbf{x}\in\mathcal{D}
\end{aligned}
\tag{12.64}
$$

with

$$\mathbf{f} = \nu\Delta\mathbf{b} + \boldsymbol{\omega}\wedge\mathbf{x}\cdot\nabla\mathbf{b} - \boldsymbol{\omega}\wedge\mathbf{b} - \mathbf{b}\cdot\nabla\mathbf{b}. \tag{12.65}$$

Let us recall the following results concerning the steady Stokes problem.

Lemma 12.11 *Let $\mathcal{D} \subset \mathbb{R}^3$ be a bounded domain with C^2 boundary. Let $(\mathbf{w}, \pi)$ be a solution to the Stokes problem*

$$
\begin{aligned}
\Delta\mathbf{w} - \nabla\pi &= g \text{ in } \mathcal{D},\\
\nabla\cdot\mathbf{w} &= 0 \text{ in } \mathcal{D},\\
\mathbf{w} &= \mathbf{0} \text{ at } \partial\mathcal{D}
\end{aligned}
\tag{12.66}
$$

with $g \in L^2(\mathcal{D})$. Then $\mathbf{w} \in H^2(\mathcal{D})$ and

$$\|\nabla^2 \mathbf{w}\| \le C_1(\|Pg\|_2 + \|\nabla \mathbf{w}\|_2)$$
$$\|\nabla \mathbf{w}\|_3 \le C_2(\|Pg\|_2^{1/2}\|\nabla \mathbf{w}\|_2^{1/2} + \|\nabla w\|_2) \tag{12.67}$$

with $C_i = C_i(\mathcal{D})$, $i = 1, 2$, depending only on the C^2 regularity of $\partial\mathcal{D}$. □

The next lemma is about the construction of a solenoidal extension in $\mathcal{D}$ of the boundary velocity $\mathbf{u}(\mathbf{x}, t)|_{\partial\mathcal{D}} = \boldsymbol{\omega} \wedge \mathbf{x}$.

Lemma 12.12 *There exists a function $\mathbf{b} \in C_0^\infty(\overline{\mathcal{D}})$ such that*

$$\mathbf{b}|_{\partial\mathcal{D}}(\mathbf{x}) = \boldsymbol{\omega} \wedge \mathbf{x} \text{ and}$$
$$\|\mathbf{b}\|_{2,q} \le c|\boldsymbol{\omega}| \text{ for all } q \in [1, \infty], \tag{12.68}$$

where $c = c(\mathcal{D}) > 0$. □

Notation

$$H(\mathcal{D}) = \{\mathbf{u} \in L^2(\mathcal{D}) \colon \nabla \cdot \mathbf{u} = 0,\ \mathbf{u} \cdot \boldsymbol{n}|_{\partial\mathcal{D}} = 0\},$$

$$V(\mathcal{D}) = \{u \in H_0^1(\mathcal{D}) \colon \nabla \cdot \mathbf{u} = 0\},\ P \text{ is projection from } L^2(\mathcal{D}) \text{ onto } H(\mathcal{D}).$$

Lemma 12.13 *Let $\mathbf{w} \in V(\mathcal{D}_R) \cap H^2(\mathcal{D}_R)$. Then,*

(i) $\boldsymbol{\omega} \wedge \mathbf{x} \cdot \nabla \mathbf{u}|_{\partial B_R} = 0$.

(ii) $(\boldsymbol{\omega} \wedge \mathbf{w} - \boldsymbol{\omega} \wedge \mathbf{x} \cdot \nabla \mathbf{w}) \in H(\mathcal{D}_R)$.

(iii)
$$\begin{aligned} &\int_{\mathcal{D}_R} (\boldsymbol{\omega} \wedge \mathbf{w} - \boldsymbol{\omega} \wedge \mathbf{w} \cdot \nabla \mathbf{w}) \cdot \nabla \mathbf{w} \\ &= \int_{\partial\mathcal{D}} [-\boldsymbol{n} \cdot \nabla \mathbf{w} \cdot (\boldsymbol{\omega} \wedge \mathbf{x} \cdot \nabla \mathbf{w}) + \frac{1}{2}|\nabla \mathbf{w}|^2 \boldsymbol{\omega} \wedge \mathbf{x} \cdot \mathbf{w}] \\ &\quad - \int_{\mathcal{D}_R} \nabla(\boldsymbol{\omega} \wedge \mathbf{w}) : \nabla \mathbf{w}. \end{aligned} \tag{12.69}$$

Now, we would like to show that the question of solvability of the initial-boundary solve problem (12.59) can be reduced to system (12.64). The goal of the next part of the subsection is to show suitable a priori estimates for approximating solutions to (12.64), where the extension $\mathbf{b}$ of the boundary data is either (i) the field given in Lemma (12.12) or (ii) $\mathbf{b} = \mathbf{u}^s$, where $\mathbf{u}^s$ verifies (12.61) and (12.60). Note that $\mathbf{b}$ has the following property:

$$\|\mathbf{b}\|_{1,\infty} + \|\nabla \mathbf{b}\|_3 \leq c|\boldsymbol{\omega}|, \tag{12.70}$$

where $c = c(\mathcal{D})$.

The property follows from the extension of Lemma 12.12, while it follows from (12.60) if $\mathbf{b} = \mathbf{u}^s$.

Further from (12.65), Lemma 12.12, (12.60), and $(12.63)_2$, we also have

$$\begin{aligned} &\|\mathbf{f}\|_q \leq c|\boldsymbol{\omega}|, \ q \geq 1 \ \text{in case (i)}, \\ &c = c(\mathcal{D}, q) > 0, \\ &f = \nabla p^s \in L^2 \ \text{in case (ii)}. \end{aligned}$$

The idea behind the proof is to construct approximating solutions in bounded domains (invading domains $\{\mathcal{D}_{R_i}\}$, $R_i > \delta(R)$, $\mathcal{D}_{R_i} \subset \mathcal{D}_{R_i+1}$, $\mathcal{D} = \bigcup_{i\in\mathbb{N}} \mathcal{D}_{R_i}$).

To find these solutions we shall apply the Faedo–Galerkin method with eigenfunctions $a_{R_{i,j}}, {}_{j\in\mathbb{N}}$ of the Stokes operator as base functions

$$P_{\Delta a_{R_{i,j}}} = -\lambda_{R_{i,j}} a_{R_{i,j}}, \ j \in \mathbb{N}, \tag{12.71}$$

where $\{\lambda_{R_{i,j}};\ j \in \mathbb{N}\}$ are corresponding eigenvalues. The corresponding eigenfunctions are chosen to be orthonormal in $L^2(\mathcal{D}_{R_i})$.

Let us begin with the following result.

Lemma 12.14 *Let* $\boldsymbol{a} \in \boldsymbol{H}^2(\mathcal{D}) \cap V(\mathcal{D})$ *with* $\boldsymbol{\omega} \wedge \mathbf{x}\nabla \boldsymbol{a} \in \mathbf{L}^2(\mathcal{D})$, *and let* $\{D_{R_i};\ i \in \mathbb{N}\}$ *be a sequence of "invading domains,"* R_1 *sufficiently large. Then there exists a sequence of function* $\{\boldsymbol{a}_{R_i};\ i \in \mathbb{N}\}$ *such that:*

(i) $\boldsymbol{\alpha}_{R_i}(\mathbf{x}) = 0$ *for all* $x \in \mathcal{D}^{R_i/2}$.
(ii) $\boldsymbol{\alpha}_{R_i} \in \boldsymbol{H}^2(\mathcal{D}_{R_i}) \cap \boldsymbol{V}(\mathcal{D}_{R_i})$.
(iii) $\lim_{R_i\to\infty} \|\boldsymbol{\alpha}_{R_i} - \boldsymbol{a}\| = 0$. □

Further, let

$$\boldsymbol{\alpha}_{R,k} = \sum_{j=1}^{k} (\boldsymbol{\alpha}_{R_i}, \boldsymbol{\alpha}_{R_j})\boldsymbol{\alpha}_{R_{ij}}$$

with $(\cdot,)$ scalar product in $L^2(\mathcal{D}_{R_i})$, and the following inequalities hold:

$$\begin{aligned} \|\boldsymbol{\alpha}_{R_i,k}\|_{2,\Omega_{R_i}} &\le c\|\boldsymbol{a}\|_2; \\ \|\nabla\boldsymbol{\alpha}_{R,k}\|_{2,\mathcal{D}_{R_i}} &\le c\|\nabla\boldsymbol{a}\|_2; \\ \|P\Delta\boldsymbol{\alpha}_{R_i,k}\|_{2,\mathcal{D}_{R_i}} &\le c\|\boldsymbol{a}\|_{2,2} \end{aligned} \tag{12.72}$$

for all $k \in \mathbb{N}$ and some $c = c(\mathcal{D}) > 0$.

Finally, there is $N = N(R_i) \in \mathbb{N}$ such that

$$\|\boldsymbol{\omega} \wedge \mathbf{x} \cdot \nabla\boldsymbol{\alpha}_{R_ik}\|_{2,\mathcal{D}_{R_i}} \le c(\|\boldsymbol{\omega} \wedge \mathbf{x} \cdot \nabla\boldsymbol{a}\|_2 + \|a\|_2) \text{ for all } k \ge N,$$

where $c = c(\mathcal{D}) > 0$.

Let us discus the proof of the existence of the strong solution of (12.59).

We begin with the local existence result.

Theorem 12.3 *Let $\mathcal{D}$ be an exterior domain with boundary $\mathcal{B}$ of class C^2. Given $\boldsymbol{\omega} \in \mathbb{R}^3$ and*

$$\mathbf{u}_0 \in \boldsymbol{H}^2(\mathcal{D}), \qquad \boldsymbol{\omega} \wedge \mathbf{x} \cdot \nabla\mathbf{u}_0 \in \boldsymbol{L}^2(\mathcal{D}) \tag{12.73}$$

with

$$\nabla \cdot \mathbf{u}_0 = 0 \text{ in } \mathcal{D}, \qquad \mathbf{u}_0/_{\sum} = \boldsymbol{\omega} \wedge \mathbf{x}, \tag{12.74}$$

there exist $T^* > 0$ and the pair of functions $(\mathbf{u}, p)$, satisfying the problem (12.59) *in $\mathcal{D} \times (0, T^*)$, such that*

$$\begin{aligned} &\partial_t\mathbf{u} \in L^\infty(0, T^*; \boldsymbol{L}^2(\mathcal{D})), \qquad \nabla\partial_t\mathbf{u},\ \nabla^2\mathbf{u} \in L^2(0, T^*; \boldsymbol{L}^2(\mathcal{D})) \\ &\mathbf{u} \in C([0, T^*); \boldsymbol{H}^1(\mathcal{D})) \cap C((0, T^*); \boldsymbol{H}^2(\mathcal{D})); \\ &\nabla p \in L^2(0, T^*; L^2(D)); \qquad \lim_{t\to 0^+} \|\mathbf{u}(t) - \mathbf{u}_0\|_{1,2} = 0. \end{aligned} \tag{12.75}$$

Proof Put $R \equiv R_i$ and

$$\mathbf{v}_{R,k}(0) = \sum_{j=1}^{k} (\boldsymbol{a}_{R_i}, \boldsymbol{a}_{R_j})\boldsymbol{a}_{R_j}.$$

Using Lemma 12.14, we obtain the estimates

$$\begin{aligned}\|\mathbf{v}_{R,k}(0)\|_{2,D_R} &\le c\|\boldsymbol{a}\|_2;\\ \|\nabla\mathbf{v}_{R,k}(0)\|_{2,D_R} &\le c\|\lambda\boldsymbol{a}\|_2;\\ \|P\Delta\mathbf{v}_{R,k}(0)\|_2 &\le c\|\lambda\boldsymbol{a}\|_{2,2};\\ \|\boldsymbol{\omega}\wedge\mathbf{x}\cdot\nabla\mathbf{v}_{R,k}(0)\|_{2,\mathcal{D}_R} &\le c(\|\boldsymbol{\omega}\wedge\mathbf{x}\cdot\nabla\boldsymbol{a}\|_2+\|\boldsymbol{a}\|_2)\end{aligned} \tag{12.76}$$

for all $k\in\mathbb{N}$. We choose extension $\boldsymbol{b}$ from Lemma 12.12. Now, since $\|\nabla\mathbf{b}\|_\infty\le c|\boldsymbol{\omega}|$, $c=c(\mathcal{D})$, and due to Lemma 12.14, we have

$$\frac{\mathrm{d}}{\mathrm{d}t}\|\mathbf{v}\|_2^2+\nu\|\nabla\mathbf{v}\|_2^2\le -2\int_{\mathcal{D}_k}\mathbf{v}\cdot\nabla\boldsymbol{b}\cdot\mathbf{v}+c(\mathcal{D},\nu)\|\mathbf{f}\|_{6/5}^2,$$

and together with $(12.76)_4$, we obtain

$$\|\mathbf{v}(\boldsymbol{b})\|_2^2+\int_0^t\|\nabla\mathbf{v}(\tau)\|_2^2\le C\|\boldsymbol{a}\|_2^2e^{c|\omega|t}+te^{c|\omega|t}c\|\mathbf{f}\|_{6/5} \tag{12.77}$$

for all $t\in[0,T_{R,k}]$, $c=c(\mathcal{D},\nu)>0$. (12.77) holds in any interval $[0,T]$, $T>0$.

Now, we will construct a global weak solution. From (12.77) and from Lemma 12.14, we obtain

$$\begin{aligned}\frac{\mathrm{d}}{\mathrm{d}t}\|\nabla\mathbf{v}\|_2^2+c_1(\nu)\|P\Delta\mathbf{v}\|_2^2&\le c(|\boldsymbol{\omega}|^2+|\boldsymbol{\omega}|)\|\nabla\mathbf{v}\|_2^2\\ &\quad+c(\|\nabla\mathbf{v}\|_2^4+\|\nabla\mathbf{v}\|_2^6)+c_2(\nu)\|\mathbf{f}\|_2^2.\end{aligned} \tag{12.78}$$

We get that there exist some interval $[0,T')$ with $T'=T'(\nu,\|\boldsymbol{a}\|_{1,2},\|\varphi\|_2)>0$ and continuous functions G and $\widehat{G}$ defined on $[0,T')$ such that

$$\|\nabla\mathbf{v}(t)\|_2\le G(t),\qquad \int_0^t\|P\Delta\mathbf{v}(t)\|_2^2\le\widehat{G}(t). \tag{12.79}$$

Finally, using the inequality $\|\nabla\boldsymbol{b}\|_\infty\le c|\boldsymbol{\omega}|$ and from Lemma 12.14,

$$\frac{\mathrm{d}}{\mathrm{d}t}\|\partial_t\mathbf{v}\|_2^2+\nu\|\nabla\partial_t\mathbf{v}\|_2^2\le c\|\nabla v\|_2^4\,\|\partial_t v\|_2^2-2\int_{\mathcal{D}_R}\partial_t\mathbf{v}\cdot\nabla\boldsymbol{b}\cdot\partial_t\mathbf{v},$$

and (12.79), we can find some interval $[0,T'']$, $0<T''\le T'$, and continuous functions H and $\widehat{H}$ defined on $[0,T'']$, such that

$$\|\partial_t\mathbf{v}(t)\|_2\le H(t),\qquad \int_0^t\|\nabla(\partial_t)(\tau)\|_2^2\le\widehat{H}(t). \tag{12.80}$$

These estimates require that the quantity $\|\partial_t v_{Rk}(0)\|_2$ is finite and bounded in terms $\boldsymbol{a}$ and $\mathbf{f}$, only at least along a subsequence. We get such a property using the assumptions on $\mathbf{u}_0$.

Let us write the construction of a sequence of approximating solutions $(\mathbf{v}_{Rk};\ k \in \mathbb{N})$ defined by $\mathbf{v}_{Rk}(x,t) = \sum_{j=1}^{k} c_{Rk_j}(t)\boldsymbol{a}_{R_j}(\mathbf{x})$, where c_{Rk_j} solve the system of ordinary differential equations

$$\begin{aligned}\frac{\mathrm{d}}{\mathrm{d}t}\int_{\mathcal{D}_R}\mathbf{v}_{Rk}\boldsymbol{a}_{R_j} &= \int_{D_R}[\nu\Delta\mathbf{v}_{Rk} - \mathbf{v}_{Rk}\cdot\nabla v_{Rk} - \boldsymbol{b}\cdot\nabla\mathbf{v}_{Rk} - \mathbf{v}_{Rk}\cdot\nabla l]\boldsymbol{a}_{R_j}\\ &+\int_{\mathcal{D}_R}[\boldsymbol{\omega}\wedge\mathbf{x}\cdot\nabla\mathbf{v}_{Rk} - \boldsymbol{\omega}\wedge\mathbf{v}_{Rk} + \mathbf{f}]\boldsymbol{a}_{R_j}\end{aligned} \tag{12.81}$$

with initial conditions $c_{Rl_j}(0) = (\boldsymbol{\alpha}_R, \boldsymbol{a}_{R_j})$, $j = 1,\ldots,k$, where $\boldsymbol{\alpha}_R$ is defined in Lemma 12.14.

We multiply (12.81) by $\frac{dc_{Rk}}{dt}$ at making $t = 0$, and using $(12.76)_4$ and (12.70), we get

$$\begin{aligned}\|\partial_t\mathbf{v}_{Rk}(0)\|_2 &\le \nu\|\Delta\mathbf{v}_{Rk}(0)\|_2 + \|\mathbf{v}_{Rk}(0)\|_\infty\|\nabla\mathbf{v}_{Rk}(0)\|_2\\ &+\|\boldsymbol{b}\|_\infty\|\nabla\mathbf{v}_{Rk}(0)\|_2 + \|\mathbf{v}_{Rk}(0)\|_2\|\nabla\boldsymbol{b}\|_\infty\\ &+\|(\boldsymbol{\omega}\wedge\mathbf{x})\cdot\nabla\mathbf{v}_{Rk}(0)\|_2 + \|\boldsymbol{\omega}\wedge\mathbf{v}_{Rk}(0)\|_2 + \|\mathbf{f}\|_2\\ &\le C(\mathcal{D},\boldsymbol{\omega},\nu)(\|\boldsymbol{a}\|_{2,2} + \|\boldsymbol{a}\|_\infty\|\nabla\boldsymbol{a}\|_2\\ &+\|(\boldsymbol{\omega}\wedge\mathbf{x})\nabla\boldsymbol{a}\|_2) + \|\mathbf{f}\|_2\end{aligned} \tag{12.82}$$

for all $k \ge N$.

From (12.73) it follows that the right-hand side (12.82) is finite. Since $\|\mathbf{u}_0\|_\infty \le c\|\mathbf{u}_0\|_{2,2}$, $c = c(\mathcal{D}) > 0$. From Lemma 12.14,

$$\begin{aligned}\frac{\mathrm{d}}{\mathrm{d}t}\|\nabla_t\mathbf{v}\|_2^2 + c(\nu)\|P\Delta\partial_t\mathbf{v}\|_2^2 &\le c_1[\|\nabla\partial_t\mathbf{v}\|_2^2(|\boldsymbol{\omega}| + |\boldsymbol{\omega}|^2 + \|\nabla\mathbf{v}\|_2^2\\ &+\|\nabla\mathbf{v}\|_2^4 + \|P\Delta\mathbf{v}\|_2^2) + |\boldsymbol{\omega}|^2\|\partial_t\mathbf{v}\|_2^2],\end{aligned} \tag{12.83}$$

where $c(\nu) > 0$, $c_1 = c_1(\mathcal{D},\nu) > 0$.

Multiplying (12.83) by $\tau > 0$, using (12.79) and (12.80), and applying the Gronwall lemma, we obtain the existence of a continuous function J over $[0, T'')$ such that

$$\int_0^t \tau\|\nabla^2\partial_\tau v(\tau)\|_2^2 \le J(t). \tag{12.84}$$

Further, for each fixed R, we will pass $k \to \infty$ along a suitable subsequence. From (12.77)–(12.80) we find an approximate solution $(\mathbf{v}_R, p_R)$ of the problem in the domain $\mathcal{D}_R$ satisfying

$$\begin{aligned} \partial_t \mathbf{v}_R + \mathbf{v}_R \cdot \nabla \mathbf{v}_R &= \nu \Delta \mathbf{v}_R - \nabla p_R - \boldsymbol{b} \cdot \nabla \mathbf{v}_R - \mathbf{v}_R \cdot \nabla \boldsymbol{b} \\ &\quad + (\boldsymbol{\omega} \wedge \mathbf{x}) \cdot \nabla \mathbf{v}_R - \boldsymbol{\omega} \wedge \mathbf{v}_R + \mathbf{f}, \qquad \mathcal{D}_R \times [0, T'); \\ \nabla \cdot \mathbf{v}_R &= 0, \qquad \mathcal{D}_R \times J_0; \\ \mathbf{v}_R &= 0 \ \text{ at } \ \partial \mathcal{D}_R \times (0, T'); \\ \mathbf{v}_R(x, 0) &= \boldsymbol{a}_R(\mathbf{x}), \qquad x \in \mathcal{D}_R \end{aligned} \tag{12.85}$$

together with estimates (12.77)–(12.79). Now, we let $R \to \infty$; it implies the existence of $\mathbf{v}$ satisfying $(12.75)_{1,3}$ with $\mathbf{u} \equiv \mathbf{v}$, $\mathbf{u}_0 \equiv \boldsymbol{a} + \boldsymbol{b}$ and (12.64) for some $p \in L^2(0, T';\ H^1_{\text{loc}}(\overline{\mathcal{D}}))$ estimate (12.77)–(12.79). For more details we refer to the article by Heywood, see [134].

Further,

$$\lim_{R \to \infty} \int_0^{T'} \|\mathbf{v}_R(t) - \mathbf{v}(t)\|_{1,2,\mathcal{D}_{R_0}} = 0 \tag{12.86}$$

for any fixed $R_0 > \delta(B)$. Using Lemma 12.13, it implies $(\boldsymbol{\omega} \wedge \mathbf{x} \cdot \nabla \mathbf{v}_R - \boldsymbol{\omega} \wedge \mathbf{v}_R) \in \boldsymbol{H}(\mathcal{D}_R)$. From $(12.85)_1$ we obtain

$$\begin{aligned} \|(\boldsymbol{\omega} \wedge \mathbf{x}) \cdot \nabla \mathbf{v}_r - \boldsymbol{\omega} \wedge \mathbf{v}_R\|_{2,\mathcal{D}_{R_0}} &\le \|\partial_t \mathbf{v}_R\|_2 + \|\mathbf{v}_R \cdot \nabla \mathbf{v}_R\|_2 + \|\boldsymbol{b} \nabla \mathbf{v}_R\|_2 \\ &\quad + \nu \|\Delta \mathbf{v}_R\| + \|\mathbf{v}_R \cdot \nabla \boldsymbol{b}\|_2 + \|\mathbf{f}\|_2; \end{aligned}$$

therefore it implies the existence of a continuous function $L(t)$ defined on $[0, T']$, such that

$$\int_0^t \|(\boldsymbol{\omega} \wedge \mathbf{x}) \cdot \nabla \mathbf{v}_R - \boldsymbol{\omega} \wedge \mathbf{v}_R\|^2_{2,\mathcal{D}_{R_0}} \le L(t)$$

for all $R \ge R_0 \ge \delta(B)$. From (12.87) it implies that $\mathbf{v} \in L^2(\Omega \times [0, T'])$, and we obtain

$$(\boldsymbol{\omega} \wedge \mathbf{x}) \cdot \nabla \mathbf{v}_R \in L^2(\Omega \times [0, T']).$$

From previous together with $(12.64)_1$ and the properties of $\mathbf{v}$ it implies that p satisfies (12.75). Since $\mathbf{u} = \mathbf{v} + \boldsymbol{b}$, it follows from (12.64) that also $(\mathbf{u}, p)$ satisfy all assertions of the theorem. □

Theorem 12.4 *Let $\mathcal{D}$ be an exterior domain with boundary $\partial\mathcal{D}$ of class C^2 and let $\boldsymbol{\omega} \in \mathbb{R}^3$. Suppose that $(\mathbf{u}_0 - \mathbf{u}^s) \in \boldsymbol{H}^2(\mathcal{D})$, with $\mathbf{u}_0$ satisfying condition* (12.74)*, and $(\mathbf{u}^s, p^s)$ is a solution of* (12.60) *satisfying* (12.61)*. Moreover, assume*

$$\boldsymbol{\omega} \wedge \mathbf{x} \cdot \nabla(\mathbf{u}_0 - \mathbf{u}^s) \in |\boldsymbol{L}^2(\mathcal{D}).$$

Then there is a constant $C = C(\mathcal{D}, \nu) > 0$ such that if

$$\|\mathbf{u}_0 - \mathbf{u}^s\|_{\boldsymbol{H}^1(\mathcal{D})} + |\boldsymbol{\omega}| < C,$$

problem (12.21) *has a global strong solution in* $[0, \infty)$*, which satisfies*

$$\partial_t \mathbf{u} \in L^\infty(0, \infty; \boldsymbol{L}^2(\mathcal{D})); \qquad \nabla\partial_t \mathbf{u}, \nabla^2 \mathbf{u} \in L^2(0, \infty; \boldsymbol{L}^2(\mathcal{D})). \tag{12.87}$$

Further, for all $T > 0$,

$$\begin{aligned} &(\mathbf{u} - \mathbf{u}^s) \in C([0, T]; \boldsymbol{L}^2(\mathcal{D})), && \nabla \mathbf{u} \in C(0, \infty; \boldsymbol{L}^2(\mathcal{D})), \\ &\nabla^2 \mathbf{u} \in C(0, \infty; \boldsymbol{L}^2(\mathcal{D})), && \nabla p \in L^2(0, T; L^2(\mathcal{D})), \end{aligned}$$

and

$$\lim_{t \to 0^+} \|\mathbf{u}(t) - \mathbf{u}_0\|_{H^1(\mathcal{D})} = 0.$$

Finally, $\mathbf{u}(t)$ converges as $t \to \infty$ in the following sense:

$$\lim_{t \to \infty} \|\nabla(\mathbf{u} - \mathbf{u}^s)\|_2 = 0.$$

Proof First let us recall the fundamental estimates from Lemma 12.14 precisely

$$\frac{\mathrm{d}}{\mathrm{d}t}\|\mathbf{v}\|_2^2 + \nu\|\nabla\mathbf{v}\|_2^2 \le -2\int_{\mathcal{D}_R} \mathbf{v} \cdot \nabla\boldsymbol{b} \cdot \mathbf{v} + C(\mathcal{D}, \nu)\|\mathbf{f}\|_{6/5} \tag{12.88}$$

$$\begin{aligned} \frac{\mathrm{d}}{\mathrm{d}t}\|\nabla\mathbf{v}\|_2^2 + C_1(\nu)\|P\Delta\mathbf{v}\|_2^2 \le C(|\boldsymbol{\omega}|^2 + |\boldsymbol{\omega}|)\|\nabla\mathbf{v}\|_2^2 \\ + C(\|\nabla\mathbf{v}\|_2^4\|\nabla\mathbf{v}\|_2^6) + C_2(\nu)\|\mathbf{f}\|_2^2, \end{aligned} \tag{12.89}$$

$C_1(\nu) > 0$, $C = C(\mathcal{D}, \nu)$.

$$\frac{\mathrm{d}}{\mathrm{d}t}\|\partial_t\mathbf{v}\|_2^2 + \nu\|\nabla(\partial_t\mathbf{v})\|_2^2 \le C\|\mathbf{v}\|_2^4\|\partial_t\mathbf{v}\|_2^2 - 2\int_{\mathcal{D}_k} \partial_t\mathbf{v} \cdot \nabla\boldsymbol{b} \cdot \partial_t\mathbf{v}, \tag{12.90}$$

$C = C(\nu) > 0$. Let us choose $\boldsymbol{b} = \mathbf{u}^s$, use Hardy's inequality, and by (12.61) we get

$$-2\int_{\Omega_R} \mathbf{w}\cdot\nabla\boldsymbol{b}\cdot\mathbf{w} \le C|\boldsymbol{\omega}|\,\|\nabla\mathbf{w}\|^2 \qquad \text{for all } \mathbf{w}\in \boldsymbol{H}_0^1(\mathcal{D}_R) \tag{12.91}$$

with $C = C(\mathcal{D})$. Taking $\boldsymbol{\omega}$ sufficiently small, from (12.88) (with $R \equiv R_i$), (12.91), and (12.75), we obtain

$$\|\mathbf{v}(t)\|_2^2 + \int_0^t \|\nabla\mathbf{v}(\tau)\|_2^2 \le C\|\boldsymbol{a}\|_2^2 \qquad \text{for all } t\in[0, T_{Rk}], \tag{12.92}$$

where $C = C(\mathcal{D})$. We can take $T_{Rk} = \infty$. Assuming that $\|\boldsymbol{a}\|_{1,2} = \|\mathbf{u}_0 - \mathbf{u}^s\|_{1,2}$ and $|\boldsymbol{\omega}|$ are small enough, we get by (12.92) and by a classical method that (12.89) has a global solution satisfying

$$\|\nabla\mathbf{v}(t)\|_2 \le C(\|\mathbf{u}_0 - \mathbf{u}^s\|_{1,2}, \omega, \nu, \mathcal{D}) \qquad \text{for all } t\in[o,\infty) \tag{12.93}$$

and

$$\int_0^\infty \|P\Delta\mathbf{v}(t)\|_2^2 \le C'(\|\mathbf{u}_0 - \mathbf{u}^s\|_{1,2}, |\boldsymbol{\omega}|, \nu, \mathcal{D}). \tag{12.94}$$

Let us consider (12.90). Using (12.88) and taking $|\boldsymbol{\omega}|$ small, we find

$$\frac{\mathrm{d}}{\mathrm{d}t}\|\partial_t\mathbf{v}\|_2^2 + C_1(\nu)\|\nabla(\partial_t\mathbf{v})\|_2^2 \le C\|\nabla\mathbf{v}\|_2^4\,\|\partial_t\mathbf{v}\|_2^2, \tag{12.95}$$

where $C_1(\nu) > 0$.

Using Gronwall's lemma we get

$$\|\partial_t\mathbf{v}(t)\|_2^2 \le \|\partial_t\mathbf{v}(0)\|_2^2 e^{C(\nu)\int_0^\infty \|\nabla\mathbf{v}(t)\|_2^4}. \tag{12.96}$$

Let us recall $\boldsymbol{a} = \mathbf{u}_0 - \mathbf{u}^s$, and then from (12.82) we obtain $\|\partial_t\mathbf{v}_{Rk}(0)\|^2$ uniformly bounded in R and k. From (12.96) and (12.93), we get

$$\|\partial_t v(t)\|_2 \le C, \tag{12.97}$$

where C depends on norms of $\mathbf{u}_0 - \mathbf{u}^s$, $\mathcal{D}$, $\boldsymbol{\omega}$, and ν. Coming back to (12.95) and taking (12.93), (12.92), and (12.97), we get

$$\int_0^\infty \|\nabla(\partial_t v)(t)\|_2^2 \le C, \tag{12.98}$$

where C depends on suitable norms of $\mathbf{u}_0 - \mathbf{u}^s$, $\mathcal{D}$, $\boldsymbol{\omega}$, and ν like (12.97).

Moreover, from (12.93) and (12.96) and from Theorem 12.3, it gives us that for all $t > 0$

$$\int_0^T t\|\nabla^2\partial_t \mathbf{v}(t)\|_2^2 \le C, \tag{12.99}$$

with $C = C(\text{data}, t)$.

From (12.92)–(12.94) and (12.97)–(12.100) and as in Theorem 12.3, it follows the existence of a global strong solution $\mathbf{v}$, p to (12.64) which we can deduce the existence of a solution $\mathbf{u} \equiv \mathbf{v} + \mathbf{u}^s$, p to (12.21) which satisfied the required properties. Moreover, we can observe

$$\int_0^\infty \|\nabla(\mathbf{u} - \mathbf{u}^s)\|_2^2 < \infty, \qquad \int_0^\infty \|\nabla(\partial_t \mathbf{u})\|_2^2 < \infty,$$

which can be concluded that

$$\lim_{t\to\infty} \|\nabla(\mathbf{u} - \mathbf{u}^s)\|_2 = 0. \tag{12.100}$$

Thus the proof of Theorem 12.4 is completed. □

12.3 Notes on Chap. 12

- Note that the global in time existence of a weak solution to the Navier–Stokes equations in a general time-varying domain (with the prescribed time variation, which includes situations when a part of the boundary strikes another part), was later proved in [180]. The application of general results from [180] to the flow around one or more rotating bodies (with the possibility that the bodies rotate about different axes) is shown in [180, Section 6 (Example 1)]. Considering the transformations described in Chap. 1, the main theorems from [180] can be reformulated so that they are valid for the problem (1.23), (1.24), (1.27), (1.30), too.
- Let us mention that the existence of the time-periodic case was shown recently, see [55].
- The existence of the weak solutions in the compressible case was investigated, see [146].
- A generalization see [118]. First result was done by Weinerger see [212], [213] and the mathematical analysis see D. Serre [192].

Part IV
Appendix

Chapter 13
Appendix

13.1 The Case When $\mathbf{u}_\infty$ Is Not Vanished or Not Parallel to the Angular Velocity $\boldsymbol{\omega}$

Let us consider a system

$$\begin{aligned} \partial_t \mathbf{v} - \nu \Delta \mathbf{v} + (\mathbf{v} \cdot \nabla)\, \mathbf{v} + \nabla q &= \tilde{\mathbf{f}} && \text{in } \mathcal{D}(t),\ t \in (0, \infty) \\ \operatorname{div} \mathbf{v} &= 0 && \text{in } \mathcal{D}(t),\ t \in (0, \infty) \\ \mathbf{v}(\mathbf{y}, t) &= \mathbf{V}(\mathbf{y}, t) && \text{on } \partial\mathcal{D}(t),\ t \in (0, \infty) \\ \mathbf{v}(\mathbf{y}, t) &\to \mathbf{0} && \text{as } |\mathbf{y}| \to \infty \end{aligned} \tag{13.1}$$

in a time-dependent exterior domain $\mathcal{D}(t) \subset \mathbb{R}^3$. We discuss the case of a time-independent angular velocity $\tilde{\boldsymbol{\omega}} = k\boldsymbol{e}_3$ and constant translational velocity $0 \neq \dot{\mathbf{y}}_C = \mathbf{u}_\infty \in \mathbb{R}^3$ so that $\mathbf{y}_C(t) = \mathbf{u}_\infty t$. For this reason we introduce the change of variables

$$\mathbf{x} = O(t)^T \big(\mathbf{y} - \mathbf{y}_C(t)\big) \tag{13.2}$$

and the new functions

$$\mathbf{u}(\mathbf{x}, t) = O(t)^T \mathbf{v}(\mathbf{y}, t), \quad p(\mathbf{x}, t) = q(\mathbf{y}, t), \quad \mathbf{f}(\mathbf{x}, t) = O(t)^T \tilde{\boldsymbol{f}}(\mathbf{y}, t)\,, \tag{13.3}$$

where the matrix of rotation is defined by

$$O(t) = \begin{pmatrix} \cos kt & -\sin kt & 0 \\ \sin kt & \cos kt & 0 \\ 0 & 0 & 1 \end{pmatrix}. \tag{13.4}$$

Š. Nečasová et al., *Mathematical Theory of a Fluid Flow Around a Rotating and Translating Body*, Springer Monographs in Mathematics,
https://doi.org/10.1007/978-3-031-84120-0_13

Then $(\mathbf{u}, p)$ satisfies—after a linearization around $\mathbf{u} = 0$—the system

$$\begin{aligned} \partial_t \mathbf{u} - \nu \Delta \mathbf{u} + \nabla p - \\ -\big[\big(\boldsymbol{\omega} \wedge \mathbf{x} + O(t)^T \mathbf{u}_\infty\big) \cdot \nabla\big]\mathbf{u} + \boldsymbol{\omega} \wedge \mathbf{u} &= \mathbf{f} \quad \text{in } \mathcal{D} \times (0, \infty) \\ \operatorname{div} \mathbf{u} &= 0 \quad \text{in } \mathcal{D} \times (0, \infty) \\ \mathbf{u} &= \mathbf{u}_{\partial \mathcal{D}} \text{ on } \partial \mathcal{D} \times (0, \infty) \\ \mathbf{u}(\mathbf{x}, t) &\to \mathbf{0} \quad \text{as } |x| \to \infty \end{aligned} \tag{13.5}$$

in a time-independent exterior domain $\mathcal{D} \subset \mathbb{R}^3$, where $\boldsymbol{\omega} = \tilde{\boldsymbol{\omega}} = k\boldsymbol{e}_3$ and $u_{\partial \mathcal{D}}(\mathbf{x}, t) = \boldsymbol{\omega} \wedge \mathbf{x} + O(t)^T \mathbf{u}_\infty$.

Note that if $\mathbf{u}_\infty$ is transversal or even orthogonal to e_3, then (13.5) contains the time-dependent term $(O(t)^T \mathbf{u}_\infty) \cdot \nabla u$ which appears in a natural way for an observer sitting on the rotating and translating obstacle and seeing the fluid flowing past him from the time-dependent direction $O(t)^T \mathbf{u}_\infty$.

To describe the general procedure leading from (13.1) to (13.5) we introduce the skew-symmetric matrix $\widetilde{\Omega(t)} \in \mathbb{R}^{3,3}$, $t \geq 0$, defined by the property $\widetilde{\Omega} \boldsymbol{a} = \tilde{\boldsymbol{\omega}} \wedge \boldsymbol{a}$ for all $\boldsymbol{a} \in \mathbb{R}^3$ and the orthogonal matrix of rotation $O(t) \in \mathbb{R}^{3,3}$ defined by the linear system of ordinary differential equations

$$\dot{O} = \widetilde{\Omega}\, O, \quad O(0) = I.$$

Note that $\dot{O}\, O^T = \widetilde{\Omega} = -O\, \dot{O}^T$. Then the domain $\mathcal{D}(t)$ occupied by the fluid at time $t \geq 0$ is given by

$$\mathcal{D}(t) = O(t)\mathcal{D} + \mathbf{y}_C(t),$$

where $\mathcal{D} = \mathbb{R}^3 \setminus \mathcal{B}$ is the given exterior domain at time $t = 0$.

Now we introduce the change of variables

$$\mathbf{x} = O(t)^T \big(\mathbf{y} - \mathbf{y}_C(t)\big) \tag{13.6}$$

and the new functions, cf. (13.3),

$$\mathbf{u}(\mathbf{x}, t) = O(t)^T \mathbf{v}(\mathbf{y}, t), \ \ p(\mathbf{x}, t) = q(\mathbf{y}, t), \ \ \mathbf{f}(\mathbf{x}, t) = O(t)^T \tilde{\mathbf{f}}(\mathbf{y}, t)\,. \tag{13.7}$$

Then $\mathbf{v}(\mathbf{y}, t) = O(t)\, \mathbf{u}(\mathbf{x}, t) = O(t)\, \mathbf{u}(O^T(\mathbf{y} - \mathbf{y}_C(t)), t)$ has the time derivative

$$\begin{aligned} \mathbf{v}_t &= \dot{O}\, \mathbf{u} + O\, \mathbf{u}_t + O\big[\big(\dot{O}^T(\mathbf{y} - \mathbf{y}_C(t)) - O^T \dot{\mathbf{y}}_C\big) \cdot \nabla_x\big]\mathbf{u} \\ &= O\big(\mathbf{u}_t + O^T \dot{O} \mathbf{u} + \big[\big(\dot{O}^T O \mathbf{x} - O^T \dot{\mathbf{y}}_C\big) \cdot \nabla_x\big]u\big). \end{aligned}$$

Moreover, a simple calculation implies that $\mathbf{v}\cdot\nabla_y\mathbf{v} = O\big(\mathbf{u}\cdot\nabla_x\mathbf{u}\big)$, $\Delta_y\mathbf{v} = O\,\Delta_x\mathbf{u}$ and $\nabla_y q = O\,\nabla_x p$. Hence from (13.1) we get that in $\mathcal{D}$

$$\mathbf{u}_t - \nu\Delta_x\mathbf{u} + u\cdot\nabla_x\mathbf{u} + \big[\big(\dot{O}^T O\mathbf{x} - O^T\dot{\mathbf{y}}_C\big)\cdot\nabla_x\big]u + O^T\dot{O}\mathbf{u} + \nabla_x p = \mathbf{f}. \tag{13.8}$$

To simplify (13.8) we define in addition to $\widetilde{\Omega}$, $\tilde{\boldsymbol{\omega}}$ the skew-symmetric matrix and angular velocity, $\Omega = \Omega(t)$, $\boldsymbol{\omega} = \boldsymbol{\omega}(t)$, $t \geq 0$, respectively, by

$$\Omega = O^T\dot{O} \quad \text{and} \quad \Omega a = \boldsymbol{\omega}\wedge a \text{ for all } a \in \mathbb{R}^3,$$

so that $\dot{O}^T O = -\Omega$, and the "new" path of the center of mass, $\mathbf{x}_C(t)$, defined by

$$\dot{\mathbf{x}}_C = O^T\dot{\mathbf{y}}_C,\ \mathbf{x}_C(0) = \mathbf{y}_C(0) = 0. \tag{13.9}$$

Then (13.8) reads

$$\mathbf{u}_t - \nu\Delta_x\mathbf{u} + \mathbf{u}\cdot\nabla_x\mathbf{u} - \big[\big(\boldsymbol{\omega}\wedge\mathbf{x} + \dot{x}_C\big)\cdot\nabla_x\big]\mathbf{u} + \boldsymbol{\omega}\wedge\mathbf{u} + \nabla_x p = f, \tag{13.10}$$

where $\boldsymbol{\omega}\wedge\mathbf{u}$ except for the factor 2 is the Coriolis force, $\dot{\mathbf{x}}_C(t)$ denotes the velocity of the center of mass in the new coordinate system attached to the rotating obstacle, and $(\boldsymbol{\omega}\wedge\mathbf{x})\cdot\nabla\mathbf{u}$ is a new term not subordinate to the Laplacian in the exterior domain $\mathcal{D}$. Note that $\operatorname{div}_x\mathbf{u} = \operatorname{div}_y\mathbf{v} = 0$.

Summarizing the previous results, we get that $(\mathbf{u},\ p)$ is a solution of the nonlinear system

$$\begin{aligned} \partial_t\mathbf{u} - \nu\Delta\mathbf{u} + (\mathbf{u}\cdot\nabla)\,\mathbf{u} + \nabla p - & \\ -\big[\big(\boldsymbol{\omega}\wedge\mathbf{x} + \dot{\mathbf{x}}_C\big)\cdot\nabla\big]\mathbf{u} + \boldsymbol{\omega}\wedge\mathbf{u} &= \mathbf{f} \quad \text{in } \mathcal{D}\times(0,\infty) \\ \operatorname{div}\mathbf{u} &= 0 \quad \text{in } \mathcal{D}\times(0,\infty) \\ \mathbf{u} &= \mathbf{u}_{\partial\mathcal{D}} \text{ on } \partial\mathcal{D}\times(0,\infty) \\ \mathbf{u}(x,t) &\to \mathbf{0} \quad \text{as } |\mathbf{x}|\to\infty, \end{aligned} \tag{13.11}$$

where $\mathbf{u}_{\partial\mathcal{D}}(\mathbf{x},t) = \boldsymbol{\omega}\wedge\mathbf{x} + \dot{\mathbf{x}}_C$. For more details on this change of coordinates including stress and inertia tensors, we refer to [96, Ch. 1].

Let us describe the main theorems from [65]. The aim was to find an explicit formula of the fundamental solution of (13.5) and to discuss the asymptotic behavior as $|x| \to \infty$. In particular, we are interested in the existence of a wake for any angular velocity $\boldsymbol{\Omega}$ and translational velocity $\mathbf{u}_\infty \neq 0$. To describe the main results, we assume, for simplicity, that $\nu = 1$ and introduce the y- and t-dependent Oseen type operator

$$\mathcal{L}v = \mathcal{L}_{\mathbf{y},t}\mathbf{v} = -\Delta\mathbf{v} - (O(t)^T\mathbf{u}_\infty + \boldsymbol{\omega}\wedge\mathbf{y})\cdot\nabla\mathbf{v} + \boldsymbol{\omega}\wedge\mathbf{v}.$$

Then the fundamental tensor of (13.5) comprises a 3×3-matrix of distributions $\Gamma(\mathbf{y},\mathbf{z},t,s)$ and a three-dimensional vector of distributions $Q(y,z,t,s)$ such that for any vector $a \in \mathbb{R}^3$ the distributions

$$\mathbf{v}_{\mathbf{z},s}(\mathbf{y},t) = \Gamma(\mathbf{y},\mathbf{z},t,s)\boldsymbol{a}, \ t \geq s, \quad \mathbf{v}_{\mathbf{z},s}(\mathbf{y},t) = 0, \ t < s,$$
$$\pi_{\mathbf{z},s}(\mathbf{y},t) = Q(\mathbf{y},\mathbf{z},t,s)\boldsymbol{a}, \ t \geq s, \quad \pi_{\mathbf{z},s}(\mathbf{y},t) = 0, \ t < s$$

solve the system

$$\begin{aligned} \frac{\partial \mathbf{v}_{\mathbf{z},s}}{\partial t} + \mathcal{L}v_{\mathbf{z},s} + \nabla\pi_{\mathbf{z},s} &= \delta_s(t)\delta_z(\mathbf{y})a \\ \operatorname{div}\mathbf{v}_{\mathbf{z},s} &= 0 \end{aligned} \tag{13.12}$$

in the sense of distributions. That is, for all test functions $\varphi \in C_0^\infty(\mathbb{R}^3 \times \mathbb{R})^3$

$$\langle \partial_t \mathbf{v}_{\mathbf{z},s} + \mathcal{L}v_{\mathbf{z},s} + \nabla\pi_{\mathbf{z},s}, \varphi\rangle = \varphi(\mathbf{z},s)\cdot a = \langle \delta_{\mathbf{z}} \otimes \delta_s, \varphi \cdot a\rangle$$

and $\operatorname{div}\mathbf{v}_{\mathbf{z},s} = 0$ for all $t > s$. Here $\delta_s(t)$, $\delta_{\mathbf{z}}(\mathbf{y})$ denote the point masses concentrated at $t = s$, $\mathbf{y} = \mathbf{z}$, respectively.

Moreover, we introduce the fundamental solution of the heat equation in $\mathbb{R}^3$,

$$K(\mathbf{x},t) = \frac{1}{(4\pi t)^{3/2}} \exp\left(-\frac{|\mathbf{x}|^2}{4t}\right).$$

Let ${}_1F_1(a,c,\cdot)$, $a,\ c > 0$, denote the Kummer function

$${}_1F_1(a,c,\lambda) = \sum_{n=0}^{\infty} \frac{(a)_n}{(c)_n}\frac{\lambda^n}{n!}, \tag{13.13}$$

where $(b)_n = \Gamma(b+n)/(\Gamma(b))$ is the Pochhammer symbol. Furthermore, let $\boldsymbol{Y}(t,s)$ be the solution of the ordinary differential equation

$$\frac{\partial \mathbf{Y}}{\partial t} + \boldsymbol{\omega} \wedge \mathbf{Y} = O(t)^T \mathbf{u}_\infty, \quad t > s, \quad \mathbf{Y}(s,s) = 0, \tag{13.14}$$

i.e., $\mathbf{Y}(t,s) = (t-s)O(t)^T\mathbf{u}_\infty$. Finally, for $\mathbf{z} \in \mathbb{R}^3$, let

$$\tilde{\mathbf{z}}(t,s,z) = O(t-s)^T\mathbf{z} - \mathbf{Y}(t,s) = O(s-t)\big(\mathbf{z} - (t-s)O(s)^T\mathbf{u}_\infty\big). \tag{13.15}$$

Theorem 13.1 *The fundamental tensor $\Gamma(\mathbf{y},\mathbf{z},t,s)$, $Q(\mathbf{y},\mathbf{z},t,s)$ of the linearized problem (13.12) can be written in the form*

$$\Gamma(\mathbf{y},\mathbf{z},t,s) = \Gamma_0(\mathbf{y} - \tilde{z}(t,s,\mathbf{z}), t-s), \quad Q(\mathbf{y},\mathbf{z},t,s) = Q_0(\mathbf{y} - \tilde{z}(t,s,\mathbf{z}), t-s),$$

where $\tilde{z}(t, s, \mathbf{z})$ *was defined in (13.15) above and*

$$\Gamma_0(\mathbf{y}, \tau) = K(\mathbf{y}, \tau)\left\{\left[I - \frac{\mathbf{y} \otimes \mathbf{y}}{|\mathbf{y}|^2}\right] - {}_1F_1\left(1, \frac{5}{2}, \frac{|\mathbf{y}|^2}{4\tau}\right)\left[\frac{1}{3}I - \frac{\mathbf{y} \otimes \mathbf{y}}{|\mathbf{y}|^2}\right]\right\}O(\tau)^T$$

$$Q_0(\mathbf{y}, \tau) = Q^*(\mathbf{y})\delta_0(\tau), \quad Q^*(\mathbf{y}) = -\frac{1}{4\pi}\nabla_y \frac{1}{|\mathbf{y}|}.$$

In particular, for every initial value $\mathbf{u}_0 \in \mathcal{S}(\mathbb{R}^3)^3$ *and* $s \in \mathbb{R}$

$$\lim_{t \to s+} \int_{\mathbb{R}^3} \Gamma_0(\mathbf{y} - \tilde{z}(t, s, \mathbf{z}), t - s)\mathbf{u}_0(\mathbf{z})\, d\mathbf{z} = P\mathbf{u}_0(\mathbf{y}), \quad \mathbf{y} \in \mathbb{R}^3,$$

where P *denotes the Helmholtz decomposition on* $\mathbb{R}^3$.

In the following we will also use cylindrical coordinates $r, \theta, x_3 \in [0, \infty) \times [0, 2\pi) \times \mathbb{R}$ for $\mathbf{x}$ such that $(\boldsymbol{\omega} \wedge \mathbf{x}) \cdot \nabla \mathbf{u} = \partial_\theta \mathbf{u}$ where ∂_θ denotes the angular derivative with respect to θ. Obviously $-\Delta$ commutes with ∂_θ. Let $\nabla' = (\partial_1, \partial_2)$.

Recall the function space $\mathcal{J}_T^{q,s}$, $1 < q, s < \infty$, of initial values with norm

$$\|\mathbf{u}_0\|_{\mathcal{J}_T^{q,s}} = \left(\int_0^T \left(\|e^{-t\mathscr{A}_0^0}\Pi_\sigma^q \mathbf{u}_0\|_q^s + \|\mathscr{A}_0^0 e^{-t\mathscr{A}_0^0}\Pi_\sigma^q \mathbf{u}_0\|_q^s\right) dt\right)^{1/s},$$

where $\Pi_\sigma^q q$ is the Helmholtz projection on $L^q(\mathbb{R}^3)$ and $\mathscr{A}_0^0 = -\Pi_\sigma^q \Delta$ is the Stokes operator. The following theorem states that the equation under consideration is well posed in this space.

Theorem 13.2 *Let* $0 < T < \infty$ *and assume that for some* $1 < q, s < \infty$ *the data* $u_0 \in L_\sigma^q(\mathbb{R}^3)^3$ *and* $\mathbf{f} \in L^s(0, T; \boldsymbol{L}^q(\mathbb{R}^3))$ *satisfy*

$$\mathbf{f},\ \partial_\theta \mathbf{f},\ t\nabla'\mathbf{f} \in L^s(0, T; \boldsymbol{L}^q(\mathbb{R}^3))$$

and $\mathbf{u}_0,\ \partial_\theta \mathbf{u}_0 \in \mathcal{J}_T^{q,s}$. *Then the unique solution* $(\mathbf{v}, \nabla p) \in L^s(0, T; (L^q(\mathbb{R}^3))^6$ *of*

$$\begin{aligned} \tfrac{\partial \mathbf{v}}{\partial t} + \mathcal{L}\mathbf{v} + \nabla p = \mathbf{f} \quad & \text{in} \quad \mathbb{R}^3 \times (0, \infty) \\ \nabla \cdot \mathbf{v} = 0 \quad & \text{in} \quad \mathbb{R}^3 \times (0, \infty) \end{aligned} \tag{13.16}$$

with initial data $\mathbf{v}(0, \mathbf{y}) = \mathbf{u}_0(\mathbf{y})$ *is given by*

$$\begin{aligned} \mathbf{v}(\mathbf{y}, t) = & \int_0^t \int_{\mathbb{R}^3} \Gamma_0(\mathbf{y} - \tilde{z}(t, s, \mathbf{z}), t - s)\mathbf{f}(\mathbf{z}, s)\, dz\, ds \\ & + \int_{\mathbb{R}^3} \Gamma_0(\mathbf{y} - \tilde{z}(t, 0, \mathbf{z}), t)\mathbf{u}_0(\mathbf{z})\, d, \end{aligned} \tag{13.17}$$

$$p(\mathbf{y},t)=\int_0^t\int_{\mathbb{R}^3} Q_0(\mathbf{y}-\tilde{z}(t,s,\mathbf{z}),t-s)\cdot\mathbf{f}(\mathbf{z},s)\,d\mathbf{z}\,ds$$

$$=\int_{\mathbb{R}^3} Q^*(\mathbf{y}-\mathbf{z})\cdot\mathbf{f}(\mathbf{z},t)\,d\mathbf{z}. \tag{13.18}$$

Moreover, $\mathbf{v},\ p$ *satisfies the a priori estimate*

$$\begin{aligned}\|\mathbf{v};\ \nabla\mathbf{v};\ \nabla^2\mathbf{v};\ \mathbf{v}_t;\ \partial_\theta\mathbf{v};\ \nabla p\|_{L^s(0,T;L^q)} &\le C(1+T)\|\mathbf{u}_0;\ \partial_\theta\mathbf{u}_0\|_{\mathcal{J}_T^{q,s}}+\\ &+C(1+T)\big[\|\mathbf{f};\ \partial_\theta\mathbf{f}\|_{L^s(0,T;L^q)}+|\boldsymbol{\Omega}\wedge\mathbf{u}_\infty|\big((1+T)\|\mathbf{f}\|_{L^s(0,T;L^q)}\\ &+\|t\nabla'\mathbf{f}\|\|_{L^s(0,T;L^q)}\big)\big],\end{aligned} \tag{13.19}$$

where the constant C *depends on* q,s *and* $\boldsymbol{\Omega}$, $\mathbf{u}_\infty$, *but not on* T.

Remark 13.1 We note that in the simpler case when $\mathbf{u}_\infty$ is parallel to $\boldsymbol{\Omega}$ and consequently $\boldsymbol{\Omega}\wedge\mathbf{u}_\infty=0$ the terms $|\boldsymbol{\Omega}\wedge\mathbf{u}_\infty|(1+T)\big((1+T)\|\mathbf{f}\|_{L^s(0,T;L^q)}+\|t\nabla'\mathbf{f}\|_{L^s(0,T;L^q)}\big)$ are not needed in (13.19). Other terms which are already present in an estimate of $\mathbf{v}_t$ and when $\boldsymbol{\Omega}||\mathbf{u}_\infty$ are due to the fact that the operator $\mathcal{L}$ does not generate an analytic semigroup and will not satisfy the standard maximal regularity estimate, see [62, 75, 116, 126].

Corollary 13.1

(i) The fundamental solution Γ *from Theorem 13.1 is unique.*
(ii) For any $\mathbf{y},\mathbf{z}\in\mathbb{R}^3$ *and* $s<\tau<t$, *one has the semigroup property*

$$\int_{\mathbb{R}^3}\Gamma(\mathbf{y},\mathbf{z}',t,\tau)\,\Gamma(\mathbf{z}',\mathbf{z},\tau,s)\,dz'=\Gamma(\mathbf{y},\mathbf{z},t,s). \tag{13.20}$$

(iii) For $u\in\mathcal{S}(\mathbb{R}^3)^3$

$$\lim_{(\mathbf{y},t)\to(\mathbf{y}^0,0^+)}\int_{\mathbb{R}^3}\Gamma_0(\mathbf{y}-\tilde{z}(t,0,\mathbf{z}),t)\mathbf{u}(\mathbf{z})\,d\mathbf{z}=\Pi_\sigma\mathbf{u}(\mathbf{y}^0)\,.$$

(iv) The (backward in time) adjoint problem

$$(-\partial_s+\mathcal{L}^*)\mathbf{w}+\nabla\pi=g,\ \ \nabla\cdot\mathbf{w}=0\quad on\ \mathbb{R}^3\times(0,T),\ \ \mathbf{w}(T)=0$$

with the operator $\mathcal{L}^*\mathbf{w}=-\Delta\mathbf{w}+(O(t)^T\cdot\mathbf{u}_\infty+\boldsymbol{\omega}\wedge\mathbf{y})\cdot\nabla\mathbf{w}-\boldsymbol{\Omega}\wedge\mathbf{w}$ *has the fundamental solution*

$$\Gamma'(\mathbf{z},\mathbf{y},s,t)=\Gamma_0(\mathbf{z}-\tilde{y}(s,t,\mathbf{y}),s-t),$$

where $\tilde{\mathbf{y}}(s,t,\mathbf{y})=O(t-s)\big(\mathbf{y}+(t-s)O(t)^T\mathbf{u}_\infty\big)$.

For more details, see [65].

13.2 Estimates of the Oseen Kernels

13.2.1 Introduction and Basic Notation

On several places of this book (Chaps. 5, 6, 10, and 11) the results are based on the asymptotic estimates of convolution integrals whose kernels are given by the Oseen fundamental tensor $\mathcal{O}(\cdot\,;\lambda)$ and its first or second gradients $\nabla\mathcal{O}(\cdot\,;\lambda)$, $\nabla^2\mathcal{O}(\cdot\,;\lambda)$ as well as by the $\nabla\mathcal{E}(\mathbf{x})$ and $\nabla^2\mathcal{E}(\mathbf{x})$, $\mathcal{E}(\mathbf{x})$ being the fundamental solution to the Laplace equation, which play the role of the fundamental pressure.

It is well known that the Oseen tensor exhibits various decay properties in various directions in $\mathbb{R}^N$, and this is the mathematical reason for dealing with anisotropically weighted spaces.

In this part of Appendix, we give the summary of these estimates in weighted L^p spaces with the following weight functions: $\nu_\beta^\alpha(\mathbf{x};\lambda) = |\mathbf{x}|^\alpha(1+s(\lambda\mathbf{x}))^\beta$, $\eta_\beta^\alpha(\mathbf{x};\lambda) = \eta_\beta^\alpha(\lambda\mathbf{x};1) = (1+\lambda|\mathbf{x}|)^\alpha(1+s(\lambda\mathbf{x}))^\beta$, $\lambda = 2Re$, in both two- and three-dimensional cases.

The derivation of the estimates is based on the technique proposed by Farwig in [57, 58], where the volume potentials $\nabla^k\mathcal{O} * f$, $k = 0, 1, 2$ are studied in anisotropically weighted L^2 spaces with the weight function given by the formula $\eta_\beta^\alpha(\mathbf{x}) = (1+|\mathbf{x}|)^\alpha(1+s(\mathbf{x}))^\beta$, $s(\mathbf{x}) = |\mathbf{x}| - x_1$. Similar results as those presented in [57] were obtained also by Kobayashi and Shibata by slightly different techniques, see [141]. The results of Kurtz and Wheeden concerning singular integrals in L^p weight spaces, see [159], are also used.

First, some L^∞-weighted estimates of a certain convolution play an essential role in the next section where the L^∞-estimates of convolutions with Oseen kernels are studied. Applying these results, we get, in the next sections, the L^p-weighted theory of Oseen potentials for both weakly singular and singular kernels.

The book deals with incompressible flow; nevertheless, such estimates can be applied to the investigation of qualitative properties of solutions of the stationary compressible Navier–Stokes equations in exterior domains using the method of decomposition, see, e.g., [182]. They can also be applied with some modifications to the case of stationary flows of certain non-Newtonian fluids, see, e.g., [185] or [183, 184].

We use the following notation:

$c, c_0, c_1, \ldots, C, C_0, C_1, \ldots$—positive constants

$s(\mathbf{x}) = |\mathbf{x}| - x_1$, $|\mathbf{x}| = (x_1 + x_2 + \cdots + x_N)^{1/2}$, $\mathbf{x} \in \mathbb{R}^N$

$\mathcal{E}(\mathbf{x})$—fundamental solution of the Laplace equation

$(\mathcal{O}(\mathbf{x};\lambda), \mathcal{P}(\mathbf{x}))$—fundamental solution of the Oseen problem

$(\mathcal{S}(\mathbf{x}), \mathcal{P}(\mathbf{x}))$—fundamental solution of the Stokes problem

$\eta_\beta^\alpha(\mathbf{x}) = (1+|\mathbf{x}|)^\alpha(1+s(\mathbf{x}))^\beta$, $\sigma_\beta^\alpha(\mathbf{x}) = |\mathbf{x}|^\alpha s(\mathbf{x})^\beta$,

$\nu_\beta^\alpha(\mathbf{x}) = |\mathbf{x}|^\alpha(1+s(\mathbf{x}))^\beta$, $\mu_\beta^{\alpha,\gamma}(\mathbf{x}) = \eta_\beta^{\alpha-\gamma}(\mathbf{x})\nu_0^\gamma(\mathbf{x})$

$\nu_\beta^\alpha(\mathbf{x};\lambda) = |x|^\alpha(1+s(\lambda x))^\beta$, $\eta_\beta^\alpha(\mathbf{x};\lambda) = (1+|\lambda\mathbf{x}|)^\alpha(1+s(\lambda\mathbf{x}))^\beta$

$L^p(\Omega; w) = \Big\{f;\ \|f\|_{p,(w),\Omega}^p = \int_\Omega |f|^p w\,\mathrm{d}\mathbf{x} < +\infty\Big\}$,

$\mathcal{L}^p(\Omega; w) = \overline{C_0^\infty(\Omega)}^{\|\cdot\|_{p,(w),\Omega}}$, $p > 1$, $w > 0$ (usually $w = \eta_\beta^\alpha, \sigma_\beta^\alpha, \nu_\beta^\alpha$), $\Omega \subseteq \mathbb{R}^N$, $N = 2, 3$

$B_r(\mathbf{a}) = \{\mathbf{x} \in \mathbb{R}^N, |\mathbf{x} - \mathbf{a}| < r\}$, $B^r(\mathbf{a}) = \{\mathbf{x} \in \mathbb{R}^N, |\mathbf{x} - \mathbf{a}| > r\}$, $\partial B_r(\mathbf{a}) = \{\mathbf{x} \in \mathbb{R}^N, |\mathbf{x} - \mathbf{a}| = r\}$, $r \in \mathbb{R}^1$, $\mathbf{a} \in \mathbb{R}^N$, $|\cdot|$—norm in $\mathbb{R}^N$.

13.2.2 Oseen Fundamental Solution

In this section we recall some basic facts about the fundamental solution to the Oseen problem. Denote by $\boldsymbol{\mathcal{O}}(\cdot\,;\lambda) = \big(\mathcal{O}_{ij}(\cdot\,;\lambda)\big)$, $\boldsymbol{\mathcal{P}} = (\mathcal{P}_i)$ its fundamental solution; it satisfies the identities

$$\begin{aligned} \partial_j \mathcal{O}_{ij} &= 0 \\ \Delta \mathcal{O}_{ij} + \partial_j \mathcal{P}_i - \lambda \partial_1 \mathcal{O}_{ij} &= \delta_{ij}\delta \end{aligned} \tag{13.21}$$

in the sense of distributions, where δ_{ij} denotes the Kronecker delta, while δ denotes the Dirac delta distribution. The latter is equivalent to

$$-\eta_i(\mathbf{x}) = \big[\partial_k \mathcal{O}_{ij}(\cdot\,;\lambda) * \partial_k \eta_j + \mathcal{P}_i * \partial_j \eta_j - \lambda \mathcal{O}_{ij}(\cdot\,;\lambda) * \partial_1 \eta_j\big](\mathbf{x})$$
$$\forall \eta \in C_0^\infty(\mathbb{R}^N).$$

In particular, it holds

$$\Delta \mathcal{O}_{ij}(\mathbf{x};\lambda) - \lambda \partial_1 \mathcal{O}_{ij}(\mathbf{x};\lambda) + \partial_j \mathcal{P}_i(\mathbf{x}) = 0$$

pointwise in $\mathbb{R}^N \setminus \{\mathbf{0}\}$.

Now we shall study separately the three- and two-dimensional situations.

In three space dimensions we can easily verify (see, e.g., [95]) that the fundamental solution can be written as

$$\mathcal{P}_i(\mathbf{x}) = \partial_i \mathcal{E}(\mathbf{x}) = \frac{1}{4\pi} \frac{x_i}{|\mathbf{x}|^3} \tag{13.22}$$

$$\mathcal{O}_{ij}(\mathbf{x};\lambda) = (\delta_{ij}\Delta - \partial_i \partial_j)\varphi_{\mathcal{O}}(\mathbf{x};\lambda)\,, \tag{13.23}$$

where

$$\varphi_{\mathcal{O}}(\mathbf{x};\lambda) = \frac{-1}{4\pi\lambda} \psi\Big(\frac{\lambda s(\mathbf{x})}{2}\Big) \tag{13.24}$$

with

$$\psi(z) = \int_0^z \frac{1-e^{-t}}{t}\mathrm{d}t = \sum_{i=1}^{\infty} \frac{(-1)^{i+1}}{i!i} z^i \tag{13.25}$$

and

$$s(\mathbf{x}) = |\mathbf{x}| - x_1 . \tag{13.26}$$

The formulas (13.24)–(13.26) yield useful rescaling property

$$\lambda \mathcal{O}(\lambda \mathbf{x};\, 1) = \mathcal{O}(\mathbf{x};\, \lambda). \tag{13.27}$$

The integral representation (13.25) implies

$$\begin{aligned} \psi'(t) &= \tfrac{1-e^{-t}}{t}\,, \quad \psi''(t) = \tfrac{-1+e^{-t}+te^{-t}}{t^2}\,,\\ \psi'''(t) &= \tfrac{2-2e^{-t}-2te^{-t}-t^2e^{-t}}{t^3}\,,\\ \psi^{(iv)}(t) &= \tfrac{-6+6e^{-t}+6te^{-t}+3t^2e^{-t}+t^3e^{-t}}{t^4}\,. \end{aligned}$$

The representation by the sum in (13.25) yields

$$\psi^{(k)}(t) = \frac{(-1)^{k+1}}{k} + O(t) \quad \text{as } t \to 0\,, \quad k = 1, 2, \dots . \tag{13.28}$$

When differentiating (13.26), we obtain

$$\frac{\partial s(\mathbf{x})}{\partial x_i} = \frac{x_i}{|\mathbf{x}|} - \delta_{1i}\,. \tag{13.29}$$

From here we get the estimates

$$\left|\frac{\partial s(\mathbf{x})}{\partial x_k}\right| \le \begin{cases} \frac{s(\mathbf{x})}{|\mathbf{x}|} & (k=1) \\ \sqrt{2}\sqrt{\frac{s(\mathbf{x})}{|\mathbf{x}|}} & (k \neq 1) \end{cases} \qquad \left|D^\alpha s(\mathbf{x})\right| \le \frac{c(\alpha)}{|\mathbf{x}|^{|\alpha|-1}}\,. \tag{13.30}$$

From (13.24)–(13.26) and (13.30) it is seen that $\mathcal{O}(\cdot\,;\,\cdot) \in C^\infty\big((\mathbb{R}^3 \setminus \{0\}) \times \mathbb{R}\big)$, and for fixed $\mathbf{x} \neq \mathbf{0}$, $\mathcal{O}(\mathbf{x};\,\cdot)$ is an analytic function.

Now we calculate the derivatives of $\varphi_{\mathcal{O}}(\cdot\,;\lambda)$ in order to establish the asymptotic behavior of $\boldsymbol{\mathcal{O}}(\cdot\,;\lambda)$ and of its first and second derivatives near zero and at infinity.

$$
\begin{aligned}
-\partial_i\varphi_{\mathcal{O}}(\mathbf{x};\lambda) &= \tfrac{\lambda}{8\pi}\psi'(\tfrac{\lambda s(\mathbf{x})}{2})\,\partial_i s(\mathbf{x})\\
-\partial_r\partial_i\varphi_{\mathcal{O}}(\mathbf{x};\lambda) &= \tfrac{1}{16\pi}\psi''(\tfrac{\lambda s(\mathbf{x})}{2})\,\partial_r s(\mathbf{x})\,\partial_i s(\mathbf{x}) + \tfrac{1}{8\pi}\psi'(\tfrac{\lambda s(\mathbf{x})}{2})\,\partial_r\partial_i s(\mathbf{x})\\
-\partial_k\partial_r\partial_i\varphi_{\mathcal{O}}(\mathbf{x};\lambda) &= \tfrac{\lambda^2}{32\pi}\psi'''(\tfrac{\lambda s(\mathbf{x})}{2})\,\partial_k s(\mathbf{x})\partial_r s(\mathbf{x})\partial_i s(\mathbf{x})\\
&\quad+\tfrac{\lambda}{16\pi}\psi''(\tfrac{\lambda s(\mathbf{x})}{2})\big[\partial_k\partial_r s(\mathbf{x})\,\partial_i s(\mathbf{x}) + \partial_k\partial_i s(\mathbf{x})\,\partial_r s(\mathbf{x})\\
&\quad+\partial_r\partial_i s(\mathbf{x})\,\partial_k s(\mathbf{x})\big] + \tfrac{1}{8\pi}\psi'\tfrac{\lambda s(\mathbf{x})}{2})\,\partial_k\partial_r\partial_i s(\mathbf{x})\\
-\partial_l\partial_k\partial_r\partial_i\varphi_{\mathcal{O}}(\mathbf{x};\lambda) &= \tfrac{\lambda^3}{64\pi}\psi^{(iv)}(\tfrac{\lambda s(\mathbf{x})}{2})\,\partial_l s(\mathbf{x})\,\partial_k s(\mathbf{x})\,\partial_r s(\mathbf{x})\,\partial_i s(\mathbf{x})\\
&\quad+\tfrac{\lambda^2}{32\pi}\psi'''(\tfrac{\lambda s(\mathbf{x})}{2})\,\big[\partial_l\partial_k s(\mathbf{x})\partial_r s(\mathbf{x})\partial_i s(\mathbf{x})\\
&\quad+\partial_l\partial_r s(\mathbf{x})\partial_k s(\mathbf{x})\partial_i s(\mathbf{x}) + \partial_l\partial_i s(\mathbf{x})\partial_k s(\mathbf{x})\partial_r s(\mathbf{x})\\
&\quad+\partial_k\partial_r s(\mathbf{x})\partial_i s(\mathbf{x})\partial_l s(\mathbf{x}) + \partial_k\partial_i s(\mathbf{x})\partial_r s(\mathbf{x})\partial_l s(\mathbf{x})\\
&\quad+\partial_r\partial_i s(\mathbf{x})\partial_k s(\mathbf{x})\partial_l s(\mathbf{x})\big] + \tfrac{\lambda}{16\pi}\psi''(\tfrac{\lambda s(\mathbf{x})}{2})\\
&\quad\cdot\big[\partial_l\partial_k\partial_r s(\mathbf{x})\,\partial_i s(\mathbf{x}) + \partial_l\partial_k\partial_i s(\mathbf{x})\,\partial_r s(\mathbf{x})\\
&\quad+\partial_l\partial_r\partial_i s(\mathbf{x})\,\partial_k s(\mathbf{x}) + \partial_k\partial_r\partial_i s(\mathbf{x})\,\partial_l s(\mathbf{x})\\
&\quad+\partial_i\partial_k s(\mathbf{x})\,\partial_r\partial_l s(\mathbf{x}) + \partial_r\partial_k s(\mathbf{x})\,\partial_i\partial_l s(\mathbf{x})\\
&\quad+\partial_i\partial_r s(\mathbf{x})\,\partial_k\partial_l s(\mathbf{x})\big] + \tfrac{1}{8\pi}\psi'(\tfrac{\lambda s(\mathbf{x})}{2})\,\partial_l\partial_k\partial_r\partial_i s(\mathbf{x}).
\end{aligned}
$$

These formulas together with (13.28), (13.30), and (13.23) yield

$$
\begin{aligned}
\boldsymbol{\mathcal{O}}(\mathbf{x};\lambda) &= \boldsymbol{\mathcal{S}}(\mathbf{x}) + \lambda O(1) \quad \text{as } \lambda|\mathbf{x}|\to 0\\
\nabla\boldsymbol{\mathcal{O}}(\mathbf{x};\lambda) &= \nabla\boldsymbol{\mathcal{S}}(\mathbf{x}) + \lambda^2 O(\tfrac{1}{\lambda|\mathbf{x}|}) \quad \text{as } \lambda|\mathbf{x}|\to 0\\
\nabla^2\boldsymbol{\mathcal{O}}(\mathbf{x};\lambda) &= \nabla^2\boldsymbol{\mathcal{S}}(\mathbf{x}) + \lambda^3 O(\tfrac{1}{\lambda^2|\mathbf{x}|^2}) \quad \text{as } \lambda|\mathbf{x}|\to 0\,,
\end{aligned}
\tag{13.31}
$$

where $(\boldsymbol{\mathcal{S}},\boldsymbol{\mathcal{P}})$ is the Stokes fundamental solution (see, e.g., [95]),

$$
\mathcal{S}_{ij}(\mathbf{x}) = \frac{-1}{8\pi}\left[\frac{\delta_{ij}}{|\mathbf{x}|} + \frac{x_i x_j}{|\mathbf{x}|^3}\right]. \tag{13.32}
$$

It can be shown (see, e.g., [98] and also Section 5) that both the second derivatives of $\boldsymbol{\mathcal{S}}$ and $\nabla\boldsymbol{\mathcal{P}}$ represent Calderón–Zygmund singular integral kernels.

In particular, for $\lambda \in (0;\lambda_0)$, $R > 0$ and $|\lambda\mathbf{x}| \le R$

$$
\left|\nabla^k\boldsymbol{\mathcal{O}}(\mathbf{x};\lambda)\right| \le \frac{c(R;\lambda_0,k)}{|\mathbf{x}|^{k+1}}. \tag{13.33}
$$

For any $\mathbf{x} \neq \mathbf{0}$ formulas (13.23) and (13.30) together with the properties of the function $s(\mathbf{x})$ give

$$\begin{aligned}
|\mathcal{O}(\mathbf{x};\lambda)| &\le \frac{c}{\lambda}\frac{1-e^{-\frac{\lambda s(\mathbf{x})}{2}}}{s(\mathbf{x})|\mathbf{x}|} \\
|\nabla\mathcal{O}(\mathbf{x};\lambda)| &\le \frac{c}{\lambda}\Big[\frac{1-e^{-\frac{\lambda s(\mathbf{x})}{2}}}{s(\mathbf{x})|\mathbf{x}|^2} + \frac{1-e^{-\frac{\lambda s(\mathbf{x})}{2}} - \frac{\lambda s(\mathbf{x})}{2}e^{-\frac{\lambda s(\mathbf{x})}{2}}}{s^{3/2}(\mathbf{x})|\mathbf{x}|^{3/2}}\Big] \\
\big|\nabla^2\mathcal{O}(\mathbf{x};\lambda)\big| &\le \frac{c}{\lambda}\Big[\frac{1-e^{-\frac{\lambda s(\mathbf{x})}{2}}}{s(\mathbf{x})|\mathbf{x}|^3} + \frac{1-e^{-\frac{\lambda s(\mathbf{x})}{2}} - \frac{\lambda s(\mathbf{x})}{2}e^{-\frac{\lambda s(\mathbf{x})}{2}}}{s^{2}(\mathbf{x})|\mathbf{x}|^{2}}\Big].
\end{aligned} \tag{13.34}$$

This yields for $|\lambda\mathbf{x}| \ge R$ and any $\kappa > 0$

$$\begin{aligned}
|\mathcal{O}(\mathbf{x};\lambda)| &\le \frac{c(\kappa,R)}{|\mathbf{x}|(\kappa + s(\lambda\mathbf{x}))} \\
|\nabla\mathcal{O}(\mathbf{x};\lambda)| &\le \frac{c(\kappa,R)\lambda^{1/2}}{|\mathbf{x}|^{3/2}\big(\kappa + s(\lambda\mathbf{x})\big)^{3/2}} \\
\big|\nabla^2\mathcal{O}(\mathbf{x};\lambda)\big| &\le \frac{c(\kappa,R)\lambda}{|\mathbf{x}|^{2}\big(\kappa + s(\lambda\mathbf{x})\big)^{2}}.
\end{aligned} \tag{13.35}$$

Formulas (13.32) and (13.35) give us in particular that $\mathcal{O}$ and $\nabla\mathcal{O}$ are analogous to $\mathcal{P}$ weakly singular kernels, while the second derivative of $\mathcal{O}$ can be written as a sum of a singular kernel ($\mathcal{S}$) and a weakly singular part.

With $(\mathcal{O}, \mathcal{P})$ at hand, we can write explicitly a C^∞-solution of the problem

$$\begin{aligned}
-\Delta u_i + \lambda\partial_1 u_i + \partial_i \Pi &= f_i \\
\nabla\cdot\boldsymbol{u} &= g
\end{aligned}$$

with $\mathbf{f}, g \in C_0^\infty(\mathbb{R}^3)$. Namely

$$\begin{aligned}
u_i &= -\mathcal{O}_{ij}(\cdot\,;\lambda) * f_j + \mathcal{P}_i * g \\
\Pi &= \mathcal{P}_j * f_j + g - \lambda\partial_1(\mathcal{E} * g) = \mathcal{P}_j * f_j + g - \lambda\mathcal{P}_1 * g\,.
\end{aligned}$$

In the case of $\mathbf{f} = \mathrm{div}\mathcal{F}$ we have

$$\begin{aligned}
u_i &= -\partial_s\mathcal{O}_{ij} * \mathcal{F}_{js} + \mathcal{P}_i * g \\
\Pi &= \partial_s\mathcal{P}_j * \mathcal{F}_{js} + g - \lambda\mathcal{P}_1 * g + c_{js}\mathcal{F}_{js}\,,
\end{aligned}$$

where c_{js}, $j, s = 1, \ldots, N$ are constants.

The two-dimensional case is a bit more complicated. Therefore, we shall study this case in more detail. Let $K_0(z)$ be the modified Bessel function, i.e., $K_0(z)$ solves the modified Bessel equation

$$z^2 K_0''(z) + zK_0'(z) - z^2 K_0(z) = 0 \tag{13.36}$$

for $z \neq 0$ and it is singular at 0, i.e.,

$$K_0(z) = -\ln(z) + \ln 2 - \gamma - (\ln \frac{z}{2}) \sum_{k=1}^{\infty} \frac{1}{k!}\Big(\frac{z}{2}\Big)^{2k} + \sum_{k=1}^{\infty} \frac{1}{(k!)^2}\Big(\sum_{j=1}^{k} \frac{1}{j} - \gamma\Big)\Big(\frac{z}{2}\Big)^{2k} \tag{13.37}$$

for $z \neq 0$ sufficiently small (see, e.g., [142]); here γ is the Euler constant. We put

$$\Phi(\mathbf{x}) = \frac{1}{4\pi\lambda} \int^{-x_1} \Big\{ \log\sqrt{\tau^2 + x_2^2} + K_0(\lambda\sqrt{\tau^2 + x_2^2}) e^{-\lambda\tau} \Big\} d\tau. \tag{13.38}$$

The problem consists of the right choice of the constants which in turn corresponds to the right choice of lower bound for the integral in (13.38). The lower bound is, at least formally, required to be equal to ∞. Unfortunately, then the integral in (13.38) does not converge, as K_0 behaves regularly at infinity (see (13.47)). We calculate formally the derivatives of (13.38) and put

$$G(\mathbf{x}; 2\lambda) = \frac{1}{4\pi\lambda}(\log\sqrt{x_1^2 + x_2^2} + K_0(\lambda\sqrt{x_1^2 + x_2^2}) e^{\lambda x_1}), \tag{13.39}$$

$$\begin{aligned} H(\mathbf{x}; 2\lambda) = \frac{1}{4\pi\lambda} \int_{\infty}^{-x_1} \Big[&\frac{\tau^2 - x_2^2}{(\tau^2 + x_2^2)^2} + \Big(K_0''(\lambda\sqrt{\tau^2 + x_2^2}) \frac{\lambda^2 x_2^2}{\tau^2 + x_2^2} + \\ &+ K_0'(\lambda\sqrt{\tau^2 + x_2^2}) \frac{\lambda\tau^2}{(\tau^2 + x_2^2)^{\frac{3}{2}}} \Big) e^{-\lambda\tau} \Big] d\tau\,, \end{aligned} \tag{13.40}$$

i.e., G is formally taken derivative of (13.38) with respect to x_1 multiplied by -1, H the second derivative of (13.38) with respect to x_2^2. First we express (13.40) in a more appropriate way (without the integrals).

We formally calculate

$$\begin{aligned} \frac{\partial^2 \Phi(\mathbf{x})}{\partial x_2^2} = \frac{1}{4\pi\lambda} \int_{\infty}^{-x_1} \Big[&\frac{\tau^2 - x_2^2}{(\tau^2 + x_2^2)^2} + \Big(K_0''(\lambda\sqrt{\tau^2 + x_2^2}) \frac{\lambda^2 x_2^2}{\tau^2 + x_2^2} \\ &+ K_0'(\lambda\sqrt{\tau^2 + x_2^2}) \frac{\lambda\tau^2}{(\tau^2 + x_2^2)^{\frac{3}{2}}} \Big) e^{-\lambda\tau} \Big] d\tau = \frac{1}{4\pi\lambda} \Big\{ \frac{x_1}{x_1^2 + x_2^2} \\ &+ \Big[K_0'(\lambda\sqrt{x_1^2 + x_2^2}) \frac{\lambda x_1}{\sqrt{x_1^2 + x_2^2}} - K_0(\lambda\sqrt{x_1^2 + x_2^2})\lambda \Big] e^{\lambda x_1} \Big\}. \end{aligned}$$

Following [95] it can be verified that $\mathcal{O}_{ij}(\mathbf{x};\lambda) = (\delta p_{ij}\Delta - \partial_i\partial_j)\Phi(\mathbf{x})$, where the derivative is calculated formally—the integral in (13.38) is not finite. Nevertheless we shall verify that the formally deduced function is really the fundamental solution being sought. We put $\left(r = \sqrt{x_1^2 + x_2^2}\right)$

$$\begin{aligned}
\mathcal{O}_{11}(\mathbf{x};2\lambda) = H(\mathbf{x};2\lambda) &= \frac{1}{4\pi\lambda}\left\{\frac{x_1}{r^2} + [K_0'(\lambda r)\frac{\lambda x_1}{r} - K_0(\lambda r)\lambda]e^{\lambda x_1}\right\}\\
\mathcal{O}_{12}(\mathbf{x};2\lambda) = \mathcal{O}_{21}(\mathbf{x};2\lambda) &= \frac{\partial}{\partial x_2}G(\mathbf{x},2\lambda) = \frac{1}{4\pi\lambda}\left[\frac{x_2}{r^2} + K_0'(\lambda r)\frac{\lambda x_2}{r}e^{\lambda x_1}\right]\\
\mathcal{O}_{22}(\mathbf{x};2\lambda) = -\frac{\partial}{\partial x_1}G(\mathbf{x};2\lambda) &= -\frac{1}{4\pi\lambda}\left\{\frac{x_1}{r^2} + [K_0'(\lambda r)\frac{\lambda x_1}{r} + K_0(\lambda r)\lambda]e^{\lambda x_1}\right\}
\end{aligned} \tag{13.41}$$

$$\mathcal{P}_i(\mathbf{x}) = \frac{1}{2\pi}\frac{x_i}{r^2} = \partial_i\mathcal{E}(\mathbf{x})\,. \tag{13.42}$$

From (13.41) and (13.37) we get for small $\lambda|\mathbf{x}|$

$$\begin{aligned}
\mathcal{O}_{ii}(\mathbf{x};2\lambda) &= \mathcal{S}_{ii}(\mathbf{x}) - \tfrac{1}{4\pi}\log\tfrac{1}{\lambda} + O(1)\\
\mathcal{O}_{ij}(\mathbf{x};2\lambda) &= \mathcal{S}_{ij}(\mathbf{x}) + O(\lambda r\log(\lambda r)) \quad i \neq j\\
\nabla\mathcal{O}_{ij}(\mathbf{x};2\lambda) &= \nabla\mathcal{S}_{ij}(\mathbf{x}) + \lambda O(\log(\lambda r))\\
\nabla^2\mathcal{O}_{ij}(\mathbf{x};2\lambda) &= \nabla^2\mathcal{S}_{ij}(\mathbf{x}) + \lambda^2 O(\tfrac{1}{\lambda r})\,,
\end{aligned} \tag{13.43}$$

where $\mathcal{S}_{ij}(\mathbf{x}) = \frac{-1}{4\pi}[\delta_{ij}\log\frac{1}{r} + \frac{x_i x_j}{r^2}]$ is the fundamental Stokes tensor (see [95]).

The explicit formulas (13.41) also imply the following homogeneity property

$$\mathcal{O}_{ij}(\mathbf{x};2\lambda) = \mathcal{O}_{ij}(2\lambda\mathbf{x};1)\,. \tag{13.44}$$

We are going to verify that $\boldsymbol{\mathcal{O}}, \boldsymbol{\mathcal{P}}$ solve (13.21) in the sense of distributions. Let us observe that

$$\begin{aligned}
\mathcal{O}_{11}(\mathbf{x};2\lambda) &= \tfrac{1}{2\lambda}\tfrac{\partial\mathcal{E}}{\partial x_1}(\mathbf{x}) - \tfrac{1}{2\lambda}\tfrac{\partial\Phi_1}{\partial x_1}(\mathbf{x};2\lambda) + \Phi_1(\mathbf{x};2\lambda)\\
\mathcal{O}_{12}(\mathbf{x};2\lambda) = \mathcal{O}_{21}(\mathbf{x};2\lambda) &= \tfrac{1}{2\lambda}\tfrac{\partial\mathcal{E}}{\partial x_2}(\mathbf{x}) - \tfrac{1}{2\lambda}\tfrac{\partial\Phi_1}{\partial x_2}(\mathbf{x};2\lambda)\\
\mathcal{O}_{22}(\mathbf{x};2\lambda) &= -\tfrac{1}{2\lambda}\tfrac{\partial\mathcal{E}}{\partial x_1}(\mathbf{x}) + \tfrac{1}{2\lambda}\tfrac{\partial\Phi_1}{\partial x_1}(\mathbf{x};2\lambda)\,,
\end{aligned} \tag{13.45}$$

where $\Phi_1(\mathbf{x}) = -\frac{1}{2\pi}K_0(\lambda|\mathbf{x}|)e^{\lambda x_1}$. We therefore easily see that $\mathcal{O}_{ij}, \mathcal{P}_j$ solves $(13.21)_1$ for $\mathbf{x} \neq \mathbf{0}$. Moreover

$$\frac{\partial\mathcal{O}_{21}}{\partial x_1} + \frac{\partial\mathcal{O}_{22}}{\partial x_2} = 0 \qquad \text{for } \mathbf{x} \neq \mathbf{0}$$

and

$$\frac{\partial \mathcal{O}_{11}}{\partial x_1} + \frac{\partial \mathcal{O}_{12}}{\partial x_2} = \frac{-1}{4\pi\lambda}(K_0''(\lambda r)\lambda^2 + K_0'(\lambda r)\frac{\lambda}{r} - \lambda^2 K_0(\lambda r))\mathrm{e}^{\lambda x_1} = 0$$

for $\mathbf{x} \neq \mathbf{0}$. Moreover from (13.43) we have that $|\nabla \boldsymbol{\mathcal{O}}(\mathbf{x}; 1)| \leq \frac{C}{|\mathbf{x}|}$ for $|\mathbf{x}|$ small and therefore

$$\frac{\partial \mathcal{O}_{ij}}{\partial x_j} = 0 \text{ in } \mathcal{D}'(\mathbb{R}^2).$$

It remains to verify that

$$(\Delta - \lambda \partial_1)\mathcal{O}_{ij} + \partial_j \mathcal{P}_i = \delta_{ij}\delta.$$

Due to the asymptotic behavior of $\boldsymbol{\mathcal{O}}$ and due to the fact that

$$\lim_{\varepsilon \to 0^+} \int_{\partial B^\varepsilon(\mathbf{x})} \mathcal{P}_i(\mathbf{x}-\mathbf{y}) n_j(\mathbf{y}) F_j(\mathbf{y})\, \mathrm{d}_y S = \frac{1}{2} F_j(\mathbf{x})\delta_{ij},$$

it is enough to verify that

$$\lim_{\varepsilon \to 0^+} \int_{\partial B^\varepsilon(\mathbf{x})} \frac{\partial O_{ij}(\mathbf{x}-\mathbf{y})}{\partial n} F_j(\mathbf{y})\, \mathrm{d}_y S = \frac{1}{2} F_j(\mathbf{x})\delta_{ij}. \tag{13.46}$$

Namely, then we easily get ($\mathbf{n}$ is the outer normal to $B^\varepsilon(\mathbf{x})$, i.e., the inner normal to $B_\varepsilon(\mathbf{x})$)

$$\begin{aligned}
&\int_{\mathbb{R}^2} \Big[\mathcal{O}_{ij}(\mathbf{x}-\mathbf{y}; \lambda)[(\Delta - \lambda \frac{\partial}{\partial y_1}) F_j(\mathbf{y})] + \mathcal{P}_i(\mathbf{x}-\mathbf{y}) \frac{\partial F_j}{\partial y_j}(\mathbf{y})\Big]\, \mathrm{d}\mathbf{y} \\
&\quad = \text{v.p.} \int_{\mathbb{R}^2} \Big[(\Delta - \lambda \frac{\partial}{\partial x_1})\mathcal{O}_{ij}(\mathbf{x}-\mathbf{y}; \lambda) + \frac{\partial \mathcal{P}_i}{\partial x_j}(\mathbf{x}-\mathbf{y})\Big] F_j(\mathbf{y})\, \mathrm{d}\mathbf{y} \\
&\qquad + \lim_{\varepsilon \to 0^+} \Big[\int_{\partial B^\varepsilon(\mathbf{x})} \frac{\partial \mathcal{O}_{ij}(\mathbf{x}-\mathbf{y}; \lambda)}{\partial \mathbf{n}} F_j(\mathbf{y})\, \mathrm{d}_y S \\
&\qquad - \lambda \int_{\partial B^\varepsilon(\mathbf{x})} \mathcal{O}_{ij}(\mathbf{x}-\mathbf{y}; \lambda) n_i(\mathbf{y}) F_j(\mathbf{y})\, \mathrm{d}_y S \\
&\qquad - \lambda \int_{\partial B^\varepsilon(\mathbf{x})} \mathcal{O}_{ij}(\mathbf{x}-\mathbf{y}; \lambda) n_i(\mathbf{y}) F_j(\mathbf{y})\, \mathrm{d}_y S \\
&\qquad + \int_{\partial B^\varepsilon(\mathbf{x})} \mathcal{P}_i(\mathbf{x}-\mathbf{y}) n_j(\mathbf{y}) F_j(\mathbf{y})\, \mathrm{d}_y S\Big] \\
&\quad = \lim_{\varepsilon \to 0^+} \int_{\partial B^\varepsilon(\mathbf{x})} \Big[\frac{\partial \mathcal{O}_{ij}}{\partial \mathbf{n}}(\mathbf{x}-\mathbf{y}; \lambda) + \mathcal{P}_i(\mathbf{x}-\mathbf{y}) n_j(\mathbf{y})\Big] F_j(\mathbf{y})\, \mathrm{d}_y S = F_j(\mathbf{x})\delta_{ij}.
\end{aligned}$$

From (13.42) and (13.37) we find after a bit tedious but straightforward calculations that

$$\begin{aligned}
\frac{\partial \mathcal{O}_{11}}{\partial n} &= -\frac{1}{4\pi}\frac{1}{r} + O(\ln r)\\
\frac{\partial \mathcal{O}_{12}}{\partial n} = \frac{\partial \mathcal{O}_{21}}{\partial n} &= O(\ln r)\\
\frac{\partial \mathcal{O}_{22}}{\partial n} &= -\frac{1}{4\pi}\frac{1}{r} + O(\ln r)
\end{aligned}$$

and therefore (13.46) follows.

The next part is devoted to the asymptotic properties when $\lambda r \to \infty$. Unlike the Stokes fundamental tensor, we get anisotropic structure. First, let us recall that for $z \to \infty$ (see, e.g., [142]):

$$\begin{aligned}
K_0(z) &= \left(\tfrac{\pi}{2z}\right)^{\frac{1}{2}} e^{-z}\Big[1 - \tfrac{1}{8z} + \tfrac{9}{2!(8z)^2} + O(z^{-3})\Big]\\
K_0'(z) &= \left(\tfrac{\pi}{2z}\right)^{\frac{1}{2}} e^{-z}\Big[-1 - \tfrac{3}{8z} + \tfrac{15}{128z^2} + O(z^{-3})\Big]\\
K_0''(z) &= \left(\tfrac{\pi}{2z}\right)^{\frac{1}{2}} e^{-z}\Big[1 + \tfrac{7}{8z} + \tfrac{57}{128z^2} + O(z^{-3})\Big]\\
K_0'''(z) &= \left(\tfrac{\pi}{2z}\right)^{\frac{1}{2}} e^{-z}\Big[-1 - \tfrac{11}{8z} - \tfrac{225}{128z^2} + O(z^{-3})\Big].
\end{aligned} \tag{13.47}$$

Using the polar coordinates $x_1 = r\cos\varphi$, $x_2 = r\sin\varphi$, we get from (13.26)

$$s(\mathbf{x}) = (r - x_1) = r(1 - \cos\varphi)$$

and (13.41) yields the following asymptotic expansion of $\boldsymbol{\mathcal{O}}$:

$$\begin{aligned}
\mathcal{O}_{11}(\mathbf{x}; 2\lambda) &= \tfrac{1}{4\pi\lambda r}\cos\varphi \;-\; \tfrac{1}{4\sqrt{2\pi\lambda r}} e^{-\lambda s}\Big[\cos\varphi\; + 1\\
&\quad + \tfrac{1}{\lambda r}\big(\tfrac{3}{8}\cos\varphi\; - \tfrac{1}{8}\big) + \nu(\lambda r)\Big]\\
\mathcal{O}_{12}(\mathbf{x}; 2\lambda) &= \tfrac{1}{4\pi\lambda r}\sin\varphi \;-\; \tfrac{1}{4\sqrt{2\pi\lambda r}} e^{-\lambda s}\sin\varphi\,\Big[1 + \tfrac{3}{8\lambda r} + \nu(\lambda r)\Big]\\
\mathcal{O}_{22}(\mathbf{x}; 2\lambda) &= \tfrac{-1}{4\pi\lambda r}\cos\varphi \;+\; \tfrac{1}{4\sqrt{2\pi\lambda r}} e^{-\lambda s}\Big[(\cos\varphi - 1)\\
&\quad + \tfrac{1}{\lambda r}\big(\tfrac{3}{8}\cos\varphi\; + \tfrac{1}{8}\big) + \nu(\lambda r)\Big].
\end{aligned} \tag{13.48}$$

Using several straightforward properties of the function $s(\mathbf{x})$

$$\frac{\partial s}{\partial x_1} = \frac{-s}{r}, \qquad \frac{\partial s}{\partial x_2} = \frac{x_2}{r} = \sin\varphi\,, \tag{13.49}$$

$$\frac{\partial \varphi}{\partial x_2} = \frac{\cos\varphi}{r}, \qquad \frac{\partial \varphi}{\partial x_1} = -\frac{\sin\varphi}{r} \tag{13.50}$$

$$s \sim r \quad \text{for} \quad x_1 \le 0 \quad \text{but} \quad s \sim \frac{(x_2)^2}{r} \quad \text{for} \quad x_1 > 0\,. \tag{13.51}$$

$$e^{-2\lambda s} \sin^2\varphi\, \lambda r = \lambda s e^{-2\lambda s}(1 + \cos\varphi) \le e^{-1}, \tag{13.52}$$

we get the following uniform behavior

$$\left.\begin{aligned} |\mathcal{O}_{12}(\mathbf{x}; 2\lambda)|, |\mathcal{O}_{22}(\mathbf{x}; 2\lambda)| &\le \tfrac{C}{\lambda r} \\ |\mathcal{O}_{11}(\mathbf{x}; 2\lambda)| &\le \tfrac{C}{\sqrt{\lambda r}} \end{aligned}\right\} \quad \text{as } \lambda r \to \infty\,. \tag{13.53}$$

Moreover, from $(13.48)_1$ we may deduce the following anisotropic structure

$$|\mathcal{O}_{11}(\mathbf{x}; 2\lambda)| \le \frac{C}{\sqrt{\lambda r} \cdot \sqrt{1 + \lambda s}}\,. \tag{13.54}$$

Now we could calculate all the derivatives and get the asymptotic expansions of them. But we are interested only in the estimates of the type (13.53)–(13.54). The formulas (13.49)–(13.52) yield

$$\begin{aligned} |\partial_2\mathcal{O}_{11}(\mathbf{x}; 2\lambda)| &\le \tfrac{C}{r(1+\lambda s)} \\ |\partial_2\mathcal{O}_{12}(\mathbf{x}; 2\lambda)|\,, \left|\partial_1\mathcal{O}_{11}(\mathbf{x}; 2\lambda)\right| &\le \tfrac{C}{r\sqrt{\lambda r(1+\lambda s)}} \\ |\partial_1\mathcal{O}_{12}(\mathbf{x}; 2\lambda)|\,, |\partial_1\mathcal{O}_{22}(\mathbf{x}; 2\lambda)|\,, |\partial_2\mathcal{O}_{22}(\mathbf{x}; 2\lambda)| &\le \tfrac{C}{\lambda r^2}\,. \end{aligned} \tag{13.55}$$

For higher derivatives we do not need such precise estimates. We therefore only observe that

$$\left|\partial_2\partial_2\mathcal{O}_{11}(\mathbf{x}; 2\lambda)\right| \le \frac{C\sqrt{\lambda}}{r^{\frac{3}{2}}(1 + \lambda s)^{\frac{3}{2}}}\,, \tag{13.56}$$

while for the other terms we have the following uniform estimate:

$$|\widetilde{D}^2\boldsymbol{\mathcal{O}}(\mathbf{x}; 2\lambda)| \le \frac{C}{r^2}\,, \tag{13.57}$$

where $\widetilde{D}^2\boldsymbol{\mathcal{O}}$ contains the second gradient of $\boldsymbol{\mathcal{O}}$ except of $\partial_2\partial_2\mathcal{O}_{11}$. For higher derivatives we get analogously

$$|\nabla^k\boldsymbol{\mathcal{O}}(\mathbf{x}; 2\lambda)| \le \frac{C\lambda^{\frac{k}{2}-\frac{1}{2}}}{r^{\frac{k}{2}+\frac{1}{2}}} \quad \text{for } k \ge 3\,. \tag{13.58}$$

The formulas (13.53)–(13.58) are valid for $\lambda r > 1$.

Let us summarize the asymptotic behavior of $\boldsymbol{\mathcal{O}}$ and its derivatives using the weights ν^α_β, $\mu^{\alpha,\gamma}_\beta$, and η^α_β, introduced in Section 0. Moreover, we assume $\lambda = 1$, and for $\lambda \neq 1$ we may use the homogeneity properties (13.27) and (13.44). Then we have for $N = 3$ and $\mathbf{x} \in \mathbb{R}^3 \setminus \{\mathbf{0}\}$

$$\begin{aligned} |\boldsymbol{\mathcal{O}}(\mathbf{x};1)| &\le C\nu_{-1}^{-1}(\mathbf{x}) \\ |\nabla\boldsymbol{\mathcal{O}}(\mathbf{x};1)| &\le C\mu_{-\frac{3}{2}}^{-\frac{3}{2},-2}(\mathbf{x}) \\ |\partial_1\boldsymbol{\mathcal{O}}(\mathbf{x};1)| &\le C\nu_{-1}^{-2}(\mathbf{x}) \\ |\nabla^2\boldsymbol{\mathcal{O}}(\mathbf{x};1) - \nabla^2\boldsymbol{\mathcal{S}}(\mathbf{x})| &\le C\nu_{-1}^{-2}(\mathbf{x}) \end{aligned} \tag{13.59}$$

and for $N = 2$ and $\mathbf{x} \in \mathbb{R}^3 \setminus \{\mathbf{0}\}$

$$\begin{aligned} |\mathcal{O}_{11}(\mathbf{x};1)| &\le C\eta_{-\frac{1}{2}}^{-\frac{1}{2}}(\mathbf{x})\log(2+\tfrac{1}{|\mathbf{x}|}) \\ |\mathcal{O}_{ij}(\mathbf{x};1)| &\le C\eta_0^{-1}(\mathbf{x})\log(2+\tfrac{1}{|\mathbf{x}|}) \quad i.j > 1 \\ |\partial_2\mathcal{O}_{11}(\mathbf{x};1)| &\le C\nu_{-1}^{-1}(\mathbf{x}) \\ \left.\begin{array}{r} |\partial_1\mathcal{O}_{11}(\mathbf{x};1)| \\ |\partial_2\mathcal{O}_{12}(\mathbf{x};1)| \\ |\partial_2\partial_2\mathcal{O}_{11}(\mathbf{x};1) - \partial_2\partial_2\mathcal{S}_{11}(\mathbf{x})| \end{array}\right\} &\le C\mu_{-\frac{1}{2}}^{-\frac{3}{2},-1}(\mathbf{x}) \\ \left.\begin{array}{r} |\partial_1\mathcal{O}_{12}(\mathbf{x};1)| \\ |\partial_1\mathcal{O}_{22}(\mathbf{x};1)| \\ |\partial_2\mathcal{O}_{22}(\mathbf{x};1)| \\ |\partial_k\partial_l\mathcal{O}_{mn}(\mathbf{x};1) - \partial_k\partial_l\mathcal{S}_{mn}(\mathbf{x})| \\ (k,l,m,n) \neq (2,2,1,1) \end{array}\right\} &\le C\mu_0^{-2,-1}(\mathbf{x})\,. \end{aligned} \tag{13.60}$$

13.2.3 L^∞-Estimates of a Convolution in $\mathbb{R}^N$

This section is devoted to the study of an auxiliary problem—the L^∞ estimates of certain convolution which will play a fundamental role in the following sections. Our aim is to give conditions on a, b, c, d, e, f such that

$$(\eta_{-b}^{-a} * \eta_{-d}^{-c})(\mathbf{x}) \le K\eta_{-f}^{-e}(\mathbf{x})\,, \quad \mathbf{x} \in \mathbb{R}^N\,. \tag{13.61}$$

We shall calculate the estimates for $N \ge 2$. Since we study the physically interesting cases $N = 2, 3$, the results will be summarized in Tables 13.1, 13.2, 13.3, and 13.4 only in these situations. Nevertheless, the calculations will be performed for general dimensions, and the results can be easily read from the integrals I_0—I_{15}.

Before calculating estimates of the type (13.61), we shall first study the asymptotic behavior of the function

$$s(\mathbf{x}) = |\mathbf{x}| - x_1 .$$

In what follows, by $f(\mathbf{x}) \sim g(\mathbf{x})$ as $|\mathbf{x}| \to A$ we mean the following: There exist $C_1, C_2 > 0$, and $U(A)$, a neighborhood of A, such that

$$C_1 f(\mathbf{x}) \leq g(\mathbf{x}) \leq C_2 f(\mathbf{x}) \tag{13.62}$$

for all $\mathbf{x}$ such that $|\mathbf{x}|$ is from $U(A)$. If we write $f(\mathbf{x}) \sim g(\mathbf{x})$, $\mathbf{x} \in M \subseteq \mathbb{R}^N$, then the inequality (13.62) holds for any $\mathbf{x} \in M$ with constants independent of $\mathbf{x}$.

Let us denote by $\mathbf{x}'$ the vector of the last $N-1$ components of $\mathbf{x}$, i.e., $\mathbf{x} = (x_1, \mathbf{x}')$. We have:

Lemma 13.1 *If* $x_1 > 0$, *then* $s(\mathbf{x}) \sim \frac{|\mathbf{x}'|^2}{|\mathbf{x}|}$; *otherwise* $s(\mathbf{x}) \sim |\mathbf{x}|$.

Proof Introducing the generalized spherical coordinates ($N \geq 3$)

$$\begin{aligned} x_1 &= R\cos\theta_1 \\ x_2 &= R\sin\theta_1 \cos\theta_2 \\ &\cdots \\ x_{N-1} &= R\sin\theta_1 \cdots \sin\theta_{N-2}\cos\theta_{N-1} \\ x_N &= R\sin\theta_1 \cdots \sin\theta_{N-2}\sin\theta_{N-1}, \end{aligned} \tag{13.63}$$

where $\theta_1, \cdots, \theta_{N-2} \in (0; \pi)$, $\theta_{N-1} \in (0; 2\pi)$, we have

$$s = R(1 - \cos\theta_1) = 2\frac{(R\sin\theta_1)^2}{R}\Big(\frac{\sin(\theta_1/2)}{\sin\theta_1}\Big)^2 .$$

For $x_1 > 0$ we have

$$\theta_1 \in (0; \pi/2), \text{ i.e. } 2\Big(\frac{\sin(\theta_1/2)}{\sin\theta_1}\Big)^2 \in (1/2; 1),$$

which implies

$$\frac{1}{2}\frac{|\mathbf{x}'|^2}{R} \leq s(\mathbf{x}) \leq \frac{|\mathbf{x}'|^2}{R} .$$

Analogously we proceed for $x_1 \leq 0$ where $\theta_1 \in [\pi/2; \pi]$ and $|\mathbf{x}| \leq s(\mathbf{x}) \leq 2|\mathbf{x}|$. If $N = 2$, we use the polar coordinates and the only change consists in the fact that $\varphi = \theta_1 \in (-\pi/2; \pi/2)$ for $x_1 > 0$ and $\varphi \in [\pi/2; 3/2\pi]$ for $x_1 \leq 0$. □

Next we study the integral of $\eta_{-b}^{-a}(\mathbf{x})$ over the sphere for sufficiently large $R = |\mathbf{x}|$.

Lemma 13.2 *Let $N \geq 2$. Then for the exponents $a, b \in \mathbb{R}$ we have*

$$\int_{\partial B_R} \eta_{-b}^{-a}(\mathbf{x})dS \sim R^{N-1-a-\min(\frac{N-1}{2},b)} \cdot (\ln R \text{ if } b = (N-1)/2) \tag{13.64}$$

as $R \to \infty$. Consequently, $\int_{\mathbb{R}^N} \eta_{-b}^{-a}(\mathbf{x})d\mathbf{x} < \infty \iff a + \min(\frac{N-1}{2}, b) > N$.

Proof Using the generalized spherical coordinates (if $N \geq 3$—see (13.63)) or the polar ones ($N = 2$), we get

$$\begin{aligned}\int_{\partial B_R} \eta_{-b}^{-a}(\mathbf{x})\mathrm{d}S &= C\int_0^{\pi} (1+R)^{-a}(1+s)^{-b}R^{N-1}(\sin\theta_1)^{N-2}\mathrm{d}\theta_1 \\ &= C\int_0^{\pi} (1+R)^{-a}(1+R(1-\cos\theta_1))^{-b}R^{N-1}\sin\theta_1^{N-2}\mathrm{d}\theta_1 \,.\end{aligned}$$

Changing the variables $s = R(1 - \cos\theta_1)$, we estimate the last integral by

$$C(1+R)^{1-a}\int_0^{2R} (1+s)^{-b}(\sqrt{2sR - s^2})^{N-3}\mathrm{d}s \,. \tag{13.65}$$

We estimate the integral (13.65) over three subintervals. Let us also note that for $N = 3$ it can be calculated explicitly. We have

$$\begin{aligned}&\int_0^1 (1+s)^{-b}(2sR - s^2)^{\frac{N-3}{2}}\mathrm{d}s \sim R^{\frac{N-3}{2}}\int_0^1 s^{\frac{N-3}{2}}\mathrm{d}s \sim R^{\frac{N-3}{2}}, \\ &\int_1^R (1+s)^{-b}(2sR - s^2)^{\frac{N-3}{2}}\mathrm{d}s \sim R^{\frac{N-3}{2}}\int_1^R s^{-b+\frac{N-3}{2}}\mathrm{d}s \\ &\qquad \sim R^{N-2-\min(b,\frac{N-1}{2})} \cdot (\ln R \text{ if } b = \tfrac{N-1}{2}), \\ &\int_R^{2R} (1+s)^{-b}(2sR - s^2)^{\frac{N-3}{2}}\mathrm{d}s \sim R^{N-2-b},\end{aligned}$$

which implies (13.64). As $\eta_{-b}^{-a}(\cdot) \in C(\mathbb{R}^N)$, the condition implying global integrability follows trivially. □

We can start to deal with the convolution (13.61). Recall that similar estimates were for the first time studied by Finn (see [82, 83]) in the three-dimensional case (but only for special values of c, d) and by Smith (see [200]) in the two-dimensional case. A generalization of their approach due to Farwig (see [58]) for the three-dimensional case was adapted to the two-dimensional case by Dutto (see [48]). We shall repeat their calculation in N dimensions where $N \geq 2$, arbitrary.

Finally note that the estimate (13.61) remains true if we replace the kernel $\eta_{-d}^{-c}(\mathbf{x} - \mathbf{y})$ by

$$K(\mathbf{z}) \sim \mu_{-d}^{-c,-\gamma}(\mathbf{z}), \quad \gamma < N, \mathbf{z} \in \mathbb{R}^N \tag{13.66}$$

(see also I_1 below).

In the sequel we shall use the following notation:

$$\begin{array}{lll} \mathbf{x} = (x_1, \mathbf{x}') & \mathbf{y} = (y_1, \mathbf{y}') & \\ R = |\mathbf{x}| & r = |\mathbf{y}| & \tilde{r} = |\mathbf{x} - \mathbf{y}| \\ s = s(\mathbf{x}) & t = y_1 & \tilde{t} = x_1 - y_1 \\ \varrho = |\mathbf{y}'| & \tilde{\varrho} = |\mathbf{x}' - \mathbf{y}'| & . \end{array} \tag{13.67}$$

In order to capture the anisotropic structure of the function $\eta^{-a}_{-b}(\cdot)$ we shall study the convolution (13.61) in four different situations:

(A) $R \le R_0$.
(B) $x_1 > 0$, $|\mathbf{x}'| \le \sqrt{x_1}$, $R > R_0$.
(C) $x_1 > 0$, $|\mathbf{x}'| = \frac{1}{2} R^{\frac{1}{2}+\sigma}$, $R > R_0$, $\sigma \in [0; \frac{1}{2}]$.
(D) $x_1 > 0$, $|\mathbf{x}'| \ge \frac{R}{2}$, $R > R_0$ or $x_1 < 0$, $R > R_0$. □

Using Lemma 13.1 we easily verify that

$$\begin{array}{l} \eta^{-a}_{-b}(\mathbf{y}) \sim \begin{cases} 1\,, & r \le 1 \\ r^{-a}\,, & r > 1\,,\, t > 0\,,\, \varrho < \sqrt{t} \\ r^{-a+b}\varrho^{-2b}\,, & r > 1\,,\, t > 0\,,\, \varrho \ge \sqrt{t} \\ r^{-a-b}\,, & r > 1\,,\ t \le 0 \end{cases} \\ \eta^{-c}_{-d}(\mathbf{x}-\mathbf{y}) \sim \begin{cases} 1\,, & \tilde{r} \le 1 \\ \tilde{r}^{-c}\,, & \tilde{r} > 1\,,\, \tilde{t} > 0\,,\, \tilde{\varrho} < \sqrt{\tilde{t}} \\ \tilde{r}^{-c+d}\tilde{\varrho}^{-2d}\,, & \tilde{r} > 1\,,\, \tilde{t} > 0\,,\, \tilde{\varrho} \ge \sqrt{\tilde{t}} \\ \tilde{r}^{-c-d}\,, & \tilde{r} > 1\,,\, \tilde{t} \le 0\,. \end{cases} \end{array} \tag{13.68}$$

For notational convenience we denote $\nu = \sigma + \frac{1}{2}$ and $b^* = \min(\frac{N-1}{2}, b)$; analogously $d^* = \min(\frac{N-1}{2}, d)$.

We start with the case (A). Applying Lemma 13.2 to the half-spaces $y_1 > 0$ and $y_1 < 0$, we find that the convolution is uniformly bounded if

$$a + b^* + c + d > N \qquad a + b + c + d^* > N\,. \tag{13.69}$$

Next we continue with the most complicated case (C). We follow Farwig (see [58]) and divide $\mathbb{R}^N$ into 16 subdomains as shown in Fig. 13.1. If $N = 2$, the subdomains are plane; otherwise they are cylindrical.

We calculate the convolutions separately on each subdomain. For the reader's convenience, the results are summarized in Tables 13.1 and 13.2 (for $N = 3$) and Tables 13.3 and 13.4 (for $N = 2$). We denote by I_k the corresponding part of the integral (13.61) over Ω_k, $k = 0, 1, \cdots, 15$. We shall get

$$I_i(\mathbf{x}) \le K R^{-e_i - 2\sigma f_i} \sim \eta^{-e_i}_{-f_i}(\mathbf{x})\,.$$

Unfortunately in many cases additional logarithmic terms will appear which will cause some losses in the weighted estimates later on.

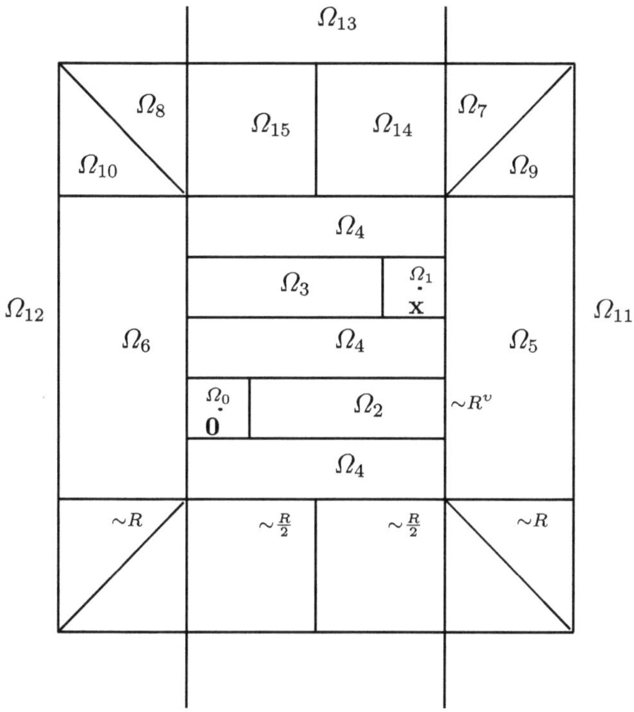

Fig. 13.1 Division of $\mathfrak{R}^N$ into 16 subdomains

Remark 13.2 Let A be a positive function. We denote

$$\ln_+ A = \max(\ln A, 1)\,.$$

We start by estimating the convolutions over the sixteen subdomains.

$\underline{\mathrm{I}_0}$ We have $\Omega_0 = \left\{\mathbf{y} \in \mathbb{R}^N : |t| \le \frac{1}{8}R^\nu; \varrho \le \frac{1}{8}R^\nu\right\}$, and therefore $\eta^{-a}_{-b}(\mathbf{y}) \sim r^{-a}(1+s(\mathbf{y}))^{-b}$, $\eta^{-c}_{-d}(\mathbf{x}-\mathbf{y}) \sim R^{-c-2d\sigma}$. Applying Lemma 13.2 we get

$$I_0 \sim R^{-c-2\sigma d}\int_0^{R^\nu}(1+r)^{N-1-a-b^*}\cdot(\ln_+ r\,,\ \text{if}\ b=\tfrac{N-1}{2})\mathrm{d}r$$
$$\sim R^{-(c+\frac{1}{2}\min(0,a+b^*-N))-2\sigma(d+\min(0,a+b^*-N))}$$
$$\cdot\begin{cases}\ln R & \text{if}\ a+b^*=N\,,\ b\neq\frac{N-1}{2}\\ & \text{or}\ a+b^*<N\,,\ b=\frac{N-1}{2}\\ \ln^2 R & \text{if}\ a+b^*=N\,,\ b=\frac{N-1}{2}\,.\end{cases}$$

The results are summarized in Tables 13.1, 13.2, 13.3, and 13.4.

Table 13.1 $N = 3$

Dom.	t	ϱ	r	$\tilde{t}$	$\tilde{\varrho}$	$\tilde{r}$	$\eta_{-b}^{-a}(\mathbf{y})$	$\eta_{-d}^{-c}(\mathbf{x}-\mathbf{y})$	e	f	*log. factors*
Ω_0	$(-\frac{1}{8}R^{\upsilon}; \frac{1}{8}R^{\upsilon})$	$(0; \frac{1}{8}R^{\upsilon})$	$(0; \frac{1}{8}R^{\upsilon})$	$\sim x_1$ R	$\sim \lvert\mathbf{x}'\rvert$ R^{υ}	$\sim \lvert\mathbf{x}\rvert$ R	$r^{-a}(1+s(\mathbf{y}))^{-b}$	$R^{-c-2\sigma d}$	$c + \frac{1}{2}\min(0, a+b^*-3)$	$d + \frac{1}{2}\min(0, a+b^*-3)$	$\ln R(b=1 \wedge a<2)\vee$ $\vee (b \neq 1 \wedge a+b^*=3)$ $\ln^2 R(b=1 \wedge a=2)$
Ω_1	$\sim x_1$ R	$\sim \lvert\mathbf{x}'\rvert$ R^{υ}	$\sim \lvert\mathbf{x}\rvert$ R	$(-\frac{1}{8}R^{\upsilon}; \frac{1}{8}R^{\upsilon})$	$(0; \frac{1}{8}R^{\upsilon})$	$(0; \frac{1}{8}R^{\upsilon})$	$R^{-a-2\sigma b}$	$\tilde{r}^{-c}(1+s(\mathbf{x}-\mathbf{y}))^{-d}$	$a + \frac{1}{2}\min(0, c+d^*-3)$	$b + \frac{1}{2}\min(0, c+d^*-3)$	$\ln R(d=1 \wedge c<2)\vee$ $\vee (d \neq 1 \wedge c+d^*=3)$ $\ln^2 R\ (c=2 \wedge d=1)$
Ω_2	$\sim r$ $(R^{\upsilon}; R)$	$(0; \frac{1}{8}R^{\upsilon})$	$\sim t$ $(R^{\upsilon}; R)$	$(-\frac{1}{8}R^{\upsilon};$ $R-R^{\upsilon})$	$\sim \lvert\mathbf{x}'\rvert$ R^{υ}	$R+R^{\upsilon}-r$	$r^{-a}\quad \varrho < \sqrt{t}$ $r^{-a+b}\varrho^{-2b}\quad \varrho > \sqrt{t}$	$\tilde{r}^{-c+d}R^{-2d\upsilon}$	$a+c-2+\frac{1}{2}\min$ $(0, 1+b^*-a, 1+d-c)$	$b^*+d-1-\frac{1}{2}\min$ $(0, 1+b^*-a, 1+d-c)$	$\ln \frac{R}{1+s}(\min(1+b^*-a,$ $1+d-c)=0 \wedge b \neq 1)$ $\ln_+ s \cdot \ln \frac{R}{1+s}$ $(b=1 \wedge 1+d=0)$ $(\ln_+ s\quad a<2)(\ln R\quad a>2)$ $\ln R\ \ln \frac{R}{1+s}\quad a=2) \wedge b=1$
Ω_3	$(-\frac{1}{8}R^{\upsilon};$ $R-R^{\upsilon})$	$\sim \lvert\mathbf{x}'\rvert$ R^{υ}	$R+R^{\upsilon}-\tilde{r}$	$\sim \tilde{r}$ $(R^{\upsilon}; R)$	$(0; \frac{1}{8}R^{\upsilon})$	$\sim \tilde{t}$ $(R^{\upsilon}; R)$	$r^{-a+b}R^{-2b\upsilon}$	$\tilde{r}^{-c}\quad \tilde{\varrho} < \sqrt{\tau}$ $\tilde{r}^{-c+d}\tilde{\varrho}^{-2d}\quad \tilde{\varrho} > \sqrt{\tau}$	$a+c-2+\frac{1}{2}\min$ $(0, 1+b-a, 1+d^*-c)$	$b+d^*-1-\frac{1}{2}\min$ $(0, 1+b-a, 1+d^*-c)$	$\ln \frac{R}{1+s}(\min(1+b-a,$ $1+d^*-c)=0 \wedge d \neq 1)$ $\ln_+ s \cdot \ln \frac{R}{1+s}(d=1\wedge$ $\wedge 1+b-a=0)$ $(\ln_+ s\ c<2)(\ln R\ c>2)$ $(\ln R \ln \frac{R}{1+s}\ c=2) \wedge d=1$
Ω_4									see Ω_2, Ω_3	see Ω_2, Ω_3	see Ω_2, Ω_3
Ω_5	$\sim r$ R	$(0; R^{\upsilon})$	$\sim t$ R	$\sim -\tilde{r}$ $(-R; -R^{\upsilon})$	$(0; R^{\upsilon})$	$\sim \lvert\tilde{t}\rvert$ $(R^{\upsilon}; R)$	$R^{-a}\quad \varrho < \sqrt{t}$ $R^{-a+b}\varrho^{-2b}\quad \varrho > \sqrt{t}$	$\lvert\tilde{t}\rvert^{-c-d}$	$a+c+d-2+\frac{1}{2}$ $\min(0, 1-c-d)$	$b^*-1-\frac{1}{2}\min$ $(0, 1-c-d)$	$(\ln_+ s\quad b=1)\cdot$ $\cdot(\ln \frac{R}{1+s}\ c+d=1)$
Ω_6	$\sim -r$ $(-R; -R^{\upsilon})$	$(0; R^{\upsilon})$	$\sim \lvert t\rvert$ $(R^{\upsilon}; R)$	$\sim \tilde{r}$ R	$(0; R^{\upsilon})$	$\sim \tilde{t}$ R	$\lvert t\rvert^{-a-b}$	$R^{-c}\quad \tilde{\varrho} < \sqrt{\tilde{t}}$ $R^{-c+d}\tilde{\varrho}^{-2d}\quad \tilde{\varrho} > \sqrt{\tilde{t}}$	$a+b+c-2+\frac{1}{2}\min$ $(0, 1-a-b)$	$d^*-1-\frac{1}{2}\min$ $(0, 1-a-b)$	$(\ln_+ s\quad d=1)\cdot$ $\cdot(\ln \frac{R}{1+s}\quad a+b=1)$
Ω_7	R	$\sim \tilde{\varrho}$ $(\lvert\tilde{t}\rvert; R)$	R	$(-R; -R^{\upsilon})$	$\sim \varrho, \tilde{r}$ $(\lvert\tilde{t}\rvert; R)$	$\sim \tilde{\varrho}$ $(\lvert\tilde{t}\rvert; R)$	$R^{-a+b}\varrho^{-2b}$	ϱ^{-c-d}	$a+b+c+d-3+$ $\frac{1}{2}\min(0, 3-2b-c-d)$	$-\frac{1}{2}\min(0, 3-2b-$ $-c-d)$	$\ln \frac{R}{1+s}\quad 2b+c+d=3$
Ω_8	$(-R; -R^{\upsilon})$	$\sim \tilde{\varrho}, r$ $(\lvert t\rvert; R)$	$\sim \varrho$ $(\lvert t\rvert; R)$	R	$\sim \varrho$ $(\lvert t\rvert; R)$	R	$\tilde{\varrho}^{-a-b}$	$R^{-c+d}\tilde{\varrho}^{-2d}$	$a+b+c+d-3+$ $\frac{1}{2}\min(0, 3-a-b-2d)$	$-\frac{1}{2}\min(0, 3-a-b-2d)$	$\ln \frac{R}{1+s}\quad a+b+2d=3$

Table 13.2 $N = 3$

Dom.	t	ϱ	r	$\tilde{t}$	$\tilde{\varrho}$	$\tilde{r}$	$\eta_{-b}^{-a}(\mathbf{y})$	$\eta_{-d}^{-c}(\mathbf{x}-\mathbf{y})$	e	f	*log. factors*
Ω_9	$\sim r$ R	$\sim \tilde{\varrho}$ $(R^\upsilon; \lvert\tilde{t}\rvert)$	$\sim \tilde{t}$ R	$\sim -\tilde{r}$ $(-R; -R^\upsilon)$	$\sim \varrho$ $(R^\upsilon; \lvert\tilde{t}\rvert)$	$\sim \lvert\tilde{t}\rvert$ $(R^\upsilon; R)$	$R^{-a+b}\varrho^{-2b}$	$\lvert\tilde{t}\rvert^{-c-d}$	$b > 1$ see Ω_5 $b < 1$ see Ω_7 $b = 1$ see Ω_5	$b > 1$ see Ω_5 $b < 1$ see Ω_7 $b = 1$ see Ω_5	$\Big((\ln \frac{R}{1+s}\ c+d<1)$ $(\ln^2 \frac{R}{1+s}\ c+d=1)\Big) \wedge b=1$
Ω_{10}	$\sim -r$ $(-R; -R^\upsilon)$	$\sim \tilde{\varrho}$ $(R^\upsilon; \lvert t\rvert)$	$\sim \lvert t\rvert$ $(R^\upsilon; R)$	$\sim \tilde{r}$ R	$\sim \varrho$ $(R^\upsilon; \lvert t\rvert)$	$\sim \tilde{t}$ R	$\lvert t\rvert^{-a-b}$	$R^{-c+d}\ \tilde{\varrho}^{-2d}$	$d > 1$ see Ω_6 $d < 1$ see Ω_8 $d = 1$ see Ω_6	$d > 1$ see Ω_6 $d < 1$ see Ω_8 $d = 1$ see Ω_6	$\Big((\ln \frac{R}{1+s}\ a+b<1)$ $(\ln^2 \frac{R}{1+s}\ a+b=1)\Big) \wedge d=1$
Ω_{11}	$(R; \infty)$	$(0; \infty)$	$\sim \tilde{r}$ $(R; \infty)$	$(-\infty; -R^\upsilon)$	$(0; \infty)$	$\sim r$ $(R; \infty)$	$(1+r)^{-a}(1+s(\mathbf{y}))^{-b}$	r^{-c-d}	$a+b^*+c+d-3>0$	0	$\ln R \quad b=1$
Ω_{12}	$(-\infty; -R^\upsilon)$	$(0; \infty)$	$\sim \tilde{r}$ $(R; \infty)$	$(R; \infty)$	$(0; \infty)$	$\sim r$ $(R; \infty)$	$\tilde{r}^{-a-b}$	$(1+\tilde{r})^{-c}(1+s(\mathbf{x}-\mathbf{y}))^{-d}$	$a+b+c+d^*-3>0$	0	$\ln R \quad d=1$
Ω_{13}		$\sim r, \tilde{r}$ $(R; \infty)$	$\sim \tilde{r}$ $(R; \infty)$		$\sim \varrho$ $(R; \infty)$	$\sim r$ $(R; \infty)$			see Ω_{11} and Ω_{12}	see Ω_{11} and Ω_{12}	see Ω_{11} and Ω_{12}
Ω_{14}	$\sim r$ R	$\sim \tilde{\varrho}$ $(R^\upsilon; R)$	$\sim t$ R	$(-\frac{1}{8}R^\upsilon; \frac{R}{2})$	$\sim \varrho$ $(R^\upsilon; R)$	$\sim \lvert\tilde{t}\rvert + \tilde{\varrho}$	$R^{-a+b}\ \varrho^{-2b}$	$(\tilde{t}+\tilde{\varrho})^{-c+d}\ \tilde{\varrho}^{-2d}\ \tilde{t}>0$ $\tilde{\varrho}^{-c-d}\ \tilde{t}<0$	see Ω_2, Ω_3 Ω_7 Ω_2 $a+b+c+d-3$	$b+d>1 \wedge 1+d-c>0$ $1+d-c<0$ $1+d-c=0 \wedge b+d>1$ 0 otherwise	$\ln \frac{R}{1+s}\ b+d=1 \wedge 1+d-c>0$ $\ln^2 \frac{R}{1+s}\ b+d=1 \wedge 1+d-c=0$
Ω_{15}	$(-R^\upsilon; \frac{R}{2})$	$\sim \tilde{\varrho}$ $(R^\upsilon; R)$	$\sim \lvert t\rvert + \varrho$	$\sim r$ R	$\sim \varrho$ $(R^\upsilon; R)$	$\sim t$ R	$(t+\varrho)^{-a+b}\ \varrho^{-2b}\quad t>0$ $\varrho^{-a-b}\quad t<0$	$R^{-c+d}\ \tilde{\varrho}^{-2d}$	see Ω_2, Ω_3 Ω_8 Ω_3 $a+b+c+d-3$	$b+d>1 \wedge 1+b-a>0$ $1+b-a<0$ $1+b-a=0 \wedge b+d>1$ 0 otherwise	$\ln \frac{R}{1+s}\ b+d=1 \wedge 1+b-a>0$ $\ln^2 \frac{R}{1+s}\ b+d=1 \wedge 1+b-a=0$

Table 13.3 $N = 2$

Dom.	t	ϱ	r	$\tilde{t}$	$\tilde{\varrho}$	$\tilde{r}$	$\eta^{-a}_{-b}(\mathbf{y})$	$\eta^{-c}_{-d}(\mathbf{x}-\mathbf{y})$	e	f	log. factors
Ω_0	$(-\frac{1}{8}R^{\upsilon};\ \frac{1}{8}R^{\upsilon})$	$(0;\ \frac{1}{8}R^{\upsilon})$	$(0;\ \frac{1}{8}R^{\upsilon})$	$\sim x_1$ R	$\sim \|x_2\|$ R^{υ}	$\sim \|\mathbf{x}\|$ R	$r^{-a}(1+s(\mathbf{y}))^{-b}$	$R^{-c-2\sigma d}$	$c + \frac{1}{2}\min(0, a + b^* - 2)$	$d + \frac{1}{2}\min(0, a + b^* - 2)$	$\ln R(b = \frac{1}{2} \wedge a < \frac{3}{2}) \vee$ $\vee (b \neq \frac{1}{2} \wedge a + b^* = 2)$ $\ln^2 R(b = \frac{1}{2} \wedge a = \frac{3}{2})$
Ω_1	$\sim x_1$ R	$\sim \|x_2\|$ R^{υ}	$\sim \|\mathbf{x}\|$ R	$(-\frac{1}{8}R^{\upsilon};\ \frac{1}{8}R^{\upsilon})$	$(0;\ \frac{1}{8}R^{\upsilon})$	$(0;\ \frac{1}{8}R^{\upsilon})$	$R^{-a-2\sigma b}$	$\tilde{r}^{-c}(1+s(\mathbf{x}-\mathbf{y}))^{-d}$	$a + \frac{1}{2}\min(0, c + d^* - 2)$	$b + \frac{1}{2}\min(0, c + d^* - 2)$	$\ln R(d = \frac{1}{2} \wedge c < \frac{3}{2}) \vee$ $\vee (d \neq \frac{1}{2} \wedge c + d^* = 2)$ $\ln^2 R\ (c = \frac{3}{2} \wedge d = \frac{1}{2})$
Ω_2	$\sim r$ $(R^{\upsilon}; R)$	 $(0;\ \frac{1}{8}R^{\upsilon})$	$\sim t$ $(R^{\upsilon}; R)$	$(-\frac{1}{8}R^{\upsilon};$ $R - R^{\upsilon})$	$\sim \|\mathbf{x}'\|$ R^{υ}	 $R + R^{\upsilon} - r$	$r^{-a} \quad \varrho < \sqrt{t}$ $r^{-a+b}\varrho^{-2b} \quad \varrho > \sqrt{t}$	 $\tilde{r}^{-c+d}R^{-2d\upsilon}$	$a + c - \frac{3}{2} + \frac{1}{2}\min$ $(0, 1 + b^* - a, 1 + d - c)$	$b^* + d - \frac{1}{2} - \frac{1}{2}\min$ $(0, 1 + b^* - a, 1 + d - c)$	$\ln \frac{R}{1+s}(\min(1 + b^* - a,$ $1 + d - c) = 0 \wedge b \neq \frac{1}{2})$ $\ln_+ s \cdot \ln \frac{R}{1+s}$ $(b = \frac{1}{2} \wedge 1 + d = c)$ $(\ln_+ s \quad a < \frac{3}{2})(\ln R\ a > \frac{3}{2})$ $(\ln R\ \ln \frac{R}{1+s} \quad a = \frac{3}{2}) \wedge b = \frac{1}{2}$
Ω_3	$(-\frac{1}{8}R^{\upsilon};$ $R - R^{\upsilon})$	$\sim \|x_2\|$ R^{υ}	 $R + R^{\upsilon} - \tilde{r}$	$\sim \tilde{r}$ $(R^{\upsilon}; R)$	 $(0;\ \frac{1}{8}R^{\upsilon})$	$\sim \tilde{t}$ $(R^{\upsilon}; R)$	 $r^{-a+b}R^{-2b\upsilon}$	$\tilde{r}^{-c} \quad \tilde{\varrho} < \sqrt{\tau}$ $\tilde{r}^{-c+d}\tilde{\varrho}^{-2d} \quad \tilde{\varrho} > \sqrt{\tau}$	$a + c - \frac{3}{2} + \frac{1}{2}\min$ $(0, 1 + b - a, 1 + d^* - c)$	$b + d^* - \frac{1}{2} - \frac{1}{2}\min$ $(0, 1 + b - a, 1 + d^* - c)$	$\ln \frac{R}{1+s}(\min(1 + b - a,$ $1 + d^* - c) = 0 \wedge d \neq 1)$ $\ln_+ s \cdot \ln \frac{R}{1+s}(d = \frac{1}{2} \wedge$ $\wedge 1 + b - a = 0)$ $(\ln_+ s\ c < \frac{3}{2})(\ln R\ c > \frac{3}{2})$ $(\ln R \ln \frac{R}{1+s}\ c = \frac{3}{2}) \wedge d = \frac{1}{2}$
Ω_4									see Ω_2, Ω_3	see Ω_2, Ω_3	see Ω_2, Ω_3
Ω_5	$\sim r$ R	 $(0; R^{\upsilon})$	$\sim t$ R	$\sim -\tilde{r}$ $(-R; -R^{\upsilon})$	 $(0; R^{\upsilon})$	$\sim \|\tilde{t}\|$ $(R^{\upsilon}; R)$	$R^{-a} \quad \varrho < \sqrt{t}$ $R^{-a+b}\varrho^{-2b} \quad \varrho > \sqrt{t}$	$\|\tilde{t}\|^{-c-d}$	$a + c + d - \frac{3}{2} + \frac{1}{2}\min$ $(0, 1 - c - d)$	$b^* - \frac{1}{2} - \frac{1}{2}\min$ $(0, 1 - c - d)$	$(\ln_+ s \quad b = \frac{1}{2})\cdot$ $\cdot (\ln \frac{R}{1+s}\ c + d = 1)$
Ω_6	$\sim -r$ $(-R; -R^{\upsilon})$	 $(0; R^{\upsilon})$	$\sim \|t\|$ $(R^{\upsilon}; R)$	$\sim \tilde{r}$ R	 $(0; R^{\upsilon})$	$\sim \tilde{t}$ R	$\|t\|^{-a-b}$	$R^{-c} \quad \tilde{\varrho} < \sqrt{\tilde{t}}$ $R^{-c+d}\tilde{\varrho}^{-2d} \quad \tilde{\varrho} > \sqrt{\tilde{t}}$	$a + b + c - \frac{3}{2} + \frac{1}{2}\min$ $(0, 1 - a - b)$	$d^* - \frac{1}{2} - \frac{1}{2}\min$ $(0, 1 - a - b)$	$(\ln_+ s\ d = \frac{1}{2})\cdot(\ln \frac{R}{1+s}\ a + b = 1)$
Ω_7	 R	$\sim \tilde{\varrho}$ $(\|\tilde{t}\|; R)$	 R	 $(-R; -R^{\upsilon})$	$\sim \varrho, \tilde{r}$ $(\|\tilde{t}\|; R)$	$\sim \tilde{\varrho}$ $(\|\tilde{t}\|; R)$	$R^{-a+b}\varrho^{-2b}$	ϱ^{-c-d}	$a + b + c + d - 2 +$ $\frac{1}{2}\min(0, 2 - 2b - c - d)$	$-\frac{1}{2}\min(0, 2 - 2b -$ $- c - d)$	$\ln \frac{R}{1+s} \quad 2b + c + d = 2$
Ω_8	 $(-R; -R^{\upsilon})$	$\sim \tilde{\varrho}, r$ $(\|t\|; R)$	$\sim \varrho$ $(\|t\|; R)$	 R	$\sim \varrho$ $(\|t\|; R)$	 R	$\tilde{\varrho}^{-a-b}$	$R^{-c+d}\tilde{\varrho}^{-2d}$	$a + b + c + d - 2 +$ $\frac{1}{2}\min(0, 2 - a - b - 2d)$	$-\frac{1}{2}\min(0, 2 - a - b - 2d)$	$\ln \frac{R}{1+s} \quad a + b + 2d = 2$

Table 13.4 $N = 2$

Dom.	t	ϱ	r	$\tilde{t}$	$\tilde{\varrho}$	$\tilde{r}$	$\eta^{-a}_{-b}(\mathbf{y})$	$\eta^{-c}_{-d}(\mathbf{x}-\mathbf{y})$	e	f	log. factors
Ω_9	$\sim r$ R	$\sim \tilde{\varrho}$ $(R^{\upsilon}; \lvert\tilde{t}\rvert)$	$\sim t$ R	$\sim -\tilde{r}$ $(-R; -R^{\upsilon})$	$\sim \varrho$ $(R^{\upsilon}; \lvert\tilde{t}\rvert)$	$\sim \lvert\tilde{t}\rvert$ $(R^{\upsilon}; R)$	$R^{-a+b}\,\varrho^{-2b}$	$\lvert\tilde{t}\rvert^{-c-d}$	$b > \frac{1}{2}$ see Ω_5 $b < \frac{1}{2}$ see Ω_7 $b = \frac{1}{2}$ see Ω_5	$b > \frac{1}{2}$ see Ω_5 $b < \frac{1}{2}$ see Ω_7 $b = \frac{1}{2}$ see Ω_5	$\Big((\ln \frac{R}{1+s}\;\; c+d<1)$ $(\ln^2 \frac{R}{1+s}\;\; c+d=1)\Big) \wedge b = \frac{1}{2}$
Ω_{10}	$\sim -r$ $(-R; -R^{\upsilon})$	$\sim \tilde{\varrho}$ $(R^{\upsilon}; \lvert t\rvert)$	$\sim \lvert t\rvert$ $(R^{\upsilon}; R)$	$\sim \tilde{r}$ R	$\sim \varrho$ $(R^{\upsilon}; \lvert t\rvert)$	$\sim \tilde{t}$ R	$\lvert t\rvert^{-a-b}$	$R^{-c+d}\,\tilde{\varrho}^{-2d}$	$d > \frac{1}{2}$ see Ω_6 $d < \frac{1}{2}$ see Ω_8 $d = \frac{1}{2}$ see Ω_6	$d > \frac{1}{2}$ see Ω_6 $d < \frac{1}{2}$ see Ω_8 $d = \frac{1}{2}$ see Ω_6	$\Big((\ln \frac{R}{1+s}\;\; a+b<1)$ $(\ln^2 \frac{R}{1+s}\;\; a+b=1)\Big) \wedge d = \frac{1}{2}$
Ω_{11}	$(R; \infty)$	$(0; \infty)$	$\sim \tilde{r}$ $(R; \infty)$	$(-\infty; -R^{\upsilon})$	$(0; \infty)$	$\sim r$ $(R; \infty)$	$(1+r)^{-a}(1+s(\mathbf{y}))^{-b}$	r^{-c-d}	$a+b^{*}+c+d-2>0$	0	$\ln R \;\; b = \frac{1}{2}$
Ω_{12}	$(-\infty; -R^{\upsilon})$	$(0; \infty)$	$\sim \tilde{r}$ $(R; \infty)$	$(R; \infty)$	$(0; \infty)$	$\sim r$ $(R; \infty)$	$\tilde{r}^{-a-b}$	$(1+\tilde{r})^{-c}(1+s(\mathbf{x}-\mathbf{y}))^{-d}$	$a+b+c+d^{*}-2>0$	0	$\ln R \;\; d = \frac{1}{2}$
Ω_{13}		$\sim r, \tilde{r}$ $(R; \infty)$	$\sim \tilde{r}$ $(R; \infty)$		$\sim \varrho$ $(R; \infty)$	$\sim r$ $(R; \infty)$			see Ω_{11} and Ω_{12}	see Ω_{11} and Ω_{12}	see Ω_{11} and Ω_{12}
Ω_{14}	$\sim r$ R	$\sim \tilde{\varrho}$ $(R^{\upsilon}; R)$	$\sim t$ R	$(-\frac{1}{8}R^{\upsilon}; \frac{R}{2})$	$\sim \varrho$ $(R^{\upsilon}; R)$	$\sim \lvert\tilde{t}\rvert + \tilde{\varrho}$	$R^{-a+b}\,\varrho^{-2b}$	$(\tilde{t}+\tilde{\varrho})^{-c+d}\,\tilde{\varrho}^{-2d}\;\; \tilde{t}>0$ $\tilde{\varrho}^{-c-d}\;\; \tilde{t}<0$	see Ω_2, Ω_3 Ω_7 Ω_2 $a+b+c+d-2$	$b+d>\frac{1}{2} \wedge 1+d-c>0$ $1+d-c<0$ $1+d-c=0 \wedge b+d>\frac{1}{2}$ 0 otherwise	$\ln \frac{R}{1+s}\;\; b+d=\frac{1}{2} \wedge 1+d-c>0$ $\ln^2 \frac{R}{1+s}\;\; b+d=\frac{1}{2} \wedge 1+d-c=0$
Ω_{15}	$(-R^{\upsilon}; \frac{R}{2})$	$\sim \tilde{\varrho}$ $(R^{\upsilon}; R)$	$\sim \lvert t\rvert + \varrho$	$\sim r$ R	$\sim \varrho$ $(R^{\upsilon}; R)$	$\sim t$ R	$(t+\varrho)^{-a+b}\,\varrho^{-2b}\;\; t>0$ $\varrho^{-a-b}\;\; t<0$	$R^{-c+d}\,\tilde{\varrho}^{-2d}$	see Ω_2, Ω_3 Ω_8 Ω_3 $a+b+c+d-2$	$b+d>\frac{1}{2} \wedge 1+b-a>0$ $1+b-a<0$ $1+b-a=0 \wedge b+d>\frac{1}{2}$ 0 otherwise	$\ln \frac{R}{1+s}\;\; b+d=\frac{1}{2} \wedge 1+b-a>0$ $\ln^2 \frac{R}{1+s}\;\; b+d=\frac{1}{2} \wedge 1+b-a=0$

$\underline{I_1}$ The integral can be estimated in the same way by exchanging a, b for c, d. Assuming the kernel (13.66) instead of η_{-d}^{-c} we have

$$\tilde{I}_1(\mathbf{x}) = \int_{B_1(\mathbf{0})} K(\mathbf{x}-\mathbf{y})\eta_{-b}^{-a}(\mathbf{x})\mathrm{d}\mathbf{x} \sim R^{-a-2\sigma b}$$

since $\gamma < N$. Again, the summarized results can be found in Tables 13.1, 13.2, 13.3, and 13.4.

$\underline{I_2}$ We have in Ω_2 that $r \sim t \in (R^\nu; R)$. So $\eta_{-b}^{-a}(\mathbf{y}) \sim r^{-a}$ for $\varrho < \sqrt{t}$ and $\eta_{-b}^{-a}(\mathbf{y}) \sim r^{-a+b}\varrho^{-2b}$ for $\varrho > \sqrt{t}$. Further $\tilde{\varrho} \sim R^\nu$ and $\tilde{r} \sim R + R^\nu - r$; therefore $\eta_{-d}^{-c}(\mathbf{x}-\mathbf{y}) \sim \tilde{r}^{-c+d}R^{-2\nu d}$. Thus

$$\begin{aligned}
I_2 \sim R^{-2d\nu} &\int_{R^\nu}^{R} \mathrm{d}r(R+R^\nu-r)^{d-c}\Big[r^{-a}\int_0^{\sqrt{r}} \varrho^{N-2}\mathrm{d}\varrho \\
&+ r^{b-a}\int_{\sqrt{r}}^{R^\nu} \varrho^{N-2-2b}\mathrm{d}\varrho\Big] = R^{-2\nu d}\int_{R^\nu}^{R} \mathrm{d}r(R+R^\nu-r)^{d-c} \\
&\cdot\Big[r^{\frac{N}{2}-\frac{1}{2}-a} + r^{b-a}(R^{\nu(N-1-2b)} - r^{\frac{N}{2}-\frac{1}{2}-b})\Big]\cdot(\ln\tfrac{R^\nu}{\sqrt{r}} \text{ if } b = \tfrac{N-1}{2}) \\
&\sim R^{-2\nu(b^*-\frac{N-1}{2}+d)}\int_{R^\nu}^{R} \mathrm{d}r(R+R^\nu-r)^{d-c}r^{b^*-a} \\
&\cdot(\ln\tfrac{R^\nu}{\sqrt{r}} \text{ if} b = \tfrac{N-1}{2}) \equiv J\,.
\end{aligned}$$

In order to verify the last equivalence, it is enough to consider first $b \geq \frac{N-1}{2}$, i.e., $b^* = \frac{N-1}{2}$, and then $b < \frac{N-1}{2}$, i.e., $b^* = b$, and estimate the integrals. Let us divide J into two parts—the integral over $(R^\nu; \frac{R}{2})$ and the integral over $(\frac{R}{2}; R)$. We estimate these two parts separately.

$$\begin{aligned}
R^{-c+d-2\nu(b^*-\frac{N-1}{2}+d)} &\int_{R^\nu}^{\frac{R}{2}} \mathrm{d}r\, r^{b^*-a}\cdot(\ln\tfrac{R^\nu}{\sqrt{r}} \text{ if } b = \tfrac{N-1}{2}) \\
&\sim R^{-(c+a-\frac{N+1}{2})-2\sigma(b^*-\frac{N-1}{2}+d)+(\sigma-\frac{1}{2})\min(0,b^*-a+1)} \\
&\cdot(\ln_+\tfrac{R}{1+s} \text{ if } b \neq \tfrac{N-1}{2},\ a = b^*+1) \\
&\cdot\Big((\ln_+ s,\ a < \tfrac{N+1}{2})\cdot(\ln R,\ a > \tfrac{N+1}{2}) \\
&\cdot(\ln R\ln_+\tfrac{R}{1+s},\ a = \tfrac{N+1}{2}) \text{ if } b = \tfrac{N-1}{2}\Big).
\end{aligned}$$

We used the fact that $s(\mathbf{x}) \sim \frac{|\mathbf{x}'|^2}{|\mathbf{x}|} \sim R^{2\sigma}$ and $\ln_+ \frac{R}{R^\nu} = \frac{1}{2}\ln_+ \frac{R}{R^{2\sigma}} \sim \ln_+ \frac{R}{1+s}$. Analogously

$$R^{-2\nu(b^*-\frac{N-1}{2}+d)+b^*-a} \int_{\frac{R}{2}}^{R} \mathrm{d}r (R+R^\nu-r)^{d-c} \cdot (\ln \tfrac{R^\nu}{\sqrt{r}} \text{ if } b=\tfrac{N-1}{2})$$
$$\sim R^{-(c+a-\frac{N+1}{2})-2\sigma(b^*-\frac{N-1}{2}+d)+(\sigma-\frac{1}{2})\min(0,d-c+1)}$$
$$\cdot(\ln_+ \tfrac{R}{1+s} \text{ if } c=d+1)\cdot(\ln_+ s \text{ if } b=\tfrac{N-1}{2}).$$

The results may again be found in Tables 13.1, 13.2, 13.3, and 13.4.

$\underline{I_3}$ We proceed analogously as for I_2 exchanging a, b for c, d.

$\underline{I_4}$ Ω_4 can be considered as a subset of Ω_2 and Ω_3. Therefore I_4 can be estimated by I_2 and I_3.

$\underline{I_5}$ We have in Ω_5 $t \sim r \sim R$, so $\eta^{-a}_{-b}(\mathbf{y}) \sim R^{-a}$ $(\varrho < \sqrt{t})$ or $\eta^{-a}_{-b}(\mathbf{y}) \sim R^{-a+b}\varrho^{-2b}$ $(\varrho > \sqrt{t})$ where ϱ varies between 0 and R^ν. Further $\tilde{r} \sim |\tilde{t}| \in (R^\nu; R)$. As $\tilde{t} < 0$ we have $\eta^{-c}_{-d}(\mathbf{x}-\mathbf{y}) \sim |\tilde{t}|^{-c-d}$.

$$I_5 \sim \int_{R^\nu}^{R} \mathrm{d}\tau\, \tau^{-c-d}\Big(R^{-a}\int_0^{\sqrt{R}} \varrho^{N-2}\mathrm{d}\varrho + R^{b-a}\int_{\sqrt{R}}^{R^\nu} \varrho^{N-2-2b}\mathrm{d}\varrho\Big)$$
$$\sim R^{1-c-d+(\sigma-\frac{1}{2})\min(0,1-c-d)}[R^{\frac{N-1}{2}-a} + (R^{(N-2-2b)\nu} - R^{\frac{N-2-2b}{2}})$$
$$\cdot R^{b-a}(\ln_+ s\,,\ \text{if } b=\tfrac{N-1}{2})]\cdot(\ln \tfrac{R}{R^\nu} \text{ if } c+d=1)$$
$$\sim R^{1-c-d+\frac{N-1}{2}-a+(\sigma-\frac{1}{2})\min(0,1-c-d)+2\sigma(\frac{N-1}{2}-b^*)}$$
$$\cdot(\ln_+ s \text{ if } b=\tfrac{N-1}{2})\cdot(\ln_+ \tfrac{R}{1+s} \text{ if } c+d=1).$$

$\underline{I_6}$ It is sufficient to interchange a, b by c, d and use the result for Ω_5.

$\underline{I_7}$ Denoting $\tau = |\tilde{t}| \in (R^\nu, R)$ we have in Ω_7 that $t \sim r \sim R$, $\varrho \sim \tilde{\varrho} \sim \tilde{r} \in (\tau; R)$. Therefore $\eta^{-a}_{-b}(\mathbf{y}) \sim R^{-a+b}\rho^{-2b}$, $\eta^{-c}_{-d}(\mathbf{x}-\mathbf{y}) \sim \rho^{-c-d}$ and

$$I_7 \sim R^{-a+b}\int_{R^\nu}^{R} \mathrm{d}\tau\Big(\int_\tau^R \varrho^{N-2-d-c-2b}\mathrm{d}\varrho\Big)$$
$$\sim R^{b-a}\int_{R^\nu}^{R} \mathrm{d}\tau (R^{N-1-c-d-2b} - \tau^{N-1-c-d-2b})$$
$$\cdot(\ln \tfrac{R}{\tau} \text{ if } c+d+2b=N-1)$$
$$\sim R^{N-a-b-c-d} + R^{N-a-b-c-d+(\sigma-\frac{1}{2})\min(0,N-c-d-2b)}$$
$$\cdot(\ln \tfrac{R}{1+s} \text{ if } c+d+2b=N)$$
$$\sim R^{N-a-b-c-d+(\sigma-\frac{1}{2})\min(0,N-c-d-2b)}\cdot(\ln \tfrac{R}{1+s} \text{ if } c+d+2b=N).$$

I8 We get the result by interchanging a, b by c, d and using the result for Ω_7.

I9 Analogously as in Ω_7 we have $t \sim r \sim R$, $\varrho \sim \tilde{\varrho} \in (R^\nu; \tau)$, $\tilde{r} \sim \tau = |\tilde{t}| \in (R^\nu, R)$; so $\eta_{-b}^{-a}(\mathbf{y}) \sim R^{-a+b}\rho^{-2b}$, $\eta_{-d}^{-c}(\mathbf{x}-\mathbf{y}) \sim \rho^{-c-d}$ and

$$I_9 \sim R^{-a+b} \int_{R^\nu}^{R} \mathrm{d}\tau\, \tau^{-d-c} \Big(\int_{R^\nu}^{\tau} \varrho^{N-2-2b} \mathrm{d}\varrho \Big) .$$

If $b > \frac{N-1}{2}$, the significant term in the inner integral will be the lower bound, and we can use I_5. If $b < \frac{N-1}{2}$, the significant term in the inner integral will be the upper bound and we can use I_7. If $b = \frac{N-1}{2}$, then

$$I_9 \sim R^{-a+\frac{N-1}{2}} \int_{R^\nu}^{R} \tau^{-d-c} \ln \frac{\tau}{R^\nu} \mathrm{d}\tau .$$

In comparison with I_5, we get some additional logarithmic factors

$$b = \tfrac{N-1}{2} \ : (\ln_+ \tfrac{R}{1+s} \text{ if } c+d < 1) \cdot (\ln_+^2 \tfrac{R}{1+s} \text{ if } c+d = 1) .$$

I10 As in I_9, we may use I_6 for $d > \frac{N-1}{2}$, I_8 for $d > \frac{N-1}{2}$ and get some additional logarithmic factors to I_8 for $d = \frac{N-1}{2}$.

I11 The domain Ω_{11} is unbounded. We have $\tilde{r} \sim r \in (R; \infty)$. Therefore $\eta_{-d}^{-c}(\mathbf{x}-\mathbf{y}) \sim r^{-c-d}$, $\eta_{-b}^{-a}(\mathbf{y}) \sim r^{-a}(1+s(\mathbf{y}))^{-b}$, and applying Lemma 13.2 under the assumption $a + b^* + c + d > N$ we get

$$\begin{aligned} I_{11} &\sim \int_R^\infty \mathrm{d}r\, r^{N-1-a-b^*-c-d} \cdot (\ln r \text{ if } b = \tfrac{N-1}{2}) \\ &\sim R^{N-a-b^*-c-d} \cdot (\ln R \text{ if } b = \tfrac{N-1}{2}) . \end{aligned}$$

I12 We proceed as in the previous case, and under the assumption $a+b+c+d^* > N$ we get

$$I_{12} \sim R^{N-a-b-c-d^*} \cdot \Big(\ln R \text{ if } d = \tfrac{N-1}{2}\Big) .$$

I13 The domain Ω_{13} can be considered as a subset of Ω_{11} and Ω_{12}. Therefore I_{13} can be bounded by I_{11} and I_{12}.

I14 In this subdomain we have $r \sim R$, $\varrho \sim \tilde{\varrho} \in (R^\nu; R)$. Moreover $\tilde{r} \sim |\tilde{t}| + \tilde{\varrho}$ where $\tilde{t} \in (-\frac{1}{8}R^\nu; \frac{R}{2})$. Then $\eta_{-b}^{-a}(\mathbf{y}) \sim R^{-a+b}\varrho^{-2b}$, $\eta_{-d}^{-c}(\mathbf{x}-\mathbf{y}) \sim (\tilde{t} + \tilde{\varrho})^{-c+d}\varrho^{-2d}$ if $\tilde{t} > 0$ and $\eta_{-d}^{-c}(\mathbf{x}-\mathbf{y}) \sim \tilde{\varrho}^{-c-d}$ if $\tilde{t} < 0$. Let us note that the strip $\tilde{t} \in (-\frac{1}{8}R; 0)$ has no influence on the asymptotic behavior since $\tilde{\varrho} > |\tilde{t}|$ there.

$$
\begin{aligned}
I_{14} &\sim R^{b-a} \int_{R^\nu}^{R} \mathrm{d}\varrho\, \varrho^{N-2-2b-2d} \int_0^{R/2} (\tilde{t}+\rho)^{d-c} \mathrm{d}\tilde{t} \\
&\sim R^{b-a} \int_{R^\nu}^{R} \mathrm{d}\varrho\, \varrho^{N-2-2b-2d} \cdot \begin{cases} R^{1+d-c} & 1+d-c>0 \\ \varrho^{1+d-c} & 1+d-c<0 \\ \ln \frac{R}{\varrho} & 1+d-c=0. \end{cases}
\end{aligned}
$$

Now we distinguish three cases:

(a) $1+d-c>0$
If $b+d \le \frac{N-1}{2}$, then

$$
I_{14} \sim R^{N-a-b-c-d} \cdot (\ln_+ \tfrac{R}{1+s} \text{ if } b+d = \tfrac{N-1}{2}),
$$

while for $b+d > \frac{N-1}{2}$ we have

$$
I_{14} \sim R^{-a-c+\frac{N+1}{2}+2\sigma(\frac{N-1}{2}-b-d)} \qquad (\text{see } I_2\,,\ I_3)\,.
$$

(b) $1+d-c<0$

$$
I_{14} \sim R^{b-a} \int_{R^\nu}^{R} \varrho^{-2b-c-d+N-1} \mathrm{d}\varrho,
$$

and the integral can be estimated by I_7.

(c) $1+d-c=0$

$$
\begin{aligned}
I_{14} &\sim R^{b-a} \int_{R^\nu}^{R} \varrho^{N-2-2b-2d} \ln \tfrac{R}{\varrho} \mathrm{d}\varrho \\
&= R^{N-a-b-c-d} \int_1^{R^{1-\nu}} z^{2b+2d-N} \ln z\, \mathrm{d}z\,.
\end{aligned}
$$

Now for $b+d \le \frac{N-1}{2}$

$$
I_{14} \sim R^{N-a-b-c-d} \cdot (\ln_+^2 \tfrac{R}{1+s} \text{ if } b+d = \tfrac{N-1}{2})
$$

and for $b+d > \frac{N-1}{2}$

$$
I_{14} \sim R^{\frac{N+1}{2}-a-c-2\sigma(b+d-\frac{N-1}{2})} \ln_+ \tfrac{R}{1+s},
$$

which can be estimated by I_2.

$\underline{I_{15}}$ Interchanging a, b by c, d we can use the results from I_{14}.

We have now completed investigation of the situation (C). The results are summarized in Tables 13.1 and 13.2 ($N = 3$) and Tables 13.3 and 13.4 ($N = 2$).

The situation (D) is almost trivial since we are left with subdomains of the type Ω_1, Ω_2, Ω_{11}, Ω_{12}, and Ω_{13}. The integrals can be estimated by the corresponding integrals in (C) taking $\sigma = \frac{1}{2}$, i.e., $\nu = 1$.

Finally in the case (B) we proceed as in case (C), but the subdomains Ω_2, Ω_3, and Ω_4 coincide. The other integrals can again be estimated by the corresponding ones from the part (C) taking $\sigma = 0$, i.e., $\nu = \frac{1}{2}$.

The study of the convolution (13.61) is therefore completed. In the next section, we shall apply the results in the study of L^∞-estimates of convolutions with the Oseen kernels.

13.2.4 L^∞-Estimates for Weakly Singular Oseen Kernels

Now we intend to get estimates for the functions ν_β^α, $\mu_\beta^{\alpha,\gamma}$ by analogy with the preceding section,

$$\nu_\beta^\alpha(\mathbf{x}) = \nu_\beta^\alpha(\mathbf{x}; 1) = |\mathbf{x}|^\alpha (1 + s(\mathbf{x}))^\beta, \qquad \mu_\beta^{\alpha,\gamma}(\mathbf{x}) = \eta_\beta^{\alpha-\gamma}(\mathbf{x})\nu_0^\gamma(\mathbf{x})\,.$$

This is the aim of Lemmas 13.3 and 13.4. We will formulate these lemmas in the case of $\mathbb{R}^N$, $N \in \mathbb{N}$, $N \geq 2$. Afterward, we apply the lemmas in the study of L^∞-estimates of convolutions with Oseen kernels. Let $I_{\alpha,\gamma}(\mathbf{x})$ denote the following integral

$$I_{\alpha,\gamma}^{(N)}(\mathbf{x}) = \int_{B_1(\mathbf{0})} \nu_0^{-\alpha}(\mathbf{y})\nu_0^{-\gamma}(\mathbf{x}-\mathbf{y})\mathrm{d}\mathbf{y} = \int_{B_1(\mathbf{x})} \nu_0^{-\alpha}(\mathbf{x}-\mathbf{y}')\nu_0^{-\gamma}(\mathbf{y}')\mathrm{d}\mathbf{y}',$$

where $\mathbf{x}, \mathbf{y}, \mathbf{y}' \in \mathbb{R}^N$. For the notational convenience we also denote

$$\ln_- |\mathbf{x}| := \max(1, -\ln|\mathbf{x}|)$$

for $\mathbf{x} \neq \mathbf{0}$.

Lemma 13.3 *For $\alpha < N$, $\gamma < N$, there exists a positive constant C_1 such that for $\mathbf{x} \in B_2(\mathbf{0}) \setminus \{\mathbf{0}\} \subset \mathbb{R}^N$*

$$I_{\alpha,\gamma}^{(N)}(\mathbf{x}) \leq C_1 \begin{cases} \nu_0^{-(\alpha+\gamma-N)}(\mathbf{x}), & \text{if } \alpha+\gamma > N \\ \ln_-|\mathbf{x}|, & \text{if } \alpha+\gamma = N \\ 1, & \text{if } \alpha+\gamma < N\,. \end{cases}$$

Moreover, there exists a positive constant C_2 such that for $\mathbf{x} \in B^2(\mathbf{0}) \subset \mathbb{R}^N$

$$I^{(N)}_{\alpha,\gamma}(\mathbf{x}) \le C_2\, \nu_0^{-\gamma}(\mathbf{x})\,.$$

Proof We divide the proof into two parts:

(a) First we assume $|\mathbf{x}| \le 2$. We will estimate integrals over sets, whose union contains unit ball $B_1(\mathbf{0})$.
$B_1(\mathbf{0}) \subset B_{|\mathbf{x}|/2}(\mathbf{0}) \cup B_{|\mathbf{x}|/2}(\mathbf{x}) \cup \big\{B_{2|\mathbf{x}|}(\mathbf{0}) \setminus \big(B_{|\mathbf{x}|/2}(\mathbf{0}) \cup B_{|\mathbf{x}|/2}(\mathbf{x})\big)\big\} \cup \big\{B_4(\mathbf{0}) \setminus B_{2|\mathbf{x}|}(\mathbf{0})\big\} \equiv M_1 \cup M_2 \cup M_3 \cup M_4$

$$\int_{M_1} \frac{1}{|\mathbf{y}|^\alpha}\frac{1}{|\mathbf{x}-\mathbf{y}|^\gamma}\,\mathrm{d}\mathbf{y} \le \frac{c_1}{|\mathbf{x}|^\gamma}\int_0^{|\mathbf{x}|/2}\frac{1}{r^\alpha}r^{N-1}\mathrm{d}r \le \frac{c_2}{|\mathbf{x}|^{\alpha+\gamma-N}},\quad \alpha < N$$

$$\int_{M_2} \frac{1}{|\mathbf{y}|^\alpha}\frac{1}{|\mathbf{x}-\mathbf{y}|^\gamma}\,\mathrm{d}\mathbf{y} \le \frac{c_3}{|\mathbf{x}|^\alpha}\int_0^{|\mathbf{x}|/2}\frac{1}{r^\gamma}r^{N-1}\mathrm{d}r \le \frac{c_4}{|\mathbf{x}|^{\alpha+\gamma-N}},\quad \gamma < N$$

$$\int_{M_3} \frac{1}{|\mathbf{y}|^\alpha}\frac{1}{|\mathbf{x}-\mathbf{y}|^\gamma}\,\mathrm{d}\mathbf{y} \le \frac{c_5}{|\mathbf{x}|^{\alpha+\gamma}}\int_0^{2|\mathbf{x}|}r^{N-1}\mathrm{d}r \le \frac{c_6}{|\mathbf{x}|^{\alpha+\gamma-N}}$$

$$\int_{M_4} \frac{1}{|\mathbf{y}|^\alpha}\frac{1}{|\mathbf{x}-\mathbf{y}|^\gamma}\,\mathrm{d}\mathbf{y} \le c_7\int_{2|\mathbf{x}|}^4 \frac{r^{-\alpha}\,r^{N-1}}{|r-|\mathbf{x}||^\gamma}\mathrm{d}r$$

$\le c_8\displaystyle\int_{2|\mathbf{x}|}^4\left(\frac{r}{|r-|\mathbf{x}||}\right)^\gamma r^{-\alpha-\gamma+N-1}\mathrm{d}r \le c_9\int_{2|\mathbf{x}|}^4 r^{-\alpha-\gamma+N-1}\mathrm{d}r$; here we use the inequality $\frac{r}{|r-|\mathbf{x}||} \le 2$.

The last integral can be estimated by $c_{10}|\mathbf{x}|^{-\alpha-\gamma+N}$ if $\alpha+\gamma > N$, by $c_{11}\ln_-|\mathbf{x}|$ if $\alpha+\gamma = N$, and by some constant if $\alpha+\gamma < N$.

(b) Now we assume $|\mathbf{x}| \ge 2$.

$$\int_{B_1(\mathbf{0})} \frac{1}{|\mathbf{y}|^\alpha}\frac{1}{|\mathbf{x}-\mathbf{y}|^\gamma}\,\mathrm{d}\mathbf{y} \le \frac{c_{11}}{|\mathbf{x}|^\gamma}\int_{B_1(\mathbf{0})}\frac{1}{|\mathbf{y}|^\alpha}\mathrm{d}\mathbf{y} \le \frac{c_{12}}{|\mathbf{x}|^\gamma}.$$

The assertion of Lemma 13.3 follows from these five estimates of convolution integrals.

□

We define for pairs of real numbers $[a,b] \le [c,d]$: $a \le c$ and $a+b \le c+d$.
It is evident that $\eta^a_b(\mathbf{x}) \le 2^{c-a}\eta^c_d(\mathbf{x})$, $\mathbf{x} \in \mathbb{R}^N$ if $[a,b] \le [c,d]$.

In the formulation of Lemma 13.4 we use functions $\mu^{\alpha,\gamma}_\beta(\mathbf{x}) = \eta^{\alpha-\gamma}_\beta(\mathbf{x})\nu^\gamma_0(\mathbf{x})$. Let us note that $\mu^{\alpha,\alpha}_\beta(\mathbf{x}) = \eta^0_\beta(\mathbf{x})\nu^\alpha_0(\mathbf{x}) = \nu^\alpha_\beta(\mathbf{x})$, $\mu^{\alpha,\gamma}_\beta(\mathbf{x}) \sim \eta^\alpha_\beta(\mathbf{x})$, $\mathbf{x} \in B^1(\mathbf{0})$, $\mu^{\alpha,\gamma}_\beta(\mathbf{x}) \sim \nu^\gamma_\beta(\mathbf{x}) \sim \nu^\gamma_0(\mathbf{x})$, $\mathbf{x} \in B_1(\mathbf{0}) \setminus \{\mathbf{0}\}$.

Lemma 13.4 *Let $a, b, c, d, e, f \in \mathbb{R}$ and positive constant C be such that for all* $\mathbf{x} \in \mathbb{R}^N$*:* $\int_{\mathbb{R}^N}\eta^{-c}_{-d}(\mathbf{x}-\mathbf{y})\,\eta^{-a}_{-b}(\mathbf{y})\,d\mathbf{y} \le C\eta^{-e}_{-f}(\mathbf{x})$, $N \in \mathbb{N}$, $N \ge 2$.

Let $g < N$, $h < N$, $[e, f] \leq [a, b]$, $[e, f] \leq [c, d]$. Then there exists a positive constant C' such that the following inequality is satisfied for $\mathbf{x} \in \mathbb{R}^N \setminus \{\mathbf{0}\}$:

$$\int_{\mathbb{R}^N} \mu_{-d}^{-c,-h}(\mathbf{x}-\mathbf{y})\, \mu_{-b}^{-a,-g}(\mathbf{y})\, d\mathbf{y} \leq C' \begin{cases} \mu_{-f}^{-e,-g-h+N}(\mathbf{x}) & g+h > N \\ \mu_{-f}^{-e,-\delta}(\mathbf{x}),\ \delta > 0 & g+h = N \\ \mu_{-f}^{-e,0}(\mathbf{x}) \equiv \eta_{-f}^{-e}(\mathbf{x}) & g+h < N. \end{cases}$$

Proof Evidently, for $\alpha, \beta \in \mathbb{R}^1$ there exist positive constants c_1, c_2, c_3, and c_4 such that

$$\begin{array}{lll} (a) & c_1\, \eta_\beta^\alpha(\mathbf{x}) \leq \nu_\beta^\alpha(\mathbf{x}) \leq c_2\, \eta_\beta^\alpha(\mathbf{x}) & \text{for all } \mathbf{x} \in B^1(\mathbf{0}) \\ (b) & c_3\, \eta_\beta^\alpha(\mathbf{x}) \leq \eta_\beta^\alpha(\mathbf{y}) \leq c_4\, \eta_\beta^\alpha(\mathbf{x}) & \text{for all } \mathbf{x} \in \mathbb{R}^N,\ \mathbf{y} \in B_1(\mathbf{x})\,. \end{array}$$

Now we will prove the assertion of Lemma 13.4:

$$(\mu_{-d}^{-c,-h} * \mu_{-b}^{-a,-g})(\mathbf{x}) = \int_{\mathbb{R}^N} \mu_{-d}^{-c,-h}(\mathbf{x}-\mathbf{y})\, \mu_{-b}^{-a,-g}(\mathbf{y}) \mathrm{d}\mathbf{y}$$

$$\leq C_1 \int_{B_1(\mathbf{x})} \nu_0^{-h}(\mathbf{x}-\mathbf{y}) \mu_{-b}^{-a,-g}(\mathbf{y})\, \mathrm{d}\mathbf{y} + C_2 \int_{B^1(\mathbf{x})} \eta_{-d}^{-c}(\mathbf{x}-\mathbf{y}) \mu_{-b}^{-a,-g}(\mathbf{y}) \mathrm{d}\mathbf{y}.$$

We will study these two integrals separately. By using Lemma 13.3 we get the estimate of $\int_{B_1(\mathbf{x})} \nu_0^{-h}(\mathbf{x}-\mathbf{y})\, \mu_{-b}^{-a,-g}(\mathbf{y})\, \mathrm{d}\mathbf{y} = \int_{B_1(\mathbf{0})} \mu_{-b}^{-a,-g}(\mathbf{x}-\mathbf{y})\, \nu_0^{-h}(\mathbf{y})\, \mathrm{d}\mathbf{y}$.

$$\int_{B_1(\mathbf{0})} \mu_{-b}^{-a,-g}(\mathbf{x}-\mathbf{y})\, \nu_0^{-h}(\mathbf{y})\, \mathrm{d}\mathbf{y}$$

$$\leq \max_{\mathbf{y} \in B_1(\mathbf{x})} \eta_{-b}^{-a+g}(\mathbf{y}) \int_{B_1(\mathbf{0})} \nu_0^{-g}(\mathbf{x}-\mathbf{y}) \nu_0^{-h}(\mathbf{y}) \mathrm{d}\mathbf{y}$$

$$\leq C_3 \eta_{-b}^{-a+g}(\mathbf{x}) \begin{cases} \left.\begin{array}{ll} \nu_0^{-g-h+N}(\mathbf{x}), & g+h > N \\ \ln_-|\mathbf{x}|, & g+h = N \\ 1, & g+h < N \end{array}\right\} & \mathbf{x} \in B_1 \setminus \{\mathbf{0}\} \\ \eta_0^{-g}(\mathbf{x}) & \mathbf{x} \in B^1 \end{cases}$$

$$\leq C_5 \begin{cases} \mu_{-b}^{-a,-g-h+N}(\mathbf{x}), & g+h > N \\ \mu_{-b}^{-a,-\delta}(\mathbf{x}),\ \delta > 0, & g+h = N \\ \mu_{-b}^{-a,0}(\mathbf{x}), & g+h < N \end{cases} \leq C_6 \begin{cases} \mu_{-f}^{-e,-g-h+N}(\mathbf{x}), & g+h > N \\ \mu_{-f}^{-e,-\delta}(\mathbf{x}),\ \delta > 0, & g+h = N \\ \mu_{-f}^{-e,0}(\mathbf{x}) \equiv \eta_{-f}^{-e}(\mathbf{x}), & g+h < N\,. \end{cases}$$

In the second inequality we use Lemma 13.3 and the relation (b). In the last inequality we take into account the assumption $[e, f] \leq [a, b]$.

We estimate the remaining integral $\int_{B^1(\mathbf{x})} \eta_{-d}^{-c}(\mathbf{x}-\mathbf{y})\mu_{-b}^{-a,-g}(\mathbf{y})\mathrm{d}\mathbf{y}$ for $\mathbf{x} \in \mathbb{R}^N \setminus \{\mathbf{0}\}$ in the following way:

$$\begin{aligned}\int_{B^1(\mathbf{x})} \eta_{-d}^{-c}(\mathbf{x}-\mathbf{y})\mu_{-b}^{-a,-g}(\mathbf{y})\mathrm{d}\mathbf{y} &\leq \int_{\mathbb{R}^N} \eta_{-d}^{-c}(\mathbf{x}-\mathbf{y})\mu_{-b}^{-a,-g}(\mathbf{y})\mathrm{d}\mathbf{y}\\ &= \int_{B_1} \eta_{-d}^{-c}(\mathbf{x}-\mathbf{y})\, \nu_0^{-g}(\mathbf{y})\mathrm{d}\mathbf{y} + \int_{B^1} \eta_{-d}^{-c}(\mathbf{x}-\mathbf{y})\eta_{-b}^{-a}(\mathbf{y})\mathrm{d}\mathbf{y}\\ &\leq C_5 \max_{\mathbf{y}\in B_1(\mathbf{x})} \eta_{-d}^{-c}(\mathbf{y}) \int_{B_1(\mathbf{0})} \nu_0^{-g}(\mathbf{y})\mathrm{d}\mathbf{y} + C_6\eta_{-f}^{-e}(\mathbf{x}) \leq C_7\eta_{-f}^{-e}(\mathbf{x})\,.\end{aligned}$$

The proof of Lemma 13.4 follows from these estimates. □

We now formulate main results of this section. We will use the following notation:

$$\overline{\eta}_F^E(\mathbf{x};\lambda) = \eta_F^E(\mathbf{x};\lambda) \quad \text{if no logarithmic factor appears, and}$$

$$\overline{\eta}_F^E(\mathbf{x};\lambda) = \eta_F^E(\mathbf{x};\lambda)\cdot \begin{cases} P(\ln_+^{-1}|\lambda\mathbf{x}|) \\ P(\ln_+^{-1} s(\lambda\mathbf{x})) \end{cases} \quad \text{if there are logarithmic factors,}$$

where function $P(\cdot)$ is a polynomial of the first or the second order, see also Remark 13.3. Similarly we define $\overline{\nu}_F^E(\cdot\,;\lambda)$. Then we have:

Theorem 13.3 *Let $A+B^* > 1$. Let $f \in L^\infty(\mathbb{R}^3, \eta_B^A(\cdot\,;\lambda))$. Then we have $|\mathcal{O}| * f \in L^\infty(\mathbb{R}^3, \overline{\eta}_F^E(\cdot\,;\lambda))$, where*

$$E = \begin{cases} A-1 & \text{for } A \leq B^*+1 \\ \frac{A+B^*-1}{2} & \text{for } A \geq B+1,\ A+B \leq 3 \\ 1 & \text{for } A+B^* \geq 3 \end{cases} \tag{i}$$

$$E+F = \begin{cases} A+B^*-1 & \text{for } A+B^* \leq 3 \\ 2 & \text{for } A+B^* \geq 3 \end{cases} \tag{ii}$$

with logarithmic factors

$$\ln_+(\lambda|\mathbf{x}|) \ \text{for} \ \begin{cases} A+B^* = 3 \\ A = B+1,\ 0 \leq B \leq 1 \end{cases} \tag{iii}$$

$$\ln_+(\lambda s(\mathbf{x})) \ \text{for} \ \ A+B < 3,\ B \leq 1 \tag{iv}$$

(see Remark 13.3). Moreover we have

$$\| |\mathcal{O}(\cdot\,;\lambda)| * f \|_{\infty,(\overline{\eta}_F^E(\cdot\,;\lambda)),\mathbb{R}^3} \le C\lambda^{-2} \| f \|_{\infty,(\eta_B^A(\cdot\,;\lambda)),\mathbb{R}^3}\,. \tag{13.70}$$

Let in addition for A, B the following conditions be satisfied

$$1 \le A < 3,\ \ B > 0,\ \ or\ \ A \le B+5,\ \ 1 < A+B \le 3,\ \ B \le 0\,. \tag{v}$$

Then for $f \in L^\infty(\mathbb{R}^3, \nu_B^A(\cdot\,;\lambda))$ *we have* $|\mathcal{O}(\cdot\,;\lambda)| * f \in L^\infty(\mathbb{R}^3, \overline{\nu}_F^E(\cdot\,;\lambda))$ *and*

$$\| |\mathcal{O}(\cdot\,;\lambda)| * f \|_{\infty,(\overline{\nu}_F^E(\cdot\,;\lambda)),\mathbb{R}^3} \le C\lambda^{-2+A-E} \| f \|_{\infty,(\nu_B^A(\cdot\,;\lambda)),\mathbb{R}^3}\,. \tag{13.71}$$

Remark 13.3 The inequalities (13.70), (13.71) must be understood in the following sense. If no logarithmic terms appear, then

$$\| |\mathcal{O}(\cdot\,;\lambda)| * f \|_{\infty,(\eta_F^E(\cdot\,;\lambda)),\mathbb{R}^3} \le C\lambda^{-2} \| f \|_{\infty,(\eta_B^A(\cdot\,;\lambda)),\mathbb{R}^3}\,.$$

Analogously for the weight $\nu_F^E(\cdot\,;\lambda)$. But for $A+B^* = 3$ or $A = B+1, 0 \le B \le 1$, we have

$$\| |\mathcal{O}(\cdot\,;\lambda)| * f \|_{\infty,(\eta_F^E(\cdot\,;\lambda)\,P(\ln_+^{-1}(\lambda|\cdot|))),\mathbb{R}^3} \le C\lambda^{-2} \| f \|_{\infty,(\eta_B^A(\cdot\,;\lambda)),\mathbb{R}^3} \tag{13.72}$$

and for $A+B < 3$, $B \le 1$

$$\| |\mathcal{O}(\cdot\,;\lambda)| * f \|_{\infty,(\eta_F^E(\cdot\,;\lambda)\,P(\ln_+^{-1}(s(\lambda\cdot)))),\mathbb{R}^3} \le C\lambda^{-2} \| f \|_{\infty,(\eta_B^A(\cdot\,;\lambda)),\mathbb{R}^3}\,, \tag{13.73}$$

where $P(\cdot)$ is a polynomial. The order of the polynomial can be traced out from the proof of Theorem 13.3 using Tables 13.1 and 13.2. Analogously for the weights $\nu_F^E(\cdot\,;\lambda)$. We can use instead of (13.72), (13.73) for $\varepsilon > 0$

$$\| |\mathcal{O}(\cdot\,;\lambda)| * f \|_{\infty,(\eta_F^{E-\varepsilon}(\cdot\,;\lambda)),\mathbb{R}^3} \le C\lambda^{-2} \| f \|_{\infty,(\eta_B^A(\cdot\,;\lambda)),\mathbb{R}^3}\,, \tag{13.72'}$$

$$\| |\mathcal{O}(\cdot\,;\lambda)| * f \|_{\infty,(\eta_{F-\varepsilon}^E(\cdot\,;\lambda)),\mathbb{R}^3} \le C\lambda^{-2} \| f \|_{\infty,(\eta_B^A(\cdot\,;\lambda)),\mathbb{R}^3}\,, \tag{13.73'}$$

respectively.

Finally, in the case of $f = 0$ in $B_{1/2}(\mathbf{0})$ (this is usually the case for $\Omega \subset \mathbb{R}^N$, exterior domain) we can get for the weight $\nu_B^A(\cdot\,;\lambda)$

$$\| |\mathcal{O}(\cdot\,;\lambda)| * f \|_{\infty,(\nu_F^{E-\varepsilon}(\cdot\,;\lambda)),\mathbb{R}^3} \le C\lambda^{-2+A-E+\varepsilon} \| f \|_{\infty,(\nu_B^A(\cdot\,;\lambda)),\mathbb{R}^3}\,.$$

□

The proof of Theorem 13.3 is similar to the proof of the other theorems in this section. We therefore give only the proof of Theorem 13.4 which is in applications the most often used one. The other theorems in both two- and three-dimensional cases can be shown by analogy with this proof.

Theorem 13.4 *Let $A + B > 1/2$, $A > -1$. Then for $f \in L^\infty(\mathbb{R}^3, \eta_B^A(\cdot\,;\lambda))$ we have $|\nabla\mathcal{O}| * f \in L^\infty(\mathbb{R}^3, \overline{\eta}_F^E(\cdot\,;\lambda))$, where*

$$E = \begin{cases} A - \frac{1}{2} & \text{for } A \le 2,\ A \le B + 1,\ B \ge 0 \\ \frac{3}{2} & \text{for } A + B^* \ge 3 \\ A + B - \frac{1}{2} & \text{for } B < 0,\ A + B \le 1 \\ \frac{A+B}{2} & \text{for } B \le A - 1,\ 1 \le A + B \le 3 \end{cases} \tag{i}$$

$$E + F = \begin{cases} A + B^* & \text{for } A \le 2,\ B \ge \frac{3}{2} \\ A + B - \frac{1}{2} & \text{for } A + B \le \frac{7}{2} \\ 3 & \text{for } A + B \ge \frac{7}{2},\ A \ge 2 \end{cases} \tag{ii}$$

with logarithmic factors

$$\ln_+(\lambda|\mathbf{x}|)\ \text{for} \begin{cases} A + B^* = 3 \\ A = B + 1,\ 0 \le B \le 1 \\ A + B = 1,\ B \le 0 \\ B = -\frac{1}{2},\ \frac{3}{4} \le A \le \frac{5}{4}\,. \end{cases} \tag{iii}$$

Moreover we have

$$\| |\nabla\mathcal{O}(\cdot\,;\lambda)| * f \|_{\infty,(\overline{\eta}_F^E(\cdot\,;\lambda)),\mathbb{R}^3} \le C\lambda^{-1} \|f\|_{\infty,(\eta_B^A(\cdot\,;\lambda)),\mathbb{R}^3}\,. \tag{13.74}$$

Let in addition for A, B the following conditions be satisfied:

$$\frac{1}{2} \le A \le \frac{5}{2},\quad B \ge -\frac{1}{2},\quad A \le B + 2. \tag{iv}$$

Then for $f \in L^\infty(\mathbb{R}^3, \nu_B^A(\cdot\,;\lambda))$ we have $|\nabla\mathcal{O}(\cdot\,;\lambda)| * f \in L^\infty(\mathbb{R}^3, \overline{\nu}_F^E(\cdot\,;\lambda))$ and

$$\| |\nabla\mathcal{O}(\cdot\,;\lambda)| * f \|_{\infty,(\overline{\nu}_F^E(\cdot\,;\lambda)),\mathbb{R}^3} \le C\lambda^{-1+A-E} \|f\|_{\infty,(\nu_B^A(\cdot\,;\lambda)),\mathbb{R}^3}\,. \tag{13.75}$$

Theorem 13.5 *Let $A + B^* > 0$. Let $R = |\nabla^2 \boldsymbol{\mathcal{O}} - \nabla^2 \boldsymbol{\mathcal{S}}|$ or $R = |\partial_1 \boldsymbol{\mathcal{O}}|$. Then for $f \in L^\infty(\mathbb{R}^3, \eta^A_B(\cdot\,;\lambda))$ we have $R * f \in L^\infty(\mathbb{R}^3, \overline{\eta}^E_F(\cdot\,;\lambda))$, where*

$$E = \begin{cases} A & \text{for } -1 < A \le 2,\ A \le B+1,\ B \ge 0 \\ 2 & \text{for } A + B^* \ge 3 \\ A + B & \text{for } B \le 0,\ 0 < A + B < 1 \\ \frac{A+B+1}{2} & \text{for } B \le A - 1,\ 1 \le A + B \le 3 \end{cases} \tag{i}$$

$$E + F = \begin{cases} A + B^* & \text{for } A + B^* \le 3 \\ 3 & \text{for } A + B^* \ge 3 \end{cases} \tag{ii}$$

with logarithmic factors

$$\ln_+(\lambda|\mathbf{x}|) \text{ for } A + B^* \le 3. \tag{iii}$$

Moreover we have

$$\Big\| |\nabla^2 \boldsymbol{\mathcal{O}}(\cdot\,;\lambda) - \nabla^2 \boldsymbol{\mathcal{S}}(\cdot)| * f \Big\|_{\infty,(\overline{\eta}^E_F(\cdot\,;\lambda)),\mathbb{R}^3} \le C \, \|f\|_{\infty,(\eta^A_B(\cdot\,;\lambda)),\mathbb{R}^3}\,, \tag{13.76}$$

$$\| |\partial_1 \boldsymbol{\mathcal{O}}(\cdot\,;\lambda)| * f \|_{\infty,(\overline{\eta}^E_F(\cdot\,;\lambda)),\mathbb{R}^3} \le C\lambda^{-1} \, \|f\|_{\infty,(\eta^A_B(\cdot\,;\lambda)),\mathbb{R}^3}\,. \tag{13.77}$$

Let in addition for A, B the following conditions be satisfied

$$0 \le A < 3, \quad B \ge -1, \quad A < B + 3. \tag{iv}$$

Then for $f \in L^\infty(\mathbb{R}^3, \nu^A_B(\cdot\,;\lambda))$ we have $R * f \in L^\infty(\mathbb{R}^3, \overline{\nu}^E_F(\cdot\,;\lambda))$ and

$$\Big\| |\nabla^2 \boldsymbol{\mathcal{O}}(\cdot\,;\lambda) - \nabla^2 \boldsymbol{\mathcal{S}}(\cdot)| * f \Big\|_{\infty,(\overline{\nu}^E_F(\cdot\,;\lambda)),\mathbb{R}^3} \le C\lambda^{A-E} \, \|f\|_{\infty,(\nu^A_B(\cdot\,;\lambda)),\mathbb{R}^3}\,, \tag{13.78}$$

$$\| |\partial_1 \boldsymbol{\mathcal{O}}(\cdot\,;\lambda)| * f \|_{\infty,(\overline{\nu}^E_F(\cdot\,;\lambda)),\mathbb{R}^3} \le C\lambda^{-1+A-E} \, \|f\|_{\infty,(\nu^A_B(\cdot\,;\lambda)),\mathbb{R}^3}\,. \tag{13.79}$$

Theorem 13.6 *Let $A + B^* > 1$. Then for $f \in L^\infty(\mathbb{R}^3, \eta^A_B(\cdot\,;\lambda))$ we have $|\boldsymbol{\mathcal{P}}| * f \in L^\infty(\mathbb{R}^3, \overline{\eta}^E_F(\cdot\,;\lambda))$, where*

$$E = \begin{cases} 2 & \text{for } A + B^* \ge 3,\ A \ge \frac{5}{2} \\ A - \frac{1}{2} & \text{for } A \le \frac{5}{2},\ B \ge \frac{1}{2} \\ A + B^* - 1 & \text{for } B \le \frac{1}{2},\ A + B^* \le 3 \end{cases} \tag{i}$$

$$E + F = \begin{cases} 2 & \text{for } A + B^* \ge 3 \\ A + B^* - 1 & \text{for } A + B^* \le 3 \end{cases} \tag{ii}$$

with logarithmic factors

$$\ln_+(\lambda|\mathbf{x}|) \ \textit{for} \ \begin{cases} B = \frac{1}{2}, \ \frac{1}{2} < A < \frac{5}{2} \\ B = 1, \ 2 \le A \le \frac{5}{2} \\ A + B^* = 3, \ A \ge \frac{5}{2} \end{cases} \tag{iii}$$

$$\ln_+(\lambda s(\mathbf{x})) \ \textit{for} \ \begin{cases} A + B^* = 3, \ 2 \le A < \frac{5}{2} \\ B = 1, \ 0 < A < 2. \end{cases} \tag{iv}$$

Moreover we have

$$\| |\mathcal{P}| * f \|_{\infty,(\overline{\eta}^E_F(\cdot\,;\lambda)),\mathbb{R}^3} \le C\lambda^{-1} \| f \|_{\infty,(\eta^A_B(\cdot\,;\lambda)),\mathbb{R}^3} \,. \tag{13.80}$$

Let in addition for A, B the following conditions be satisfied

$$\frac{1}{2} \le A < 3, \quad B \ge 0. \tag{v}$$

Then for $f \in L^\infty(\mathbb{R}^3, \nu^A_B(\cdot\,;\lambda))$ *we have* $|\mathcal{P}| * f \in L^\infty(\mathbb{R}^3, \overline{\nu}^E_F(\cdot\,;\lambda))$ *and*

$$\| |\mathcal{P}| * f \|_{\infty,(\overline{\nu}^E_F(\cdot\,;\lambda)),\mathbb{R}^3} \le C\lambda^{-1+A-E} \| f \|_{\infty,(\nu^A_B(\cdot\,;\lambda)),\mathbb{R}^3} \,. \tag{13.81}$$

In the next part of this section we give formulations of the theorems in the two-dimensional case:

Theorem 13.7 *Let* $A + B^* > 1$. *Then for* $f \in L^\infty(\mathbb{R}^2, \eta^A_B(\cdot\,;\lambda))$ *we have* $|\mathcal{O}_{11}| * f \in L^\infty(\mathbb{R}^2, \overline{\eta}^E_F(\cdot\,;\lambda))$, *where*

$$E = \begin{cases} A - 1 & \textit{for } A \le B^* + 1 \\ \frac{1}{2}(A + B - 1) & \textit{for } A \ge B + 1 \ A + B \le 2 \\ \frac{1}{2} & \textit{for } A + B^* \ge 2 \end{cases} \tag{i}$$

$$E + F = \begin{cases} 1 & \textit{for } A + B^* \ge 2 \\ A + B^* - 1 & \textit{for } A + B^* \le 2 \end{cases} \tag{ii}$$

with logarithmic factors

$$\ln_+(\lambda|\mathbf{x}|) \ \textit{for} \ \begin{cases} A + B^* = 2 \\ A = B + 1, \ 0 < B \le \frac{1}{2} \end{cases} \tag{iii}$$

$$\ln_+(\lambda s(\mathbf{x})) \ \textit{for} \ \left\{ A + B < 2, \ B \le \tfrac{1}{2}. \right. \tag{iv}$$

Moreover we have

$$\| |\mathcal{O}_{11}(\cdot\,;\lambda)| * f \|_{\infty,(\overline{\eta}_F^E(\cdot\,;\lambda)),\mathbb{R}^2} \le C\lambda^{-2} \|f\|_{\infty,(\eta_B^A(\cdot\,;\lambda)),\mathbb{R}^2} . \tag{13.82}$$

Let in addition for A *the following conditions be satisfied*

$$1 \le A < 2. \tag{v}$$

Then for $f \in L^\infty(\mathbb{R}^2, \nu_B^A(\cdot\,;\lambda))$ *we have* $|\mathcal{O}_{11}(\cdot\,;\lambda)| * f \in L^\infty(\mathbb{R}^2, \overline{\nu}_F^E(\cdot\,;\lambda))$ *and*

$$\| |\mathcal{O}_{11}(\cdot\,;\lambda)| * f \|_{\infty,(\overline{\nu}_F^E(\cdot\,;\lambda)),\mathbb{R}^2} \le C\lambda^{-2+A-E} \|f\|_{\infty,(\nu_B^A(\cdot\,;\lambda)),\mathbb{R}^2} . \tag{13.83}$$

Theorem 13.8 *Let* $A + B^* > 1$, $i, j = 1, 2$, $i \cdot j \ne 1$, $R = |\mathcal{O}_{ij}|$, *or* $R = |\mathcal{P}|$. *Then for* $f \in L^\infty(\mathbb{R}^2, \eta_B^A(\cdot\,;\lambda))$ *we have* $R * f \in L^\infty(\mathbb{R}^2, \overline{\eta}_F^E(\cdot\,;\lambda))$, *where*

$$E = E + F = \begin{cases} 1 & \text{for } A + B^* \ge 2 \\ A + B^* - 1 & \text{for } A + B^* \le 2 \end{cases} \tag{i}$$

with logarithmic factors

$$\ln_+(\lambda|\mathbf{x}|) \ \text{for} \ \begin{cases} A + B^* = 2 \\ A + B^* \le 2, \ B \ge \frac{1}{2}. \end{cases} \tag{ii}$$

Moreover, we have

$$\left\| |\mathcal{O}_{ij}(\cdot\,;\lambda)| * f \right\|_{\infty,(\overline{\eta}_F^E(\cdot\,;\lambda)),\mathbb{R}^2} \le C\lambda^{-2} \|f\|_{\infty,(\eta_B^A(\cdot\,;\lambda)),\mathbb{R}^2} , \tag{13.84}$$

$$\| |\mathcal{P}| * f \|_{\infty,(\overline{\eta}_F^E(\cdot\,;\lambda)),\mathbb{R}^2} \le C\lambda^{-1} \|f\|_{\infty,(\eta_B^A(\cdot\,;\lambda)),\mathbb{R}^2} . \tag{13.85}$$

Let in addition for A, B *the following conditions be satisfied*

$$A < 2, \ B^* \ge 0 \ (\text{ if } R = |\mathcal{P}|) \quad \text{or} \quad A < 2, \ (\text{ if } R = |\mathcal{O}_{ij}|). \tag{iii}$$

Then for $f \in L^\infty(\mathbb{R}^2, \nu_B^A(\cdot\,;\lambda))$ *we have* $R * f \in L^\infty(\mathbb{R}^2, \overline{\nu}_F^E(\cdot\,;\lambda))$ *and*

$$\left\| |\mathcal{O}_{ij}(\cdot\,;\lambda)| * f \right\|_{\infty,(\overline{\nu}_F^E(\cdot\,;\lambda)),\mathbb{R}^2} \le C\lambda^{-2+A-E} \|f\|_{\infty,(\nu_B^A(\cdot\,;\lambda)),\mathbb{R}^2} , \tag{13.86}$$

$$\| |\mathcal{P}| * f \|_{\infty,(\overline{\nu}_F^E(\cdot\,;\lambda)),\mathbb{R}^2} \le C\lambda^{-1+A-E} \|f\|_{\infty,(\nu_B^A(\cdot\,;\lambda)),\mathbb{R}^2} . \tag{13.87}$$

Theorem 13.9 *Let $A + B^* > 0$ and $A + B > \frac{1}{2}$. Then for $f \in L^\infty(\mathbb{R}^2, \eta_B^A(\cdot\,;\lambda))$ we have $|\partial_2 \mathcal{O}_{11}(\cdot\,;\lambda)| * f \in L^\infty(\mathbb{R}^2, \overline{\eta}_F^E(\cdot\,;\lambda))$, where*

$$E = \begin{cases} 1 & \text{for } A + B^* \geq 2 \\ A - \frac{1}{2} & \text{for } -\frac{1}{2} < A \leq \frac{3}{2},\ B \geq 0,\ A \leq B + 1 \\ \frac{A+B^*}{2} & \text{for } 1 \leq A + B \leq 2,\ A \geq B + 1 \\ A + B - \frac{1}{2} & \text{for } B \leq 0,\ A + B \leq 1 \end{cases} \tag{i}$$

$$E + F = \begin{cases} 2 & \text{for } A + B \geq \frac{5}{2},\ A \geq \frac{3}{2} \\ A + B^* & \text{for } -\frac{1}{2} < A \leq \frac{3}{2},\ B \geq 1 \\ A + B - \frac{1}{2} & \text{for } A + B \leq \frac{5}{2},\ B \leq 1 \end{cases} \tag{ii}$$

with logarithmic factors

$$\ln_+(\lambda|\mathbf{x}|) \ \text{for} \ \begin{cases} A + B^* = 2 \\ B = A - 1,\ 0 \leq B \leq \frac{1}{2} \\ A + B = 1,\ B \leq 0 \end{cases} \tag{iii}$$

$$\ln_+(\lambda s(\mathbf{x})) \ \text{for} \ A + B = 1,\ 0 < B \leq 1. \tag{iv}$$

Moreover, we have

$$\||\partial_2 \mathcal{O}_{11}(\cdot\,;\lambda)| * f\|_{\infty,(\overline{\eta}_F^E(\cdot\,;\lambda)),\mathbb{R}^2} \leq C\lambda^{-1} \|f\|_{\infty,(\eta_B^A(\cdot\,;\lambda)),\mathbb{R}^2}\,. \tag{13.88}$$

Let in addition for A, B the following conditions be satisfied

$$\frac{1}{2} < A < 2,\ B \geq -\frac{1}{2},\ B \geq A - 2. \tag{v}$$

Then for $f \in L^\infty(\mathbb{R}^2, \nu_B^A(\cdot\,;\lambda))$ we have $|\partial_2 \mathcal{O}_{11}(\cdot\,;\lambda)| * f \in L^\infty(\mathbb{R}^2, \overline{\nu}_F^E(\cdot\,;\lambda))$ and

$$\||\partial_2 \mathcal{O}_{11}(\cdot\,;\lambda)| * f\|_{\infty,(\overline{\nu}_F^E(\cdot\,;\lambda)),\mathbb{R}^2} \leq C\lambda^{-1+A-E} \|f\|_{\infty,(\nu_B^A(\cdot\,;\lambda)),\mathbb{R}^2}\,. \tag{13.89}$$

Theorem 13.10 *Let $A + B^* > 0$ and $R = |\nabla^2 \boldsymbol{\mathcal{O}} - \nabla^2 \boldsymbol{\mathcal{S}}|$ or $R = |\partial_k \mathcal{O}_{ij}|$, $i, j, k = 1, 2$, except $|\partial_2 \mathcal{O}_{11}|$. Then for $f \in L^\infty(\mathbb{R}^2, \eta_B^A(\cdot\,;\lambda))$ we have $R * f \in L^\infty(\mathbb{R}^2, \overline{\eta}_F^E(\cdot\,;\lambda))$, where*

$$E = \begin{cases} \frac{3}{2} & \text{for } A + B^* \geq 2 \\ A & \text{for } A \leq \frac{3}{2},\ B \geq 0,\ A \leq B + 1 \\ \frac{A+B+1}{2} & \text{for } 1 \leq A + B \leq 2,\ A \geq B + 1 \\ A + B & \text{for } B \leq 0,\ A + B \leq 1 \end{cases} \tag{i}$$

$$E + F = \begin{cases} 2 & \text{for } A + B^* \geq 2 \\ A + B^* & \text{for } A + B^* \leq 2, \end{cases} \tag{ii}$$

with logarithmic factors

$$\ln_+(\lambda|\mathbf{x}|) \ \text{for} \ \ A + B^* \leq 2. \tag{iii}$$

Moreover, we have

$$\Big\| |\nabla^2 \boldsymbol{\mathcal{O}}(\cdot\,;\lambda) - \nabla^2 \boldsymbol{\mathcal{S}}(\cdot)| * f \Big\|_{\infty,(\overline{\eta}_F^E(\cdot\,;\lambda)),\mathbb{R}^2} \leq C\, \|f\|_{\infty,(\eta_B^A(\cdot\,;\lambda)),\mathbb{R}^2}\,, \tag{13.90}$$

$$\big\| |\partial_k \mathcal{O}_{ij}(\cdot\,;\lambda)| * f \big\|_{\infty,(\overline{\eta}_F^E(\cdot\,;\lambda)),\mathbb{R}^2} \leq C\lambda^{-1}\, \|f\|_{\infty,(\eta_B^A(\cdot\,;\lambda)),\mathbb{R}^2}\,. \tag{13.91}$$

Let in addition for A, B the following conditions be satisfied

$$0 \leq A < 2,\ B \geq -1. \tag{iv}$$

Then for $f \in L^\infty(\mathbb{R}^2, \nu_B^A(\cdot\,;\lambda))$ we have $R * f \in L^\infty(\mathbb{R}^2, \overline{\nu}_F^E(\cdot\,;\lambda))$ and

$$\Big\| |\nabla^2 \boldsymbol{\mathcal{O}}(\cdot\,;\lambda) - \nabla^2 \boldsymbol{\mathcal{S}}(\cdot)| * f \Big\|_{\infty,(\overline{\nu}_F^E(\cdot\,;\lambda)),\mathbb{R}^2} \leq C\lambda^{A-E}\, \|f\|_{\infty,(\nu_B^A(\cdot\,;\lambda)),\mathbb{R}^2}\,, \tag{13.92}$$

$$\big\| |\partial_k \mathcal{O}_{ij}(\cdot\,;\lambda)| * f \big\|_{\infty,(\overline{\nu}_F^E(\cdot\,;\lambda)),\mathbb{R}^2} \leq C\lambda^{-1+A-E}\, \|f\|_{\infty,(\nu_B^A(\cdot\,;\lambda)),\mathbb{R}^2}\,. \tag{13.93}$$

Proof of Theorem 13.4 Let $f \in L^\infty(\mathbb{R}^3, \eta_B^A(\cdot\,;1))$. Recalling that $|\nabla \boldsymbol{\mathcal{O}}(\mathbf{x} - \mathbf{y}; 1)| \leq C_0 \mu_{-\frac{3}{2}}^{-\frac{3}{2},-2}(\mathbf{x} - \mathbf{y}; 1)$ we have

$$\begin{aligned} ||\nabla \boldsymbol{\mathcal{O}}(\cdot\,;1)| * f(\mathbf{x})| &\leq C_1\, \mu_{-\frac{3}{2}}^{-\frac{3}{2},-2}(\cdot; 1) * \eta_{-B}^{-A}(\cdot; 1)(\mathbf{x}) \\ &\leq C_2\, \eta_{-\frac{3}{2}}^{-\frac{3}{2}}(\cdot; 1) * \eta_{-B}^{-A}(\cdot; 1)(\mathbf{x})\,. \end{aligned} \tag{13.94}$$

We have therefore to study the convolution (13.94); we apply Tables 13.1 and 13.2 with $c = d = \frac{3}{2}$, $a = A$, $b = B$, and we get, under the condition $A + B^* > 1$, that (we skip the logarithmic factors, for the moment)

$$(\eta^{-\frac{3}{2}}_{-\frac{3}{2}}(\cdot; 1) * \eta^{-A}_{-B}(\cdot; 1))(\mathbf{x}) \le C\eta^{-E}_{-F}(\mathbf{x}; 1)$$

with

$$E \le \min\left(\tfrac{3}{2},\ \tfrac{A+B^*}{2},\ A - \tfrac{1}{4},\ A - \tfrac{1}{2},\ A + B - \tfrac{1}{2},\ \tfrac{A+B}{2},\ A + B^*\right)$$

$$= \min\left(\tfrac{3}{2},\ \tfrac{A+B^*}{2},\ A - \tfrac{1}{2},\ A + B - \tfrac{1}{2}\right) \tag{13.95}$$

$$E + F \le \min\left(3,\ A + B^*,\ A + B - \frac{1}{2}\right).$$

We therefore easily get (i) and (ii), see Figs. 13.2 and 13.3:
Let us now regard the logarithmic factors. From Ω_0 we have $\ln_+(\lambda|\mathbf{x}|)$ whenever $B = 1$, $A \le 2$ or $A + B^* = 3$ and $e_0 = \frac{3}{2} + \frac{1}{2}\min(0,\ A + B^* - 3)$, $e_0 + f_0 = 3 + \min(0,\ A + B^* - 3)$. Therefore, if $A + B^* = 3$, the factor $\ln_+(\lambda|\mathbf{x}|)$ must be taken into account. But for $B = 1$, $-1 < A < 2$ we have $e_0 = \frac{A+1}{2} > A - \frac{1}{2}$, $e_0 + f_0 = A + B^* > A + B - \frac{1}{2}$, and therefore, there are in fact no logarithmic factors.

Next in Ω_3 we have $\ln_+ \frac{\lambda|\mathbf{x}|}{1+\lambda s(\mathbf{x})}$ for $A = 1 + B$. But if $B \ge 0$, then $e_3 = A - \frac{1}{2}$ and $e_3 + f_3 = A + B - \frac{1}{2}$, and then for $0 \le B \le 1$ the factor $\ln_+(\lambda|\mathbf{x}|)$ must be

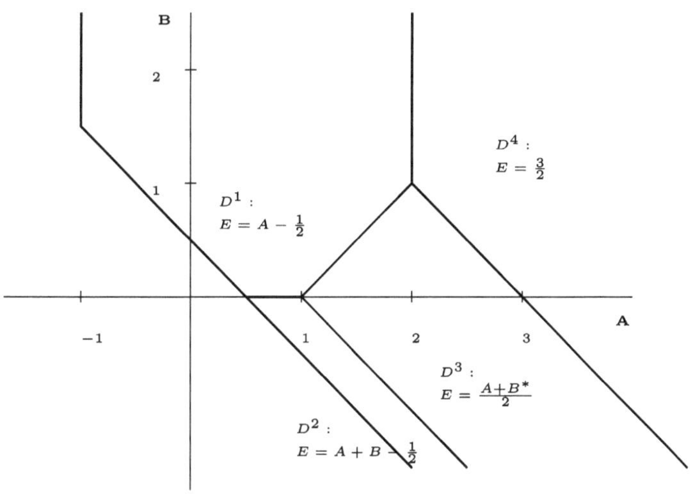

Fig. 13.2 Dependence of E on A, B, Theorem 13.4(i)

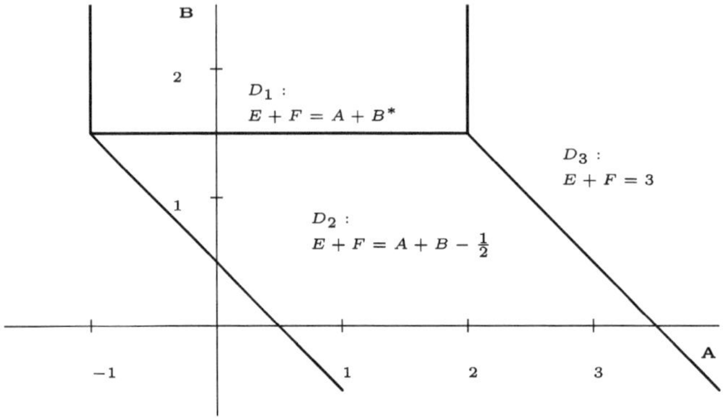

Fig. 13.3 Dependence of E+F on A, B, Theorem 13.4(ii)

taken into account. But for $B<0$ we have $e_3>A+B-\frac{1}{2}$, and for $B>1$ we have $e_3>\frac{3}{2}$. Thus (as if $s(\mathbf{x})\sim|\mathbf{x}|$ then $\frac{\lambda|\mathbf{x}|}{1+\lambda s(\mathbf{x})}\sim 1$) we can disregard the logarithmic factors in these situations.

Analogously we proceed in other subdomains and get (iii). The estimate (13.74) for $\lambda=1$ is therefore shown. In order to show (13.74) for $\lambda\neq 1$, let us recall the homogeneity property of $\mathcal{O}_{ij}(\mathbf{x}-\mathbf{y};\lambda)$. Namely, for $N=3$ we have $\mathcal{O}_{ij}(\mathbf{x}-\mathbf{y};\lambda)=\lambda\mathcal{O}_{ij}(\lambda(\mathbf{x}-\mathbf{y});1)$ and therefore

$$\begin{aligned}&\Big|\int_{\mathbb{R}^3}|\nabla\boldsymbol{\mathcal{O}}(\mathbf{x}-\mathbf{y};\lambda)|\,f(\mathbf{y})\,\mathrm{d}\mathbf{y}\Big|=\lambda^{-1}\Big|\int_{\mathbb{R}^3}|\nabla\boldsymbol{\mathcal{O}}(\lambda\mathbf{x}-\mathbf{z};1)|\,f(\frac{\mathbf{z}}{\lambda})\,\mathrm{d}\mathbf{z}\Big|\\&\le\lambda^{-1}\sup_{\mathbf{y}\in\mathbb{R}^3}\Big|f(\mathbf{y})\,\eta_B^A(\lambda\mathbf{y};1)\Big|\,\eta_{-F}^{-E}(\lambda\mathbf{x};1)\,P_1(\ln_+(\lambda|\mathbf{x}|))\,P_2(\ln_+ s(\lambda\mathbf{x}))\end{aligned}$$

and so, as $\eta_A^1(B;1)\lambda\mathbf{x}=\eta_B^A(\mathbf{x};\lambda)$, we have (13.74).

Let us study the weight $\nu_B^A(\mathbf{x};\lambda)$. From Lemma 13.4 we have the following conditions $E\ge\max(0,A-1)$ and $A<3$ and therefore we get on D^1 that $A\ge\frac{1}{2}$, on D^2 that $A+B-\frac{1}{2}\ge A-1$, i.e., $B\ge-\frac{1}{2}$, on D^3 that $\frac{A+B}{2}\ge A-1$, i.e., $A\le B+2$ and on D^4 that $A\le\frac{5}{2}$.

Finally, to show (13.75) we proceed as in the case of the estimate (13.74). Evidently, (13.75) holds for $\lambda=1$. Therefore

$$\begin{aligned}&\Big|\int_{\mathbb{R}^3}|\nabla\boldsymbol{\mathcal{O}}(\mathbf{x}-\mathbf{y};\lambda)|\,f(\mathbf{y})\mathrm{d}\mathbf{y}\Big|\\&\le\lambda^{-1}\sup_{\mathbf{y}\in\mathbb{R}^3}\Big|f(\mathbf{y})\nu_B^A(\lambda\,\mathbf{y};1)\Big|\nu_{-F}^{-E}(\lambda\,\mathbf{x};1)\,P_1(\ln_+(\lambda|\mathbf{x}|))\,P_2(\ln_+ s(\lambda\mathbf{x}))\\&=\lambda^{-1+A-E}\,\|f\|_{\infty,(\nu_B^A(\cdot;\lambda)),\mathbb{R}^3}\,\nu_{-F}^{-E}(\mathbf{x};\lambda)\,P_1(\ln_+(\lambda|\mathbf{x}|))P_2(\ln_+ s(\lambda\mathbf{x}))\,.\end{aligned}$$

□

13.2.5 L^p-Estimates for Weakly Singular Oseen Kernels

This section is devoted to the L^p-estimates of convolutions with Oseen kernels. Here we shall use the results from the previous section, i.e., the L^∞-theory. The proofs are again similar each to other, and we give it at the end of this section only for Theorem 13.13 for $R = |\nabla^2\mathcal{O} - \nabla^2\mathcal{S}|$ which is in application the most often used case.

Theorem 13.11 *Let T be an integral operator with the kernel $|\mathcal{O}|$, $T : f \mapsto |\mathcal{O}| * f$ and and let $1 < p < \infty$. Then T is a well-defined continuous operator:*

(a) $L^p(\mathbb{R}^3; \eta_\beta^{\alpha+p/2}(\cdot\,;\lambda)) \longmapsto L^p(\mathbb{R}^3; \eta_\beta^{\alpha-p/2-\varepsilon}(\cdot\,;\lambda))$
for $-\varepsilon(p-1)/p < \beta < p-1+\varepsilon(p-1)/p$, $p/2-1-\varepsilon/p < \alpha+\beta < 5p/2-3+\varepsilon$, $\alpha-\beta < p/2-1+\varepsilon+\varepsilon(p-1)/p$, $-p/2-\varepsilon/p < \alpha < 3p/2-2+\varepsilon$, $0 < \varepsilon \le p$.
(b) $L^p(\mathbb{R}^3; \nu_\beta^{\alpha+p/2}(\cdot\,;\lambda)) \longmapsto L^p(\mathbb{R}^3; \nu_\beta^{\alpha-p/2-\varepsilon}(\cdot\,;\lambda))$

for $-\varepsilon(p-1)/p < \beta < p-1+\varepsilon(p-1)/p$, $p/2-1-\varepsilon/p < \alpha+\beta < 5p/2-3+\varepsilon$, $\alpha - \beta < p/2 - 1 + \varepsilon + \varepsilon(p-1)/p$, $\max\{-p/2-\varepsilon/p, p/2-3+\varepsilon\} < \alpha < \min\{3p/2-2+\varepsilon, 5p/2-3\}$, $0 < \varepsilon \le p$.

Moreover we have for α, β specified in (a) and (b), respectively:

ad a)

$$\| |\mathcal{O}(\cdot\,;\lambda)| * f \|_{p,(\eta_\beta^{\alpha-p/2-\varepsilon}(\cdot\,;\lambda)),\mathbb{R}^3} \le C\,\lambda^{-2}\, \|f\|_{p,(\eta_\beta^{\alpha+p/2}(\cdot\,;\lambda)),\mathbb{R}^3}\,. \tag{13.96}$$

ad b)

$$\| |\mathcal{O}(\cdot\,;\lambda)| * f \|_{p,(\nu_\beta^{\alpha-p/2-\varepsilon}(\cdot\,;\lambda)),\mathbb{R}^3} \le C\,\lambda^{\frac{\varepsilon}{p}-1}\, \|f\|_{p,(\nu_\beta^{\alpha+p/2}(\cdot\,;\lambda)),\mathbb{R}^3}\,. \tag{13.97}$$

Theorem 13.12 *Let T be an integral operator with the kernel $|\nabla\mathcal{O}|$, $T : f \mapsto |\nabla\mathcal{O}| * f$, and let $1 < p < \infty$. Then T is a well-defined continuous operator:*

(a) $L^p(\mathbb{R}^3; \eta_\beta^{\alpha+p/2}(\cdot\,;\lambda)) \longmapsto L^p(\mathbb{R}^3; \eta_\beta^{\alpha}(\cdot\,;\lambda))$
for $0 < \beta < 3p/2-3/2$, $-1 < \alpha+\beta$, $\alpha < 3p/2-2$, $\alpha-\beta < p/2-1$.
(b) $L^p(\mathbb{R}^3; \nu_\beta^{\alpha+p/2}(\cdot\,;\lambda)) \longmapsto L^p(\mathbb{R}^3; \nu_\beta^{\alpha}(\cdot\,;\lambda))$

for $0 < \beta < 3p/2-3/2$, $-1 < \alpha+\beta$, $-3 < \alpha < 3p/2-2$, $\alpha-\beta < p/2-1$.

Moreover we have for α, β specified in (a), and (b), respectively:

ad a)

$$\| |\nabla\mathcal{O}(\cdot\,;\lambda)| * f \|_{p,(\eta_\beta^{\alpha}(\cdot\,;\lambda)),\mathbb{R}^3} \le C\lambda^{-1}\, \|f\|_{p,(\eta_\beta^{\alpha+p/2}(\cdot\,;\lambda)),\mathbb{R}^3}\,. \tag{13.98}$$

ad b)

$$\| |\nabla \mathcal{O}(\cdot\,;\lambda)| * f \|_{p,(\nu_\beta^\alpha(\cdot\,;\lambda)),\mathbb{R}^3} \le C\lambda^{-\frac{1}{2}} \|f\|_{p,(\nu_\beta^{\alpha+p/2}(\cdot\,;\lambda)),\mathbb{R}^3} . \tag{13.99}$$

Theorem 13.13 *Let* $R = |\nabla^2\mathcal{O} - \nabla^2\mathcal{S}|$ *or* $R = |\partial_1\mathcal{O}|$. *Let* T *be an integral operator with the kernel* R, $T : f \mapsto R * f$, *and let* $1 < p < \infty$. *Then* T *is a well-defined continuous operator:*

(a) $L^p(\mathbb{R}^3; \eta_\beta^{\alpha+p/2}(\cdot\,;\lambda)) \longmapsto L^p(\mathbb{R}^3; \eta_\beta^{\alpha+p/2-\varepsilon}(\cdot\,;\lambda))$
for $-\varepsilon(p-1)/p < \beta < p-1+\varepsilon(p-1)/p,\ -p/2-1-\varepsilon/p < \alpha+\beta < 5p/2-3+\varepsilon,\ \alpha-\beta < p/2-1+\varepsilon+\varepsilon(p-1)/p,\ -3p/2-\varepsilon/p < \alpha < 3p/2-2+\varepsilon,$ $0 < \varepsilon \le p$.
(b) $L^p(\mathbb{R}^3; \nu_\beta^{\alpha+p/2}(\cdot\,;\lambda)) \longmapsto L^p(\mathbb{R}^3; \nu_\beta^{\alpha+p/2-\varepsilon}(\cdot\,;\lambda))$ □

for $-\varepsilon(p-1)/p < \beta < p-1+\varepsilon(p-1)/p,\ -p/2-1-\varepsilon/p < \alpha+\beta < 5p/2-3+\varepsilon,$ $\alpha-\beta < p/2-1+\varepsilon+\varepsilon(p-1)/p,\ \max\{-3p/2-\varepsilon/p, -p/2-3+\varepsilon\} < \alpha < \min\{3p/2-2+\varepsilon, 5p/2-3\},\ 0 < \varepsilon \le p.$

Moreover we have for α, β *specified in* (a) *and* (b), *respectively:*

ad a)

$$\left\| |\nabla^2\mathcal{O}(\cdot\,;\lambda) - \nabla^2\mathcal{S}(\cdot)| * f \right\|_{p,(\eta_\beta^{\alpha+p/2-\varepsilon}(\cdot\,;\lambda)),\mathbb{R}^3} \le C\,\|f\|_{p,(\eta_\beta^{\alpha+p/2}(\cdot\,;\lambda)),\mathbb{R}^3} , \tag{13.100}$$

$$\| |\partial_1\mathcal{O}(\cdot\,;\lambda)| * f \|_{p,(\eta_\beta^{\alpha+p/2-\varepsilon}(\cdot\,;\lambda)),\mathbb{R}^3} \le C\,\lambda^{-1}\,\|f\|_{p,(\eta_\beta^{\alpha+p/2}(\cdot\,;\lambda)),\mathbb{R}^3} . \tag{13.101}$$

ad b)

$$\left\| |\nabla^2\mathcal{O}(\cdot\,;\lambda) - \nabla^2\mathcal{S}(\cdot)| * f \right\|_{p,(\nu_\beta^{\alpha+p/2-\varepsilon}(\cdot\,;\lambda)),\mathbb{R}^3} \le C\,\lambda^{\frac{\varepsilon}{p}}\,\|f\|_{p,(\nu_\beta^{\alpha+p/2}(\cdot\,;\lambda)),\mathbb{R}^3} \tag{13.102}$$

$$\| |\partial_1\mathcal{O}(\cdot\,;\lambda)| * f \|_{p,(\nu_\beta^{\alpha+p/2-\varepsilon}(\cdot\,;\lambda)),\mathbb{R}^3} \le C\,\lambda^{\frac{\varepsilon}{p}-1}\,\|f\|_{p,(\nu_\beta^{\alpha+p/2}(\cdot\,;\lambda)),\mathbb{R}^3} . \tag{13.103}$$

Theorem 13.14 *Let* T *be an integral operator with the kernel* $|\mathcal{P}|$, $T : f \mapsto |\mathcal{P}| * f$, $i = 1, 2, 3$, *and let* $1 < p < \infty$. *Then* T *is a well-defined continuous operator:*

(a) $L^p(\mathbb{R}^3; \eta_\beta^{\alpha+p/2}(\cdot\,;\lambda)) \longmapsto L^p(\mathbb{R}^3; \eta_\beta^{\alpha-p/2}(\cdot\,;\lambda))$
for $0 < \beta < p-1,\ \ p/2-3 < \alpha+\beta < 5p/2-3.$
(b) $L^p(\mathbb{R}^3; \nu_\beta^{\alpha+p/2}(\cdot\,;\lambda)) \longmapsto L^p(\mathbb{R}^3; \nu_\beta^{\alpha-p/2}(\cdot\,;\lambda))$ □

for $0 < \beta < p-1,\ p/2-3 < \alpha+\beta < 5p/2-3,\ p/2-3 < \alpha < 5p/2-3.$

Moreover we have for α, β *specified in* (*a*) *and* (*b*), *respectively:*

ad a)

$$\| |\mathcal{P}| * f \|_{p,(\eta_\beta^{\alpha-p/2}(\cdot\,;\lambda)),\mathbb{R}^3} \leq C\lambda^{-1} \|f\|_{p,(\eta_\beta^{\alpha+p/2}(\cdot\,;\lambda)),\mathbb{R}^3} \,. \tag{13.104}$$

ad b)

$$\| |\mathcal{P}| * f \|_{p,(\nu_\beta^{\alpha-p/2}(\cdot\,;\lambda)),\mathbb{R}^3} \leq C \, \|f\|_{p,(\nu_\beta^{\alpha+p/2}(\cdot\,;\lambda)),\mathbb{R}^3} \,. \tag{13.105}$$

Next we formulate analogous results also in the two-dimensional case.

Theorem 13.15 *Let* T *be an integral operator with the kernel* $|\mathcal{O}_{11}|$ $T : f \mapsto |\mathcal{O}_{11}| * f$, *and let* $1 < p < \infty$. *Then* T *is a well-defined continuous operator:*

(a) $$L^p(\mathbb{R}^2; \eta_\beta^{\alpha+p/2}(\cdot\,;\lambda)) \longmapsto L^p(\mathbb{R}^2; \eta_\beta^{\alpha-p/2-\varepsilon}(\cdot\,;\lambda))$$
for $-\varepsilon(p-1)/p < \beta < p/2 - 1/2 + \varepsilon(p-1)/p$, $p/2 - 1 - \varepsilon/p < \alpha + \beta < 3p/2 - 2 + \varepsilon$, $\alpha - \beta < p/2 - 1 + \varepsilon + \varepsilon(p-1)/p$, $-1/2 - \varepsilon/p < \alpha < p - 3/2 + \varepsilon$, $0 < \varepsilon \leq p/2$.

(b) $$L^p(\mathbb{R}^2; \nu_\beta^{\alpha+p/2}(\cdot\,;\lambda)) \longmapsto L^p(\mathbb{R}^2; \nu_\beta^{\alpha-p/2-\varepsilon}(\cdot\,;\lambda))$$

for $-\varepsilon(p-1)/p < \beta < p/2 - 1/2 + \varepsilon(p-1)/p$, $p/2 - 1 - \varepsilon/p < \alpha + \beta < 3p/2 - 2 + \varepsilon$, $\alpha - \beta < p/2 - 1 + \varepsilon + \varepsilon(p-1)/p$, $\max\{-1/2 - \varepsilon/p, p/2 - 2 + \varepsilon\} < \alpha < \min\{p - 3/2 + \varepsilon, 3p/2 - 2\}$, $0 < \varepsilon \leq p/2$.

Moreover we have for α, β *specified in* (*a*) *and* (*b*), *respectively:*

ad a)

$$\| |\mathcal{O}_{11}(\cdot\,;\lambda)| * f \|_{p,(\eta_\beta^{\alpha-p/2-\varepsilon}(\cdot\,;\lambda)),\mathbb{R}^2} \leq C\,\lambda^{-2} \|f\|_{p,(\eta_\beta^{\alpha+p/2}(\cdot\,;\lambda)),\mathbb{R}^2} \,. \tag{13.106}$$

ad b)

$$\| |\mathcal{O}_{11}(\cdot\,;\lambda)| * f \|_{p,(\nu_\beta^{\alpha-p/2-\varepsilon}(\cdot\,;\lambda)),\mathbb{R}^2} \leq C\,\lambda^{\frac{\varepsilon}{p}-1} \|f\|_{p,(\nu_\beta^{\alpha+p/2}(\cdot\,;\lambda)),\mathbb{R}^2} \,. \tag{13.107}$$

Theorem 13.16 *Let* T *be an integral operator with the kernel* $R = |\mathcal{O}_{ij}|$, $i, j = 1, 2$, $i \cdot j \neq 1$ *or* $R = |\mathcal{P}|$; $T : f \mapsto R * f$, *and let* $1 < p < \infty$. *Then* T *is a well-defined continuous operator:*

(a) $$L^p(\mathbb{R}^2; \eta_\beta^{\alpha+p/2}(\cdot\,;\lambda)) \longmapsto L^p(\mathbb{R}^2; \eta_\beta^{\alpha-p/2}(\cdot\,;\lambda))$$
for $0 < \beta < p/2 - 1/2$, $p/2 - 2 < \alpha + \beta < 3p/2 - 2$.

(b) $$L^p(\mathbb{R}^2; \nu_\beta^{\alpha+p/2}(\cdot\,;\lambda)) \longmapsto L^p(\mathbb{R}^2; \nu_\beta^{\alpha-p/2}(\cdot\,;\lambda)) \quad \square$$

for $0 < \beta < p/2 - 1/2$, $p/2 - 2 < \alpha + \beta < 3p/2 - 2$, $p/2 - 2 < \alpha < 3p/2 - 2$.

Moreover we have for α, β specified in (a) and (b), respectively:

ad a)

$$\big\| |\mathcal{O}_{ij}(\cdot\,;\lambda)| * f \big\|_{p,(\eta_\beta^{\alpha-p/2}(\cdot\,;\lambda)),\mathbb{R}^2} \le C\,\lambda^{-2}\,\|f\|_{p,(\eta_\beta^{\alpha+p/2}(\cdot\,;\lambda)),\mathbb{R}^2}, \tag{13.108}$$

$$\||\mathcal{P}| * f\|_{p,(\eta_\beta^{\alpha-p/2}(\cdot\,;\lambda)),\mathbb{R}^2} \le C\,\lambda^{-1}\,\|f\|_{p,(\eta_\beta^{\alpha+p/2}(\cdot\,;\lambda)),\mathbb{R}^2}. \tag{13.109}$$

ad b)

$$\big\| |\mathcal{O}_{ij}(\cdot\,;\lambda)| * f \big\|_{p,(\nu_\beta^{\alpha-p/2}(\cdot\,;\lambda)),\mathbb{R}^2} \le C\,\lambda^{-1}\,\|f\|_{p,(\nu_\beta^{\alpha+p/2}(\cdot\,;\lambda)),\mathbb{R}^2}, \tag{13.110}$$

$$\||\mathcal{P}| * f\|_{p,(\nu_\beta^{\alpha-p/2}(\cdot\,;\lambda)),\mathbb{R}^2} \le C\,\|f\|_{p,(\nu_\beta^{\alpha+p/2}(\cdot\,;\lambda)),\mathbb{R}^2}. \tag{13.111}$$

Theorem 13.17 *Let T be an integral operator with the kernel $|\partial_2\mathcal{O}_{11}|$, $T : f \mapsto |\partial_2\mathcal{O}_{11}| * f$, and let $1 < p < \infty$. Then T is a well-defined continuous operator:*

(a) $$L^p(\mathbb{R}^2;\eta_\beta^{\alpha+p/2}(\cdot\,;\lambda)) \longmapsto L^p(\mathbb{R}^2;\eta_\beta^{\alpha}(\cdot\,;\lambda))$$
for $0 < \beta < p-1,\ -1 < \alpha+\beta,\ \alpha < p-3/2,\ \alpha-\beta < p/2-1.$

(b) $$L^p(\mathbb{R}^2;\nu_\beta^{\alpha+p/2}(\cdot\,;\lambda)) \longmapsto L^p(\mathbb{R}^2;\nu_\beta^{\alpha-p/2}(\cdot\,;\lambda)) \quad \square$$

for $0 < \beta < p-1,\ -1 < \alpha+\beta,\ -2 < \alpha < p-3/2,\ \alpha-\beta < p/2-1.$

Moreover we have for α, β specified in (a) and (b), respectively:

ad a)

$$\||\partial_2\mathcal{O}_{11}(\cdot\,;\lambda)| * f\|_{p,(\eta_\beta^{\alpha}(\cdot\,;\lambda)),\mathbb{R}^2} \le C\,\lambda^{-1}\,\|f\|_{p,(\eta_\beta^{\alpha+p/2}(\cdot\,;\lambda)),\mathbb{R}^2}. \tag{13.112}$$

ad b)

$$\||\partial_2\mathcal{O}_{11}(\cdot\,;\lambda)| * f\|_{p,(\nu_\beta^{\alpha}(\cdot\,;\lambda)),\mathbb{R}^2} \le C\,\lambda^{-\frac{1}{2}}\,\|f\|_{p,(\nu_\beta^{\alpha+p/2}(\cdot\,;\lambda)),\mathbb{R}^2}. \tag{13.113}$$

Theorem 13.18 *Let T be an integral operator with the kernel $R = |\nabla^2\boldsymbol{\mathcal{O}} - \nabla^2\boldsymbol{\mathcal{S}}|$, or $R = |\partial_k\mathcal{O}_{ij}|$, $i, j, k = 1, 2$, except $|\partial_2\mathcal{O}_{11}|$; $T : f \mapsto R * f$, and let $1 < p < \infty$. Then T is a well-defined continuous operator:*

(a) $$L^p(\mathbb{R}^2;\eta_\beta^{\alpha+p/2}(\cdot\,;\lambda)) \longmapsto L^p(\mathbb{R}^2;\eta_\beta^{\alpha+p/2-\varepsilon}(\cdot\,;\lambda))$$
for $-\varepsilon(p-1)/p < \beta < p/2-1/2+\varepsilon(p-1)/p,\ -p/2-1-\varepsilon/p < \alpha+\beta < 3p/2-2+\varepsilon,\ \alpha-\beta < p/2-1+\varepsilon+\varepsilon(p-1)/p,\ -p-1/2-\varepsilon/p < \alpha < p-3/2+\varepsilon,\ 0 < \varepsilon \le p/2.$

(b) $$L^p(\mathbb{R}^2;\nu_\beta^{\alpha+p/2}(\cdot\,;\lambda)) \longmapsto L^p(\mathbb{R}^2;\nu_\beta^{\alpha+p/2-\varepsilon}(\cdot\,;\lambda))$$

$\square$

for $-\varepsilon(p-1)/p < \beta < p/2 - 1/2 + \varepsilon(p-1)/p$, $-p/2 - 1 - \varepsilon/p < \alpha + \beta < 3p/2 - 2 + \varepsilon$, $\alpha - \beta < p/2 - 1 + \varepsilon + \varepsilon(p-1)/p$, $\max\{-p - 1/2 - \varepsilon/p, -p/2 - 2 + \varepsilon\} < \alpha < \min\{p - 3/2 + \varepsilon, 3p/2 - 2\}$, $0 < \varepsilon \le p/2$.

Moreover we have for α, β *specified in* (*a*), *and* (*b*), *respectively:*

ad a)

$$\Big\| |\nabla^2 \boldsymbol{\mathcal{O}}(\cdot\,;\lambda) - \nabla^2 \boldsymbol{\mathcal{S}}(\cdot)| * f \Big\|_{p,(\eta_\beta^{\alpha+p/2-\varepsilon}(\cdot\,;\lambda)),\mathbb{R}^2} \le C\, \|f\|_{p,(\eta_\beta^{\alpha+p/2}(\cdot\,;\lambda)),\mathbb{R}^2}, \tag{13.114}$$

$$\big\| |\partial_k \mathcal{O}_{ij}(\cdot\,;\lambda)| * f \big\|_{p,(\eta_\beta^{\alpha+p/2-\varepsilon}(\cdot\,;\lambda)),\mathbb{R}^2} \le C\,\lambda^{-1}\, \|f\|_{p,(\eta_\beta^{\alpha+p/2}(\cdot\,;\lambda)),\mathbb{R}^2}. \tag{13.115}$$

ad b)

$$\Big\| |\nabla^2 \boldsymbol{\mathcal{O}}(\cdot\,;\lambda) - \nabla^2 \boldsymbol{\mathcal{S}}(\cdot)| * f \Big\|_{p,(\nu_\beta^{\alpha+p/2-\varepsilon}(\cdot\,;\lambda)),\mathbb{R}^2} \le C\,\lambda^{\frac{\varepsilon}{p}}\, \|f\|_{p,(\nu_\beta^{\alpha+p/2}(\cdot\,;\lambda)),\mathbb{R}^2}, \tag{13.116}$$

$$\big\| |\partial_k \mathcal{O}_{ij}(\cdot\,;\lambda)| * f \big\|_{p,(\nu_\beta^{\alpha+p/2-\varepsilon}(\cdot\,;\lambda)),\mathbb{R}^2} \le C\,\lambda^{\frac{\varepsilon}{p}-1}\, \|f\|_{p,(\nu_\beta^{\alpha+p/2}(\cdot\,;\lambda)),\mathbb{R}^2}. \tag{13.117}$$

Proof of Theorem 13.13 We show only the case $R = |\nabla^2 \boldsymbol{\mathcal{O}} - \nabla^2 \boldsymbol{\mathcal{S}}|$; the other case differs only very slightly. We proceed similarly as in the case of L^∞-weighted estimates. Studying first $\eta_\beta^\alpha(\cdot\,;\lambda)$ weights, we show (13.100) for $\lambda = 1$. Applying the homogeneity properties of $\mathcal{O}_{ij}(\cdot;\lambda)$ we get (13.100) in the general situation $\lambda \ne 1$. Next, using the results from (a) together with Lemma 13.4 we show (13.102).

Let us denote

$$\begin{aligned} K(\mathbf{x},\mathbf{y}) &= R(\mathbf{x}-\mathbf{y})\left(\eta_\beta^{\alpha+p/2-\varepsilon}(\mathbf{x})\right)^{1/p}\left(\eta_\beta^{\alpha+p/2}(\mathbf{y})\right)^{-1/p}, \\ F(\mathbf{y}) &= f(\mathbf{y})\left(\eta_\beta^{\alpha+p/2}(\mathbf{y})\right)^{1/p}. \end{aligned}$$

We easily observe that, in order to verify (13.100) with $\lambda = 1$, it is sufficient to show that there exists $C > 0$, independent of f, such that

$$\Big\| \int_{\mathbb{R}^3} K(\cdot\,,\mathbf{y})\,F(\mathbf{y})\,\mathrm{d}\mathbf{y} \Big\|_p \le C\|F\|_p. \tag{13.118}$$

Let $L(\cdot)$ and $M(\cdot)$ be nonnegative functions defined on $\mathbb{R}^3$ such that for all $\mathbf{x}$, $\mathbf{y} \in \mathbb{R}^3$

$$J_0(\mathbf{x}) = \int_{\mathbb{R}^3} |K(\mathbf{x},\mathbf{y})|\ L(\mathbf{y})^q\,\mathrm{d}\mathbf{y} \le C^q\ M(\mathbf{x})^q, \tag{13.119}$$

$$J_1(\mathbf{y}) := \int_{\mathbb{R}^3} |K(\mathbf{x},\mathbf{y})|\ M(\mathbf{x})^p\,\mathrm{d}\mathbf{x} \le C^p\ L(\mathbf{y})^p, \tag{13.120}$$

where $C > 0$, $1 < p < \infty$, and $p^{-1} + q^{-1} = 1$. Then relation (13.118) is satisfied. Indeed,

$$\begin{gathered}
\Big\| \int_{\mathbb{R}^3} K(\cdot\,,\mathbf{y})\,F(\mathbf{y})\,\mathrm{d}\mathbf{y}\Big\|_p^p \le \\
\int_{\mathbb{R}^3} \Big\{\Big(\int_{\mathbb{R}^3} |K(\mathbf{x},\mathbf{y})|\,|F(\mathbf{y})|^p L(\mathbf{y})^{-p}\,\mathrm{d}\mathbf{y}\Big)^{\frac{1}{p}} J_0(\mathbf{x})^{\frac{1}{q}}\Big\}^p \mathrm{d}\mathbf{x} \le \\
C^p \int_{\mathbb{R}^3} M(\mathbf{x})^p \int_{\mathbb{R}^3} |K(\mathbf{x},\mathbf{y})|\,|F(\mathbf{y})|^p L(\mathbf{y})^{-p}\mathrm{d}\mathbf{y}\,\mathrm{d}\mathbf{x} = \\
C^p \int_{\mathbb{R}^3} |F(\mathbf{y})|^p J_1(\mathbf{y}) L(\mathbf{y})^{-p}\mathrm{d}\mathbf{y} \le C^{2p}\,\|F\|_p^p\,,
\end{gathered}$$

i.e., we get (13.118). We shall suppose the functions $L(\cdot)$, $M(\cdot)$ in the form $L(\mathbf{x}) = M(\mathbf{x}) = \eta^{-A}_{-\frac{\beta}{p^2}}(\mathbf{x})$, $A \in \mathbb{R}^1$. Denoting

$$\begin{array}{ll}
a_0 = qA + \frac{\alpha}{p} + \frac{1}{2} & a_1 = pA - \frac{\alpha}{p} - \frac{1}{2} + \frac{\varepsilon}{p} \\
b_0 = \frac{\beta}{p-1} & b_1 = 0
\end{array} \tag{13.121}$$

we get that in order to verify (13.119) we have to find such a_0 b_0 that

$$\int_{\mathbb{R}^3} R(\mathbf{x}-\mathbf{y})\,\eta_{-b_0}^{-a_0}(\mathbf{y})\,\mathrm{d}\mathbf{y} \le C\,\eta_{-b_0}^{-a_0+\frac{\varepsilon}{p}}(\mathbf{x})$$

for all $\mathbf{x} \in \mathbb{R}^3$, and in order to verify (13.120) we have to find a_1 such that

$$\int_{\mathbb{R}^3} R(-\mathbf{x}-\mathbf{y})\,\eta_0^{-a_1}(-\mathbf{x})\,\mathrm{d}\mathbf{x} \le C\,\eta_0^{-a_1+\frac{\varepsilon}{p}}(\mathbf{y})$$

for all $\mathbf{y} \in \mathbb{R}^3$.

Applying Theorem 13.5 with $f = \eta_{-b_0}^{-a_0}(\cdot)$ we get for the first inequality the following set of conditions

$$\begin{gathered}
-\varepsilon/p < b_0 < 1 + \varepsilon/p, \qquad a_0 < b_0 + 1 + 2\varepsilon/p, \qquad a_0 + b_0^* > 0, \\
a_0 + b_0 < 3 + \varepsilon/p, \qquad a_0 < 2 + \varepsilon/p,
\end{gathered} \tag{13.122}$$

while for the second inequality we directly apply Tables 13.1 and 13.2 and get[1]

$$0 < a_1 < 1 + 2\varepsilon/p\,, \tag{13.123}$$

a_i, b_i defined in (13.121). The conditions on a_i, b_i can be satisfied for some $A \in \mathbb{R}^1$ if we have for $0 < \varepsilon \le p$: $-\varepsilon(p-1)/p < \beta < p - 1 + \varepsilon(p-1)/p$, $-p/2 - 1 - \varepsilon/p < \alpha + \beta < 5p/2 - 3 + \varepsilon$, $\alpha - \beta < p/2 - 1 + \varepsilon + \varepsilon(p-1)/p$, $-3p/2 - \varepsilon/p < \alpha < 3p/2 - 2 + \varepsilon$. Thus (13.100) is proved for $\lambda = 1$.

Next let $\lambda \neq 1$. As $|\nabla^2 \boldsymbol{\mathcal{O}}(\mathbf{x}-\mathbf{y};\lambda) - \nabla^2 \boldsymbol{\mathcal{S}}(\mathbf{x})| = \lambda^3 |\nabla^2 \boldsymbol{\mathcal{O}}(\lambda(\mathbf{x}-\mathbf{y});1) - \nabla^2 \boldsymbol{\mathcal{S}}(\lambda\mathbf{x})|$, we easily have

$$\begin{aligned}
&\int_{\mathbb{R}^3} \Big| \int_{\mathbb{R}^3} R(\mathbf{x}-\mathbf{y};\lambda) f(\mathbf{y}) \mathrm{d}\mathbf{y} \Big|^p \eta_\beta^{\alpha+p/2-\varepsilon}(\mathbf{x};\lambda) \mathrm{d}\mathbf{x} \\
&= \int_{\mathbb{R}^3} \Big| \int_{\mathbb{R}^3} R(\lambda\mathbf{x}-\mathbf{z};1) f(\mathbf{z}/\lambda) \mathrm{d}\mathbf{z} \Big|^p \eta_\beta^{\alpha+p/2-\varepsilon}(\lambda\mathbf{x};1) \mathrm{d}\mathbf{x} \\
&\qquad \le C\lambda^{-3} \int_{\mathbb{R}^3} \Big| f(\frac{\mathbf{z}}{\lambda}) \Big|^p \eta_\beta^{\alpha+p/2}(\mathbf{z};1) \mathrm{d}\mathbf{z} \\
&\qquad = C\lambda^{-2p} \int_{\mathbb{R}^3} |f(\mathbf{y})|^p \eta_\beta^{\alpha+p/2}(\mathbf{y};\lambda) \mathrm{d}\mathbf{y}
\end{aligned}$$

and we have (13.100) with $\lambda \neq 1$.

In order to prove (13.102), we redefine functions $K(\cdot,\cdot)$ and $F(\cdot)$:

$$\begin{aligned}
K(\mathbf{x},\mathbf{y}) &= R(\mathbf{x}-\mathbf{y}) \left(\nu_\beta^{\alpha+p/2-\varepsilon}(\mathbf{x})\right)^{1/p} \left(\nu_\beta^{\alpha+p/2}(\mathbf{y})\right)^{-1/p}, \\
F(\mathbf{y}) &= f(\mathbf{y}) \left(\nu_\beta^{\alpha+p/2}(\mathbf{y})\right)^{1/p}.
\end{aligned}$$

We will now proceed as in the first part of the proof, but now we search the functions $L(\cdot)$, $M(\cdot)$ in the form $L(\mathbf{x}) = \mu_{-\frac{\beta}{p^2}}^{-A,-G}(\mathbf{x})$, $M(\mathbf{x}) = \mu_{-\frac{\beta}{p^2}}^{-A,-H}(\mathbf{x})$. Denoting

$$\begin{aligned}
c_0 &= qG + \tfrac{\alpha}{p} + \tfrac{1}{2} & c_1 &= pH - \tfrac{\alpha}{p} - \tfrac{1}{2} + \tfrac{\varepsilon}{p} \\
d_0 &= qH + \tfrac{\alpha}{p} + \tfrac{1}{2} - \tfrac{\varepsilon}{p} & d_1 &= pG - \tfrac{\alpha}{p} - \tfrac{1}{2}
\end{aligned} \tag{13.124}$$

we see that in order to verify (13.119) and (13.120) we have to find such a_i, b_i, see (13.121), (13.122), and c_i, d_i such that

$$\begin{aligned}
&\int_{\mathbb{R}^3} R(\mathbf{x}-\mathbf{y}) \mu_{-b_0}^{-a_0,-c_0}(\mathbf{y}) \mathrm{d}\mathbf{y} \le C\mu_{-b_0}^{-a_0+\frac{\varepsilon}{p},-d_0}(\mathbf{x})\,,\ \mathbf{x} \in \mathbb{R}^3 \setminus \{\mathbf{0}\} \\
&\int_{\mathbb{R}^3} R(-\mathbf{x}-\mathbf{y}) \mu_0^{-a_1,-c_1}(-\mathbf{x}) d\mathbf{x} \le C\mu_0^{-a_1+\frac{\varepsilon}{p},-d_1}(\mathbf{y}),\ \mathbf{y} \in \mathbb{R}^3 \setminus \{\mathbf{0}\}.
\end{aligned}$$

---

[1] The convolution on the left-hand side of (13.120) is again the same convolution as treated in Section 3, but at the point $-\mathbf{y}$.

Recalling that $|R(\mathbf{x}-\mathbf{y})| \leq C_0 \nu_{-1}^{-2}(\mathbf{x}-\mathbf{y};\, 1)$ we get from Lemma 13.4 the following two possible sets of conditions for c_i, d_i:

$$(i) \qquad \begin{aligned} c_i &< 3 \\ c_i + 2 &> 3 \\ d_i &\geq c_i - 1 \end{aligned} \qquad\qquad (ii) \qquad \begin{aligned} c_i &< 1 \\ d_i &\geq 0\,, \end{aligned}$$

where in both cases $i = 0, 1$. Conditions for a_i, b_i are the same as in the first part of this proof. From the conditions (i) we get the following additional condition:

$$-\,p/2 - 3 + \varepsilon < \alpha < 5p/2 - 3.$$

Case (ii) gives more restrictive conditions on α, no extension of the result. So, (13.102) is proved in the case $\lambda = 1$.

Finally to get (13.102) with $\lambda \neq 1$ we proceed as in the case of the weights $\eta_B^A(\cdot\,;\lambda)$. We have

$$\begin{aligned} &\int_{\mathbb{R}^3} \Big| \int_{\mathbb{R}^3} R(\mathbf{x}-\mathbf{y};\, \lambda) f(\mathbf{y}) \mathrm{d}\mathbf{y} \Big|^p \nu_\beta^{\alpha+p/2-\varepsilon}(\mathbf{x};\, \lambda) \mathrm{d}\mathbf{x} \\ = \lambda^{-\alpha-p/2+\varepsilon} &\int_{\mathbb{R}^3} \Big| \int_{\mathbb{R}^3} R(\lambda\mathbf{x}-\mathbf{z};\, 1) f(\mathbf{z}/\lambda) \mathrm{d}\mathbf{z} \Big|^p \nu_\beta^{\alpha+p/2-\varepsilon}(\lambda\mathbf{x};\, 1) \mathrm{d}\mathbf{x} \\ \leq C\lambda^{-\alpha-p/2+\varepsilon-3} &\int_{\mathbb{R}^3} \Big| f(\mathbf{z}/\lambda) \Big|^p \nu_\beta^{\alpha+p/2}(\mathbf{z};\, 1) \mathrm{d}\mathbf{z} \\ = C\,\lambda^\varepsilon &\int_{\mathbb{R}^3} |f(\mathbf{y})|^p \nu_\beta^{\alpha+p/2}(\mathbf{y};\, \lambda) \mathrm{d}\mathbf{y}, \end{aligned}$$

and we have (13.102) with $\lambda \neq 1$. This completes our proof. □

13.3 Singular Integrals

The aim of this section is to present some results concerning L^p-estimates of certain singular operators and finally to apply them on the convolutions (defined in the sense of principal value) of the type $\nabla \boldsymbol{\mathcal{P}} * f$, $\nabla^2 \boldsymbol{\mathcal{S}} * f$, and $\nabla^2 \boldsymbol{\mathcal{O}} * f$.

We shall use the idea of Farwig (see [58]). Before formulating the result from [159], we need to define some notation.

Definition The weight w, $w \geq 0$ belongs to the Muckenhoupt class A_p, $1 \leq p < +\infty$ if there is a constant C such that

$$\begin{aligned} &\sup_Q \left[\left(\tfrac{1}{|Q|} \int_Q w(\mathbf{x}) \mathrm{d}\mathbf{x} \right) \left(\tfrac{1}{|Q|} \int_Q w(\mathbf{x})^{-\frac{1}{p-1}} \mathrm{d}\mathbf{x} \right)^{p-1} \right] \leq C < +\infty \\ &\sup_Q \tfrac{1}{|Q|} \int_Q w(\mathbf{x}) \mathrm{d}\mathbf{x} \leq C w(\mathbf{x}_0), \quad \forall \mathbf{x}_0 \in \mathbb{R}^N \end{aligned} \tag{13.125}$$

for $p \in (1;\infty)$ and $p = 1$, respectively. In the first case, the supremum is taken over all cubes Q in $\mathbb{R}^N$, in the second case only over those cubes which contain $\mathbf{x}_0$; $|Q|$ denotes the Lebesgue measure of Q. The constant does not depend on $\mathbf{x}_0$.

Remark 13.4

(a) For $p = 1$, the condition $(13.125)_2$ can be replaced by

$$Mw(\mathbf{x}) \leq Cw(\mathbf{x}) \text{ for a.a. } \mathbf{x} \in \mathbb{R}^N, \tag{13.126}$$

where $Mg(\mathbf{x})$ is the Hardy–Littlewood maximal function which is defined by the left-hand side in $(13.125)_2$.

(b) In (13.125) it is enough to take the supremum over all cubes with edges parallel to an arbitrary chosen Cartesian system X.

To show this, let X be a Cartesian system in $\mathbb{R}^N$ and X' another one arising from X by any rotation. Then we have

$$\frac{1}{N^{\frac{N}{2}}}\frac{1}{|Q_1|}\int_{Q_1} w(\mathbf{x})\mathrm{d}\mathbf{x} \leq \frac{1}{|Q'|}\int_{Q'} w(\mathbf{x})\mathrm{d}\mathbf{x} \leq \frac{N^{\frac{N}{2}}}{|Q_2|}\int_{Q_2} w(\mathbf{x})\mathrm{d}\mathbf{x} \tag{13.127}$$

for any locally integrable function $w \geq 0$. In (13.127) Q' is a cube with edges parallel to the axes of X', Q_1 is the greatest cube with edges parallel to the axes of X such that $Q_1 \subset Q'$,, and Q_2 is the smallest cube with edges parallel to the axes of X such that $Q' \subset Q_2$.

Next part is devoted to the investigation under what condition the weights defined in Section 1 belong to A_p for some $1 \leq p < +\infty$. First we recall several general results:

Lemma 13.5

(a) *If $w_1, w_2 \in A_1$, then for any $1 \leq p < +\infty$ the weight $w \equiv w_1\, w_2^{1-p} \in A_p$.*
(b) *If $w_1,\ w_2 \in A_p$ for some $1 \leq p < +\infty$, then for any $\hbar \in [0;1]$, $w_\hbar \equiv w_1^\hbar w_2^{1-\hbar} \in A_p$.*
(c) *If $w \in A_p$ for some $1 \leq p < +\infty$, then for any $\hbar \in [0;1]$, $w^\hbar \in A_p$.*

Proof It follows directly from the definition of A_p and in case (b) from the Hölder inequality. □

Definition Let μ be a nonnegative Borel measure. We define the maximal function

$$M\mu(x) = \sup_Q \frac{1}{|Q|}\int_Q \mathrm{d}\mu(y), \tag{13.128}$$

where the supremum is taken over all cubes Q such that $x \in Q$. Analogously we define $Mf(x)$ for $f \in L^1_{loc}(\mathbb{R}^N)$, replacing $d\mu(y)$ by $|f|dy$. (See also Remark 13.4 (a) and (b).)

The proof of the following lemma can be found in [211] (Theorems IX.5.5 and IX.3.4).

Lemma 13.6

(a) *If* $M\mu$ *is finite for a.a.* $\mathbf{x} \in \mathbb{R}^N$, *then, for any* $\hbar \in [0; 1)$, $(M\mu)^\hbar \in A_1$.
(b) *Let* $w \in A_1$. *Then there exists a function* $f \in L^1_{loc}(\mathbb{R}^N)$ *such that* $w \sim (Mf)^\hbar$ *for some* $\hbar \in [0; 1)$.

Using Lemma 13.6 we can easily show:

Lemma 13.7 *The weights* $|\mathbf{x}|^{-a}$ *and* $(1+|\mathbf{x}|)^{-a}$ *satisfy the* A_1*-condition on* $\mathbb{R}^N$ *for each* $a \in [0; N)$.

Proof We have that for $\mu = \delta_0$ the maximal function $M\mu(x) \sim |\mathbf{x}|^{-N}$ and so $|\mathbf{x}|^{-N\hbar} \in A_1, \forall \hbar \in [0; 1)$. Further, if we define $\mu(A) = |A \cap B_1(\mathbf{0})|$, then $M\mu(x) \sim (1 + |x|)^{-N}$ and again Lemma 13.6(a) furnishes the result. □

Now we shall concentrate on the two-dimensional case. We have:

Lemma 13.8 *For* $b \in (-1; 1]$ *and* $\hbar \in [0; 1)$ *the function* $w_0^{(2)}(\mathbf{x}) = \left(\frac{|\mathbf{x}|^{b-\frac{1}{2}}}{s(\mathbf{x})^{\frac{1}{2}}}\right)^\hbar$, $\mathbf{x} \in \mathbb{R}^2$ *is a weight of the class* A_1 *in* $\mathbb{R}^2$.

Proof Let $b \in (-1; 1]$. We define the measure μ by

$$\mu(A) \equiv \int_{A^+} x_1^b \mathrm{d}x_1$$

where $A^+ = \{x_1 \in \mathbb{R},\ x_1 > 0 :\ (x_1, 0) \in A\}$ for $A \subset \mathbb{R}^2$, measurable. Evidently μ is a nonnegative Borel measure on $\mathbb{R}^2$. We shall show that $M\mu \sim \frac{r^{b-\frac{1}{2}}}{s^{\frac{1}{2}}}$; then the assertion of the lemma follows from Lemma 13.6(a).

Let Q_a be a closed square containing $\mathbf{x}$ with sides parallel to the axes (see Remark 13.4 (*b*)) with the side length $a > |x_2|$, $q(a) = \frac{\mu(Q_a)}{|Q_a|}$ and $|\mathbf{x}|$ denoting the maximum norm of $\mathbf{x} = (x_1, x_2) \in \mathbb{R}^2$. We have to distinguish several cases:

(A) Let $x_1 > 0$, $|x_2| \le x_1$, and $b \in (-1; 0]$. We first consider squares Q_a with $|x_2| \le a \le x_1$ and $Q_a^+ = [x_1 - a; x_1]$. Then we have

$$q(a) = \frac{1}{b+1} \frac{x_1^{b+1} - (x_1 - a)^{b+1}}{a^2} = \frac{1}{b+1} x_1^{b-1} \varphi\left(\frac{a}{x_1}\right)$$

with

$$\varphi(y) = \frac{1 - (1 - y)^{b+1}}{y^2} \qquad \frac{|x_2|}{x_1} \le y \le 1.$$

Let $b < 0$. Then $\varphi(0^+) = +\infty$, $\varphi(1) = 1$, $\varphi'(1^-) = \infty$. It is possible to show that the function φ has exactly one local minimum on $(0; 1)$ with a value less than 1. Therefore, there exists exactly one point $y_b \in (0, 1)$ such that $\varphi(y_b) = 1$. If $|x_2| \le y_b \cdot x_1$, then φ achieves its maximum on $\left[\frac{|x_2|}{x_1}, 1\right)$ at the point $\frac{|x_2|}{x_1}$ and

$$\max\{q(a); |x_2| \le a \le x_1\} = q(|x_2|) \sim \frac{x_1^b}{|x_2|} \sim \frac{r^{b-\frac{1}{2}}}{s^{\frac{1}{2}}}.$$

If $|x_2| > y_b \cdot x_1$, then φ is maximal at the point 1 what yields

$$\max\{q(a); |x_2| \le a \le x_1\} = q(x_1) \sim x_1^{b-1} \sim \frac{r^b}{|x_2|} \qquad \text{as } x_1 \sim |x_2|.$$

Now let us consider squares Q_a with $a \ge x_1$ and $Q_a^+ = [0; a)$. Then

$$q(a) = \frac{a^{b-1}}{b+1}$$

is strictly decreasing in $a \ge x_1$. Therefore

$$\max\{q(a); x_1 \le a\} = \frac{x_1^{b-1}}{b+1} \le c\frac{r^b}{|x_2|}.$$

Combining this with the fact that $s(\mathbf{x}) \sim \frac{|x_2|^2}{r}$ we get

$$M\mu \sim \frac{r^{b-\frac{1}{2}}}{s^{\frac{1}{2}}}.$$

The case $b = 0$ is trivial.

(B) Let $x_1 > 0$, $|x_2| > x_1$, and $b \in (-1; 0]$. It suffices to consider squares Q_a with $a \ge |\mathbf{x}'|$ and $Q_a^+ = [0, a]$. But then obviously

$$\max\{q(a); a \ge |x_2|\} \sim q(|x_2|) \sim |x_2|^{b-1} \sim \frac{r^{b-\frac{1}{2}}}{s^{\frac{1}{2}}} \qquad \text{as } s(\mathbf{x}) \sim |x_2| \sim r.$$

(C) Let $x_1 > 0$ and $b \in (0; 1]$. It suffices to consider only squares Q_a such that $a \ge |x_2|$ and $Q_a^+ = [x_1; x_1 + a]$. Therefore

$$q(a) = \frac{1}{b+1}\frac{(x_1+a)^{b+1} - x_1^{b+1}}{a^2},$$

and since $q(a)$ is evidently decreasing, we get $\max\{q(a); a \geq |x_2|\} = q(|x_2|)$.

Now, if $|x_2| < x_1$, then $q(|x_2|) \sim \frac{x_1^b}{|x_2|} \sim \frac{r^{b-\frac{1}{2}}}{s^{\frac{1}{2}}}$. On the other side, if $|x_2| > x_1$, then $q(|x_2|) \sim |x_2|^{b-1} \sim \frac{r^{b-\frac{1}{2}}}{s^{\frac{1}{2}}}$.

(D) Let $x_1 < 0$. It suffices to consider only squares Q_a with $a \geq \max(|x_2|, |x_1|)$ and $Q_a^+ = [0; a - |x_1|]$. Then

$$q(a) = \frac{|x_1|^{b-1}}{b+1}\varphi\left(\frac{a}{|x_1|}\right) \quad \text{with } \varphi(y) = \frac{(y-1)^{b+1}}{y^2}.$$

The function $\varphi(y)$ vanishes at the point 1 and, if $b < 1$, at infinity. Thus for $b < 1$ there is a point $y_b > 1$ such that φ is maximal in y_b for $y \geq 1$. If $|x_2| \leq y_b \cdot |x_1|$, then

$$\max\{q(a) : a \geq |x_2|\} = q(y_b \cdot |x_1|) \sim x_1^{b-1} \sim \frac{r^{b-\frac{1}{2}}}{s^{\frac{1}{2}}} \quad \text{as } s \sim r \quad \text{for } x_1 < 0.$$

But if $|x_2| > y_b \cdot |x_1|$, then

$$\max\{q(a); a \geq |x_2|\} \simeq q(|x_2|) \sim |x_2|^{b-1} \sim \frac{r^{b-\frac{1}{2}}}{s^{\frac{1}{2}}}.$$

The case $b = 1$ is trivial since q is maximal for $a \to \infty$ yielding

$$\max\{q(a); a \geq |x_2|\} \sim 1 \sim \frac{r^{\frac{1}{2}}}{s^{\frac{1}{2}}}.$$

□

Using this result, we can now show the following:

Lemma 13.9 *For $b \in (-1; \frac{1}{2}]$ and $\hbar \in [0; 1)$ the function $w_1^{(2)} = \left(\nu_{-\frac{1}{2}}^{b-\frac{1}{2}}\right)^{\hbar}$ is a weight of class A_1 in $\mathbb{R}^2$.*

Proof We have to verify that $(Mw_1^{(2)})(\mathbf{x}) \leq Cw_1^{(2)}(\mathbf{x})$ a.e. in $\mathbb{R}^2$. Let Q_a denote, similarly as in the previous lemma, a closed square with sides parallel to the axes, and with the side length a; R will be a sufficiently large constant. We again distinguish several cases:

(A) $s(\mathbf{x}) \leq 1, \quad r = |\mathbf{x}| \geq R$.

α) $\quad a \leq \frac{1}{2} r^{\frac{1}{2}}$

Then for all $\mathbf{y} \in Q_a$ we have that $w_1^{(2)}(\mathbf{y}) \sim w_1^{(2)}(\mathbf{x})$ and

$$\int_{Q_a} w_1^{(2)}(\mathbf{y})\,\mathrm{d}\mathbf{y} \le C|Q_a|w_1^{(2)}(\mathbf{x}).$$

$\beta)$ $a = \frac{1}{2}r^{\frac{1}{2}+\sigma}$, $\sigma \in (0; \frac{1}{2}]$
Now $Q_a \subset \{\mathbf{y} \in \mathbb{R}^2;\ ||\mathbf{y}| - r| \le cr^{\frac{1}{2}+\sigma};\ s(\mathbf{y}) \le cr^{2\sigma}\}$, and we may use

$$\int_{Q_a} w_1^{(2)}(\mathbf{y})\,\mathrm{d}\mathbf{y} \le C\int_{r-cr^{\frac{1}{2}+\sigma}}^{r+cr^{\frac{1}{2}+\sigma}} \mathrm{d}\varrho\, \varrho^{\frac{1}{2}+(b-\frac{1}{2})\hbar} \int_0^{Cr^{2\sigma}} \frac{\mathrm{d}s}{s^{\frac{1}{2}}(1+s)^{\hbar}}$$

$$\le C|Q_a|r^{(b-\frac{1}{2})\hbar-\sigma\hbar} \le C|Q_a|w_1^{(2)}(\mathbf{x}) \quad \text{as } r = |\mathbf{x}| \ge R \gg 1,\ s(\mathbf{x}) \le 1.$$

$\gamma)$ $a \ge \frac{r}{2}$
In this case Q_a is contained in the ball B_{3a}. We use the following

$$\int_{\partial B_R} \nu_{-\beta}^{-\alpha}\,\mathrm{d}S \sim R^{1-\alpha}(R+1)^{-\min(\beta,\frac{1}{2})}\begin{cases}\ln R, & \beta = \frac{1}{2} \\ 1, & \beta \ne \frac{1}{2}\end{cases},\quad R \in (0; +\infty).$$

So we get

$$\int_{S_\tau} w_1^{(2)}\,\mathrm{d}S \sim \tau^{1+(b-\frac{1}{2})\hbar}(\tau+1)^{-\frac{\hbar}{2}},\ \tau \in (0; +\infty). \tag{13.129}$$

This implies for $-1 < b \le 1$, $R \le r \le 2a$

$$\int_{B_{3a}} w_1^{(2)}(\mathbf{y})\,\mathrm{d}S(\mathbf{y}) \le C\,|Q_a|a^{(b-1)\hbar} \le C\,|Q_a|r^{(b-1)\hbar} \le C\,|Q_a|r^{(b-\frac{1}{2})\hbar}.$$

(B) $s(\mathbf{x}) \ge 1$, $r \ge R$
Because $\hbar \ge 0$ we have

$$w_1^{(2)}(\mathbf{x}) = \left(\frac{r^{b-\frac{1}{2}}}{(1+s)^{\frac{1}{2}}}\right)^{\hbar} \le \left(\frac{r^{b-\frac{1}{2}}}{s^{\frac{1}{2}}}\right)^{\hbar} = w_0^{(2)}(\mathbf{x}).$$

Since $w_0^{(2)} \in A_1$, we have

$$\int_{Q_a} w_1^{(2)}(\mathbf{y})\,\mathrm{d}\mathbf{y} \le C\int_{Q_a} w_0^{(2)}(\mathbf{y})\,\mathrm{d}y \le C\,|Q_a|\,w_0^{(2)}(\mathbf{x}).$$

As $s(\mathbf{x}) \ge 1$, we have $w_0^{(2)}(\mathbf{x}) \le Cw_1^{(2)}(\mathbf{x})$.

(C) $r \leq R$:

α) If $\frac{a}{2} \leq R$, then easily

$$\int_{Q_a} w_1^{(2)}(\mathbf{y})\, \mathrm{d}\mathbf{y} \leq C|Q_a|\, w_1^{(2)}(\mathbf{x}).$$

Here we use the condition $b \leq \frac{1}{2}$ so $w_1^{(2)} \geq c' > 0$.

β) If $\frac{a}{2} > R$, then $Q_a \subset B_{3a}$; using (13.129) we have

$$\int_{B_{3a}} w_1^{(2)}(\mathbf{y})\, \mathrm{d}\mathbf{y} \leq C\, |Q_a|\, a^{(b-\frac{1}{2})\hbar}(1+a)^{-\frac{1}{2}\hbar}$$
$$\leq C_1\, |Q_a|\, r^{(b-\frac{1}{2})\hbar}(1+2R)^{-\frac{1}{2}\hbar} \leq$$
$$\leq C_1\, |Q_a|\, r^{(b-\frac{1}{2})\hbar}(1+s)^{-\frac{1}{2}\hbar}.$$

In the last two inequalities we used the conditions $b \leq \frac{1}{2}$, $\frac{a}{2} > R \geq r$.

□

Lemma 13.10 *For $b \in (-1; 1]$ and $\hbar \in [0; 1)$ the function $w_2^{(2)} = \left(\eta_{-\frac{1}{2}}^{b-\frac{1}{2}}\right)^{\hbar}$ is a weight of class A_1 in $\mathbb{R}^2$.*

Proof We proceed as in Lemma 13.9. The part (A) remains the same. In the part (B) we use that for some $C > 0$

$$\left(1+\frac{1}{r}\right)^{b-\frac{1}{2}} \leq \left(1+\frac{1}{r}\right)^{\frac{1}{2}} \leq C\left(1+\frac{1}{s}\right)^{\frac{1}{2}}$$

for all $\mathbf{y} \in \mathbb{R}^2$ such that $s(\mathbf{y}) \geq 1$, $r(\mathbf{y}) \geq R$, if $b \leq 1$. In the part Cα) the condition $w_2^{(2)} \geq C > 0$ for $|\mathbf{x}| \leq R$ is satisfied without the necessity of the additional condition $b \leq \frac{1}{2}$ unlike in Lemma 13.9. In the part Cβ) we proceed similarly as in Lemma 13.9, but we use

$$\int_{S_\tau} w_2^{(2)}\, \mathrm{d}S \sim \tau\, (\tau+1)^{1+(b-1)\hbar}, \quad \tau \in (0; +\infty)$$

instead of (13.129). Also here we do not need the condition $b \leq \frac{1}{2}$. This is why we get Lemma 13.10 under less restrictive condition on b than in Lemma 13.9. □

Combining Lemmas 13.8, 13.9, 13.10, and 1.6 with the properties of Muckenhoupt classes A_p we can show the following:

Theorem 13.19 *Let $-\frac{1}{2} < \beta < \frac{1}{2}(p-1)$, $-2 < \alpha+\beta < 2(p-1)$. Then the weights $\eta_\beta^\alpha(\mathbf{x})$ and $\sigma_\beta^\alpha(\mathbf{x})$ are A_p-weights in $\mathbb{R}^2$ for $p \in (1; \infty)$.*

Let $-\frac{1}{2} < \beta < \frac{1}{2}(p-1)$, $-2 < \alpha + \beta < 2(p-1)$, $-2 < \alpha < 2(p-1)$. *Then the weight* $\nu_\beta^\alpha(\mathbf{x})$ *is* A_p*-weight in* $\mathbb{R}^2$ *for* $p \in (1; \infty)$.

Now we shall concentrate on the three-dimensional case.

Lemma 13.11 *For* $b \in (-1; 2]$ *and* $\hbar \in [0; 1)$ *the function* $w_0^{(3)}(\mathbf{x}) = \left(\frac{|\mathbf{x}|^{b-1}}{s(\mathbf{x})}\right)^\hbar$, $\mathbf{x} \in \mathbb{R}^3$ *is a weight of the class* A_1 *in* $\mathbb{R}^3$.

The proof is analogous to the proof of Lemma 13.8. For the proof see also [58].

Lemma 13.12 *For* $b \in (-1; 1]$ *and* $\hbar \in [0; 1)$ *the function* $w_1^{(3)} = \left(\nu_{-1}^{b-1}\right)^\hbar$ *is a weight of class* A_1 *in* $\mathbb{R}^3$.

The proof of the lemma is analogous to the proof of Lemma 13.10. But in the part Aβ) and Cβ) the estimates are slightly less technical than in Lemma 13.10.

Lemma 13.13 *For* $b \in (-1; 2]$ *and* $\hbar \in [0; 1)$ *the function* $w_2^{(3)} = \left(\eta_{-1}^{b-1}\right)^\hbar$ *is a weight of class* A_1 *in* $\mathbb{R}^3$.

Proof We proceed as in Lemma 13.9. Part (A) remains the same. In part (B) we use the fact that for some $C > 0$

$$\left(1 + \frac{1}{r}\right)^{b-1} \leq \left(1 + \frac{1}{r}\right) \leq C\left(1 + \frac{1}{s}\right)$$

for all $\mathbf{y} \in \mathbb{R}^3$ such that $s(\mathbf{y}) \geq 1$, $r(\mathbf{y}) \geq R$, if $b \leq 2$. In part Cα) the condition $w_2^{(3)} \geq C > 0$ for $|\mathbf{x}| \leq R$ is satisfied without the necessity of any additional condition on b. In part Cβ) we proceed as in Lemma 13.9, but we use

$$\int_{S_\tau} w_2^{(3)}\,\mathrm{d}S \sim \tau^2\,(\tau + 1)^{1+(b-2)\hbar}.$$

□

Combining Lemmas 13.11, 13.12, and 13.13 with the properties of Muckenhoupt classes A_p, we can show the following:

Theorem 13.20 *Let* $-1 < \beta < p-1$, $-3 < \alpha+\beta < 3(p-1)$. *Then the weights* $\eta_\beta^\alpha(\mathbf{x})$ *and* $\sigma_\beta^\alpha(\mathbf{x})$ *are* A_p*-weights in* $\mathbb{R}^3$ *for* $p \in (1; \infty)$.

Let $-1 < \beta < p-1$, $-3 < \alpha + \beta < 3(p-1)$, $-3 < \alpha < 3(p-1)$. *Then the weight* $\nu_\beta^\alpha(\mathbf{x})$ *is* A_p*-weight in* $\mathbb{R}^3$ *for* $p \in (1; \infty)$.

In order to formulate the fundamental theorem used in this section, we need to define the so-called L^∞-Dini condition. We will use the following notation in the definition.

ρ - rotation of $\partial B_1(\mathbf{0})$ with magnitude $|\rho| = \sup\limits_{\mathbf{x}\in\partial B_1(\mathbf{0})} |\rho\mathbf{x} - \mathbf{x}|$.

Definition Let Ω be a function defined on $\partial B_1(\mathbf{0})$, $\Omega \in L^\infty(\partial B_1(\mathbf{0}))$. We say that function Ω satisfies the L^∞-Dini condition ($\Omega \in L^\infty$-Dini) if

$$\int_0^1 \theta_\infty(\delta)\frac{1}{\delta}\mathrm{d}\delta < +\infty\,, \quad \text{where } \theta_\infty(\delta) := \sup_{|\rho|<\delta} \|\Omega(\rho\mathbf{x}) - \Omega(\mathbf{x})\|_{L^\infty(\partial B_1(\mathbf{0}))}.$$

Theorem 13.21 *Let $N \in \mathbb{N}$, $N \geq 2$, $\Omega \in L^\infty(\partial B_1(\mathbf{0}))$, $\Omega \in L^\infty$-Dini, $\int_{\partial B_1(\mathbf{0})} \Omega dS = 0$, Ω is a positively homogeneous function of degree zero, $R(\mathbf{x}) := \Omega(\mathbf{x}')/|\mathbf{x}|^N$, $\mathbf{x}' = \mathbf{x}/|\mathbf{x}|$.*
Let T be an operator with the kernel R, i.e., $Tf(\mathbf{x}) = (R * f)(\mathbf{x})$ in the principal-value sense and $w \in A_p$ in $\mathbb{R}^N$, $p > 1$. Then T is a continuous operator $\mathcal{L}^p(\mathbb{R}^N; w) \mapsto L^p(\mathbb{R}^N; w)$.

For a proof of the theorem, see [159].

Remark 13.5 The fact that $R(\cdot)$ satisfies the conditions formulated in Theorem 13.21 means that $R(\cdot)$ represents a Calderón–Zygmund singular integral kernel. We will write in this case $R(\cdot) \in CZ$. It is known that $\nabla^2\boldsymbol{\mathcal{S}}$, $\nabla\boldsymbol{\mathcal{P}} \in CZ$ in both the two- and three-dimensional cases.

Remark 13.6 If w is a weight such that $f \in L^p(\mathbb{R}^N; w) \Rightarrow f \in L^1_{loc}(\mathbb{R}^N)$ and $1 \leq p < +\infty$, then it is an easy matter to see that $\mathcal{L}^p(\mathbb{R}^N; w)$ and $L^p(\mathbb{R}^N; w)$ coincide. The weights η^α_β have this property. For the other weights we will distinguish spaces $\mathcal{L}^p(\mathbb{R}^N; w)$ and $L^p(\mathbb{R}^N; w)$.

As the corollary of Theorem 13.21 we get:

Corollary *Let $N = 2, 3$, R be either $\partial_i\partial_j\mathcal{S}_{rs}$ or $\partial_i\mathcal{P}_r$, $i, j, r, s = 1, 2, ...N$ and $f \in \mathcal{L}^p(\mathbb{R}^N; w)$, where w stands for η^α_β, σ^α_β, or ν^α_β with $1 < p < \infty$ and let α, β be such that the corresponding weights are A_p-weights. Then v.p.$(R*f) \in L^p(\mathbb{R}^N; w)$ and*

$$\|v.p.\ (R * f)\|_{p,(w),\mathbb{R}^N} \leq C\|f\|_{p,(w),\mathbb{R}^N}.$$

We will now formulate two theorems about $\nabla\boldsymbol{\mathcal{P}}$ both of which follow from Theorem 13.21.

Theorem 13.22 $\nabla\boldsymbol{\mathcal{P}} : f \mapsto \nabla\boldsymbol{\mathcal{P}} * f$ *defines continuous operators:*

$$(a) \qquad L^p(\mathbb{R}^3; \eta_\beta^{\alpha+p/2}) \longmapsto L^p(\mathbb{R}^3; \eta_\beta^{\alpha+p/2})$$

for $1 < p < \infty,\ -1 < \beta < p-1,\ -3-p/2 < \alpha+\beta < 5p/2-3$

(*b*) $$\mathcal{L}^p(\mathbb{R}^3;\nu_\beta^{\alpha+p/2}(\cdot\,;\lambda)) \longmapsto L^p(\mathbb{R}^3;\nu_\beta^{\alpha+p/2})$$

for $1 < p < \infty,\ -1 < \beta < p-1,\ -3-p/2 < \alpha < 5p/2-3,\ -3-p/2 < \alpha+\beta < 5p/2-3$.

Theorem 13.23 $\nabla\mathcal{P} : f \mapsto \nabla\mathcal{P} * f$ *defines continuous operators:*

(a)

$$L^p(\mathbb{R}^2;\eta_\beta^{\alpha+p/2}) \longmapsto L^p(\mathbb{R}^2;\eta_\beta^{\alpha+p/2}) \tag{13.130}$$

for $1 < p < \infty,\ -\frac{1}{2} < \beta < \frac{1}{2}(p-1),\ -2-p/2 < \alpha+\beta < 3p/2-2$.

(b)

$$\mathcal{L}^p(\mathbb{R}^2;\nu_\beta^{\alpha+p/2}(\cdot\,;\lambda)) \longmapsto L^p(\mathbb{R}^2;\nu_\beta^{\alpha+p/2}) \tag{13.131}$$

for $1 < p < \infty,\ -\frac{1}{2} < \beta < \frac{1}{2}(p-1),\ -2-p/2 < \alpha < 3p/2-2,\ -2-p/2 < \alpha+\beta < 3p/2-2$.

The next two theorems follow from Theorems 13.13, 13.18, 13.19, 13.20, and Corollary 13.3.

Theorem 13.24 *Let* T *be an integral operator in the principal-value sense with the kernel* $\partial_k\partial_l\mathcal{O}_{ij}(\cdot\,;\lambda)$, $i,j,k,l = 1,2,3$, $T : f \mapsto R * f$, *and let* $1 < p < \infty$. *Then* T *is a well-defined continuous operator:*

(*a*) $$L^p(\mathbb{R}^3;\eta_\beta^{\alpha+p/2}(\cdot\,;\lambda)) \longmapsto L^p(\mathbb{R}^3;\eta_\beta^{\alpha+p/2-\varepsilon}(\cdot\,;\lambda))$$
for $\max\{-1,-\varepsilon(p-1)/p\} < \beta < p-1,\ -p/2-1-\varepsilon/p < \alpha+\beta < 5p/2-3$, $\alpha-\beta < p/2-1+\varepsilon+\varepsilon(p-1)/p,\ -3p/2-\varepsilon/p < \alpha < 3p/2-2+\varepsilon$, $0 < \varepsilon \le p$.

(*b*) $$\mathcal{L}^p(\mathbb{R}^3;\nu_\beta^{\alpha+p/2}(\cdot\,;\lambda)) \longmapsto L^p(\mathbb{R}^3;\mu_\beta^{\alpha+p/2-\varepsilon,\alpha+p/2}(\cdot\,;\lambda)).$$

$L^p(\mathbb{R}^3\setminus\Omega;\nu_\beta^{\alpha+p/2}(\cdot\,;\lambda)) \longmapsto L^p(\mathbb{R}^3;\nu_\beta^{\alpha+p/2-\varepsilon}(\cdot\,;\lambda))$ *for* $\max\{-1,-\varepsilon(p-1)/p\} < \beta < p-1,\ -p/2-1-\varepsilon/p < \alpha+\beta < 5p/2-3$, $\alpha-\beta < p/2-1+\varepsilon+\varepsilon(p-1)/p$, $\max\{-p/2-3,-3p/2-\varepsilon/p\} < \alpha < \min\{5p/2-3, 3p/2-2+\varepsilon\}$, $0 < \varepsilon \le p$, $\Omega \subset \mathbb{R}^3$*—an arbitrary domain,* $\mathbf{0} \in \Omega$.

Moreover, we have for α, β *specified in* (*a*) *and* (*b*), *respectively:*

ad a)

$$\left\| v.p.\ (\partial_k\partial_l\mathcal{O}_{ij}(\cdot\,;\lambda) * f)\right\|_{p,(\eta_\beta^{\alpha+p/2-\varepsilon}(\cdot\,;\lambda)),\mathbb{R}^3} \le C\ \|f\|_{p,(\eta_\beta^{\alpha+p/2}(\cdot\,;\lambda)),\mathbb{R}^3}. \tag{13.132}$$

ad b)

$$\left\| v.p.\, (\partial_k \partial_l \mathcal{O}_{ij}(\cdot\,;\lambda) * f) \right\|_{p,(\nu_\beta^{\alpha+p/2-\varepsilon}(\cdot\,;\lambda)),\mathbb{R}^3} \le C\, \|f\|_{p,(\mu_\beta^{\alpha+p/2-\varepsilon,\alpha+p/2}(\cdot\,;\lambda)),\mathbb{R}^3} \tag{13.133}$$

$$\|v.p.\, (\partial_k \partial_l \mathcal{O}_{ij}(\cdot\,;\lambda) * f)\|_{p,(\nu_\beta^{\alpha+p/2-\varepsilon}(\cdot;\lambda)),\mathbb{R}^3\setminus\Omega} \le C\, \lambda^{\frac{\varepsilon}{p}}\, \|f\|_{p,(\nu_\beta^{\alpha+p/2}(\cdot\,;\lambda)),\mathbb{R}^3}\,. \tag{13.134}$$

Theorem 13.25 *Let T be an integral operator in the value-principal sense with the kernel $\partial_k \partial_l \mathcal{O}_{ij}(\cdot\,;\lambda)$, $i, j, k, l = 1, 2$, $T : f \mapsto R * f$, and let $1 < p < \infty$. Then T is a well-defined continuous operator:*

(a) $$L^p(\mathbb{R}^2; \eta_\beta^{\alpha+p/2}(\cdot\,;\lambda)) \longmapsto L^p(\mathbb{R}^2; \eta_\beta^{\alpha+p/2-\varepsilon}(\cdot\,;\lambda))$$
for $\max\{-1/2, -\varepsilon(p-1)/p\} < \beta < p/2-1/2$, $-p/2-1-\varepsilon/p < \alpha+\beta < 3p/2-2$, $\alpha-\beta < p/2-1+\varepsilon+\varepsilon(p-1)/p$, $-p-1/2-\varepsilon/p < \alpha < p-3/2+\varepsilon$, $0 < \varepsilon \le p/2$.

(b) $$\mathcal{L}^p(\mathbb{R}^2; \nu_\beta^{\alpha+p/2}(\cdot\,;\lambda)) \longmapsto L^p(\mathbb{R}^2; \mu_\beta^{\alpha+p/2-\varepsilon,\alpha+p/2}(\cdot\,;\lambda))$$

$$L^p(\mathbb{R}^2 \setminus \Omega; \nu_\beta^{\alpha+p/2}(\cdot\,;\lambda)) \longmapsto L^p(\mathbb{R}^2; \nu_\beta^{\alpha+p/2-\varepsilon}(\cdot\,;\lambda))$$

for $\max\{-1/2, -\varepsilon(p-1)/p\} < \beta < p/2-1/2$, $-p/2-1-\varepsilon/p < \alpha+\beta < 3p/2-2$, $\alpha-\beta < p/2-1+\varepsilon+\varepsilon(p-1)/p$, $\max\{-p/2-2, -p-1/2-\varepsilon/p\} < \alpha < \min\{3p/2-2, p-3/2+\varepsilon\}$, $0 < \varepsilon \le p/2$, $\Omega \subset \mathbb{R}^2$*—an arbitrary domain,* $\mathbf{0} \in \Omega$.

Moreover, we have for α, β specified in (a) and (b), respectively:

ad a)

$$\left\| v.p.\, (\partial_k \partial_l \mathcal{O}_{ij}(\cdot\,;\lambda) * f) \right\|_{p,(\eta_\beta^{\alpha+p/2-\varepsilon}(\cdot\,;\lambda)),\mathbb{R}^2} \le C\, \|f\|_{p,(\eta_\beta^{\alpha+p/2}(\cdot\,;\lambda)),\mathbb{R}^2}\,. \tag{13.135}$$

ad b)

$$\left\| v.p.\, (\partial_k \partial_l \mathcal{O}_{ij}(\cdot\,;\lambda) * f) \right\|_{p,(\nu_\beta^{\alpha+p/2-\varepsilon}(\cdot\,;\lambda)),\mathbb{R}^2} \le C\, \|f\|_{p,(\mu_\beta^{\alpha+p/2-\varepsilon,\alpha+p/2}(\cdot\,;\lambda)),\mathbb{R}^2}, \tag{13.136}$$

$$\|v.p.\, (\partial_k \partial_l \mathcal{O}_{ij}(\cdot\,;\lambda) * f)\|_{p,(\nu_\beta^{\alpha+p/2-\varepsilon}(\cdot;\lambda)),\mathbb{R}^2\setminus\Omega} \le C\, \lambda^{\frac{\varepsilon}{p}}\, \|f\|_{p,(\nu_\beta^{\alpha+p/2}(\cdot\,;\lambda)),\mathbb{R}^2}\,. \tag{13.137}$$

References

1. Abada, N., Boulmezaoud,T.Z., Kerdid, N.: The Stokes flow around a rotating body in the whole space. J. Math. Soc. Japan **65**, 607–632 (2013)
2. Adams, R.A.: Sobolev spaces. J. Pure Appl. Math. **65**, xviii+268 pp. (1975). Academic Press [A subsidiary of Harcourt Brace Jovanovich, Publishers], New York-London
3. Alt, H.W.: Lineare Funktionalanalysis. Springer, Heidelberg (1985)
4. Aldoss, T.K., Abou-Arab, T.W.: Experimental study of the flow around a rotating cylinder in crossflow. Exp. Thermal Fluid Sci. **3**(3), 316–322 (1990)
5. Amann, H.: On the strong solvability of the Navier-Stokes equations. J. Math. Fluid Mech. **2**, 1–98 (2000)
6. Amann, H.: Nonhomogeneous Navier-Stokes equations with integrable low- regularity data. In: Nonlinear Problems in Mathematical Physics and Related Topics, II, pp. 1–28. Kluwer/Plenum, New York. The International Mathematical Series (2002)
7. Amrouche, C., Girault, V.: Decomposition of vector spaces and application to the Stokes problem in arbitrary dimension. Czech. Math. J. **44**, 109–140 (1994)
8. Amrouche, C., Consiglieri, L.: On the stationary Oseen equations in $\mathbb{R}^3$. Comm. Math. Anal. **10**, 5–29 (2010)
9. Amrouche, C., Rodriguez-Bellido, M.A.: Stationary Stokes, Oseen and Navier-Stokes equations with singular data. Arch. Ration. Mech. Anal. **199**, 597–651 (2011)
10. Rosa da Silva, A., Aristeu da Silveira, N., Marcos G. de Lima, A., Alves Rade, D.: Numerical simulations of flows over a rotating circular cylinder using the immersed boundary method. J. Braz. Soc. Mech. Sci. Eng. **33**(1), (2011). Rio de Janeiro Jan./Mar.
11. Babenko, K.I.: On stationary solutions of the problem of flow past a body of a viscous incompressible fluid. Mat. Sb. **91**, 3–26 (1973) (Russian); English translation: Math. USSR-Sbornik **20**, 1–25 (1973)
12. Babenko, K.I.: Spectrum of the linearized problem of flow of a viscous incompressible liquid round a body. Sov. Phys. Dokl. **27**, 25–27 (1982)
13. Bensoussan, A., Da Prato, G., Delfour, M., Mitter, S.K.: Representation and control of infinite-dimensional systems. In: Systems & Control: Foundations & Applications, vol. II. Birkhäuser Boston, Inc., Boston (1993)
14. Bergh, J., Löfström, J.: Interpolation Spaces. Springer, New York (1976)
15. Bogovskii, M.E.: Solution of the first boundary value problem for the equation of continuity of an incompressible medium. Sov. Math. Dokl. **20**, 1094–1098 (1979)

Š. Nečasová et al., *Mathematical Theory of a Fluid Flow Around a Rotating and Translating Body*, Springer Monographs in Mathematics,
https://doi.org/10.1007/978-3-031-84120-0

16. Bogovskii, M.E.: Solution of some vector analysis problems connected with operators div and grad. (In Russian.) Trudy Seminar S.L. Sobolev **1**, 80. Akademia Nauk SSR, Sibirskoe Otdelenie Matematiki, Novosibirsk, 5–40 (1980)
17. Borchers, W.: Zur Stabilität und Faktorisienrungsmethode für die Navier-Stokes Gleichungen inkompressibler viskoser Flüssigkeiten. University of Paderborn, Habilitationsschrift (1992)
18. Borchers, W., Sohr, H.: On the equations rot $v = g$ and $\operatorname{div} v = f$ with zero boundary conditions. Hokkaido Math. J. **19**, 67–87 (1990)
19. Boyer, F., Fabrie, P.: Outflow boundary conditions for the incompressible non-homogeneous Navier–Stokes equations. DCDS—Series B **7**, 219–250 (2007)
20. Braack, M., Mucha, P.B.: Directional do-nothing condition for the Navier–Stokes equations. J. Comp. Math. **32**, 507–521 (2014)
21. Bruhat, F.: Distributions sur un groupe localement compact et applications l'étude des représentations des groupes P-adiques. Bull. Soc. Math. France **89**, 43–75 (1961)
22. Cattabriga, L.: Su un problema al contorno relativo al sistema di equazioni di Stokes. (Italian) Rend. Sem. Mat. Univ. Padova **31**, 308–340 (1961)
23. Chang, I.,Finn, R.: On the solutions of a class of equations occurring in continuum mechanics, with application to the Stokes paradox. Arch. Rational Mech. Anal. **7**, 388–401 (1961)
24. Clark, D.C.: The vorticity at infinity for solutions of the stationary Navier-Stokes equations in exterior domains. Indiana Univ. Math. J. **20**, 633–654 (1970/71)
25. Chemetov, N.V., Nečasová, Š., Muha, B.: Weak-strong uniqueness for fluid-rigid body interaction problem with slip boundary condition. J. Math. Phys. **60**(1) (2019)
26. Cumsille, P.,Tucsnak, M.: Wellposedness for the Navier-Stokes flow in the exterior of a rotating obstacle. Math. Method Appl. Sci. **29**, 595–623 (2006)
27. Da Prato, G., Lunardi, A.: On the Ornstein-Uhlenbeck operator in spaces of continuous functions. J. Funct. Anal. **131**, 94–114 (1995)
28. de Leeuw K.: On L_p multipliers. Ann. Math. **2**, 364–274 (1965)
29. Deuring, P.: Pointwise spatial decay of time-dependent Oseen flows: the case of data with noncompact support. Discrete Contin. Dyn. Syst. **33**(7), 2757–2776 (2013)
30. Deuring, P.: Finite element methods for the Stokes system in three-dimensional exterior domains. Math. Methods Appl. Sci. **20**, 245–269 (1997)
31. Deuring, P.: Spatial decay of time-dependent incompressible Navier-Stokes flows with nonzero velocity at infinity. SIAM J. Math. Anal. **45**(3), 1388–1421 (2013)
32. Deuring, P.: The single-layer potential associated with the time-dependent Oseen system. In: Proceedings of the 2006 IASME/WSEAS International Conference on Continuum Mechanics. Chalkida, Greece, May 11–13, pp. 117–125 (2006)
33. Deuring, P., Kračmar, S.: Artificial boundary conditions for the Oseen system in 3D exterior domains. Analysis **20**, 65–90 (2000)
34. Deuring, P., Kračmar, S.: Exterior stationary Navier-Stokes flows in 3D with non-zero velocity at infinity: approximation by flows in bounded domains. Math. Nachr. **269–270**, 86–115 (2004)
35. Deuring, P., Kračmar, S., Nečasová, Š.: A representation formula for linearized stationary incompressible viscous flows around rotating and translating bodies. Discrete Contin. Dyn. Syst., Ser. S **3**, 237–253 (2010)
36. Deuring, P., Kračmar, S., Nečasová, Š.: On pointwise decay of linearized stationary incompressible viscous flow around rotating and translating bodies. SIAM J. Math. Anal. **43**, 705–738 (2011)
37. Deuring, P. Kračmar, S., Nečasová, Š.: A linearized system describing stationary incompressible viscous flow around rotating and translating bodies: improved decay estimates of the velocity and its gradient. DCDS, Supplement, pp. 351–361 (2011)
38. Deuring, P., Kračmar, S., Nečasová, Š.: Linearized stationary incompressible flow around rotating and translating bodies: asymptotic profile of the velocity gradient and decay estimate of the second derivatives of the velocity. J. Differ. Equations **252**, 459–476 (2012)

39. Deuring, P., Kračmar, S., Nečasová, Š.: Pointwise decay of stationary rotational viscous incompressible flows with nonzero velocity at infinity. J. Differ. Equations **255**, 1576–1606 (2013)
40. Deuring, P., Kračmar, S., Nečasová, Š.: Linearized stationary incompressible flow around rotating and translating bodies—Leray solution. Discrete Contin. Dyn. Syst. Ser. S **7**, 967–979 (2014)
41. Deuring, P., Kračmar, S., Nečasová, Š.: Leading terms of the velocity and its gradients of the stationary rotational viscous incompressible flows with nonzero velocity at infinity. Discrete Contin. Dyn. Syst. **37**(3), 1389–1409 (2017)
42. Deuring, P., Kračmar, S., Nečasová, Š.: Asymptotic structure of viscous incompressible flow around a rotating body, with nonvanishing flow field at infinity. Z. Angew. Math. Phys. **68**(1), Art. 16, 15 pp. (2017)
43. Deuring, P., Kračmar, S., Nečasová, Š.: Artificial boundary conditions for linearized stationary incompressible viscous flow around rotating and translating body. Math. Nachr. **294**(1), 56–73 (2021)
44. Deuring, P., Kračmar, S., Nečasová, Š.,: Note to the problem of asymptotic behavior of viscous incompressible flow around a rotating body. C.R. Math. **354**(8), 794–798 (2016)
45. Deuring, P., Varnhorn, W.: On Oseen resolvent estimates. Differ. Integr. Equat. **23**, 1139–1149 (2010)
46. Deuring, P.: Exterior stationary Navier-Stokes flows in 3D with nonzero velocity at infinity: asymptotic behavior of the second derivatives of the velocity. Commun. Partial Differ. Equations **30**(7–9), 987–1020 (2005)
47. Dintelmann, E., Geissert, M., Hieber, M.: Strong L^p solutions to the Navier-Stokes flow past moving obstacles: the case of several obstacles and time dependent velocity. Trans. Amer. Math. Soc. **361**, 653–669 (2009)
48. Dutto, P.: Solutions Physiquement Raisonnables des équations de Navier–Stokes Compressibles Stationnaires dans un domaine extérieur du Plan. Ph.D. Thesis, University of Toulon, France (1998)
49. Edwards, R., Gaudry, G.: Littlewood-Paley and Multiplier Theory. Springer, Berlin (1977)
50. Eidelman, S.D.: On fundamental solutions of parabolic systems II. Mat. Zbornik **53**, 73–136 (1961) (Russian)
51. Eiter, T., Kyed, M.: Estimates of time-periodic fundamental solutions to the linearized Navier-Stokes equations. J. Math. Fluid Mech. **20**(2), 517–529 (2018)
52. Eiter, T., Kyed, M.: Time-periodic linearized Navier-Stokes equations: an approach based on Fourier multipliers. Particles in flows, 77–137, Adv. Math. Fluid Mech., Birkhaäuser/Springer, Cham (2017)
53. Eiter, T.: On the Oseen-type resolvent problem associated with time-periodic flow past a rotating body. SIAM J. Math. Anal. **54**(4), 4987–5012 (2022)
54. Eiter, T.: On the Stokes-type resolvent problem associated with time-periodic flow around a rotating obstacle. J. Math. Fluid Mech. **24**(2), Paper No. 52, 17 pp. (2022)
55. Eiter, T., Kyed, M.: Viscous flow around a rigid body performing a time-periodic motion. J. Math. Fluid Mech. **23**(1), Paper No. 28, 23 pp. (2021)
56. Engel, K.J., Nagel, R.: One-Parameter Semigroups for Linear Evolution Equations. Texts in Mathematics, vol. 194. Springer, Berlin (2000)
57. Farwig R.: Das stationäre Außenraumproblem der Navier-Stokes-Gleichungen bei nicht-verschwindender Anströmgeschwindigkeit in anisotrop gewichteten Sobolevräumen. SFB 256 preprint no. **110** (Habilitationsschrift), University of Bonn, Bonn (1990)
58. Farwig, R.: The stationary exterior 3D-problem of Oseen and Navier-Stokes equations in anisotropically weighted Sobolev spaces. Math. Z. **211**, 409–447 (1992)
59. Farwig, R.: The stationary Navier-Stokes equations in a 3D—exterior domain. Recent Topics on Mathematical Theory of Viscous Incompressible Fluid, Lecture Notes in Num. Appl. Anal. **16**, 53–115 (1998)

60. Farwig, R.: Estimates of lower order derivatives of viscous fluid flow past a rotating obstacle. Regularity and other aspects of the Navier-Stokes equations. Banach Center Warsaw **70**, 73–84 (2005). Polish Academy of Sciences, Warsaw
61. Farwig, R.: An L^q-analysis of viscous fluid flow past a rotating obstacle. Tôhoku Math. J. **920**, 129–147 (2006)
62. Farwig, R., Neustupa, J.: On the spectrum of a Stokes-type operator arising from flow around a rotating body. Manuscr. Math. **122**, 419–437 (2007)
63. Farwig, R., Galdi, G.P., Kyed, M.: Asymptotic structure of a Leray solution to the Navier-Stokes flow around a rotating body. Pacific J. Math. **253**, 367–382 (2011)
64. Farwig, R., Galdi G.P., Sohr, H.: A new class of very weak solutions of the Navier-Stokes equations with nonhomogeneous data. J. Math. Fluid Mech. **8**, 423–444 (2006)
65. Farwig, R., Guenther, R.B., Nečasová, Š., Thomann, E.A.: The fundamental solution of the linearized instationary Navier-Stokes equations of motion around a rotating and translating body. Discrete Contin. Dyn. Syst. **34**, 511–529 (2014)
66. Farwig, R., Hishida, T.: Stationary Navier-Stokes flow around a rotating obstacles. Funkcial. Ekvac. **50**, 371–403 (2007)
67. Farwig, R., Hishida, T.: Asymptotic profiles of steady Stokes and Navier-Stokes flows around a rotating obstacle. Ann. Univ. Ferrara, Sez. VII **55**, 263–277 (2009)
68. Farwig, R., Hishida, T.: Asymptotic profile of steady Stokes flow around a rotating obstacle. Manuscripta Mathematica **136**, 315–338 (2011)
69. Farwig, R., Hishida, T.: Leading term at infinity of steady Navier-Stokes flow around a rotating obstacle. Math. Nachr. **284**, 2065–2077 (2011)
70. Farwig, R., Hishida, T., Müller, D.: L^q-Theory of a singular "winding" integral operator arising from fluid dynamics. Pacific J. Math. **215**, 297–312 (2004)
71. Farwig, R., Kozono, H., Sohr, H.: Very weak solutions of the Navier-Stokes equations in exterior domains with nonhomogeneous data. J. Math. Soc. Japan **59**, 127–150 (2007)
72. Farwig, R., Krbec, M., Nečasová, Š.: A weighted L^q approach to Oseen flow around a rotating body. Math. Methods Appl. Sci. **31**, 551–574 (2008)
73. Farwig, R., Krbec, M., Nečasová, Š.: A weighted L^q approach to Stokes flow around a rotating body. Annali di Ferrara **54**, 61–84 (2008)
74. Farwig, R., Nečasová, Š., Neustupa, J.: On the essential spectrum of a Stokes-type operator arising from flow around a rotating body in the L^q-framework. Kyoto Conference on the Navier-Stokes Equations and their Applications, pp. 93–105 (2007). RIMS Kukyuroku Bessatsu, B1, Research Institute for Mathematical Sciences (RIMS), Kyoto.
75. Farwig, R., Nečasová, Š., Neustupa, J.: Spectral analysis of a Stokes-type operator arising from flow around a rotating body. J. Math. Soc. Japan **63**, 163–194 (2011)
76. Farwig, R., Neustupa, J.: On the spectrum of an Oseen-type operator arising from flow past a rotating body. Integral Equations Oper. Theory **62**, 169–189 (2008)
77. Farwig, R., Neustupa, J.: Spectral properties in L^q of an Oseen operator modelling fluid flow past a rotating body. Tohoku Math. J. **62**, 287–309 (2010)
78. Farwig, R., Sohr, H.: An approach to resolvent estimates for the Stokes system in L^q spaces. Lect. Notes Math. **1530**, 97–110 (1992)
79. Farwig, R., Sohr, H.: Generalized resolvent estimates for the Stokes system in bounded and unbounded domains. J. Math. Soc. Japan **48**, 607–643 (1994)
80. Farwig, R., Sohr, H.: Weighted estimates for the Oseen equations and the Navier-Stokes equations in exterior domains. In: Theory of the Navier-Stokes Equations. Series on Advances in Mathematics for Applied Sciences, vol. 47, pp. 11–30. World Scientific Publishing, River Edge, NJ (1998)
81. Farwig, R., Sohr, H.: The stationary and nonstationary Stokes system in exterior domains with nonzero divergence and nonzero boundary values. Math. Methods Appl. Sci. **17**(4), 269–291 (1994)
82. Finn, R.: Estimates at infinity for stationary solution of Navier-Stokes equations. Bult. Math. de la Soc. Sci. Math. de la R.P.R. Tome 3 **53**, 387–418 (1959)

83. Finn, R.: On the exterior stationary problem for the Navier-Stokes equations, and associated perturbation problems. Arch. Rational Mech. Anal. **19**, 363–406 (1965)
84. Finn, R.: Estimates at infinity for stationary solutions of the Navier-Stokes equations. Bull. Math. Soc. Sci. Math. Phys. R. P. Roumaine (N.S.) **3**(51), 387–418 (1959)
85. Finn, R.: Estimates at infinity for steady state solutions of the Navier-Stokes equations. In: Proceedings of Symposia in Pure Mathematics, vol. IV, pp. 143–148. American Mathematical Society, Providence, R.I. (1961)
86. Feng, J., Hu, H.H., Joseph, D.D.: Direct simulation of initial value problems for the motion of solid bodies in a Newtonian Fluid. Part 2. Couette and Poiseuille flows. J. Fluid Mech. **277**, 271–301 (1994)
87. Finn, R.: Estimates at infinity for stationary solutions of the Navier-Stokes equations. Bull. Math. Soc. Sci. Math. Phys. R. P. Roumaine **3**, 387–418 (1959)
88. Fröhlich A.: Solutions of the Navier-Stokes initial value problem in weighted L^q-spaces. Math. Nachr. **150–166**, 269–270 (2004)
89. Fröhlich, A.: The Stokes operator in weighted L^q-spaces II: Weighted resolvent estimates and maximal L^p-regularity. Math. Ann. **339**, 287–316 (2007)
90. Fujita, H., Kato, T.: On the Navier-Stokes initial value problem. I Arch. Rational Mech. Anal. **16**, 269–315 (1964)
91. Galdi, G.P., Neustupa, J.: Spectral instability of a steady flow of an incompressible fluid past a rotating obstacle. Math. Annal. **382**(1-2), 357–382 (2022)
92. Galdi, G.P.: Existence, uniqueness and asymptotic behavior of regular time-periodic solutions to the Navier-Stokes equations around a moving body: rotational case. Indiana Univ. Math. J. **71**(6), 2259–2281 (2022)
93. Galdi, G.P.: An introduction to the mathematical theory of the Navier-Stokes equations: Linearised steady problems. In: Springer Tracts in Natural Philosophy, vol. 38, 2nd edn. Springer, Berlin (1998)
94. Galdi, G.P.: An introduction to the mathematical theory of the Navier-Stokes equations. In: Nonlinear Steady Problems, vol. II. Springer, New York e.a. (1994)
95. Galdi, G.P.: An Introduction to the mathematical theory of the Navier-Stokes equations. In: Linearized Steady Problems, vol. I (rev. ed.). Springer, New York e.a. (1998)
96. Galdi, G.P.: On the motion of a rigid body in a viscous liquid: A mathematical analysis with applications. In: Friedlander, S., Serre, D. (eds.) Handbook of Mathematical Fluid Dynamics, vol. 1. Elsevier, Amsterdam (2002)
97. Galdi, G.P.: Steady flow of a Navier-Stokes fluid around a rotating obstacle. Essays and papers dedicated to the memory of Clifford Ambrose Truesdell III, II. J. Elasticity **71**, 1–31 (2003)
98. Galdi, G.P.: An Introduction to the Mathematical Theory of the Navier-Stokes Equations. Steady-State Problems, 2nd edn. Springer Monographs in Mathematics. Springer, New York (2011)
99. Galdi, G.P.: On the asymptotic structure of D-solutions to steady Navier-Stokes equations in exterior domains. Mathematical problems relating to the Navier-Stokes equation. In: Series on Advances in Mathematics for Applied Sciences, vol. 11, pp. 81–104. World Scientific Publishing, River Edge, NJ (1992)
100. Galdi, G.P., Kyed, M.: Asymptotic behavior of a leray solution around a rotating obstacle. Progr. Nonlinear Differential Equations Appl. **60**, 251–266 (2011)
101. Galdi, G.P., Kyed, M.: Steady-state Navier-Stokes flows past a rotating body: Leray solutions are physically reasonable. Arch. Rational Mech. **200**, 21–58 (2011)
102. Galdi, G.P., Kyed, M.: A simple proof of L^q-estimates for the steady-state Oseen and Stokes equations in a rotating frame. Part I: strong solutions. Proc. Am. Math. Soc. **141**, 573–583 (2013)
103. Galdi, G.P., Kyed, M.: A simple proof of L^q-estimates for the steady-state Oseen and Stokes equations in a rotating frame. Part II: Weak solutions. Proc. Am. Math. Soc. **141**, 1313–1322 (2013)

104. Galdi, G.P., Kyed, M.: Time periodic solutions to the Navier-Stokes equations. In: Handbook of Mathematical Analysis in Mechanics of Viscous Fluids, pp. 509–578. Springer, Cham (2018)
105. Galdi, G.P., Neustupa, J.: Stability of Steady Flow past a Rotating Body. In: Suzuki, Y., Shibata, Y., (eds.) Mathematical Fluid Dynamics, Present and Future, Springer Proceedings in Mathematics & Statistics, vol. 183, pp. 71–94. Springer Japan, Tokyo (2016)
106. Galdi, G.P., Neustupa, J.: Steady flows around moving bodies. In: Giga, Y., Novotny, A. (eds.) Handbook of Mathematical Analysis in Mechanics of Viscous Fluids, pp. 341–417. Springer, Cham (2018)
107. Galdi, G.P., Padula, M.: A new approach to energy theory in the stability of fluid motion. Arch. Rational Mech. Anal. **110**, 187–286 (1990)
108. Galdi, G.P., Silvestre, A.S.: Strong solutions to the Navier-Stokes equations around a rotating obstacle. Arch. Rat. Mech. Appl. **176**, 331–350 (2005)
109. Galdi, G.P., Silvestre A.L.: On the steady motion of a Navier-Stokes liquid around a rigid body. Arch. Rat. Mech. Anal. **184**, 371–400 (2006)
110. Galdi, G.P., Silvestre, A.L.: Further results on steady-state flow of a Navier-Stokes liquid around a rigid body. Existence of the wake. In: Kyoto Conference on the Navier-Stokes Equations and Their Applications, pp. 127–143. RIMS Kôkyûroku Bessatsu, B1, Research Institute for Mathematical Sciences (RIMS), Kyoto (2007)
111. Galdi, G.P., Silvestre, S.A.: The steady motion of a Navier-Stokes liquid around a rigid body. Arch. Rat. Mech. Appl. **184**, 371–400 (2007)
112. Galdi, G.P., Simader, C., Sohr, H.: A class of solutions to stationary Stokes and Navier-Stokes equations with boundary data in $W^{-1/q,q}$. Math. Ann. **331**, 41–74 (2005)
113. Galdi, G.P.: Steady-state Navier-Stokes problem past a rotating body: geometric-functional properties and related questions. In: Topics in Mathematical Fluid Mechanics. Lecture Notes in Mathematics, vol. 2073, pp. 109–197. Fond. CIME/CIME Found. Subser. Springer, Heidelberg (2013)
114. Galdi, G.P., Neustupa, J.: Steady-state Navier-Stokes flow around a moving body. In: Handbook of Mathematical Analysis in Mechanics of Viscous Fluids, pp. 341–417. Springer, Cham (2018)
115. Garcia-Cuerva, J., Rubio de Francia, J.L.: Weighted Norm Inequalities and Related Topics. North Holland, Amsterdam (1985)
116. Geissert, M., Heck, H., Hieber, M.: L^p theory of the Navier-Stokes flow in the exterior of a moving or rotating obstacle. J. Reine Angew. Math. **596**, 45–62 (2006)
117. Geissert, M., Heck, H., Hieber, M.: On the equation div $= 0$ and Bogovskii's operator in Sobolev spaces of negative order. In: Partial Differential Equations and Functional Analysis. Operator Theory: Advances and Applications. Birkhauser, Basel, vol. 168, pp. 113–121 (2006)
118. Geissert, M., Hansel, T.: A non-autonomous model problem for the Oseen-Navier-Stokes flow with rotating effects. J. Math. Soc. Japan **63**(3), 1027–1037 (2011)
119. Kračmar, S., Novotný, A., Pokorný, M.: Estimates of Oseen kernels in weighted L^p spaces. J. Math. Soc. Japan **53**, 59–111 (2001)
120. Giga, Y.: Analyticity of the semigroup generated by the Stokes operator in L_p-spaces. Math. Z. **178**, 287–329 (1981)
121. Girault, V., Raviart, P.A.: Finite Element Approximation of the Navier-Stokes Equations. In: Lecture Notes in Mathematics, vol. 749. Springer, Berlin (1979)
122. Glazman, I.M.: Direct Methods of Qualitative Spectral Analysis of Singular Differential Operators. Moscow 1963 (In Russian). English version: Israel Progress for Science Translation, Jerusalem (1965)
123. Grafakos, L.: Classical Fourier Analysis. 3rd edn. Springer, Berlin (2014)
124. Guenther, R.B., Thomann, E.A.: The fundamental solution of the linearized Navier-Stokes equations for spinning bodies in three spatial dimensions—time dependent case. J. Math. Fluid Mech. **8**, 77–98 (2006)

125. Hishida, T.: An existence theorem for the Navier-Stokes flow in the exterior of a rotating obstacle. Arch. Rat. Mech. Anal. **150**, 307–348 (1999)
126. Hishida, T.: The Stokes operator with rotating effect in exterior domains. Analysis **19**, 51–67 (1999)
127. Hishida, T.: L^q estimates of weak solutions to the stationary Stokes equations around a rotating body. J. Math. Soc. Japan **58**, 743–767 (2006)
128. Hishida, T.: Asymptotic structure of steady Stokes flow around a rotating obstacle in two dimensions. In: Mathematical Fluid Dynamics, Present and Future, Springer Process Mathematical Statistics, vol. 183, pp. 95–137. Springer, Tokyo (2016)
129. Hishida, T., Shibata, Y.: $L_p - L_q$ estimate of Stokes operator and Navier-Stokes flows in the exterior of a rotating obstacle. Arch. Ration. Mech. Anal. **193**, 339–421 (2009)
130. Hishida, T., Shibata, Y.: Decay estimates of the Stokes flow around a rotating obstacle. RIMS Kôkyûroku Bessatsu **B1**, 167–186 (2007)
131. Heck, H., Kim, H., Kozono, H.: On the stationary Navier-Stokes flow around a rotating body. Manus. Math. **138**, 315–345 (2012)
132. Henry, D.: Geometric Theory of Semilinear Parabolic Equations. Lecture Notes in Mathematics, vol. 840. Springer, Berlin (1981)
133. Heuser, H.G.: Functional Analysis. Wiley, New York (1982)
134. Heywood, J.G.: The Navier-Stokes equations: on the existence, regularity and decay of solutions. Indiana Univ. Math. J. **29**(5), 639–681 (1980)
135. Inoue, A., Wakimoto, M.: On existence of solutions of the Navier-Stokes equation in a time dependent domain. J. Fac. Sci. Univ. Tokyo Sect. IA Math. **24**, 303–319 (1977)
136. Kato, T.: Growth properties of solutions of the reduced wave equation with a variable coefficient. Comm. Pure Appl. Math. **XII**, 403–425 (1959)
137. Kato, T.: Perturbation Theory for Linear Operators. Springer, Berlin (1966)
138. Kim, H., Kozono, H.: A removable isolated singularity theorem for the stationary Navier-Stokes equations. J. Differ. Equ. **220**, 68–84 (2006)
139. Kim, D.: Lq-estimates for the stationary Oseen equations on the exterior of a rotating obstacle. Math. Methods Appl. Sci. **41**(12), 4506–4527 (2018)
140. Kim, D.: Lr-results of the stationary Navier-Stokes flows around a rotating obstacle. Z. Angew. Math. Phys. **74**(5), (187), 21. pp (2023)
141. Kobayashi, T., Shibata, Y.: On the Oseen equation in the three-dimensional exterior domains. Math. Ann. **310**, 1–45 (1998)
142. Korn G.A., Korn T.M.: Mathematical Handbook for Scientists and Engineers. Mc Graw–Hill, New York (1961)
143. Korolev, A., Šverák, V.: On the large-distance asymptotics of steady-state solutions of the Navier-Stokes equations in 3D exterior domains. Ann. Inst. Henri Poincaré, Anal. Non Linéaire **28**, 303–313 (2011)
144. Kozono H., Sohr H.: New a priori estimates for the Stokes equations in exterior domains. Indiana Univ. Math. J. **41**, 1–27 (1991)
145. Kračmar, S., Nečasová: Navier-Stokes flow around a rotating obstacle: Mathematical analysis of its asymptotic behavior. In: Atlantis Briefs in Differential Equations, vol. 3, x+96 pp. Atlantis Press, Paris (2016). ISBN: 978-94-6239-230-4; 978-94-6239-231-1 76-02
146. Kračmar,S., Nečasová, Novotný, A.: The motion of a compressible viscous fluid around rotating body. Ann. Univ. Ferrara Sez. VII Sci. Mat. **60**, 189–208 (2014)
147. Kračmar, S., Nečasová, Š., Penel, P.: Estimates of weak solutions in anisotropically weighted Sobolev spaces to the stationary rotating Oseen equations. IASME Trans. **2**, 854–861 (2005)
148. Kračmar, S., Nečasová, Š., Penel, P.: Anisotropic L^2 estimates of weak solutions to the stationary Oseen type equations in R^3 for a rotating body. RIMS Kôkyûroku Bessatsu **B1**, 219–235 (2007)
149. Kračmar, S., Nečasová, Š., Penel, P.: L^q approach of weak solutions of Oseen Flow around a rotating body. In: Parabolic and Navier-Stokes equations, Banach Centrum Publications, vol. 81, pp. 259–276 (2008)

150. Kračmar, S., Nečasová, Š., Penel, P.: Remarks on the non-homogeneous Oseen problem arising from modeling of the fluid around a rotating body. In: S. Benzoni-Gavage, D. Serre (eds.) Hyperbolic Problems: Theory, Numerics, Applications. Proceedings of the Eleventh International Conference on Hyperbolic Problem held in Ecole Normale Supériore, Lyon, July 17–21, (2006), pp. 776–782. Springer, Berlin (2008)
151. Kračmar, S., Nečasová, Š., Penel, P.: L^q—approach of weak solutions of Oseen flow around a rotating body. Quart. J. Math. **LXVIII**(3), 421–437 (2010)
152. Kračmar, S., Krbec, M., Nečasová, Š., Penel, P., Schumacher, K.: Very weak solutions to the rotating Stokes, Oseen and Navier-Stokes problems in weighted space. Math. Nachr. **289**(11–12), 1466–1487 (2016)
153. Kračmar, S., Krbec, M., Nečasová, Š., Penel, P., Schumacher, K.: An L^q approach with generalized anisotropic weights of the weak solution of the Oseen flow around a rotating body in the whole space. Nonlinear Anal. **71**, e2940–e2957 (2009)
154. Kračmar, S., Nečasová, Š., Penel, P.: Anisotropic L^2-estimates of weak solutions to the stationary Oseen-type equations in 3D-exterior domain for a rotating body. J. Math. Soc. Japan **62**, 239–268 (2010)
155. Kračmar, S., Novotný, A., Pokorný, M.: Estimates of three dimensional Oseen kernels in weighted L^p spaces. In: Applied Nonlinear Analysis, pp. 281–316. Kluwer Academic, London (1999)
156. Kračmar, S., Penel, P.: Variational properties of a generic model equation in exterior 3D domains. Funkcialaj Ekvacioj **47**, 499–523 (2004)
157. Kračmar, S., Penel, P.: New regularity results for a generic model equation in exterior 3D domains. Banach Center Publications Warsaw **70**, 139–155 (2005)
158. Kurtz, D.S.: Littlewood-Paley and multiplier theorems on weighted L^p spaces. Trans. Am. Math. Soc. **259**, 235–254 (1980)
159. Kurtz D.S., Wheeden R.L.: Results on weighted norm inequalities for multipliers. Trans. Am. Math. Soc. **255**, 343–364 (1979)
160. Kyed, M.: Asymptotic profile of a linearized flow past a rotating body. Quart. Appl. Math. **71**, 489–500 (2013)
161. Kyed, M.: Periodic solutions to the Navier-Stokes equations. Habilitation thesis, Technische Universität Darmstadt, Darmstadt (2012)
162. Kyed, M.: On a mapping property of the Oseen operator with rotation. Discrete Contin. Dynam. Syst., Ser. S **6**, 1315–1322 (2013)
163. Kyed, M.: On the asymptotic structure of a Navier-Stokes flow past a rotating body. J. Math. Soc. Japan **66**, 1–16 (2014)
164. Kyed, M.: Maximal regularity of the time-periodic linearized Navier-Stokes system. J. Math. Fluid Mech. **16**(3), 523–538 (2014)
165. Ladyzhenskaya, O.A.: The Mathematical Theory of Viscous Incompressible Flow. Second English edition, revised and enlarged. Translated from the Russian by Richard A. Silverman, John Chu. Mathematics and its Applications, vol. 2. Gordon and Breach, Science Publishers, New York (1969)
166. Ladyzhenskaya, O.A.: Investigation of the Navier-Stokes equation for stationary motion of an incompressible fluid. (Russian) Uspehi Mat. Nauk **14**(3), 75–97 (1959)
167. Leis, R.: Initial Boundary Value Problems in Mathematical Physics. B. G. Teubner, Chichester (1986)
168. Lemarié-Rieusset, P.G.: Recent Developments in the Navier-Stokes Problem. In: Research Notes in Mathematics Series. Chapman & Hall/CRC, New York (2002)
169. Lions, J.L., Magenes, E.: Problèmes aus limites non-homogènes et applications, vol. 1 & 2. Dunod, Paris (1968)
170. Lizorkin, P.I.: On multipliers of Fourier integrals in the spaces $L_{p,\vartheta}$. Trudy Mat. Inst. Stekl. **89**, 231–248 (1967) (Russian); English translation: Proc. Steklov Inst. Math. **89**, 269–290 (1968)

171. Magnus, W., Oberhettinger, F., Soni, R.P.: Formulas and Theorems for the Special Functions of Mathematical Physics, 3rd edn. Die Grundlehren der mathematischen Wissenschaften, Band 52. Springer, New York (1966)
172. Miyakawa, T.: On nonstationary solutions of the Navier–Stokes equations in an exterior domain. Hiroshima Math. J. **12**, 115–140 (1982)
173. Muckenhoupt, B.: Weighted norm inequalities for the Hardy maximal function. Trans. Am. Math. Soc. **165**, 207–226 (1972)
174. Nečas, J.: Direct methods in the theory of elliptic equations. Translated from the 1967 French original by Gerard Tronel and Alois Kufner. In: Simader, C.G. (ed.) Editorial coordination and preface by Šárka Nečasová and a contribution. Springer Monographs in Mathematics. Springer, Heidelberg (2012)
175. Nečasová, Š.: Asymptotic properties of the steady fall of a body in viscous fluids. Math. Meth. Appl. Sci. **27**, 1969–1995 (2004)
176. Nečasová, Š.: On the problem of the Stokes flow and Oseen flow in $\mathbb{R}^3$ with Coriolis force arising from fluid dynamics. IASME Transaction **2**, 1262–1270 (2005)
177. Nečasová, Š., Schumacher, K.: Strong solutions to the Stokes equations of a flow around a rotating body in weighted L^q-spaces. Math. Nachr. **284**, 1701–1714 (2011)
178. Neri, U.: Singular Integrals. In: Lecture Notes in Mathematics, vol. 200. Springer, Berlin (1971)
179. van Neerven, J.: The Asymptotic Behaviour of Semigroups of Linear Operators. Birkhäuser, Basel-Boston-Berlin (1996)
180. Neustupa, J.: Existence of a weak solution to the Navier-Stokes equation in a general time-varying domain by the Rothe method. Math. Methods Appl. Sci. **32**, 653–683 (2009)
181. Neustupa, J., Penel, P.: A weak solvability of the Navier-Stokes equation with Navier's boundary condition around a ball striking the wall. In: Advances in Mathematical Fluid Mechanics, pp. 385–407. Springer, Berlin (2010)
182. Novotný A., Padula M.: Physically reasonable solutions to steady compressible Navier–Stokes equations in 3D–exterior domains ($v_\infty \neq 0$). Math. Ann. **370**, 439–489 (1997)
183. Novotný, A., Pokorný, M.: Three-dimensional steady flow of viscoelastic fluid past an obstacle. J. Math. Fluid Dynamic **2**, 294–314 (2000)
184. Novotný A., Pokorný M.: Steady plane flow of viscoelastic fluid past an obstacle. Appl Math **47**, 231–254 (2002)
185. Pokorný M.: Asymptotic behaviour of some equations describing the flow of fluids in unbounded domains. Ph.D. thesis, Charles University Prague & University of Toulon, France (1999)
186. Rychkov, V.: Littlewood-Paley theory and function spaces with A_p^{loc} weights. Math. Nachr. **224**, 145–180 (2001)
187. Sawyer, E.: Weighted inequalities for the one-sided Hardy-Little-wood maximal function. Trans. Am. Math. Soc. **297**, 53–61 (1986)
188. Schechter M.: On the essential spectrum of an arbitrary operator I. J. Math. Anal. Appl. **13**, 205–215 (1966)
189. Schumacher, K.: Solutions to the equation $\operatorname{div} u = f$ in weighted Sobolev spaces. In: Parabolic and Navier-Stokes equations. Part 2, 433–440 (2008). Banach Center Publications, vol. 81. Part 2, Polish Academy of Sciences Institute Mathematical, Warsaw
190. Schumacher, K.: Very weak solutions to the Stokes and Stokes resolvent problem in weighted function spaces. Ann. Univ. Ferrara Sez. VII Sci. Mat. **54**, 123–144 (2008)
191. Schumacher, K.: Stationary Stokes and Navier-Stokes equations in weighted Bessel-potential spaces. J. Math. Soc. Japan **61**, 1–38 (2009)
192. Serre, D.: Chute libre d'un solide dans un fluide visqueux incompressible. Existence. (French) [Free fall of a rigid body in an incompressible viscous fluid. Existence] Japan J. Appl. Math. **4**(1), 99–110 (1987)
193. Shibata, Y.: A Stability Theorem of the Navier-Stokes Flow Past a Rotating Body. In: Parabolic and Navier-Stokes equations. Part 2, Banach Center Publications, Polish Academy of Sciences Institute Mathematical, Warsaw, vol. 81 Part 2, pp. 441–455 (2008)

194. Shibata, Y. On the Oseen semigroup with rotating effect. In: Functional Analysis and Evolution Equations, pp. 595–611. Birkhauser, Basel (2008)
195. Shibata, Y.: On a C^0 semigroup associated with a modified Oseen equation with rotating effect. In: Advances in Mathematical Fluid Mechanics, pp. 513–551. Springer, Berlin (2010)
196. Silvestre, A.L.: On the existence of steady flows of a Navier-Stokes liquid around a moving rigid body. Math. Appl. Sci **27**, 1399–1409 (2004)
197. Silvestre, A.L.: The fundamental solution of the steady 2D generalized Oseen equations. Appl. Math. Lett. **124**, Paper No. 107613, 6 pp (2022)
198. Simader, C.G., Sohr, H.: The Dirichlet problem for the Laplacian in bounded and unbounded domain. In: Pitman Research Notes in Mathematics, vol. 360. Addison Wesley Longman, Harlow (1996)
199. Simader, Ch., Sohr, H.: A new approach to the Helmholtz decomposition and the Neumann problem in L^q–spaces for bounded and exterior domains. In: Galdi, G.P. (ed.) Mathematics Problems Relating to the Navier–Stokes Equations. Series: Advances in Mathematics for Applied Sciences, vol. 11, pp. 1–35. World Scientific Publishing, River Edge, NJ (1992)
200. Smith D.: Estimates at infinity for stationary solutions of the N.S. equations in two dimensions. Arch. Rat. Mech. Anal. **20**, 341–372 (1965)
201. Sohr, H.: The Navier-Stokes Equations: An Elementary Functional Analytic Approach. Birkhäuser, Berlin (2001)
202. Solonnikov, V.A.: A priori estimates for second order parabolic equations. Trudy Mat. Inst. Steklov. **70**, 133–212 (1964). (In Russian). English translation: AMS Translations **65**, 51–137 (1967)
203. Solonnikov, V.A.: Estimates of the solutions of a nonstationary linearized system of Navier-Stokes equations. Trudy Mat. Inst. Steklov. **70**, 213–317 (1964) (Russian); English translation : AMS Translations **75**, 1–116 (1968)
204. Stein, E.M.: Singular Integrals and Differentiability Properties of Functions. Princeton University Press, Princeton (1970)
205. Stein, E.M.: Harmonic Analysis: Real-Variable Methods, Orthogonality, and Oscillatory Integrals. Princeton University Press, Princeton, N.J. (1993)
206. Strömberg, J. O., Torchinsky, A.: Weighted hardy spaces. In: Lecture Notes in Mathematics, vol. 1381. Springer, Berlin (1989)
207. Takahashi, T.: Existence of strong solution for the problem of a rigid—fluid system. C. R. Acad. Sci. Paris, Ser. I **1336**, 453–458 (2003)
208. Takahashi, T., Tucsnak, M.: Global Strong solutions for the two-dimensional motion of an infinite cylinder in a viscous fluid. J. Math. Fluid Mech. **6**, 53–77 (2004)
209. Takahahsi, T.: Anisotropically weighted $L^q - L^r$ estimates of the Oseen semigroup in exterior domains, with applications to the Navier-Stokes flow past a rigid body. arXiv:2208.14584
210. Temam, R.: Navier-Stokes equations. Theory and numerical analysis. With an appendix by F. Thomasset, 3rd edn. Studies in Mathematics and its Applications, vol. 2. North-Holland Publishing Co., Amsterdam (1984)
211. Torchinski A.: Real Variable Methods in Harmonic Analysis. Academic Press, Orlando (1986)
212. Weinberger, H.F.: Variational properties of steady fall in Stokes flow. J. Fluid Mech. **52**, 321–344 (1972)
213. Weinberger, H.F.: On the steady fall of a body in a Navier-Stokes fluid. Partial differential equations. In: Proceedings of Symposia in Pure Mathematics, vol. XXIII, pp. 421–439. University of California, California (1971). American Mathematical Society, Providence, R.I. (1973)

Index

A

A_q-consistently decreasing mapping, 21
A_q-consistently increasing mapping, 21
A_q^+ - constant of w, 209, 211
Adjoint problem, 31
Algebraic multiplicity, 366
Analytic semigroup, 416
Angular maximal operator, 68
Approximate deficiency of operator $\mathscr{T}$, 365
Approximate nullity of operator $\mathscr{T}$, 365
Artificial boundary conditions, 307
Associated ordinary angular differential equation, 59
Associated time periodic Stokes problem, 57

B

Bogovskii operator, 22

C

Centered Hardy–Littlewood maximal operator, 21
Christoffel's symbol, 524
Classical Hardy–Littlewood maximal operator, 68
Constant A_q-consistency, 21
Continuous extension operator $\mathcal{E}_p$, 288, 290
Continuous spectrum, 366
C_0-semigroup, 416
Cylindrical domain, 523

D

Deficiency of operator $\mathscr{T}$, 365
Doubling property, 20

E

Effective oscillatory velocity part, 57
Essential growth bound of the semigroup $\mathrm{e}^{\mathscr{T}t}$, 419
Essential spectral radius of the semigroup $\mathrm{e}^{\mathscr{T}t}$ at time t, 419

F

Fourier transform, 21
Fredholm operator, 365
Fundamental solution of the heat equation, 29, 448
Fundamental solution of the Oseen system, 489
Fundamental solution of the Poisson equation, 448
Fundamental solution of the scalar Oseen equation, 488
Fundamental solution of the stationary Stokes type problem $\boldsymbol{\Sigma}_{st}$, 65
Fundamental solution of the time-dependent Stokes system, 492

G

Gamma function, 29, 448
Generalized eigenfunction, 366
Generalized eigenvector, 366
Green's formulas, 248, 250
G-Schwartz-Brulat space, 48

Š. Nečasová et al., *Mathematical Theory of a Fluid Flow Around a Rotating and Translating Body*, Springer Monographs in Mathematics,
https://doi.org/10.1007/978-3-031-84120-0

H
Haar measure, 48
Helmholtz decomposition, 351
Helmholtz projection, 29, 30, 352
Homogeneous Sobolev space, 18
Homogeneous weighted Sobolev space, 20
Hörmander–Mikhlin multiplier, 24

I
Index of $\mathscr{T}$, 365

K
Kernel of the Newton potential, 448
Kummer function, 29, 448

L
Lebesgue space, 17
Leray solution, 298, 300, 444
Linearized problem in a truncation domain, 307
Littlewood-Paley decomposition, 27
Locally compact Abelian group, 48
The local transformation, 521
Lorentz space $L_{q,r}(\mathcal{D})$, 438

M
Marcinkiewic's multiplier theorem, 75
Maximal operator, 68
Metric contravariant tensor, 524
Metric covariant tensor, 524
Muckenhoupt class A_q, 19
Muckenhoupt class $A_q(\mathcal{C})$, 208, 210
Muckenhoupt class $A_q(\mathcal{R})$, 208, 210
Multi–index, 17
Multiplicity of an eigenvalue, 365

N
Newton potential, 248, 250
Normal operator, 360
Nullity of operator $\mathscr{T}$, 365

O
One-dimensional anisotropic weight function, 213, 215
One-dimensional Muckenhoupt condition, 98
One-sided maximal operator, 209, 211
One-sided Muckenhoupt weight, 209, 211
Ornstein–Uhlenbeck operators, 417
Oseen operator, 352
Oseen–type operator, 352
Oseen-type operator $\mathcal{L}$, 28

P
Pochhammer symbol, 32
Point spectrum, 365
Pointwise Hörmander-Mikhlin multiplier condition, 212, 214
Poisson equation, 248, 250

R
Residual spectrum, 366
Resolvent set, 365

S
Schwartz space, 21
Semi-Fredholm operator, 365
Sobolev space, 17
Space of regularity $\mathfrak{M}_p$, 282, 284
Spectral bound of the operator $\mathscr{T}$, 419
Spectrum, 365
Star-shaped domain, 22
Stokes operator, 30, 352
Stokes–type operator, 352
Strong centered Hardy-Littlewood maximal operator, 211, 213

T
Truncating boundary ∂B_R, 307
Truncation error estimate, 307
T-time periodic function, 47

V
Volume potential, 271, 273

W
Weighted Littlewood-Paley decomposition of L^q_w, 220, 222
Weighted Sobolev space, 20

MIX
Papier aus verantwortungsvollen Quellen
Paper from responsible sources
FSC® C105338

If you have any concerns about our products,
you can contact us on
ProductSafety@springernature.com

In case Publisher is established outside the EU,
the EU authorized representative is:
Springer Nature Customer Service Center GmbH
Europaplatz 3, 69115 Heidelberg, Germany

Printed by Libri Plureos GmbH
in Hamburg, Germany